Table of Atomic Masses

	Symbol	Atomic Number	Atomic Mass		Symbol	Atomic Number	Atomic Mass
Actinium	Ac	89	(227)†	Mercury	Hg	80	200.6
Aluminum	Al	13	26.98	Molybdenum	Mo	42	95.94
Americium	Am	95	(243)	Neodymium	Nd	60	144.2
Antimony	Sb	51	121.8	Neon	Ne	10	20.18
Argon	Ar	18	39.95	Neptunium	Np	93	(237)
Arsenic	As	33	74.92	Nickel	Ni	28	58.70
Astatine	At	85	(210)	Niobium	Nb	41	92.91
Barium	Ba	56	137.3	Nitrogen	N	7	14.01
Berkelium	Bk	97	(247)	Nobelium	No	102	(259)
Beryllium	Be	4	9.012	Osmium	Os	76	190.2
Bismuth	Bi	83	209.0	Oxygen	O	8	16.00
Boron	B	5	10.81	Palladium	Pd	46	106.4
Bromine	Br	35	79.90	Phosphorus	P	15	30.97
Cadmium	Cd	48	112.4	Platinum	Pt	78	195.1
Calcium	Ca	20	40.08	Plutonium	Pu	94	(244)
Californium	Cf	98	(251)	Polonium	Po	84	(209)
Carbon	C	6	12.01	Potassium	K	19	39.10
Cerium	Ce	58	140.1	Praseodymium	Pr	59	140.9
Cesium	Cs	55	132.9	Promethium	Pm	61	(145)
Chlorine	Cl	17	35.45	Protactinium	Pa	91	(231)
Chromium	Cr	24	52.00	Radium	Ra	88	226.0
Cobalt	Co	27	58.93	Radon	Rn	86	(222)
Copper	Cu	29	63.55	Rhenium	Re	75	186.2
Curium	Cm	96	(247)	Rhodium	Rh	45	102.9
Dysprosium	Dy	66	162.5	Rubidium	Rb	37	85.47
Einsteinium	Es	99	(252)	Ruthenium	Ru	44	101.1
Erbium	Er	68	167.3	Samarium	Sm	62	150.4
Europium	Eu	63	152.0	Scandium	Sc	21	44.96
Fermium	Fm	100	(257)	Selenium	Se	34	78.96
Fluorine	F	9	19.00	Silicon	Si	14	28.09
Francium	Fr	87	(223)	Silver	Ag	47	107.9
Gadolinium	Gd	64	157.3	Sodium	Na	11	22.99
Gallium	Ga	31	69.72	Strontium	Sr	38	87.62
Germanium	Ge	32	72.59	Sulfur	S	16	32.06
Gold	Au	79	197.0	Tantalum	Ta	73	180.9
Hafnium	Hf	72	178.5	Technetium	Tc	43	(98)
Helium	He	2	4.003	Tellurium	Te	52	127.6
Holmium	Ho	67	164.9	Terbium	Tb	65	[illegible]
Hydrogen	H	1	1.008	Thallium	Tl	81	[illegible]
Indium	In	49	114.8	Thorium	Th	90	[illegible]
Iodine	I	53	126.9	Thulium	Tm	69	[illegible]
Iridium	Ir	77	192.2	Tin	Sn	50	[illegible]
Iron	Fe	26	55.85	Titanium	Ti	22	[illegible]
Krypton	Kr	36	83.80	Tungsten	W	74	[illegible]
Lanthanum	La	57	138.9	Uranium	U	92	[illegible]
Lawrencium	Lr	103	(260)	Vanadium	V	23	[illegible]
Lead	Pb	82	207.2	Xenon	Xe	54	[illegible]
Lithium	Li	3	6.941	Ytterbium	Yb	70	[illegible]
Lutetium	Lu	71	175.0	Yttrium	Y	39	[illegible]
Magnesium	Mg	12	24.31	Zinc	Zn	30	[illegible]
Manganese	Mn	25	54.94	Zirconium	Zr	40	[illegible]
Mendelevium	Md	101	(258)				

*The values given here are to four significant figures. A table of more accurate atomic masses is given in Appendix E.

†A value given in parentheses denotes the mass of the longest-lived isotope.

GENERAL CHEMISTRY

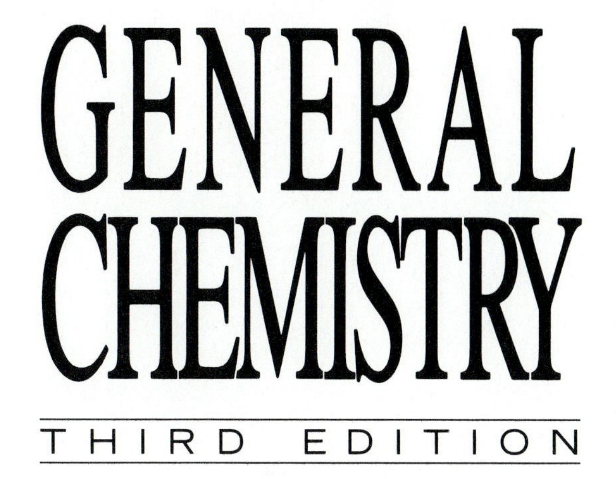

GENERAL CHEMISTRY

THIRD EDITION

Donald A. McQuarrie
University of California, Davis

Peter A. Rock
University of California, Davis

W. H. Freeman and Company
New York

COVER IMAGE: Xenon flash lamps (Figure 26-38).

Photograph by Russell Schlepman, EG&G Company

Library of Congress Cataloging-in-Publication Data

McQuarrie, Donald A. (Donald Allan)
General chemistry / Donald A. McQuarrie, Peter A. Rock.—3d ed.
p. cm
Includes index.
ISBN 0-7167-2132-5; ISBN 0-7167-2169-4 (Int. stud. ed.)
1. Chemistry. I. Rock, Peter A., 1939- . II. Title.
QD31.2.M388 1991
540—dc20 90-3706
CIP

Printed in the United States of America

1 2 3 4 5 6 7 8 9 0 KP 9 9 8 7 6 5 4 3 2 1

CONTENTS IN BRIEF

CONTENTS

CHAPTER 10
IONIC BONDS AND COMPOUNDS 354

CHAPTER 11
LEWIS FORMULAS 375

CHAPTER 12
PREDICTION OF MOLECULAR GEOMETRIES 409

CHAPTER 13
COVALENT BONDING 438

CHAPTER 14
LIQUIDS AND SOLIDS 473

CHAPTER 15
COLLIGATIVE PROPERTIES OF SOLUTIONS 518

CHAPTER 21
OXIDATION-REDUCTION REACTIONS 752

CHAPTER 22
ENTROPY, GIBBS FREE ENERGY, AND CHEMICAL REACTIVITY 774

CHAPTER 23
ELECTROCHEMISTRY 812

CHAPTER 24
NUCLEAR CHEMISTRY AND RADIOCHEMISTRY 850

CHAPTER 25
THE CHEMISTRY OF THE MAIN-GROUP ELEMENTS I 886

CHAPTER 26
THE CHEMISTRY OF THE MAIN-GROUP ELEMENTS II 926

PREFACE

We offer this third edition to you with great enthusiasm. The opportunity to revise two successful previous editions is a great luxury, and we sincerely believe that this new edition is exceptional.

As in the previous two editions, we have taken what we call an experimental approach to chemistry. We believe that students should be exposed to chemistry as it is practiced and applied. In most cases we introduce experiments and discuss the data before developing the theory to tie these observations together. We make a great effort to use actual chemical compounds and real data in formulating in-chapter examples and end-of-chapter problems. We avoid discussions of unspecified substances A and B or reactions such as $A + B \rightarrow C$; rather, we use actual chemical species—their properties, their reactions, and their important uses—to make the presentation real and vivid. With this approach, students not only learn chemical principles more easily but also understand the nature of chemistry as a laboratory science and enjoy its study more.

COLOR IS USED TO TEACH CHEMISTRY

In keeping with the experimental approach, ours was the first text to show chemistry as the colorful subject that it is. Even the most casual look at any other general chemistry text shows that this idea was long overdue and eagerly adopted. In this third edition, we have greatly expanded the use of color throughout, but we do not use color frivolously. Many elements, compounds, and reactions can be fully appreciated only when seen in color. With the help of a superb photo researcher and a scientific photographer for the Smithsonian Institution, we have assembled a collection of photographs that illustrate the beauty and excitement of chemistry.

THERE ARE SEVERAL IMPROVEMENTS IN THE THIRD EDITION

Expanded Coverage of the Basics

The first six chapters have been reorganized and expanded to ensure that students have a firm grasp of the basics. They also allow the early introduction of important classes of chemical reactions and a detailed discussion of stoichiometric calculations in solution. We believe that the changes will accommodate a variety of laboratory schedules and strengthen our experimental approach.

Chapter 1, "Chemistry and the Scientific Method," presents a thorough discussion of the scientific method and quantitative measurements, the metric system and SI units, precision and significant figures, dimensional analysis, and the Guggenheim notation. Chapter 2, "Atoms and Molecules," presents the atomic theory, elementary chemical nomenclature, atomic and molecular mass, the discovery of the atomic nucleus and atomic structure, and isotopes. Chapter 3, "The Periodic Table and Chemical Periodicity," develops the group properties of the elements and the periodic table. Chapter 4, "Chemical Reactivity," is new to this edition. It introduces several classifications of chemical reactions, leading in a natural way to acids and bases, more chemical nomenclature (including that of oxyacids and oxyanions), and the relative reactivities of metals. In this chapter we introduce oxidation states and their use in balancing oxidation-reduction reactions. Chapter 5, "Chemical Calculations," is now a more thorough treatment of stoichiometry, including several examples involving mixtures. Chapter 6, "Chemical Calculations for Solutions," another new chapter, presents molarity, solubility rules, precipitation reactions, acid-base titrations, and oxidation-reduction titrations.

As in the previous editions, an early introduction to descriptive chemistry allows us to integrate it throughout the text. When a compound is used to illustrate a principle or a type of calculation, we often comment on one or more of its chemical or physical properties or its industrial or commercial uses.

New Chapters on Descriptive Chemistry

The early and frequent introduction of chemical properties and reactions was a strong feature of our first two editions. We now have included three chapters (Chapters 25–27), a total of over 100 pages, that deal formally with the chemistry of the elements.

Interchapters

Because of these three new chapters, interchapters no longer need serve as a formal introduction to descriptive chemistry. We can instead introduce timely new interchapters to apply and reinforce the chemical principles that students have learned. The thirteen fully optional interchapters now present applications of chemistry to industry, to the environment, to chemical research, and in our everyday lives. Although shorter than interchapters in our previous editions, each is long enough for a careful discussion of such topics as "The Chemical Industry," "The Atmosphere," "Lasers," "High-Temperature Superconductivity," "Chemistry and Photography," and "Natural and Synthetic Gemstones." We hope that their titles and striking new design will invite students to read and enjoy them.

Three interchapters benefit from the wisdom and expertise of guest authors: F. E. Bailey and J. V. Koleske of Union Carbide, Warren S. Warren of Princeton University, and James H. Swinehart of the University of California, Davis. Their contributions greatly enrich our third edition.

Extended Coverage

We have included several topics that did not appear in the previous editions. In Chapter 7 we introduce the root-mean-square speed of molecules and partially derive the fundamental equation of the kinetic theory of gases. In Chapter 8, we carefully introduce the concept of energy, using simple mechanical models, before discussing energy changes associated with chemical reactions. In Chapter 13, we have expanded our treatment of covalent bonding, using the positive and negative values of atomic orbitals in the construction of molecular orbitals. We also introduce nonbonding molecular orbitals by way of a discussion of the bonding in ozone. One of our reviewers called this chapter the best treatment of bonding that he has ever seen at this level.

Our chapter on chemical kinetics (Chapter 16) still precedes our chapter on chemical equilibrium (Chapter 17). The relative order of these two chapters enjoys no consensus, there being about the same degree of preference for either approach. To add flexibility for the instructor, we introduce the basic ideas of both rate and equilibrium (particularly the concept of dynamic equilibrium) in our discussion of vapor pressure in Chapter 14, "Liquids and Solids," and of the dissolution of a solid in a liquid in Chapter 15, "Colligative Properties of Solutions." Chapter 14 now includes a section on the physical properties of liquids, including viscosity, surface tension, and dielectric constants.

The last section of Chapter 19 involves a calculation of the concentrations of all the species in an aqueous solution of a polyprotic acid, such as phosphoric acid. We have also extended our treatment of the second law of thermodynamics in Chapter 22, and we present a derivation of the Bragg equation of X-ray crystallography in Interchapter M. Finally, we have switched from using common logarithms to natural logarithms in our treatment of first-order kinetics, the Arrhenius equation, radioactive decay, and the van't Hoff equation.

In most cases, we have placed advanced material in the final section of a chapter, where it may be passed over without interrupting the discussion of a primary topic. In fact, as always, we purposefully isolate topics of optional interest as often as possible to better accommodate a wide range of course levels and emphases.

Practice Problems

A new feature of the third edition that has received an enthusiastic response from our many reviewers is the inclusion of new exercises, called practice problems, after each worked example. They provide the reader with additional, independent practice and reinforcement of skills. Answers to all the practice problems appear in the text, and their complete solutions appear in the *Instructor's Manual.*

Chapter-Ending Problems

There are over 500 new problems in this edition. In keeping with the format of the first two editions, many are grouped by topic in the order covered in the chapter. Most of these are arranged in matched pairs, each member dealing with the same principle or operation. This gives students who have difficulty with a particular problem a chance to test their understanding by working a similar one. As in the second edition, we then include additional problems that are not identified by topic, that are not paired, and that often are more challenging. We have roughly doubled the number of these additional problems, to about 40 per chapter. As a checkpoint for students, the answers to odd-numbered problems appear in Appendix K at the end of the text.

Expanded Tables of Physical and Chemical Properties

We have included rather large tables of solubility-product constants, thermodynamic data, acid-dissociation constants, base-protonation constants, and standard reduction voltages in the appendixes to offer the instructor a great variety of data for additional problems and testing.

WELL-RECEIVED FEATURES HAVE BEEN RETAINED

The following still are noteworthy emphases of our text:

- We order the material with the companion laboratory course in mind. We introduce stoichiometry, solutions, oxidation-reduction reactions, and the properties of acids and bases early to accommodate a variety of laboratory schedules.
- We think that first-year students should become proficient in writing Lewis formulas, and in developing Chapter 11 we have taken special care to teach this skill. Rules for writing Lewis formulas are especially clearcut in this edition. This chapter can be covered in two lectures.
- Chapter 12, "Prediction of Molecular Geometries" (formerly entitled "VSEPR Theory") develops the valence-shell electron-pair repulsion theory. We think VSEPR theory is easy to understand, easy to apply, and amazingly reliable. It both reinforces the writing of Lewis formulas and introduces first-year students to a large number of compounds. In our experience, students enjoy VSEPR theory because of its simplicity and predictive power. For the instructor who does not wish to present VSEPR, most, if not all, of the chapter can be omitted.
- Our section headings take the form of declarative sentences rather than brief terms or phrases. These headings focus on the underlying principle or primary objective of each section. Simply reading the section headings gives the students a good overview of any chapter.
- Each chapter contains numerous detailed worked examples that reinforce the concepts presented. We have increased the number of worked examples by about 25% in this edition, to nearly 350. No other general chemistry text offers so extensive a development of problem-solving skills.
- Each chapter ends with a summary, a list of terms students should know (cross-referenced to the page numbers on which the terms are introduced), and a list of equations that students should know how to use.
- We use SI units almost exclusively. Thus, we use joules instead of calories and picometers instead of angstroms. Authors of textbooks today face a dilemma: although SI units are endorsed by numerous organizations and journals, many instructors are reluctant to change. However, we generally express pressure in units of atmospheres or torr (mmHg), although we include a separate section of gas-law problems involving pascals for instructors who have made a complete transition to SI units.
- We use the Guggenheim "slash" notation to label headings in tables and the axes of graphs in figures. This notation, which is endorsed by the International Union of Pure and Applied Chemistry (IUPAC), is explained in Section 1-8. It is so much more convenient and less ambiguous than other notations that its use is expanding rapidly.

A COMPLETE INSTRUCTIONAL PACKAGE IS AVAILABLE

Study Guide/Solutions Manual

As in our first two editions, we firmly believe the *Study Guide/Solutions Manual,* by Carole McQuarrie and us, to be of real benefit to the student. The fact that it is the most used study guide on a per text basis—in fact, probably on any basis—bears us out. Consequently, its format remains unchanged. We continue to be sensitive to the difficulty many students have with numerical problems, and thus we give considerable emphasis to problem-solving skills.

For each chapter in the text, the *Study Guide/Solutions Manual* provides

- an outline of the chapter (section headings and short descriptive sentences)
- a self-test (about 50 fill-in-the-blank questions, not computational problems)
- a list of calculations that students should know how to do
- detailed solutions to the odd-numbered problems (unquestionably the most valuable feature to the student)
- the answers to the self-test

A glossary at the end of the manual is cross-referenced to the text.

Laboratory Manuals and Separates

The accompanying laboratory manual, *General Chemistry in the Laboratory,* second edition, by Julian L. Roberts, Jr., J. Leland Hollenberg, and James M. Postma, is derived from the popular Frantz-Malm series and provides suggestions for using the 42 experiments in conjunction with our text. Like the Frantz-Malm series, each experiment is also available separately as a lab separate. The accompanying *Instructor's Manual* contains filled-in report forms for all the experiments in the manual.

Test Bank and Instructor's Manual

For the instructor, our test bank, by Robert J. Balahura, allows selection from among more than 1300 questions, including both multiple-choice items and a variety of short-answer questions, crossword puzzles, and brain teasers. A computerized version of the test bank is available for use with IBM personal computers. Our *Instructor's Manual* contains detailed solutions to all even-numbered problems and practice problems in the text. The carefully selected set of 150 overhead transparencies in full color provides a useful lecture aid.

ACKNOWLEDGMENTS

We begin by thanking the 150 teachers who either answered our survey or were interviewed by phone and the scores of others who sent in unsolicited comments: your encouragement and suggestions very much influenced the course we took with this edition.

Many individuals deserve special recognition for their detailed reviews as we prepared our present work:

Robert J. Balahura, *University of Guelph*
David Becker, *Oakland University*
John Bellama, *University of Maryland*
Scott Briggs, *Western Washington University*
Joan Buillion, *University of Southern California*
Betty Deroski, *Suffolk Community College*
William H. Fink, *University of California, Davis*
John I. Gelder, *Oklahoma State University*
Daniel A. Geselowitz, *Haverford College*
Anthony V. Guzzo, *University of Wyoming*
Alton Hassell, *Baylor University*
Timothy Kling, *Lakeland Community College*
Robert M. Kren, *University of Michigan, Flint*
Joseph E. Ledbetter, *Contra Costa College*
Joel T. Mague, *Tulane University*
David Malik, *Indiana University-Purdue University at Indianapolis*
Lawrence J. Sacks, *Christopher Newport College*
Charles W. J. Scaife, *Union College*
Neil H. Schore, *University of California, Davis*
Maurice Schwartz, *University of Notre Dame*
Spencer L. Seager, *Weber State College*
Henry Shanfield, *University of Houston*
John Sharp, *University of Vermont*
Elizabeth Swiger, *Fairmont State College*
Harold Swofford, *University of Minnesota*
Warren S. Warren, *Princeton University*
Larry Westmoreland, *Central State University*
William Wilk, *California State University, Dominguez Hills*
Cary Willard, *University of New Hampshire*
Noel Zaugg, *Ricks College.*

In addition, we also wish to remember by name the more than 40 reviewers of the first two editions, whose counsel not only led to the success of that work but has had an ongoing effect on our writing: David L. Adams, *North Shore Community College;* Robert C. Atkins, *James Madison University;* Robert J. Balahura, *University of Guelph;* Otto T. Benfey, *Guilford College;* Larry E. Bennett, *San Diego State University;* David W. Brooks, *University of Nebraska–Lincoln;* Bruce W. Brown, *Portland State University;* George Brubaker, *Illinois Institute of Technology;* Ian S. Butler, *McGill University;* Harvey F. Carroll, *Kingsborough Community College, CUNY;* Ronald J. Clark, *Florida State University;* John M. D'Auria, *Simon Fraser University;* Derek A. Davenport, *Purdue University;* Daniel R. Decious, *California State University, Sacramento;* Robert Desiderato, *North Texas State University;* Timothy C. Donnelly, *University of California, Davis;* Frank J. Gomba, *United States Naval Academy;* Charles G. Haas, Jr., *Pennsylvania State University;* Edward D. Harris, *Texas A&M University;* Henry M. Hellman, *New York University;* Forrest C. Hentz, Jr., *North Carolina State University;* Earl S. Huyser, *University of Kansas;* Joseph E. Ledbetter, *Contra Costa College;* Edward C. Lingafelter, *University of Washington;* William M. Litchman, *University of New Mexico;* Saundra Y. McGuire, *University of Tennessee;* Arlene M. McPherson, *Tulane University;* Charles P. Nash, *University of California, Davis;* John M. Newey, *American River College;* Dennis G. Peters, *Indiana University;* Grace S. Petrie, *Nassau Community College;* Henry Po, *California State University, Long Beach;* James M. Postma, *California State University, Chico;* W. H. Reinmuth, *Columbia University;* Randall J. Remmel, *University of Alabama in Birmingham;* Don Roach, *Miami-Dade Community College;* Charles B. Rose, *University of Nevada;* Barbara Sawrey, *San Diego State University;* William M. Scovell, *Bowling Green State University;* Donald Showalter, *University of Wisconsin;* R. T. Smedberg, *American River College;* James C. Thompson, *University of Toronto;* Russell F. Trimble, *Southern Illinois University;* Carl Trindle, *University of Virginia;* Carl A. von Frankenberg, *University of Delaware;* E. J. Wells, *Simon Frazer University;* and Helmut Wieser, *University of Calgary.*

We had the tremendous advantage of having two very different but excellent editors throughout the preparation of the manuscript. One of these editors, Nancy White, had majored in philosophy and was a novice to chemistry, but she turned out to be one of the most perceptive, critical, and enthusiastic students that we have ever had. The other, John Haber, a former physics major, was exceedingly helpful and insightful; he contributed tremendously to the scientific quality of the text. This unusual editorial team helped us make this third edition as sound and literate as any text could be.

They are among many at W. H. Freeman and Company to whom we give special thanks: Linda Chaput, president, for always giving us her enthusiastic support and encouragement; Philip McCaffrey for overseeing the project from manuscript to bound book; Jodi Simpson for her superb copyediting and indexing; Nancy Singer for design; Howard Johnson for layout; Susan Stetzer for coordinating production; Mara Kasler for coordinating illustrations; Travis Amos for outstanding photo research and numerous excellent suggestions; and Chip Clark for his brilliant photography. Finally, we thank Carole McQuarrie for her generous and invaluable scientific help and Elaine Rock for helping to prepare the manuscript.

Donald A. McQuarrie
Peter A. Rock

GENERAL CHEMISTRY

1 CHEMISTRY AND THE SCIENTIFIC METHOD

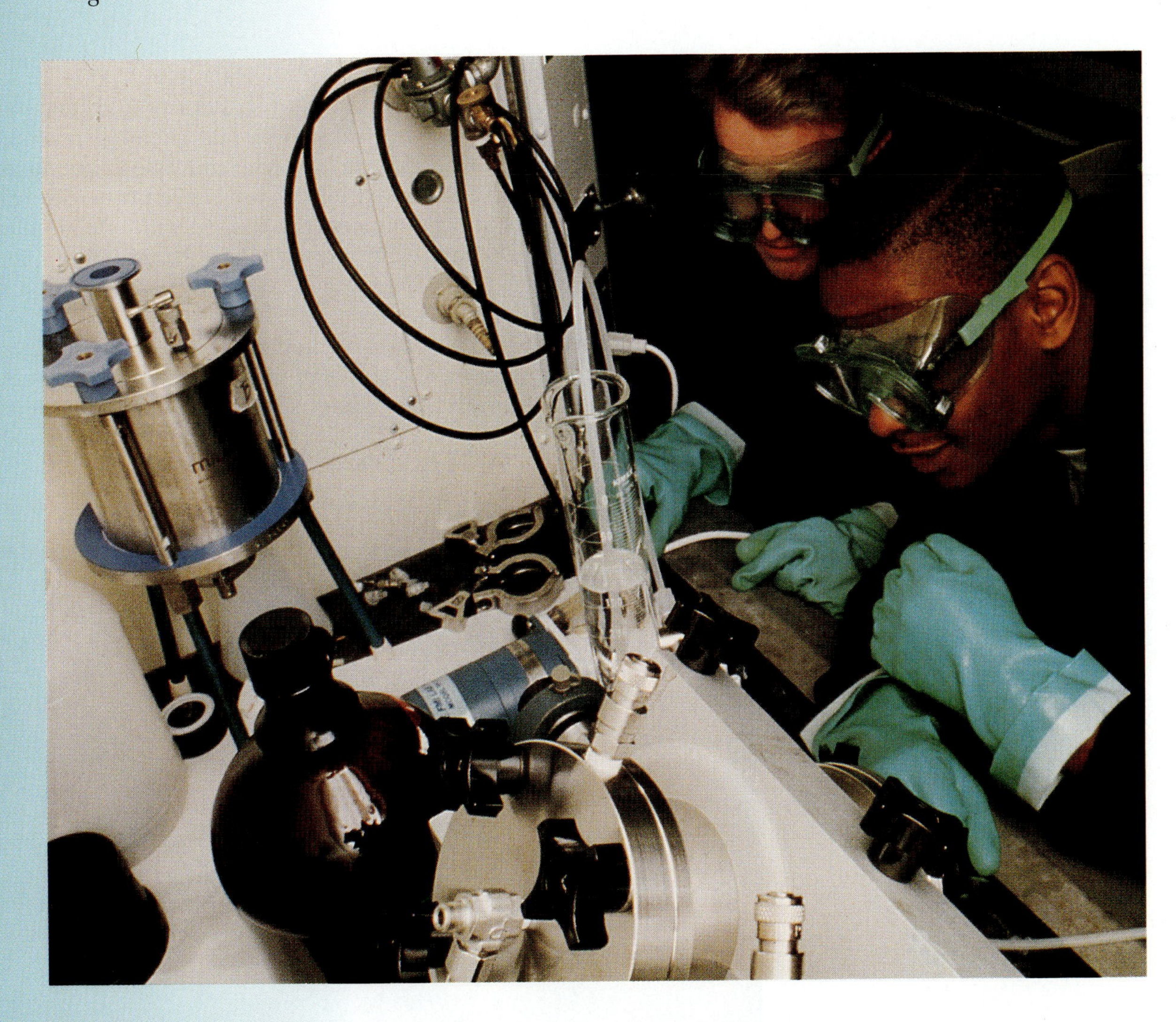

Chemistry is based firmly on the results of experiments. Carefully planned experiments are an endless source of fascination, excitement, and challenge.

You and about 400,000 other students in the United States and Canada are about to begin your first college course in chemistry. Most of you do not plan to become professional chemists; probably only about 10,000 of you will graduate with a bachelor's degree in chemistry. Whatever your chosen field of study, however, there is a very good chance that you will need a knowledge of elementary chemistry.

Chemists use the scientific method to describe the immense variety of the world's substances, from a grain of sand to the components of the human body. As you shall see, they can do this because chemistry is a quantitative science, based on experimental measurements and scientific calculations. You must therefore begin with a clear understanding of the methods scientists use to measure and calculate physical quantities. This chapter gives you these foundations.

1-1. WHY SHOULD YOU STUDY CHEMISTRY?

Chemistry is the study of the properties of substances and how they react with one another. Chemical substances and chemical reactions pervade all aspects of the world—in fact, the universe—around us. Usually, the new substances formed in reactions have properties very different from those of the substances that reacted with one another, properties that chemists can predict and put to use. Hundreds of materials that we use every day, directly and indirectly, are products of chemical research (Figure 1-1).

The examples of useful products of chemical reactions are limitless. The development of fertilizers, one of the major focuses of the chemical industry, has profoundly affected agricultural production in devel-

Figure 1-1 A modern chemical research laboratory.

oped countries. Equally important is the pharmaceutical industry. Who among us has not taken an antibiotic to cure an infection or used a drug to alleviate the pain associated with dental work, an accident, or surgery? Modern medicine, which rests firmly upon chemistry, has increased our life expectancy by about 15 years since the 1920s. It is hard to believe that, little over a century ago, many people died from simple infections.

Perhaps the chemical products most familiar to all of us are plastics. About 50 percent of industrial chemists are involved with the development and production of plastics. Every year the United States produces over 10 billion pounds of synthetic fibers, or over 40 pounds per person. Names such as nylon, polyethylene, Formica, Saran, Teflon, Hollofil, Gore-Tex, polyester, and silicone are familiar to us in our homes, our clothing, and the activities of our daily life. Chemistry also underlies the products of materials science—from computer chips to goods in paper and wood. Metals such as steel and lightweight alloys of titanium and aluminum make possible modern ships and aircraft.

It is remarkable that all these substances are built up from only about 100 different basic units, called atoms. Atomic theory pictures substances as atoms, or groups of atoms, joined together into units called molecules and ions. You will explore atomic theory in Chapter 2 and then go on to study chemical reactions and chemical calculations. You will learn to make predictions about what reactions take place, when they take place, and how quickly they take place; what substances are produced in these reactions; and what the structure, properties, and behavior of these substances will be. You will learn the chemistry behind many of the materials and processes we have already mentioned. We are confident that you will find your study of chemistry both interesting and enjoyable.

1-2. CHEMISTRY IS AN EXPERIMENTAL SCIENCE

Chemistry is an experimental science based on the scientific method. The essence of the **scientific method** is the use of carefully controlled experiments to answer scientific questions.

As an example, consider the statement, "Hot, boiled water freezes faster than cold water." This statement seems incorrect to many people, because, as they argue, hot water first has to get cold before it can freeze. Therefore, they reason that hot water must take longer to freeze than cold water. However, an argument is not an experiment. There are no guarantees that arguments supported only by intuition or by what seems to be commonsense reasoning are correct.

To use the scientific method, we must first define our goal, that is, we must first formulate the question we wish to answer. In this case, the question is, "Does hot or cold water freeze faster?" Let's go through the subsequent steps in our inquiry using the scientific method and then return to see how we would apply those steps to this question.

After defining our goal, our next step is to collect information or data about the subject under consideration. The data we collect will be of two sorts—**qualitative data,** consisting of descriptive observations; and **quantitative data,** consisting of numbers obtained by measurement. If we gather enough data about our subject, then we will be able to form a hypothesis. A **hypothesis** is a proposition put forth as the possible explanation for, or prediction of, an observation or a phenomenon. If hypotheses are supported by a sufficient number of experimental observations obtained under a wide variety of conditions, then they evolve into scientific theories.

To test a hypothesis, we perform experiments. If the experiments do not show that the hypothesis is incorrect, then we perform further experiments to see whether our results are reproducible. In particular, we wish to see whether the hypothesis stands up under a variety of experimental conditions. After many experiments, a pattern may emerge in the form of a constant relationship among phenomena under the same conditions. A concise statement of this relationship is called a **natural law** or a **scientific law.** Note that a law summarizes the relationship but does not explain it.

Once a law has been formulated, scientists try to develop a **theory,** or unifying principle, that explains the law based on the experimental observations. Eventually the theory also will be tested, and perhaps modified or rejected as the result of further experimentation. Let's turn back, now, to the question of whether hot or cold water freezes faster and apply the scientific method to the question.

Goal: To establish whether hot or cold water freezes faster.

Data:

1. Ordinary tap water contains dissolved air.
2. Dissolved air is expelled during boiling.
3. Dissolved air is expelled during freezing.
4. Ice formed from boiled water is clear, whereas ice formed from tap water is opaque because of trapped air pockets (Figure 1-2).
5. Water freezes from the top down.
6. Unexpelled air from ordinary tap water accumulates in a layer between the ice that has already formed and the liquid remaining underneath.
7. This trapped air layer impedes the transfer of water as vapor from the liquid to the ice layer above.

Hypothesis: Because the dissolved air in boiled water has already been expelled, there will be no air layer to impede the transfer of water from liquid to solid. Therefore, boiled water will freeze faster than ordinary tap water.

Experiment: Place a tray of boiled water and a tray of ordinary tap water side by side in the freezer. Check the trays at 1-hour intervals to see which one freezes completely. We leave the discovery of the result to you as an exercise in the application of the scientific method.

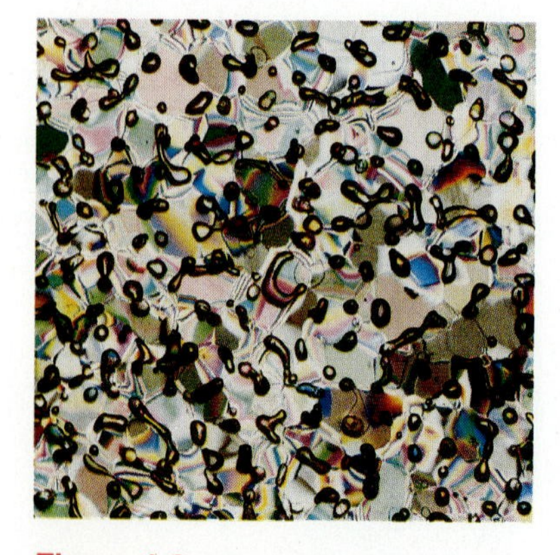

Figure 1-2 When ice is formed from water containing dissolved air, the air is expelled from the water and trapped in tiny pockets within the ice.

The scientific method underlies all of chemistry. When we study atomic theory in Chapter 2, we will see how the results of a large body of experiments led to the discovery of several important laws, which were in turn explained by a unifying atomic theory of matter.

It is important to remember that no theory can ever be proved correct by experiment. Experimental results can provide supporting data for a theory, but no matter how many experiments yield results consistent with a theory, the possibility always remains that additional experiments will demonstrate a flaw in the theory. This is the primary reason why experiments should be designed to disprove a hypothesis or theory rather than simply to provide additional support for the theory. The role of experiments, hypotheses, laws, and theories in the scientific method is outlined in Figure 1-3.

As Figure 1-3 shows, scientific theories are subject to ongoing revision, and most theories in use have known limitations. An imperfect theory is often useful, however, even though we cannot have complete confidence in its theoretical predictions. For example, a theory that correctly predicts the result, say, 90 percent of the time is very useful. Because scientific theories produce a unification of ideas, imperfect theories generally are not abandoned until a better theory is developed.

We conclude this section with a short story that illustrates the importance of carefully defining the quantities we use in experimental work.

A certain retired sea captain made his home in a secluded spot on the island of Zanzibar. As a sentimental reminder of his seafaring career he still had his ship's chronometer and religiously kept it

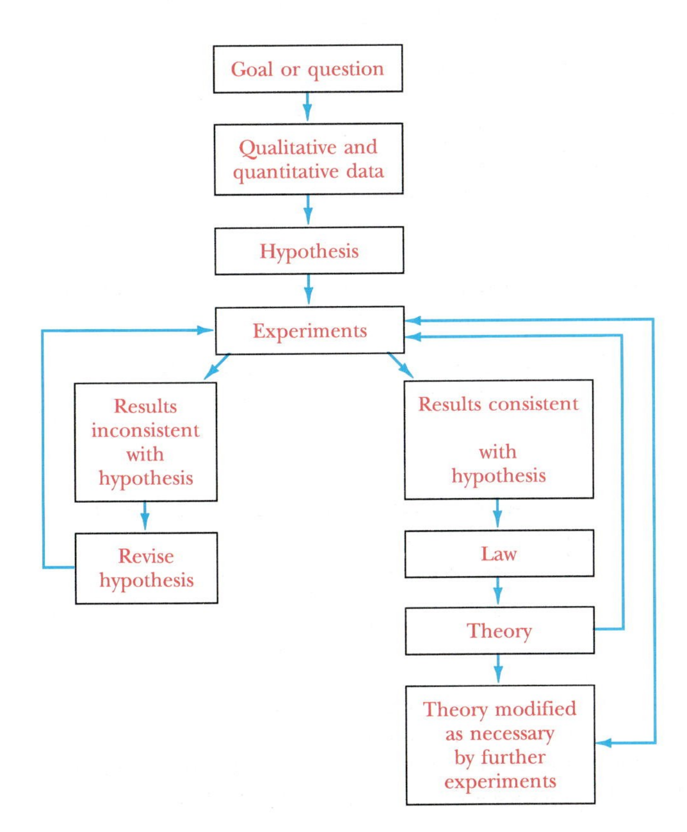

Figure 1-3 The interactive role of experiments, hypotheses, laws, and theories in the scientific method.

wound and in good operating condition. Every day exactly at noon, as indicated on his chronometer, he performed the ritual of firing off a volley from a small cannon. On one rare occasion, he received a visit from an old friend who inquired how the captain verified the correctness of his chronometer. "Oh," he replied, "there is a horologist over there in the town of Zanzibar where I go whenever I lay in supplies. He has very reliable time and as I have fairly frequent occasion to go that way I almost always walk past his window and check my time against his." After his visit was over the visitor dropped into the horologist's shop and inquired how the horologist checked his time. "Oh," replied he, "there's an old sea captain over on the other end of the island who, I am told, is quite a fanatic about accurate time and who shoots off a gun every day exactly at noon, so I always check my time and correct it by his."*

Because the sea captain defined accurate time in terms of the horologist's clock and the horologist defined accurate time in terms of the sea captain's chronometer, neither had an adequate definition. What each needed was an independent standard for accuracy that was universally agreed upon.

1-3. CHEMISTRY IS BASED ON QUANTITATIVE MEASUREMENTS

Figure 1-4 The French chemist Antoine Lavoisier and his wife and colleague, Marie-Anne Pierrette. Marie-Anne assisted Antoine in much of his work and illustrated and helped write his famous book, *Elementary Treatise on Chemistry.* Because of his financial connection with a much hated tax-collecting firm, Lavoisier was denounced, arrested, and guillotined in 1794 by supporters of the French Revolution.

Chemistry began to develop as a science in the eighteenth century. The early chemists were concerned primarily with qualitative observations of the general characteristics of substances, such as color. The modern physical sciences, however, are based firmly on **quantitative measurements,** measurements in which the result is expressed as a number. For example, the determinations that the mass of 1.00 cubic centimeter (cm^3) of gold is 19.3 grams (g) and that 1.25 g of calcium reacts with 1.00 g of sulfur express the results of quantitative measurements. Compare these determinations with **qualitative observations,** where we note general characteristics, such as color, odor, taste, and the tendency to undergo chemical change in the presence of other substances. An example of a qualitative statement is that lead is much denser than aluminum. As we shall see later in this chapter, the corresponding quantitative statement is that the mass of 1.00 cm^3 of lead is 11.3 g, whereas the mass of 1.00 cm^3 of aluminum is 2.70 g.

The French scientist Antoine Lavoisier (Figure 1-4) was the first chemist to fully appreciate the importance of carrying out quantitative chemical measurements. Lavoisier designed special balances that were more accurate than any devised before, and he used these balances to discover the **law of conservation of mass:** in a chemical reaction, the total mass of the reacting substances is equal to the total mass of the

*This story comes from E. R. Cohen, K. M. Crowe, and J. W. M. Drummond, *Fundamental Constants of Physics,* John Wiley Interscience, New York, 1957; the authors attribute the story to Professor George Harrison. Used by permission of John Wiley Interscience.

products formed. In other words, by careful quantitative measurements Lavoisier was able to show that mass is conserved in chemical reactions. Lavoisier's influence on the development of chemistry as a modern science cannot be overstated. In 1789, he published his *Elementary Treatise on Chemistry,* in which he presented a unified picture of chemical knowledge. The *Elementary Treatise on Chemistry* (Figure 1-5) was translated into many languages and was the first textbook of chemistry based on quantitative experiments.

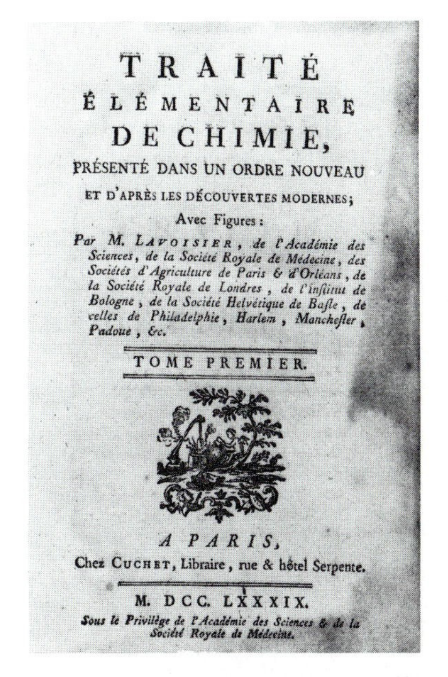

TRAITÉ
ÉLÉMENTAIRE
DE CHIMIE,
PRÉSENTÉ DANS UN ORDRE NOUVEAU
ET D'APRÈS LES DÉCOUVERTES MODERNES;
Avec Figures:
Par M. LAVOISIER, de l'Académie des Sciences, de la Société Royale de Médecine, des Sociétés d'Agriculture de Paris & d'Orléans, de la Société Royale de Londres, de l'institut de Bologne, de la Société Helvétique de Basle, de celles de Philadelphie, Harlem, Manchester, Padoue, &c.

TOME PREMIER.

A PARIS,
Chez CUCHET, Libraire, rue & hôtel Serpente.

M. DCC. LXXXIX.
Sous le Privilège de l'Académie des Sciences & de la Société Royale de Médecine.

Figure 1-5 The title page to Lavoisier's textbook of chemistry.

1-4. THE METRIC SYSTEM OF UNITS AND STANDARDS IS USED IN SCIENTIFIC WORK

With every number that represents a measurement, the units of that measurement must be indicated. If we measure the thickness of a wire and find it to be 1.35 millimeters (mm), then we express the result as 1.35 mm. To say that the thickness of the wire is 1.35 would be meaningless.

The preferred system of units used in scientific work is the **metric system.** Several sets of units make up the metric system, but in recent years there has been a worldwide movement to express all measurements in terms of just one set of metric units called **SI units** (for *S*ystème *I*nternational). We introduce here the three basic SI units representing the dimensions of length, mass, and temperature. The **basic SI units** are those from which all other units needed for measurement may be derived. Examples of derived units that we discuss are those for volume and density. The **derived SI units** are defined in terms of the basic SI units. For example, volume is measured in the unit of length cubed, and density is given by the mass divided by volume. The other basic SI units shown in Table 1-1, such as the unit of time (second), and other derived SI units, such as those representing pressure and concentration, will be discussed when we first encounter the need for them in the text. Appendix B contains a complete description of the metric system and SI units.

The basic SI unit of **length** (i.e., the distance between two points) is the **meter** (m). Prior to the advent of SI units in 1960, the meter was defined as the length of a special platinum rod maintained in a repository in Paris, France. However, this standard for the meter is not precise enough for modern scientific work of the highest accuracy, primar-

Table 1-1 Basic SI units

What is measured	Unit of measurement	Symbol
length	meter	m
mass	kilogram	kg
temperature	kelvin	K
time	second	s
amount of substance	mole	mol
charge	coulomb	C

Figure 1-6 Comparison of a meter stick with a yardstick (the meter stick is about 10 percent longer than the yardstick); a liter with a quart (the volume of a liter is about 6 percent larger than the volume of a quart); and a kilogram (the mass of a kilogram is about 2.2 times larger than the mass of a pound on earth).

ily because of the variation in the length of the "meter rod" with temperature. The SI definition of the meter is now given in terms of the speed of light in a vacuum.

If we denote the **speed of light** in meters per second by the symbol c, then a meter is defined as the distance traveled by light in a vacuum in the time equal to the reciprocal of the speed of light, that is, $1/c$. You can more easily comprehend this definition by imagining a person who walks at a constant speed of exactly 4 miles per hour; you could then define a mile as the distance that person can walk in exactly 1/4 hour. The speed of light in a vacuum is 2.9979×10^8 meters per second. Therefore, a meter is the distance light travels in a vacuum in $1/2.9979 \times 10^8$ second. The important point to grasp is that the speed of light in a vacuum is a fundamental constant of nature. Furthermore, it is not a physical object; and therefore neither dependent on temperature nor subject to mechanical damage or loss, as is a platinum rod. In more familiar terms, a meter is equivalent to 1.094 yards, or to 39.37 inches. Thus, a meter stick is 3.37 inches longer than a yardstick (36.00 inches) (Figure 1-6).

The SI system of units uses a series of prefixes to indicate factors-often multiples and fractions of SI units. For example, 1000 meters is called a **kilometer** (km), which is equivalent to 0.621 miles and is the unit used to express distances between towns and cities on maps and road signs in most countries throughout the world (Figure 1-7). A **centimeter** (cm) is one one-hundredth of a meter—there are 2.54 cm in one inch. Common SI-unit prefixes are given in Table 1-2.

Figure 1-7 Road sign showing distances in miles and kilometers.

EXAMPLE 1-1: Using the prefixes given in Table 1-2, explain what is meant by (a) a microsecond; (b) a milligram; and (c) 100 picometers.

Solution: (a) Table 1-2 shows that the prefix micro- means 10^{-6}, so a microsecond (μs) is 10^{-6} seconds or one one-millionth (1/1,000,000) of one second

(1 s). Events that take place in microseconds are common in scientific experiments. (b) The prefix milli- means 10^{-3}, so a milligram is 10^{-3} grams or one one-thousandth (1/1000) of a gram. (c) The prefix pico- means 10^{-12}, so a picometer is 10^{-12} meters or one-millionth of one-millionth of a meter (10^{-12} is equal to $10^{-6} \times 10^{-6}$ and 10^{-6} is one one-millionth). Thus, 100 picometers, or 100 pm, is equal to $100 \times 1 \times 10^{-12}$ m or 1.00×10^{-10} m. We shall see that the picometer is a convenient unit of length when discussing the sizes of atoms and molecules.

PRACTICE PROBLEM 1-1: Explain what is meant by (a) 400 nm; and (b) 20 ps.

Answer: (a) 4.00×10^{-7} m; (b) 2.0×10^{-11} s

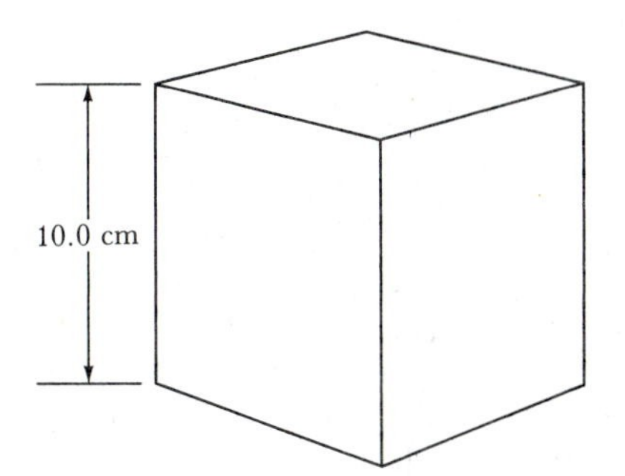

Figure 1-8 Each edge of a one-liter cube is 10.0 cm in length. A cubic meter (1 m^3) contains 1000 times ($10 \times 10 \times 10$) the volume of a liter (1 L).

Measures of **volume,** denoted by the symbol V, are derived from the basic SI unit for length, which is the meter. A cubic meter (1 m^3) is the volume of a cube that is 1 meter on each edge. However, a more convenient measure of volume for laboratory work is the liter. A **liter** (1 L) is equal to the volume of a cube that is 10 centimeters (or one-tenth of a meter) on each edge (Figure 1-8). The volume of a cube is equal to the cube (third power) of the length of an edge of the cube; thus,

$$1 \text{ L} = 10 \text{ cm} \times 10 \text{ cm} \times 10 \text{ cm} = (10 \text{ cm})^3 = 1000 \text{ cm}^3$$

A **milliliter,** 1 mL, is one one-thousandth (1/1000) of a liter; in other words, there are 1000 mL in 1 L. Because 1000 mL and 1000 cm^3 are both equal to 1 L, we conclude that 1 mL = 1 cm^3; that is, a milliliter (1 mL) and a **cubic centimeter** (1 cm^3) are equal to each other. One liter is equal to 1.057 liquid U.S. quarts (see the inside of the back cover); thus, the volume of a liter is 5.7 percent larger than the volume of a quart (Figure 1-6). A U.S. gallon contains exactly 4 U.S. quarts and corresponds to 3.78 L.

The basic SI unit of **mass** is the **kilogram.** The mass of a cylinder of a platinum-iridium alloy kept by the International Bureau of Weights

Table 1-2 **Common prefixes for SI units**

Prefix	Symbol	Multiple	Example
mega-	M	10^6 or 1,000,000	1 megameter, 1 Mm = 1×10^6 m
kilo-	k	10^3 or 1000	1 kilometer, 1 km = 1×10^3 m
centi-	c	10^{-2} or 1/100	1 centimeter, 1 cm = 1×10^{-2} m
milli-	m	10^{-3} or 1/1000	1 millimeter, 1 mm = 1×10^{-3} m
micro-	μ	10^{-6} or 1/1,000,000	1 micrometer, 1 μm = 1×10^{-6} m
nano-	n	10^{-9} or 1/1,000,000,000	1 nanometer, 1 nm = 1×10^{-9} m
pico-	p	10^{-12} or 1/1,000,000,000,000	1 picometer, 1 pm = 1×10^{-12} pm

Figure 1-9 An automatic analytical balance and a laboratory beam balance; both are used to determine mass. The beam balance requires the placement of standard masses on one of the pans to achieve a mass balance.

and Measures in Sèvres, France, represents the standard kilogram, which is the only basic SI unit still defined by an artifact. One kilogram is equal to 1000 **grams** (1 kg = 1000 g).

The mass of an object is determined in the laboratory by balancing its weight against the weight of a reference set of masses (Figure 1-9). The mass values of the set of reference masses are fixed by comparison with the standard kilogram.

Because a mass is determined by balancing its weight against reference masses, the terms *mass* and *weight* are often used interchangeably (for example, "the sample weighs 28 grams"). However, these terms are not the same. The **weight** (w) of an object is equal to its mass (m) times the gravitational acceleration (g), that is, $w = mg$. Weight is thus the force of attraction of the object to a large body, such as the earth or the moon. An object on the moon weighs about one-sixth as much as it does on the earth, but its mass is the same in both places (Figure 1-10). To avoid such ambiguities, we generally use the term *mass* rather than the term *weight* throughout this book. A 1-kg mass weighs 2.205 pounds on earth; therefore, 1 pound is equivalent to 453.6 g (Figure 1-6).

Figure 1-10 An object on the moon weighs about one-sixth as much as it does on the earth.

A modern analytical balance (shown on the left in Figure 1-9) has the standard weights enclosed by the balance housing. The weights are controlled by an internal set of movable levers. The basic principle of operation is the same as that for the beam balance shown on the right in Figure 1-9 except that the balance point is detected optically with a light beam rather than visually with the naked eye.

Temperature is a property represented by another basic SI unit and constitutes a quantitative measure of the relative tendency of heat to escape from a body. The higher the temperature of a body, the greater is the tendency of heat to escape from the body. When we say that water is "hot" to the touch, we mean that heat flows readily from the water to our fingers, which are at a lower temperature than the water. When the water is "cold," the flow of heat is from our fingers to the water, which is at a lower temperature than our fingers. Numerical **temperature scales** are established by assigning temperatures to two reference systems. For example, a temperature scale can be established by assigning a temperature of exactly zero degrees to the freezing point of water and exactly one hundred degrees to the boiling point of water at 1.00 atmospheric pressure (i.e., the pressure of air at sea level on a clear day).

A **thermometer** is a device used to measure temperature. A thermometer contains a substance whose properties change in a reproducible way with changes in temperature. For example, the property might be the volume of a certain liquid, such as mercury. Because the increase in volume of an enclosed sample of mercury increases with increasing temperature, it can be used to measure temperature. We simply mark on the glass rod the position of the mercury column when the thermometer is in contact with an ice-water mixture; we label this position 0.0. Then we mark the position of the mercury column, as 100.0, when the thermometer is in contact with boiling water at 1 atmosphere. These two marks are **calibration points.** The temperature scale is then determined by marking off the thermometer scale linearly between the two calibration points.

We discuss temperature in more detail in Chapter 7. It is sufficient for our purposes at this stage to recognize that there are three different (but interrelated) temperature scales in common use (Figure 1-11): the Celsius temperature scale, the Fahrenheit temperature scale, and the Kelvin temperature scale.

The **Kelvin temperature** scale defines the SI unit for temperature, which is called the **kelvin** (K). The lowest possible temperature of the Kelvin scale is zero kelvin (0 K); consequently, this scale is called an absolute scale. Note that the degree sign is omitted, and all temperatures on the Kelvin scale are positive. **Celsius temperatures** (denoted by t) are related to Kelvin temperatures (denoted by T) by the equation

$$T \text{ (in K)} = t \text{ (in °C)} + 273.15 \qquad (1\text{-}1)$$

Thus, a Kelvin temperature of 373.15 K corresponds to 373.15 − 273.15 = 100.00°C. In other words, 0°C is 273.15 kelvins higher than 0 K (see Figure 1-11). One degree on the Celsius temperature scale corresponds to the same temperature interval as one degree on the Kelvin temperature scale. The two scales differ only in their zero points.

Celsius temperatures are related to **Fahrenheit temperatures** by the equation

$$t \text{ (in °C)} = (5/9)\ \{t \text{ (in °F)} - 32.0\} \qquad (1\text{-}2)$$

Thus, a Fahrenheit temperature of 98.6°F corresponds to a Celsius temperature of

$$t = (5/9)(98.6 - 32.0)\text{°C} = 37.0\text{°C}$$

and a Kelvin temperature of

$$T = (273.15 + 37.0)\text{K} = 310.2 \text{ K}$$

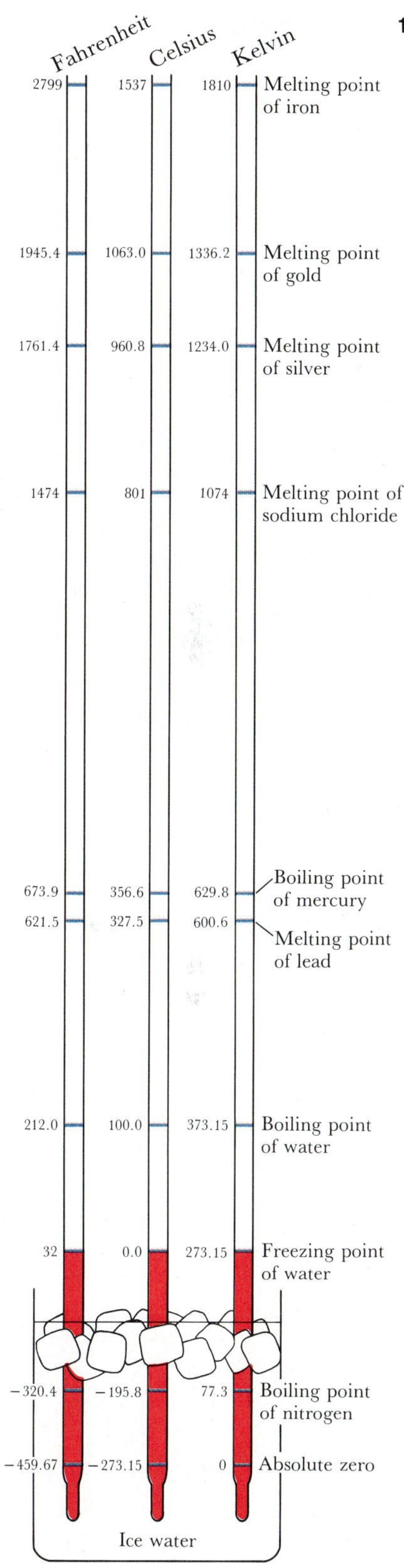

Figure 1-11 Comparison of the Fahrenheit, Celsius, and Kelvin temperature scales.

EXAMPLE 1-2: Derive Equation (1-2) by considering the information in Figure 1-11.

Solution: We first note that there are 100°C between the boiling and freezing points of water on the Celsius scale. Comparison with the Fahrenheit scale shows that there are 212°F − 32°F = 180°F between the same two points on the Fahrenheit scale. Thus, a degree Fahrenheit is five-ninths (100°C/180°F = 5°C/9°F) of a degree Celsius. We also note the 0°C corresponds to 32°F; thus, there is a (5/9) × 32-degree shift in the zero point on going from the Fahrenheit to the Celsius scale. Combination of these two results yields

$$t\ (\text{in }^\circ\text{C}) = \left(\frac{5}{9}\right)t\ (\text{in }^\circ\text{F}) - \left(\frac{5}{9}\right)(32.0^\circ\text{F})$$

or

$$t\ (\text{in }^\circ\text{C}) = \left(\frac{5}{9}\right)\{t\ (\text{in }^\circ\text{F}) - 32\}$$

PRACTICE PROBLEM 1-2: (a) Convert −90°F to degrees Celsius and to kelvins. (b) Find the one value of the temperature at which the temperatures on the Celsius and Fahrenheit scales coincide numerically.

Answer: (a) −68°C, 205 K; (b) −40

Density is an example of a property expressed in **compound units** because it involves both the unit of mass and the unit of volume. **Density** is defined as the mass per unit volume of a substance

$$\text{density} = \frac{\text{mass}}{\text{volume}}$$

or in symbols

$$d = \frac{m}{V} \tag{1-3}$$

To find the density of a material, we simply determine the volume of a known mass of the material and then use Equation (1-3).

The units of density are said to be derived units in the sense that they involve combinations of the basic SI units for mass and length. The concept of density allows us to compare the masses of equal volumes of materials. The old joke that asks, "Which is heavier, a pound of lead or a pound of feathers?" plays upon our intuition regarding density. Often, when we say that a material is heavy or light, we really mean that its density is high or low, respectively.

Equation (1-3) indicates that the **dimensions** of density are mass per unit volume, which can be expressed in a variety of units. If we express the mass in grams and the volume in cubic centimeters, then the units of density are grams per cubic centimeter. For example, the density of ice is 0.92 g/cm^3, where the slash denotes "per." In other words, 0.92 grams of ice occupies a volume of 1 cm^3 (the "unit volume" in this case).

From algebra, we know that

$$\frac{1}{a^n} = a^{-n}$$

where n is an exponent. Thus,

$$\frac{1}{\text{cm}^3} = \text{cm}^{-3}$$

Therefore, we also can express density as g·cm^{-3} instead of g/cm^3. The use of centered dots in compound units is an SI convention (Appendix B) and is used to avoid ambiguities. For example, m·s denotes meter-second, whereas ms (without the dot) denotes millisecond.

EXAMPLE 1-3: Calculate the density of gold, given that, at 20°C, 5.00 cm^3 of gold has a mass of 96.5 g.

Solution: Using Equation (1-3), we find that

$$d = \frac{m}{V} = \frac{96.5\ \text{g}}{5.00\ \text{cm}^3} = 19.3\ \text{g·cm}^{-3}$$

PRACTICE PROBLEM 1-3: Mercury is the only metal that is a liquid at 25°C. Given that 1.667 mL of mercury has a mass of 22.60 g at 25°C, calculate the density of mercury in g·mL^{-1} and g·cm^{-3}.

Answer: 13.56 g·mL^{-1} = 13.56 g·cm^{-3}

The density of a substance depends on the temperature, because the volume of a given mass of a substance depends on the temperature. For most substances, the density decreases as the temperature increases because the volume of most substances increases when their temperature increases.

Both density and temperature are examples of the **intensive properties** of a substance, which are properties whose values are independent of the amount of a substance. For example, the density of gold at 20°C is 19.3 g·cm^{-3}, whether the gold sample consists of 5.00 g or 5.00 kg. In contrast to intensive properties, **extensive properties** are directly proportional to the amount of a substance. Thus, mass and volume are both extensive properties. If we double the amount of a substance, then the mass and the volume also double, but the density remains the same.

1-5. THE PRECISION OF A MEASURED QUANTITY IS INDICATED BY THE NUMBER OF SIGNIFICANT FIGURES

The counting of objects is the only type of experiment that can be carried out with complete accuracy, that is, without any inherent error. Let's consider the problem of determining how many pennies there are in a jar. We can determine the exact number of pennies simply by counting them. Now suppose that there are 1542 pennies in the jar. The number 1542 is exact; there is no uncertainty associated with it. It is a different matter, however, when we determine the mass of 1542 pennies with a balance (Figure 1-9). Suppose that the balance we use is capable of measuring the mass of an object to the nearest one-tenth of a gram and that we use it to determine the mass of the 1542 pennies as 4776.2 ± 0.1 g. Because our balance was capable of measuring only to the nearest one-tenth of a gram, we know that the .2 in 4776.2 is not an exact number. It could actually be any number between .1 and .3; for example, it could be .13 or .26. The ±0.1 indicates the uncertainty in the last digit in the result 4776.2 g.

Now suppose we had a more sensitive balance capable of measuring the mass of the pennies to the nearest hundredth of a gram. Our result might be 4776.23 ± 0.01, meaning any number in the range 4776.22 to 4776.24. We would need a still more sensitive balance to determine the mass of the pennies to the nearest milligram (±0.001 g; 1 mg = 1/1000g).

In scientific work we need to distinguish between the precision of a result and the accuracy of a result. **Accuracy** refers to how close our result is to the actual value. To clarify the distinction between precision and accuracy, let's return to our measurement of the mass of the 1542 pennies. The result 4776.2 ± 0.1 g for the mass of the pennies may differ significantly from the actual mass, because perhaps the balance was improperly calibrated, or we misread the result displayed by the balance, or we failed to put all 1542 pennies back in the jar, or we forgot to correct for the mass of the jar. Each of these potential sources of error could have a significant effect on the accuracy of our result. The **precision** of a result conveys both how well repeated measurements of a quantity give results that agree with one another and how sensitive a measuring instrument was used. But high precision is no guarantee of high accuracy, because the same error source may be present in each and every measurement (Figure 1-12).

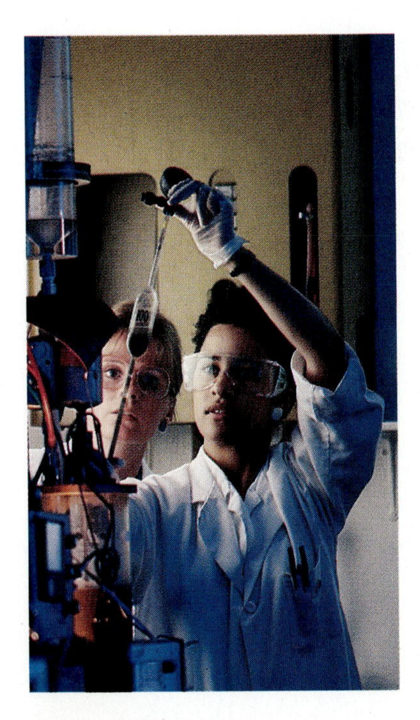

Figure 1-12 Accurate measurements are essential to obtain reproducible experimental results.

Denoting the uncertainty of a measured quantity by the ± notation is desirable in describing scientific results, but such a notation is too cumbersome for our purposes. We indicate the precision of measured quantities by the number of **significant figures** used to express a result. All digits in a numerical result are significant if only the last digit has some uncertainty. A result expressed as 4776.2 g means that there is an uncertainty of at least one unit in the last digit and that there are five significant figures in the result.

Zeros are considered to be significant figures if they are not included just to locate the decimal point; thus, both zeros in 1001.2 are significant figures. Zeros are not considered to be significant figures if their presence serves only to position the decimal point (as in 0.00125). In certain cases, it is not clear just how many significant figures are implied. Consider the number 100. With the number presented as 100, we might mean that the value of the number is exactly 100. But the two zeros might not be significant and we might mean that the value is approximately 100—say, 100 ± 10. The number of significant figures in such a case is uncertain. Usually the number of significant figures can be deduced from the statement of the problem. It is the writer's obligation to indicate clearly the number of significant figures by appropriate use of decimal points and zeros. Thus, 100.0 has four significant figures and 35×10^3 has two significant figures.

The rules for determining the significant figures in a measured quantity can be summarized as follows:

1. All nonzero digits or zeros between nonzero digits are significant figures. For example, 4023 mL has four significant figures.

2. Zeros used solely to position the decimal point are not significant figures. For example, 0.000206 L has three significant figures. The zeros to the left of the 2 are not significant figures.

3. If a numerical result ends in a zero or zeros to the right of the decimal point, then those zeros are significant figures. For example, 2.200 g has four significant figures.

4. If a numerical result ends in zeros that are not to the right of a decimal point, then those zeros may or may not be significant figures. In such cases we must deduce the number of significant figures from the statement of the problem. For example, the statement "350,000 spectators lined the parade route" involves a number that probably has only two significant figures at best; we infer this because it is obvious that no one actually counted the spectators.

5. A useful rule of thumb to use for determining whether or not zeros are significant figures is that zeros are not significant figures if the zeros disappear when scientific notation is used. For example, in the number 0.0197 ($= 1.97 \times 10^{-2}$) the zeros are not significant figures; in the number 0.01090 ($= 1.090 \times 10^{-2}$) the first two zeros are not significant and the second two are. (See Appendix A if you need to review scientific notation for numbers.)

EXAMPLE 1-4: State the number of significant figures in each of the following numbers:
(a) 0.0312; (b) 0.03120; (c) 3.1200×10^5

Solution:
(a) 0.0312 has three significant figures: 3, 1, and 2.
(b) 0.03120 has four significant figures: 3, 1, 2, and 0.
(c) 3.1200×10^5 has five significant figures: 3, 1, 2, 0, and 0.

PRACTICE PROBLEM 1-4: Determine the number of significant figures in the following numbers: (a) The human population of the United States is about 252,000,000; (b) 30,006; (c) 0.0029060.

Answer: (a) only 3; (b) 5; (c) 5

1-6. CALCULATED NUMERICAL RESULTS SHOULD SHOW THE CORRECT NUMBER OF SIGNIFICANT FIGURES

In scientific calculations we must express the final numerical result with the correct number of significant figures; otherwise, an incorrect impression of the precision of the results is conveyed.

In multiplication and division, the calculated result should not be expressed to more significant figures than the factor in the calculation with the least number of significant figures. For example, if we perform the multiplication 8.3143×298.2 on a hand calculator, then the following result comes up on the calculator display: 2479.3243. However, not all the figures in this result are significant. The correct result is 2479 because the factor 298.2 has only four significant figures; thus, the result cannot have more than four significant figures. The extra figures are not significant and should be discarded.

EXAMPLE 1-5: Determine the result to the correct number of significant figures:

$$y = \frac{3.00 \times 0.08205 \times 298}{0.93}$$

Solution: Using a hand calculator, we obtain

$$y = 78.873871$$

The factor 0.93 has the least number of significant figures—only two. Thus, the calculated result should be rounded off to two significant figures. The correct result is $y = 79$.

PRACTICE PROBLEM 1-5: Determine the result of the following calculation to the correct number of significant figures:

$$y = \frac{8.314 \times 298.15}{96{,}487.2}$$

Answer: 0.02569

In addition or subtraction, the calculated result should have no more figures after the decimal point than the least number of figures after the decimal point in any of the numbers that are being added or subtracted. Consider the sum

$$\begin{array}{r} 6.939 \\ +1.00797 \\ \hline 7.94697 \end{array} \quad \text{(round off to 7.947)}$$

The last two digits in 7.94697 are not significant, because we know the value of the first number in the addition, 6.939, to only three digits beyond the decimal place. Thus, the result cannot be accurate to more than three digits past the decimal. Therefore, our result expressed to the correct number of significant figures is 7.947.

In discarding insignificant figures, we use the following convention: If the figure following the last figure retained is a 5, 6, 7, 8, or 9, then the preceding figure should be increased by 1; otherwise (i.e., for 0, 1, 2, 3, and 4), the preceding figure should be left unchanged. Thus, rounding off the following numbers to three significant figures, we obtain $27.35 \rightarrow 27.4$ and $27.34 \rightarrow 27.3$. Note that in half of the cases (0,1,2,3,4) we discard the insignificant digit and in the other half (5,6,7,8,9) we increase the preceding digit by 1 when we discard the insignificant digits.

EXAMPLE 1-6: Determine the result of the following calculation to the correct number of significant figures:

$$y = 2796.8 - 2795$$

Solution: Subtraction yields 1.8, which we must round off to 2, because the second number (2795) has no digits to the right of the decimal point; therefore, the correct result should have no digits to the right of the decimal point.

PRACTICE PROBLEM 1-6: Determine the result of the following calculation to the correct number of significant figures:

$$y = \frac{7.2960}{8.9000} - 132.0$$

Answer: -131.2

1-7. DIMENSIONAL ANALYSIS IS USED TO SIMPLIFY MANY TYPES OF CHEMICAL CALCULATIONS

Dimensional analysis is a particularly useful method for calculations that involve quantities with units. The basic idea involved in **dimensional analysis** is to treat the units of the various quantities involved in the calculations as quantities that follow the rules of algebra. The calculation is set up in such a way that the undesired units cancel out and the numerical answer is obtained in the desired units. If the answer obtained does not have the desired units, then we know that the procedure used to obtain the result is incorrect.

Let's consider some specific examples of the use of dimensional analysis in calculations. Only numbers that have the same units can be added or subtracted. If we add 2.12 cm and 4.73 cm, we obtain 6.85 cm. If we wish to add 76.4 cm to 1.19 m, we must first convert 76.4 cm to meters or 1.19 m to centimeters. We convert from one unit to another by using a **unit conversion factor.** Suppose we want to convert meters to centimeters. In Appendix B we find that

$$1 \text{ m} = 100 \text{ cm} \tag{1-4}$$

Equation (1-4) is a definition and therefore exact; there is no limit to the number of significant figures on either side of the equation. If we divide both sides of Equation (1-4) by 1 m, we get

$$1 = \frac{100 \text{ cm}}{1 \text{ m}} \tag{1-5}$$

Equation (1-5) is called a unit conversion factor, because we can use it to convert a number in meters to a number in centimeters by multiplying the number in meters by the unit conversion factor in Equation (1-5). A unit conversion factor, as shown in Equation (1-5), is equal to unity; thus, we can multiply any quantity by a unit conversion factor without changing its intrinsic value. If we multiply 1.19 m by the unit conversion factor in Equation (1-5), then we obtain

$$(1.19 \cancel{\text{m}})\left(\frac{100 \text{ cm}}{1 \cancel{\text{m}}}\right) = 119 \text{ cm}$$

Notice that the unit of meter cancels, giving the final result in centimeters. To convert centimeters to meters, we use the reciprocal of Equation (1-5):

$$(76.4 \cancel{\text{cm}})\left(\frac{1 \text{ m}}{100 \cancel{\text{cm}}}\right) = 0.764 \text{ m}$$

Notice in this case that the unit of centimeter cancels out, giving the final result in meters. From these results, we see that the sum of 76.4 cm and 1.19 m is

$$76.4 \text{ cm} + (1.19 \cancel{\text{m}})\left(\frac{100 \text{ cm}}{1 \cancel{\text{m}}}\right) = 195 \text{ cm}$$

or

$$(76.4 \cancel{\text{cm}})\left(\frac{1 \text{ m}}{100 \cancel{\text{cm}}}\right) + 1.19 \text{ m} = 1.95 \text{ m}$$

Notice that both results are given to three significant figures.

As another example of converting from one set of units to another, let's convert 55 miles per hour (mph) to kilometers (km) per hour (h). From the inside of the back cover, we find that

$$1 \text{ mile} = 1.61 \text{ km} \tag{1-6}$$

Dividing both sides of Equation (1-6) by 1 mile yields the unit conversion factor:

$$1 = \frac{1.61 \text{ km}}{1 \text{ mile}} \tag{1-7}$$

Equation (1-7) is the unit conversion factor that is used to convert a speed given in miles per hour to a speed in kilometers per hour. Thus,

$$\left(\frac{55 \cancel{\text{miles}}}{1 \text{ h}}\right)\left(\frac{1.61 \text{ km}}{1 \cancel{\text{mile}}}\right) = 89 \text{ km}\cdot\text{h}^{-1}$$

Note that the use of the proper units for each quantity provides an internal check on the correctness of the calculation. We must multiply 55 mph by 1.61 km/1 mile to obtain the result in the desired units of $\text{km}\cdot\text{h}^{-1}$. Also note that if we had used the conversion 1 km = 0.62 mile from the inside of the back cover, then the unit conversion would be

$$\left(\frac{55\ \text{miles}}{1\ \text{h}}\right)\left(\frac{1\ \text{km}}{0.62\ \text{mile}}\right) = 89\ \text{km}\cdot\text{h}^{-1},$$

and the unit of mile once again cancels out.

EXAMPLE 1-7: Given that the unit conversion factor for the conversion of U.S. quarts to liters is 0.946 L per quart (qt), compute the number of liters in 1.00 U.S. gallon (gal).

Solution: There are 4 quarts per U.S. gallon. Thus, the number of liters of gasoline in one gallon of gasoline is

$$\left(1.00\ \text{gal}\right)\left(\frac{4\ \text{qt}}{1\ \text{gal}}\right)\left(\frac{0.946\ \text{L}}{1\ \text{qt}}\right) = 3.78\ \text{L}$$

Note that we wrote a sequence of conversions, working on each unit conversion separately.

PRACTICE PROBLEM 1-7: Suppose you are driving in France, and your speedometer reads $80\ \text{km}\cdot\text{h}^{-1}$. (a) If the next town is 72 km away, then how far away is it in miles? (b) If the elevation of the town is 1250 m, then what is its elevation in feet? (c) How fast are you going in miles per hour?

Answer: (a) 45 miles; (b) 4101 ft; (c) 50 mph

EXAMPLE 1-8: In most countries meat is sold in the market by the kilogram. Suppose the price of a certain cut of beef is 1400 pesos per kilogram and the exchange rate is 124 pesos to the U.S. dollar. What is the cost of the meat in dollars per pound (lb)?

Solution: The cost of the meat per pound obtained by using dimensional analysis is

$$\left(\frac{1400\ \text{peso}}{1\ \text{kg}}\right)\left(\frac{1\ \text{dollar}}{124\ \text{peso}}\right)\left(\frac{1\ \text{kg}}{2.20\ \text{lb}}\right) = 5.13\ \text{dollar}\cdot\text{lb}^{-1}$$

PRACTICE PROBLEM 1-8: Determine the number of grams in one U.S. ounce.

Answer: $28.35\ \text{g}\cdot\text{oz}^{-1}$

We now consider an example with several unit conversions involving quantities in compound units. It has been estimated that all the gold that has ever been mined would occupy a cube 17 m on a side. Given that the density of gold is 19.3 $g\cdot cm^{-3}$, let's calculate the mass of all this gold. The volume of a cube 17 m on a side is

$$\text{volume} = (17\ \text{m})^3 = 4913\ \text{m}^3 \qquad (1\text{-}8)$$

Although the volume calculated in Equation (1-8) is good to only two significant figures (because the value 17 m is good to only two significant figures), we shall carry extra significant figures through the calculation and then round off the final result to two significant figures. This procedure minimizes accumulation of roundoff errors in the calculation. The mass of the gold is obtained by multiplying the density by the volume; but before doing this, we must convert cubic meters to cubic centimeters, because the density is given in $g\cdot cm^{-3}$. From Appendix B, we find that 1 m = 100 cm. By cubing both sides of this expression, we obtain $1\ m^3 = 10^6\ cm^3$; so the unit conversion factor is $1 = 10^6\ cm^3/1 m^3$. Thus, the volume in cubic centimeters is equal to

$$\text{volume} = \left(4913\ \text{m}^3\right)\left(\frac{10^6\ \text{cm}^3}{1\ \text{m}^3}\right) = 4.913 \times 10^9\ \text{cm}^3$$

If we multiply the volume of the gold by the density of gold, then we obtain the mass of the gold:

$$\begin{aligned}\text{mass} &= \text{volume} \times \text{density}\\ &= (4.913 \times 10^9\ \text{cm}^3)(19.3\ \text{g}\cdot\text{cm}^{-3}) = 9.5 \times 10^{10}\ \text{g}\end{aligned}$$

The result is rounded off to two significant figures, because, as we mentioned before, the side of the cube (17 m) is given to only two significant figures. In obtaining this result, we have used the fact that $(cm^3)(cm^{-3}) = 1$.

Let's see what this mass of gold would be worth at \$450 per troy ounce (oz). (Gold is sold by the troy ounce, which is about 10 percent heavier than the avoirdupois ounce, the unit used for foods.) There are 31.1 g in 1 troy oz, so the unit conversion factor is 1 = 1 troy oz per 31.1 g. The mass of gold in troy ounces is

$$\begin{aligned}\text{mass} &= \left(9.5 \times 10^{10}\ \text{g}\right)\left(\frac{1\ \text{troy oz}}{31.1\ \text{g}}\right)\\ &= 3.1 \times 10^9\ \text{troy oz}\end{aligned}$$

At \$450 per troy ounce, the value in U.S. dollars of all the gold ever mined is

$$\begin{aligned}\text{value} &= (3.1 \times 10^9\ \text{troy oz})\left(\frac{\$450}{1\ \text{troy oz}}\right)\\ &= \$1.4 \times 10^{12} = 1.4\ \text{trillion U.S. dollars}\end{aligned}$$

Gold bars from the Echo Bay Mines in Edmonton, Alberta, Canada.

EXAMPLE 1-9: An old rule of thumb states that "A pint is a pound, the world around." Take the density of the liquid as 1.00 $g\cdot cm^{-3}$ and compute the number of pounds in a pint (pt).

Solution: There are 2 pints in 1 quart, 1 quart contains 0.946 liter, and there are 453.6 grams in 1 pound (inside the back cover of the text). Thus, we have

$$\left(\frac{1\text{ qt}}{2\text{ pt}}\right)\left(\frac{0.946\text{ L}}{1\text{ qt}}\right)\left(\frac{1000\text{ ml}}{1\text{ L}}\right)\left(\frac{1\text{ cm}^3}{1\text{ mL}}\right)\left(\frac{1.00\text{ g}}{1\text{ cm}^3}\right)\left(\frac{1\text{ lb}}{453.6\text{ g}}\right) = 1.04\text{ lb}\cdot\text{pt}^{-1}$$

which tells us that the old rule of thumb is good to within 4 percent, provided the density of the liquid is 1.00 $g \cdot cm^{-3}$, which is the density of water at room temperature (20°C).

PRACTICE PROBLEM 1-9: Gold prospectors used to carry mercury to extract gold flakes that were stuck to sand particles. The gold dissolves in the liquid mercury and can be recovered by distillation (see Chapter 2). Compute the weight in pounds of 3.00 L of mercury, given that the density of mercury is 13.6 $g \cdot cm^{-3}$.

Answer: 89.9 lb

Scientific calculations involve ways of handling numbers and units that may be new to you. If, however, you carefully set up the necessary conversion factors and make certain that the appropriate units cancel to give the units required for the answer—that is, if you use the dimensional analysis approach—then making unit conversions becomes a straightforward procedure.

1-8. THE GUGGENHEIM NOTATION IS USED TO LABEL TABLE HEADINGS AND THE AXES OF GRAPHS

In presenting tables of quantities with units, it is convenient to list the numerical values without their units and to use a column heading to specify the units. The least ambiguous way to do this is to write the name or symbol of the quantity followed by a slash and the symbol for the units. For example, the heading "Distance/m" indicates that the units of the numerical entries are distances expressed in meters. This notation is called the **Guggenheim notation** after E. A. Guggenheim, the British chemist who proposed its use

To see how Guggenheim notation works, let's suppose that we wish to tabulate a number of masses, such as 1.604 g, 2.763 g, and 3.006 g. We use the Guggenheim notation in the heading—"Mass/g"—and list the masses as numbers without units, as shown in Table 1-3, column (a). Now suppose that later we wish to retrieve the values with their units. The heading indicates that the numbers in the column are masses divided by grams, so we write, for example, mass/g = 1.604. We can multiply both sides of the equation by g to obtain

$$\text{g} \times \text{mass/g} = 1.604 \times \text{g}$$

or

$$\text{mass} = 1.604\text{ g}$$

Table 1-3 Tabulated data with headings in the Guggenheim notation

(a) Mass/g	(b) Mass/10^{-4} g
1.604	1.29
2.763	3.58
3.006	7.16

In the resulting expression we can easily recognize each component of the data:

$$\text{mass} = 1.604 \text{ g}$$
$$\text{property} = \text{value} \times \text{unit of measure}$$

Note that the heading is treated as an algebraic quantity and that we retrieve the data through an algebraic process.

The Guggenheim notation is particularly convenient when the values to be tabulated are expressed in scientific notation. Suppose that we wish to tabulate the masses 1.29×10^{-4} g, 3.58×10^{-4} g, and 7.16×10^{-4} g. In this case we can use the heading "Mass/10^{-4} g" to simplify the tabulated data, as shown in Table 1-3, column (b). To retrieve the data from the table, we write, for example, mass/10^{-4} g = 1.29, from which we get mass = 1.29×10^{-4} g. Notice that the Guggenheim notation enables us to list tabular entries as unitless numbers.

EXAMPLE 1-10: Consider the following tabulated data:

Time/10^{-5} s	**Speed/10^{5} m·s^{-1}**
1.00	3.061
1.50	4.153
2.00	6.302
2.50	8.999

Retrieve the actual data—the values and their units for time and speed, respectively—as four data pairs.

Solution: To find the actual times, we use, for example, time/10^{-5} s = 1.00, from which we obtain time = 1.00×10^{-5} s. The corresponding speed is given by speed/10^{5} m·s^{-1} = 3.061, or speed = 3.061×10^{5} m·s^{-1}. The other data pairs are 1.50×10^{-5} s, 4.153×10^{5} m·s^{-1}; 2.00×10^{-5} s, 6.302×10^{5} m·s^{-1}; and 2.50×10^{-5} s, 8.999×10^{5} m·s^{-1}.

PRACTICE PROBLEM 1-10: The SI unit for the quantity of electrical charge is a **coulomb** (C). Use the Guggenheim notation and tabulate the following data: 7.05×10^{-15} C, 3.24×10^{-15} C, and 9.86×10^{-16} C.

Answer:

Charge/10^{-15} C
7.05
3.24
0.986

Looking through this book, you will see that the axes of graphs in figures are labeled like the column headings of tabulated data and that the numbers on the axes are unitless. For example, the vertical axis in Figure 7-7 is labeled *V*/L (volume divided by liters), and the numbers on the axis are 1.0, 2.0, and 3.0. For a point half way between 1.0 and 2.0, we would have *V*/L = 1.5 or *V* = 1.5 L. The Guggenheim notation is especially useful for the graphical presentation of data, because only dimensionless quantities can be graphed and the retrieval of quantities with appropriate units from the graph is unambiguous when the Guggenheim notation is used on the graph axes.

EXAMPLE 1-11: Use the Guggenheim notation to label the graph axes and plot the following data:

$v/\text{ft}\cdot\text{s}^{-1}$	t/s
0	0
16	1.0
64	2.0
144	3.0
256	4.0

Solution: On a suitable piece of graph paper (Figure 1-13) we mark off the vertical axis ($v/\text{ft}\cdot\text{s}^{-1}$) and the horizontal axis (t/s). The data pairs are the coordinates of the various points on the graph and thus are used to position the points on the graph. Once the points are located, we draw a smooth curve through the points.

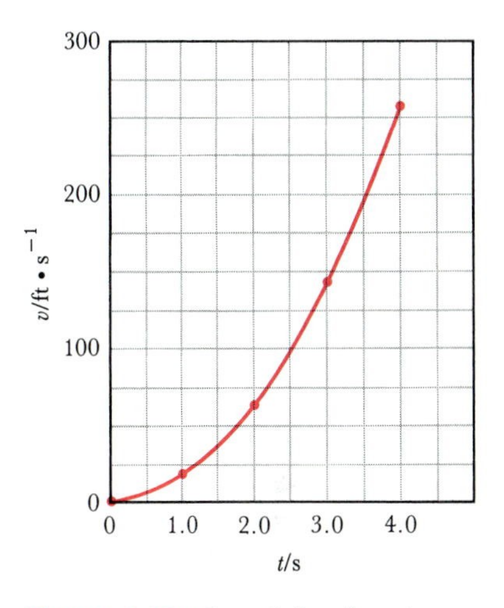

Figure 1-13 Plot of the data in Example 1-11 using the Guggenheim notation to label the axes.

PRACTICE PROBLEM 1-11: Replot the data given in Example 1-11 in the form v versus t^2. Label the axes according to the Guggenheim notation. What does your result tell you about the mathematical dependence of v on t? If you encounter difficulty with Example 1-11 and Practice Problem 1-11, you should study Appendix A, which contains a review of numerical calculation procedures.

Answer: Plot is a straight line, therefore v is proportional to t^2.

The Guggenheim notation has been adopted by the International Union of Pure and Applied Chemistry (IUPAC),* so it is the officially recommended notation for labeling columns of data in tables and axes of graphs. Although older, less convenient, and often ambiguous, alternative conventions are still in use, they are being phased out in the scientific literature.

This section concludes our discussion of units and various aspects of scientific calculations, including dimensional analysis and significant figures. The next chapter is devoted to developing an understanding of the nature of elements and chemical compounds within the framework of Dalton's atomic theory.

*See the *IUPAC Manual of Symbols and Terminology for Physicochemical Quantities and Units,* 1979. For a thorough discussion of the advantages of the Guggenheim notation, see E. A. Guggenheim, "Notations in Physics and Chemistry," *Journal of Chemical Education* **35:**606 (1958).

SUMMARY

Chemistry is an experimental science based on the scientific method. Scientific questions are answered by carrying out appropriate experiments. Scientific laws are concise summaries of large numbers of experimental observations. Scientific theories are designed to provide explanations for the laws and observations, and are subject to ongoing refinement based on new observations and results.

Modern chemistry began in the late eighteenth century when Antoine Lavoisier, considered to be the founder of modern chemistry, introduced quantitative measurements into chemical research.

The internationally sanctioned units for scientific measurements are the SI units. The basic SI unit for length is the meter, the basic SI unit for mass is the kilogram, and the basic SI unit for temperature is the kelvin. Derived SI units are combinations of basic SI units.

To do scientific calculations, you should understand significant figures, units of measurement, and unit conversion factors. The number of significant figures indicates the precision of experimental results. There are rules for determining the correct number of significant figures in a calculated result. Units must always be included with numbers that represent the results of measurements; otherwise, the numbers are meaningless. Similar quantities must be converted to the same type of units before they can be used together in calculations. Unit conversions are made with unit conversion factors.

The general procedure for setting up a scientific calculation in such a way that all the units but those desired in the final result cancel out is called dimensional analysis.

Guggenheim notation is used to label table entries and graph axes so that we can retrieve without ambiguity the entries in a table or a graph and their appropriate units.

TERMS YOU SHOULD KNOW*

scientific method 3
qualitative data 3
quantitative data 3
hypothesis 4
natural or scientific law 4
theory 4
quantitative measurement 6
qualitative observation 6
law of conservation of mass 6
metric system 7
SI units 7
basic SI units 7
derived SI units 7
length, l 7
meter, m 7
speed of light, *c* 8
kilometer, km 8
centimeter, cm 8
mega-, M 9
kilo-, k 9
centi-, c 9
milli-, m 9
micro-, μ 9
nano-, n 9
pico-, p 9
volume, *V* 9
liter, L 9
milliliter, mL 9
cubic centimeter, cm^3 9
mass, *m* 9
kilogram, kg 9
gram, g 10
weight, *w* 10
temperature 10
temperature scale 11
thermometer 11
calibration point 11
Kelvin temperature, K 11
kelvin 11
Celsius temperature, °C 11
Fahrenheit temperature, °F 11
compound units 12
density, *d* 12
dimensions 12
intensive property 13
extensive property 13
accuracy 14
precision 14
significant figures 14
dimensional analysis 17
unit conversion factor 18
Guggenheim notation 21
coulomb, C 22

EQUATIONS YOU SHOULD KNOW HOW TO USE

$T \text{ (in K)} = t \text{ (in °C)} + 273.15$ (1-1) (relation between Kelvin and Celsius temperature scales)

$t \text{ (in °C)} = \left(\frac{5}{9}\right)\{t \text{ (in °F)} - 32\}$ (1-2) (relation between Celsius and Fahrenheit temperature scales)

$d = \frac{m}{V}$ (1-3) (definition of density)

*Terms are listed in the order in which they appear in the text. The numbers refer to the pages on which the terms are introduced. A complete glossary of these terms is in the *Study Guide/Solutions Manual* by C. A. McQuarrie, D. A. McQuarrie, and P. A. Rock, W. H. Freeman and Company, New York, 1991.

PROBLEMS*

UNITS

1-1. Arrange the following quantities in order of increasing length:

(a) 100 nm (b) 1.0 km
(c) 1.0×10^3 cm (d) 100 pm
(e) 1.00×10^3 nm (f) 1000 m

1-2. Arrange the following quantities in order of increasing volume:

(a) 10 L (b) 100 mL
(c) 0.10 ML (d) 1.0×10^3 μL
(e) 20 cL (f) 1.0×10^4 nL

1-3. The volume of a sphere is given by $V = (4/3)\pi r^3$, where r is the radius. Compute the volume of a sphere with a radius of 100 pm.

1-4. The volume of a cube is given by $V = l^3$, where l is the length of an edge of the cube. Compute the volume of a cube with $l = 200$ pm.

1-5. A 20.4-g mass of a substance has a volume of 1.50 cm^3. Compute the density of the substance in $g{\cdot}cm^{-3}$.

1-6. The density of gold is 19.3 $g{\cdot}cm^{-3}$. Compute the volume of 31.1 g (1 troy ounce) of gold.

SIGNIFICANT FIGURES

1-7. Determine the number of significant figures in each of the following:

(a) 0.0390 (b) 6.022×10^{23}
(c) 3.652×10^{-5}
(d) 1,200,000,000, which is the 1990 population of China
(e) the number 16

1-8. Determine the number of significant figures in each of the following:

(a) 578 (b) 0.000578
(c) There are 1000 m in 1 km.
(d) The distance from the earth to the sun is 93,000,000 miles.

1-9. Calculate z to the correct number of significant figures in each part:

(a) $656.29 - 654 = z$; (b) $(27.5)^3 = z$
(c) $51/18.02 = z$; (d) $(6.022 \times 10^{23})(5.6 \times 10^{-2}) = z$

1-10. Calculate z to the correct number of significant figures in each part:

(a) $213.3642 + 17.54 + 32978 = z$
(b) $373.26 - 119 = z$
(c) $(6.626196 \times 10^{-34})(2.997925 \times 10^{9})/(1.38062 \times 10^{-23}) = z$
(d) $(9.1109558 \times 10^{-31} + 1.67252 \times 10^{-27} - 1.67482 \times 10^{-27})(2.997925 \times 10^{8})^2 = z$

UNIT CONVERSIONS AND DIMENSIONAL ANALYSIS

1-11. Use the information from the inside of the back cover to make the following unit conversions, expressing your results to the correct number of significant figures:

(a) 1.00 liter to quarts
(b) 186,000 miles per second to meters per second
(c) 8.314 $J{\cdot}K^{-1}{\cdot}mol^{-1}$ to $cal{\cdot}K^{-1}{\cdot}mol^{-1}$

1-12. Use the information from the inside of the back cover to make the following unit conversions, expressing your results to the correct number of significant figures:

(a) 325 feet to meters
(b) 1.54 angstroms (Å) to picometers and to nanometers
(c) 175 pounds to kilograms

1-13. A light-year is the distance light travels in 1 year. The speed of light is 3.00×10^8 $m{\cdot}s^{-1}$. Compute in meters and in miles the distance that light travels in 1 year.

1-14. In older U.S. cars, total cylinder volume is expressed in cubic inches. Compute the total cylinder volume in liters of a 454-cubic-inch engine.

1-15. You are shopping for groceries and see that one bottle of soda has a volume of 2.00 liters and costs \$1.49, whereas a six-pack of 16-oz bottles costs \$3.50. Which is the better buy?

1-16. Compute the speed in meters per second of a 90-mph fastball. The pitcher's mound on a regulation baseball diamond is 60 ft, 6 in. from home plate. Compute the time in seconds that it takes a 90-mph fastball to travel from the pitcher's mound to home plate.

*Problems grouped by subject area are arranged such that both usually involve the same principle or operation. Answers to odd-numbered problems are given in Appendix K. Detailed solutions to odd-numbered problems can be found in the *Study Guide/Solutions Manual.* Problems given under the heading "Additional Problems" are not identified by subject area and are not paired.

1-17. In an experiment to determine the relationship between the temperature of a gas and its volume, the following data were determined: at 0.0°C, the volume of the gas was 1.000 L; at 100°C, the volume was 1.37 L; at 200°C, the volume was 1.73 L; and at 300°C, the volume was 2.10 L. Make a table of the data using the Guggenheim notation.

1-18. In an experiment to determine the acceleration of a car, the following data were determined: at 0 s, the car was at rest; at 2.0 s, the distance traveled was 51 ft; at 4.0 s, the distance traveled was 204 ft; and at 6.0 s, the distance traveled was 459 ft. Make a table of the data using the Guggenheim notation.

1-19. Plot the following data on an 8.5-inch by 11-inch piece of graph paper using the Guggenheim notation to label the axes:

$[N_2O_5]/10^{-2}$ M	t/min
1.24	0
0.62	23
0.31	46
0.16	69
0.080	92

1-20. Consider the following tabulated data:

height/ft	time/s
0.87	6.09
1.40	11.65
1.99	18.11
4.26	30.41

Convert the height to centimeters and construct a new table.

ADDITIONAL PROBLEMS

1-21. Given that "room temperature" is 68°F, compute the corresponding room temperatures in degrees Celsius and in kelvins.

1-22. The U.S. wine and spirits industries readily adopted the metric system of units. One of the reasons for this is related to the results of the following comparison. Compare the number of milliliters in a 750-mL bottle of wine or spirits with the number of milliliters in a "fifth," which has a volume equal to four-fifths of a quart.

1-23. Sulfuric acid sold for laboratory use consists of 96.7 percent by mass sulfuric acid; the rest is water. The density of the solution is 1.845 $g\cdot cm^{-3}$. Compute the number of kilograms and pounds of sulfuric acid in a 2.20-L bottle of laboratory sulfuric acid.

1-24. What volume of benzene contains 55 g of benzene, given that the density of benzene is 0.879 $g\cdot cm^{-3}$?

1-25. According to the official major league baseball rules, a regulation baseball "shall weigh not less than five nor more than 5-1/4 ounces avoirdupois and measure not less than nine nor more than 9-1/4 inches in circumference." Compute the allowed range in density of an official major league baseball.

1-26. Compute the piston volume in liters of a 183-cubic-inch displacement automobile engine.

1-27. Given that a woman weighs 135 pounds on earth and is 5 feet 7 inches in height, compute her mass in kilograms and her height in centimeters.

1-28. If a person can run the 100-yard dash in 10 s, what is his average speed in miles per hour during the run?

1-29. The volume of a right circular cylinder is equal to the area of the base (πr^2) times the height (h); thus $V = \pi r^2 h$. Calculate the volume in milliliters of a can with a diameter of 2.50 inches and a height of 4.75 inches.

1-30. Compute the time in seconds that it takes for light to travel a distance of exactly 1 m in space.

1-31. The velocity of sound at sea level is about 770 mph. Calculate the time it takes for sound to travel one mile at sea level and compare that value with the time it takes light to travel the same distance.

1-32. In the context of the scientific method, comment on the statement, "The theory of evolution is a fact."

1-33. In the context of the scientific method, comment on the statement, "No two snowflakes are alike."

1-34. How would you use the scientific method to check the statement, "Dead fish float belly up in the water"?

1-35. The density of air at sea level and 98°F is about 1.20 $g\cdot L^{-1}$. The volume of the adult human lungs in the expanded state is about 6.0 L. Given that air is 20 percent oxygen by mass, compute the number of grams of oxygen in your lungs.

1-36. Using the number of significant figures given for the various physical constants on the inside of the back cover, compute the result to the correct number of significant figures:

(a) hc/k
(b) RT/F at $T = 298.1500$ K (Note: 1 J = 1 V·C)
(c) $2m_ec^2$
(d) eF (e is the charge on a proton)

1-37. What volume of acetone has the same mass as 10.0 mL of mercury? Take the densities of acetone and mercury to be 0.792 $g \cdot cm^{-3}$ and 13.56 $g \cdot cm^{-3}$, respectively.

1-38. A typical U.S.-made beer has an ethyl alcohol content of about 4.0 percent by volume, wine is about 12 percent alcohol by volume, and many hard liquors are about 40 percent (80 proof) alcohol by volume. Given that the densities of alcohol and water are 0.80 and 1.00 $g \cdot cm^{-3}$, respectively, compare the alcohol content in the following amounts of alcoholic beverages: (a) a 12.0-fluid-ounce can of beer (b) a 6.0-fluid-ounce glass of wine (c) a mixed drink with 1.25 fluid ounces of hard liquor.

1-39. You are sending a recipe to a friend who uses SI units exclusively. If your recipe calls for 2 cups of milk and 1 tablespoon of baking soda, and cooks at 350°F, how would you write your recipe in metric units? (Note: 1 cup = 8 fluid ounces, 1 tablespoon = 0.50 fluid ounce.)

1-40. The density of pure gold is 19.3 $g \cdot cm^{-3}$. A quantity of what appears to be gold has a mass of 465 grams and a volume of 26.5 mL. Is the substance likely to be pure gold?

1-41. Calculate the Celsius and Kelvin temperatures corresponding to 104°F, which is the body temperature of a typical bird.

1-42. A simple laboratory gas burner (Bunsen burner) has a flame temperature of about 1200°C. Calculate the corresponding Fahrenheit and Kelvin temperatures.

1-43. Derive the following relationship between the Kelvin and Fahrenheit temperature scales:

$$T \text{ (in K)} = 255.37 + (5/9)t \text{ (in °F)}$$

1-44. What is the Fahrenheit temperature that corresponds to absolute zero on the Kelvin temperature scale?

1-45. A container that can hold 6780 g of mercury can hold only 797 g of carbon tetrachloride. Given that the density of mercury is 13.6 $g \cdot mL^{-1}$, calculate the density of carbon tetrachloride.

1-46. An aspirin tablet contains 325 mg of aspirin. How many grams of aspirin are required to produce a 64-bottle box of aspirin, if each bottle contains 250 tablets?

1-47. Which is longer, a 400-m race or a 440-yard race?

1-48. One metric ton is 1000 kilograms. What is the corresponding weight in pounds?

1-49. A bottle that holds 250 g of water holds only 175 grams of gasoline. What is the density of the gasoline? Take the density of water to be 1.00 $g \cdot mL^{-1}$.

1-50. The U.S. recommended daily allowance (RDA) of vitamin B_2 (riboflavin) is 1.7 mg. How many vitamin pills would contain one g of vitamin B_2, assuming that each tablet contains the RDA of vitamin B_2?

1-51. Suppose that a runner runs 100 m in 9.80 s. What is his/her average speed in miles per hour?

1-52. Given that a weight lifter lifts 160 kilograms, compute the corresponding number of pounds.

1-53. The metal with the highest melting point (3410°C) is tungsten. What is the melting point of tungsten in °F?

1-54. Iridium is a metal with the greatest density, 22.65 $g \cdot cm^{-3}$. What is the volume of 192.2 g of iridium?

1-55. A swimming pool has average dimensions of 40 feet by 20 feet by 6 feet. What is the capacity of the pool in liters? Given that the density of water is 1.0 $g \cdot mL^{-1}$, how many kilograms of water is this?

1-56. Devise an experiment to determine the total volume of your own body and your density.

1-57. Air at normal temperatures and pressures has a density of 1.184×10^{-3} $g \cdot cm^{-3}$.

(a) How much does the air weigh (in pounds) in a room measuring 30.0 feet and 7.0 inches by 41.0 feet by 9.00 feet?
(b) What is the volume of 1.0 g of air?

1-58. What is the difference in volume (in cubic centimeters) between a sphere of radius 6.25 cm and a cylinder 10.0 cm high with a radius of 3.25 cm?

1-59. When you follow the manufacturer's specifications, one gallon of a latex, gloss enamel paint supposedly covers 350 square feet of a sealed surface. What is the average thickness (in millimeters) of a coat of this paint?

1-60. Suppose that you are driving in a country whose monetary unit is called a peso (124 pesos = 1 U.S. dollar). If you pay 53.6 pesos per liter of gasoline, then what is the cost in U.S. dollars per gallon of gasoline?

1-61. How many minutes does it take for light to travel from the sun to the earth, given that the sun is about 93 million miles from the earth?

INTERCHAPTER A

The Chemical Industry

F. E. Bailey, Jr., and J. V. Koleske, **Union Carbide Chemical and Plastics Company**

Chemistry is unique among the sciences in its close relationship to engineering and industry. There is no "physics industry," or "biology industry," but there is a huge and well-established chemical industry. The products of chemical research are a vital part of both national economies and international trade.

A-1. THE PRODUCTS OF CHEMICAL RESEARCH

Chemical research has led to advances not only in chemistry but also in almost every field of science, especially biology, medicine, and agriculture. Chemical products, from commercial fertilizers to pharmaceuticals now being tested, and from microelectronic circuits to the clothes we wear have a profound effect on the way we live.

Take a close look at a typical household room. The chair cushion is stuffed with polyurethane. The TV set, housed in a fire-resistant, stain-resistant cabinet, contains electronic circuits that use a variety of chemicals. The carpet, which may be made of nylon, polyester, or acrylic, is backed by a synthetic foam pad. Plywood floors are bonded with a polymeric adhesive, while the walls are covered with latex paint, and electrical outlets are made of molded plastic. Behind the walls, electrical circuits and water pipes are sealed with polymers. A glance at the bathroom cabinet would reveal a still greater variety of chemical products, from shampoos to prescription medicines. Not all the chemicals we have mentioned will be familiar yet, but their uses are all around us.

A research chemist in the fermentation laboratory of the Merck Chemical Manufacturing Division in Elaton, Virginia. The research involves using fermentation to manufacture chemicals.

All these everyday items and many more are manufactured from chemical intermediates, chemicals that are synthesized from more basic substances such as coal and petroleum. As the production moves from the basic substances, to chemical intermediates, to the end-products, value is added. There is more of a higher technology component in the product as its uses become more specific and specialized. In other words, chemistry is used to transform basic resources into more specialized and more valuable products. This addition of value is what makes chemical products so important to business. Not surprisingly, the more advanced a country's economy, the more highly developed is its chemical industry.

A-2. THE BUSINESS OF CHEMISTRY

How large is the chemical industry? The size of an industry can be described either in terms of the tonnage of material produced or the dollar value of the products, just as the size of the economy of an entire country can be described in terms of its "gross national product," the value of all the goods and services produced in a given year. In either of these terms, the chemical industry is *big*. Sales of chemicals and related products came to just under $300 billion for 1989 in the United States alone. The largest chemical producers—Du Pont and Dow in the United States, BASF and Bayer in Europe, and Asahi and Mitsubishi in Japan—had sales in recent years ranging from about $4 billion to more than $22 billion (Table A-1). Each of these companies is a highly complex business, employing tens of thousands of people.

The chemical industry has been an outstanding demonstration that the United States can compete in high-technology industries. World trade in chemicals has more than doubled in the last ten years, with the United States's share holding at about 15 percent. Much is being said these days about the "balance of trade," and the United States had a trade deficit of over $100 billion each year in the late 1980s. This means that a greater value of goods was imported from other countries than was made and sold overseas. In brief, the United States has been building up a huge debt that is only partially offset by foreign investment.

In those same years, however, the chemical industry contributed a trade *surplus* of over $15 billion. The success of the chemical industry will be more significant than ever in the years ahead if the United States is to compete in a global marketplace.

A-3. RESEARCH OPPORTUNITIES FOR CHEMISTS

As other countries increase their production of basic chemicals and raw materials, the United States is likely to focus instead on later stages in the manufacturing process—that is, on goods with more "value added." That means an emphasis on innovation and on developing sophisticated manufacturing processes for new products, both for other industries and for everyday use. We can expect challenging opportunities for chemists in such areas as biochemistry, plastics, agriculture, microelectronics, pharmaceuticals, and chemical engineering.

Table A-1 Dollar value of chemical sales of the largest chemical companies

	1989 sales/billion $
United States	
Du Pont	15.2
Dow Chemical	14.2
Exxon	10.6
Western Europe	
BASF	25.3
Hoechst	24.4
Bayer	23.3
Japan	
Asahi	6.4
Mitsubishi Kasei	5.3
Sumitomo	4.4

Electrolytic refining facility to produce nickel at the Sudbury Nickel Mine (Inco) in Ontario, Canada. Ontario is one of the largest producers of nickel in the world.

Environmental chemistry will also grow more important. The sheer size of modern industry means that proper waste disposal is critical. The best way to solve the problem, of course, is not to create the waste in the first place. Chemical research can help, by creating new processes that are environmentally acceptable. Research can also show how to put waste products to use as raw materials.

The chemical industry will remain vital to the world economy and the environment. It will stay closely related to the science of chemistry. Both basic research and practical innovation are needed in our ever more complex society. Both should continue to offer challenges—and jobs—for chemists in the years ahead.

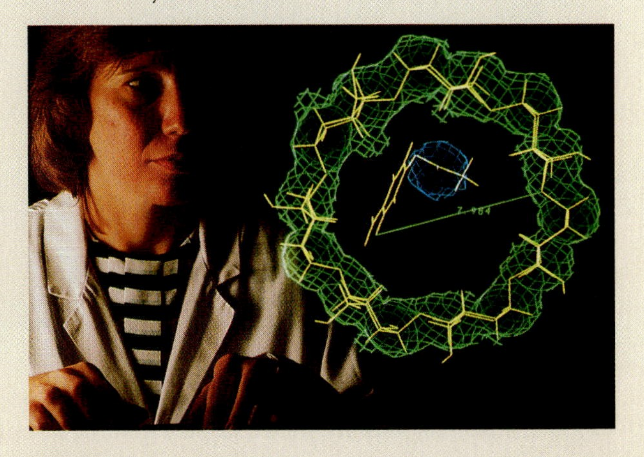

Many chemical research laboratories use computers to design new pharmaceuticals.

2 ATOMS AND MOLECULES

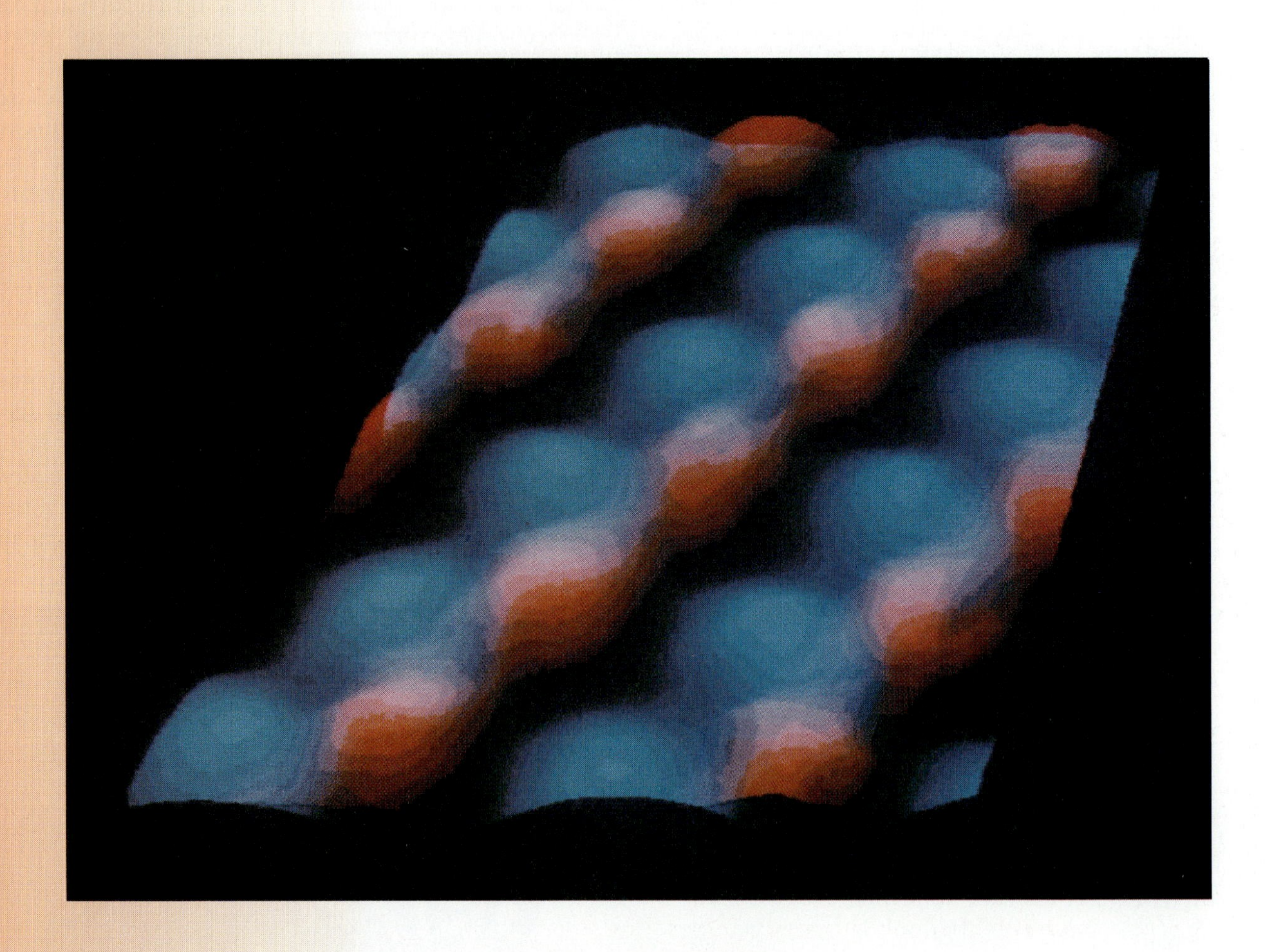

Scanning tunneling micrograph of the surface of a gallium arsenide crystal. The gallium ions are shown in pink and the arsenide ions are shown in blue.

The millions of chemicals known today are made up of little more than 100 elementary components—the chemical elements. In the early 1800s, an English schoolteacher named John Dalton drew on the idea of the elements and their chemical combinations to propose the atomic theory. Dalton began with the idea of structureless, solid spheres, which he called atoms. In Dalton's atomic theory, all substances consist either of atoms, of molecules (which are groups of atoms joined together), or of ions (which are atoms or groups of atoms having a positive or a negative charge). In this chapter you will see how Dalton's theory gives a simple picture of chemical reactions and explains countless chemical observations. You will also learn how experiments in the last years of the nineteenth and the early years of the twentieth centuries led to a dramatically new view of the atom—the nuclear model. With the atomic theory and the nuclear model, we begin to develop the concepts that enable us to understand the properties of matter.

2-1. ELEMENTS ARE THE SIMPLEST SUBSTANCES

Most substances in nature are **mixtures,** in which the component substances exist together without combining chemically. A familiar example of a mixture is air, which consists of 78 percent nitrogen and 21 percent oxygen with small amounts of argon, water vapor, and carbon dioxide. Another example is the mixture of iron, sugar, sand, and gold that we discuss in Section 2-12. Not only most naturally occurring substances but also many laboratory preparations consist of mixtures; and it is usually necessary to separate a mixture into its **pure,** or unmixed, components in order to study their chemical properties.

Almost all the millions of different chemicals known today can be broken down into simpler substances. Any substance that cannot be broken down into simpler substances is called an **element.** A pure substance that can be broken down into elements is called a **compound.** Before the early 1800s, many substances were classified incorrectly as elements because methods to break them down had not yet been developed, but these errors were rectified over the years. Although our definition of an element given above is a satisfactory working definition at this stage, we will soon learn the modern definition of an element: an element is a substance that consists only of atoms with the same nuclear charge.

There are 108 known chemical elements. Some are very rare; about 40 of them constitute 99.99 percent of all substances. Table 2-1 lists the most common elements found in the earth's crust, the oceans, and the atmosphere. Note that only 10 elements make up over 99 percent of the total mass. Oxygen and silicon are the most common elements; they are the major constituents of sand, soil, and rocks. Oxygen also occurs as a free element in the atmosphere and in combination with hydrogen in water. Table 2-2 lists the most common elements found in the human body. Note that only 10 elements constitute over 99.8 percent of the total mass of the human body. Because about 70 percent of the mass of the human body is water, much of this mass is oxygen and hydrogen.

Table 2-1 Elemental composition of the earth's surface, which includes the crust, oceans, and atmosphere

Element	Percent by mass
oxygen	49.1
silicon	26.1
aluminum	7.5
iron	4.7
calcium	3.4
sodium	2.6
potassium	2.4
magnesium	1.9
hydrogen	0.88
titanium	0.58
chlorine	0.19
carbon	0.09
all others	0.56

2-2. ABOUT THREE FOURTHS OF THE ELEMENTS ARE METALS

The elements are divided into two broad classes: **metals** and **nonmetals.** We are all familiar with the properties of solid metals. They have a characteristic luster, can be cast into various shapes, and are usually good conductors of electricity and heat. In addition, they are **malleable,** a term that means they can be rolled or hammered into sheets, and **ductile,** a term that means they can be drawn into wires.

About three fourths of the elements are metals. All the metals except mercury are solids at room temperature (about 20°C). Mercury is a shiny, silver-colored liquid at room temperature (Figure 2-1). Mercury used to be called quicksilver because of its silvery luster and the tendency of drops of mercury to roll rapidly on nonlevel surfaces.

Table 2-3 lists some common metals and their chemical symbols, and Figure 2-2 shows some of these metals. **Chemical symbols** are abbreviations used to designate the elements; usually they consist of the first one or two letters in the name of the element. But some chemical symbols do not seem to correspond at all to the elements' names; these symbols are derived from the Latin names of the elements (Table 2-4). It is essential for you to memorize the chemical symbols of the more common elements, because we shall be using them throughout this book. For an interesting discussion of the origins of the names of the elements, see Interchapter B.

Unlike metals, nonmetals vary greatly in their appearance. Over half of the nonmetals are gases at room temperature; the others are solids, except for bromine, which is a red-brown, corrosive liquid (Figure 2-1). In contrast to metals, nonmetals are poor conductors of electricity and heat, they cannot be rolled into sheets or drawn into wires, and they do

Table 2-2 Elemental composition of the human body

Element	Percent by mass
oxygen	64.6
carbon	18.0
hydrogen	10.0
nitrogen	3.1
calcium	1.9
phosphorus	1.1
chlorine	0.40
potassium	0.36
sulfur	0.25
sodium	0.11
magnesium	0.03
iron	0.005
zinc	0.002
copper	0.0004
tin	0.0001
manganese	0.0001
iodine	0.0001

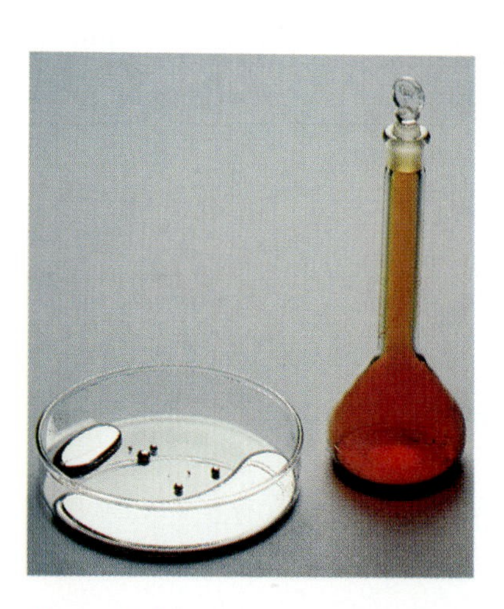

Figure 2-1 Mercury and bromine are the only elements that are liquids at room temperature (20°C). The red-brown vapor above the liquid bromine is bromine gas.

Table 2-3 Common metals and their chemical symbols

Element	Symbol	Element	Symbol
aluminum	Al	mercury	Hg
barium	Ba	nickel	Ni
cadmium	Cd	platinum	Pt
calcium	Ca	potassium	K
chromium	Cr	silver	Ag
cobalt	Co	sodium	Na
copper	Cu	strontium	Sr
gold	Au	tin	Sn
iron	Fe	titanium	Ti
lead	Pb	tungsten	W
lithium	Li	uranium	U
magnesium	Mg	zinc	Zn
manganese	Mn		

Figure 2-2 Common metals. Clockwise starting with the cylinder at top center are titanium, nickel, copper, aluminum, iron, and zinc.

not have a characteristic luster. Table 2-5 lists several common nonmetals, their chemical symbols, and their appearances (see also Figure 2-3). Note that several of the symbols of the nonmetallic elements in Table 2-5 have a subscript 2. This number indicates that these elements—hydrogen (H_2), nitrogen (N_2), oxygen (O_2), fluorine (F_2), chlorine (Cl_2), bromine (Br_2), and iodine (I_2)—exist in nature as a unit consisting of two smaller basic units called **atoms** that are joined together. The unit consisting of two or more atoms that are joined together is called a

Table 2-4 Elements whose symbol corresponds to the Latin name

Element	Symbol	Latin name
antimony	Sb	stibium
copper	Cu	cuprum
gold	Au	aurum
iron	Fe	ferrum
lead	Pb	plumbum
mercury	Hg	hydrargyrum
potassium	K	kalium
silver	Ag	argentum
sodium	Na	natrium
tin	Sn	stannum

Figure 2-3 Common nonmetals. Top row (*left to right*): arsenic, iodine, and selenium; bottom row (*left to right*): sulfur, carbon, boron, and phosphorus.

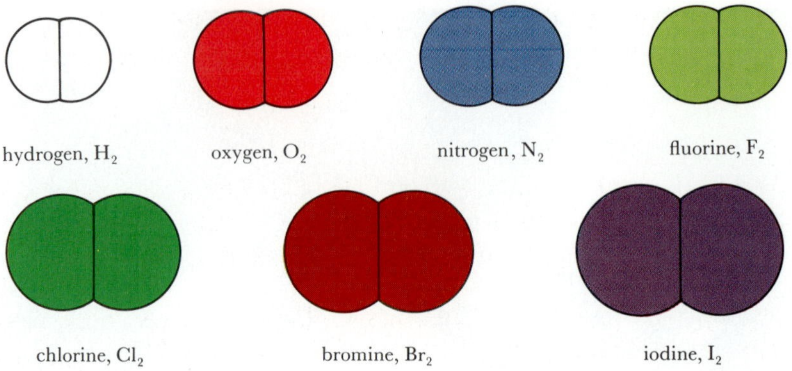

hydrogen, H_2 oxygen, O_2 nitrogen, N_2 fluorine, F_2

chlorine, Cl_2 bromine, Br_2 iodine, I_2

Figure 2-4 Scale models of molecules of hydrogen, oxygen, nitrogen, fluorine, chlorine, bromine, and iodine. These substances exist as diatomic molecules in their natural states but are still classified as elements, because their molecules consist of identical atoms.

Table 2-5 Some common nonmetals and their appearances at room temperature

Element	Symbol*	Appearance
Gases		
hydrogen	H_2	colorless
helium	He	colorless
nitrogen	N_2	colorless
oxygen	O_2	colorless
fluorine	F_2	pale yellow
neon	Ne	colorless
chlorine	Cl_2	green-yellow
argon	Ar	colorless
krypton	Kr	colorless
xenon	Xe	colorless
Liquids		
bromine	Br_2	red-brown
Solids		
carbon	C	black (in the form of coal or graphite)
phosphorus	P	pale yellow or red
sulfur	S	lemon yellow
iodine	I_2	violet-black

*The subscript 2 tells us that, at room temperature, the element exists as a diatomic molecule, that is, a molecule consisting of two atoms.

molecule. A molecule consisting of just two atoms is called a **diatomic molecule.** Scale models of diatomic molecules are shown in Figure 2-4.

2-3. THE LAW OF CONSTANT COMPOSITION STATES THAT THE RELATIVE AMOUNT OF EACH ELEMENT IN A COMPOUND IS ALWAYS THE SAME

The quantitative approach pioneered by Lavoisier was used in the chemical analysis of compounds. The quantitative chemical analysis of a great many compounds led to the **law of constant composition:** the relative amount of each element in a particular compound is always the same, regardless of the source of the compound or how the compound is prepared. For example, if calcium metal is heated with sulfur in the absence of water and oxygen, then the compound called calcium sulfide, which is used in fluorescent paints, is formed (Figure 2-5). We can specify the relative amounts of calcium and sulfur in calcium sulfide as the **mass percentage** of each element. The mass percentages of calcium and sulfur in calcium sulfide are defined as

$$\left(\begin{array}{c}\text{mass percentage of calcium}\\ \text{in calcium sulfide}\end{array}\right) = \left(\frac{\text{mass of calcium}}{\text{mass of calcium sulfide}}\right) \times 100$$

where the factor of 100 is necessary to convert the ratio of masses to a percentage. Similarly

$$\left(\begin{array}{c}\text{mass percentage of sulfur}\\ \text{in calcium sulfide}\end{array}\right) = \left(\frac{\text{mass of sulfur}}{\text{mass of calcium sulfide}}\right) \times 100$$

Suppose we analyze 1.630 g of calcium sulfide and find that it consists of 0.906 g of calcium and 0.724 g of sulfur. Then the mass percentages of calcium and sulfur in calcium sulfide are

$$\left(\begin{array}{c}\text{mass percentage of calcium}\\ \text{in calcium sulfide}\end{array}\right) = \left(\frac{\text{mass of calcium}}{\text{mass of calcium sulfide}}\right) \times 100$$

$$= \left(\frac{0.906\text{ g}}{1.630\text{ g}}\right) \times 100 = 55.6\%$$

$$\begin{pmatrix}\text{mass percentage of sulfur}\\\text{in calcium sulfide}\end{pmatrix} = \left(\frac{\text{mass of sulfur}}{\text{mass of calcium sulfide}}\right) \times 100$$

$$= \left(\frac{0.724\text{ g}}{1.630\text{ g}}\right) \times 100 = 44.4\%$$

Figure 2-5 The reaction between the elements calcium and sulfur yields calcium sulfide. The calcium sulfide is being formed at high temperature and is emitting light (fluorescing).

Note that, because calcium and sulfur are the only two elements present in calcium sulfide, the sum of the mass percentages of calcium and sulfur must add up to one hundred percent (55.6% + 44.4% = 100.0%).

The law of constant composition says that the mass percentage of calcium in pure calcium sulfide is always 55.6 percent. It does not matter whether the calcium sulfide is prepared by heating a large amount of calcium with a small amount of sulfur or by heating a small amount of calcium with a large amount of sulfur. Similarly, the mass percentage of sulfur in calcium sulfide is always 44.4 percent. Any excess of calcium or of sulfur simply does not react to form calcium sulfide. If calcium is in excess, then, in addition to the reaction product calcium sulfide, we have unreacted calcium metal remaining in the reaction vessel. If sulfur is in excess, then, in addition to the reaction product, unreacted sulfur remains.

EXAMPLE 2-1: Suppose we analyze 2.83 g of a compound of lead and sulfur and find that it consists of 2.45 g of lead and 0.380 g of sulfur. Calculate the mass percentages of lead and sulfur in the compound, which is called lead sulfide.

Solution: The mass percentage of lead in lead sulfide is

$$\begin{pmatrix}\text{mass percentage of lead}\\\text{in lead sulfide}\end{pmatrix} = \left(\frac{\text{mass of lead}}{\text{mass of lead sulfide}}\right) \times 100$$

$$= \left(\frac{2.45\text{ g}}{2.83\text{ g}}\right) \times 100 = 86.6\%$$

The mass percentage of sulfur in lead sulfide is

$$\begin{pmatrix}\text{mass percentage of sulfur}\\\text{in lead sulfide}\end{pmatrix} = \left(\frac{\text{mass of sulfur}}{\text{mass of lead sulfide}}\right) \times 100$$

$$= \left(\frac{0.380\text{ g}}{2.83\text{ g}}\right) \times 100 = 13.4\%$$

The law of constant composition assures us that the mass percentage of lead in lead sulfide is independent of the source of the lead sulfide. The principal natural source of lead sulfide is an ore known as *galena* (Figure 2-6).

Figure 2-6 Metal sulfides are valuable ores of metals. Shown here is galena, the principal ore of lead.

PRACTICE PROBLEM 2-1: A 5.650-g sample of a compound containing the elements potassium, nitrogen, and oxygen was found to contain 38.67% K and 13.86% N. Calculate the number of grams of each element in the sample. (Hint: Recall that the sum of the mass percentages of all the elements in a compound must total one hundred percent.)

Answer: 2.185 g of K, 0.7831 g of N, and 2.682 g of O

2-4. DALTON'S ATOMIC THEORY EXPLAINS THE LAW OF CONSTANT COMPOSITION

Figure 2-7 John Dalton (1766–1844). English schoolteacher, chemist, and physicist.

By the end of the eighteenth century, scientists had analyzed many compounds and had amassed a large amount of experimental data. But they lacked a theory that could bring all these data into a single framework. In 1803 John Dalton, an English schoolteacher, proposed an **atomic theory** (Figure 2-7). His theory provided a simple and beautiful explanation of both the law of constant composition and the law of conservation of mass. We can express the postulates of Dalton's atomic theory in modern terms as follows:

1. Matter is composed of small, indivisible particles called atoms.
2. The atoms of a given element all have the same mass and are identical in all respects, including chemical behavior.
3. The atoms of different elements differ in mass and in chemical behavior.
4. Chemical compounds are composed of two or more atoms of different elements joined together. The particle that results when two or more atoms join together is called a molecule. But the atoms in a molecule do not necessarily have to be different. If the atoms are the same, then the molecule is that of an element. If the atoms are different, then the molecule is that of a chemical compound.
5. In a chemical reaction, the atoms involved are rearranged, separated, or recombined to form different molecules. No atoms are created or destroyed, and the atoms themselves are not changed.

As we shall see, some of these postulates were later modified, but the main features of Dalton's atomic theory still are accepted today.

The law of conservation of mass in chemical reactions follows directly from Dalton's postulate that atoms are neither created nor destroyed in chemical reactions; rather, they are simply rearranged to form new substances. The law of constant composition follows from Dalton's postulate that atoms are indivisible and that compounds are formed by joining together different types of atoms. That is, compounds have constant composition because they contain fixed ratios of the different types of atoms. For example, suppose calcium sulfide is formed when calcium and sulfur are combined in a one-to-one ratio. In such a case, the ratio of calcium atoms to sulfur atoms is one-to-one, no matter how the sample is prepared. Although Dalton's application of his theory was complicated by several incorrect guesses about the relative numbers of atoms in compounds (Figure 2-8), these errors were eventually resolved. Meanwhile, the atomic theory gained wide acceptance and is now universally accepted.

Dalton's atomic theory enables us to set up a scale of relative atomic masses. Consider calcium sulfide, which we know consists of 55.6 percent calcium by mass and 44.4 percent sulfur by mass. Suppose there is one calcium atom for each sulfur atom in calcium sulfide. Because we know that the mass of a calcium atom relative to that of a sulfur atom must be the same as the mass percentages in calcium sulfide, we know that the ratio of the mass of a calcium atom to that of a sulfur atom is

$$\left(\frac{\text{mass of a calcium atom}}{\text{mass of a sulfur atom}}\right) = \left(\frac{55.6}{44.4}\right) = 1.25$$

or

$$(\text{mass of a calcium atom}) = 1.25 \times (\text{mass of a sulfur atom})$$

Thus, even though we cannot easily determine the mass of any individual atom, we can use the quantitative results of chemical analyses to determine the *relative* masses of atoms. Of course, we have based our result for calcium and sulfur on the assumption that there is one atom of calcium for each atom of sulfur in calcium sulfide.

Let's consider another compound, hydrogen chloride. Quantitative chemical analysis shows that the mass percentages of hydrogen and chlorine in hydrogen chloride are 2.76 percent and 97.24 percent, respectively. Once again, assuming (correctly, it turns out) that one atom of hydrogen is combined with one atom of chlorine, we find that

$$\left(\frac{\text{mass of a chlorine atom}}{\text{mass of a hydrogen atom}}\right) = \left(\frac{97.24}{2.76}\right) = 35.2$$

or

$$(\text{mass of a chlorine atom}) = 35.2 \times (\text{mass of a hydrogen atom})$$

By continuing in this manner with other compounds, it is possible to build up a table of relative atomic masses. We define a quantity called **atomic mass ratio** (usually referred to simply as the **atomic mass**), which is the ratio of the mass of a given atom to the mass of some particular reference atom. At one time the mass of hydrogen, the lightest atom, was arbitrarily given the value of exactly 1 and used as the reference in terms of which all other atomic masses were expressed. As discussed in Section 2-10, however, a form of carbon is now used as the standard. Thus, today the atomic mass of hydrogen is 1.008 instead of exactly 1. The presently accepted atomic masses of the elements are given on the inside of the front cover of the text.

Because "atomic masses" are actually ratios of masses, they have no units. Nevertheless, it is useful to assign to atomic masses a unit called the **atomic mass unit (amu).** Thus, for example, we can say that the atomic mass of carbon is 12.01 or 12.01 amu; both statements are correct. Although we will refer to atomic mass ratios for elements as simply atomic masses, it is important to recognize that atomic masses are actually relative, dimensionless quantities. The particular values assigned to atomic masses (but not their ratios) depend on the reference atom chosen to set up the scale.

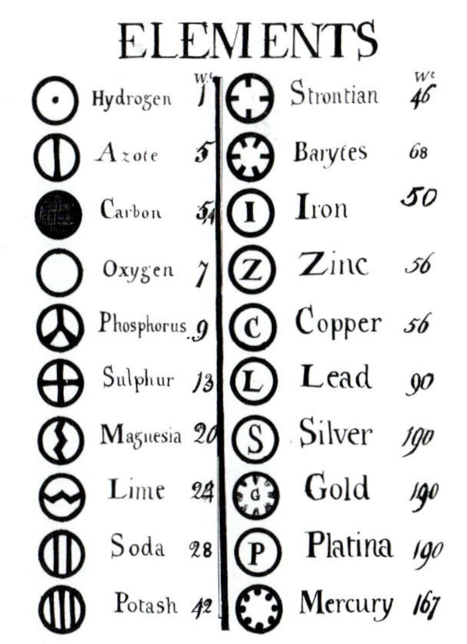

Figure 2-8 Dalton's symbols for chemical elements. Some of these "elements" are now known to be compounds, not elements (e.g., lime, soda, and potash). Some of Dalton's atomic masses were in error because of incorrect assumptions regarding the relative numbers of atoms in compounds.

EXAMPLE 2-2: Suppose the atomic mass of hydrogen were set at exactly 1. What would the atomic mass of carbon be?

Solution: Because the ratio of the masses of two atoms is independent of the value of the mass chosen for the reference element, we have in the present system

$$\frac{(\text{mass of C})}{(\text{mass of H})} = \frac{12.01}{1.008} = 11.91$$

Thus, for the revised (H = 1 exactly) system, we would have

$$\frac{(\text{mass of C})}{(\text{mass of H})} = \frac{11.91}{1}$$

and the atomic mass of carbon on the H = 1 scale would be 11.91 rather than 12.01.

PRACTICE PROBLEM 2-2: Prior to the adoption of the present carbon-based scale of atomic mass, the atomic mass of oxygen was set equal to exactly 16 and oxygen was used as the standard. Use the table of atomic masses given in Appendix E to calculate the atomic mass of carbon to five significant figures on the O = 16 scale.

Answer: 12.011. Difference is not significant to five significant figures.

2-5. MOLECULES ARE GROUPS OF ATOMS JOINED TOGETHER

Dalton postulated in the original version of his atomic theory that an element is a substance consisting of identical atoms and that a compound is a substance consisting of identical molecules. Although Dalton did not realize it at the time, some of the elements occur naturally as molecules containing more than one of the same kind of atoms. This distinction is accounted for in our statement of the fourth postulate of Dalton's atomic theory. As we already noted in Table 2-5, the elements hydrogen, nitrogen, oxygen, fluorine, chlorine, bromine, and iodine exist as diatomic molecules of the same kind of atoms (see Figure 2-4). Consequently, these substances are classified as elements. Compounds, on the other hand, are made up of molecules containing different kinds of atoms. Examples of the chemical formulas for molecules of chemical compounds are

These formulas indicate how the atoms are joined together in the molecules. We shall learn how to write such formulas in Chapter 11.

H—Cl
hydrogen chloride, HCl

water, H_2O

ammonia, NH_3

methyl alcohol, CH_3OH (wood alcohol)

methane, CH_3 (principal constituent of natural gas)

Scale models of these molecules are shown in Figure 2-9.

Dalton's atomic theory provides a nice pictorial view of chemical reactions. Recall that Dalton proposed that, in a chemical reaction, the atoms in the reactant molecules are separated and then rearranged into product molecules. According to this view, the chemical reaction between hydrogen and oxygen to form water involves the rearrangement:

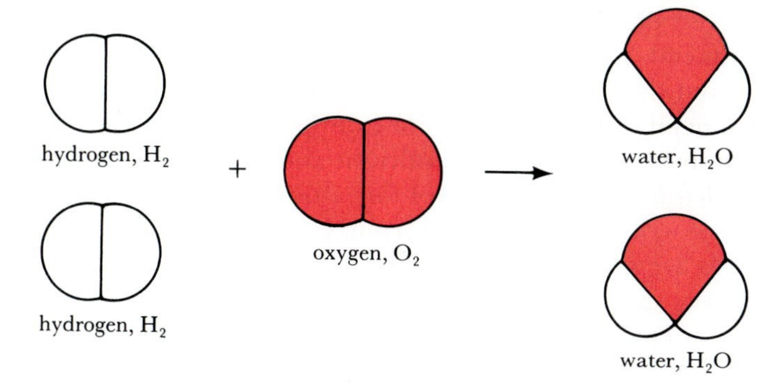

Note that completely different molecules, and hence completely different substances, are formed in a chemical reaction. Hydrogen and oxygen are gases, whereas water is a liquid at room temperature.

As another example, consider the burning of carbon in oxygen to form carbon dioxide:

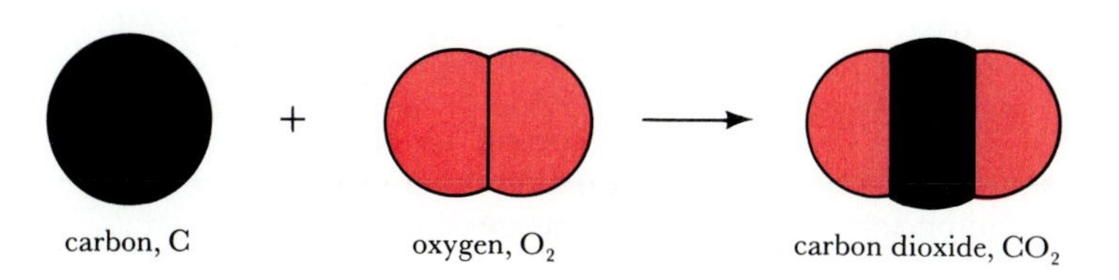

Once again, note that a completely new substance is formed. Carbon is a black solid; the product, carbon dioxide, is a colorless gas.

As a final example, consider the reaction between steam (hot gaseous water) and red-hot carbon to form hydrogen and carbon monoxide:

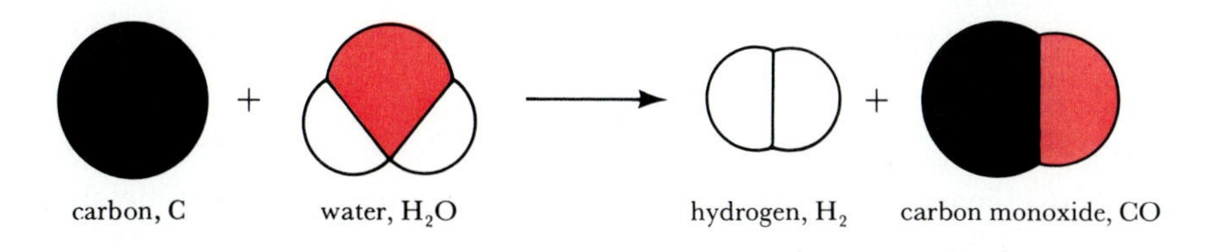

Notice that in each of the three reactions pictured above, the numbers of each kind of atoms do not change. Atoms are neither created nor destroyed in chemical reactions; they are simply rearranged into new molecules, in accordance with the conservation of atoms and of mass in chemical reactions.

2-6. COMPOUNDS ARE NAMED BY AN ORDERLY SYSTEM OF CHEMICAL NOMENCLATURE

The system for the assignment of names to compounds is called **chemical nomenclature.** In this chapter we discuss only the system of naming compounds consisting of two elements, that is, **binary compounds.**

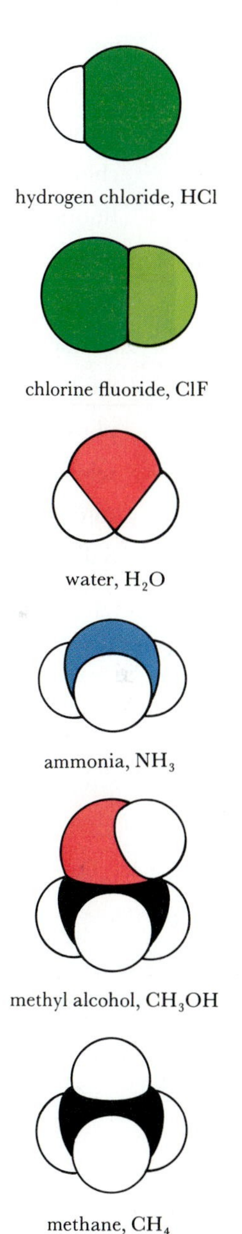

Figure 2-9 Scale models of molecules of hydrogen chloride, chlorine fluoride, water, ammonia, methyl alcohol, and methane.

When the two elements that make up a binary compound are a metal and a nonmetal, we name the compound by first writing the name of the metal and then that of the nonmetal, with the ending of the name of the nonmetal changed to *-ide*. For example, we saw that the name of the compound formed between calcium and sulfur is calcium sulfide. Because calcium sulfide consists of one atom of calcium for each atom of sulfur, we write the **chemical formula** of calcium sulfide as CaS; in other words, we simply join the chemical symbols of the two elements. In a different case, calcium combines with two atoms of chlorine to form calcium chloride; thus, the formula of calcium chloride is $CaCl_2$. Note that the number of atoms is indicated by a subscript. The subscript 2 in $CaCl_2$ means that there are two chlorine atoms per calcium atom in calcium chloride. Table 2-6 lists the *-ide* nomenclature for common nonmetals.

Table 2-6 The *-ide* nomenclature of the nonmetals

Element	*-ide* Nomenclature
arsenic	arsenide
bromine	bromide
carbon	carbide
chlorine	chloride
fluorine	fluoride
hydrogen	hydride
iodine	iodide
nitrogen	nitride
oxygen	oxide
phosphorus	phosphide
selenium	selenide
sulfur	sulfide
tellurium	telluride

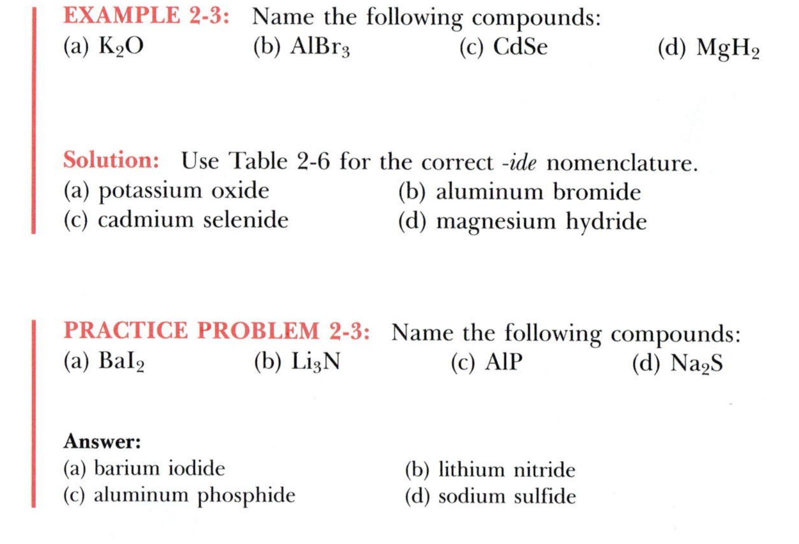

EXAMPLE 2-3: Name the following compounds:
(a) K_2O (b) $AlBr_3$ (c) CdSe (d) MgH_2

Solution: Use Table 2-6 for the correct *-ide* nomenclature.
(a) potassium oxide (b) aluminum bromide
(c) cadmium selenide (d) magnesium hydride

PRACTICE PROBLEM 2-3: Name the following compounds:
(a) BaI_2 (b) Li_3N (c) AlP (d) Na_2S

Answer:
(a) barium iodide (b) lithium nitride
(c) aluminum phosphide (d) sodium sulfide

Many binary compounds involve combinations of two nonmetals. Because more than one binary compound may result from the combination of the same two nonmetallic elements (e.g., CO and CO_2), we need to distinguish the various possibilities; we do so by means of Greek numerical prefixes (Table 2-7). For example,

CO carbon *mono*xide CO_2 carbon *di*oxide

Some other examples are

SO_2 sulfur *di*oxide SO_3 sulfur *tri*oxide

SF_4 sulfur *tetra*fluoride SF_6 sulfur *hexa*fluoride

PCl_3 phosphorus *tri*chloride PCl_5 phosphorus *penta*chloride

Ball and stick models of these compounds are shown in Figure 2-10.

Table 2-7 Greek prefixes used to indicate the number of atoms of a given type in a molecule

Number	Prefix*	Example
1	*mono-, mon-*	carbon monoxide, CO
2	*di-*	carbon dioxide, CO_2
3	*tri-*	sulfur trioxide, SO_3
4	*tetra-, tetr-*	carbon tetrachloride, CCl_4
5	*penta-, pent-*	phosphorus pentachloride, PCl_5
6	*hexa-*	sulfur hexafluoride, SF_6

*The final *a* or *o* is dropped from the prefix when it is combined with a name beginning with a vowel. For example, we say *pentachloride* but *pentoxide* and *monohydride* but *monoxide*.

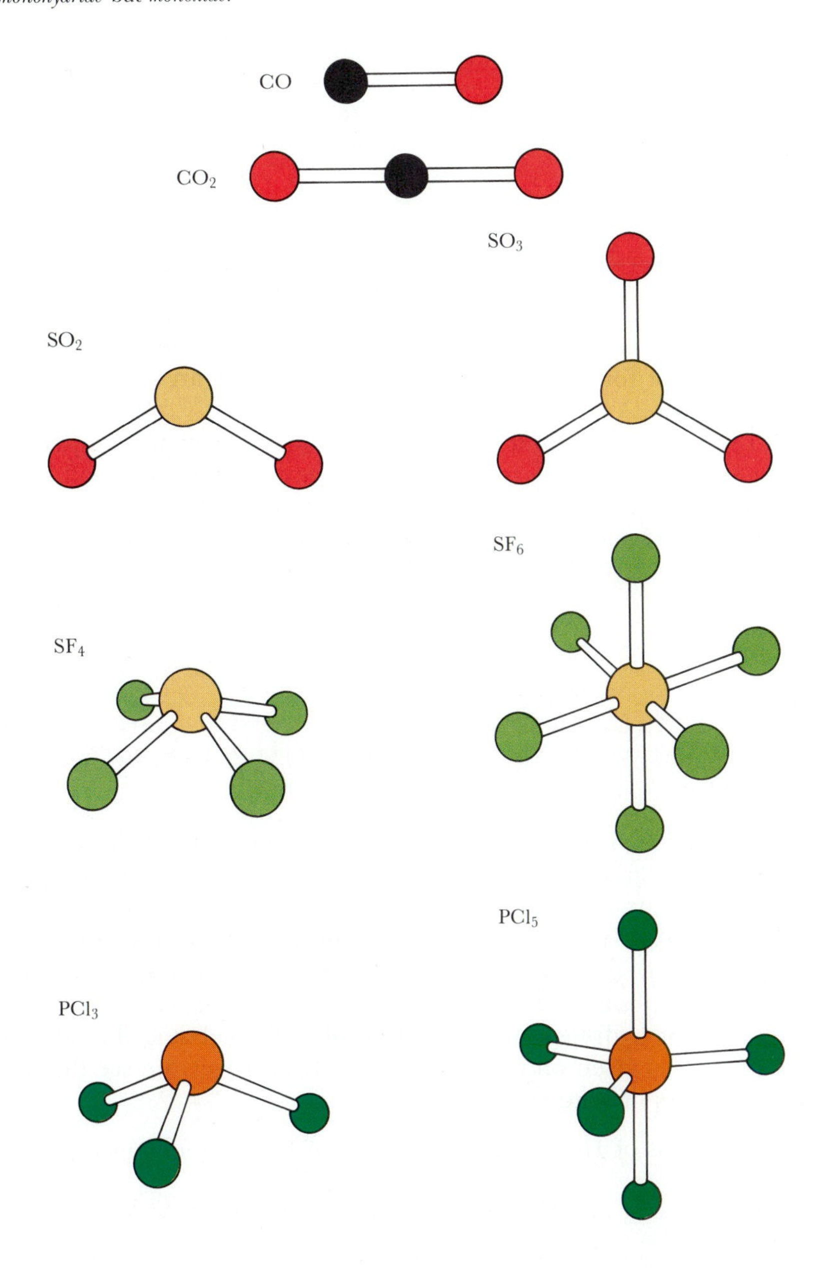

Figure 2-10 Structures of several binary nonmetallic compounds.

EXAMPLE 2-4: Name the following binary nonmetallic compounds:
(a) BrF_5 (b) XeF_4 (c) N_2O_4

Solution: Because these compounds involve two nonmetallic elements, we must denote the relative numbers of the two types of atoms in the name. (a) Bromine is written first in the formula; thus we name the compound bromine pentafluoride (the prefix mono- is often dropped, especially on the first named element); (b) xenon tetrafluoride; (c) dinitrogen tetroxide.

PRACTICE PROBLEM 2-4: Name the following compounds:
(a) N_2O (b) NO
(c) N_2O_3 (d) N_2O_5
(e) NO_2

Answer:
(a) dinitrogen oxide (nitrous oxide)
(b) nitrogen oxide (or nitrogen monoxide)
(c) dinitrogen trioxide
(d) dinitrogen pentoxide
(e) nitrogen dioxide

The compound N_2O (common name, nitrous oxide) was the first known general anesthetic (laughing gas) and is still used in dentistry. It is also used as a propellant for canned whipped cream and shaving cream. Except for N_2O_3, all of the nitrogen oxides are gases at room temperature.

At this point you should understand how to name binary compounds when you are given the formula. In the next chapter, we shall learn how to write a correct formula from the name. In Chapter 10 we will explain why elements are given in a particular order in the chemical formula (e.g., N_2O_5 rather than O_5N_2).

2-7. MOLECULAR MASS IS THE SUM OF THE ATOMIC MASSES OF THE ATOMS IN A MOLECULE

Now that we can distinguish different compounds by their chemical formulas, we can make our explanation of the law of constant composition still clearer and more powerful by introducing the idea of molecular or formula mass. The sum of the atomic masses of the atoms in a molecule or a compound is called the **molecular mass,** or **formula mass,** of the substance. For example, a water molecule, H_2O, consists of two atoms of hydrogen and one atom of oxygen. Using the table of atomic masses given on the inside of the front cover, we see that the formula mass of water is

$$\begin{aligned}(\text{formula mass of } H_2O) &= 2(\text{atomic mass of H}) + (\text{atomic mass of O}) \\ &= 2(1.008) + (16.00) \\ &= 18.02\end{aligned}$$

Using the table of atomic masses given on the inside of the front cover, we see that the formula mass of dinitrogen pentoxide, N_2O_5, is

$$\begin{aligned}\text{(formula mass of } N_2O_5) &= 2(\text{atomic mass of N}) + 5(\text{atomic mass of O}) \\ &= 2(14.01) + 5(16.00) \\ &= 108.02\end{aligned}$$

The following example shows how to use atomic and molecular (or formula) masses to calculate the mass percentage composition of compounds.

EXAMPLE 2-5: Using the fact that the atomic mass of lead is 207.2 and that of sulfur is 32.06, calculate the mass percentages of lead and sulfur in the compound lead sulfide, PbS.

Solution: As the formula PbS indicates, lead sulfide consists of one atom of lead for each atom of sulfur. The formula mass of lead sulfide is

$$\begin{aligned}\begin{pmatrix}\text{formula mass of}\\ \text{lead sulfide}\end{pmatrix} &= \begin{pmatrix}\text{atomic mass}\\ \text{of lead}\end{pmatrix} + \begin{pmatrix}\text{atomic mass}\\ \text{of sulfur}\end{pmatrix} \\ &= 207.2 + 32.06 \\ &= 239.3\end{aligned}$$

The mass percentages of lead and sulfur in lead sulfide are

$$\begin{aligned}\text{(mass percentage of lead)} &= \left(\frac{\text{atomic mass of lead}}{\text{molecular mass of lead sulfide}}\right) \times 100 \\ &= \left(\frac{207.2}{239.3}\right) \times 100 = 86.59\%\end{aligned}$$

$$\begin{aligned}\text{(mass percentage of sulfur)} &= \left(\frac{\text{atomic mass of sulfur}}{\text{molecular mass of lead sulfide}}\right) \times 100 \\ &= \left(\frac{32.06}{239.3}\right) \times 100 = 13.40\%\end{aligned}$$

Note that this result is the same as that calculated in Example 2-1. The table of atomic masses must be consistent with experimental values of mass percentages. The two mass percentages in this example do not add up to exactly 100.00 percent because of a slight round-off error.

PRACTICE PROBLEM 2-5: Calculate the mass percentages of bromine and fluorine in BrF_5.

Answer: 45.68 percent Br and 54.32 percent F

One of the great advantages of Dalton's atomic theory was that he was able to use it to devise a table of atomic masses that could then be used in chemical calculations like those in Example 2-5. What Dalton did not know, however, is that not all atoms of a given element have the same atomic mass. This discovery, which was made in the twentieth century, required a new model of the atom.

Figure 2-11 Sir Joseph John Thomson (1856–1940). English physicist, awarded the Nobel prize in physics in 1906.

2-8. MOST OF THE MASS OF AN ATOM IS CONCENTRATED IN ITS NUCLEUS

For most of the nineteenth century, atoms were considered to be indivisible, stable particles, as proposed by Dalton. Toward the end of the century, however, new experiments indicated that the atom is composed of even smaller **subatomic particles.**

One of the first experiments on subatomic particles was carried out by the English physicist J. J. Thomson in 1897 (Figure 2-11). Some years earlier, it had been discovered that an electric discharge (glowing current) flows between metallic electrodes that are sealed in a partially evacuated glass tube, as shown in Figures 2-12 and 2-13. Using an apparatus of the type depicted in Figure 2-12, Thomson deflected the electric discharge with electric and magnetic fields and showed that it was actually a stream of identical, negatively charged particles and that the mass of each particle was only 1/1837 that of a hydrogen atom. Because the hydrogen atom is the lightest atom, he correctly reasoned that these particles, which are now called **electrons,** are constituents of atoms. The electron was the first subatomic particle to be discovered.

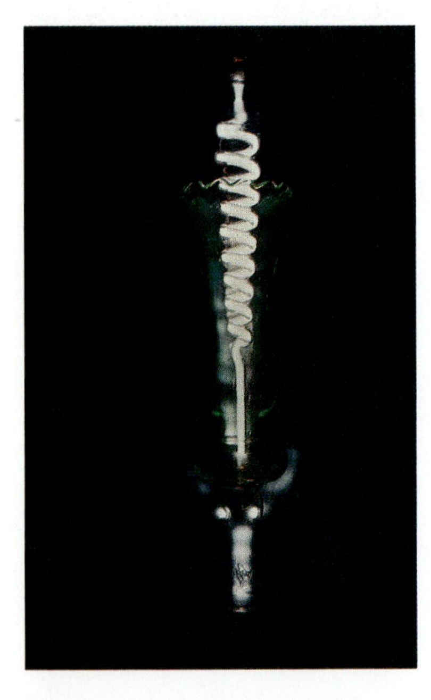

Figure 2-12 A discharge tube apparatus like the one Thomson used to discover the electron.

If an atom contains electrons, which are negatively charged particles, then it also must contain positively charged particles, because atoms are electrically neutral. The total amount of negative charge in a neutral atom must be balanced by an equal amount of positive charge. The question is, How are the positively charged particles and electrons arranged within an atom? The first person to answer this question was the New Zealand–born physicist Ernest Rutherford (Figure 2-14).

Before we explain Rutherford's experiment, we must first mention another discovery of the late 1890s—radioactivity. About the same time

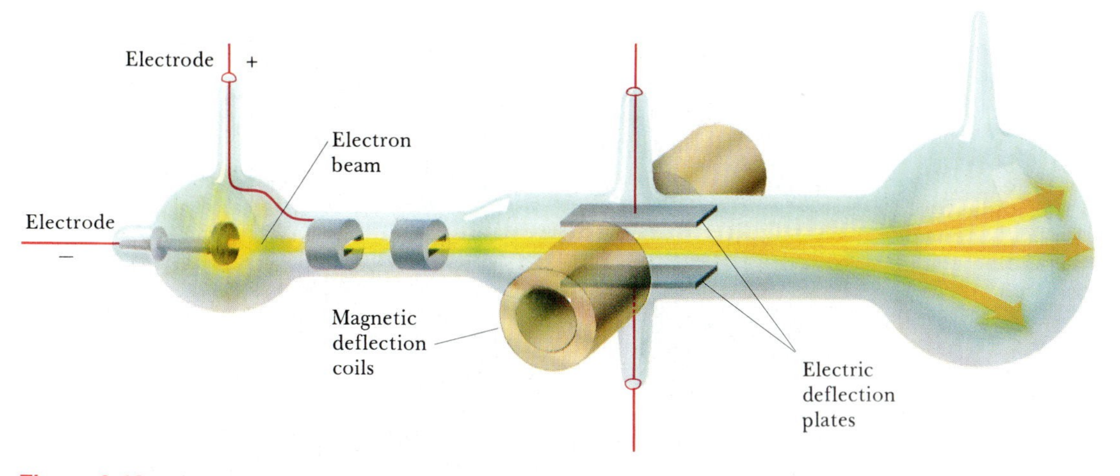

Figure 2-13 Schematic of a discharge tube. When a voltage is applied across electrodes that are sealed in a partially evacuated glass tube, the space between the electrodes glows.

that Thomson discovered the electron, the French scientist Antoine-Henri Becquerel discovered **radioactivity,** the process by which certain atoms spontaneously break apart. Becquerel showed that uranium atoms are **radioactive.** Shortly after Becquerel's discovery, Marie and Pierre Curie, working in Paris, discovered other radioactive elements such as radium (so named because it emits rays) and polonium (named for Poland, Marie Curie's native country). Other scientists discovered that the radiation emitted by radioactive substances consists of three types, which are now called ***α*-particles** (alpha-particles), ***β*-particles** (beta-particles), and ***γ*-rays** (gamma-rays). Experiments by a number of researchers showed that α-particles have a charge equal in magnitude to that of two electrons but of opposite (i.e., positive) sign and a mass equal to the mass of a helium atom (4.00 amu); β-particles are simply electrons that result from radioactive disintegrations; and γ-rays are very similar to X-rays; X-rays constitute a very penetrating form of radiation that was known to be emitted by certain metals at high temperatures. Table 2-8 summarizes the properties of these three common emissions from radioactive substances.

Figure 2-14 Sir Ernest Rutherford (1871–1937), awarded the Nobel prize in chemistry in 1908.

Table 2-8 Properties of the three radioactive emissions discovered by Rutherford

Original name	Modern name	Mass*	Charge†
α-ray	α-particle	4.00	+2
β-ray	β-particle (electron)	5.49×10^{-4}	−1
γ-ray	γ-ray	0	0

*In atomic mass units.
†Relative to the charge on an electron, which is defined as −1. The actual charge on an electron is -1.602×10^{-19} coulomb. The charges given here are, in effect, in units of the charge on the electron.

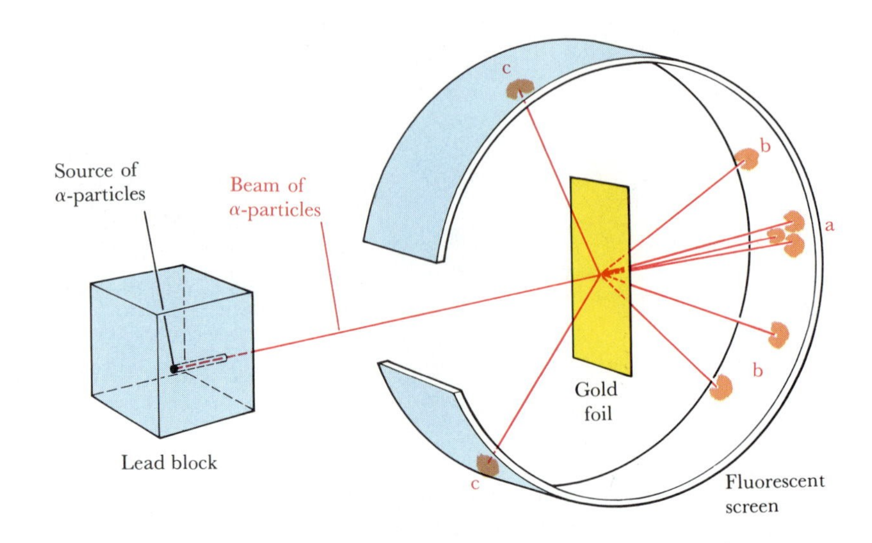

Figure 2-15 In 1911, Rutherford and Marsden set up an experiment in which a thin gold foil was bombarded with α-particles. Most of the particles passed through the foil (pathway a). Some were deflected only slightly (pathway b) when they passed near a gold nucleus in the foil, and a few were deflected backward (pathway c) when they collided with a nucleus.

Rutherford became intrigued with the idea of using α-particles as subatomic projectiles. In a now-famous experiment, Ernest Marsden, one of Rutherford's students, took a piece of gold and rolled it into an extremely thin foil (gold is very malleable). He then directed a beam of α-particles at the gold foil and observed the paths of the α-particles by watching for flashes as they struck a fluorescent screen surrounding the foil (Figure 2-15). Contrary to expectations, most of the particles passed straight through the foil, but a few were deflected through large angles (pathway c in Figure 2-15). Rutherford interpreted this unexpected result to mean that an atom is mostly empty space and that all the positive charge and essentially all the mass of an atom are concentrated in a very small volume in the center of the atom, which he called the **nucleus** (Figure 2-16). Most α-particles passed through the gold foil; the deflection of a few through large angles was the result of collisions of positively charged α-particles with positively charged gold nuclei.

Those α-particles that were deflected through intermediate angles (pathway b in Figure 2-15) had passed near gold nuclei and were repelled. Because α-particles are positively charged, we would expect them to be repelled by a positively charged nucleus, because like charges repel each other. By counting the number of α-particles deflected in various directions, Rutherford was able to show that the diameter of a nucleus is about 1/100,000 times the diameter of an atom.

Rutherford subsequently discovered that the positive charge of an atom is due to particles that he named **protons.** These subatomic particles have a positive charge equal in magnitude to that of an electron but opposite in sign. The mass of a proton is almost the same as the mass of a hydrogen atom, about 1836 times the mass of an electron.

The size of the nucleus relative to the size of the whole atom can be grasped from the following analogy. If an atom could be enlarged so that its nucleus were the size of a pea, then the entire atom would be about the size of the Metrodome in Minneapolis. The electrons in an atom are located throughout the space surrounding the nucleus (Figure 2-16). Just how the electrons are arranged in an atom is taken up in later chapters.

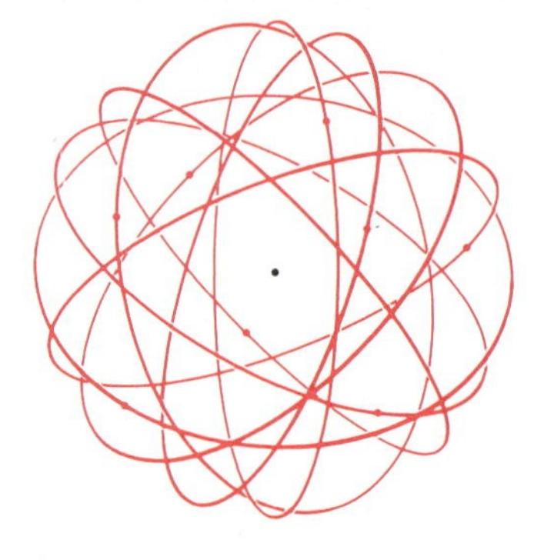

Figure 2-16 Nuclear model of the atom. The nucleus is very small and located at the center. The electrons are located in the space around the nucleus. In fact, electrons do *not* travel around a nucleus in well-defined orbits, as depicted here.

2-9. ATOMS CONSIST OF PROTONS, NEUTRONS, AND ELECTRONS

Our picture of the atom is not yet complete. Experiments suggested that the mass of a nucleus cannot be attributed to the protons alone. Scientists hypothesized in the 1920s, and in 1932 James Chadwick verified, that there is another type of particle in the nucleus. This particle has a slightly greater mass than a proton and is called a **neutron,** because it is electrically neutral.

The modern picture of an atom, then, consists of three types of particles—electrons, protons, and neutrons. The properties of these three subatomic particles are

Particle	Charge*	Mass/amu	Where located
proton	+1	1.0073	in nucleus
neutron	0	1.0087	in nucleus
electron	−1	5.49×10^{-4}	outside nucleus

*Relative to the charge on a proton. The actual charge on a proton is 1.602×10^{-19} coulomb.

The number of protons in an atom is called the **atomic number** of that atom and is denoted by *Z*. In a **neutral atom,** the number of electrons is equal to the number of protons. The differences between atoms are a result of the different atomic numbers, and each element is characterized by a unique atomic number. In other words, no two elements have the same atomic number. For example, hydrogen has an atomic number of 1 (1 proton in the nucleus), helium has an atomic number of 2 (2 protons in the nucleus), and uranium has an atomic number of 92 (92 protons in the nucleus). The table of the elements given on the inside of the front cover of this book lists the atomic numbers of all the known elements. The total number of protons and neutrons in an atom is called the **mass number** of that atom and is denoted by *A*.

EXAMPLE 2-6: Use the mass data given in the table above to calculate the percentage of the mass of a hydrogen atom that is located in the nucleus. Assume that the hydrogen nucleus consists of a single proton.

Solution: The mass percentage in the nucleus of a hydrogen atom is given by the ratio of the mass of a proton to the mass of a proton plus the mass of an electron times 100 to convert the result to a percentage

$$\left(\frac{1.0073}{1.0073 + 0.000549}\right) \times 100 = 99.946\%$$

Note that a hydrogen atom has a lower mass percentage in its nucleus than any other atom.

PRACTICE PROBLEM 2-6: Given that the diameter of a nucleus is roughly 1×10^{-5} times the diameter of an atom, calculate the percentage by volume of an atom that is occupied by the nucleus. (Hint: Recall that the volume of a sphere is given by $V = 4\pi r^3/3$).

Answer: $1 \times 10^{-13}\%$

2-10. MOST ELEMENTS OCCUR IN NATURE AS MIXTURES OF ISOTOPES

Nuclei are made up of protons and neutrons, each of which has a mass of approximately 1 amu; therefore, you might expect atomic masses to be approximately equal to whole numbers. Although many atomic masses are approximately whole numbers (for example, the atomic mass of carbon is 12.0 and the atomic mass of fluorine is 19.0), many others are not. Chlorine ($Z = 17$) has an atomic mass of 35.45, magnesium ($Z = 12$) has an atomic mass of 24.31, and copper ($Z = 29$) has an atomic mass of 63.55. The explanation for these variations lies in the fact that many elements consist of two or more **isotopes,** which are atoms of one element that contain the same number of protons but different numbers of neutrons. Recall that it is the number of *protons* (the atomic number) that characterizes a particular element, but nuclei of the same element may have different numbers of neutrons. For example, the most common isotope of the simplest element, hydrogen, contains one proton and one electron, but no neutrons. Another less common isotope of hydrogen contains one proton, one neutron, and one electron. These two hydrogen isotopes both undergo the same chemical reactions. The heavier isotope is called heavy hydrogen—or, more commonly, **deuterium**—and is often denoted by the special symbol D. Water that is made from deuterium is called **heavy water** and is usually denoted by D_2O. Heavy water is produced by distillation. Its major use is in heavy-water nuclear reactors, like those in Canadian nuclear power plants (Figure 2-17).

Figure 2-17 Ontario Hydro's Bruce Nuclear Power Development heavy-water plant near Twerton on the shores of Lake Huron.

An isotope is specified by its atomic number and its mass number. The notation used to designate isotopes is the chemical symbol of the element written with its atomic number as a left subscript and its mass number as a left superscript:

$$\text{mass number} \rightarrow {}^{A}_{Z}\text{X} \leftarrow \text{chemical symbol}$$
$$\text{atomic number} \rightarrow Z$$

For example, an ordinary hydrogen atom is denoted ${}^{1}_{1}\text{H}$ and a deuterium atom is denoted ${}^{2}_{1}\text{H}$. The number of neutrons, N, in an atom is equal to

$$N = A - Z \qquad (2\text{-}1)$$

EXAMPLE 2-7: Fill in the blanks.

Symbol	Atomic number	Number of neutrons	Mass number
(a) ____	22	____	48
(b) ____	____	110	184
(c) ${}^{?}_{?}\text{Co}$	____	____	60

Solution:
(a) The number of neutrons is the mass number minus the atomic number (Equation 2-1), or $48 - 22 = 26$ neutrons. The element with atomic number 22 is titanium (see the inside of the front cover), so the symbol of this isotope is ${}^{48}_{22}\text{Ti}$. It is called titanium-48.
(b) The number of protons is the mass number minus the number of neutrons, or $184 - 110 = 74$ protons. The element with atomic number 74 is tungsten, so the symbol is ${}^{184}_{74}\text{W}$.
(c) According to the symbol, the element is cobalt, whose atomic number is 27. The symbol of the particular isotope is ${}^{60}_{27}\text{Co}$, and the isotope has $60 - 27 = 33$ neutrons. Cobalt-60 is used as a γ-radiation source for the treatment of cancer (radiation therapy).

PRACTICE PROBLEM 2-7: The radioactive phosphorus isotope phosphorus-32 is used extensively in biochemistry and medicine to monitor chemical reactions. Give the number of protons, neutrons, and electrons in a neutral ${}^{32}\text{P}$ atom.

Answer: 15, 17, 15, respectively

Although one of the postulates of Dalton's atomic theory was that all the atoms of a given element have the same mass, we now see that this is not usually so. Isotopes of the same element have different masses, but all atoms of a given isotope have the same mass. Several common natural isotopes and their corresponding masses are given in Table 2-9. Note that the isotopic mass of carbon-12 is exactly 12. The modern atomic mass scale is based on this convention. All atomic masses are given relative to the mass of the carbon-12 isotope, which is defined by international convention to be *exactly* 12.

Table 2-9 also shows two isotopes for helium: helium-3 and helium-4. The atomic number of helium is 2; in other words, a helium nucleus has two protons and a nuclear charge of +2. A helium-4 nucleus has a charge of +2 and an atomic mass of 4, the same as an alpha-particle (Table 2-8). In fact, we now know that an α-particle is simply the nucleus of a helium-4 isotope.

As Table 2-9 implies, many elements occur in nature as mixtures of isotopes. The naturally occurring percentages of the isotopes of a particular element are referred to as the **natural abundances** of that element. Naturally occurring chlorine consists of two isotopes: 75.77 percent chlorine-35 and 24.23 percent chlorine-37. These proportions are nearly independent of the natural source of the chlorine. In other words, chlorine obtained from, say, salt deposits in Africa or Australia or North America has nearly the same isotopic composition as that given in Table 2-9. The atomic mass of chlorine is the sum of the masses of each isotope, each multiplied by its natural abundance mass fraction. When we use the isotopic masses and natural abundances of chlorine given in Table 2-9, we obtain

$$\begin{aligned}\text{(atomic mass of chlorine)} &= (34.97)\left(\frac{75.77}{100}\right) + (36.97)\left(\frac{24.23}{100}\right) \\ &= 35.45\end{aligned}$$

Table 2-9 Naturally occurring isotopes of some common elements*

Element	Isotope	Natural abundance/%	Isotopic mass/amu	Protons	Neutrons	Mass number
hydrogen	$^{1}_{1}H$	99.985	1.0078	1	0	1
	$^{2}_{1}H$	0.015	2.0141	1	1	2
	$^{3}_{1}H$	trace	3.0160	1	2	3
helium	$^{3}_{2}He$	1.4×10^{-4}	3.0160	2	1	3
	$^{4}_{2}He$	99.99986	4.0026	2	2	4
carbon	$^{12}_{6}C$	98.89	12.0000	6	6	12
	$^{13}_{6}C$	1.11	13.0034	6	7	13
	$^{14}_{6}C$	trace	14.0032	6	8	14
oxygen	$^{16}_{8}O$	99.758	15.9949	8	8	16
	$^{17}_{8}O$	0.038	16.9991	8	9	17
	$^{18}_{8}O$	0.204	17.9992	8	10	18
fluorine	$^{19}_{9}F$	100.0	18.9984	9	10	19
magnesium	$^{24}_{12}Mg$	78.99	23.9850	12	12	24
	$^{25}_{12}Mg$	10.00	24.9858	12	13	25
	$^{26}_{12}Mg$	11.01	25.9826	12	14	26
chlorine	$^{35}_{17}Cl$	75.77	34.9689	17	18	35
	$^{37}_{17}Cl$	24.23	36.9659	17	20	37

*Data like these are available for all the naturally occurring elements.

This value is the atomic mass of chlorine given on the inside of the front cover of the book. The factors (75.77/100) and (24.23/100) must be included in order to take into account the relative natural abundance of each isotope. Thus, tables of atomic masses of the elements contain the relative average masses of atoms. The average mass is related to the masses of the individual isotopes of the element in the manner just illustrated for chlorine.

EXAMPLE 2-8: Naturally occurring chromium is a mixture of four isotopes with the following isotopic masses and natural abundances:

Mass number	Isotopic mass	Natural abundance/%
50	49.946	4.35
52	51.941	83.79
53	52.941	9.50
54	53.939	2.36

Calculate the atomic mass of chromium.

Solution: The atomic mass is the sum of the masses of the four isotopes each weighted by their respective abundances:

$$\begin{pmatrix}\text{atomic mass}\\ \text{of naturally}\\ \text{occurring chromium}\end{pmatrix} = (49.946)\left(\frac{4.35}{100}\right) + (51.941)\left(\frac{83.79}{100}\right) + (52.941)\left(\frac{9.50}{100}\right) + (53.939)\left(\frac{2.36}{100}\right) = 52.00$$

PRACTICE PROBLEM 2-8: Naturally occurring lithium is composed of two isotopes, lithium-6 (6.0169 amu) and lithium-7 (7.0182 amu). Given that the atomic mass of lithium is 6.941 amu, compute the percentage natural abundances of ^{6}Li and ^{7}Li. (Hint: If we denote the percentage of ^{7}Li in naturally occurring lithium by x, then the percentage of ^{6}Li is $100 - x$, and the calculation proceeds like that in Example 2-8, except that now we know the atomic mass of naturally occurring lithium and we seek the mass percentages of the two isotopes.)

Answer: 7.71 percent ^{6}Li and 92.29 percent ^{7}Li

Small variations in natural abundances of the isotopic compositions of the elements limit the precision with which atomic masses can be specified. The masses of individual isotopes are known much more accurately than atomic masses given in the periodic table.

2-11. ISOTOPIC MASSES ARE DETERMINED WITH A MASS SPECTROMETER

The mass and percentage of each isotope of an element can be determined by using a **mass spectrometer.** When a gas consisting of atoms is bombarded with electrons from an external source, the bombarding electrons knock other electrons out of the neutral atoms in the gas, producing positively charged, atomic-sized particles called **ions.** Atomic ions are atoms that have either a deficiency of electrons (in which case the ions are positively charged and are called **cations**) or extra electrons (in which case the ions are negatively charged and are called **anions**). The positively charged gas ions produced by electron bombardment in the mass spectrometer are accelerated by an electric field and pass through slits to form a narrow, well-focused beam (Figures 2-18 and 2-19). The beam of ions then passes through a magnetic field, which deflects each ion by an amount proportional to its mass. Thus, the original beam of ions splits into several separate beams, one for each isotope of the gas. The intensities of the separated ion beams, which can be determined experimentally, are a direct measure of the number of ions in each beam. In this manner, we can determine not only the mass of each isotope of any element (by the amount of deflection of each beam) but also the percentage of each isotope (by the intensity of each beam). Also note that the mass spectrometer separates the mixture of isotopes into isotopically pure components; therefore, we can use it to prepare isotopically pure samples of elements. We also can study in a mass spectrometer elements that are solids or liquids by heating them to a temperature high enough to vaporize them.

We will encounter ions throughout our study of chemistry, so we introduce a notation for them here. An atom that has lost one electron has a net charge of +1; one that has lost two electrons has a charge of

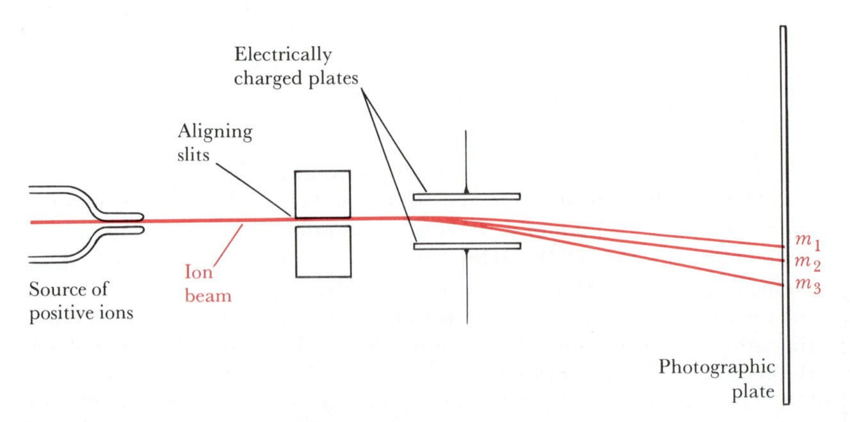

Figure 2-18 Schematic diagram of a mass spectrometer. A gas is bombarded with electrons that knock other electrons out of the gas atoms or molecules and produce positively charged ions. These ions are then accelerated by an electric field and form a narrow beam as they pass through two aligning slits. The beam then passes through a magnetic field. The ions of different masses (m) are deflected to different extents, and the beam is split into several separate beams. The particles of different masses strike a detector, such as a photographic plate, at different places, and the amount of exposure at various places is proportional to the number of particles having a particular mass. Here, $m_1 > m_2 > m_3$.

Figure 2-19 A mass spectrometer like the one shown here can be used to separate the isotopes of an element.

+2; one that has gained an electron has a charge of −1; and so on. We denote an ion by the chemical symbol of the element with a right-hand superscript to indicate its charge:

K^+	singly charged potassium ion
Mg^{2+}	doubly charged magnesium ion
Cl^-	singly charged chloride ion
S^{2-}	doubly charged sulfide ion

The K^+ ion has 18 electrons ($19 - 1 = 18$), and the Cl^- ion also has 18 electrons ($17 + 1 = 18$). Species that contain the same number of electrons are said to be **isoelectronic.** Note from these examples that the names of anions have the *-ide* ending characteristic of the second-named element in binary compounds.

EXAMPLE 2-9: How many electrons are there in the Mg^{2+} cation and the S^{2-} anion?

Solution: From the table on the inside of the front cover, we see that the atomic number of magnesium is 12. The Mg^{2+} ion is a magnesium atom that has lost two electrons, so Mg^{2+} has 10 electrons. The atomic number of sulfur is 16. The S^{2-} ion has two more electrons than a sulfur atom, so S^{2-} has 18 electrons.

PRACTICE PROBLEM 2-9: Give an example of a cation that is isoelectronic with the oxide ion, O^{2-}.

Answer: Na^+ or Mg^{2+} or Al^{3+}

In this chapter we have discussed atoms, molecules, elements, and compounds within the framework of Dalton's atomic theory. Before going on to discuss the chemical properties of the elements in Chapter 3, we shall discuss some of the many techniques that chemists use to separate mixtures of compounds into pure substances.

2-12. HETEROGENEOUS MIXTURES ARE NOT UNIFORM IN COMPOSITION FROM POINT TO POINT

When determining the physical properties (e.g., density, melting point, or molecular mass) or the chemical properties (e.g., the ability to react chemically with other substances) of an element or a compound, chemists must be certain that the substance is reasonably pure, otherwise the

(a) The components of the mixture cannot be determined by casual inspection.

(b) The pure, separated components of the mixture.

(c) A microscopic view of the mixture. Note that the mixture is heterogeneous (i.e., not uniform from point to point) and that each of the four components is clearly distinguishable.

(d) A magnet can be used to separate iron filings from the mixture. The iron filings are attracted by the magnet, but the other three components are not.

Figure 2-20 A mixture of sugar, sand, iron filings, and gold dust.

results have very limited applicability in that they are restricted to the *particular* impure sample that was studied. Most substances in nature are mixtures of compounds and/or elements in which the constituent substances exist together without combining chemically. Thus, it is often necessary for chemists to separate the various pure components from a mixture. As we have noted, a mass spectrometer is one example of a device that is used to separate the constituents of a mixture, namely, a mixture of isotopes. But chemists also use other techniques to separate the components of mixtures and thereby obtain pure substances.

Let's consider the problem of separating a mixture composed of sugar, sand, iron filings, and gold dust into its four components (Figure 2-20). The first thing to recognize about the mixture is that it is **heterogeneous;** that is, it is not uniform from point to point. The heterogeneity of the mixture can be seen clearly with the aid of a microscope (Figure 2-20c). We could separate the four components of the mixture by using a tweezers, a microscope, and a lot of time and patience; however, we can achieve much more rapid separations with other methods. We can separate the iron filings from the mixture by using a magnet (Figure 2-20d), which attracts the magnetic iron particles but has no effect on the other three components. The same technique is used on a much larger scale in waste recycling to separate ferrous metals (iron, steel, nickel) from nonferrous refuse (e.g., aluminum, glass, paper, and plastics).

After the iron has been removed from our mixture, the sugar can be separated from the remaining components by adding water. We call this process **dissolution.** Only the sugar dissolves in the water to form a solution of sugar in water. A **solution** is a **homogeneous** (i.e., uniform from point to point) mixture of two or more components. The formation of the sugar-water solution leaves the sand and gold particles at the bottom of the container. The heterogeneous mixture of gold, sand, and sugar-water solution can then be separated by **filtration** (Figure 2-21). The sugar-water solution passes readily through the small pores in the filter paper, but the solid particles are too large to pass through and are trapped on the paper. The sugar can be recovered from the sugar-water solution by evaporating the water, a process that leaves the recrystallized sugar in the container. Salts are separated from seawater and brines on a commercial scale by the **evaporation** of the water from the brines (Figure 2-22).

The sand and gold dust can be separated by panning or by sluice-box techniques, which rely on the differences in density of the two solids to achieve a separation. In simple panning, water is added to the mixture of sand and gold and the slurry is swirled in a shallow, saucer-shaped metal pan. The dense (19.3 $g{\cdot}cm^{-3}$) gold particles collect near the center of the pan, whereas the less dense sand particles (2 to 3 $g{\cdot}cm^{-3}$) swirl out of the pan. In the sluice-box technique, running water is passed over an agitated sand–gold mixture; the less dense sand particles, which rise higher in the water than the gold, are swept away in the stream of water.

When fine gold particles are firmly attached to sand particles, the gold can be separated by shaking the mixture with liquid mercury, in which the gold dissolves. The sand, which floats on the mercury, is removed. The solution of gold in mercury is then separated by **distilla-**

Figure 2-21 Filtration can be used to separate a liquid from a solid. The liquid passes through filter paper, but the solid particles are too large to do so. Filter paper is available in a wide range of pore sizes, down to pores small enough (2.5×10^{-8} m) to remove bacteria (the smallest bacteria are about 1×10^{-7} m in diameter). Micropore filters can be used in place of pasteurization to produce bacteria-free liquids such as canned draft beer and bottled water.

2-3. Name the elements with the following chemical symbols:

(a) Ge (b) Sc (c) Ir
(d) Cs (e) Sr (f) Am
(g) Mo (h) In
(i) Pu (j) Xe

2-4. Name the elements with the following chemical symbols:

(a) Pt (b) Te (c) Pb
(d) Ta (e) Ba (f) Ti
(g) Re (h) La
(i) Eu (j) Pr

MASS PERCENTAGES IN COMPOUNDS

2-5. A 1.659-g sample of a compound of sodium and oxygen contains 0.978 g of sodium and 0.681 g of oxygen. Calculate the mass percentages of sodium and oxygen in the compound.

2-6. The compound lanthanum oxide is used in the production of optical glass and the fluorescent phosphors used to coat television screens. An 8.29-g sample is found to contain 7.08 g of lanthanum and 1.21 g of oxygen. Calculate the mass percentages of lanthanum and oxygen in lanthanum oxide.

2-7. A 1.28-g sample of copper is heated with sulfur to produce 1.60 g of a copper sulfide compound. Calculate the mass percentages of copper and sulfur in the compound.

2-8. Stannous fluoride, which is an active ingredient in some toothpastes and helps to prevent cavities, contains tin and fluorine. A 1.793-g sample was found to contain 1.358 g of tin. Calculate the mass percentages of tin and fluorine in stannous fluoride.

2-9. Potassium cyanide is used in extracting gold and silver from their ores. A 12.63-mg sample is found to contain 7.58 mg of potassium, 2.33 mg of carbon, and 2.72 mg of nitrogen. Calculate the mass percentages of potassium, carbon, and nitrogen in potassium cyanide.

2-10. Ethyl alcohol, the alcohol in alcoholic beverages, is a compound of carbon, hydrogen, and oxygen. A 3.70-g sample of ethyl alcohol contains 1.93 g of carbon and 0.49 g of hydrogen. Calculate the mass percentages of carbon, hydrogen, and oxygen in ethyl alcohol.

NOMENCLATURE

2-11. Name the following binary compounds:

(a) Li_2S (b) BaO
(c) Mg_3P_2 (d) $CsBr$

2-12. Name the following binary compounds:

(a) BaF_2 (b) Mg_3N_2
(c) $CsCl$ (d) CaS

2-13. Name the following binary compounds:

(a) SiC (b) GaP
(c) Al_2O_3 (d) $BeCl_2$

2-14. Name the following binary compounds:

(a) MgF_2 (b) AlN
(c) $MgSe$ (d) Li_3P

2-15. Name the following pairs of compounds:

(a) ClF_3 and ClF_5 (b) SF_4 and SF_6
(c) KrF_2 and KrF_4 (d) BrO and BrO_2

2-16. Name the following pairs of compounds:

(a) $SbCl_3$ and $SbCl_5$ (b) ICl_3 and ICl_5
(c) SeO_2 and SeO_3 (d) CS and CS_2

MOLECULAR MASSES

2-17. Calculate the molecular mass for each of the following oxides:

(a) TiO_2 (white pigment)
(b) Fe_2O_3 (rust)
(c) V_2O_5 (a catalyst)
(d) P_4O_{10} (dehydrating agent)

2-18. Calculate the molecular mass for each of the following ores:

(a) $CaWO_4$ (scheelite, an ore of tungsten)
(b) Fe_3O_4 (magnetite)
(c) Na_3AlF_6 (cryolite)
(d) $Be_3Al_2Si_6O_{18}$ (beryl)
(e) Zn_2SiO_4 (willemite)

2-19. Calculate the molecular mass for each of the following halogen compounds:

(a) BrN_3 (explosive)
(b) $NaIO_3$ (antiseptic)
(c) CCl_2F_2 (refrigerant)
(d) $C_{14}H_9Cl_6$ (DDT)

2-20. Calculate the molecular mass for each of the following vitamins:

(a) $C_{20}H_{30}O$ (vitamin A)
(b) $C_{12}H_{17}ClN_4OS$ (vitamin B_1, thiamine)
(c) $C_{17}H_{20}N_4O_6$ (vitamin B_2, riboflavin)
(d) $C_{56}H_{88}O_2$ (vitamin D_1)
(e) $C_6H_8O_6$ (vitamin C, ascorbic acid)

MASS PERCENTAGES AND ATOMIC MASSES

2-21. Use the atomic masses given on the inside of the front cover of the text to calculate the mass percentages of chlorine and fluorine in chlorine trifluoride.

2-22. Use the atomic masses given on the inside of the front cover of the text to calculate the mass percentages of nitrogen and oxygen in dinitrogen oxide.

2-23. Ordinary table sugar, whose common chemical name is sucrose, has the chemical formula $C_{12}H_{22}O_{11}$. Calculate the mass percentages of carbon, hydrogen, and oxygen in sucrose.

2-24. A key compound in the production of aluminum metal is cryolite, Na_3AlF_6. Calculate the mass percentages of sodium, aluminum, and fluorine in this compound.

2-25. Calculate the number of grams of xenon in 2.000 g of the compound xenon tetrafluoride.

2-26. Calculate the number of grams of sulfur in 5.585 g of the compound sulfur trioxide.

PROTONS, NEUTRONS, AND ELECTRONS

2-27. The following isotopes are used widely in medicine or industry:

(a) iodine-131 (b) cobalt-60
(c) potassium-43 (d) indium-113

How many protons, neutrons, and electrons are there in a neutral atom of each of these isotopes?

2-28. The following isotopes do not occur naturally but are produced in nuclear reactors:

(a) phosphorus-30 (b) technetium-97
(c) iron-55 (d) americium-240

How many protons, neutrons, and electrons are there in a neutral atom of each of these isotopes?

2-29. Fill in the blanks in the following table:

Symbol	Atomic number	Number of neutrons	Mass number
$^{14}_{6}C$	___	___	___
$^{?}_{?}Am$	___	___	241
___	53	___	123
___	___	10	18

2-30. Fill in the blanks in the following table:

Symbol	Atomic number	Number of neutrons	Mass number
$^{?}_{?}Ca$	___	___	48
___	40	___	90
___	___	78	131
$^{?}_{?}Mo$	___	57	___

2-31. Fill in the blanks in the following table:

Symbol	Atomic number	Number of neutrons	Mass number
___	31	36	___
___	___	8	15
___	27	___	58
$^{?}_{?}Xe$	___	___	133

2-32. Fill in the blanks in the following table:

Symbol	Atomic number	Number of neutrons	Mass number
$^{39}_{19}K$	___	___	___
$^{?}_{?}Fe$	___	___	56
___	36	___	84
___	___	70	120

ISOTOPIC COMPOSITION

2-33. Naturally occurring hydrogen consists of three isotopes with the atomic masses and abundances given in Table 2-9. Calculate the atomic mass of naturally occurring hydrogen.

2-34. Naturally occurring magnesium consists of three isotopes with the atomic masses and abundances given in Table 2-9. Calculate the atomic mass of naturally occurring magnesium.

2-35. Naturally occurring neon is a mixture of three isotopes with the following atomic masses and abundances:

Mass number	Atomic mass	Abundance/%
20	19.99	90.51
21	20.99	0.27
22	21.99	9.22

Calculate the atomic mass of naturally occurring neon.

2-36. Naturally occurring silicon consists of three isotopes with the following atomic masses and abundances:

Mass number	Atomic mass	Abundance/%
28	27.977	92.23
29	28.977	4.67
30	29.974	3.10

Calculate the atomic mass of naturally occurring silicon.

2-37. Naturally occurring bromine consists of two isotopes, ^{79}Br and ^{81}Br, whose atomic masses are 78.9183 and 80.9163, respectively. Given that the observed atomic mass of bromine is 79.904, calculate the percentages of ^{79}Br and ^{81}Br in naturally occurring bromine.

2-38. Naturally occurring boron consists of two isotopes with the atomic masses 10.013 and 11.009. The observed atomic mass of boron is 10.811. Calculate the abundance of each isotope.

2-39. Nitrogen has two naturally occurring isotopes, ^{14}N and ^{15}N, whose atomic masses are 14.0031 and 15.0001, respectively. The atomic mass of nitrogen is 14.0067. Use these data to compute the percentage of ^{15}N in naturally occurring nitrogen.

2-40. Naturally occurring europium consists of two isotopes, ^{151}Eu and ^{153}Eu, whose atomic masses are 150.9199 and 152.9212, respectively. Given that the atomic mass of europium is 151.96, calculate its isotopic percentage composition.

IONS

2-41. How many electrons are there in the following ions?

(a) Cs^+ (b) I^-
(c) Se^{2-} (d) N^{3-}

2-42. How many electrons are there in the following ions?

(a) Br^- (b) P^{3-}
(c) Ag^+ (d) Pb^{4+}

2-43. Determine the number of electrons in the following ions:

(a) Ba^{2+} (b) Tl^{3+}
(c) Fe^{2+} (d) Ti^{4+}

2-44. Determine the number of electrons in the following ions:

(a) Te^{2-} (b) La^{3+}
(c) Au^+ (d) Ir^{3+}

2-45. Give three ions that are isoelectronic with each of the following:

(a) K^+ (b) Kr
(c) N^{3-} (d) I^-

2-46. Give three ions that are isoelectronic with each of the following:

(a) F^- (b) Se^{2-}
(c) Ba^{2+} (d) La^{3+}

2-47. Use the atomic masses given in Appendix E to compute formula masses of the following ions to the greatest number of significant figures justified by the data:

(a) OH^- (b) H_3O^+
(c) AlF_6^{3-} (d) PCl_4^+

2-48. Use the atomic masses given in Appendix E to compute formula masses of the following ions to the greatest number of significant figures justified by the data:

(a) NH_4^+ (b) HO_2^-
(c) $AgCl_2^-$ (d) PCl_6^-

SEPARATIONS

2-49. Explain how you could separate iron filings from aluminum powder.

2-50. Explain how you could separate table salt, NaCl, from sand.

2-51. Describe how filtration can be used in separations.

2-52. Describe how flotation can be used in separations.

2-53. What is the role of the condenser in distillation?

2-54. What is meant when a liquid is said to be volatile?

2-55. What is the role of the carrier gas in gas-liquid chromatography?

2-56. Describe how the components of a gas mixture can be separated by gas-liquid chromatography.

2-57. What is the origin of the term "chromatography"?

2-58. Explain how paper chromatography works.

2-59. Describe how evaporation can be used to separate certain components in a solution.

2-60. What is distillation used for?

2-61. What are the contrasting physical properties of gold and sand on which panning of gold depends?

2-62. What is meant by a heterogeneous mixture?

ADDITIONAL PROBLEMS

2-63. Give an example of a separation technique that can be used to separate the components of a heterogeneous solid mixture but cannot be used to separate the components of a solution.

2-64. Rank the following fertilizers in decreasing order of mass percentage of nitrogen:

(a) NH_4NO_3 (b) NH_3
(c) $(NH_4)_2SO_4$ (d) $(NH_4)_2HPO_4$
(e) $(NH_4)H_2PO_4$ (f) KNO_3

2-65. Suppose we decide to establish an atomic mass scale by setting the atomic mass of ^{12}C exactly equal to one. What would be the atomic masses of naturally occurring hydrogen and oxygen on this scale?

2-66. Boron consists of two isotopes: boron-10 (10.013 amu) and boron-11 (11.009 amu). Use this information to determine the percent abundance of each isotope of boron in a nonnatural sample of boron with an atomic mass of 10.600.

2-67. A separation technique that is used extensively is adsorption. In adsorption, a component or components of a gaseous mixture or a solution is removed by preferential binding to the surface of a solid. Give two or more examples based on your personal experience of adsorption-based separations. (Hints: baking soda, carbon canister.)

2-68. Give an example of a method discussed in this chapter that could be used to separate ^6Li from ^7Li in a sample of the metal.

2-69. An isotope of iodine used to treat hyperactive thyroids is iodine-131. It forms the iodide ion I^-.

(a) How many protons are there in a nucleus of I and in a nucleus of I^-?
(b) How many neutrons are there in a nucleus of I and in a nucleus of I^-?
(c) How many electrons are there in an iodine atom and in an iodide ion?

INTERCHAPTER B

Elemental Etymology: What's in a Name?

David W. Ball, Rice University

Adapted with permission from the *Journal of Chemical Education,* volume 62 (September 1985), pages 787–788.

Like any discipline of science, chemistry has its own nomenclature, or system of naming. Mastery of this nomenclature is essential for the survival of any chemistry student. Initially, students may be scared witless by the seemingly complex organic and inorganic names. Soon, though, they are rattling off polysyllabic words with the ease of an expert.

The names of the elements provide the root of chemical naming, from simple salts to coordination compounds, from simple acids and bases to complicated organic species. While chemistry students use the names of the elements to name compounds, many of them have no idea of how the elements *themselves* got their names. For example, many students perceive the connection between the name "hydrogen" and the prefix "hydro-," meaning "water," but most do not know the meanings behind other element names. Study of element names can be used as an important tool for recognizing certain properties of particular elements, and the origins themselves can make excellent mnemonic devices in remembering an element's properties, symbol, and uses. Let us, then, take a close look at the origin of the names, or etymologies, of the chemical elements.

Alchemist in Search of the Philosopher's Stone Discovers Phosphorus, 1795, by Joseph Wright of Derby

Most of the available etymological information can be obtained from a dictionary or a chemical handbook, but neither points out the fact that there are many similarities and trends among the names. For example, many people immediately notice that a few elements are named after famous scientists. What they may not realize, however, is that a few elements are named after prominent mythological figures, too. To illustrate these patterns better (and to increase the pedagogical value of this paper), I have grouped the elements in six categories according to the origins of their names.

It is generally accepted by scientists that the discoverer of an element has the honor of naming it. However, the International Union of Pure and Applied Chemistry (IUPAC) reserves the right to select an approved name, as well as an approved symbol, regardless of the priority of discovery. The only IUPAC rule governing the naming process is the inclusion of the *-ium* suffix in the name of any new metallic element. However, because many elements were discovered before the existence of this

Table B-1 Elements with names of obscure origin

Element	Origin*
gold	Sanskrit, *jval;* Ang.-Sax., *gold;* ME, *guld*
iron	Ang.-Sax., *iron;* ME, *iren*
lead	Ang.-Sax., *lead;* ME, *leed*
silver	Ang.-Sax., *seolfor, sylfer*
sulfur	Sanskrit, *sulvere;* ME, *sulphre*
tin	Ang.-Sax., *tin;* ME, *tin*
zinc	Ang.-Sax., *zinc*

*Key to all tables: Ang.-Sax. = Anglo-Saxon, Eng. = English, Ger. = German, Gr. = Greek, L. = Latin, ME = Middle English, Sp. = Spanish, Swed. = Swedish.

rule, many elements (nonmetals as well as metals) do not have this suffix.

Table B-1 lists elements of ancient or even prehistoric discovery and whose names are of obscure origins. These seven elements, six of them metals, have been known and used by mankind for thousands of years; a few of these names are among the oldest words in any language. Unlike some of the other groups of elements, these ancient names follow no standard form and have no common or distinguishing syllables.

Table B-2 lists elements that are named because of the color of the element or its compounds and properties. For example, since the salts of iridium are of various colors, it seems appropriate that the element name is derived from a word meaning "rainbow." Chlorine and iodine are named for their respective colors, whereas rubidium, a silvery-white metal, is named for the intense ruby line in its atomic spectrum. The other elements listed here have similar etymologies.

Table B-2 Elements named for colors

Element	Origin
bismuth	Ger., *weisse Masse,* white mass
cesium	L., *caesius,* sky blue
chlorine	Gr., *chloros,* greenish yellow
chromium	Gr., *chroma,* color
indium	the color indigo
iodine	Gr., *iodes,* violet
iridium	L., *iris,* rainbow
praesodymium	Gr., *prasios* + *didymos,* green twin
rubidium	L., *rubidos,* deepest red
zirconium	Arabic, *zargun,* gold color

Table B-3 gives the elements that are named after real or imaginary personages. About half of these elements are named after famed scientists; the other half are named after various mythological figures. . . . Note that all elements discovered since 1952 (beginning with einsteinium, element 99) have been named after people.

Table B-4 lists elements whose names are derived from some geographic location. At least five countries are represented, one of them twice (gallium comes from *Gallia,* the Latin name for France). A total of *four* elements are named after Ytterby, a small Swedish town about 10 miles north of Göteborg, Sweden's second largest city. There is

Table B-3 Elements named after people (real or mythical)

Element	Origin
curium	Pierre and Marie Curie, discoverers of radium
einsteinium	Albert Einstein, originator of the theories of relativity
fermium	Enrico Fermi, pioneer in nuclear physics
gadolinium	Johann Gadolin, a Finnish chemist who discovered yttrium
lawrencium	Ernest O. Lawrence, developer of the cyclotron
mendelevium	Dmitri Mendeleev, developer of the periodic table
niobium	Niobe, an evil and blasphemous daughter of Tantalos (see below)
nobelium	Alfred Nobel, founder of the Nobel Prizes and inventor of dynamite
promethium	Prometheus, the Greek god who gave mankind fire
tantalum	Tantalos, the Greek mythical figure banished to a tantalizing fate in Hades
thorium	Thor, the Norse god of thunder
titanium	the Titans, Greek gods
vanadium	Vanadis, a "Wise Woman" in Scandinavian mythology

3-9) arranged the elements in order of increasing atomic mass and was able to show that the chemical properties of the elements exhibit repetitive patterns in chemical behavior. To illustrate the idea of Mendeleev's observation, let's start with the element lithium and arrange the succeeding elements in order of increasing atomic mass, as shown in Table 3-1. If we examine Table 3-1 carefully, then we see that the chemical properties of the elements show a remarkably repetitive, or periodic, pattern. The variations in the properties of the elements increasing in atomic number from lithium to neon are repeated in the properties of the elements from sodium to argon. The repeating pattern (or periodicity) is seen more clearly if we arrange the elements horizontally in two rows, or **periods:**

Figure 3-9 Dmitri Ivanovich Mendeleev (1834–1907), Russian chemist who played the major role in the development of the periodic table of the elements.

First row (or period) of elements:

Li	Be	B	C	N	O	F	Ne
lithium	beryllium	boron	carbon	nitrogen	oxygen	fluorine	neon

Second row (or period) of elements:

Na	Mg	Al	Si	P	S	Cl	Ar
sodium	magnesium	aluminum	silicon	phosphorus	sulfur	chlorine	argon

Sodium is placed below lithium to start a new period (row), because the chemical properties of sodium are similar to those of lithium. This placement of sodium leads, by continuation of the listing by atomic mass order, to the placement of Mg below Be and Cl below F (Cl_2 and F_2 have similar chemical behaviors). Mendeleev did not know about the existence of the nobel gases (see Section 3-5). He then began a new period by placing K in the same group as Na and Li, because K is very similar to Na in chemical behavior. He continued on, listing the elements in order of increasing atomic mass, making certain in each case that the elements in the same group (column) had similar chemical properties.

Mendeleev's genius was not in the arrangement of elements in order of increasing atomic mass. His genius lay in the **periodic arrangement** of the elements by atomic mass and in his realization that apparent gaps in the periodic arrangement, that is, in the **periodicity,** must correspond to missing elements. As we will see, he used the concept of periodicity to predict many of the chemical and physical properties of these, at that time unknown, elements.

Figure 3-10 presents a modern version of Mendeleev's **periodic table of the elements.** This version is more complicated than the version with which we began, because it contains more elements than just the 16 listed in Table 3-1, and many more elements than were known to Mendeleev. In the modern periodic table in Figure 3-10, the elements are arranged in order of increasing atomic number instead of increasing atomic mass. With a few exceptions, the order is the same in both cases. The idea of atomic number was not developed until the early 1900s, about 40 years after Mendeleev's first periodic table was published.

Figure 3-10 (*facing page*) A modern version of the periodic table of the elements. In this version the elements are ordered according to atomic number, rather than atomic mass. The chemical properties of the elements show a periodic pattern, that is, elements in a column have similar chemical properties.

Table 3-1 **The chemical properties of sixteen elements, listed in order of increasing atomic mass**

Atomic mass	Element	Symbol	Properties	Formula of element	Formula of halogen compound*
6.9	lithium	Li	very reactive metal	Li(*s*)	LiX
9.0	beryllium	Be	reactive metal	Be(*s*)	BeX_2
10.8	boron	B	semimetal†	B(*s*)	BX_3
12.0	carbon	C	nonmetallic solid	C(*s*)	CX_4
14.0	nitrogen	N	nonmetallic diatomic gas	$N_2(g)$	NX_3
16.0	oxygen	O	nonmetallic, moderately reactive diatomic gas	$O_{2(g)}$	OX_2
19.0	fluorine	F	very reactive diatomic gas	$F_2(g)$	FX
20.2	neon	Ne	very unreactive monatomic gas	Ne(*g*)	none
23.0	sodium	Na	very reactive metal	Na(*s*)	NaX
24.3	magnesium	Mg	reactive metal	Mg(*s*)	MgX_2
27.0	aluminum	Al	metal	Al(*s*)	AlX_3
28.1	silicon	Si	semimetal†	Si(*s*)	SiX_4
31.0	phosphorus	P	nonmetallic solid‡	$P_4(s)$	PX_3
32.1	sulfur	S	nonmetallic solid§	$S_8(s)$	SX_2
35.5	chlorine	Cl	very reactive diatomic gas	$Cl_2(g)$	ClX
39.9	argon	Ar	very unreactive monatomic gas	Ar(*g*)	none

*X stands for F, Cl, Br, or I.
†Boron and silicon are called semimetals because they have properties that are intermediate between those of the metals and those of the nonmetals.
‡Phosphorus exists as P_4 units in the solid state and is often denoted by $P_4(s)$ rather than P(*s*).
§Sulfur exist as S_8 units in the solid state and is sometimes denoted as $S_8(s)$.

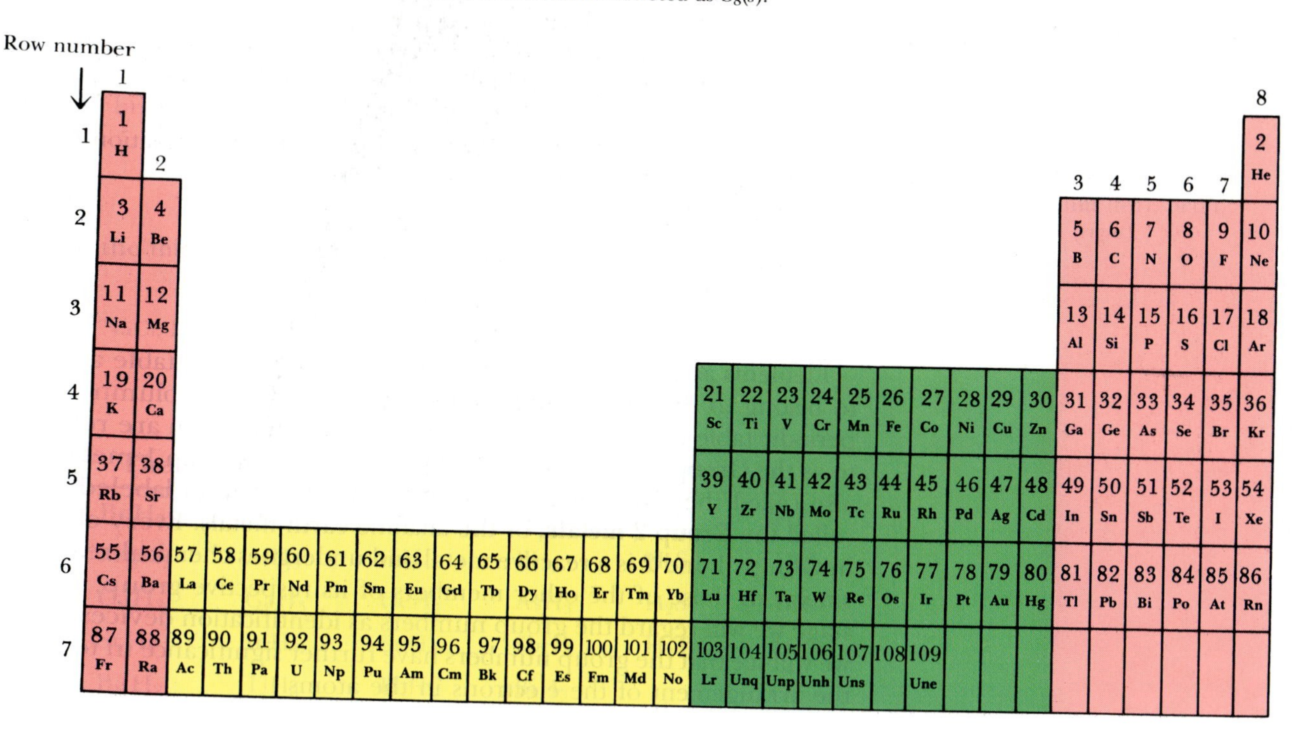

All chemical reactions can be assigned to one of two classes: reactions in which electrons are transferred from one chemical **species** (i.e., an element, a compound, or an ion) to another and reactions in which electrons are not transferred. Reactions in which electrons are transferred from one species to another are called oxidation-reduction ("redox") reactions or electron-transfer reactions. To understand oxidation-reduction reactions, we need to understand the concept of oxidation states in chemical species. Changes in the oxidation state of one or more elements occur during a redox reaction as a result of a transfer of electrons. We determine the oxidation state of an element according to a set of rules and use that information to identify oxidation-reduction reactions and to balance them in a systematic way.

Another useful classification scheme for chemical reactions has four categories: (1) combination reactions, (2) decomposition reactions, (3) single-replacement reactions, and (4) double-replacement reactions. Although not all chemical reactions fall into these four categories, many do; and it is helpful when first learning about chemical reactions to try to classify them into these types. In this chapter, we describe examples of each of these four classes of reactions. And a discussion of reaction types leads naturally to chemical nomenclature, to relative reactivities of metals, and to an introduction to the chemistry of acids and bases.

4-1. OXIDATION-REDUCTION REACTIONS INVOLVE THE TRANSFER OF ELECTRONS BETWEEN SPECIES

The simplest example of an **oxidation-reduction ("redox") reaction,** or an **electron-transfer reaction,** is the reaction between a metal and a nonmetal. For example, the reaction of sodium metal with sulfur produces the ionic compound sodium sulfide (Figure 4-1):

$$2Na(s) + S(s) \rightarrow Na_2S(s)$$

and the reaction of calcium metal with oxygen gas produces the ionic compound calcium oxide (Figure 4-2):

$$2Ca(s) + O_2(g) \rightarrow 2CaO(s)$$

In both of these reactions, electrons are transferred from the metal atoms to the nonmetal atoms; the transfer produces an ionic compound in which the ions have a noble-gas arrangement.

The transfer of electrons between species necessarily involves a change in the charge on each species. For example, consider again the reaction between sodium and sulfur to give sodium sulfide, Na_2S:

$$2Na(s) + S(s) \rightarrow Na_2S(s)$$

A neutral sodium atom has no net charge (a charge of 0). The sodium ion in Na_2S has a charge of +1, so the charge on sodium changes from 0 to +1 in the reaction. Similarly, the charge on sulfur changes in the reaction from 0 to −2. Each sodium atom loses one electron as it

Figure 4-1 The alkali metal sodium reacts with the nonmetal sulfur to produce sodium sulfide.

(a)

(b)

Figure 4-2 Calcium is a reactive Group 2 metal that reacts with oxygen and water vapor in air. The freshly exposed calcium metal surfaces shown in (a) are rapidly oxidized to calcium oxide, CaO(*s*), when exposed to air, as shown in (b).

changes from Na to Na^+, so the two sodium atoms lose a total of two electrons; and each sulfur atom gains two electrons as it changes from S to S^{2-}. As this example shows, the electron transfer between species in a reaction must be balanced; that is, in any oxidation-reduction reaction the total number of electrons lost always equals the total number of electrons gained.

We can use the idea of the conservation of electrons to determine the balancing coefficients in chemical equations for oxidation-reduction reactions. Recall that in Section 3-9 we used the idea of balancing ionic charges on monatomic species to write the chemical formula for a binary compound consisting of a metal and a nonmetal. The concept of an ionic charge on an element within a polyatomic species is so convenient in balancing chemical equations for redox reactions that it has been generalized, and this type of ionic charge is called an **oxidation number** or an **oxidation state.**

We can assign an oxidation state to each atom in a chemical species on the basis of a set of rules. The rules originate from a consideration of the number of electrons in a neutral atom of an element relative to the number of electrons that we assign to that element when it is incorporated in a molecule or ion. In the case of monatomic ions (e.g., Na^+, Ca^{2+}, O^{2-}, and Cl^-), the assigned oxidation state is simply equal to the charge on the atomic ion. For chemical species involving two or more atoms, the assigned oxidation states are not, in general, equal to the actual charges on the atoms. In such species, the assignment of oxidation states is, in essence, a bookkeeping device that is useful for identifying and balancing redox reactions.

The general procedure for assigning oxidation states to elements in chemical species containing two or more atoms is given by the following set of rules, *which take priority in the order given,* in the final determination of the oxidation states of the elements in a species.

Rules for Assigning Oxidation States

1. Free elements are assigned an oxidation state of 0.
2. The sum of the oxidation states of all the atoms in a species must be equal to the net charge on the species.
3. The alkali metals (Li, Na, K, Rb, and Cs) in compounds are always assigned an oxidation state of +1.
4. Fluorine in compounds is always assigned an oxidation state of −1.
5. The alkaline earth metals (Be, Mg, Ca, Sr, Ba, and Ra) and also Zn and Cd in compounds are always assigned an oxidation state of +2.
6. Hydrogen in compounds is assigned an oxidation state of +1.
7. Oxygen in compounds is assigned an oxidation state of −2. (By applying these rules in the order given, we eliminate the need for memorizing numerous qualifications and exceptions such as oxygen is −1 in peroxides and hydrogen is −1 in hydrides and so on.)

These seven rules are easy to apply. We have already used rules 1, 2, 3, and 5 in Chapter 3, when we discussed ionic charges and their use in writing chemical formulas for binary ionic compounds. The rules given above cover most, but not all, cases. For compounds not covered by the rules, we assign oxidation states by analogy with similar compounds derived from elements in the same group in the periodic table.

The +1 oxidation state of alkali metals (Group 1 metals) in compounds corresponds to the ionic charge of the alkali metal ions. The +1 state represents the loss of an electron from the neutral atom. The +2 oxidation state of the alkaline earth metals (Group 2 metals) corresponds to the ionic charge of the alkaline earth metal ions. The +2 state represents the loss of two electrons from the neutral atoms.

The ionic charges of the metal ions discussed in Section 3-9 (Figure 3-22) correspond to the oxidation states of those elements (Table 4-1). For example, aluminum is always assigned an oxidation state of +3 in its compounds; and because gallium is in the same group as aluminum, gallium, by analogy with aluminum, also is assigned an oxidation state of +3 in its compounds.

Table 4-1 Ionic charges and oxidation states for metal ions

Group	Ionic charge	Oxidation state
Group 1: alkali metal ions (Li^+, Na^+, K^+, Cs^+, Rb^+)	+1	+1
Group 2: alkaline earth metal ions (Be^{2+}, Mg^{2+}, Ca^{2+}, Sr^{2+}, Ba^{2+})	+2	+2
Group 3 ions (Al^{3+}, Ga^{3+})	+3	+3

EXAMPLE 4-1: Assign an oxidation state to each atom in the following compounds:
(a) CsF (b) NO_2 (c) $HClO_3$ (d) H_2O_2 (e) NaH

Solution: (a) We assign to cesium an oxidation state of +1 (rule 3); thus, fluorine is assigned an oxidation state of −1 (rule 2), because CsF is a neutral species and +1 − 1 = 0.
(b) We assign to oxygen an oxidation state of −2 (rule 7). The oxidation state of nitrogen in NO_2, represented by x, is thus (rule 2)

$$x + 2(-2) = 0$$
$$x = +4$$

The oxidation state of nitrogen in NO_2 is +4.
(c) We assign to hydrogen an oxidation state of +1 (rule 6) and to oxygen an oxidation state of −2 (rule 7). So the oxidation state x of chlorine is (rule 2)

$$+1 + x + 3(-2) = 0$$
$$x = +5$$

The oxidation state of chlorine in $HClO_3$ is +5.
(d) We assign to hydrogen an oxidation state of +1 (rule 6), so the oxidation state x of oxygen is (rule 2)

$$2(+1) + 2x = 0$$
$$x = -1$$

Thus, the oxidation state of oxygen in H_2O_2 is −1, which is characteristic of peroxides. Although this result seems to contradict rule 7, we assign to oxygen in peroxides on oxidation state of −1 because rules 2 and 6 take precedence.
(e) We assign to sodium an oxidation state of +1 (rule 3); and, according to rule 2, the oxidation state of hydrogen in NaH is −1, which is characteristic of hydrides. Although this result seems to violate rule 6, rules 2 and 3 take precedence over all subsequent rules.

PRACTICE PROBLEM 4-1: Assign an oxidation state to each atom in the following compounds:
(a) $HClO_4$ (b) KNO_3 (c) Mn_2O_7

Answer: (a) H(+1), O(−2), Cl(+7) (b) K(+1), O(−2), N(+5) (c) O(−2), Mn(+7)

Example 4-1 involves only neutral compounds, whose net charge must be 0. For ionic species, the sum of the oxidation states of each atom must equal the net charge on the ion.

EXAMPLE 4-2: Assign an oxidation state to each atom in the following ions:
(a) CrO_4^{2-} (b) HF_2^- (c) NH_4^+

Solution: (a) We assign to oxygen an oxidation state of −2 (rule 7), and the charge on the ion is −2. Thus, the oxidation state x of chromium is (rule 2)

$$x + 4(-2) = -2$$
$$x = +6$$

The oxidation state of chromium in CrO_4^{2-} is +6.
(b) We assign to fluorine an oxidation state of −1 (rule 4), and the charge on the ion is −1. Thus, the oxidation state x of hydrogen is (rule 2)

$$x + 2(-1) = -1$$
$$x = +1$$

The oxidation state of H in HF_2^- is +1.
(c) We assign to hydrogen an oxidation state of +1 (rule 6), and the charge on the ion is +1. Thus, the oxidation state x of nitrogen is (rule 2)

$$x + 4(+1) = +1$$
$$x = -3$$

PRACTICE PROBLEM 4-2: Assign an oxidation state to each atom in the following ions:
(a) $Cr_2O_7^{2-}$ (b) BF_4^- (c) PCl_4^+

Answer: (a) O(−2), Cr(+6) (b) F(−1), B(+3) (c) Cl(−1), P(+5)

In Section 4-3 we present the chemical formulas for common polyatomic ions. That information will enable us to assign oxidation states to the elements in compounds such as $NH_4BF_4(s)$.

When the oxidation state of an atom increases (i.e., becomes more positive), we say the atom is **oxidized.** The term **oxidation** denotes a loss of electrons. When the oxidation state of an atom decreases, we say the atom is **reduced.** The term **reduction** denotes a gain of electrons.

Recall that in the reaction between sodium and sulfur

$$2Na(s) + S(s) \rightarrow Na_2S(s)$$

each sodium atom loses one electron and each sulfur atom gains two electrons; thus, it requires two sodium atoms to react with each sulfur atom. Sodium loses an electron and is oxidized, whereas sulfur gains two electrons and is reduced. For any oxidation-reduction reaction, the total number of electrons lost by the element that is oxidized must equal the total number of electrons gained by the element that is reduced (conservation of electrons).

In an oxidation-reduction reaction, the species that loses electrons is called the **reducing agent** and the species that gains electrons is called the **oxidizing agent.** Thus, in the equation

$$2Na(s) + S(s) \rightarrow Na_2S(s)$$

Na, which loses electrons (its oxidation state increases from 0 to +1), is the reducing agent, and S, which gains electrons (its oxidation state decreases from 0 to −2), is the oxidizing agent. Note that the reducing agent, which loses electrons, is oxidized and that the oxidizing agent, which gains electrons, is reduced.

Figure 4-3 Magnesium burns in air to produce magnesium oxide.

EXAMPLE 4-3: The combustion of magnesium in oxygen (Figure 4-3) is described by the equation

$$2Mg(s) + O_2(g) \rightarrow 2MgO(s)$$

In this reaction, which element is oxidized and which element is reduced? How many electrons are transferred per formula unit of MgO formed? Identify the oxidizing and reducing agents in the reaction.

Solution: A magnesium atom is neutral, so we assign to it an oxidation state of 0. An oxygen molecule is neutral, so we assign to each oxygen atom in O_2 an oxidation state of 0. In MgO the oxidation states of magnesium and oxygen are +2 and −2, respectively. The oxidation state of magnesium increases from 0 to +2, and the oxidation state of oxygen decreases from 0 to −2. Thus, magnesium is oxidized (loss of electrons) and oxygen is reduced (gain of electrons) in this reaction. Each of the two oxygen atoms gains two electrons for a total of four electrons gained. Because each Mg atom that is oxidized to Mg^{2+} supplies two electrons, two Mg atoms are required to reduce one O_2 molecule to two O^{2-} ions. The transfer of four electrons yields two MgO formula units; thus, two electrons are transferred per MgO formula unit produced. Because Mg supplies electrons, it is the reducing agent; and, because O_2 takes up electrons, it is the oxidizing agent.

Figure 4-4 Iron metal and chlorine gas react to form $FeCl_3(s)$.

PRACTICE PROBLEM 4-3: For the reaction between iron metal and chlorine gas (Figure 4-4), described by the equation

$$2Fe(s) + 3Cl_2(g) \rightarrow 2FeCl_3(s)$$

indicate which species is oxidized and which species is reduced. Indicate the reducing agent and the oxidizing agent. How many electrons are transferred for each $FeCl_3$ formula unit produced?

Answer: Fe is oxidized and is the reducing agent; Cl_2 is reduced and is the oxidizing agent; there are three electrons transferred per $FeCl_3$ formula unit produced.

4-2. OXIDATION STATES CAN BE USED TO BALANCE EQUATIONS FOR OXIDATION-REDUCTION REACTIONS

We now consider the problem of balancing oxidation-reduction equations. We use the fact that, for any oxidation-reduction reaction, the total number of electrons lost (donated) by the reducing agent must equal the total number of electrons gained (accepted) by the oxidizing agent (conservation of electrons). We call the method of balancing oxidation-reduction equations on the basis of changes of oxidation states coupled with the conservation of electrons the **oxidation-state method of balancing oxidation-reduction equations.** As an example, consider the unbalanced oxidation-reduction equation

$$Fe(s) + Cl_2(g) \rightarrow FeCl_3(s) \qquad (4\text{-}1)$$

First, note that Equation (4-1) is an oxidation-reduction equation because it involves changes in oxidation states (Fe goes from 0 to +3 and Cl goes from 0 in Cl_2 to −1 in $FeCl_3$). We diagram the changes in oxidation state in the reaction as follows (where the numbers next to the vertical lines represent oxidation states)

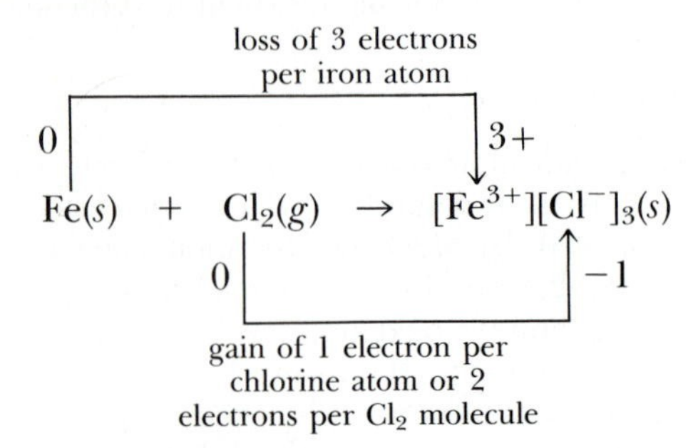

Each Fe atom that is oxidized loses three electrons and each Cl_2 molecule that is reduced gains two electrons (one electron per chlorine atom in Cl_2). The number of electrons donated by the reducing agent (Fe in this case) must be equal to the number of electrons accepted by the oxidizing agent (Cl_2 in this case). One Cl_2 molecule can accept only two electrons, whereas each Fe atom denotes three electrons. To balance the number of electrons donated and accepted, we note that

$$\underset{\text{(from Fe)}}{2 \times 3e^-} = \underset{\text{(to } Cl_2)}{3 \times 2e^-}$$

The multiplication factors (2 for Fe and 3 for Cl_2) are the balancing coefficients for those reactant species in the balanced chemical equation. Thus, we have the (still unbalanced) equation

$$2Fe(s) + 3Cl_2(g) \rightarrow FeCl_3(s) \tag{4-2}$$

Noting that the use of two Fe atoms must produce two $FeCl_3$ formula units (conservation of Fe atoms), we place a 2 in front of $FeCl_3$

$$2Fe(s) + 3Cl_2(g) \rightarrow 2FeCl_3(s) \tag{4-3}$$

a coefficient that also balances the Cl atoms. Now Equation (4-3) is balanced.

We also could have balanced Equation (4-1) by the method of inspection discussed in Chapter 3. However, the method that we have used here is more general, and it can be used to balance redox equations that would be very difficult to balance by the method of inspection.

EXAMPLE 4-4: The equation

$$SO_2(g) + H_2S(g) \rightarrow S(s) + H_2O(g)$$

describes a reaction that produces the sulfur deposits around hot springs (Figure 4-5). Use the oxidation-state method to balance this equation.

Solution: First, diagram the changes in oxidation states:

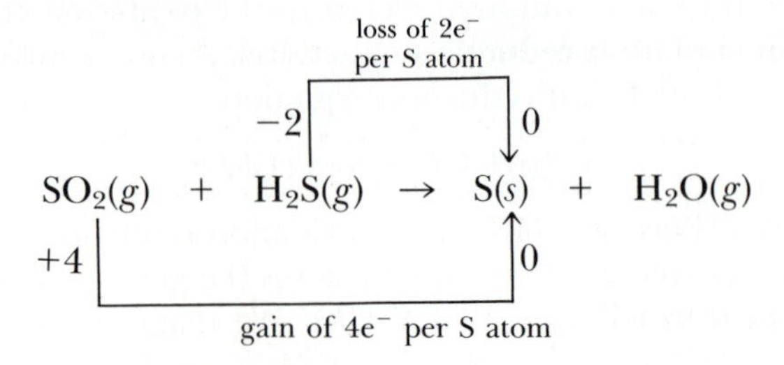

Figure 4-5 The sulfur deposits that occur around hot springs usually result from the oxidation-reduction reaction between sulfur dioxide and hydrogen sulfide.

Note that in this reaction some sulfur atoms are reduced ($SO_2 \rightarrow S$) and some are oxidized ($H_2S \rightarrow S$). The oxidizing agent (**electron acceptor;** gains electrons) is SO_2 and the reducing agent (**electron donor;** loses electrons) is H_2S. To balance the number of electrons donated and accepted, we note that

$$\underset{(\text{to } SO_2)}{1 \times 4e^-} = \underset{(\text{from } H_2S)}{2 \times 2e^-}$$

Using the multiplication factors as the balancing coefficients for the respective species gives the (not yet completely balanced) equation

$$SO_2(g) + 2H_2S(g) \rightarrow S(s) + H_2O(g) \qquad \text{(unbalanced)}$$

Now note that there is a total of three S atoms on the left; thus, we need three S atoms on the right

$$SO_2 + 2H_2S(g) \rightarrow 3S(s) + H_2O(g) \qquad \text{(unbalanced)}$$

The S atoms are now balanced, but the H and the O atoms are not. Because there are two O atoms on the left, we can balance the O atoms by placing a 2 in front of H_2O on the right, thereby simultaneously balancing the H atoms (four on each side)

$$SO_2(g) + 2H_2S(g) \rightarrow 3S(s) + 2H_2O(g) \qquad \text{(balanced)}$$

The equation is now balanced with respect to all the different types of atoms.

PRACTICE PROBLEM 4-4: Use the oxidation-state method to balance the equation

$$Fe_2O_3(s) + CO(g) \rightarrow CO_2(g) + Fe(l)$$

which represents one of the reactions occurring in the blast furnaces that convert iron ores to iron metal.

Answer: $Fe_2O_3(s) + 3CO(g) \rightarrow 3CO_2(g) + 2Fe(l)$

pounds. Our first example is the reaction between sodium oxide and carbon dioxide to form the ionic compound sodium carbonate:

$$Na_2O(s) + CO_2(g) \rightarrow \underset{\text{sodium carbonate}}{Na_2CO_3(s)}$$

This combination reaction can be used to remove CO_2 from air.

When the gases ammonia and hydrogen chloride are mixed (Figure 4-7), they combine to form the ionic compound ammonium chloride

$$NH_3(g) + HCl(g) \rightarrow \underset{\text{ammonium chloride}}{NH_4Cl(s)}$$

Another example of a combination reaction involving compounds as reactants is the reaction of sulfur trioxide with magnesium oxide to form the ionic compound magnesium sulfate

$$MgO(s) + SO_3(g) \rightarrow \underset{\text{magnesium sulfate}}{MgSO_4(s)}$$

This combination reaction can be used to remove sulfur trioxide from air.

The three combination reactions illustrated above all involve the formation of polyatomic ions. For example, when magnesium sulfate is dissolved in water, only two ions per $MgSO_4$ formula unit are produced:

$$\underset{\text{magnesium sulfate}}{MgSO_4(s)} \xrightarrow[H_2O(l)]{} \underset{\text{magnesium ion}}{Mg^{2+}(aq)} + \underset{\text{sulfate ion}}{SO_4^{2-}(aq)}$$

Figure 4-7 Ammonia and hydrogen chloride are colorless gases. They react in a combination reaction to produce the solid white compound ammonium chloride. The white cloud in the picture consists of small ammonium chloride particles formed where the gaseous NH_3 from one bottle comes into contact with the gaseous HCl from the other.

The ion SO_4^{2-} remains intact as a unit (a polyatomic ion) in solution. Magnesium sulfate, $MgSO_4$, consists of the ions Mg^{2+} and SO_4^{2-}. A **polyatomic ion** is an ion that contains more than one atom. There are numerous combinations of atoms, such as SO_4^{2-}, that form stable polyatomic ions. The polyatomic ion SO_4^{2-} is called a sulfate ion; similarly, the name for $MgSO_4$ is magnesium sulfate. Table 4-2 lists twenty common polyatomic ions, their names, and their charges. Compounds containing the ions in Table 4-2 are named according to the rules for naming binary compounds. For example, NaOH is called sodium hydroxide, KCN is called potassium cyanide, and NH_4Cl is called ammonium chloride. You should learn the formulas and names of all the polyatomic ions in Table 4-2.

Table 4-2 Common polyatomic ions

Positive ions	
ammonium	NH_4^+
mercury(I)*	Hg_2^{2+}
Negative ions	
acetate	$C_2H_3O_2^-$
carbonate	CO_3^{2-}
chlorate	ClO_3^-
chromate	CrO_4^{2-}
cyanide	CN^-
dichromate	$Cr_2O_7^{2-}$
hydrogen carbonate ("bicarbonate")	HCO_3^-
hydrogen sulfate ("bisulfate")	HSO_4^-
hydroxide	OH^-
hypochlorite	ClO^-
nitrate	NO_3^-
nitrite	NO_2^-
perchlorate	ClO_4^-
permanganate	MnO_4^-
phosphate	PO_4^{3-}
sulfate	SO_4^{2-}
sulfite	SO_3^{2-}
thiosulfate	$S_2O_3^{2-}$

*Note that Hg_2^{2+} contains two Hg^+ ions joined together.

EXAMPLE 4-6: Name the following ionic compounds:
(a) $RbMnO_4(s)$ (b) $Co(NO_2)_2(s)$ (c) $CrPO_4(s)$

Solution: (a) From Table 4-2, we note that MnO_4^- is called the permanganate ion, so $RbMnO_4$ is called rubidium permanganate.
(b) The NO_2^- ion is called the nitrite ion, so $Co(NO_2)_2$ is called cobalt(II) nitrite (see Table 3-5).
(c) The PO_4^{3-} ion is called the phosphate ion, so $CrPO_4$ is called chromium(III) phosphate (see Table 3-5).

PRACTICE PROBLEM 4-6: Name the following ionic compounds:
(a) $NH_4C_2H_3O_2(s)$ (b) $PbCrO_4(s)$ (c) $K_2Cr_2O_7(s)$

Answer: (a) ammonium acetate (b) lead(II) chromate (c) potassium dichromate

EXAMPLE 4-7: Write the chemical formula for sodium thiosulfate, copper(II) perchlorate, calcium hydroxide.

Solution: The formula for sodium thiosulfate is written by combining Na^+ and $S_2O_3^{2-}$. Because $S_2O_3^{2-}$ has an ionic charge of -2, it requires two Na^+ for each $S_2O_3^{2-}$; so $Na_2S_2O_3$ is the formula for sodium thiosulfate.

Copper(II) perchlorate requires two ClO_4^- for each Cu^{2+}, so the formula of the compound is $Cu(ClO_4)_2$. Note that the entire perchlorate ion, which exists as a unit, is enclosed in parentheses and that the subscript 2 lies outside the parentheses. One formula unit of $Cu(ClO_4)_2$ contains one copper atom, two chlorine atoms, and eight oxygen atoms.

Calcium hydroxide involves the ions Ca^{2+} and OH^-; thus, it has the formula $Ca(OH)_2$. Once again, note the use of parentheses.

PRACTICE PROBLEM 4-7: Write the chemical formula for
(a) sodium phosphate (b) mercury(I) nitrate
(c) nickel(II) sulfite

Answer: (a) Na_3PO_4 (b) $Hg_2(NO_3)_2$ (c) $NiSO_3$

Example 5-1 illustrates an important point. *In order to calculate the number of moles in a given mass of a chemical compound, it is necessary to know the chemical formula of the compound. A mole of any compound is defined only in terms of its chemical formula.* If a substance (such as coal or wood) cannot be represented by a single chemical formula, then we can give only the mass of the substance.

5-2. ONE MOLE OF ANY SUBSTANCE CONTAINS AVOGADRO'S NUMBER OF FORMULA UNITS

Figure 5-2 The Italian scientist Amedeo Avogadro (1776–1856) was one of the first scientists to propose a clear distinction between atoms and molecules.

It has been determined experimentally that one mole of any substance contains 6.022×10^{23} formula units. This number is called **Avogadro's number** after Amedeo Avogadro, who was an Italian scientist and one of the earliest proponents of the atomic theory (Figure 5-2). We say not only that one mole of any substance contains Avogadro's number of formula units but also that one mole is that mass of a substance containing Avogadro's number of formula units, or "elementary entities." For example, the atomic mass of carbon-12 is taken to be exactly 12, so 12.00 g of carbon-12 contains 6.022×10^{23} atoms. Likewise, the molecular mass of water is 18.02, so 18.02 g of water contains 6.022×10^{23} molecules. A mole is, in essence, a shorthand term for a certain number of "things" (that is, formula units), just as a dozen stands for 12 things. But instead of the number 12 implied by the term dozen, the number of things in a mole is 6.022×10^{23}. A molecular mass or formula mass is thus the mass in grams of 6.022×10^{23} molecules or formula units.

Avogadro's number is an enormous number, probably much larger than any number you have encountered. In order to appreciate the magnitude of Avogadro's number, let's compute how many years it would take to spend Avogadro's number of dollars at the rate of one million dollars per second. There are 3.15×10^7 s in 1 year; thus, the number of years required to spend 6.022×10^{23} dollars is

$$(6.022 \times 10^{23} \text{ dollars})\left(\frac{1 \text{ s}}{10^6 \text{ dollars}}\right)\left(\frac{1 \text{ year}}{3.15 \times 10^7 \text{ s}}\right) = 1.91 \times 10^{10} \text{ years}$$

or 19.1 billion years (1 billion = 10^9). This interval is over four times longer than the estimated age of the earth (4.6 billion years) and roughly equal to the estimated age of the universe (13 to 20 billion years). This calculation illustrates how large Avogadro's number is and, consequently, how small atoms and molecules are. Look again at the Frontispiece to this chapter. Each of those samples contains 6.022×10^{23} formula units of the indicated substance.

We can use Avogadro's number to calculate the mass of one atom or molecule.

EXAMPLE 5-2: Using Avogadro's number, calculate the mass of one nitrogen molecule.

Solution: Recall that nitrogen occurs as a diatomic molecule. The formula of molecular nitrogen is N_2, so its formula mass, or molecular mass, is 28.02. Thus, there are 6.022×10^{23} molecules of nitrogen in 28.02 g of nitrogen.

The mass of one nitrogen molecule is

$$\begin{aligned}\begin{pmatrix}\text{mass of one}\\ \text{nitrogen molecule}\end{pmatrix} &= \frac{\text{mass of Avogadro's number of nitrogen molecules}}{\text{Avogadro's number}}\\ &= \frac{\text{mass corresponding to 1 mol N}_2}{\text{Avogadro's number}}\\ &= \frac{28.02\text{ g N}_2\cdot\text{mol}^{-1}}{6.022\times10^{23}\text{ N}_2\text{ molecules}\cdot\text{mol}^{-1}}\\ &= 4.653\times10^{-23}\text{ g}\cdot\text{molecule}^{-1}\end{aligned}$$

PRACTICE PROBLEM 5-2: In Chapter 7 when we study gases, we will calculate the masses of molecules in kilograms. Calculate the mass of a CO_2 molecule and of an SF_6 molecule in kilograms.

Answer: CO_2, 7.308×10^{-26} kg; SF_6, 2.426×10^{-25} kg

Avogadro's number also can be used to calculate the number of atoms or molecules in a given mass of a substance. The next Example illustrates this type of calculation.

EXAMPLE 5-3: Calculate how many methane molecules and how many hydrogen and carbon atoms there are in 25.0 g of methane.

Solution: The formula mass of CH_4 is $12.0 + (4 \times 1.0) = 16.0$, so 25.0 g of CH_4 corresponds to

$$\text{moles of CH}_4 = (25.0\text{ g CH}_4)\left(\frac{1\text{ mol CH}_4}{16.0\text{ g CH}_4}\right) = 1.56\text{ mol CH}_4$$

Because 1 mol of CH_4 contains 6.022×10^{23} molecules, 1.56 mol must contain

$$\begin{aligned}\text{number of CH}_4\text{ molecules} &= (1.56\text{ mol CH}_4)\left(\frac{6.022\times10^{23}\text{ molecules}}{1\text{ mol}}\right)\\ &= 9.39\times10^{23}\text{ CH}_4\text{ molecules}\end{aligned}$$

Each molecule of methane contains one carbon atom and four hydrogen atoms, so

$$\begin{aligned}\text{number of C atoms} &= (9.39\times10^{23}\text{ CH}_4\text{ molecules})\left(\frac{1\text{ C atom}}{1\text{ CH}_4\text{ molecule}}\right)\\ &= 9.39\times10^{23}\text{ C atoms}\end{aligned}$$

$$\begin{aligned}\text{number of H atoms} &= (9.39\times10^{23}\text{ CH}_4\text{ molecules})\left(\frac{4\text{ H atoms}}{1\text{ CH}_4\text{ molecule}}\right)\\ &= 3.76\times10^{24}\text{ H atoms}\end{aligned}$$

PRACTICE PROBLEM 5-3: Refer to Practice Problem 5-1. Calculate how many molecules of each of the top five chemicals are produced annually in the United States. How many hydrogen atoms are contained in the produced quantities of each of these five chemicals?

Answer: H_2SO_4, 2.417×10^{35} molecules (4.834×10^{35} H atoms); N_2, 5.242×10^{35} molecules; O_2, 3.222×10^{35} molecules; C_2H_4, 3.404×10^{35} molecules (1.362×10^{36} H atoms); NH_3, 5.415×10^{35} molecules (1.625×10^{36} H atoms)

Table 5-1 summarizes the relationships between molar quantities. In reading the table, recall that sodium chloride is an ionic compound, with the formula unit consisting of one sodium ion, Na^+, and one chloride ion, Cl^-. Similarly, the formula unit of barium fluoride consists of one barium ion, Ba^{2+}, and *two* fluoride ions, F^-.

We conclude this section with the official SI definition of mole: "The mole is the amount of substance of a system that contains as many elementary entities as there are atoms in exactly 0.012 kg of carbon-12. When the mole is used, the elementary entities must be specified; they may be atoms, molecules, ions, electrons, other particles, or specified groups of such particles." Note that because the atomic mass of ^{12}C is exactly 12 by definition, a mole of ^{12}C contains exactly 12 g (= 0.012 kg) of carbon. This SI definition of a mole is equivalent to the other definitions given in this section.

The concept of a mole is one of the most important concepts, if not the most important, in all of chemistry. We will be using moles not only throughout this chapter, but in almost every chapter in the rest of the book. It is very important that you understand what a mole is and how to use it to do chemical calculations. If you are not comfortable with the mole concept at this point, then be sure to do as many as needed of the first 12 problems at the end of the chapter to guarantee that you understand the concept.

Table 5-1 **Representative relationships between molar quantities**

Substance	Formula	Formula mass	Molar mass/ $g \cdot mol^{-1}$	Number of particles in 1 mol	Moles
atomic chlorine	Cl	35.45	35.45	6.022×10^{23} chlorine atoms	1 mol Cl atoms
chlorine gas	Cl_2	70.90	70.90	6.022×10^{23} chlorine molecules	1 mol Cl_2 molecules
				12.044×10^{23} chlorine atoms	2 mol Cl atoms
water	H_2O	18.02	18.02	6.022×10^{23} water molecules	1 mol H_2O molecules
				12.044×10^{23} hydrogen atoms	2 mol H atoms
				6.022×10^{23} oxygen atoms	1 mol O atoms
sodium chloride	$NaCl$	58.44	58.44	6.022×10^{23} NaCl formula units	1 mol NaCl formula units
				6.022×10^{23} sodium ions	1 mol Na^+ ions
				6.022×10^{23} chloride ions	1 mol Cl^- ions
barium fluoride	BaF_2	175.3	175.3	6.022×10^{23} BaF_2 formula units	1 mol BaF_2 formula units
				6.022×10^{23} barium ions	1 mol Ba^{2+} ions
				12.044×10^{23} fluoride ions	2 mol F^- ions
nitrate ion	NO_3^-	62.01	62.01	6.022×10^{23} nitrate ions	1 mol NO_3^- ions
				6.022×10^{23} nitrogen atoms	1 mol N atoms
				18.066×10^{23} oxygen atoms	3 mol O atoms

5-3. SIMPLEST FORMULAS CAN BE DETERMINED BY CHEMICAL ANALYSIS

Stoichiometry (stoi′ke om′i tre) is the calculation of the quantities of elements or compounds involved in chemical reactions. The word stoichiometry is derived from the Greek words *stoicheio,* meaning "simplest components or parts," and *metrein,* meaning "to measure." The concept of a mole is central to carrying out stoichiometric calculations. For example, we can use the concept of a mole to determine the simplest chemical formula of a substance. Zinc oxide is found by chemical analysis to be 80.3 percent (by mass) zinc and 19.7 percent (by mass) oxygen. When working with mass percentages in chemical calculations, it is convenient to consider a 100-g sample so that the mass percentages can be converted easily to grams. For example, a 100-g sample of zinc oxide contains 80.3 g of zinc and 19.7 g of oxygen. We can write this schematically as

$$80.3 \text{ g Zn} \mathrel{\hat{=}} 19.7 \text{ g O}$$

where the symbol $\mathrel{\hat{=}}$ means "is **stoichiometrically equivalent to**" or, in this case, "**combines with.**" If we divide 80.3 g Zn by the atomic mass of zinc (65.38) and 19.7 g O by the atomic mass of oxygen (16.00), then we find that

$$\text{moles of Zn} = (80.3 \text{ g Zn})\left(\frac{1 \text{ mol Zn}}{65.38 \text{ g Zn}}\right) = 1.23 \text{ mol Zn}$$

and

$$\text{moles of O} = (19.7 \text{ g O})\left(\frac{1 \text{ mol O}}{16.00 \text{ g O}}\right) = 1.23 \text{ mol O}$$

Thus, we have

$$1.23 \text{ mol Zn} \mathrel{\hat{=}} 1.23 \text{ mol O}$$

or, dividing both sides of this expression by 1.23,

$$1.00 \text{ mol Zn} \mathrel{\hat{=}} 1.00 \text{ mol O}$$

Because 1.00 mole corresponds to Avogadro's number of atoms, we can divide both sides by Avogadro's number and get

$$1.00 \text{ atom of Zn} \mathrel{\hat{=}} 1.00 \text{ atom of O}$$

This expression says that one atom of zinc combines with one atom of oxygen and that the **simplest formula** of zinc oxide is therefore ZnO. We call ZnO the simplest chemical formula of zinc oxide because chemical analysis provides us with only the *ratios* of atoms in a chemical formula and not the actual number of atoms. Mass percentages alone cannot be used to distinguish among ZnO, Zn_2O_2, Zn_3O_3, or any other multiple of ZnO. For this reason, simplest formulas are often called **empirical formulas.** The following Example illustrates another calculation of a simplest, or empirical, formula.

EXAMPLE 5-4: The chemical name for rubbing alcohol is isopropyl alcohol. Chemical analysis shows that pure isopropyl alcohol is 60.0 percent carbon, 13.4 percent hydrogen, and 26.6 percent oxygen by mass. Determine the empirical formula of isopropyl alcohol.

Solution: As usual, we take a 100-g sample and write

$$60.0\text{ g C} \Bumpeq 13.4\text{ g H} \Bumpeq 26.6\text{ g O}$$

We divide each value by the corresponding atomic mass and get

$$(60.0\text{ g C})\left(\frac{1\text{ mol C}}{12.01\text{ g C}}\right) = 5.00\text{ mol C}$$
$$\Bumpeq (13.4\text{ g H})\left(\frac{1\text{ mol H}}{1.008\text{ g H}}\right) = 13.3\text{ mol H}$$
$$\Bumpeq (26.6\text{ g O})\left(\frac{1\text{ mol O}}{16.00\text{ g O}}\right) = 1.66\text{ mol O}$$

or

$$5.00\text{ mol C} \Bumpeq 13.3\text{ mol H} \Bumpeq 1.66\text{ mol O}$$

To find a simple, whole-number relationship for these values, we divide through by the smallest value (1.66) and get

$$3.0\text{ mol C} \Bumpeq 8.0\text{ mol H} \Bumpeq 1.0\text{ mol O}$$

Therefore, we find that the empirical formula of an isopropyl alcohol molecule consists of three carbon atoms, eight hydrogen atoms, and one oxygen atom, or that the empirical formula is C_3H_8O.

PRACTICE PROBLEM 5-4: Ethyl alcohol, the alcohol in alcoholic beverages, is 52.1 percent carbon, 13.2 percent hydrogen, and 34.7 percent oxygen. Determine the empirical formula of ethyl alcohol.

Answer: C_2H_6O

The next Example illustrates an experimental procedure for determining empirical formulas.

Figure 5-3 Magnesium nitride is produced when magnesium is burned in an atmosphere of nitrogen.

EXAMPLE 5-5: A 0.450-g sample of magnesium metal is reacted completely with nitrogen to produce 0.623 g of magnesium nitride (Figure 5-3). Use these data to determine the empirical formula of magnesium nitride.

Solution: The 0.623 g of magnesium nitride contains 0.450 g of magnesium, so the mass of nitrogen in the product is

$$\text{mass of N atoms in product} = 0.623\text{ g} - 0.450\text{ g} = 0.173\text{ g N}$$

We can convert 0.450 g of Mg to moles of Mg by dividing by its atomic mass (24.31):

$$\text{moles of Mg} = (0.450\text{ g Mg})\left(\frac{1\text{ mol Mg}}{24.31\text{ g Mg}}\right) = 0.0185\text{ mol Mg}$$

Similarly, by dividing the mass of the nitrogen atoms by the atomic mass of nitrogen (14.01), we obtain

$$\text{moles of N} = (0.173\text{ g N})\left(\frac{1\text{ mol N}}{14.01\text{ g N}}\right) = 0.0123\text{ mol N}$$

Thus, we have the relation

$$0.0185\text{ mol Mg} \Bumpeq 0.0123\text{ mol N}$$

When we divide both quantities by the smaller number (0.0123), we obtain

$$1.50 \text{ mol Mg} \Leftrightarrow 1.00 \text{ mol N}$$

Because we are seeking the simplest whole-number relationship, we multiply both sides by 2 to get

$$3.00 \text{ mol Mg} \Leftrightarrow 2.00 \text{ mol N}$$

Thus, we see that 3.00 mol of magnesium atoms combine with 2.00 mol of nitrogen atoms. Thus, the empirical formula for magnesium nitride is Mg_3N_2.

PRACTICE PROBLEM 5-5: A 2.18-g sample of scandium is burned in oxygen to produce 3.34 g of scandium oxide. Use these data to determine the empirical formula of scandium oxide.

Answer: Sc_2O_3

5-4. EMPIRICAL FORMULAS CAN BE USED TO DETERMINE AN UNKNOWN ATOMIC MASS

If we know the empirical formula of a compound, then we can determine the atomic mass of one of the elements in the compound if the atomic masses of the others are known. This determination is a standard experiment in many general chemistry laboratory courses.

EXAMPLE 5-6: The empirical formula of a metal oxide is MO, where M stands for the chemical symbol of the metal. A weighed quantity of metal, 0.490 g, is burned in oxygen, and the metal oxide produced is found to have a mass of 0.813 g. Given that the atomic mass of oxygen is 16.00, determine the atomic mass of the metal.

Solution: The mass of oxygen in the metal oxide is

$$\text{mass of O in sample} = 0.813 \text{ g MO} - 0.490 \text{ g M} = 0.323 \text{ g O}$$

The number of moles of O is

$$\text{moles of O} = (0.323 \text{ g O})\left(\frac{1 \text{ mol O}}{16.00 \text{ g O}}\right) = 0.0202 \text{ mol O}$$

The empirical formula MO tells us that 0.0202 mol of M is combined with 0.0202 mol of O, so we have

$$0.490 \text{ g M} \Leftrightarrow 0.0202 \text{ mol M}$$

The atomic mass of the metal can be obtained if we determine how many grams of the metal correspond to 1.00 mol. To determine this, we divide the stoichiometric correspondence by 0.0202 to get

$$24.3 \text{ g M} \Leftrightarrow 1.00 \text{ mol M}$$

In other words, the atomic mass of the metal is 24.3. By consulting a table of atomic masses, we see that the metal is magnesium.

PRACTICE PROBLEM 5-6: The empirical formula of a metal oxide is M_2O_3. A 3.058-g sample of the metal is burned in oxygen, and the M_2O_3 produced is found to have a mass of 4.111 g. Given that the atomic mass of oxygen is 16.00, determine the atomic mass of the metal. What is the metal?

Answer: 69.70; gallium

You can see from these examples that it is possible to determine the empirical formula of a compound if the atomic masses are known and that it is possible to determine the atomic mass of an element if the empirical formula of one of its compounds and the atomic masses of the other elements that make up the compound are known. We are faced with a dilemma here. Atomic masses can be determined if empirical formulas are known (Example 5-6), but atomic masses must be known to determine empirical formulas (Examples 5-4 and 5-5). This was a serious problem in the early 1800s, shortly after Dalton formulated his atomic theory, because it was necessary to guess the empirical formulas of compounds—an incorrect guess led to an incorrect atomic mass. We show in Chapter 7 that it was the quantitative study of gases and of reactions between gases that provided the key for resolving the difficulty in determining reliable values of atomic masses. There are now experimental techniques for the *direct* determination of atomic masses, so ambiguities no longer exist in the table of atomic masses.

5-5. AN EMPIRICAL FORMULA ALONG WITH THE MOLECULAR MASS DETERMINES THE MOLECULAR FORMULA

Suppose that the chemical analysis of a compound gives 85.7 percent carbon and 14.3 percent hydrogen by mass. We then have

$$\begin{aligned}&85.7\text{ g C} \approx 14.3\text{ g H}\\&7.14\text{ mol C} \approx 14.2\text{ mol H}\\&1\text{ mol C} \approx 2\text{ mol H}\end{aligned}$$

and conclude that the empirical formula is CH_2. However, the actual formula, the **molecular formula,** might be C_2H_4, C_3H_6, or, generally, C_nH_{2n} for any whole number n. The chemical analysis gives us only ratios of numbers of atoms. If we know the molecular mass from another experiment, however, then we can determine the molecular formula unambiguously. For example, suppose we know that the molecular mass of our compound, as determined from other experiments, is 42. By listing the various possible formulas, all of which have the empirical formula CH_2,

Formula	**Formula mass**
CH_2	14
C_2H_4	28
C_3H_6	42
C_4H_8	56

we see that C_3H_6 is the molecular formula that has a formula mass of 42. The molecular formula of the compound, therefore, is C_3H_6. You can understand now why the formula deduced from chemical analysis is called the simplest formula. It must be supplemented by molecular mass data to determine the molecular formula.

Note that the molecular formula in the above Example is "three times" the empirical formula. This factor of 3 can be obtained directly by using the following relationship:

$$\text{"factor"} = \frac{\text{molecular mass}}{\text{empirical formula mass}} = \frac{42}{14} = 3$$

EXAMPLE 5-7: There are many compounds, called **hydrocarbons,** that consist of only carbon and hydrogen. Gasoline typically is a mixture of over 100 different hydrocarbons. Chemical analysis of one of the constituents of gasoline yields 92.30 percent carbon and 7.70 percent hydrogen by mass. (a) Determine the simplest formula of this compound. (b) Given that its molecular mass is 78, determine its molecular formula.

Solution: (a) The determination of the simplest formula can be summarized by

$$\begin{aligned}&92.30\text{ g C} \Leftrightarrow 7.70\text{ g H}\\&7.69\text{ mol C} \Leftrightarrow 7.64\text{ mol H}\\&1\text{ mol C} \Leftrightarrow 1\text{ mol H}\end{aligned}$$

The simplest formula is CH, which has a formula mass of 13.
(b) The molecular mass is 78 and the empirical formula mass is 13, so the molecular formula must be 78/13 = 6 times the empirical formula. Thus, the molecular formula is C_6H_6.

PRACTICE PROBLEM 5-7: Ethylene dichloride is listed as fifteenth in amount produced annually in the United States. Chemical analysis shows that ethylene dichloride is 24.27 percent carbon, 4.075 percent hydrogen, and 71.65 percent chlorine by mass. Given that the molecular mass of ethylene dichloride is 98.95, determine its chemical formula.

Answer: $C_2H_4Cl_2$

5-6. THE COEFFICIENTS IN CHEMICAL EQUATIONS CAN BE INTERPRETED AS NUMBERS OF MOLES

A subject of great practical importance in chemistry is the determination of what quantity of product can be obtained from a given quantity of reactants. For example, the reaction between hydrogen and nitrogen to produce ammonia, NH_3, is described by the equation

$$3H_2(g) + N_2(g) \rightarrow 2NH_3(g)$$

where the coefficients in the equation are called **balancing coefficients** or **stoichiometric coefficients.** We might wish to know how much NH_3

is produced when 100 g of H_2 reacts with excess N_2. To answer questions like this, we interpret the equation in terms of moles rather than molecules.

The molecular interpretation of the hydrogen-nitrogen reaction is

$$3 \text{ molecules of hydrogen} + 1 \text{ molecule of nitrogen} \rightarrow 2 \text{ molecules of ammonia}$$

If we multiply both sides of this equation by Avogadro's number, then we obtain

$$3(6.022) \times 10^{23}) \text{ molecules of hydrogen} + 6.022 \times 10^{23} \text{ molecules of nitrogen} \rightarrow 2(6.022 \times 10^{23}) \text{ molecules of ammonia}$$

If we use the fact that 6.022×10^{23} molecules corresponds to 1 mol, then we also have

$$3 \text{ mol } H_2 + 1 \text{ mol } N_2 \rightarrow 2 \text{ mol } NH_3$$

This result is important. It tells us that the stoichiometric or balancing coefficients are the relative numbers of moles of each substance in a balanced chemical equation.

We can also interpret the hydrogen-nitrogen reaction in terms of masses. If we convert moles to masses by multiplying by the appropriate molar masses, then we get

$$6.05 \text{ g } H_2 + 28.02 \text{ g } N_2 \rightarrow 34.07 \text{ g } NH_3$$

Note that the total mass is the same on the two sides of the equation, in accord with the law of conservation of mass. Table 5-2 summarizes the various interpretations of the hydrogen-nitrogen reaction as well as those of the sodium-chlorine reaction.

We are now ready to calculate how much ammonia is produced when a given quantity of nitrogen or hydrogen is used.

Let's calculate how many moles of $NH_3(g)$ can be produced from 10.0 mol of $N_2(g)$, assuming that an excess amount of $H_2(g)$ is available. According to the balanced chemical equation,

$$3H_2(g) + N_2(g) \rightarrow 2NH_3(g)$$

2 mol of $NH_3(g)$ are produced from each mole of $N_2(g)$. We can express this relationship as a **stoichiometric unit conversion factor**

$$1 \text{ mol } N_2 \approx 2\text{mol } NH_3$$

or

$$1 = \frac{2 \text{ mol } NH_3}{1 \text{ mol } N_2}$$

Therefore, 10.0 mol of $N_2(g)$ yields

$$(10.0 \text{ mol } N_2)\left(\frac{2 \text{ mol } NH_3}{1 \text{ mol } N_2}\right) = 20.0 \text{ mol } NH_3$$

We can also calculate the number of grams of $NH_3(g)$ produced by using the fact that 17.0 g of NH_3 corresponds to 1 mol of NH_3.

$$\text{mass of } NH_3 \text{ produced} = (20.0 \text{ mol } NH_3)\left(\frac{17.0 \text{ g } NH_3}{1 \text{ mol } NH_3}\right)$$

$$= 340 \text{ g } NH_3$$

Table 5-2 The various interpretations of two chemical equations

Interpretation	$3H_2$	$+ N_2$	$\rightarrow 2NH_3$
molecular	3 molecules	+ 1 molecule	→ 2 molecules
molar	3 mol	+ 1 mol	→ 2 mol
mass	6.05 g	+ 28.02 g	→ 34.07 g
Interpretation	**2Na**	**+ Cl_2**	**→ 2NaCl**
molecular	2 atoms	+ 1 molecule	→ 2 ion pairs or 2 formula units
molar	2 mol	+ 1 mol	→ 2 mol
mass	45.98 g	+ 70.90 g	→ 116.88 g

EXAMPLE 5-8: How many grams of NH_3 can be produced from 8.50 g of $H_2(g)$, assuming that an excess amount of $N_2(g)$ is available? How many grams of $N_2(g)$ are required?

Solution: The number of moles of $H_2(g)$ corresponding to 8.50 g is

$$\text{moles of } H_2 = (8.50 \text{ g } H_2)\left(\frac{1 \text{ mol } H_2}{2.016 \text{ g } H_2}\right) = 4.22 \text{ mol } H_2$$

The number of moles of NH_3 is obtained by using the stoichiometric unit conversion factor between $NH_3(g)$ and $H_2(g)$

$$1 = \frac{2 \text{ mol } NH_3}{3 \text{ mol } H_2}$$

which we obtain directly from the balanced chemical equation. Therefore,

$$\text{moles of } NH_3(g) = (4.22 \text{ mol } H_2)\left(\frac{2 \text{ mol } NH_3}{3 \text{ mol } H_2}\right) = 2.81 \text{ mol } NH_3$$

The mass of NH_3 is given by

$$\text{mass of } NH_3 = (2.81 \text{ mol } NH_3)\left(\frac{17.0 \text{ g } NH_3}{1 \text{ mol } NH_3}\right) = 47.8 \text{ g } NH_3$$

To calculate the mass of $N_2(g)$ required to react completely with 8.50 g, or 4.22 mol of $H_2(g)$, we first calculate the number of moles of $N_2(g)$ required:

$$\text{moles of } N_2 = (4.22 \text{ mol } H_2)\left(\frac{1 \text{ mol } N_2}{3 \text{ mol } H_2}\right) = 1.41 \text{ mol } N_2$$

where, as always, the stoichiometric unit conversion factor is obtained from the balanced chemical equation. The mass of $N_2(g)$ is

$$\text{mass of } N_2 = (1.41 \text{ mol } N_2)\left(\frac{28.02 \text{ g } N_2}{1 \text{ mol } N_2}\right) = 39.5 \text{ g } N_2$$

Solution: Because all the chloride in the mixture was precipitated as AgCl(*s*), we have

$$\begin{pmatrix}\text{number of}\\ \text{chloride ions}\\ \text{in NaCl}\end{pmatrix} + \begin{pmatrix}\text{number of}\\ \text{chloride ions}\\ \text{in KCl}\end{pmatrix} = \begin{pmatrix}\text{number of}\\ \text{chloride ions}\\ \text{in AgCl}\end{pmatrix}$$

or in terms of moles,

$$\text{mol NaCl} + \text{mol KCl} = \text{mol AgCl}$$

If we let x be the number of grams of NaCl in the mixture, then $1.25 - x$ is the number of grams of KCl. The number of moles of NaCl, KCl, and AgCl are given by

$$\text{moles of NaCl} = (x\ \text{g NaCl})\left(\frac{1\ \text{mol NaCl}}{58.44\ \text{g NaCl}}\right)$$

$$\text{moles of KCl} = [(1.25\ \text{g} - x)\ \text{KCl}]\left(\frac{1\ \text{mol KCl}}{74.55\ \text{g KCl}}\right)$$

$$\text{moles of AgCl} = (2.50\ \text{g AgCl})\left(\frac{1\ \text{mol AgCl}}{143.3\ \text{g AgCl}}\right)$$
$$= 0.01744\ \text{mol}$$

Using the relation

$$\text{mol NaCl} + \text{mol KCl} = \text{mol AgCl}$$

gives

$$\frac{x}{58.44} + \frac{(1.25 - x)}{74.55} = 0.01744$$

or

$$0.0171x + 0.0168 - 0.0134x = 0.01744$$

Solving for x, we find

$$3.7 \times 10^{-3}\ x = 6.4 \times 10^{-4}$$

or

$$x = 0.17\ \text{g}$$

Thus, the mass percentages of NaCl and KCl in the mixture are

$$\text{mass \% NaCl} = \left(\frac{0.17\ \text{g}}{1.25\ \text{g}}\right) \times 100 = 14\ \text{percent NaCl}$$

$$\text{mass \% KCl} = \left(\frac{1.25\ \text{g} - 0.17\ \text{g}}{1.25\ \text{g}}\right) \times 100 = 86\ \text{percent KCl}$$

PRACTICE PROBLEM 5-13: A mixture of NaCl and $BaCl_2$ weighing 2.86 g is dissolved in water; when $AgNO_3$ is added to the solution, a precipitate of AgCl which weighs 4.81 g is obtained. Calculate the mass percentages of NaCl and of $BaCl_2$ in the sample.

Answer: 28.3 percent NaCl and 71.7 percent $BaCl_2$

5-9. WHEN TWO OR MORE SUBSTANCES REACT, THE MASS OF THE PRODUCT IS DETERMINED BY THE LIMITING REACTANT

If you look back over the Examples in this chapter, you will notice that in no case did we start out with the masses of more than one of the reactants given. We have always assumed that there was sufficient material present to react with all that reactant whose quantity was given. Let's now consider an example in which we do have quantities of two reactants given. Cadmium sulfide, which is used in light meters, solar cells, and other light-sensitive devices, can be made by the direct combination of the two elements:

$$\text{Cd}(s) + \text{S}(s) \rightarrow \text{CdS}(s) \qquad (5\text{-}1)$$

How much CdS is produced if we start out with 2.00 g of cadmium and 2.00 g of sulfur? As in all stoichiometric calculations, we first determine the number of moles of each reactant:

$$\text{moles of Cd} = (2.00\text{ g Cd})\left(\frac{1\text{ mol Cd}}{112.4\text{ g Cd}}\right) = 0.0178\text{ mol Cd}$$

$$\text{moles of S} = (2.00\text{ g S})\left(\frac{1\text{ mol S}}{32.06\text{ g S}}\right) = 0.0624\text{ mol S}$$

We know from inspection of the balanced chemical equation that 1 mol of cadmium requires 1 mol of sulfur, so the 0.0178 mol of cadmium requires 0.0178 mol of sulfur. Thus, there is excess sulfur; only 0.0178 mol of sulfur reacts and (0.0624 − 0.0178) mol = 0.0446 mol of sulfur remains after the reaction is completed. The cadmium reacts completely, and the moles of cadmium consumed determine how much CdS is produced. The reactant that is consumed completely and thereby limits the amount of product formed is called the **limiting reactant** (Cd in the Example), and any other reactants are called **excess reactants** (S in the Example). The initial mass of limiting reactant must be used to calculate how much product is formed.

In Equation (5-1), 0.0178 mol of cadmium reacts with 0.0178 mol of sulfur to produce 0.0178 mol of CdS. The mass of CdS produced is

$$\text{mass of CdS} = (0.0178\text{ mol CdS})\left(\frac{144.5\text{ g CdS}}{1\text{ mol CdS}}\right) = 2.57\text{ g CdS}$$

The unused sulfur (0.0446 mol) has a mass of

$$\text{mass of unused S} = (0.0446\text{ mol S})\left(\frac{32.06\text{ g S}}{1\text{ mol S}}\right) = 1.43\text{ g S}$$

Note that before the reaction there are 2.00 g of cadmium and 2.00 g of sulfur, or 4.00 g of reactants. After the reaction, there are 2.57 g of cadmium sulfide and 1.43 g of sulfur, or a total of 4.00 g, as required by the law of conservation of mass.

When the masses of two or more reactants are given in a problem, we must check to see which, if either, is a limiting reactant. If one of the reactants is limiting, then it is the one to be used in the calculation of the mass of product obtained.

EXAMPLE 5-14: A mixture is prepared from 25.0 g of aluminum and 85.0 g of Fe_2O_3. The reaction that occurs is described by the equation

$$Fe_2O_3(s) + 2Al(s) \rightarrow Al_2O_3(s) + 2Fe(l)$$

How much iron is produced in the reaction?

Solution: Because the masses of both reactants are given, we must check to see which, if either, is a limiting reactant. The number of moles of Al and Fe_2O_3 available is

$$\text{moles of Al} = (25.0 \text{ g Al})\left(\frac{1 \text{ mol Al}}{26.98 \text{ g Al}}\right) = 0.927 \text{ mol Al}$$

$$\text{moles of } Fe_2O_3 = (85.0 \text{ g } Fe_2O_3)\left(\frac{1 \text{ mol } Fe_2O_3}{159.7 \text{ g } Fe_2O_3}\right) = 0.532 \text{ mol } Fe_2O_3$$

From the balanced chemical equation, we see that 0.927 mol of aluminum consumes only

$$(0.927 \text{ mol Al})\left(\frac{1 \text{ mol } Fe_2O_3}{2 \text{ mol Al}}\right) = 0.464 \text{ mol } Fe_2O_3$$

Thus, we see that the Fe_2O_3 is in excess. The amount of Fe_2O_3 in excess is

$$\text{mol excess } Fe_2O_3 = (0.532 - 0.464) \text{ mol } Fe_2O_3 = 0.068 \text{ mol } Fe_2O_3$$

The aluminum is the limiting reactant and the one to use in calculating how much iron is produced. From the balanced equation,

$$2 \text{ mol Al} \Leftrightarrow 2 \text{ mol Fe}$$

or

$$0.927 \text{ mol Al} \Leftrightarrow 0.927 \text{ mol Fe}$$

The mass of iron corresponding to 0.927 mol is

$$\text{grams of Fe} = (0.927 \text{ mol Fe})\left(\frac{55.85 \text{ g Fe}}{1 \text{ mol Fe}}\right) = 51.8 \text{ g Fe}$$

The reaction of aluminum metal with a metal oxide is called a **thermite reaction** and has numerous applications (Figure 5-8). Thermite reactions were once used to weld railroad rails and are used in thermite grenades, which are employed by the military to destroy heavy equipment. In a thermite reaction the reaction temperature can exceed 3500°C.

PRACTICE PROBLEM 5-14: Calcium sulfide, which is used in luminous paints and as a depilatory, can be made by heating calcium sulfate with charcoal at a high temperature. The unbalanced equation is

$$CaSO_4(s) + C(s) \rightarrow CaS(s) + CO(g) \qquad \text{(unbalanced)}$$

How many grams of $CaS(s)$ can be prepared from 100 g each of $CaSO_4(s)$ and $C(s)$?

Answer: 53.0 g

Figure 5-8 Thermite reaction. A spectacular example of a single-replacement reaction is the reaction between powdered aluminum metal and iron(III) oxide

$$2Al(s) + Fe_2O_3(s) \rightarrow 2Fe(l) + Al_2O_3(s)$$

Once this reaction is initiated by a heat source such as a burning magnesium ribbon, it proceeds vigorously, producing so much heat that the iron is formed as a liquid.

An easy way to see which reactant is the limiting reactant is to compare the number of moles given for each reactant divided by the corresponding stoichiometric coefficient; the smallest resulting number of moles identifies the limiting reactant. Thus, in Example 5-14 we have

$$\frac{\text{mol Al}}{2} = \frac{0.927\ \text{mol}}{2} = 0.464\ \text{mol}$$

and

$$\frac{\text{mol Fe}_2\text{O}_3}{1} = 0.532\ \text{mol}$$

This result tells us that the Al(*s*) is the limiting reactant. This procedure is equivalent to that used in Example 5-14.

There are many instances in which it is important to add reactants in stoichiometric proportions so as not to have any reactants left over. The propulsion of rockets and space vehicles serves as a good example (Figure 5-9). The Lunar Lander rocket engines were powered by a reaction similar to

$$\underset{\text{dinitrogen tetroxide}}{N_2O_4(l)} + \underset{\text{hydrazine}}{2N_2H_4(l)} \rightarrow 3N_2(g) + 4H_2O(g)$$

Dinitrogen tetroxide and hydrazine react explosively when brought into contact. These two reactants were kept in separate tanks and pumped through pipes into the rocket engines, where they reacted. The gases produced (H_2O is a gas at the exhaust temperatures of the rocket engines) exited through the exhaust chamber of the engine and propelled the rocket forward. The cost of carrying materials into space is enormous, and the two fuels must be combined in the correct proportions. It would be wasteful to carry any excess reactant.

Figure 5-9 Apollo Lunar Lander.

5-10. FOR MANY CHEMICAL REACTIONS THE AMOUNT OF THE DESIRED PRODUCT OBTAINED IS LESS THAN THE THEORETICAL AMOUNT

In all the Examples considered up to now, we have assumed that the amount of products produced can be calculated from the complete reaction of the limiting reactant. However, it frequently happens that less than this **theoretical yield** of product is obtained because (1) the reaction may fail to go to completion; (2) there may be side reactions that give rise to undesired products; or (3) some of the desired product may not be readily recoverable or may be lost in the purification process. In these instances, the mass of the product that actually is obtained is called the **actual yield** and the efficiency of conversion of reactants into recovered products can be expressed as the **percentage yield** (% yield). The percentage yield is defined as

$$\%\ \text{yield} = \left(\frac{\text{actual yield}}{\text{theoretical yield}}\right) \times 100 \qquad (5\text{-}2)$$

The industrial production of methyl alcohol, $CH_3OH(l)$, from the high-pressure reaction

$$CO(g) + 2H_2(g) \rightarrow CH_3OH(l)$$

serves to illustrate the difference between the theoretical yield and the actual yield of a reaction. For a variety of reasons, this reaction does not give a 100 percent yield. Suppose that 5.12 metric tons of $CH_3OH(l)$ is obtained from 1.00 metric ton of $H_2(g)$ reacting with an excess of $CO(g)$. Let's calculate the percentage yield of $CH_3OH(l)$. The theoretical yield is

$$\text{theoretical yield} = (1 \text{ metric ton } H_2)\left(\frac{10^6 \text{ g}}{1 \text{ metric ton}}\right)\left(\frac{1 \text{ mol } H_2}{2.016 \text{ g } H_2}\right) \times \left(\frac{1 \text{ mol } CH_3OH}{2 \text{ mol } H_2}\right)\left(\frac{32.03 \text{ g } CH_3OH}{1 \text{ mol } CH_3OH}\right)$$

$$= 7.94 \times 10^6 \text{ g } CH_3OH = 7.94 \text{ metric tons } CH_3OH$$

The percentage yield is given by Equation (5-2):

$$\% \text{ yield} = \left(\frac{\text{actual yield}}{\text{theoretical yield}}\right) \times 100 = \left(\frac{5.12 \text{ metric tons}}{7.94 \text{ metric tons}}\right) \times 100 = 64.5\%$$

EXAMPLE 5-15: A 0.473-g sample of phosphorus is reacted with an excess of chlorine, and 2.12 g of phosphorus pentachloride, PCl_5, is collected. The equation for the reaction is

$$2P(s) + 5Cl_2(g) \rightarrow 2PCl_5(s)$$

What is the percentage yield of PCl_5?

Solution: The theoretical yield of PCl_5 is

$$\text{theoretical yield} = (0.473 \text{ g P})\left(\frac{1 \text{ mol P}}{30.97 \text{ g P}}\right)\left(\frac{1 \text{ mol } PCl_5}{1 \text{ mol P}}\right) \times \left(\frac{208.2 \text{ g } PCl_5}{1 \text{ mol } PCl_5}\right)$$

$$= 3.18 \text{ g } PCl_5$$

The percentage yield is

$$\% \text{ yield} = \left(\frac{\text{actual yield}}{\text{theoretical yield}}\right) \times 100 = \left(\frac{2.12 \text{ g}}{3.18 \text{ g}}\right) \times 100 = 66.7\%$$

PRACTICE PROBLEM 5-15: Tin(IV) chloride can be made by heating tin in an atmosphere of dry chlorine

$$Sn(s) + 2Cl_2(g) \rightarrow SnCl_4(l)$$

If the percentage yield of this process is 64.3 percent, then how many grams of tin are required to produce 0.106 g of $SnCl_4(l)$?

Answer: 0.0751 g

Most reactions take place in solution, so in the next chapter we will discuss how to express quantitatively the concentrations of solutions and how to calculate reaction quantities involving reactions that take place in solutions.

SUMMARY

Stoichiometric calculations are based on the concept of a mole. The quantity of a substance that is numerically equal to its formula mass in grams is called a mole of that substance. Thus, in order to calculate the number of moles in a given mass of a substance, it is necessary to know its chemical formula. Another definition of a mole is that mass of a substance that contains Avogadro's number (6.022×10^{23}) of formula units or elementary entities. By using the concept of a mole and Avogadro's number, we can carry out a large variety of chemical calculations. For example, we can calculate the masses of individual atoms and molecules, or calculate how many atoms and molecules there are in a given mass of a substance. We can also use the idea of a mole to determine chemical formulas from chemical analysis. By interpreting the balancing coefficients in chemical equations in terms of moles, we can use chemical equations to calculate quantities of substances involved in chemical reactions. For example, we can calculate how much product can be obtained from a given amount of reactant, or how much reactant to use to obtain a given amount of product.

TERMS YOU SHOULD KNOW

atomic substance 140
molecular substance 140
formula mass 140
formula unit 140
mole (mol) 140
molar mass 140
Avogadro's number (6.022×10^{23} formula units per mole) 142
stoichiometry 145
$\Leftrightarrow$ ("stoichiometrically equivalent to" or "combines with") 145
simplest formula 145
empirical formula 145
molecular formula 148
hydrocarbon 149
balancing coefficient 149
stoichiometric coefficient 149
stoichiometric unit conversion factor 150
metric ton 156
limiting reactant 159
excess reactant 159
thermite reaction 160
theoretical yield 161
actual yield 161
percentage yield (% yield) 161

AN EQUATION YOU SHOULD KNOW HOW TO USE

$$\% \text{ yield} = \left(\frac{\text{actual yield}}{\text{theoretical yield}}\right) \times 100 \qquad (5\text{-}2) \qquad \text{(calculation of percentage yield)}$$

PROBLEMS

NUMBER OF MOLES

5-1. Calculate the number of moles in

(a) 28.0 g of H_2O (1 ounce)
(b) 200 mg of diamond (C) (1 carat)
(c) 454 g of NaCl (1 pound)
(d) 1000 kg of CaO (1 metric ton)

5-2. Calculate the number of moles in the recommended daily allowance of the following substances:

(a) 15 mg of zinc
(b) 60 mg of vitamin C, $C_6H_8O_6$
(c) 1.5 mg of vitamin A, $C_{20}H_{30}O$
(d) 6.0 μg of vitamin B_{12}, $C_{63}H_{88}CoN_{14}O_{14}P$

5-3. Calculate the number of moles in

(a) 1.00 kg of malathion, $C_{10}H_{19}O_6PS_2$
(b) 75.0 g of aluminum sulfate, $Al_2(SO_4)_3$
(c) 50.0 mg of oil of peppermint, $C_{10}H_{20}O$
(d) 2.756 g of potassium dichromate, $K_2Cr_2O_7$

5-4. Calculate the number of moles in

(a) 2.00 kg of parathion, $C_{10}H_{14}NO_5PS$
(b) 250.0 g of ammonium hydrogen phosphate, $(NH_4)_2HPO_4$
(c) 75.0 mg of oil of cinnamon, $C_{18}H_{14}O_3$
(d) 150.0 g of Epsom salt, $MgSO_4{\cdot}7H_2O$

5-5. A baseball has a mass of 142 g. Calculate the mass of Avogadro's number of baseballs (1 mol) and compare your result with the mass of the earth, 6.0×10^{24} kg.

5-6. The U.S. population is about 255 million. If Avogadro's number of dollars were distributed equally among the population, how many dollars would each person receive?

5-7. Compute the mass in grams of

(a) one CO_2 molecule
(b) one $C_6H_{12}O_6$ (glucose) molecule
(c) one $CaCl_2$ formula unit

5-8. Compute the mass in grams of

(a) one O_2 molecule
(b) one $FeSO_4$ formula unit
(c) one $C_{12}H_{22}O_{11}$ (sucrose) molecule

5-9. Calculate the mass in grams of

(a) 200 iron atoms
(b) 1.0×10^{16} water molecules
(c) 1.0×10^{6} oxygen atoms
(d) 1.0×10^{6} oxygen molecules (O_2)

5-10. Calculate the mass in grams of

(a) 100 molecules of nitroglycerin, $C_3H_5N_3O_9$
(b) 5000 molecules of TNT, $C_7H_5N_3O_6$
(c) 10^{10} molecules of octane, C_8H_{18}
(d) 10 molecules of ozone, O_3

5-11. A 50.0-g sample of H_2O contains ____ moles of H_2O, ____ molecules of H_2O, and a total of ____ atoms.

5-12. A 450-g sample of CH_3OH contains ____ moles of CH_3OH, ____ molecules of CH_3OH, and a total of ____ atoms.

EMPIRICAL FORMULA

5-13. Calcium carbide produces acetylene when water is added to it. The acetylene evolved is burned to provide the light source on spelunkers' helmets. Chemical analysis shows that calcium carbide is 62.5 percent (by mass) calcium and 37.5 percent (by mass) carbon. Determine the empirical formula of calcium carbide.

5-14. Rust occurs when iron metal reacts with the oxygen in the air. Chemical analysis shows that dry rust is 69.9 percent iron and 30.1 percent oxygen by mass. Determine the empirical formula of rust.

5-15. A 2.46-g sample of copper metal is reacted completely with chlorine gas to produce 5.22 g of copper chloride. Determine the empirical formula for this chloride.

5-16. A 3.78-g sample of iron metal is reacted with sulfur to produce 5.95 g of iron sulfide. Determine the empirical formula of this compound.

5-17. A 28.1-g sample of cobalt metal was reacted completely with excess chlorine gas. The mass of the compound formed was 61.9 g. Determine its empirical formula.

5-18. A 5.00-g sample of aluminum metal is burned in an oxygen atmosphere to provide 9.45 g of aluminum oxide. Use these data to determine the empirical formula of aluminum oxide.

5-19. Given the following mass percentages of the elements in certain compounds, determine the empirical formulas in each case.

(a) 46.45% Li   53.55% O
(b) 59.78% Li   40.22% N
(c) 14.17% Li   85.83% N
(d) 36.11% Ca   63.89% Cl

5-20. Given the following mass percentages of the elements in certain compounds, determine the empirical formula in each case.

(a) 71.89% Tl   28.11% Br
(b) 74.51% Pb   25.49% Cl
(c) 82.24% N   17.76% H
(d) 72.24% Mg   27.76% N

DETERMINATION OF ATOMIC MASS

5-21. A 1.443-g sample of metal is reacted with excess oxygen to yield 1.683 g of the oxide M_2O_3. Calculate the atomic mass of the element M.

5-22. An element forms a chloride whose formula is XCl_4, which is known to consist of 75.0 percent chlorine by mass. Calculate the atomic mass of X and identify it.

5-23. A sample of a compound with the formula $MCl_2 \cdot 2H_2O$ has a mass of 0.642 g. When the compound is heated to remove the water of hydration (represented by $\cdot\ 2H_2O$ in the formula), 0.0949 g of water is collected. What element is M?

5-24. The formula of an acid is only partially known as HXO_3. The mass of 0.0133 mol of this acid is 1.123 g. Find the atomic mass of X and identify the element represented by X.

MOLECULAR FORMULAS

5-25. Acetone is an important chemical solvent; a familiar home use is as a nail polish remover. Chemical analysis shows that acetone is 62.0 percent carbon, 10.4 percent hydrogen, and 27.5 percent oxygen by mass. Determine the empirical formula of acetone. In a separate experiment, the molecular mass is found to be 58.1. What is the molecular formula of acetone?

5-26. Glucose, one of the main sources of energy used by living organisms, has a molecular mass of 180.2. Chemical analysis shows that glucose is 40.0 percent carbon, 6.71 percent hydrogen, and 53.3 percent oxygen by mass. Determine its molecular formula.

5-27. A class of compounds called sodium metaphosphates were used as additives to detergents to improve cleaning ability. One of them has a molecular mass of 612. Chemical analysis shows that this sodium metaphosphate consists of 22.5 percent sodium, 30.4 percent phosphorus, and 47.1 percent oxygen by mass. Determine the molecular formula of this compound.

5-28. A hemoglobin sample was found to be 0.373 percent (by mass) iron. Given that there are four iron atoms per hemoglobin molecule, determine the molecular mass of hemoglobin.

CALCULATIONS INVOLVING CHEMICAL REACTIONS

5-29. The combustion of propane occurs via the reaction

$$C_3H_8(g) + 5O_2(g) \rightarrow 3CO_2(g) + 4H_2O(g)$$

How many grams of oxygen are required to burn completely 10.0 g of propane?

5-30. Iodine is prepared both in the laboratory and commercially by adding $Cl_2(g)$ to an aqueous solution containing sodium iodide according to

$$2NaI(aq) + Cl_2(g) \rightarrow I_2(s) + 2NaCl(aq)$$

How many grams of sodium iodide must be used to produce 50.0 g of iodine?

5-31. Small quantities of chlorine can be prepared in the laboratory by the reaction

$$MnO_2(s) + 4HCl(aq) \rightarrow MnCl_2(aq) + Cl_2(g) + 2H_2O(l)$$

How many grams of chlorine can be prepared from 100 g of manganese dioxide?

5-32. Small quantities of oxygen can be prepared in the laboratory by heating potassium chlorate, $KClO_3$. The equation for the reaction is

$$2KClO_3(s) \rightarrow 2KCl(s) + 3O_2(g)$$

Calculate how many grams of O_2 can be produced from heating 10.0 g of $KClO_3$.

5-33. Lithium nitride reacts with water to produce ammonia and lithium hydroxide:

$$Li_3N(s) + 3H_2O(l) \rightarrow NH_3(g) + 3LiOH(aq)$$

Heavy water is water with the isotope deuterium in place of ordinary hydrogen, and its formula is D_2O. The above reaction can be used to produce heavy ammonia, ND_3.

$$Li_3N(s) + 3D_2O(l) \rightarrow ND_3(g) + 3LiOD(aq)$$

Calculate how many grams of heavy water are required to produce 200 mg of ND_3. The atomic mass of deuterium is 2.014.

5-34. A common natural source of phosphorus is phosphate rock, an ore found in extensive deposits in areas that were originally ocean floor. The formula of one type of phosphate rock is $Ca_{10}(OH)_2(PO_4)_6$. Phosphate rock is converted to phosphoric acid by the reaction

$$Ca_{10}(OH)_2(PO_4)_6(s) + 10H_2SO_4(l) \rightarrow 6H_3PO_4(l) + 10CaSO_4(s) + 2H_2O(l)$$

Calculate how many metric tons of phosphoric acid can be produced from 100 metric tons of phosphate rock (1 metric ton = 1000 kg).

5-35. Zinc is produced from its principal ore, sphalerite (ZnS), by the two-step process

$$2ZnS(s) + 3O_2(g) \rightarrow 2ZnO(s) + 2SO_2(g)$$

$$ZnO(s) + C(s) \rightarrow Zn(s) + CO(g)$$

How many kilograms of zinc can be produced from 2.00×10^5 kg of ZnS?

5-36. Titanium is produced from its principal ore, rutile (TiO_2), by the two-step process

$$TiO_2(s) + 2Cl_2(g) + 2C(s) \rightarrow TiCl_4(g) + 2CO(g)$$

$$TiCl_4(g) + 2Mg(s) \rightarrow Ti(s) + 2MgCl_2(s)$$

How many kilograms of titanium can be produced from 4.10×10^3 kg of TiO_2?

CALCULATIONS WITHOUT THE CHEMICAL EQUATION

5-37. Hydrazine, $N_2H_4(l)$, is produced commercially by the reaction of ammonia, $NH_3(g)$, with NaOCl (Raschig synthesis). Assuming that all the nitrogen in ammonia ends up in hydrazine, calculate how many metric tons of hydrazine can be produced from 10.0 metric tons of $NH_3(g)$.

5-38. Sulfur is obtained in large quantities from the hydrogen sulfide in so-called sour natural gas resulting from the removal of sulfur from petroleum. How many metric tons of sulfur can be obtained from 10.0 metric tons of hydrogen sulfide?

5-39. The thallium in a 12.76-g sample of an insecticide was precipitated as thallium(I) iodide, TlI(*s*). Calculate the mass percentage of $Tl_2SO_4(s)$ in the insecticide that would yield 0.6112 g of precipitated TlI from the sample.

The most common type of solution is a solid dissolved in a liquid. The solid that is dissolved is called the **solute,** and the liquid in which it is dissolved is called the **solvent.** The terms solvent and solute are merely terms of convenience, because all the components of a solution are uniformly dispersed throughout the solution. When NaCl(*s*) is dissolved in water, NaCl(*s*) is the solute and $H_2O(l)$ is the solvent. The process of dissolving NaCl(*s*) in water is represented by the equation

$$\mathrm{NaCl}(s) \xrightarrow[\mathrm{H_2O}(l)]{} \mathrm{Na}^+(aq) + \mathrm{Cl}^-(aq)$$

where $H_2O(l)$ under the arrow tells us that water is the solvent. The species $Na^+(aq)$ and $Cl^-(aq)$ represent a sodium ion and a chloride ion in an aqueous solution. As Figure 6-1 illustrates, these ions are solvated by water molecules; that is, they are surrounded by a loosely bound shell of water molecules.

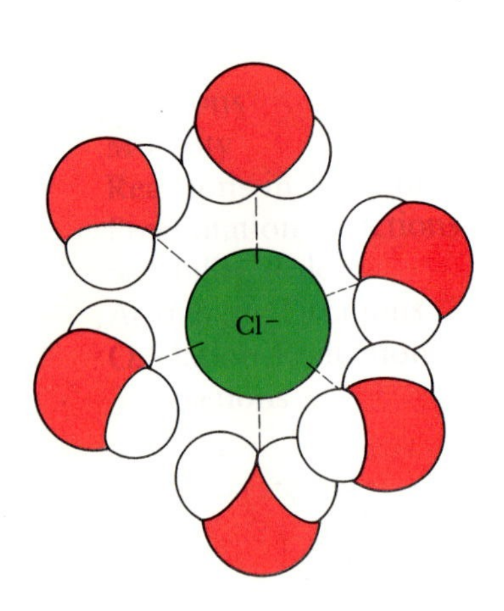

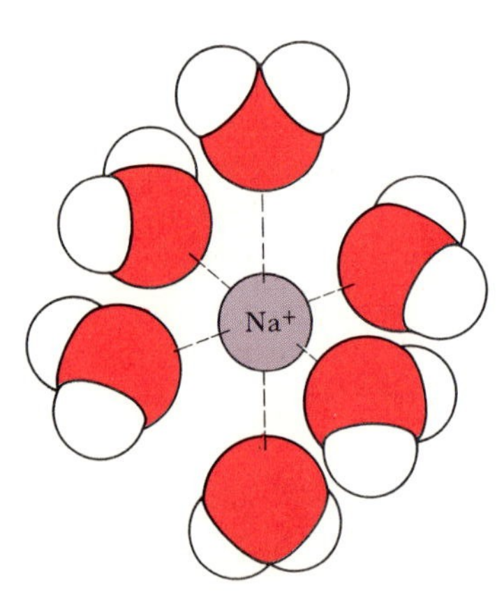

Figure 6-1 Ions in aqueous solutions are surrounded by a loosely bound shell of water molecules. Such ions are said to be solvated.

When a small quantity of sodium chloride is added to water, the sodium chloride dissolves completely, leaving no crystals at the bottom of the beaker. As more and more sodium chloride is added, we reach a point where no more sodium chloride can dissolve, so any further sodium chloride crystals that we add simply remain at the bottom of the beaker. Such a solution is called a **saturated solution,** and the maximum quantity of solute dissolved is called the **solubility** of that solute. Solubility can be expressed in a variety of units, but more commonly it is expressed as grams of solute per 100 g of solvent. For example, we can say that the solubility of KCl(*s*) in water at 20°C is about 32 g per 100 g of $H_2O(l)$.

It is important to realize that the solubility of a substance is the maximum quantity that can be dissolved in a saturated solution at a particular temperature. The solubility of NaCl(*s*) at 20°C is about 36 g per 100 g of $H_2O(l)$. If we add 50 g of NaCl(*s*) to 100 g of $H_2O(l)$ at 20°C, then 36 g dissolves and 14 g is left as undissolved NaCl(*s*). The solution is saturated. If we add 25 g of NaCl(*s*) to 100 g of $H_2O(l)$, then all the NaCl(*s*) dissolves to form what is called an **unsaturated solution;** that is, a solution in which we can dissolve more of a particular solute.

In most cases the solubility of a substance depends on temperature. The effect of temperature on the solubility of several salts in water is shown in Figure 6-2. Almost all salts become more soluble in water as the temperature increases. For example, potassium nitrate is about five times more soluble in water at 40°C than at 0°C.

6-2. MOLARITY IS THE MOST COMMON UNIT OF CONCENTRATION

The **concentration** of solute in a solution describes the quantity of solute dissolved in a given quantity of solvent or a given quantity of solution. The most common method of expressing the concentration of a solute is **molarity,** which is denoted by the symbol M. Molarity is defined as the number of moles of solute per liter of solution:

$$\text{molarity} = \frac{\text{moles of solute}}{\text{liters of solution}} \qquad (6\text{-}1)$$

Equation (6-1) can be expressed symbolically as

$$M = \frac{n}{V} \tag{6-2}$$

where M represents the molarity of the solution, n is the number of moles of solute dissolved in the solution, and V is the volume of the solution in liters. To see how to use Equation (6-2), let's calculate the molarity of a solution prepared by dissolving 62.3 g of sucrose, $C_{12}H_{22}O_{11}(s)$, in enough water to form 0.500 L of solution. The formula mass of sucrose is 342, so 62.3 g corresponds to

$$\text{moles of sucrose} = (6.23\ \text{g sucrose})\left(\frac{1\ \text{mol sucrose}}{342\ \text{g sucrose}}\right) = 0.182\ \text{mol}$$

The molarity of the solution is given by

$$M = \frac{n}{V} = \left(\frac{0.182\ \text{mol}}{0.500\ \text{L}}\right) = 0.364\ \text{mol·L}^{-1} = 0.364\ \text{M}$$

We say that the concentration of sucrose in the solution is 0.364 **molar** or 0.364 M.

The definition of molarity involves the volume of the solution, *not* the volume of the solvent. Suppose we wish to prepare one liter of a 0.100 M aqueous solution of potassium dichromate, $K_2Cr_2O_7$. We would prepare the solution by weighing out 0.100 mol (29.4 g) of $K_2Cr_2O_7(s)$, dissolving it in less than one liter of water, say, about 500 mL, and then adding water while stirring until the final volume of the solution is precisely one liter (Figure 6-3). As shown in Figure 6-3, we use a **volumetric flask,** which is a precision piece of glassware used to prepare precise volumes. It would be incorrect to add 0.100 mol of $K_2Cr_2O_7(s)$ to one liter of water; the final volume of such a solution is not precisely one liter because the added $K_2Cr_2O_7(s)$ changes the volume from 1.00 L to 1.02 L. The following example illustrates the procedure for making up a solution of a specified molarity.

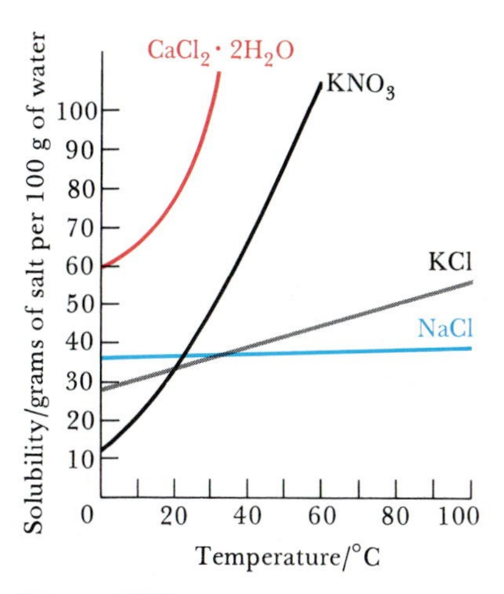

Figure 6-2 The solubility of most salts is a function of temperature. Usually solubility increases with increasing temperature.

Figure 6-3 The procedure used to prepare one liter of a solution of a specific molarity, such as 0.100 M $K_2Cr_2O_7$. (a) The 0.100 mol of $K_2Cr_2O_7$ (29.4 g) is weighed out and (b) added to a 1-L volumetric flask that is only partially filled with water. (c) The $K_2Cr_2O_7$ is dissolved, and then more water is added to bring the final volume up to the 1-L mark on the flask. The solution is swirled to ensure uniform mixing.

EXAMPLE 6-1: Potassium bromide, KBr(*s*), is used as a sedative and as an anticonvulsive agent. Explain how you would prepare 250 mL of a 0.600 M aqueous KBr solution.

Solution: From Equation (6-2) and the specified volume and concentration, we can calculate the number of moles of KBr(*s*) required. Equation (6-2) can be written as

$$n = MV \tag{6-3}$$

so

$$\begin{aligned}\text{moles of KBr} &= (0.600\ \text{M})(0.250\ \text{L}) = (0.600\ \text{mol}\cdot\text{L}^{-1})(0.250\ \text{L}) \\ &= 0.150\ \text{mol}\end{aligned}$$

We can convert moles to grams by multiplying by the formula mass of KBr (119.0):

$$\text{grams of KBr} = (0.150\ \text{mol KBr})\left(\frac{119.0\ \text{g KBr}}{1\ \text{mol KBr}}\right) = 17.9\ \text{g}$$

To prepare the solution, we add 17.9 g of KBr(*s*) to a 250-mL volumetric flask that is partially filled with distilled water. We swirl the flask until the salt is dissolved and then dilute the solution to the 250-mL mark on the flask and shake it again to assure uniformity. Do not add the KBr(*s*) to 250 mL of water, because the volume of the resulting solution will not necessarily be 250 mL.

PRACTICE PROBLEM 6-1: Ammonium selenate, $(NH_4)_2SeO_4(s)$, is used as a mothproofing agent. Describe how you would prepare 500 mL of a 0.155 M aqueous solution of ammonium selenate.

Answer: Dissolve 13.9 g of ammonium selenate in less than 500 mL of water and then dilute to 500 mL by using a volumetric flask.

Occasionally the concentration of a solution is given as the mass percentage of the solute. For example, commercial sulfuric acid is sold as a solution that is 96.7 percent $H_2SO_4(l)$ and 3.3 percent $H_2O(l)$ by mass. If you know the density of such a solution, you can calculate its molarity. The density of the commercial sulfuric acid solution is 1.84 $\text{g}\cdot\text{mL}^{-1}$. The mass of $H_2SO_4(l)$ in one liter of solution is given by

$$\begin{aligned}\left(\begin{array}{c}\text{mass of } H_2SO_4(l)\\ \text{per liter of solution}\end{array}\right) &= (96.7\ \text{percent})(\text{mass of 1 L of solution}) \\ &= (0.967)(\text{mass of 1 L of solution}) \\ &= (0.967)(1.84\ \text{g}\cdot\text{mL}^{-1})\left(\frac{1000\ \text{mL}}{1\ \text{L}}\right) \\ &= 1780\ \text{g } H_2SO_4(l) \text{ per liter of solution}\end{aligned}$$

and the number of moles of $H_2SO_4(l)$ per liter of solution—or, in other words, the molarity—is given by

$$\begin{aligned}\text{molarity of } H_2SO_4(aq) &= (1780\ \text{g}\cdot\text{L}^{-1})\left(\frac{1\ \text{mol } H_2SO_4}{98.08\ \text{g } H_2SO_4}\right) \\ &= 18.1\ \text{M}\end{aligned}$$

EXAMPLE 6-2: Ammonia is sold as an aqueous solution that is 28.0 percent NH_3 by mass and has a density of $0.90\ g{\cdot}mL^{-1}$. Calculate the molarity of this solution.

Solution: The mass of one liter of the solution is

$$\text{mass of 1 L of solution} = (0.90\ g{\cdot}mL^{-1})\left(\frac{1000\ mL}{1\ L}\right) = 900\ g{\cdot}L^{-1}$$

and the mass of NH_3 in one liter of solution is

$$\left(\begin{matrix}\text{mass of } NH_3 \\ \text{per liter of solution}\end{matrix}\right) = (28.0\ \text{percent})(900\ g{\cdot}L^{-1}) = (0.280)(900\ g{\cdot}L^{-1})$$

$$= 252\ g\ NH_3 \text{ per liter of solution}$$

The molarity is given by

$$\text{molarity of } NH_3(aq) = (252\ g{\cdot}L^{-1})\left(\frac{1\ mol\ NH_3}{17.03\ g\ NH_3}\right) = 14.8\ M$$

PRACTICE PROBLEM 6-2: An aqueous sodium hydroxide solution is 30.0 percent NaOH by mass and has a density of $1.33\ g{\cdot}mL^{-1}$. Calculate the molarity of the solution.

Answer: 9.98 M

It is often necessary in laboratory work to prepare a more dilute solution from a more concentrated stock solution. In such cases, a certain volume of a solution of known molarity is diluted with a certain volume of pure solvent to produce the final solution with the desired molarity. The key point to recognize in carrying out such **dilution** calculations is that the number of moles of solute does not change on dilution with solvent. Thus, from Equation (6-3), we have

$$\text{moles of solute} = n = M_1V_1 = M_2V_2$$

or

$$M_1V_1 = M_2V_2 \qquad \text{(dilution)} \tag{6-4}$$

where M_1 and V_1 are the initial molarity and initial volume, and M_2 and V_2 are the final molarity and final volume of the solution. The following Example illustrates a dilution calculation.

EXAMPLE 6-3: Compute the volume of 6.00 M $H_2SO_4(aq)$ required to produce 500 mL of 0.30 M $H_2SO_4(aq)$.

Solution: From Equation (6-4), we have

$$M_1V_1 = M_2V_2$$

$$(6.00\ mol{\cdot}L^{-1})V_1 = (0.30\ mol{\cdot}L^{-1})(0.500\ L)$$

Thus,

$$V_1 = \frac{(0.500 \text{ L})(0.30 \text{ mol·L}^{-1})}{6.00 \text{ mol·L}^{-1}} = 0.025 \text{ L}$$

or $V_1 = 25$ mL. Thus, we add 25 mL of 6.00 M $H_2SO_4(aq)$ to a 500-mL volumetric flask that is about half-filled with water, swirl the solution, and dilute with water to the 500-mL mark on the flask. Finally, we swirl again to make the new solution homogeneous.

PRACTICE PROBLEM 6-3: Commercial nitric acid is a 15.9 M aqueous solution. How would you prepare one liter of 6.00 M $HNO_3(aq)$ solution?

Answer: Dilute 377 mL of the 15.9 M $HNO_3(aq)$ to one liter using a volumetric flask.

6-3. MOLARITY IS USED IN STOICHIOMETRIC CALCULATIONS FOR REACTIONS THAT TAKE PLACE IN SOLUTION

The concept of molarity allows us to apply the techniques we discussed in Chapter 5 to reactions that take place in solution. For example, we can calculate the number of moles of a reactant dissolved in a given volume of a solution of known concentration. Suppose we wish to calculate how many moles of $AgNO_3(aq)$ there are in 35.6 mL of 0.135 M $AgNO_3(aq)$ solution. Using Equation (6-3) and remembering that the volume must be expressed in liters, we write

$$\begin{aligned}\text{moles of AgNO}_3 &= (0.135 \text{ M})(35.6 \text{ mL})\left(\frac{1 \text{ L}}{1000 \text{ mL}}\right) \\ &= (0.135 \text{ mol·L}^{-1})(0.0356 \text{ L}) \\ &= 4.81 \times 10^{-3} \text{ mol}\end{aligned}$$

This type of calculation usually is an intermediate step in a calculation involving a chemical reaction. For example, a standard laboratory preparation of small quantities of bromine involves the reaction

$$MnO_2(s) + 4HBr(aq) \rightarrow MnBr_2(aq) + Br_2(l) + 2H_2O(l)$$

What volume of 8.84 M $HBr(aq)$ solution would be required to react completely with 3.62 g of $MnO_2(s)$? How many grams of $Br_2(l)$ will be produced? From the chemical equation, we see that 1 mol of MnO_2 requires 4 mol of HBr to react completely, or that

$$1 \text{ mol MnO}_2 \Leftrightarrow 4 \text{ mol HBr}$$

The number of moles of $MnO_2(s)$ is given by

$$\begin{aligned}\text{moles of MnO}_2 &= (3.62 \text{ g MnO}_2)\left(\frac{1 \text{ mol MnO}_2}{86.94 \text{ g MnO}_2}\right) \\ &= 0.0416 \text{ mol}\end{aligned}$$

and the corresponding number of moles of HBr required is

$$\text{moles of HBr} = (0.0416 \text{ mol MnO}_2)\left(\frac{4 \text{ mol HBr}}{1 \text{ mol MnO}_2}\right)$$
$$= 0.166 \text{ mol}$$

We use a rearranged version of Equation (6-2) to calculate the volume of the 8.84 M HBr(*aq*) solution that is required

$$V = \frac{n}{M} = \frac{0.166 \text{ mol}}{8.84 \text{ M}} = \frac{0.166 \text{ mol}}{8.84 \text{ mol·L}^{-1}}$$
$$= 0.0188 \text{ L} = 18.8 \text{ mL}$$

From the chemical equation, we see that the number of moles of bromine produced is equal to the number of moles of $MnO_2(s)$ reacted, so

$$\text{mol Br}_2 = \text{mol MnO}_2 = 0.0416 \text{ mol}$$

The mass of bromine is given by

$$\text{mass of Br}_2 = (0.0416 \text{ mol Br}_2)\left(\frac{159.8 \text{ g Br}_2}{1 \text{ mol Br}_2}\right)$$
$$= 6.65 \text{ g}$$

The following Example illustrates another calculation involving a reaction between a solution and a solid.

Figure 6-4 Zinc metal reacts with an aqueous solution of hydrochloric acid. The bubbles are hydrogen gas escaping from the solution.

EXAMPLE 6-4: Zinc reacts with hydrochloric acid, HCl(*aq*) (Figure 6-4):

$$Zn(s) + 2HCl(aq) \rightarrow ZnCl_2(aq) + H_2(g)$$

Calculate how many grams of zinc react with 50.0 mL of 6.00 M HCl(*aq*).

Solution: The equation for the reaction indicates that 1 mol of Zn reacts with 2 mol of HCl(*aq*). We can use Equation (6-3) to calculate how many moles of HCl there are in 50.0 mL of a 6.00 M HCl solution:

$$\text{moles of HCl} = MV = (6.00 \text{ M})(50.0 \text{ mL})\left(\frac{1 \text{ L}}{1000 \text{ mL}}\right)$$
$$= 0.300 \text{ mol}$$

Thus, the number of moles of zinc that react is given by

$$\text{moles of Zn} = (0.300 \text{ mol HCl})\left(\frac{1 \text{ mol Zn}}{2 \text{ mol HCl}}\right) = 0.150 \text{ mol}$$

The mass that corresponds to 0.150 mol of Zn (atomic mass 65.38) is

$$\text{grams of Zn} = (0.150 \text{ mol Zn})\left(\frac{65.38 \text{ g Zn}}{1 \text{ mol Zn}}\right)$$
$$= 9.81 \text{ g}$$

Figure 6-5 Drāno consists of a mixture of pieces of aluminum and NaOH(*s*). When Drāno is added to water, the aluminum reacts with the NaOH(*aq*) to produce hydrogen gas.

PRACTICE PROBLEM 6-4: Aluminum reacts with a moderately concentrated sodium hydroxide solution (Figure 6-5):

$$2Al(s) + 2NaOH(aq) + 6H_2O(l) \rightarrow 2NaAl(OH)_4(aq) + 3H_2(g)$$

In this chapter, we examine the properties of gases. Many chemical reactions involve gases as reactants or products, or both, so we must determine how the properties of gases depend on conditions such as temperature, pressure, volume, and number of moles. We show how gases respond to changes in pressure and temperature and then discuss how the pressure, temperature, and volume of a gas are related to one another. After presenting a number of experimental observations concerning gases, we discuss the kinetic theory of gases, which gives us insight into the molecular nature of gases.

7-1. MOST OF THE VOLUME OF A GAS IS EMPTY SPACE

Before we discuss the nature of gases, we must first consider the three physical states of matter: solid, liquid, and gas. A **solid** has a fixed volume and shape. A **liquid,** however, has a fixed volume but assumes the shape of the container into which it is poured. A **gas** has neither a fixed volume nor a fixed shape; it always expands to occupy the entire volume of any closed container into which it is placed.

The molecular picture of a solid is that of an ordered array, called a lattice, of particles (atoms, molecules, or ions), as shown in Figure 7-1a. The individual particles vibrate around fixed lattice positions but are not free to move about (Figure 7-2a). The restrictions on the motion of the particles of a solid are reflected by the fixed volume and shape that characterize a solid.

A molecular view of liquid (Figure 7-1b) shows the particles in continuous contact with one another but free to move about throughout the liquid. There is no orderly, fixed arrangement of particles in a liquid as there is in a solid. When a solid melts and becomes a liquid, the lattice breaks down and the constituent particles are no longer held in fixed positions (Figure 7-2b). The fact that the densities of the solid phase and the liquid phase of any substance do not differ greatly from each other indicates that the distance between particles is similar in the two

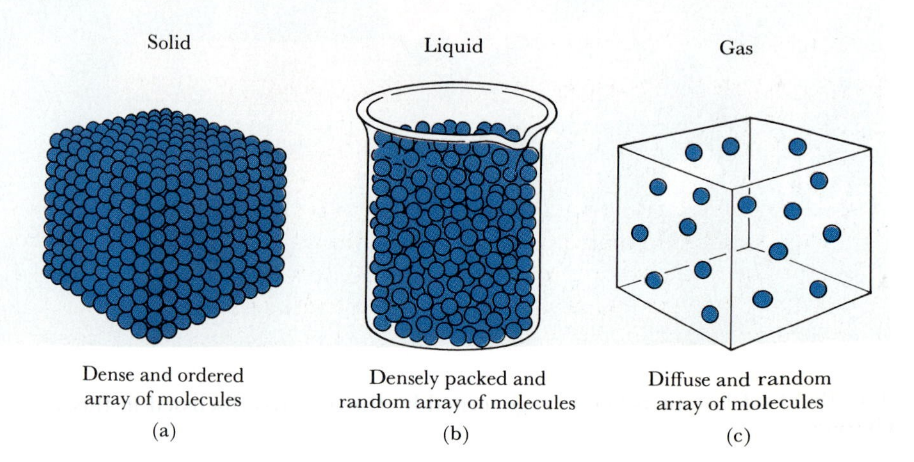

Figure 7-1 Molecular views of (a) a solid, (b) a liquid, and (c) a gas.

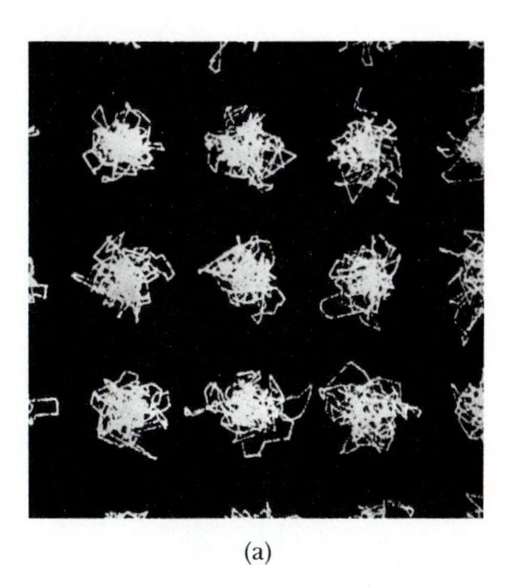
(a)

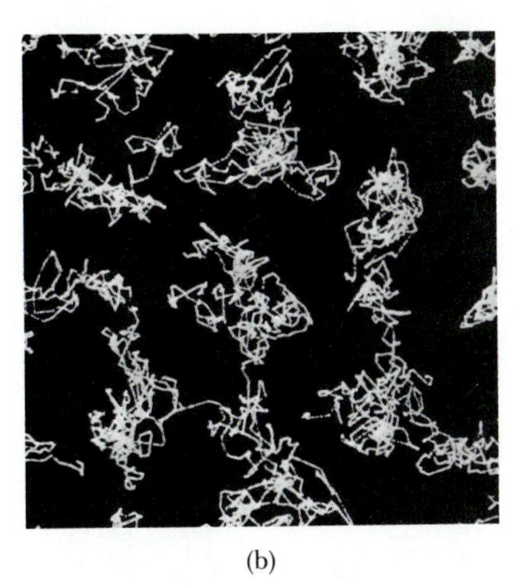
(b)

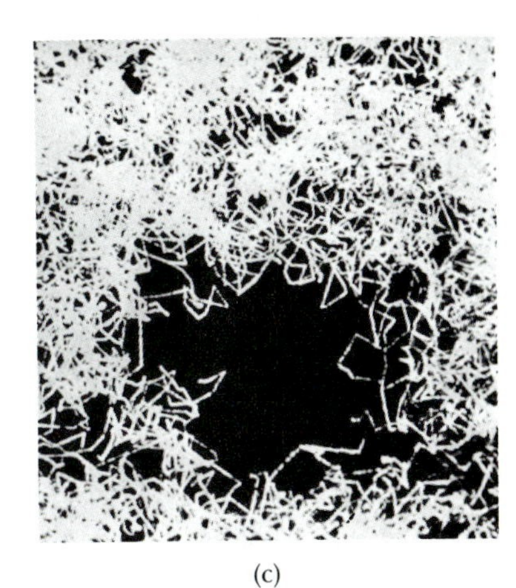
(c)

Figure 7-2 Computer-calculated paths of particles appear as bright lines on the face of a cathode-ray tube coupled to a computer. (a) Motion of atoms in an atomic crystal (note that the atoms move only about fixed positions). (b) A crystal in the process of melting (note the breakdown of the ordered array). (c) A liquid and its vapor (the dark area represents a gas bubble surrounded by particles whose motions characterize a liquid).

phases. Furthermore, the solid and liquid phases of a substance have similar, small **compressibilities,** meaning that their volumes do not change appreciably with increasing pressure. The similar compressibilities of the solid and liquid phases of a substance are further evidence that the particles in the two phases have similar separations.

When a given mass of liquid is vaporized, there is a huge increase in its volume. For example, 1 mol of liquid water occupies 17.3 mL at 100°C, whereas 1 mol of water vapor occupies over 30,000 mL under the same conditions. Upon vaporization, the molecules of a substance become widely separated, as indicated in Figure 7-1c. The picture of a gas as a substance with widely separated particles accounts nicely for the relative ease with which gases can be compressed. The particles take up only a small fraction of the total space occupied by a gas; most of the volume of a gas is empty space. As we note later in this chapter, the volume of a gas decreases markedly as the pressure increases; in other words, gases have large compressibilities.

7-2. A MANOMETER IS USED TO MEASURE THE PRESSURE OF A GAS

A gas is mostly empty space, with the molecules widely separated from one another. The molecules are in constant motion, traveling about at high speeds and colliding with one another and with the walls of the container. It is the force of these incessant, numerous collisions with the walls of the container that is responsible for the **pressure** exerted by a gas.

A common laboratory setup used to measure the pressure exerted by a gas is a **manometer,** which is a glass U-shaped tube partially filled with

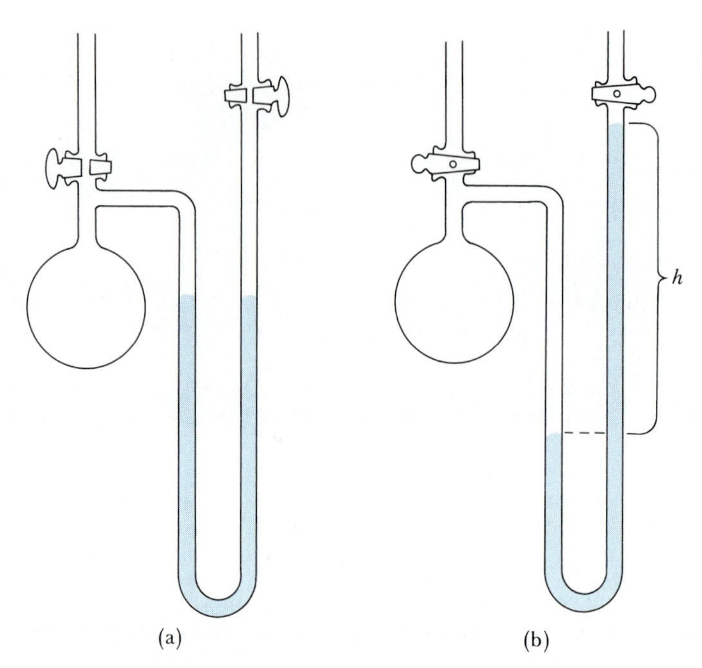

Figure 7-3 A mercury manometer. (a) Both stopcocks are open to the atmosphere, so both columns are exposed to atmospheric pressure. Both columns are at the same height because the pressure is the same on both surfaces. (b) The two stopcocks are closed, and the air in the right-hand column has been evacuated so that the pressure on the top of the right-hand column of mercury is essentially zero. The heights of the columns are no longer the same. The difference in heights is a direct measure of the pressure of the gas in the flask.

a liquid (Figure 7-3). Mercury is commonly used as the liquid in a manometer because it has a high density and is fairly unreactive. Figure 7-3 illustrates the measurement of gas pressure with a manometer. The height h of the column of mercury that is supported by the gas in the flask is directly proportional to the pressure of the gas. Because of this direct proportionality, it is convenient to express pressure in terms of the height of a column of mercury that the gas will support. This height is usually measured in millimeters, so pressure is expressed in terms of millimeters of mercury (mm Hg). The pressure unit mm Hg is called a **torr,** after the Italian scientist Evangelista Torricelli. He invented the **barometer,** which is used to measure atmospheric pressure (Figure 7-4). Thus, we say, for example, that the pressure of a gas is 600 torr.

Although mercury is most often used as the liquid in a manometer, other liquids can be used. The height of the column of liquid that can be supported by a gas is inversely proportional to the density of the liquid; that is, the less dense the liquid, the taller the column will be.

EXAMPLE 7-1: Calculate the height of a column of water that will be supported by a pressure of 1 atm. Take the density of mercury to be 13.6 $g \cdot mL^{-1}$ and that of water to be 1.00 $g \cdot mL^{-1}$.

Solution: A pressure of 1 atm corresponds to 760 torr, or a column of mercury 760 mm high. Because mercury is 13.6 times more dense than water, the column of water supported will be 13.6 times higher than that of mercury, or

$$\text{height of } H_2O \text{ column} = (13.6)(\text{height of mercury column})$$
$$= (13.6)(760 \text{ mm})$$
$$= 1.03 \times 10^4 \text{ mm} = 33.9 \text{ ft}$$

For a liquid to be useful as a manometric fluid, typical heights supported must be large enough to be accurately measured, but small enough to avoid the necessity for holes in the ceiling. Mercury is the only liquid that is dense enough to avoid the latter problem for measurements of atmospheric pressure at room temperature. Various organic liquids, such as certain silicone oils and di-*n*-butyl phthalate, $C_6H_4(COOC_4H_9)_2$, are occasionally used for measurements of low pressures.

PRACTICE PROBLEM 7-1: Di-*n*-butyl phthalate, $C_6H_4(COOC_4H_9)_2$, is an oily, unreactive liquid with a density of $1.046 \text{ g}\cdot\text{mL}^{-1}$ at 20°C. Calculate the height of a column of di-*n*-butyl phthalate that will be supported by a pressure of 2.00 torr.

Answer: 26.0 mm

Figure 7-4 The pressure exerted by the atmosphere can support a column of mercury that is about 760 mm high, as seen in the central tube of the barometer shown here. This barometer is at the National Maritime Museum in Greenwich, England.

7-3. A STANDARD ATMOSPHERE IS 760 TORR

The atmosphere surrounding the earth is a gas that exerts a pressure. The manometer pictured in Figure 7-3 can be used to demonstrate this pressure. If the flask is open to the atmosphere and the air in the right-hand side is evacuated, a column of mercury will be supported by the atmospheric (barometric) pressure. The height of the mercury column depends on elevation above sea level, temperature, and climatic conditions, but at sea level on a clear day it is about 760 mm.

Several units are used to express pressure. One **standard atmosphere** (atm) is defined as a pressure of 760 torr. It is common to express pressure in terms of standard atmospheres or, more simply, atmospheres. Strictly speaking, however, torr and atmosphere are not units of pressure but rather are quantities that are directly proportional to pressure. Pressure is defined as a *force per unit area.* The SI unit of pressure is the **pascal** (Pa). The precise definition of a pascal is given in Appendix B; however, a pascal is defined simply and operationally as the pressure exerted on a one-square-meter surface by a mass of 102 g. More important for us is the relation between torr, atmosphere, and pascal:

$$760 \text{ torr} = 1 \text{ atm} = 1.013 \times 10^5 \text{ Pa}$$

The common units for expressing pressure are summarized in Table 7-1. The units torr and atmosphere are so widely used by chemists that their replacement by the pascal will be extremely slow. Consequently, we will use torr or atmosphere in this text, but a section of problems using SI pressure units is included at the end of this chapter.

Table 7-1 Various units for expressing pressure

Unit	Definition
SI unit	
1 pascal (Pa)	= pressure (force per unit area) exerted by a mass of 102 g on a 1-m^2 surface (see Appendix B for precise definition)
"convenience" unit	
	height of a column of mercury supported by the pressure; commonly expressed as torr (1 torr = 1 mm Hg)
defined unit	
1 standard atmosphere	$= 1.013 \times 10^5$ Pa
	= 101.3 kPa
	= 760 torr
	$= 14.7\ \text{lb}\cdot\text{in.}^{-2}$
meterological unit	
1 bar	$= 10^5$ Pa
1 atm	= 1.013 bar = 1013 mbar (the bar is derived from *bar*ometer)

EXAMPLE 7-2: In a recent issue of the journal *Science,* a research group discussed experiments in which they determined the structure of cesium iodide crystals at a pressure of 302 gigapascals (GPa). How many atmospheres is this pressure?

Solution: We refer to Table 7-1 and see that 1 atm is equivalent to 1.013×10^5 Pa, so we have the unit conversion factor

$$\frac{1\ \text{atm}}{1.013 \times 10^5\ \text{Pa}} = 1$$

Using the fact that the prefix *giga-* stands for 10^9 (see the inside of the back cover), we write

$$(302\ \text{GPa})\left(\frac{1 \times 10^9\ \text{Pa}}{1\ \text{GPa}}\right)\left(\frac{1\ \text{atm}}{1.013 \times 10^5\ \text{Pa}}\right) = 2.98 \times 10^6\ \text{atm}$$

or almost three million atmospheres!

Figure 7-5 The English scientist Robert Boyle (1627–1691) was one of the first to use the modern scientific method. Boyle studied the properties of gases and formulated the law we now call Boyle's law.

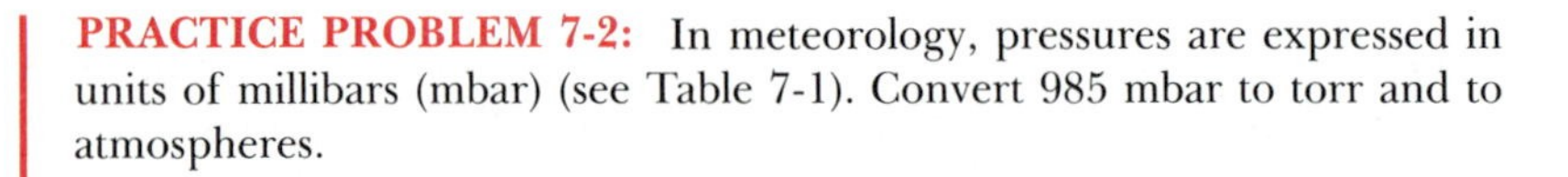

PRACTICE PROBLEM 7-2: In meteorology, pressures are expressed in units of millibars (mbar) (see Table 7-1). Convert 985 mbar to torr and to atmospheres.

Answer: 739 torr; 0.972 atm

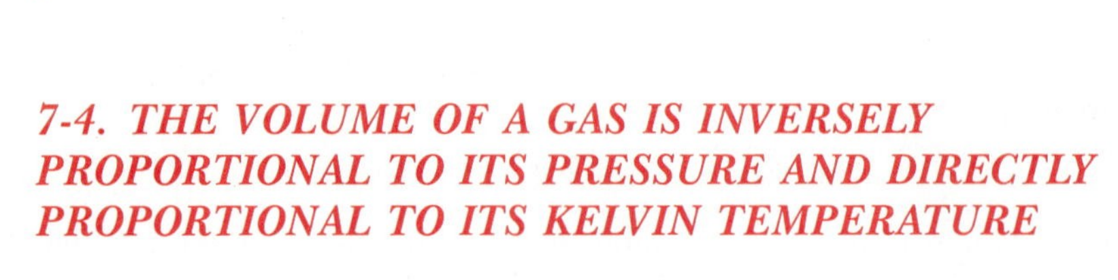

7-4. THE VOLUME OF A GAS IS INVERSELY PROPORTIONAL TO ITS PRESSURE AND DIRECTLY PROPORTIONAL TO ITS KELVIN TEMPERATURE

The first systematic study of the behavior of gases under different applied pressures was carried out in the 1660s by the English scientist Robert Boyle (Figure 7-5). Boyle was able to show that, at constant temperature, the volume of a given sample of gas is inversely proportional to the pressure:

$$V \propto \frac{1}{P}$$

We can also write

$$V = \frac{c}{P} \quad \text{(constant temperature)} \qquad (7\text{-}1)$$

where c denotes a proportionality constant whose value in each case is determined by the amount and the temperature of the gas. The relationship between pressure and volume expressed in Equation (7-1) is known as **Boyle's law.** Equation (7-1) is plotted in Figure 7-6. Note that the greater the pressure on a gas, the smaller is the volume at constant temperature. If we double the pressure on a gas, then its volume decreases by a factor of 2.

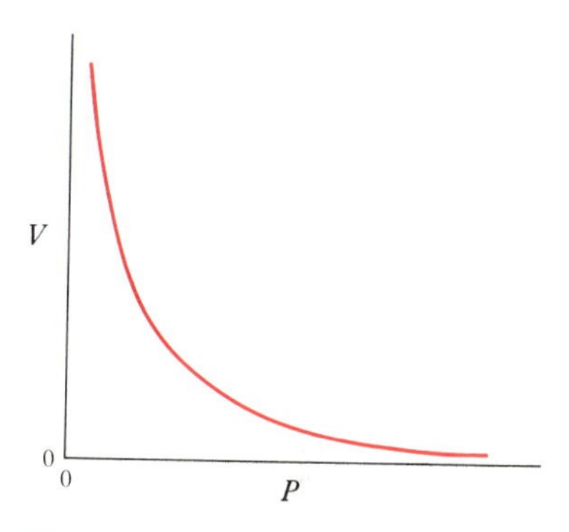

Figure 7-6 Boyle's law states that the volume of a gas at constant temperature is inversely proportional to its pressure.

Jacques Charles, a French scientist and adventurer, showed that there is a linear relationship between the volume of a gas and its temperature at constant pressure. Typical experimental data are plotted as volume versus temperature in Figure 7-7. Notice in Figure 7-7 that all three sets of experimental data extrapolate to the same point. Careful experimental measurements show that this point corresponds to a temperature of −273.15°C. The curves in Figure 7-7 suggest that if we add 273.15 to the temperatures expressed on the Celsius scale, then all three of the lines in Figure 7-7 will meet at the origin, as shown in Figure 7-8. This new temperature scale that we have proposed—adding 273.15 to the Celsius temperatures—turns out to be a fundamental temperature scale. We call it the **absolute** or the **Kelvin temper-**

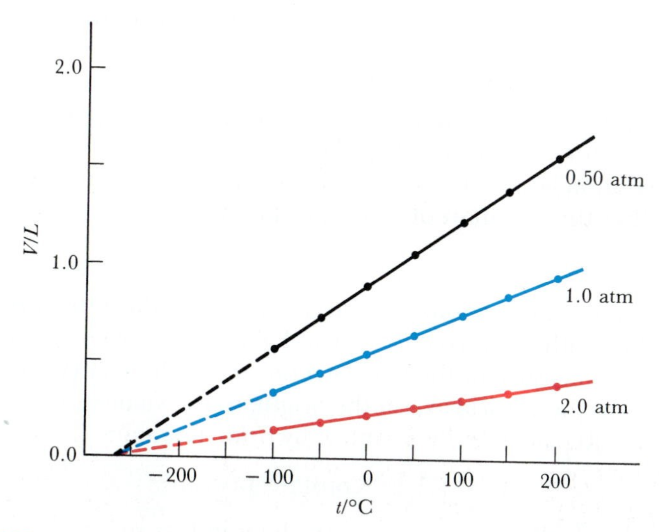

Figure 7-7 The volume (V) of 0.580 g of air plotted as a function of temperature (t) at three different pressures. Note that all three curves extrapolate to $V = 0$ at −273.15°C. These plots suggest that we can define a temperature scale that reflects the basic relationship of volume and temperature by adding 273.15 to the Celsius scale.

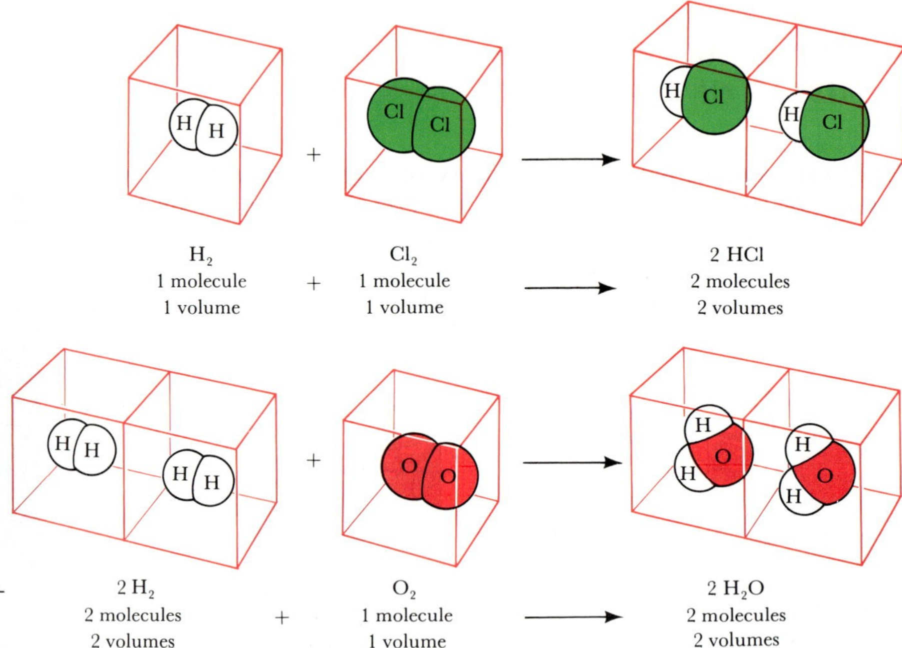

Figure 7-13 An illustration of Avogadro's explanation of Gay-Lussac's law of combining volumes.

of chlorine to produce *two molecules* of hydrogen chloride (Figure 7-13). If one molecule of hydrogen can form two molecules of hydrogen chloride, then a hydrogen molecule must consist of two atoms (at least) of hydrogen. Avogadro pictured both hydrogen and chlorine as diatomic gases and was able to represent the reaction between them as

$$H_2(g) + Cl_2(g) \rightarrow 2HCl(g)$$
$$1 \text{ molecule} + 1 \text{ molecule} \rightarrow 2 \text{ molecules}$$
$$1 \text{ volume} + 1 \text{ volume} \rightarrow 2 \text{ volumes}$$

Prior to Avogadro's explanation, it was difficult to see how one volume of hydrogen could produce two volumes of hydrogen chloride. If hydrogen occurred simply as atoms, there would be no way to explain Gay-Lussac's law of combining volumes. Another example of Avogadro's law is illustrated by

$$3H_2(g) + N_2(g) \rightarrow 2NH_3(g)$$
$$3 \text{ molecules} + 1 \text{ molecule} \rightarrow 2 \text{ molecules}$$
$$3 \text{ volumes} + 1 \text{ volume} \rightarrow 2 \text{ volumes}$$

It is interesting to note that, in spite of the beautiful simplicity of Avogadro's explanation of these reactions, his work was largely ignored and chemists continued to confuse atoms and molecules and to use many incorrect chemical formulas. It was not until the mid-1800s, after Avogadro's death, that his hypothesis was finally appreciated and generally accepted.

7-6. THE IDEAL-GAS EQUATION IS A COMBINATION OF BOYLE'S, CHARLES'S, AND AVOGADRO'S LAWS

Avogadro postulated that equal volumes of gases at the same pressure and temperature contain the same number of molecules. This statement implies that equal volumes of gases at the same pressure and temperature contain equal numbers of moles, n. Thus, we can write Avogadro's law as

$$V \propto n \qquad \text{(fixed } P \text{ and } T\text{)}$$

Boyle's law and Charles's law are, respectively,

$$V \propto \frac{1}{P} \qquad \text{(fixed } T \text{ and } n\text{)}$$

$$V \propto T \qquad \text{(fixed } P \text{ and } n\text{)}$$

We can combine these three proportionality statements for V into one by writing

$$V \propto \frac{nT}{P}$$

Note how the three individual statements for V are all included in the combined statement. For example, if P and n are fixed, then only T can vary; and we see that $V \propto T$, which is Charles's law. If P and T are fixed, then only n can vary; and we have Avogadro's law, $V \propto n$. Last, if T and n are fixed, then we have Boyle's law, $V \propto 1/P$.

We can convert the combined proportionality statement for V to an equation by introducing a proportionality constant R and writing the equation as

$$PV = nRT \tag{7-6}$$

This expression is called the **ideal-gas law** or **ideal-gas equation.** It is based on Boyle's law, Charles's law, and Avogadro's law. Boyle's law and Charles's law are valid only at low pressures (less than a few atmospheres, say), so Equation (7-6) is valid only at low pressures. It turns out, however, that most gases do not deviate from Equation (7-6) by more than a few percent up to tens of atmospheres, so Equation (7-6) is very useful. Gases that satisfy the ideal-gas equation are said to behave ideally, or to be **ideal gases.**

Before we can use Equation (7-6), we must determine the value of R, which is called the **molar gas constant.** It has been determined experimentally that 1 mol of an ideal gas at 0°C and 1.00 atm occupies 22.4 L. The volume 22.4 L, shown in Figure 7-14, is called the **molar volume** of an ideal gas at 0°C and 1.00 atm. If we solve Equation (7-6) for R and substitute this information into the resulting equation, then we find that

$$R = \frac{PV}{nT} = \frac{(1.00\ \text{atm})(22.4\ \text{L})}{(1.00\ \text{mol})(273\ \text{K})} \tag{7-7}$$
$$= 0.0821\ \text{L}\cdot\text{atm}\cdot\text{mol}^{-1}\cdot\text{K}^{-1}$$

You should pay careful attention to the units of R. When the value $R = 0.0821\ \text{L}\cdot\text{atm}\cdot\text{mol}^{-1}\cdot\text{K}^{-1}$ is used in Equation (7-6), P must be expressed in atmospheres, V in liters, n in moles, and T in kelvins. (Note that we say that T is "in kelvins," not "in degrees kelvin.")

Now that we have determined the value of the molar gas constant, we can use Equation (7-6) in many ways.

Figure 7-14 The volume of 1 mol of an ideal gas at 0°C and 1 atm. A volume of 22.4 L is represented by this cube whose edges measure 28.2 cm. A basketball in its carton is shown for comparison.

EXAMPLE 7-4: A spherical balloon is inflated with helium to a diameter of 30 m. If the pressure is 740 torr and the temperature is 27°C, then what is the mass of helium in the balloon?

Solution: The volume of the (spherical) balloon is

$$V = \frac{4\pi r^3}{3} = \frac{(4\pi)\left(\frac{30\text{ m}}{2}\right)^3}{3} = 1.41 \times 10^4\text{ m}^3$$

Note that we know the value of three variables—V, P, T—and we wish to calculate the fourth—n. To use the ideal-gas equation with $R = 0.0821$ L·atm·mol^{-1}·K^{-1}, we must express V in liters, P in atmospheres, and T in kelvins:

$$V = (1.41 \times 10^4\text{ m}^3)\left(\frac{100\text{ cm}}{1\text{ m}}\right)^3\left(\frac{1\text{ mL}}{1\text{ cm}^3}\right)\left(\frac{1\text{ L}}{1000\text{ mL}}\right) = 1.41 \times 10^7\text{ L}$$

$$P = (740\text{ torr})\left(\frac{1\text{ atm}}{760\text{ torr}}\right) = 0.974\text{ atm}$$

$$T = 27°\text{C} + 273 = 300\text{ K}$$

Solving Equation (7-6) for n gives

$$n = \frac{PV}{RT} = \frac{(0.974\text{ atm})(1.41 \times 10^7\text{ L})}{(0.0821\text{ L·atm·mol}^{-1}\text{·K}^{-1})(300\text{ K})}$$
$$= 5.58 \times 10^5\text{ mol}$$

The number of grams of helium is

$$\text{mass of helium} = (5.58 \times 10^5\text{ mol He})\left(\frac{4.003\text{ g He}}{1\text{ mol He}}\right)$$
$$= 2.23 \times 10^6\text{ g He}$$
$$= 2.23\text{ metric tons He}$$

PRACTICE PROBLEM 7-4: The pressure in a 10.0-L gas cylinder containing nitrogen is 4.15 atm at 20°C. What is the mass of nitrogen in the cylinder?

Answer: 48.3 g

EXAMPLE 7-5: Ammonia can be produced by heating ammonium chloride with calcium hydroxide according to

$$2NH_4Cl(s) + Ca(OH)_2(s) \rightarrow CaCl_2(s) + 2H_2O(l) + 2NH_3(g)$$

If all the ammonia produced by reacting 2.50 g of $NH_4Cl(s)$ with excess $Ca(OH)_2(s)$ occupies a 500-mL container at 15°C, then what will the pressure of the ammonia be?

Solution: The number of moles of $NH_3(g)$ generated is

$$\text{moles of NH}_3 = (2.50\text{ g NH}_4\text{Cl})\left(\frac{1\text{ mol NH}_4\text{Cl}}{53.49\text{ g NH}_4\text{Cl}}\right)\left(\frac{2\text{ mol NH}_3}{2\text{ mol NH}_4\text{Cl}}\right)$$
$$= 4.67 \times 10^{-2}\text{ mol}$$

The pressure is calculated from Equation (7-6):

$$P = \frac{nRT}{V} = \frac{(4.67 \times 10^{-2}\text{ mol})(0.0821\text{ L·atm·mol}^{-1}\text{·K}^{-1})(288\text{ K})}{0.500\text{ L}}$$
$$= 2.21\text{ atm}$$

PRACTICE PROBLEM 7-5: If all the oxygen generated by the thermal decomposition of 1.34 g of potassium chlorate, $KClO_3(s)$, occupies 250 mL at 20°C, what will the pressure be?

Answer: 1.58 atm

EXAMPLE 7-6: How many milliliters of $H_2(g)$ at 18°C and 736 torr will be generated by reacting 0.914 g of zinc with 50.0 mL of 0.650 M $HCl(aq)$?

Solution: The relevant equation is

$$Zn(s) + 2HCl(aq) \rightarrow ZnCl_2(aq) + H_2(g)$$

We are given quantities of both reactants, so we first must see whether one of them acts as a limiting reactant. The number of millimoles of HCl is given by

$$\text{millimoles of HCl} = MV = (0.650\ \text{M})(50.0\ \text{mL}) = 32.5\ \text{mmol}$$

and the number of millimoles of zinc is given by

$$\begin{aligned}\text{millimoles of Zn} &= (0.914\ \text{g Zn})\left(\frac{1\ \text{mol Zn}}{65.38\ \text{g Zn}}\right)\left(\frac{1\ \text{mmol}}{10^{-3}\ \text{mol}}\right)\\ &= 14.0\ \text{mmol}\end{aligned}$$

From the reaction stoichiometry, we see that

$$1\ \text{mol Zn} \approx 2\ \text{mol HCl}$$

Therefore,

$$14.0\ \text{mmol Zn} \approx 28.0\ \text{mmol HCl}$$

so we see that HCl is in excess (4.5 mmol in excess) and that zinc is the limiting reactant. Because

$$1\ \text{mol Zn} \approx 1\ \text{mol}\ H_2$$

we can write

$$\text{millimoles of}\ H_2 = \text{millimoles of Zn} = 14.0\ \text{mmol}$$

The volume of $H_2(g)$ is given by

$$\begin{aligned}V &= \frac{nRT}{P}\\ &= \frac{(14.0\ \text{mmol})\left(\frac{10^{-3}\ \text{mol}}{1\ \text{mmol}}\right)(0.0821\ \text{L·atm·mol}^{-1}\text{·K}^{-1})(291\ \text{K})}{(736\ \text{torr})\left(\frac{1\ \text{atm}}{760\ \text{torr}}\right)}\\ &= 0.345\ \text{L} = 345\ \text{mL}\end{aligned}$$

PRACTICE PROBLEM 7-6: Powdered charcoal, $C(s)$, ignites spontaneously in $F_2(g)$ at room temperature and forms carbon tetrafluoride, $CF_4(g)$. What volume of $CF_4(g)$ at 0°C and 760 torr will be produced by reacting 250 mL of $F_2(g)$ at 20°C and 1200 torr with 26.2 mg of charcoal?

Answer: 48.9 mL

In Examples 7-4 through 7-6 we were given three quantities and had to calculate a fourth. Another type of application of the ideal-gas equation involves changes from one set of conditions to another.

EXAMPLE 7-7: One mole of O_2 gas occupies 22.4 L at 0°C and 1.00 atm. What volume does it occupy at 100°C and 4.00 atm?

Solution: Note that in this problem we are given V at one set of conditions (T and P) and asked to calculate V under another set of conditions (that is, a different T and P). Because R is a constant and because n is a constant in this problem, we can write the ideal-gas equation as

$$\frac{PV}{T} = nR = \text{constant}$$

This equation says that the ratio PV/T remains constant, so we can write

$$\frac{P_i V_i}{T_i} = \frac{P_f V_f}{T_f}$$

where the subscripts i and f denote *initial* and *final*, respectively. Recall that we wish to calculate a final volume, so we solve this equation for V_f.

$$V_f = V_i\left(\frac{P_i}{P_f}\right)\left(\frac{T_f}{T_i}\right) \tag{7-8}$$

If we substitute the given quantities into this equation, then we obtain

$$V_f = (22.4\ \text{L})\left(\frac{1.00\ \text{atm}}{4.00\ \text{atm}}\right)\left(\frac{373\ \text{K}}{273\ \text{K}}\right)$$
$$= 7.65\ \text{L}$$

Note that the increase in pressure (from 1.00 atm to 4.00 atm) decreases the gas volume (the gas is compressed), whereas the increase in temperature increases it. The pressure increases by a factor of 4.00, whereas the temperature increases only by a factor of 373/273 = 1.37; thus the *net* effect is a decrease in the volume of the gas.

PRACTICE PROBLEM 7-7: A 2-L gas cylinder has a pressure of 2.50 atm at 0°C. If the cylinder cannot withstand pressures in excess of 5.0 atm, will the cylinder burst if it is heated to 300°C?

Answer: Yes (5.25 atm)

Note that in Example 7-7 we multiplied the initial volume (22.4 L) by a pressure ratio and a temperature ratio. The pressure increased from 1.00 atm to 4.00 atm, and the pressure ratio used was 1.00/4.00; this multiplication yields a smaller volume, as you would expect. Similarly, the temperature increased from 273 K to 373 K, and the temperature ratio used was 373/273; this multiplication yields a larger volume. A "common-sense" method of solving this problem is to write

$$V_f = V_i \times \text{pressure ratio} \times \text{temperature ratio}$$

and to decide by simple reasoning whether each ratio to be used is greater or less than unity.

7-7. THE IDEAL-GAS EQUATION CAN BE USED TO CALCULATE THE MOLECULAR MASSES OF GASES

One of the most important applications of the ideal-gas equation involves the determination of the molecular mass of a gas. Let's see how we know that chlorine is a diatomic species. Suppose we find that a 0.286-g sample of chlorine occupies 250 mL at 300 torr and 25°C. We are given V, P, and T, so we can use Equation (7-6) to calculate the number of moles, n:

$$n = \frac{PV}{RT} = \frac{(300\ \text{torr})\left(\dfrac{1\ \text{atm}}{760\ \text{torr}}\right)(0.250\ \text{L})}{(0.0821\ \text{L·atm·mol}^{-1}\text{·K}^{-1})(298\ \text{K})}$$
$$= 4.03 \times 10^{-3}\ \text{mol}$$

Thus, 0.286 g of chlorine gas corresponds to 4.03×10^{-3} mol:

$$0.286\ \text{g} \approx 4.03 \times 10^{-3}\ \text{mol}$$

We would like to have this stoichiometric correspondence read

$$\text{a certain number of grams} \approx 1.00\ \text{mol}$$

We can achieve this by dividing both sides by 4.03×10^{-3} to obtain

$$\frac{0.286\ \text{g}}{4.03 \times 10^{-3}} \approx \frac{4.03 \times 10^{-3}\ \text{mol}}{4.03 \times 10^{-3}}$$

or

$$71.0\ \text{g} \approx 1.00\ \text{mol}$$

Thus, we find that the molecular mass of chlorine is 71.0. This result implies that chlorine is a diatomic gas (Cl_2) because the atomic mass of chlorine is 35.45.

EXAMPLE 7-8: The density of dry air at 1.00 atm and 20°C is 1.205 g·L^{-1}. Calculate the effective or average molecular mass of air; in other words, calculate the molecular mass of air as if it were a pure gas.

Solution: The density of the air in moles per liter is given by

$$\frac{n}{V} = \frac{P}{RT} = \frac{1.00\ \text{atm}}{(0.0821\ \text{L·atm·mol}^{-1}\text{·K}^{-1})(293\ \text{K})}$$
$$= 0.0416\ \text{mol·L}^{-1}$$

Taking a one-liter sample, we see that

$$0.0416\ \text{mol} \approx 1.205\ \text{g}$$

Dividing through by 0.0416, we obtain

$$1\ \text{mol} \approx 29.0\ \text{g}$$

so the effective molecular mass of dry air is 29.0.

PRACTICE PROBLEM 7-8: Meteorites contain small quantities of trapped argon, which originates from radioactive processes (see Chapter 24). The density of argon from a particular meteorite is found to be 1.504 g per liter at 15°C and 700 torr. Determine the atomic mass of argon from this meteorite sample.

Answer: 40.0

We can combine a calculation in which we determine the molecular mass with a determination of the empirical (simplest) formula of the compound from chemical analysis to determine the molecular formula.

EXAMPLE 7-9: Chemical analysis shows that the gas acetylene is 92.3 percent carbon and 7.70 percent hydrogen by mass. It has a density of 0.711 $g\cdot L^{-1}$ at 20°C and 500 torr. Use these data to determine the molecular formula of acetylene.

Solution: The determination of the empirical formula from chemical analysis is explained in Section 5-3. Following the procedure given there, we write

$$92.3 \text{ g C} \approx 7.70 \text{ g H}$$

Dividing the left side by 12.01 g C/mol C and the right by 1.008 g H/mol H gives

$$7.69 \text{ mol C} \approx 7.64 \text{ mol H}$$

Dividing both sides by 7.64 and rounding yield

$$1 \text{ mol C} \approx 1 \text{ mol H}$$

so the empirical formula of acetylene is CH.

We use the density data to determine the molecular mass of acetylene. Solving Equation (7-6) for n/V, we have

$$\frac{n}{V} = \frac{P}{RT} = \frac{(500 \text{ torr})\left(\frac{1 \text{ atm}}{760 \text{ torr}}\right)}{(0.0821 \text{ L}\cdot\text{atm}\cdot\text{mol}^{-1}\cdot\text{K}^{-1})(293 \text{ K})}$$
$$= 0.0273 \text{ mol}\cdot\text{L}^{-1}$$

Taking a one-liter sample, we see that

$$0.0269 \text{ mol} \approx 0.711 \text{ g}$$

Dividing through by 0.0269, we obtain

$$1 \text{ mol} \approx 26.0 \text{ g}$$

so the molecular mass of acetylene is 26.0. Its empirical formula is CH and the empirical formula mass is 13.0, so the molecular formula of acetylene must be C_2H_2 (formula mass = 26.0).

PRACTICE PROBLEM 7-9: Chemical analysis shows that the gas propene is 85.6 percent carbon and 14.4 percent hydrogen. Given that it has a density of 1.55 $g \cdot L^{-1}$ at 40°C and 720 torr, determine the molecular formula of propene.

Answer: C_3H_6

We can also use the ideal-gas equation to calculate the densities of gases. If we solve Equation (7-6) for n/V, then we get

$$\frac{n}{V} = \frac{P}{RT}$$

The ratio n/V is equal to gas density in the units mole per liter. We can convert from moles per liter to grams per liter by multiplying both sides of the equation by the number of grams per mole, or the **molar mass,** M. (Molar masses are numerically equal to formula masses but have the units of $g \cdot mol^{-1}$. For example, the formula mass of H_2O is 18.0 and its molar mass is 18.0 $g \cdot mol^{-1}$.) If we denote the density in grams per liter by the symbol ρ (the Greek letter rho), then we can write

$$\rho = \frac{Mn}{V} = \frac{MP}{RT} \tag{7-9}$$

Note that gas density increases as pressure increases and as temperature decreases.

EXAMPLE 7-10: Calculate the density of nitrogen dioxide gas at 0°C and 1.00 atm.

Solution: The molar mass of NO_2 is 46.01 $g \cdot mol^{-1}$. Using Equation (7-9) with the appropriate units gives

$$\rho = \frac{MP}{RT} = \frac{(46.01\ g \cdot mol^{-1})(1.00\ atm)}{(0.0821\ L \cdot atm \cdot mol^{-1} \cdot K^{-1})(273\ K)}$$
$$= 2.05\ g \cdot L^{-1}$$

Nitrogen dioxide is more dense than air, so it can be poured from one container to another, as shown in Figure 7-15.

PRACTICE PROBLEM 7-10: The density of CO_2 was carefully determined to be 1.7192 $g \cdot L^{-1}$ at the temperature and pressure at which the density of oxygen gas was determined to be 1.2500 $g \cdot L^{-1}$. Using the known atomic mass of oxygen (15.9994), calculate the atomic mass of carbon to five significant figures.

Answer: 12.011

Figure 7-15 Like liquids, gases have flow properties and can be poured from one container to another if they are denser than air. This photo shows $NO_2(g)$ being poured.

7-8. THE TOTAL PRESSURE OF A MIXTURE OF IDEAL GASES IS THE SUM OF THE PARTIAL PRESSURES OF ALL THE GASES IN THE MIXTURE

Up to this point we have not considered explicitly mixtures of gases, and yet mixtures of gases are of great importance. For example, air is a mixture of 78 percent nitrogen, 21 percent oxygen, and 1 percent argon by volume with lesser amounts of other gases, such as carbon dioxide. Many industrial processes involve gaseous mixtures. For example, the commercial production of ammonia involves the reaction

$$3H_2(g) + N_2(g) \xrightarrow[500°C]{300 \text{ atm}} 2NH_3(g)$$

thus, the reaction vessel contains a mixture of N_2, H_2, and NH_3.

In a mixture of ideal gases, each gas exerts a pressure as if it were present alone in the container. For a mixture of two ideal gases, we have

$$P_{total} = P_1 + P_2 \tag{7-10}$$

The pressure exerted by each gas is called its **partial pressure,** and Equation (7-10) is known as **Dalton's law of partial pressures.** Each of the gases obeys the ideal-gas equation, so

$$P_1 = \frac{n_1RT}{V} \qquad P_2 = \frac{n_2RT}{V} \tag{7-11}$$

Notice that the volume occupied by each gas is V, because each gas in a mixture occupies the entire container. If the partial pressures P_1 and P_2 are substituted into Equation (7-10), then we get for our two-gas mixture,

$$\begin{aligned} P_{total} &= \frac{n_1RT}{V} + \frac{n_2RT}{V} \\ &= (n_1 + n_2)\frac{RT}{V} \\ &= n_{total}\frac{RT}{V} \end{aligned} \tag{7-12}$$

The total pressure exerted by a mixture of gases is determined by the total number of moles of gas in the mixture.

If we divide each of the Equations (7-11) by Equation (7-12), then we obtain

$$\frac{P_1}{P_{total}} = \frac{n_1}{n_1 + n_2} \qquad \frac{P_2}{P_{total}} = \frac{n_2}{n_1 + n_2} \tag{7-13}$$

Let us write

$$X_1 = \frac{n_1}{n_1 + n_2} \qquad X_2 = \frac{n_2}{n_1 + n_2} \tag{7-14}$$

where X_1 is the mole fraction of gas 1 in the mixture and X_2 is the mole fraction of gas 2. Notice that a **mole fraction** is unitless (it is a fraction) and that

$$X_1 + X_2 = 1$$

Equation (7-13) can be written as

$$P_1 = X_1P_{total} \qquad P_2 = X_2P_{total} \tag{7-15}$$

which expresses the partial pressure of the ith species in terms of its mole fraction and the total pressure.

If we multiply both the numerators and the denominators in Equations (7-14) by Avogadro's number, then X_1 and X_2 will be expressed as molecular fractions,

$$X_1 = \frac{N_1}{N_1 + N_2} \qquad X_2 = \frac{N_2}{N_1 + N_2} \tag{7-16}$$

where N_1 and N_2 are the number of molecules of gases 1 and 2, respectively. Thus, X_1, for example, is the fraction of molecules in a mixture that are gas 1 molecules. With this in mind, we can give a simple molecular interpretation of Dalton's law of partial pressures. The pressure exerted by a gas or a gaseous mixture is due to the incessant collisions of the molecules of the gas with the walls of the container. According to the ideal-gas law, at a fixed temperature and volume, the pressure exerted by a gas is proportional only to the number of molecules of the gas. Thus, in a mixture of gases, the partial pressure of each gas is the total pressure multiplied by the fraction of molecules of each gas.

EXAMPLE 7-11: A 0.428-g mixture of gases contained in a vessel at 1.75 atm is found to be 15.6 percent $N_2(g)$, 46.0 percent $N_2O(g)$, and 38.4 percent $CO_2(g)$ by mass. What is the partial pressure of each gas in the mixture?

Solution: The masses of each of the three gases in the mixture are

$$\text{mass of } N_2 = (15.6\%)(0.428 \text{ g}) = 0.0668 \text{ g } N_2$$
$$\text{mass of } N_2O = (46.0\%)(0.428 \text{ g}) = 0.197 \text{ g } N_2O$$
$$\text{mass of } CO_2 = (38.4\%)(0.428 \text{ g}) = 0.164 \text{ g } CO_2$$

and the numbers of moles of each gas are

$$\text{moles of } N_2 = (0.0668 \text{ g } N_2)\left(\frac{1 \text{ mol } N_2}{28.06 \text{ g } N_2}\right) = 2.38 \times 10^{-3} \text{ mol}$$
$$\text{moles of } N_2O = (0.197 \text{ g } N_2O)\left(\frac{1 \text{ mol } N_2O}{44.02 \text{ g } N_2O}\right) = 4.47 \times 10^{-3} \text{ mol}$$
$$\text{moles of } CO_2 = (0.164 \text{ g } CO_2)\left(\frac{1 \text{ mol } CO_2}{44.01 \text{ g } CO_2}\right) = 3.73 \times 10^{-3} \text{ mol}$$

The total number of moles is 10.58×10^{-3} mol and the various mole fractions are

$$X_{N_2} = \frac{2.38 \times 10^{-3} \text{ mol}}{10.58 \times 10^{-3} \text{ mol}} = 0.225$$
$$X_{N_2O} = \frac{4.47 \times 10^{-3} \text{ mol}}{10.58 \times 10^{-3} \text{ mol}} = 0.422$$
$$X_{CO_2} = \frac{3.73 \times 10^{-3} \text{ mol}}{10.58 \times 10^{-3} \text{ mol}} = 0.353$$

and the partial pressures are

$$P_{N_2} = X_{N_2}P_{\text{total}} = (0.225)(1.75 \text{ atm}) = 0.394 \text{ atm}$$
$$P_{N_2O} = X_{N_2O}P_{\text{total}} = (0.422)(1.75 \text{ atm}) = 0.738 \text{ atm}$$
$$P_{CO_2} = X_{CO_2}P_{\text{total}} = (0.353)(1.75 \text{ atm}) = 0.618 \text{ atm}$$

Figure 7-16 Chemists can manipulate gases at various pressures using an apparatus called a vacuum rack. This photo shows a vacuum rack in a research laboratory at the Scripps Institution in San Diego.

PRACTICE PROBLEM 7-11: Consider two flasks connected by a stopcock (Figure 7-16). One flask has a volume of 500 mL and contains $N_2(g)$ at a pressure of 700 torr, and the other flask has a volume of 400 mL and contains $O_2(g)$ at a pressure of 950 torr. If the stopcock is opened so that the two gases mix completely, calculate the partial pressures of $N_2(g)$ and $O_2(g)$ and the total pressure of the resultant mixture.

Answer: $P_{O_2} = 422$ torr; $P_{N_2} = 389$ torr; $P_{\text{total}} = 811$ torr

EXAMPLE 7-12: Suppose that we have a mixture of $N_2(g)$ and $O_2(g)$ of unknown composition whose total pressure is 385 torr. If all the $O_2(g)$ is removed from the mixture by reaction with phosphorus, which does not react directly with nitrogen, then the new pressure is 250 torr. Calculate the mole fraction of each gas in the original mixture.

Solution: The pressure before the removal of the $O_2(g)$ is

$$P_{\text{total}} = 385 \text{ torr} = P_{N_2} + P_{O_2}$$

After removal of the $O_2(g)$

$$P_{N_2} = 250 \text{ torr}$$

Therefore, the pressure due to $O_2(g)$ initially is

$$P_{O_2} = P_{\text{total}} - P_{N_2} = 385 \text{ torr} - 250 \text{ torr} = 135 \text{ torr}$$

Now let X_{N_2} be the mole fraction of $N_2(g)$ in the mixture and $X_{O_2} = 1 - X_{N_2}$ be the mole fraction of the $O_2(g)$. Using Equation (7-15), the mole fraction of $N_2(g)$ in the mixture is

$$X_{N_2} = \frac{P_{N_2}}{P_{\text{total}}} = \frac{250 \text{ torr}}{385 \text{ torr}} = 0.649$$

and the mole fraction of $O_2(g)$ is

$$X_{O_2} = \frac{P_{O_2}}{P_{\text{total}}} = \frac{135 \text{ torr}}{385 \text{ torr}} = 0.351$$

or, alternatively,

$$X_{O_2} = 1 - X_{N_2} = 1.000 - 0.649 = 0.351$$

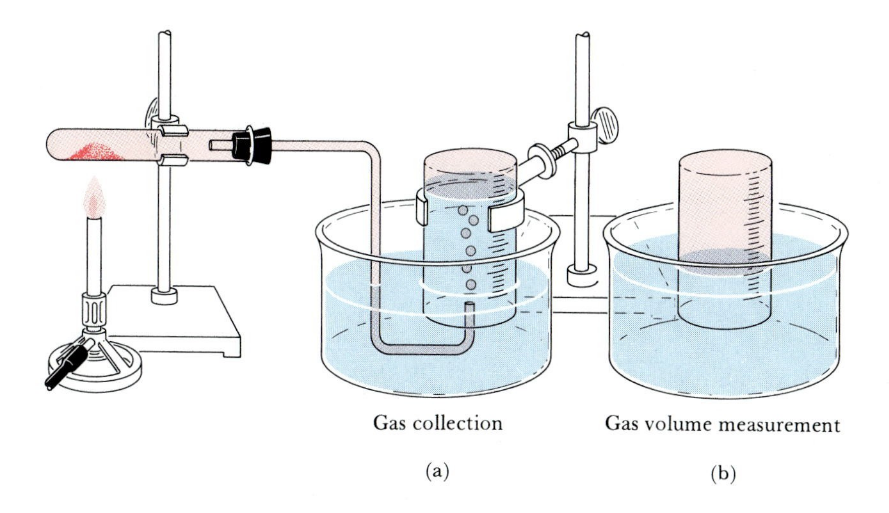

Figure 7-17 (a) The collection of a gas over water. (b) When the water levels inside and outside the container are equal, the pressure inside the container and the atmospheric pressure must be equal.

PRACTICE PROBLEM 7-12: A mixture of $N_2(g)$ and $H_2(g)$ has mole fractions 0.40 and 0.60, respectively. Determine the density of the mixture at 0°C and 1 atm.

Answer: $0.554\ \text{g·L}^{-1}$

Practical applications of Dalton's law of partial pressures arise often in the laboratory. A standard method for determining the quantity of a water-insoluble gas evolved in a chemical reaction is diagramed in Figure 7-17. The gas displaces the water from an inverted beaker that is initially filled with water. When the reaction is completed, the beaker is raised or lowered until the water levels inside and outside are the same. When the two levels are the same, the pressure inside the beaker is equal to the atmospheric pressure. The pressure inside the beaker, however, is not due just to the gas collected; there is also water vapor present. Thus, the pressure inside the beaker is

$$P_{\text{total}} = P_{\text{gas}} + P_{H_2O} = P_{\text{atmospheric}} \tag{7-17}$$

The vapor pressure of H_2O, P_{H_2O}, depends only on the temperature; Table 14-5 gives the vapor pressure of H_2O at various temperatures. We will study the pressure due to water vapor more fully in Chapter 14.

EXAMPLE 7-13: The reaction

$$2KClO_3(s) \rightarrow 2KCl(s) + 3O_2(g)$$

represents a common laboratory procedure for producing small quantities of pure oxygen. A 0.250-L flask is filled with oxygen that has been collected over water at an atmospheric pressure of 729 torr (see Figure 7-17). The temperature of the water and the gas is 14°C. Calculate the molar volume of oxygen at 0°C and 760 torr. The vapor pressure of H_2O at 14°C is 12.0 torr.

Solution: To calculate the molar volume of oxygen, we must first determine its partial pressure. The atmospheric pressure is 729 torr, and the vapor pressure of H_2O at 14°C is 12.0 torr; therefore,

$$P_{O_2} = P_{\text{total}} - P_{H_2O} = (729 - 12.0)\ \text{torr} = 717\ \text{torr}$$

We calculate the number of moles of O_2 produced by using the ideal-gas equation:

$$n = \frac{PV}{RT} = \frac{(717\ \text{torr})\left(\frac{1\ \text{atm}}{760\ \text{torr}}\right)(0.250\ \text{L})}{(0.0821\ \text{L·atm·mol}^{-1}\text{·K}^{-1})(287\ \text{K})}$$
$$= 0.0100\ \text{mol}$$

We wish to calculate the molar volume of O_2 at 0°C and 760 torr, so we use Equation (7-8):

$$V_f = V_i\left(\frac{P_i}{P_f}\right)\left(\frac{T_f}{T_i}\right) = (0.250\ \text{L})\left(\frac{717\ \text{torr}}{760\ \text{torr}}\right)\left(\frac{273\ \text{K}}{287\ \text{K}}\right)$$
$$= 0.224\ \text{L}$$

The molar volume is the volume occupied by 1 mol of gas. Thus,

$$\text{molar volume of } O_2 = \frac{0.224\ \text{L}}{0.0100\ \text{mol}} = 22.4\ \text{L·mol}^{-1}$$

PRACTICE PROBLEM 7-13: The hydrogen generated in a reaction is collected over water and occupies 425 mL at 18°C and 742 torr. What is the volume of the dry $H_2(g)$ at 0°C and 760 torr. The vapor pressure of water at 18°C is 15.5 torr.

Answer: 381 mL

7-9. THE MOLECULES OF A GAS HAVE A DISTRIBUTION OF SPEEDS

The fact that the ideal-gas equation can be used for *all* gases at low enough pressures suggests that this law reflects the fundamental nature of gases. As we have seen, the molecules in a gas are widely separated and in constant motion, and exert a pressure as they collide with the walls of the container. By applying the laws of physics to the motion of the molecules, it is possible to calculate the pressure exerted by the molecules as a result of their collisions with the walls of the container. Because this approach focuses on the motion of the molecules, it is called the **kinetic theory of gases.** Before we can discuss the kinetic theory of gases, we must introduce the concept of kinetic energy.

A body in motion has energy by virtue of the fact that it is in motion. The energy associated with the motion of a body, called **kinetic energy** (E_k), is given by the formula

$$E_k = \tfrac{1}{2}mv^2 \qquad (7\text{-}18)$$

where m is the mass of the body and v is its speed. If m is expressed in kilograms and v in meters per second, then E_k has the units kg·m^2·s^{-2}. This combination of units is called a **joule** (J), which is the SI unit of energy: 1 J = 1 kg·m^2·s^{-2}. (See Appendix B.)

To get an idea of the size of a joule, let's calculate the kinetic energy of a golf ball (mass of 45.9 g) that is traveling at 250 km per hour (about 150 miles per hour), which is the speed of a well-driven golf ball. To use

Equation (7-18), we must first convert the mass to kilograms and the speed to meters per second:

$$m = (45.9\ \text{g})\left(\frac{1\ \text{kg}}{1000\ \text{g}}\right) = 0.0459\ \text{kg}$$

and

$$v = (250\ \text{km·h}^{-1})\left(\frac{1\ \text{h}}{3600\ \text{s}}\right)\left(\frac{1000\ \text{m}}{1\ \text{km}}\right) = 69.4\ \text{m·s}^{-1}$$

The kinetic energy is

$$E_\text{k} = \tfrac{1}{2}mv^2 = \tfrac{1}{2}(0.0459\ \text{kg})(69.4\ \text{m·s}^{-1})^2$$
$$= 111\ \text{kg·m}^2\text{·s}^{-2} = 111\ \text{J}$$

Of course, atomic and molecular energies are much smaller. We shall see in Section 7-11 that the typical speed of a nitrogen molecule at room temperature is about 500 m·s^{-1}, or about 1100 miles per hour (mph), so its kinetic energy is about 6×10^{-21} joules. The kinetic energy per mole of molecules, each traveling with a speed, v, is given by

$$E_\text{k} = \tfrac{1}{2}M_\text{kg}v^2 \tag{7-19}$$

where M_kg is the molar mass in kilograms. Note that the molar mass must be expressed in kilograms per mole (kg·mol^{-1}) so that the units of E_k will be joules per mole (J·mol^{-1}). Thus, the kinetic energy per mole of nitrogen molecules would be

$$E_\text{k} = (\tfrac{1}{2})(0.0280\ \text{kg·mol}^{-1})(500\ \text{m·s}^{-1})^2$$
$$= 3500\ \text{J·mol}^{-1} = 3.50\ \text{kJ·mol}^{-1}$$

if each of the nitrogen molecules had a speed of 500 m·s^{-1}.

Although, for simplicity, we have just implied that all the molecules in a gas travel at the same speed, in fact they do not. Figure 7-18 illustrates an experimental apparatus that can be used to determine the

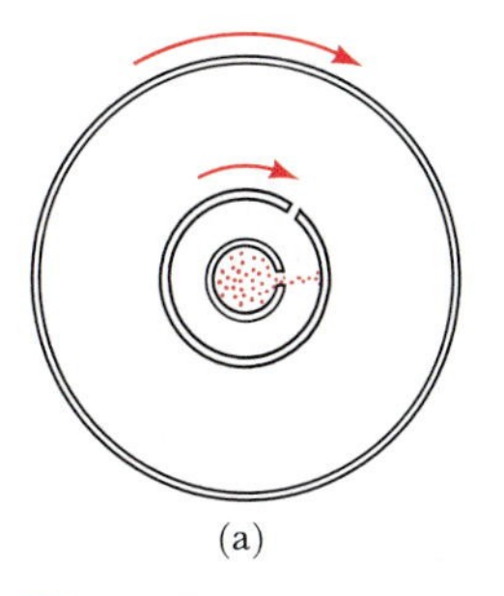

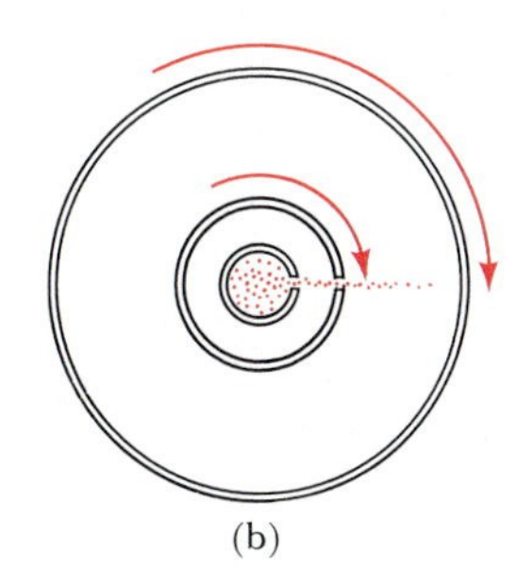

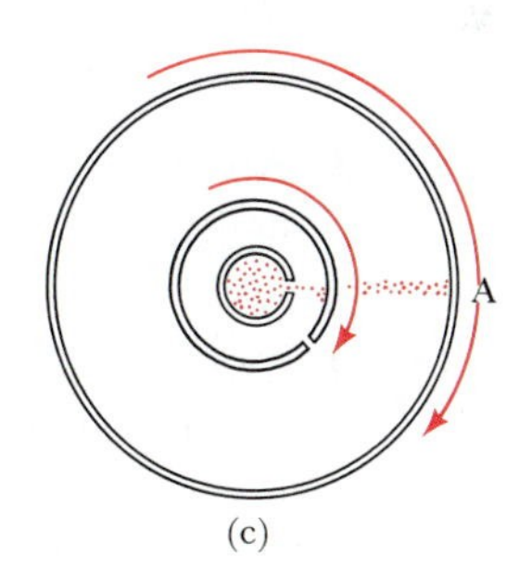

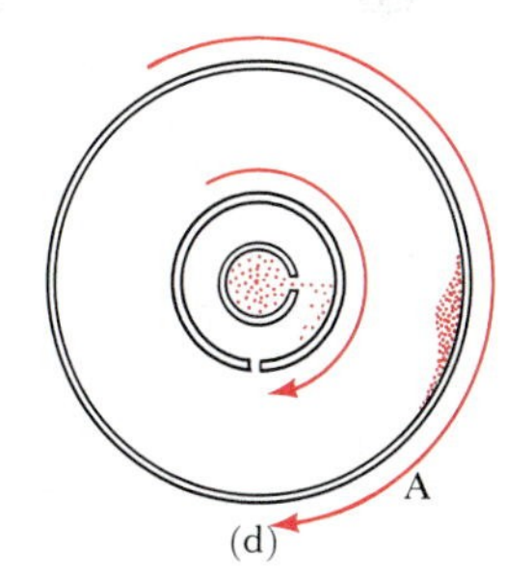

Figure 7-18 An experimental setup that was used not only to demonstrate that there is a distribution of molecular speeds in a gas but also to measure that distribution. (a) The apparatus consists of three concentric evacuated cylindrical drums. The two outermost drums rotate together at the same angular speed, and the innermost drum is stationary. The innermost drum contains a gas or vapor such as silver vapor. The two innermost drums have small slits, and when these slits momentarily line up, as shown in (b), a beam of silver atoms is directed to the inner surface of the outermost drum. The silver atoms with the greatest speed reach the outermost drum first, at point A in (c). By the time the slower silver atoms reach the outer drum, it will have rotated further, so the deposit of silver atoms will be spread out, as seen in (d). The thickness of the silver deposit is proportional to the number of silver atoms with a certain speed, so the variation in thickness represents the actual distribution of speeds of the silver atoms in the vapor.

distribution of the speeds of the molecules in a gas, and Figure 7-19 shows the results of such an experiment. In Figure 7-19 we plot the fraction of molecules that have a speed v versus v at two different temperatures. Notice that both curves start at the origin, rise to a maximum, and then fall off to zero as the speed increases. Also notice that more molecules travel at higher speeds at higher temperatures. The distribution of molecular speeds in a gas is called a **Maxwell-Boltzmann distribution,** after James Clerk Maxwell and Ludwig Boltzmann, two scientists who developed the kinetic theory of gases in the latter half of the nineteenth century.

Because the molecules in a gas travel at different speeds, they also have different kinetic energies. Thus, there is a distribution of molecular kinetic energies in a gas. This distribution is shown in Figure 7-20, where the fraction of molecules with kinetic energy E_k is plotted against E_k. Note that the curves are somewhat similar to those in Figure 7-20, in the sense that they rise through a maximum and then fall off to zero as v increases.

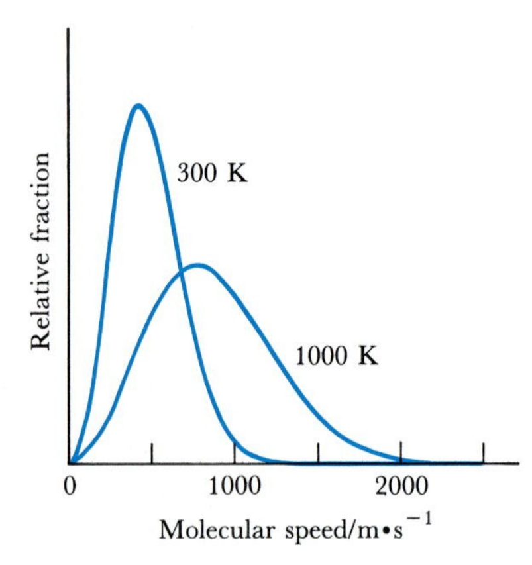

Figure 7-19 The distribution of speeds for nitrogen molecules at 300 K and 1000 K. The distribution is represented by the fraction of nitrogen molecules that have speed v plotted against that speed. Note, for example, that the fraction of molecules with a speed of 1 km·s^{-1} is greater at 1000 K than at 300 K.

7-10. THE AVERAGE KINETIC ENERGY OF A GAS IS PROPORTIONAL TO ITS KELVIN TEMPERATURE

We are now ready to discuss the kinetic theory of gases. The postulates of the kinetic theory of gases are

1. The molecules in a gas are incessantly in motion. They collide randomly with one another and with the walls of the container.
2. The average distance between the molecules in a gas is much larger than the sizes of the molecules. In other words, the gas consists mostly of empty space.
3. Any interactions between the molecules in a gas are negligible. We assume that the molecules in a gas neither repel nor attract one another.
4. The average kinetic energy of the molecules in a gas is proportional to the kelvin temperature of the gas.

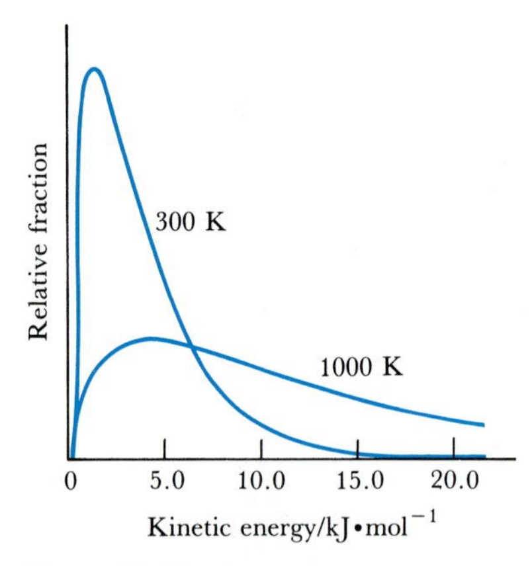

Figure 7-20 The distribution of kinetic energies for nitrogen molecules at 300 K and 1000 K. The distribution is represented by the fraction of nitrogen molecules that have kinetic energy E_k plotted against that kinetic energy.

This last postulate involves the ***average* kinetic energy** because there is a distribution of molecular kinetic energies. It is common notation to denote the average of a quantity by placing a bar over the symbol for that quantity, so, using Equation (7-19), we can write the average kinetic energy per mole as

$$\overline{E_k} = \tfrac{1}{2}M_{kg}\overline{v^2} \tag{7-20}$$

Our fourth postulate of the kinetic theory of gases states that

$$\overline{E_k} = \tfrac{1}{2}M_{kg}\overline{v^2} \propto T \tag{7-21}$$

Note that the average on the right-hand side of Equation (7-20) is the average of v^2. The **average** of any set of N values is obtained by adding all the values and then dividing by N. For example, the average of the four speeds $v_1 = 1.50 \times 10^3\ \text{m·s}^{-1}$, $v_2 = 2.10 \times 10^3\ \text{m·s}^{-1}$, $v_3 = 2.75 \times 10^3\ \text{m·s}^{-1}$, $v_4 = 3.25 \times 10^3\ \text{m·s}^{-1}$ is given by

$$\begin{aligned}\overline{v} &= \frac{v_1 + v_2 + v_3 + v_4}{4} \\ &= \frac{1}{4}\Big(1.50 \times 10^3\ \text{m·s}^{-1} + 2.10 \times 10^3\ \text{m·s}^{-1} \\ &\qquad + 2.75 \times 10^3\ \text{m·s}^{-1} + 3.25 \times 10^3\ \text{m·s}^{-1}\Big) \\ &= 2.40 \times 10^3\ \text{m·s}^{-1}\end{aligned}$$

The quantity in Equation (7-20) is $\overline{v^2}$, however, which is given by

$$\begin{aligned}\overline{v^2} &= \frac{v_1^2 + v_2^2 + v_3^2 + v_4^2}{4} \\ &= (\tfrac{1}{4})[(1.50 \times 10^3\ \text{m·s}^{-1})^2 + (2.10 \times 10^3\ \text{m·s}^{-1})^2 \\ &\qquad + (2.75 \times 10^3\ \text{m·s}^{-1})^2 + (3.25 \times 10^3\ \text{m·s}^{-1})^2] \\ &= 6.20 \times 10^6\ \text{m}^2\text{·s}^{-2}\end{aligned}$$

Notice that $\overline{v^2} = 6.20 \times 10^6\ \text{m}^2\text{·s}^{-2} \neq (\overline{v})^2 = (2.40 \times 10^3\ \text{m·s}^{-1})^2 = 5.76 \times 10^6\ \text{m}^2\text{·s}^{-2}$. In other words, the average of v^2 does *not* equal the average of v squared.

Although you may be familiar with the idea of an average of a quantity (such as the average on an exam), you may not be familiar with the average of the square of a quantity. It turns out, however, that such an average is used often in statistics, in particular to calculate a quantity called the standard deviation, which is a measure of how much the values cluster around the average.

EXAMPLE 7-14: The mass m of an object was determined by five separate students with the results 0.1065 g, 0.1060 g, 0.1059 g, 0.1071 g, and 0.1066 g. Calculate the average and the average of the squares of these results.

Solution: Let m denote mass; so we can write

$$\begin{aligned}\overline{m} &= \frac{0.1065\ \text{g} + 0.1060\ \text{g} + 0.1059\ \text{g} + 0.1071\ \text{g} + 0.1066\ \text{g}}{5} \\ &= 0.1064\ \text{g}\end{aligned}$$

Similarly,

$$\begin{aligned}\overline{m^2} &= \frac{(0.1065\ \text{g})^2 + (0.1060\ \text{g})^2 + (0.1059\ \text{g})^2 + (0.1071\ \text{g})^2 + (0.1066\ \text{g})^2}{5} \\ &= 0.01133\ \text{g}^2\end{aligned}$$

Note once again that $\overline{m^2} \neq (\overline{m})^2$.

PRACTICE PROBLEM 7-14: The pressure of a gas is measured in a series of experiments with the results 410 torr, 432 torr, 385 torr, 417 torr, 372 torr, and 451 torr. Calculate $\overline{P}$ and $\overline{P^2}$.

Answer: 411 torr; 170,000 torr^2

7-11. THE KINETIC THEORY OF GASES ALLOWS US TO CALCULATE THE ROOT-MEAN-SQUARE SPEED OF A MOLECULE

The first three postulates of the kinetic theory of gases describe the model of a gas that we have already introduced, and we will see the importance of the final postulate shortly. From these postulates, we can calculate the properties of gases in terms of molecular quantities. For example, we can calculate an expression for the pressure of a gas in terms of the speeds of its constituent molecules. The pressure exerted by the gas is due to collisions of the molecules with the walls of the container, and it depends on three factors:

1. As the number of particles per unit volume increases, so should the number of collisions. We call the number of particles per unit volume, N/V, the **number density.** Thus, we can write one factor as $P \propto (N/V)$.
2. The greater the speed of the molecules, the more frequently they will collide with the walls, so we have as another factor $P \propto v$.
3. The greater the momentum of the gas molecules, the greater the pressure they exert from their collisions. The momentum of a molecule is due to its mass as well as its speed and is given by the product mv. Thus, we have the third factor $P \propto mv$.

If we combine these three factors in one proportionality statement, then we obtain the expression

$$P \propto \left(\frac{N}{V}\right)(v)(mv)$$

or

$$PV \propto Nmv^2$$

If we take into account that the molecules in a gas travel at different speeds, then the right-hand side of this expression is replaced by its average value, or

$$PV \propto Nm\overline{v^2} \tag{7-22}$$

A detailed derivation gives a proportionality constant of 1/3, so Expression (7-22) becomes

$$PV = \left(\frac{1}{3}\right)Nm\overline{v^2} \tag{7-23}$$

This equation is called the **fundamental equation of the kinetic theory of gases.** For 1 mol of gas, Equation (7-23) becomes

$$PV = \left(\frac{1}{3}\right)N_0m\overline{v^2} \tag{7-24}$$

where N_0 is Avogadro's number. If we compare Equation (7-24) with the ideal-gas equation for 1 mol, $PV = RT$, then we obtain

$$\left(\frac{1}{3}\right)N_0m\overline{v^2} = RT$$

If we multiply both sides of this equation by 3/2 and recognize that N_0m is the mass of one mole of molecules, then we obtain

$$\left(\frac{1}{2}\right)M_{kg}\overline{v^2} = \left(\frac{3}{2}\right)RT \quad (7\text{-}25)$$

Remember that M_{kg} is the molar mass in units of $kg{\cdot}mol^{-1}$.

The left-hand side of Equation (7-25) is the average kinetic energy of one mole of the gas given by Equation (7-20), so we can write

$$\overline{E_k} = \left(\frac{3}{2}\right)RT \quad (7\text{-}26)$$

which is a fundamental result of the kinetic theory of gases. Equation (7-26) states that the temperature of a gas is directly related to its average kinetic energy, which is the idea behind the fourth postulate of the kinetic theory of gases. It tells us the meaning of temperature, an *observed* quantity, in terms of the average value of a *molecular* property.

Because $\overline{E_k}$ is expressed in units of $joule{\cdot}mole^{-1}$, we must express R in units of $joule{\cdot}mole^{-1}{\cdot}kelvin^{-1}$ when we use Equation (7-26). It is shown in Appendix B that $R = 8.314\ J{\cdot}mol^{-1}{\cdot}K^{-1}$. We can use Equation (7-25) to calculate an important measure of the speed at which the molecules in a gas travel. We first solve Equation (7-25) for the average value of v^2

$$\overline{v^2} = \frac{3RT}{M_{kg}}$$

and then take the square root of both sides:

$$(\overline{v^2})^{1/2} = \left(\frac{3RT}{M_{kg}}\right)^{1/2} \quad (7\text{-}27)$$

The quantity in the left-hand side of this equation has the units of speed ($m{\cdot}s^{-1}$). It is called the **root-mean-square (rms) speed** because it is the square *root* of the *mean* (or average) of the *square* of the molecular speeds. We will denote the root-mean-square speed by v_{rms}, so Equation (7-27) can be written as

$$v_{rms} = \left(\frac{3RT}{M_{kg}}\right)^{1/2} \quad (7\text{-}28)$$

The root-mean-square speed is a very good measure of the average molecular speed in a gas (Problem 7-95).

EXAMPLE 7-15: Calculate the root-mean-square speed for N_2 at 20°C.

Solution: The molar mass of N_2 in kilograms per mole is

$$M_{kg} = \frac{28.0\ g{\cdot}mol^{-1}}{1000\ g{\cdot}kg^{-1}} = 0.0280\ kg{\cdot}mol^{-1}$$

Thus, using Equation (7-28), we calculate

$$\begin{aligned} v_{rms} &= \left(\frac{3RT}{M_{kg}}\right)^{1/2} \\ &= \left[\frac{(3)(8.314\ J{\cdot}mol^{-1}{\cdot}K^{-1})(293\ K)}{0.028\ kg{\cdot}mol^{-1}}\right]^{1/2} \\ &= (2.61 \times 10^5\ J{\cdot}kg^{-1})^{1/2} \\ &= (2.61 \times 10^5\ m^2{\cdot}s^{-2})^{1/2} = 511\ m{\cdot}s^{-1} \end{aligned}$$

Table 7-2 Values of v_{rms} for four gases at 20°C and 1000°C

Molecule	Formula mass	$v_{rms}/m \cdot s^{-1}$	
		t = 20°C	t = 1000°C
H_2	2.0	1900	4000
N_2	28.0	510	1060
O_2	32.0	480	1000
CO_2	44.0	410	850

A speed of 511 $m \cdot s^{-1}$ is comparable to the muzzle velocity of a high-speed rifle bullet.

Values of v_{rms} for several gases are given in Table 7-2. Note that v_{rms} decreases with increasing molecular mass at constant temperature, as is required by Equation (7-28).

A sound wave is a pressure disturbance that travels through a substance. The speed with which a sound wave travels through a gas depends on the speeds of the molecules in the gas. It can be shown from the kinetic theory of gases that the speed of sound through a gas is about 0.7 v_{rms}. The speed of sound in air at 20°C and 1 atm is about 770 mph, or 340 $m \cdot s^{-1}$.

PRACTICE PROBLEM 7-15: The speed of sound in an ideal diatomic gas is given by

$$v_{sound} = \left(\frac{7RT}{5M_{kg}}\right)^{1/2}$$

Calculate the speed of sound through nitrogen at 0°C. Compare this value to the speed of sound in air at 0°C.

Answer: 337 $m \cdot s^{-1}$; the speed of sound in air at 0°C is 331 $m \cdot s^{-1}$ (740 mph) at 1 atm (sea level).

7-12. GASES CAN BE SEPARATED BY EFFUSION

We can use Equation (7-28) to derive a formula for the relative rates at which gases leak from a container through a small hole, a process called **effusion.** For two gases at the same pressure and temperature, the rate of effusion is directly proportional to the root-mean-square speed of the molecules. We let $v_{rms,A}$ and $v_{rms,B}$ be the root-mean-square speeds of two gases A and B, and we use Equation (7-28) to write

$$v_{rms,A} = \left(\frac{3RT}{M_{kg,A}}\right)^{1/2} \quad \text{and} \quad v_{rms,B} = \left(\frac{3RT}{M_{kg,B}}\right)^{1/2}$$

The temperature does not have a subscript because both gases are at the same temperature. If we divide $v_{rms,A}$ by $v_{rms,B}$, then we obtain

$$\frac{v_{rms,A}}{v_{rms,B}} = \left(\frac{M_{kg,B}}{M_{kg,A}}\right)^{1/2} = \left(\frac{M_B}{M_A}\right)^{1/2}$$

where M_A and M_B are the molar masses of gases A and B. The rate of effusion is directly proportional to v_{rms}, so

$$\frac{\text{rate}_A}{\text{rate}_B} = \left(\frac{M_B}{M_A}\right)^{1/2} \tag{7-29}$$

This relation was observed experimentally by the British scientist Thomas Graham in the 1840s and is called **Graham's law of effusion.**

EXAMPLE 7-16: A porous container is filled with $N_2(g)$. It is observed that the pressure within the container decreased by 20 percent in 6.0 hours. If the same container is filled with He(g) under the same conditions, then how long would it take for the pressure to decrease by 20 percent?

Solution: We must first recognize that the rate and time are inversely proportional to each other; the shorter the time for a certain amount of effusion to occur, the greater is the rate of effusion. Thus, Equation (7-29) can be written in the form

$$\frac{\text{rate}_A}{\text{rate}_B} = \frac{t_B}{t_A} = \left(\frac{M_B}{M_A}\right)^{1/2} \tag{7-30}$$

where t is effusion time. If we let A denote $N_2(g)$ and B denote He(g), then Equation (7-30) becomes

$$t_{He} = t_{N_2}\left(\frac{M_{He}}{M_{N_2}}\right)^{1/2} = (6.0 \text{ h})\left(\frac{4.003}{28.02}\right)^{1/2} = 2.3 \text{ h}$$

Note that helium takes less time to escape because, being lighter, its molecules travel at higher average speeds at a given temperature.

PRACTICE PROBLEM 7-16: A porous container is filled with equal amounts of $N_2(g)$ and an unknown gas. The $N_2(g)$ escaped 2.3 times faster than the unknown gas did. What is the molecular mass of the unknown gas?

Answer: 148

Naturally occurring uranium consists primarily of two isotopes, U-238 (99.275 percent) and U-235 (0.720 percent). Only the U-235 isotope can be used in the construction of an atomic bomb (see Chapter 24), so the separation of the two isotopes, or at least enriching uranium in U-235, became a large-scale technological problem during the development of the atomic bomb in World War II. The isotopes U-235 and U-238 undergo the very same chemical reactions and cannot be separated by chemical means. They can, however, be separated by effusion. Uranium is a solid, but uranium hexafluoride, UF_6, is easily vaporized. Furthermore, there is only one naturally occurring isotope of fluorine (fluorine-19), and so U-235 and U-238 form UF_6 molecules of only formula masses 349 and 352, respectively. According to Graham's law of effusion, the lighter species will effuse faster than the heavier one, so the effusing mixture will be slightly enriched in U-235. Using Equation (7-29), we find that

$$\frac{\text{rate}_{235}}{\text{rate}_{238}} = \left(\frac{M_{238\ UF_6}}{M_{235\ UF_6}}\right)^{1/2} = \left(\frac{352}{349}\right)^{1/2} = 1.004$$

Figure 7-21 Isotopes of uranium can be separated by making the gaseous compound UF_6 and the using Graham's law of effusion, which says that the lighter isotopic compound will effuse more quickly than the heavier one. This photo shows the large process equipment in which effusion is carried out in stages to achieve isotopic enrichment.

so the mixture is 1.004 times richer in $^{235}UF_6$ than in $^{238}UF_6$. This figure certainly is not impressive, but recycling this mixture through a number of stages gives a significant enrichment (Figure 7-21).

7-13. THE AVERAGE DISTANCE A MOLECULE TRAVELS BETWEEN COLLISIONS IS CALLED THE MEAN FREE PATH

Although the molecules in a gas at 1 atm and 20°C travel with speeds of hundreds of meters per second, they do not travel any appreciable distances that rapidly. We all have observed that it may take several minutes for an odor to spread through a draft-free room. The explanation for this observation lies in the fact that the molecules in a gas undergo many collisions, so their actual path is a chaotic, zigzag path like that shown in Figure 7-22. Between collisions, gas molecules travel with speeds of hundreds of meters per second, but their net progress is quite slow. The average distance traveled between collisions is called the **mean free path** (l).

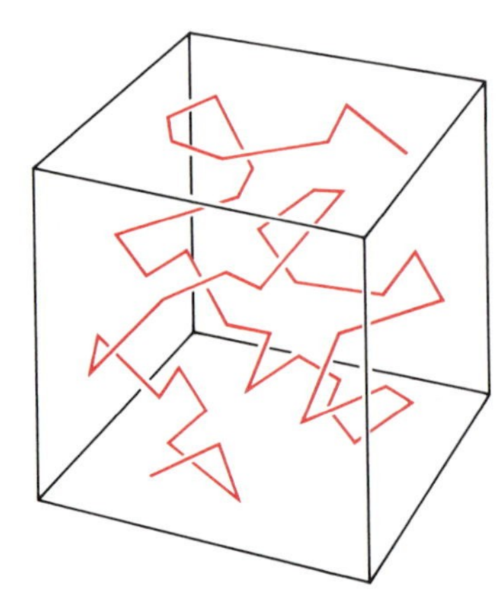

Figure 7-22 A typical path followed by a gas molecule. The molecule travels in a straight line until it collides with another molecule, at which point its direction is changed in an almost random manner. At 0°C and 1 atm, a molecule undergoes about 10^{10} collisions per second.

We can estimate the mean free path by reasoning that (1) the greater the density of molecules, the more often a molecule will collide; so the mean free path will be shorter. Therefore, the mean free path should be inversely proportional to the number density of molecules, N/V. (2) The greater the cross-sectional area of a molecule, the greater a target it will present for other molecules; and again the mean free path will be shorter. If a molecule has a diameter equal to σ (the Greek letter sigma), then its cross section will be $\pi(\sigma/2)^2$. Therefore, the mean free path should be inversely proportional to the *square* of the diameter of the molecule. If we let l be the mean free path, then we write

$$l \propto \left(\frac{1}{N/V}\right)\left(\frac{1}{\sigma^2}\right)$$

If we introduce a proportionality constant c, then we can write this proportionality statement as an equation

$$l = \frac{c}{\sigma^2(N/V)}$$

The value of the proportionality constant c depends on the units of the **molecular diameter, σ,** the units of the gas density, N/V, and the units of the mean free path, l. Molecular diameters are obviously very small and are commonly expressed in units of picometers (pm), 1 pm = 10^{-12} meters (Table 7-3). If l and σ are expressed in picometers, and the gas density in moles per liter, then our relationship becomes

$$l = \frac{3.7 \times 10^8 \text{ pm}^3\cdot\text{mol}\cdot\text{L}^{-1}}{\sigma^2(n/V)} \tag{7-31}$$

We can use Equation (7-31) to calculate the mean free path in $N_2(g)$ at 1.00 atm and 20°C. The gas density under these conditions is

$$\frac{n}{V} = \frac{P}{RT} = \frac{1.00 \text{ atm}}{(0.0821 \text{ L}\cdot\text{atm}\cdot\text{mol}^{-1}\cdot\text{K}^{-1})(293 \text{ K})}$$
$$= 0.0416 \text{ mol}\cdot\text{L}^{-1}$$

For N_2, σ = 370 pm (Table 7-3), so

$$l = \frac{3.7 \times 10^8 \text{ pm}^3\cdot\text{mol}\cdot\text{L}^{-1}}{(370 \text{ pm})^2(0.0416 \text{ mol}\cdot\text{L}^{-1})} = 6.6 \times 10^4 \text{ pm}$$
$$= (6.6 \times 10^4 \text{ pm})\left(\frac{10^{-12} \text{ m}}{1 \text{ pm}}\right) = 6.6 \times 10^{-8} \text{ m}$$

and

$$\frac{l}{\sigma} = \frac{6.6 \times 10^4 \text{ pm}}{370 \text{ pm}} = 178$$

The second expression means that at 1.00 atm and 20°C, a nitrogen molecule travels an average distance of almost 200 molecular diameters between collisions.

We can also derive a formula for the **collision frequency** of a molecule, that is, the number of collisions that a molecule undergoes per second. If we denote the collision frequency by z, then

$$z = \text{number of collisions per second}$$
$$= \frac{\text{distance traveled per second}}{\text{distance traveled per collision}} \tag{7-32}$$
$$= \frac{\text{collisions}}{\text{second}} = \frac{v_{\text{rms}}}{l}$$

Let's use this result to calculate the collision frequency of a nitrogen molecule at 1.00 atm and 20°C. From Example 7-15 we see that v_{rms} = 511 $\text{m}\cdot\text{s}^{-1}$, and we saw in this section that $l = 6.6 \times 10^{-8}$ m under the same conditions. Thus, for N_2 at 20°C and 1.00 atm, we get

$$z = \frac{v_{\text{rms}}}{l} = \frac{511 \text{ m}\cdot\text{s}^{-1}}{6.6 \times 10^{-8} \text{ m}\cdot\text{collision}^{-1}}$$
$$= 7.7 \times 10^9 \text{ collisions}\cdot\text{s}^{-1}$$

Thus, we see that one nitrogen molecule undergoes about eight billion collisions per second at 20°C and 1 atm.

Table 7-3 Atomic or molecular diameters for eight gases

Gas	Diameter/pm
He	210
Ne	250
Ar	360
Kr	410
Xe	470
H_2	280
N_2	370
O_2	370

(c) The pressure of a hydrogen gas sample is 3.75 atm. Convert this pressure to pascals and to kilopascals.
(d) The pressure of an automobile tire is 39 psi. Convert this pressure to atmospheres and to kilopascals.

7-3. A gas bubble has a volume of 0.650 mL at the bottom of a lake, where the pressure is 3.46 atm. What is the volume of the bubble at the surface of the lake, where the pressure is 1.00 atm? What are the diameters of the bubble at the two depths? Assume that the temperature is constant and that the bubble is spherical.

7-4. Suppose we wish to inflate a weather balloon with helium. The balloon will have a volume of 100 m^3 when inflated to a pressure of 0.10 atm. If we use 50-L cylinders of compressed helium gas at a pressure of 100 atm, how many cylinders do we need? Assume that the temperature remains constant.

TEMPERATURE AND CHARLES'S LAW

7-5. Convert the following temperatures to the Kelvin scale:

(a) body temperature, 37°C
(b) room temperature, 20°C
(c) the freezing point of hydrogen, −259°C
(d) the boiling point of ethylene glycol, 199°C

7-6. Convert the following temperatures to the Kelvin scale:

(a) −183°C (the melting point of oxygen)
(b) 6000°C (temperature at the surface of the sun)
(c) −269°C (the boiling point of helium)
(d) 800°C (the melting point of sodium chloride)

7-7. Suppose that in a gas thermometer the gas occupies 14.7 mL at 0°C. The thermometer is immersed in boiling water (100°C). What is the volume of the gas at 100°C?

7-8. Suppose that in a gas thermometer the gas occupies 12.6 mL at 20°C. The thermometer is immersed in a container of solid carbon dioxide chips ("Dry Ice"); the gas then occupies 8.4 mL. What is the temperature of the Dry Ice?

GAY-LUSSAC'S LAW

7-9. Methane burns according to the equation

$$CH_4(g) + 2O_2(g) \rightarrow CO_2(g) + 2H_2O(g)$$

What volume of air, which is 20 percent oxygen by volume, is required to burn 5.0 L of methane when both are the same temperature and pressure?

7-10. Hydrogen and oxygen react violently with each other once the reaction is initiated. For example, a spark can set off the reaction and cause the mixture to explode. What volume of oxygen will react with 0.55 L of hydrogen if both are at 300°C and 1 atm? What volume of water will be produced at 300°C and 1 atm?

IDEAL-GAS LAW

7-11. Calculate the volume that 0.65 mol of ammonia gas occupies at 37°C and 600 torr.

7-12. Calculate the number of grams of propane, C_3H_8, in a 50-L container at a pressure of 7.5 atm and a temperature of 25°C.

7-13. Calculate the pressure exerted by 18 g of steam (H_2O) confined to a volume of 18 L at 100°C. What volume would the water occupy if the steam were condensed to liquid water at 25°C? The density of liquid water is 1.00 $g \cdot mL^{-1}$ at 25°C.

7-14. Calculate the volume in liters occupied by 0.55 kg of dimethyl ether, C_2H_6O, at 950 torr and 15°C.

7-15. Calculate the number of atoms of helium in 1.0 L at −200°C and 0.0010 atm. Compare this value with the number in 1.0 L at 0°C and 1.0 atm.

7-16. Calculate the number of molecules of SO_3 in 100 mL at 100°C and 1.00 atm.

7-17. The ozone molecules in the stratosphere absorb much of the ultraviolet radiation from the sun. The temperature of the stratosphere is −23°C, and the pressure due to the ozone is 1.4×10^{-7} atm. Calculate the number of ozone molecules present in 1.0 mL.

7-18. A low pressure of 1.0×10^{-3} torr is readily obtained in the laboratory by means of a vacuum pump. Calculate the number of molecules in 1.00 mL of gas at this pressure and 20°C.

IDEAL-GAS LAW AND CHEMICAL REACTIONS

7-19. Acetylene is prepared by the reaction of calcium carbide with water:

$$CaC_2(s) + 2H_2O(l) \rightarrow Ca(OH)_2(s) + C_2H_2(g)$$

What volume of acetylene can be obtained from 100 g of calcium carbide and 100 g of water at 0°C and 1.00 atm? What volume results when the temperature is 120°C and the pressure is 1.00 atm?

7-20. Lithium metal reacts with nitrogen at room temperature (20°C) according to the equation

$$6Li(s) + N_2(g) \rightarrow 2Li_3N(s)$$

A sample of lithium metal was placed under a nitrogen atmosphere in a sealed 1.00-L container at a pressure of 1.23 atm. One hour later the pressure dropped to 0.92 atm. Calculate the number of grams of nitrogen that reacted with the lithium metal. Assuming that all the lithium reacted, calculate the mass of lithium originally present.

7-21. Cellular respiration occurs according to the overall equation

$$\underset{\text{glucose}}{C_6H_{12}O_6(s)} + 6O_2(g) \rightarrow 6CO_2(g) + 6H_2O(l)$$

Calculate the volume of $CO_2(g)$ produced at 37°C (body temperature) and 1.00 atm when 1.00 g of glucose is metabolized.

7-22. Chlorine is produced by the electrolysis of an aqueous solution of sodium chloride:

$$2NaCl(aq) + 2H_2O(l) \xrightarrow{\text{electrolysis}} 2NaOH(aq) + H_2(g) + Cl_2(g)$$

The hydrogen gas and chlorine gas are collected separately at 10.0 atm and 25°C. What volume of each can be obtained from 2.50 kg of sodium chloride?

7-23. Chlorine gas can be prepared in the laboratory by the reaction of manganese dioxide with hydrochloric acid

$$MnO_2(s) + 4HCl(aq) \rightarrow MnCl_2(aq) + 2H_2O(l) + Cl_2(g)$$

How much MnO_2 should be added to excess HCl to obtain 500 mL of chlorine gas at 25°C and 750 torr?

7-24. About 50 percent of U.S. and most Canadian sulfur is produced by the Claus process, in which sulfur is obtained from the $H_2S(g)$ that occurs in natural gas deposits or is produced when sulfur is removed from petroleum. The reactions are described by the equations

$$2H_2S(g) + 3O_2(g) \rightarrow 2SO_2(g) + 2H_2O(g)$$
$$SO_2(g) + 2H_2S(g) \rightarrow 3S(l) + 2H_2O(g)$$

How many metric tons of sulfur can be produced from 2.00 million liters of $H_2S(g)$ at 6.00 atm and 200°C?

GAS DENSITY

7-25. Calculate the density of water in the gas phase at 100°C and 1.00 atm. Compare this value with the density of liquid water at 100°C and 1.00 atm (0.958 $g{\cdot}mL^{-1}$).

7-26. Calculate the density of the gas CF_2Cl_2 at 0°C and 1.00 atm.

7-27. A 0.271-g sample of an unknown vapor occupies 294 mL at 100°C and 765 torr. The simplest formula of the compound is CH_2. What is the molecular formula of the compound?

7-28. Upon chemical analysis, a gaseous hydrocarbon is found to contain 88.82 percent C and 11.18 percent H by mass. A 62.6-mg sample of the gas occupies 34.9 mL at 772 torr and 100°C. Determine the molecular formula of the hydrocarbon.

7-29. Ethylene is a gas produced in petroleum cracking and is used to synthesize a variety of important chemicals, such as polyethylene and polyvinylchloride. Chemical analysis shows that ethylene is 85.60 percent carbon and 14.40 percent hydrogen by mass. It has a density of 0.9588 $g{\cdot}L^{-1}$ at 25°C and 635 torr. Use these data to determine the molecular formula of ethylene.

7-30. Benzene is the sixteenth most widely used chemical in the United States, and its principal source is petroleum. Benzene has a wide range of uses, including its use as a solvent and in the synthesis of nylon and detergents. Chemical analysis shows that benzene is 92.24 percent carbon and 7.76 percent hydrogen by mass. A 2.334-g sample of benzene was vaporized in a sealed 500-mL container at 100°C, producing a pressure of 1.83 atm. Use these data to determine the molecular formula of benzene.

PARTIAL PRESSURES

7-31. A gaseous mixture consisting of 0.513 g H_2 and 16.1 g N_2 occupies 10.0 L at 20°C. Calculate the partial pressures of H_2 and N_2 in the mixture.

7-32. A 2.0-L sample of H_2 at 1.00 atm, an 8.0-L sample of N_2 at 3.00 atm, and a 4.0-L sample of Kr at 0.50 atm are transferred to a 10.0-L container. Calculate the partial pressure of each gas and the total final pressure. Assume constant temperature conditions.

7-33. A mixture of O_2 and N_2 is reacted with white phosphorus, which removes the oxygen. If the volume of the mixture decreases from 50.0 mL to 35.0 mL, calculate the partial pressures of O_2 and N_2 in the mixture. Assume that the total pressure remains constant at 740 torr.

7-34. A gaseous mixture of three volumes of carbon dioxide and one volume of water vapor at 200°C and 2.00 atm is cooled to 10°C, thereby condensing the water vapor. If the total volume remains the same, what is the pressure of the carbon dioxide at 10°C? Assume that there is no water vapor present after condensation.

7-35. Nitroglycerin decomposes according to the equation

$$4C_3H_5(NO_3)_3(s) \rightarrow 12CO_2(g) + 10H_2O(l) + 6N_2(g) + O_2(g)$$

What is the total volume of gases produced when collected at 1.0 atm and 25°C from 10 g of nitroglycerin? What pressure is produced if the reaction is confined to a volume of 0.50 L at 25°C? Assume that you can use the ideal-gas equation. Neglect any pressure due to water vapor.

7-36. Explosions occur when a substance decomposes very rapidly with the production of a large volume of gases. When detonated, TNT (trinitrotoluene) decomposes according to the equation

$$2C_7H_5(NO_2)_3(s) \rightarrow 12CO(g) + 2C(s) + 5H_2(g) + 3N_2(g)$$

What is the total volume of gases produced from 1.00 kg of TNT if collected at 0°C and 1.0 atm? What pressure is produced if the reaction is confined to a 50-L container at 500°C? Assume that you can use the ideal-gas equation.

MOLECULAR SPEEDS

7-37. Calculate the root-mean-square speed, v_{rms}, of a fluorine molecule at 25°C.

7-38. Calculate the root-mean-square speed, v_{rms}, for N_2O at 20°C, 200°C, and 2000°C.

7-39. If the temperature of a gas is doubled, how much is the root-mean-square speed of the molecules increased?

7-40. The speed of sound in air at sea level at 20°C is about 760 mph. Compare this value with the root-mean-square speed of N_2 and O_2 gas molecules at 20°C (see Table 7-2).

7-41. Arrange the following gas molecules in order of increasing root-mean-square speed at the same temperatures: O_2, N_2, H_2O, CO_2, NO_2, $^{235}UF_6$, and $^{238}UF_6$.

7-42. Consider a mixture of $H_2(g)$ and $I_2(g)$. Calculate the ratio of the root-mean-square speeds of H_2 and I_2 molecules in the reaction mixture.

MEAN FREE PATH

7-43. Interstellar space has an average temperature of about 10 K and an average density of hydrogen atoms of about one hydrogen atom per cubic meter. Calculate the mean free path of hydrogen atoms in interstellar space. Take $\sigma = 100$ pm.

7-44. Calculate the pressures at which the mean free path of a hydrogen molecule will be 1.00 μm, 1.00 mm, and 1.00 m at 20°C.

7-45. Calculate the number of collisions per second of one hydrogen molecule at 20°C and 1.0 atm.

7-46. Calculate the number of collisions per second that one molecule of nitrogen undergoes at 20°C and 1.0×10^{-3} torr.

GRAHAM'S LAW OF EFFUSION

7-47. Two identical balloons are filled, one with helium and one with nitrogen, at the same temperature and pressure. If the nitrogen leaks out from its balloon at the rate of 75 $mL \cdot h^{-1}$, then what will be the rate of leakage from the helium-filled balloon?

7-48. Two identical porous containers are filled, one with hydrogen and one with carbon dioxide, at the same temperature and pressure. After one day, 1.50 mL of carbon dioxide has leaked out of its container. How much hydrogen has leaked out in one day?

7-49. It takes 145 s for 1.00 mL of N_2 to effuse from a certain porous container. Given that it takes 230 s for 1.00 mL of an unknown gas to effuse under the same temperature and pressure, calculate the molecular mass of the unknown gas.

7-50. Suppose that it takes 175 s for 1.00 mL of N_2 to effuse from a porous container under a certain temperature and pressure and that it takes 200 s for 1.00 mL of a CO-CO_2 mixture to effuse under the same conditions. What is the volume percentage of CO in the mixture?

VAN DER WAALS EQUATION

7-51. Use the van der Waals equation to calculate the pressure exerted by 24.5 g of NH_3 confined to a 2.15-L container at 300 K. Compare your answer with the pressure calculated using the ideal-gas equation.

7-52. Use the van der Waals equation to calculate the pressure exerted by 45 g of propane, C_3H_8, confined to a 2.2-L container at 300°C. Compare your answer with the pressure calculated using the ideal-gas equation.

IDEAL-GAS LAW USING PASCALS

7-53. Calculate the number of Cl_2 gas molecules in a volume of 5.00 mL at 40°C and 2.15×10^4 Pa.

7-54. Calculate the pressure in pascals that is exerted by 6.15 mg of CO_2 occupying 2.10 mL at 75°C.

7-55. A sample of radon occupies 7.12 μL at 22°C and 8.72×10^4 Pa. Calculate the volume at 0°C and 1.013×10^5 Pa. What is the mass of the radon?

7-56. Calculate the mass of $N_2O(g)$ that occupies 2.10 L at a pressure of 4.50×10^4 Pa and a temperature of 15°C.

7-57. Calculate the density of $ND_3(g)$ at 0°C and 2.00×10^3 Pa..Take $M = 20.06$ $g \cdot mol^{-1}$.

7-58. Using the fact that 1 atm = 1.013×10^5 Pa, calculate the value of the gas constant in SI units, $J \cdot mol^{-1} \cdot K^{-1}$.

ADDITIONAL PROBLEMS

7-59. A pressure of exactly one atmosphere is defined to be the pressure that supports a column of mercury 760.0

mm high. Pressure is force per unit area, and the force, F, that a mass, m, exerts on a surface is

$$F = mg$$

where $g = 9.806\ \text{m}\cdot\text{s}^{-2}$ is a constant called the acceleration of gravity. Given that the density of mercury is $13.59\ \text{g}\cdot\text{cm}^{-3}$, show that 1 atm $= 1.013 \times 10^5\ \text{N}\cdot\text{m}^{-2}$, or 1.013×10^5 Pa.

7-60. Gallium metal can be used as a manometer fluid at high temperatures because of its wide liquid range (30 to 2400°C). Compute the height of the liquid gallium column in a gallium manometer when the temperature is 850°C and the pressure is 1300 torr. Take the density of liquid gallium to be $6.0\ \text{g}\cdot\text{mL}^{-1}$.

7-61. Given below are pressure-volume data for a sample of 0.28 g of N_2 at 25°C. Verify Boyle's law for these data. Plot the data so that a straight line is obtained.

P/atm	*V*/L	*P*/atm	*V*/L
0.26	0.938	2.10	0.116
0.41	0.595	2.63	0.093
0.83	0.294	3.14	0.078
1.20	0.203		

7-62. A container of carbon dioxide has a volume of 50 L. Originally the carbon dioxide was at a pressure of 10.0 atm at 25°C. After the tank had been used for a month at 25°C, the pressure dropped to 4.7 atm. Calculate the number of grams of carbon dioxide used during this period of time.

7-63. Recent measurements have shown that the atmosphere of Venus is mostly carbon dioxide. At the surface, the temperature is about 800°C and the pressure is about 75 atm. In the unlikely chance that a resident of Venus defined a standard temperature and pressure and took those values as standard conditions, what value would he find for the volume of a mole of ideal gas at Venusian standard conditions?

7-64. The organic compound di-*n*-butyl phthalate, $C_{16}H_{22}O_4$, is sometimes used as a low-density ($1.046\ \text{g}\cdot\text{mL}^{-1}$) manometer fluid. Compute the pressure in torr of a gas that supports a 500-mm column of di-*n*-butyl phthalate.

7-65. Several television commercials state that it requires 10,000 gallons of air to burn 1 gal of gasoline. Using octane, C_8H_{18}, as the chemical formula of gasoline and using the fact that air is 20 percent oxygen by volume, calculate the volume of air at 0°C and 1.0 atm that is required to burn 1 gal of gasoline. The density of octane is $0.70\ \text{g}\cdot\text{mL}^{-1}$.

7-66. Lactic acid is produced by the muscles when insufficient oxygen is available and is responsible for muscle cramps during vigorous exercising. It also provides the acidity found in dairy products. Chemical analysis shows that lactic acid is 39.99 percent carbon, 6.73 percent hydrogen, and 53.28 percent oxygen by mass. A 0.3338-g sample of lactic acid was vaporized in a sealed 300-mL container at 150°C, producing a pressure of 326 torr. Use these data to determine the molecular formula of lactic acid.

7-67. A 0.428-g sample of a mixture of KCl and $KClO_3$ is heated; 80.7 mL of O_2 is collected over water at 18°C and 756 torr. Calculate the mass percentage of $KClO_3$ in the mixture. ($P_{H_2O} = 15.5$ torr at 18°C.)

7-68. Nitrous oxide, N_2O, sometimes called laughing gas, is used as an anesthetic. It is prepared by the decomposition of ammonium nitrate with heat

$$NH_4NO_3(s) \xrightarrow{\text{high T}} N_2O(g) + 2H_2O(g)$$

What volume of nitrous oxide will be produced when 10.0 g of ammonium nitrate is heated to 200°C at 1.00 atm? The gases are then cooled to 0°C to condense $H_2O(g)$ to water. What volume will the nitrous oxide occupy at 0°C and 1.00 atm? Neglect any pressure due to water vapor.

7-69. Calculate the volume of 0.200 M NaOH required to prepare 150 mL of $H_2(g)$ at 10°C and 750 torr from the reaction

$$2Al(s) + 2NaOH(aq) + 2H_2O(l) \rightarrow 2NaAlO_2(aq) + 3H_2(g)$$

7-70. Calculate the temperature at which a carbon dioxide molecule would have the same root-mean-square speed as a neon atom at 100°C.

7-71. Ammonia gas reacts with hydrogen chloride gas as described by the equation

$$NH_3(g) + HCl(g) \rightarrow NH_4Cl(s)$$

Suppose 5.0 g of NH_3 is reacted with 10.0 g of $HCl(g)$ in a 1.00-L vessel at 75°C. Compute the final pressure (in atm) of gas in the vessel.

7-72. In many everyday situations, atmospheric pressure is reported in "inches of mercury." Calculate the pressure of a standard atmosphere in inches of mercury.

7-73. Compare the mass of oxygen that you utilize to the mass of solid food that you consume each day. Assume that you breathe in 0.5 L of air with each breath, that your respiratory rate is 14 times per minute, that you utilize about 25 percent of the oxygen inhaled, and that you eat 2 pounds of solid food daily.

7-74. Using the data given in Table 7-4, predict which compound of those listed shows the largest deviation from ideal-gas behavior at 1000 atm.

7-75. What must be the temperature in order that CO_2 molecules have a root-mean-square speed of $1000\ \text{m}\cdot\text{s}^{-1}$?

7-76. Carbon dioxide can be prepared in the laboratory by the reaction of dilute hydrochloric acid with marble chips [primarily $CaCO_3(s)$]:

$$CaCO_3(s) + 2HCl(aq) \rightarrow CaCl_2(aq) + H_2O(l) + CO_2(g)$$

How many milliliters of 0.250 M $HCl(aq)$ are required to produce 500 mL of $CO_2(g)$ at 14°C and 756 torr by means of the reaction above?

7-77. How many milliliters of 0.620 M $HCl(aq)$ must be added to excess zinc to produce 250 mL of hydrogen measured at 20°C and 745 torr?

7-78. Dry ammonia can be prepared by gently heating a mixture of ammonium chloride and calcium hydroxide

$$2NH_4Cl(s) + Ca(OH)_2(s) \rightarrow CaCl_2(s) + 2H_2O(g) + 2NH_3(g)$$

and passing the gaseous products through a column packed with calcium oxide to remove the water vapor according to

$$CaO(s) + H_2O(g) \rightarrow Ca(OH)_2(s)$$

How many milliliters of ammonia, measured at 15°C and 750 torr, can be produced by heating 1.00 gram each of ammonium chloride and calcium hydroxide?

7-79. Sodium peroxide is used in self-contained breathing devices to absorb exhaled carbon dioxide and simultaneously produce oxygen. The equation for the reaction is

$$2Na_2O_2(s) + 2CO_2(g) \rightarrow 2Na_2CO_3(s) + O_2(g)$$

How many liters of $CO_2(g)$, measured at 0°C and 1 atmosphere, can be absorbed by 1 kg of $Na_2O_2(s)$? How many liters of $O_2(g)$ will be produced under the same conditions?

7-80. Nitrous oxide, N_2O, can be prepared by gently heating equimolar quantities of potassium nitrate and ammonium chloride:

$$NH_4Cl(s) + KNO_3(s) \rightarrow KCl(s) + 2H_2O(l) + N_2O(g)$$

Ammonium nitrate also can be used to produce N_2O

$$NH_4NO_3(s) \rightarrow N_2O(g) + 2H_2O(l)$$

but NH_4NO_3 is dangerously explosive. How many grams of $NH_4Cl(s)$ and $KNO_3(s)$ are required to produce 800 mL of $N_2O(g)$ at 20°C and 748 torr?

7-81. Acetylene, $C_2H_2(g)$, can be prepared by the reaction

$$CaC_2(s) + 2H_2O(l) \rightarrow C_2H_2(g) + Ca(OH)_2(s)$$

How many grams of calcium carbide are required to fill a 500-mL canister with acetylene at 18°C and 3.00 atm?

7-82. Gunpowder is a mixture of potassium nitrate, charcoal, and sulfur in the approximate proportions of 6:1:1 by mass. The explosive character of gunpowder is due to the sudden production of a large volume of gaseous products and heat generated upon detonation. The equation for the principal reaction is

$$2KNO_3(s) + S(s) + 3C(s) \rightarrow K_2S(s) + N_2(g) + 3CO_2(g)$$

Use the reaction stoichiometry to explain why the approximate proportions are 6:1:1 by mass. Suppose that 1 kg of gunpowder is detonated in a 2-L container. Assuming a final temperature of 300°C, calculate the pressure generated. [Assume that the kilogram of gunpowder consists of 748 g of $KNO_3(s)$, 119 g of $S(s)$, and 133 g of $C(s)$.]

7-83. The pressure in a ceramic vessel that contained nitrogen dropped from 1850 torr to 915 torr in 30.0 min. When the same vessel was filled with another gas, the pressure dropped from 1850 torr to 915 in 54.3 min. Calculate the molecular mass of the second gas assuming that the gases effuse from the container.

7-84. A mixture of neon and argon has a density of $1.64\ g\cdot L^{-1}$ at 0°C and 800 torr. Compute the ratio of the number of moles of neon to the number of moles of argon in the mixture.

7-85. On a hot, humid day the partial pressure of water vapor in the atmosphere is typically 30 to 40 torr. Suppose that the partial pressure of water vapor is 35 torr and that the temperature is 35°C. If all the water vapor in a room that measures 3.0 m by 5.0 m by 6.0 m were condensed, how many milliliters of $H_2O(l)$ would be obtained?

7-86. A certain gaseous hydrocarbon was determined to be 82.66 percent carbon and 17.34 percent hydrogen by mass. A 6.09-g sample of the gas occupied 2.48 L at 1.00 atm and 15°C. Determine the molecular formula of the gas.

7-87. Calculate the volume of $Cl_2(g)$ produced at 815 torr and 15°C if 6.75 g of $KMnO_4(s)$ are added to 300 mL of 0.1150 M $HCl(aq)$. The equation for the reaction is

$$2KMnO_4(s) + 16HCl(aq) \rightarrow 2MnCl_2(aq) + 2KCl(aq) + 5Cl_2(g) + 8H_2O(l)$$

7-88. A vessel with a very small hole in one of its walls is filled with $N_2(g)$ at 750 torr and 15°C. In 1 min, 2.75 mL of $N_2(g)$ effuses from the vessel. The vessel is evacuated and then filled with an unknown gas. In 1 min, 1.50 mL of the unknown gas effuses from the vessel. Calculate the molecular mass of the unknown gas.

7-89. Two glass bulbs are connected by a valve. One bulb has a volume of 650 mL and is occupied by $N_2(g)$ at 825 torr. The other has a volume of 500 mL and is occupied by $O_2(g)$ at 730 torr. The valve is opened and the two gases mix. Calculate the total pressure and the partial pressure of $N_2(g)$ and $O_2(g)$ of the resulting mixture.

7-90. Calculate the value of absolute zero on the Fahrenheit scale.

Table C-I Composition of atmosphere below 100 km

Major constituents	Content in f molecules (a
nitrogen (N_2)	0.7808 (75.51
oxygen (O_2)	0.2095 (23.14
argon	0.0093 (1.28%
water vapor	0–0.04

Minor constituents	Content in p per million
carbon dioxide	325 ppm
neon	18 ppm
helium	5 ppm
methane	2 ppm
krypton	1 ppm
hydrogen (H_2)	0.5 ppm
dinitrogen oxide	0.5 ppm
xenon	0.1 ppm

*The unit ppm denotes parts per m
325 ppm of carbon dioxide means
molecules are CO_2.

for more than 80 percent of t
all the gaseous water and cl
and ice crystals) in the eart
characterized by strong vertic
ple, in clear air, a molecule ca
level to the top of the tropos
during severe thunderstorr
travel the same distance i
weather takes place in the tr
The *stratosphere* lies above t
troposphere and stratosphere
99.9 percent of the mass of
stratosphere is calm; it is char
vertical mixing. Debris fron
and dust from volcanic erupti
stratosphere for years before
osphere. Above the stratosp
called the *mesosphere* (literally
and the *ionosphere*. The iono
and electrons, which are prod
high-energy solar radiation
Contemporary geological t
much of the earth's atmosph
gases discharged from the ea
volcanic activity (Figure C-4).
sition is fairly uniform up to
C-1) and is roughly 78 perc
percent oxygen (by volume),

7-91. When 300 mL of an aqueous solution of hydrochloric acid is reacted with an excess of zinc, 1.65 L of $H_2(g)$ at 20°C and 752 torr are evolved. What is the molarity of the HCl(*aq*) solution?

7-92. A mixture of zinc and aluminum with a mass of 5.62 g reacts completely with hydrochloric acid, and 2.67 L of hydrogen gas at 23°C and 773 torr are liberated. Calculate the mass percentage of zinc in the mixture.

7-93. Both sodium hydride, NaH(*s*), and calcium hydride, $CaH_2(s)$, react with water to produce $H_2(g)$ and the respective hydroxides. Suppose that a 3.75-g sample of a mixture of NaH(*s*) and $CaH_2(s)$ is added to water and the evolved $H_2(g)$ is collected. The $H_2(g)$ occupies a volume of 4.12 L at 742 torr and 17°C. Calculate the mass percentages of NaH(*s*) and $CaH_2(s)$ in the mixture.

7-94. The speed of sound in an ideal monatomic gas is given by

$$v_{\text{sound}} = \left(\frac{5RT}{3M_{\text{kg}}}\right)^{1/2}$$

Derive an equation for the ratio $v_{\text{rms}}/v_{\text{sound}}$. Calculate v_{rms} for an argon atom at 20°C and compare your result to the speed of sound in argon.

7-95. The root-mean-square speed is defined as

$$v_{\text{rms}} = (\overline{v^2})^{1/2} = \left(\frac{v_1^2 + v_2^2 + v_3^2 + v_4^2 + \ \ldots}{N}\right)^{1/2}$$

where the notation + . . . means "and so on." We have learned in this chapter that $v_{\text{rms}} = (3RT/M_{\text{kg}})^{1/2}$. It is possible to use the kinetic theory of gases to derive a formula for simply the average molecular speed, $\overline{v}$

$$\overline{v} = \frac{v_1 + v_2 + v_3 + v_4 + \ \ldots}{N}$$

This result turns out to be $\overline{v} = (8RT/\pi M_{\text{kg}})^{1/2}$. Show that

$$\frac{v_{\text{rms}}}{\overline{v}} = 1.085$$

Notice that the average speed and the root-mean-square speed differ by less than 10 percent.

EXAMPLE 8-1: Calculate the maximum height that a bullet of mass 25 g will reach if it is shot straight upward with an initial speed of 400 $\text{m}\cdot\text{s}^{-1}$. (Ignore the effects of air resistance.)

Solution: Ignoring the height of the gun barrel above the ground, all the energy of the bullet on exit from the barrel is kinetic energy, so we have (Equation 8-1)

$$E_{\text{k}}(\text{initial}) = mv_0^2/2 = (0.025\text{ kg})(400\text{ m}\cdot\text{s}^{-1})^2/2$$
$$= 2000\text{ kg}\cdot\text{m}^2\cdot\text{s}^{-2} = 2000\text{ J}$$

and the total energy is

$$E_{\text{total}}(\text{initial}) = E_{\text{k}}(\text{initial}) + E_{\text{p}}(\text{initial})$$
$$= 2000\text{ J} + 0\text{ J} = 2000\text{ J}$$

At the apex of the bullet's flight, the kinetic energy will be zero (because its speed is zero) and all its energy will be potential energy. Thus, we have

$$E_{\text{total}}(\text{apex}) = E_{\text{k}}(\text{apex}) + E_{\text{p}}(\text{apex}) = 0\text{ J} + 2000\text{ J} = 2000\text{ J}$$

We use Equation (8-2) to solve for the maximum height

$$E_{\text{p}}(\text{apex}) = 2000\text{ J} = gmh(\text{apex})$$

Thus, we have for $h(\text{apex})$

$$h(\text{apex}) = \frac{E_{\text{p}}(\text{apex})}{gm} = \frac{2000\text{ J}}{(9.81\text{ m}\cdot\text{s}^{-2})(0.025\text{ kg})}$$
$$= \frac{2000\text{ kg}\cdot\text{m}^2\cdot\text{s}^{-2}}{(9.81\text{ m}\cdot\text{s}^{-2})(0.025\text{ kg})}$$
$$= 8200\text{ m}$$

PRACTICE PROBLEM 8-1: Electricity is generated at hydroelectric facilities by converting the potential energy of water in a lake to kinetic energy as the water falls down large pipes that are connected to turbines (Figure 8-3). The turbines convert the kinetic energy of the falling water into electricity. Given that the water falls 200 m, calculate the kinetic energy available at the turbine input connection per kilogram of water.

Answer: 1.96×10^3 J

Figure 8-3 Hoover Dam at Lake Mead. The potential energy of the water in Lake Mead is converted to electrical energy when it drops from near the lake surface to the bottom of the dam. The falling water drives turbines that are within the dam and generate electricity.

Figure C-1 The earth fror clearly visible.

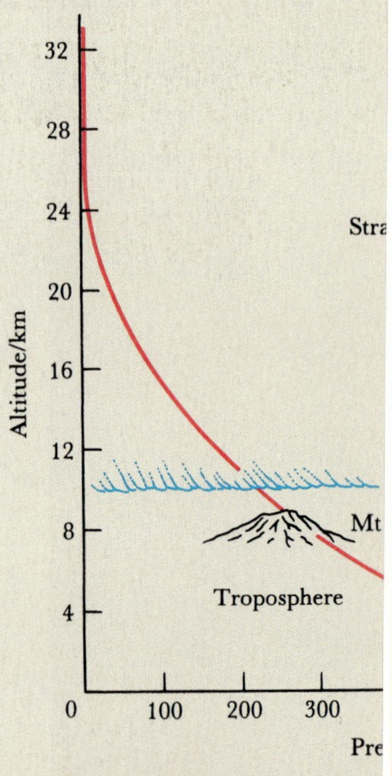

Figure C-2 The pressure function of altitude above decreases rapidly with inc 380 torr (0.5 atm) at abou

The rate at which energy is produced or utilized is called **power.** The SI unit of power is a **watt,** which is defined as exactly one joule per second; that is, $1\ \mathrm{W} = 1\ \mathrm{J{\cdot}s^{-1}}$. For example, a 40-watt fluorescent light bulb emits energy in the form of light (20 percent) and heat (80 percent) at a rate of $40\ \mathrm{J{\cdot}s^{-1}}$. Home energy consumption is often expressed in units of **kilowatt-hours.** A kilowatt-hour (kW·h) is the energy used by a one-kilowatt device operating for 1 hour, Thus,

$$\begin{aligned} 1 \text{ kilowatt-hour} &= (1\ \mathrm{kJ{\cdot}s^{-1}})(1\ \mathrm{h}) \\ &= (1\ \mathrm{kJ{\cdot}s^{-1}})(1\ \mathrm{h})\left(\frac{3600\ \mathrm{s}}{1\ \mathrm{h}}\right) \\ &= 3600\ \mathrm{kJ} \end{aligned}$$

EXAMPLE 8-2: The energy requirements of an average U.S. home are about 30 kW·h per day. A major energy source that remains largely untapped, at least by humans, is sunlight. The sunlight that reaches the surface of the United States on a clear day has a power level of about $1.0\ \mathrm{kW{\cdot}m^{-2}}$. If 30 percent of the incident solar energy could be collected and used, then show that for 8 hours of sunlight in a day, all the household energy requirements could be satisfied by about $12\ \mathrm{m^2}$ of collector surface.

Solution: The total power level in the sunlight incident on a $12\ \mathrm{m^2}$ collector surface is

$$(1.0\ \mathrm{kW{\cdot}m^{-2}})(12\ \mathrm{m^2}) = 12\ \mathrm{kW}$$

In 8.0 h the number of kilowatt-hours available is

$$(12\ \mathrm{kW})(8.0\ \mathrm{h}) = 96\ \mathrm{kW{\cdot}h}$$

Because the solar collector can only convert 30 percent of the incident energy to electricity, the available energy level is only 30 percent of 96 kW·h or

$$(96\ \mathrm{kW{\cdot}h})(0.30) = 29\ \mathrm{kW{\cdot}h}$$

Thus, the converter can generate at least 29 kW·h per day, provided the sun shines for at least 8 h.

PRACTICE PROBLEM 8-2: Although power company electricity rates vary widely, a rate of 10 U.S. cents per kilowatt-hour is representative. Modern electric clothes dryers use about 7.5 kW when in the heating mode. Assuming it takes 50 min to dry a large load of clothes, calculate the cost of the electricity used to dry a load of clothes.

Answer: $0.625

8-2. THE TRANSFER OF ENERGY BETWEEN A REACTION SYSTEM AND ITS SURROUNDINGS OCCURS AS WORK OR HEAT

Chemical reactions almost always involve a transfer of energy between the reaction system and its surroundings. The reaction system includes all the chemicals involved in the reaction. The surroundings consist of everything else that can exchange heat or work with the reaction system, for example, containers, water baths, and the atmosphere. **Ther-**

energy transferred as heat by the symbol q. It should be recognized that a system does not contain heat or work. Rather, heat and work are ways in which energy is transferred, as the following analogy illustrates. If a container is left outside in the rain, the question, "How much rain is in the container?" is answered correctly as, "There is no rain in the container, only water." Rain is the means by which water was transferred from the clouds to the container.

In thermodynamics the energy of a system is denoted by the symbol U. We denote the **energy change** of a system by ΔU, where

$$\Delta U = U_{\mathrm{f}} - U_{\mathrm{i}}$$

As we noted earlier, the energy of a system can change as a result of a transfer of energy as heat (q) or as work (w). Application of the law of conservation of energy to a reaction system yields

$$\Delta U = U_{\mathrm{f}} - U_{\mathrm{i}} = q + w \tag{8-7}$$

where q is the energy transferred as heat to or from the system and w is the energy transferred as work to or from the system. Equation (8-7), which constitutes a mathematical expression of the first law of thermodynamics, is an energy balance equation. It says that in going from some particular initial state to some particular final state of a system, the energy transferred as work and the energy transferred as heat must always add up to give the same value of ΔU for the system. Because ΔU is equal to $U_{\mathrm{final}} - U_{\mathrm{initial}}$, the value of ΔU depends *only* on the initial and final states of the system. The value of ΔU does not depend on how the system goes from the initial to the final state. Functions that depend only on the state of a system and not on how that state is achieved are called **state functions** in thermodynamics. To get a better idea of what we mean by a state function, let's consider the temperature. There are many ways in which we could take a system from initial temperature T_1 to some final temperature T_2. For example, we could take a metal rod at T_1 and drop it into a large bath of hot water at T_2. The final temperature of the rod would be T_2, and the difference in temperature between the initial and final states would be $T_2 - T_1$. Alternatively, we could apply a flame to the rod until enough energy has been transferred to the rod to raise its final temperature to T_2. In either case, the change in temperature is simply $T_2 - T_1$. It makes no difference whatsoever to the value of $T_2 - T_1$ how the change in temperature was achieved. This is the essential requirement for a state function. Other important state functions are pressure and volume.

Work and heat are not state functions, but **energy transfer functions.** Their values depend on how a process is carried out. To illustrate this distinction, consider the following analogy. Two rock climbers of equal mass scale the same cliff face. One climber gets a significant way up the face, slips part of the way down, and then climbs to the top. The other climber reaches the summit with no backsliding. Obviously, the first climber has performed more work (and generated more heat) than the second, even though their potential energies at the summit are the same.

State functions are expressed as uppercase letters (P for pressure, V for volume, T for temperature, and U for energy), whereas energy transfer functions are expressed as lowercase letters, for example, w for work and q for heat.

If we apply the first law of thermodynamics to a chemical reaction, then Equation (8-7) becomes

$$\Delta U_{\text{rxn}} = q + w \tag{8-8}$$

where the subscript rxn stands for *reaction.* The value of ΔU_{rxn} is the **energy change of the reaction.**

For a reaction that occurs at constant pressure, we can substitute Equation (8-6) into Equation (8-8) to obtain

$$\Delta U_{\text{rxn}} = q - P\Delta V \tag{8-9}$$

If a reaction occurs in a rigid, closed container, then there is no change in volume, no energy is transferred as work, and thus $\Delta V = 0$ in Equation (8-9). In this case we have

$$\Delta U_{\text{rxn}} = q_{\text{V}} = \begin{pmatrix}\text{heat evolved or absorbed}\\ \text{at constant volume}\end{pmatrix} \tag{8-10}$$

The V subscript on q emphasizes that q_{V} is the heat evolved or absorbed when the volume of the reaction system is constant.

For reactions that occur at constant pressure, it is convenient to introduce a thermodynamic function H, called the **enthalpy,** for the reaction system, which is defined by the equation

$$H = U + PV \tag{8-11}$$

where U is the energy, P is the pressure, and V is the volume of the reaction system. Application of Equation (8-11) to a chemical reaction yields

$$\Delta H_{\text{rxn}} = H_{\text{f}} - H_{\text{i}} = U_{\text{f}} - U_{\text{i}} + P_{\text{f}}\,V_{\text{f}} - P_{\text{i}}\,V_{\text{i}} \tag{8-12}$$

where ΔH_{rxn} is the **enthalpy change** for the chemical reaction. If the reaction occurs at constant pressure, then $P_{\text{f}} = P_{\text{i}}$, and Equation (8-12) becomes ($P_{\text{f}} = P_{\text{i}} = P$)

$$\Delta H_{\text{rxn}} = \Delta U_{\text{rxn}} + P\Delta V_{\text{rxn}} \tag{8-13}$$

The difference between a constant-pressure process and a constant-volume process can be seen by considering a reaction in which the volume of the reaction mixture increases during the course of the reaction. In this case, the mixture has to expend energy to push back the surrounding atmosphere as it expands. Consequently, the energy evolved or absorbed as heat in a constant-pressure reaction is not the same as that for the same reaction run at constant volume. The difference between ΔU_{rxn} and ΔH_{rxn} is just the energy that is required for the system to expand against a constant atmospheric pressure. Substitution of Equation (8-9) into (8-13) yields

$$\Delta H_{\text{rxn}} = q_{\text{P}} - P\Delta V + P\Delta V = q_{\text{P}}$$

where the P subscript on q emphasizes that q_{P} is the heat transferred when the reaction takes place at constant pressure. Thus, we see that the enthalpy change for a reaction has the useful property that

$$\Delta H_{\text{rxn}} = q_{\text{P}} = \begin{pmatrix}\text{heat evolved or absorbed}\\ \text{at constant pressure}\end{pmatrix} \tag{8-14}$$

Equation (8-14) is the constant-pressure analogue of Equation (8-10). The difference between ΔU_{rxn} and ΔH_{rxn} is usually small, as we will

If we cancel $CO(g)$ from both sides, then we get

$$(3) \quad C(s) + O_2(g) \rightarrow CO_2(g)$$

The additive property ΔH°_{rxn} tells us that ΔH°_{rxn} for Equation (3) is simply

$$\begin{aligned}\Delta H^\circ_{rxn}(3) &= \Delta H^\circ_{rxn}(1) + \Delta H^\circ_{rxn}(2) \\ &= -110.5\ \text{kJ} + (-283.0\ \text{kJ}) = -393.5\ \text{kJ}\end{aligned} \tag{8-16}$$

In effect we can imagine Equations (1) and (2) as representing a two-step process with the same initial and final steps as Equation (3). The total enthalpy change for the two equations together must, therefore, be the same as if the reaction proceeded in a single step.

The additivity property of ΔH°_{rxn} values is known as **Hess's law.** If two or more chemical equations are added together, then the value of ΔH°_{rxn} for the resulting equation is equal to the sum of the ΔH°_{rxn} values for the separate equations. Thus, if the values of $\Delta H^\circ_{rxn}(1)$ and $\Delta H^\circ_{rxn}(2)$ are known from experiment, then we need not independently determine the experimental value of $\Delta H^\circ_{rxn}(3)$, because its value is equal to the value of the sum $\Delta H^\circ_{rxn}(1) + \Delta H^\circ_{rxn}(2)$.

Suppose that we add a chemical equation to itself, for example,

$$\begin{array}{lll} & SO_2(g) \rightarrow S(s) + O_2(g) & \Delta H^\circ_{rxn}(1) \\ & SO_2(g) \rightarrow S(s) + O_2(g) & \Delta H^\circ_{rxn}(1) \\ \hline \text{sum} & 2SO_2(g) \rightarrow 2S(s) + 2O_2(g) & \Delta H^\circ_{rxn}(3) \end{array}$$

In this case

$$\Delta H^\circ_{rxn}(3) = \Delta H^\circ_{rxn}(1) + \Delta H^\circ_{rxn}(1) = 2\Delta H^\circ_{rxn}(1) \tag{8-17}$$

Notice that adding a chemical equation to itself is equivalent to multiplying both sides of the chemical equation by 2, that is,

$$2[SO_2(g) \rightarrow S(s) + O_2(g)]$$

becomes

$$2SO_2(g) \rightarrow 2S(s) + 2O_2(g)$$

Equation (8-17) can be generalized to cover multiplication of a chemical equation by any numerical factor n. For example,

$$n[SO_2(g) \rightarrow S(s) + O_2(g)]$$

becomes

$$nSO_2(g) \rightarrow nS(s) + nO_2(g)$$

and the value of ΔH°_{rxn} for the resulting equation is

$$\Delta H^\circ_{rxn} = n\Delta H^\circ_{rxn}(1) \tag{8-18}$$

Equation (8-18) follows directly from the fact that multiplication of the chemical equation through by n is equivalent to writing the equation out n times and adding the n equations together.

Now let's consider the following combination of chemical equations

$$\begin{array}{lll} (1) & SO_2(g) \rightarrow S(s) + O_2(g) & \Delta H^\circ_{rxn}(1) \\ (2) & S(s) + O_2(g) \rightarrow SO_2(g) & \Delta H^\circ_{rxn}(2) \end{array}$$

By Equation (8-16), the value of ΔH°_{rxn} for the sum of these two equations is

$$\Delta H^\circ_{rxn}(3) = \Delta H^\circ_{rxn}(1) + \Delta H^\circ_{rxn}(2)$$

But the addition of these two equations yields no net reactants and no net products or, in other words, no net chemical change whatsoever. Because there is no net change, the value of $\Delta H^\circ_{rxn}(3)$ must be 0 and we conclude that

$$\Delta H^\circ_{rxn}(2) = -\Delta H^\circ_{rxn}(1) \qquad (8\text{-}19)$$

Because Equation (2) is simply the reverse of Equation (1), we conclude from Hess's law that

$$\Delta H^\circ_{rxn}(\text{reverse}) = -\Delta H^\circ_{rxn}(\text{forward}) \qquad (8\text{-}20)$$

Equation (8-20) is easy to apply. If we reverse a chemical equation, then the reactants become the products and the products become the reactants, and the sign of ΔH°_{rxn} changes. The value of ΔH°_{rxn} for the reaction described by the chemical equation

$$(1) \quad CO_2(g) \rightarrow C(s) + O_2(g)$$

which is the reverse of the equation

$$(2) \quad C(s) + O_2(g) \rightarrow CO_2(g) \qquad \Delta H_{rxn} = -393.5 \text{ kJ}$$

is

$$\Delta H^\circ_{rxn}(1) = -\Delta H^\circ_{rxn}(2)$$
$$= -(-393.5 \text{ kJ}) = +393.5 \text{ kJ}$$

EXAMPLE 8-5: Given the following ΔH°_{rxn} values

$$(1) \quad SO_2(g) \rightarrow S(s) + O_2(g) \qquad \Delta H^\circ_{rxn}(1) = +296.8 \text{ kJ}$$
$$(2) \quad 2S(s) + 3O_2(g) \rightarrow 2SO_3(g) \qquad \Delta H^\circ_{rxn}(2) = -791.4 \text{ kJ}$$

calculate the value of ΔH°_{rxn} for the equation

$$(3) \quad 2SO_2(g) + O_2(g) \rightarrow 2SO_3(g)$$

Solution: To obtain Equation (3) from Equations (1) and (2), it is first necessary to multiply Equation (1) through by 2, because Equation (3) involves 2 mol of $SO_2(g)$ as a reactant. For the equation

$$(4) \quad 2SO_2(g) \rightarrow 2S(s) + 2O_2(g)$$

we have from Equation (8-18) with $n = 2$

$$\Delta H^\circ_{rxn}(4) = 2\Delta H^\circ_{rxn}(1) = 2 \times 296.8 \text{ kJ} = +593.6 \text{ kJ}$$

Addition of Equations (2) and (4) yields

$$2S(s) + 3O_2(g) + 2SO_2(g) \rightarrow 2S(s) + 2O_2(g) + 2SO_3(g)$$

If we cancel $2S(s)$ and $2O_2(g)$ from both sides, then we get Equation (3):

$$2SO_2(g) + O_2(g) \rightarrow 2SO_3(g)$$

The corresponding value of ΔH°_{rxn} is

$$\Delta H^\circ_{rxn}(3) = 2\Delta H^\circ_{rxn}(1) + \Delta H^\circ_{rxn}(2)$$
$$= +593.6 \text{ kJ} - 791.4 \text{ kJ}$$
$$= -197.8 \text{ kJ}$$

EXAMPLE 8-6: The hydrocarbon butane (C_4H_{10}) exists as two structurally different isomers called *n*-butane and isobutane. The standard molar heats of combustion at 25°C of *n*-butane and isobutane are $-2878\ kJ \cdot mol^{-1}$ and $-2871\ kJ \cdot mol^{-1}$, respectively. Calculate ΔH°_{rxn} for the conversion of *n*-butane to isobutane

$$n\text{-butane} \rightarrow \text{isobutane}$$

or

$$n\text{-}C_4H_{10}(g) \rightarrow i\text{-}C_4H_{10}(g)$$

The equations for the two combustion reactions are

$$(1)\quad n\text{-}C_4H_{10}(g) + \tfrac{13}{2}O_2(g) \rightarrow 4CO_2(g) + 5H_2O(l) \qquad \Delta H^\circ_{rxn}(1) = -2878\ kJ$$

$$(2)\quad i\text{-}C_4H_{10}(g) + \tfrac{13}{2}O_2(g) \rightarrow 4CO_2(g) + 5H_2O(l) \qquad \Delta H^\circ_{rxn}(2) = -2871\ kJ$$

Solution: If we reverse the second equation and add the result to the first equation, then we obtain the desired equation

$$(3)\quad n\text{-}C_4H_{10} \rightarrow i\text{-}C_4H_{10}$$

From Equation (8-20), we see that the value of ΔH°_{rxn} for the reverse of Equation (2) is $\Delta H^\circ_{rxn}[\text{reverse of (2)}] = -\Delta H^\circ_{rxn}(2)$; combination of this result with Equation (8-16) yields

$$\begin{aligned} \Delta H^\circ_{rxn}(3) &= \Delta H^\circ_{rxn}(1) - \Delta H^\circ_{rxn}(2) \\ &= -2878\ kJ - (-2871\ kJ) \\ &= -7\ kJ \end{aligned}$$

PRACTICE PROBLEM 8-6: The chemical equation for the combustion of magnesium in sulfur dioxide is

$$3Mg(s) + SO_2(g) \rightarrow MgS(s) + 2MgO(s)$$

In this reaction magnesium is the fuel and sulfur dioxide is the oxidizer. Calculate the value of ΔH°_{rxn} for the reaction, given the following ΔH°_{rxn} values:

$$(1)\quad Mg(s) + \tfrac{1}{2}O_2(g) \rightarrow MgO(s) \qquad \Delta H^\circ_{rxn}(1) = -601.7\ kJ$$

$$(2)\quad Mg(s) + S(s) \rightarrow MgS(s) \qquad \Delta H^\circ_{rxn}(2) = -598.0\ kJ$$

$$(3)\quad S(s) + O_2(g) \rightarrow SO_2(g) \qquad \Delta H^\circ_{rxn}(3) = -296.8\ kJ$$

Answer: -1504.6 kJ

8-5. HEATS OF REACTIONS CAN BE CALCULATED FROM TABULATED HEATS OF FORMATION

The value of ΔH°_{rxn} for the reaction of carbon with oxygen is

$$\begin{aligned} C(s) + O_2(g) \rightarrow CO_2(g) \qquad \Delta H^\circ_{rxn} &= -393.5\ kJ \\ &= \Delta H^\circ_f[CO_2(g)] \end{aligned}$$

We refer to this value of ΔH°_{rxn} as the standard (molar) enthalpy of formation for $CO_2(g)$ and denote it by the symbol $\Delta H^\circ_f[CO_2(g)]$. The

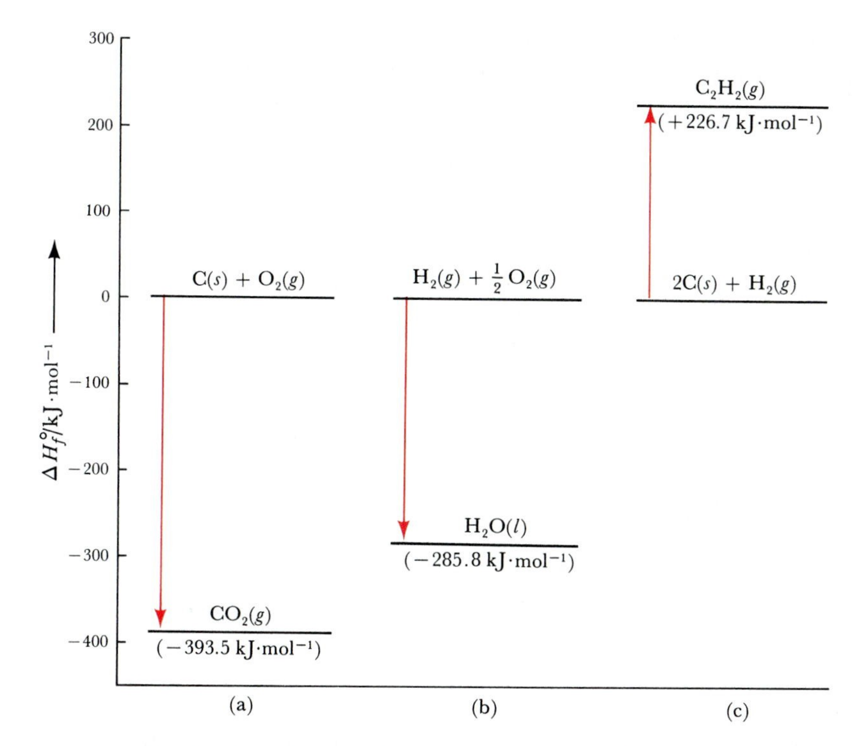

Figure 8-9 Enthalpy changes involved in the formation of $CO_2(g)$, $H_2O(l)$, and $C_2H_2(g)$ from their elements. Note that (a) $CO_2(g)$ lies 393.5 $kJ \cdot mol^{-1}$ (on the enthalpy scale) below the elements; (b) $H_2O(l)$ lies 285.8 $kJ \cdot mol^{-1}$ below the elements; and (c) $C_2H_2(g)$ lies 226.7 $kJ \cdot mol^{-1}$ above the elements.

quantity ΔH_f° also is called the **heat of formation.** The **standard molar enthalpy of formation** is defined as the enthalpy change for the equation in which one mole of a compound is formed in its one-atmosphere standard state from its constituent elements in their one-atmosphere standard states at the temperature of interest.

The subscript f on ΔH_f° stands for formation from the elements and also indicates that the value is for 1 mol. A $\Delta H_f^\circ[CO_2(g)]$ value of -393.5 kJ tells us that 1 mol of $CO_2(g)$ lies 393.5 kJ "downhill" on the enthalpy scale relative to its constituent elements (Figure 8-9a).

The standard molar enthalpy of formation of water from its elements is equal to the ΔH_{rxn}° value for the reaction in which one mole of $H_2O(l)$ is formed from its elements (Figure 8-9b):

$$H_2(g) + \tfrac{1}{2}O_2(g) \rightarrow H_2O(l)$$

$$\Delta H_{rxn}^\circ = \Delta H_f^\circ[H_2O(l)] = -285.8\ kJ \cdot mol^{-1}$$

Most compounds cannot be formed directly from their elements. For example, an attempt to make the hydrocarbon acetylene (C_2H_2) by reaction with hydrogen

$$2C(s) + H_2(g) \rightarrow C_2H_2(g)$$

yields not just C_2H_2 but a complex mixture of various hydrocarbons such as C_2H_4 and C_2H_6, among others. Nevertheless, we can determine the value of ΔH_f° for acetylene by using Hess's law, together with the available ΔH_{rxn}° data on combustion reactions. All three species in the above chemical equation burn in oxygen:

(1) $C(s) + O_2(g) \rightarrow CO_2(g)$ $\qquad \Delta H_{rxn}^\circ(1) = -393.5$ kJ

(2) $H_2(g) + \tfrac{1}{2}O_2(g) \rightarrow H_2O(l)$ $\qquad \Delta H_{rxn}^\circ(2) = -285.8$ kJ

(3) $C_2H_2(g) + \tfrac{5}{2}O_2(g) \rightarrow 2CO_2(g) + H_2O(l)$ $\qquad \Delta H_{rxn}^\circ(3) = -1299.5$ kJ

If we multiply Equation (1) by 2, reverse Equation (3), and add the results to Equation (2), then we obtain

$$(4)\quad 2C(s) + H_2(g) \rightarrow C_2H_2(g)$$

From Hess's law, we know that the value of ΔH°_{rxn} for Equation (4) is

$$\begin{aligned}\Delta H^\circ_{rxn}(4) &= 2\Delta H^\circ_{rxn}(1) + \Delta H^\circ_{rxn}(2) - \Delta H^\circ_{rxn}(3) \\ &= (2)(-393.5\text{ kJ}) + (-285.8\text{ kJ}) - (-1299.5\text{ kJ}) \\ &= +226.7\text{ kJ}\end{aligned}$$

Because Equation (4) represents the formation of one mole of $C_2H_2(g)$ from its elements, we conclude that $\Delta H^\circ_{rxn}[C_2H_2(g)] = +226.7\text{ kJ}\cdot\text{mol}^{-1}$ (Figure 8-9c). Thus, we see that it is possible to obtain values of ΔH°_f even if the compound cannot be formed directly from its elements.

EXAMPLE 8-7: Given that the heats of combustion of $C(s)$, $H_2(g)$, and $CH_4(g)$ are -393.5 kJ, -285.8 kJ, and -890.2 kJ, respectively, calculate the standard molar enthalpy of formation of methane, $CH_4(g)$.

Solution: The chemical equations for the three combustion reactions are

(1) $C(s) + O_2(g) \rightarrow CO_2(g)$ $\quad \Delta H^\circ_{rxn}(1) = -393.5$ kJ

(2) $H_2(g) + \frac{1}{2}O_2(g) \rightarrow H_2O(l)$ $\quad \Delta H^\circ_{rxn}(2) = -285.8$ kJ

(3) $CH_4(g) + 2O_2(g) \rightarrow CO_2(g) + 2H_2O(l)$ $\quad \Delta H^\circ_{rxn}(3) = -890.2$ kJ

If we reverse Equation (3), multiply Equation (2) by 2, and add the results to Equation (1), then we obtain the equation for the formation of $CH_4(g)$ from the elements

$$(4)\quad C(s) + 2H_2(g) \rightarrow CH_4(g) \qquad \Delta H^\circ_{rxn}(4) = \Delta H^\circ_f[CH_4(g)]$$

along with

$$\begin{aligned}\Delta H^\circ_{rxn}(4) &= \Delta H^\circ_{rxn}(1) + 2\Delta H^\circ_{rxn}(2) - \Delta H^\circ_{rxn}(3) \\ &= (-393.5\text{ kJ}) + (2)(-285.8\text{ kJ}) - (-890.2\text{ kJ}) \\ &= -74.9\text{ kJ}\end{aligned}$$

Because Equation (4) represents the formation of one mole of $CH_4(g)$ directly from its elements, we have $\Delta H^\circ_f[CH_4(g)] = -74.9\text{ kJ}\cdot\text{mol}^{-1}$.

PRACTICE PROBLEM 8-7: Diborane, B_2H_6, cannot be made directly from boron and hydrogen, but its standard molar enthalpy of formation can be determined from the following ΔH°_{rxn} data for combustion reactions:

(1) $4B(s) + 3O_2(g) \rightarrow 2B_2O_3(s)$ $\quad \Delta H^\circ_{rxn}(1) = -2509.1$ kJ

(2) $2H_2(g) + O_2(g) \rightarrow 2H_2O(l)$ $\quad \Delta H^\circ_{rxn}(2) = -571.6$ kJ

(3) $B_2H_6(g) + 3O_2(g) \rightarrow B_2O_3(s) + 3H_2O(l)$ $\quad \Delta H^\circ_{rxn}(3) = -2147.5$ kJ

Calculate $\Delta H^\circ_f[B_2H_6(g)]$ from these data.

Answer: $+35.6\text{ kJ}\cdot\text{mol}^{-1}$

Table 8-1 Representative elemental forms for which we take $\Delta H^\circ_f = 0$ at 25°C

Element	Formula
hydrogen	$H_2(g)$
oxygen	$O_2(g)$
nitrogen	$N_2(g)$
chlorine	$Cl_2(g)$
fluorine	$F_2(g)$
bromine	$Br_2(l)$
mercury	$Hg(l)$
sodium	$Na(s)$
magnesium	$Mg(s)$
carbon (graphite)*	$C(s)$
sulfur (rhombic)	$S(s)$
iron	$Fe(s)$

*Elemental carbon occurs both as diamond and graphite at 25°C. We take $\Delta H^\circ_f = 0$ for graphite, because it is the more stable form at 25°C and 1 atm. Energy is required to convert graphite to diamond (see Table 8-2).

As suggested by Figure 8-9, we can set up a table of ΔH_f° values for compounds by setting the ΔH_f° values for the elements equal to zero. That is, for each element in its *normal* physical state at 1 atm at the temperature of interest (usually, but by no means always, 25°C), we set ΔH_f° equal to zero, as illustrated below (see also Table 8-1):

$$\Delta H_f^\circ[\mathrm{Fe}(s)] = 0 \qquad \Delta H_f^\circ[\mathrm{O_2}(g)] = 0$$

Thus, standard molar enthalpies of formation of compounds are given relative to the elements in their normal physical states at 1 atm. Table 8-2 lists values of ΔH_f° at 25°C for a number of substances. If you look at Table 8-2, you will see that $\Delta H_f^\circ[\mathrm{C(diamond)}] = +1.897\ \mathrm{kJ{\cdot}mol^{-1}}$, $\Delta H_f^\circ[\mathrm{Br_2}(g)] = +30.91\ \mathrm{kJ{\cdot}mol^{-1}}$, and $\Delta H_f^\circ[\mathrm{I_2}(g)] = +62.4\ \mathrm{kJ{\cdot}mol^{-1}}$.

Table 8-2 Standard enthalpies of formation, ΔH_f°, for various substances at 25°C

Substance	Formula	$\Delta H_f^\circ/\mathrm{kJ{\cdot}mol^{-1}}$
acetylene	$C_2H_2(g)$	+226.7
ammonia	$NH_3(g)$	−46.19
ammonium dichromate	$(NH_4)_2Cr_2O_7(s)$	−1806
benzene	$C_6H_6(l)$	+49.03
benzoic acid	$HC_7H_5O_2(s)$	−385.1
bromine vapor	$Br_2(g)$	+30.91
butane	$C_4H_{10}(g)$	−126.1
carbon dioxide	$CO_2(g)$	−393.5
carbon monoxide	$CO(g)$	−110.5
carbon tetrachloride	$CCl_4(l)$	−135.4
	$CCl_4(g)$	−103.0
chromium (III) oxide	$Cr_2O_3(s)$	−1135
cyclohexane	$C_6H_{12}(l)$	−156.2
diamond	$C(s)$	+1.897
ethane	$C_2H_6(g)$	−84.68
ethanol (ethyl alcohol)	$C_2H_5OH(l)$	−277.7
ethene (ethylene)	$C_2H_4(g)$	+52.28
freon-12	$CF_2Cl_2(g)$	−493.3
glucose	$C_6H_{12}O_6(s)$	−1260
graphite	$C(s)$	0
hexane	$C_6H_{14}(l)$	−198.8
hydrazine	$N_2H_4(l)$	+50.6
	$N_2H_4(g)$	+95.40
hydrogen bromide	$HBr(g)$	−36.4
hydrogen chloride	$HCl(g)$	−92.31
hydrogen fluoride	$HF(g)$	−271.1
hydrogen iodide	$HI(g)$	+26.1
hydrogen peroxide	$H_2O_2(l)$	−187.8
iodine vapor	$I_2(g)$	+62.4
magnesium carbonate	$MgCO_3(s)$	−1096
magnesium oxide	$MgO(s)$	−601.7
magnesium sulfide	$MgS(s)$	−346
methane	$CH_4(g)$	−74.86
methanol (methyl alcohol)	$CH_3OH(l)$	−238.7
	$CH_3OH(g)$	−200.7
nitrogen oxide	$NO(g)$	+90.37
nitrogen dioxide	$NO_2(g)$	+33.85
dinitrogen tetroxide	$N_2O_4(g)$	+9.66
	$N_2O_4(l)$	−19.5
octane	$C_8H_{18}(l)$	−250.0
pentane	$C_5H_{12}(l)$	−146.4
propane	$C_3H_8(g)$	−103.8
sodium carbonate	$Na_2CO_3(s)$	−1131
sodium oxide	$Na_2O(s)$	−418.0
sucrose	$C_{12}H_{22}O_{11}(s)$	−2220
sulfur dioxide	$SO_2(g)$	−296.8
sulfur trioxide	$SO_3(g)$	−395.7
water	$H_2O(l)$	−285.8
	$H_2O(g)$	−241.8

The ΔH_f° values for these forms of the elements are not equal to zero because C(diamond), $Br_2(g)$, and $I_2(g)$ are not the normal physical states of these elements at 25°C and 1 atm. The normal physical states of these elements at 25°C are C(graphite), $Br_2(l)$, and $I_2(s)$.

Once we have tabulated heats of formation, we can then use Hess's law to calculate the enthalpy change of any chemical equation for which we know the ΔH_f° values of all the products and reactants. For example, let's consider one of the reactions that is used in the production of tin from its principal ore, *cassiterite*, SnO_2:

$$\text{(1)} \quad SnO_2(s) + C(s) \rightarrow Sn(s) + CO_2(g)$$

We can write this equation as the sum of the two equations

$$C(s) + O_2(g) \rightarrow CO_2(g)$$

$$SnO_2(s) \rightarrow Sn(s) + O_2(g)$$

Notice that ΔH_{rxn}° for the first equation is equal to $\Delta H_f^\circ[CO_2(g)]$. Moreover, ΔH_{rxn}° for the second equation is equal to $-\Delta H_f^\circ[SnO_2(s)]$, because it is the equation for the reverse of the formation of $SnO_2(s)$ directly from its elements. If we add these two equations along with their corresponding values of ΔH_{rxn}°, then we obtain

$$SnO_2(s) + C(s) \rightarrow Sn(s) + CO_2(g)$$
$$\Delta H_{rxn}^\circ = \Delta H_f^\circ[CO_2(g)] - \Delta H_f^\circ[SnO_2(s)]$$

Because, by our convention for ΔH_f° values of elements, we have $\Delta H_f^\circ[C(s)] = 0$ and $\Delta H_f^\circ[Sn(s)] = 0$, we can write ΔH_{rxn}° for Equation (1) as

$$\Delta H_{rxn}^\circ = \{\Delta H_f^\circ[Sn(s)] + \Delta H_f^\circ[CO_2(g)]\} - \{\Delta H_f^\circ[SnO_2(s)] + \Delta H_f^\circ[C(s)]\}$$

This equation has the form

$$\Delta H_{rxn}^\circ = \Delta H_f^\circ[\text{all products}] - \Delta H_f^\circ[\text{all reactants}] \quad (8\text{-}21)$$

The following example suggests that Equation (8-21) is general.

EXAMPLE 8-8: Express ΔH_{rxn}° for

$$\text{(1)} \quad NH_3(g) + 3Cl_2(g) \rightarrow NCl_3(l) + 3HCl(g)$$

in terms of the standard molar enthalpies of formation of the reactants and the products.

Solution: The equations for the formations of one mole of the reactant and product compounds directly from their elements are

$$\text{(2)} \quad \tfrac{1}{2}N_2(g) + \tfrac{3}{2}H_2(g) \rightarrow NH_3(g) \qquad \Delta H_{rxn}^\circ = \Delta H_f^\circ[NH_3(g)]$$

$$\text{(3)} \quad \tfrac{1}{2}N_2(g) + \tfrac{3}{2}Cl_2(g) \rightarrow NCl_3(l) \qquad \Delta H_{rxn}^\circ = \Delta H_f^\circ[NCl_3(l)]$$

$$\text{(4)} \quad \tfrac{1}{2}H_2(g) + \tfrac{1}{2}Cl_2(g) \rightarrow HCl(g) \qquad \Delta H_{rxn}^\circ = \Delta H_f^\circ[HCl(g)]$$

Equation (1) can be obtained by combining Equations (2), (3), and (4) as follows:

$$\text{Equation (1)} = \text{Equation (3)} + 3 \times \text{Equation (4)} - \text{Equation (2)}$$

so

$$\Delta H^\circ_{rxn}(1) = \Delta H^\circ_{rxn}(3) + 3\Delta H^\circ_{rxn}(4) - \Delta H^\circ_{rxn}(2)$$

or

$$\Delta H^\circ_{rxn}(1) = \Delta H^\circ_f[NCl_3(l)] + 3\Delta H^\circ_f[HCl(g)] - \Delta H^\circ_f[NH_3(g)]$$

Notice that this result is in the form of Equation (8-21) and that $\Delta H^\circ_f[Cl_2(g)]$ does not appear explicitly, because it is equal to zero by definition.

PRACTICE PROBLEM 8-8: Express ΔH°_{rxn} for the equation

$$CS_2(l) + 3O_2(g) \rightarrow CO_2(g) + 2SO_2(g)$$

in terms of the standard molar enthalpies of formation of the reactants and products.

Answer: $\Delta H^\circ_{rxn} = \Delta H^\circ_f[CO_2(g)] + 2\Delta H^\circ_f[SO_2(g)] - \Delta H^\circ_f[CS_2(l)]$

When using Equation (8-21), it is necessary to specify whether each substance is a gas, a liquid, or a solid, because the value of ΔH°_f depends on the physical state of the substance. To further illustrate the use of Equation (8-21), let's calculate the heat of reaction (i.e., the value of ΔH°_{rxn}) for the decomposition of $MgCO_3(s)$ into solid magnesium oxide and gaseous carbon dioxide:

$$MgCO_3(s) \rightarrow MgO(s) + CO_2(g)$$

According to Equation (8-21), we can determine the value of ΔH°_{rxn} from

$$\Delta H^\circ_{rxn} = \Delta H^\circ_f[\text{products}] - \Delta H^\circ_f[\text{reactants}]$$

Thus, we have

$$\Delta H^\circ_{rxn} = \Delta H^\circ_f[MgO(s)] + \Delta H^\circ_f[CO_2(g)] - \Delta H^\circ_f[MgCO_3(s)]$$

Using the ΔH°_f values given in Table 8-2, we obtain

$$\begin{aligned}\Delta H^\circ_{rxn} &= (1\ \text{mol})(-601.7\ \text{kJ}\cdot\text{mol}^{-1}) + (1\ \text{mol})(-393.5\ \text{kJ}\cdot\text{mol}^{-1}) \\ &\qquad - (1\ \text{mol})(-1096\ \text{kJ}\cdot\text{mol}^{-1}) \\ &= +101\ \text{kJ}\end{aligned}$$

The decomposition reaction of $MgCO_3(s)$ is endothermic; that is, it takes energy as heat to decompose magnesium carbonate into magnesium oxide and carbon dioxide.

EXAMPLE 8-9: Use the ΔH°_f data in Table 8-2 to calculate ΔH°_{rxn} for the combustion of liquid ethyl alcohol, $C_2H_5OH(l)$, at 25°C:

$$C_2H_5OH(l) + 3O_2(g) \rightarrow 2CO_2(g) + 3H_2O(l)$$

Solution: Referring to Tables 8-1 and 8-2, we find that $\Delta H^\circ_f[CO_2(g)] = -393.5\ \text{kJ}\cdot\text{mol}^{-1}$; $\Delta H^\circ_f[H_2O(l)] = -285.8\ \text{kJ}\cdot\text{mol}^{-1}$; $\Delta H^\circ_f[O_2(g)] = 0$; and $\Delta H^\circ_f[C_2H_5OH(l)] = -277.7\ \text{kJ}\cdot\text{mol}^{-1}$. Application of Equation (8-21) yields

$$\begin{aligned}\Delta H^\circ_{rxn} &= 2\Delta H^\circ_f[CO_2(g)] + 3\Delta H^\circ_f[H_2O(l)] - \Delta H^\circ_f[C_2H_5OH(l)] - 3\Delta H^\circ_f[O_2(g)] \\ &= (2\ \text{mol})(-393.5\ \text{kJ}\cdot\text{mol}^{-1}) + (3\ \text{mol})(-285.8\ \text{kJ}\cdot\text{mol}^{-1}) \\ &\qquad - (1\ \text{mol})(-277.7\ \text{kJ}\cdot\text{mol}^{-1}) - (3\ \text{mol})(0) \\ &= -1366.7\ \text{kJ}\end{aligned}$$

The ethyl alcohol combustion reaction is highly exothermic and is used extensively in alcohol burners of various types to keep food warm in chafing dishes.

PRACTICE PROBLEM 8-9: Use the data in Table 8-2 to calculate the heat of combustion of cyclohexane, $C_6H_{12}(l)$, and glucose, $C_6H_{12}O_6(s)$, at 25°C. (Hint: First write out and balance the combustion reactions, given that the only products in each case are $CO_2(g)$ and $H_2O(l)$ at 25°C.)

Answer: $\Delta H^\circ_{rxn} = -3919.6$ kJ for cyclohexane; $\Delta H^\circ_{rxn} = -2816$ kJ for glucose

A Sterno burner in operation. Sterno is a mixture of ethyl and methyl alcohol in a gelling agent, which makes the fuel a solidlike material.

8-6. HEAT CAPACITY MEASURES THE ABILITY OF A SUBSTANCE TO TAKE UP ENERGY AS HEAT

The ΔH°_f values given in Table 8-2 are derived from experimentally determined heats of reaction. We will see how to measure heats of reaction in Section 8-8, but first we must introduce a quantity called heat capacity.

The **heat capacity** of a sample of a substance is defined as the heat required to raise the temperature of the sample by one degree Celsius, or equivalently, by one kelvin. If the substance is heated at constant pressure, then the heat capacity is denoted by c_P, where the subscript P denotes constant pressure. We can write an equation for the definition of the heat capacity at constant pressure:

$$c_P = \frac{q_P}{\Delta T} \tag{8-22}$$

Because $q_P = \Delta H$ [Equation (8-14)], Equation (8-22) can also be written as

$$c_P = \frac{\Delta H}{\Delta T} \tag{8-23}$$

Thus,

$$q_P = \Delta H = c_P \Delta T \tag{8-24}$$

where q_P is the energy added as heat and ΔT is the increase in temperature of the substance arising from the heat input. All that is needed to determine the heat capacity of a substance is to add a known quantity of energy as heat and then measure the resulting increase in temperature. The heat capacity of a substance is always positive. The following Example illustrates the use of Equation (8-22).

EXAMPLE 8-10: When 421.2 J of heat is added to 36.0 g of liquid water, the temperature of the water increases from 10.000°C to 12.800°C. Calculate the heat capacity of the 36.0 g of $H_2O(l)$.

Solution: We start with Equation (8-22):

$$c_P = \frac{q_P}{\Delta T}$$

The value of ΔT is equal to 12.800°C − 10.000°C = 2.800°C. Because a kelvin and a Celsius degree are the same size, we have

$$c_P = \frac{421.2\ \text{J}}{2.800\ \text{K}} = 150.4\ \text{J}\cdot\text{K}^{-1}$$

for the heat capacity of 36.0 g of $H_2O(l)$.

PRACTICE PROBLEM 8-10: The heat capacity of 18.0 g of ice is 37.7 $\text{J}\cdot\text{K}^{-1}$. Calculate the final temperature of 18.0 g of ice, initially at −20°C, that results when 200 J of heat is absorbed by the ice.

Answer: −14.7°C

We denote the heat capacity per mole, or the **molar heat capacity,** of any substance by C_P. (Note that a capital C is used to denote molar heat capacity.) The heat capacity per mole of water can be computed from the c_P value of 150.4 $\text{J}\cdot\text{K}^{-1}$ for 36.0 g of water calculated in Example 8-10. Because 1.00 mol of water has a mass of 18.0 g, 36.0 g of water contains 2.00 mol of water. Thus, the value of C_P for liquid water is

$$C_P = \frac{150.4\ \text{J}\cdot\text{K}^{-1}}{2.00\ \text{mol}} = 75.2\ \text{J}\cdot\text{K}^{-1}\cdot\text{mol}^{-1}$$

The molar heat capacities for a variety of substances are given in Table 8-3.

Table 8-3 Molar heat capacities at constant pressure for various substances at 25°C

Name	Formula	$C_P/\text{J}\cdot\text{K}^{-1}\cdot\text{mol}^{-1}$	Name	Formula	$C_P/\text{J}\cdot\text{K}^{-1}\cdot\text{mol}^{-1}$
acetylene	$C_2H_2(g)$	43.9	mercury	$Hg(g)$	20.8
aluminum	$Al(s)$	24.3		$Hg(l)$	28.0
ammonia	$NH_3(g)$	35.1		$Hg(s)$	28.3
argon	$Ar(g)$	20.8	methane	$CH_4(g)$	35.7
carbon dioxide	$CO_2(g)$	37.1	neon	$Ne(g)$	20.8
carbon monoxide	$CO(g)$	29.2	nitrogen	$N_2(g)$	29.2
copper	$Cu(s)$	24.5	oxygen	$O_2(g)$	29.4
ethane	$C_2H_6(g)$	52.5	silver	$Ag(s)$	25.4
ethylene	$C_2H_4(g)$	42.0	sodium	$Na(s)$	28.2
gold	$Au(s)$	25.4		$Na(l)$	32.7
helium	$He(g)$	20.8	tin	$Sn(s)$	27.0
hydrogen	$H_2(g)$	28.8	water	$H_2O(s)$	37.7
iron	$Fe(s)$	25.1		$H_2O(l)$	75.2
lithium	$Li(s)$	24.8		$H_2O(g)$	33.6
	$Li(l)$	31.3			

8-8. A CALORIMETER IS A DEVICE USED TO MEASURE THE AMOUNT OF HEAT EVOLVED OR ABSORBED IN A REACTION

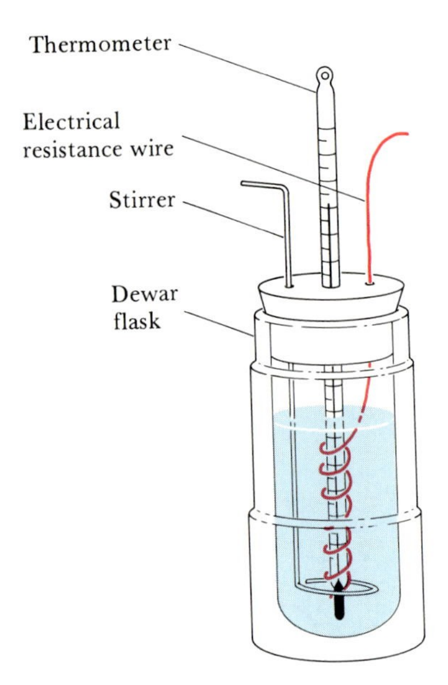

Figure 8-11 A simple calorimeter, consisting of a Dewar flask and its cover (which prevents a significant loss or gain of heat from the surroundings); a high-precision thermometer (which gives the temperature to within ±0.001 K); a simple ring-type stirrer; and an electrical resistance heater. One reactant is placed in the Dewar flask and then the other reactant, at the same temperature, is added. As the reaction mixture is stirred, the change in temperature is measured.

The value of ΔH°_{rxn} for a chemical reaction can be measured in a device called a **calorimeter.** A simple calorimeter, consisting of a **Dewar flask** ("thermos bottle") equipped with a high-precision thermometer, is shown in Figure 8-11. A calorimeter works on the principle that the total energy is always conserved.

For a chemical reaction occurring at fixed pressure, the value of ΔH_{rxn} is equal to the heat evolved or absorbed in the process [Equation (8-14)]:

$$\Delta H_{rxn} = q_P$$

Consider an exothermic reaction run in a Dewar flask. The heat evolved by the reaction cannot escape from the flask, so it is absorbed by the calorimeter contents (reaction mixture, thermometer, stirrer, and so on). The absorption of the heat evolved by the calorimeter contents leads to an increase in the temperature of the calorimeter contents. Because all the heat evolved by the reaction is absorbed by the calorimeter contents, we can write

$$\Delta H_{rxn} = -\Delta H_{calorimeter} \tag{8-28}$$

From the definition of heat capacity [Equation (8-24)], we get

$$q_P = c_{P,calorimeter}\Delta T = \Delta H_{calorimeter} \tag{8-29}$$

where ΔT is the observed temperature change. Substitution of Equation (8-29) into Equation (8-28) yields

$$\Delta H_{rxn} = -c_{P,calorimeter}\Delta T \tag{8-30}$$

Equation (8-30) tells us that if we run a chemical reaction in a calorimeter with a known heat capacity ($c_{P,calorimeter}$) and determine the temperature change, then the value of ΔH_{rxn} can be calculated. The value of $c_{P,calorimeter}$ can be determined by electrical resistance heating (a process analogous to that described in Example 8-10). The value of ΔH_{rxn} for reactions involving solutions depends on concentration, but this effect is usually not large. We will ignore the effect of concentration on ΔH_{rxn} values and assume that $\Delta H_{rxn} \simeq \Delta H^\circ_{rxn}$.

EXAMPLE 8-14: A 0.500-L sample of 0.200 M NaCl(*aq*) is added to 0.500 L of 0.200 M $AgNO_3$(*aq*) in a calorimeter with a known total heat capacity equal to 4.60×10^3 J·K^{-1}. The observed ΔT is +1.423 K. Calculate the value of ΔH°_{rxn} for the equation

$$AgNO_3(aq) + NaCl(aq) \rightarrow AgCl(s) + NaNO_3(aq)$$

Solution: The observed increase in temperature arises from the formation of the precipitate AgCl(*s*). The number of moles of AgCl(*s*) formed in the reaction is given by

$$\text{moles of AgCl}(s)\text{ formed} = (0.500\text{ L})(0.200\text{ mol·L}^{-1}) = 0.100\text{ mol}$$

The value of ΔH°_{rxn} for the formation of 0.100 mol of AgCl(*s*) is

$$\Delta H^\circ_{rxn} = -c_{P,\text{calorimeter}}\Delta T$$
$$= -(4.60 \times 10^3\ \text{J}\cdot\text{K}^{-1})(1.423\ \text{K}) = -6550\ \text{J}$$

For the formation of 1.00 mol of AgCl(*s*), the value of ΔH°_{rxn} is 10 times the value for the formation of 0.100 mol; thus,

$$\Delta H^\circ_{rxn} = \left(\frac{-6550\ \text{J}}{0.100\ \text{mol}}\right)(1.00\ \text{mol})$$
$$= -65.5 \times 10^3\ \text{J} = -65.5\ \text{kJ}$$

PRACTICE PROBLEM 8-14: A 100-mL sample of 0.100 M HCl(*aq*) solution is mixed with a 100-mL sample of 0.100 M NaOH(*aq*) solution in a calorimeter with a total heat capacity of 1710 $\text{J}\cdot\text{K}^{-1}$. If the observed temperature increase is 0.326°C, then calculate the value of ΔH°_{rxn} for the neutralization of a strong acid by a strong base.

Answer: $-55.7\ \text{kJ}\cdot\text{mol}^{-1}$

The heat of combustion of a substance can be determined in a **bomb calorimeter** like that shown in Figure 8-12. A known mass of the sub-

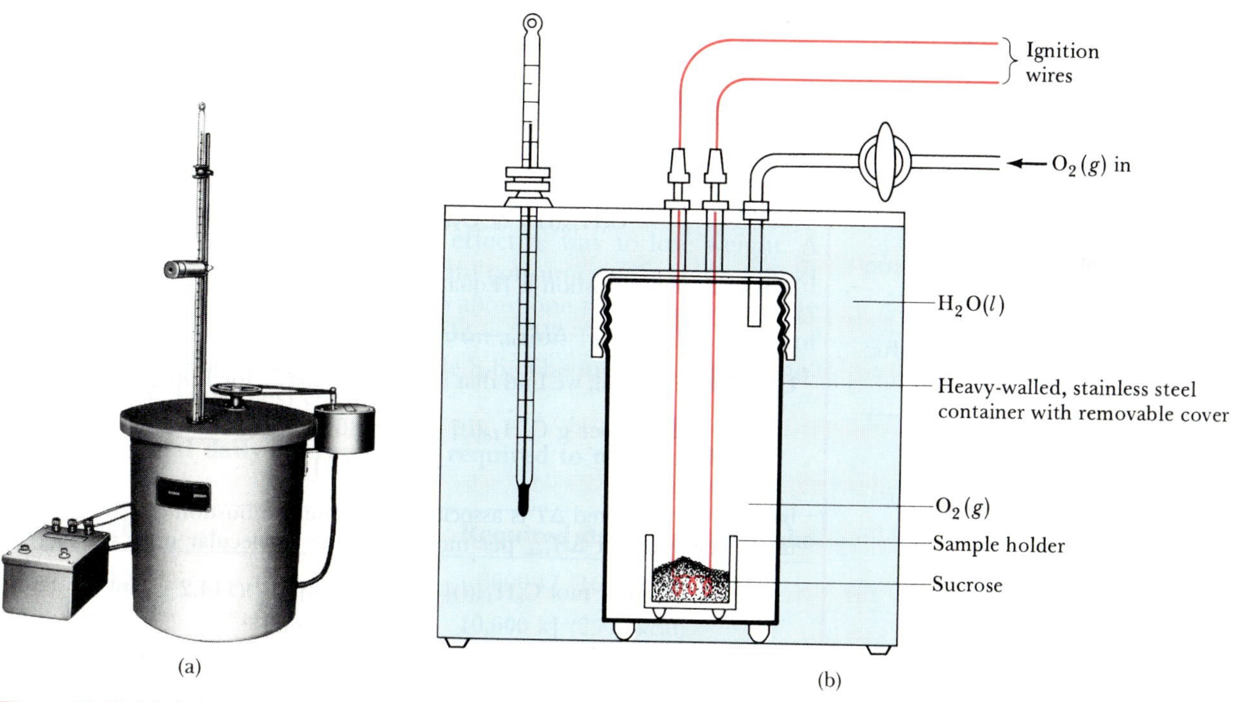

Figure 8-12 (a) A bomb calorimeter and (b) its cross section. The inner container ("bomb"), within which the combustion reaction occurs, is placed in the water filling the outer container, as shown here. The thermometer, which is equipped with a small telescopic eyepiece for more precise reading, sticks through the top of the outer container and into the water surrounding the bomb. The stirrer motor is the device in (a) that is attached to the drive belt; it powers stirrer blades immersed in the water. The small box to the left in (a) is the battery-powered ignition device that ignites the reactants.

8-11. ENERGY IS OBTAINED FROM FOSSIL FUELS BY COMBUSTION REACTIONS

Table 8-8 Heats of combustion of fuels*

Fuel	$\Delta H^\circ_{rxn}/kJ \cdot g^{-1}$
hydrogen, $H_2(g)$	−131
methane, $CH_4(g)$	−50
propane, $C_3H_8(l)$	−48
carbon, $C(s)$	−33
ethyl alcohol, $C_2H_5OH(l)$	−27
refined heating oil	−44
gasoline, kerosene, or diesel fuel	−48
coal	−28
dry seasoned wood	−25

*Enthalpy changes refer to the reactions where gaseous products, $CO_2(g)$ and $H_2O(g)$, are formed.

Fossil fuels are essentially mixtures of hydrocarbons. Thermal energy is obtained from fuels by burning them in air. An important criterion for the quality of a substance that is used as a fuel in a combustion reaction is the quantity of heat evolved per gram of fuel when the fuel is burned. The heats of combustion in kilojoules per gram for a variety of fuels are given in Table 8-8. The more energy evolved as heat per gram of fuel, the greater the quality of the fuel as an energy source. Other important criteria for fuels are cost, ease of transport, and hazards associated with use.

Natural gas is primarily methane (CH_4), with traces of other hydrocarbons, and is second only to hydrogen in the energy released per gram of fuel burned. Propane (C_3H_8), which is used as a fuel in areas not serviced by natural gas mains, is stored on-site as a liquid in metal tanks. The vapor over the liquid propane has a sufficiently high pressure to flow out of the tank to the combustion region when a valve is opened to the atmosphere. Conventional liquid fuels such as gasoline, jet fuel, diesel fuel, and heating oils are complex mixtures of hydrocarbons. For example, gasolines consist of mixtures of over 100 different hydrocarbon compounds in variable proportions. The hydrocarbons have from 4 to 14 carbon atoms per formula unit. Various blends are produced, depending on the environmental conditions of use and the quality of the gasoline. Diesel fuel and heating oils contain various hydrocarbons that have from 10 to 20 carbon atoms per formula unit.

Gasohol is a mixture of gasoline and ethyl alcohol (C_2H_5OH). The most common blend is 90 percent conventional gasoline plus 10 percent ethyl alcohol. Ethyl alcohol is produced by fermentation of sugars from plants using yeasts as the fermentation agent; for example,

$$\underset{\substack{\text{glucose} \\ \text{(a sugar)}}}{C_6H_{12}O_6(aq)} \xrightarrow{\text{yeast}} \underset{\text{ethyl alcohol}}{2C_2H_5OH(aq)} + 2CO_2(g)$$

The heat of combustion per gram for ethyl alcohol is significantly less than that for a hydrocarbon, because ethyl alcohol is already partially oxidized relative to a hydrocarbon:

$$\underset{\text{ethane}}{C_2H_6(g)} + \tfrac{1}{2}O_2(g) \rightarrow C_2H_5OH(l) \qquad \Delta H^\circ_{rxn} = -193 \text{ kJ}$$

As a consequence, an automobile running on pure ethyl alcohol fuel uses about 75 percent more fuel per mile than does an automobile running on gasoline. However, the fraction of unburned fuel is less for gasohol than for gasoline; thus, gasohol is a cleaner burning fuel.

Coal is a complex substance that contains highly variable amounts of many different elements. Sulfur is a major impurity in many coals, and the SO_2 produced when the coal is burned in a power plant must be removed before the combustion gases exit the plant. The need to remove a wide range of environmental pollutants from coal combustion gases has increased greatly the costs of building coal-fired power plants.

Coal is not as convenient or versatile an energy source as petroleum liquids because it is a solid. Coal is much more expensive to mine and to transport than oil, which is moved through pipelines, so coal is most economically used on a large scale close to where it is mined. Because petroleum liquids are in short supply relative to coal and also because petroleum liquids are much more valuable as transportation fuels, many oil-fired power plants have been converted to coal-fired plants in the last two decades.

8-12. ROCKETS ARE POWERED BY HIGHLY EXOTHERMIC REACTIONS WITH GASEOUS PRODUCTS

The heat-of-combustion data in Table 8-8 show that the most energy-rich fuel on a mass basis is hydrogen, which has an energy content per gram of well over twice that of the next-best fuel. Because of its unusually high energy content per gram, liquid hydrogen was used in the first stage of the Apollo series spaceships that traveled to the moon (the fuel must be lifted as part of the space vehicle). The main disadvantages of liquid hydrogen as a fuel are that H_2 can be maintained as a liquid only at very low temperatures (about 20 K at 1 atm) and that hydrogen readily forms explosive mixtures with air. The second and third stages of the Apollo spaceships were powered by the reaction between kerosene and **liquid oxygen (LOX),** both of which were stored on the spaceship.

The power used to propel the Apollo Lunar Lander spaceships to and from the surface of the moon was the energy released by the reaction of *N,N*-dimethylhydrazine, $H_2NN(CH_3)_2(l)$, with dinitrogen tetroxide, $N_2O_4(l)$:

$$H_2NN(CH_3)_2(l) + 2N_2O_4(l) \rightarrow 3N_2(g) + 2CO_2(g) + 4H_2O(g)$$
$$\Delta H^\circ_{rxn} = -29 \text{ kJ per gram of fuel}$$

This reaction is especially suitable for a lunar escape vehicle, because the reaction starts spontaneously on mixing. No battery or spark plugs, with associated electrical circuitry, are required.

The principal solid fuel used in the space shuttle booster rockets, Minuteman ICBMs (intercontinental ballistic missiles), Polaris and Trident submarine missiles (Figure 8-15), and the airplane-to-airplane Sidewinder missiles typically consists of a mixture of 70 percent ammonium perchlorate, 18 percent aluminum metal powder, and 12 percent binder. Ammonium perchlorate is a self-contained solid fuel—the fuel NH_4^+ and the oxidizer ClO_4^- are together in the solid. The overall equation for the NH_4ClO_4 decomposition reaction is

$$2NH_4ClO_4(s) \rightarrow N_2(g) + 2HCl(g) + 3H_2O(g) + \tfrac{5}{2}O_2(g)$$

Note that the NH_4ClO_4 is oxygen rich; that is, O_2 is a reaction product. To utilize this available oxygen and thereby to provide more rocket thrust, aluminum powder and a binder are added. The aluminum is oxidized to aluminum oxide:

$$4Al(s) + 3O_2(g) \rightarrow 2Al_2O_3(s)$$

and the binder is oxidized to CO_2 and H_2O. The aluminum powder also promotes a more rapid and even decomposition of the NH_4ClO_4.

Figure 8-15 Underwater firing of a Polaris A-3 test missile from a nuclear submarine.

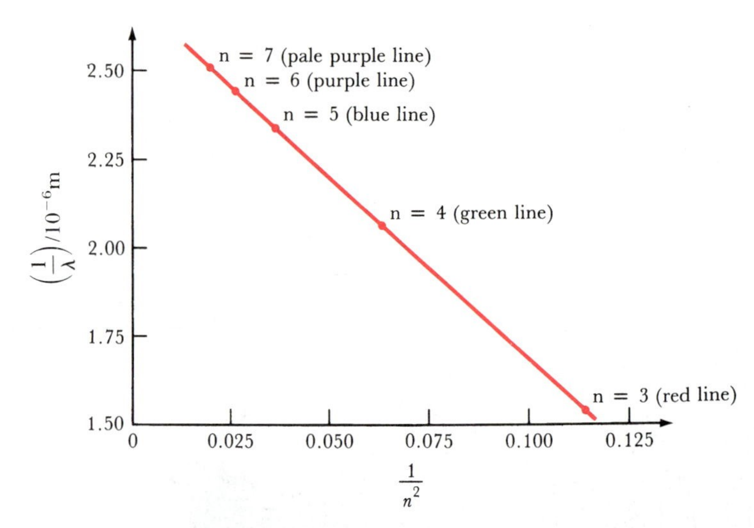

Figure 9-8 Plot of $1/\lambda$ versus $1/n^2$ for the lines in the visible spectrum of atomic hydrogen.

the ladder correspond to the discrete frequencies attainable and the ladder itself corresponds to the continuous spectrum in which any value of the frequency is allowed.

For many years scientists tried to find some pattern in the wavelengths or frequencies corresponding to the lines in the atomic spectrum of hydrogen. Finally, in 1885, Johann Balmer, an amateur Swiss scientist, showed that a plot of the reciprocal of the wavelength ($1/\lambda$) corresponding to the lines shown in Figure 9-7 versus $1/n^2$ (where n is an integer, $n = 3$ is the red line, $n = 4$ is the green line, and so on) is a straight line (Figure 9-8). Balmer's discovery was extended by Johannes Rydberg, a Swedish physicist, who gave the following empirical equation for the wavelengths of lines in the visible spectrum of hydrogen:

$$\frac{1}{\lambda} = (1.097 \times 10^7 \text{ m}^{-1})\left(\frac{1}{4} - \frac{1}{n^2}\right), \qquad n = 3,4,5,\ldots \qquad (9\text{-}2)$$

Equation (9-2) is known as the **Rydberg-Balmer equation** and the constant $1.097 \times 10^7 \text{ m}^{-1}$ is known as the **Rydberg constant.** The Rydberg-Balmer equation accurately predicts the wavelengths of the lines in the visible spectrum of hydrogen (Table 9-4) and is the equation for the line shown in Figure 9-8.

The atomic emission spectra of all elements consist of series of lines similar to the Balmer series for hydrogen (see Figure 9-9a). However,

Table 9-4 Comparison of the experimentally observed wavelengths corresponding to the lines in the visible region of the hydrogen atomic emission spectrum and those calculated using Equation (9-2)

Line color	n	Observed λ/nm	Calculated λ/nm
red	3	656.3	656.1
green	4	486.1	486.0
blue	5	434.0	433.9
indigo	6	410.1	410.1
violet	7	396.9	397.0

Si
di
sh
sil
D

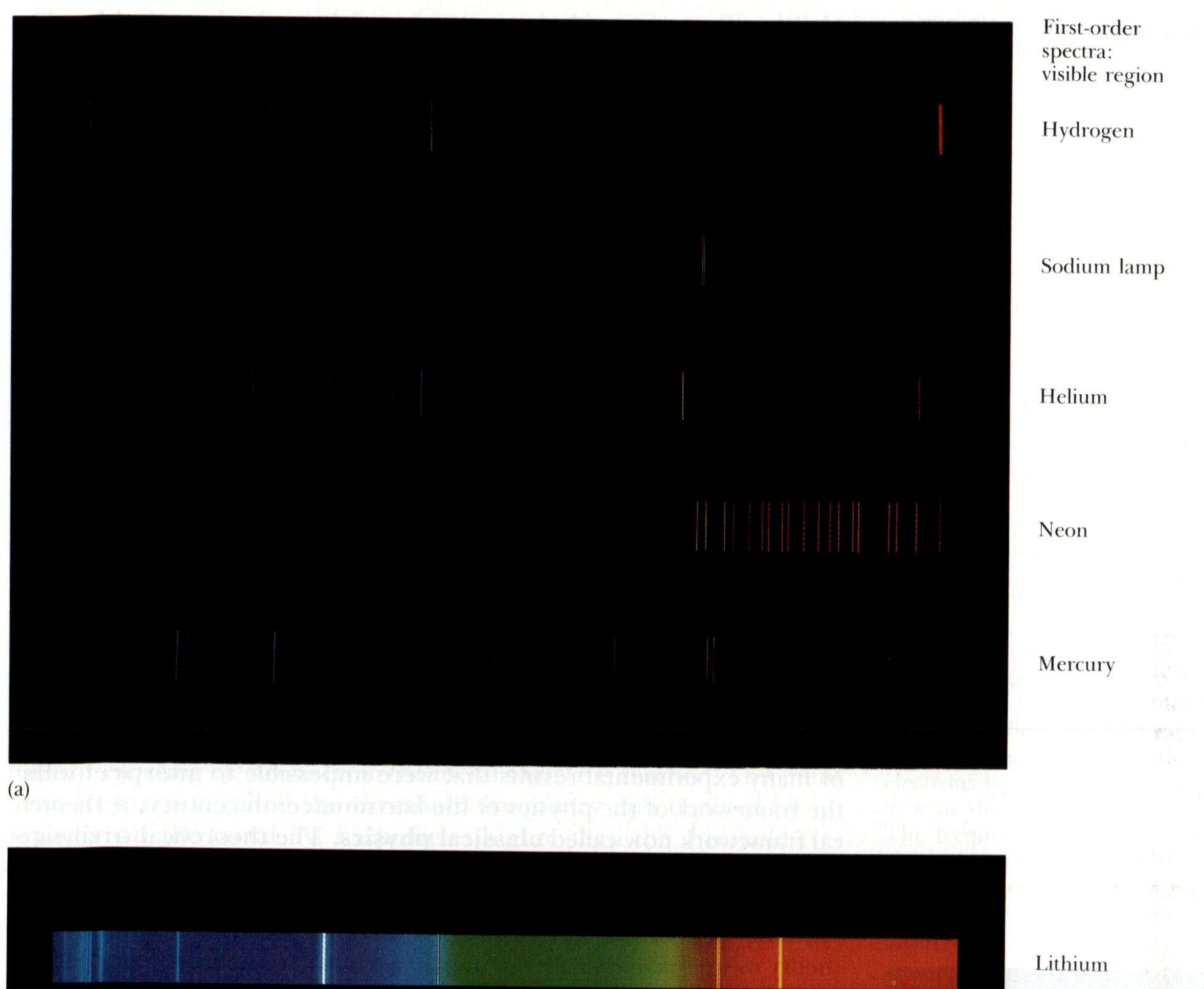

(a)

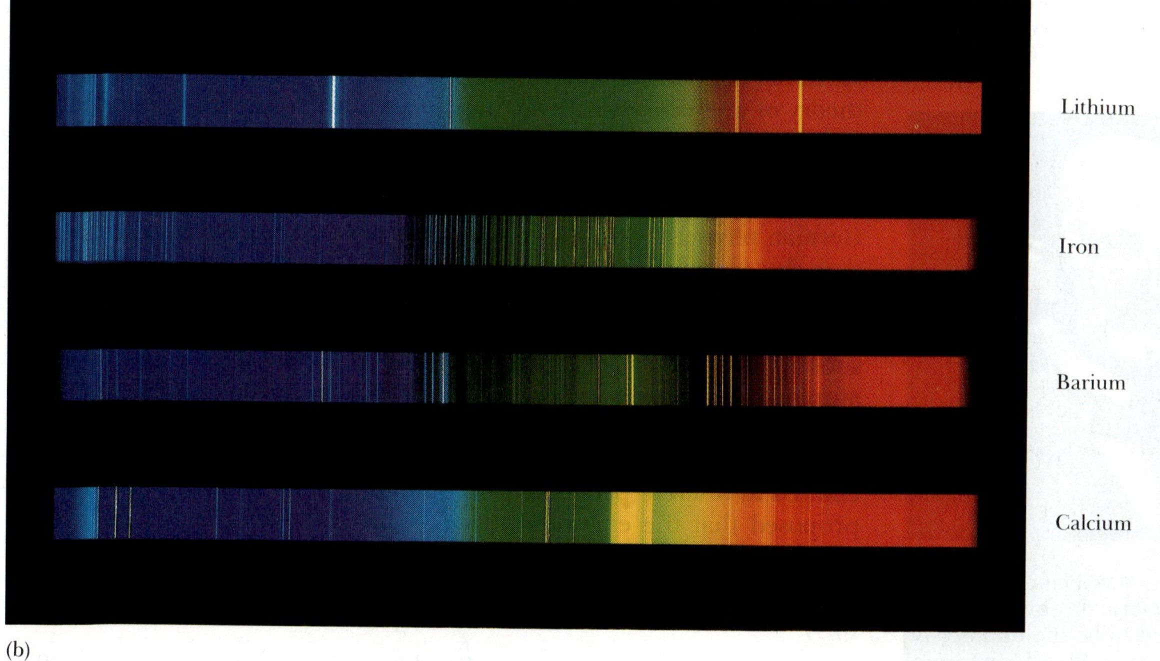

(b)

Figure 9-9 Atomic spectra. (a) The emission spectra of atomic hydrogen, sodium, helium, neon, and mercury. (b) The absorption spectra of lithium, iron, barium, and calcium. Absorption spectra of atoms also consist of series of lines that occur at the same wavelengths as those found in the emission spectrum of the element.

upward represents $m_s = +\frac{1}{2}$, and the arrow pointing downward represents $m_s = -\frac{1}{2}$. This pictorial representation is so ingrained that chemists often use the terms **spin up** and **spin down** to refer to electrons with $m_s = +\frac{1}{2}$ and $m_s = -\frac{1}{2}$, respectively. When two electrons occupy an orbital, they are said to have their electron spins **paired.** A single electron in an orbital has its electron spin **unpaired** and is said to be a **spin-unpaired electron.** According to the Pauli exclusion principle, the spin quantum numbers of the electrons in a given orbital cannot be the same; if they were, the electrons would have the same set of four quantum numbers. Thus, the representations (↑↑) and (↓↓) are not allowed; that is, they are forbidden.

The $n = 1$ level is complete with two electrons, because there are only two possible sets of four quantum numbers with $n = 1$. When $n = 2$, there are two possible values of l, namely, 0 and 1. The $l = 0$ value corresponds to a 2*s* orbital, which can hold two electrons of opposite spins. The $l = 1$ value corresponds to three 2*p* orbitals ($m_l = 1$, $m_l = 0$, $m_l = -1$), each of which can hold two electrons of opposite spins, giving a total of six electrons in the three *p* orbitals. The $n = 2$ level, then, can hold eight electrons (two in the 2*s* orbital and six in the 2*p* orbitals). Note that no more than two electrons can occupy any orbital:

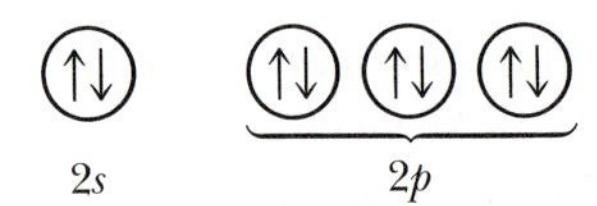

For historical reasons, the levels designated by n are called **shells.** The $n = 1$ shell is called the K shell, the $n = 2$ shell is called the L shell, the $n = 3$ shell is called the M shell, and so forth (see Table 9-9). The groups of orbitals designated by different l values within these shells are called **subshells.** For $n = 2$, there are two subshells: the *s* subshell, which can contain a maximum of two electrons, and the *p* subshell, which can contain a maximum of six electrons.

For $n = 3$, we have 3*s*, 3*p*, and 3*d* subshells. The only new feature here is the *d* subshell. Because each *d* subshell contains five *d* orbitals and each orbital can contain only two electrons with opposite spins, the *d* subshell can contain up to 10 electrons. Thus, as Table 9-9 shows, the $n = 3$ level, or M shell, can contain 18 (= 2 + 6 + 10) electrons. The only new feature for $n = 4$ is the *f* subshell. Because there are seven *f* orbitals and each one can contain only two electrons with opposite spins, the *f* subshell can contain up to 14 electrons, giving a total capacity of 32 (= 2 + 6 + 10 + 14) electrons for the $n = 4$ level.

9-18. ELECTRON CONFIGURATIONS DESIGNATE THE OCCUPANCY OF ELECTRONS IN ATOMIC ORBITALS

We are now ready to use Table 9-9 to interpret in terms of electronic structure some of the principal features of the periodic table. Consider first the helium atom. The lowest energy state of the helium atom is achieved by placing both electrons in the 1*s* orbital, because this orbital has the lowest energy. Thus, we can represent the **ground electronic**

state (the allowed electronic state of lowest energy) in helium by (↑↓) or by $1s^2$. The latter notation is standard. The $1s$ means that we are considering a $1s$ orbital, and the superscript denotes the two electrons in the $1s$ orbital. It is understood that the electrons have different spin quantum numbers, or opposite spins. If we are depicting five electrons in the $3p$ orbitals, then we write $3p^5$. The arrangement of electrons in the orbitals is called the **electron configuration** of the atom. Thus, we say that the electron configuration of the ground state of helium is $1s^2$.

Let's go on now and consider the case of lithium with its three electrons. It is not possible to place three electrons in a $1s$ orbital without violating the Pauli exclusion principle, because two of the electrons would have the same set of four quantum numbers. The $1s$ orbital is completely filled by two of the electrons, so the third electron must be assigned to the next available orbital, the $2s$ orbital. The electron in the $2s$ orbital can have $m_s = +\frac{1}{2}$ or $-\frac{1}{2}$, so we can represent the lithium atom by

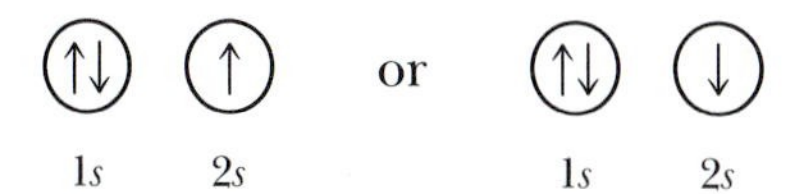

The direction of the arrow in the $2s$ orbital is not important here, but it is customary to use the spin-up picture. The standard notation is $1s^2 2s^1$.

We used the experimental values of the ionization energies for lithium given in Table 9-1 to argue that lithium can be represented as a helium core with one outer electron. In Table 9-2, we represented the lithium atom by the electron-dot formula [He]· or Li·, which shows one valence electron. Now, we see that this same conclusion follows naturally from the quantum theory.

The ground state of beryllium ($Z = 4$) is obtained by placing the fourth electron in the $2s$ orbital such that the two electrons there have opposite spins. Pictorially, we have

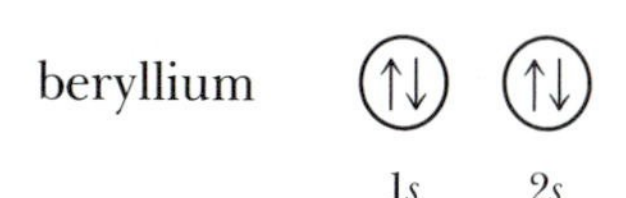

and the standard notation for this ground-state electron configuration is $1s^2 2s^2$.

In boron ($Z = 5$) both the $1s$ and $2s$ orbitals are filled, so we must use the $2p$ orbitals. Thus, we have for boron

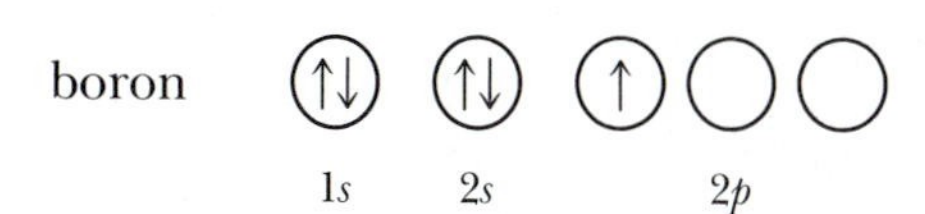

The three p orbitals have the same energy in the absence of any external electric or magnetic fields, so it does not matter into which of the three p orbitals we place the electron. The ground-state electron configuration of boron is written $1s^2 2s^2 2p^1$.

EXAMPLE 9-13: The ground-state electron configuration of ions can be described by the same notation that we have discussed for atoms. What is the ground-state electron configuration of B^+?

Solution: A boron atom has five electrons ($Z = 5$), and the B^+ ion has one fewer electron than a neutral boron atom; therefore, B^+ has $5 - 1 = 4$ electrons. The ground electronic state is obtained by placing two of these electrons in the 1*s* orbital and two in the 2*s* orbital:

B^+ (↑↓) (↑↓)

1*s* 2*s*

or $1s^2 2s^2$.

PRACTICE PROBLEM 9-13: Give the ground-state electron configuration for the F^- ion.

Answer: $1s^2 2s^2 2p^6$ [10 electrons (9 + 1) total]

9-19. HUND'S RULE IS USED TO PREDICT GROUND-STATE ELECTRON CONFIGURATIONS

For a carbon atom ($Z = 6$), we have three distinct choices for the placement of the two 2*p* electrons. The three configurations that obey the Pauli exclusion principle are

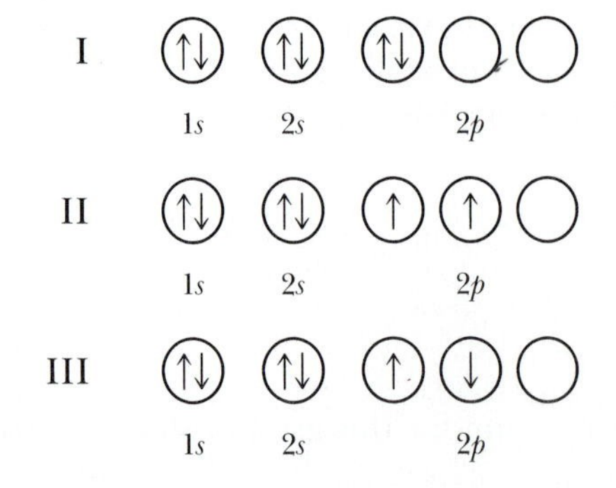

There are, however, small differences in the energies of these three configurations. In configuration I, both electrons are in the same *p* orbital and hence are restricted, on the average, to the same region in space. In the other two cases, the two electrons are in different *p* orbitals, so they are, on the average, in different regions of space. Because electrons have the same charge and so repel each other, the placement of the two electrons into different *p* orbitals and hence different regions of space minimizes the repulsion between the electrons. Thus, we conclude that configurations II and III have lower energies and are favored over configuration I. It has been determined experimentally that the configuration in which the two *p* electrons are placed in different *p* orbitals with **parallel spins** leads to the lowest energy, or ground-state, configuration. Therefore, the ground-state electron configuration of the carbon atom is

carbon (↑↓) (↑↓) (↑)(↑)( )

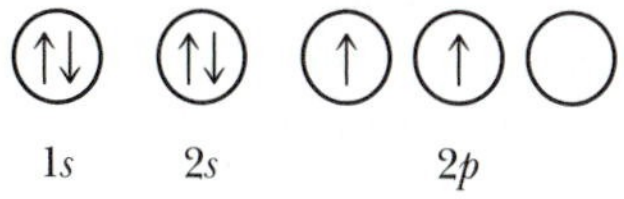

1s 2s 2p

The arguments given above can be generalized to give what is called **Hund's rule:** for any set of orbitals of the same energy, that is, for any subshell, the ground-state electron configuration is obtained by placing the electrons in different orbitals of this set with parallel spins. No orbital in the subshell contains two electrons until each orbital contains one electron. Using Hund's rule, we write for nitrogen ($Z = 7$)

nitrogen (↑↓) (↑↓) (↑)(↑)(↑)

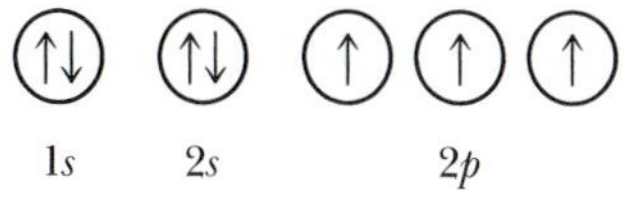

1s 2s 2p

The standard notation is $1s^2 2s^2 2p_x^1 2p_y^1 2p_z^1$. This notation often is condensed to $1s^2 2s^2 2p^3$. In both cases the reader is assumed to know that the three $2p$ electrons have parallel spins in the ground state.

For an oxygen atom ($Z = 8$), we begin to pair up the p electrons and obtain

oxygen (↑↓) (↑↓) (↑↓)(↑)(↑)

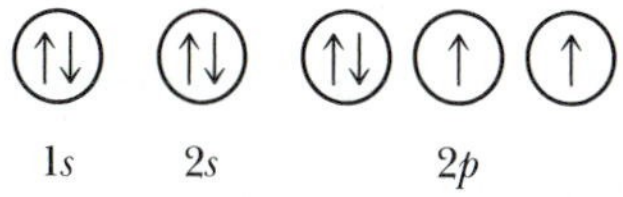

1s 2s 2p

or $1s^2 2s^2 2p_x^2 2p_y^1 2p_z^1$, or simply $1s^2 2s^2 2p^4$. It does not matter into which p orbital we place the paired electrons. The electron configurations $1s^2 2s^2 2p_x^1 2p_y^2 2p_z^1$ and $1s^2 2s^2 2p_x^1 2p_y^1 2p_z^2$ are equivalent to each other and to $1s^2 2s^2 2p_x^2 2p_y^1 2p_z^1$.

EXAMPLE 9-14: What is the ground-state electron configuration of O^+?

Solution: The O^+ ion has seven electrons (for oxygen, $Z = 8$; for O^+ we have $8 - 1 = 7$ electrons). Four of the electrons are in the $1s$ and $2s$ orbitals. The other three are in the $2p$ orbitals. According to Hund's rule, the three $2p$ electrons are in different $2p$ orbitals and all have the same spin. The ground-state electron configuration is $1s^2 2s^2 2p_x^1 2p_y^1 2p_z^1$, or simply $1s^2 2s^2 2p^3$.

PRACTICE PROBLEM 9-14: What is the ground-state electron configuration of O^{2-}?

Answer: $1s^2 2s^2 2p^6$ [10 electrons (8 + 2) total]

The ground-state electron configurations of the first 10 elements are shown in Table 9-10. Note that helium has a filled $n = 1$ shell and that neon has a filled $n = 2$ shell. The ground-state electron configuration is obtained by filling up the atomic orbitals of lowest energy in accord with the Pauli exclusion principle and Hund's rule.

Table 9-10 Ground-state electron configurations of the first ten elements

Element	Ground-state electron configuration
hydrogen	$1s^1$
helium	$1s^2$
lithium	$1s^2 2s^1$
beryllium	$1s^2 2s^2$
boron	$1s^2 2s^2 2p^1$
carbon	$1s^2 2s^2 2p^2$
nitrogen	$1s^2 2s^2 2p^3$
oxygen	$1s^2 2s^2 2p^4$
fluorine	$1s^2 2s^2 2p^5$
neon	$1s^2 2s^2 2p^6$

We saw in Section 9-10 that an atom can absorb electromagnetic radiation. In this process an electron is promoted to an orbital of higher energy, and the atom is said to be in an excited state. For example, a lithium atom absorbs electromagnetic radiation of wavelength 671 nm and undergoes the electronic transition

$$\text{Li}(1s^2 2s^1) + h\nu \rightarrow \text{Li}(1s^2 2p^1)$$

We see that the electron in the 2*s* orbital is promoted to a 2*p* orbital in the process. The resulting lithium atom is in an excited state, and its excited-state electron configuration is $1s^2 2p^1$. The first excited state is obtained by promoting the electron of highest energy in the ground state to the next available orbital. We are interested primarily in ground electronic states, but we should realize that the ground state is just the lowest of a set of allowed atomic energy states.

EXAMPLE 9-15: What is the electron configuration of the first excited state of neon?

Solution: The ground-state electron configuration of neon is $1s^2 2s^2 2p^6$. The electron of highest energy is any one of the 2*p* electrons. The next available orbital is the 3*s* orbital, so

$$\text{Ne* (first excited state): } 1s^2 2s^2 2p^5 3s^1$$

The asterisk indicates an excited state.

PRACTICE PROBLEM 9-15: What is the electronic configuration of the first excited state of the O^{2-} ion?

Answer: First excited state is $1s^2 2s^2 2p^5 3s^1$.

9-20. ELEMENTS IN THE SAME COLUMN OF THE PERIODIC TABLE HAVE SIMILAR VALENCE-ELECTRON CONFIGURATIONS

According to either Figure 9-32b or 9-33, after neon we use the 3*s* and 3*p* orbitals and obtain the next row of the periodic table. In this series of elements we are filling up the 3*s* and 3*p* orbitals outside a neon inner-shell structure. It is therefore common practice to use the abbreviated form of the electron configurations shown in the right-hand column of Table 9-11.

Note how the ground-state electron configurations of the elements correlate with the Lewis electron-dot formulas (Tables 9-2 and 9-12.) In each case, the number of dots displayed in the Lewis electron-dot formula is the same as the total number of electrons in the outer (valence) shell, as indicated in the electron configuration.

Recall that electrons in the outermost occupied shell (highest *n*-value shell with an electron in it) of a neutral atom or a monatomic ion of a main-group element are called **valence electrons.** A main-group atomic *cation* for which the outermost occupied shell is completely filled (e.g., Na^+ with a neon configuration) has no valence electrons. A main-group atomic *anion* for which the outermost *ns* and *np* subshells are

Table 9-11 Ground-state electron configurations of third-row elements

Element	Ground-state electron configuration	Abbreviated form of ground-state electron configuration
sodium	$1s^22s^22p^63s^1$	$[Ne]3s^1$
magnesium	$1s^22s^22p^63s^2$	$[Ne]3s^2$
aluminum	$1s^22s^22p^63s^23p^1$	$[Ne]3s^23p^1$
silicon	$1s^22s^22p^63s^23p^2$	$[Ne]3s^23p^2$
phosphorus	$1s^22s^22p^63s^23p^3$	$[Ne]3s^23p^3$
sulfur	$1s^22s^22p^63s^23p^4$	$[Ne]3s^23p^4$
chlorine	$1s^22s^22p^63s^23p^5$	$[Ne]3s^23p^5$
argon	$1s^22s^22p^63s^23p^6$	$[Ne]3s^23p^6$

filled completely have eight (ns^2np^6 and $2 + 6 = 8$) valence electrons. Also recall that, by definition, **isoelectronic** ions have the same *total* (as opposed to valence) number of electrons (e.g., K^+ and Cl^- or Mg^{2+} and F^-).

EXAMPLE 9-16: How many valence electrons are there in (a) the ion O^{2-} and (b) the ion Ne^+?

Solution: (a) The ground-state electron configuration of O^{2-} is $1s^22s^22p^6$, so there are eight valence electrons in O^{2-}. The Lewis electron-dot formula of O^{2-} is $:\ddot{\underset{..}{O}}:^{2-}$. (b) The ground-state electron configuration of Ne^+ is $1s^22s^22p^5$, so there are seven valence electrons in Ne^+. The Lewis electron-dot formula of Ne^+ is $:\ddot{\underset{..}{Ne}}^+$.

PRACTICE PROBLEM 9-16: How many valence electrons are there in (a) Al^{3+}, (b) Mg^{2+}, and (c) Cl^-?

Answer: (a) 0; (b) 0; (c) 8

If we compare the electron configurations of sodium through argon (Table 9-11) with those of lithium through neon (Table 9-10), then we see why these two series of elements have a periodic correlation in

Table 9-12 Comparison of Lewis electron-dot formula and ground-state electron configuration

Element	Lewis electron-dot formula	Ground-state electron configuration
carbon	$\cdot\dot{C}:$	$1s^22s^22p^2$
fluorine	$:\ddot{\underset{\cdot}{F}}:$	$1s^22s^22p^5$
neon	$:\ddot{\underset{..}{Ne}}:$	$1s^22s^22p^6$
sodium	$Na\cdot$	$[Ne]3s^1$
chlorine	$:\ddot{\underset{\cdot}{Cl}}:$	$[Ne]3s^23p^5$

chemical properties. Their valence electron configurations range from ns^1 to ns^2np^6 ($n = 2$ and $n = 3$, respectively) in the same manner.

Figure 9-32b shows that after argon, the next available orbital is the 4*s* orbital. Thus, the electron configurations of the next two elements after argon are

potassium $[Ar]4s^1$ calcium $[Ar]4s^2$

where [Ar] denotes the ground-state electron configuration of an argon atom. If we consider the ground-state electron configurations of lithium, sodium, and potassium, then we see why they fall naturally into the same column of the periodic table. Each has an ns^1 configuration outside a noble-gas configuration, that is,

lithium $[He]2s^1$ sodium $[Ne]3s^1$ potassium $[Ar]4s^1$

Also note that the principal quantum number of the outer *s* orbital coincides with the number of the row of the periodic table (Figure 9-34). Each row starts off with an alkali metal, whose electron configu-

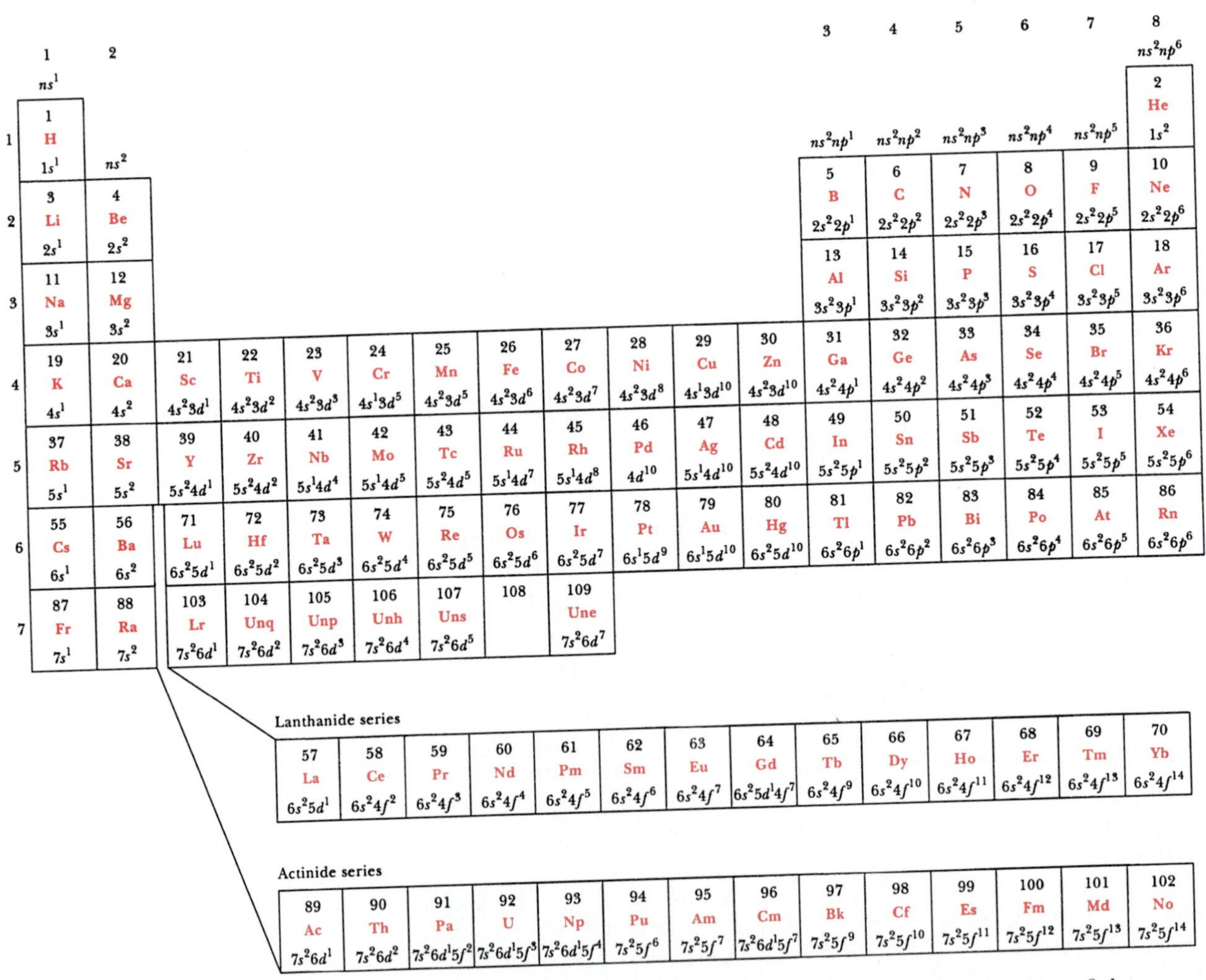

Figure 9-34 A periodic table showing the ground-state electron configurations of the outer electrons of the elements. The general valence-electron configurations of the main-group elements are given above each group. Thus the alkali metals have the valence-electron configuration ns^1, the alkaline earths ns^2, and so on.

ration is [noble gas]ns^1. For example, cesium, which follows xenon and begins the sixth row of the table, has the electron configuration

cesium [Xe]$6s^1$

The same type of observation can be used to explain why the alkaline earths all occur in the second column in the periodic table. The electron configuration of an alkaline earth metal is [noble gas]ns^2 (Table 9-13).

Table 9-13 Ground-state electron configurations of the alkaline earth metals

Element	Ground-state electron configuration
beryllium	[He]$2s^2$
magnesium	[Ne]$3s^2$
calcium	[Ar]$4s^2$
strontium	[Kr]$5s^2$
barium	[Xe]$6s^2$
radium	[Rn]$7s^2$

9-21. THE OCCUPIED ORBITALS OF HIGHEST ENERGY ARE *d* ORBITALS FOR NEUTRAL TRANSITION-METAL ATOMS AND *f* ORBITALS FOR LANTHANIDES AND ACTINIDES

Once we reach calcium ($Z = 20$), the $4s$ orbital is completely filled. Figure 9-32b shows that the next available orbitals are the five $3d$ orbitals. Each of these orbitals can be occupied by two electrons of opposite spins, giving a maximum of 10 electrons in all. Note that this number corresponds perfectly with the 10 transition metals that occur between calcium and gallium in the periodic table. Thus, in the first set of transition metals, we see the sequential filling of the five $3d$ orbitals. Because of this, the first set of transition metals is called the **3*d* transition-metal series.** You may think that the ground-state electron configurations of these 10 elements go smoothly from [Ar]$3d^1 4s^2$ to [Ar]$3d^{10}4s^2$, but this is not so. The actual ground-state electron configurations of the $3d$ transition metals are as shown in Table 9-14. We see that chromium and copper have only one $4s$ electron. Note that in each case an electron has been taken from the $4s$ orbital in order to either half-fill or completely fill all of the $3d$ orbitals. This filling pattern results because an extra stability is realized by the electron configurations

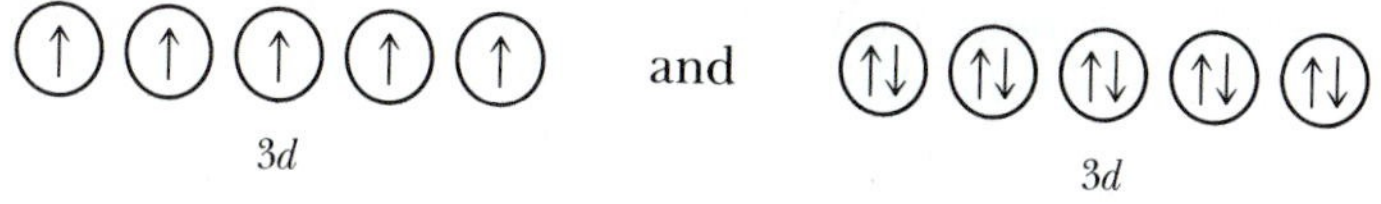

relative to the *incorrect* $4s^2 3d^4$ and $4s^2 3d^9$ ground-state configurations for the neutral gaseous atoms of these elements. It so happens that the energies of the electrons in the $4s$ and $3d$ orbitals are very similar (Figure 9-32b), and deviations from the regular filling order of these two orbitals—as well as for the $5s$ and $4d$ and $6s$ and $5d$ orbitals—are found, especially when such deviations result in a half or completely filled nd subshell (see Figure 9-34). Note that there are even more irregularities in the filling order of the $4d$ series [at niobium ($5s^1 4d^4$), molybdenum ($5s^1 4d^5$), ruthenium ($5s^1 4d^7$), rhodium ($5s^1 4d^8$), palladium ($4d^{10}$), and silver ($5s^1 4d^{10}$)] than for the $3d$ series. In the $5d$ series there are only two irregularities in the filling order (can you identify the two elements involved?).

After the $3d$ orbitals are filled, the next available orbitals are the $4p$ orbitals, which fill up as shown in Table 9-15. For these six elements, the $4p$ orbitals are sequentially filled, and these elements fall naturally into the fourth row of the periodic table under the sequence of elements boron through neon and aluminum through argon, which fill the $2p$ and $3p$ orbitals, respectively (Figure 9-34).

Table 9-14 Ground-state electron configurations of the 3*d* transition metals

Element	Ground-state electron configuration
scandium	[Ar]$4s^2 3d^1$
titanium	[Ar]$4s^2 3d^2$
vanadium	[Ar]$4s^2 3d^3$
chromium	[Ar]$4s^1 3d^5$
manganese	[Ar]$4s^2 3d^5$
iron	[Ar]$4s^2 3d^6$
cobalt	[Ar]$4s^2 3d^7$
nickel	[Ar]$4s^2 3d^8$
copper	[Ar]$4s^1 3d^{10}$
zinc	[Ar]$4s^2 3d^{10}$

Table 9-15 Ground-state electron configurations of the *p*-block fourth-row elements

Element	Ground-state electron configuration
gallium	$[Ar]4s^23d^{10}4p^1$
germanium	$[Ar]4s^23d^{10}4p^2$
arsenic	$[Ar]4s^23d^{10}4p^3$
selenium	$[Ar]4s^23d^{10}4p^4$
bromine	$[Ar]4s^23d^{10}4p^5$
krypton	$[Ar]4s^23d^{10}4p^6$

Krypton, like all the noble gases, has a completely filled set of *p* orbitals whose principal quantum number corresponds to the row in which it is located in the periodic table. Figure 9-32b shows that the 5*s* orbital follows the 4*p* orbital, so we are back to the left-hand column of the periodic table with the alkali metal rubidium followed by the alkaline earth metal strontium. These two metals have the ground-state electron configurations $[Kr]5s^1$ and $[Kr]5s^2$, respectively. The next available orbitals are the 4*d* orbitals, which lead to the **4*d* transition-metal series,** yttrium through cadmium. The ground-state electron configurations of the outer electrons of these 10 metals (Figure 9-34) show irregularities like those found in the 3*d* transition-metal series. After cadmium, $[Kr]5s^24d^{10}$, the 5*p* orbitals are filled to give the six elements indium through the noble gas xenon, which has the ground-state electron configuration $[Kr]5s^24d^{10}5p^6$. As before, the completion of a set of *p* orbitals leads to a noble gas located in the far right-hand column of the periodic table. The two reactive metals cesium and barium follow xenon by filling the 6*s* orbital to give the ground-state electron configurations $[Xe]6s^1$ and $[Xe]6s^2$, respectively.

After filling the 6*s* orbital, we begin to fill the seven 4*f* orbitals. Because each of these seven orbitals can hold two electrons of opposite spin, we expect that the next fourteen elements should involve the filling of the 4*f* orbitals. The elements lanthanum ($Z = 57$) through ytterbium ($Z = 70$) constitute what is called the **lanthanide series,** because the series begins with the element lanthanum in the periodic table. Figure 9-34 shows that except for a few irregularities like those found for the *d* transition-metal series, the lanthanides involve a sequential filling of the seven 4*f* orbitals. The chemistry of these elements is so similar that for many years it proved very difficult to separate them from the naturally occurring mixtures. However, separations are now achieved using ion-exchange and chromatographic methods.

If we consider that the lanthanides differ only in the number of electrons in the 4*f* subshells, with the 6*s* and 5*p* subshells already filled, the reason for their chemical similarity becomes clear. According to the quantum theory, the average distance of an electron from a nucleus depends on both the principal quantum number n and the azimuthal quantum number l. Although the average distance of an electron from the nucleus increases with n, it increases less as l increases than as n increases. For this reason, the average distance of 4*f* ($n = 4, l = 3$) electrons from the nucleus is less than that of 6*s* ($n = 6, l = 0$) or 5*p* ($n = 5, l = 1$) electrons. For the 4*f* electrons, not only is n smaller but l is larger than for 6*s* or 5*p* electrons. The 4*f* electron density, then, is concentrated in the interior of the atom, so 4*f* electrons have little effect on the chemical activity of the atom, which is dominated by the outer electrons. For this reason, the lanthanides are also called **inner transition metals.** The outer electron configuration, which plays a principal role in determining chemical activity, is the same for all the lanthanides ($5p^66s^2$) and accounts for their similar chemical properties.

Following the lanthanides is a third transition-metal series (the 5*d* transition-metal series) consisting of the elements lutetium ($Z = 71$) through mercury ($Z = 80$). This series, in which the 5*d* orbitals are filled, is followed by the six elements thallium ($Z = 81$) through radon ($Z = 86$). Radon, a radioactive noble gas with the ground-state electron configuration $[Xe]6s^24f^{14}5d^{10}6p^6$, finishes the sixth row of the table.

The next two elements, the radioactive metals francium, [Rn]$7s^1$, and radium, [Rn]$7s^2$, are followed by another inner transition-metal series in which the $5f$ orbitals are filled. This series begins with actinium (Z = 89) and ends with nobelium (Z = 102); it is called the **actinide series.** All the elements in this series are radioactive. In fact, with the exception of trace quantities of plutonium, the elements beyond uranium (Z = 92), called the **transuranium elements,** have not been found in nature. They are synthesized in nuclear reactors (Chapter 24).

EXAMPLE 9-17: Predict the ground-state electron configuration of a neodymium atom (Z = 60).

Solution: Neodymium occurs in the sixth row of the periodic table. The noble gas preceding this row is xenon. The ground-state electron configuration of barium, the element that precedes the lanthanides, is [Xe]$6s^2$. Neodymium is the fourth member of the lanthanides, so we predict that it has four $4f$ electrons. The predicted ground-state electron configuration is

$$\text{Nd} \qquad [\text{Xe}]6s^2 4f^4$$

Figure 9-34 shows that this result is correct. Notice, however, that there are several irregularities in the electron configurations of the lanthanides.

PRACTICE PROBLEM 9-17: Write out the complete ground-state electron configuration of plutonium.

Answer: $1s^2 2s^2 2p^6 3s^2 3p^6 4s^2 3d^{10} 4p^6 5s^2 4d^{10} 5p^6 6s^2 4f^{14} 5d^{10} 6p^6 7s^2 5f^6$

Figure 9-35 shows a periodic table that indicates which orbitals are being filled by electrons as we move through the various regions of the table. These regions are referred to as the ***s*-block elements** (Groups 1 and 2), the ***p*-block elements** (Groups 3 through 8), the ***d*-block elements** (the transition metals), and the ***f*-block elements** (the inner transition metals).

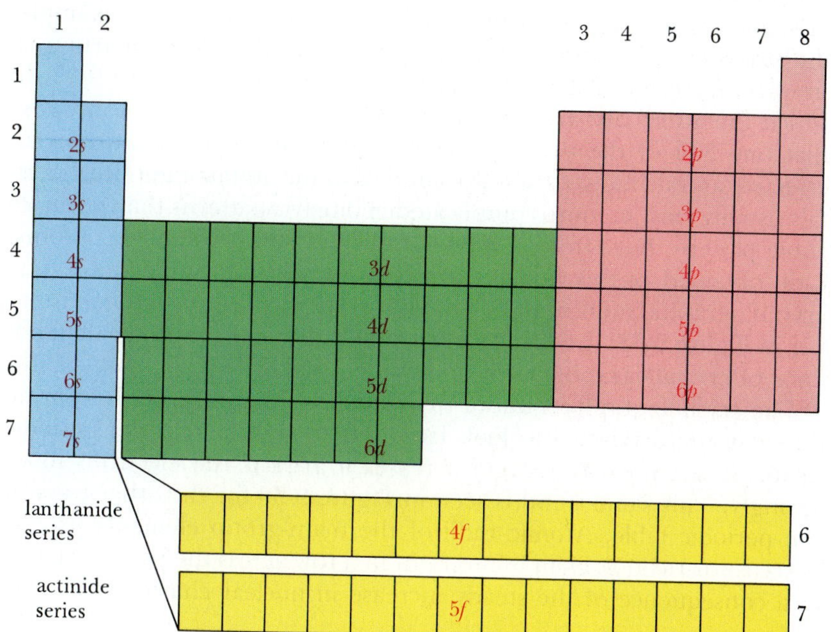

Figure 9-35 A periodic table indicating which orbitals are filling up with electrons as we move through various regions of the table.

As we will see in the next four chapters, the concept of valence electrons is especially useful in understanding the formation of chemical bonds. For main-group elements, the number of valence electrons in the neutral atom is simply equal to the group number. For the transition metals, the situation is not so simple, because of the inverted order of filling of the two outermost occupied orbitals (recall, for example, that 4*s* fills before 3*d*) and the closeness in energy of these orbitals. For transition-series metals we have, in effect, more than one possibility for the number of valence electrons. Our definition of the number of valence electrons in a transition-metal *ion* is the number equal to the oxidation state of the transition metal in the compound or species. Thus, the number of valence electrons for zinc in $ZnCl_2$ is 2 and for titanium in $TiCl_4$ it is 4.

9-22. ATOMIC RADIUS IS A PERIODIC PROPERTY

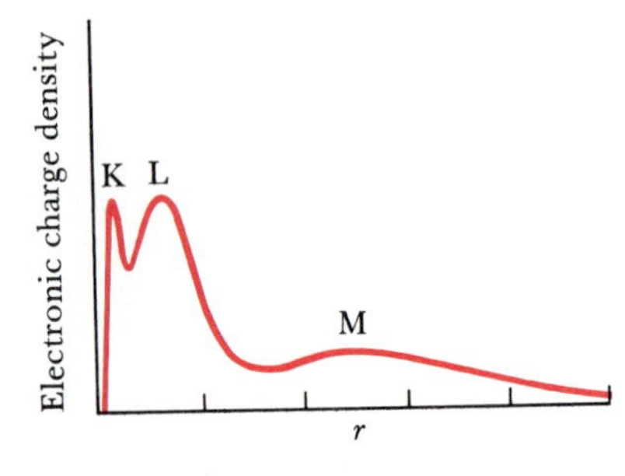

Figure 9-36 The distribution of electronic charge density versus distance from the nucleus for an argon atom can be obtained by solving the Schrödinger equation with a computer. Note that there appear to be three shells. Two of these are well defined and close to the nucleus (the K and L shells). The third, outermost shell (the M shell) is more diffuse.

As can be seen from Figures 9-26 and 9-27, the probability of finding an electron at some distance, r, from the nucleus decreases with increasing r. Even though the decrease occurs fairly rapidly with increasing r, it never actually becomes zero, even at very large distances from the nucleus. As a consequence, it is not possible to define unambiguously an outer "edge" of an atom, that is, a distance beyond which there is zero probability of finding an electron associated with the nucleus. In other words, an atom has no sharp boundary. Although the Schrödinger equation is complicated for multielectron atoms, it can be solved with a computer. The results of such a calculation for argon are sketched in Figure 9-36. We can discern clearly three shells: the inner two shells, the K shell and the L shell, are relatively well defined. The third, outermost shell, the M shell, is much more diffuse.

Even though atoms do not have well-defined edges, we can propose practical definitions for **atomic radii** based on models. For example, the atoms in a crystal of an element are arranged in ordered arrays. A simple version of such an ordered array is shown in Figure 9-37, in which the atoms are arranged in a simple cubic array. If we propose that one half of the distance between adjacent nuclei constitutes an effective atomic radius, then we can determine atomic radii. Real crystals usually exist in more complicated geometric patterns than a simple cubic pattern, but effective atomic radii can still be deduced. Atomic radii obtained in this manner are called **crystallographic radii.** The crystallographic radii of the elements are plotted against atomic number in Figure 9-38; the resulting patterns indicate the periodic dependence of crystallographic radii on atomic number.

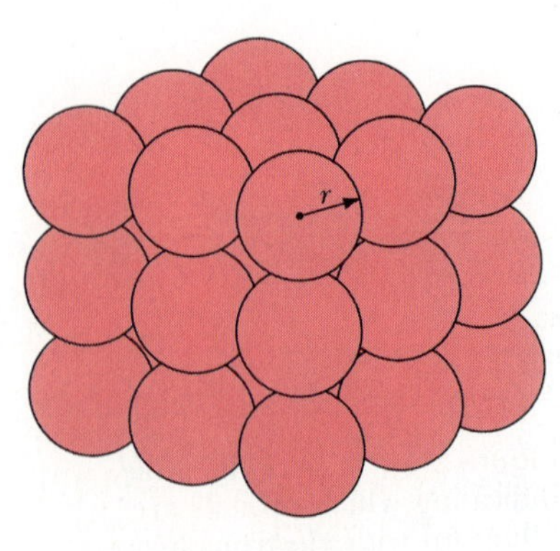

Figure 9-37 A simple cubic arrangement of atoms in a crystal.

The crystallographic radii of the elements lithium through fluorine decrease uniformly as we look from left to right across the periodic table. As the nuclear charge increases it attracts the electrons more strongly. This same trend is seen in Figure 9-38 for the other rows of the periodic table. Atomic radii of the main-group elements usually decrease as we look from left to right in a row across the periodic table, as a consequence of the steady increase in nuclear charge within the row.

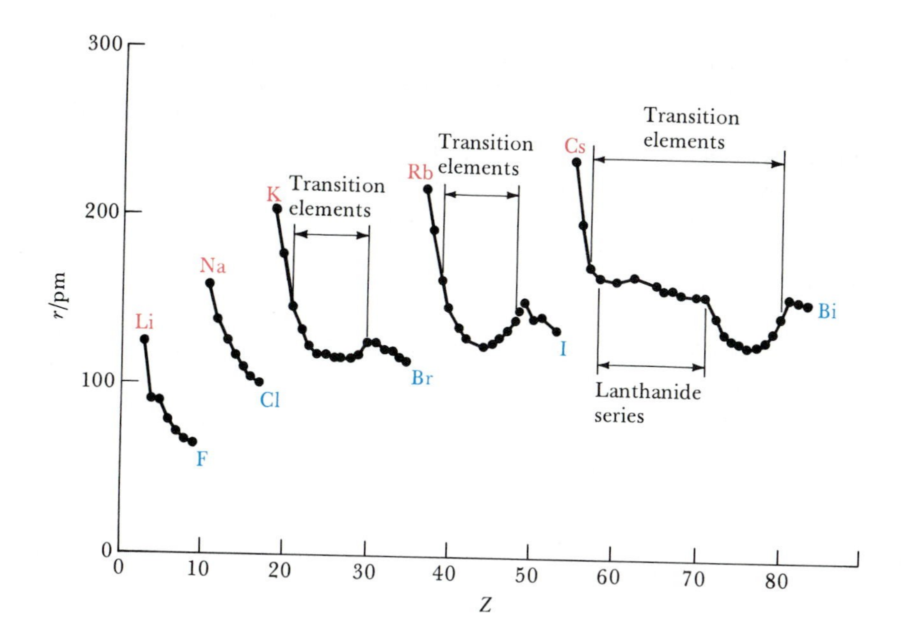

Figure 9-38 Crystallographic radii of the elements versus atomic number. Note that atomic radius is a periodic property.

The crystallographic radii of the alkali metal group also increase as we look down the periodic table within the group. Although the nuclear charge increases, the outermost electrons begin new shells, and, as shown in Figure 9-39, this effect outweighs the increased nuclear attraction. Similar behavior is found for other groups in the periodic table.

The reasoning we have just used to explain the variation of atomic radii in the periodic table also can be used to explain variations in first ionization energies (Figure 9-1). Atomic radius increases as we go down a group in the periodic table. The farther the electron is from the nucleus, the less the nuclear attraction, so the more easily the electron is removed. Therefore, first ionization energies are seen to decrease as we look down the periodic table within a group. Similarly, the decrease in atomic size as we move from left to right across a row of the periodic table is reflected in the corresponding increase in first ionization energies. Trends in atomic radii and ionization energies are thus seen to follow directly from quantum theory.

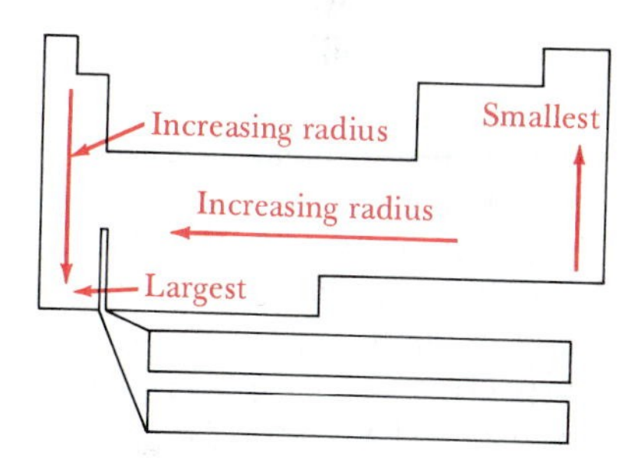

Figure 9-39 The trend of atomic radii in the periodic table.

SUMMARY

First ionization energy is a periodic property (Figure 9-1). The values of successive ionization energies suggest that electrons in atoms are arranged in shell structures (Figure 9-3), as depicted in Lewis electron-dot formulas given in Table 9-2.

In 1900 Planck initiated quantum theory by postulating that electromagnetic radiation is emitted from heated bodies only in quanta, or little packets, with energies given by $E = h\nu$. Five years later Einstein used the same ideas to describe the photoelectric effect. Photons, electrons, and other small particles exhibit the phenomenon of wave-particle duality, and a particle of mass m has a de Broglie wavelength $\lambda = h/mv$.

In 1911 Bohr developed a model of the hydrogen atom that was able to account for the atomic spectrum of hydrogen. It was later shown, however, that the Bohr theory could not be used to explain the atomic spectra of multielectron atoms and also was inconsistent with the Heisenberg uncertainty principle. In 1925 Schrödinger proposed the central equation of the quantum theory that describes the motion of electrons, atoms, and molecules. One consequence of the Schrödinger equation is that the electrons in atoms and molecules can have only certain discrete, or quantized, energies. In addition, Schrödinger showed that an electron in an atom or molecule must be described by a wave function, or orbital, which is obtained by solving the Schrödinger equation. The square of a wave function gives the probability density associated with finding the electron in some region of space. The hydrogen atom wave functions serve as the prototype wave functions for all other atoms. The wave functions are specified by three quantum numbers: n, the principal quantum number; l, the azimuthal quantum number; and m_l, the magnetic quantum number. Orbitals with $l = 0$ are called s orbitals; orbitals with $l = 1$ are called p orbitals; orbitals with $l = 2$ are called d orbitals; and orbitals with $l = 3$ are called f orbitals. For a given value of n, there is one s orbital, three p orbitals (for $n \geq 2$), five d orbitals (for $n \geq 3$), and seven f orbitals (for $n \geq 4$) (see Table 9-9).

To explain certain fine details in atomic spectra, Uhlenbeck, Goudsmit, and Pauli introduced a fourth quantum number, the spin quantum number, m_s, which specifies the intrinsic spin of an electron. The spin quantum number can have the value of $+\frac{1}{2}$ or $-\frac{1}{2}$.

The energy states of the hydrogen atom depend only on the principal quantum number n; for all other atoms, the energy states depend on both n and the azimuthal quantum number l. According to the Pauli exclusion principle, no two electrons in an atom can have the same set of four quantum numbers (n, l, m_l, m_s). Using this principle and the order of the energy states given in Figures 9-32b and 9-33, together with Hund's rule, we are able to write ground-state electron configurations and correlate these with the periodic table. For main-group elements, the number of valence electrons is equal to the group number. Electron configurations enable us to understand the trends of atomic radii and ionization energies within the periodic table.

TERMS YOU SHOULD KNOW

electronic structure 291
electron arrangement 291
ionization energy, I 291
first ionization energy, I_1 291
second ionization energy, I_2 291
shell 294
outer electron 295
valence electron 295
Lewis electron-dot formula 295
electromagnetic theory of radiation 296
wavelength, λ 297
frequency, ν 297
speed of light, c 297
hertz (Hz) 298
electromagnetic spectrum 298
visible region 298
continuous spectrum 299
line spectrum 299
atomic emission spectrum 299
Rydberg-Balmer equation 300
Rydberg constant ($1.097 \times 10^7\ m^{-1}$) 300
atomic absorption spectrum 302
atomic spectroscopy 302
classical physics 302
blackbody radiation 302
quantum 302
Planck's constant, h (6.626×10^{-34} J·s) 302
photoelectric effect 303
threshold frequency, ν_0 303
photon 303
Einstein 304
work function, Φ 304
quantization of energy 306
quantum theory 306
wave-particle duality 306
de Broglie wavelength, λ 306
momentum, p 306
X-ray diffraction 307
electron diffraction 308
electron microscope 308
quantum condition 310
quantized 310
energy state 310
stationary state 311
ground state 311
excited state 311
first excited state 311
second excited state 311
Lyman series 312
Balmer series 313
emission spectrum 314
flame test 314
absorption spectrum 314
Heisenberg uncertainty principle 316
Schrödinger equation 318
wave function, ψ 318
orbital 318
probability density 318
quantum number 318
principal quantum number, n 319
spherically symmetric 319
azimuthal quantum number, l 320
angular momentum quantum number, l 320
s orbital 320
p orbital 320
d orbital 320
f orbital 320
nodal surface 322
cylindrically symmetric 323
planar node 323
spherical node 323
magnetic quantum number, m_l 323
intrinsic electron spin 325
spin quantum number, m_s 325
Pauli exclusion principle 329
spin up 330
spin down 330
spin-paired electron 330
spin-unpaired electron 330
shell 330
subshell 330
ground electronic state 330
electron configuration 331
parallel spins 332
Hund's rule 333
valence electron 334
isoelectronic 335
$3d$ transition-metal series 337
$4d$ transition-metal series 338
lanthanide series 338
inner transition metal 338
actinide series 339
transuranium elements 339
s-block elements 339
p-block elements 339
d-block elements 339
f-block elements 339
atomic radius 340
crystallographic radius 340

$\lambda\nu = c$ (9-1) (relation between wavelength and frequency)

$\frac{1}{\lambda} = (1.097 \times 10^7\ \mathrm{m}^{-1})\left(\frac{1}{4} - \frac{1}{n^2}\right),\quad n = 3,4,5, \ldots$ (9-2) (wavelengths of lines in atomic hydrogen spectrum in the visible region)

$E = h\nu$ (9-3) (energy of a photon in terms of frequency)

$E = \frac{hc}{\lambda}$ (9-4) (energy of a photon in terms of wavelength)

$\Phi = h\nu_0$ (9-5) (work function in terms of threshold frequency)

$\text{K.E.} = h\nu - \Phi$ (9-6) (photoelectric effect)

$\lambda = \frac{h}{mv}$ (9-7) (de Broglie wavelength)

$E_n = \frac{-2.18 \times 10^{-18}\ \mathrm{J}}{n^2},\quad n = 1,2,3, \ldots$ (9-9) (energies of the electron in a hydrogen atom)

$\nu_{n_1} = (3.29 \times 10^{15}\ \mathrm{s}^{-1})\left(1 - \frac{1}{n^2}\right),\quad n = 2,3,4, \ldots$ (9-13) (Lyman series frequencies)

$(\Delta p)(\Delta x) \simeq h$ (9-17) (Heisenberg uncertainty principle)

$$\left.\begin{aligned} n &= 1,2, \ldots \\ l &= 0,1,2, \ldots, n-1 \\ m_l &= l, l-1, l-2, \ldots, -1, -2, \ldots, -l \\ m_s &= +\tfrac{1}{2} \text{ or } -\tfrac{1}{2} \end{aligned}\right\} \text{quantum numbers}$$

PROBLEMS

IONIZATION ENERGIES

9-1. Arrange the following species in order of increasing first ionization energy:

He Be Kr Ne

9-2. Arrange the following species in order of increasing first ionization energy:

Ca Mg Ba Sr

9-3. Use the data in Table 9-1 to plot the logarithms of the ionization energies of the boron atom versus the number of electrons removed. What does the plot suggest about the electronic structure of boron?

9-4. Use the data in Table 9-1 to plot the logarithms of the ionization energies of beryllium versus the number of electrons removed. Compare your plot to Figure 9-3.

LEWIS ELECTRON-DOT FORMULAS

9-5. Write Lewis electron-dot formulas for all the alkali metal atoms and for all the halogen atoms. What is the similarity in all the alkali metal atom formulas and in all the halogen formulas?

9-6. Write the Lewis electron-dot formulas for the Group 6 elements. Comment on the similarities in the valence-electron configurations.

9-7. Write the Lewis electron-dot formula for

Ar S S^{2-} Al^{3+} Cl^-

9-8. Write the Lewis electron-dot formula for

B^+ N^{3-} F^- O^{2-} Na^+

ELECTROMAGNETIC RADIATION

9-9. A helium-neon laser produces light of wavelength 633 nm. What is the frequency of this light?

9-10. The radiation given off by a sodium lamp, which is used in streetlights, has a wavelength of 589.2 nm. What is the frequency of this radiation?

9-11. Assume the first ionization energy of potassium is 419 kJ·mol^{-1}. What is the wavelength of light that is just sufficient to ionize one potassium atom?

9-12. Assume the first ionization energy of argon is 1.52 MJ·mol^{-1}. Do X-rays with a wavelength of 80 nm have sufficient energy to ionize argon?

9-13. The human eye is able to detect as little as 2.35×10^{-18} J of green light of wavelength 510 nm. Calculate the minimum number of photons that can be detected by the human eye.

9-14. Calculate the energy of 1.00 mol of X-ray photons of wavelength 210 pm.

PHOTOELECTRIC EFFECT

9-15. The work function of gold metal is 7.7×10^{-19} J. Will ultraviolet radiation of wavelength 200 nm eject electrons from the surface of metallic gold?

9-16. Photocells that are used in "electric eye" door openers are applications of the photoelectric effect. A beam of light strikes a metal surface, from which electrons are emitted, producing an electric current. When the beam of light is blocked by a person walking through the beam, the electric circuit is broken, thereby opening the door. If the source of light is a sodium vapor lamp that emits light at a wavelength of 589 nm, would copper be a satisfactory metal to use in the photocell? The work function of copper is 6.69×10^{-19} J.

9-17. Given that the work function of cesium metal is 2.90×10^{-19} J, calculate the kinetic energy of an electron ejected from the surface of cesium metal when it is irradiated with light of wavelength 400 nm.

9-18. The work function of a metal can be determined from measurements of the speed of the ejected electrons. Electrons were ejected from a metal with a speed of 5.00×10^5 m·s^{-1} when irradiated by light having a wavelength of 390 nm. Find the work function of this metal and the threshold frequency.

DE BROGLIE WAVELENGTH

9-19. Calculate the de Broglie wavelength of a proton traveling at a speed of 1.00×10^5 m·s^{-1}. The mass of a proton is 1.67×10^{-27} kg.

9-20. A certain rifle bullet has a mass of 5.00 g. Calculate the de Broglie wavelength of the bullet traveling at 1200 mph.

9-21. Calculate the de Broglie wavelength of a hydrogen molecule traveling with a speed of 2000 m·s^{-1}.

9-22. The de Broglie wavelength of electrons used in an experiment utilizing an electron microscope is 96.0 pm. What is the speed of one of these electrons?

HYDROGEN ATOMIC SPECTRUM

9-23. How much energy is required for an electron in a hydrogen atom to make a transition from the $n = 2$ state to the $n = 3$ state? What is the wavelength of a photon having this energy?

9-24. A line in the Lyman series of hydrogen has a wavelength of 1.03×10^{-7} m. Find the original energy level of the electron.

9-25. A ground-state hydrogen atom absorbs a photon of light having a wavelength of 97.2 nm. It then gives off a photon having a wavelength of 486 nm. What is the final state of the hydrogen atom?

9-26. Use Equation (9-9) to compute the ionization energy of a hydrogen atom in its first excited state.

9-27. The energy levels of one-electron ions, such as He^+ and Li^{2+}, are given by the equation

$$E_n = -\frac{(2.18 \times 10^{-18}\ \text{J})Z^2}{n^2}$$

where Z is the atomic number. Compare the measured ionization energies (Table 9-1) for He^+, Li^{2+}, and Be^{3+} ions with the values calculated from this equation.

9-28. A helium ion is called hydrogenlike because it consists of one electron and one nucleus. The Schrödinger equation can be applied to He^+, and the result that corresponds to Equation (9-9) is

$$E_n = -\frac{8.72 \times 10^{-18}\ \text{J}}{n^2}$$

Show that the spectrum of He^+ consists of a number of separate series, just as the spectrum of atomic hydrogen does. Consider the $n > 1$ to $n = 1$ series of transitions.

QUANTUM NUMBERS AND ORBITALS

9-29. Indicate which of the following atomic orbital designations are impossible:

(a) 7*s* (b) 1*p* (c) 5*d*
(d) 2*d* (e) 4*f*

9-30. Give all the possible sets of four quantum numbers for an electron in a 5*d* orbital.

9-31. Give the corresponding atomic orbital designations (that is, 1*s*, 3*p*, and so on) for electrons with the following sets of quantum numbers:

	n	l	m_l	m_s
(a)	4	1	0	$-\frac{1}{2}$
(b)	3	2	0	$+\frac{1}{2}$
(c)	4	2	-1	$-\frac{1}{2}$
(d)	2	0	0	$-\frac{1}{2}$

9-32. Give the corresponding atomic orbital designations for electrons with the following sets of quantum numbers:

	n	l	m_l	m_s
(a)	3	1	-1	$+\frac{1}{2}$
(b)	5	0	0	$+\frac{1}{2}$
(c)	2	1	0	$+\frac{1}{2}$
(d)	4	3	-2	$+\frac{1}{2}$

9-33. If $l = 2$, what can you deduce about n? If $m_l = 3$, what can you say about l?

9-34. Indicate which of the following sets of quantum numbers are allowed (that is, possible) for an electron in an atom:

	n	l	m_l	m_s
(a)	2	1	0	$+\frac{1}{2}$
(b)	3	0	$+1$	$-\frac{1}{2}$
(c)	3	2	-2	$-\frac{1}{2}$
(d)	1	1	0	$+\frac{1}{2}$
(e)	2	1	0	0

ORBITALS AND ELECTRONS

9-35. Give all the possible sets of four quantum numbers for an electron in a 3*d* orbital.

9-36. Give all the possible sets of four quantum numbers for an electron in a 4*f* orbital.

9-37. Without referring to the text, deduce the maximum number of electrons that can occupy an *s* orbital, a subshell of *p* orbitals, a subshell of *d* orbitals, and a subshell of *f* orbitals.

9-38. Without referring to the text, deduce the maximum number of electrons that can occupy a K shell, an L shell, an M shell, and an N shell.

9-39. Explain why there are 10 members of each *d* transition series.

9-40. Explain why there are 14 members of each *f* transition series.

ELECTRON CONFIGURATIONS OF ATOMS

9-41. Indicate which of the following electron configurations are ruled out by the Pauli exclusion principle:

(a) $1s^22s^22p^7$ (b) $1s^22s^22p^63s^3$
(c) $1s^22s^22p^63s^23p^64s^23d^{12}$ (d) $1s^22s^22p^63s^23p^6$

9-42. Explain why the following ground-state electron configurations are not possible:

(a) $1s^22s^32p^3$ (b) $1s^22s^22p^33s^6$
(c) $1s^22s^22p^73s^23p^8$ (d) $1s^22s^22p^63s^23p^14s^23d^{14}$

9-43. Write the corresponding electron configuration for each of the following pictorial representations. Name the element, assuming that the configuration describes a neutral atom:

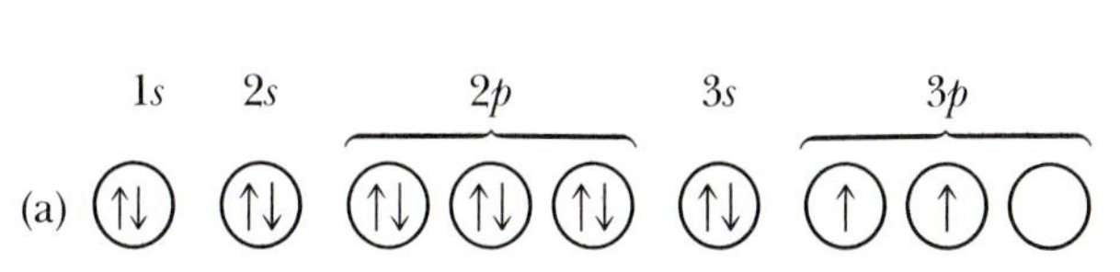

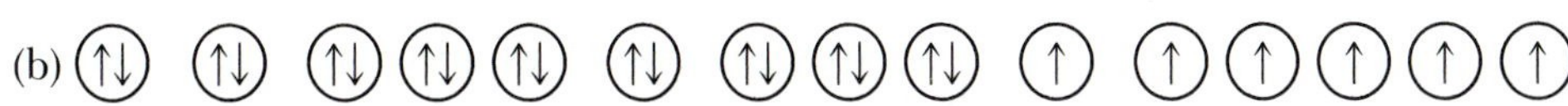

(c)

(d)

(e)

9-43

9-72. For a particle moving in a circular orbit, the quantity mvr (mass × velocity × radius of the orbit) is a fundamental quantity called the angular momentum of the particle. Show that Equation (9-8) is equivalent to the condition that the angular momentum of the electron in a hydrogen atom must be an integral multiple of $h/2\pi$.

9-73. Make a graph of frequency versus $1/n^2$ for the lines in the Lyman series of atomic hydrogen.

9-74. Compute the energy necessary to completely remove an electron from the $n = 2$ level of an He^+ ion (see Problem 9-27).

9-75. Estimate the value of ΔH_{rxn} for the following reactions using the data given in Table 9-1.

(a) $Li(g) + Na^+(g) \rightarrow Li^+(g) + Na(g)$

(b) $Mg^{2+}(g) + Mg(g) \rightarrow 2Mg^+(g)$

(c) $Al^{3+}(g) + 3e^- \rightarrow Al(g)$

9-76. Without counting the total number of electrons, determine the neutral atom whose ground-state electron configuration is

(a) $1s^22s^22p^63s^23p^64s^23d^8$

(b) $1s^22s^22p^63s^23p^64s^23d^{10}4p^65s^14d^{10}$

(c) $1s^22s^22p^63s^23p^4$

(d) $1s^22s^22p^63s^23p^64s^23d^{10}4p^65s^24d^{10}5p^66s^24f^{14}5d^{10}6p^2$

9-77. Without looking at a periodic table, deduce the atomic numbers of the other elements that are in the same family as the element with atomic number (a) 16 and (b) 11.

9-78. Name each of the atoms with the following ground-state electron configuration for its valence electrons:

(a) $3s^23p^1$ (b) $2s^22p^4$
(c) $4s^23d^{10}$ (d) $4s^24p^6$

9-79. For elements of atomic number (a) 15, (b) 26, and (c) 32 in their ground states, answer the following questions without reference to the text or to a periodic table:

How many d electrons?
How many electrons having quantum number $l = 1$?
How many unpaired electrons?

9-80. Show what the periodic table would look like if the order of the energies of atomic orbitals were regular; that is, if the order were $1s < 2s < 2p < 3s < 3p < 3d < 4s$, and so on.

9-81. How would the ground-state electron configurations of the elements in the second row of the periodic table differ if the $2s$ and $2p$ orbitals had the same energy, as they do for a hydrogen atom?

9-82. The order of the orbitals given in Figure 9-32b can be deduced by the following argument. The energy of an orbital increases with the sum $n + l$. For orbitals with the same value of $n + l$, those with the smaller value of n have lower energies. This observation is also known as Hund's rule (there are several Hund's rules). Show that this rule is consistent with the order given in Figure 9-32b.

9-83. Given that we know the position of a hydrogen atom to within 1.00 pm, compute the uncertainty in the velocity of the atom.

9-84. The atomic spectrum of hydrogen consists of several series of lines, one of which is the Balmer series. Recall that the Balmer series corresponds to electronic transitions from levels with $n > 2$ to $n = 2$. Other series of lines in the hydrogen spectrum are

Series	Name
$n > 1$ to $n = 1$	Lyman
$n > 2$ to $n = 2$	Balmer
$n > 3$ to $n = 3$	Paschen
$n > 4$ to $n = 4$	Brackett
$n > 5$ to $n = 5$	Pfund

Write the equations, analogous to Equation (9-2), that predict the wavelengths of the lines in each of these series. Use these equations to calculate the shortest wavelength line in each series and use your results to determine the region of the electromagnetic spectrum (for example, ultraviolet or infrared) in which each series occurs.

9-85. Determine the number of valence electrons of the transition metal in each of the following species:

(a) $HgCl_2(s)$ (b) $KMnO_4(s)$ (c) $K_2Cr_2O_7(s)$
(d) $AuCl_2^-(aq)$ (e) $NiBr_4^{2-}(aq)$

9-86. Determine the number of valence electrons of the transition metal in each of the following species:

(a) $HgI_4^{2-}(aq)$ (b) $PtBr_4(s)$ (c) $CuI_2^-(aq)$
(d) $V_2O_5(s)$ (e) $MnO_2(s)$

9-87. Ionization energies are sometimes tabulated in electron volts (eV). (An electron volt is the energy of an electron subject to an electric potential of one volt.) Show that $1\ eV = 96.9\ kJ\cdot mol^{-1}$ given that $1\ J = 1\ C\cdot V$, where C stands for coulomb, the SI unit of electric charge and V stands for volt, the SI unit of electric potential. The charge on an electron is 1.61×10^{-19} C.

9-88. Compare the work function of a sodium metal (Example 9-4) with the first ionization energy of $Na(g)$ (Table 9-1). Note that $\Phi < I$, because of interactions between Na atoms in the metal.

9-89. Instead of wavelength or frequency, spectroscopists often refer to the *wave number*, $\bar{\nu}$, of a particular photon. This quantity is defined by

$$\bar{\nu} = \frac{1}{\lambda} \qquad \text{(reciprocal wavelength)}$$

and is often tabulated in cm^{-1} units. Rewrite Equation (9-2) using wave numbers and deduce the Rydberg constant in wave numbers.

9-90. Use the data in Table 9-1 to calculate the energy required to remove two electrons from a magnesium atom.

9-91. The heat capacity of water is $4.18\ J{\cdot}K^{-1}{\cdot}g^{-1}$. Calculate the number of infrared photons with a wavelength of 900 nm that is required to raise the temperature of 1.00 L of water by 1.00°C.

9-92. A watt is a unit of energy per unit time, and one watt (W) is equal to one joule per second ($J{\cdot}s^{-1}$). A 100-W light bulb produces about 4 percent of its energy as visible light. Assuming that the light has an average wavelength of 510 nm, calculate how many such photons are emitted per second by a 100-W light bulb.

9-93. Data for the photoelectric effect of silver are given in the following table:

Frequency of incident radiation/$10^{15}\ s^{-1}$	Kinetic energy of ejected electrons/10^{-19} J
2.00	5.90
2.50	9.21
3.00	12.52
3.50	15.84
4.00	19.15

Plot the kinetic energy of the ejected electrons versus the frequency of the incident radiation and determine the value of Planck's constant and the threshold frequency.

9-94. Sunlight reaches the earth's surface at Madison, Wisconsin, with an average power of approximately $1.0\ kJ{\cdot}s^{-1}{\cdot}m^{-2}$. If the sunlight consists of photons with an average wavelength of 510 nm, how many photons strike a $1.0\text{-}cm^2$ area per second?

INTERCHAPTER E

Lasers

Warren S. Warren, **Princeton University**

To most people the word laser conjures up visions of star wars and death rays. Some lasers do emit pulses or beams of light that can bore holes in steel or burn bricks (Figure E-1), and if these laser beams are invisible they seem particularly dramatic. But many less powerful types of lasers play more benign roles in our everyday lives. Lasers are used to scan the bar codes on goods that we buy in the supermarket, to read compact disks in compact disk players, to perform surgery on the retina of the eye and other forms of microsurgery, to align tunnels and other construction projects, to measure long distances very accurately, and to carry thousands of telephone signals through thin glass fibers. In addition, lasers play a central role in modern chemical research.

Although virtually everyone has heard of lasers, neither the device nor the very word itself existed before 1960. We can get an idea of how lasers work from what we have just learned in Chapter 9, because most laser action is based on the transitions that take place between atomic (or molecular) energy levels when atoms absorb or emit electromagnetic radiation (photons). Just as a hydrogen atom has infinitely many energy levels [all integral values of n in Equation (9-9) are possible], so do multielectron atoms and molecules. Nonetheless, we can understand the fundamental processes involved in lasers by considering only two energy levels, as shown in Figure E-2.

Figure E-1 An ultraviolet laser producing 100 W of power in a small (1-cm^2) spot can burn a brick. Although the laser light is visible in the photograph (because the film used is sensitive to ultraviolet light), it cannot be seen by the naked eye.

Ordinarily (that is, at equilibrium), most of the molecules in a sample will be in the lower level, as indicated by the relative number of dots in Figure E-2a. In that case, if the sample is irradiated with photons that have exactly the same energy as the difference between the two levels, energy is absorbed. (The light waves should be pictured as traveling from left to right. Note that the height of the light waves decreases as photons are used to excite the molecules to the higher energy.) Every time one photon is absorbed, one molecule moves to the excited state.

We can represent the absorption of a photon by the equation

$$Nh\nu + A \rightarrow (N - 1)h\nu + A^* \qquad \text{(E-1)}$$

This equation says that N photons, each with energy $h\nu$, interact with one ground-state molecule A by exciting the molecule and losing one photon. If we turn off the light and wait long enough, eventually A^* returns to the ground state A. Most often, the molecule emits a photon to return to the lower state in Figure E-2b:

$$A^* \rightarrow h\nu + A \qquad \text{(E-2)}$$

This process is called *spontaneous emission*; if the emitted light is strong, it is also often called *fluorescence.* For example, if a hydrogen atom is excited from a 1s to a 2p orbital, it will fluoresce back to the 1s orbital in about 10^{-9} s, giving back the photon it absorbed. The direction of this emission is random, as shown in Figure E-2b.

Excited molecules often spontaneously emit photons with lower energy (longer wavelength) than the ones they absorbed. The molecule has absorbed more energy than it has lost, so by conservation of energy, some energy must stay in the molecule, often as heat. New T-shirts look very

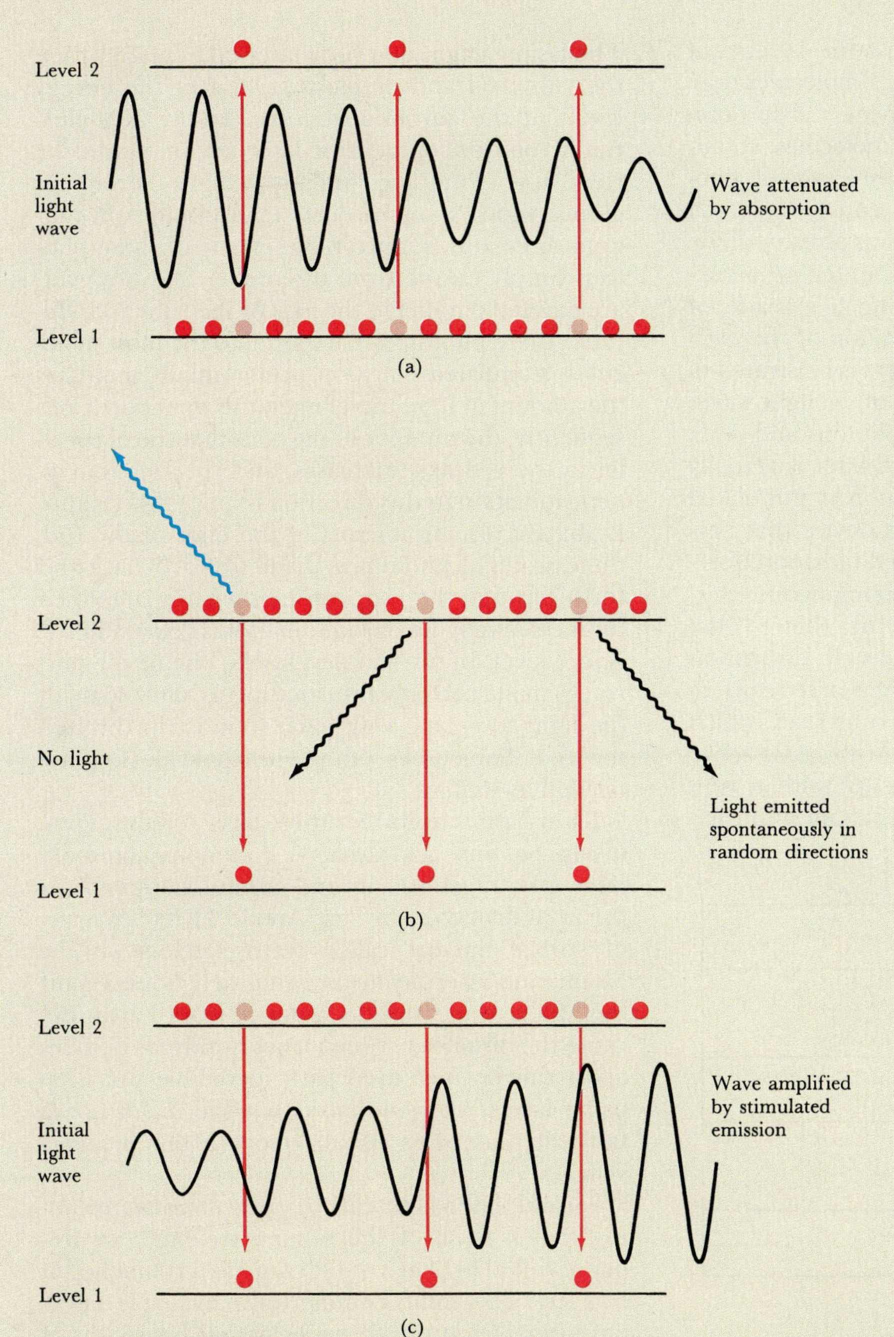

Figure E-2 Two atomic energy levels separated by an energy E. (a) Normally the number of atoms in the lower level is much greater than the number in the upper level. The dots indicate the relative population of the two levels. If light is applied with an energy per photon equal to E, the light is absorbed, thus exciting molecules to the upper energy level. (b) Excited molecules can relax to a lower level by emitting light spontaneously. This spontaneous emission is random in direction. (c) If light (with an energy per photon equal to E) interacts with excited molecules, the molecules can be stimulated to emit more light. Note that the height of the light waves increases as the molecules emit photons and relax back to the lower level. This amplification is the basis of laser action.

white because added dye molecules (whiteners) absorb ultraviolet light from the sun but emit visible light. The molecules in fluorescent ink glow by exactly the same process, which is why fluorescent marking pens that use fluorescent ink make brighter lines than pens that use normal ink.

Another process is possible if there is a so-called *population inversion,* where there are many more molecules in the upper level than the lower level. This population inversion can be achieved in some cases by "pumping" the molecules with a flash lamp similar to the flash on a camera. Once a population inversion is achieved, the molecules begin to fall back to the lower level. In the process, they emit photons of energy $E = h\nu$, just as with spontaneous emission. If the excited molecule encounters *many* photons whose energies just match the energy difference between the excited state and a lower state, it can be stimulated to emit light, which adds to the electromagnetic field. This *stimulated emission* can be expressed by the equation

$$(N - 1)h\nu + A^* \rightarrow Nh\nu + A \qquad \text{(E-3)}$$

Notice that Equation (E-3) is just the reverse of Equation (E-1). If there are more A molecules than A^* molecules (as usually happens), absorption dominates; if there are more A^* molecules, stimulated emission dominates. The direction of photons produced in stimulated emission is *not* random—the light adds to what was already there. These photons in turn can encounter other excited-state atoms, thereby causing a cascade of emitted photons and an amplification of the electromagnetic radiation. This process is illustrated in Figure E-2c. Note that the height of the light waves increases as the molecules emit photons and relax back to the lower level. The word laser is actually an acronym, from *l*ight *a*mplification by *s*timulated *e*mission of *r*adiation. A laser is a device that harnesses and channels this stimulated radiation, forcing it to be much stronger than spontaneous emission. It was Albert Einstein who first showed that stimulated emission must also exist if absorption exists—but he did not live to see the laser realized.

The first successful laser was a ruby laser, which uses a cylindrical crystalline rod of ruby as its active medium. Ruby is primarily $Al_2O_3(s)$ with a very small amount of atomic chromium as an impurity.

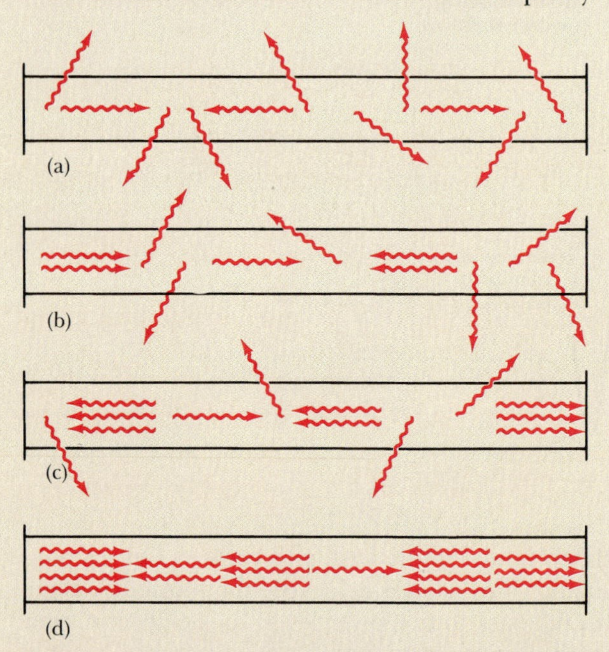

Figure E-3 The development of laser action. (a) After pumping, photons are emitted spontaneously in random directions, with some emitted in a direction parallel to the axis of the figure. (b) Those photons emitted parallel to the axis stimulate the emission of more photons, which also travel parallel to the axis. (c) The stimulated emission continues to build up as the photons are reflected back and forth, resulting in (d), a high density of photons traveling back and forth.

The lasing action of a ruby laser takes place within the ruby rod and employs several of the energy levels of the chromium atoms. Figure E-3 illustrates the development of laser action within the ruby rod. Pumping initially excites chromium atoms, which spontaneously emit photons in random directions (Figure E-3a). Many of these photons simply escape from the rod. A few, however, are emitted parallel to the axis of the ruby rod. Because these photons are exposed to the most molecules, stimulated emission preferentially increases the amount of light traveling in this direction. Consequently, the number of photons that travel parallel to the rod axis increases, and the electromagnetic radiation in this direction is amplified (Figure E-3b). By placing mirrors at the ends of the rod, the cascade of photons is made to travel back and forth through the rod, constantly being amplified until essentially no chromium atoms remain in the upper level. In the simplest lasers, one of the mirrors is made partially transmitting to allow some of the light to escape with each round trip through the rod. Sometimes other electronic devices are needed as well.

Laser light is useful for three basic reasons. First, it can be very *directional.* A common laboratory laser can travel 100 m and produce a spot only 0.1 m in diameter. In fact, Apollo 11 left an array of special mirrors called retroreflectors on the Moon, and laser beams are routinely bounced off the Moon to measure its distance from Earth. Because the distance between cities hundreds of miles apart can be measured with incredible precision using lasers, we now know that the major active faults in California usually move a few inches a year.

Second, laser light can be *monochromatic,* meaning that it is all of the same wavelength or frequency. Red light at $\lambda = 600$ nm has a frequency of 5×10^{14} Hz; some commercially available lasers have tunable output frequencies and linewidths of only 10^6 Hz, meaning that the frequency fluctuates by less than a millionth of a percent! Even if such a laser produces only a few milliwatts of power at this frequency, the effective brightness is far higher than that of conventional light sources, which distribute their energy over a broad range of frequencies and in all directions. This selectivity is crucial in scientific experiments, because many molecules absorb only at well-defined frequencies, and excited molecules can undergo new kinds of chemical reactions. Because of this monochromaticity, laser radiation is exactly in phase, meaning that the maxima and minima of each wave superimpose.

Finally, lasers can be very *intense.* Laser nuclear fusion experiments are being attempted at the Lawrence Livermore National Laboratory in California and in other places around the world. At Livermore, the largest laser system uses neodymium ions (atomic number 60) in an inert host such as garnet as the lasing medium. This laser can produce single pulses of infrared light ($\lambda = 1.06\ \mu m$) that last for about 10^{-9} s but have a total energy of 10^6 J; its output can also be efficiently converted to green light ($\lambda = 0.53\ \mu m$) at exactly twice the frequency. The peak power of such a laser is $10^6\ J/10^{-9}\ s = 10^{15}$ W. This beam is focused very tightly on tiny samples of deuterium (^2H) or tritium (^3H). The intensity is so high that the photons themselves exert enormous pressures on the sample, fusing the nuclei into helium nuclei and releasing energy. Unfortunately, these lasers can currently give only a few pulses a day.

Table E-1 lists a variety of common lasers used in chemical research and their typical characteristics. Chemical reactions are often very rapid, occurring within 10^{-14}–10^{-12} s after two molecules come in contact. Such reactions are often studied with pulsed laser systems. The shortest laser pulses generated to date last only 6×10^{-15}s. Since light travels $3 \times 10^8\ m \cdot s^{-1}$, such pulses are less than 2 μm long. For optical communications, lasers can be modulated to carry information, and this modulation can be detected at a great distance. In research laboratories, modulation with 10^{-13}s resolution or better has been demonstrated. This could eventually mean that a tremendous number of "bits" of information per second could be transmitted over a single optical fiber.

Many of the industrial or medical applications of lasers depend on the color of the output. For example, long-wavelength lasers such as carbon dioxide lasers ($\lambda \approx 10\ \mu m$) are frequently used for welding metals, because shorter wavelengths would be reflected off shiny metallic surfaces instead of being absorbed, and of course reflected light does not heat the metal. Carbon dioxide lasers are also used in surgical procedures. Biological tissue consists primarily of water, which can be vaporized instantly by a carbon dioxide laser beam. Tight focusing of the laser beam permits precise cutting, and the heat deposited in the surrounding tissue cauterizes the wound and prevents the bleeding normally associated with conventional surgery. But many other kinds of lasers are used in different surgical procedures. Excimer lasers can produce light in the ultraviolet, which has much more energy per photon than visible lasers. It has been shown that 193-nm light can break chemical bonds, with enough energy per photon left over to vaporize the target tissue, thus making an extremely clean cut without scarring or cauterizing (Figure E-4).

There are other types of lasers, but they all work by the same principle. The key ingredients of any laser are the production of a population inversion followed by the stimulated emission of electromagnetic radiation.

Table E-1 Some common lasers*

Gases	Pulse (peak power)
helium-neon (632.8 nm)	continuous (1–50 mW)
argon ion, Ar^+ (514.5 nm)	continuous (20 W)
(488.0 nm)	or 100 ps (1000 W)
excimers XeCl (308 nm)	10 ns (10^8 W)
KrF (248 nm)	
ArF (193 nm)	
carbon dioxide (10,600 nm)	continuous (10,000 W) down to 10^{-6} s (10^9 W)
Liquids	
dye lasers (tunable 400–1000 nm)	continuous (1 W) down to 6×10^{-15} s (10^9 W)
Solids	
Nd^{3+} in yttrium aluminum garnet, Nd:YAG (1064 nm)	continuous (100 W) down to 10^{-9} s (Livermore: 10^{15} W)

*All the combinations of parameters listed here are approximate. For example, many different kinds of dye lasers with quite different characteristic pulse lengths and pulse energies are currently used.

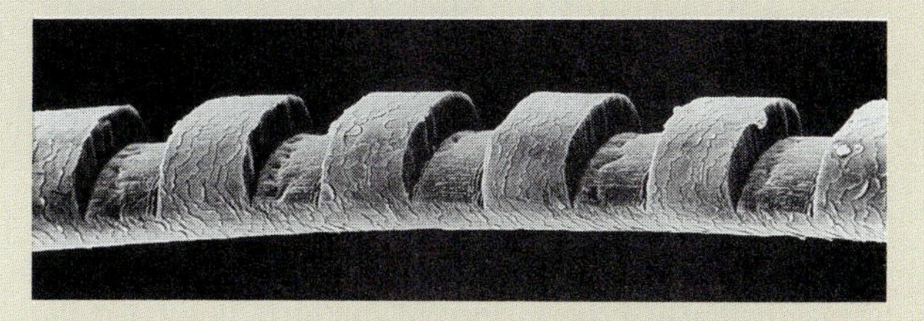

Figure E-4 Cut made in a human hair by an excimer laser. The energy per photon is so high that the tissue is actually blown out of the cut.

10 IONIC BONDS AND COMPOUNDS

Like charges repel and opposite charges attract. In this photo the negative charges produced by an electrostatic generator are transferred by hand contact to the surface of the girl's body and hair. The charges on the hair fibers repel and cause the hair fibers to spread out. The girl is standing on a rubber mat, which prevents the charge from leaving her body.

In Chapter 9 we showed how the electronic structure of atoms underlies the structure of the periodic table. Thus, it seems reasonable to suppose that an understanding of atomic structure should help us understand the chemical bonding that occurs between atoms in molecules. For example, we can use the electron configurations of the sodium atom and the chlorine atom to understand why one atom of sodium combines with just one atom of chlorine to form sodium chloride. We can understand why the result is NaCl, instead of $NaCl_2$ or Na_2Cl. We also can understand why sodium chloride is an ionic compound, capable of conducting an electric current when it is dissolved in water or melted. Similarly, we will learn why carbon and hydrogen combine to form the stable molecule methane, whose formula is CH_4 instead of CH or CH_2, and why nitrogen is a diatomic gas, N_2, at room temperature. We also will learn why sodium chloride is a solid with a high melting point and why a solution of sucrose dissolved in water is a poor conductor of electric current. All these observations relate to the bonding that occurs between atoms. In this and the next few chapters we will develop an understanding of these observations in terms of chemical bonds.

10-1. SOLUTIONS THAT CONTAIN IONS CONDUCT AN ELECTRIC CURRENT

As our first step toward understanding ionic bonding, we discuss an important experimental property of ionic compounds in aqueous solution. When most ionic compounds dissolve in water, the crystals break up into mobile ions rather than neutral molecules. For example, an aqueous solution of sodium chloride consists of $Na^+(aq)$ and $Cl^-(aq)$ ions that move throughout the water. If electrodes (for example, strips of an inert metal like platinum) connected to the poles of a battery are dipped into a solution containing ions, then the positive ions are attracted to the negative electrode and the negative ions are attracted to the positive electrode (Figure 10-1). The movement of the ions toward the respective electrodes constitutes an electric current through the solution.

Compounds that yield neutral molecules when they dissolve in water are very poor conductors of an electric current, because no charge carriers like those in solutions of ionic compounds are present. For example, an aqueous solution of sucrose (table sugar, $C_{12}H_{22}O_{11}$) contains neutral sucrose molecules, so it does not conduct an electric current

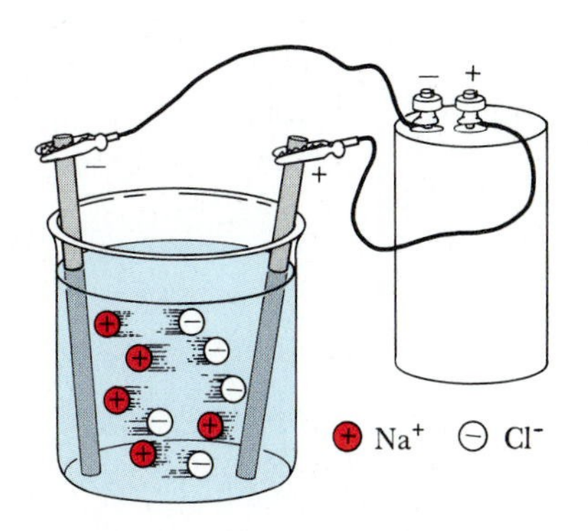

Figure 10-1 An aqueous solution of NaCl conducts an electric current. An electric voltage is applied by dipping metal strips (electrodes) attached to poles of a battery into the solution. Like the poles of a battery, one of the electrodes is positive and the other is negative. The positively charged sodium ions are attracted to the negative electrode, and the negatively charged chloride ions are attracted to the positive electrode. Thus, the Na^+ ions migrate to the left in the figure and the Cl^- ions migrate to the right. The migration of the ions constitutes an electric current through the solution.

Figure 10-2 Comparison of the currents (as measured using ammeters) through 1 M NaCl(*aq*) and 1 M sucrose (*aq*). Note that the solution of strong electrolyte (NaCl) is a much better conductor of electricity than the solution of nonelectrolyte (sucrose).

(Figure 10-2). Substances like sodium chloride or calcium chloride, whose aqueous solutions conduct an electric current, are called **electrolytes.** Substances like sucrose, whose aqueous solutions do not conduct an electric current, are called **nonelectrolytes.**

Not all solutions of electrolytes conduct an electric current to the same extent. For example, a 0.10 M $HgCl_2(aq)$ solution is a much poorer conductor of electricity than is a 0.10 M $CaCl_2(aq)$ solution. For this reason, we call calcium chloride a **strong electrolyte** and mercury(II) chloride a **weak electrolyte.** When a strong electrolyte such as calcium chloride dissolves in water, essentially all the calcium chloride formula units **dissociate** (break up) in solution into free ions, which are available to conduct an electric current. However, when a weak electrolyte such as mercury(II) chloride dissolves in water, only a small fraction of the mercury(II) chloride formula units dissociate into ions; most exist as molecular mercury(II) chloride units. Because a $HgCl_2(aq)$ solution contains far fewer ions to conduct a current than does a $CaCl_2(aq)$ solution of the same concentration, a $HgCl_2(aq)$ solution is a poorer conductor than is a $CaCl_2(aq)$ solution at the same concentration.

The following simple rules (which should be memorized) can be used to determine whether a substance is a strong electrolyte, a weak electrolyte, or a nonelectrolyte.

1. The acids HCl, HBr, HI, HNO_3, H_2SO_4, and $HClO_4$ are strong electrolytes. Most other acids are weak electrolytes. In other words, if an acid is not on this short list of strong electrolytes, then it is a weak electrolyte.

2. The soluble hydroxides of the Group 1 and 2 metals are strong electrolytes. Most other bases, and particularly ammonia, are weak electrolytes.

3. Most salts are strong electrolytes in aqueous solution.

4. The halides and cyanides of the "heavy metals" (i.e., those with high atomic numbers), for example, mercury and lead, are often weak electrolytes.

5. Most organic compounds, that is, compounds that consist of carbon, hydrogen, and possibly other atoms, are nonelectrolytes. Notable exceptions are organic acids and bases, which are usually weak electrolytes.

EXAMPLE 10-1: Classify each of the following compounds as either a strong electrolyte, a weak electrolyte, or a nonelectrolyte in aqueous solution: (a) $NaNO_3$; (b) C_2H_5OH (ethyl alcohol); (c) $Ba(OH)_2$; (d) $AuCl_3$.

Solution: (a) Sodium nitrate is a water-soluble salt and a strong electrolyte. (b) Ethyl alcohol is an organic compound and a nonelectrolyte. (c) Barium hydroxide is a water-soluble Group 2 hydroxide and a strong electrolyte. (d) Gold(III) chloride is a heavy metal halide; thus, we predict that it is a weak electrolyte in aqueous solution. This prediction is correct.

PRACTICE PROBLEM 10-1: Classify each of the following compounds either as a strong electrolyte, a weak electrolyte, or a nonelectrolyte in aqueous solution: (a) $KClO_3$; (b) $(CH_3)_2CO$ (acetone); (c) H_2SO_3; (d) $Hg(CN)_2$.

Answer: (a)$KClO_3$ (strong); (b) $(CH_3)_2CO$ (nonelectrolyte); (c) H_2SO_3 (weak); (d) $Hg(CN)_2$ (weak)

Recall that a weak electrolyte differs from a strong electrolyte in the extent to which the formula units of the compound dissociate into ions when it dissolves. For example, when sufficient $CaCl_2(s)$ is dissolved in water to form, say, a 0.10 M solution, essentially all the calcium chloride in the solution exists as the ions $Ca^{2+}(aq)$ and $Cl^-(aq)$. When $HgCl_2(s)$ dissolves in water, most of it exists as undissociated $HgCl_2(aq)$ units, with only traces of the ions $HgCl^+(aq)$, $Hg^{2+}(aq)$, and $Cl^-(aq)$. For this situation, we can write the chemical equations

$$CaCl_2(s) \xrightarrow[H_2O(l)]{100\%} Ca^{2+}(aq) + 2Cl^-(aq)$$

$$HgCl_2(s) \xrightarrow{H_2O(l)} \begin{cases} \xrightarrow{99.8\%} HgCl_2(aq) \\ \xrightarrow{0.18\%} HgCl^+(aq) + Cl^-(aq) \\ \xrightarrow{2 \times 10^{-4}\%} Hg^{2+}(aq) + 2Cl^-(aq) \end{cases}$$

where the percentages refer to the extent of reaction in the 0.10 M solution.

The extent to which a compound dissociates into ions in solution is called the **degree of dissociation.** The degree of dissociation of a dissolved compound is obtained by determining the **electrical conductance** of the solution. At a particular concentration of salt, the greater the degree of dissociation, the more ions there will be in the solution, so the greater will be the electrical conductance, because the ions conduct the current in the solution. We use the **molar conductance** to compare the conductivities of salts on a per mole basis—the molar conductance is the electrical conductivity of the solution per mole of the dissolved compound. Weak electrolytes have much lower molar conductances than strong electrolytes have (Table 10-1).

The key question is, What is the origin of the difference in the degrees of dissociation between strong electrolytes (nearly 100 percent dissociated) and weak electrolytes (usually less than 10 percent dissociated)? To answer this question, we must develop an understanding of chemical bonding, which is the main focus of the remainder of this chapter and of Chapters 11, 12, and 13.

Table 10-1 Molar conductances of strong and weak electrolytes (25°C, 0.10 M aqueous solutions)

Compound	Molar conductance/ $ohm^{-1} \cdot cm^2 \cdot mol^{-1}$
Strong electrolytes	
$HCl(aq)$	391
$KCl(aq)$	129
$NaOH(aq)$	221
$AgNO_3(aq)$	109
$BaCl_2(aq)$	210
$NaC_2H_3O_2(aq)$ (sodium acetate)	73
Weak electrolytes	
$HC_2H_3O_2(aq)$ (acetic acid)	5.2
$NH_3(aq)$	3.5
$HgCl_2(aq)$	2

10-2. THE ELECTROSTATIC FORCE THAT BINDS OPPOSITELY CHARGED IONS TOGETHER IS CALLED AN IONIC BOND

To understand ionic bonds, let's first consider the reaction between a sodium atom and a chlorine atom. The ground-state electron configurations of the sodium and chlorine atoms are

$$\text{Na} \quad [\text{Ne}]3s^1 \qquad \text{Cl} \quad [\text{Ne}]3s^23p^5$$

Note that the electron configuration of a sodium atom consists of a neonlike inner core with a 3*s* electron outside the core. If the sodium atom loses the 3*s* electron, then the resultant species is a sodium ion, with an electron configuration like that of the noble gas neon. We can describe the ionization process by the equation

$$\text{Na}([\text{Ne}]3s^1) \rightarrow \text{Na}^+([\text{Ne}]) + e^-$$

Once a sodium atom loses its 3*s* electron, the resultant sodium ion has a neonlike electron configuration and is relatively stable to further ionization.

If a chlorine atom accepts an electron, then the resultant species is a chloride ion, with an electron configuration like that of the noble gas argon. We write this as

$$\text{Cl}([\text{Ne}]3s^23p^5) + e^- \rightarrow \text{Cl}^-([\text{Ar}])$$

Thus, we see that both a sodium atom and a chlorine atom can simultaneously achieve noble-gas electron configurations through the transfer of an electron from the sodium atom to the chlorine atom. We can describe the electron transfer by the equation

$$\text{Na}([\text{Ne}]3s^1) + \text{Cl}([\text{Ne}]3s^23p^5) \rightarrow \text{Na}^+([\text{Ne}]) + \text{Cl}^-([\text{Ar}])$$

or, in terms of Lewis electron-dot formulas,

$$\text{Na}\cdot + \cdot\ddot{\underset{\cdot\cdot}{\text{Cl}}}: \longrightarrow \underbrace{\text{Na}^+ + :\ddot{\underset{\cdot\cdot}{\text{Cl}}}:^-}_{\text{Na}^+\text{Cl}^-}$$

The sodium ion and the chloride ion have opposite charges, so they attract each other. This **electrostatic force** binds the ions together and is called an **ionic bond.**

We have seen that noble-gas electron configurations are relatively stable to the gain or loss of additional electrons. Because both sodium ions and chloride ions achieve a noble-gas electron configuration, the above reaction occurs easily, and there is no tendency for additional electron transfer. Because a sodium atom readily loses one and only one electron, whereas a chlorine atom readily gains one and only one electron, when a sodium atom reacts with a chlorine atom, the transfer of an electron from the sodium atom to the chlorine atom results in one sodium ion and one chloride ion. The compound, sodium chloride, like all chemical compounds, is electrically neutral. Therefore, the chemical formula of sodium chloride must be NaCl and not $NaCl_2$, Na_2Cl, or anything other than NaCl. Furthermore, sodium chloride is an **ionic compound,** that is, a compound composed of ions.

The reaction between sodium and chlorine is an example of a reaction between a reactive metal and a reactive nonmetal that produces ions with noble-gas electron configurations. We shall see repeatedly that a noble-gas electron configuration is a particularly stable electron arrangement and that there is a strong tendency for this configuration to occur. This latter characteristic is especially true for the elements in the first two rows of the periodic table. Figure 10-3 shows some of the common atoms that lose or gain electrons to achieve a noble-gas electron configuration. All the ions in Figure 10-3 have an outer electron configuration of $ns^2\,np^6$, except for Li^+ and Be^{2+}, which have a helium-like electron configuration of $1s^2$.

Note that metallic elements lose electrons to become positively charged ions (called **cations**) and nonmetallic elements gain electrons to become negatively charged ions (called **anions**). Also note that the charges on these ions correspond exactly to the ionic charges discussed in Chapter 3. In fact, the rules developed there for writing correct formulas for simple chemical compounds ensure that the group of ions indicated by the chemical formula has no net electrical charge.

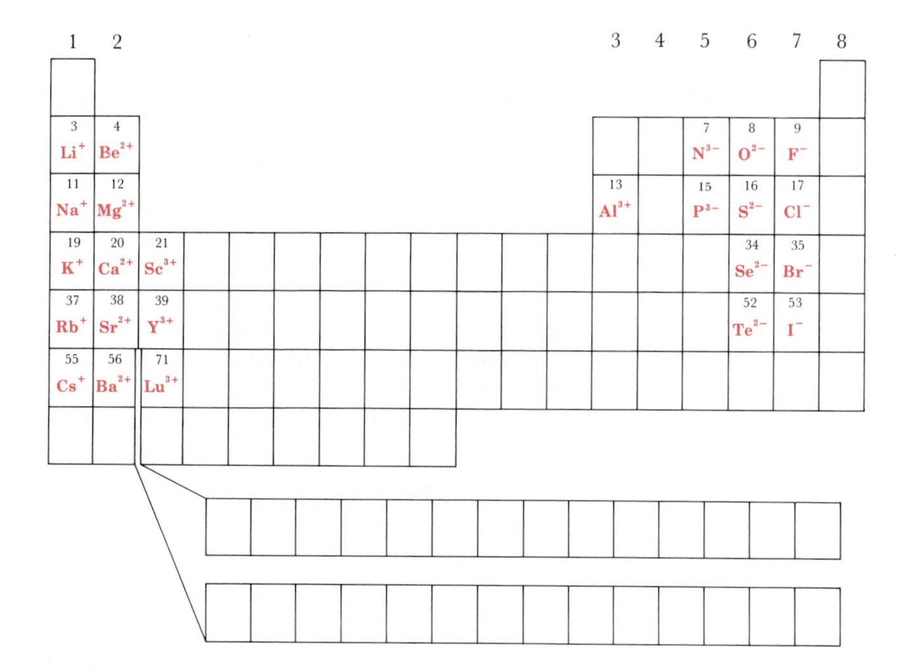

Figure 10-3 Ions with a noble-gas outer electron configuration.

EXAMPLE 10-2: Write the ground-state electron configurations of Ca^{2+} and Se^{2-} and predict the formula for calcium selenide.

Solution: The ground-state electron configurations of Ca and Se are

$$\text{Ca: } 1s^2 2s^2 2p^6 3s^2 3p^6 4s^2$$
$$\text{Se: } 1s^2 2s^2 2p^6 3s^2 3p^6 4s^2 3d^{10} 4p^4$$

To form Ca^{2+}, a calcium atom loses its two $4s$ electrons; and to form Se^{2-}, a selenium atom gains two $4p$ electrons. Thus, the electron configurations of these ions are

$$Ca^{2+}\text{: } 1s^2 2s^2 2p^6 3s^2 3p^6 \quad \text{or} \quad [\text{Ar}]$$
$$Se^{2-}\text{: } 1s^2 2s^2 2p^6 3s^2 3p^6 4s^2 3d^{10} 4p^6 \quad \text{or} \quad [\text{Kr}]$$

In each case, the resultant ion has a noble-gas electron configuration. Because the $+2$ charge on one calcium ion exactly balances the -2 charge on one selenide ion, the formula for calcium selenide is CaSe.

PRACTICE PROBLEM 10-2: Write the ground-state electron configurations for Mg^{2+} and N^{3-} and predict the formula for magnesium nitride.

Answer: $Mg^{2+}(1s^2 2s^2 2p^6)$; $N^{3-}(1s^2 2s^2 2p^6)$; $Mg_3N_2(s)$

10-3. THE COMMON OXIDATION STATES OF TRANSITION-METAL IONS CAN BE UNDERSTOOD IN TERMS OF ELECTRON CONFIGURATIONS

There are many metal ions not listed in Figure 10-3. For example, let's consider silver ($Z = 47$), which has the ground-state electron configuration $[\text{Kr}]5s^1 4d^{10}$. The silver atom would have to lose 11 electrons or gain 7 electrons to achieve a noble-gas electron configuration. A review of the data in Table 9-1 shows that the first of these alternatives would require an enormous amount of energy. The energy required for the addition of seven electrons is also prohibitively large. Each successive electron would have to overcome a larger and larger repulsion as the negative charge on the ion increased. Consequently, atomic ions with a charge greater than three are rare, but not unknown.

Although a silver atom cannot reasonably be expected to achieve a noble-gas configuration, its outer electron configuration will be $4s^2 4p^6 4d^{10}$ if it loses its $5s$ electron. This configuration, with 18 electrons in the outer shell, is a relatively stable electron configuration and is sometimes called an **18-outer electron configuration.** The unusual stability of the $ns^2 np^6 nd^{10}$ outer electron configuration is often referred to simply as the **18-electron rule.** Silver forms an 18-electron-rule ion, Ag^+, by loss of an electron

$$Ag([\text{Kr}]5s^1 4d^{10}) \rightarrow Ag^+([\text{Kr}]4d^{10}) + e^-$$
$$\text{or}$$
$$Ag^+(1s^2 2s^2 2p^6 3s^2 3p^6 3d^{10} 4s^2 4p^6 4d^{10}) + e^-$$

EXAMPLE 10-3: Use the 18-electron rule to predict the electron configuration and the charge of a zinc ion.

Solution: A zinc atom has a $1s^2 2s^2 2p^6 3s^2 3p^6 4s^2 3d^{10}$ ground-state electron configuration. The zinc atom can achieve the 18-outer electron configuration $3s^2 3p^6 3d^{10}$ by losing its two $4s$ electrons. Thus, we predict that the electron configuration of the zinc ion is

$$Zn^{2+}(1s^2 2s^2 2p^6 3s^2 3p^6 3d^{10}) \quad \text{or} \quad Zn^{2+}([Ar]3d^{10})$$

and its charge is +2, which is correct. Note that Zn^{2+} has a completely filled $n = 3$ (that is, M) shell.

PRACTICE PROBLEM 10-3: Use the 18-electron rule to predict a possible charge on an indium ion.

Answer: In^{3+}

Some of the other metals that form 18-electron ions are shown in Figure 10-4. Note that these metals occur near the ends of the *d* transition series and that the charge on the ions increases by one unit as we look left to right along a row of the periodic table.

Another outer electron configuration that is often found in ions is illustrated by the element thallium ($Z = 81$). The electron configuration of a thallium atom is $[Xe]6s^2 5d^{10} 6p^1$. Loss of the $6p$ electron yields the electron configuration $[Xe]5d^{10}6s^2$ for the ion Tl^+. Although a thallium(I) ion does not have a noble-gas electron configuration or an 18-electron-rule configuration, it does have all its subshells completely filled, and this type of electron configuration is also relatively stable. Other elements that behave like thallium are shown in Figure 10-5. Note that these elements are in Groups 3, 4, and 5. The unipositive ions in Figure 10-5 can lose two more electrons to achieve an 18-electron-rule configuration. For example, if Tl^+ loses its two $6s$ electrons, then the resulting configuration for Tl^{3+} is $[Xe]5d^{10}$, which appears in Figure 10-4. Thus, we see that thallium and indium have two possible ionic charges, +1 and +3; tin and lead also have two possible ionic charges, +2 and +4.

The ground-state electron configurations of transition-metal ions are relatively easy to deduce. We learned in Chapter 9 that the $3d$ orbitals are filled after the $4s$ orbital in neutral atoms. This order of filling occurs because the order of energy levels *for neutral atoms* is that shown in Figures 9-32b and 9-33. However, when electrons are lost from a neutral atom, the charge on the ion alters the order of the orbital energies such that in the transition-metal ions, the energy of the $3d$ orbitals is less than that of the $4s$ orbital. A similar situation occurs for the $4d$ and $5s$ orbitals and the $5d$ and $6s$ orbitals. Therefore, the filling order of the orbitals for transition-metal ions is regular, that is, the order of orbital energies is

$$1s < 2s < 2p < 3s < 3p < 3d < 4s < 4p < 4d < 4f < \ldots$$

Thus, although the ground-state electron configuration of a nickel atom is $[Ar]4s^2 3d^8$, that for a nickel ion, Ni^{2+}, is $[Ar]3d^8$ and not $[Ar]4s^2 3d^6$.

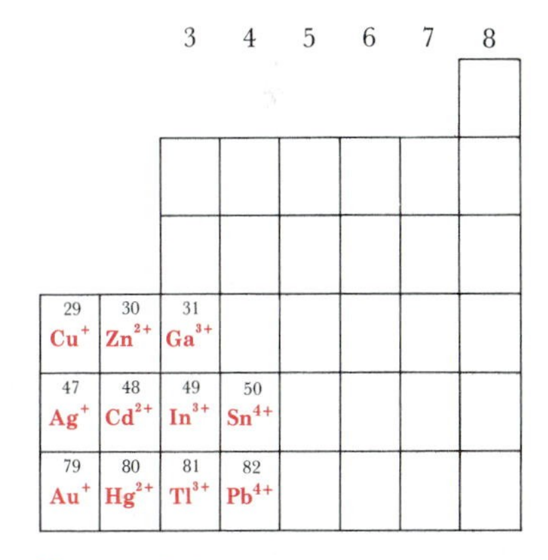

Figure 10-4 Metal ions with an 18-outer electron configuration, $ns^2 np^6 nd^{10}$.

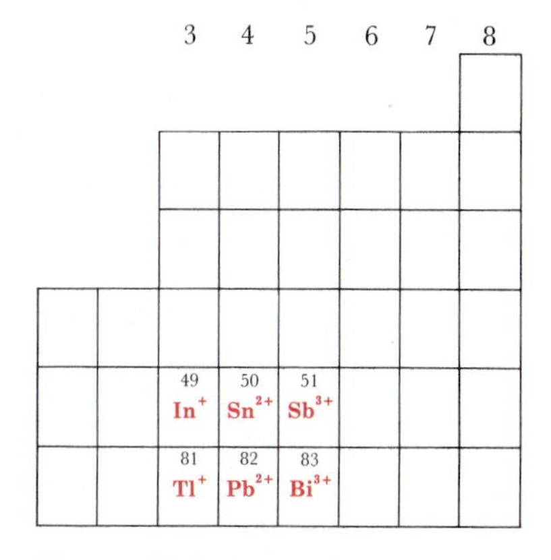

Figure 10-5 Ions with the outer electron configuration [noble gas]$nd^{10}(n + 1)s^2$.

EXAMPLE 10-4: Predict the ground-state electron configuration of Ti^{3+}.

Solution: The ground-state electron configuration of the titanium atom ($Z = 22$) is

$$[Ar]4s^2 3d^2$$

Although the 4*s* orbital has a lower energy than the 3*d* orbitals have in neutral atoms, this situation is not true for ions. In Ti^{3+} the 3*d* orbitals are lower in energy than the 4*s* orbital is. Therefore, the ground-state electron configuration of Ti^{3+}, which has three fewer electrons than the neutral titanium atom has, is

$$[Ar]3d^1$$

Note that the electron configuration of Ti^{3+} is regular in the sense that it is

$$1s^2 2s^2 2p^6 3s^2 3p^6 3d^1$$

The orbitals of transition-metal ions are filled in the regular order.

PRACTICE PROBLEM 10-4: Predict the ground-state electron configurations of Pd^{2+}.

Answer: Pd^{2+}: $[Kr]4d^8$

EXAMPLE 10-5: Referring only to the periodic table, predict the ground-state electron configurations of the ions Cu^+, Cu^{2+}, and Cu^{3+}.

Solution: The atomic number of copper is 29, so Cu^+ has 28 electrons, Cu^{2+} has 27 electrons, and Cu^{3+} has 26 electrons. The order of filling the orbitals is regular, so the electron configurations are

$$Cu^+: 1s^2 2s^2 2p^6 3s^2 3p^6 3d^{10}$$
$$Cu^{2+}: 1s^2 2s^2 2p^6 3s^2 3p^6 3d^9$$
$$Cu^{3+}: 1s^2 2s^2 2p^6 3s^2 3p^6 3d^8$$

The Cu(II) oxidation state is the most common oxidation state of copper ions.

PRACTICE PROBLEM 10-5: Predict the ground-state electron configurations of Cr(II), Cr(III), and Cr(VI).

Answers: Cr(II): $[Ar]3d^4$; Cr(III): $[Ar]3d^3$; Cr(VI): [Ar]

10-4. CATIONS ARE SMALLER AND ANIONS ARE LARGER THAN THEIR PARENT NEUTRAL ATOMS

Because atoms and ions are different species, we should expect atomic radii and **ionic radii** to have different values. For example, the average distance from the nucleus of the 3*s* electron in a sodium atom is greater than that of the 1*s*, 2*s*, and 2*p* electrons, because the 3*s* electron is in the $n = 3$ (M) shell. When a sodium atom loses its 3*s* electron, only the K and L shells are occupied, so a sodium ion is smaller than a sodium atom. In addition, the excess positive charge draws the remaining electrons toward the nucleus and causes the electron cloud to contract. Positive atomic ions are always smaller than their corresponding neutral atoms for this reason.

The relative sizes of the alkali metal atoms and ions are shown in the first column of Figure 10-6. The Group 2 metals lose two outer *s* electrons in becoming M^{2+} ions. The excess positive charge of +2 contracts the remaining electron shells even more than in the case of the Group 1 metals, as you can see by comparing the sizes in the first and second columns of Figure 10-6.

The atoms of nonmetals gain electrons in becoming ions. The addition of an extra electron increases the electron-electron repulsion and causes the electron cloud to expand. Negative ions are always larger than their corresponding neutral atoms for this reason. The relative atomic and ionic sizes of the halogen atoms and halide ions are shown in the column headed by 7 in Figure 10-6. Numerical values of ionic radii are given in Table 10-2.

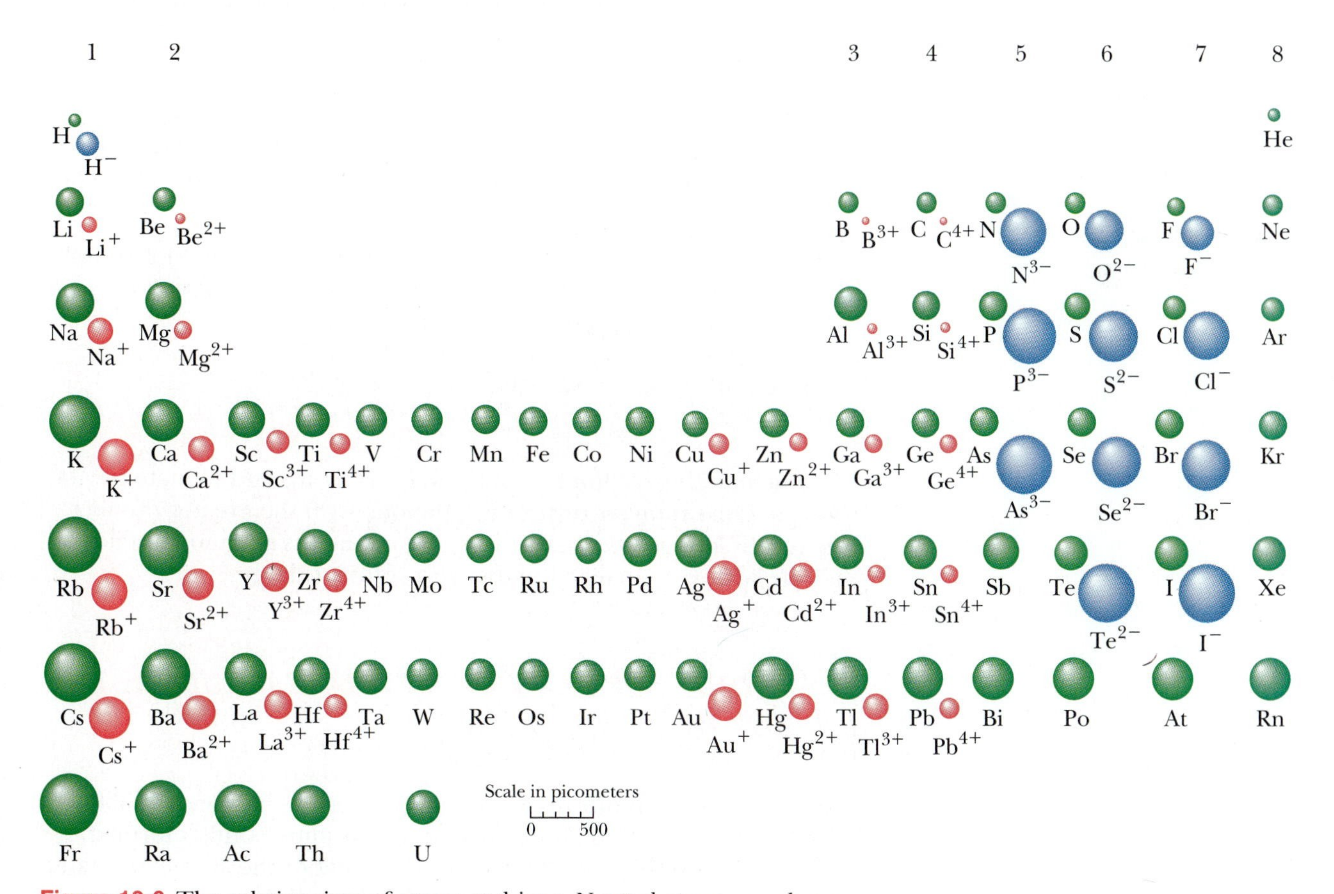

Figure 10-6 The relative sizes of atoms and ions. Neutral atoms are shown in green, cations in red, and anions in blue. Note that cations are smaller and anions are larger than the respective neutral parent atoms.

Table 10-2 **Ionic radii in picometers (1 pm = 10^{-12} m)**

Ion	Radius	Ion	Radius	Ion	Radius	Ion	Radius	Ion	Radius	Ion	Radius	Ion	Radius
Cations								**Anions**					
Ag^+	126	Ba^{2+}	135	Al^{3+}	50	Ce^{4+}	101	Br^-	195	O^{2-}	140	N^{3-}	171
Cs^+	169	Ca^{2+}	99	B^{3+}	20	Ti^{4+}	68	Cl^-	181	S^{2-}	184	P^{3-}	212
Cu^+	96	Cd^{2+}	97	Cr^{3+}	65	U^{4+}	97	F^-	136	Se^{2-}	196		
K^+	133	Co^{2+}	82	Fe^{3+}	67	Zr^{4+}	80	H^-	154	Te^{2-}	221		
Li^+	60	Cu^{2+}	70	Ga^{3+}	62			I^-	216				
Na^+	95	Fe^{2+}	78	In^{3+}	81								
NH_4^+	148	Mg^{2+}	65	La^{3+}	115								
Rb^+	148	Ni^{2+}	69	Tl^{3+}	95								
Tl^+	144	Sr^{2+}	113	Y^{3+}	93								
		Zn^{2+}	74										

EXAMPLE 10-6: Without reference to Table 10-2 or to Figure 10-6, predict which of the isoelectronic species is the larger ion, K^+ or Cl^-.

Solution: The ground-state electron configuration of both K^+ and Cl^- is $1s^2 2s^2 2p^6 3s^2 3p^6$. The excess positive charge of K^+ contracts the K and L and M shells, and the excess negative charge of Cl^- leads to an enlargement of the shells, so we predict that Cl^- is larger than K^+. Note also that both ions have the same number of electrons (18) (that is, they are **isoelectronic**), but potassium has a nuclear charge of +19 and chlorine has a nuclear charge of only +17. The radius of K^+ is 133 pm and that of Cl^- is 181 pm.

PRACTICE PROBLEM 10-6: Without reference to Table 10-2 or to Figure 10-6, arrange the following ions in order of increasing size:

$$Mg^{2+} \quad Na^+ \quad I^- \quad Br^- \quad Al^{3+}$$

Answer: $Al^{3+} < Mg^{2+} < Na^+ < Br^- < I^-$

We conclude this section by noting that for atoms and monatomic ions with the same number of protons, the one with the greater number of electrons is larger, whereas for isoelectronic atoms and monatomic ions, the one with the smaller number of protons is larger.

10-5. COULOMB'S LAW IS USED TO CALCULATE THE ENERGY OF AN ION-PAIR

Prior to this section our discussion of ionic bonds has been qualitative. We now show by calculations that, when an ionic bond is formed, the energy of the ionic products is lower than that of the atomic reactants. Let's consider the reaction described by the equation

$$Na(g) + Cl(g) \rightarrow Na^+Cl^-(g)$$

Figure 10-7 An atomic and molecular beam apparatus, with its designer, Professor Yuan T. Lee of the University of California, Berkeley, who was awarded the 1988 Nobel Prize in chemistry for his studies of reactions in molecular beams.

Reactions of this type can be studied by firing a beam of sodium atoms at a beam of chlorine atoms and observing the gaseous products formed (Figure 10-7). The net energy change for the reaction can be calculated by breaking it down into three separate steps and applying Hess's law (page 256) to the sum of the three steps:

1. The electron is removed from the sodium atom (ionization). The energy required to ionize a mole of sodium atoms is 496 $kJ \cdot mol^{-1}$.
2. The electron removed from the sodium atom is added to the chlorine atom. Energy is released in the process, and this energy is called the **electron affinity** of chlorine. (We will discuss the idea of electron affinity below.) The electron affinity of atomic chlorine is -348 $kJ \cdot mol^{-1}$.
3. The sodium ion and chloride ion are brought together as shown in Figure 10-8. From Tables 10-2 we see that the radius of a sodium ion is 95 pm and that of a chloride ion is 181 pm. Thus, the centers of the two ions are 181 + 95 = 276 pm apart when the two ions are just touching, assuming that the ions behave like hard spheres. We refer to the distance between the centers of an ionically bonded ion-pair as the **equilibrium ion-pair separation,** denoted by d_{eq}.

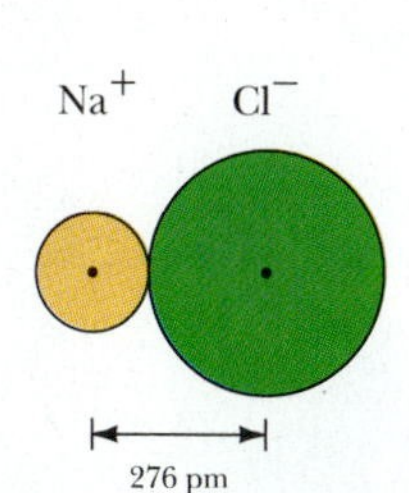

Figure 10-8 A solid-sphere representation of the ion-pair NaCl. According to Table 10-2, the sodium ion can be represented as a sphere of radius 95 pm and the chloride ion can be represented as a sphere of radius 181 pm. Because the two ions have opposite charges, they draw together until they touch, with a distance between their centers of 95 + 181 = 276 pm. They are bound together at this distance in an ionic bond.

Step 1 involves ionization energy, which we discussed in Chapter 9. We now need to discuss the energy associated with steps 2 and 3. Consider the process of adding an electron to a gaseous atom:

$$A(g) + e^- \rightarrow A^-(g)$$

The energy associated with this process is called the **first electron affinity,** EA_1, of the atom $A(g)$. For example, for chlorine we have

$$Cl(g) + e^- \rightarrow Cl^-(g) \qquad EA_1 = -348\ kJ \cdot mol^{-1}$$

The value of EA_1 in this case is negative, because energy is released in the process. Notice that the electron affinity of an atom is the negative of the first ionization energy of the ion, $A^-(g)$. In an equation, we have

$$A^-(g) \rightarrow A(g) + e^- \qquad I_1 = -EA_1$$

Table 10-3 Electron affinities of the atoms of some reactive nonmetals

Atom	EA/kJ·mol^{-1}
H	−72
F	−333
Cl	−348
Br	−324
I	−295
O	−136
	+780 (EA_2)
S	−200
	+590 (EA_2)
Se	−210
	+420 (EA_2)
N	−58
	+800 (EA_2)
	+1300 (EA_3)

Thus, we see that the first ionization energy of an isolated chloride ion, $Cl^-(g)$, is +348 kJ·mol^{-1}. Just as we define successive ionization energies, we define successive electron affinities. For example, the first two electron affinities for an oxygen atom are defined by

$$O(g) + e^- \rightarrow O^-(g) \qquad EA_1 = -136 \text{ kJ·mol}^{-1}$$

$$O^-(g) + e^- \rightarrow O^{2-}(g) \qquad EA_2 = +780 \text{ kJ·mol}^{-1}$$

Notice that the value of the **second electron affinity,** EA_2, is positive; it requires energy to overcome the repulsion of the negatively charged ion O^- (g) and the electron. The most important electron affinities for our purposes are those of the reactive nonmetals (Table 10-3).

So far then, we can write for steps 1 and 2

$$\text{step 1: } Na(g) \rightarrow Na^+(g) + e^- \qquad I_1 = 496 \text{ kJ·mol}^{-1}$$

$$\text{step 2: } Cl(g) + e^- \rightarrow Cl^-(g) \qquad EA_1 = -348 \text{ kJ·mol}^{-1}$$

If we add these two equations, then we find that

$$\text{step 1 + step 2: } \quad Na(g) + Cl(g) \rightarrow Na^+(g) + Cl^-(g)$$

$$\Delta H^\circ_{rxn} \simeq \Delta U^\circ_{rxn} = I_1 + EA_1$$

$$= +148 \text{ kJ·mol}^{-1}$$

Notice that an energy input of 148 kJ·mol^{-1} is required to drive this reaction.

Each of the species in this process is in the gas phase; in other words, $Na^+(g)$ and $Cl^-(g)$ are so far apart that they are effectively isolated entities. We now must calculate the energy change involved in bringing the two widely separated ions (where their energy of interaction is taken to be zero) to their equilibrium ion-pair separation of 276 pm (Figure 10-8). To calculate this energy, we use a relation discovered by Augustin Coulomb (Figure 10-9), that is, Coulomb's law.

Figure 10-9 Portrait by Émile Lecomte of Charles-Augustin de Coulomb (1736–1806), French physicist and discoverer of what is now called Coulomb's law, stated in Equation (10-1).

Coulomb's law says that the energy of interaction of two ions is directly proportional to the product of their electrical charges and is inversely proportional to the distance between their centers. Thus, we have

$$E = \frac{kZ_1Z_2}{d}$$

where Z_1 and Z_2 are the charges of the two ions, d is the distance between the centers of the two ions, and k is a proportionality constant. The value of the proportionality constant depends on the units of the charges and of the distance d; if the charges are measured in units of the charge on an electron (+1 for Na^+, −1 for Cl^-, and so on); if d is measured in picometers (1 pm = 10^{-12} m), and E is expressed in joules, then E is given by (Figure 10-10)

$$E = \frac{(2.31 \times 10^{-16} \text{ J·pm})Z_1Z_2}{d_{eq}} \qquad (10\text{-}1)$$

Note from Equation (10-1) that as d, the ion-pair separation, gets very large, the electrostatic energy of interaction goes to zero. If the charges of the ions have the same sign, then E in Equation (10-1) is positive. If the ions are oppositely charged, then E is negative. For Na^+Cl^-, $Z_1 = +1$ (Na^+), $Z_2 = -1$ (Cl^-), and $d_{eq} = 276$ pm (= 95 pm + 181 pm),

$$E = \frac{(2.31 \times 10^{-16}\ \text{J}\cdot\text{pm})(+1)(-1)}{276\ \text{pm}}$$
$$= -8.37 \times 10^{-19}\ \text{J}$$

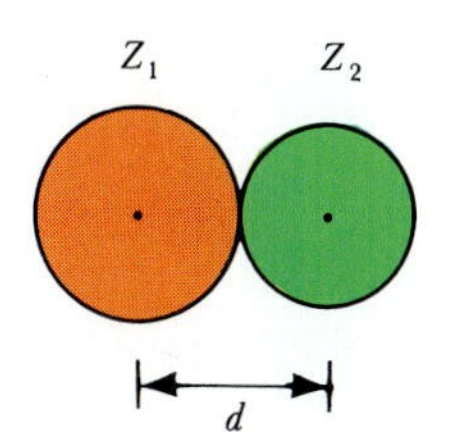

Figure 10-10 Two ions separated by a distance d_{eq}. The charges of the ions are Z_1 and Z_2. The energy of interaction of two ions is given by Coulomb's law [Equation (10-1)].

The minus sign means that the ions attract each other and that energy is released when the ions are brought together. Thus, the energy of the pair of ions at 276 pm is less than it is when the ions are very far from each other. The quantity -8.37×10^{-19} J is for the formation of one **ion-pair.** For the formation of 1 mol of sodium chloride ion-pairs, the energy released is

$$E = \left(\frac{-8.37 \times 10^{-19}\ \text{J}}{1\ \text{ion-pair}}\right)\left(\frac{6.02 \times 10^{23}\ \text{ion-pair}}{1\ \text{mol}}\right)$$
$$= -5.04 \times 10^{5}\ \text{J}\cdot\text{mol}^{-1}$$
$$= -5.04\ \text{kJ}\cdot\text{mol}^{-1}$$

We can express ion-pair formation as a chemical equation:

$$\text{step 3: } Na^+(g) + Cl^-(g) \rightarrow Na^+Cl^-(g) \qquad d_{eq} = 276\ \text{pm}$$

with

$$\Delta H_3^\circ = -504\ \text{kJ}\cdot\text{mol}^{-1}$$

If we add the chemical equation for step 3 to the sum of steps 1 and 2, then we obtain (Hess's law) for the equation

$$Na(g) + Cl(g) \rightarrow Na^+Cl^-(g) \qquad d_{eq} = 276\ \text{pm}$$

$$\Delta H_{rxn}^\circ = \Delta H_3^\circ + I_1 + EA_1 = -504\ \text{kJ}\cdot\text{mol}^{-1} + 148\ \text{kJ}\cdot\text{mol}^{-1}$$
$$= -356\ \text{kJ}\cdot\text{mol}^{-1}$$

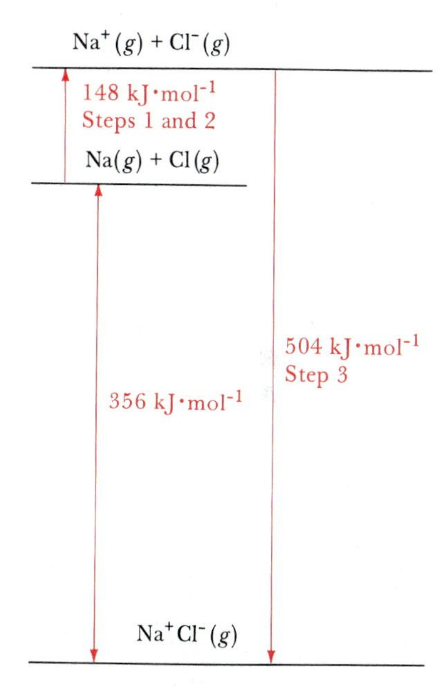

Figure 10-11 Steps used to calculate the energy evolved in the process $Na(g) + Cl(g) \rightarrow Na^+Cl^-(g)$. First the atoms are converted to ions (steps 1 and 2), and then the two ions are brought together to a distance equal to the sum of their ionic radii (step 3). This first step uses the ionization energy of sodium and the electron affinity of chlorine. The second step uses Coulomb's law to calculate the energy involved in bringing the two isolated ions together.

Our calculations show that 356 kJ of energy is released in the formation of 1 mol of sodium chloride ion-pairs from 1 mol of sodium atoms and 1 mol of chlorine atoms. The fact that energy is released in the process means that the energy of the ion-pair is lower than that of the two separated atoms. Because the ion-pair has a lower energy, it is stable with respect to the atoms. The overall process is illustrated in Figure 10-11.

EXAMPLE 10-7: Calculate the energy released in the reaction described by the equation

$$Cs(g) + Cl(g) \rightarrow Cs^+Cl^-(g)$$

given that the first ionization energy of cesium is 376 $\text{kJ}\cdot\text{mol}^{-1}$.

Solution: We must break this reaction down into three steps: (1) the ionization of Cs; (2) the addition of an electron to Cl; and (3) the bringing together of Cs^+ and Cl^- to a distance equal to their equilibrium ion-pair separation.

1. From the first ionization energy of cesium, we write

$$Cs(g) \rightarrow Cs^{+}(g) + e^{-} \qquad I_1 = +376\ kJ{\cdot}mol^{-1}$$

2. From Table 10-3 we see that the first electron affinity of Cl is −348 $kJ{\cdot}mol^{-1}$, so

$$Cl(g) + e^{-} \rightarrow Cl^{-}(g) \qquad EA_1 = -348\ kJ{\cdot}mol^{-1}$$

If we add the results of steps 1 and 2, then we get

$$Cs\ (g) + Cl(g) \rightarrow Cs^{+}(g) + Cl^{-}(g) \qquad \Delta H^{\circ}_{rxn} = +28\ kJ{\cdot}mol^{-1}$$

3. We now calculate the energy involved in bringing Cs^{+} and Cl^{-} to a distance equal to their equilibrium ion-pair separation. According to Table 10-2, the radius of Cs^{+} is 169 pm and that of Cl^{-} is 181 pm. Thus, their equilibrium ion-pair separation is 350 pm. We now use Equation (10-1):

$$\begin{aligned} E &= \frac{(2.31 \times 10^{-16}\ J{\cdot}pm)Z_1Z_2}{d_{eq}} \\ &= \frac{(2.31 \times 10^{-16}\ J{\cdot}pm)(+1)(-1)}{350\ pm} \\ &= -6.60 \times 10^{-19}\ J \end{aligned}$$

This is the energy released by the formation of one ion-pair. For 1 mol of ion pairs, we multiply this result by Avogadro's number:

$$\begin{aligned} E &= \left(\frac{-6.60 \times 10^{-19}\ J}{1\ \text{ion-pair}}\right)\left(\frac{6.02 \times 10^{23}\ \text{ion-pair}}{1\ mol}\right) \\ &= -397\ kJ{\cdot}mol^{-1} \end{aligned}$$

The minus sign indicates that energy is released in the process. We can express this result in the form of an equation:

$$Cs^{+}(g) + Cl^{-}(g) \rightarrow Cs^{+}Cl^{-}(g) \qquad \Delta H^{\circ}_{rxn} = -397\ kJ{\cdot}mol^{-1}$$
$$d_{eq} = 350\ pm$$

If we combine this result with the net result of steps 1 and 2, then for the equation

$$Cs(g) + Cl(g) \rightarrow Cs^{+}Cl^{-}(g)$$

we get

$$\Delta H^{\circ}_{rxn} = 28\ kJ{\cdot}mol^{-1} - 397\ kJ{\cdot}mol^{-1} = -369\ kJ{\cdot}mol^{-1}$$

Thus, 369 kJ of energy is evolved in the formation of 1 mol of CsCl ion-pairs from 1 mol of cesium atoms and 1 mol of chlorine atoms.

PRACTICE PROBLEM 10-7: Calculate the value of ΔH°_{rxn} for the reaction

$$Ca(g) + O(g) \rightarrow CaO(g)$$

The first and second ionization energies of Ca(g) are 590 $kJ{\cdot}mol^{-1}$ and 1140 $kJ{\cdot}mol^{-1}$, respectively.

Answer: $\Delta H^{\circ}_{rxn} = +44\ kJ{\cdot}mol^{-1}$

Purely ionic chemical bonds are the simplest type of chemical bonds. They are the result of an electrostatic attraction (Coulomb's law) between oppositely charged ions. If we know the ionic charges involved and their equilibrium ion-pair separation, then we can use Equation (10-1) to calculate the energy released when the ionic bond is formed.

The negative of this energy value is the energy that must be supplied to break the ionic bond when separating the ions.

The Coulomb's law calculations carried out in this section apply only to widely separated gaseous ion-pairs; we have yet to consider the energy released during crystal formation. For example, the energy released in the formation of crystalline sodium chloride is greater than for gaseous sodium chloride, because each ion in crystalline sodium chloride is surrounded by six ions of opposite charge, thus giving an additional stability. We discuss the energy released upon crystal formation next.

10-6. THE FORMATION OF IONIC SOLIDS FROM THE ELEMENTS IS AN EXOTHERMIC PROCESS

The reactions that we have considered so far in this chapter have been simplified in the sense that we have discussed only reactions between gaseous atoms to form gaseous ion-pairs. At room temperature chlorine exists as a diatomic molecule, and sodium and sodium chloride are solids. The direct reaction between the elements sodium and chlorine at 25°C and 1 atm is described by the equation

$$\mathrm{Na}(s) + \tfrac{1}{2}\mathrm{Cl}_2(g) \rightarrow \mathrm{NaCl}(s)$$

rather than by the equation

$$\mathrm{Na}(g) + \mathrm{Cl}(g) \rightarrow \mathrm{NaCl}(g)$$

We can break down the first of these two equations into five steps:

1. Vaporize 1.0 mol of sodium metal so that the sodium atoms are far apart and effectively isolated from one another. The energy required for this step is the energy of vaporization of sodium, which is 93 $\mathrm{kJ \cdot mol^{-1}}$ at room temperature. We can write this process as

$$\mathrm{Na}(s) \rightarrow \mathrm{Na}(g) \qquad \Delta H^\circ_{\mathrm{vap}} = +93\ \mathrm{kJ \cdot mol^{-1}}$$

2. Dissociate 0.50 mol of $\mathrm{Cl_2}$ (g) into 1.0 mol of chlorine atoms. The energy required for this process is 122 $\mathrm{kJ \cdot mol^{-1}}$, so we write

$$\tfrac{1}{2}\mathrm{Cl}_2(g) \rightarrow \mathrm{Cl}(g) \qquad \Delta H^\circ_{\mathrm{diss}} = +122\ \mathrm{kJ \cdot mol^{-1}}$$

3. Ionize the mole of Na(g). The energy required is the first ionization energy of sodium, which is 496 $\mathrm{kJ \cdot mol^{-1}}$. Therefore, we write

$$\mathrm{Na}(g) \rightarrow \mathrm{Na}^+(g) + \mathrm{e}^- \qquad I_1 = +496\ \mathrm{kJ \cdot mol^{-1}}$$

4. Add the mole of electrons generated in step 3 to a mole of chlorine atoms (Table 10-3)

$$\mathrm{Cl}(g) + \mathrm{e}^- \rightarrow \mathrm{Cl}^-(g) \qquad EA_1 = -348\ \mathrm{kJ \cdot mol^{-1}}$$

5. Bring the mole of isolated sodium ions and the mole of isolated chloride ions together to form one mole of the sodium chloride crystal (a portion of which is shown in Figure 10-12). Energy is released in this step. This energy, called the **lattice energy,** is known to be

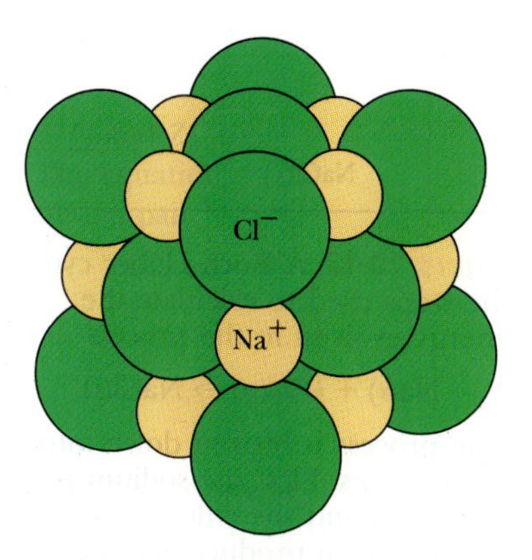

Figure 10-12 The crystalline structure of NaCl. Each $\mathrm{Na^+}$ ion is surrounded by six $\mathrm{Cl^-}$ ions, and each $\mathrm{Cl^-}$ ion is surrounded by six $\mathrm{Na^+}$ ions. This drawing shows the ions as spheres drawn to scale and illustrates the packing that occurs in the crystal.

10-4. Predict the products of the following reactions from a consideration of the Lewis electron-dot formulas of the reactants and the achievement of noble-gas electron configurations in the product ions:

(a) calcium and nitrogen (as N)

(b) aluminum and chlorine (as Cl)

(c) lithium and oxygen (as O)

ELECTRON CONFIGURATIONS OF IONS

10-5. Predict the ground-state electron configuration of

(a) Cr^{2+} (b) Cu^{2+} (c) Co^{3+} (d) Mn^{2+}

10-6. Predict the ground-state electron configuration of

(a) Ru^{2+} (b) W^{3+} (c) Pd^{2+} (d) Ag^{2+}

10-7. Which *d* transition-metal ions with a +2 charge have

(a) six *d* electrons (b) ten *d* electrons

(c) one *d* electron (d) five *d* electrons

Do you see a connection between the number of *d* electrons in the +2 ion and the position of the ion in its transition-metal series?

10-8. How many *d* electrons are there in

(a) Fe^{2+} (b) Zn^{2+} (c) V^{2+} (d) Ni^{2+}

Can you see a pattern between the number of *d* electrons and the position of these ions in the first transition-metal series?

10-9. Using only a periodic table, predict the ground-state 18-outer electron configuration and the charge of

(a) cadmium ion (b) indium(III) ion

(c) thallium(III) ion (d) zinc ion

10-10. Predict the ground-state 18-outer electron configuration and the charge of

(a) copper(I) ion (b) gallium ion
(c) mercury(II) ion (d) gold(I) ion

10-11. Determine which of the following salts are composed of isoelectronic cations and anions:

(a) LiF (b) NaF

(c) KBr (d) KCl

(e) BaI_2 (f) AlF_3

10-12. Determine which of the following salts are composed of isoelectronic cations and anions:

(a) NaCl (b) RbBr
(c) $SrCl_2$ (d) $SrBr_2$
(e) MgF_2 (f) KI

CHEMICAL FORMULAS OF IONIC COMPOUNDS

10-13. Write the chemical formula for

(a) yttrium sulfide (b) lanthanum bromide
(c) magnesium telluride (d) rubidium nitride
(e) aluminum selenide (f) calcium oxide

10-14. Write the chemical formula for

(a) thallium chloride (b) cadmium sulfide
(c) zinc nitride (d) copper(I) bromide
(e) gallium oxide (f) tin(II) fluoride

IONIC RADII

10-15. The following pairs of ions are isoelectronic. Predict which is the larger ion in each pair:

(a) K^+ and Cl^- (b) Ag^+ and Cd^{2+}
(c) Cu^+ and Zn^{2+} (d) F^- and O^{2-}

10-16. List the following ions in order of increasing radius:

Na^+ O^{2-} Mg^{2+} F^- Al^{3+}

IONIZATION ENERGIES AND ELECTRON AFFINITIES

10-17. List the following atoms in order of the ease with which they gain an electron to form an anion:

Br I H Cl

10-18. List the following atoms in order of the ease with which they lose electron(s) to form cations:

Ca K Na Al Li

10-19. Use the electron affinity data for Cl, Br, and I in Table 10-3, together with your knowledge of atomic periodicity trends to estimate the electron affinity of astatine.

10-20. Explain why the magnitude of the first electron affinity of a chlorine atom is greater than that of a sulfur atom.

10-21. Using Tables 9-1 and 10-3, calculate ΔH°_{rxn} for

(a) $Li(g) + Br(g) \rightarrow Li^+(g) + Br^-(g)$
(b) $I^-(g) + Cl(g) \rightarrow I(g) + Cl^-(g)$
(c) $Na(g) + H(g) \rightarrow Na^+(g) + H^-(g)$

10-22. Using Tables 9-1 and 10-3 calculate ΔH°_{rxn} for

(a) $2Na(g) + S(g) \rightarrow 2Na^+(g) + S^{2-}(g)$
(b) $Mg(g) + O(g) \rightarrow Mg^{2+}(g) + O^{2-}(g)$
(c) $Mg(g) + 2Br(g) \rightarrow Mg^{2+}(g) + 2Br^-(g)$

CALCULATIONS INVOLVING COULOMB'S LAW

10-23. Use Coulomb's law to calculate the energy of a zinc ion and an oxide ion that are just touching.

10-24. Use Coulomb's law to calculate the energy of a sodium ion and a fluoride ion that are just touching.

10-25. Calculate the energy released (in $kJ \cdot mol^{-1}$) in the reaction

$$K(g) + Br(g) \rightarrow K^+Br^-(g)$$

The first ionization energy of potassium is 419 $kJ \cdot mol^{-1}$.

10-26. Calculate the energy change in $kJ \cdot mol^{-1}$ for the reaction

$$Mg(g) + O(g) \rightarrow Mg^{2+}O^{2-}(g)$$

10-27. Calculate the energy change in $kJ \cdot mol^{-1}$ for the reaction

$$Na(g) + H(g) \rightarrow Na^+H^-(g)$$

10-28. Calculate ΔH°_{rxn} for the following equation:

$$Zn(g) + S(g) \rightarrow Zn^{2+}S^{2-}(g)$$

The ionization energy for the process

$$Zn(g) \rightarrow Zn^{2+}(g) + 2e^-$$

is 2,640 $kJ \cdot mol^{-1}$.

10-29. Construct a diagram like that shown in Figure 10-8 for $Li^+F^-(g)$ (see Tables 9-1, 10-2, and 10-3 for the necessary data).

10-30. Construct a diagram like that shown in Figure 10-8 for $Na^+H^-(g)$ (see Tables 9-1, 10-2, and 10-3 for the necessary data).

LATTICE ENERGIES

10-31. Calculate the energy released (in $kJ \cdot mol^{-1}$) in the reaction

$$Na(s) + \tfrac{1}{2}F_2(g) \rightarrow NaF(s)$$

The heat of vaporization of Na is 93 $kJ \cdot mol^{-1}$, the dissociation energy of one mole of F_2 is 155 $kJ \cdot mol^{-1}$, and the lattice energy of NaF is -919 $kJ \cdot mol^{-1}$.

10-32. Calculate the energy released (in $kJ \cdot mol^{-1}$) in the reaction

$$K(s) + \tfrac{1}{2}Br_2(l) \rightarrow KBr(s)$$

The heat of vaporization of potassium is 89 $kJ \cdot mol^{-1}$, the first ionization energy of potassium is 419 $kJ \cdot mol^{-1}$, the sum of the dissociation and vaporization energies for $Br_2(l)$ is 223 $kJ \cdot mol^{-1}$, and the lattice energy of KBr is -688 $kJ \cdot mol^{-1}$.

10-33. Calculate the energy released (in $kJ \cdot mol^{-1}$) when $NaI(s)$ is formed in the reaction

$$Na(s) + \tfrac{1}{2}I_2(s) \rightarrow NaI(s)$$

The energy of vaporization of $Na(s)$ is 93 $kJ \cdot mol^{-1}$. The sum of the heats of dissociation and vaporization of $I_2(s)$ is 214 $kJ \cdot mol^{-1}$, and the lattice energy of NaI is -704 $kJ \cdot mol^{-1}$.

10-34. Calculate the energy released (in $kJ \cdot mol^{-1}$) when $LiH(s)$ is formed in the reaction

$$Li(s) + \tfrac{1}{2}H_2(g) \rightarrow LiH(s)$$

The heat of vaporization of lithium is 161 $kJ \cdot mol^{-1}$, the dissociation energy of H_2 is 436 $kJ \cdot mol^{-1}$, and the lattice energy of LiH is -917 $kJ \cdot mol^{-1}$.

10-35. Calculate the energy released in the reaction

$$Ca(s) + Cl_2(g) \rightarrow CaCl_2(s)$$

The heat of vaporization of $Ca(s)$ is 193 $kJ \cdot mol^{-1}$, the dissociation energy of Cl_2 is 244 $kJ \cdot mol^{-1}$, the lattice energy of $CaCl_2(s)$ is -2266 $kJ \cdot mol^{-1}$, the first ionization energy of $Ca(g)$ is 590 $kJ \cdot mol^{-1}$, and the second ionization energy of $Ca(g)$ is 1140 $kJ \cdot mol^{-1}$.

10-36. Given the following data

$K(s) + \tfrac{1}{2}Br_2(l) \rightarrow KBr(s)$ $\quad \Delta H^\circ_{rxn} = -392\ kJ \cdot mol^{-1}$

$K(g) + Br(g) \rightarrow KBr(g)$ $\quad \Delta H^\circ_{rxn} = -329\ kJ \cdot mol^{-1}$

calculate the heat of vaporization for $KBr(s)$ (see Problem 10-32 for additional data).

ADDITIONAL PROBLEMS

10-37. List three ions that are isoelectronic with F^-.

10-38. Predict the charge on (a) a lutetium ion; (b) a lawrencium ion.

10-39. The ionic radius of K^+ is 133 pm, while the ionic radius of Cu^+ is 96 pm. Explain why the radius of Cu^+ is smaller than that of K^+.

10-40. Calculate the lattice energy of $LiCl(s)$ given the following data:

chlorine-chlorine bond energy: 244 $kJ \cdot mol^{-1}$

first ionization energy of $Li(g)$: 519 $kJ \cdot mol^{-1}$

the heat of vaporization of $Li(s)$: 161 $kJ \cdot mol^{-1}$

the electron affinity of Cl: -348 $kJ \cdot mol^{-1}$

heat of formation of $LiCl(s)$: -408 $kJ \cdot mol^{-1}$.

In the last chapter, we discussed how metals react with nonmetals to form ionic compounds. Outer-shell electrons are transferred completely from one atom to another, a process resulting in electrostatic attraction—the ionic bond. In 1916, the American chemist G. N. Lewis postulated another kind of chemical bond in which two atoms share a pair of electrons—the covalent bond. Lewis published his results nearly a decade before the birth of the quantum theory, which was to give his idea of electron pairs, or covalent bonding, a firm theoretical basis.

Covalent bonds make possible an enormous class of compounds. The compounds are poor conductors of an electric current, and many are gases or liquids at room temperature. In this chapter, we investigate covalent bonding by studying the method of writing molecular formulas that Lewis introduced. We also learn how to characterize bonds that are intermediate between covalent and ionic bonds. Although a full understanding of covalent bonding depends on the quantum theory, which we consider in Chapter 13, Lewis formulas remain one of the most useful concepts in chemistry.

11-1. A COVALENT BOND CAN BE DESCRIBED AS A PAIR OF ELECTRONS SHARED BY TWO ATOMS

Consider the chlorine molecule, Cl_2. The Lewis electron-dot formula for a chlorine atom is

$$:\ddot{\underset{\cdot\cdot}{\mathrm{Cl}}}\cdot$$

Figure 11-1 The regular arrangement of the chlorine molecules in crystalline Cl_2. This pattern is repeated throughout the crystal. The chlorine molecules are neutral, so they do not attract one another as strongly as neighboring ions in an ionic lattice. Consequently, molecular crystals like Cl_2 usually have lower melting points than ionic crystals. The melting point of chlorine is −101°C. For comparison, the melting point of NaCl(*s*) is 800°C.

Recall that a Lewis electron-dot formula for an atom shows only the **valence electrons** (the outer-shell electrons) and that the number of valence electrons in a main-group element is equal to the group number of that element. The chlorine atom is one electron short of having eight electrons in its outer shell and achieving an argonlike electron configuration. A chlorine atom in Cl_2 could get this electron from the other chlorine atom, but certainly that chlorine atom does not wish to give up an electron. In a sense there is a stalemate with respect to electron transfer. Both atoms have the same driving force to gain an electron. From a more quantitative point of view, although a chlorine atom has a large electron affinity ($-348\ \mathrm{kJ\cdot mol^{-1}}$), it has a much higher ionization energy ($1260\ \mathrm{kJ\cdot mol^{-1}}$) and does not lose an electron very easily. Ionic bonds result in binary compounds only when one atomic reactant is a metal with a relatively low ionization energy and the other is a nonmetal with a relatively high electron affinity.

Although we have ruled out the formation of an ionic bond in Cl_2, there is a way for the two chlorine atoms to achieve an argonlike electron configuration *simultaneously*. If the two chlorine atoms *share* a pair of electrons between them, then the resulting distribution of valence electrons can be pictured as

$$:\ddot{\underset{\cdot\cdot}{\mathrm{Cl}}}\cdot + \cdot\ddot{\underset{\cdot\cdot}{\mathrm{Cl}}}: \rightarrow :\ddot{\underset{\cdot\cdot}{\mathrm{Cl}}}:\ddot{\underset{\cdot\cdot}{\mathrm{Cl}}}:$$

Notice that *each* chlorine atom has eight electrons in its outer shell:

Thus by sharing a pair of electrons, each chlorine atom is able to achieve the stable argonlike outer electron configuration of eight electrons. According to Lewis's picture, the shared electron pair is responsible for holding the two chlorine atoms together as a chlorine molecule. The bond formed between two atoms by a shared electron pair is called a **covalent bond.**

The electron-dot formula depicted for Cl_2 is called a **Lewis formula.** It is conventional to indicate the electron-pair bond as a line joining the two atoms and the other electrons as pairs of dots surrounding the atoms:

$$:\ddot{\underset{\cdot\cdot}{Cl}}—\ddot{\underset{\cdot\cdot}{Cl}}:$$

In general, the halogens have the Lewis formula $:\ddot{\underset{\cdot\cdot}{X}}—\ddot{\underset{\cdot\cdot}{X}}:$, where X is F, Cl, Br, or I. The pairs of electrons that are not shared between the chlorine atoms are called **lone electron pairs,** or simply **lone pairs.** A Lewis formula correctly depicts a covalent bond as a pair of electrons shared between two atoms.

When Cl_2 is solidified (its freezing point is −101°C), it forms a **molecular crystal** (Figure 11-1). In contrast to an ionic crystal, the constituent particles of a molecular crystal are molecules, in this case chlorine molecules. The low melting point of molecular chlorine indicates that the attraction between the molecules is weak relative to the attraction between ions in an ionic crystal. The chlorine molecules are neutral; thus, there is no net electrostatic attraction between them in the crystal. The interactions of (neutral) molecules are discussed in Section 14-4.

Figure 11-2, which shows molecular models of the halogen molecules, can be used to help define **bond length.** The drawings suggest that the nuclei of the two halogen atoms are held a fixed distance apart. We refer to this internuclear distance as bond length. In a real molecule the atoms vibrate about these positions, but they vibrate about a well-defined average bond length. Table 11-1 shows the average bond lengths of the halogens. Note that the bond lengths in the diatomic halogen molecules increase as atomic number increases.

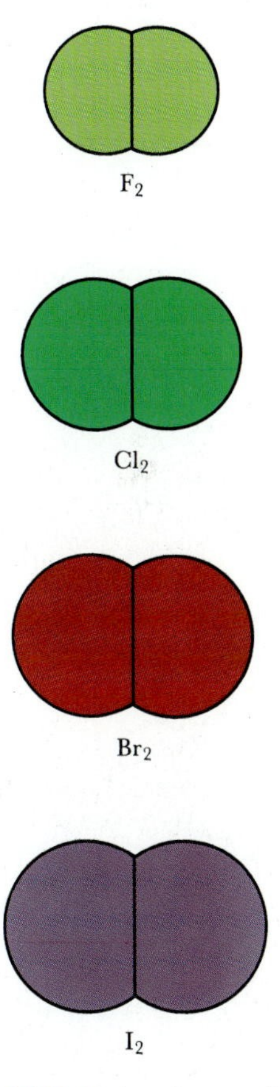

Figure 11-2 Molecular models of the halogen molecules, drawn to scale to indicate the relative sizes of the atoms. Note that the halogens become larger as we go down the group in the periodic table.

Table 11-1 The bond lengths and bond energies of the halogen molecules

Molecule	Bond length/pm	Bond energy/$kJ\cdot mol^{-1}$
F_2	128	155
Cl_2	198	243
Br_2	228	192
I_2	266	152

11-2. WE ALWAYS TRY TO SATISFY THE OCTET RULE WHEN WRITING LEWIS FORMULAS

When writing Lewis formulas, we always try to satisfy the **octet rule:** each element forms covalent bonds such that eight electrons occupy its outer shell. The octet rule has its origin in the special stability of the noble-gas electron configuration. Thus, for example, carbon, nitrogen, oxygen, and fluorine achieve a neonlike electron configuration when they are surrounded by eight valence electrons. We will show that, although there are exceptions to the octet rule, it is still useful because of the large number of compounds that do obey it. We do not violate the octet rule in writing Lewis formulas unless there is a good reason to do so. The following example illustrates how to apply the octet rule.

When fluorine gas is bubbled through an aqueous solution of sodium hydroxide, the pale yellow gas oxygen difluoride, OF_2, is formed. We can deduce the Lewis formula for OF_2 by first writing the Lewis electron-dot formulas for the atoms:

$$:\ddot{\underset{\cdot\cdot}{F}}\cdot \quad \cdot\ddot{\underset{\cdot\cdot}{O}}\cdot \quad \cdot\ddot{\underset{\cdot\cdot}{F}}:$$

We wish to join these three atoms such that each has eight electrons in its outer shell. Pictorially, we wish to join these atoms such that each one can be written with eight valence electrons surrounding the nucleus. By bringing the fluorine atoms in toward the oxygen atom, we see that the electron-dot formula $:\ddot{\underset{\cdot\cdot}{F}}:\ddot{\underset{\cdot\cdot}{O}}:\ddot{\underset{\cdot\cdot}{F}}:$ allows each of the three atoms to be surrounded simultaneously by eight electrons. Thus, we conclude that a satisfactory Lewis formula for OF_2 is

$$:\ddot{\underset{\cdot\cdot}{F}}—\ddot{\underset{\cdot\cdot}{O}}—\ddot{\underset{\cdot\cdot}{F}}:$$

As a final check of this formula, note that there are 20 valence electrons indicated in the Lewis formula for the molecule and that there is a total of 20 valence electrons $[(2 \times 7) + 6]$ in the Lewis electron-dot formulas for the individual atoms.

Note that the Lewis formula for oxygen difluoride depicts the two fluorine atoms attached to a central oxygen atom. One great utility of Lewis formulas is that they suggest which atoms are actually bonded to each other in a molecule.

We can write Lewis formulas in a systematic manner by using the following four-step procedure:

1. Arrange the symbols of the atoms that are bonded together in the molecule next to one another. For oxygen difluoride we would write

   $$\text{F} \quad \text{O} \quad \text{F}$$

   Although this may seem like a difficult step to you at this stage, you will become more confident with experience. Often, if there is only one atom of a particular element, it is a good first try to assume that this atom is the central atom (as in OF_2) and that the other atoms are bonded to it. Sometimes, however, the correct arrangement can only be found by trial and error.

2. Compute the total number of valence electrons in the molecule by adding the numbers of valence electrons for all the atoms in the

molecule. If the species is an ion rather than a molecule, then you must take the charge of the ion into account by adding electrons if it is a negative ion or subtracting electrons if it is a positive ion. For example,

Ion	Total number of valence electrons
NH_2^-	N 2H
	$(1 \times 5) + (2 \times 1) + 1 = 8$
NH_4^+	N 4H
	$(1 \times 5) + (4 \times 1) - 1 = 8$

3. Represent a two-electron covalent bond by placing a line between the atoms that are assumed to be bonded to each other. For oxygen difluoride, we have

F—O—F

4. Now arrange the remaining valence electrons as lone pairs about each atom so that the octet rule is satisfied for each one:

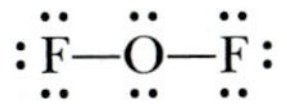

$$:\ddot{\underset{\cdot\cdot}{F}}—\ddot{\underset{\cdot\cdot}{O}}—\ddot{\underset{\cdot\cdot}{F}}:$$

The use of this procedure is illustrated in the following Examples.

EXAMPLE 11-1: Write the Lewis formula for carbon tetrachloride, CCl_4.

Solution: Because carbon is the unique atom in this molecule, we will assume that it is central and that each chlorine atom is attached to it:

```
     Cl
Cl   C   Cl
     Cl
```

The total number of valence electrons is $(1 \times 4) + (4 \times 7) = 32$. We use eight of these electrons to form carbon-chlorine bonds, an arrangement that satisfies the octet rule about the carbon atom. We place the remaining 24 valence electrons as lone pairs on the chlorine atoms to satisfy the octet rule on the chlorine atoms. The Lewis formula is

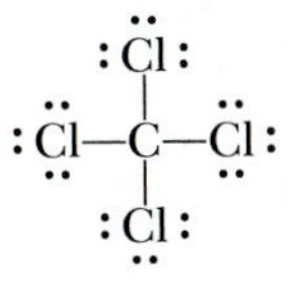

```
       ..
      :Cl:
  ..   |   ..
 :Cl—C—Cl:
  ..   |   ..
      :Cl:
       ..
```

PRACTICE PROBLEM 11-1: Silicon tetrachloride is a colorless, fuming liquid that forms a dense and persistent cloud when exposed to moist air. Because of this property, silicon tetrachloride is used to produce smoke screens. The equation for the reaction is

$$SiCl_4(l) + 2H_2O(l) \rightarrow SiO_2(s) + 4HCl(g)$$

The white cloud consists of highly dispersed particles of silicon dioxide. Write the Lewis formula of silicon tetrachloride, which is a covalent compound.

Answer:

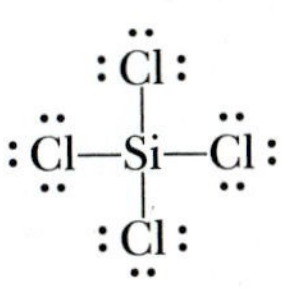

```
       ..
      :Cl:
  ..   |   ..
 :Cl—Si—Cl:
  ..   |   ..
      :Cl:
       ..
```

EXAMPLE 11-2: Write the Lewis formula for nitrogen trifluoride, NF_3.

Solution: Because nitrogen is the unique atom in this molecule, we will assume that it is the central atom and that each fluorine atom is attached to it:

F N F

F

The total number of valence electrons is $(1 \times 5) + (3 \times 7) = 26$. We use six of these valence electrons to form nitrogen-fluorine bonds. We now place valence electrons as lone pairs on each fluorine atom (accounting for 18 of the 20 remaining valence electrons) and the remaining two valence electrons as a lone pair on the nitrogen atom. The completed Lewis formula is

$$:\ddot{\underset{..}{F}}-\ddot{\underset{|}{N}}-\ddot{\underset{..}{F}}:$$
$$:\underset{..}{F}:$$

Notice that the octet rule is satisfied for all four atoms.

PRACTICE PROBLEM 11-2: Phosphorus pentachloride is a solid at room temperature and consists of ion-pairs of the type $PCl_4^+PCl_6^-$. Write the Lewis formula of the PCl_4^+ ion.

Answer:

$$\left[\begin{array}{c} :\ddot{Cl}: \\ | \\ :\ddot{\underset{..}{Cl}}-P-\ddot{\underset{..}{Cl}}: \\ | \\ :\underset{..}{Cl}: \end{array}\right]^{\oplus}$$

11-3. HYDROGEN ATOMS ARE ALMOST ALWAYS TERMINAL ATOMS IN LEWIS FORMULAS

We have said that there are exceptions to the octet rule. One important exception is the hydrogen atom. The noble gas closest to hydrogen in the periodic table is helium. We might expect, then, that hydrogen needs only two electrons in order to attain a noble-gas electron configuration. For example, let's consider H_2 itself. The electron-dot formula for hydrogen is $H\cdot$. Each hydrogen atom can be surrounded by two electrons if the two hydrogen atoms share electrons:

$$H\cdot + \cdot H \rightarrow H:H \qquad \text{or} \qquad H-H$$

In this way each hydrogen atom achieves a heliumlike electron configuration.

The Lewis formulas for the hydrogen halides are obtained directly from the electron-dot formulas of the individual atoms. If we let X be F, Cl, Br, or I, we can write

$$H\cdot + \cdot\ddot{\underset{..}{X}}: \rightarrow H-\ddot{\underset{..}{X}}:$$

The hydrogen atom has two electrons surrounding it, and the halogen atom has eight. Molecular models of the hydrogen halides are shown in Figure 11-3. The bond length and bond energies of the hydrogen halides are given in Table 11-2.

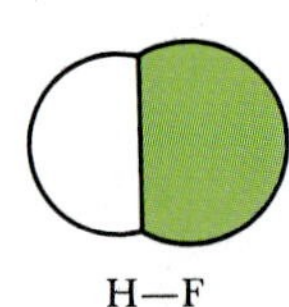

H—F

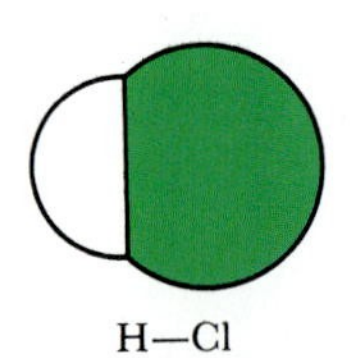

H—Cl

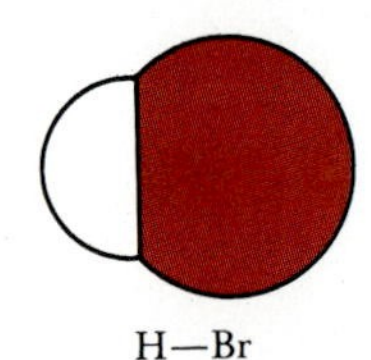

H—Br

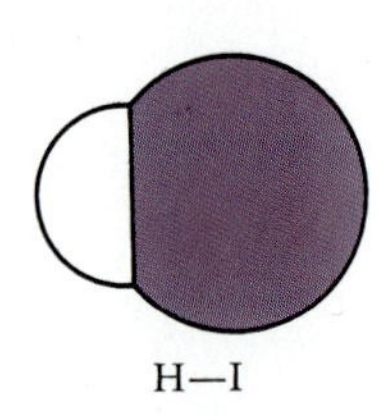

H—I

Figure 11-3 Molecular models of the hydrogen halides.

Table 11-2 The bond lengths and bond energies of the hydrogen halides

Compound	Bond length/pm	Bond energy/$kJ \cdot mol^{-1}$
HF	92	565
HCl	127	431
HBr	141	368
HI	161	300

EXAMPLE 11-3: Write the Lewis formula for the ammonium ion, NH_4^+.

Solution: We first arrange the atoms as

$$\begin{array}{ccc} & H & \\ H & N & H \\ & H & \end{array}$$

There is a total of $(4 \times 1) + 5 - 1 = 8$ valence electrons. We use all eight of these to form the hydrogen-nitrogen bonds:

$$\begin{array}{ccccc} & & H & & \\ & & | & & \\ H & - & N & - & H \\ & & | & & \\ & & H & & \end{array}$$

We indicate that this species has a charge of +1 by writing

$$\left[\begin{array}{ccccc} & & H & & \\ & & | & & \\ H & - & N & - & H \\ & & | & & \\ & & H & & \end{array}\right]^{\oplus}$$

Note that the nitrogen atom has eight electrons around it and each hydrogen atom has two.

PRACTICE PROBLEM 11-3: Phosphine, $PH_3(g)$, is a colorless, toxic gas with a garliclike odor. It is used as a doping agent for semiconductor production. Write the Lewis formula of phosphine.

Answer:

$$\begin{array}{ccccc} & & \cdot\cdot & & \\ H & - & P & - & H \\ & & | & & \\ & & H & & \end{array}$$

Because a hydrogen atom completes its valence shell with a total of two electrons, hydrogen atoms almost always form a covalent bond to only one other atom and so almost always are terminal atoms in Lewis formulas.

EXAMPLE 11-4: Write the Lewis formula for chloroform, $HCCl_3$.

Solution: The chloroform molecule has three different types of atoms; thus, the first step is to decide how to arrange them in the Lewis formula. Hydrogen is almost always a terminal atom in a Lewis formula. Of the re-

and in formula II are

$$\text{formal charge on N} = 5 - 2 - \tfrac{1}{2}(6) = 0$$

$$\text{formal charge on O} = 6 - 4 - \tfrac{1}{2}(4) = 0$$

If we write the Lewis formulas with the calculated formal charges, then we have

```
    H
    |⊕ ..⊖            ..  ..
H—N—O:            H—N—O—H
    |  ..              |  ..
    H                  H
    I                  II
```

Because II has zero formal charges everywhere, we predict (correctly) that II represents the actual structure of hydroxylamine. The chemical formula of hydroxylamine is usually written as NH_2OH to reflect its Lewis formula.

PRACTICE PROBLEM 11-7: Use formal charges to choose which of the two Lewis formulas best represents the structure of hydrogen peroxide, H_2O_2.

```
H
 \ ..  ..              ..  ..
  O—O:            H—O—O—H
 /     ..              ..  ..
H
    I                   II
```

Answer: Formula II has a formal charge of 0 on each atom and is the preferred Lewis formula.

11-5. IT IS NOT ALWAYS POSSIBLE TO SATISFY THE OCTET RULE BY USING ONLY SINGLE BONDS

In all the molecules that we have discussed so far, exactly the correct number of valence electrons remained after step 3 (page 379) to be used in step 4. In this section we consider the case in which there are not enough electrons to satisfy the octet rule for each atom by using only single bonds. The molecule ethylene, C_2H_4, serves as a good example. Using the same reasoning that we did for hydrazine, we arrange the atoms as

```
H  C  C  H
   H  H
```

There is a total of $(4 \times 1) + (2 \times 4) = 12$ valence electrons. We use 10 of them to join the atoms:

```
H—C—C—H
  |  |
  H  H
```

If we use only single bonds, it is not possible to satisfy the octet rule for each carbon atom with only the two remaining valence electrons. We are short two electrons. When this situation occurs, we add one more bond for each two electrons that we are short. In the case of ethylene, we add another bond between the carbon atoms to get

$$\begin{array}{ccc} H & & H \\ & C{=}C & \\ H & & H \end{array}$$

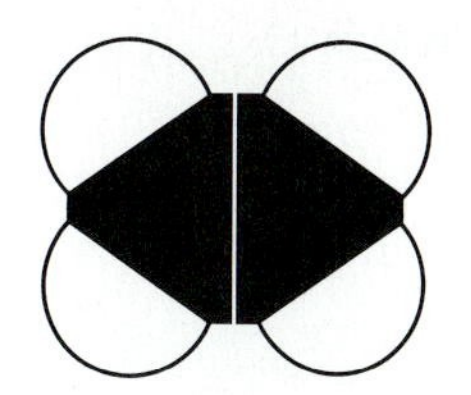

Molecular model of ethylene, C_2H_4

Notice that now the octet rule is satisfied for each carbon atom. When two atoms are joined by two pairs of electrons, we say that there is a **double bond** between the atoms. A double bond between two atoms is shorter and stronger than a single bond between the same two atoms. The carbon-carbon bond in C_2H_4 is indeed shorter and stronger than, for example, the carbon-carbon single bond in ethane, C_2H_6. Table 11-3 gives typical bond lengths and bond energies for various single and double bonds.

It is also possible to have a **triple bond,** as we now show for N_2. There are 10 valence electrons in N_2. If we add one bond and then try to satisfy the octet rule for each nitrogen atom, we find that we are four electrons short. For example,

$$\ddot{\underset{..}{N}}-\ddot{\underset{..}{N}} \quad \text{(violates the octet rule)}$$

Thus, we add two more bonds to obtain

$$N{\equiv}N$$

The four remaining valence electrons are now added according to step 4 to obtain

$$:N{\equiv}N:$$

Table 11-3 Average bond lengths and bond energies of single, double, and triple bonds

Bond	Average bond length/pm	Average bond energy/ $kJ\cdot mol^{-1}$
C—O	143	350
C=O	120	730
C—C	154	350
C=C	134	615
C≡C	120	810
N—N	145	160
N=N	125	420
N≡N	110	950

EXAMPLE 11-8: Write a Lewis formula for CO_2.

Solution: We arrange the atoms with the carbon atom in the center:

$$O \quad C \quad O$$

There is a total of $(1 \times 4) + (2 \times 6) = 16$ valence electrons. If we add one bond between each oxygen atom and the carbon atom and try to satisfy the octet rule for each atom, then we find that we are four electrons short. For example,

$$:\ddot{O}-\ddot{\underset{..}{C}}-\ddot{O}: \quad \text{(violates the octet rule about the oxygen atoms)}$$

Thus, we go back to step 3 and add two more bonds:

$$O{=}C{=}O$$

Now we use step 4 and arrange the remaining eight valence electrons as lone pairs to satisfy the octet rule for each atom:

$$\dot{\underset{.}{}}O{=}C{=}O\dot{\underset{.}{}}$$

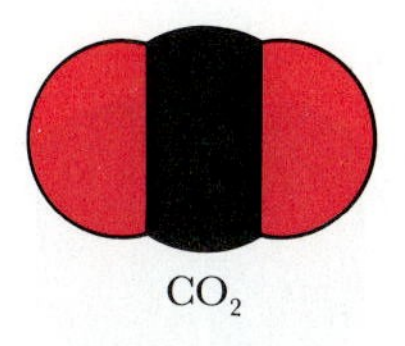

Molecular model of carbon dioxide, CO_2

The Lewis formula of CO_2 shows two carbon-oxygen double bonds. Incidentally, we could also have made one single bond and one triple bond instead of two double bonds. This would, however, produce a Lewis formula with formal charge separation $^{\oplus}:O{\equiv}C-\ddot{\underset{..}{O}}:^{\ominus}$. The Lewis formula with two double bonds gives no charge separation; therefore, it is the preferred one.

PRACTICE PROBLEM 11-8: (a) Formaldehyde, H_2CO, is a gas with a pungent, characteristic odor. An aqueous solution of formaldehyde is called formalin and is used to preserve biological specimens. Formaldehyde is used extensively in the production of certain plastics (bakelite, melamine). Write the Lewis formula of formaldehyde.

Answer:

```
H
 \
  C=O:
 /
H
```

(b) If one of the hydrogen atoms in formaldehyde is replaced by a methyl group (CH_3—), the compound is called acetaldehyde, a colorless liquid with a pungent, fruity odor. Write the Lewis formula of acetaldehyde.

Answer:

```
   H  H
   |  |
H—C—C=O:
   |
   H
```

EXAMPLE 11-9: Write the Lewis formula for the gas hydrogen cyanide, HCN.

Solution: Either the carbon atom or the nitrogen atom might be the central atom in this case. When in doubt, arrange the atoms as the formula is written:

H C N

Use 4 of the 10 valence electrons to write

H—C—N

We are four electrons short of satisfying the octet rule for both the carbon atom and the nitrogen atom, so we add two more bonds. The hydrogen atom already has two electrons around it, so in this case we must form a triple bond between the carbon atom and the nitrogen atom:

H—C≡N

The remaining two valence electrons are placed on the nitrogen atom as a lone pair so that both the carbon atom and the nitrogen atom satisfy the octet rule. The Lewis formula is

H—C≡N:

PRACTICE PROBLEM 11-9: Acetylene, C_2H_2, is a colorless gas that burns in oxygen to give a relatively high flame temperature. Acetylene is the fuel in an oxyacetylene torch, but it finds even greater use as a raw material in the plastics industry. Write the Lewis formula of acetylene.

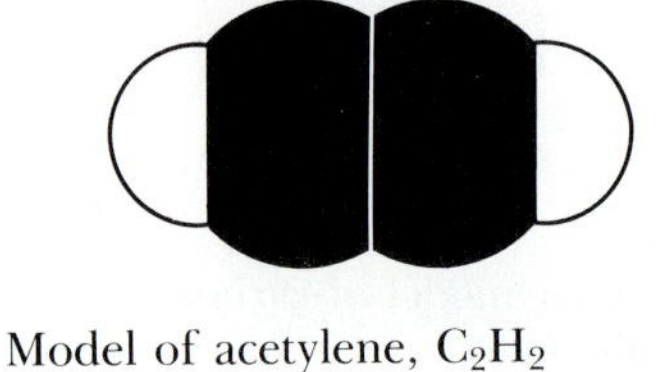

Model of acetylene, C_2H_2

Answer:

H—C≡C—H

11-6. A RESONANCE HYBRID IS A SUPERPOSITION OF LEWIS FORMULAS

For many molecules and ions, we can write two or more equally satisfactory Lewis formulas. For example, let's consider the nitrite ion, NO_2^-. One Lewis formula for NO_2^- is

$$:\ddot{O}=\ddot{N}-\ddot{\underset{\cdot\cdot}{O}}:^{\ominus}$$

One of the oxygen atoms, the right-hand one as written, has a formal charge of -1, having three lone pairs and one bond. Another equally acceptable Lewis formula for NO_2^- is

$$^{\ominus}:\ddot{\underset{\cdot\cdot}{O}}-\ddot{N}=\ddot{O}:$$

In this case the negative formal charge is on the other oxygen atom. Both of these Lewis formulas satisfy the octet rule. When it is possible to write two or more satisfactory Lewis formulas *without altering the positions of the nuclei,* the actual formula is viewed as an average or as a superposition of the individual formulas. Each of the individual Lewis formulas is said to be a **resonance form,** and the use of multiple Lewis formulas is called **resonance.** We indicate resonance forms by means of a two-headed arrow, as in

$$:\ddot{O}=\ddot{N}-\ddot{\underset{\cdot\cdot}{O}}:^{\ominus} \longleftrightarrow {}^{\ominus}:\ddot{\underset{\cdot\cdot}{O}}-\ddot{N}=\ddot{O}:$$

Neither of the individual Lewis formulas taken separately accurately reflects the actual bonding. Two Lewis formulas taken together are necessary to describe the bonding in NO_2^-.

There is no generally accepted way to represent resonance pictorially, but one way is to write the formula of NO_2^- as

$$\left[O\overset{\cdot\cdot}{\,{-}\!{-}\,N\,{-}\!{-}\,}O\right]^{\ominus}$$

where the two dashed lines taken together represent a pair of bonding electrons which spread over the two bonds. Such a superimposed formula is called a **resonance hybrid,** because it is a hybrid of the various resonance forms. Each of the nitrogen-oxygen bonds in NO_2^- can be thought of as an *average* of a single bond and a double bond. The superimposed Lewis formulas for NO_2^- suggest that the two nitrogen-oxygen bonds in NO_2^- are equivalent; this idea is in accord with experimental observation: the two bonds do have exactly the same length, 113 pm. Notice that using only one Lewis formula suggests that the two nitrogen-oxygen bonds are not equivalent, one being a single bond and the other being a double bond.

The resonance hybrid for the nitrite ion suggests that the -1 charge of the ion is shared equally by the two oxygen atoms, rather than being completely on either one of two oxygen atoms as suggested by either of the two resonance forms. In such as case, we say that the charge is **delocalized.** Resonance hybrids with delocalized charges have lower energies than their (hypothetical) individual resonance forms. This difference in energy is called **resonance energy.**

Another example of the need to use resonance forms to obtain a satisfactory picture of the distribution of electronic charge occurs in the nitrate ion, NO_3^-. Three equally satisfactory Lewis formulas for NO_3^- are

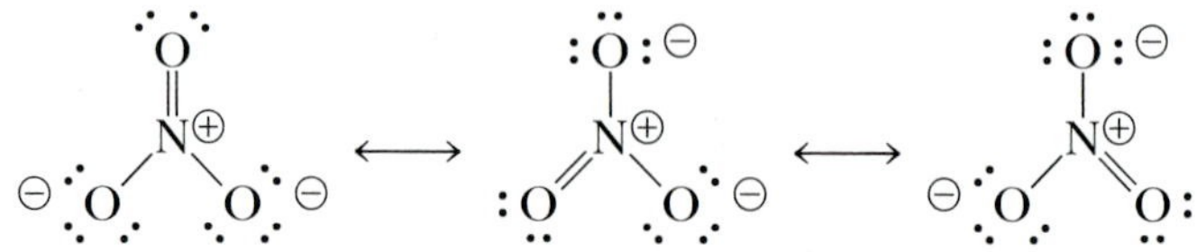

Because each of these formulas is equally satisfactory, the actual structure is viewed as a superposition or an average of the three formulas and can be represented pictorially as the resonance hybrid.

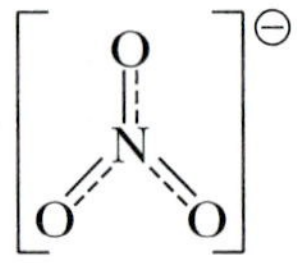

where the three dashed lines taken together represent a pair of bonding electrons which spread over the three bonds. In this case each of the nitrogen-oxygen bonds is an average of a double bond and two single bonds. As the superimposed representation suggests, the three nitrogen-oxygen bonds are equivalent, which is in agreement with experimental observations (each nitrogen-oxygen bond is 122 pm in length). Furthermore, there are no known chemical reactions that can be used to distinguish one oxygen atom from another in a nitrate ion, another observation suggesting that all three oxygen atoms are bonded equivalently to the nitrogen atom.

The need for resonance forms arises from the fact that Lewis formulas involve electron-pair bonds. If the species involves a bond intermediate between a single and a double bond, then we need to write two or more Lewis formulas to describe the bonding in the molecule. Resonance is in no sense a real phenomenon. It is a device that enables us to give a more realistic picture of the electron distribution in a species by the use of Lewis formulas.

The following Example is particularly important because it involves several of the concepts that we have discussed in this chapter.

EXAMPLE 11-10: Write Lewis formulas for the two resonance forms of sulfur dioxide, SO_2. Indicate formal charges and discuss the bonding in this molecule.

Solution: We arrange the atoms as

S

O O

The molecule has 18 valence electrons. The two resonance forms are

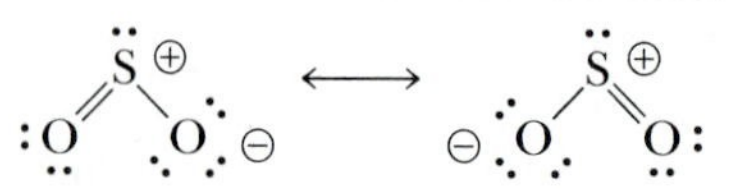

The formal charges indicated are calculated by using Equation (11-1):

$$\text{formal charge on S in } SO_2 = 6 - 2 - \tfrac{1}{2}(6) = +1$$

$$\text{formal charge on singly bonded O in } SO_2 = 6 - 6 - \tfrac{1}{2}(2) = -1$$

$$\text{formal charge on doubly bonded O in } SO_2 = 6 - 4 - \tfrac{1}{2}(4) = 0$$

These two Lewis formulas constitute equivalent resonance forms, so the actual formula is an average of the two, which can be represented by the resonance hybrid

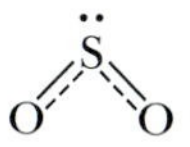

This formula suggests that the two sulfur-oxygen bonds in SO_2 are equivalent, or that the two sulfur-oxygen bond lengths in SO_2 are equal. This prediction is in agreement with experimental results.

PRACTICE PROBLEM 11-10: Sodium carbonate is an ionic solid that is used in the manufacture of glass. Write Lewis formulas for the resonance forms of the carbonate ion. Indicate formal charges and discuss the bonding in this ion.

Answer:

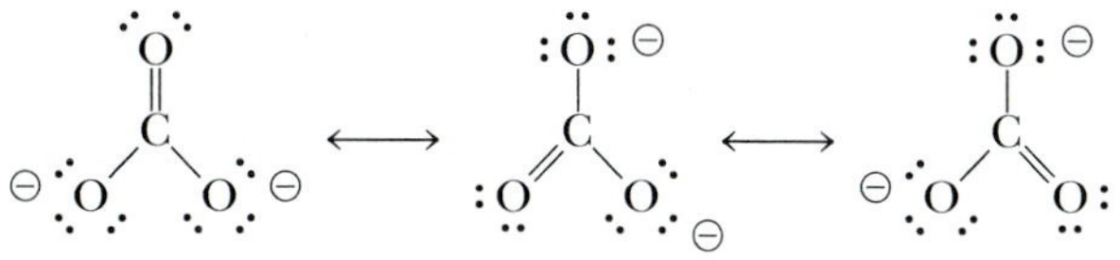

The three carbon-oxygen bonds are equivalent.

An important example of resonance and its consequences is provided by benzene, C_6H_6, a clear, colorless, highly flammable liquid with a characteristic odor. Benzene is obtained from petroleum and coal tar and has many chemical uses. The benzene molecule has two principal resonance forms:

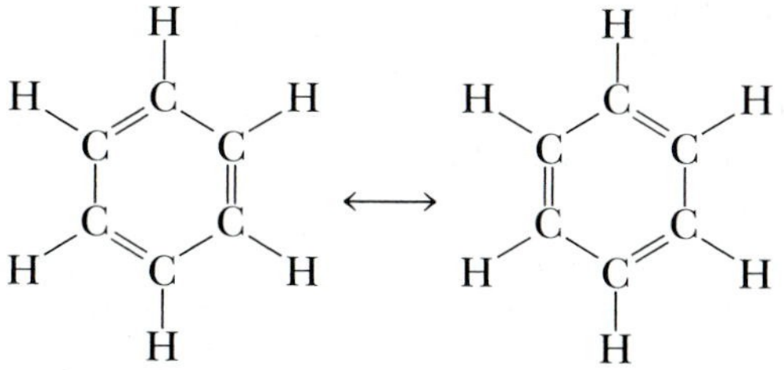

We depict the superposition or average of these two Lewis formulas by the resonance hybrid

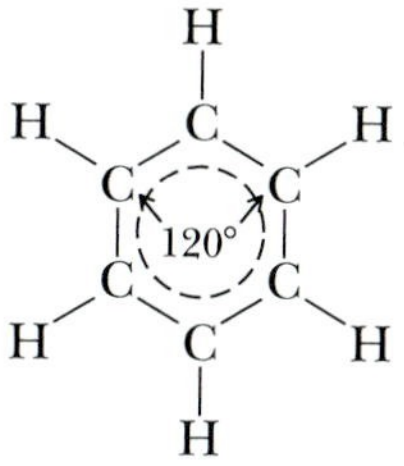

This formula predicts that all the carbon-carbon bonds in benzene are equivalent, as observed experimentally, and that each is the average of a single bond and a double bond. The carbon-carbon bond distance in

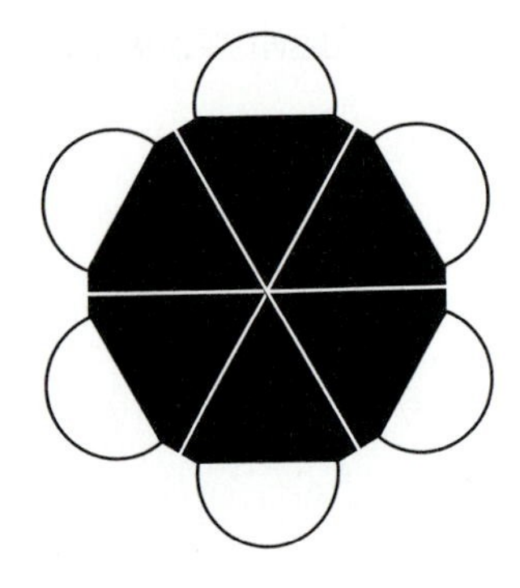

Molecular model of benzene, C_6H_6

benzene is 140 pm, which is intermediate between the usual carbon-carbon single-bond (154 pm) and double-bond (134 pm) distances. Benzene is a **planar molecule** (all the atoms lie in the same plane), with the ring of carbon atoms constituting a perfect hexagon with 120° interior carbon-carbon bond angles. A commonly used representation of the benzene molecule is simply

where each vertex represents a carbon atom attached to a hydrogen atom. The benzene ring is part of the chemical formulas of a great many organic compounds. We will learn in Chapter 29 that benzene behaves chemically as a substance with no localized double bonds and is a relatively unreactive molecule. The unusual stability of a benzene molecule is ascribed to what chemists call **resonance stabilization:** the energy of the actual molecule, represented by a superposition of Lewis formulas, is lower than the energy of any of its (hypothetical) individual Lewis formulas.

11-7. A SPECIES WITH ONE OR MORE UNPAIRED ELECTRONS IS CALLED A FREE RADICAL

As useful as the octet rule is, in some cases it cannot be satisfied. In particular, the octet rule cannot be satisfied by each atom in a species with an odd total number of electrons. For example, consider the molecule nitrogen oxide, NO, which has 5 + 6 = 11 valence electrons. The electron-dot formulas of nitrogen and oxygen are

$$\cdot\dot{\underset{\cdot\cdot}{\text{N}}}\cdot \qquad \text{and} \qquad \cdot\ddot{\underset{\cdot\cdot}{\text{O}}}\cdot$$

If we try to write a Lewis formula for NO, we find that it is not possible to satisfy the octet rule. The best that we can do is

$$\dot{\text{N}}\text{=}\ddot{\text{O}} \qquad \text{or} \qquad {}^{\ominus}\dot{\text{N}}\text{=}\ddot{\text{O}}^{\oplus}$$

The difficulty here is that the total number of valence electrons is an odd number (11), so it is impossible to pair up all the electrons as we have been doing. A species that has one or more unpaired electrons is called a **free radical.** Because of the unpaired electron(s), free radicals are usually very reactive, for these species have a tendency to form additional bonds so that all the atoms can satisfy the octet rule.

Another example of a free radical is chlorine dioxide, ClO_2, a yellow to reddish yellow gas with an unpleasant odor similar to that of chlorine; chlorine dioxide reacts explosively with many substances. The chlorine atom has seven valence electrons, and each oxygen atom has six valence electrons. Therefore, ClO_2 has an odd number (19) of valence electrons. Two resonance forms for ClO_2 are

$$^{\ominus}:\ddot{\underset{\cdot\cdot}{\text{O}}}-\overset{\oplus}{\ddot{\underset{\cdot\cdot}{\text{Cl}}}}-\ddot{\underset{\cdot\cdot}{\text{O}}}\cdot \qquad \text{and} \qquad \cdot\ddot{\underset{\cdot\cdot}{\text{O}}}-\overset{\oplus}{\ddot{\underset{\cdot\cdot}{\text{Cl}}}}-\ddot{\underset{\cdot\cdot}{\text{O}}}:^{\ominus}$$

The ClO_2 free radical is viewed as a hybrid of the resonance forms, and the two chlorine-oxygen bonds in ClO_2 are equivalent.

The molecules NO and ClO_2 are free radicals. They have an odd number of electrons, so they cannot satisfy the octet rule. Other compounds, called **electron-deficient compounds,** have an even number of outer electrons but do not have enough electrons to form octets about each atom. Compounds of beryllium and boron serve as good examples of electron-deficient compounds. Consider the molecule beryllium hydride, BeH_2. The electron-dot formulas for the beryllium atom and the hydrogen atom are

$$\text{H}\cdot \qquad \text{and} \qquad \cdot\text{Be}\cdot$$

A Lewis formula for BeH_2 is

$$\text{H—Be—H}$$

The beryllium atom is four electrons short of satisfying the octet rule. Electron-deficient compounds, like free radicals, generally are highly reactive.

EXAMPLE 11-11: Suggest a Lewis formula for boron fluoride, BF_3, that satisfies the octet rule. Give a reason why the electron-deficient formula is preferred.

Solution: Boron trifluoride is an electron-deficient molecule. Each fluorine atom has 7 valence electrons and the boron atom has 3, for a total of 24. A Lewis formula for BF_3 using 12 electron pairs is

$$:\ddot{\underset{..}{F}}—\overset{\ominus}{B}—\ddot{\underset{..}{F}}: \quad (\text{B}=\overset{\oplus}{\text{F}}\text{ double bond below})$$

The formal charge of +1 on the very reactive nonmetallic fluorine atom can be used to decide that this Lewis formula is less favorable than the formula

$$:\ddot{\underset{..}{F}}—\underset{|\atop :\underset{..}{F}:}{B}—\ddot{\underset{..}{F}}:$$

for which the formal charges are all 0.

PRACTICE PROBLEM 11-11: Nitrogen oxide can be prepared by reacting copper with nitric acid:

$$3Cu(s) + 8HNO_3(aq) \rightarrow 3Cu(NO_3)_2(aq) + 4H_2O(l) + 2NO(g)$$

Notice that nitric acid acts as an oxidizing agent in this reaction; this reaction is not of the type, metal + acid → salt + hydrogen, in that H^+ is not reduced to H_2. Although $NO(g)$ is colorless, if the reaction is run in a vessel open to the atmosphere, then the gas evolved is reddish brown because of the reaction

$$\underset{\text{colorless}}{NO(g)} + \tfrac{1}{2}O_2(g) \rightarrow \underset{\text{reddish brown}}{NO_2(g)}$$

Write the Lewis formula for NO_2.

Answer: Two resonance forms are

$$:\underset{..}{O}=\overset{\cdot\,\oplus}{N}—\underset{..}{\ddot{O}}:^{\ominus} \longleftrightarrow {}^{\ominus}:\underset{..}{\ddot{O}}—\overset{\cdot\,\oplus}{N}=\underset{..}{O}:$$

As noted, electron-deficient compounds are usually highly reactive species. For example, the electron-deficient compound BF_3 readily reacts with NH_3 to form H_3NBF_3:

```
                  ..                  ..
    H            :F:            H    :F:
    |             |   ..        ⊕|    |⊖..
 H—N: +          B—F:   →    H—N—B—F:
    |             |   ..         |    |  ..
    H            :F:            H    :F:
                  ..                  ..
```

The lone electron pair in NH_3 can be shared between the nitrogen atom and the boron atom so that the octet rule is now satisfied for each atom. A covalent bond that is formed when one atom contributes both electrons is called a **coordinate-covalent bond.** The product of the foregoing reaction, H_3NBF_3, is called a **donor-acceptor complex.**

11-8. ATOMS OF ELEMENTS BELOW CARBON THROUGH NEON IN THE PERIODIC TABLE CAN EXPAND THEIR VALENCE SHELLS

We have not yet considered the case in which there are more valence electrons than are needed to satisfy the octet rule on each atom. Such a case can happen when one of the elements in the compound lies below the second-row elements carbon, nitrogen, oxygen, and fluorine in the periodic table. This element will usually be the central atom and its valence electrons have $n > 3$ as the principal quantum number. In this case, we assign the "extra" electrons as lone pairs to that element, which we say has an **expanded valence shell.**

As an example, let's write the Lewis formula for sulfur tetrafluoride, SF_4. First we arrange the atoms as

```
   F
F  S  F
   F
```

Of the $6 + (4 \times 7) = 34$ valence electrons, we use 8 electrons to form four sulfur-fluorine bonds. We can satisfy the octet rule for each atom by using only 24 of the remaining valence electrons:

```
      ..
     :F:
 ..   |   ..
:F—S—F:        (two valence electrons unassigned)
 ..   |   ..
     :F:
      ..
```

Two valence electrons are still to be accounted for. Sulfur lies in the third row of the periodic table, so we add these as a lone pair to the sulfur atom. Thus, the Lewis formula is

```
      ..
     :F:
 ..   |·· ..
:F—S—F:
 ..   |   ..
     :F:
      ..
```

The exact position of the lone pair on the sulfur atom is not important; for example, we could just as well have placed the pair at the upper left of the sulfur atom. Notice that the formal charges are 0 on all the atoms in the Lewis formula.

In sulfur tetrafluoride, SF_4, sulfur expands its valence shell by using its *d* orbitals. It is not possible for the atoms of elements in the second row of the periodic table to expand their valence shells beyond eight electrons, because second-row elements complete the L shell, which does not contain *d* orbitals, when they satisfy the octet rule. Second-row elements would have to use orbitals in the M shell to accommodate more electrons, and the energies of the orbitals in the M shell are much higher than those of the orbitals in the L shell. Thus, although SF_4 has been synthesized, OF_4 has never been observed.

EXAMPLE 11-12: Xenon difluoride, XeF_2, was one of the first xenon compounds to be prepared. Write the Lewis formula for the XeF_2 molecule.

Solution: We arrange the atoms as

F Xe F

Of the $8 + (2 \times 7) = 22$ valence electrons, 4 are used to form the two xenon-fluorine bonds. We can use 12 of the remaining 18 valence electrons to satisfy the octet rule on each fluorine atom:

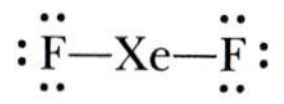

:F̤̈—Xe—F̤̈: (six valence electrons unassigned)

The "extra" six valence electrons are placed on the xenon atom as three lone pairs. The Lewis formula is

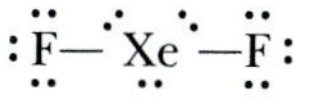

:F̤̈—Ẍ̤e—F̤̈: (with three lone pairs on Xe)

which has 0 formal charges on all the atoms.

PRACTICE PROBLEM 11-12: Phosphorus oxychloride, $POCl_3(l)$, is a colorless, clear, strongly fuming liquid with a pungent odor. It is used as a chlorinating agent, especially to replace oxygen atoms with chlorine atoms in organic compounds. Write a Lewis formula of phosphorus oxychloride that has no formal charges.

Answer:

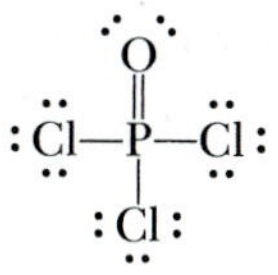

```
        ˙O˙
        ||
 :Cl—P—Cl:
        |
       :Cl:
```

Other examples in which using step 4 leads to more than eight electrons around the central atom are

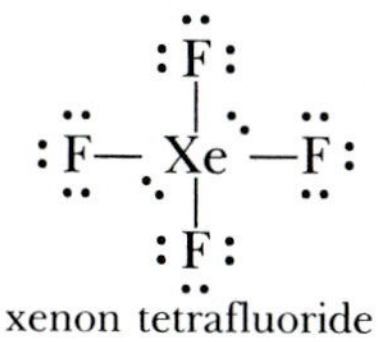

```
       :F:                          
        |                           
 :F—  Xe  —F:              :F—  Br  —F:
        |                           |
       :F:                         :F:
 xenon tetrafluoride       bromine trifluoride

        ⊖                          :F:
 :I—  I  —I:            ⊖       ⊕ |        ⊖
                          :O—  I  —O:
                                  |
                                 :F:
 triiodide ion            iodine dioxide difluoride ion
```

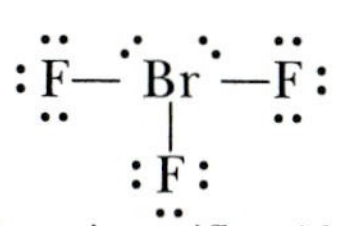

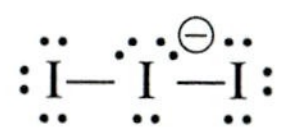

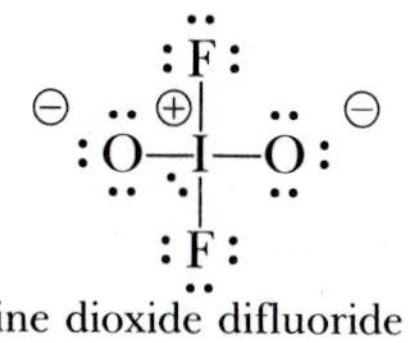

Because the atoms of elements below the second row of the periodic table can accommodate more than eight electrons in the valence shells, they are able to bond to more than four atoms. Some examples are

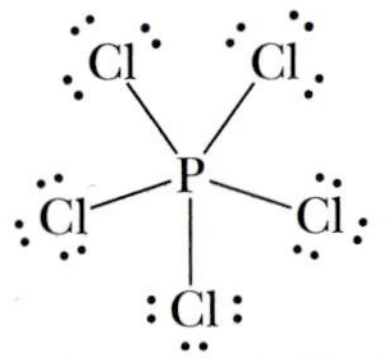

phosphorus pentachloride

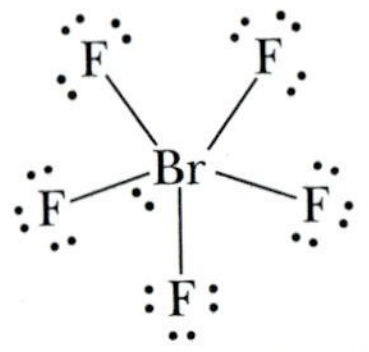

bromine pentafluoride

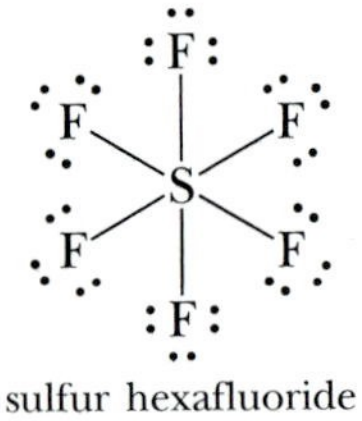

sulfur hexafluoride

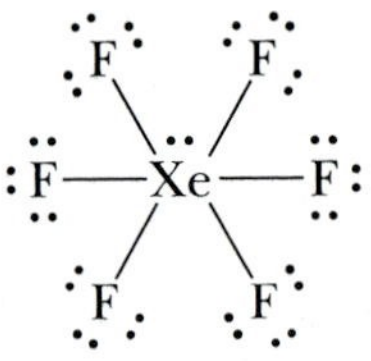

xenon hexafluoride

EXAMPLE 11-13: Tellurium forms the pentafluorotellurate ion, TeF_5^-, when KF(*s*) and TeO_2(*s*) are dissolved in HF(*aq*). Write the Lewis formula of the TeF_5^- ion.

Solution: A tellurium atom has six valence electrons, and each fluorine atom has seven valence electrons. The negative charge on the ion gives a total of 42 valence electrons. Of these 42 valence electrons, 10 are used to form bonds between the five fluorine atoms and the central tellurium atom, 30 more are used as lone pairs on the fluorine atoms to satisfy the octet rule, and the remaining 2 valence electrons constitute a lone pair on the tellurium atom. Thus, the Lewis formula of TeF_5^- is

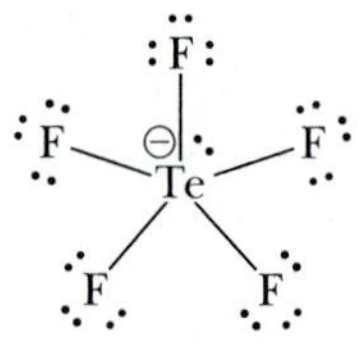

The formal charge on the tellurium atom is given by Equation (11-1):

$$\text{formal charge on Te} = 6 - 2 - \tfrac{1}{2}(10) = -1$$

PRACTICE PROBLEM 11-13: Write the Lewis formula of thionyl tetrafluoride, SOF_4.

Answer:

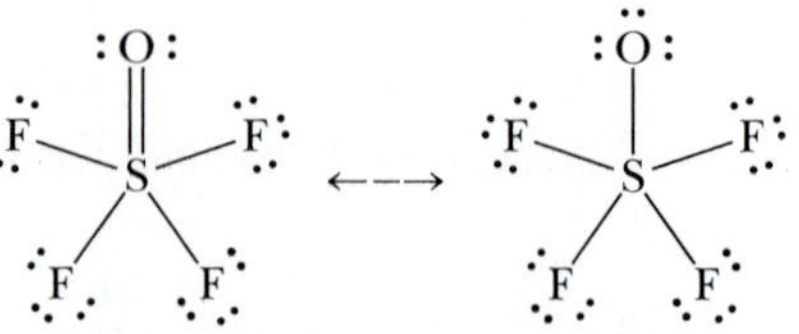

The fact that atoms of elements below the second row of the periodic table can expand their valence shells leads to additional resonance formulas for many of the compounds involving those atoms. For example, consider sulfuryl chloride, SO_2Cl_2. According to the rules that we have presented, the Lewis formula for SO_2Cl_2 is

This Lewis formula, however, displays a large formal charge separation that can be reduced by writing additional Lewis formulas, which recognize the ability of sulfur to expand its valence shell:

and $\longleftrightarrow$

All four of these Lewis formulas are resonance forms of SO_2Cl_2, so we have

EXAMPLE 11-14: Write Lewis formulas for the various resonance forms of sulfur trioxide, SO_3, including those in which the sulfur atom has an expanded valence shell. (There are seven resonance forms.) Indicate formal charges and discuss the bonding in SO_3.

Solution: A sulfur atom has 6 valence electrons and each oxygen atom has 6 valence electrons for a total of 24 valence electrons. The various resonance forms of SO_3 are

three of these involving one S=O bond

three of these involving two S=O bonds

one of these

A superposition of all seven resonance hybrids suggests (correctly) that the three sulfur-oxygen bonds in SO_3 are equivalent.

PRACTICE PROBLEM 11-14: Write Lewis formulas for the various resonance forms of a phosphate ion, including those in which the phosphate atom has an expanded valence shell. Indicate formal charges and discuss the bonding in a phosphate ion.

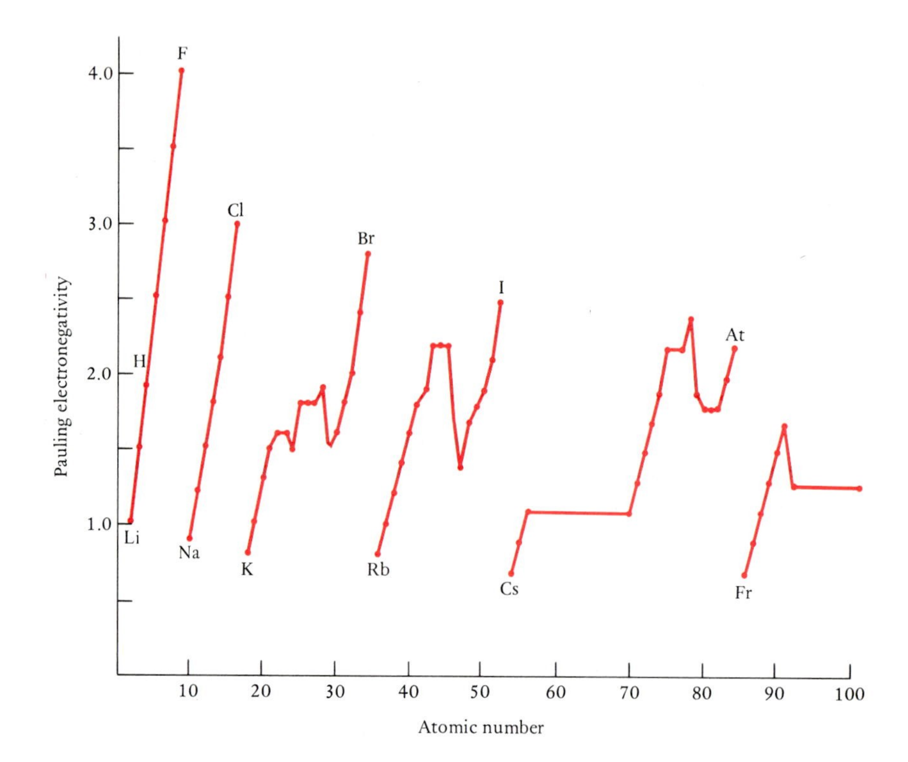

Figure 11-4 Pauling electronegativities plotted against atomic number.

ativity scale is based on the differences in the bond energies of a **heteronuclear molecule** (a molecule consisting of two unlike atoms) AB and the **homonuclear molecules** (molecules consisting of identical atoms) A_2 and B_2. Pauling's formula can be written as

$$|X_A - X_B| = 0.102[E_{AB} - (E_{A_2}E_{B_2})^{1/2}] \qquad (11\text{-}2)$$

Here X_A and X_B are the electronegativities of atoms A and B, respectively, and E_{AB}, E_{A_2}, and E_{B_2} are the bond energies of molecules AB, A_2, and B_2. The numerical factor of 0.102 is simply due to the units used. The vertical lines enclosing $X_A - X_B$ indicate that the absolute value of the difference between X_A and X_B is taken; in other words, the difference is always taken to have a positive sign. The important point to notice is that Equation (11-2) involves only the *differences* in the electronegativities of A and B. To determine individual electronegativities, one atom must be assigned a specific value by convention, and then all other values can be related to that one. Pauling chose the value of the most electronegative element, fluorine, to be 4.0.

In Figure 11-4, the Pauling electronegativities are plotted against atomic number. Figure 11-4 shows clearly that electronegativity is a periodic property. Note that electronegativities increase from left to right across the short (second and third) rows of the periodic table, as the elements become increasingly nonmetallic. Note also that electronegativities decrease in going down a column (Figure 11-5), because the nuclear attraction of the outer electrons decreases as the size of the atom increases. Note that fluorine is the most electronegative atom; the least electronegative are cesium and francium. The order for the generally most useful electronegativities is (you should learn this)

$$\underset{4.0}{F} > \underset{3.5}{O} > N \simeq \underset{3.0}{} Cl > C \simeq \underset{2.5}{} S > H \simeq \underset{2.1}{} P$$

1 H 2.1																	2 He -
3 Li 1.0	4 Be 1.5											5 B 1.9	6 C 2.5	7 N 3.0	8 O 3.5	9 F 4.0	10 Ne -
11 Na 0.9	12 Mg 1.2											13 Al 1.5	14 Si 1.8	15 P 2.1	16 S 2.5	17 Cl 3.0	18 Ar -
19 K 0.8	20 Ca 1.0	21 Sc 1.3	22 Ti 1.5	23 V 1.6	24 Cr 1.6	25 Mn 1.5	26 Fe 1.8	27 Co 1.8	28 Ni 1.8	29 Cu 1.9	30 Zn 1.5	31 Ga 1.6	32 Ge 1.8	33 As 2.0	34 Se 2.4	35 Br 2.8	36 Kr -
37 Rb 0.8	38 Sr 1.0	39 Y 1.2	40 Zr 1.4	41 Nb 1.6	42 Mo 1.8	43 Tc 1.9	44 Ru 2.2	45 Rh 2.2	46 Pd 2.2	47 Ag 1.7	48 Cd 1.4	49 In 1.7	50 Sn 1.8	51 Sb 1.9	52 Te 2.1	53 I 2.5	54 Xe -
55 Cs 0.7	56 Ba 0.9	57-71 1.1-1.2	72 Hf 1.3	73 Ta 1.5	74 W 1.7	75 Re 1.9	76 Os 2.2	77 Ir 2.2	78 Pt 2.2	79 Au 2.4	80 Hg 1.9	81 Tl 1.8	82 Pb 1.8	83 Bi 1.8	84 Po 2.0	85 At 2.2	86 Rn -
87 Fr 0.7	88 Ra 0.9	89 Ac 1.1	90 Th 1.3	91 Pa 1.5	92 U 1.7	93-103 Np-Lr 1.3											

Figure 11-5 Electronegativities of the elements as calculated by Linus Pauling. Note that the electronegativities of the elements in the second and third rows increase from left to right, and that they increase from bottom to top in a given column.

Remember that only *differences* in electronegativities are meaningful. It is significant that the electronegativity difference between fluorine and hydrogen is 1.9, but the electronegativity of fluorine is twice that of hydrogen only because of the scale we are using. A comparison of Figures 11-5 and 11-6 shows the inverse relationship between metallic character and electronegativity.

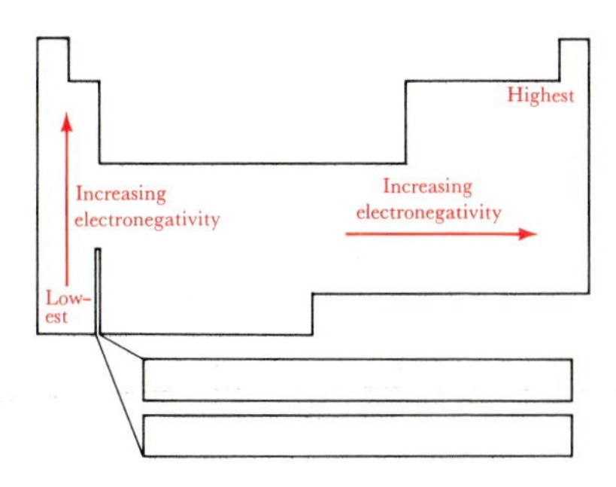

Figure 11-6 The trends of electronegativities in the periodic table.

11-11. WE CAN USE ELECTRONEGATIVITY TO PREDICT THE POLARITY OF CHEMICAL BONDS

It is the *difference* in electronegativities of the two atoms in a covalent bond that determines how the electrons in the bond are shared. If the electronegativities are nearly the same, then the electrons in the bond are shared equally, and the bond is called a **pure covalent bond,** or a **nonpolar bond.** Equal sharing of bonding electrons occurs in homonuclear diatomic molecules. If the electronegativities of the two atoms differ, then electrons in the bond are not shared equally and the bond is said to be a **polar bond.** The extreme case of a polar bond occurs when the difference in electronegativities is large, say, greater than about 2.0. For such a case, the electron pair is completely on the more electronegative atom, and the bond is a **pure ionic bond.**

A polar bond can be illustrated by HCl. According to Figure 11-5, the electronegativity of a hydrogen atom is 2.1 and that of a chlorine atom

11-3. Pure hydrogen peroxide, H_2O_2, is a colorless liquid that is caustic to the skin, but a 3 percent aqueous solution is a mild bleaching agent. Write the Lewis formula for H_2O_2.

11-4. Tetrafluorohydrazine, N_2F_4, is a colorless liquid that is used as rocket fuel. Write the Lewis formula for tetrafluorohydrazine.

11-5. Write the Lewis formula for

(a) methane, CH_4
(b) fluoromethane, CH_3F
(c) aminomethane, CH_3NH_2

11-6. Write the Lewis formula for

(a) methyl mercaptan, CH_3SH
(b) dimethyl ether, CH_3OCH_3
(c) trimethyl amine, $N(CH_3)_3$

MULTIPLE BONDS

11-7. Write the Lewis formula for

(a) acetylene, C_2H_2
(b) diazine, N_2H_2
(c) phosgene, $COCl_2$

11-8. Write the Lewis formula for

(a) nitrous acid, HNO_2
(b) silicon dioxide, SiO_2
(c) propylene, CH_3CHCH_2

11-9. Formic acid is a colorless liquid with a penetrating odor. It is the irritating ingredient in the bite of ants. Its chemical formula is HCOOH. Write the Lewis formula for formic acid.

11-10. Hydrazoic acid, HN_3, is a dangerously explosive, colorless liquid. Azides of heavy metals explode when struck sharply and are used in detonation caps. Write the Lewis formula for hydrazoic acid (HN_3) and for the azide ion (N_3^-).

11-11. Vinyl chloride is an important industrial chemical used in the manufacture of polyvinyl chloride. Its chemical formula is C_2H_3Cl. Write the Lewis formula for vinyl chloride.

11-12. Acetone is an organic compound widely used in the chemcial industry as a solvent, for example, in paints and varnishes. You may be familiar with its sweet odor because it is used as a fingernail-polish remover. Its chemical formula is CH_3COCH_3. Write the Lewis formula for acetone.

FORMAL CHARGE

11-13. Use formal charge considerations to rule out the Lewis formula for NF_3 in which the nitrogen atom and the three fluorine atoms are connected in a row.

11-14. Use formal charge considerations to predict the arrangement of the atoms in NOCl.

11-15. Laughing gas, an anesthetic and a propellent in whipped-cream-dispensing cans, has the chemical formula N_2O. Use Lewis formulas and formal charge considerations to predict which structure, NNO or NON, is the more likely.

11-16. Use formal charge considerations to rule out the Lewis formula for NO_2^- in which the arrangement of the atoms is O-O-N.

RESONANCE

11-17. Write Lewis formulas for the resonance forms of the formate ion, $HCOO^-$. Indicate formal charges and discuss the bonding of this ion.

11-18. Write Lewis formulas for the resonance forms of the acetate ion, CH_3COO^-. Indicate formal charges and discuss the bonding in this ion.

11-19. Write Lewis formulas for the resonance forms of the thiosulfate ion, $S_2O_3^{2-}$. Indicate formal charges and discuss the bonding in this ion.

11-20. Write Lewis formulas for the resonance forms of ozone, O_3. Indicate formal charges and discuss the bonding in this molecule.

OCTET RULE VIOLATIONS

11-21. Which of the following species contain an odd number of electrons?

(a) NO_2 (b) CO (c) O_3^- (d) O_2^-

Write a Lewis formula for each of these species.

11-22. Which of the following species contain an odd number of electrons?

(a) BrO_3 (b) SO_3 (c) HNO (d) HO_2

Write a Lewis formula for each of these species.

11-23. Nitrosamines are carcinogens that are found in tobacco smoke. They can also be formed in the body from the nitrites and nitrates used to preserve processed meats, especially bacon and sausage. The simplest nitrosamine is methylnitrosamine, H_3CNNO. Write the Lewis formula for this molecule. Is it a free radical?

11-24. Many free radicals combine to form molecules that do not contain any unpaired electrons. The driving force for the radical-radical combination reaction is the formation of a new electron-pair bond. Write Lewis formulas for the reactant and product species in the following reactions:

(a) $CH_3(g) + CH_3(g) \rightarrow H_3CCH_3(g)$
(b) $N(g) + NO(g) \rightarrow NNO(g)$
(c) $2OH(g) \rightarrow H_2O_2(g)$

11-25. Write the Lewis formula for

(a) PCl_6^- (b) I_3^- (c) SiF_6^{2-}

11-26. Write a Lewis formula for each of the following compounds of xenon:

Compound	Form at 25°C	Melting point/°C
XeF_2	colorless crystals	129
XeF_4	colorless crystals	117
XeF_6	colorless crystals	50
$XeOF_4$	colorless liquid	−46
XeO_2F_2	colorless crystals	31

11-27. Write the Lewis formulas of the interhalogen compounds IF_3 and IF_5.

11-28. Write the Lewis formulas of the interhalogen ions ICl_4^-, IF_4^+, and IF_2^-.

11-29. Write Lewis formulas for sulfur tetrafluoride and sulfur hexafluoride.

11-30. Write Lewis formulas for sulfinyl fluoride, SOF_2, and sulfonyl fluoride, SO_2F_2.

COVALENT TRANSITION-METAL COMPOUNDS

11-31. Write Lewis formulas for the covalent ions

(a) VO_2^+ (b) VO_3^{2-}

11-32. Write Lewis formulas of the covalent mercury ions

(a) $HgCl_4^{2-}$ (b) $HgCl_3^-$

11-33. Write Lewis formulas for the covalent titanium ions

(a) TiF_6^{2-} (b) $TiBr_5^-$

11-34. Write Lewis formulas for the covalent species

(a) MoF_6 (b) CrO_3 (c) $VOCl_3$

DIPOLE MOMENTS

11-35. Write the Lewis formula for bromine chloride and indicate its dipole moment.

11-36. Arrange the following groups of molecules in order of increasing dipole moment:

(a) HCl HF HI HBr
(b) PH_3 NH_3 AsH_3 (tripod-shaped molecules)
(c) Cl_2O F_2O H_2O (bent molecules)
(d) ClF_3 BrF_3 IF_3 (T-shaped molecules)
(e) H_2O H_2S H_2Te H_2Se (bent molecules)

11-37. Describe the charge distribution in

(a) nitrogen trifluoride, NF_3
(b) oxygen difluoride, OF_2
(c) oxygen dibromide, OBr_2

11-38. Describe the charge distribution in

(a) hydrogen fluoride, HF
(b) phosphine, PH_3
(c) hydrogen sulfide, H_2S

ADDITIONAL PROBLEMS

11-39. Write the Lewis formula for

(a) the sulfate ion, SO_4^{2-}
(b) the phosphate ion, PO_4^{3-}
(c) the acetate ion, $CH_3CO_2^-$

Show the formal charges and discuss the bonding in each ion.

11-40. The halogens form a number of interhalogen compounds. For example, chlorine pentafluoride, ClF_5, can be prepared by the reaction

$$KCl(s) + 3F_2(g) \rightarrow \underset{\text{colorless}}{ClF_5(g)} + KF(s)$$

The halogen fluorides are very reactive, combining explosively with water, for example. Write the Lewis formula for each of the following halogen fluoride species:

(a) ClF_5 (b) IF_3 (c) IF_7 (d) IF_4^+

11-41. Write the Lewis formula for

(a) tetrafluoroammonium ion, NF_4^+
(b) tetrafluorochlorinium ion, ClF_4^+
(c) phosphonium ion, PH_4^+
(d) hexafluoroarsenate ion, AsF_6^-
(e) tetrafluorobromate ion, BrF_4^-

11-42. Write the Lewis formula for each of the following oxychlorine species:

(a) perchlorate ion, ClO_4^-
(b) chlorine oxide, ClO
(c) chlorate ion, ClO_3^-
(d) chlorine dioxide, ClO_2
(e) hypochlorite ion, ClO^-

11-43. Write the Lewis formula for each of the following acids:

(a) $HClO_3$ (b) HNO_2
(c) HIO_4 (d) $HBrO_2$

11-44. Write the Lewis formula for each of the following oxyacids of sulfur:

(a) sulfuric acid, H_2SO_4
(b) thiosulfuric acid, $H_2S_2O_3$
(c) disulfuric acid, $H_2S_2O_7$ (has an S—O—S bond)
(d) dithionic acid, $H_2S_2O_6$ (has an S—S bond)
(e) peroxydisulfuric acid, $H_2S_2O_8$ (has an O—O bond)

The shapes of molecules play a major role in determining a wide variety of chemical properties including chemical reactivity, odor, taste, and drug action. In this chapter we devise a set of simple, systematic rules that allow us to predict the shapes of thousands of molecules. These rules are based on the Lewis formulas that we developed in Chapter 11 and are collectively called the valence-shell electron-pair repulsion (VSEPR) theory. In spite of its rather imposing name, VSEPR theory is easy to understand, easy to apply, and remarkably reliable.

12-1. LEWIS FORMULAS DO NOT GIVE US THE SHAPES OF MOLECULES

Lewis formulas show the bonding relationships among atoms in a molecule but do not indicate the molecule's actual shape. That is, a Lewis formula does not tell us the geometrical arrangement of the nuclei in the molecule. As an example, consider the molecule dichloromethane, CH_2Cl_2. One Lewis formula for dichloromethane is

$$\begin{array}{ccccc} & & \mathrm{H} & & \\ & & | & & \\ :\ddot{\underset{..}{\mathrm{Cl}}}- & & \mathrm{C} & & -\ddot{\underset{..}{\mathrm{Cl}}}: \\ & & | & & \\ & & \mathrm{H} & & \end{array}$$

I

If we infer this from Lewis formula that dichloromethane is flat, or **planar,** then we must conclude that the Lewis formula

$$\begin{array}{ccc} & :\ddot{\mathrm{Cl}}: & \\ & | & \\ \mathrm{H}- & \mathrm{C} & -\ddot{\underset{..}{\mathrm{Cl}}}: \\ & | & \\ & \mathrm{H} & \end{array}$$

II

represents a different geometry for dichloromethane. In formula I the two chlorine atoms lie 180° apart, whereas in formula II they lie 90° apart. Molecules that have the same chemical formula (CH_2Cl_2, in this case) but different geometric arrangements of the atoms are called **geometric isomers.** Geometric isomers are different molecular species and therefore have different chemical and physical properties. For example, their boiling points differ, so they can be separated by distillation.

Two isomers of dichloromethane have never been observed; there is only one kind of dichloromethane molecule. This finding suggests that the four bonds are oriented such that there is only one distinct way of bonding the two hydrogen atoms and the two chlorine atoms to the central carbon atom. Therefore, our assumption that dichloromethane is a planar molecule is incorrect.

A geometric arrangement that shows why there are no geometric isomers of dichloromethane was reported independently in 1874 by the Dutch chemist Jacobus van't Hoff and by the French chemist Joseph Le Bel. They proposed that the four bonds about a central carbon atom in a molecule such as dichloromethane, CH_2Cl_2, or methane, CH_4, are directed toward the vertices of a **tetrahedron** (Figure 12-1a). A regular

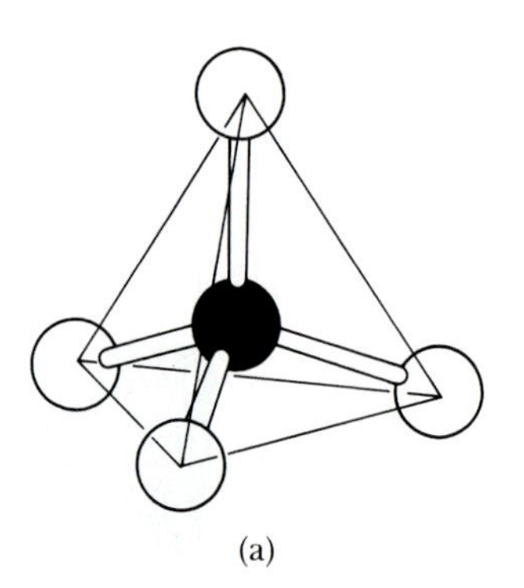

(a)

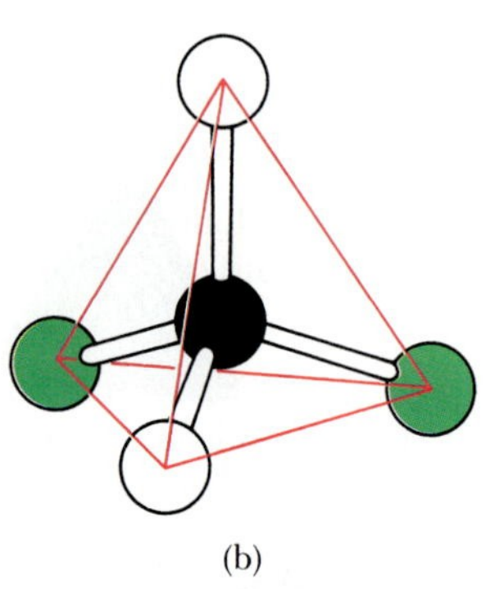

(b)

Figure 12-1 (a) Each carbon-hydrogen bond in a methane molecule points toward the vertex of a regular tetrahedron. The positions of all four hydrogen atoms in CH_4 are equivalent by symmetry. All the H—C—H bond angles are the same, 109.5°. (b) Dichloromethane, CH_2Cl_2. Note that it makes no difference at which two vertices we place the two chlorine atoms; in each case, exactly the same molecule results.

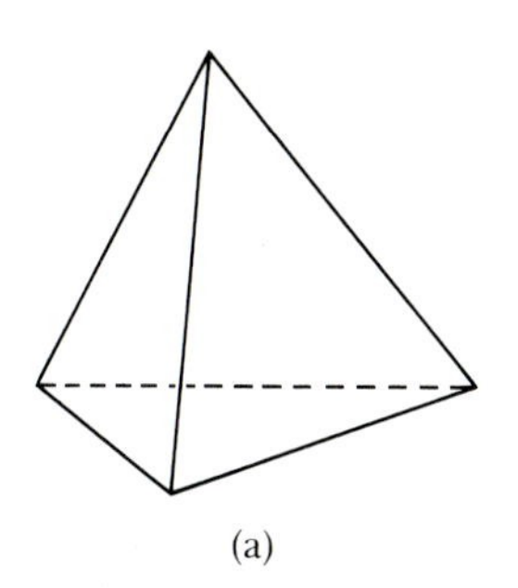

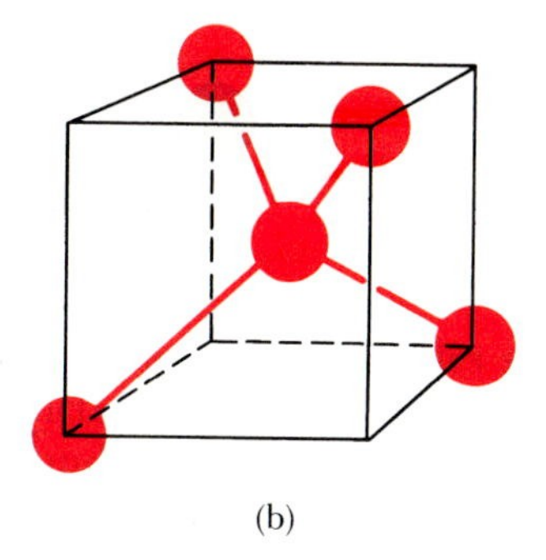

Figure 12-2 (a) A regular tetrahedron is a symmetric body consisting of four equivalent vertices and four equivalent faces. Each face is an equilateral triangle. Note that a tetrahedron differs from the more familiar square pyramid which has a square base and four triangular sides. (b) A tetrahedron also can be viewed as derived from a cube by placing atoms at four of the eight vertices as shown, and then placing an atom in the center of the cube. From this diagram you can derive the tetrahedral angle of 109.5°.

tetrahedron is a four-sided figure that has four equivalent vertices and four identical faces, each of which is an equilateral triangle (Figure 12-2). Appendix C gives instructions for building a regular tetrahedron out of cardboard.

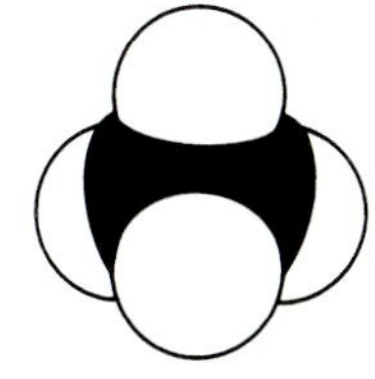

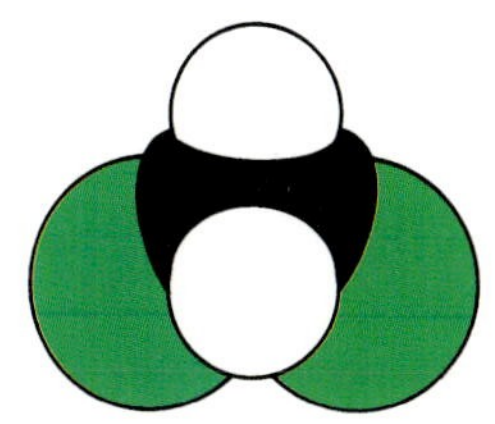

Figure 12-3 Space-filling molecular models of methane and dichloromethane.

12-2. ALL FOUR VERTICES OF A REGULAR TETRAHEDRON ARE EQUIVALENT

You can verify from a model or from Figures 12-1a and 12-2 that the four vertices of a tetrahedron are equivalent and there is only one way of bonding two hydrogen atoms and two chlorine atoms directly to a central carbon atom (Figure 12-1b). The tetrahedral model for CH_2Cl_2 is thus in accord with the experimental fact that dichloromethane has no isomers.

A molecular model of the type shown in Figure 12-3 is called a **space-filling molecular model.** Such models give fairly accurate representations of the angles between bonds and of the relative sizes of the atoms in molecules. A less realistic molecular model, but one in which the geometry is usually easier to see, is the **ball-and-stick molecular model** shown in Figure 12-1.

In a tetrahedral molecule like methane, CH_4, all the H—C—H bond angles are equal to 109.5°, which is called the **tetrahedral bond angle.** The tetrahedral bond angle of 109.5° is a direct consequence of the geometrical properties of a regular tetrahedron. It is the angle between any two vertices of the tetrahedron and a point located exactly in the center of the tetrahedron (Figure 12-4).

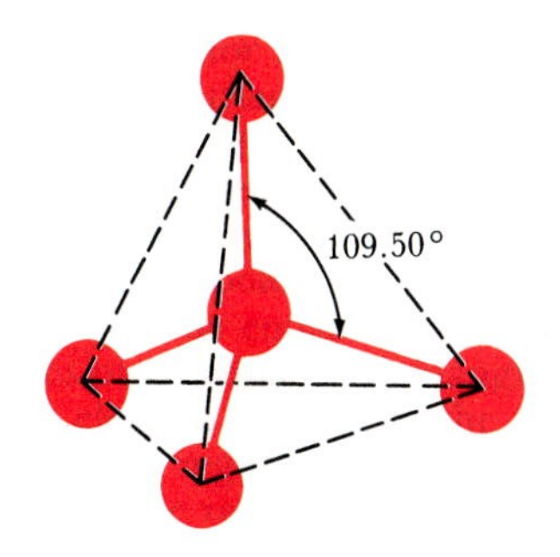

Figure 12-4 A tetrahedron showing the six equivalent tetrahedral bond angles.

A carbon atom that is bonded to four other atoms is called a **tetravalent** carbon atom. The hypothesis of van't Hoff and Le Bel that the bonds of a tetravalent carbon atom are tetrahedrally oriented was the beginning of what is called **structural chemistry,** the area of chemistry in which the shapes and sizes of molecules are studied. Many experimental methods have been developed to determine molecular geometries. Most of the methods involve the interaction of electromagnetic radiation or electrons with molecules. Using such methods, we can central sulfur atom. These six electron pairs mutually repel one another. The mutual repulsion of the six pairs is minimized if the six electron pairs point toward the vertices of a regular **octahedron** (Figure 12-8). This geometric figure has six vertices and eight faces. All eight faces are identical equilateral triangles. An important property of a regular octahedron is that all six vertices are equivalent. This characteristic can be seen by building an octahedron using the directions given in Appendix C. We see, then, that SF_6 is octahedral and that the six fluorine atoms are geometrically equivalent. There is no way, by either chemical or physical methods, to distinguish among the six sulfur-fluorine bonds in SF_6. All the F—S—F bond angles that involve adjacent fluorine atoms in SF_6 are 90°.

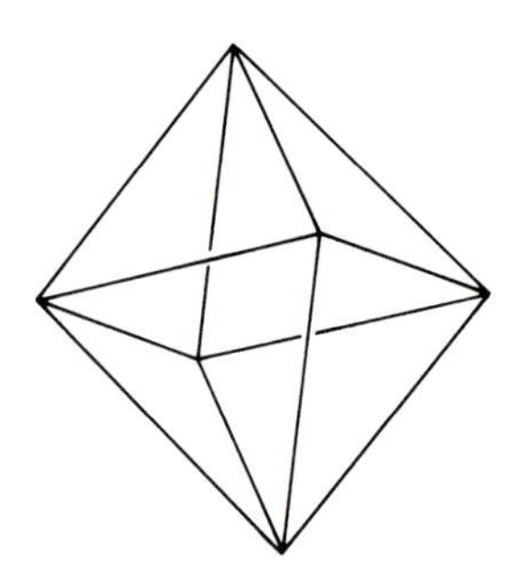

Figure 12-8 A regular octahedron is a symmetric body consisting of six equivalent vertices and eight identical faces that are equilateral triangles.

Sulfur hexafluoride is a very inert gas that is used in high-quality tennis balls, in which the enclosed gas is about 50 percent SF_6. The SF_6 leaks out of the rubber tennis ball casing much less rapidly than air, so these balls keep their "bounce" longer. The slower rate of escape of SF_6, relative to that of O_2 and N_2, is a result of the slower speed and larger size of SF_6 molecules. The ball is not completely filled with SF_6, however, because, when it is, it then does not bounce as high as the 50-50 mixture of SF_6 and air.

Table 12-1 The bond angles associated with shapes shown in Figure 12-6

Shape	Structure
180°	linear
120°	trigonal planar
109.5°	tetrahedral
90°, 120°	trigonal bipyramidal
90°, 90°	octahedral

EXAMPLE 12-3: Predict the shape of the hexachlorophosphate ion, PCl_6^-.

Solution: The Lewis formula for PCl_6^- is

$$\left[\begin{array}{c} :\ddot{Cl}: \quad :\ddot{Cl}: \quad :\ddot{Cl}: \\ \diagdown \ | \ \diagup \\ P \\ \diagup \ | \ \diagdown \\ :\ddot{Cl}: \quad :\ddot{Cl}: \quad :\ddot{Cl}: \end{array}\right]^{\ominus}$$

The six covalent bonds are directed toward the vertices of a regular octahedron, and we predict correctly that the PCl_6^- ion is octahedral. Solid PCl_5, however, unlike gaseous PCl_5, is composed of PCl_4^+ (tetrahedral) cations and PCl_6^- (octahedral) anions. There are no PCl_5 molecules in $PCl_5(s)$.

PRACTICE PROBLEM 12-3: Predict the geometry and the F—Al—F bond angles involving adjacent fluorine atoms in the AlF_6^{3-} ion.

Answer: AlF_6^{3-} is octahedral with F—Al—F angles involving adjacent fluorine atoms of 90°

Table 12-1 shows the bond angles associated with the molecular shapes we have discussed thus far.

12-5. LONE ELECTRON PAIRS IN THE VALENCE SHELL AFFECT THE SHAPES OF MOLECULES

In each case we have discussed so far, all the electron pairs in the valence shell of the central atom have been in covalent bonds. Now let's consider cases in which there are lone pairs of electrons as well as cova-

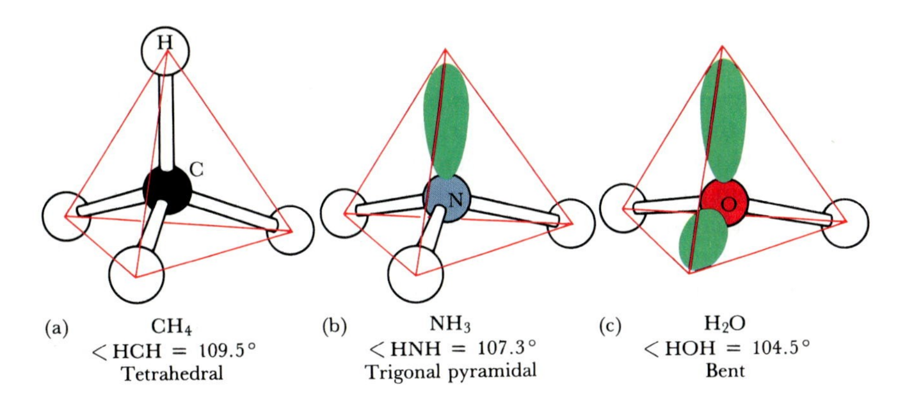

Figure 12-9 The role of bond pairs and lone pairs of electrons in determining molecular geometry.

lent bonds in the valence shell of the central atom. As an example, consider the ammonia molecule. The Lewis formula for NH_3 is

$$\mathrm{H{-}\ddot{N}{-}H} \\ \quad | \\ \quad \mathrm{H}$$

There are four electron pairs in the valence shell of the nitrogen atom. Three of them are in covalent bonds, and one is a lone pair. These four valence-shell electron pairs mutually repel one another and therefore are directed toward the corners of a tetrahedron (Figure 12-9b). The three hydrogen atoms form an equilateral triangle, and the nitrogen atom sits directly above the center of the plane of the triangle. Such a structure is called a triangular pyramid or **trigonal pyramid.** The ammonia molecule is shaped like a **tripod,** with the three N—H bonds forming the legs of the tripod. A space-filling molecular model of NH_3 is shown in Figure 12-10. It is important to keep in mind that what we mean by the shape of a molecule is defined by the positions of the nuclei in the molecule, because it is only the positions of these nuclei that can be located in most methods used to experimentally determine molecular structures. The lone-pair electrons, on the other hand, are relatively diffuse and are generally not located in structure determinations.

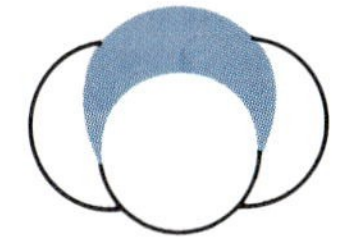

Figure 12-10 A space-filling molecular model of ammonia, NH_3.

However, even when lone pairs exist in the valence shell of the central atom, we still use the concept of minimization of repulsions between electron pairs to deduce the geometry of the electron pairs around the central atom. The geometrical arrangement of the electron pairs around the central atom determines the spatial arrangement of the atoms bonded to the central atom. If all the electron pairs around the central atom are bonding pairs, then the geometrical arrangement of the electron pairs and the geometrical arrangement of the nuclei in the molecule are the same. However, if one or more of the electron pairs around the central atom are lone pairs, then the geometrical arrangements of the electron pairs and of the nuclei are not the same. In such cases, the number of atoms bonded to the central atom is less than the number of electron pairs around the central atom, and the structure of the molecule is *derived from* the geometrical arrangement of all the electron pairs. We want to predict the geometrical arrangement of the nuclei around the central atom, not the geometrical arrangement of the electron pairs, but we must use the geometrical arrangement of the electron pairs to make our prediction of the geometrical arrangement of the nuclei.

EXAMPLE 12-7: Although elemental iodine, I_2, is not very soluble in water, it is very soluble in aqueous solutions of potassium iodide. The increased solubility is due to the formation of the triiodide ion, I_3^-, through the reaction

$$I_2(aq) + I^-(aq) \rightarrow I_3^-(aq)$$

Predict the geometry of the triiodide ion.

Solution: The Lewis formula for I_3^- is

$$:\ddot{\underset{..}{I}}-\overset{\ominus}{\ddot{\underset{..}{I}}}-\ddot{\underset{..}{I}}:$$

This Lewis formula shows that I_3^- belongs to the class AX_2E_3. Figure 12-14c indicates that the three lone pairs occupy equatorial positions in a trigonal bipyramid; thus, I_3^- is a linear ion (Figure 12-17).

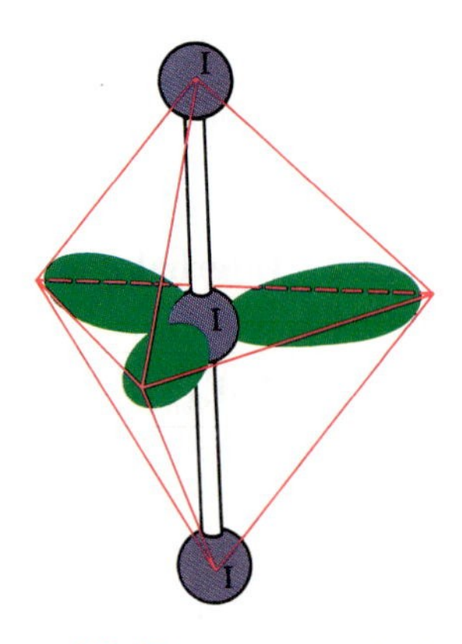

Figure 12-17 The ion I_3^- belongs to the class AX_2E_3. The three lone pairs occupy the equatorial vertices of a trigonal bipyramid. The two iodine atoms occupy the axial positions, so I_3^- is a linear ion.

PRACTICE PROBLEM 12-7: Predict the shape of xenon difluoride.

Answer: Linear

One of the impressive successes of VSEPR theory is its correct prediction of the structures of noble-gas compounds. Some noble-gas compounds that have been synthesized are the xenon fluorides (XeF_2, XeF_4, and XeF_6), xenon oxyfluorides ($XeOF_4$, XeO_2F_2), xenon oxides (XeO_3, XeO_4), and krypton fluorides (KrF_2, KrF_4).

12-8. TWO LONE ELECTRON PAIRS OCCUPY OPPOSITE VERTICES OF AN OCTAHEDRON

The octahedral class (that is, AX_mE_n species with $m + n = 6$) involves species of the type AX_6, AX_5E, AX_4E_2, and AX_3E_3. The corresponding structures are shown in Figure 12-18. Because all six vertices of a regular octahedron are equivalent, so are all six possible positions of the lone pair in AX_5E. To minimize the lone-pair–lone-pair repulsion in an AX_4E_2 molecule, however, the two lone pairs must be placed at opposite vertices, as shown in Figure 12-18c. No AX_3E_3 molecules are known, but Figure 12-18d predicts that they would be T-shaped.

An example of an AX_6 molecule is sulfur hexafluoride, SF_6, which has the predicted octahedral shape (Figure 12-18a). The Lewis formula for the interhalogen compound bromine pentafluoride, BrF_5, is

:F:
F | F
Br
F   ..   F

This Lewis formula shows that BrF_5 belongs to the class AX_5E. According to Figure 12-18b, then, we predict that BrF_5 has a **square pyra-**

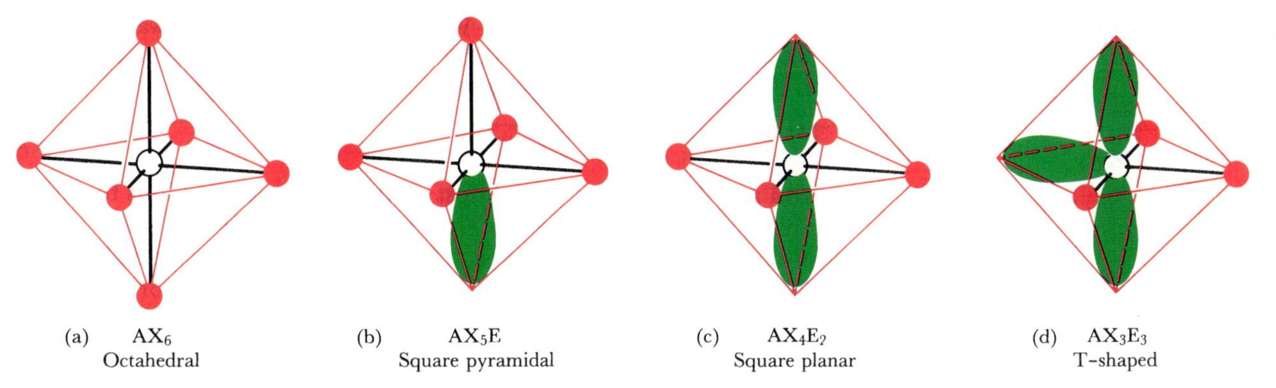

Figure 12-18 The ideal shapes associated with the classes (a) AX_6, (b) AX_5E, (c) AX_4E_2, and (d) AX_3E_3. In (c) and (d), two of the lone electron pairs occupy opposite vertices because this placement minimizes the relatively strong lone-pair–lone-pair electron repulsions.

midal shape, as shown in Figure 12-19. The F—Br—F bond angles are slightly less than the ideal of 90°, because of the lone pair sitting at one vertex. The following example shows that an AX_4E_2 molecule has a **square planar** geometry.

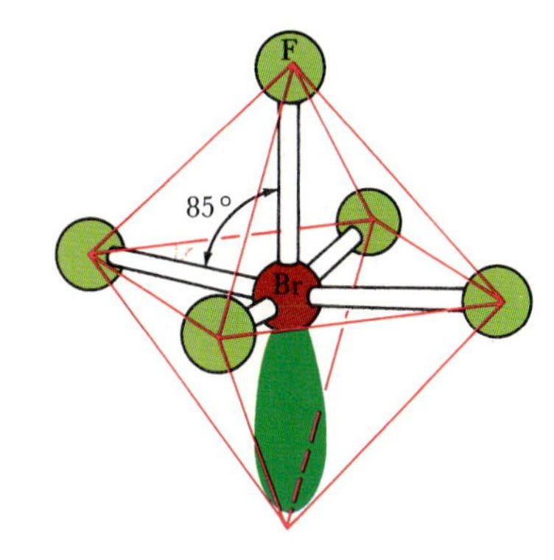

Figure 12-19 The shape of the interhalogen molecule bromine pentafluoride, BrF_5. The lone electron pair repels the bromine-fluorine bonds, causing the bromine atom to lie slightly below the plane formed by four of the fluorine atoms. The BrF_5 molecule has a shape somewhat like an opened umbrella with its handle pointing upward.

EXAMPLE 12-8: Xenon tetrafluoride is prepared by heating Xe and F_2 at 400°C at a pressure of 6 atm in a nickel container. The equation is

$$\mathrm{Xe}(g) + 2\mathrm{F_2}(g) \rightarrow \mathrm{XeF_4}(s)$$

Predict the shape of XeF_4.

Solution: The Lewis formula for XeF_4 is

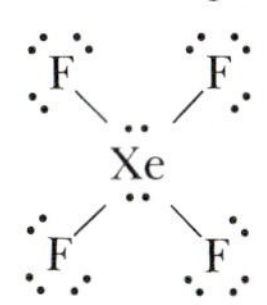

This molecule belongs to the class AX_4E_2; so according to Figure 12-18c, we predict that XeF_4 is a square planar molecule. This shape is indeed that observed for xenon tetrafluoride, as shown in Figure 12-20.

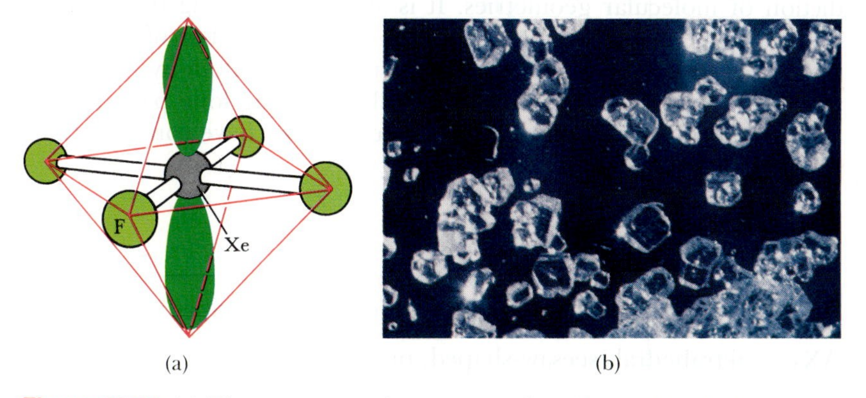

Figure 12-20 (a) The geometry of xenon tetrafluoride, XeF_4, which belongs to the class AX_4E_2. The two lone electron pairs occupy opposite vertices of the octahedron, so XeF_4 is a planar molecule. (b) Crystals of xenon tetrafluoride, one of the first noble-gas compounds synthesized.

SUMMARY

The valence-shell electron-pair repulsion (VSEPR) theory is used to predict the shapes of molecules and ions in which there is a central atom bonded to ligands. Molecular shapes, are defined by the arrangements of the nuclei. However, VSEPR theory takes into account all the valence-shell electrons, including lone pairs, to predict the geometrical arrangement of the ligands around the central atom. The theory is based on the premise that the valence-shell electron pairs around a central atom arrange themselves to minimize their mutual repulsion. The procedure for using VSEPR theory to predict molecular shapes can be summarized as follows:

1. Use the Lewis formula to determine the class, AX_mE_n, to which the molecule or ion belongs. (A represents a central atom, X represents a ligand atom, and E represents a lone pair of electrons.)
2. Given the class AX_mE_n to which the molecule or ion belongs, use Table 12-3 to predict its shape and polarity.

Lone pairs of electrons and multiple bonds have a larger spatial requirement than bond pairs and produce small distortions of the molecular geometry from regular geometric shapes.

TERMS YOU SHOULD KNOW

planar 410
geometric isomer 410
tetrahedron, tetrahedral 410
space-filling molecular model 411
ball-and-stick molecular model 411
tetrahedral bond angle (109.5°) 411
tetravalent 411
structural chemistry 411
valence-shell electron-pair repulsion (VSEPR) theory 413
linear 413
trigonal planar 414
trigonal bipyramid, trigonal bipyramidal 415
equatorial vertex 415
axial vertex 415
octahedron, octahedral 416
trigonal pyramid, trigonal pyramidal 417
tripod 417
bent or V-shaped 418
ligand 419
AX_mE_n 419
seesaw 422
T-shaped 422
interhalogen compound 423
square pyramid, square pyramidal 424
square planar 425
nonpolar molecule 428
polar molecule 429

AN EQUATION YOU SHOULD KNOW HOW TO USE

lone-pair–lone-pair repulsion >
lone-pair–bond-pair repulsion >
bond-pair–bond-pair repulsion (12-1) (relative magnitudes of electron-pair repulsions)

PROBLEMS

MOLECULES AND IONS INVOLVING ONLY SINGLE BONDS

12-1. Which of the following molecules have bond angles of 90°?

(a) TeF_6 (b) $AsBr_5$ (c) GaI_3 (d) XeF_4

12-2. Which of the following species have bond angles of 90°?

(a) NH_4^+ (b) PF_5 (c) AlF_6^{3-} (d) $SiCl_4$

12-3. Which of the following species have 120° bond angles?

(a) ClF_3 (b) $SbBr_6^-$ (c) $SbCl_5$ (d) $InCl_3$

12-4. Which of the following species have 180° bond angles?

(a) SeF_6 (b) BrF_2^- (c) SCl_2 (d) $SiCl_4$

12-5. Which of the following triatomic molecules are linear? Which are bent?

(a) TeF_2 (b) $SnBr_2$ (c) KrF_2 (d) OF_2

12-6. Which of the following triatomic ions are linear? Which are bent?

(a) NH_2^- (b) PF_2^+ (c) IF_2^+ (d) Br_3^-

12-7. Which of the following molecules are tetrahedral?

(a) XeF_4 (b) XeO_2F_2 (c) NF_3O (d) SeF_4

12-8. Which of the following molecules are trigonal pyramidal?

(a) NH_2Cl (b) ClF_3 (c) PF_3 (d) BF_3

12-9. Which of the following species are trigonal bipyramidal?

(a) BrF_5 (b) $SbCl_5$ (c) GeF_5^- (d) SF_5^-

12-10. Name the geometry (tetrahedral, seesaw, or square planar) that describes the shape of each of the following ions:

(a) IBr_4^- (b) PCl_4^+ (c) BF_4^- (d) IF_4^+

12-11. Indicate which, if any, of the listed bond angles occur in the following species:

(a) TeF_6 (1) 90°
(b) $SbCl_5$ (2) 109.5°
(c) ICl_4^- (3) 120°
(d) $InBr_3$

12-12. Indicate which, if any, of the listed bond angles occur in the following ions:

(a) BF_4^- (1) 90°
(b) SiF_6^{2-} (2) 109.5°
(c) SiF_3^+ (3) 120°
(d) $SnCl_6^{2-}$

12-13. Indicate which, if any, of the listed bond angles occur in the following ions:

(a) AlH_4^- (1) 90°
(b) SbF_6^- (2) 109.5°
(c) BrF_4^- (3) 120°
(d) $AsCl_4^+$

12-14. From the accompanying list, select the appropriate description(s) of the bond angles that occur in each of the following molecules:

(a) $GeCl_4$
(b) $SbCl_3$
(c) TeF_6
(d) SF_4

(1) exactly 90°
(2) slightly less than 90°
(3) exactly 109.5°
(4) slightly less than 109.5°
(5) exactly 120°
(6) slightly less than 120°
(7) slightly greater than 120°
(8) exactly 180°
(9) slightly less than 180°

12-15. VSEPR theory has been successful in predicting the geometry of interhalogen molecules and ions. Predict the shapes of the interhalogen molecules given in Table 12-2.

12-16. Predict the shapes of the following iodofluorine ions:

(a) IF_2^+ (b) IF_6^+ (c) IF_4^+ (d) IF_4^-

MOLECULES OR IONS THAT MAY INVOLVE MULTIPLE BONDS

12-17. Write a Lewis formula for each of the following molecules and predict their shapes:

(a) $SeOCl_2$ (b) SO_2Cl_2 (c) SOF_4 (d) ClO_3F

12-18. Write a Lewis formula for each of the following species and predict their shapes:

(a) $XeOF_4$ (b) IOF_5 (c) $PO_2F_2^-$ (d) PO_3F^{2-}

12-19. Write a Lewis formula for each of the following species and predict their shapes:

(a) CCl_2O b) NSF_3 (c) N_3^- (d) SbOCl

12-20. Write a Lewis formula for each of the following species and predict their shapes:

(a) $IO_2F_2^-$ (b) ClO_2^- (c) NOCl (d) NO_2Cl

12-21. Predict the shapes of the following ions:

(a) BrO_2^- (b) TeF_5^- (c) SO_3Cl^- (d) SF_3^+

12-22. Predict the shapes of the following molecules:

(a) NF_3O (b) GeO_2 (c) $AsOCl_3$ (d) XeO_2

12-23. Write Lewis formulas for the following molecules and predict their shapes:

(a) XeO_2F_4 (b) IO_2F_3 (c) IO_2F (d) IO_3F

12-24. Write Lewis formulas for the following ions and predict their shapes:

(a) TlF_4^- (b) IO_2^- (c) CS_3^{2-} (d) BrO_3^-

12-25. The species NO_2^+ and NO_2^- have O—N—O bond angles of 180° and 115°, respectively. Use VSEPR theory to explain the difference in bond angles.

12-26. Compare the shapes of the oxynitrogen ions

(a) NO_2^- (b) NO_3^- (c) NO_2^+ (d) NO_4^{3-}

MOLECULAR SHAPES AND DIPOLE MOMENTS

12-27. For each of the following molecules, write a Lewis formula and predict the shape. Indicate which ones have a dipole moment.

(a) XeF_2 (b) AsF_5 (c) $TeCl_4$ (d) Cl_2O

12-28. For each of the following molecules, write a Lewis formula and predict the shape. Indicate which ones have a dipole moment.

(a) $GeCl_4$ (b) SCl_2 (c) PoF_6 (d) BrF_3

12-29. For each of the following molecules, write a Lewis formula and predict the shape. Indicate which ones have a dipole moment.

(a) $GaCl_3$ (b) $TeCl_2$ (c) TeF_4 (d) $SbCl_5$

12-30. For each of the following molecules, write a Lewis formula and predict the shape. Indicate which ones have a dipole moment.

(a) TeF_6 (b) ClF_5 (c) $SiCl_4$ (d) $SeCl_2$

12-31. Predict which of these molecules are polar:

(a) CF_4 (b) AsF_3 (c) XeF_4 (d) SeF_4

12-32. Predict which of the following molecules are polar:

(a) $TeCl_4$ (b) BCl_3 (c) SF_6 (d) PCl_5

12-33. Describe the bond polarities in the following molecules

(a) nitrogen trifluoride, NF_3
(b) oxygen difluoride, OF_2
(c) oxygen dibromide, OBr_2

12-34. Describe the bond polarities in the following molecules.

(a) CCl_4 (b) PCl_3 (c) ClF_3

GEOMETRIC ISOMERS

12-35. Describe the possible geometric isomers of

(a) a tetrahedral molecule AX_3Y
(b) a tetrahedral molecule AX_2YZ
(c) a square planar molecule AX_3Y
(d) a square planar molecule AX_2Y_2

where X, Y, and Z are different ligands.

12-36. Describe the possible geometric isomers of a trigonal bipyramidal molecule whose formula is

(a) AX_4Y (b) AX_3Y_2 (c) AX_2Y_3

12-37. Describe the possible geometric isomers of an octahedral molecule whose formula is

(a) AX_5Y (b) AX_4Y_2 (c) AX_3Y_3

12-38. Describe the possible geometric isomers of $C_2H_4Cl_2$.

12-39. Describe the possible geometric isomers of $C_2H_2F_2Cl_2$.

12-40. Describe the possible geometric isomers of $N_2H_2F_2$.

ADDITIONAL PROBLEMS

12-41. Name the geometry (tetrahedral, seesaw, or square planar) that describes the shape of each of the following halides:

(a) SF_4 (b) KrF_4 (c) CF_4 (d) $GeCl_4$

12-42. From the accompanying list, select the appropriate description(s) of the bond angles that occur in each of the following fluorides:

(a) SeF_6
(b) GeF_4
(c) BrF_3
(d) IF_5

(1) exactly 90°
(2) slightly less than 90°
(3) exactly 109.5°
(4) slightly less than 109.5°
(5) exactly 120°
(6) slightly less than 120°
(7) slightly greater than 120°
(8) exactly 180°
(9) slightly less than 180°

12-43. Give one example of each of the following:

(a) bent molecule
(b) bent ion
(c) tetrahedral ion
(d) octahedral molecule

12-44. Give one example of each of the following:

(a) trigonal planar molecule
(b) trigonal pyramidal molecule
(c) T-shaped molecule
(d) octahedral ion

12-45. VSEPR theory has been successful in predicting the molecular geometry of noble-gas compounds. Predict the molecular geometry of each of the following xenon species:

(a) XeO_3 (b) XeO_4 (c) XeO_2F_2 (d) XeO_6^{4-}

12-46. Predict the geometry of each of the following phosphorus-containing species:

(a) POF_3 (b) $POCl$ (c) PH_2^- (d) PCl_4^+

12-47. Compare the shapes of the following oxysulfur ions:

(a) sulfoxylate ion, SO_2^{2-}
(b) sulfite ion, SO_3^{2-}
(c) sulfate ion, SO_4^{2-}

12-48. Compare the shapes of the following oxychloro ions:

(a) chlorite, ClO_2^-
(b) chlorate, ClO_3^-
(c) perchlorate, ClO_4^-

12-49. Arrange the following groups of molecules in order of increasing dipole moment:

(a) HCl HF HI HBr
(b) PH_3 NH_3 AsH_3
(c) ClF_3 BrF_3 IF_3
(d) H_2O H_2S H_2Te H_2Se

12-50. Which of the following molecules is trigonal pyramidal?

(a) SOF_2 (b) ClF_3 (c) NO_2Cl (d) BF_3

12-51. Which of the following fluorides has 90° bond angles?

(a) XeF_4 (b) CF_4 (c) SF_2 (d) XeF_2

12-52. Which of the following molecules is linear?

(a) $CdCl_2$ (b) O_3 (c) OCl_2 (d) NOF

12-53. Predict the shapes of the following species involving transition metals:

(a) TiF_6^{2-} (b) VO_2^+ (c) $VOCl_3$ (d) CrO_4^{2-}

12-54. Predict the shapes of the following species involving mercury:

(a) $HgCl_2$ (b) $HgCl_4^{2-}$ (c) $HgCl_3^-$

12-55. Predict the shapes of the following species involving transition metals:

(a) $TiBr_5^-$ (b) MoF_6 (c) VF_6^- (d) MnO_4^-

12-56. Predict the shapes of the following bromofluoride ions:

(a) BrF_2^- (b) BrF_4^- (c) BrF_2^+ (d) BrF_4^+

12-57. Predict the shapes of the following oxyfluoro compounds of sulfur:

(a) SOF_4 (b) SOF_2 (c) SO_2F_2

12-58. Predict the geometries of the following species:

(a) HgI_4^{2-} (b) $NiCl_4^{2-}$ (c) $AuCl_4^-$

12-59. Predict the geometries of the following species:

(a) OsO_4 (b) $NiBr_4^{2-}$ (c) $FeCl_4^-$

12-60. Predict the geometries of the following species:

(a) CrF_6^{3-} (b) $CoCl_4^{2-}$ (c) SnF_2

12-61. Use Figure 12-2b to show that the tetrahedral angle in a regular tetrahedron is 109.5°. (Hint: If we let the edge of the cube be of length a, then the diagonal on a face of the cube has a length $\sqrt{2}a$, by the Pythagorean theorem. From this information, you can determine the distance in terms of a from the center of the cube to a vertex from which, in turn, you can determine the tetrahedral angle.)

Proton Magnetic Resonance

Beginning students of chemistry often ask just how the structure of a molecule is determined experimentally. For example, how can we determine whether the two chlorine atoms in a sample whose empirical formula is $C_2H_4Cl_2$ are on the same carbon atom or on different carbon atoms? Chemists have developed a number of ways to determine molecular structures. In *spectroscopic techniques,* the molecules are brought to interact with various types of electromagnetic radiation. Some of these techniques—such as infrared spectroscopy, ultraviolet spectroscopy, and microwave spectroscopy—are named after the particular region of the electromagnetic spectrum used. Here we shall learn about one particular type of spectroscopy, *nuclear magnetic resonance* (NMR) *spectroscopy.* NMR spectroscopy, developed in the 1950s, has become one of the most useful and most powerful tools for the determination of molecular structures. Hardly a chemistry department or a chemical research laboratory in the world is without at least one NMR spectrometer (Figure F-1).

Before we begin our discussion of NMR spectroscopy, let's see what some NMR spectra look like. Figure F-2 shows the NMR spectra of the two *isomers,* 1,2-dichloroethane and 1,1-dichloroethane, whose Lewis formulas are shown with the spectra. These compounds have the same empirical formula, $C_2H_4Cl_2$, and are closely related; yet their NMR spectra are very different. This comparison shows that NMR spectra are very sensitive to the structures of molecules and so can be used to determine molecular structures.

We shall limit our discussion to NMR spectra that are due to observation of only the hydrogen nuclei, or protons, in a molecule. Nuclear magnetic resonance spectroscopy due to protons only is called *proton magnetic resonance* (PMR). We shall see

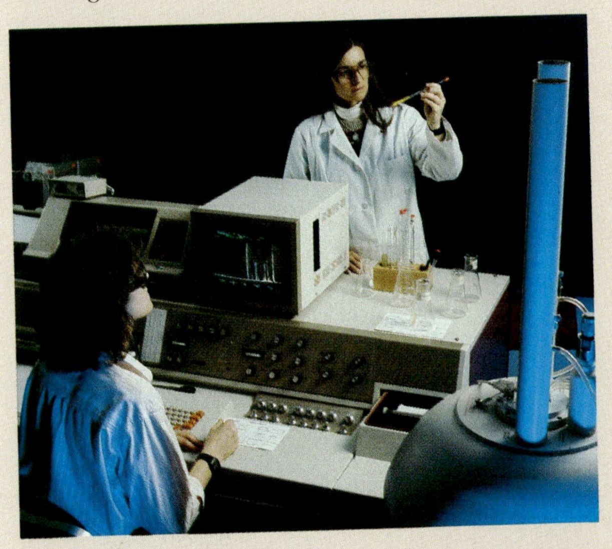

Figure F-1 An NMR spectrometer. Note the spectrum appearing on the console screen.

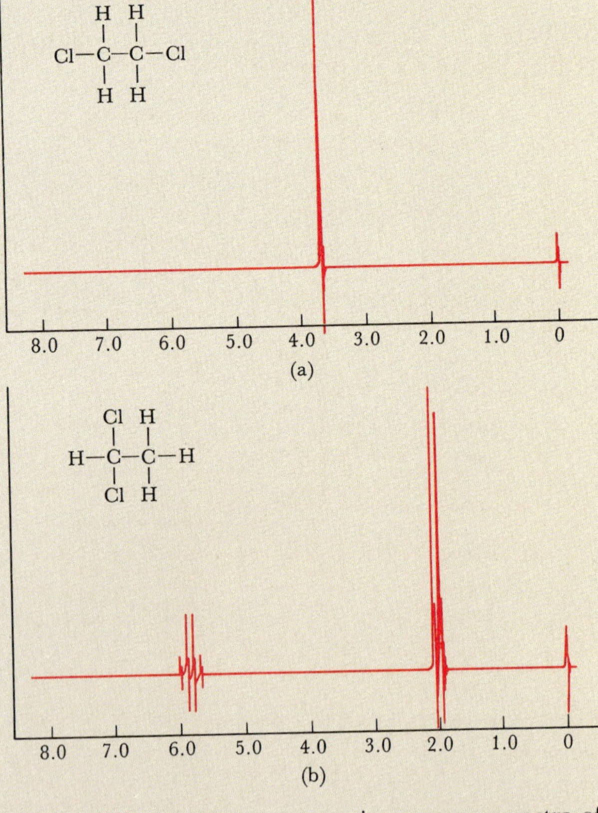

Figure F-2 The nuclear magnetic resonance spectra of (a) 1,2-dichloroethane and (b) 1,1-dichloroethane. These two isomers of dichloroethane have very different NMR spectra. This difference shows that the NMR spectrum of a molecule is a sensitive probe into the structure of a molecule.

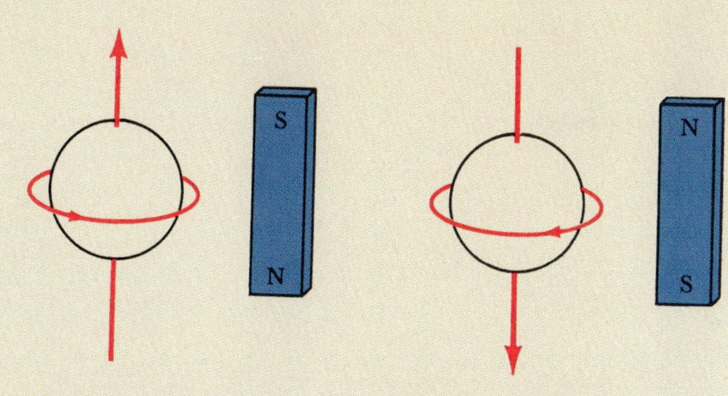

Figure F-3 A spinning charge creates a magnetic field, so a proton behaves like a tiny magnet. The two possible directions of the spin correspond to two different orientations of the magnet, as shown.

that PMR spectroscopy is a sensitive probe into the arrangement of hydrogen atoms within molecules.

Recall that the spin quantum number of an electron can have only two possible values, $+\frac{1}{2}$ or $-\frac{1}{2}$. Although the spin quantum number is strictly a quantum mechanical quantity, we can imagine that an electron spins about its axis like a top; in this simplified interpretation, the two possible values of $+\frac{1}{2}$ and $-\frac{1}{2}$ represent the two possible directions of the spin about the axis. Protons, neutrons, and even entire nuclei also have spins, and in proton magnetic resonance we use the spins of protons. Protons have only two values of the spin quantum number, $+\frac{1}{2}$ and $-\frac{1}{2}$, and so can be pictured as spinning about their axes like electrons.

A spinning charge creates a magnetic field, and so a proton acts like a tiny magnet. The $\pm\frac{1}{2}$ spins of a proton correspond to two different orientations of the magnet, as shown in Figure F-3. When an external magnetic field is applied to a molecule containing hydrogen, quantum mechanics tells us that the protons (the hydrogen nuclei) must align their spins either with the field or against the field. These are the only two orientations that the protons can have in an externally applied magnetic field.

It takes energy to align a proton *against* a magnetic field, and so a proton that is aligned *with* the magnetic field has a lower energy. If such a proton is irradiated with electromagnetic radiation of frequency, ν, given by $E = h\nu$, then the radiation will cause the proton to make a transition from the lower-energy state to the higher-energy state, as shown in Figure F-4.

The difference between the two possible values of the energy for a proton in an externally applied magnetic field turns out to be directly proportional to the strength of the magnetic field, as shown in Figure F-5a. Thus, for a fixed magnetic field strength, one can vary the frequency of the electro-

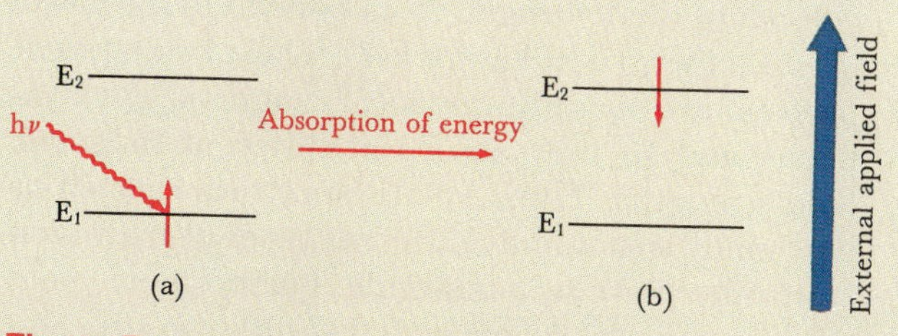

Figure F-4 A proton that is lined up with an externally applied magnetic field (a) has a lower energy than a proton that is aligned against an externally applied magnetic field (b). The energy difference is given as $E_2 - E_1$ in the figure. When a proton aligned with the magnetic field is irradiated with electromagnetic radiation whose frequency, ν, is given by $E_2 - E_1 = h\nu$, the proton will absorb the radiation and become aligned against the magnetic field.

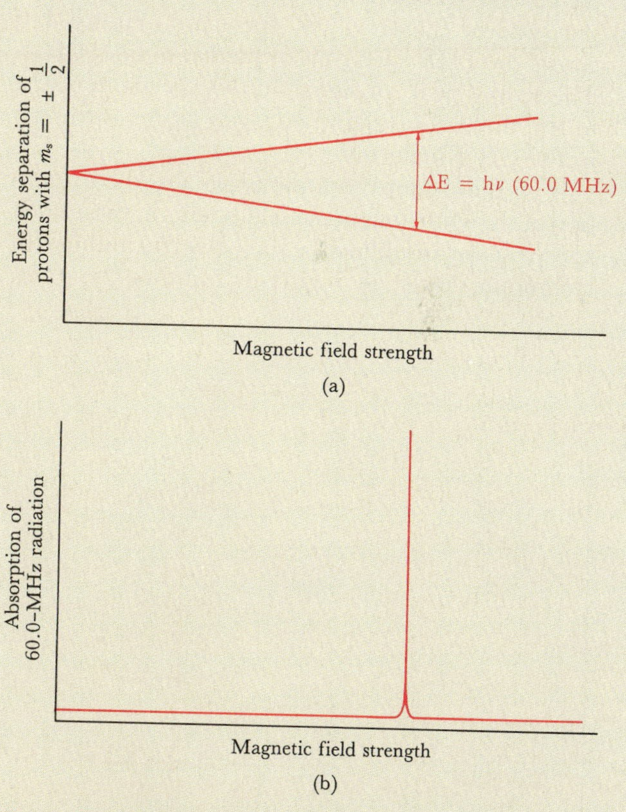

Figure F-5 (a) The energy separation of a proton aligned with or against an applied magnetic field increases with the strength of the magnetic field. (b) When the strength of the magnetic field is such that the separation just matches the energy of the 60.0-MHz radiation that irradiates the sample, the sample will absorb the 60.0-MHz radiation and give the PMR spectrum shown.

magnetic radiation until absorption occurs; or, conversely, one can fix the frequency of the radiation and vary the strength of the magnetic field until absorption occurs. Many PMR spectrometers fix the frequency at 60.0 MHz (megahertz) and vary the magnetic field.

In a PMR experiment, a compound containing hydrogen (protons) is placed between the poles of a strong magnet whose field strength can be varied. The sample is irradiated by 60.0-MHz radiation, and the amount that is absorbed by the sample is detected and recorded. When the magnetic field strength is just enough to cause protons to make transitions, as shown in Figure F-5a, the 60.0-MHz radiation is absorbed by the sample, as illustrated in Figure F-5b.

The absorption of radiation is called *resonance.* When the resonance is due to protons in a magnetic field, it is called proton magnetic resonance. The absorption of the 60.0-MHz radiation is detected and recorded as a *proton magnetic resonance spectrum,* or simply PMR spectrum. Figure F-6 shows the PMR spectrum of methyl iodide, CH_3I. The strong peak, or signal, in this spectrum is due to the three equivalent hydrogen atoms (protons) in the CH_3I molecule.

If all the protons in a molecule absorbed radiation at the same magnetic field strength, then PMR spectroscopy would not be the powerful and useful technique that it is. Figure F-7 shows the PMR spectrum of methyl formate, $HCOOCH_3$. Note that there are two signals in the spectrum of methyl formate, whereas there is only one signal in the spectrum of methyl iodide (Figure F-6). The explanation for this difference lies in the number of structurally different types of hydrogen atoms in the two molecules. All three hydrogen atoms in methyl iodide are chemically equivalent; each one experiences the same magnetic environment as the other. Not all the hydrogen atoms in methyl formate are chemically and magnetically equivalent, however; there are *two* structurally different types of hydrogen atoms in methyl formate. One type consists of the three hydrogen atoms in the CH_3 group (labeled b in Figure F-7), and the other type consists of the hydrogen atom bonded to the C=O group (labeled a in Figure F-7). These two types of protons are in different local magnetic environments and so they absorb at different magnetic field strengths. Thus, a PMR spectrum can tell us how many structurally different types of hydrogen atoms there are in a molecule. Furthermore, note that the two signals in Figure F-7 differ in size. Signal b is about three times as large as signal a because there are three equivalent protons in set b and only one in set a, as shown in the Lewis formula. Thus, a PMR spectrum tells us not only how many structurally different protons there are in a molecule, but how many there are of each kind.

One of the most powerful applications of proton magnetic resonance spectroscopy is the identifica-

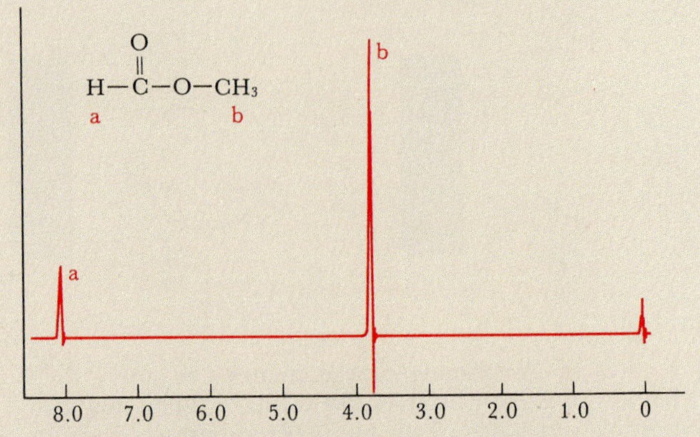

Figure F-7 The PMR spectrum of methyl formate. The small signal at 0 on the horizontal scale is simply a reference signal. The signals at 3.8 and 8.1 are due to the two sets of equivalent protons in methyl formate. Note that the signal due to the three protons labeled b is about three times as great as the signal due to the single proton labeled a. The area under each peak is a measure of the number of protons producing the signal.

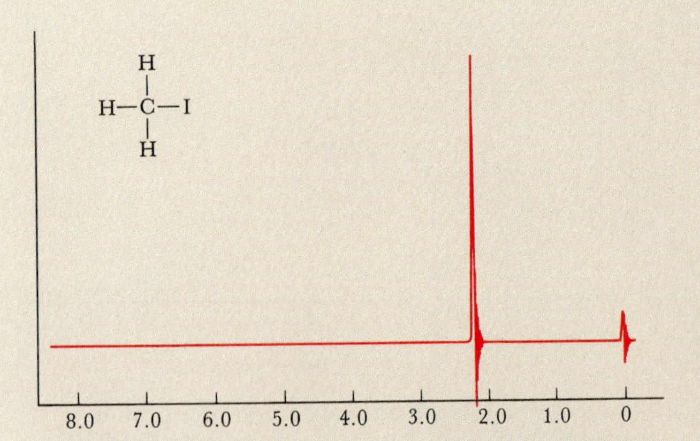

Figure F-6 The PMR spectrum of methyl iodide, CH_3I. There is a strong signal at around 2.2 on the horizontal axis, which is due to the absorption by the three equivalent protons in CH_3I. (The scale of the horizontal axis is somewhat arbitrary. Similarly, the small signal at 0 is a reference peak.)

tion of an unknown molecule. Suppose that we have a compound that we know is either methyl acetate or ethyl formate. The Lewis formulas of these two compounds, which have the same empirical formula, C_3H_6O, are

$$\underset{a}{CH_3}-O-\overset{\overset{\displaystyle O}{\|}}{C}-\underset{b}{CH_3} \qquad \underset{a}{CH_3}\underset{b}{CH_2}-O-\overset{\overset{\displaystyle O}{\|}}{C}-\underset{c}{H}$$

methyl acetate — ethyl formate

The PMR spectrum of our unknown compound is shown in Figure F-8. Is this compound methyl acetate or ethyl formate? From the Lewis formula of methyl acetate, we see that methyl acetate has two methyl groups. One methyl group is attached to an oxygen atom, the other to a carbon atom. Consequently, these two methyl groups are not equivalent, and so methyl acetate has two sets of equivalent protons, each set containing three protons. The Lewis formula of ethyl formate suggests that there are three different sets of protons in this molecule. There are only two signals observed in the PMR spectrum, so we conclude that the unknown compound is methyl acetate. In addition, the two signals are the same size, indicating that the number of protons in each set is the same, as in methyl acetate.

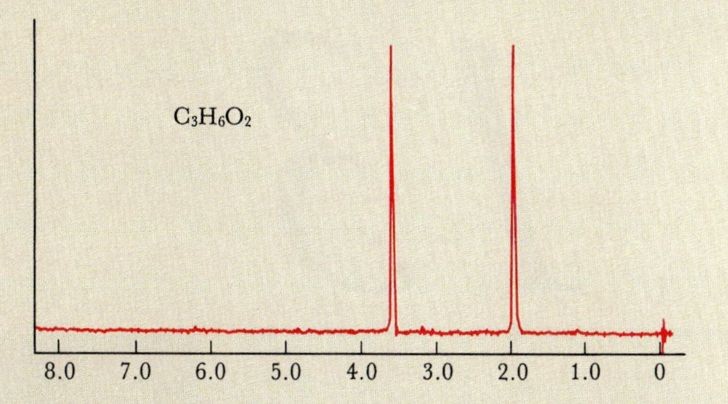

Figure F-8 The PMR spectrum of a compound whose empirical formula is $C_3H_6O_2$. Using this spectrum, it is possible to decide whether the compound is methyl acetate, CH_3COOCH_3, or ethyl formate, $HCOOCH_2CH_3$.

You can see from this very brief introduction how powerful a tool magnetic resonance spectroscopy can be. Many nuclei other than hydrogen can be probed. Carbon-13, which also has spins of $\pm\frac{1}{2}$, has proved to be an exceedingly valuable nucleus to probe in organic compounds. Using various spectroscopic techniques, chemists can identify substances and determine molecular structures.

13 COVALENT BONDING

Linus Pauling, one of America's greatest chemists, was a pioneer in the development of the theory and understanding of chemical bonding. His book, *The Nature of the Chemical Bond,* first published in 1939, is one of the most influential chemistry texts of the twentieth century. During the 1950s, Pauling was in the forefront of the fight against nuclear bomb testing. He was awarded the Nobel Prize for chemistry in 1954 and the Nobel Peace Prize in 1963.

The VSEPR theory allows us to predict how bonds are arranged, but not how and why bonds form in the first place. Why, for example, does hydrogen exist as a diatomic molecule and helium as a monatomic gas? Why has the species He_2 never been observed? To answer these and similar questions, we need a more fundamental theory of covalent bonding. We showed in Chapter 9 how the quantum theory describes the electrons in atoms in terms of atomic orbitals. Now we will discuss how molecular orbitals are formed from overlapping atomic orbitals from different atoms. In the first few sections, we apply a theory called molecular orbital theory to homonuclear diatomic molecules (diatomic molecules in which the two atoms are the same), describing how to write electron configurations for these molecules and how to predict relative bond lengths and bond energies. Then after applying molecular orbital theory to heteronuclear diatomic molecules (diatomic molecules in which the two atoms are different), we go on to discuss polyatomic molecules. A **polyatomic molecule** can be pictured as a group of atoms held together by covalent bonds. The bonding in polyatomic molecules can be described using molecular orbitals in a manner similar to that used for diatomic molecules. In many instances, however, a simpler picture can be obtained if the covalent bonds are considered to be localized between adjacent atoms. To describe these localized covalent bonds, we introduce the idea of hybrid orbitals, which are combinations of atomic orbitals of the same atom. This approach is called valence-bond theory and is mathematically equivalent to the more elaborate molecular orbital theory. In those instances where a localized bond model is inappropriate, the molecular orbital theory will be used.

13-1. A MOLECULAR ORBITAL IS A COMBINATION OF ATOMIC ORBITALS ON DIFFERENT ATOMS

The simplest neutral molecule is diatomic hydrogen, H_2, which has only two electrons. The Schrödinger equation (Section 9-12) that describes the motion of the electrons in H_2 can be solved with a computer to a high degree of accuracy. The results are valuable because they are similar to the results for more complicated molecules. Let's look, therefore, at the approach of quantum theory in more detail. As a first step in setting up the Schrödinger equation for H_2, the two nuclei are fixed at some given separation. Then the two electrons are included, and the equation is solved to give the wave functions and energies that describe the two electrons. The wave function that corresponds to the lowest energy, the **ground-state wave function,** can be used to compute contour diagrams, much like the maps used to show peaks and valleys in hilly terrain. These diagrams show the distribution of electron density around the two nuclei.

Figure 13-1 shows contour diagrams of the ground-state electron density as a function of the separation of the two nuclei of the hydrogen atoms. Note that at large separations the two atoms hardly interact, so the electron density is just that of two electrons, each in a 1*s* orbital about each of the hydrogen atoms. As the separation decreases, however, the two 1*s* orbitals overlap, combining into one orbital that is distributed around both nuclei. Such an orbital is called a **molecular orbital,** because it extends over both nuclei in the molecule. Throughout this chapter, we will build molecular orbitals by overlapping atomic orbitals on different atoms. The buildup of electron density between the nuclei results in a covalent bond. Note how the detailed quantum theoretical results shown in Figure 13-1 correspond to the Lewis formula; both approaches picture a covalent bond as the sharing of an electron pair between two nuclei.

The lower part of Figure 13-1 shows the energies that correspond to the electron densities. Notice that interaction energies are negative for any distances at which the atoms attract each other. These negative values mean that energy is released when the H_2 bond is formed. The graph shows that, for H_2, the interaction energy has a minimum at the internuclear separation $R = 74$ pm. This value of R is the predicted length of a bond and is in excellent agreement with the experimental value.

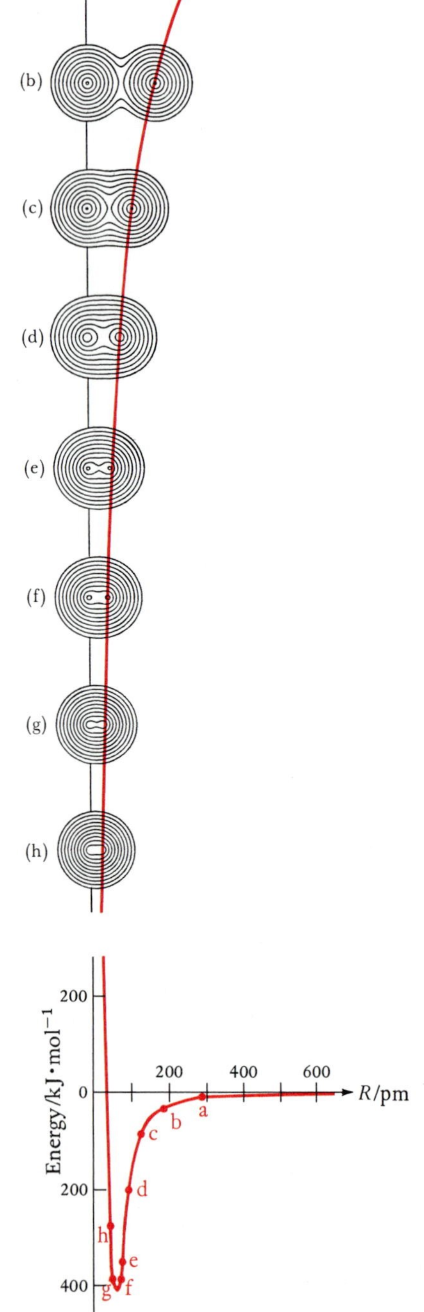

13-2. THE HYDROGEN MOLECULAR ION H_2^+ IS THE SIMPLEST DIATOMIC SPECIES

In this section we discuss a theory of bonding, called **molecular orbital theory,** that gives us insight into why, for example, two hydrogen atoms join to form a stable molecule whereas two helium atoms do not. This theory was developed in the 1930s by the German scientist Friedrich Hund and the American scientist Robert Mullikin. Although molecular orbital theory can be applied to all molecules, for simplicity we will consider only diatomic molecules.

Recall that we describe the electronic structure of atoms in terms of atomic orbitals, which are based on the set of orbitals that were given for a hydrogen atom. Because a hydrogen atom has only one electron, its atomic orbitals are relatively simple to calculate from the Schrödinger equation and serve as approximate orbitals for more complicated atoms. A one-electron system that applies to homonuclear diatomic molecules is the **hydrogen molecular ion,** H_2^+, which consists of two protons and one electron. The H_2^+ ion is stable relative to a separated H and H^+; its bond length is 106 pm and its bond energy is 255 $kJ \cdot mol^{-1}$.

Figure 13-1 Electron density contour diagrams of two hydrogen atoms as a function of their separation (upper part). At large separations, as in (a), the two orbitals appear simply as those of two separate atoms. As the atoms come together, the two separate atomic orbitals combine into one molecular orbital encompassing both nuclei, as in (b) through (h).

The lower part of the figure shows the energy of two hydrogen atoms as a function of their separation R. The labels (a) through (h) correspond to those in the upper part of the figure. At large distances (a), the two hydrogen atoms do not interact, so their interaction energy is zero. As the two atoms come together, they attract each other, and so their interaction energy becomes negative. When they are less than 74 pm apart, the interaction energy increases and they repel each other. The bond length of H_2 is the distance at which the energy is a minimum, that is, 74 pm. The energy at this distance is -436 $kJ \cdot mol^{-1}$, which is the energy required to dissociate the H_2 molecule into two separate hydrogen atoms.

The Schrödinger equation for H_2^+, like that for a hydrogen atom, is relatively easy to solve, and we obtain a set of wave functions, or orbitals, and a corresponding set of energies. As noted earlier, these orbitals extend over both nuclei in H_2^+ and therefore are called molecular orbitals. In Chapter 9 we discussed the shapes of the various hydrogen atomic orbitals and then used them to build up the electronic structures of more complicated atoms. In just the same way we now use the various H_2^+ molecular orbitals to build up the electronic structures of more complicated diatomic molecules.

Figure 13-2 shows the shapes of the first several molecular orbitals of H_2^+. Each shape represents the three-dimensional surface that encloses

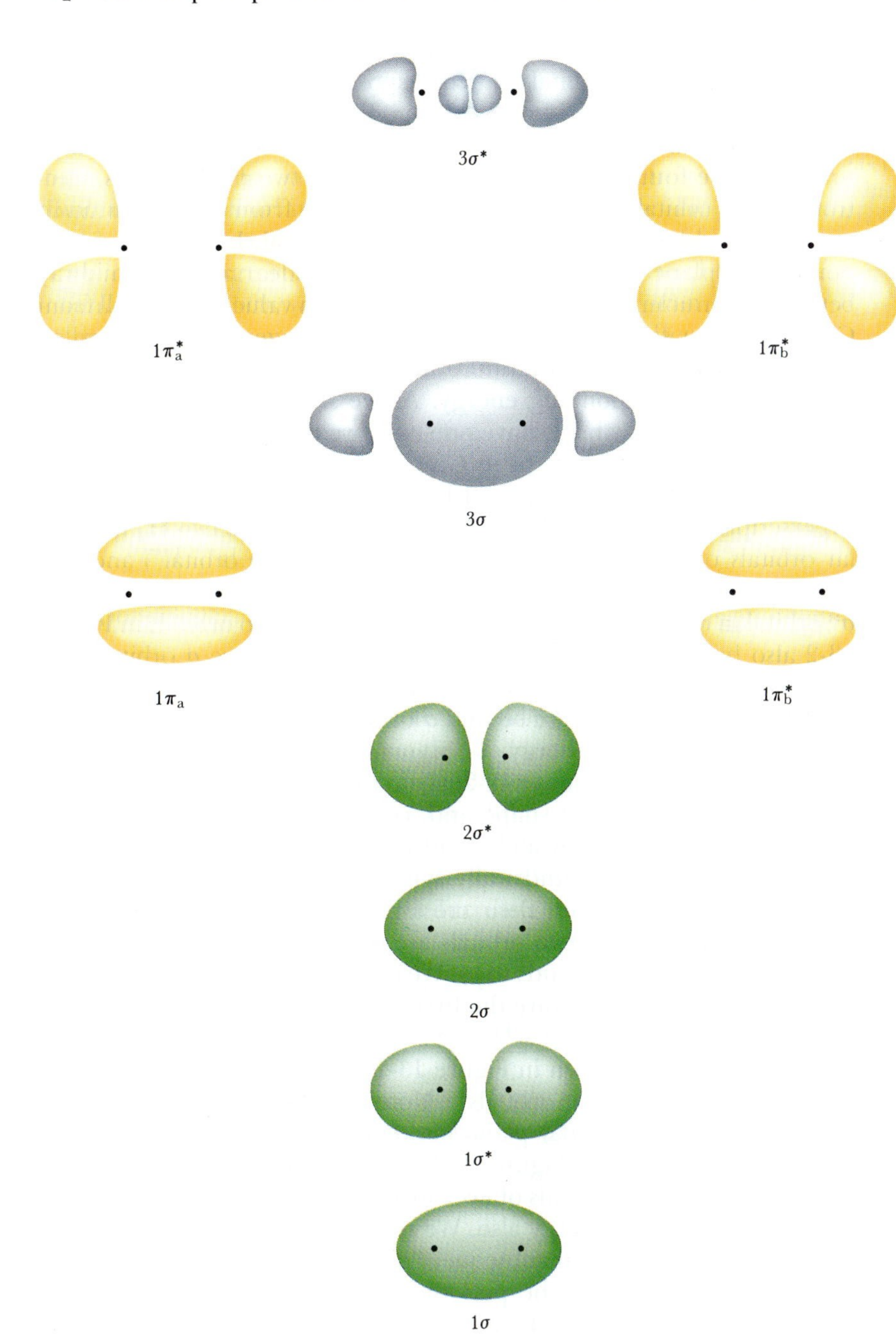

Figure 13-2 The three-dimensional surfaces that depict the shapes but not the relative sizes of the first few H_2^+ molecular orbitals. The orbitals are listed in order of increasing energy. Note that some molecular orbitals have nodal planes between the two nuclei, which are shown as heavy black dots. Also note that the two molecular orbitals designated by $1\pi_a$ and $1\pi_b$ have the same energy and that the two designated by $1\pi_a^*$ and $1\pi_b^*$ have the same energy.

13-3. MOLECULAR ORBITAL THEORY PREDICTS MOLECULAR ELECTRON CONFIGURATIONS

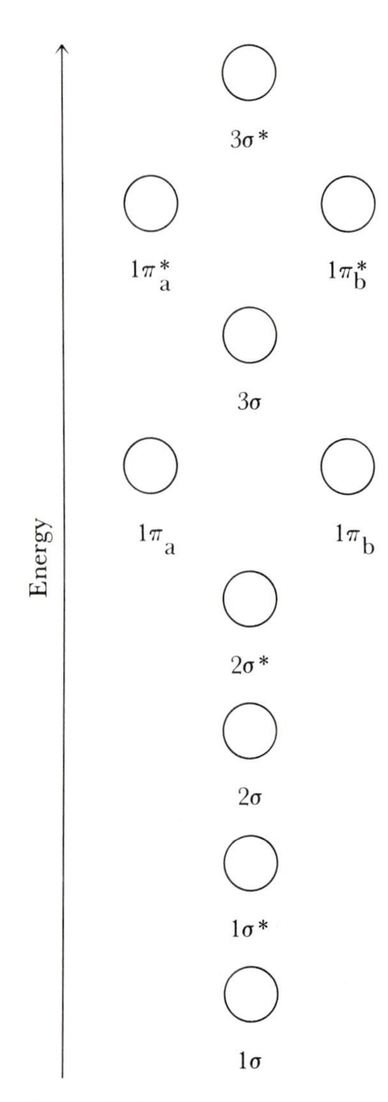

Figure 13-5 An energy-level diagram to be used for the homonuclear diatomic molecules H_2 through Ne_2. The orbitals are listed in order of increasing energy, $1\sigma < 1\sigma^* < 2\sigma < 2\sigma^* < 1\pi_a = 1\pi_b < 3\sigma < 1\pi_a^* = 1\pi_b^* < 3\sigma^*$. Electrons occupy these orbitals in accord with the Pauli exclusion principle.

Figure 13-5 lists the molecular orbitals 1σ to $3\sigma^*$ listed in order of increasing energy. We can use Figure 13-5 to write ground-state electron configurations for the homonuclear diatomic molecules Li_2 through Ne_2.

Lithium vapor contains diatomic lithium molecules, Li_2. A lithium atom has three electrons, so Li_2 has a total of six electrons. In the ground state of Li_2, the six electrons occupy the lowest three molecular orbitals in Figure 13-5, in accord with the Pauli exclusion principle. The ground-state electron configuration of Li_2 is $(1\sigma)^2(1\sigma^*)^2(2\sigma)^2$. There is a net of two bonding electrons, so the bond order is 1. Thus, we predict that Li_2 is more stable than two separated lithium atoms. Table 13-2 shows that Li_2 has a bond length of 267 pm and a bond energy of 110 $kJ \cdot mol^{-1}$. The process

$$Li_2(g) \rightarrow 2Li(g) \qquad \Delta H^\circ_{rxn} = 110 \text{ kJ}$$

is endothermic.

EXAMPLE 13-1: Use Figure 13-5 to write the ground-state electron configuration of N_2. Calculate the bond order of N_2 and compare your result with the Lewis formula for N_2.

Solution: There are 14 electrons in N_2. Using Figure 13-5, we see that its ground-state electron configuration is $(1\sigma)^2(1\sigma^*)^2(2\sigma)^2(2\sigma^*)^2(1\pi)^4(3\sigma)^2$. According to Equation (13-1), the bond order in N_2 is

$$\text{bond order} = \frac{10 - 4}{2} = 3$$

The Lewis formula for N_2, $:N{\equiv}N:$, is thus in agreement with molecular orbital theory. The triple bond in N_2 accounts for its short bond length (110 pm) and its unusually large bond energy (941 $kJ \cdot mol^{-1}$). The bond in N_2 is one of the strongest known bonds.

PRACTICE PROBLEM 13-1: Use molecular orbital theory to explain why neon does not form a stable diatomic molecule.

Answer: Neon's ground-state electron configuration is $(1\sigma)^2(1\sigma^*)^2(2\sigma)^2(2\sigma^*)^2(1\pi)^4(3\sigma)^2(1\pi^*)^4(3\sigma^*)^2$ giving a bond order of $(10 - 10)/2 = 0$.

One of the most impressive aspects of molecular orbital theory is its ability to predict that oxygen molecules will be **paramagnetic.** This property means that oxygen is attracted to a region between the poles of a magnet (Figure 13-6). Most substances are **diamagnetic,** meaning that they are slightly repelled by a magnetic field. Let's see how the paramagnetism of O_2 is related to its electron structure.

Each oxygen atom has 8 electrons; thus, O_2 has a total of 16 electrons. When the 16 electrons are placed according to the molecular orbital diagram given in Figure 13-5, the last 2 go into the $1\pi^*$ orbitals. As in the atomic case, we apply Hund's rule (Section 9-19), because the two $1\pi^*$ orbitals have the same energy. We, therefore, place one electron in each $1\pi^*$ orbital such that the two electrons have unpaired spins as shown in Figure 13-7. The ground-state electron configuration of O_2

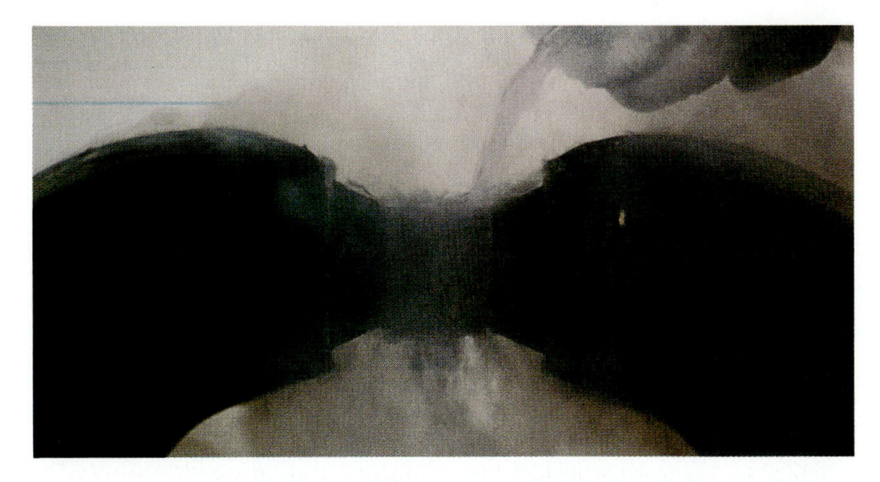

Figure 13-6 Liquid oxygen is attracted to the magnetic field between the poles of a magnet because oxygen is paramagnetic.

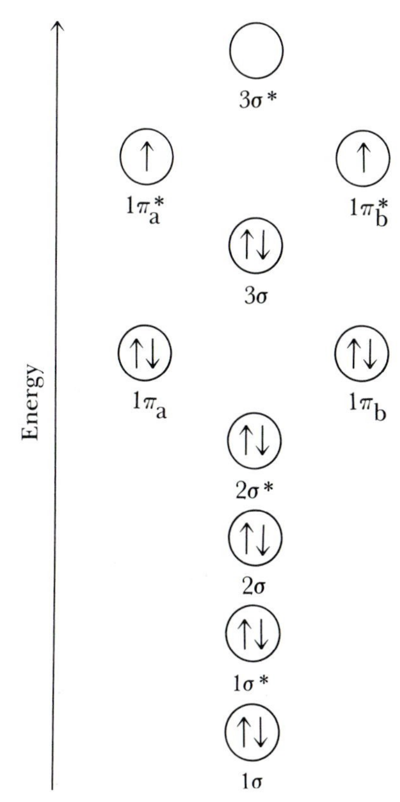

Figure 13-7 The ground-state electron configuration of O_2. There are 16 electrons in O_2, and they occupy the molecular orbitals as shown. Note that two of the electrons occupy the $1\pi^*$ orbitals in accord with Hund's rule, being placed in separate orbitals with unpaired spins. The molecule itself has a net electron spin, so it acts like a tiny magnet.

is $(1\sigma)^2(1\sigma^*)^2(2\sigma)^2(2\sigma^*)^2(1\pi)^4(3\sigma)^2(1\pi_a{}^*)^1(1\pi_b{}^*)^1$. Because each $1\pi^*$ orbital is occupied by one electron and the spins are unpaired, an oxygen molecule has a net electron spin; so it acts like a tiny magnet. Thus, O_2 is attracted into a region between the poles of a magnet.

The amount of oxygen in air can be monitored by measuring its paramagnetism. Because oxygen is the only major component in air that is paramagnetic, the measured paramagnetism of air is directly proportional to the amount of oxygen present. Linus Pauling developed a method using the paramagnetism of oxygen to monitor oxygen levels in submarines and airplanes in World War II. A similar method is still used by physicians to monitor the oxygen content in blood during anesthesia.

The Lewis formula of O_2 does not account for the paramagnetism of O_2. According to the octet rule, we should write the Lewis formula of O_2 as :Ö=Ö:, but this formula incorrectly implies that all the electrons are paired. The oxygen molecule is an exception to the utility of Lewis formulas, whereas the more fundamental molecular orbital theory is able to account successfully for the distribution of the electrons in O_2.

Table 13-2 gives the ground-state electron configurations of the homonuclear diatomic molecules Li_2 through Ne_2.

Table 13-2 Properties of the homonuclear diatomic molecules of the second-row elements

Species	Ground-state electron configuration	Bond order	Bond length/pm	Bond energy/kJ·mol^{-1}
Li_2	$(1\sigma)^2(1\sigma^*)^2(2\sigma)^2$	1	267	110
Be_2	$(1\sigma)^2(1\sigma^*)^2(2\sigma)^2(2\sigma^*)^2$	0	not observed	not observed
B_2	$(1\sigma)^2(1\sigma^*)^2(2\sigma)^2(2\sigma^*)^2(1\pi_a)^1(1\pi_b)^1$	1	159	289
C_2	$(1\sigma)^2(1\sigma^*)^2(2\sigma)^2(2\sigma^*)^2(1\pi)^4$	2	124	599
N_2	$(1\sigma)^2(1\sigma^*)^2(2\sigma)^2(2\sigma^*)^2(1\pi)^4(3\sigma)^2$	3	110	941
O_2	$(1\sigma)^2(1\sigma^*)^2(2\sigma)^2(2\sigma^*)^2(1\pi)^4(3\sigma)^2(1\pi_a^*)^1(1\pi_b^*)^1$	2	121	494
F_2	$(1\sigma)^2(1\sigma^*)^2(2\sigma)^2(2\sigma^*)^2(1\pi)^4(3\sigma)^2(1\pi^*)^4$	1	142	154
Ne_2	$(1\sigma)^2(1\sigma^*)^2(2\sigma)^2(2\sigma^*)^2(1\pi)^4(3\sigma)^2(1\pi^*)^4(3\sigma^*)^2$	0	not observed	not observed

EXAMPLE 13-2: Use Figure 13-5 to determine which species has the greater bond length, F_2 or F_2^-.

Solution: The ground-state electron configurations of F_2 and F_2^- are

$$F_2\text{: } (1\sigma)^2(1\sigma^*)^2(2\sigma)^2(2\sigma^*)^2(1\pi)^4(3\sigma)^2(1\pi^*)^4$$
$$F_2^-\text{: } (1\sigma)^2(1\sigma^*)^2(2\sigma)^2(2\sigma^*)^2(1\pi)^4(3\sigma)^2(1\pi^*)^4(3\sigma^*)^1$$

The bond orders are

$$\text{bond order } F_2 = \frac{10-8}{2} = 1$$

$$\text{bond order } F_2^- = \frac{10-9}{2} = \frac{1}{2}$$

Thus, we predict that F_2^- has a longer bond length than F_2.

PRACTICE PROBLEM 13-2: An excited state of O_2 has the electron configuration $(1\sigma)^2(1\sigma^*)^2(2\sigma)^2(2\sigma^*)^2(1\pi)^4(3\sigma)^1(1\pi^*)^3$. Compare the bond length of O_2 in this excited state to the bond length of O_2 in its ground state.

Answer: The bond length of O_2 is shorter in its ground state.

Molecular orbital theory can also be applied to heteronuclear diatomic molecules. The energy-level scheme in Figure 13-5 can be used if the atomic numbers of the two atoms in the molecule differ by only one or two atomic numbers.

EXAMPLE 13-3: Which of the following species would you expect to have the shortest bond length, CO^+, CO, or CO^-?

Solution: Using Figure 13-5, we see that the ground-state electron configurations of these three species are

$$CO^+\text{: } (1\sigma)^2(1\sigma^*)^2(2\sigma)^2(2\sigma^*)^2(1\pi)^4(3\sigma)^1$$
$$CO\text{ : } (1\sigma)^2(1\sigma^*)^2(2\sigma)^2(2\sigma^*)^2(1\pi)^4(3\sigma)^2$$
$$CO^-\text{: } (1\sigma)^2(1\sigma^*)^2(2\sigma)^2(2\sigma^*)^2(1\pi)^4(3\sigma)^2(1\pi^*)^1$$

with bond orders $2\frac{1}{2}$, 3, and $2\frac{1}{2}$, respectively. Thus, we predict that CO has the shortest bond, which is correct.

PRACTICE PROBLEM 13-3: Predict which of the following species has the largest bond energy, CN^+, CN, or CN^-?

Answer: CN^-

13-4. THE BONDING IN POLYATOMIC MOLECULES CAN BE DESCRIBED IN TERMS OF BOND ORBITALS

Molecular orbital theory can also be applied to polyatomic molecules. In doing so we construct molecular orbitals, this time by combining atomic orbitals from all the atoms in the molecule. Then, using the corresponding energy-level diagram, we place electrons into the molecular orbitals in accord with the Pauli exclusion principle. Because molecular orbitals are combinations of atomic orbitals from all the atoms in a polyatomic molecule, they are often spread over the entire molecule.

There is an alternative theory, called valence-bond theory, that is mathematically equivalent to molecular orbital theory. It recognizes that many chemical bonds have properties such as bond lengths and bond energies that are fairly constant from molecule to molecule. For example, the carbon-hydrogen bond lengths in many molecules are about 110 pm and their bond energies are a little greater than 400 $kJ \cdot mol^{-1}$. A large amount of experimental data such as these suggests that the bonding in many polyatomic molecules can be analyzed in terms of orbitals that are localized between pairs of bonded atoms.

Consider the methane molecule, CH_4, in which each hydrogen atom is joined to the central carbon atom by a covalent bond. As Figure 13-8 suggests, the bonding electrons, and hence the orbitals that describe them, are localized along the line joining the carbon and the hydrogen atoms. These electrons are said to occupy **localized bond orbitals,** and the two electrons that occupy a localized bond orbital are said to constitute a **localized covalent bond.** Note the similarity between this bonding picture in CH_4 and the Lewis formula for CH_4:

```
    H
    |
H—C—H
    |
    H
```

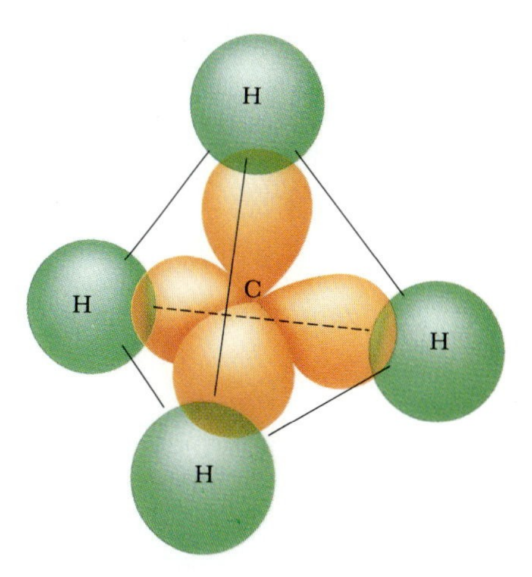

Figure 13-8 The bond orbitals in a methane molecule can be pictured as four carbon-hydrogen bond orbitals, directed toward the vertices of a tetrahedron. A localized bond orbital that is occupied by two electrons with opposite spins constitutes a covalent bond localized between two atoms.

The localized bond orbital approach that we are describing is a simplified version of the **valence-bond theory,** which Linus Pauling developed in the 1930s. The valence-bond theory makes it possible to translate Lewis formulas into the mathematical formulas of the quantum theory. Although molecular orbital theory and valence-bond theory appear to be quite different, they lead to essentially the same results. In our treatment of polyatomic molecules, we use a mixture of both molecular orbital theory (delocalized bonds) and valence-bond theory (localized bonds). We use valence-bond theory to treat σ bonds and molecular orbital theory for π bonds. We use this mixed approach because of its close connection with Lewis formulas and because in a great many molecules the σ bonds are localized and the π bonds are delocalized.

13-5. HYBRID ORBITALS ARE COMBINATIONS OF ATOMIC ORBITALS ON THE SAME ATOM

The simplest neutral polyatomic molecule is BeH_2. Beryllium hydride is an electron-deficient compound. Its Lewis formula,

H—Be—H

does not satisfy the octet rule. According to VSEPR theory, beryllium hydride is a symmetric linear molecule; the two Be—H bonds are 180° apart and are equivalent. Therefore, according to valence-bond theory, we must form two equivalent bond orbitals that are localized along the H—Be—H axis.

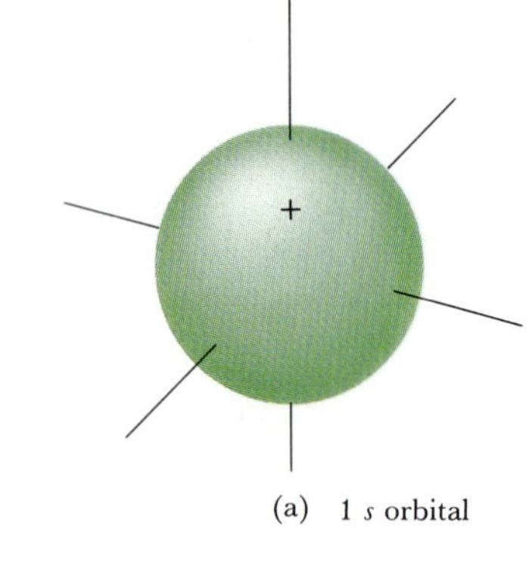

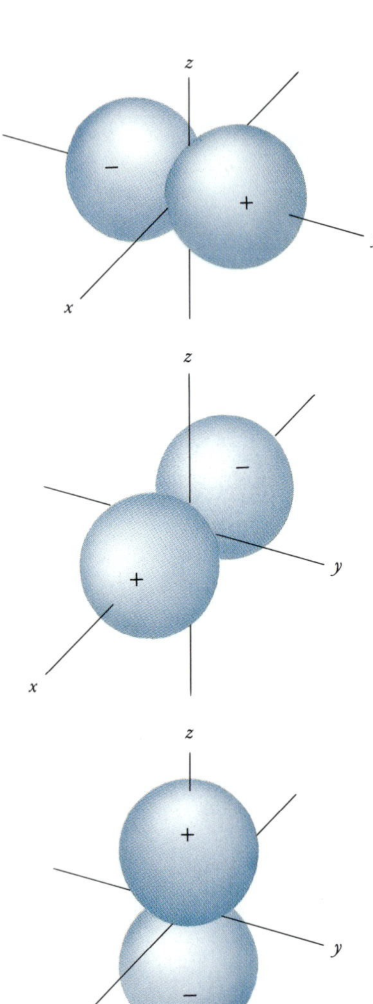

Figure 13-9 The hydrogen 1*s* and 2*p* orbitals showing the regions in which they have positive values and regions in which they have negative values. (a) The 1*s* orbital has a positive value everywhere, (b) but the 2*p* orbitals have a positive value in one lobe and a negative value in the other.

In order to describe the bonding in BeH_2 and many other polyatomic molecules, we introduce the idea of a hybrid orbital. We define a **hybrid orbital** as a combination of atomic orbitals *of the same atom*. In the case of BeH_2, we consider combinations of the valence atomic orbitals on a beryllium atom, namely, the 2*s* and the 2*p* orbitals. Because BeH_2 has two equivalent bonds 180° apart, we wish to construct two hybrid orbitals of the same symmetry. It turns out that it is possible to combine the 2*s* and *one* of the 2*p* orbitals to produce such orbitals.

Before we can discuss the formation of hybrid orbitals from atomic orbitals, we must consider a property of atomic orbitals that we did not discuss in Chapter 9: namely, atomic orbitals, which are described by certain mathematical formulas, have both positive and negative values. For example, as Figure 13-9 shows, a 1*s* orbital has positive values everywhere, but each 2*p* orbital has a positive value in one lobe and a negative value in the other. The values are an algebraic property of the orbitals and should not be confused with the sign of the electrical charge of an electron, which is always negative.

Figure 13-10 illustrates the formation of two hybrid orbitals from a 2*s* and a 2*p* orbital. Note that there is a buildup of the hybrid orbital where the values of the overlapping 2*s* and $2p_z$ orbitals have the same signs (+) and partial cancellation where they have opposite signs. Generally, when atomic orbitals overlap, they reinforce in regions where they have positive values and partially cancel in regions where their values have opposite signs. Figure 13-10 shows that one of our hybrid orbitals is directed toward the right (as drawn) and that the other has the same shape but is directed toward the left (as drawn).

These two hybrid orbitals on the beryllium atom are called ***sp* orbitals** because they are formed from the 2*s* orbital and one of the 2*p* orbitals. The *sp* orbitals have two important features: (1) each one provides a large region of positive sign to combine with a hydrogen 1*s* orbital, and (2) they are 180° apart. The two empty 2*p* orbitals that are not used to form the hybrid *sp* orbitals are perpendicular to each other and to the line formed by the *sp* orbitals (Figure 13-11).

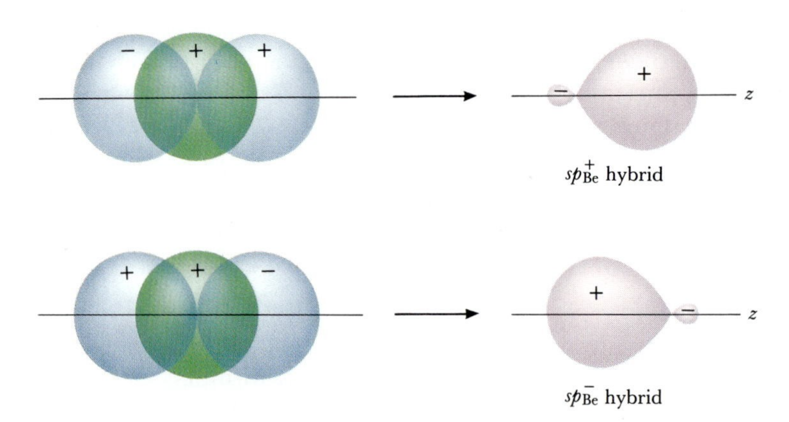

Figure 13-10 Formation of two *sp* hybrid orbitals from a 2*s* and a $2p_z$ atomic orbital. Note that the resulting *sp* orbitals are 180° apart. The *s* and *p* orbitals reinforce each other in regions where they have values with the same sign and partially cancel each other in regions where they have values of opposite sign. Consequently, each *sp* orbital consists of a large lobe of positive value and a small lobe of negative value. For simplicity, we often shall omit the little lobes of negative value.

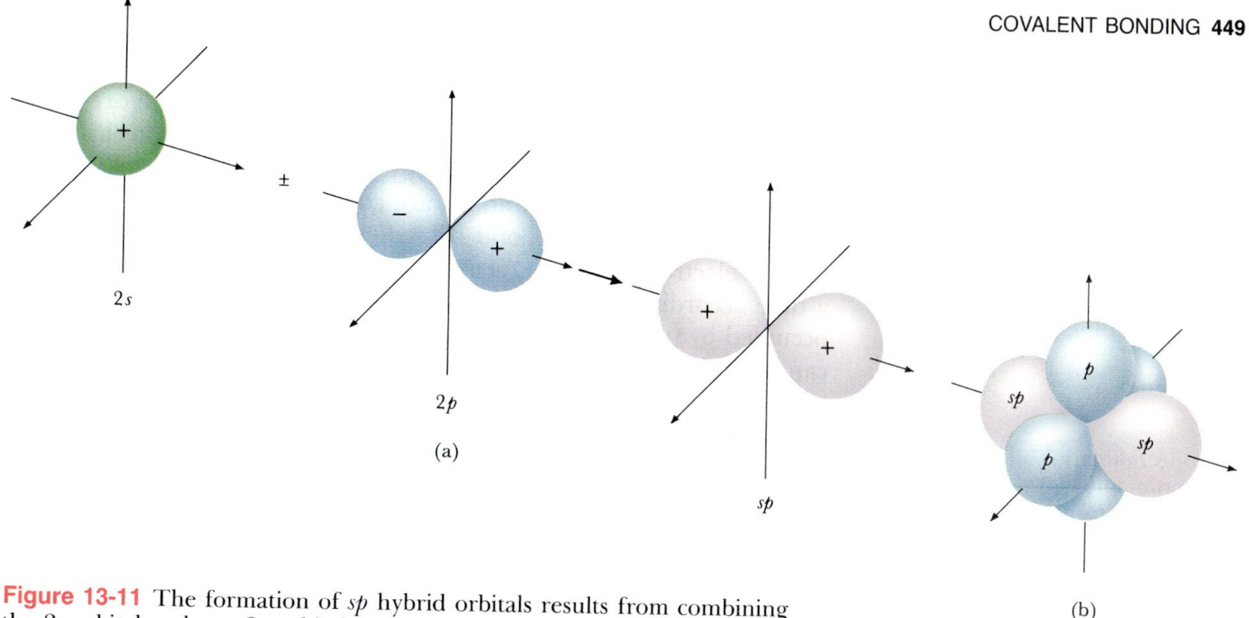

Figure 13-11 The formation of *sp* hybrid orbitals results from combining the 2*s* orbital and one 2*p* orbital on a single atom. The two *sp* orbitals are equivalent and are 180° apart. In (a), for simplicity, only the 2*p* orbital that is combined with the 2*s* orbital is shown. In (b), all the orbitals are shown. The two 2*p* orbitals that are not combined with the 2*s* orbital are perpendicular to each other and to the line formed by the *sp* orbitals. The little lobes of negative sign cannot be seen in this figure.

We now form two covalent bond orbitals by combining or overlapping each *sp* orbital with a hydrogen atomic 1*s* orbital. As shown in Figure 13-12, these are σ bond orbitals. Each is localized between the beryllium atom and a hydrogen atom, and they are 180° apart. Recall from Section 13-2 that the H_2^+ molecular orbitals occur as bonding and antibonding pairs. Similarly, when each beryllium *sp* overlaps with a

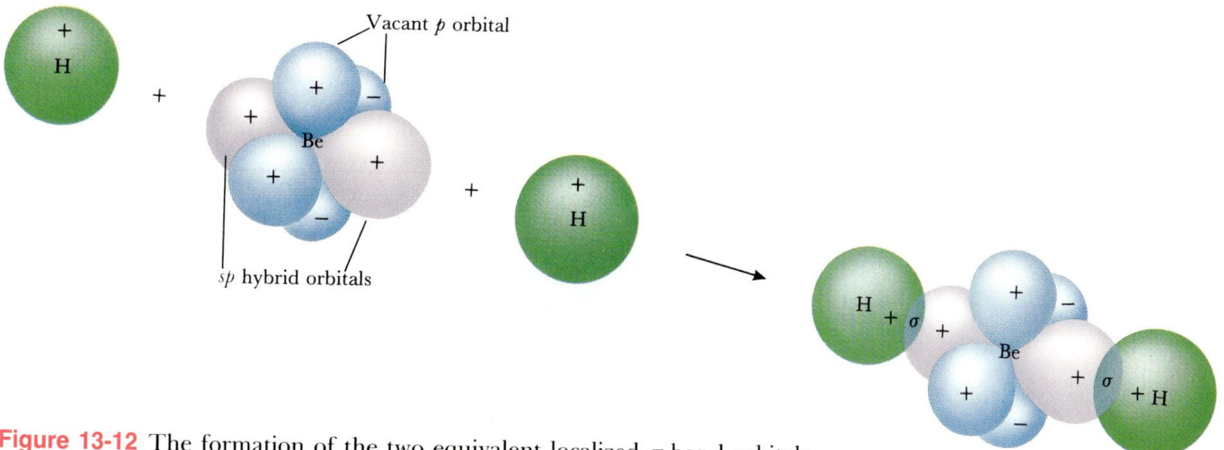

Figure 13-12 The formation of the two equivalent localized σ bond orbitals in BeH_2. Each bond orbital is formed by the overlap of a beryllium *sp* orbital and a hydrogen 1*s* orbital. There are four valence electrons in BeH_2: two from the beryllium atom and one from each of the two hydrogen atoms. The four valence electrons occupy the two localized bond orbitals, forming the two localized beryllium-hydrogen bonds in BeH_2.

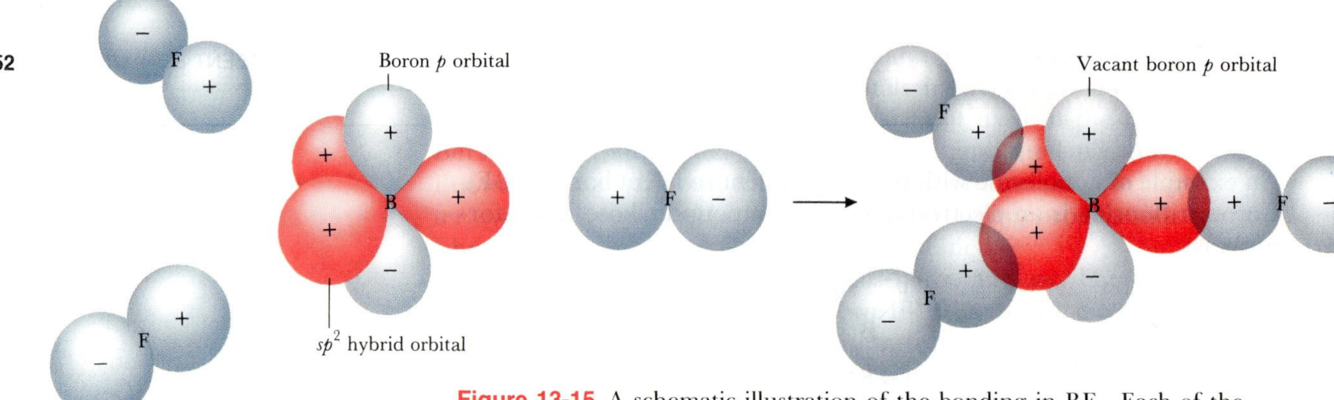

Figure 13-15 A schematic illustration of the bonding in BF_3. Each of the three boron-fluorine bond orbitals is formed by the overlap of a boron sp^2 orbital and a fluorine $2p$ orbital. The three localized boron-fluorine bond orbitals $B(sp^2) + F(2p)$ are occupied by six of the valence electrons and constitute the three covalent boron-fluorine bonds in BF_3.

resulting bond orbitals are cylindrically symmetric, so they are σ orbitals. Each localized σ bond orbital is occupied by two electrons of opposite spin to form a σ bond. The bonding in BF_3 is shown in Figure 13-15.

Note that when we combine a $2s$ orbital with one $2p$ orbital, we get two sp hybrid orbitals; and when we combine a $2s$ orbital with two $2p$ orbitals, we get three sp^2 hybrid orbitals. These two results are an example of the **principle of conservation of orbitals:**

1. If we combine atomic orbitals *on the same atom* to form hybrid orbitals, then the number of resulting hybrid orbitals is equal to the number of atomic orbitals combined.

2. If we overlap atomic orbitals *from different atoms,* then the number of resulting molecular orbitals is equal to the number of atomic orbitals combined. (Recall that we shall ignore the antibonding orbitals that are formed.)

In every case involving different atoms that we have considered so far, we have combined one atomic orbital from each of two atoms; the result has been one bonding molecular orbital and one antibonding molecular orbital. In later sections, however, we will show that when we combine atomic orbitals from more than two atoms, we get more than two molecular orbitals.

13-7. sp^3 HYBRID ORBITALS POINT TOWARD THE VERTICES OF A TETRAHEDRON

As predicted by VSEPR theory, methane (CH_4) is a tetrahedral molecule, and its four carbon-hydrogen bonds are equivalent. Thus, in order to describe the bonding in methane with valence-bond theory, we must construct four equivalent bond orbitals on the carbon atom. If we combine the $2s$ orbital and all three $2p$ orbitals on the carbon atom, then we get four equivalent hybrid orbitals, each pointing to one vertex of a tetrahedron (Figure 13-16). Because the four equivalent hybrid orbitals result from combining the $2s$ and all three $2p$ orbitals on the carbon atom, they are called ***sp*^{3} orbitals.**

Four equivalent localized σ bond orbitals in CH_4 are formed by overlapping each sp^3 orbital with a hydrogen $1s$ orbital (Figure 13-17). (Re-

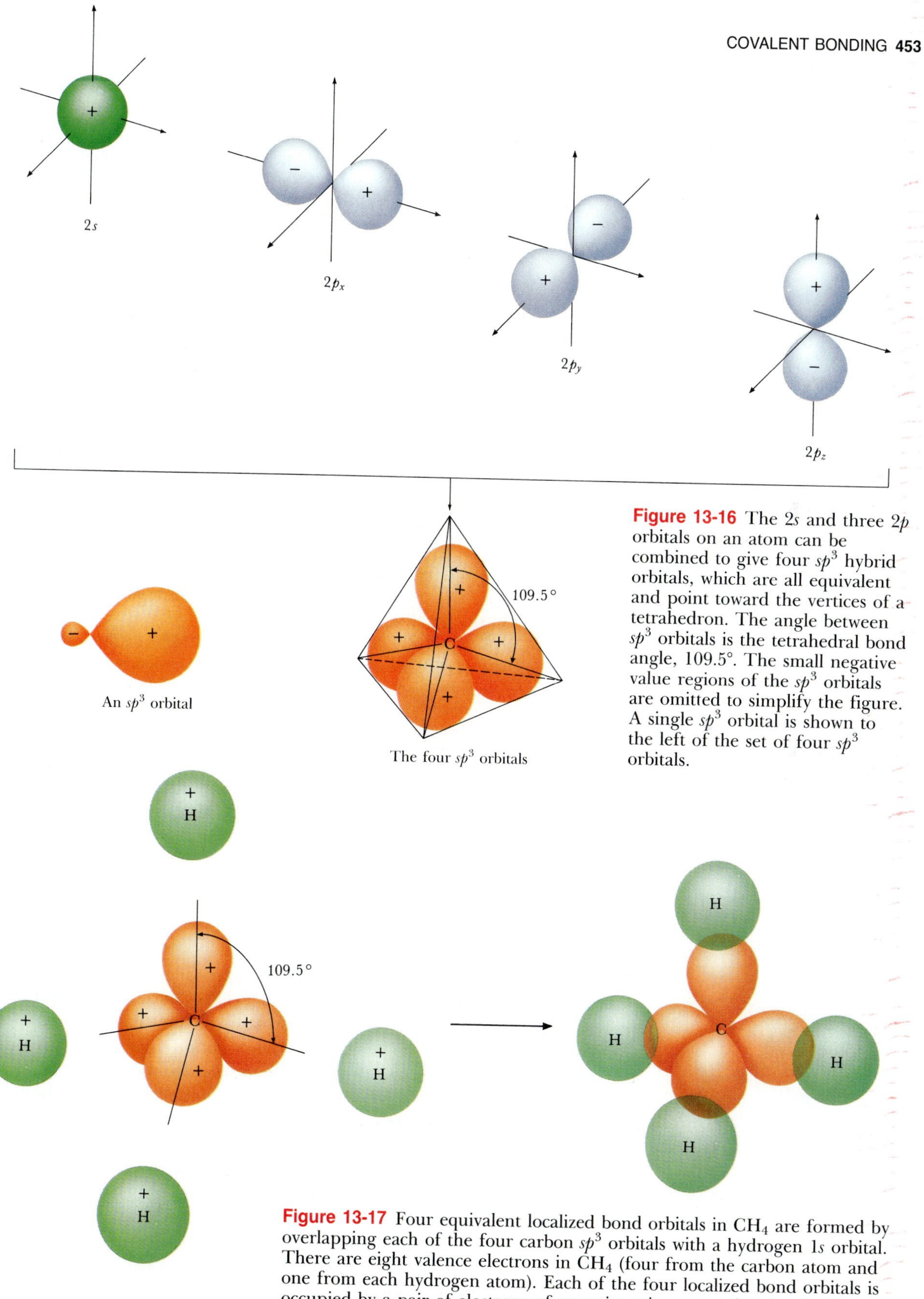

Figure 13-16 The 2*s* and three 2*p* orbitals on an atom can be combined to give four sp^3 hybrid orbitals, which are all equivalent and point toward the vertices of a tetrahedron. The angle between sp^3 orbitals is the tetrahedral bond angle, 109.5°. The small negative value regions of the sp^3 orbitals are omitted to simplify the figure. A single sp^3 orbital is shown to the left of the set of four sp^3 orbitals.

Figure 13-17 Four equivalent localized bond orbitals in CH_4 are formed by overlapping each of the four carbon sp^3 orbitals with a hydrogen 1*s* orbital. There are eight valence electrons in CH_4 (four from the carbon atom and one from each hydrogen atom). Each of the four localized bond orbitals is occupied by a pair of electrons of opposite spin, accounting for the four localized carbon-hydrogen bonds in CH_4.

call that we are ignoring the four σ^* orbitals also formed.) There are $4 + (4 \times 1) = 8$ valence electrons in CH_4. Each of the four localized bond orbitals is occupied by a pair of electrons of opposite spin, accounting for the four localized covalent bonds in CH_4.

EXAMPLE 13-5: Describe the bonding in the ammonium ion, NH_4^+, whose Lewis formula is

```
        H
        | ⊕
    H—N—H
        |
        H
```

Solution: We learned in Chapter 12 that NH_4^+ is tetrahedral. Thus, we wish to form four localized bond orbitals that point toward the vertices of a tetrahedron. It is appropriate to do this by forming sp^3 hybrid orbitals on the nitrogen atom by combining the nitrogen $2s$ orbital and all three of the nitrogen $2p$ orbitals. The resulting sp^3 hybrid orbitals are similar to the carbon atom sp^3 orbitals shown in Figure 13-17. We next form four equivalent localized bond orbitals by overlapping each sp^3 orbital on the nitrogen atom with a hydrogen $1s$ orbital. There are $5 + (4 \times 1) - 1 = 8$ valence electrons in NH_4^+. Two valence electrons of opposite spins occupy each of the four bond orbitals, thereby forming the four covalent bonds in NH_4^+. The bonding and the shape of the ammonium ion are similar to that shown for methane in Figure 13-17.

PRACTICE PROBLEM 13-5: Use localized bond orbitals to describe the bonding in the tetrafluoroborate ion, BF_4^-.

Answer: The BF_4^- has tetrahedral geometry, so it is appropriate to use sp^3 hybrid orbitals on the boron atom and $2p$ orbitals on the fluorine atoms. We overlap each of the boron sp^3 orbitals with a fluorine p orbital to form four $B(sp^3) + F(2p)$ localized bond orbitals. The eight valence electrons in BF_4^- occupy these four bond orbitals in accord with the Pauli exclusion principle.

We can also use sp^3 orbitals to describe the bonding in molecules that have no single central atom. An example is ethane, C_2H_6, whose Lewis formula is

```
       H  H
       |  |
    H—C—C—H
       |  |
       H  H
```

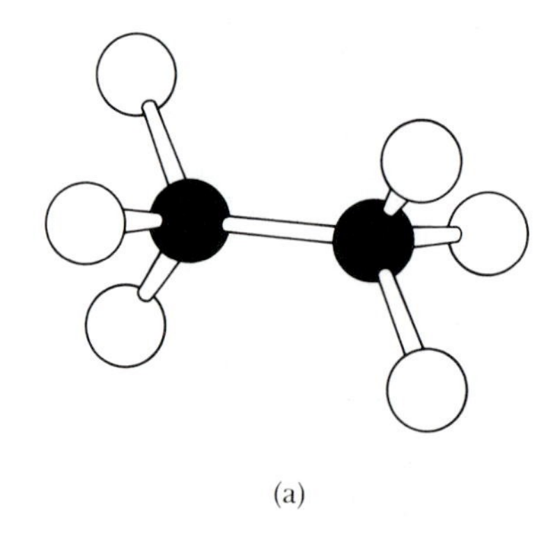

(a)

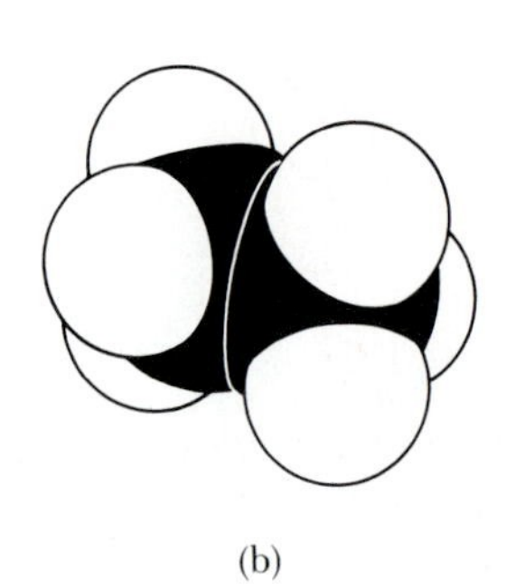

(b)

Figure 13-18 Molecular models of ethane. (a) Ball-and-stick model. (b) Space-filling model. Note that the bonds about each carbon atom are tetrahedrally oriented.

Figure 13-18 shows ball-and-stick and space-filling models of ethane. The disposition of the bonds about each carbon atom is tetrahedral, so it is appropriate to use sp^3 hybrid orbitals on the carbon atoms to describe the bonding. The carbon-carbon bond orbital in ethane is formed by the overlap of two sp^3 orbitals, one from each carbon atom; the six carbon-hydrogen bond orbitals in ethane result from the overlap of the three remaining sp^3 orbitals on each carbon atom with the hydrogen $1s$ atomic orbitals. Note that there are seven bond orbitals in ethane, and $(6 \times 1) + (2 \times 4) = 14$ valence electrons. The 14 valence electrons occupy the seven bond orbitals in ethane such that each bond orbital has 2 electrons of opposite spins. The resulting bonding in ethane is shown in Figure 13-19.

We should emphasize at this point that hybrid orbitals are "after the fact" constructions. The geometry associated with a set of hybrid orbit-

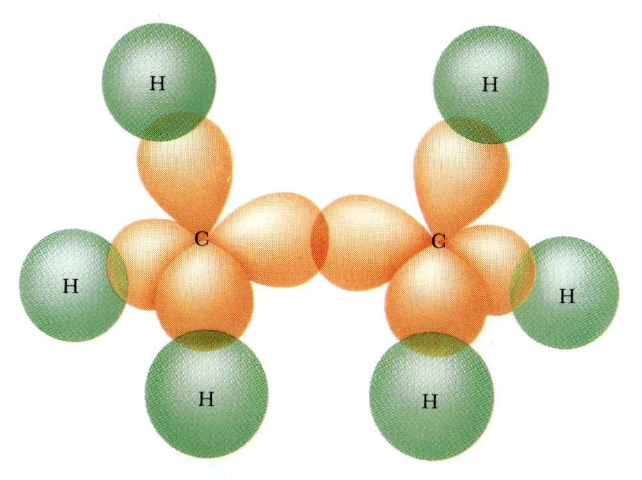

Figure 13-19 The six carbon-hydrogen bond orbitals in ethane result from the overlap of sp^3 orbitals on the carbon atoms and $1s$ orbitals on the hydrogen atoms. The carbon-carbon bond orbital results from the overlap of two sp^3 orbitals, one from each carbon atom. There are $(6 \times 1) + (2 \times 4) = 14$ valence electrons in ethane. Each of the seven bond orbitals is occupied by two valence electrons of opposite spins, accounting for the seven bonds in ethane.

als does *not* determine the geometry of the molecule; rather, the geometry of the molecule determines which type of hybrid orbitals are appropriate to describe the bonding in that molecule. For example, methane is not tetrahedral because the valence orbitals of the carbon atom are sp^3. The molecule is tetrahedral because that shape gives methane its lowest-possible energy.

13-8. WE CAN USE sp^3 ORBITALS TO DESCRIBE THE BONDING IN MOLECULES THAT HAVE FOUR ELECTRON PAIRS ABOUT THE CENTRAL ATOM

None of the molecules that we have considered up to now has had a lone pair of electrons. So let's consider H_2O, which has two lone pairs:

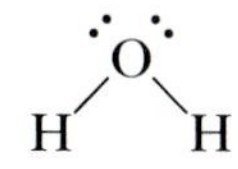

The oxygen atom in a water molecule is surrounded by four pairs of electrons: two in covalent bonds and two lone pairs. We expect from VSEPR theory that the four pairs of electrons will be tetrahedrally disposed. Thus, it is appropriate to use sp^3 hybrid orbitals on the oxygen atom to describe the bonding in H_2O. Each hydrogen $1s$ orbital overlaps with one of the sp^3 orbitals on the oxygen atom to produce an oxygen-hydrogen bond orbital (Figure 13-20). Next, we account for the eight valence electrons in H_2O. Four of the valence electrons occupy the two σ bond orbitals. The other four occupy the two nonbonded sp^3 orbitals, constituting the two lone electron pairs on the oxygen atom.

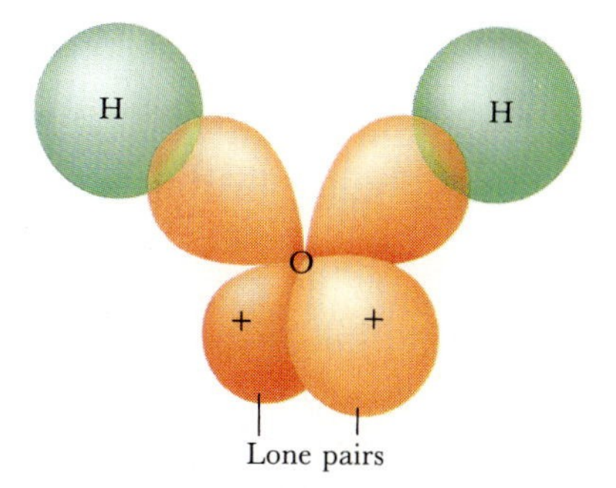

Figure 13-20 Bonding in the water molecule. Two of the oxygen sp^3 orbitals overlap with hydrogen $1s$ orbitals to form two equivalent localized bond orbitals. Of the eight valence electrons, four occupy the two bond orbitals and four occupy the two nonbonded sp^3 orbitals on the oxygen atom. The latter are lone electron pairs.

On the basis of this bonding, we predict that the H—O—H bond angle will be 109.5°. This prediction differs from the experimental value of 104.5°, because the four orbitals surrounding the oxygen atom are not used in the same way. Two are used to form bonds with the hydrogen atoms, and two are used for the lone-pair electrons. Recall from our discussion of VSEPR theory that we predicted that the H—O—H bond angle in H_2O will be somewhat less than the tetrahedral value of 109.5° because the lone electron pairs repel the two hydrogen-oxygen bonds.

EXAMPLE 13-6: Use hybrid orbitals to describe the bonding in ammonia, NH_3. The Lewis formula for NH_3 is

```
   ..
   N
  /|\
 H H H
```

Solution: Ammonia has three covalent bonds and one lone pair of electrons. We know from VSEPR theory that the four electron pairs in the valence shell of the nitrogen atom point toward the vertices of a tetrahedron. Therefore, it is appropriate to use sp^3 hybrid orbitals on the nitrogen atom. Three of these sp^3 orbitals form localized bond orbitals by overlapping with the hydrogen 1*s* orbitals. Thus, we can describe the bonding in an ammonia molecule in terms of three localized bond orbitals and a lone-pair (nonbonded) sp^3 orbital on the nitrogen atom.

There are eight valence electrons in NH_3. Six of them occupy the three localized bond orbitals and two occupy the nonbonded sp^3 orbital. The three occupied bond orbitals describe the three covalent bonds in NH_3; the occupied nonbonded sp^3 orbital on the nitrogen atom describes the lone pair of electrons in NH_3 (Figure 13-21). The use of sp^3 orbitals implies that the H—N—H bond angles are 109.5°. The four valence orbitals in NH_3 are not used equivalently, however, because one describes a lone pair. Thus, we should expect to find small deviations from a regular tetrahedral shape; and, in fact, the observed H—N—H bond angles in NH_3 are 107.3°.

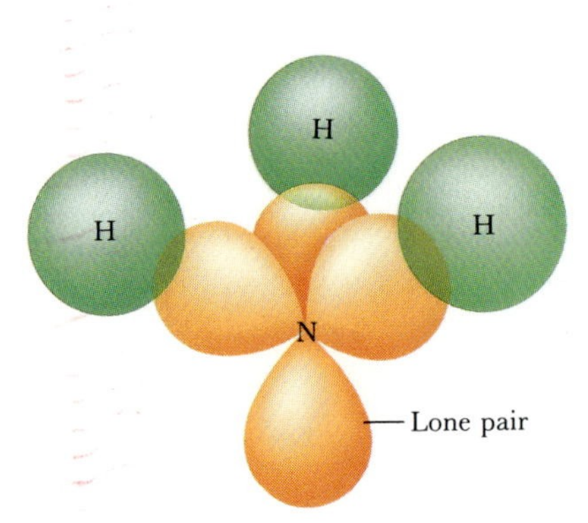

Figure 13-21 The use of sp^3 hybrid orbitals on nitrogen to describe the bonding in NH_3. Three of the nitrogen sp^3 orbitals overlap with hydrogen 1*s* orbitals to form three equivalent localized bond orbitals. The fourth nitrogen sp^3 orbital is a nonbonded orbital and is occupied by the lone pair of electrons in ammonia.

PRACTICE PROBLEM 13-6: Use localized bond orbitals to describe the bonding in NH_2^-.

Answer: NH_2^- is isoelectronic with H_2O, and the bonding in NH_2^- is similar to that in H_2O.

The sp^3 orbitals are appropriate to describe the bonds on an oxygen atom in alcohols, which are organic compounds involving an —OH group bonded to a carbon atom. The simplest alcohol is

```
    H
    |  ..
 H—C—O—H
    |  ..
    H
```

methanol
CH_3OH

The bonding in methanol is illustrated in Figure 13-22. Both the carbon atom and the oxygen atom are surrounded by four pairs of electrons, which according to VSEPR theory, we expect to be tetrahedrally disposed. Therefore, it is appropriate to use sp^3 hybrid orbitals on both the carbon atom and the oxygen atom to describe the bonding in a methanol molecule. The carbon-oxygen bond orbital results from the

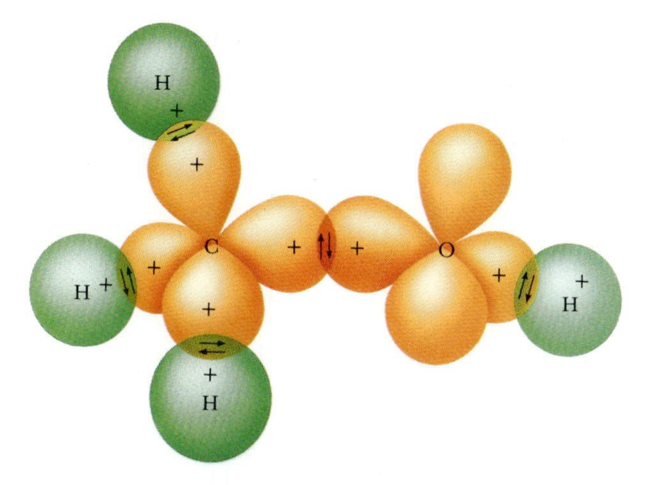

Figure 13-22 A schematic representation of the bonding orbitals in methanol. We use sp^3 orbitals on both the carbon atom and the oxygen atom.

overlap of sp^3 orbitals from each atom. The carbon-hydrogen and oxygen-hydrogen bonds result from the overlap of sp^3 and hydrogen 1*s* orbitals.

Note that there are five bond orbitals in CH_3OH. Ten of $4 + (4 \times 1) + 6 = 14$ valence electrons in CH_3OH occupy these five bond orbitals, and the other four valence electrons occupy the two remaining sp^3 orbitals as lone pairs on the oxygen atom.

EXAMPLE 13-7: Describe the bonding in a hydrogen peroxide molecule, H_2O_2. The Lewis formula for H_2O_2 is

$$\mathrm{H{-}\ddot{\underset{..}{O}}{-}\ddot{\underset{..}{O}}{-}H}$$

Solution: Each oxygen atom has four pairs of electrons, which we predict on the basis of VSEPR theory will be tetrahedrally disposed. Therefore, it is appropriate to use sp^3 hybrid orbitals on the oxygen atoms. The oxygen-oxygen bond orbital is formed by the overlap of an sp^3 orbital on each oxygen atom. Each hydrogen-oxygen bond orbital is formed by the overlap of an oxygen sp^3 orbital and a hydrogen 1*s* orbital.

There are $(2 \times 1) + (2 \times 6) = 14$ valence electrons in H_2O_2. Six of the valence electrons occupy the three bonding orbitals, forming the three bonds in the Lewis formula for H_2O_2. The other eight valence electrons occupy the four sp^3 orbitals, two on each oxygen atom, forming the two lone electron pairs shown on each oxygen atom in the Lewis formula. Figure 13-23 illustrates the bonding in H_2O_2.

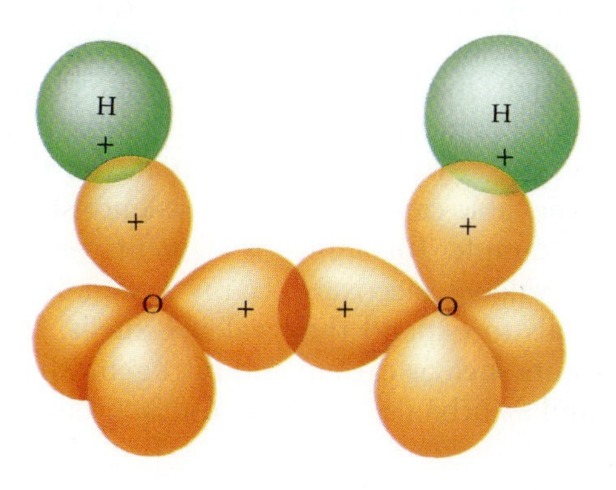

Figure 13-23 A schematic representation of the bonding orbitals in hydrogen peroxide, H_2O_2. The oxygen-oxygen bond orbital is formed by the overlap of an sp^3 orbital from each oxygen atom; each hydrogen-oxygen bond orbital is formed by the overlap of an oxygen sp^3 orbital and a hydrogen 1*s* orbital.

PRACTICE PROBLEM 13-7: Use localized bond orbitals to describe the bonding in hydrazine, N_2N_4.

Answer: The N—H bonds are N(sp^3) + H(1*s*) localized bonding orbitals. The N—N bond is a N(sp^3) + N(sp^3) localized bond orbital, and the lone pairs occupy the remaining N(sp^3) orbitals.

13-9. HYBRID ATOMIC ORBITALS CAN INVOLVE d *ORBITALS*

In Chapter 12 we learned about molecules that are trigonal bipyramidal (for example, phosphorus pentachloride, PCl_5) and octahedral (for example, sulfur hexafluoride, SF_6). The central atom in each of these molecules has an expanded valence shell, and one way of describing the bonding in such molecules is to include *d* orbitals in the construction of hybrid orbitals.

A combination of a 3*s* orbital, three 3*p* orbitals, and one 3*d* orbital gives five hybrid atomic orbitals that have trigonal bipyramidal symmetry. These five ***sp*³*d* orbitals** are interesting because they are not equivalent to one another. In fact, there are two sets of orbitals: a set of three equivalent equatorial orbitals and a set of two equivalent axial orbitals. This conclusion is consistent with the experimental fact that the five chlorine atoms in PCl_5 are not equivalent (Section 12-4). The five phosphorus-chlorine bond orbitals are formed by overlapping each phosphorus sp^3d hybrid orbital with a chlorine 3*p* orbital. Ten of the valence electrons (five from the phosphorus atom and one from each of the chlorine atoms) occupy the five localized bond orbitals (two electrons in each orbital) to form the five localized covalent bonds.

To describe the bonding in the octahedral SF_6 molecule, we need six equivalent hybrid orbitals on the sulfur atom, which point toward the vertices of an octahedron. This arrangement can be achieved by combining the 3*s* orbital, three 3*p* orbitals, and two 3*d* orbitals on the sulfur atom. The resulting six ***sp*³*d*² orbitals** point toward the vertices of a regular octahedron. The six sulfur-fluorine bond orbitals in SF_6 are formed by overlapping each sulfur sp^3d^2 orbital with a fluorine 2*p* orbital. Twelve of the valence electrons (six from the sulfur atom and one from each of the fluorine atoms) occupy the six localized bond orbitals to form the six localized covalent bonds.

For PCl_5 and SF_6, we use 3*s*, 3*p*, and some of the 3*d* orbitals on the central atom to form hybrid atomic orbitals. Quantum theory tells us that only orbitals of similar energy combine effectively. In other words, we can combine 3*s*, 3*p*, and 3*d* orbitals because they have similar energies. The combination of 3*d* orbitals with 2*s* and 2*p* orbitals does not produce hybrid orbitals that are effective in forming bonds because the 3*d* orbitals are much higher in energy than the 2*s* or 2*p* orbitals are. This restriction explains why only elements in the third and higher rows of the periodic table can expand their valence shells, as we saw in Chapter 11. For example, for atoms of elements such as phosphorus and sulfur, whose valence electrons occupy 3*s* and 3*p* orbitals, we can use their 3*d* orbitals to expand their valence shells, but for atoms of second-row elements such as carbon and nitrogen, whose valence electrons occupy 2*s* and 2*p* orbitals, we do not use 3*d* orbitals to form hybrid orbitals.

Table 13-3 Properties of hybrid orbitals

Hybrid	Number	Orbital geometry	Orbital angle	Examples	Hybridization required
sp	2	linear	180°	BeH_2, BeF_2	180°
sp^2	3	trigonal planar	120°	BF_3	120°
sp^3	4	tetrahedral	109.5°	CH_4, BF_4^-, NH_4^+	109.5°
sp^3d	5	trigonal bipyramidal	90°, 120°	PCl_5	90°, 120°
sp^3d^2	6	octahedral	90°	SF_6, AlF_6^{3-}	90°, 90°

Table 13-3 summarizes the hybrid atomic orbitals that we have introduced. Note that in each case the number of resulting hybrid orbitals is equal to the number of atomic orbitals used to construct them, in accord with the principle of the conservation of orbitals.

13-10. A DOUBLE BOND CAN BE REPRESENTED BY A σ BOND AND A π BOND

All the molecules that we have discussed so far in this chapter have only single bonds. One of the simplest molecules in which there is a double bond is ethene, C_2H_4, which is commonly known as ethylene. Its Lewis formula is

```
H       H
 \     /
  C═C
 /     \
H       H
```

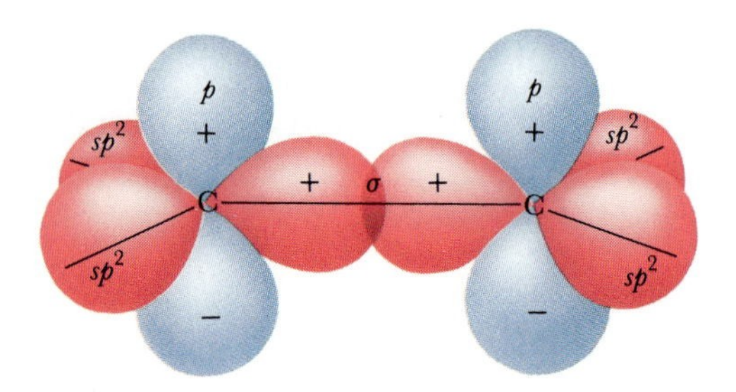

Figure 13-24 Two carbon atoms joined by the overlap of an sp^2 orbital from each. The resulting bond orbital is cylindrically symmetric around the carbon-carbon axis and therefore is a σ bond orbital. The carbon-carbon σ bond orbital constitutes part of the double bond in ethylene.

The geometry of an ethene molecule is quite different from that of ethane (Section 13-7). All six atoms in an ethene molecule lie in one plane, and each carbon atom is bonded to three other atoms. The geometry about each carbon atom is trigonal planar, and we saw in Section 13-6 that sp^2 hybrid orbitals are appropriate for describing this geometry. Therefore, we will describe the bonding in ethene using sp^2 orbitals on each carbon atom.

The first step is to join the two carbon atoms by overlapping an sp^2 orbital from each, as shown in Figure 13-24. The resulting carbon-carbon bond orbital is a σ bond orbital. There are $(2 \times 4) + (4 \times 1) = 12$ valence electrons in ethene. Two of these valence electrons occupy the carbon-carbon σ bond orbital to form a carbon-carbon σ bond. The four hydrogen atoms are bonded, two to each carbon atom, by overlapping the hydrogen 1*s* orbitals with the four remaining sp^2 orbitals on the carbon atoms, as shown in Figure 13-25. These four carbon-hydrogen σ bond orbitals are occupied by eight of the valence electrons to form four carbon-hydrogen σ bonds. All five bonds formed so far are σ bonds, and Figure 13-25 shows the **σ-bond framework** in ethene.

Recall that there is an unused 2*p* orbital on each carbon atom perpendicular to each H—C—H plane (Figure 13-24). If the two ends of the molecule are oriented such that these two 2*p* orbitals are parallel, then their overlap is maximized and the π orbital in Figure 13-26 results. This π orbital is occupied by the remaining two valence electrons to form a **π bond.** The double bond in ethene is described by the σ bond *and* the π bond in Figure 13-26. Note that the bond order due to the σ

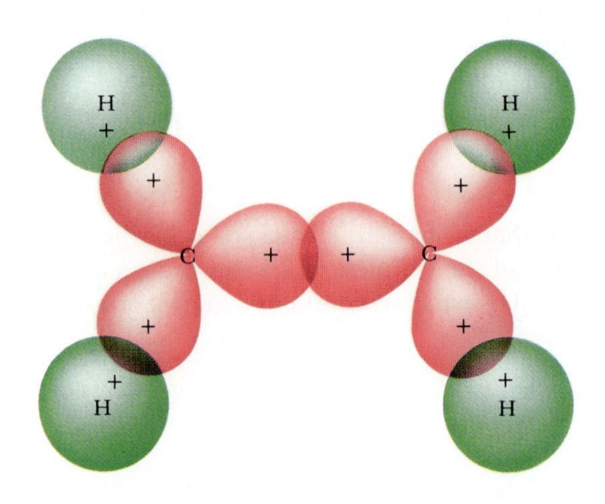

Figure 13-25 The σ-bond framework in ethene. The carbon-carbon bond orbital results from the overlap of two sp^2 orbitals, one from each carbon atom. The four carbon-hydrogen bond orbitals result from the overlap of carbon sp^2 orbitals and hydrogen 1*s* orbitals. The remaining *p* orbitals on the carbon atoms are not shown but are perpendicular to the page.

bond is one and that due to the π bond is one, for a total bond order of two. A σ bond and a π bond do not have the same energy, so a double bond, although much stronger than a single bond, is not twice as strong as a single bond. Carbon-carbon single-bond energies are about 350 $kJ \cdot mol^{-1}$, whereas carbon-carbon double-bond energies are about 600 $kJ \cdot mol^{-1}$.

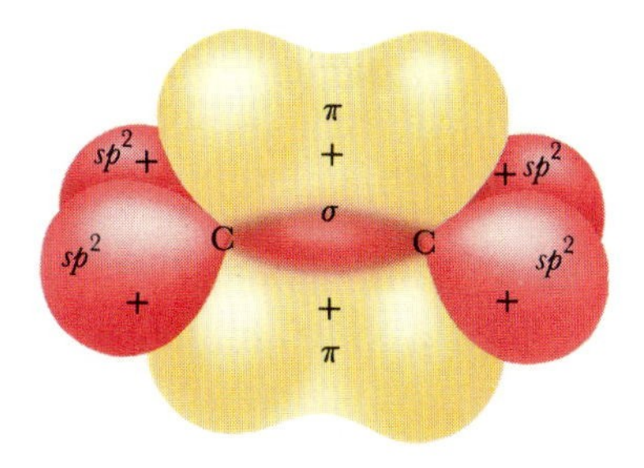

Figure 13-26 A double bond consists of a σ bond and a π bond. The σ bond results from the overlap of two sp^2 orbitals, one from each atom. The π bond results from the overlap of two p orbitals, one from each atom. The π orbital maintains the σ-bond framework in a planar shape and prevents rotation about the double bond.

EXAMPLE 13-8: Describe the bonding in the formaldehyde molecule, which has the Lewis formula

$$\begin{array}{l} H \\ \quad \diagdown \\ \quad\; C{=}\ddot{O}\!: \\ \quad \diagup \\ H \end{array}$$

Solution: From VSEPR theory we conclude that the formaldehyde molecule is planar with a trigonal geometry around the carbon atom. Because the bond angles are about 120°, it is appropriate to use sp^2 hybrid orbitals on the carbon atom. Further, because there are also three groups of electrons around the oxygen atom (the double bond and two lone pairs), it is appropriate to use sp^2 hybrid orbitals on the oxygen atom.

First, then, we overlap an sp^2 orbital on the carbon atom with an sp^2 orbital on the oxygen atom to form a carbon-oxygen σ bond orbital. The remaining two sp^2 orbitals on the carbon atom overlap with the 1s orbitals on the hydrogen atoms to form the two carbon-hydrogen σ bonds orbitals. There are $4 + 6 + (2 \times 1) = 12$ valence electrons in formaldehyde. Two of these valence electrons occupy the carbon-oxygen σ bond orbital to form a carbon-oxygen σ bond, and four of the valence electrons occupy the two carbon-hydrogen σ bond orbitals to form two carbon-hydrogen σ bonds. The remaining two sp^2 orbitals on the oxygen atom are occupied by four of the valence electrons, constituting the two lone pairs on the oxygen atom.

The remaining p orbital on the carbon atom and the remaining p orbital on the oxygen atom, both of which are perpendicular to the plane of the molecule, are now overlapped to form a carbon-oxygen π bond orbital, which is occupied by the two remaining valence electrons. Thus, the carbon-oxygen double bond is composed of a σ bond and a π bond. The bonding in formaldehyde is shown in Figure 13-27.

PRACTICE PROBLEM 13-8: Use localized bond orbitals to describe the bonding in phosgene, Cl_2CO.

Answer: As in formaldehyde, it is appropriate to use sp^2 hybrid orbitals on both the carbon atom and the oxygen atom. The resulting bonding can be summarized by a $C(sp^2) + O(sp^2)$ σ bond, two $Cl(3p_z) + C(sp^2)$ σ bonds, two lone pairs occupying two of the sp^2 orbitals on the oxygen atom, and a $C(2p_x) + O(2p_x)$ π bond.

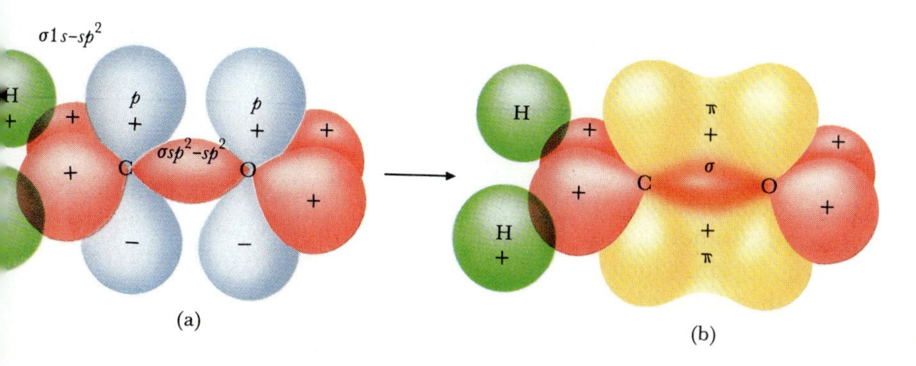

Figure 13-27 The bonding in formaldehyde, H_2CO. (a) The σ-bond framework, showing the unused 2p orbitals that are perpendicular to the plane formed by the four atoms. These two 2p orbitals combine to form a π bond orbital, which is occupied by two of the valence electrons. (b) The carbon-oxygen double bond consists of one σ bond and one π bond.

13-11. THERE IS LIMITED ROTATION ABOUT A DOUBLE BOND

The double bond in ethene consists of a σ bond and a π bond. The π bond locks the molecule into a planar shape (Figure 13-26). A significant amount of energy is required to break a π bond, so essentially no rotation occurs about a double bond at room temperature.

To see the consequences of the lack of rotation about double bonds, consider the molecule called 1,2-dichloroethene, ClCH=CHCl. (The 1,2- designation tells us that the chlorine atoms are attached to different carbon atoms.) Because there is essentially no rotation about the carbon-carbon double bond, there are two distinct structural forms of 1,2- dichloroethene:

trans isomer   *cis* isomer

The first of these is called *trans*-1,2-dichloroethene because the chlorine atoms lie across (*trans* means "across") the double bond from each other. The other is called *cis*-1,2-dichloroethene because the chlorine atoms lie on the same side (*cis* means "on the same side").

Molecules with the same atom-to-atom bonding but different spatial arrangements are called **stereoisomers.** The particular type of stereoisomerism that is displayed by 1,2-dichloroethene is called **cis-trans isomerism.** Stereoisomers, and cis-trans isomers in particular, have different physical properties. Although both the trans and cis isomers have polar bonds, the trans isomer has no net dipole moment, but the cis isomer does. We will show in the next chapter why the cis isomer, with its net dipole moment, has a higher boiling point than the nonpolar trans isomer. The boiling point of the trans isomer of 1,2-dichloroethene is 48°C and that of the cis isomer is 60°C.

An example of the importance of cis-trans isomerism occurs in the chemistry of vision. Although we have stated that there is no rotation allowed about double bonds in the ground state of a molecule, cis and trans isomers can interconvert if the molecule is supplied with sufficient energy in the form of heat or light:

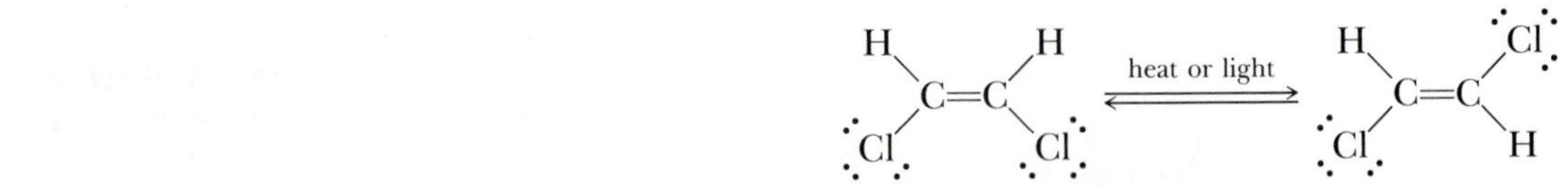

heat or light

It was determined in the 1950s that the chemistry of vision involves cis to trans isomerization. The retina of the eye contains a substance called rhodopsin, which consists of a molecule called 11-*cis*-retinal combined with a protein called opsin. When 11-*cis*-retinal is struck by a photon of visible light, it isomerizes at the cis double bond to give 11-*trans*-retinal:

11-*cis*-retinal $\xrightarrow{\text{light}}$ 11-*trans*-retinal

The numbers in this drawing refer to carbon atoms that are not specifically shown; recall the shorthand representation of benzene in Chapter 11. The shading in these formulas represents a planar region in the molecule. The shapes of the cis and trans isomers are significantly different, and the light-induced change in shape triggers a response in the optic nerve cells that is transmitted to the brain and perceived as vision. The vision response occurs through a sequence of processes that has been investigated thoroughly. The primary event, however, is the conversion of the cis to the trans isomer of retinal.

13-12. A TRIPLE BOND CAN BE REPRESENTED BY ONE σ BOND AND TWO π BONDS

Let us next consider a molecule that contains a triple bond. A good example is ethyne, C_2H_2, which is commonly called acetylene. The Lewis formula for the acetylene molecule is

$$\text{H—C}\equiv\text{C—H}$$

The acetylene molecule is linear, with each carbon atom bonded to only two other atoms. We saw in Section 13-5 that *sp* hybrid orbitals are appropriate to describe the bonding of an atom that forms two bonds separated by 180° (Figure 13-11).

We build a σ-bond framework for the acetylene molecule in two steps. We first form a carbon-carbon bond orbital by overlapping two *sp* orbitals, one from each carbon atom. Then we form the carbon-hydrogen bond orbitals by overlapping a hydrogen 1*s* orbital with the remaining *sp* orbital on each carbon atom. These three σ orbitals are occupied by six of the $(2 \times 4) + (2 \times 1) = 10$ valence electrons in acetylene to form the three σ bonds shown in Figure 13-27. The σ-bond framework of acetylene is shown in Figure 13-28.

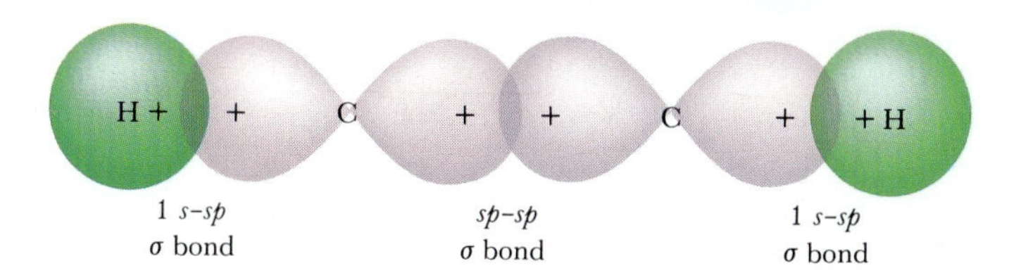

Figure 13-28 The σ-bond framework of acetylene. The carbon-carbon σ bond orbital results from overlapping two *sp* orbitals, one from each carbon atom. Each of the two carbon-hydrogen bond orbitals results from overlapping a carbon *sp* orbital and a hydrogen 1*s* orbital.

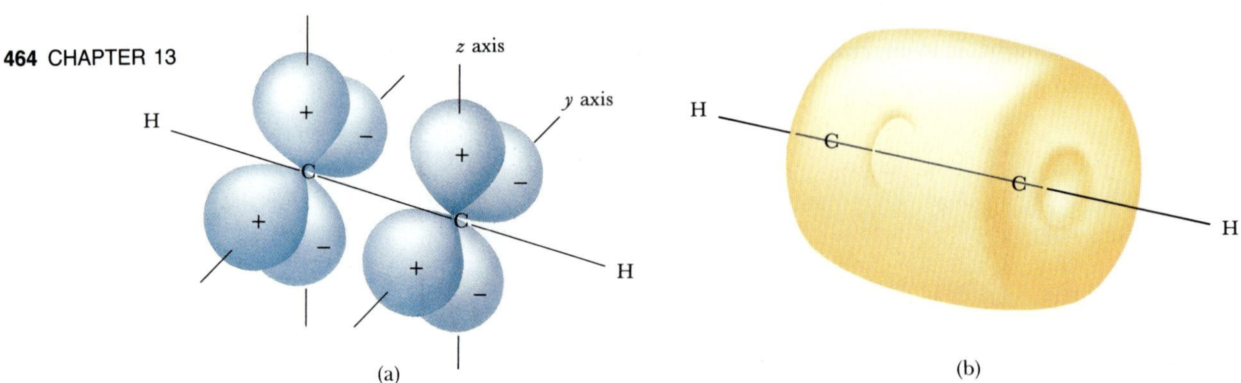

Figure 13-29 (a) The $2p$ orbitals on the carbon atoms in acetylene. The $2p$ orbitals that are directed along the z axis combine to form one π bond orbital, and the $2p$ orbitals directed along the y axis overlap to form another π bond orbital. (b) The two π bonds in acetylene constitute a barrel-shaped distribution of electron density in the bond region.

The remaining carbon $2p$ orbitals are perpendicular to the H—C—C—H axis, as shown in Figure 13-29a. These orbitals can combine to produce two π bond orbitals. These two bond orbitals are occupied by the four remaining valence electrons to form two π bonds. Note that the total bond order is three; one due to the σ bond and two due to the π bonds. The carbon-carbon triple bond consists of one σ bond and two π bonds (Figure 13-29b).

EXAMPLE 13-9: Compare the bonding in acetylene with that in hydrogen cyanide, HCN.

Solution: The Lewis formula for HCN is

H—C≡N:

Because VSEPR theory predicts that the molecule is linear, it is appropriate to use sp orbitals on both the carbon atom and the nitrogen atom in HCN. The σ-bond framework of HCN is shown in Figure 13-30; it is similar to that of C_2H_2 (Figures 13-28 and 13-29). The unused $2p$ orbitals of the carbon and nitrogen atoms combine to form the two π bond orbitals. There are four bond orbitals in HCN; two are σ bond orbitals and two are π bond orbitals. There are 10 valence electrons in HCN: 8 occupy the four bond orbitals, and 2 occupy the nitrogen sp orbital and constitute a lone electron pair on the nitrogen atom.

PRACTICE PROBLEM 13-9: Use localized bond orbitals to describe the bonding in CO.

Answer: The Lewis formula of CO is :C≡O:. The triple bond can be described as a C(sp) + O(sp) σ orbital, a C($2p_x$) + O($2p_x$) π orbital, and a C($2p_y$) + O($2p_y$) π orbital. The lone pairs occupy sp orbitals.

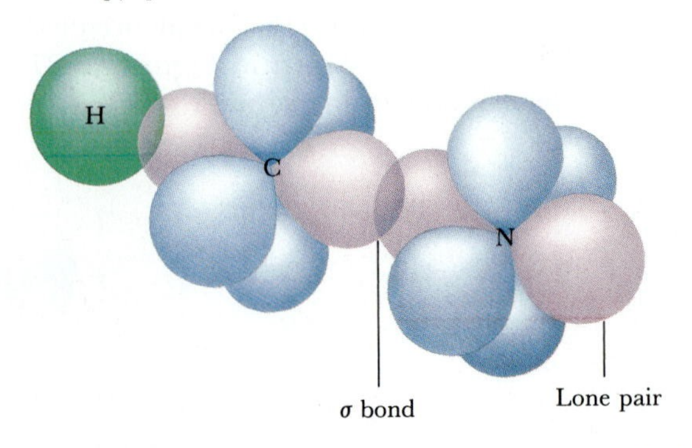

Figure 13-30 The σ-bond framework of hydrogen cyanide, HCN.

13-13. THE π ELECTRONS IN BENZENE ARE DELOCALIZED

In many molecules and ions, there are orbitals that extend over more than two adjacent atoms. One of the most important examples is benzene, C_6H_6. Recall that the Lewis formula representation of benzene (Section 11-6) involves two resonance forms. We expressed the superposition of these two resonance forms as

or

We can describe the bonding in benzene in terms of σ bonds and π bonds. Benzene is a planar molecule in the shape of a regular hexagon. The angles in a regular hexagon are 120°, so the three bonds surrounding each carbon atom lie in a plane at an angle of 120°. Thus, it is appropriate to use sp^2 hybrid orbitals on the carbon atoms to describe the bonding in benzene. This depiction leads directly to the σ-bond framework shown in Figure 13-31. Note that there are 12 σ bond orbitals.

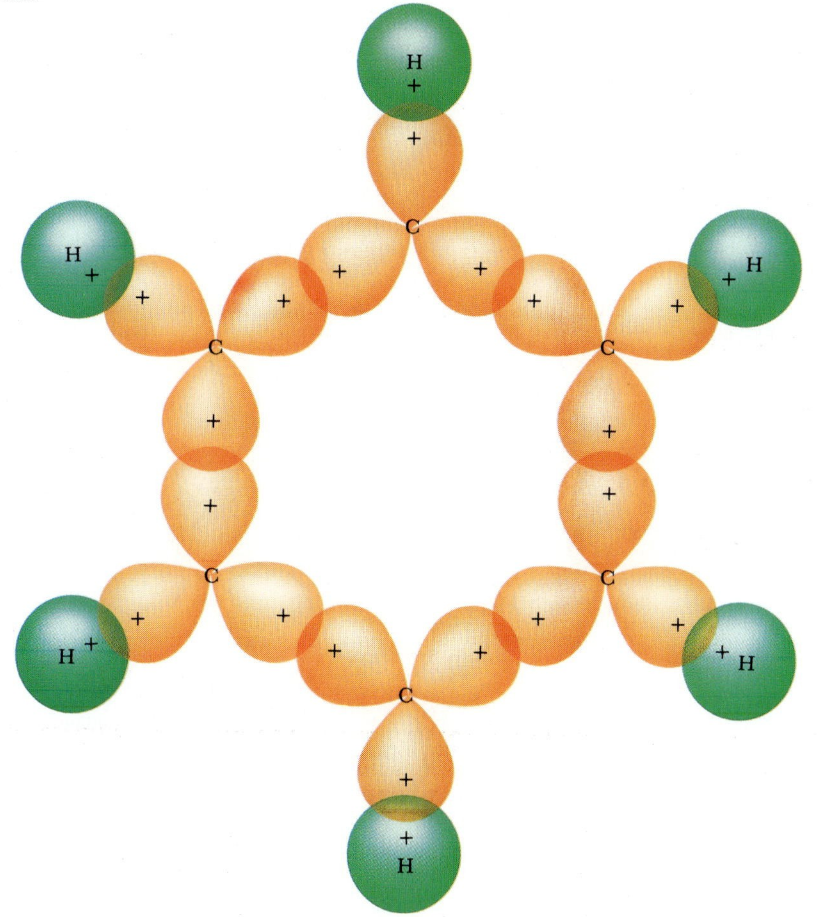

Figure 13-31 The σ-bond framework in a benzene molecule. Each carbon-carbon bond orbital results from the overlap of sp^2 orbitals, and each carbon-hydrogen bond orbital results from the overlap of a carbon sp^2 orbital and a hydrogen 1*s* orbital. All 12 atoms lie in a single plane, so benzene is a planar molecule. The six carbon atoms form a regular hexagon. Not shown are the six vacant *p* orbitals, one on each carbon atom, that are perpendicular to the hexagonal plane. (See Figure 13-32.)

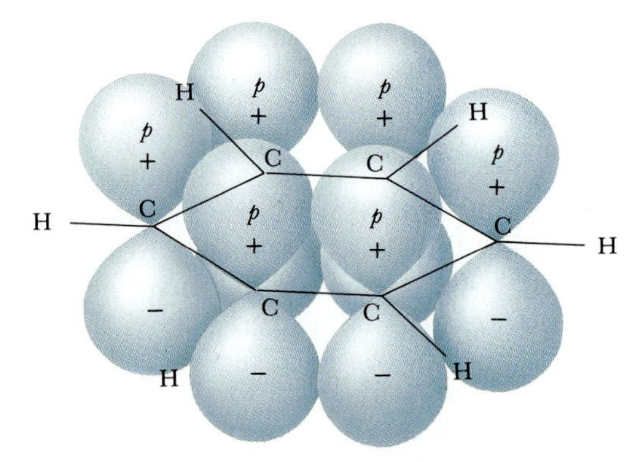

Figure 13-32 The individual $2p$ orbitals on the carbon atoms in a benzene ring. Note that they are perpendicular to the ring.

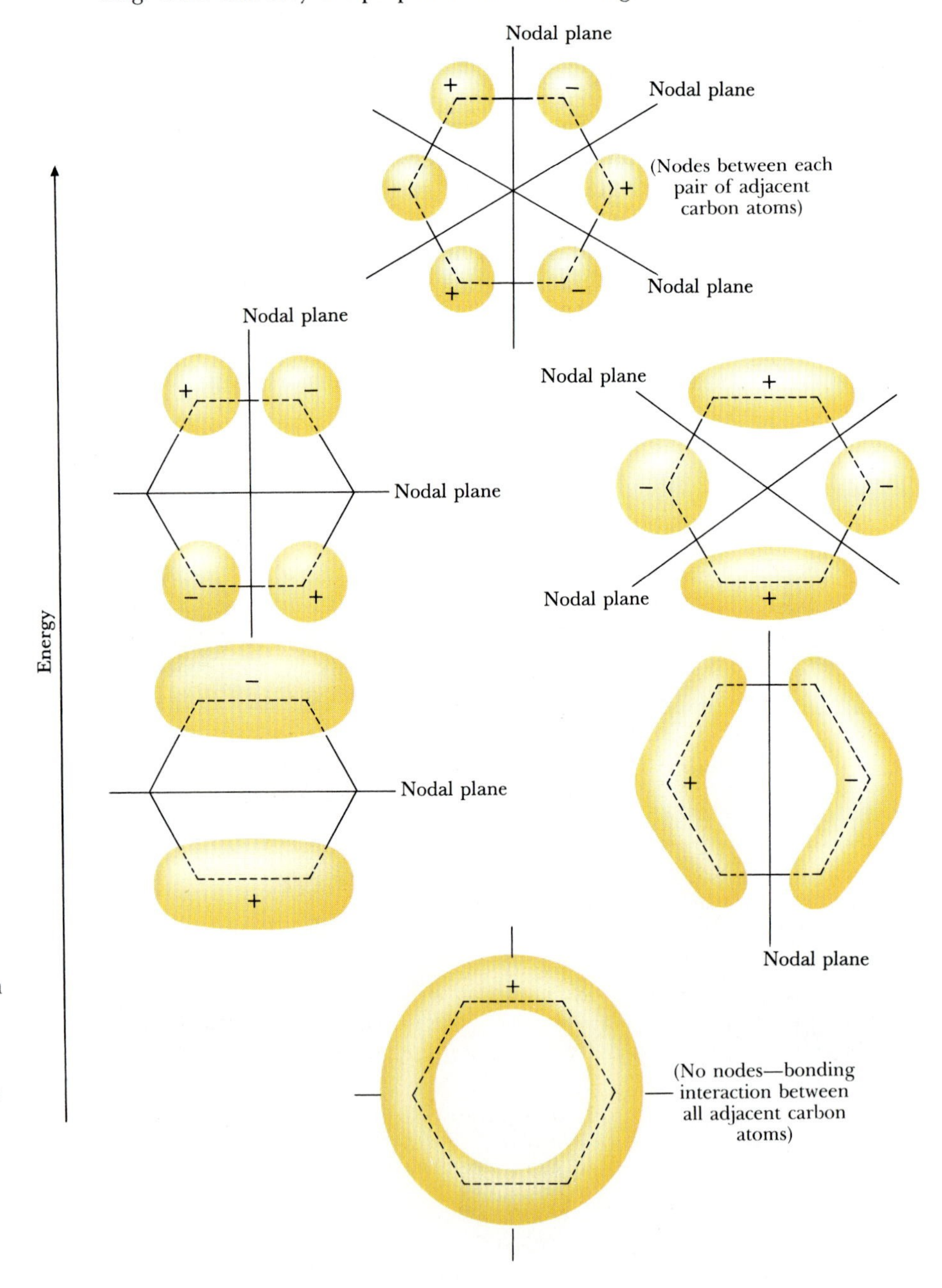

Figure 13-33 The six π molecular orbitals of benzene that result from combining the $2p$ orbitals on each of the carbon atoms. The three π orbitals of lowest energy are bonding orbitals, and the other three are antibonding orbitals. Note that the energies of the orbitals increase as the number of nodal planes increases.

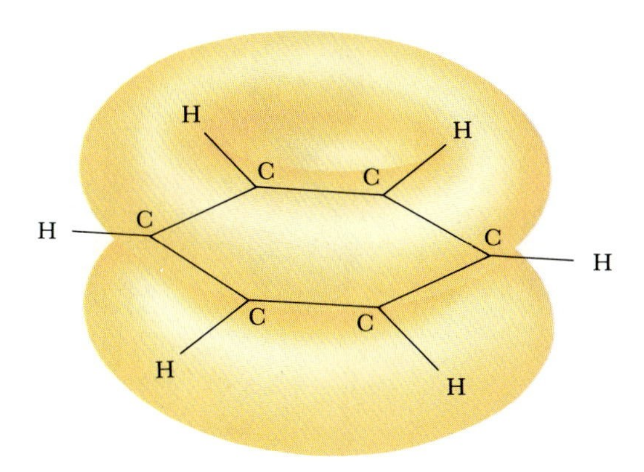

Figure 13-34 The total π-electron density of benzene.

Each carbon atom also has a $2p$ orbital that is perpendicular to the hexagonal plane (Figure 13-32). These six p orbitals combine to give a total of six π orbitals (conservation of orbitals), which are depicted in Figure 13-33. The three π orbitals of lowest energy are bonding orbitals, whereas the three π orbitals of highest energy are antibonding orbitals. Note that the energies of these π orbitals increase as the number of nodal planes increases.

There are $(6 \times 4) + (6 \times 1) = 30$ valence electrons in benzene, and 24 of these valence electrons occupy the 12 σ bond orbitals. The remaining 6 valence electrons occupy the three bonding π orbitals. The total π electron charge density is depicted in Figure 13-34. The bond order of each carbon-carbon bond due to the σ bond is 1, and the total bond order due to the π electrons is 3, or $\frac{1}{2}$ for each carbon-carbon bond. Thus, the total, average carbon-carbon bond order in benzene is $1\frac{1}{2}$.

The bonding π orbitals in benzene are not associated with any particular pair of carbon atoms, so they are said to be **delocalized orbitals,** and electrons in these orbitals are said to be **delocalized.** The delocalization of the electrons around the benzene ring is an example of **charge delocalization.** Charge delocalization in our quantum theoretical description is what resonance in the Lewis formula description attempts to show. The use of delocalized π molecular orbitals provides a more satisfying and clearer picture than do the Lewis formula resonance forms. Charge delocalization confers a greater degree of stability on a molecule relative to the hypothetical species with localized π bonds.

13-14. SOME π ORBITALS ARE NONBONDING ORBITALS

When we combine p orbitals from more than two atoms, the resulting π orbitals depend not only on the number of p orbitals but also on the geometry of the molecule. Consider an ozone molecule, O_3, which is described by two Lewis formula resonance forms

$$:\underset{..}{O}=\overset{\oplus}{\ddot{O}}-\underset{..}{\ddot{O}}:^{\ominus} \longleftrightarrow {}^{\ominus}:\underset{..}{\ddot{O}}-\overset{\oplus}{\ddot{O}}=\underset{..}{O}:$$

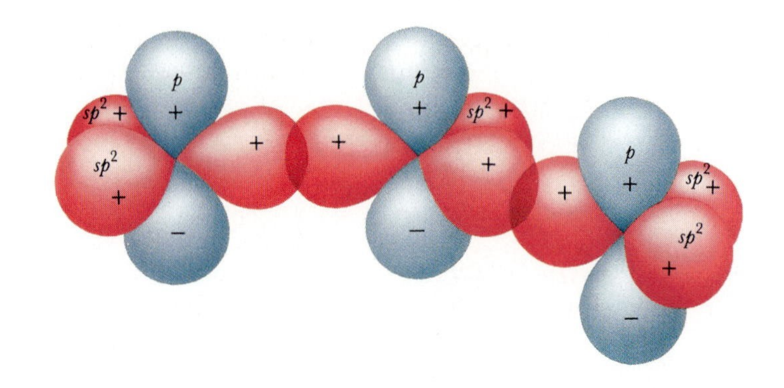

Figure 13-35 The σ-bond framework of a bent homonuclear triatomic molecule such as O_3. The σ-bond framework is formed from sp^2 hybrid orbitals on each of the three atoms.

The central oxygen atom is surrounded by three sets of electrons, so VSEPR theory tells us that O_3 is bent. Thus, it is appropriate to use sp^2 hybrid orbitals to describe the bonding in O_3. We will use sp^2 hybrid orbitals on all three atoms. A σ-bond framework can be formed, where each oxygen-oxygen bond orbital results from overlapping an sp^2 orbital from each atom (Figure 13-35). Four of the $3 \times 6 = 18$ valence electrons occupy the two oxygen-oxygen σ bond orbitals to form the two oxygen-oxygen σ bonds, and ten of them occupy the remaining sp^2 orbitals as lone-pair electrons (Figure 13-35).

The three $2p$ orbitals (one on each oxygen atom) combine to give the three π molecular orbitals shown in Figure 13-36. The orbitals of lowest and highest energies are a bonding and an antibonding π orbital, respectively. The middle π orbital is a **nonbonding orbital.** Electrons in

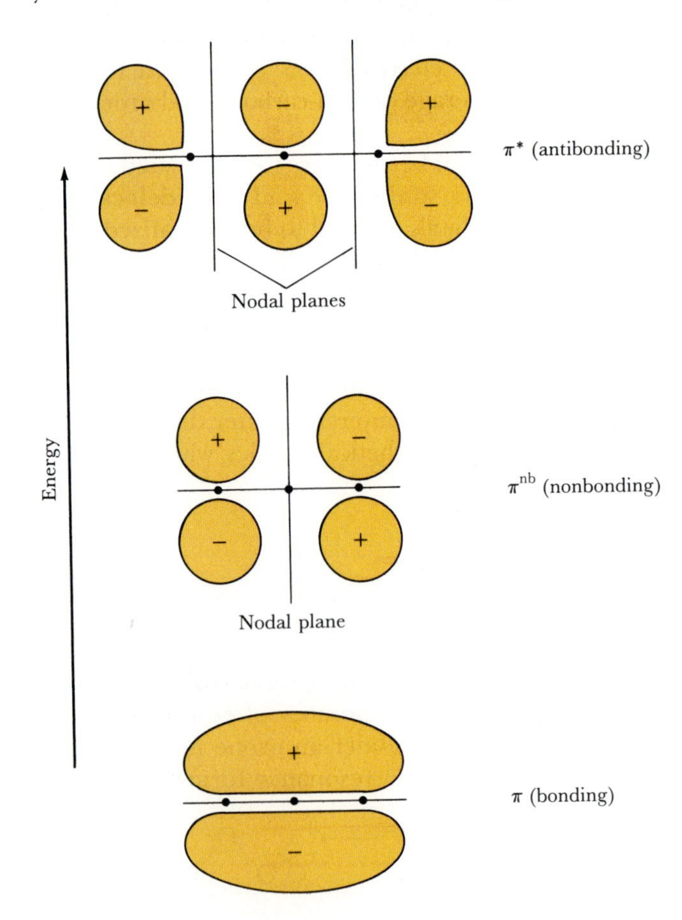

Figure 13-36 The three π orbitals of O_3 that result from combining the $2p$ orbitals on each of the oxygen atoms. The π orbital of lowest energy is a bonding orbital; the orbital of highest energy is an antibonding orbital; and the middle orbital is a nonbonding orbital. Electrons in nonbonding molecular orbitals neither contribute to nor detract from bonding.

nonbonding molecular orbitals neither contribute to nor detract from bonding. **Nonbonding electrons** (electrons in nonbonding molecular orbitals) are not included in a calculation of bond order (Equation 13-1). As also noted in Figure 13-33 for benzene, the energies of these orbitals increase as the number of nodal planes increases.

Up to now we have accounted for 14 of the valence electrons in O_3. The four remaining valence electrons occupy the two π orbitals of lowest energy. Two of these electrons occupy the π bonding orbital and the other two occupy the nonbonding orbital. The total bond order in O_3 is three: two due to the localized oxygen-oxygen σ bonds and one due to the one occupied delocalized π bond. The average oxygen-oxygen bond order is $1\frac{1}{2}$, and therefore, the oxygen-oxygen bond length should be between those for oxygen-oxygen single and double bonds. This prediction is found to be true experimentally.

SUMMARY

In Chapter 9 we discussed the atomic orbitals of the simplest atom, hydrogen, and then used them to write electron configurations for other atoms. In this chapter, we presented the molecular orbitals of the simplest diatomic species, H_2^+, and then used them to discuss the bonding in other diatomic molecules. The bonding properties of diatomic molecules depend on the number of electrons in bonding and antibonding orbitals. Molecular orbital theory correctly predicts that the diatomic molecules He_2 and Ne_2 do not exist and that O_2 is paramagnetic.

In our treatment of polyatomic molecules, we use a mixture of valence-bond theory and molecular orbital theory. We use valence-bond theory to treat the localized bonds, usually in terms of hybrid orbitals, and we use molecular orbital theory to treat delocalized bonds. Once the geometry of a molecule is known, we can choose a hybrid orbital to describe the bonding. Table 13-3 summarizes the associated geometries of various hybrid atomic orbitals.

Bond orbitals for single bonds are cylindrically symmetric about the bond axis and are called σ bond orbitals. A σ bond orbital occupied by two electrons of opposite spins constitutes a σ bond.

Double bonds can be represented by one σ bond and one π bond. The π bond is formed by the combination of p orbitals on adjacent atoms. The atoms bonded directly to double-bonded atoms all lie in one plane. Because a molecule in its ground state cannot rotate about a double bond, *cis* and *trans* isomers result. Triple bonds can be represented by one σ bond and two π bonds.

In some molecules, such as benzene, the π bond orbitals are spread uniformly over many atoms and are said to be delocalized. The electrons that occupy delocalized orbitals are also delocalized, giving what is called charge delocalization. Charge delocalization confers an extra degree of stability on a molecule or ion and accounts for the relative stability of benzene.

TERMS YOU SHOULD KNOW

polyatomic molecule 439
ground-state wave function 439
molecular orbital 440
molecular orbital theory 440
hydrogen molecular ion, H_2^+ 440
internuclear axis 442
σ orbital 442
bonding orbital 442
antibonding orbital 442
nodal surface 442
σ^* orbital 442
π orbital 442
π^* orbital 442
bond order 443
paramagnetic 444
diamagnetic 444
localized bonding orbital 447
localized covalent bond 447
valence-bond theory 447
hybrid orbital 448
sp (hybrid) orbital 448
sp^2 (hybrid) orbital 451
principle of conservation of orbitals 452
sp^3 (hybrid) orbital 452
sp^3d (hybrid) orbital 458
sp^3d^2 (hybrid) orbital 458
σ-bond framework 460
π bond 460
stereoisomer 462
cis- and *trans*-isomers 462
cis-trans isomerism 462
delocalized orbital 467
delocalized electron 467
charge delocalization 467
nonbonding orbital 468
nonbonding electrons 469

AN EQUATION YOU SHOULD KNOW HOW TO USE

$$\text{bond order} = \frac{\begin{pmatrix}\text{number of}\\ \text{electrons in}\\ \text{bonding orbitals}\end{pmatrix} - \begin{pmatrix}\text{number of}\\ \text{electrons in}\\ \text{antibonding orbitals}\end{pmatrix}}{2} \qquad (13\text{-}1)$$

(definition of bond order)

PROBLEMS

DIATOMIC MOLECULES

13-1. Use molecular orbital theory to explain why diatomic beryllium does not exist.

13-2. Use molecular orbital theory to predict whether or not diatomic boron is paramagnetic.

13-3. Use molecular orbital theory to explain why the bond energy of N_2 is greater than that of N_2^+, but the bond energy of O_2 is less than that of O_2^+.

13-4. Use molecular orbital theory to predict the relative bond energies and bond lengths of F_2 and F_2^+.

13-5. Use molecular orbital theory to predict the relative bond energies and bond lengths of diatomic carbon, C_2, and the acetylide ion, C_2^{2-}.

13-6. Use molecular orbital theory to determine the ground-state electron configurations and bond orders of NF, NF^+, and NF^-. Which of these species do you predict to be paramagnetic?

13-7. Write the ground-state electron configurations and determine the bond orders of the following ions:

(a) O_2^{2-} (b) C_2^+ (c) Be_2^+ (d) Ne_2^+

13-8. For each of the following molecular electron configurations, decide whether it describes a ground electronic state or an excited electronic state.

(a) $(1\sigma)^2(1\sigma^*)^2(2\sigma^*)^1$
(b) $(1\sigma)^2(1\sigma^*)^2(2\sigma)^2(2\sigma^*)^1$
(c) $(1\sigma)^2(1\sigma^*)^2(2\sigma)^2(2\sigma^*)^2(1\pi)^3(3\sigma)^1$
(d) $(1\sigma)^2(1\sigma^*)^2(2\sigma)^2(2\sigma^*)^2(1\pi)^4(3\sigma)^2$

13-9. In some cases the removal of an electron from a species can result in a stronger net bonding (e.g., O_2^+ versus O_2). Give an example in which the addition of an electron to a species produces a stronger net bonding.

13-10. One of the excited states of C_2 has the electron configuration $(1\sigma)^2(1\sigma^*)^2(2\sigma)^2(2\sigma^*)^1(1\pi)^4(3\sigma)^1$. Would you expect the bond length in this excited state to be longer or shorter than that in the ground state?

13-11. Which of the following species are paramagnetic?

(a) C_2 (b) B_2 (c) C_2^{2+} (d) F_2^{2+}

13-12. The energy-level diagram in Figure 13-5 can be continued to higher energies. The next few orbitals in order of increasing energy are 4σ, $4\sigma^*$, 2π, 5σ, $2\pi^*$, and $5\sigma^*$; in other words, the 2σ to $3\sigma^*$ pattern is repeated. Use this extended energy-level diagram to predict whether or not Mg_2 is stable.

POLYATOMIC MOLECULES

13-13. How many valence electrons are there in BH_3? Describe the bonding in BH_3 in terms of hybrid orbitals.

13-14. How many valence electrons are there in $HgCl_2$? Describe the covalent bonding in $HgCl_2$ in terms of hybrid orbitals.

13-15. How many valence electrons are there in CF_4? Use hybrid orbitals to describe the bonding in CF_4.

13-16. How many valence electrons are there in PCl_6^-? Use hybrid orbitals to describe the bonding in PCl_6^-.

13-17. How many valence electrons are there in chloromethane, CH_3Cl. Describe the bonding in CH_3Cl.

13-18. How many valence electrons are there in chloroform, $HCCl_3$? Describe the bonding in the chloroform molecule.

13-19. The hydronium ion, H_3O^+, is trigonal pyramidal with H—O—H bond angles of 110°. Describe the bonding in H_3O^+.

13-20. Use the hybrid orbitals to describe the bonding in OF_2.

13-21. Use the hybrid orbitals to describe the bonding in NF_3.

13-22. How many valence electrons are there in SF_4? Use hybrid orbitals to describe the bonding in SF_4.

13-23. How many valence electrons are there in PCl_3? Use hybrid orbitals to describe the bonding in PCl_3.

13-24. How many valence electrons are there in $SnCl_2$? Use hybrid orbitals to describe the bonding in $SnCl_2$.

MOLECULES WITH NO UNIQUE CENTRAL ATOM

13-25. Use hybrid orbitals to describe the bonding in hydroxylamine, $HONH_2$.

13-26. A class of organic compounds called alcohols may be viewed as derived from HOH by replacing one of the hydrogen atoms by an alkyl group, which is a hydrocarbon group, such as $—CH_3$ (methyl) or $—CH_2CH_3$ (ethyl). A simple alcohol is ethyl alcohol, whose Lewis formula is

```
    H  H
    |  |   ..
H—C—C—O—H
    |  |   ..
    H  H
```

Describe the bonding and shape around both carbon atoms and the oxygen atom in ethyl alcohol.

13-27. A class of organic compounds called amines may be viewed as derived from NH_3 with one or more hydrogen atoms replaced by alkyl groups (see Problem 13-26). Examples of amines are

CH_3NH_2	$(CH_3)_2NH$	$(CH_3)_3N$
methylamine	dimethylamine	trimethylamine

Describe the bonding and the shape of methylamine.

13-28. Describe the bonding and shape of dimethylamine (Problem 13-27). How many σ bonds are there? How many lone pairs of electrons? How many valence-shell electrons are there in the constituent atoms?

13-29. If both hydrogen atoms in HOH are replaced by alkyl groups (see Problem 13-26), the result is an ether, ROR′, where R and R′ are alkyl groups that may or may not be different. The simplest ether is dimethyl ether

```
    H       H
    |   ..  |
H—C—O—C—H
    |   ..  |
    H       H
```

dimethyl ether

Describe the bonding around both carbon atoms and the oxygen atom in dimethyl ether.

13-30. Describe the bonding in ethyl methyl ether, whose Lewis formula is

```
    H  H       H
    |  |   ..  |
H—C—C—O—C—H
    |  |   ..  |
    H  H       H
```

MULTIPLE BONDS

13-31. How many σ bonds and π bonds are there in each of the following molecules?

(a) $Cl_2C{=}CH_2$
(b) $H_2C{=}CHCH{=}CH_2$
(c) CH_3COOH
(d)

```
      CH2
    /     \
  HC       CH
  ||       ||
  HC       CH
    \     /
      CH2
```

13-32. How many σ bonds and π bonds are there in each of the following molecules?

(a) $F_2C{=}CF_2$
(b) $HOOC{—}COOH$
(c) $H_2C{=}C{=}CCl_2$
(d)

```
H       H
  \   /
   C=C
   |  |
   C=C
  /   \
H       H
```

13-33. How many σ bonds are there in ethylacetylene, $CH_3CH_2C{\equiv}CH$? How many π bonds? How many valence electrons?

13-34. How many σ bonds and π bonds are there in methyl cyanide, CH_3CN? How many valence electrons?

13-35. Describe the bonding in carbon monoxide, CO.

13-36. Describe the bonding in the acetylide ion, C_2^{2-}.

DELOCALIZED BONDS

13-37. Write the complete Lewis formula for and describe the bonding in naphthalene, $C_{10}H_8$,

a white, crystalline solid with an odor characteristic of mothballs.

13-38. Write the complete Lewis formula for and describe the bonding in anthracene, $C_{14}H_{10}$,

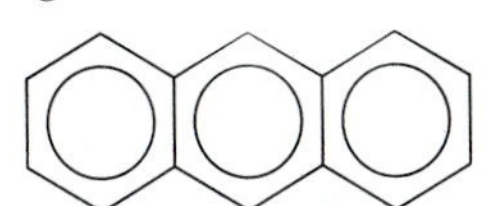

a yellow, crystalline solid found in coal tar.

13-39. The π orbitals in NO_2^- are similar to those illustrated in Figure 13-36 for O_3. Using Figure 13-36, describe the bonding in NO_2^-.

13-40. Describe the bonding in a formate ion, CHO_2^-, using the π orbitals illustrated in Figure 13-36 for O_3. How many delocalized π electrons are there in CHO_2^-? Calculate the carbon-oxygen bond orders in CHO_2^-. Can you rationalize your result with the Lewis formula resonance forms of CHO_2^-?

ADDITIONAL PROBLEMS

13-41. Formamide, H_2NCHO, is known to be planar. Describe the bonding in formamide.

13-42. The H—As—H bond angles in AsH_3 are about 90°. What atomic orbitals would you use to form the localized bond orbitals in this molecule?

13-43. Aldehydes are organic compounds that have the general Lewis formula

R
C=O:
H

where R is either a hydrogen atom (giving formaldehyde) or an alkyl group such as $—CH_3$ (methyl) or $—CH_2CH_3$ (ethyl). Describe the bonding in acetaldehyde, CH_3CHO.

13-44. Ketones are organic compounds with the general Lewis formula

R′
C=O:
R

where R and R′ are alkyl groups that may or may not be different. The simplest ketone is acetone, $(CH_3)_2CO$, one of the most important solvents. Describe the bonding and shape of acetone.

13-45. Describe the bonding and shape of methyl-acetylene, $CH_3C{\equiv}CH$. How many σ and π bonds are there? How many valence electrons?

13-46. Describe the bonding and shape of ethyl cyanide, CH_3CH_2CN. How many σ and π bonds are there? How many valence electrons?

13-47. The bond angle in H_2Te is about 90°. What atomic orbitals would you use to form the localized bond orbitals in this molecule?

13-48. Write the ground-state electron configuration of $Na_2(g)$. (See Problem 13-12.)

13-49. Predict whether Cl_2 or Cl_2^+ has the longer bond length. (See Problem 13-12.)

13-50. Use molecular orbital theory to determine the relative bond lengths and bond energies of NO and NO^+.

13-51. Use molecular orbital theory to determine the relative bond lengths and bond energies of CO and CO^+.

13-52. Determine the type of hybrid orbitals that would be appropriate to use for the valence orbitals on the central atom(s) in the following molecules:

(a) $C_2F_4(g)$ (b) $CS_2(g)$

13-53. One of the excited states of N_2 has the electron configuration $(1\sigma)^2(1\sigma^*)^2(2\sigma)^2(2\sigma^*)^2(1\pi)^4(3\sigma)^1(1\pi^*)^1$. Compare the bond length in this excited state to the bond length in the ground state of N_2.

13-54. Use molecular orbital theory to determine the bond order in BO.

13-55. How many σ bonds and π bonds are there in the following molecules?

(a) CH_3CHCH_2 (b) CH_3CHO
(c) CH_3CH_2CN (d) CH_3OCH_3

13-56. Use molecular orbital theory to determine the ground-state electron configuration of the cyanide ion, CN^-. What other species do you know that have the same bond order as that of CN^-?

13-57. How do you account for the fact that there are two distinct 1,2-dibromoethene species. One of them has no dipole moment, but the other does.

13-58. Describe the bonding in the tetrahydroborate ion, BH_4^-.

13-59. Describe the bonding in a methyl cation, CH_3^+.

13-60. The π energy-level diagram for a trigonal planar species such as a nitrate ion, NO_3^-, is (π^{nb} designates a nonbonding π orbital)

energy
π^*
π^{nb} π^{nb}
π

Use this π energy-level diagram to describe the bonding in NO_3^-. What is the total bond order in NO_3^-? Can you rationalize your result with Lewis formulas for NO_3^-?

13-61. Use the π energy-level diagram shown in Problem 13-60 to describe the bonding in a carbonate ion, CO_3^{2-}. Calculate the total bond order. Can you rationalize your results with the Lewis formulas for CO_3^{2-}?

13-62. Use the π energy-level diagram in Figure 13-36 to describe the bonding in N_3^-. Calculate the total bond order. Is the Lewis formula for N_3^- consistent with your result?

13-63. Use the π energy-level diagram in Figure 13-36 to describe the bonding in NO_2^+. Calculate the total bond order. Is the Lewis formula for NO_2^+ consistent with your result?

14 LIQUIDS AND SOLIDS

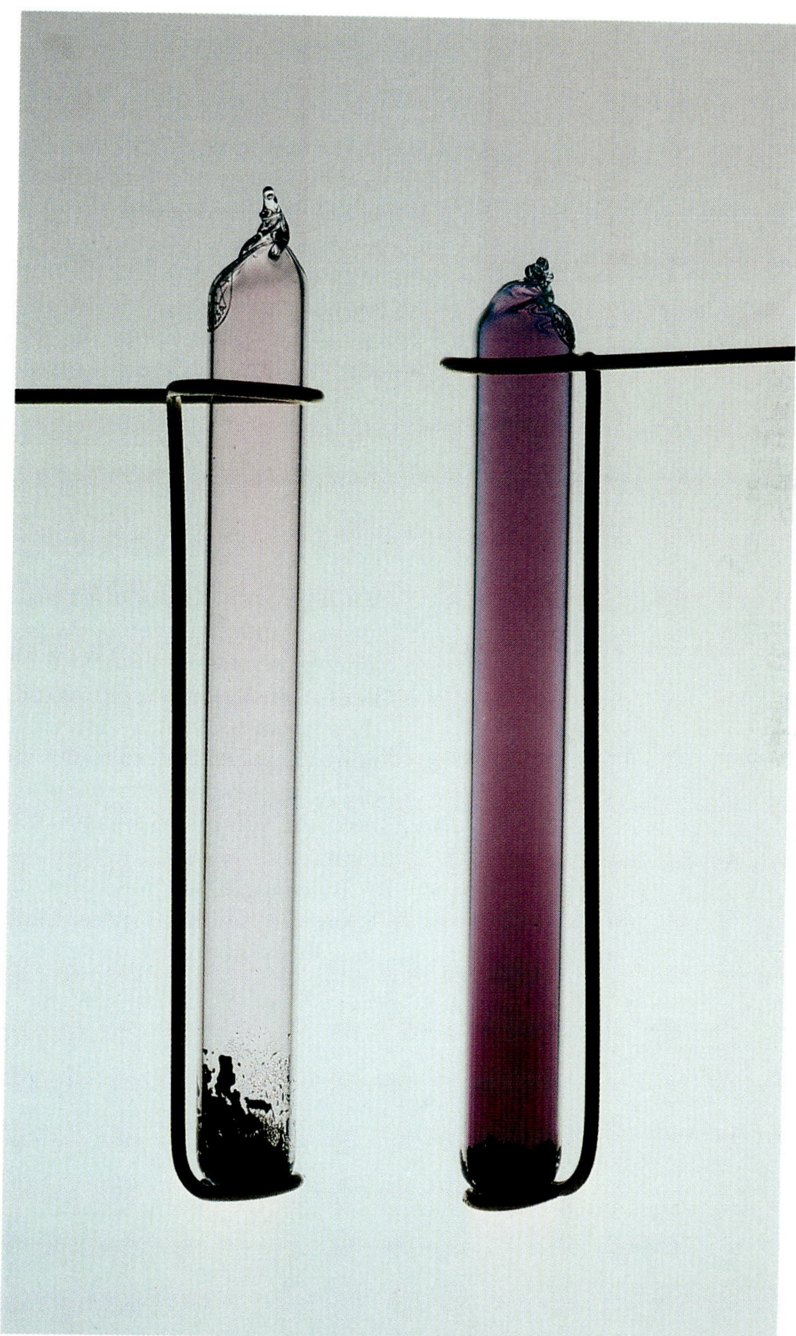

Solid iodine in equilibrium with gaseous iodine at 40°C (lighter-colored vapor) and at 90°C (darker-colored vapor). The equilibrium vapor pressure increases with increasing temperature, as shown by the more intense color.

Your study of chemistry has taken you from calculations involving chemical equations to a fundamental understanding of how atoms combine to form compounds. Now we begin to reverse the process—to build on our picture of molecular structure in order to predict chemical change and the properties of matter. In this chapter, we discuss how the behavior of liquids and solids depends on the attraction between molecules.

If molecules did not interact with, or attract, one another, then gases would not condense into liquids. Without an orderly arrangement of atoms, a diamond would not retain its hardness and the "lead" in pencils its softness. We discuss these and other properties of liquids and solids in terms of intermolecular forces. We examine the processes of melting, freezing, and vaporization and then introduce phase diagrams, which give the temperatures and pressures where a substance occurs as a solid, a liquid, or a gas. We discuss how various types of crystal structures are described and how these contrast with other solids, such as glasses. First, however, we consider what properties distinguish solids and liquids from gases.

14-1. THE MOLECULES IN SOLIDS AND LIQUIDS ARE IN CONTACT WITH ONE ANOTHER

We learned in Chapter 7 that the molecules in a gas are in constant chaotic motion, traveling distances more than a hundred molecular diameters between collisions (at 1 atm). In addition to traveling in straight lines **(translational motion)**, the molecules rotate and vibrate freely as they travel between collisions (Figure 14-1). Because the average distances between the molecules are so large, gases have low densities and high compressibilities; that is, they can be readily compressed. In addition, the volume of a gas is always the same as the volume of its container.

By contrast, the molecules of a solid are close together and restricted to fixed positions in space. They vibrate about fixed positions, but they usually do not rotate or move easily, or translate, from site to site (Fig-

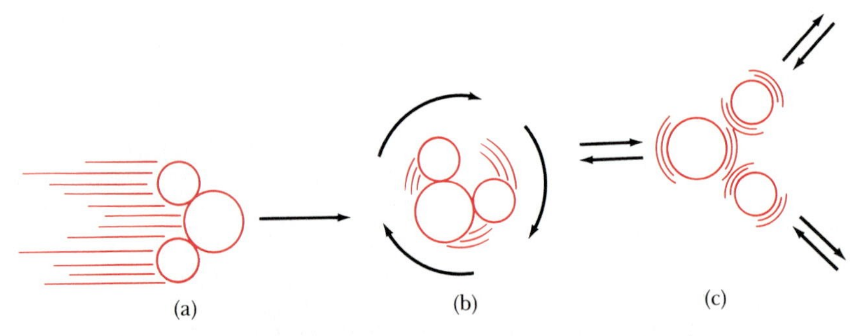

Figure 14-1 Molecular motion may be translational, rotational, or vibrational. (a) Translation is movement of the entire molecule through space. (b) Rotation is the spinning of a molecule in space. (c) Vibration is the back-and-forth movement of nuclei about relative positions.

Table 14-1 Characteristics of solids, liquids, and gases

Phase	Translation	Rotation	Vibration	Average distance between particles
solid	none	none or hindered	about fixed positions in space	less than 1 molecular diameter
liquid	hindered	hindered	free	less than 1 molecular diameter
gas	free	free	free	about 10 molecular diameters at 1 atm

ure 14-1). The molecules in a liquid are also close together, but they are not restricted to fixed positions; they translate and rotate throughout the fluid. Molecules in a liquid are essentially in constant interaction with one another, and the distance that a molecule travels between collisions is about one molecular diameter or less. The translational and rotational motion of molecules in a liquid is hindered by frequent collisions between the molecules. The molecular distinctions between a solid, a liquid, and a gas are summarized in Table 14-1.

Because the particles in solids are fixed in space and are not able to translate, a solid body has a fixed shape, and the volume and shape of a solid are independent of its container. Although the molecules in a liquid are essentially in continual contact with one another, they are able to move past one another and move throughout the liquid. Thus, unlike a solid body, a liquid can be poured from one container to another. Furthermore, a liquid has a fixed volume, because the molecules are constantly attracting one another. But, unlike a solid, a liquid assumes the shape of its container, because its constituent molecules are not held in fixed positions in space.

Liquids and solids have densities that are about a thousand times greater than the densities of gases at 1 atm. The densities of gases are often expressed in units of grams per liter, whereas the densities of liquids and solids are expressed in units of grams per milliliter or grams per cubic centimeter. The relative densities of gases, liquids, and solids imply that the distances between molecules in liquids and solids are much smaller than they are in gases.

The molar volume of a substance is a direct indication of the average separation between the molecules. The molar volume of the liquid phase of a compound is approximately equal to the molar volume of the solid phase of the same compound. Yet the molar volume of a gas is much larger than that of a liquid or a solid at the same pressure and temperature. The molar volumes for solid, liquid, and gaseous water at 0°C and 1 atm and for 100°C and 1 atm are shown in Table 14-2. Note that the molar volume of water increases by a factor of 1630 (30,600 mL/18.8 mL) upon vaporization at 100°C. The average separation between the molecules on going from the liquid to the gas phase at 1 atm increases approximately 10-fold.

Table 14-2 Molar volume of solid, liquid, and gaseous water at 0°C and 100°C and 1 atm

Phase of H_2O	Volume at 0°C/ $mL \cdot mol^{-1}$	Volume at 100°C/ $mL \cdot mol^{-1}$
solid	19.8	—
liquid	18.0	18.8
gas	—	30,600

14-2. THE PROCESSES OF MELTING AND BOILING APPEAR AS HORIZONTAL LINES ON A HEATING CURVE

Let's consider an experiment in which a pure substance is converted from a solid to a liquid to a gas by the application of heat at a constant rate at a pressure of 1 atm. Figure 14-2 shows how the temperature of the substance—water, in this case—varies with time when it is heated at a constant rate. Such a plot is called a **heating curve.** Initially the water is at a temperature of $-10°C$, so it is in the form of ice. As heat is added, the temperature of the ice increases until 0°C, the melting point of ice, is reached. At 0°C, the temperature remains constant for 60 min even though the ice is being heated at a constant rate of $100\ J\cdot min^{-1}$. The heat being added at 0°C melts the ice to liquid water. The temperature of the ice-water mixture remains at 0°C until all the ice is melted. The energy that is absorbed as heat and is required to melt one mole of any substance is called the **molar enthalpy of fusion** or the **molar heat of fusion;** it is denoted by ΔH_{fus}. The experimental data plotted in Figure 14-2 show that 60 min is required to melt 1 mol of ice when heat is added at the rate of $100\ J\cdot min^{-1}$; so the enthalpy of fusion for ice is given by

$$\begin{aligned}\Delta H_{fus} &= (60\ min\cdot mol^{-1})(100\ J\cdot min^{-1}) \\ &= 6000\ J\cdot mol^{-1} \\ &= 6.0\ kJ\cdot mol^{-1}\end{aligned}$$

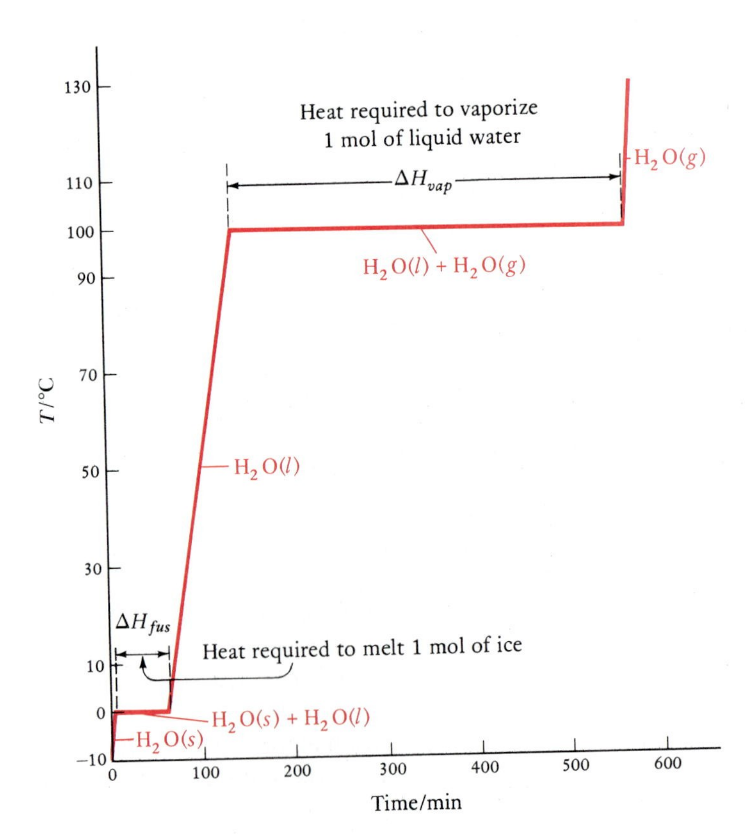

Figure 14-2 The heating curve for one mole of water starting with ice at $-10°C$. The energy is added as heat at a constant rate of $100\ J\cdot min^{-1}$. The most noteworthy features of the heating curve are the horizontal portions, which represent the heat of fusion, ΔH_{fus}, and the heat of vaporization, ΔH_{vap}. Note that the heat of vaporization is much larger than the heat of fusion.

After all the ice is melted, and not until then, the temperature increases from 0°C. The temperature continues to increase until 100°C, the boiling point of water, is reached. At 100°C, the temperature remains constant for 407 min even though heat is being added at a rate of 100 $J \cdot min^{-1}$. The heat being absorbed at 100°C is vaporizing the liquid water to water vapor. The temperature of the liquid-vapor mixture remains at 100°C until all the water is vaporized. The energy that is absorbed as heat and is required to vaporize one mole of any substance is called the **molar enthalpy of vaporization** or the **molar heat of vaporization** and is denoted by ΔH_{vap}. According to Figure 14-2, ΔH_{vap} for water is

$$\begin{aligned}\Delta H_{vap} &= (407\ \text{min}\cdot\text{mol}^{-1})(100\ \text{J}\cdot\text{min}^{-1}) \\ &= 40{,}700\ \text{J}\cdot\text{mol}^{-1} \\ &= 40.7\ \text{kJ}\cdot\text{mol}^{-1}\end{aligned}$$

Once all the water is vaporized, the temperature increases from 100°C, as shown in Figure 14-2.

The horizontal portions of the heating curve of a pure substance represent the heat of fusion and the heat of vaporization. The other regions represent pure phases being heated at a constant rate. We learned in Chapter 8 that it requires energy to raise the temperature of a substance. The heat absorbed in raising the temperature of a substance from T_1 to T_2 without a change in phase is given by [see Equation (8-24)]

$$q_P = nC_P(T_2 - T_1) \qquad (14\text{-}1)$$

where C_P is the molar heat capacity at constant pressure and n is the number of moles. The heat capacity is the measure of the ability of a substance to take up energy as heat. Different rates of temperature increase for $H_2O(s)$, $H_2O(l)$, and $H_2O(g)$ appear in Figure 14-2 (steep segments of the heating curve) because

$$C_P[H_2O(l)] > C_P[H_2O(s)] > C_P[H_2O(g)]$$

14-3. ENERGY IS REQUIRED TO MELT A SOLID AND TO VAPORIZE A LIQUID

The melting, or fusion, of one mole of ice can be represented by

$$H_2O(s) \rightarrow H_2O(l) \qquad \Delta H_{fus} = 6.0\ \text{kJ}\cdot\text{mol}^{-1}$$

or, in general, we write

$$X(s) \rightarrow X(l) \qquad \Delta H_{fus}$$

where X stands for any element or compound. The energy as heat that is required to melt n moles of a substance is given by

$$q_{fus} = n\Delta H_{fus} \qquad (14\text{-}2)$$

The enthalpy of fusion is necessarily positive because it requires energy to break up the crystal lattice. Recall that a positive value of ΔH means that heat is absorbed in the process.

The vaporization of one mole of water can be represented by

$$H_2O(l) \rightarrow H_2O(g) \qquad \Delta H_{vap} = 40.7\ \text{kJ}\cdot\text{mol}^{-1}$$

or, in general,

$$X(l) \rightarrow X(g) \qquad \Delta H_{vap}$$

The energy that is absorbed as heat and is required to vaporize n moles of a substance is given by

$$q_{vap} = n\Delta H_{vap} \tag{14-3}$$

The value of ΔH_{vap} is always positive, because energy is required to separate the molecules in a liquid from one another. Gas-phase molecules are so far apart from one another that they interact only very weakly relative to liquid-phase molecules. Essentially all the energy put in as heat of vaporization is required to separate the molecules of the liquid from one another. Because the temperature does not change during vaporization, the average kinetic energy of the molecules does not change. Table 14-3 gives the molar enthalpies of fusion and vaporization of several substances.

Table 14-3 Melting points, boiling points, and molar enthalpies of vaporization and fusion

Compound	Chemical formula	Melting point/K	Boiling point/K	ΔH_{fus}/kJ·mol^{-1}	ΔH_{vap}/kJ·mol^{-1}
ammonia	NH_3	195	240	5.65	23.4
argon	Ar	84	87	1.17	6.52
bromine	Br_2	266	332	10.6	29.5
carbon dioxide	CO_2	217	(195) sublimes	8.33	(25.2) sublimes
chlorine	Cl_2	172	239	6.41	20.4
chloromethane	CH_3Cl	139	249	6.4	21.5
formaldehyde	H_2CO	181	252	—	24.5
helium	He	—	4.2	0.014	0.081
hydrogen	H_2	14	20	0.12	0.90
hydrogen bromide	HBr	186	206	2.4	17.6
hydrogen chloride	HCl	159	188	2.0	16.2
iodine	I_2	387	458	15.5	41.9
krypton	Kr	116	121	1.63	9.03
lithium bromide	LiBr	823	1583	17.6	148.1
mercury	Hg	234	630	2.30	59.1
methane	CH_4	91	112	0.94	8.17
neon	Ne	24	27	0.33	1.76
nitrogen	N_2	63	77	0.72	5.58
oxygen	O_2	54	90	0.44	6.82
sulfur dioxide	SO_2	198	263	7.4	24.9
water	H_2O	273	373	6.01	40.7
xenon	Xe	160	166	2.30	12.63

EXAMPLE 14-1: How long would it take to convert 50.0 g of ice at 0°C to water at 100°C if the heating rate is 250 $J \cdot min^{-1}$? The molar heat capacity of $H_2O(l)$ is 75.2 $J \cdot mol^{-1} \cdot K^{-1}$.

Solution: Fifty grams of H_2O corresponds to

$$\text{moles of } H_2O = (50.0 \text{ g } H_2O)\left(\frac{1 \text{ mol } H_2O}{18.02 \text{ g } H_2O}\right) = 2.77 \text{ mol}$$

First we calculate how long it takes to melt the ice at 0°C and then calculate how long it takes to heat the water from 0°C to 100°C. The energy required to melt 2.77 mol of ice is given by Equation (14-2)

$$q_{fus} = n\Delta H_{fus} = (2.77 \text{ mol})(6.0 \text{ kJ} \cdot \text{mol}^{-1}) = 16.6 \text{ kJ}$$

The time it would take to add this much energy is given by q_P divided by the rate of heating, or

$$t = \frac{16.6 \text{ kJ}}{250 \text{ J} \cdot \text{min}^{-1}} = \frac{16.6 \times 10^3 \text{ J}}{250 \text{ J} \cdot \text{min}^{-1}} = 66.6 \text{ min}$$

The energy required to heat the water from 0°C to 100°C is [Equation (14-1)]

$$q_P = nC_P(T_2 - T_1) = (2.77 \text{ mol})(75.2 \text{ J} \cdot \text{mol}^{-1} \cdot \text{K}^{-1})(100 \text{ K})$$
$$= 2.08 \times 10^4 \text{ J}$$

The time it takes to add this much energy as heat is given by

$$t = \frac{2.08 \times 10^4 \text{ J}}{250 \text{ J} \cdot \text{min}^{-1}} = 83.3 \text{ min}$$

The total time it would take to convert 50.0 g of ice at 0°C to water at 100° C is

$$t = 66.6 \text{ min} + 83.3 \text{ min} = 150 \text{ min}$$

PRACTICE PROBLEM 14-1: How long would it take to convert 100 g of solid sodium at 20°C to sodium vapor at 1000°C if the heating rate was 8.0 $kJ \cdot min^{-1}$? The melting point of sodium is 98°C; its boiling point is 881°C; its molar enthalpy of fusion is 2.60 $kJ \cdot mol^{-1}$; its molar enthalpy of vaporization is 97.4 $kJ \cdot mol^{-1}$; and the heat capacities of solid, liquid, and gaseous sodium are 28.1 $J \cdot mol^{-1} \cdot K^{-1}$, 32.7 $J \cdot mol^{-1} \cdot K^{-1}$, and 20.8 $J \cdot mol^{-1} \cdot K^{-1}$, respectively.

Answer: 71 min

EXAMPLE 14-2: Compute the energy released as heat when 28 g of liquid water at 18°C is converted to ice at 0°C. (An ice cube contains about 1 oz of water, and 1 oz is equivalent to 28 g.) The molar heat capacity of $H_2O(l)$ is $C_P = 75.2$ $J \cdot K^{-1} \cdot mol^{-1}$, and $\Delta H_{fus} = 6.0$ $kJ \cdot mol^{-1}$ for ice.

Solution: The overall process must be broken down into two steps. We must first bring the $H_2O(l)$ from 18°C to 0°C (the freezing point of water) and then consider the process $H_2O(l) \rightarrow H_2O(s)$ at 0°C:

$$\underset{\text{at } 18°\text{C}}{28 \text{ g } H_2O(l)} \xrightarrow{\text{step 1}} \underset{\text{at } 0°\text{C}}{28 \text{ g } H_2O(l)} \xrightarrow{\text{step 2}} \underset{\text{at } 0°\text{C}}{28 \text{ g } H_2O(s)}$$

For step 1 we have, from Equation (14-1), where n is the number of moles of water,

$$q_P = nC_P(T_2 - T_1)$$
$$= \left[(28\text{ g})\left(\frac{1\text{ mol H}_2\text{O}}{18\text{ g H}_2\text{O}}\right)\right]\left[(75.2\text{ J}\cdot\text{K}^{-1}\cdot\text{mol}^{-1})\left(\frac{1\text{ kJ}}{1000\text{ J}}\right)\right](273\text{ K} - 291\text{ K})$$
$$= -2.1\text{ kJ}$$

The negative sign for q_P reflects the fact that energy must be removed to lower the temperature of the water. For step 2, where n is the number of moles of H_2O,

$$q_P = n(-\Delta H_{fus}) = (28\text{ g})\left(\frac{1\text{ mol H}_2\text{O}}{18\text{ g H}_2\text{O}}\right)(-6.0\text{ kJ}\cdot\text{mol}^{-1}) = -9.3\text{ kJ}$$

where the minus sign in front of ΔH_{fus} arises because freezing is the reverse of fusion, that is, $\Delta H_{freezing} = -\Delta H_{fus}$. The total amount of energy that must be *removed* as heat from the 28 g of water is 2.1 kJ + 9.3 kJ = 11.4 kJ.

PRACTICE PROBLEM 14-2: Using the data in Tables 14-3 and 8-3, compute the energy released as heat when 25.0 g of liquid mercury at 300°C is converted to solid mercury at −60°C.

Answer: 1.55 kJ

Figure 14-3 Naphthalene is one of a number of substances that readily sublime at room temperature. Here we see crystals of napthalene condensing directly from vapor to crystals on the surface of the cold tube.

It is possible for a solid to be converted directly to a gas without passing through the liquid phase. This process is called **sublimation.** The **molar enthalpy of sublimation,** ΔH_{sub}, or the **molar heat of sublimation** is the energy absorbed as heat when one mole of a solid is sublimed at constant pressure. Essentially all the energy put in as heat in the sublimation process is used to separate the molecules in the solid from one another. The larger the value of ΔH_{sub}, the stronger the intermolecular attractions in the solid are.

The best-known example of sublimation is the conversion of Dry Ice, $CO_2(g)$, to carbon dioxide gas:

$$CO_2(s) \rightarrow CO_2(g) \qquad \Delta H_{sub} = 25.2\text{ kJ}\cdot\text{mol}^{-1}$$

The name Dry Ice is used because the CO_2 does not become a liquid at 1 atm pressure. Dry Ice at 1 atm has a temperature of −78°C and is widely used as a one-time, low-temperature refrigerant. The sublimation of 44 g (one mole) of Dry Ice requires 25.2 kJ of heat.

Ice sublimes at temperatures below its melting point (0°C). For the sublimation of ice,

$$H_2O(s) \rightarrow H_2O(g) \qquad \Delta H_{sub} = 46.7\text{ kJ}\cdot\text{mol}^{-1}$$

Snow often sublimes, and so does the ice in the freezer compartment of your refrigerator. Perhaps you've noticed that ice cubes left in the freezer get smaller as time passes. Another substance that sublimes at temperatures below its melting point is naphthalene, a principal component of one type of mothball (Figure 14-3).

14-4. VAN DER WAALS FORCES ARE ATTRACTIVE FORCES BETWEEN MOLECULES

In the process of vaporization or sublimation, the molecules of the liquid or solid, which are in contact with one another, become separated and widely dispersed. The value of ΔH_{vap} or ΔH_{sub} reflects how strongly the molecules attract one another in the liquid or solid. The more strongly the molecules attract one another, the greater the value of ΔH_{vap} or ΔH_{sub}.

The simplest force at the atomic or molecular level is that between ions. We studied this force in Chapter 10 when we discussed ionic compounds. Two ions with opposite charges attract each other; ions with like charges repel each other. The force between ions is relatively strong; it requires a relatively large amount of energy to separate ions of opposite charge. Therefore, the enthalpies of vaporization of ionic compounds are much larger than those of nonionic compounds. Enthalpies of vaporization of ionic compounds are typically at least 100 $kJ \cdot mol^{-1}$. The boiling points of ionic compounds also are higher than those of nonionic compounds.

Most of the compounds listed in Table 14-3 are molecular compounds. In Chapter 11 we learned that some molecules have dipole moments; in other words, they are polar. An example of a polar molecule is formaldehyde, whose Lewis formula is

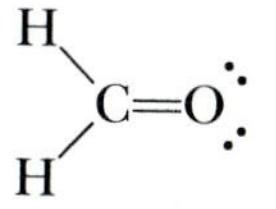

H
C=O:
H

The bond moments in a formaldehyde molecule can be illustrated by

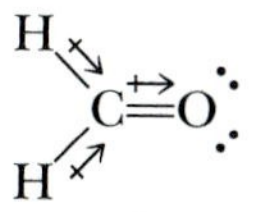

H
C=O:
H

because an oxygen atom is more electronegative than a carbon atom, which in turn is more electronegative than a hydrogen atom. The net dipole moment is illustrated by

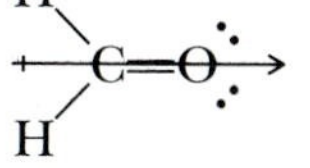

H
C=O:
H

Thus, even though a formaldehyde molecule is electrically neutral overall, it has a positively charged end and a negatively charged end and therefore is a polar molecule.

EXAMPLE 14-3: Recall from Section 12-9 that we discussed how to use the symmetry of a molecule to predict whether or not it has a dipole moment. Use VSEPR theory to predict which of the following molecules has a dipole moment:

(a) SF_6 (b) SF_4 (c) PF_5

Solution: (a) Sulfur hexafluoride, SF_6, is an AX_6 molecule; so it is octahedral and has no dipole moment (Figure 12-18). (b) Sulfur tetrafluoride, SF_4, is an AX_4E molecule; so it is seesaw-shaped (Figure 12-15) and has a dipole moment. (c) Phosphorus pentafluoride, PF_5, is an AX_5 molecule; so it is trigonal bipyramidal and has no dipole moment (Figure 12-7).

PRACTICE PROBLEM 14-3: Which of the following molecules is polar: C_2Cl_4, PCl_3, SO_2, CH_3Cl, CCl_4?

Answer: PCl_3, SO_2, and CH_3Cl are polar.

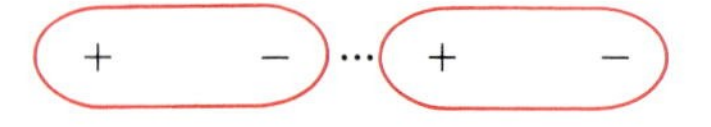

Figure 14-4 Although polar molecules are electrically neutral overall, they attract one another by a dipole-dipole force. The molecules orient themselves as shown because the positive end of one attracts the negative end of another. The dipoles are said to be oriented head-to-tail.

Polar molecules attract one another (Figure 14-4), and this attraction is called a **dipole-dipole attraction.** The charge separation in polar molecules is considerably smaller than the full electronic charges on ions, so dipole-dipole forces are weaker than ion-ion forces. Enthalpies of vaporization for many polar compounds are around 20 $kJ \cdot mol^{-1}$. We must emphasize here that the interaction energies *between* neutral molecules are much weaker than the covalent bonds *within* molecules, which have energies of hundreds of kilojoules per mole.

A particularly important dipole-dipole attraction occurs when one or more hydrogen atoms are bonded to a highly electronegative atom, as in the case of water and ammonia. Let's consider the molar enthalpies of vaporization of water, ammonia, and methane:

H_2O	NH_3	CH_4
40.7 $kJ \cdot mol^{-1}$	23.4 $kJ \cdot mol^{-1}$	8.2 $kJ \cdot mol^{-1}$

These three compounds have approximately the same molecular mass (18, 17, 16, respectively), but the amounts of energy required to separate the molecules of these three liquids are very different.

We can represent a water molecule by

$$\overset{2\delta-}{O}(H^{\delta+})_2 \quad \text{(bent: } {}^{\delta+}H - O - H^{\delta+}\text{)}$$

Water molecules in liquid water attract one another through the electrostatic interaction between a hydrogen atom and the oxygen atom on a different molecule:

$$H^{\delta+} - \overset{2\delta-}{O} - H^{\delta+} \cdots \overset{2\delta-}{O}(H^{\delta+})_2$$

hydrogen bond

The electrostatic attraction that occurs between molecules in which a hydrogen atom is covalently bonded to a highly electronegative atom, such as O, N, or F, is called **hydrogen bonding.** Because a hydrogen atom is so small, the charge on it is highly concentrated, so it strongly attracts electronegative atoms in neighboring molecules.

Hydrogen bonds are a particularly strong form of dipole-dipole attraction. The pattern shown in Figure 14-5 extends throughout liquid water and gives water its large value of ΔH_{vap} (40.7 $kJ \cdot mol^{-1}$) and its high boiling point. We will see that hydrogen bonding gives water many special properties.

Hydrogen bonding greatly affects the structure of ice (Figure 14-6), which is described as an *open structure* because of the significant amount of empty space between the molecules. The open structure of ice is a direct consequence of the fact that each hydrogen atom is hydrogen-bonded to the oxygen atom of an adjacent molecule. Note the tetrahedral arrangement of the oxygen atoms in ice. Every oxygen atom sits in the center of a tetrahedron formed by four other oxygen atoms.

The structure of liquid water is less open than the structure of ice, because when ice melts the total number of hydrogen bonds decreases. Unlike most other substances, water *increases* in density on going from solid to liquid because of a partial breakdown of the hydrogen-bonded structure. The extent of hydrogen bonding in liquid water is only about

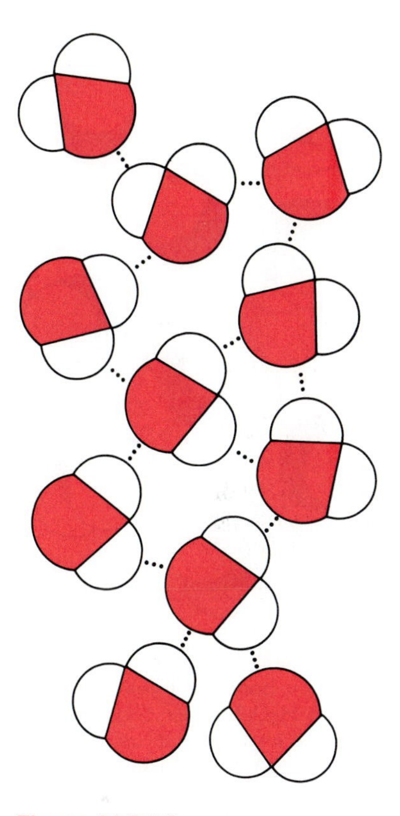

Figure 14-5 There are many hydrogen bonds in liquid water; each oxygen atom can form two hydrogen bonds because each oxygen atom has two lone pairs of electrons. Thus each water molecule has the ability to form four hydrogen bonds. At 25°C about 80 percent of the hydrogen atoms in water are hydrogen bonded.

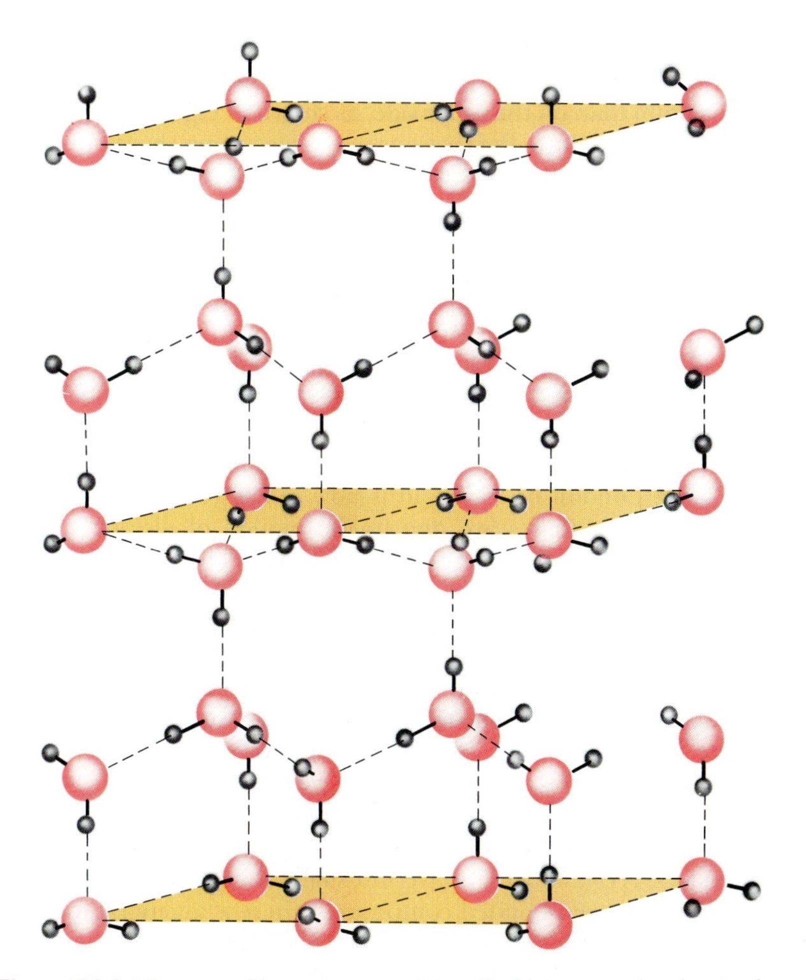

Figure 14-6 The crystalline structure of ice. Each water molecule can form four hydrogen bonds. Each oxygen atom is located in the center of a tetrahedron formed by four other oxygen atoms. The entire structure is held together by hydrogen bonds.

80 percent, whereas in ice 100 percent of the oxygen atoms are hydrogen-bonded. The extent of hydrogen bonding in water decreases as the temperature increases.

Hydrogen bonding also occurs in liquid ammonia, but the individual hydrogen bonds are weaker than those in water because nitrogen is less electronegative than oxygen and the fractional charges on the nitrogen and hydrogen atoms in NH_3 are less than those on the oxygen and hydrogen atoms in H_2O. Furthermore, there are fewer hydrogen bonds in NH_3 because each nitrogen atom can form only one hydrogen bond (Figure 14-7). Consequently, ΔH_{vap} for ammonia is only about half that for water. Methane is a nonpolar molecule because it is tetrahedral and the four bond moments cancel. Thus, of H_2O, NH_3, and CH_4, CH_4 has the lowest value of ΔH_{vap}.

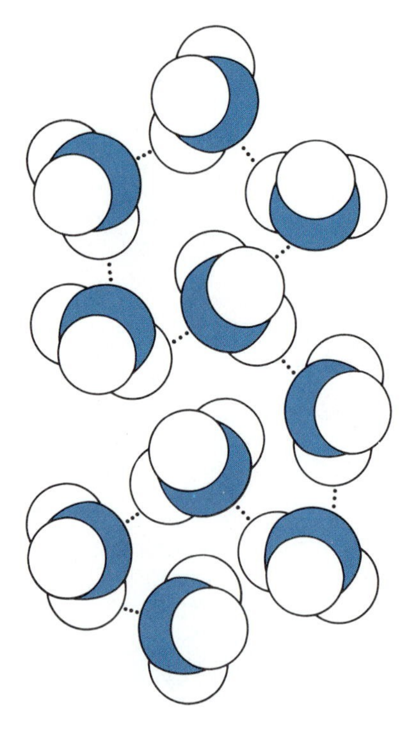

Figure 14-7 There are fewer hydrogen bonds in liquid NH_3 than in H_2O because NH_3 has only one lone pair, so each nitrogen atom can form only one hydrogen bond.

We must consider one other attractive force. Even though methane is nonpolar, it liquefies at 91 K and has an enthalpy of vaporization of 8.2 $kJ \cdot mol^{-1}$; therefore, methane molecules must attract one another. Even the noble gases, which consist of single, spherical atoms, can be liquefied. How neutral, nonpolar molecules attract one another was not understood until quantum theory was developed. Let's first consider a single argon atom. As the electrons move about in the atom, there will be instants when the distribution of the total charge will not be perfectly symmetric. For example, perhaps there will be more electrons on one side of the atom than on the other side, as shown in Figure 14-8. Over a period of time these little fluctuations in the electronic charge distribution will average out, thereby yielding a perfectly symmetric distribution. But at any instant, there will be an instantaneous *asymmetry* in the electronic charge distribution resulting in an instantaneous dipole moment (Figure 14-8). Now let's consider two argon atoms separated by a small distance. The motion of the electrons in one atom influences the motion of the electrons in the other atom in such a way that the instantaneous dipole moments line up head-to-tail (Figure 14-9). This synchronized motion of the electrons leads to an effective dipole-dipole attraction, which accounts for the attractive force between atoms as well as nonpolar molecules. This force was first explained by the German physicist Fritz London in 1930 and is now called a **London force.**

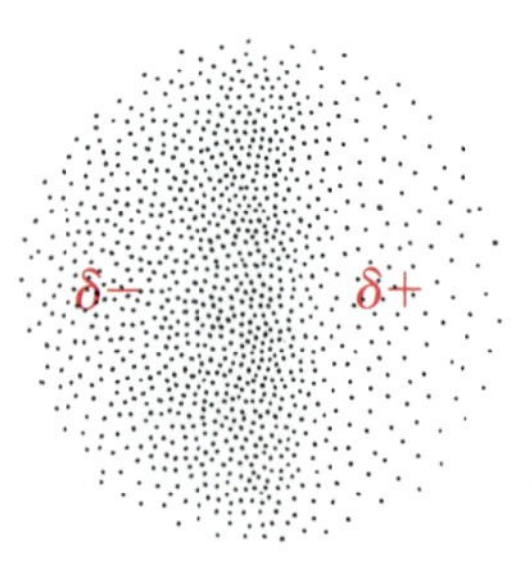

Figure 14-8 If it were possible to take an instantaneous view of an atom, it might look like this drawing. The instantaneous position of the electrons leads to an instantaneous dipole moment. The negative charge is due to a greater-than-average electronic-charge density, and the positive charge is due to a less-than-average electronic-charge density. As the electrons move around, at different times the dipole moment points in all directions and averages to zero.

Because London forces are due to the motion of electrons, their strength depends on the number of electrons. The more electrons there are in the two interacting molecules, the stronger their London attraction for each other. Therefore, for purely nonpolar substances, which attract each other only by way of London forces, we expect the value of ΔH_{vap} to increase with the number of electrons, or even with the size of the molecules. The data in Table 14-4 support this prediction. Note that, within each group, the value of ΔH_{vap} increases with the number of electrons.

Figure 14-10 shows plots of the boiling points of the noble gases and the hydrides of the nonmetallic elements. The hydrogen-bonded compounds (H_2O, NH_3, and HF) have unusually high boiling points. Except for the hydrogen-bonded compounds, there is a general increase of boiling point with increasing molecular mass. This increase is due to the increase in London forces with an increase in the number of electrons in a molecule. The attractive forces between molecules, be they dipole-dipole forces or London forces, are collectively called **van der**

Table 14-4 Relation between number of electrons in some atoms and nonpolar molecules and their heats of vaporization

Substance	Number of electrons	$\Delta H_{vap}/kJ \cdot mol^{-1}$
He	2	0.08
Ne	10	1.76
Ar	18	6.52
Kr	36	9.03
Xe	54	12.63
F_2	18	6.5
Cl_2	34	20.4
Br_2	70	29.5
I_2	106	41.9

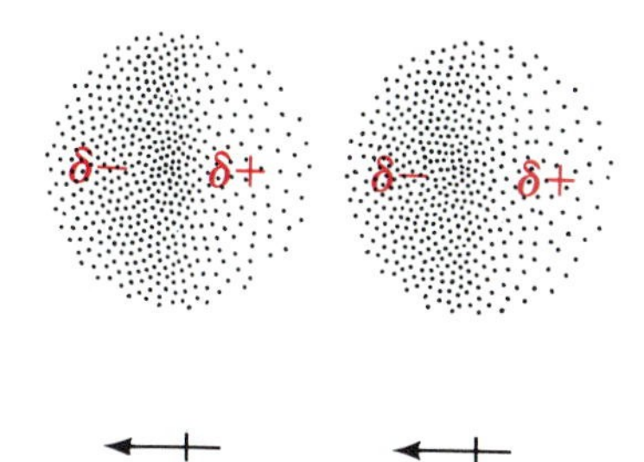

Figure 14-9 When two atoms are near each other, the motions of the electrons in the two atoms affect each other so that the instantaneous dipole moments are aligned head-to-tail. This effect leads to an instantaneous dipole-dipole attraction between the two atoms.

Waals forces. Table 14-5 compares the various forces that we have discussed in this section.

EXAMPLE 14-4: Without referring to any sources, rank the following substances in order of increasing enthalpies of vaporization and boiling points:

$$NaCl \qquad C_2H_4 \qquad CH_3OH$$

Solution: The only ionic compound listed is NaCl. Thus, we predict that NaCl has the largest value of ΔH_{vap} and the highest boiling point. The Lewis formula for CH_3OH, (see margin), shows that there is a hydrogen atom bonded to an oxygen atom in CH_3OH. Thus, we predict that in $CH_3OH(l)$ there will be hydrogen bonding between molecules. The Lewis formula for C_2H_4 (margin) indicates that C_2H_4 is a nonpolar molecule, so we predict that

$$\Delta H_{vap}[NaCl] > \Delta H_{vap}[CH_3OH] > \Delta H_{vap}[C_2H_4]$$

The boiling points are in the same order. The actual values of ΔH_{vap} are 170 $kJ \cdot mol^{-1}$, 35.3 $kJ \cdot mol^{-1}$, and 13.5 $kJ \cdot mol^{-1}$, respectively; and the boiling points are 1690 K, 337 K, and 170 K, respectively.

CH_3OH

C_2H_4

Table 14-5 Various types of attractive forces between ions and molecules

Type	Examples	Typical value of ΔH_{vap}
ion-ion	NaCl, KBr	$\sim 100\ kJ \cdot mol^{-1}$
dipole-dipole	H_2CO, HBr	$\sim 20\ kJ \cdot mol^{-1}$
hydrogen bonding	H_2O, NH_3	20–40 $kJ \cdot mol^{-1}$
London	Ar, CH_4	5–20 $kJ \cdot mol^{-1}$

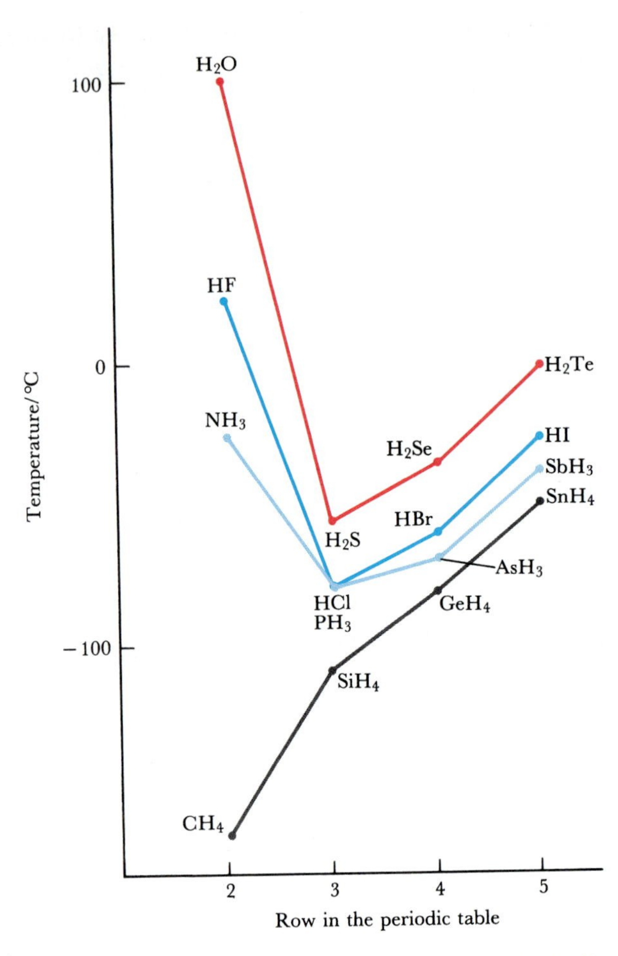

Figure 14-10 Boiling points of the hydrides of the nonmetallic elements. Note the abnormally high values for hydrogen fluoride, water, and ammonia, which are the result of hydrogen bonds in the liquid and solid phases.

PRACTICE PROBLEM 14-4: Explain the trends of the boiling points for each group of molecules in Figure 14-10.

Answer: Generally the boiling point increases with increasing molecular mass because of London forces; H_2O, HF, and NH_3, however, are anomalous because of hydrogen bonding.

14-5. VISCOSITY, SURFACE TENSION, AND DIELECTRIC CONSTANT ARE PROPERTIES OF LIQUIDS

Several properties of liquids are of practical importance to chemists and depend on intermolecular forces. The **viscosity** of a liquid is a measure of its resistance to flow. Substances that flow very slowly, such as molasses, are said to be viscous. Viscosity can be measured experimentally by timing how long it takes a certain volume of a liquid to flow through a thin tube. Many engineering processes involve the flow of liquids through pipes, so engineering design is often concerned with the viscosities of liquids and liquid mixtures. Viscosity decreases with increasing temperature, because, as the temperature increases, the molecules have greater kinetic energy and therefore can more readily overcome the intermolecular interactions. The viscosities of a variety of compounds are given in Table 14-6.

Table 14-6 Physical properties of a variety of liquids at 25°C

Compound	Formula	Viscosity (relative to water)	Surface tension/ $mJ \cdot m^{-2}$	Dielectric constant
acetone	$(CH_3)_2CO$	0.38	26	21
benzene	C_6H_6	0.68	29	2.3
bromine	Br_2	1.2	46	3.5
ethanol	C_2H_5OH	1.2	24	26
glycerol	$C_3H_8O_3$	950	63	42
hexane	C_6H_{14}	0.35	20	1.9
methanol	CH_3OH	0.61	24	33
octane	C_8H_{18}	0.61	24	2.0
sulfuric acid	H_2SO_4	27.6	—	100
water	H_2O	1.00	72	78

EXAMPLE 14-5: The time it takes for a given volume of a liquid to flow through a capillary is directly proportional to the viscosity of the liquid and inversely proportional to its density. Suppose that the time it takes a given volume of H_2O to flow through a certain capillary at 20°C is 37 s. The time required for an equal volume of carbon tetrachloride to flow through the capillary is 56 s. Given that the densities of water and carbon tetrachloride at 20°C are 1.00 $g \cdot mL^{-1}$ and 1.58 $g \cdot mL^{-1}$, respectively, calculate the viscosity of carbon tetrachloride relative to that of water.

Solution: The basic relation is

$$\text{time} \propto \frac{\text{viscosity}}{\text{density}}$$

or

$$\text{viscosity} \propto \text{time} \times \text{density}$$

or if we denote viscosity by η (Greek letter eta), then we can write

$$\eta = ctd$$

where c is a proportionality constant. If we solve this expression for c, then

$$c = \frac{\eta}{td}$$

or

$$\frac{\eta_1}{t_1 d_1} = \frac{\eta_2}{t_2 d_2} \qquad (14\text{-}4)$$

We let 1 stand for water and 2 stand for carbon tetrachloride, and solve for η_2/η_1:

$$\frac{\eta_2}{\eta_1} = \frac{t_2 d_2}{t_1 d_1} = \frac{(56\text{ s})(1.58\text{ g}\cdot\text{mL}^{-1})}{(37\text{ s})(1.00\text{ g}\cdot\text{mL}^{-1})} = 2.39$$

PRACTICE PROBLEM 14-5: Use the data in Table 14-6 to calculate how long it will take a given volume of benzene to flow through a certain capillary at 25°C if it takes 1 min for the same volume of water. The densities of water and benzene at 25°C are 1.00 $g \cdot mL^{-1}$ and 0.879 $g \cdot mL^{-1}$, respectively.

Answer: 0.77 min

Figure 14-12 Molecules in the interior of a liquid are attracted in all directions, but the molecules at the surface experience a net inward attraction that minimizes the surface area of the liquid and results in surface tension.

Figure 14-11 A metal paper clip floats on water because of the distribution of its weight and the surface tension of the water. Water has a high surface tension because the surface molecules form fewer hydrogen bonds than the interior molecules. The surface of water is like an elastic skin that resists penetration.

If a paper clip is carefully placed on a water surface, it floats even though the density of the paper clip is greater than the density of water (Figure 14-11). The clip is held up by surface tension. Water striders and some other insects can walk on water, being supported by the surface tension of the water, which resists penetration of the surface.

What is the cause of the surface tension of a liquid? A molecule in the body of a liquid is subject to attractive forces in all directions, but a molecule at the surface experiences a net attractive force toward the interior of the liquid (Figure 14-12). Thus, molecules at the surface of a liquid experience a net inward force. This force tends to minimize the number of molecules at the surface and so minimize the surface area of the liquid. This force is the **surface tension.** Any liquid whose molecules attract one another strongly has a high surface tension; water is a good example. The surface tension of a liquid tends to hold a drop of liquid in a spherical shape (Figure 14-13) because a sphere is the shape that has the smallest surface area for a given volume. The higher the surface tension is, the more nearly spherical the drop is (Figure 14-14).

Figure 14-13 Surface tension causes freely falling drops of liquid to assume a spherical shape because a sphere is the shape with the smallest surface area for a given volume.

Figure 14-14 Shapes of equal volumes of Hg, H_2O, $(CH_3)_2SO_3$ (dimethylsulfoxide), and CH_3COCH_3 (acetone), from left to right. Surface tension holds the drops in a spherical shape and gravity flattens them. The effect of gravity is the same for all the drops; thus the higher the surface tension, the more nearly spherical is the drop. Surface tension has units of millijoules per meter squared ($mJ \cdot m^{-2}$).

Figure 14-15 Capillary action is shown here as colored water rises in these glass tubes and a celery stalk. The cells in a celery stalk form a capillary structure.

Table 14-6 lists the surface tension for a variety of liquids. Note that surface tension has units of energy per unit area. Surface tension can be thought of as the energy that it takes to create an area of surface.

Certain compounds, such as sodium dodecylsulfate (SDS), $NaC_{12}H_{25}SO_3$, lower the surface tension of a liquid by concentrating at the liquid surface. Such molecules are called **surfactants** (surface active agents). A 0.1 percent SDS solution has a surface tension of 20 $mJ \cdot m^{-2}$, whereas pure water has a surface tension of 72 $mJ \cdot m^{-2}$. Reduction of the surface tension of water by surfactants is the basis of detergent action. In a detergent solution the surface tension of the water is comparable to that of the grime (oils) on clothing, for example, thereby allowing the solution to wet the grime.

Closely related to surface tension is **capillary action,** the rise of a liquid in a thin tube. Capillary action occurs when the adhesive forces between the molecules of the capillary wall surface and the molecules of the liquid are sufficiently great that the liquid adheres to, or wets, the solid surface. The adhesive force pulls the liquid up into the capillary. The liquid column rises until the upward adhesive force is balanced by the downward gravitational force. Capillary action plays a major role in the movement of water in plants, animals, and soil. The water is pulled by the capillary action up into and through living structures (Figure 14-15).

One important consequence of surface tension and capillary action is the formation of a **meniscus,** the surface of a liquid in a capillary (Figure 14-16). For a liquid that wets glass (for example, water), the liquid rises in a glass capillary and the meniscus is concave. For a liquid that does not wet glass (for example, mercury), the liquid is lower where it contacts the glass and the meniscus is convex.

One final property of liquids that we shall discuss is called the **dielectric constant.** We can define a dielectric constant by referring to Coulomb's law (Section 10-5), which says that two charged bodies separated

Figure 14-16 A meniscus formed by water *(left)* in a glass tube is concave, whereas the meniscus formed by mercury *(right)* is convex.

by a distance d *in a vacuum* have an interaction energy given by [Equation (10-1)]

$$E = (2.31 \times 10^{-16}\ \mathrm{J{\cdot}pm})\frac{Z_1 Z_2}{d}$$

where Z_1 and Z_2 are the charges on the two ions and d is their separation. If these same two charges are separated by a distance d in a material medium such as a liquid, then their energy of interaction is given by

$$E = (2.31 \times 10^{-16}\ \mathrm{J{\cdot}pm})\frac{Z_1 Z_2}{Dd} \qquad (14\text{-}5)$$

where D, a unitless quantity, is the dielectric constant of the liquid. It turns out that D is greater than one for all materials, so Equation (14-5) says that the interaction of the two charges is always less by a factor of D in a material medium that it is in a vacuum.

The dielectric constant of a liquid depends in a complicated way on the dipole moments of the molecules in the liquid. The dielectric constants of nonpolar liquids are around 2 to 3, whereas those of polar liquids are significantly greater. The dielectric constant of water, a very polar substance, is about 80 at 25°C. According to Equation (14-5), two ions attract each other with a force (energy of interaction) that is 80 times less in water than in a vacuum. Consequently, water and other liquids with relatively large dielectric constants (say, >40) are good solvents for ionic compounds. Liquids with low dielectric constants, such as hydrocarbons ($D \sim 3$), are poor solvents for ionic compounds.

Table 14-6 lists the relative viscosities, the surface tensions, and the dielectric constants of a variety of compounds. The most important thing to notice is the relative values of these quantities.

14-6. A LIQUID HAS A UNIQUE EQUILIBRIUM VAPOR PRESSURE AT EACH TEMPERATURE

Figure 14-17 When a liquid is placed in a closed container that has been evacuated, the pressure of the vapor above the liquid eventually reaches a constant value that depends upon the particular liquid and the temperature.

Let's look more closely at the process of vaporization. Suppose that we cover a beaker containing a liquid with a bell jar, as shown in Figure 14-17, and maintain a constant temperature. Now suppose that we evacuate all the gas, so that the space above the beaker and its contents is a vacuum. The molecules in the liquid are in constant motion; some at the surface will break away from the liquid and form a vapor phase above it. The pressure of the vapor is observed to increase rapidly at first and then progressively more slowly until a constant pressure is reached. Let's see why.

In order to break free of the liquid, a molecule at the surface must have enough kinetic energy to overcome the attractive force of its neighbors and be moving in the right direction. The number of molecules that leave the surface is proportional to the surface area of the liquid. Because the surface area is constant, the rate of evaporation is constant (Figure 14-18). There are no molecules in the vapor phase initially, so there is no condensation from the vapor phase to the liquid phase. As the concentration of molecules in the vapor phase increases, the pressure of the vapor increases and the number of vapor-phase molecules that collide with the liquid surface increases. As a result, the rate of condensation of the vapor increases. Eventually a state is

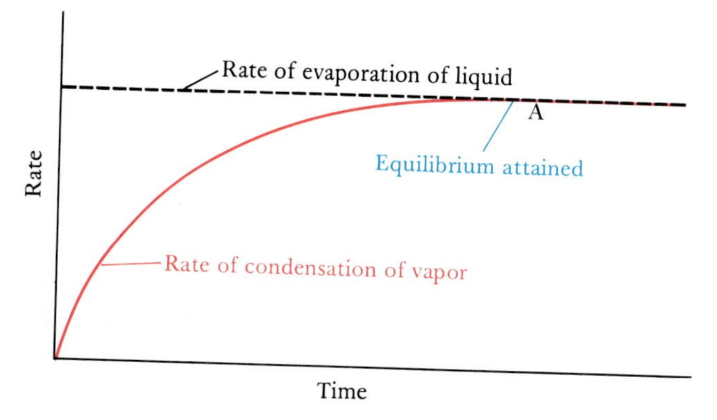

Figure 14-18 Equilibrium is attained (at A) when the rate of evaporation of the liquid equals the rate of condensation of the vapor. At equilibrium the pressure of the vapor is constant and is called the equilibrium vapor pressure.

reached where the rate of evaporation from the liquid surface is equal to the rate of condensation from the vapor phase. The pressure of the vapor no longer increases but takes on a constant value. The evaporation-condensation process appears to have stopped, and we say that the system is at **equilibrium,** meaning that no change appears to be taking place (Figure 14-19). The pressure of the vapor is now a constant value.

The equilibrium between the liquid and the vapor is a **dynamic equilibrium;** that is, the liquid continues to evaporate and the vapor continues to condense, but the rate of evaporation is exactly equal to the rate of condensation and thus there is no *net* change. In other words, we have the equilibrium condition

$$\text{rate of evaporation} = \text{rate of condensation}$$

The pressure of the vapor at equilibrium is called the **equilibrium vapor pressure.** We will show that the value of the equilibrium vapor pressure depends on the particular liquid and the temperature.

Let's consider the approach to a dynamic liquid-vapor equilibrium at two different temperatures. The higher the temperature, the more rapidly the molecules in the liquid phase move and the higher the rate of evaporation is. Figure 14-20 shows that, because the rate of evaporation at T_2 is greater than the rate of evaporation at T_1 (given that $T_2 > T_1$),

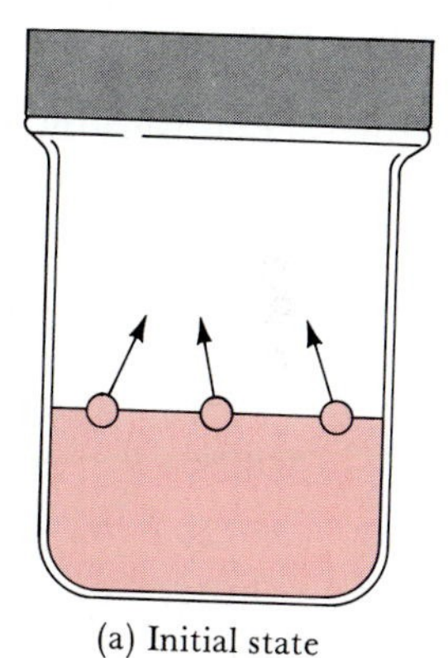

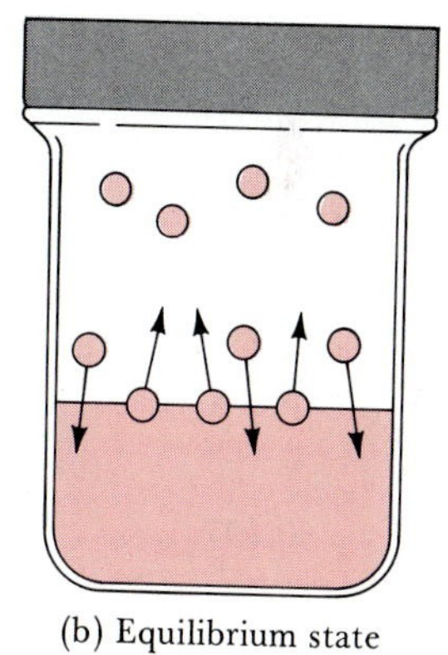

Figure 14-19 When a liquid is placed in a closed container, (a) the rate at which molecules escape from the surface is constant, but the rate at which molecules enter the liquid from the vapor is proportional to the number of molecules in the vapor. (b) When the number of molecules in the vapor is such that the rate of escape from the surface is equal to the rate of condensation from the vapor, the liquid and vapor are in equilibrium with each other.

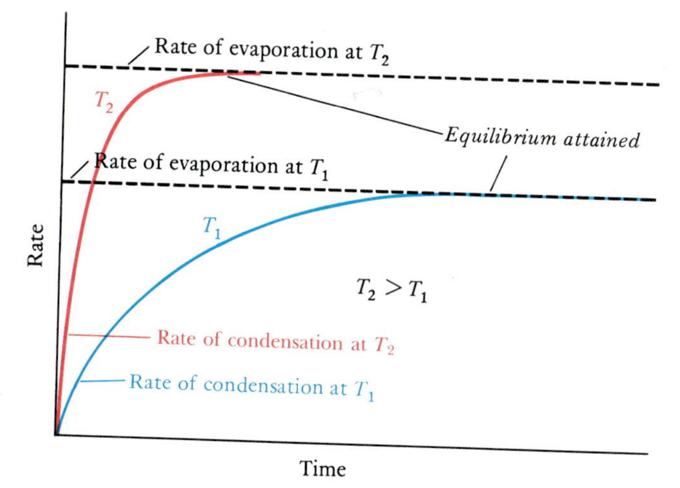

Figure 14-20 The change in the rate of condensation with time for the vapor over as it approaches equilibrium. Because the rate of evaporation increases with increasing temperature, the equilibrium vapor pressure increases with increasing temperature ($T_2 > T_1$).

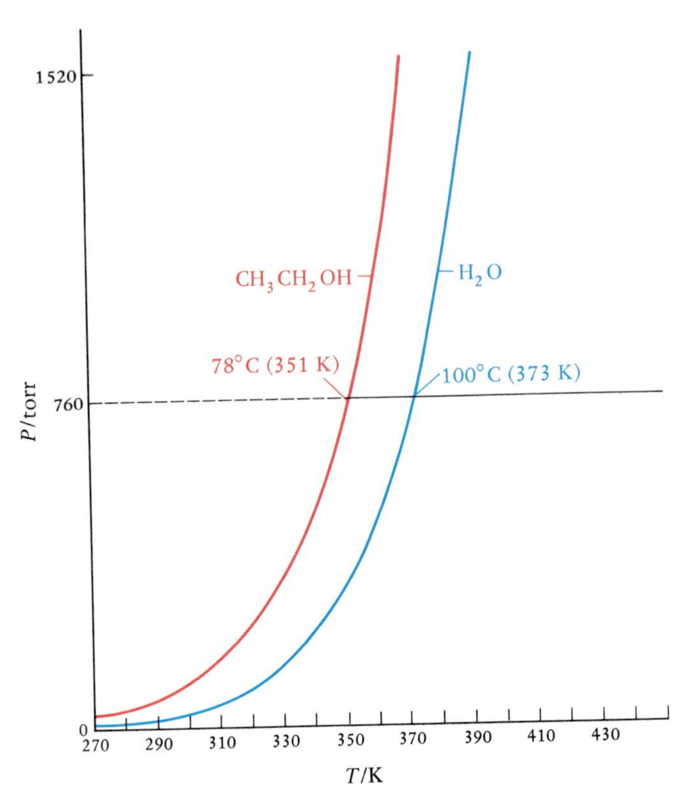

Figure 14-21 The equilibrium vapor pressure curves for water and ethanol over the temperature range 270 to 430 K (−3° to 157°C). Note the very rapid increase in vapor pressure with increasing temperature. The equilibrium vapor pressure curve for ethanol lies above that for water because ethanol has a higher equilibrium vapor pressure than water at the same temperature.

Table 14-7 Equilibrium vapor pressure of water as a function of temperature

t/°C	P/atm	P/torr
0	0.0060	4.6
5	0.0086	6.5
10	0.0121	9.2
15	0.0168	12.8
20	0.0230	17.4
25	0.0313	23.8
30	0.0418	31.6
35	0.0555	42.2
40	0.0728	55.3
45	0.0946	71.9
50	0.122	92.5
55	0.155	118.0
60	0.197	149.4
65	0.247	187.5
70	0.308	233.7
75	0.380	289.1
80	0.467	355.1
85	0.571	433.6
90	0.692	525.8
95	0.834	633.9
100	1.000	760.0
105	1.192	906.1
110	1.414	1074.6
120	2.00	1520

the equilibrium vapor pressure at T_2 is greater than that at T_1. Thus, we see that the value of the equilibrium vapor pressure of a liquid increases with increasing temperature. At each temperature, a liquid has a definite equilibrium vapor pressure. The equilibrium vapor pressure of water plotted as a function of the temperature is called a **vapor pressure curve** (Figure 14-21 and Table 14-7). Figure 14-21 also shows the equilibrium vapor pressure curve for ethanol, CH_3CH_2OH. In Chapter 22 we discuss a mathematical expression that describes the equilibrium vapor pressure as a function of absolute temperature.

EXAMPLE 14-6: A 0.0896-g sample of water is placed in a 250-mL container. Is there any liquid present when the temperature is held at 70°C?

Solution: First, we calculate the pressure assuming that all the H_2O is gaseous. If this pressure is greater than the equilibrium vapor pressure of H_2O at 70°C, then H_2O will condense until the pressure is equal to the equilibrium vapor pressure. If the pressure is less than the equilibrium vapor pressure, then no liquid will be present. The pressure is given by

$$P = \frac{nRT}{V} = \frac{\left(\dfrac{0.0896\text{ g}}{18.02\text{ g}\cdot\text{mol}^{-1}}\right)(0.0821\text{ L}\cdot\text{atm}\cdot\text{mol}^{-1}\cdot\text{K}^{-1})(343\text{ K})}{0.250\text{ L}}$$

$$= 0.560\text{ atm} = 425\text{ torr}$$

According to Table 14-7, the equilibrium vapor pressure of water is 233.7 torr at 70°C, so some of the water vapor will condense until its pressure is 233.7 torr.

PRACTICE PROBLEM 14-6: What mass of water will condense in Example 14-6?

Answer: 0.0404 g

The vapor pressure of a liquid depends on the attractive forces between its constituent molecules. Substances with relatively strong intermolecular attractions will have relatively low vapor pressures. At a given temperature, relatively few of the molecules of such a substance will have sufficient kinetic energy to overcome the attractive forces of the other molecules and enter the vapor phase.

The boiling point of a liquid is the temperature at which its vapor pressure equals the external pressure. The **normal boiling point** (the boiling point at exactly 1 atm) of water is 373 K (100°C). If the external pressure is less than 1.00 atm, then the temperature at which the vapor pressure of liquid water equals this pressure is less than 100°C. For example, the elevation at Vail, Colorado, is about 8000 feet. The atmospheric pressure at this elevation is about 0.75 atm, so water boils at 92°C. In a pressure cooker, on the other hand, when the pressure is 2.0 atm, water boils at 120°C. Because atmospheric pressure decreases with increasing elevation and because the rate at which food cooks depends on the temperature, it requires a significantly longer time to cook food by boiling in an open container at high elevations than at sea level. An egg must be boiled for about 5 min at 9200 feet in order to be cooked to the same extent as one boiled for 3 min at sea level.

14-7. RELATIVE HUMIDITY IS BASED ON THE VAPOR PRESSURE OF WATER

The vapor pressure of water in the atmosphere is expressed in terms of relative humidity. **Relative humidity** is the ratio of the partial pressure of the water vapor in the atmosphere to the equilibrium vapor pressure of water at the same temperature times 100. An equation expressing this relation is

$$\text{relative humidity} = \left(\frac{P_{H_2O}}{P^{\circ}_{H_2O}}\right) \times 100 \qquad (14\text{-}6)$$

where P_{H_2O} is the partial pressure of the water vapor in the air and $P^{\circ}_{H_2O}$ is the equilibrium vapor pressure of water at the same temperature. At 20°C, the equilibrium vapor pressure of water is 17.4 torr. If the partial pressure of the water vapor in the air is 11.2 torr, then the relative humidity is

$$\text{relative humidity} = \left(\frac{11.2\text{ torr}}{17.4\text{ torr}}\right) \times 100 = 64.4\%$$

If the temperature of the air is lowered to 13°C, where the equilibrium vapor pressure of water is 11.2 torr, then the relative humidity is

$$\text{relative humidity} = \left(\frac{11.2\ \text{torr}}{11.2\ \text{torr}}\right) \times 100 = 100\%$$

At 13°C, air that contains water vapor at a partial pressure of 11.2 torr is saturated with water vapor. At this temperature, the water vapor begins to condense as dew or fog, which consists of small droplets of water. The air temperature at which the relative humidity reaches 100 percent is called the **dew point.** Most people begin to feel uncomfortable when the dew point rises above 20°C, and air with a dew point above 24°C is generally regarded as extremely humid or muggy.

EXAMPLE 14-7: Calculate the relative humidity and the dew point for a day when the partial pressure of water vapor in the air is 22.2 torr and the temperature of the air is 30°C. The equilibrium vapor pressure of water at 30°C is 31.6 torr (Table 14-7).

Solution: The relative humidity, given by Equation (14-6), is

$$\text{relative humidity} = \left(\frac{P_{H_2O}}{P^\circ_{H_2O}}\right) \times 100 = \left(\frac{22.2\ \text{torr}}{31.6\ \text{torr}}\right) \times 100 = 70.3\%$$

The dew point is the temperature at which the equilibrium vapor pressure of water is equal to 22.2 torr. According to Table 14-7, this is about 24°C. Such a day would be considered very uncomfortable.

PRACTICE PROBLEM 14-7: Calculate the dew point for a day when the relative humidity is 78 percent at 20°C.

Answer: approximately 17°C

14-8. A PHASE DIAGRAM DISPLAYS THE REGIONS OF ALL THE PHASES OF A PURE SUBSTANCE SIMULTANEOUSLY

The vapor pressure curve of a pure substance can be combined with two other useful quantities, the sublimation pressure curve and the melting point curve, into a single diagram called a **phase diagram.** The phase diagram of water is shown in Figure 14-22.

Along the vapor pressure curve (the red curve in Figure 14-22), liquid and vapor exist together at equilibrium. To the left of this curve, at lower temperatures, the water exists as a liquid. To the right of the curve, at higher temperatures, the water exists as a vapor. The equilibrium vapor pressure of a liquid increases with temperature up to the **critical point** (Figure 14-22), where the vapor pressure curve terminates abruptly. Above the **critical temperature,** the gas and the liquid phases become indistinguishable. A gas above its critical temperature cannot be liquefied no matter how high a pressure is applied. The critical point for water occurs at 218 atm and 647 K. Water vapor above 647 K cannot be liquefied by the application of pressure.

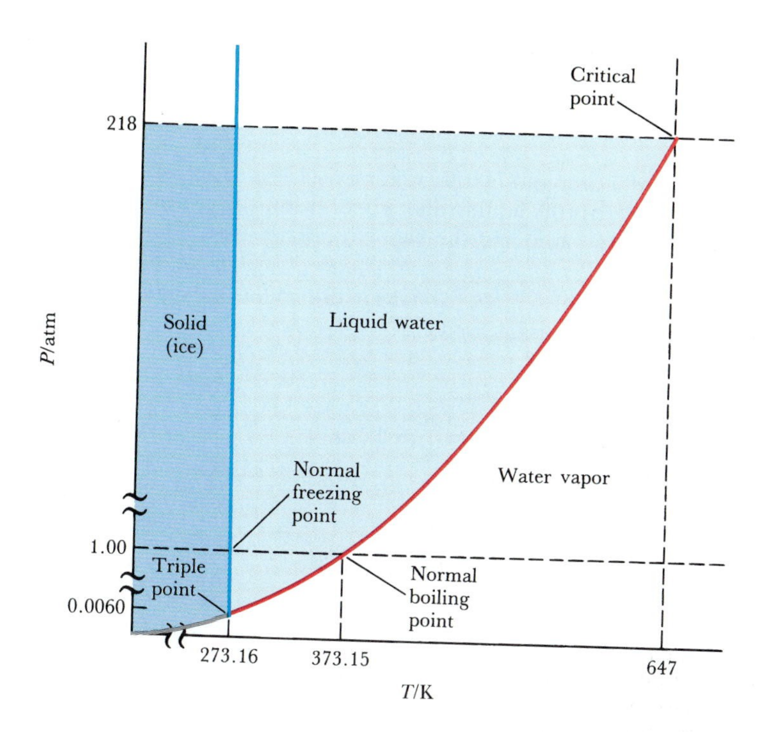

Figure 14-22 The phase diagram of water (not to scale; note the breaks in both the horizontal and vertical axes), which displays simultaneously the sublimation pressure curve (gray), the vapor pressure curve (red), and the melting point curve (blue). The triple point, the critical point, the normal boiling point, and the normal freezing point are indicated in the figure. The phase equilibrium lines are the boundaries between the regions of stability of the solid, liquid, and vapor phases, which are labeled solid (ice), liquid water, and water vapor.

Along the **sublimation pressure curve** (gray curve in Figure 14-22), solid and vapor exist together at equilibrium. To the left of this curve, at lower temperatures, water exists as a solid (ice). To the right of this curve, at higher temperatures, water exists as a vapor.

Along the **melting point curve** (blue curve in Figure 14-22), solid and liquid exist together in equilibrium. To the left of this curve, water exists as ice, and to the right of the curve, water exists as liquid. The melting point at a pressure of one atmosphere is called the **normal melting point.** However, melting points are only weakly dependent on pressure, so a melting point curve is an almost vertical line (Figure 14-22). For almost all substances, the melting point increases with increasing pressure at a rate of 0.01 to 0.03 $K \cdot atm^{-1}$. Water is anomalous because its melting point decreases with increasing pressure. The melting point of ice decreases by about 0.01 K per atmosphere of applied pressure. Consequently, unlike most other solids, ice can be melted by the application of pressure. We discuss why this is so in Section 17-7.

Notice that the three curves in Figure 14-22 separate regions in which water exists as a solid, a liquid, or a vapor. Let's use Figure 14-22 to follow the behavior of water as it is heated from −50°C to 200°C at a constant pressure of 1 atm. At −50°C (223 K) and 1 atm, water exists as ice. As we heat the ice at 1 atm, we move horizontally from left to right along the dashed line in Figure 14-22. At 0°C, we cross the melting point curve and pass from the solid region into the liquid region. At 100°C, we cross the equilibrium vapor pressure curve and pass from the liquid region into the vapor region.

The three curves in Figure 14-22 intersect at a point called the **triple point.** At the triple point, and only at the triple point, all three phases—solid, liquid, and gas—coexist in equilibrium. The triple point for water occurs at 4.58 torr and 273.16 K. Notice that if we heat ice at a constant pressure less than 4.58 torr (0.0060 atm), then the ice sublimes rather than melts.

EXAMPLE 14-8: Use the phase diagram of water given in Figure 14-22 to predict the result of increasing the pressure of water vapor initially at 1 atm and 500 K, keeping the temperature constant.

Solution: At 1 atm and 500 K, water exists as a vapor. As the pressure is increased, we cross the liquid-vapor curve below the critical point at a pressure of about 150 atm, and the vapor condenses to a liquid.

PRACTICE PROBLEM 14-8: Given the following data for iodine, I_2, sketch the phase diagram for I_2.

triple point	113°C	0.12 atm
critical point	512°C	116 atm
normal melting point	114°C	1 atm
normal boiling point	184°C	1 atm
density of solid > density of liquid	—	—

EXAMPLE 14-9: The vapor pressures (in torr) of solid and liquid argon are given by

$$\log P = 7.506 - \frac{399.1\ \text{K}}{T} \qquad \text{(solid)}$$

$$\log P = 6.617 - \frac{304.2\ \text{K}}{T} \qquad \text{(liquid)}$$

where T is the Kelvin temperature. Calculate the temperature and pressure of the triple point of argon.

Solution: The vapor pressures of the solid and the liquid are equal at the triple point, so we have

$$7.506 - \frac{399.1\ \text{K}}{T} = 6.617 - \frac{304.2\ \text{K}}{T}$$

Solving for T gives $T = 107$ K. Substituting this value into either of the vapor pressure equations gives

$$\log P = 7.506 - \frac{399.1\ \text{K}}{107\ \text{K}} = 3.776$$

or

$$P = 5970\ \text{torr} = 7.86\ \text{atm}$$

PRACTICE PROBLEM 14-9: The vapor pressures (in torr) of solid and liquid bromine are given by

$$\log P = 9.721 - \frac{2041\ \text{K}}{T} \qquad \text{(solid)}$$

$$\log P = 6.688 - \frac{1120\ \text{K}}{T} \qquad \text{(liquid)}$$

where T is the Kelvin temperature. Calculate the temperature and pressure at the triple point of bromine.

Answer: $T = 304$ K; $P = 1.34$ atm

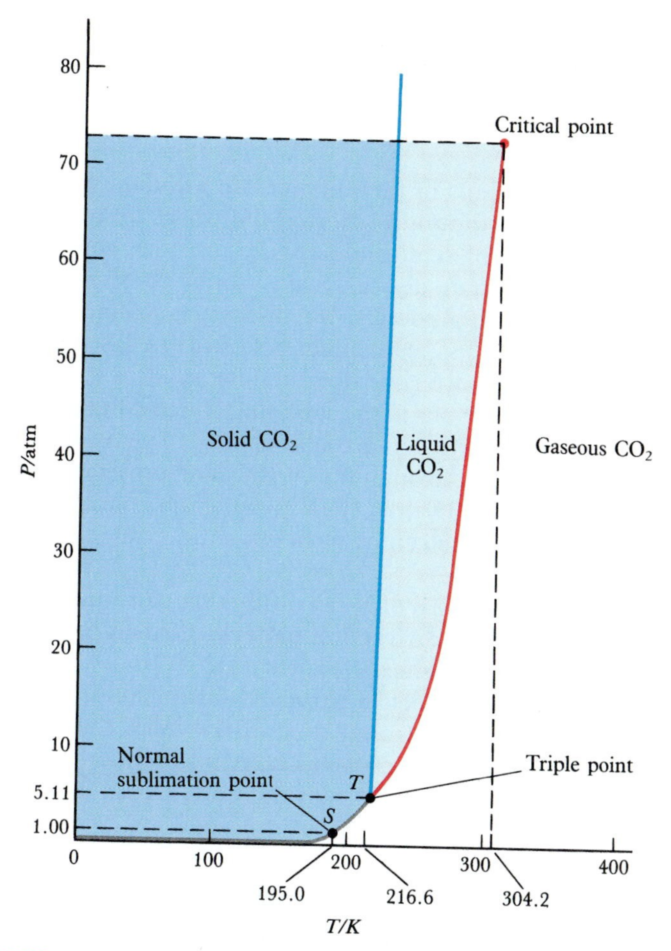

Figure 14-23 The phase diagram of carbon dioxide. The point C is the critical point, and the point T is the triple point. Note that the triple point lies above 1 atm and thus $CO_2(s)$ at 1 atm does not melt—it sublimes. The point S is the normal (1 atm) sublimation point of CO_2.

The phase diagram of carbon dioxide is shown in Figure 14-23. Although it looks similar to that of water, there are several important differences. The melting point curve of carbon dioxide goes up and to the right, a direction indicating that the melting point of CO_2 increases with increasing pressure. Recall that the melting point curve of water points up and slightly to the left, a direction indicating that the melting point of H_2O decreases with increasing pressure.

Another difference between Figures 14-22 and 14-23 are the positions of the triple points. The triple point for CO_2 occurs at 5.11 atm and 216.6 K. Because the pressure at the triple point is greater than 1 atm, CO_2 does not melt when it is heated at 1 atm as water does. Instead, CO_2 sublimes. The **normal sublimation point** of CO_2 is 195 K (−78°C), which is the temperature of solid CO_2 at 1 atm. Liquid CO_2 can be obtained by compressing $CO_2(g)$ at a temperature below its critical point (31°C). A pressure of about 60 atm is required to liquefy CO_2 at 25°C. A carbon dioxide-filled fire extinguisher at 25°C contains liquid CO_2 at a pressure of about 60 atm.

14-9. X-RAY DIFFRACTION PATTERNS YIELD INFORMATION ABOUT THE STRUCTURES OF CRYSTALLINE SOLIDS

Figure 14-24 The X-ray diffraction pattern produced by a crystal of sodium chloride. The symmetry and spacing of the dots carry detailed information regarding the arrangement of atoms in the crystals.

In the remaining sections of this chapter, we will discuss solids. Solid-state chemistry has profoundly affected the world we live in through the creation and use of semiconductors, transistors, computer chips, and many other solid-state devices. Solids can be classified into crystalline solids and amorphous solids. We discuss crystalline solids in this section and briefly discuss amorphous solids in the last section.

A distinguishing characteristic of crystalline solids is the ordered nature of the molecules or ions in the solid state, an arrangement we refer to as a **crystal lattice.** We can actually obtain a "picture" of a crystal lattice by passing X-rays through the crystal. The presence of a definite ordered array of atoms in the crystal produces a characteristic **X-ray diffraction pattern,** which can be recorded as an array of spots (Figure 14-24).

To get a feel for the origin of X-ray diffraction patterns from crystals, let's examine the **optical diffraction patterns** formed by light passing through holes in opaque sheets (Figure 14-25). Notice that each size and arrangement of holes yield a particular diffraction pattern that can be used to determine the arrangement of the holes that produced it. Just as the optical diffraction patterns carry information regarding the relative positions and the spacings of the holes, X-ray diffraction pat-

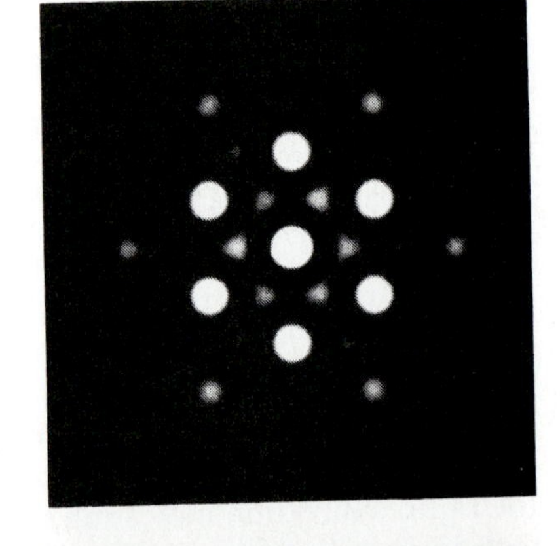

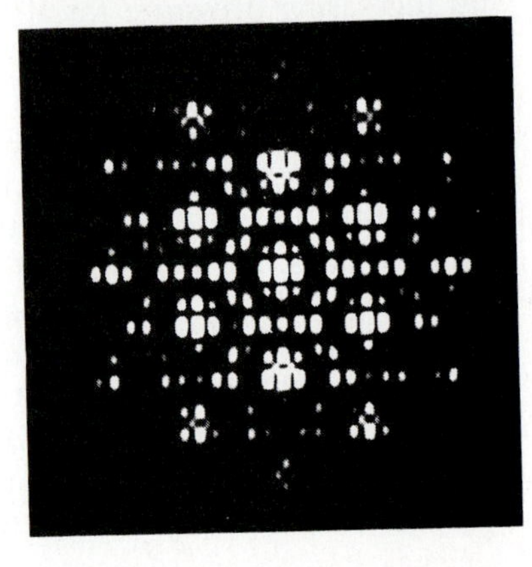

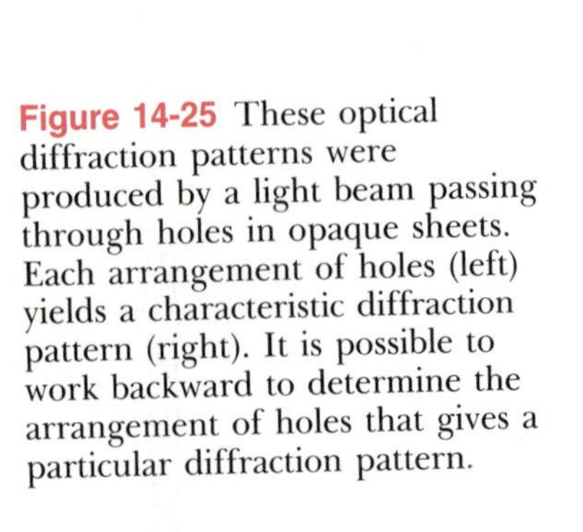

Figure 14-25 These optical diffraction patterns were produced by a light beam passing through holes in opaque sheets. Each arrangement of holes (left) yields a characteristic diffraction pattern (right). It is possible to work backward to determine the arrangement of holes that gives a particular diffraction pattern.

terns provide information about the arrangement of the atoms in crystals.

The smallest subunit of a crystal lattice that can be used to generate the entire lattice is called a **unit cell.** A crystal lattice is thus a repeating pattern of unit cells. Figure 14-26 illustrates in two dimensions how a unit cell can generate a crystal lattice. Three-dimensional unit cells are classified according to the relative lengths of the sides of the cell and the angles between the sides (Figure 14-27). Seven such types of unit cell are observed in crystalline solids, but, for simplicity, we will discuss only cubic unit cells because most metallic elements have one of three kinds of cubic unit cell (Figure 14-28).

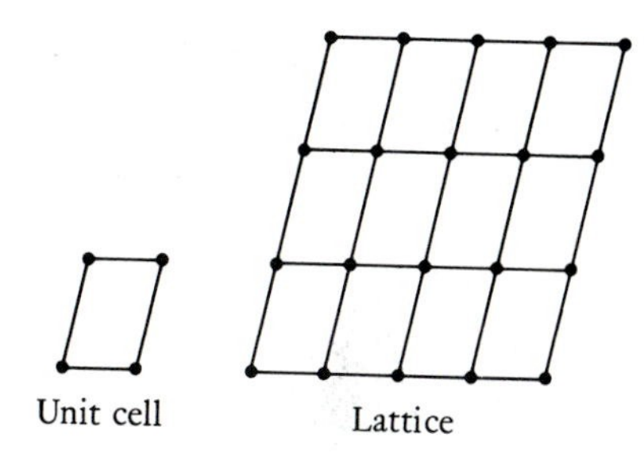

Figure 14-26 A two-dimensional illustration of the generation of a crystal lattice by a unit cell. Note that only a portion of each dot lies within the unit cell, but whole dots are generated when several unit cells are "stacked" together.

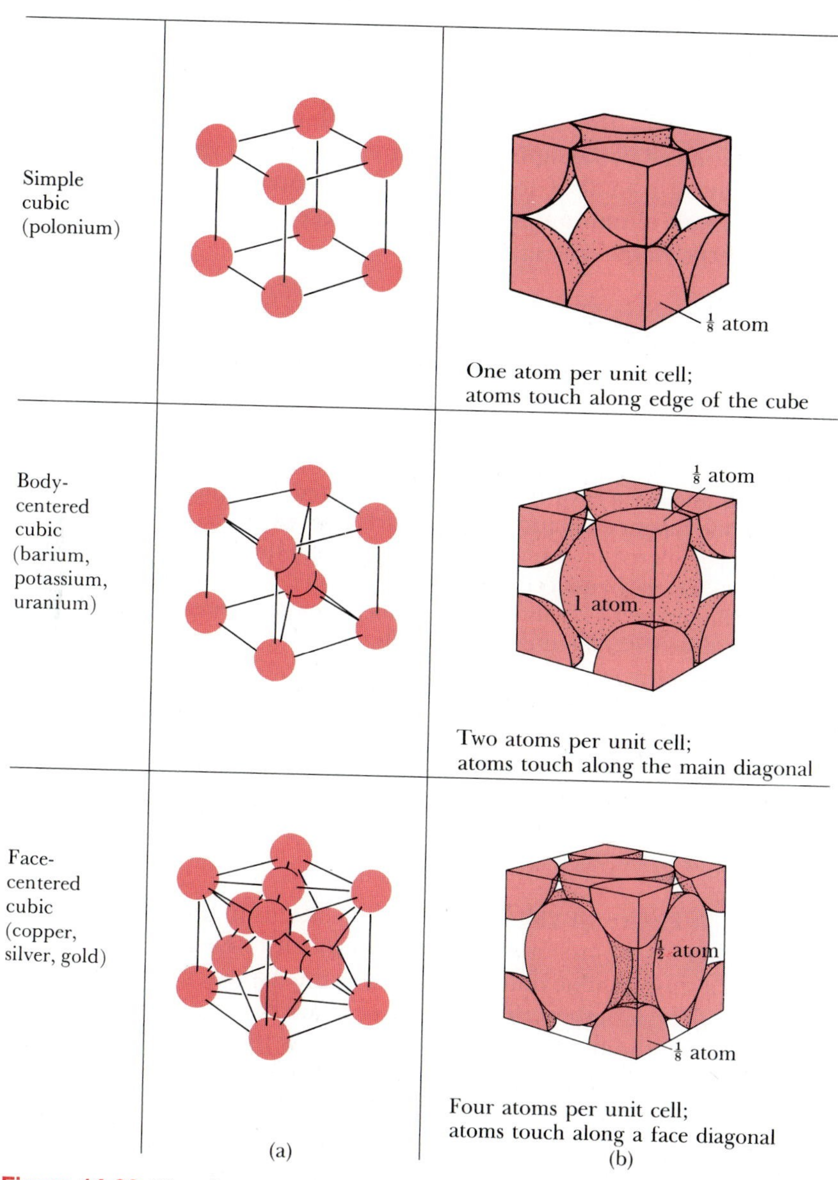

Figure 14-28 The three cubic unit cells: simple cubic, body-centered cubic, and face-centered cubic. (a) An open perspective of the unit cells. Note that the body-centered cubic unit cell has an atom at the center of the unit cell and that a face-centered unit cell has an atom at the center of each face. (b) The sharing of atoms by adjacent unit cells.

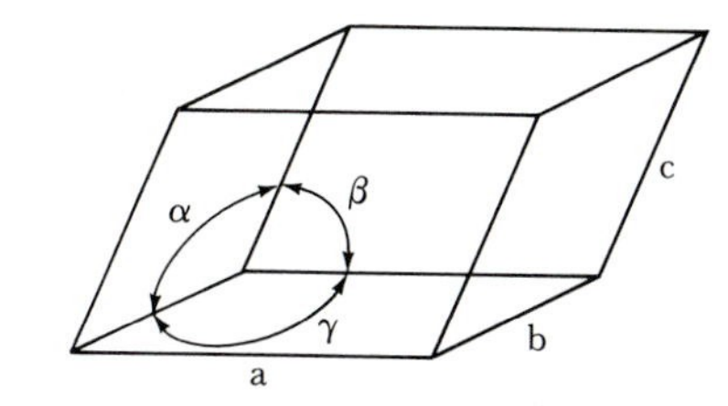

Figure 14-27 The axes and angles used to describe the seven simple unit cells.

The **simple cubic unit cell** has atoms only at the vertices of the cube, and each one of these atoms is shared by a total of eight unit cells. Because each of the eight atoms of a simple cubic unit cell is shared by a total of eight unit cells, we can assign one atom to a simple cubic unit cell. Only one metal, polonium, occurs as a simple cubic lattice. A **body-centered cubic unit cell** is similar to a simple cubic unit cell, except that there is an atom (or, in general, a particle) at the center of the unit cell. This atom is shared by no other unit cells, so we can assign a total of two atoms to a body-centered unit cell. Some metals that occur as a body-centered cubic lattice are barium, cesium, potassium, lithium, molybdinum, tantalum, uranium, and vanadium. A **face-centered cubic unit cell** has an atom at each vertex, but also one at each face of the unit cube. An atom at one of the faces is shared by two unit cells. Because there are six faces on a cube and each of these atoms is shared by two unit cells, we assign a total of three atoms to the unit cell from the faces. In addition, as in the case of a simple cubic unit cell and a body-centered unit cell, we assign one atom to this unit cell from those at the vertices, giving a total of four atoms. Face-centered cubic lattices are found in silver, aluminum, gold, copper, nickel, lead, strontium, platinum, and the noble gases.

EXAMPLE 14-10: Copper exists as a face-centered cubic lattice. How many copper atoms are there in a unit cell?

Solution: Reference to Figure 14-28 shows that each of the eight copper atoms at the corners of the unit cell are shared by eight unit cells, so we assign one copper atom ($8 \times \frac{1}{8} = 1$) to each unit cell. Each of the six atoms at the faces is shared by two unit cells, so we assign three more copper atoms ($6 \times \frac{1}{2} = 3$) to each unit cell, thereby giving a total of four copper atoms in a unit cell. Figure 14-28 also illustrates the counting process for the simple cubic and body-centered cubic unit cells.

PRACTICE PROBLEM 14-10: Cesium exists as a body-centered cubic lattice. How many cesium atoms are there in a unit cell?

Answer: 2

Note that Figure 14-28b implies that the atoms in the unit cell touch one another. If we assume that they do touch, then we find that the length of a simple cubic unit cell is equal to the diameter of the constituent atoms. The length of the diagonal of a face in a face-centered cubic lattice is twice the diameter of the atoms, as is the length of the main diagonal (in other words, a line that goes from one vertex to another and passes through the center of the cube) of a body-centered unit cell. We noted in Chapter 9 that atoms do not have well-defined radii, but we can use the suggestion of Figure 14-28 to calculate effective radii, or **crystallographic radii,** from the dimensions of atomic crystals.

If we know the volume of the unit cell, we can calculate the length of an edge or a diagonal, and hence the crystallographic radius of the constituent atoms. We can determine the volume of a unit cell of an atomic crystalline solid if we know its density and its crystal structure. Let V_{mol} be the molar volume (volume per mole) of the substance and let $V_{unit\ cell}$ be the volume of its unit cell. If n is the number of atoms per unit cell (1 for a simple cubic lattice, 2 for a body-centered cubic lattice, and 4 for a face-centered cubic lattice), then $V_{unit\ cell}/n$ is the volume per atom; and if we multiply this quantity by Avogadro's number, N_0, then we have the molar volume:

$$V_{mol} = \left(\frac{V_{unit\ cell}}{n}\right) \times N_0$$

Next we solve this equation for $V_{unit\ cell}$:

$$V_{unit\ cell} = \frac{nV_{mol}}{N_0} \qquad (14\text{-}7)$$

Finally, we can calculate the molar volume in terms of the density by taking the formula for the density,

$$d = \frac{m}{V} \qquad (14\text{-}8)$$

and replacing m by the molar mass, M. In this case, V in Equation (14-8) is the molar volume, V_{mol}, so we have

$$d = \frac{M}{V_{mol}}$$

or

$$V_{mol} = \frac{M}{d} \qquad (14\text{-}9)$$

We get our final equation by substituting this expression for V_{mol} into Equation (14-7):

$$V_{unit\ cell} = \frac{nM}{dN_0} \qquad (14\text{-}10)$$

The following Example illustrates the use of Equation (14-10).

EXAMPLE 14-11: Copper, which crystallizes as a face-centered cubic lattice, has a density of 8.930 $g{\cdot}cm^{-3}$ at 20°C. Calculate the crystallographic radius of a copper atom.

Solution: From Equation (14-10) we can calculate the volume of the unit cell. A face-centered cubic lattice has four atoms per unit cell, so $n = 4$ in Equation (14-10). Thus, we write

$$V_{unit\ cell} = \frac{(4\ \text{atom}{\cdot}\text{unit cell}^{-1})(63.55\ \text{g}{\cdot}\text{mol}^{-1})}{(8.930\ \text{g}{\cdot}\text{cm}^{-3})(6.022 \times 10^{23}\ \text{atom}{\cdot}\text{mol}^{-1})}$$
$$= 4.727 \times 10^{-23}\ \text{cm}^3{\cdot}\text{unit cell}^{-1}$$

Because the unit cell is cubic, the length of an edge is given by the cube root of $V_{unit\ cell}$

$$l = (V_{unit\ cell})^{1/3} = (4.727 \times 10^{-23}\ \text{cm}^3)^{1/3} = 3.616 \times 10^{-8}\ \text{cm}$$
$$= 361.6\ \text{pm}$$

Figure 14-28 shows that the effective radius of an atom in a face-centered cubic lattice is given by one fourth of the length of the diagonal of a face. The length of a diagonal is given by

$$\text{diagonal} = \sqrt{2}l = \sqrt{2}(361.6\ \text{pm}) = 511.4\ \text{pm}$$

so the crystallographic radius of a copper atom is (margin)

$$\text{radius} = \frac{511.4\ \text{pm}}{4} = 127.8\ \text{pm}$$

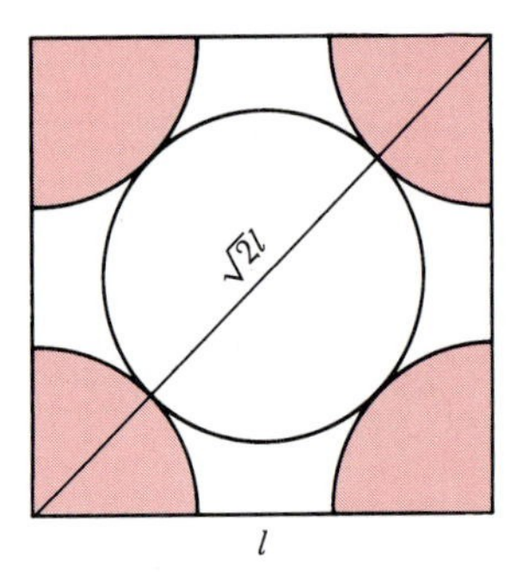

PRACTICE PROBLEM 14-11: Europium, which crystallizes in a body-centered cubic lattice, has a density of $5.243\ \text{g}\cdot\text{cm}^{-3}$ at 20°C. Calculate the crystallographic radius of a europium atom.

Answer: 162.0 pm

If the density of an atomic solid and the dimensions of its unit cell are known from X-ray analysis, then we can use Equation (14-10) to determine Avogadro's number. The following example illustrates this type of calculation.

EXAMPLE 14-12: Potassium crystallizes in a body-centered cubic lattice, and the length of a unit cell is 533.3 pm. Given that the density of potassium is $0.8560\ \text{g}\cdot\text{cm}^{-3}$, calculate Avogadro's number.

Solution: We solve Equation (14-10) for N_0, giving

$$N_0 = \frac{nM}{dV_{\text{unit cell}}}$$

Because the unit cell is cubic,

$$\begin{aligned} V_{\text{unit cell}} &= l^3 = (533.3 \times 10^{-12}\ \text{m})^3 \\ &= (533.2 \times 10^{-10}\ \text{cm})^3 = 1.517 \times 10^{-22}\ \text{cm}^3 \end{aligned}$$

The unit cell is body-centered cubic, and so $n = 2$. Thus, N_0 is given by

$$N_0 = \frac{(2\ \text{atom}\cdot\text{unit cell}^{-1})(39.10\ \text{g}\cdot\text{mol}^{-1})}{(0.8560\ \text{g}\cdot\text{cm}^{-3})(1.517 \times 10^{-22}\ \text{cm}^3)} = 6.022 \times 10^{23}\ \text{atom}\cdot\text{mol}^{-1}$$

In 1976, scientists at the National Bureau of Standards used very precise X-ray measurements on ultra pure silicon to obtain a value of $6.022098 \pm 6 \times 10^{23}$ for Avogadro's number. The ±6 indicates the uncertainty in the last digit.

PRACTICE PROBLEM 14-12: Cerium crystallizes in a face-centered cubic lattice, and the length of the unit cell is 516.0 pm. Given that the density of cerium is $6.773\ \text{g}\cdot\text{cm}^{-3}$, calculate Avogadro's number.

Answer: 6.022×10^{23}

14-10. CRYSTALS CAN BE CLASSIFIED ACCORDING TO THE FORCES BETWEEN THE CONSTITUENT PARTICLES

Crystal structures are determined by the size of the atoms, ions, or molecules making up the lattice and by the nature of the forces that act between these particles. The examples that we discussed in the previous section consisted of particles of the same size. Such crystals are called **atomic crystals,** a good example being crystals of the noble gases, which crystallize as face-centered cubic crystals. In this section we discuss ionic crystals, molecular crystals, covalent network crystals, and the conduction properties of metals.

Ionic crystals are held together by the electrostatic attraction between ions of opposite charge. The crystalline structure of ionic crystals often depends on how the cations and anions can be packed together to form a lattice. Consequently, the difference in the sizes of the cation and the anion plays a key role. For example, the unit cells of the salts sodium chloride and cesium chloride are shown in Figure 14-29. Notice that each sodium ion in NaCl is surrounded by an octahedral arrangement of chloride ions and each cesium ion in CsCl is surrounded by a cubic arrangement of chloride ions. The different packing arrangements for NaCl and CsCl are a direct consequence of the fact that cesium ions are larger than sodium ions.

Each ion in an ionic crystal is surrounded by ions of opposite charge. The total electrostatic interaction energy of the lattice accounts for much of the lattice energy, a quantity that we used in Section 10-6. Because electrostatic interactions are relatively strong, ionic solids usually have high melting points and low vapor pressures.

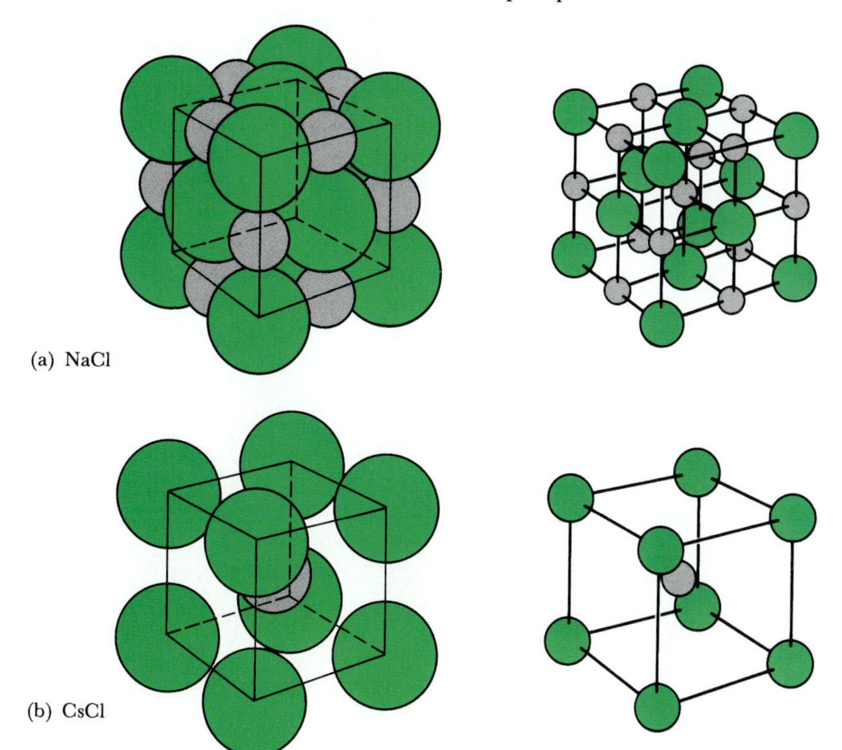

Figure 14-29 Space-filling and ball-and-stick representations of the unit cells of (a) NaCl and (b) CsCl. The different crystalline structures in the two cases are a direct consequence of the relative sizes of the cations and the anions. Recall that cations are positively charged ions and that anions are negatively charged ions.

Crystals composed of neutral molecules are called **molecular crystals.** We discussed the forces that hold the molecules together in molecular crystals in Section 14-4. These forces range in strength from the weak London attraction between small nonpolar molecules, such as H_2 and CH_4, to the relatively strong dipole-dipole attraction between large, polar molecules. Consequently, the melting points of molecular crystals range from 14 K for H_2 to over 500 K for large polar molecules. Generally, molecular crystals have lower melting points and higher vapor pressures than ionic crystals do.

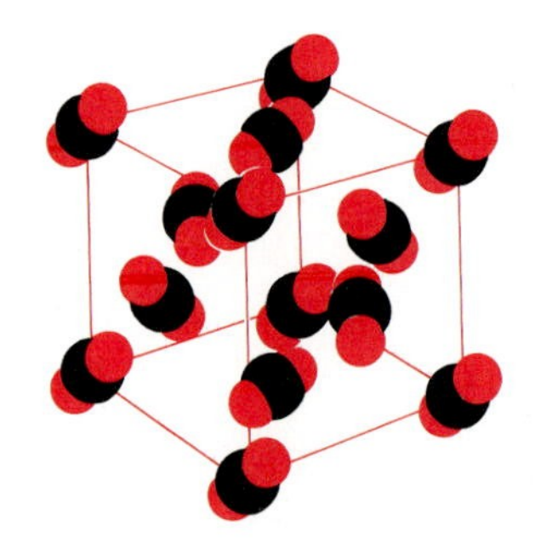

Figure 14-30 The unit cell of crystalline CO_2. The molecules have been reduced in size for clarity.

Molecular crystals come in a great variety of types. For example, methane crystallizes in a face-centered cubic structure, as do the noble gases; carbon dioxide crystals have the unit cell shown in Figure 14-30. When the positions of all the atoms in the unit cell of a molecular crystal are determined by **X-ray crystallography,** the positions of all the atoms within an individual molecule are determined as well. Thus, the determination of the crystalline structure of a molecular solid is equivalent to a determination of the structure of a single molecule.

The electron density contour map of benzoic acid is shown in Figure 14-31. The general outline of the planar molecule

```
      H       H
       \     /
        C — C           ..
       /     \       // O :
H — C         C — C
       \     /       \  ..
        C — C          O — H
       /     \         ..
      H       H
```

is clearly discernible. The characteristic shape of a benzene ring (page 465) is also evident. Such information is typical of that obtainable with X-ray crystallographic techniques. Precise details of structure such as

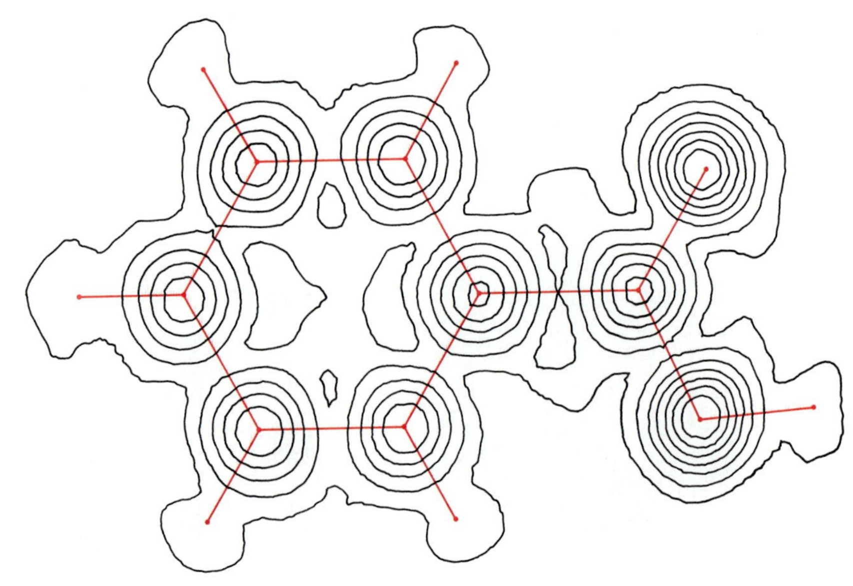

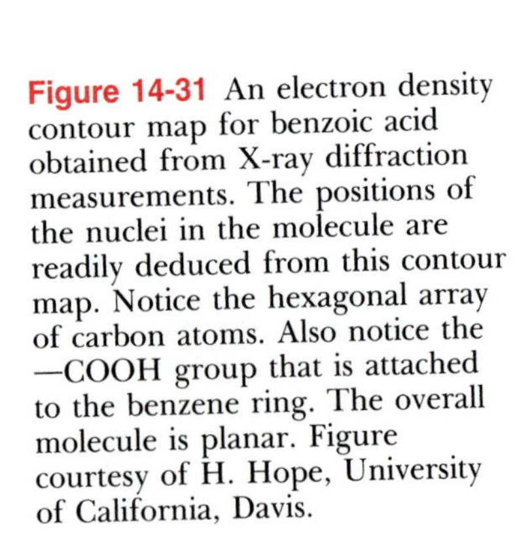

Figure 14-31 An electron density contour map for benzoic acid obtained from X-ray diffraction measurements. The positions of the nuclei in the molecule are readily deduced from this contour map. Notice the hexagonal array of carbon atoms. Also notice the —COOH group that is attached to the benzene ring. The overall molecule is planar. Figure courtesy of H. Hope, University of California, Davis.

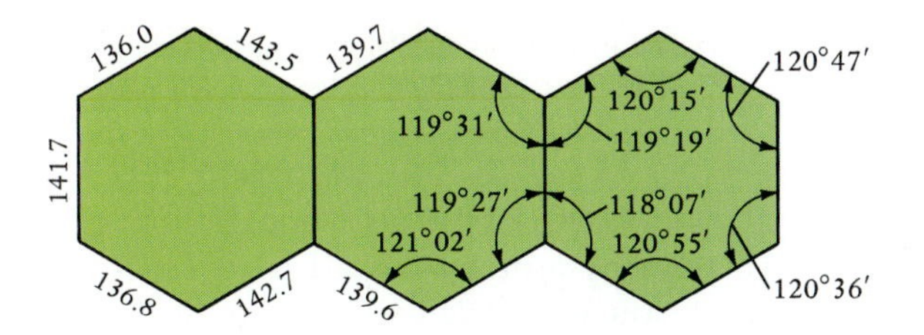

Figure 14-32 The dimensions of an individual anthracene molecule. This type of information can be obtained from X-ray diffraction experiments. The distances given are in picometers.

those shown in Figure 14-32 for the anthracene molecule are also readily determined. X-ray crystallography is one of the most powerful methods available for the determination of molecular structure and is used extensively by chemists.

A few substances form **covalent network crystals,** in which the constituent particles are held together by covalent bonds. Carbon, which in pure form can exist as diamond or graphite, is a good example of such a substance. Diamond has an extended, covalently bonded tetrahedral structure. Each carbon atom lies at the center of a tetrahedron formed by four other carbon atoms (Figure 14-33). The carbon-carbon bond distance is 154 pm, which is the same as the carbon-carbon bond distance in ethane. The diamond crystal is, in effect, a gigantic molecule. The hardness of diamond is due to the fact that each carbon atom throughout the crystal is covalently bonded to four others; thus, many strong covalent bonds must be broken in order to cleave a diamond. Graphite has the unusual layered structure shown in Figure 14-34. The carbon-carbon bond distance within a layer is 139 pm, which is close to the carbon-carbon bond distance in benzene. The distance between layers is about 340 pm. The bonding within a layer is covalent, but the interaction between layers is weak. Therefore, the layers easily slip past one another, thereby producing the molecular basis for the lubricating action of graphite. The "lead" in lead pencils is actually graphite. Layers of the graphite slide from the pencil onto the paper.

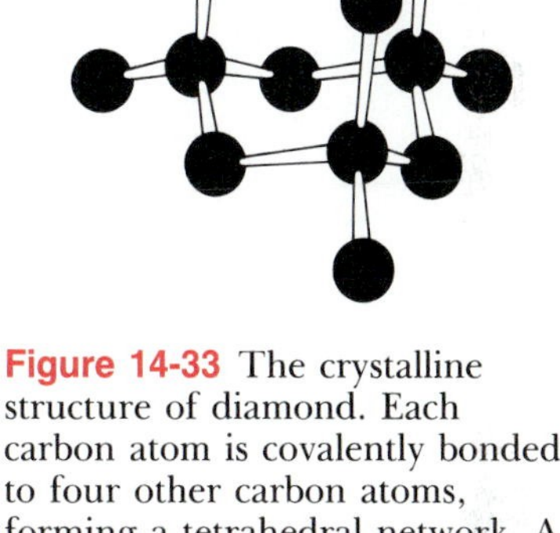

Figure 14-33 The crystalline structure of diamond. Each carbon atom is covalently bonded to four other carbon atoms, forming a tetrahedral network. A diamond crystal is essentially one gigantic molecule. (All four C—C bonds are shown for only four of the atoms displayed in this segment of the crystal).

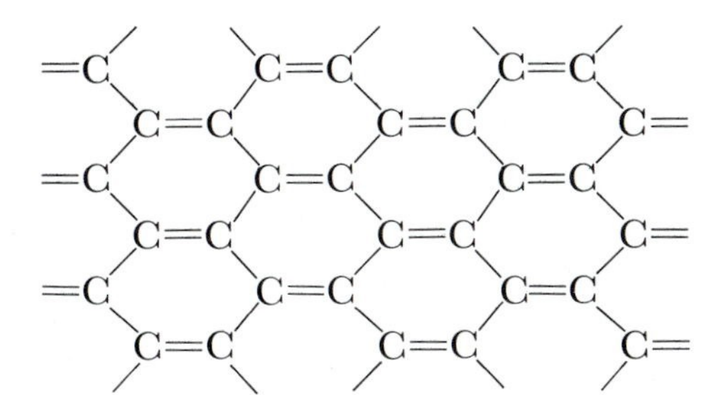

Figure 14-34 The layered structure of graphite. Each layer resembles a network of benzene rings joined together. The bonding within a layer is covalent and strong; however, the interaction between layers is due only to London forces, so is relatively weak. Consequently, the layers, which are separated by 340 pm, easily slip past each other, giving graphite its slippery feel and its use as a lubricant.

14-11. THE ELECTRONS IN METALS ARE DELOCALIZED THROUGHOUT THE CRYSTAL

The high electrical conductivity of metals is one of their most characteristic properties. To see why a metal has a high electrical conductivity and an insulator has a low electrical conductivity, we must discuss the electronic energy levels in atomic crystals.

For simplicity, imagine how we might form a crystal of metallic sodium. We could, at least ideally, bring a mole of widely separated sodium atoms to their positions in the crystalline lattice. In Chapter 13, we imagined bringing two hydrogen atoms together; here we are dealing with Avogadro's number of atoms instead of just two. Instead of getting two molecular orbitals (one bonding orbital and one antibonding orbital) as we do when we bring two hydrogen atoms together, we get Avogadro's number of molecular orbitals when we bring Avogadro's number of sodium atoms together. The energies of these orbitals are so close to one another that essentially they form a continuum of energy levels, from the bonding orbital of lowest energy to the antibonding orbital of highest energy. These molecular orbitals are delocalized over all the Avogadro's number of sodium atoms. Thus, we picture a **metallic crystal** as a lattice occupied by ions, which are formed by removing the valence electrons from the atoms of the metal; the valence electrons are delocalized over the entire crystal lattice.

A sodium atom has a neonlike core with an outer $3s$ electron. Therefore, only one half of the delocalized orbitals of the crystalline lattice are occupied by the valence electrons of the sodium atoms. Of the remaining unfilled orbitals, many are so close in energy to the higher energy filled orbitals (valence orbitals) that relatively little energy is required to excite some electrons into the unfilled orbitals. Once in these orbitals, the electrons are easily displaced by an applied electric field, a property accounting for the high electrical conductivity of the metal.

Now let's consider an insulator, such as diamond (carbon). Its extended covalent network can be described in terms of sp^3 hybrid orbitals on each carbon atom. Again, imagine bringing Avogadro's number of carbon atoms together. In this case, however, there is a relatively large gap in energy between the band of delocalized bonding orbitals (called the **valence band**) and the band of delocalized antibonding orbitals (called the **conduction band**) (Figure 14-35). The valence electrons of the carbon atoms occupy only the valence band, and the conduction band is empty. Furthermore, the energy of the gap between the two bands (called the **band gap**) is so large that the electrons in the valence band are not easily promoted into the conduction band. Consequently, no electrons are available to conduct an electric current, so diamond is an insulator. In contrast, in a good conductor, such as sodium metal, there is no gap between the energies of the bonding orbitals and the antibonding orbitals; in other words, the band gap is zero in a good conductor.

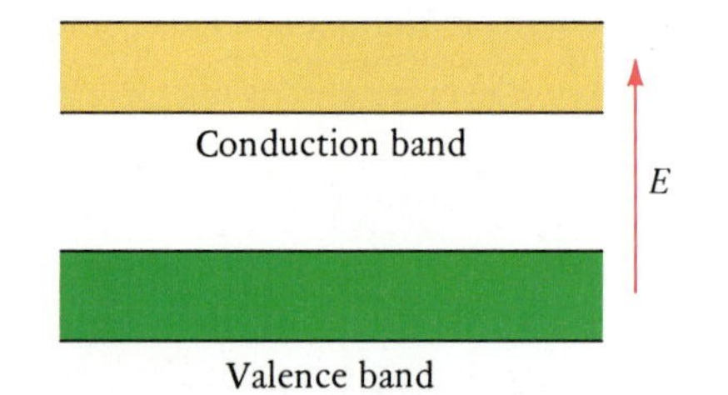

Figure 14-35 When the atoms of a crystal are brought together to form the crystal lattice, the valence orbitals of the atoms combine to form two sets of energy levels called the valence band and the conduction band.

Some substances, such as silicon and germanium, have relatively small band gaps. These substances have properties that are intermediate between conductors and insulators; therefore, they are called **semiconductors.** Figure 14-36 summarizes the band structure of metals,

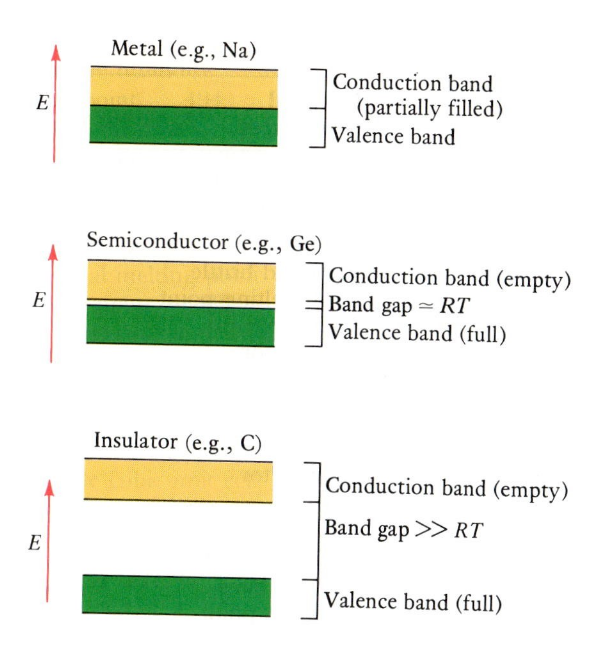

Figure 14-36 A comparison of the energy separations between the valence bands and conduction bands of metals, semiconductors, and insulators. Metals have no band gap, semiconductors have a small band gap, and insulators have a large band gap.

semiconductors, and insulators. As the temperature of a semiconductor is increased, more electrons acquire sufficient energy to be promoted from the valence band to the conduction band, so the electrical conductivity of semiconductors increases with increasing temperature. This temperature dependence is the opposite of that of metals, whose electrical conductivities decrease with increasing temperature.

14-12. AMORPHOUS SOLIDS DO NOT HAVE A CRYSTALLINE STRUCTURE

Not all solids are crystalline. When liquids of high viscosity are cooled rapidly, a rigid structure forms before the molecules have time to orient properly to form a crystalline lattice. The molecules of these materials lack the high degree of spatial order of crystalline substances and have more of a "frozen-in" liquid structure. Such materials are called **amorphous solids.** A wide variety of materials form amorphous solids under the proper conditions, but the most familiar examples are materials with large molecules such as plastics, rubber (see Chapter 30), and glasses. Amorphous solids are distinguished by the lack of a sharp melting point. Crystalline solids melt at a precise temperature, but amorphous solids gradually soften over a wide temperature range.

Amorphous solids are now an active area of research in chemistry, physics, and engineering. Research in fiber optics has enabled millions of bits of information to be transmitted through glass fibers with diameters less than that of a hair (Figure 14-37). Another recent development is amorphous silicon, which has the potential of producing a significant reduction in the cost of converting sunlight to electricity. Table 14-8 summarizes the various types of solids that we have discussed.

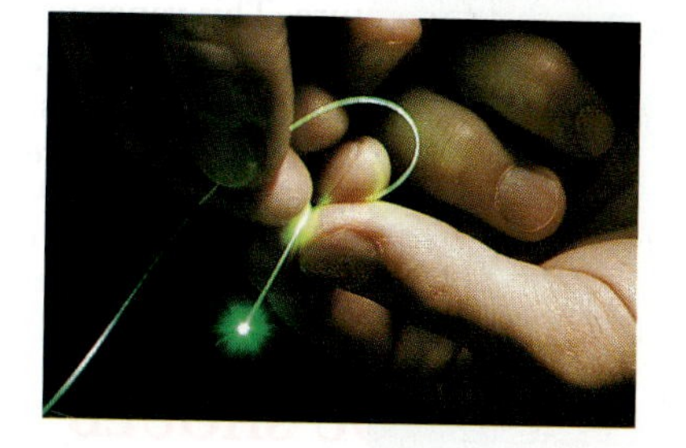

Figure 14-37 Fiber optics. Light is transmitted through thin strands of silica glass that form an optical-waveguide fiber bundle that acts as an optical pathway for the light.

Table 14-8 Ty

Type of solid

metallic crystal

ionic crystal

molecular crystal

network crystal

amorphous

SUMMARY

All molecules at
teractions are m
gases. The quar
the strength of
respectively. Th
lectively called v
polar, then the
forces. The pol
their dipole mc
minimizes their
each other beca
the instantaneo
head-to-tail; the
force.

Pure liquids h
at each tempera
increasing temp
sure, at which p
point of a liqui
pressure of the

TERMS YC

translational m
heating curve
molar enthalpy
ΔH_{fus} 476

14-6. The heat of vaporization of einsteinium was determined to be 128 $kJ \cdot mol^{-1}$ using only a 100-μg sample. How much heat is required to vaporize 100 μg of einsteinium?

14-7. Calculate the heat absorbed by the sublimation of 100.0 g of solid carbon dioxide.

14-8. Calculate the number of moles of water at 0°C that can be frozen by 1.00 mol of solid carbon dioxide. See Table 14-3 for necessary data.

HEATING CURVES

14-9. Sketch a heating curve for 7.50 g of mercury from 200 to 800 K using a heat input rate of 100 $J \cdot min^{-1}$. Refer to Table 14-3 for some of the necessary data for mercury. The molar heat capacities of solid, liquid, and gaseous mercury are 27.2 $J \cdot K^{-1} \cdot mol^{-1}$, 28.0 $J \cdot K^{-1} \cdot mol^{-1}$, and 20.8 $J \cdot K^{-1} \cdot mol^{-1}$, respectively.

14-10. What would take longer, heating 10.0 g of water at 50.0°C to 100.0°C or vaporizing the 10.0 g if the rate of heating in both cases is 5 $J \cdot s^{-1}$?

14-11. Heat was added to 25.0 g of solid sodium chloride, NaCl, at the rate of 3.00 $kJ \cdot min^{-1}$. The temperature remained constant at 801°C, the melting point of NaCl, for 250 s. Calculate the molar enthalpy of fusion of NaCl.

14-12. Heat was added to a 45.0-g sample of liquid propane, $C_3H_8(l)$, at the rate of 500 $J \cdot min^{-1}$. The temperature remained constant at −42.1°C, the normal boiling point of propane, for 38.4 min. Calculate the molar enthalpy of vaporization of propane.

VAN DER WAALS FORCES

14-13. Which of the following molecules have polar interactions?

Cl_2 ClF NF_3 F_2

14-14. Which of the following molecules can hydrogen bond?

H_2 HF CH_4 CH_3OH

14-15. Arrange the following compounds in order of increasing boiling point:

KBr C_2H_5OH C_2H_6 He

14-16. Arrange the following compounds in order of increasing boiling point:

Ar NH_3 PH_3 KCl

14-17. Arrange the following molecules in order of increasing molar enthalpy of vaporization:

CH_4 C_2H_6 CH_3OH C_2H_5OH

14-18. Arrange the following molecules in order of increasing molar enthalpy of vaporization:

CCl_4 $SiCl_4$ CH_4 $SiBr_4$

VAPOR PRESSURE

14-19. A 0.75-g sample of ethyl alcohol is placed in a sealed 400-mL container. Is there any liquid present when the temperature is held at 60°C?

14-20. Mexico City lies at an elevation of 7400 ft (2300 m). Water boils at 93°C in Mexico City. What is the normal atmospheric pressure there?

14-21. A sample of ethyl alcohol vapor in a vessel of constant volume exerts a pressure of 300 torr at 75.0°C. Use the ideal-gas law to plot pressure versus temperature of the vapor between 75.0°C and 40.0°C. Assume no condensation. Compare your result with the vapor pressure curve for ethyl alcohol shown in Figure 14-21. Estimate the temperature at which condensation occurs upon cooling from 75.0°C.

14-22. Atmospheric pressure decreases with altitude. Plot the following data to obtain the relationship between pressure and altitude:

Altitude/ft	Atmospheric pressure/atm
5000	0.83
10,000	0.70
15,000	0.58
20,000	0.47

Using your plot and the vapor pressure curve of water (Figure 14-21), estimate the boiling point of water at the following locations:

Location	Altitude/ft
Denver	5280
Mount Kilimanjaro	19,340
Mt. Washington	6290
The Matterhorn	14,690

14-23. Compare the dew points of two days with the same relative humidity of 70 percent but with temperatures of 20°C and 30°C, respectively.

14-24. The relative humidity in a greenhouse at 40°C is 92 percent. Calculate the vapor pressure of water vapor in the greenhouse.

PROPERTIES OF LIQUIDS

14-25. The surface tension of water is 72 $mJ\cdot m^{-2}$. What is the energy required to change a spherical drop of water with a diameter of 2 mm to two smaller spherical drops of equal size? The surface area of a sphere of radius r is $4\pi r^2$ and the volume is $4\pi r^3/3$.

14-26. The surface tension of water is 72 $mJ\cdot m^{-2}$. Calculate the amount of energy required to disperse one spherical drop of radius 3.0 mm into spherical drops of radius 3.0×10^{-3} mm. The surface area of a sphere of radius r is $4\pi r^2$ and the volume is $4\pi r^3/3$.

14-27. Suppose that the time it takes a given volume of H_2O to flow through a certain capillary at 20°C is 71 s. The time required for an equal volume of cyclohexanol to flow through the capillary is 2700 s. Given that the densities of water and cyclohexanol at 20°C are 0.998 $g\cdot mL^{-1}$ and 0.942 $g\cdot mL^{-1}$, respectively, calculate the viscosity of cyclohexanol relative to that of water.

14-28. Use the data in Table 14-6 to calculate how long it will take a given volume of methanol to flow through a certain capillary at 25°C if it takes 126 s for the same volume of water. The densities of water and methanol at 25°C are 0.998 $g\cdot mL^{-1}$ and 0.787 $g\cdot mL^{-1}$, respectively.

14-29. Which liquid do you think will have the higher dielectric constant at 25°C, *trans*-1,2-dichloroethene or *cis*-1,2-dichloroethene?

14-30. Sulfur melts at 119°C to a thin, pale-yellow liquid consisting of S_8 rings

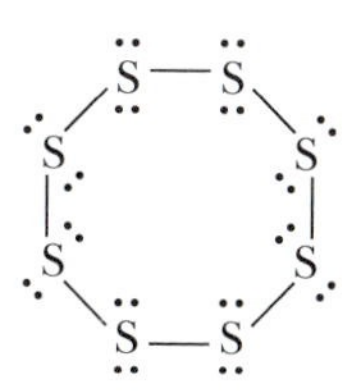

As sulfur is heated to 150°C and higher, it becomes so viscous that it hardly pours. Explain why in terms of the breaking of the S_8 rings.

PHASE DIAGRAMS

14-31. Determine whether water is a solid, liquid, or gas at the following pressure and temperature combinations (use Figure 14-22):

(a) 373 K, 0.70 atm (b) −5°C, 0.006 atm
(c) 400 K, 200 atm (d) 0°C, 300 atm

14-32. Referring to Figure 14-23, state the phase of CO_2 under the following conditions:

(a) 127°C, 8 atm (b) −50°C, 40 atm
(c) 50°C, 1 atm (d) −80°C, 5 atm

14-33. Sketch the phase diagram for oxygen using the following data:

	Triple point	Critical point
temperature/K	54.3	154.6
pressure/torr	1.14	37,823

The normal melting point and normal boiling point of oxygen are −218.4°C and −182.9°C. Does oxygen melt under an applied pressure as water does?

14-34. Sketch the phase diagram for nitrogen given the following data:

triple point, 63.156 K and 139 torr
normal melting point, 63.29 K
normal boiling point, 77.395 K
critical point, 126.1 K and 33.49 atm

CRYSTAL STRUCTURES

14-35. Potassium exists as a body-centered cubic lattice. How many potassium atoms are there per unit cell?

14-36. Crystalline potassium fluoride has the NaCl-type structure shown in Figure 14-29. How many potassium ions and fluoride ions are there per unit cell?

14-37. The density of silver is 10.50 $g\cdot cm^{-3}$ at 20°C. Given that the unit cell of silver is face-centered cubic, calculate the length of an edge of a unit cell.

14-38. The density of tantalum is 16.69 $g\cdot cm^{-3}$ at 20°C. Given that the unit cell of tantalum is body-centered cubic, calculate the length of an edge of a unit cell.

14-39. Copper crystallizes in a face-centered cubic lattice with a density of 8.93 $g\cdot cm^{-3}$. Given that the length of an edge of a unit cell is 361.6 pm, calculate Avogadro's number.

14-40. Chromium crystallizes in a body-centered cubic lattice with a density of 7.20 $g\cdot cm^{-3}$. Given that the length of an edge of a unit cell is 288.4 pm, calculate Avogadro's number.

14-41. Crystalline potassium fluoride has the NaCl-type structure shown in Figure 14-29. Given that the density of KF is 2.481 $g\cdot cm^{-3}$ at 20°C, calculate the unit cell length and the nearest-neighbor distance in KF. (The nearest-neighbor distance is the shortest distance between the centers of any two adjacent ions in the lattice.)

14-42. Crystalline cesium bromide has the CsCl-type structure shown in Figure 14-29. Given that the density of CsBr is 4.44 $g \cdot cm^{-3}$ at 20°C, calculate the unit cell length and the nearest-neighbor distance (see Problem 14-41) in CsBr.

ADDITIONAL PROBLEMS

14-43. Suppose you are stranded in a mountain cabin by a snowstorm. You have some food and fuel, but you wish to conserve them as long as possible. You remember reading that you should melt snow to get water to drink and not eat the snow directly because the body expends energy when it has to melt the snow. Explain why this is so and estimate how much energy is used per gram of snow that is melted.

14-44. Trouton's rule states that the molar enthalpy of vaporization of a liquid that does not involve strong molecular interactions such as hydrogen bonding or ion-ion attractions is given by

$$\Delta H_{vap} = (85\ J \cdot K^{-1} \cdot mol^{-1})T_b$$

where T_b is the normal boiling point of the liquid in kelvins. Use Trouton's rule to estimate ΔH_{vap} for argon, given that T_b is 87 K.

14-45. Apply Trouton's rule, given in Problem 14-44, to water and suggest a molecular explanation for any discrepancy with the value of ΔH_{vap} given in Table 14-3.

14-46. What is the minimum amount of data necessary to make a rough sketch of the phase diagram of a substance exhibiting (a) a single solid phase? (b) two solid phases?

14-47. What is the relationship between ΔH_{fus}, ΔH_{vap}, and ΔH_{sub} in the vicinity of the triple point?

14-48. Why is H_2S a gas at -10°C whereas H_2O is a solid at this temperature?

14-49. Explain why silicon carbide, SiC, and boron nitride, BN, are about as hard as diamond.

14-50. Although the temperature may not exceed 0°C, the amount of ice on a sidewalk decreases owing to sublimation. A source of heat for the sublimation is solar radiation. The average daily solar radiation in February for Boston is 8.1 $MJ \cdot m^{-2}$. Calculate how much ice will disappear from a 1.0-m^2 area in one day assuming that all the radiation is used to sublime the ice. The density of ice is 0.917 $g \cdot cm^{-3}$.

14-51. British surveyors were prevented from extending their survey of India into the Himalayas because entry into Tibet was banned by the Chinese emperor. In 1865, the Indian Nain Singh secretly entered Lhasa, the capital city of Tibet, and determined its correct location for map placement. Singh was not able to bring instruments for measuring altitude with him, but he did have a thermometer. He estimated that Lhasa was 3420 m above sea level. (Its true elevation is 3540 m.) Describe how he was able to estimate the altitude from a measurement of the boiling point of water.

14-52. A 0.677-g sample of zinc reacts completely with sulfuric acid:

$$Zn(s) + H_2SO_4(aq) \rightarrow ZnSO_4(aq) + H_2(g)$$

A volume of 263 mL of hydrogen is collected over water; the water level in the collecting vessel is the same as the outside level. Atmospheric pressure is 756 torr and the temperature is 25°C. Calculate the atomic mass of zinc.

14-53. The vapor pressures (in torr) of solid and liquid chlorine are given by

$$\log P_s = 10.560 - \frac{1640\ K}{T} \qquad \text{(solid)}$$

$$\log P_l = 7.769 - \frac{1159\ K}{T} \qquad \text{(liquid)}$$

where T is the absolute temperature. Calculate the temperature and pressure at the triple point of chlorine.

14-54. Moisture often forms on the outside of a glass containing a mixture of ice and water. Use the principles developed in this chapter to explain this phenomenon.

14-55. The relative humidity is 65 percent on a certain day on which the temperature is 30°C. As the air cools during the night, what will be the dew point?

14-56. Given that the density of KBr is 2.75 $g \cdot cm^{-3}$ and that the length of an edge of a unit cell is 654 pm, determine how many formula units of KBr there are in a unit cell. Does the unit cell have a NaCl or a CsCl structure? (See Figure 14-29.)

14-57. Given that the density of CaO is 3.25 $g \cdot cm^{-3}$ and that the length of an edge of a unit cell is 486 pm, determine how many formula units of CaO there are in a unit cell. Does the unit cell have a NaCl or a CsCl structure? (See Figure 14-29.)

14-58. Sodium chloride has the crystal structure shown in Figure 14-29. By X-ray diffraction, it is determined that the distance between a sodium ion and a chloride ion is 282 pm. Using the fact that the density of sodium chloride is 2.163 $g \cdot cm^{-3}$, calculate Avogadro's number.

14-59. Cesium chloride has the crystal structure shown in Figure 14-29. The length of a side of a unit cell is determined by X-ray diffraction to be 412.1 pm. What is the density of cesium chloride?

14-60. Commercial refrigeration units in the United States are rated in tons. A 1-ton unit is capable of removing, during 24 hours of operation, an amount of heat equal to that released when 1.00 ton of water at 0°C is converted to ice. Calculate the number of kilojoules of heat per hour that can be removed by a 4-ton home air conditioner.

14-61. Polonium is the only metal that exists as a simple cubic lattice. Given that the length of a side of the unit cell of polonium is 334.7 pm at 25°C, calculate the density of polonium.

14-62. The unit cell of lithium is body-centered cubic, and the length of an edge of a unit cell is 351 pm at 20°C. Calculate the density of lithium at 20°C.

14-63. Calculate the concentration in $mol \cdot L^{-1}$ of water vapor in air saturated with water vapor at 25°C.

14-64. Arrange the following substances in order of increasing dielectric constant:

CCl_4 $CHCl_3$ CH_2Cl_2

14-65. Suppose that the time it takes a given volume of H_2O to flow through a certain capillary at 20°C is 123 s. The time required for an equal volume of acetone to flow through the capillary is 59 s. Given that the densities of water and acetone at 20°C are $0.998\ g \cdot mL^{-1}$ and $0.791\ g \cdot mL^{-1}$, respectively, calculate the viscosity of acetone relative to that of water.

14-66. Use the data in Table 14-6 to calculate how long it will take a given volume of glycerol to flow through a certain capillary at 25°C if it takes one minute for the same volume of water. The densities of water and glycerol at 25°C are $0.998\ g \cdot mL^{-1}$ and $1.26\ g \cdot mL^{-1}$, respectively.

14-67. The phase diagram for sulfur is

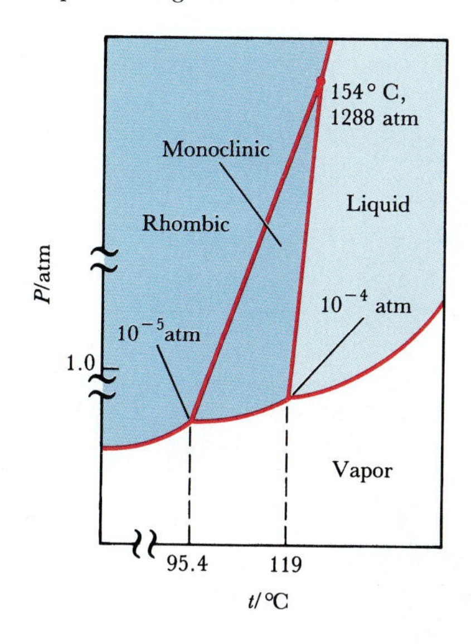

The regions labeled rhombic and monoclinic indicate two different crystalline forms of sulfur. How many triple points are there? Describe what happens if sulfur is heated from 40°C at 1 atm to 200°C at 1 atm. Below what pressure will sublimation occur?

14-68. Discuss why the vapor pressure of ethanol is greater than the vapor pressure of water at a given temperature (Figure 14-21).

14-69. Use hybrid orbitals to describe the bonding in diamond.

14-70. Use hybrid orbitals to describe the bonding in one of the molecular layers in graphite.

14-71. The vapor pressures (in torr) of solid and liquid uranium hexafluoride are given by

$$\log P_s = 10.646 - \frac{2559.1\ K}{T} \qquad \text{(solid)}$$

$$\log P_l = 7.538 - \frac{1511\ K}{T} \qquad \text{(liquid)}$$

where T is the absolute temperature. Calculate the temperature and pressure at the triple point of UF_6.

INTERCHAPTER G

High-Temperature Superconductors

It is not often that a scientific discovery causes intense interest in the general news media as well as excitement throughout the entire scientific world community, but the year 1986 saw such a discovery. In September of that year Georg Bednorz, a German scientist, and K. Alex Müller, a Swiss scientist, working at IBM's Zurich research lab (Figure G-1), announced the discovery of a substance that is a superconductor at a temperature as high as 35 K (−238°C). The very next year they were awarded the Nobel Prize in physics, an extraordinarily short time between the discovery and the award. Let's see what superconductivity is and why the work of Bednorz and Müller is so important. What can be so earth-shaking about a discovery that takes place at 35 K, certainly a very low temperature?

The study of superconductivity goes back to 1911, when the Dutch physicist Heike Kammerlingh Onnes, who pioneered in the production of extremely low temperatures, discovered that the electrical resistance of mercury falls to zero at temperatures below 4.2 K (Figure G-2). Although it

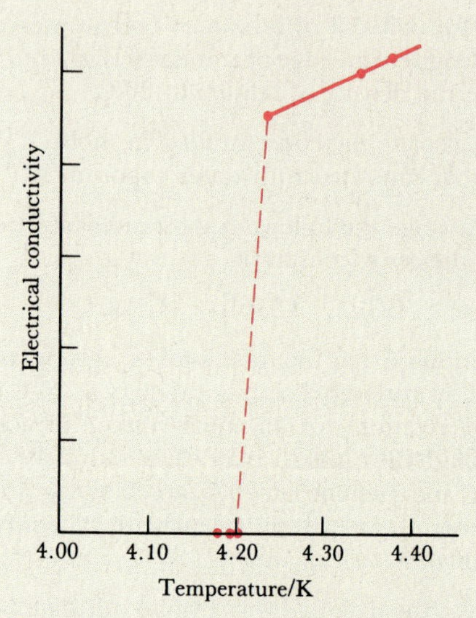

Figure G-2 The electrical conductivity of mercury, showing the onset of superconductivity at 4.2 K.

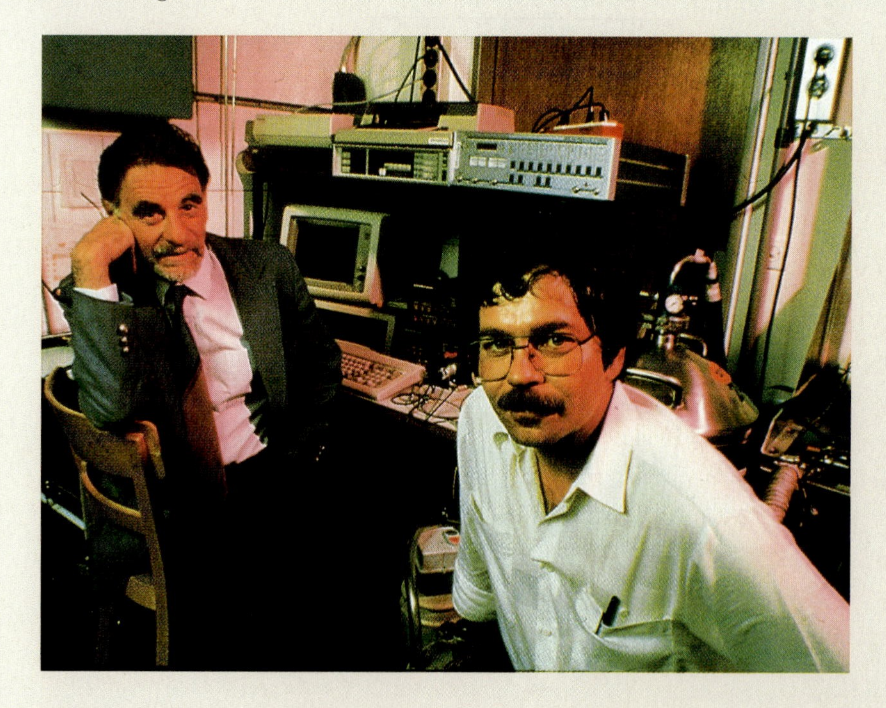

Figure G-1 Georg Bednorz and Alex Müller of IBM's Zurich laboratory were the first to synthesize a superconducting ceramic material. They received the Nobel Prize in physics in 1987.

was well known at the time that electrical resistance decreases with decreasing temperature, a sudden drop to zero resistance was totally unexpected. Electrical resistance is the opposition to a flow of electrical current, so zero electrical resistance means no opposition at all—a perfect conductor, or a *superconductor*. Once started, an electric current will flow forever in such a substance.

In addition to having zero electrical resistance, a superconductor is repelled by a magnetic field. A substance that is repelled by a magnetic field is called *diamagnetic,* whereas one that is attracted into a magnetic field is called *paramagnetic*. The strong diamagnetism of a superconductor can be used to perform a neat demonstration. If you place a small magnet directly over a superconductor, it will "levitate" above the superconductor (Figure G-3). The strong diamagnetism of the superconductor repels the magnet, causing it to remain suspended in the air.

Figure G-3 Because of the strong diamagnetism associated with a superconductor, a small magnet will "levitate" above a pellet of a superconductor.

The technological implications of superconductivity are immense. For example, the lack of electrical resistance could be used by power companies to transmit and store electricity without any loss, thus allowing them to build huge, efficient power stations at isolated locations; also computer chips could be designed that would transmit information almost instantaneously. The diamagnetism of superconductors could be used in designing trains that would hover over magnetized tracks and cars that would hover over magnetic freeways and glide along almost effortlessly; and extremely powerful electromagnets with applications in medical imaging, among many others, could be produced.

All these wondrous applications are dreams, however, if superconductivity is restricted to temperatures as low as 4.2 K. Temperatures below 20 K or so can be attained only with the use of liquid helium, which requires complex, expensive equipment to produce and maintain it. At such temperatures, the only viable applications are specialized ones that can justify the expense and the inconvenience of handling liquid helium. Current devices that routinely use superconductors are magnets for nuclear magnetic resonance spectrometers (see Interchapter F) and for storage rings in the particle accelerators used in high-energy physics. Consequently, ever since Kammerlingh Onnes's original discovery in 1911, the search has been on for materials that superconduct at higher, more convenient temperatures. Although room temperature and higher is the obvious dream, it so happens that the discovery of a substance that superconducts even at 77 K (−196°C) would be a major breakthrough, because at this temperature liquid nitrogen, which is plentiful, cheap, and relatively easy to handle, could be used as a coolant instead of liquid helium. Thus 77 K is a magic temperature in the search for "high-temperature" superconductivity. Let's go back now to 1911 and follow this search.

After the discovery of superconductive mercury at 4.2 K, many other metals were studied. Over the next 20 years, about 30 other metals were shown to be superconductors, but in 1930 the highest transition temperature known, that for niobium, was a dismal 9.2 K. Research in superconductivity stopped during World War II, but the search resumed afterward. Many alloys and intermetallic compounds were studied, and by 1973 the record was held by the intermetallic compound Nb_3Ge, with a transition temperature of 23.2 K. This

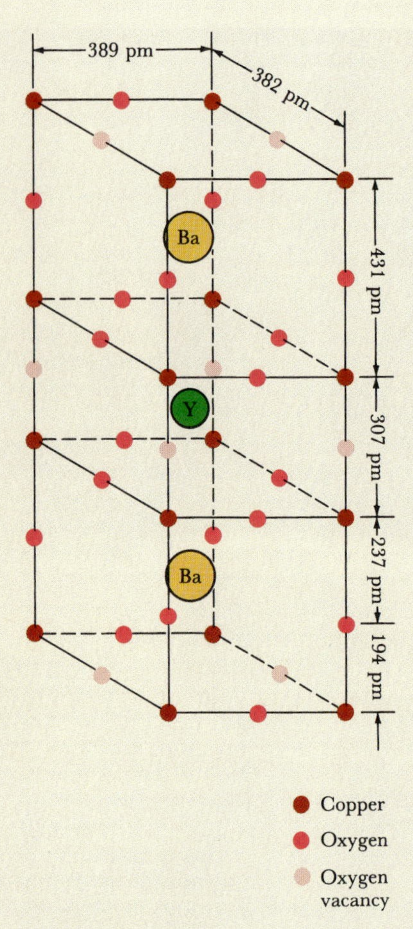

Figure G-4 The unit cell of a 123 compound, $YBa_2Cu_3O_7$. The linear arrays and planes containing the copper and oxygen atoms are believed to be the key ingredients of the 123 superconductors.

gloomy record stood until 1986, when Bednorz and Müller published their epic paper announcing a superconductor with a transition temperature of 35 K.

The substance on which Bednorz and Müller reported was not an alloy or an intermetallic compound, but a complex oxide of lanthanum, barium, and copper with the empirical formula $Ba_xLa_{2-x}CuO_4$, where x is a small, variable quantity around 0.1. This compound belongs to the family of substances called *ceramics*, and it can be made by heating the proper proportions of lanthanum oxide, $La_2O_3(s)$, barium oxide, $BaO(s)$, and copper(II) oxide, $CuO(s)$, in a simple, open crucible in a furnace to about 900° or 1000°C. Ceramics, unlike metals and their alloys, are normally poor electrical conductors, so the fact that a ceramic could not only become superconducting, but could do so at the relatively high temperature of 35 K, came as a great surprise to most scientists.

Bednorz and Müller had opened up a whole new class of materials to study, and the search for superconductors with even higher transition temperatures became almost frantic. Just four months after the Bednorz-Müller announcement, Paul Chu of the University of Houston and Maw-Kuen Wu of the University of Alabama jointly announced a new ceramic superconductor with a transition temperature above 90 K. This oxide has the empirical formula $YBa_2Cu_3O_{7-x}$, where x is from around 0.1 to 0.3. Because of the stoichiometry of the metal constituents, this compound has become known as the "123" compound. The current (1990) record transition temperature is 125 K for a thalium-calcium-barium-copper oxide, and the search continues.

Although there has been a successful theory of superconductivity in metallic superconductors since 1957, this theory does not seem to apply completely to the new ceramic superconductors. At present no generally accepted theory can explain all the properties of ceramic superconductors. In order to try to understand how these ceramic materials superconduct and to aid in the development of a general theory, the determination of the crystal structures of the superconducting ceramic is of utmost importance. As you learned in Chapter 14, crystal structures can be determined by X-ray diffraction studies. The now-famous 123 compound serves as a good example. The unit cell of the compound $YBa_2Cu_3O_7$ is shown in Figure G-4. (Can you see that the unit cell in this figure contains one yttrium atom, two barium atoms, three copper atoms, and seven oxygen atoms?) The crystal structure contains linear arrays and planes that contain copper and oxygen atoms. The present belief is that it is the electrons from the copper-oxygen linear arrays and planes that are responsible for superconductivity, while the yttrium and barium atoms serve simply to keep the copper and oxygen atoms in the right places. Because the optimal dimensions are not known, there is hope that this superconductor can be improved upon by using atoms other than yttrium and barium.

In spite of the tremendous progress that has been made in just the past few years, enormous technological problems must be overcome before ceramic superconductors become widespread. For example, the ceramic materials are powders and do not readily lend themselves to the fabrication of the

Figure G-5 A tape of high-temperature superconducting ceramic, made at Argonne National Laboratory, is flexible enough to be shaped before firing, after which it becomes extremely brittle. Argonne scientists are working on ways to form the material into practical superconducting objects of various shapes.

wires and coils necessary for transmission lines, transformers, and other electrical devices, although applications in the form of thin films seem to be promising (Figure G-5). Furthermore, in most applications, the superconductors must be made to carry large electrical currents—much larger than those that have been produced so far in ceramic superconductors. Whether or not these and many other problems will be overcome remains to be seen, but what is certain is that high-temperature superconductivity has fueled the imaginations of many people.

15 COLLIGATIVE PROPERTIES OF SOLUTIONS

A 50-50 mixture of water and automobile antifreeze at −13.3°C. Note that the water-antifreeze mixture remains as a liquid at −13.3°C, whereas pure water would be frozen solid.

Why can some salts melt ice? Why does the same antifreeze added to the coolant water in a car radiator prevent both overheating in summer and freezing in winter? What factors operate to keep biological cells from collapsing? Why do divers get the bends when they ascend too quickly to the water surface? These and many other questions can be answered when the properties of solutions are understood from a molecular point of view. The major emphasis of this chapter is on the colligative properties of solutions. Colligative properties depend primarily on the ratio of the number of solute particles to the number of solvent particles in the solution rather than on the chemical nature of the solute. The colligative properties of solutions are vapor-pressure lowering, boiling-point elevation, freezing-point depression, and osmotic pressure.

We showed in Chapter 14 that pure water has a unique equilibrium vapor pressure at each temperature. We show in this chapter that the equilibrium vapor pressure of pure water, or any other solvent, always decreases when a substance is dissolved in it. The vapor-pressure lowering effect is the key to understanding the colligative properties of solutions.

15-1. SOLUTES AFFECT THE PROPERTIES OF THE SOLVENT

Recall from Chapter 6 that when a solute is dissolved in a solvent, the solution formed usually is homogeneous throughout, right down to the molecular level. We also noted in Chapter 6 that solute molecules or ions interact with solvent molecules. The electrostatic interactions between dissolved ions and polar solvent molecules is especially strong. As a result, ions (especially cations) become **solvated** in the solution, that is, attached to solvent molecules (see Figure 6-2). Because water has a large diectric constant, it dissolves a wide variety of electrolytes (e.g., sodium chloride and magnesium sulfate) and polar nonelectrolytes (e.g., sucrose and ethanol).

The strong interactions between solute particles (molecules or ions) and solvent molecules affect dramatically many of the properties of the solvent. Such properties, called **colligative properties,** depend primarily on the *ratio* of the number of solute particles to the number of solvent molecules in a solution. Consequently, it is important to understand the different ways in which this ratio is described. The four commonly used measures of solute concentration are molarity (denoted by M), mole fraction (denoted by X), molality (denoted by m, and to be defined below), and percentage by mass (% mass).

We are already familiar with the concentration unit molarity (Section 6-3). Molarity is defined as the number of moles of solute per liter of solution. Molarity is the most widely used measure of solute concentration, because it is the easiest to use. We need only weigh out the appropriate amount of solute, place it in the appropriate volumetric flask, add solvent to dissolve the solute, and dilute up to the mark, making certain by swirling that the solution is homogeneous (Figure 6-3). However, the molarity has a serious disadvantage as a measure of concentration in that the molarity of a given solution changes as the temperature changes, because the volume of the solution changes as the temperature changes. Usually the volume of the solution increases as the temperature increases; thus, the molarity of the solution decreases as the temperature increases (same mass of solute in a larger volume of solution). For example, suppose that the volume of the solution increases by 0.10 percent per degree Celsius. If such a solution is 1.00 M at 20°C, then at 60°C the molarity of the solution will be

$$1.00 \text{ M} - (1.00 \text{ M})(0.0010°\text{C}^{-1})(40°\text{C}) = 0.96 \text{ M}$$

In contrast to molarity, solute concentrations expressed as mole fraction, molality, and % mass are all temperature independent, because they are defined in terms of mass ratios; mass-to-mass ratios, unlike mass-to-volume ratios, do not depend on the temperature.

Recall from Chapter 7 that the **mole fraction** of component 1 in a solution, X_1, is defined as the ratio of the number of moles of component 1 to the total number of moles of all components in the solution:

$$X_1 = \frac{n_1}{n_{\text{tot}}} \tag{15-1}$$

If there are only two components (call them 1 and 2 to denote solvent and solute, respectively) in the solution, then the mole fraction of component 1 is given by

$$X_1 = \frac{n_1}{n_1 + n_2} \tag{15-2}$$

Similarly, the mole fraction of component 2 in the solution is

$$X_2 = \frac{n_2}{n_1 + n_2} \tag{15-3}$$

Because mole fractions are defined in terms of mole ratios, they are dimensionless quantities. Also note from Equations (15-2) and (15-3) that the sum of the mole fractions is equal to one

$$X_1 + X_2 = \frac{n_1}{n_1 + n_2} + \frac{n_2}{n_1 + n_2} = 1 \tag{15-4}$$

In general, the sum of all the mole fractions of all the components in a solution must equal one (unity).

Solute concentrations are most commonly expressed as **% mass** (percentage by mass) when either we do not know the chemical formula of the solute (e.g., with an unknown) or the substance cannot be described by a single chemical formula, because it is a mixture of substances (e.g., a plant extract or a cup of coffee or tea). For these cases, we can characterize the solute concentration by simply stating the ratio of the number of grams of dissolved solute to the total mass of the solute plus the solvent; we then multiply the ratio by 100 to get the % mass of solute, that is,

$$\% \text{ mass} = \left(\frac{\text{mass of solute}}{\text{mass of solute} + \text{mass of solvent}}\right) \times 100 \tag{15-5}$$

For example, if we dissolve 5.85 g of an unknown solute in 100.0 g of water, then the % mass of solute in the solution is

$$\% \text{ mass} = \left(\frac{5.85\text{ g}}{5.85\text{ g} + 100.0\text{ g}}\right) \times 100 = 5.53\%$$

The mole fraction of the solvent in a dilute solution is close to one because in such a case $n_2 \ll n_1$, and $X_1 \simeq n_1/n_1 = 1$. Consequently, mole fraction is not a convenient measure of concentration for dilute solutions. In dilute solutions, usually we find it more convenient to use a concentration unit called molality, which is directly proportional to the mole fraction of solute in a dilute solution. We define the **molality,** m, of a solute as the number of moles of solute per 1000 g of solvent:

$$\begin{aligned}\text{molality} &= \frac{\text{moles of solute}}{1000\text{ g of solvent}} \\ &= \frac{\text{moles of solute}}{\text{kilogram of solvent}}\end{aligned} \qquad (15\text{-}6)$$

For example, a solution prepared by dissolving 0.100 mol of sodium chloride (5.844 g) in 1.00 kg of water is 0.100 molal (0.100 m) in NaCl.

EXAMPLE 15-1: Calculate (a) the molality and (b) the mole fractions of a solution prepared by dissolving 20.0 g of sucrose, $C_{12}H_{22}O_{11}$, in 500 g of water.

Solution: (a) The number of moles of sucrose in 20.0 g is

$$(20.0\text{ g})\left(\frac{1\text{ mol sucrose}}{342.3\text{ g sucrose}}\right) = 0.0584\text{ mol}$$

When 0.0584 mol of sucrose is dissolved in 0.500 kg of water, the molality of sucrose in the resulting solution is given by Equation (15-6):

$$m = \frac{0.0584\text{ mol}}{0.500\text{ kg}} = 0.117\text{ mol}\cdot\text{kg}^{-1} = 0.117\text{ m}$$

(b) To calculate the mole fractions of sucrose and water in the solution, we need to know both the moles of sucrose and the moles of water. From part (a), we know that there are 0.0584 mol of sucrose in the solution. The number of moles of water in the solution is

$$(500\text{ g } H_2O)\left(\frac{1\text{ mol } H_2O}{18.02\text{ g } H_2O}\right) = 27.7\text{ mol } H_2O$$

The mole fractions of sucrose, X_s, and of water, X_w, in the solution are

$$X_s = \frac{0.0584\text{ mol}}{(27.7\text{ mol} + 0.0584\text{ mol})} = 2.10 \times 10^{-3}$$

$$X_w = \frac{27.7\text{ mol}}{(27.7\text{ mol} + 0.0584\text{ mol})} = 0.998$$

Note that $X_s + X_w = 0.00210 + 0.998 = 1.000$ and also that $X_w \simeq 1.00$.

PRACTICE PROBLEM 15-1: Calculate the molality of a solution prepared by dissolving 5.25 g of $KMnO_4(s)$ in 250 g of water.

Answer: 0.133 m

The molality of a solute, which is denoted by m, is not the same as the molarity, denoted by M (Section 6-3). To prepare 500 mL of a 0.117 M solution of sucrose in water, we dissolve 0.0585 mol of sucrose in less than 500 mL of water and dilute the resulting solution with enough water to yield exactly 500 mL of *solution*. Compare this procedure with the procedure described in Example 15-1 for the preparation of a 0.117 m of sucrose solution.

It is essential for the understanding of colligative properties to realize that it is the ratio of the number of solute particles to the number of solvent particles that determines the magnitude of a colligative effect. A 0.10 m aqueous NaCl solution has twice as many solute particles per mole of water as a 0.10 m aqueous sucrose solution, because NaCl is a strong electrolyte and dissociates completely in water to $Na^{+}(aq)$ and $Cl^{-}(aq)$, whereas sucrose exists in solution as intact $C_{12}H_{22}O_{11}$ molecules. We can express this result as an equation by writing

$$m_c = im \qquad (15\text{-}7)$$

where i is the number of solute particles produced per formula unit when the solute is dissolved in the solvent. We thus distinguish between the molality, denoted by m, and the **colligative molality,** denoted by m_c, of a solute. The distinction is illustrated numerically in Table 15-1.

EXAMPLE 15-2: What is the colligative molality of a 0.20 m $K_2SO_4(aq)$ solution?

Solution: Because one formula unit of K_2SO_4 produces one $SO_4^{2-}(aq)$ and two $K^{+}(aq)$ ions in aqueous solution, the colligative molality is three times the molality, that is, $m_c = im = 3 \times 0.20\ \text{m} = 0.60\ \text{m}_c$.

PRACTICE PROBLEM 15-2: A 1.00-mol sample of each of the following substances is dissolved in 500 g of water:
(a) $CH_3OH(l)$, methanol, an organic alcohol (b) $AgNO_3(s)$ (c) $Ca(ClO_4)_2(s)$
Determine the colligative molality of each of the resulting solutions.

Answer: (a) 2.00 m_c; (b) 4.00 m_c; (c) 6.00 m_c

In the following sections of this chapter we discuss the colligative properties of solutions, namely, vapor-pressure lowering, boiling-point elevation, freezing-point depression, and osmotic pressure. In each case we will see that the magnitude of the effect on the properties of the solvent is directly proportional to the colligative molality of the solute.

Table 15-1 Comparison of molality and colligative molality of aqueous solutions

Solute	Solute molality/m	Solute particles per formula unit	Colligative molality/m_c
$C_6H_{12}O_6$	0.10	1	0.10
NaCl	0.10	2	0.20
$CaCl_2$	0.10	3	0.30

15-2. THE EQUILIBRIUM PARTIAL PRESSURE OF A PURE LIQUID ALWAYS DECREASES WHEN A SUBSTANCE IS DISSOLVED IN THE LIQUID

Consider a solution of a nonvolatile solute such as sucrose dissolved in a volatile solvent such as water. As Figure 15-1 suggests, the vapor pressure of the solvent over a solution will be less than the vapor pressure of pure solvent at the same temperature.

The equilibrium vapor pressure results when the rate of evaporation of the solvent from the solution is equal to the rate of condensation of the solvent from the vapor (Section 14-4). The rate of evaporation of the *solvent* from a solution is less than that of the pure solvent, because the presence of solute molecules at the surface of the solution decreases the number of solvent molecules per unit area of surface (Figure 15-1). The rate of condensation is directly proportional to the number of molecules per unit volume in the vapor, which in turn is proportional to the vapor pressure. The lower rate of evaporation, then, is balanced by a lower rate of condensation; the result is a lower equilibrium vapor pressure (Figure 15-2).

As Figure 15-1 implies, the vapor pressure of the solvent over a solution is directly proportional to the fraction of solvent particles at the surface. Because the number of moles is proportional to the number of particles, the mole fraction of the solvent [Equation (15-2)] can be considered a particle fraction. Therefore, we can write the relation between the equilibrium vapor pressure of the solvent P_1 and the mole fraction of solvent X_1 as

$$P_1 \propto X_1$$

By introducing a proportionality constant, k, we can write

$$P_1 = kX_1 \tag{15-8}$$

The value of the proportionality constant is determined as follows. When the mole fraction of the solvent is unity, that is, when we have

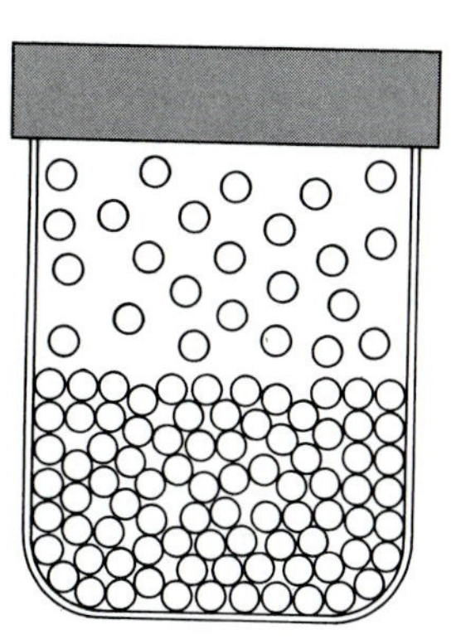

Pure solvent

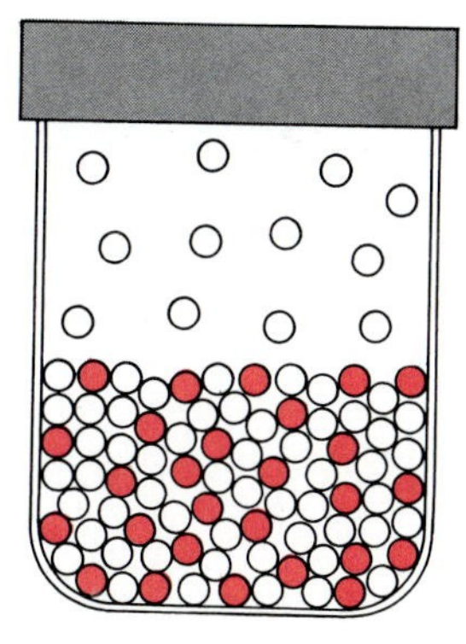

Solution with a nonvolatile solute

Figure 15-1 The effect of a nonvolatile solute on the equilibrium vapor pressure of a solvent at a fixed temperature. The solute molecules lower the equilibrium vapor pressure of the solvent relative to that of the pure solvent by partially blocking the escape of solvent molecules from the surface of the solution.

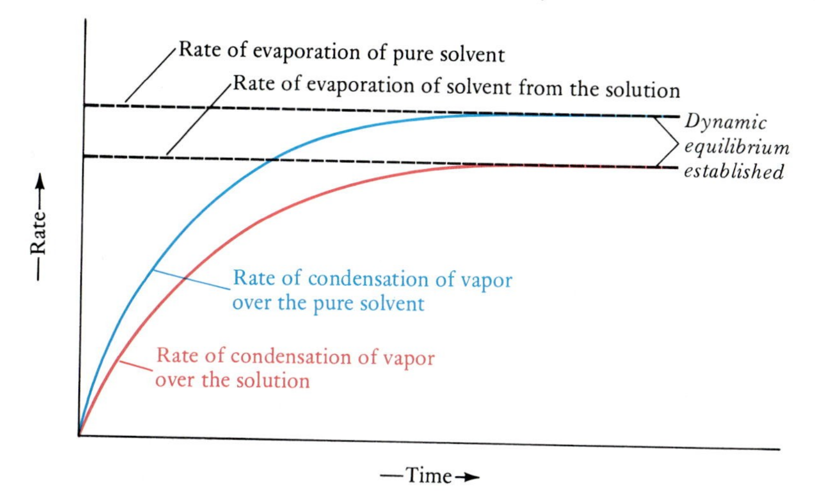

Figure 15-2 The effect of a nonvolatile solute on the solvent evaporation rate. The solute molecules lead to a lower rate of evaporation of the solvent, which in turn leads to a lower equilibrium vapor pressure of the solvent.

pure solvent, then P_1 is equal to the vapor pressure of the pure solvent. Thus, when $X_1 = 1$ in Equation (15-8), we have $P_1 = P_1^\circ$. If we substitute $X_1 = 1$ and $P_1 = P_1^\circ$ into Equation (15-8), then we see that $k = P_1^\circ$. Using this definition of k, we can then rewrite Equation (15-8) as

$$P_1 = X_1 P_1^\circ \tag{15-9}$$

Equation (15-9) is known as **Raoult's law** and was discovered by the French chemist F. M. Raoult. Raoult's law tells us that the equilibrium vapor pressure of the solvent over a solution is directly proportional to the mole fraction of the solvent in the solution.

The solvent always obeys Raoult's law if the solution is sufficiently dilute. Solutions for which all constituents obey Raoult's law for *all* concentrations are called *ideal solutions.* We discuss ideal solutions in Section 15-6, where we treat solutions consisting of two liquids.

The amount by which the vapor pressure of a solution is less than the vapor pressure of the pure solvent, that is, $P_1^\circ - P_1$, is called the **vapor-pressure lowering.** Using Equation (15-9), we can express the vapor-pressure lowering, ΔP_1, as

$$\Delta P_1 = P_1^\circ - P_1 = P_1^\circ - X_1 P_1^\circ = (1 - X_1) P_1^\circ \tag{15-10}$$

or

$$\Delta P_1 = X_2 P_1^\circ \tag{15-11}$$

where we have used the fact that $X_1 + X_2 = 1$ and where X_2 is the mole fraction of the solute.

EXAMPLE 15-3: The vapor pressure of water at 80°C is 355 torr. Use Raoult's law to calculate the vapor pressure of a solution made by dissolving 56.0 g of the nonvolatile solid sucrose, $C_{12}H_{22}O_{11}$, in 100.0 g water. Also calculate the vapor-pressure lowering of water.

Solution: The mole fraction of water in the solution, X_w, is

$$X_w = \frac{n_w}{n_w + n_s}$$

$$= \frac{(100.0\text{ g})\left(\dfrac{1\text{ mol water}}{18.02\text{ g water}}\right)}{(100.0\text{ g})\left(\dfrac{1\text{ mol water}}{18.02\text{ g water}}\right) + (56.0\text{ g})\left(\dfrac{1\text{ mol sucrose}}{342.3\text{ g sucrose}}\right)}$$

$$= 0.971$$

The equilibrium vapor pressure of water over the solution is given by Raoult's law [Equation (15-9)]:

$$P_w = X_w P_w^\circ = (0.971)(355\text{ torr}) = 345\text{ torr}$$

The vapor-pressure lowering of water produced by the dissolved sucrose is

$$\Delta P_w = P_w^\circ - P_w = 355\text{ torr} - 345\text{ torr} = 10\text{ torr}$$

which corresponds to a vapor pressure decrease of water of 2.8 percent.

PRACTICE PROBLEM 15-3: A 41.3-g sample of a solution of naphthalene, $C_{10}H_8$, in benzene, C_6H_6, has a vapor pressure at 30°C of 103 torr. The vapor pressure of pure benzene at 30°C is 120 torr. Assume that naphthalene has a negligible vapor pressure over the solution and use the vapor-pressure lowering to calculate the mole fraction of naphthalene and the number of grams of naphthalene in the solution.

Answer: $X_{naphth} = 0.142$; 8.82 g

15-3. NONVOLATILE SOLUTES INCREASE THE BOILING POINT OF A LIQUID

Recall that the boiling point is the temperature at which the equilibrium vapor pressure over the liquid phase equals the atmospheric pressure. We know from Section 15-2 that the equilibrium vapor pressure of the solvent over a solution containing a nonvolatile solute is less than that for the pure solvent at the same temperature. Therefore, the temperature at which the equilibrium vapor pressure reaches atmospheric pressure is higher for the solution than for the pure solvent (Figure 15-3). In other words, the boiling point of the solution, T_b, is higher than the boiling point of the pure solvent, T_b°. The amount by which the boiling point of the solution exceeds the boiling point of the pure liquid, that is, $T_b - T_b^\circ$, is called the **boiling-point elevation.**

In solutions with colligative molalities less than about 1.0 m_c, the mole fraction of the solute is directly proportional to the colligative molality of the solute, that is, $X_2 \propto m_c$ (see Problem 15-74). Because the vapor-pressure lowering of the solvent, $P_1^\circ - P_1$, is directly proportional to the mole fraction of the solute X_2 [Equation (15-11)] and $X_2 \propto m_c$, the vapor-pressure lowering is also proportional to the colligative molality.

$$(P_1^\circ - P_1) \propto m_c \tag{15-12}$$

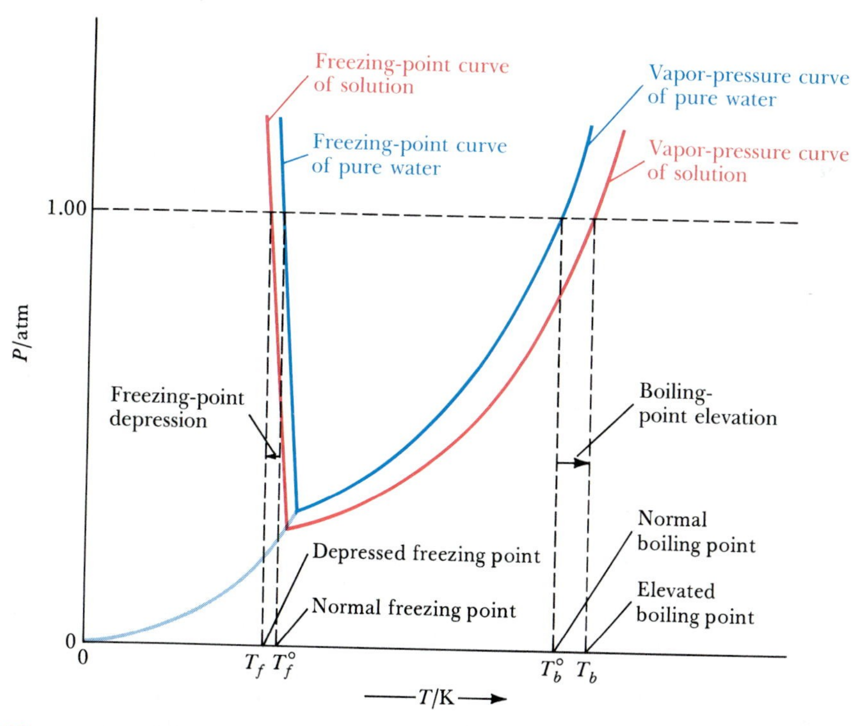

Figure 15-3 Phase diagrams for pure water (blue lines) and for water containing a nonvolatile solute (red lines). The presence of the solute lowers the vapor pressure of the solvent. The reduced vapor pressure of the solvent results in an *increase* in the boiling point of the solution relative to that of the pure solvent and a *decrease* in the freezing point of the solution relative to that of the pure solvent.

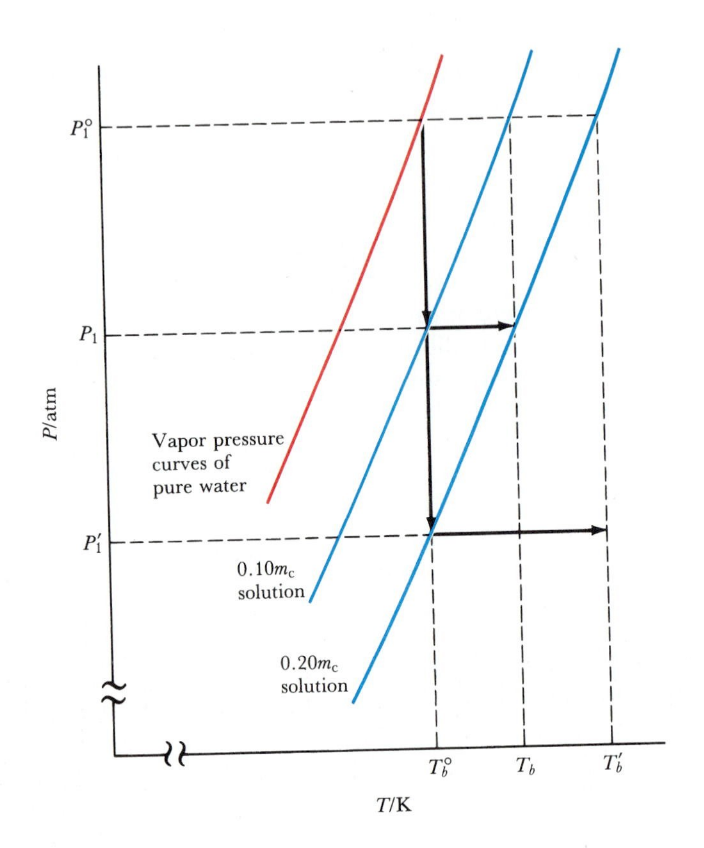

Figure 15-4 Expanded version of Figure 15-3 in the region where the vapor-pressure curves intersect the $P = 1.00$ atm line (denoted P_1° here). Because the vapor-pressure lowering is small unless m_c is large, the vapor-pressure curves are essentially parallel straight lines over the temperature range T_b° to T_b', and therefore, $(P_1^\circ - P_1) \propto (T_b - T_b^\circ)$. The blue line is a portion of the vapor-pressure curve for pure water and the red lines are the vapor-pressure curves for two different concentrations of a nonvolatile solute in water.

Figure 15-4 shows an enlarged version of the region of Figure 15-3 where several vapor-pressure curves intersect the horizontal line at $P_1^\circ = 1.00$ atm. Because the vapor-pressure lowering is small for most solutions, the vapor-pressure curves are essentially straight lines over a small temperature range. Consequently, the vapor-pressure lowering $(P_1^\circ - P_1)$ is proportional to the boiling-point elevation, $T_b - T_b^\circ$. Combination of the result, $(P_1^\circ - P_1) \propto (T_b - T_b^\circ)$, with Equation (15-12) yields the result that the boiling-point elevation, $(T_b - T_b^\circ)$, is directly proportional to the colligative molality, m_c

$$(T_b - T_b^\circ) \propto m_c \tag{15-13}$$

Equation (15-13) can be written as an equality by introducing the proportionality constant K_b

$$T_b - T_b^\circ = K_b m_c \tag{15-14}$$

Equation (15-14) tells us that the increase in the boiling point of a solution containing a nonvolatile solute is directly proportional to the colligative molality. That is, the boiling-point elevation is directly proportional to the solute (particle) concentration. The value of K_b depends only on the solvent and is called the **boiling-point-elevation constant.** Values of K_b for several solvents are given in Table 15-2.

The magnitude of the boiling-point elevation is small unless the solute concentration is high. For example, we see from Equation (15-14) and Table 15-2 that $T_b - T_b^\circ$ for a 1.0 m_c solution of glucose in water is

$$T_b - T_b^\circ = (0.52\ \text{K}\cdot\text{m}_c^{-1})(1.0\ \text{m}_c) = 0.52\ \text{K}$$

Table 15-2 Boiling-point elevation constants (K_b) and freezing-point depression constants (K_f) for various solvents

Solvent	Boiling point/°C	K_b/K·m_c^{-1}	Freezing point/°C	K_f/K·m_c^{-1}
water	100.00	0.52	0.00	1.86
acetic acid	117.9	2.93	16.6	3.90
benzene	80.0	2.53	5.50	5.10
chloroform	61.2	3.63	−63.5	4.68
cyclohexane	80.7	2.79	6.5	20.2
nitrobenzene	210.8	5.24	5.7	6.87
camphor	(208.0)*	5.95	179.8	40.0

*Sublimation point.

Figure 15-5 Pure water open to the atmosphere boils at 100°C, whereas a 50-50 mixture of water and automobile antifreeze (ethylene glycol) boils at about 109°C. The pure water and the antifreeze mixture in this photo are at the same temperature (100°C), but only the pure water is boiling.

EXAMPLE 15-4: Calculate the boiling point of an aqueous solution of ethylene glycol, $HOCH_2CH_2OH$ (automobile antifreeze), that contains 50.0 g of water and 55.0 g of ethylene glycol. Assume that the vapor pressure of ethylene glycol over the solution is negligible.

Solution: Ethylene glycol does not dissociate in water; thus, the colligative molality of the solution is

$$m_c = \left(\frac{55.0\text{ g}}{62.07\text{ g}\cdot\text{mol}^{-1}}\right)\left(\frac{1}{0.0500\text{ kg}}\right) = 17.7\text{ m}_c$$

We use Equation (15-14) to compute the boiling-point elevation

$$T_b - T_b^\circ = (0.52\text{ K}\cdot\text{m}_c^{-1})(17.7\text{ m}_c) = 9.2\text{ K}$$

The magnitude of a degree Celsius is equal to that of a degree Kelvin and $T_b^\circ = 373.2$ K or 100.0°C. Thus, the boiling point of the solution in degrees Celsius is 109.2°C (Figure 15-5).

PRACTICE PROBLEM 15-4: Seawater contains about 3.5 percent by mass dissolved solids. As a rough approximation, assume that the dissolved solids are predominantly NaCl (that is, 3.5 g of NaCl in 96.5 g of H_2O) and estimate the normal boiling point of seawater.

Answer: 100.64°C

Figure 15-6 Even though they are formed directly from seawater, icebergs that break off from the arctic ice shelves are composed of pure water. Because the density of ice is only about 91 percent of the density of liquid water, about 91 percent of the total mass of an iceberg is below the surface of the water.

15-4. SOLUTES DECREASE THE FREEZING POINT OF A LIQUID

When an aqueous solution begins to freeze, the solid that separates out is usually pure ice. For example, ice formed from seawater is free of salt, and freezing is one method of preparing fresh water from seawater to provide pure drinking water (Figure 15-6). When an aqueous solution such as NaCl (*aq*) or seawater begins to freeze, we have pure ice in equilibrium with the solution.

Pure ice and the salt solution can coexist in equilibrium only if their vapor pressures are equal. But the vapor pressure of water over the NaCl (*aq*) solution is less than that of pure water, so the vapor pressure of the ice must be less than if it were in equilibrium with pure water. Consequently, the temperature of the ice must be less than 0°C, and the freezing point of the solution must be less than that of pure water. Thus, we see that the lowering of the vapor pressure of a solvent by a solute leads to a lowering of the freezing point of the solution relative to that of the pure solvent. This effect is called the **freezing-point depression.**

By arguments analogous to those used to obtain the boiling-point elevation equation, the magnitude of the freezing-point depression produced by a solute is found to be proportional to its colligative molality m_c:

$$T_f^\circ - T_f = K_f m_c \qquad (15\text{-}15)$$

where T_f° is the freezing point of the pure solvent, T_f is the freezing point of the solution in kelvins ($T_f^\circ > T_f$), and K_f is the **freezing-point-depression constant.** The value of K_f depends only on the solvent.

The value of the freezing-point-depression constant of water is 1.86 $\text{K}\cdot\text{m}_c^{-1}$ (Table 15-2). Thus, we predict that an aqueous solution with a colligative molality of 0.50 m_c has a freezing-point depression of

$$T_f^\circ - T_f = K_f m_c = (1.86\ \text{K}\cdot\text{m}_c^{-1})(0.50\ \text{m}_c) = 0.93\ \text{K}$$

and the freezing point is $T_f = -0.93\ \text{K} + 273.15\ \text{K} = 272.22\ \text{K}$ or -0.93°C.

The freezing-point depression due to a dissolved substance is the basis of the action of **antifreezes.** The most commonly used antifreeze is ethylene glycol,

```
     H   H
     |   |
HO—C—C—OH
     |   |
     H   H
```

whose boiling point is 197°C and whose freezing point is −17.4°C. The addition of ethylene glycol to water depresses the freezing point and elevates the boiling point of the solution relative to that of pure water. An ethylene glycol-water solution in which ethylene glycol is 50 percent by volume has a freezing point of about −36°C.

EXAMPLE 15-5: Estimate the freezing point of a 5.0 m_c solution of ethylene glycol in water.

Solution: The freezing point of the solution is computed from Equation (15-15):

$$\begin{aligned} T_f^\circ - T_f &= K_f m_c \\ &= (1.86\ \text{K}\cdot\text{m}_c^{-1})(5.0\ \text{m}_c) = 9.3\ \text{K} \end{aligned}$$

The freezing point of the solution is 263.9 K, or −9.3°C.

PRACTICE PROBLEM 15-5: Compute the molality (m) of a $CaCl_2(aq)$ solution that freezes at −5.25°C.

Answer: 0.941 m

Figure 15-7 shows the freezing point of aqueous ethylene glycol solutions as a function of the molality of ethylene glycol. The effectiveness of ethylene glycol as an antifreeze is a result of its high boiling point, its chemical stability, and the tendency of the ice that freezes out of the solution to form a slushy mass rather than a solid block. In the absence of a sufficient amount of antifreeze, the 9 percent volume expansion of water on freezing can generate a force of 30,000 $lb \cdot in.^{-2}$, if there is no room available to accommodate the increase in volume. Such a force is more than sufficient to rupture a radiator or even a metal engine block.

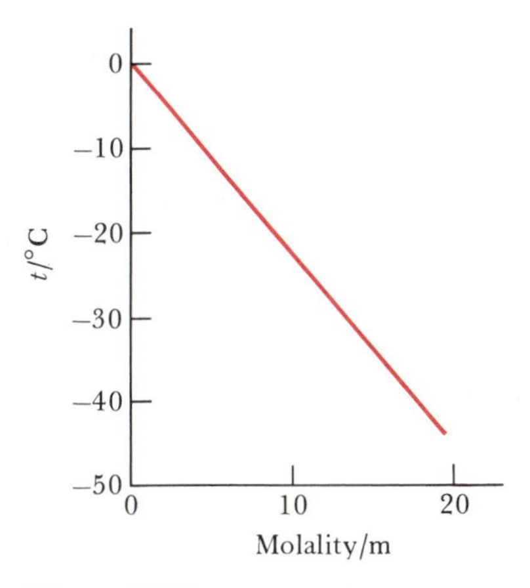

Figure 15-7 Freezing point versus molality for aqueous ethylene glycol (antifreeze) solutions. Note that $m_c = m$ for ethylene glycol.

The freezing point of seawater is about −1.85°C; therefore this temperature is also that of the seawater surrounding an arctic ice shelf. At least two species of fishes can live in the cold waters of the Ross Sea of Antarctica near the sea ice (Figure 15-8). On the basis of the total concentration of solutes dissolved in the blood serum of these fishes, the freezing point of the serum should be −1.46°C. So the blood of these fishes should freeze in the −1.85°C water, but it does not because the fishes are protected from freezing by "antifreeze" proteins in their blood. These proteins have an enhanced capacity to lower the freezing point of water. In fact, the freezing-point depression of solutions containing the antifreeze proteins is much greater than that predicted using Equation (15-15). The freezing point of the fish blood serum, after removal of all the salts (but not the antifreeze proteins), is −0.60°C. The measured concentration of antifreeze proteins in the fish blood is about 3×10^{-4} m_c. According to Equation (15-15), the predicted freezing point depression of a 3×10^{-4} m_c solution of the antifreeze protein is only

Figure 15-8 *Champeocephalus gunneri,* one of the Antarctic species of fishes that have "antifreeze" proteins in their blood.

$$T_f^\circ - T_f = (1.86\ K \cdot m_c^{-1})(3 \times 10^{-4}\ m_c) = 0.0006\ K$$

Such a solution should freeze at −0.0006°C, but it actually freezes at −0.60°C. The observed freezing point of the antifreeze protein solution is thus 1000 times (−0.60°C/−0.0006°C) greater than that predicted by Equation (15-15). How the antifreeze proteins work is not entirely clear. The currently accepted hypothesis is that they are adsorbed onto the surfaces of ice crystal nuclei, thereby stopping further growth of the crystals.

The freezing-point depression Equation (15-15) can be combined with Equation (15-7) to obtain the equation

$$T_f^\circ - T_f = K_f i m \tag{15-16}$$

The i in Equation (15-16) is called the **van't Hoff *i*-factor,** and it represents the number of solute particles produced per formula unit that dissolves. Recall that for strong electrolytes such as NaCl, which yield two ions per formula unit that dissolves in water, $i = 2$. If we measure the freezing point of a 0.10 m NaCl solution in water, then we expect the freezing-point depression to be about 0.37 K ($0.20\ m_c \times 1.86\ K \cdot m_c^{-1}$), and the value of i is

$$i = \frac{T_f^\circ - T_f}{K_f m} = \frac{0.37\ K}{(1.86\ K \cdot m_c^{-1})(0.10\ m)} = 2.0$$

or 2.0 ions per formula unit.

Equation (15-16) can be used to determine the percentage of dissociation of a weak electrolyte in aqueous solution around 0°C. To do that we measure the freezing point of a solution of the weak electrolyte in a solution of known molality. For example, consider the weak electrolyte acetic acid, $HC_2H_3O_2(aq)$, which partially dissociates in aqueous solution, as shown by the equation

$$HC_2H_3O_2(aq) + H_2O(l) \rightarrow C_2H_3O_2^-(aq) + H_3O^+(aq)$$

Only a small percentage ($< 5\%$) of the $HC_2H_3O_2(aq)$ dissociates in solution. The partial dissociation gives rise to a van't Hoff i-factor in the range $1.00 < i < 2.00$. Suppose that the freezing point of a 0.0500 m aqueous solution of acetic acid is found to be $-0.095°C$; what is the value of i? Rearranging Equation (15-16), we find that

$$i = \frac{T_f^\circ - T_f}{K_f m} = \frac{0.095\ \text{K}}{(1.86\ \text{K}\cdot\text{m}^{-1})(0.0500\ \text{m})} = 1.02$$

Because each acetic acid molecule that dissociates yields two solute particles, a value of $i = 1.02$ means that about 2 percent [i.e., $(1.02 - 1.00) \times 100 = 2$ percent] of the acetic acid molecules are dissociated in an aqueous solution that is 0.0500 m in acetic acid.

EXAMPLE 15-6: The freezing-point depression of a 0.10 m $HF(aq)$ solution is $-2.01°C$. Calculate the percentage of the $HF(aq)$ molecules in the solution that are dissociated into ions

$$HF(aq) + H_2O(l) \rightarrow H_3O^+(aq) + F^-(aq)$$

Solution: Rearranging Equation (15-16), we find that

$$i = \frac{T_f^\circ - T_f}{K_f m} = \frac{2.01\ \text{K}}{(1.86\ \text{K}\cdot\text{m}^{-1})(0.10\ \text{m})} = 1.08$$

Thus, about 8 percent of the HF molecules are dissociated.

PRACTICE PROBLEM 15-6: A 1.00 m solution of acetic acid, $HC_2H_3O_2$, in benzene has a freezing-point depression of 2.25 K. Calculate the value of i and suggest an explanation for the unusual result. (Hint: If i is less than 1.0, then each formula unit that dissolves yields less than one solute particle, an outcome suggesting aggregation of solute particles.)

Answer: $i = 0.50$; formation of dimers of composition $(HC_2H_3O_2)_2$

One method used to melt ice on streets and sidewalks is to spread rock salt (NaCl) crystals on the ice. The solubility of NaCl in liquid water around 0°C is 4.8 m. Sodium chloride in water completely dissociates into $Na^+(aq)$ and $Cl^-(aq)$, and so the colligative molality is twice the molality. Thus the freezing-point depression of a saturated aqueous NaCl solution is

$$T_f^\circ - T_f = K_f m_c = (1.86\ \text{K}\cdot\text{m}_c^{-1})(9.6\ \text{m}_c) = 18\ \text{K}$$

The freezing point of the solution, therefore, is −18°C. On contact with ice, salt dissolves to form a saturated aqueous solution with a freezing point of −18°C. Above −18°C spreading salt on ice causes the ice to melt as it forms a concentrated aqueous salt solution. When the ice temperature is below −18°C, spreading salt on it will not melt the ice because ice freezes out from the saturated NaCl(*aq*) solution.

An 18°C temperature decrease can be produced by the addition of salt to ice. This effect is used in the preparation of homemade ice cream to obtain a low enough temperature to solidify cream. The temperature of 0°F on the Fahrenheit temperature scale was chosen as the freezing point of a saturated solution of sodium chloride in water and thus 0°F corresponds to −18°C, whereas the freezing point of pure water on the Fahrenheit temperature scale is 32°F.

15-5. OSMOTIC PRESSURE REQUIRES A SEMIPERMEABLE MEMBRANE

Suppose we place pure water in one beaker and an equal volume of seawater in another beaker and then place both beakers under a bell jar, as shown in Figure 15-9a. We observe that as time passes the volume of pure water decreases and the volume of seawater increases (Figure 15-9b). The pure water has a higher equilibrium vapor pressure than the seawater; thus, the rate of condensation of the water into the seawater is greater than the rate of evaporation of the water from the seawater. The net effect is the transfer of water, via the vapor phase, from the beaker with pure water to the beaker with seawater. This transfer continues until no pure liquid water remains, and the seawater ends up diluted.

If pure water and seawater are separated by a membrane that is permeable to water but not to the ions in seawater, then the water passes directly through the membrane from the pure-water side of the mem-

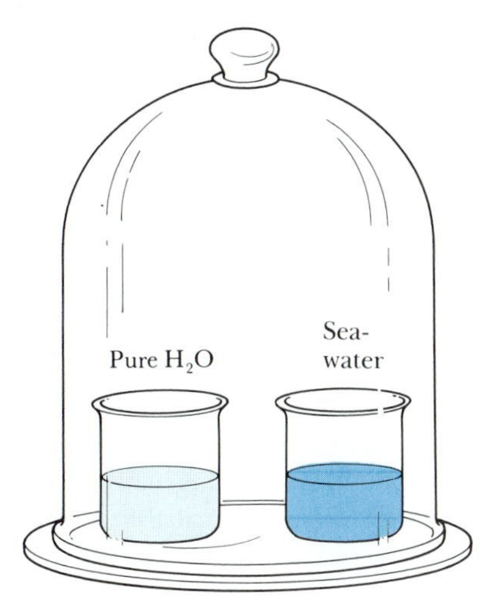

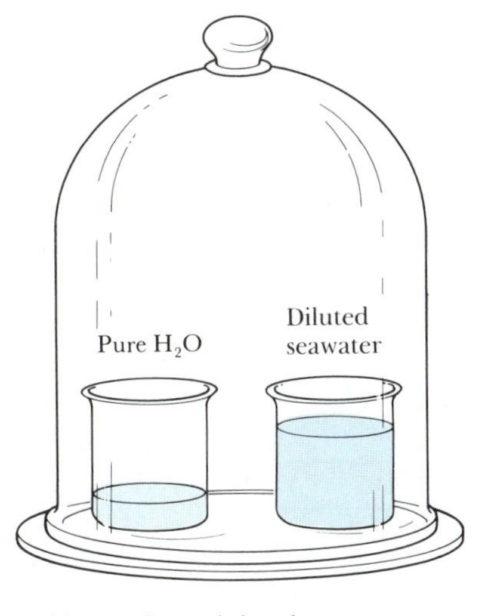

Figure 15-9 One beaker contains pure water, and the other contains seawater. (a) The equilibrium vapor pressure over the pure water is greater than that over the seawater solution. (b) As time passes, pure water is transferred via the vapor phase from the beaker containing pure water to the beaker containing seawater, thereby diluting the seawater. If we wait long enough, all the pure liquid water will transfer to the seawater beaker.

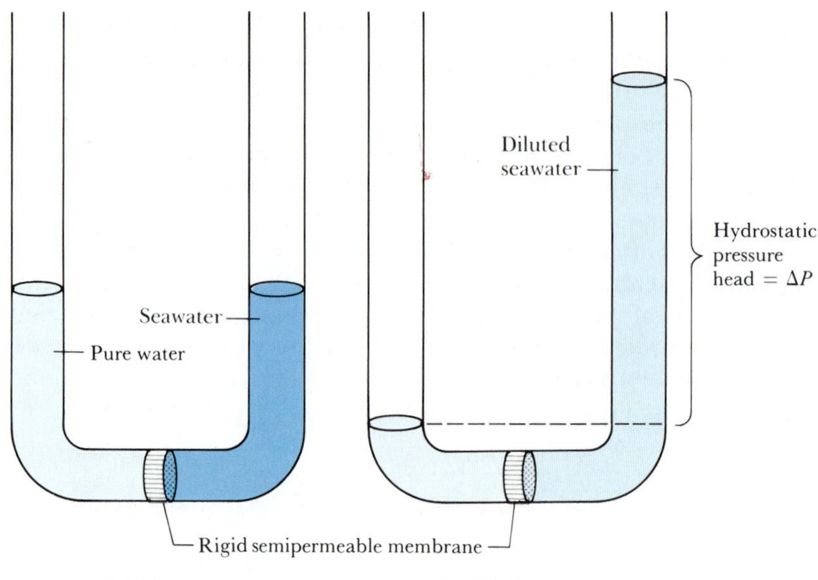

Figure 15-10 Passage of water through a rigid semipermeable membrane separating pure water from seawater. The water passes through the membrane until the escaping tendency of the water from the seawater equals the escaping tendency of the pure water. The escaping tendency of water from the seawater increases as the seawater is diluted and also as the hydrostatic pressure above the seawater column increases.

brane to the seawater side (Figure 15-10). The membrane is composed of polymer molecules intertwined like a mass of cooked spaghetti. The small neutral water molecules can move between cavities in the strands, whereas the solvated ions such as $Na^{+}(aq)$ and $Cl^{-}(aq)$ are much too large to move through the membrane channels. Such a membrane is called a **semipermeable membrane.** The tendency of the water to pass through the membrane is called the **escaping tendency** and is directly proportional to the vapor pressure of water over the solution. The escaping tendency of water from pure water is greater than the escaping tendency of water from seawater, because pure water has a higher vapor pressure than seawater. As water passes through the membrane to the seawater side, the escaping tendency of the water in the seawater increases, not only because the seawater is being diluted, but also because of the increased pressure on the seawater side of the membrane. This pressure increase arises from the pressure due to the height of the water column that develops. Recall from Chapter 7 that a column of liquid exerts a pressure proportional to the height of the column of liquid. The column of seawater rises until the escaping tendency of the water in the seawater is equal to the escaping tendency of the pure water. When this condition is attained, an equilibrium exists and the column of seawater no longer rises. The pressure of the liquid column produced in this process is called the **osmotic pressure** (Figure 15-10). The spontaneous passage of solvent through a semipermeable membrane from one solution to a more concentrated solution is called **osmosis.**

The osmotic pressure, π, of a solution is directly proportional to the solute concentration in the solution and is given by the equation

$$\pi = RTM_c \qquad (15\text{-}17)$$

where R is the gas constant (0.0821 $L \cdot atm \cdot K^{-1} \cdot mol^{-1}$), T is the absolute temperature, and M_c is the **colligative molarity.** The value of M_c is given by $M_c = iM$, where i is the number of solute particles produced per formula unit that is dissolved. Colligative molality is analogous to colligative molarity. In dilute aqueous solutions $M_c \simeq m_c$.

We have shown that we can increase the escaping tendency of water from a solution either by decreasing the solute concentration or by increasing the temperature of the solution. The escaping tendency of water from a solution also can be increased by applying pressure to the solution. The increase in pressure increases the average energy of the solvent molecules and thereby increases their escaping tendency. This effect is the basis of the osmotic pressure effect. The osmotic pressure is the pressure that must be applied to the solution to increase the vapor pressure (escaping tendency) of the solvent to a value equal to the vapor pressure of the pure solvent at that temperature.

EXAMPLE 15-7: As a rough approximation, seawater can be regarded as a 0.55 M aqueous NaCl solution. Estimate the osmotic pressure of seawater at 15°C.

Solution: The colligative molarity of seawater is approximately 2 × 0.55 M, because NaCl dissociates in water to yield two ions per formula unit. From Equation (15-17), we calculate the osmotic pressure of seawater at 15°C as:

$$\begin{aligned}\pi &= RTM_c \\ &= (0.0821\ L \cdot atm \cdot K^{-1} \cdot mol^{-1})(288\ K)(2 \times 0.55\ mol \cdot L^{-1}) \\ &= 26\ atm\end{aligned}$$

The osmotic pressure of seawater is 26 times higher than atmospheric pressure. The osmotic-pressure effect is by far the largest in magnitude of all of the colligative property effects.

PRACTICE PROBLEM 15-7: Estimate the osmotic pressure at 25°C of a 0.10 molar $CaCl_2(aq)$ solution.

Answer: 7.3 atm

If a pressure in excess of 26 atm is applied to seawater at 15°C (Example 15-7), then the escaping tendency of the water in the seawater will exceed that of pure water. Consequently, pure water can be obtained from seawater by using a rigid semipermeable membrane and an applied pressure in excess of the osmotic pressure of 26 atm. This process is known as **reverse osmosis** (Figure 15-11). Reverse osmosis units are commercially available and are used to obtain fresh water from salt

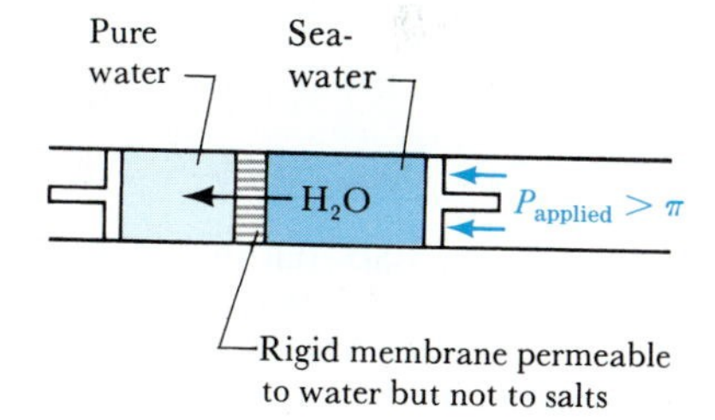

Figure 15-11 Reverse osmosis. A rigid semipermeable membrane separates pure water from seawater. A pressure in excess of the osmotic pressure of seawater (26 atm at 15°C) is applied to the seawater, and this increases the escaping tendency of water from the seawater to a value above that of pure water. Under these conditions, the net flow of water is from the seawater side through the semipermeable membrane to the pure water side, thus producing fresh water from seawater.

Figure 15-12 A commercial reverse osmosis unit. Suspended solids, including bacteria, are blocked by mechanical exclusion (Figure 15-13), and dissolved salts are chemically repulsed by the membrane. The unit shown reduces the salt concentration to less than 5 percent of the initial concentration.

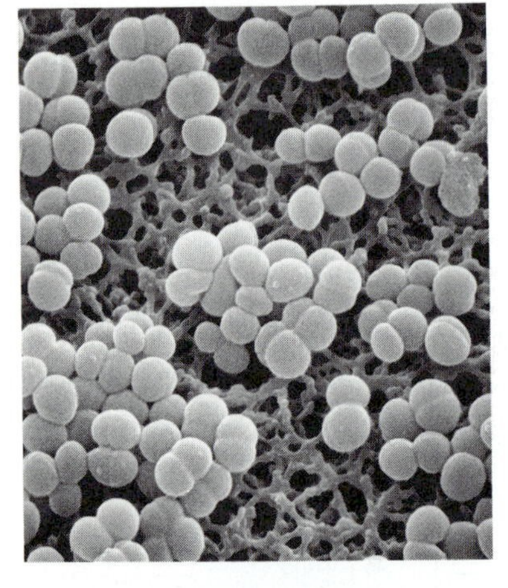

Figure 15-13 Bacterial cells on a membrane filter. A microporous membrane filter has pores so small that bacteria are trapped and thereby removed from the water. Similar membranes are used to produce "cold-filtered" bacteria-free beverages such as beer, thereby eliminating the need for pasteurization (heating) to kill bacteria.

water using a variety of semipermeable membranes, the most common of which is cellulose acetate (Figures 15-12 and 15-13).

The magnitude of the osmotic pressure effect makes osmotic pressure measurements an especially powerful method for the determination of the molecular masses of proteins. Proteins have large molecular masses and thus yield a relatively small number of solute particles for a given dissolved mass. In fact, osmotic pressure is the only colligative effect sufficiently sensitive to provide useful molecular information about proteins.

EXAMPLE 15-8: A 4.00-g sample of human hemoglobin (nondissociating in solution) was dissolved in water to make 0.100 L of solution. The osmotic pressure of the solution at 7°C was found to be 10.0 torr, or 0.0132 atm. Calculate the molecular mass of the hemoglobin.

Solution: The concentration of the hemoglobin in the aqueous solution is [Equation (15-17)]

$$M_c = \frac{\pi}{RT} = \frac{(0.0132\ \text{atm})}{(0.0821\ \text{L·atm·K}^{-1}\text{·mol}^{-1})(280\ \text{K})}$$
$$= 5.74 \times 10^{-4}\ \text{mol·L}^{-1}$$

The molecular mass can be calculated from the concentration of the protein because the dissolved mass is known. We have the correspondence

$$5.74 \times 10^{-4}\ \text{mol·L}^{-1} \approx \frac{4.00\ \text{g}}{0.100\ \text{L}} = 40.0\ \text{g·L}^{-1}$$

and therefore

$$5.74 \times 10^{-4}\ \text{mol} \approx 40.0\ \text{g}$$

By dividing both sides of this stoichiometric correspondence by 5.74×10^{-4}, we find that

$$1\ \text{mol} \approx 69{,}700\ \text{g}$$

The molecular mass of the hemoglobin is 69,700 (Figure 15-14). Protein molecular masses can be as large as 1,000,000.

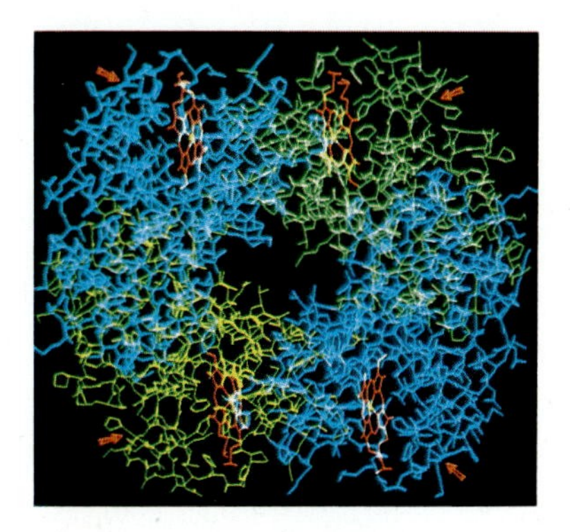

Figure 15-14 A computer-generated model of human hemoglobin, with a molecular mass of about 70,000.

PRACTICE PROBLEM 15-8: A 0.550-g sample of a nondissociating unknown material was dissolved in water at 25°C to yield 50.0 mL of solution. The osmotic pressure of the solution was found to be 0.0250 atm. Calculate the molar mass of the unknown.

Answer: $M = 10.8 \times 10^3\ \text{g}\cdot\text{mol}^{-1}$

Both plant and animal cells have membranes that are permeable to water but not, for example, to sucrose. The colligative concentration of the solution inside a typical biological cell is approximately 0.3 M_c. Most biological cells have about the same internal colligative molarity as the extracellular fluid in which the cells reside.

Water passes spontaneously through a biological cell membrane from the side with the lower colligative molarity (higher water escaping tendency) to the side with the higher colligative molarity (lower water escaping tendency). The entry of water into a cell causes the cell to expand, and the exit of water from the cell causes the cell to contract. The cell assumes its normal volume when it is placed in a solution with a colligative molarity of 0.3 M_c (Figure 15-15). More concentrated solutions cause the cell to contract, and less concentrated solutions cause it to expand. When cells are placed in distilled water at 27°C, equilibrium should be achieved at an internal cell pressure equal to the osmotic pressure of a 0.30 M_c solution, that is,

$$\pi = RTM_c = (0.0821\ \text{L}\cdot\text{atm}\cdot\text{K}^{-1}\cdot\text{mol}^{-1})(300\ \text{K})(0.30\ \text{mol}\cdot\text{L}^{-1}) = 7.4\ \text{atm}$$

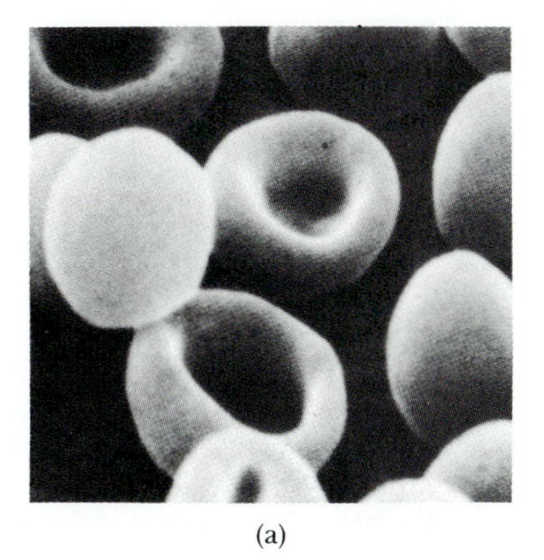

(a)

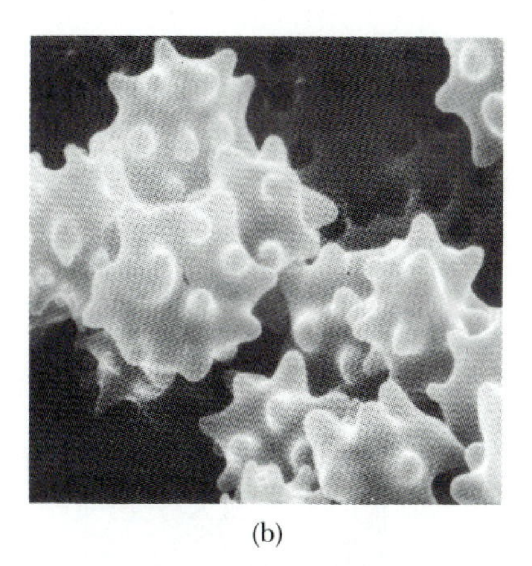

(b)

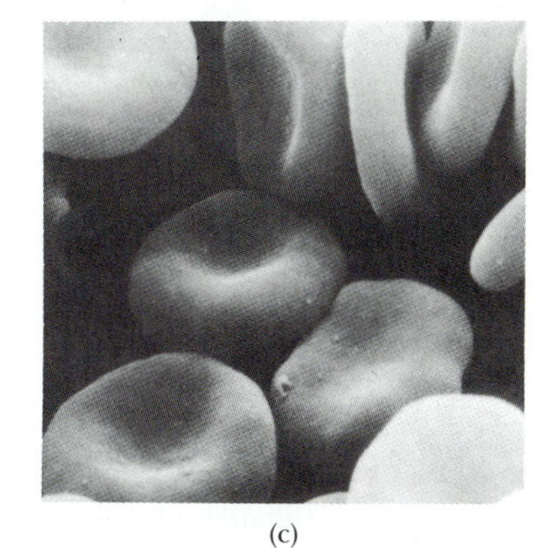

(c)

Figure 15-15 Osmosis in blood cells. (a) Blood cells placed in a solution whose colligative molarity is equal to that of the solution inside the cell (isotonic solution) neither contract nor expand. (b) Blood cells contract as a result of water loss when placed in a solution with a colligative molarity greater than 0.3 M_c (hypertonic solution). (c) Blood cells expand when placed in a solution with a colligative molarity less than 0.3 M_c (hypotonic solution).

Because a pressure of 7.4 atm cannot be sustained by animal cell membranes, the cells burst. Plant cell walls, in contrast, are rigid and can tolerate a pressure of 7.4 atm. In fact, the entry of water into plant cells gives nonwoody plants the rigidity required to stand erect.

15-6. THE COMPONENTS OF AN IDEAL SOLUTION OBEY RAOULT'S LAW

In the preceding sections of this chapter we have discussed solutions of nonvolatile solutes in volatile solvents. In this section we will discuss solutions consisting of two volatile liquids. We define an **ideal solution** as follows: a solution of two components, A and B, is said to be ideal if the interactions between an A and a B molecule are the same as those between two A molecules or two B molecules. In an ideal solution the A and B molecules are randomly distributed throughout the solution, including the region near the surface (Figure 15-16). When the molecules of the two components are very similar, then the solution is essentially ideal. For example, benzene and toluene, which have similar shapes and charge distributions (see margin) form ideal solutions.

benzene

CH_3

toluene

In an ideal solution of two liquids, the vapor pressure of each component is given by Raoult's law, so

$$P_A = X_A P_A^\circ \qquad \text{and} \qquad P_B = X_B P_B^\circ \tag{15-18}$$

The vapor over such a solution consists of both components, so the total vapor pressure is

$$P_{total} = P_A + P_B \tag{15-19}$$

Substituting Equation (15-18) into Equation (15-19) gives

$$P_{total} = X_A P_A^\circ + X_B P_B^\circ \tag{15-20}$$

Because $X_A + X_B = 1$, we can set $X_B = 1 - X_A$ and write Equation (15-20) as

$$P_{total} = P_B^\circ + X_A(P_A^\circ - P_B^\circ) \tag{15-21}$$

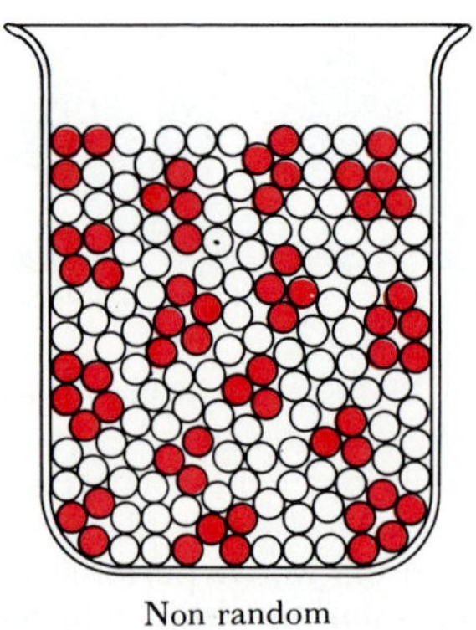

Figure 15-16 Random and nonrandom distribution of two types of molecules, A (open circles) and B (solid circles), in solution. Note in the nonrandom case that AAA . . . and BBB . . . clusters form in the solution. The random distribution represents an ideal solution.

Notice that for pure B, $X_A = 0$, so $P_{total} = P_B^\circ$, and that when $X_A = 1$ (pure A), $P_{total} = P_A^\circ$.

If we plot P_{total} versus X_A for an ideal solution, then we obtain a straight line with an intercept of P_B° and a slope of $P_A^\circ - P_B^\circ$ (Appendix A-6). The total equilibrium vapor pressure of a benzene-toluene solution is plotted against the mole fraction of benzene in Figure 15-17. These data confirm that a benzene-toluene solution is an ideal solution. Note also from Figure 15-17 that because the total vapor pressure over an ideal solution is always between the vapor pressures of the two components, the solution will boil at a temperature between the boiling points of the two components.

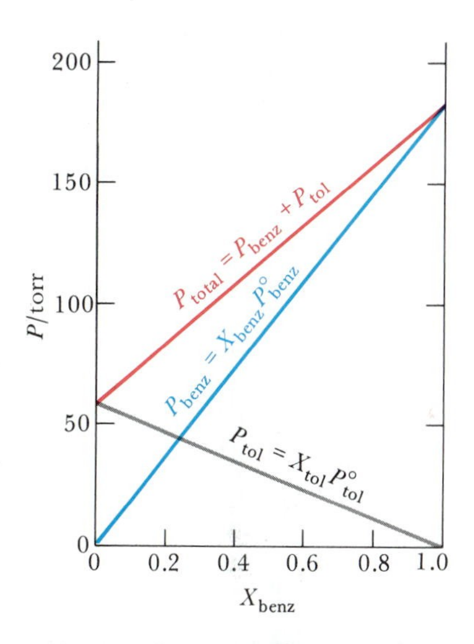

Figure 15-17 Equilibrium vapor pressure at 40°C versus the mole fraction of benzene in solutions of benzene and toluene. The solutions are essentially ideal, and the equilibrium vapor pressures of benzene and toluene are given by Raoult's law [Equation (15-18)]. The total pressure is directly proportional to X_{benz} when Raoult's law holds for both components, that is, when the solution is ideal [Equation (15-21)].

EXAMPLE 15-9: Given that the equilibrium vapor pressures of pure benzene and pure toluene are 183 torr and 59.2 torr, respectively, calculate (a) the total vapor pressure over a $X_{benz} = X_{tol} = 0.500$ solution, and (b) the mole fraction of benzene in the vapor.

Solution: (a) The partial pressures of benzene and toluene over the solution are given by Raoult's law:

$$P_{benz} = X_{benz}P_{benz}^\circ = (0.500)(183 \text{ torr}) = 91.5 \text{ torr}$$

$$P_{tol} = X_{tol}P_{tol}^\circ = (0.500)(59.2 \text{ torr}) = 29.6 \text{ torr}$$

The total vapor pressure is the sum of the partial pressures:

$$P_{total} = P_{benz} + P_{tol} = 91.5 \text{ torr} + 29.6 \text{ torr} = 121.1 \text{ torr}$$

(b) The pressure of a gas is directly proportional to the number of moles of the gas, so the mole fraction of benzene in the vapor, Y_{benz}, is given by

$$Y_{benz} = \frac{n_{benz}}{n_{benz} + n_{tol}} = \frac{P_{benz}}{P_{benz} + P_{tol}} = \frac{91.5 \text{ torr}}{121.1 \text{ torr}} = 0.756$$

We have used Y for the mole fraction in the vapor to distinguish it from the mole fraction, X, in solution.

PRACTICE PROBLEM 15-9: Given that the total pressure over a solution of benzene and toluene is 100.0 torr, calculate the mole fraction of benzene in the liquid and the vapor phases. (Take $P_{tol}^\circ = 59.2$ torr and $P_{benz}^\circ = 183$ torr.)

Answer: $X_b = 0.330$; $Y_b = 0.604$

From Example 15-9 we see that the vapor over a benzene-toluene solution is richer in benzene, the more volatile component, than is the solution. If this vapor is condensed and then reevaporated, the resulting vapor will be even richer in benzene. If this condensation-evaporation process is repeated many times, then a separation of the benzene and toluene is achieved. Such a process is called **fractional distillation** and is carried out automatically in a distillation column of the type shown in Figure 15-18. A fractional distillation column differs from an ordinary distillation column in that the former is packed with glass beads, glass rings, or glass wool. The packing material provides a large surface area for the repeated condensation-evaporation process.

Remarkable separations can be achieved with elaborate fractional distillation units. Fractional distillation techniques are used to separate heavy water (D_2O) from regular water. The heavy water is used on a large scale as a coolant in heavy-water nuclear power plants and in innumerable chemical investigations ranging from spectroscopy to thermodynamics to chemical synthesis. Regular water has a normal boiling point of 100.00°C, whereas heavy water has a normal boiling point of 101.42°C. Only 0.015 percent of the hydrogen atoms in regular water are the deuterium isotope. Nonetheless, a modern heavy-water distillation plant produces almost pure D_2O from regular water at a total cost of around $500 per kilogram of D_2O. These distillation plants have over 300 successive distillation stages and require an input of over 1 metric ton of water per gram of D_2O produced. Canada, which uses D_2O extensively in its nuclear reactors, has two heavy-water plants, with a combined D_2O output capability of 1600 tons per year.

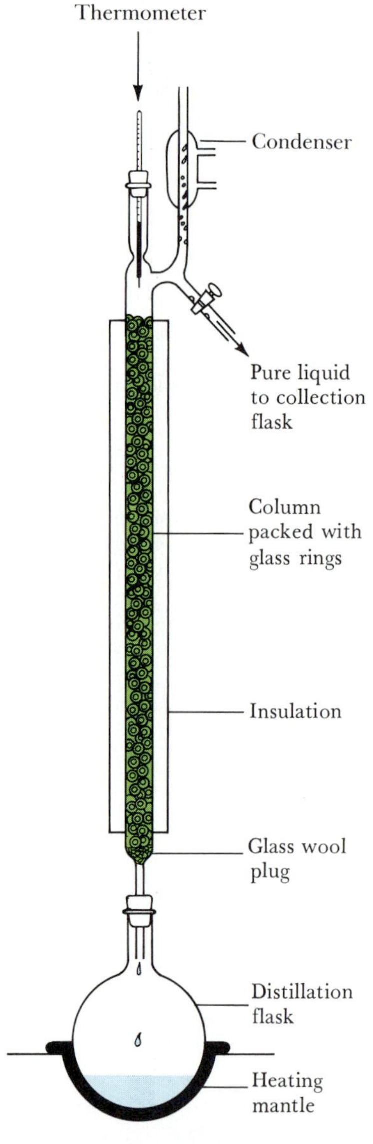

Figure 15-18 A simple fractional distillation column. Because repeated condensation and reevaporation occur along the entire column, the vapor becomes progressively richer in the more volatile component as it moves up the column.

When a two-component solution is not ideal, two different cases arise. If the attractions between an A and a B molecule are greater than those between two A molecules or two B molecules, then the A and B molecules prefer to be in solution and the total vapor pressure will be less than that calculated from Equation (15-21). An example of such a solution is a carbon disulfide-chloromethane (CS_2-CH_3Cl) solution, whose total vapor pressure is plotted in Figure 15-19a. The deviations from Raoult's law, shown as colored lines in Figure 15-19a, are called **negative deviations** from ideal behavior because the P_{total} curve lies below that for the corresponding ideal solution (dashed line). If the attractions between A and B molecules are less than those between A molecules or between B molecules, then the A and B molecules prefer not to be in solution and the total vapor pressure will be greater than that calculated from Equation (15-21). An example is an ethyl alcohol-water (C_2H_5OH-H_2O) solution, whose total vapor pressure is plotted in Figure 15-19b. In this case, we see **positive deviations** from ideal behavior. Plots like those in Figure 15-19 give us insight into intermolecular interactions.

Ethyl alcohol-water solutions provide an especially interesting example of nonideal solution behavior. If we attempt to separate alcohol from water by distillation, an amazing result is obtained. No matter how effective a distillation column we use, the distillate issuing from the column has a maximum alcohol content of 95 percent; water is 5 percent of the distillate. A 95 percent alcohol-plus-water solution distills as if it were a pure liquid. A solution that distills without change in composition is called an **azeotrope.**

Alcohol-water solutions obtained by fermentation have a maximum alcohol content of about 12 percent. Yeasts that produce alcohol as a waste product cannot survive in solutions with more than 12 percent alcohol. In effect, the yeasts are poisoned by their own waste products. The **proof** of an alcohol-water solution is defined as two times the percentage of alcohol by volume in the solution. An 80 proof liquor is 40 percent alcohol. The proof can be increased by simple distillation up to 190 (95 percent). **Absolute alcohol** (100 percent ethanol) can be pre-

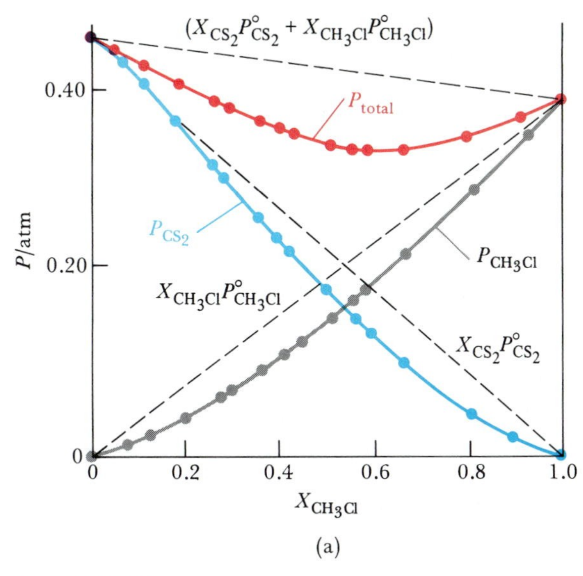

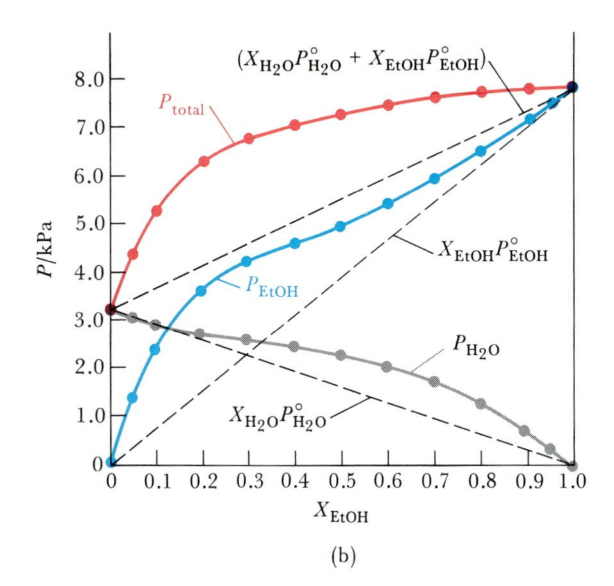

Figure 15-19 Vapor-pressure curves for nonideal solutions. In both cases the dashed lines represent the behavior expected if Raoult's law holds. (a) Equilibrium vapor pressures of carbon disulfide-chloromethane solutions at 25°C. These curves illustrate negative deviations from ideal behavior. (b) Equilibrium vapor pressures of ethyl alcohol-water solutions at 25°C. These curves illustrate positive deviations from ideal behavior.

pared from 95 percent alcohol by dehydration with calcium oxide, which removes the water from the solution by the reaction

$$CaO(s) + H_2O(soln) \rightarrow Ca(OH)_2(s)$$

The $Ca(OH)_2(s)$ is filtered off to leave absolute alcohol.

15-7. THE SOLUBILITY OF A GAS IN A LIQUID IS DIRECTLY PROPORTIONAL TO THE PRESSURE OF THE GAS OVER THE LIQUID

The solubility of a gas in a liquid is directly proportional to the partial pressure of the gas in contact with the liquid. If we express the solubility as the molarity of the dissolved gas, M_{gas}, and the partial pressure of the gas as P_{gas}, then we can write

$$P_{gas} = k_h M_{gas} \qquad (15\text{-}22)$$

Equation (15-22) is called **Henry's law.** The proportionality constant k_h, whose value depends on the gas, the solvent, and the temperature, is called the **Henry's law constant.** Henry's law tells us that if we double the pressure of, say, O_2 gas over liquid water, then the concentration of oxygen dissolved in the water also doubles. A doubling of the pressure of the gas over the solution doubles the concentration of the gas and thus doubles the rate at which the gas molecules enter the solution. This outcome requires a doubling of the concentration of the gas in the solution so that the rate of escape of the dissolved molecules from the solution balances the rate of entry of the gas molecules to the solution.

EXAMPLE 15-10: Calculate the concentration of O_2 in water that is in equilibrium with air at 25°C. The Henry's law constant for O_2 in water at 25°C is 780 atm·M^{-1}.

Solution: The partial pressure of O_2 in the atmosphere is 0.20 atm; thus, from Equation (15-22),

$$M_{O_2} = \frac{P_{O_2}}{k_h} = \frac{0.20 \text{ atm}}{780 \text{ atm}\cdot\text{M}^{-1}} = 2.6 \times 10^{-4} \text{ M}$$

PRACTICE PROBLEM 15-10: The Henry's law constant for H_2S dissolving in water at 25°C is 10 atm·M^{-1}. Calculate the concentration of dissolved H_2S in an aqueous solution in equilibrium with H_2S gas at a pressure of 1.00 atm.

Answer: 0.10 M

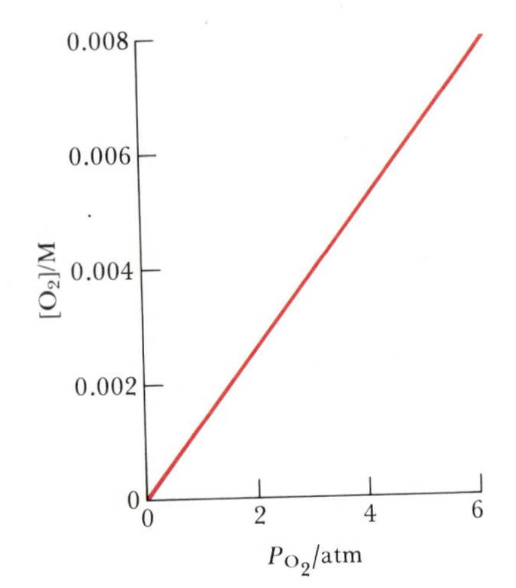

Figure 15-20 The solubility of oxygen in water at 25°C plotted against the pressure of the oxygen in contact with the water. The linear plot confirms that Henry's law holds for oxygen in water over the range of 0 to 6 atm $O_2(g)$.

Figure 15-20 shows the solubility of oxygen in water as a function of the pressure of the oxygen in contact with the water. The resulting straight line is in accord with Henry's law. Henry's law constants for several common gases are given in Table 15-3. The smaller the value of this constant for a gas, the greater the solubility of the gas, because $M_{gas} = P_{gas}/k_h$.

Carbonated beverages are pressurized with CO_2 gas at a pressure above 1 atm; sodas are pressurized with CO_2 at about 2 atm and champagne is pressurized at 4 to 5 atm. The CO_2 pressure is responsible for the rush of escaping gas that causes the "pop" when the carbonated drink container is opened. The loss of CO_2 from the solution occurs because the average atmospheric partial pressure of CO_2 is only 3×10^{-4} atm. The bubbles that form in the liquid are mostly CO_2 plus some water vapor at about 1 atm total pressure.

The air breathed by a diver under water is significantly above atmospheric pressure, because the diver must exhale the air into an environment that has a pressure greater than atmospheric pressure. For example, at a depth of 100 ft, the pressure is about 4.0 atm, so the diver must breathe air at 4.0 atm. At this pressure the solubilities of N_2 and O_2 in the blood are four times greater than they are at sea level. If a diver ascends too rapidly, then the sudden pressure drop causes the dissolved nitrogen to form numerous small gas bubbles in the blood. This phenomenon, which is extremely painful and can result in death, is called the *bends*, because it causes the diver to bend over in pain. Oxygen, which is readily metabolized, does not accumulate as bubbles in the blood. However, pure oxygen cannot be used for breathing for extended periods, because at high oxygen pressures the need to breathe is greatly reduced. Thus, CO_2 accumulates in the bloodstream and causes CO_2 asphyxiation. The solution to these problems was proposed by the chemist Joel Hildebrand, who was a professor of chemistry at the University of California for about 70 years. He was fully active in chemical research well into his nineties, over 20 years after his formal retirement. Hildebrand's solution to the problem consists of substituting helium for nitrogen. Helium is only about half as soluble in blood as nitrogen;

Table 15-3 **Henry's law constants for gases in water at 25°C**

Gas	k_h/atm·M^{-1}
He	2.7×10^3
N_2	1.6×10^3
O_2	7.8×10^2
CO_2	29
H_2S	10

thus, the magnitude of the problem is cut in half. Divers' "air" tanks contain a mixture of He and O_2 adjusted so that the pressure of O_2 is about 0.20 atm at maximum dive depth, thereby avoiding the CO_2 asphyxiation problem also.

The solubility of gases in liquids decreases with increasing temperature, because of the increase in escaping tendency of the solute with increasing temperature. Cold water in equilibrium with air has a higher concentration of dissolved oxygen than warm water. The decreased solubility of oxygen with increased temperature is the reason why most fishes, especially the more active ones, such as trout and tuna, prefer cooler water.

SUMMARY

The concentration of a solute can be expressed in various concentration units, such as molarity, molality, and mole fraction.

Vapor-pressure lowering, boiling-point elevation, freezing-point depression, and osmotic pressure are colligative properties. The key point in understanding the colligative properties of solutions is realizing that equilibrium vapor pressure of a solvent is reduced when a solute is dissolved in it. If a solution is sufficiently dilute, then the equilibrium vapor pressure of the solvent is given by Raoult's law. Colligative properties of solutions depend only on the solute particle concentration and are independent of the chemical nature of the solute.

Osmotic pressure is the largest of the colligative effects and can be used to determine the molecular mass of proteins. Osmotic pressure effects are important in biological systems in that osmotic pressure keeps biological cells inflated.

Raoult's law also applies to ideal solutions consisting of two volatile liquids. Ideal solutions result when the interactions between unlike molecules in the solution are the same as the interactions between like molecules. If this is not the case, then the solution is not ideal and deviations from Raoult's law are observed.

Henry's law states that the solubility of a gas in a liquid is directly proportional to the pressure of the gas over the solution. The solubility of gases in liquids decreases with increasing temperature.

TERMS YOU SHOULD KNOW

solvated 519
colligative property 519
mole fraction, X 520
% mass (percentage by mass) 520
molality, m 521
colligative molality, m_c 522
Raoult's law 524
vapor-pressure lowering 524
boiling-point elevation 525
boiling-point elevation constant, K_b 526
freezing-point depression 528
freezing-point depression constant, K_f 528
antifreeze 528
van't Hoff i-factor 529
semipermeable membrane 532
escaping tendency 532
osmotic pressure, π 532
osmosis 532
colligative molarity, M_c 533
reverse osmosis 533
ideal solution 536
fractional distillation 537
negative deviation from Raoult's law 538
positive deviation from Raoult's law 538
azeotrope 538
proof (alcohol) 538
absolute alcohol 538
Henry's law 539
Henry's law constant 539

EQUATIONS YOU SHOULD KNOW HOW TO USE

$X_i = \dfrac{n_i}{n_{tot}}$ (i = 1, 2) (15-1) (mole fraction)

$m = \dfrac{\text{moles of solute}}{\text{kilograms of solvent}}$ (15-6) (definition of molality)

$m_c = im$ (15-7) (colligative molality)

$P_1 = X_1 P_1^\circ$ (15-9) (Raoult's law)

$\Delta P_1 = X_2 P_1^\circ$ (15-11) (vapor-pressure lowering)

$T_b - T_b^\circ = K_b m_c$ (15-14) (boiling-point elevation)

$T_f^\circ - T_f = K_f m_c$ (15-15) (freezing-point depression)

$T_f^\circ - T_f = K_f im$ (15-16) (van't Hoff i-factor equation)

$\pi = RTM_c$ (15-17) (osmotic pressure)

$P_{total} = X_A P_A^\circ + X_B P_B^\circ$ (15-20)

$P_{total} = P_B^\circ + X_A(P_A^\circ - P_B^\circ)$ (15-21)

(Raoult's law for a 2-component solution)

$P_{gas} = k_h M_{gas}$ (15-22) (Henry's law)

PROBLEMS

MOLE FRACTION

15-1. Calculate the mole fractions in a solution that is made up of 20.0 g of ethyl alcohol, C_2H_5OH, and 80.0 g of water.

15-2. A solution of 40 percent formaldehyde, H_2CO, 10 percent methyl alcohol, CH_3OH, and 50 percent water by mass is called formalin. Calculate the mole fractions of formaldehyde, methyl alcohol, and water in formalin. Formalin is used to disinfect dwellings, ships, storage houses, and so forth.

15-3. Describe how you would prepare 1.00 kg of an aqueous solution of acetone, CH_3COCH_3, in water in which the mole fraction of acetone is 0.19.

15-4. Describe how you would prepare 500 g of a solution of sucrose, $C_{12}H_{22}O_{11}$, in water in which the mole fraction of sucrose is 0.125.

15-5. Calculate the mole fraction of isopropyl alcohol, C_3H_7OH, in a solution that is 70.0 percent isopropyl alcohol and 30.0 percent water by volume. Take the density of water as $1.00\ g\cdot cm^{-3}$ and the density of isopropyl alcohol as $0.785\ g\cdot cm^{-3}$.

15-6. Some forms of so-called maintenance-free car batteries use a lead-calcium alloy (solid solution). Given that the alloy is 95 percent lead and 5.0 percent calcium by mass, calculate the mole fraction of calcium in the solid solution.

MOLALITY

15-7. Describe how you would prepare a solution of formic acid in acetone, $(CH_3)_2CO$, that is 2.50 m in formic acid, $HCHO_2$.

15-8. Describe how you would prepare an aqueous solution that is 1.75 m in $Ba(NO_3)_2$.

15-9. The solubility of iodine in carbon tetrachloride is 2.603 g per 100.0 g of CCl_4 at 35°C. Calculate the molality of iodine in a saturated solution.

15-10. How many kilograms of water would have to be added to 18.0 g of oxalic acid, $H_2C_2O_4$, to prepare a 0.050 m solution?

15-11. A 1.0-mol sample of each of the following substances is dissolved in 1000 g of water. Determine the colligative molality of the substance in each case.

(a) $MgSO_4$ (b) $Cu(NO_3)_2$
(c) C_2H_5OH (d) $Al_2(SO_4)_3$

15-12. A 1.0-mol sample of each of the following substances is dissolved in 1000 g of water. Determine the colligative molality of the substance in each case.

(a) CH_3OH (b) $Al(NO_3)_3$
(c) $Fe(NO_3)_2$ (d) $(NH_4)_2Cr_2O_7$

VAN'T HOFF FACTOR

15-13. Predict the van't Hoff i-factor for each of the following salts dissolved in water:

(a) $AgNO_3$ (b) $MgCl_2$ (c) K_2SO_4

15-14. Predict the van't Hoff i-factor for each of the following salts dissolved in water:

(a) Na_2CO_3 (b) $(NH_4)_2SO_4$ (c) $Pb(NO_3)_2$

15-15. Which of the following compounds would you expect to have the largest van't Hoff i-factor?

(a) $HNO_3(aq)$ (b) $H_2SO_4(aq)$ (c) $HClO_4(aq)$

15-16. Which of these compounds would you expect to have the largest van't Hoff i-factor?

(a) $HCl(aq)$ (b) $HClO_2(aq)$ (c) $HClO(aq)$

RAOULT'S LAW

15-17. The vapor pressure of pure water at 37°C is 47.1 torr. Use Raoult's law to estimate the vapor pressure of an aqueous solution at 37°C containing 20.0 g of glucose, $C_6H_{12}O_6$, dissolved in 500.0 g of water. Also compute the vapor-pressure lowering.

15-18. Water at 37°C has a vapor pressure of 47.1 torr. Calculate the vapor pressure of water if 50.0 g of glycerin, $C_3H_8O_3$, is added to 100.0 mL of water. The density of water at 37°C is 0.993 $g \cdot mL^{-1}$. Also calculate the vapor-pressure lowering.

15-19. The vapor pressure of pure water at 25°C is 23.76 torr. Use Raoult's law to estimate the vapor pressure of an aqueous solution at 25°C containing 20.0 g of sucrose, $C_{12}H_{22}O_{11}$, dissolved in 195 g of water. Also calculate the vapor-pressure lowering.

15-20. The vapor pressure of pure water at 100°C is 1.00 atm. Use Raoult's law to estimate the vapor pressure of water over an aqueous solution at 100°C containing 50.0 g of ethylene glycol, $C_2H_6O_2$, dissolved in 100.0 g of water. Also calculate the vapor-pressure lowering for water.

15-21. Calculate the vapor pressure of an ethyl alcohol solution at 25°C containing 20.0 g of the nonvolatile solute urea, $(NH_2)_2CO$, dissolved in 100.0 g of ethyl alcohol, C_2H_5OH. The vapor pressure of ethyl alcohol at 25°C is 59.2 torr. Also calculate the vapor-pressure lowering.

15-22. Calculate the vapor pressure of ethyl alcohol, C_2H_5OH, over 80 proof (40 percent by volume) vodka at 19°C. The vapor pressure of pure ethyl alcohol at 19°C is 40 torr. The density of ethyl alcohol is 0.79 $g \cdot mL^{-1}$ and the density of water is 1.00 $g \cdot mL^{-1}$.

15-23. Given that the vapor pressure of water is 17.54 torr at 20°C, calculate the vapor-pressure lowering of aqueous solutions that are 0.25 m in

(a) NaCl
(b) $CaCl_2$
(c) sucrose, $C_{12}H_{22}O_{11}$
(d) $Al(ClO_4)_3$

15-24. The colligative molality of the contents of a typical human cell is about 0.30 m_c. Compute the equilibrium vapor pressure of water at 37°C for the cell solution. Take ($P^{\circ}_{H_2O}$ = 0.0313 atm):

VAPOR-PRESSURE LOWERING

15-25. Calculate the vapor-pressure lowering of the following aqueous solutions at 25°C ($P^{\circ}_{H_2O}$ = 0.0313 atm):

(a) 2.00 m $C_{12}H_{22}O_{11}$ (sucrose)
(b) 2.00 m NaCl(*aq*)
(c) 2.00 m $CaCl_2$(*aq*)

15-26. Calculate the vapor-pressure lowering of the following solutions at 25°C:

(a) 1.50 m C_2H_5OH(*aq*)
(b) 0.50 m $TlCl_3$(*aq*)
(c) 0.250 m K_2SO_4(*aq*)

15-27. The observed vapor-pressure lowering of an aqueous sucrose ($C_{12}H_{22}O_{11}$) solution is 3.56 torr at 25.0°C. Calculate the concentration of sucrose in the solution.

15-28. The observed vapor-pressure lowering of a $CaCl_2$(*aq*) solution is 5.00 torr at 20.0°C. Calculate the concentration of $CaCl_2$(*aq*) in the solution.

15-29. Given that an aqueous solution of ethylene glycol in water has a vapor-pressure lowering of 12.0 torr at 90.0°C, calculate the mole fraction of ethylene glycol in the solution.

15-30. Calculate the concentrations of the following electrolytes that will produce a vapor-pressure lowering of 2.00 torr at 20°C:

(a) NaCl(*aq*)
(b) $CaCl_2$(*aq*)
(c) K_2SO_4(*aq*)

BOILING-POINT ELEVATION

15-31. Calculate the boiling point of a 2.0 m aqueous solution of $Sc(ClO_4)_3$.

15-32. How much NaCl would have to be dissolved in 1000 g of water in order to raise the boiling point by 1.0°C?

15-33. Calculate the boiling point of a solution of 10.0 g of picric acid, $C_6H_2(OH)(NO_2)_3$, dissolved in 100.0 g of cyclohexane, C_6H_{12}. Assume that the colligative molality and the molality are the same for picric acid in cyclohexane.

15-34. The colligative molality of seawater is approximately 1.10 m_c. Calculate the boiling point of seawater at 1.00 atm and its vapor pressure at 15°C. The vapor pressure of pure water at 15°C is 12.79 torr.

15-35. Calculate the boiling point of a solution containing 25.0 g of urea, $(H_2N)_2CO$, dissolved in 1500 g of nitrobenzene, $C_6H_5NO_2$.

15-36. Calculate the boiling point of a solution of 25.0 g of urea, $(H_2N)_2CO$, plus 25.0 g of thiourea, $(H_2N)_2CS$, in 500 g of chloroform, $CHCl_3$.

FREEZING-POINT DEPRESSION

15-37. Calculate the freezing point of an aqueous solution of 60.0 g of glucose, $C_6H_{12}O_6$, dissolved in 200.0 g of water.

15-38. Calculate the freezing point of a 0.15 m aqueous solution of NaCl.

15-39. Calculate the freezing point of a solution of 22.0 g of carbon tetrachloride dissolved in 800 g of benzene.

15-40. Calculate the freezing point of an aqueous solution of 20.0 g of $Ca(NO_3)_2$ dissolved in 500 g of water.

15-41. Calculate the freezing point of a solution of 5.00 g of diphenyl, $C_{12}H_{10}$, and 7.50 g of naphthalene, $C_{10}H_8$, dissolved in 200 g of benzene.

15-42. Quinine is a natural product extracted from the bark of the cinchona tree, which is native to South America. Quinine is used as an antimalarial agent. When 1.00 g of quinine is dissolved in 10.0 g of cyclohexane, the freezing point is lowered 6.23 K. Calculate the molecular mass of quinine.

15-43. Vitamin K is involved in normal blood clotting. When 0.500 g of vitamin K is dissolved in 10.0 g of camphor, the freezing point of the solution is lowered by 4.43 K. Calculate the molecular mass of vitamin K.

15-44. Polychlorinated biphenyls (PCBs) are highly resistant to decomposition and have been used as coolants in transformers; however, PCBs are carcinogenic and are being phased out of use. A 0.100-g sample of a PCB dissolved in 10.0 g of camphor depressed the freezing point 1.22 K. Calculate the molecular mass of the compound.

15-45. Don Juan Pond in the Wright Valley of Antarctica freezes at −57°C. The major solute in the pond is $CaCl_2$. Estimate the concentration of $CaCl_2$ in the pond water.

15-46. Menthol is a crystalline substance with a peppermint taste and odor. A solution of 6.54 g of menthol per 100.0 g of cyclohexane freezes at −1.95°C. Determine the formula mass of menthol.

15-47. A solution of mercury(II) chloride, $HgCl_2$, is a poor conductor of electricity. A 40.7-g sample of $HgCl_2$ is dissolved in 100.0 g of water, and the freezing point of the solution is found to be −2.83°C. Explain why $HgCl_2$ in solution is a poor conductor of electricity.

15-48. Mayer's reagent, K_2HgI_4, is used in analytical chemistry. In order to determine its extent of dissociation in water, its effect on the freezing point of water is investigated. A 0.25 m aqueous solution is prepared, and its freezing point is found to be −1.41°C. Suggest a possible dissociation reaction that takes place when K_2HgI_4 is dissolved in water.

OSMOTIC PRESSURE

15-49. Calculate the osmotic pressure of a 0.25 M aqueous solution of sucrose, $C_{12}H_{22}O_{11}$, at 37°C.

15-50. Calculate the osmotic pressure of seawater at 37°C. Take $M_c = 1.10\ mol{\cdot}L^{-1}$ for seawater.

15-51. Insulin is a small protein hormone that regulates carbohydrate metabolism by decreasing blood glucose levels. A deficiency of insulin leads to diabetes. A 20.0-mg sample of insulin is dissolved in enough water to make 10.0 mL of solution, and the osmotic pressure of the solution at 25°C is found to be 6.48 torr. Calculate the molecular mass of insulin.

15-52. Pepsin is the principal digestive enzyme of gastric juice. A 3.00-mg sample of pepsin is dissolved in enough water to make the 10.0 mL of solution, and the osmotic pressure of the solution at 25°C is found to be 0.162 torr. Calculate the molecular mass of pepsin.

15-53. In reverse osmosis, water flows out of a salt solution until the osmotic pressure of the solution equals the applied pressure. If a pressure of 100 atm is applied to seawater, what will be the final concentration of the seawater at 20°C when reverse osmosis stops? Given that seawater is a 1.1 M_c solution of NaCl(*aq*), calculate how many liters of seawater are required to produce 10 L of fresh water at 20°C with an applied pressure of 100 atm.

15-54. What is the minimum pressure that must be applied at 25°C to obtain pure water by reverse osmosis from water that is 0.15 M in NaCl and 0.015 M in $MgSO_4$?

RAOULT'S LAW FOR TWO COMPONENTS

15-55. Given that the equilibrium vapor pressures of benzene and toluene at 81°C are 768 torr and 293 torr, respectively, (a) calculate the total vapor pressure at 81°C over a benzene-toluene solution with $X_{benz} = 0.250$, and (b) calculate the mole fraction of benzene in the vapor phase over the solution.

15-56. Propyl alcohol, $CH_3CH_2CH_2OH$, and isopropyl alcohol, $CH_3CHOHCH_3$, form ideal solutions in all proportions. Calculate the partial pressure of each component in equilibrium at 25°C with a solution of compositions $X_{prop} = 0.25$, 0.50, and 0.75, given that $P^\circ_{prop} =$ 20.9 torr and $P^\circ_{iso} = 45.2$ torr at 25°C. Calculate the composition of the vapor phase also.

HENRY'S LAW

15-57. Calculate the concentration of nitrogen in water at a nitrogen gas pressure of 0.79 atm and a temperature of 25°C.

15-58. Of the gases N_2, O_2, and CO_2, which has the highest concentration in water at 25°C when each gas has a pressure of 1.0 atm?

15-59. The Henry's law constant for CO_2 in water at 25°C is 29 atm·M^{-1}. Estimate the concentration of dissolved CO_2 in a carbonated soft drink pressurized with 2.0 atm of CO_2 gas.

15-60. Calculate the masses of oxygen and nitrogen that are dissolved in 1.00 L of aqueous solution in equilibrium with air at 25°C and 760 torr. Assume that air is 20 percent oxygen and 79 percent nitrogen by volume.

15-61. Every 33 ft under water the pressure increases by 1 atm. How many feet would a diver have to descend before the blood concentration of oxygen reaches 1.28×10^{-3} M, assuming that compressed air is used (20 percent O_2 by volume)?

15-62. Compare the concentrations of helium and nitrogen in water when each gas is at a pressure of 4.0 atm at 25°C.

ADDITIONAL PROBLEMS

15-63. Immunoglobulin G, formerly called gamma globulin, is the principal antibody in blood serum. A 0.50-g sample of immunoglobulin G is dissolved in enough water to make 0.100 L of solution, and the osmotic pressure of the solution at 25°C is found to be 0.619 torr. Calculate the molecular mass of immunoglobulin G.

15-64. Most wines are about 12 percent ethyl alcohol by volume, and many hard liquors are about 80 proof. Assuming that the only major nonaqueous constituent of wine and vodka is ethyl alcohol, calculate the freezing points of wine and vodka. Take the density of ethyl alcohol as 0.79 g·mL^{-1} and the density of water as 1.00 g·mL^{-1}. The formula of ethyl alcohol is C_2H_5OH.

15-65. The boiling point of ethylene glycol is 197°C, whereas the boiling point of ethyl alcohol is 78°C. Ethylene glycol is called a "permanent" antifreeze and ethyl alcohol a "temporary" one. Explain the difference between "permanent" and "temporary" antifreezes.

15-66. What volume percent of oxygen should be used in a diver's air tanks to make the partial pressure of oxygen in the air supply 0.20 atm at a water depth of 90 ft? (See Problem 15-61.)

15-67. Calculate the molality of sucrose, $C_{12}H_{22}O_{11}$, in a solution prepared by dissolving 2.00 teaspoons of sucrose in 1.00 cup of water (3 teaspoons = 1 tablespoon = 0.50 ounce = 14 g and 4 cups = 1 quart = 0.946 L). Take the density of water as 1.00 g·mL^{-1}.

15-68. Suppose that an aqueous solution is observed to begin freezing when the temperature is +2.0°C. What can you say about the composition of the solid that freezes out of the solution?

15-69. An aqueous solution of ethylene glycol in which ethylene glycol is 50 percent by volume has a freezing point of −36°C. What is the molality of ethylene glycol? Calculate the boiling point of this solution. Take the density of ethylene glycol to be 1.116 g·mL^{-1} and the density of water to be 1.00 g·mL^{-1}.

15-70. Recently scientists have discovered that some insects produce an antifreeze in cold weather; the antifreeze is glycerol, $HOCH_2CHOHCH_2OH$. How much glycerol must an insect produce per gram of body fluid (taken to be water) to survive at −5.0°C?

15-71. Calculate the mole fractions and molalities of methyl alcohol, CH_3OH, and ethyl alcohol, CH_3CH_2OH, if 305 mg of CH_3OH and 275 mg of CH_3CH_2OH are dissolved in 10.0 g H_2O.

15-72. Let the mole fraction of component A in the vapor over a two-component ideal solution be Y_A. Show that Y_A is given by

$$Y_A = \frac{X_A P_A^\circ}{X_A(P_A^\circ - P_B^\circ) + P_B^\circ}$$

Show that $Y_A > X_A$ if $P_A^\circ > P_B^\circ$; in other words, show that the vapor phase is richer than the liquid phase in the more volatile component.

15-73. A semipermeable membrane separates two aqueous solutions at 20°C. For each of the following cases, name the solution into which a net flow of water (if any) will occur:

(a) 0.10 M $NaCl(aq)$ and 0.10 M $KBr(aq)$

(b) 0.10 M $Al(NO_3)_3(aq)$ and 0.20 M $NaNO_3(aq)$

(c) 0.10 M $CaCl_2(aq)$ and 0.50 M $CaCl_2(aq)$

15-74. Show that the relation between the mole fraction of solute and the molality of the solution is given by

$$X_2 = \frac{(M_1 m/1000)}{\left(1 + \frac{M_1 m}{1000}\right)}$$

where M_1 is the molar mass of the solvent and m is the molality. Now argue that

$$X_2 \simeq \frac{M_1 m}{1000}$$

if the solution is dilute.

15-75. The density of a glycerol-water solution that is 40.0 percent glycerol by mass is 1.101 g·mL^{-1} at 20°C. Calculate the molality and the molarity of glycerol in the solution at 20°C. What is the molality at 0°C? The formula of glycerol is $C_3H_8O_3$.

15-76. When 2.87 g of an organic compound, which is known to be 39.12 percent carbon, 8.76 percent hydrogen, and 52.12 percent oxygen, is dissolved in 65.3 g of camphor, the freezing point of the solution is 160.7°C. Determine the molecular formula of the compound.

15-77. Calculate the molality, the colligative molality, the freezing point, and the boiling point for each of the following solutions:

(a) 5.00 g of K_2SO_4 in 250 g of water
(b) 5.00 g of ethyl alcohol, C_2H_5OH, in 250 g of water
(c) 1.00 g of aniline, $C_6H_5NH_2$, in 50.0 g of camphor

15-78. Calculate the vapor pressures of carbon tetrachloride, CCl_4, and ethyl acetate, $C_4H_8O_2$, in a solution at 50°C containing 25.0 g of carbon tetrachloride dissolved in 100.0 g of ethyl acetate. The vapor pressures of pure CCl_4 and $C_4H_8O_2$ at 50°C are 306 torr and 280 torr, respectively.

15-79. When 2.74 g of phosphorus is dissolved in 100 mL of carbon disulfide, the boiling point is 46.71°C. Given that the normal boiling point of pure carbon disulfide is 46.30°C, that its density is 1.261 $g\cdot mL^{-1}$, and that its boiling-point elevation constant is $K_b = 2.34\ K\cdot m_c^{-1}$, determine the molecular formula of phosphorus.

15-80. A 2.0-g sample of the polymer polyisobutylene, $[CH_2C(CH_3)_2]_x$, is dissolved in enough cyclohexane to make 10.0 mL of solution at 20°C and produces an osmotic pressure of 2.0×10^{-2} atm. Determine the formula mass and the number of units (x) in the polymer.

15-81. It is possible to convert from molality to molarity if the density of a solution is known. The density of a 2.00 m NaOH aqueous solution is 1.22 $g\cdot mL^{-1}$. Calculate the molarity of this solution.

15-82. In many fields outside chemistry, solution concentrations are expressed in mass percent. Calculate the molality of an aqueous solution that is 24.0 percent potassium chromate, K_2CrO_4, by mass. Given that the density of the solution is 1.21 $g\cdot mL^{-1}$, calculate the molarity.

15-83. Radiator antifreeze also provides "antiboiling" protection for automobile cooling systems. Calculate the boiling point of a solution composed of 50.0 g of water and 50.0 g of ethylene glycol. Assume that the vapor pressure of ethylene glycol is negligible at 100°C. The formula of ethylene glycol is $C_2H_6O_2$.

15-84. Use Figure 15-7 to determine the molality of ethylene glycol, $C_2H_6O_2$, in water that is necessary to give antifreeze protection down to −40°C. Compare the result obtained from Figure 15-7 with that calculated from the Equation (15-15).

15-85. Concentrated sulfuric acid, H_2SO_4, is sold as a solution that is 98 percent sulfuric acid and 2.0 percent water by mass. Given that the density is 1.84 $g\cdot mL^{-1}$, calculate the molarity of concentrated H_2SO_4.

15-86. Concentrated phosphoric acid is sold as a solution of 85 percent phosphoric acid and 15 percent water by mass. Given that its molarity is 15 M, calculate the density of concentrated phosphoric acid.

15-87. Show that in an aqueous solution the mole fraction of the solute is given by

$$X_2 = \frac{m_c}{m_c + 55.5\ mol\cdot kg^{-1}} \simeq \frac{m_c}{55.5\ mol\cdot kg^{-1}}$$

where the approximation holds to within about 2 percent when $m_c \leq 1.00\ m_c$.

15-88. Show, using the properties of similar triangles and the curves in Figure 15-4, that

$$(P_1^\circ - P_1) \propto (T_b - T_b^\circ)$$

15-89. Given that the freezing-point depression of a 1.00 m solution of $H_2SO_4(aq)$ is 3.74°C, determine the number of solute particles per formula unit of $H_2SO_4(aq)$.

15-90. Given the following freezing-point depression data, determine the number of ions produced per formula unit when the indicated substance is dissolved in aqueous solution to produce a 1.00 m solution.

Formula	**ΔT/K**
$PtCl_2\cdot 4NH_3$	5.58
$PtCl_2\cdot 3NH_3$	3.72
$PtCl_2\cdot 2NH_3$	1.86
$KPtCl_3\cdot NH_3$	3.72
K_2PtCl_4	5.58

16 RATES AND MECHANISMS OF CHEMICAL REACTIONS

When hydrogen flows through a platinum gauze, it reacts with oxygen without the need of a spark or any other initiator because the platinum acts as a catalyst.

Different chemical reactions take place at different rates. Some reactions, such as the reaction between $AgNO_3(aq)$ and $KCl(aq)$ to produce a precipitate of $AgCl(s)$, seem to occur almost instantaneously, whereas other reactions, such as the reaction between $H_2(g)$ and $N_2(g)$ to produce $NH_3(g)$, occur very slowly. Chemists study the rates of chemical reactions to determine the conditions under which reactions can be made to proceed at favorable rates. Certainly, if a chemist desires to produce a particular product, then the reaction used must take place at an appreciable rate. This condition is particularly important for reactions used for the commercial production of chemicals.

The reaction rate law, which must be determined by experiment, tells us how the rate of a reaction depends on the concentration of the reactants and of other added substances, such as catalysts. The reaction rate law provides the most important clue to the reaction mechanism, which is the sequence of steps by which the reactants are converted to products. An understanding of reaction mechanisms may enable us to adjust reaction conditions in order to produce a desired reaction rate and increases our understanding of how chemical reactions occur at the molecular level. Such an understanding enables us to control chemical reactions better and to manufacture chemical products in the most economical way.

16-1. A REACTION RATE TELLS US HOW FAST A QUANTITY OF REACTANT OR PRODUCT IS CHANGING WITH TIME

Let's consider the decomposition of dinitrogen pentoxide, $N_2O_5(g)$, to nitrogen dioxide, $NO_2(g)$, and oxygen, $O_2(g)$:

$$\underset{\text{colorless}}{N_2O_5(g)} \rightarrow \underset{\text{brown}}{2NO_2(g)} + \underset{\text{colorless}}{\tfrac{1}{2}O_2(g)}$$

Because $NO_2(g)$ is brown and the other species are colorless, the rate of this reaction can be determined from the increase in the intensity of brown color of the reaction mixture as a function of time. Figure 16-1 shows the concentrations of N_2O_5, NO_2, and O_2 versus time in a reaction system at 45°C. We will denote the concentration of a species by enclosing its chemical formula in square brackets. Thus, for example, $[N_2O_5]$ represents the concentration of N_2O_5. Figure 16-1 shows that $[N_2O_5]$ decreases with time while $[NO_2]$ and $[O_2]$ increase. Note also that $[NO_2]$ increases more rapidly than $[O_2]$. Because two moles of $NO_2(g)$ are produced for every 1/2 mole of $O_2(g)$, the rate of production of $NO_2(g)$ is actually four times that of $O_2(g)$.

We can study the rates of other chemical reactions in other ways. Consider the decomposition of hydrogen peroxide in aqueous solution:

$$2H_2O_2(aq) \rightarrow 2H_2O(l) + O_2(g)$$

Because a gaseous product $O_2(g)$ forms, we can determine how its concentration increases with time by measuring the increase in pressure

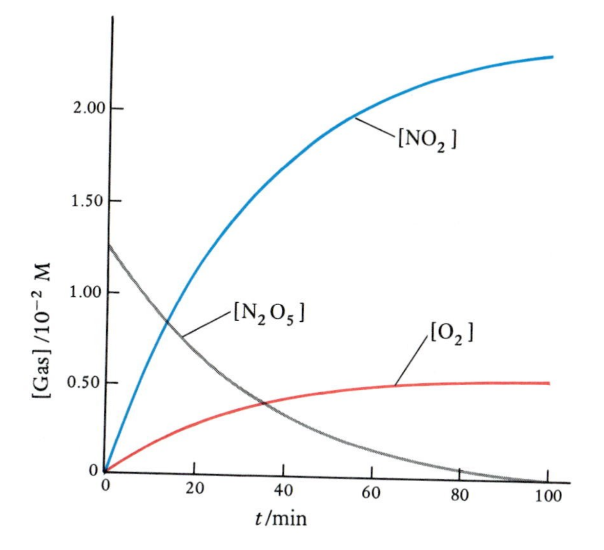

Figure 16-1 The change in concentration of $N_2O_5(g)$, $NO_2(g)$, and $O_2(g)$ as a function of time for the reaction

$$N_2O_5(g) \rightarrow 2NO_2(g) + \tfrac{1}{2}O_2(g)$$

at 45°C. The data are given in Table 16-1. Note that both $[NO_2]$ and $[O_2]$ increase with time because NO_2 and O_2 are reaction products.

above the reaction mixture. Let us see how we can use this measurement to define a reaction rate.

Suppose that we know the concentration of the oxygen gas at some time to be $[O_2]_1$, and suppose we find that at some later time it is $[O_2]_2$. We can write the change in its concentration as

$$\Delta[O_2] = [O_2]_2 - [O_2]_1$$

Here the Δ (delta) stands for change, or later value minus earlier value. Because the concentration of the product increases with time, we know that $\Delta[O_2]$ must be positive. We can now define a **reaction rate** as the average rate at which oxygen is produced, or the change in concentration divided by the elapsed time:

$$\frac{\text{change in concentration of the product}}{\text{elapsed time}} = \frac{\Delta[O_2]}{\Delta t}$$

Note that the rate of a reaction is a positive quantity. Its units are moles per liter per second ($mol{\cdot}L^{-1}{\cdot}s^{-1}$ or $M{\cdot}s^{-1}$), or, more generally, moles per liter per unit time.

We can also define a reaction rate by looking at how quickly the reactant is consumed. Now we are interested in the average number of moles per liter *reacted* per second. Because the concentration of H_2O_2 is decreasing, $\Delta[H_2O_2]$ is a negative quantity, but we can insert a minus sign to make the reaction rate positive:

$$-\frac{\text{change in concentration of the reactant}}{\text{elapsed time}} = -\frac{\Delta[H_2O_2]}{\Delta t}$$

The relation between the two equations depends on the equation for the reaction. In this case, we know the rate at which O_2 is produced is equal to one half the rate at which H_2O_2 is consumed, because two H_2O_2 molecules react for each O_2 molecule produced. Thus, we have

$$\text{rate} = \frac{\Delta[O_2]}{\Delta t} = -\left(\frac{1}{2}\right)\left(\frac{\Delta[H_2O_2]}{\Delta t}\right) \qquad (16\text{-}1)$$

EXAMPLE 16-1: Let the rate of the reaction

$$N_2O_5(g) \rightarrow 2NO_2(g) + \tfrac{1}{2}O_2(g)$$

be expressed as $-\Delta[N_2O_5]/\Delta t$. Express the rate of the reaction in terms of the concentration of each product.

Solution: The rate of loss of N_2O_5 is twice as great as the rate of production of O_2 because two N_2O_5 molecules are consumed for each O_2 molecule produced. Thus, we have

$$\text{rate} = -\frac{\Delta[N_2O_5]}{\Delta t} = \frac{2\Delta[O_2]}{\Delta t}$$

The rate of loss of N_2O_5 is one half the rate of production of NO_2 because two NO_2 molecules are formed for each N_2O_5 molecule that decomposes. Thus,

$$\text{rate} = -\frac{\Delta[N_2O_5]}{\Delta t} = \left(\frac{1}{2}\right)\left(\frac{\Delta[NO_2]}{\Delta t}\right)$$

PRACTICE PROBLEM 16-1: Let the rate of the reaction

$$H_2(g) + Br_2(g) \rightarrow 2HBr(g)$$

be expressed as $-\Delta[H_2]/\Delta t$. Express the rate of the reaction in terms of the other reactants or products.

Answer:

$$\text{rate} = -\frac{\Delta[Br_2]}{\Delta t} = +\left(\frac{1}{2}\right)\left(\frac{\Delta[HBr]}{\Delta t}\right)$$

The rate of the reaction for the decomposition of $N_2O_5(g)$ in Example 16-1 can be expressed as

$$\text{rate} = -\left(\frac{1}{1}\right)\left(\frac{\Delta[N_2O_5]}{\Delta t}\right) = \left(\frac{1}{2}\right)\left(\frac{\Delta[NO_2]}{\Delta t}\right) = \left(\frac{1}{1/2}\right)\left(\frac{\Delta[O_2]}{\Delta t}\right) \tag{16-2}$$

Notice that the numerical factor in the denominator of each expression is the same as the balancing coefficient of the species in the chemical equation.

Thus, for the reaction

$$2O_3(g) \rightarrow 3O_2(g)$$

we define the rate to be

$$\text{rate} = -\left(\frac{1}{2}\right)\left(\frac{\Delta[O_3]}{\Delta t}\right) = \left(\frac{1}{3}\right)\left(\frac{\Delta[O_2]}{\Delta t}\right)$$

Recall that the negative sign here ensures that the rate is a positive quantity.

EXAMPLE 16-2: Express the rate of the following reaction in terms of the rate of change for each of the three species involved.

$$2PH_3(g) \rightarrow 2P(g) + 3H_2(g)$$

Solution: Using the general ideas discussed above, we write

$$\text{rate} = -\left(\frac{1}{2}\right)\left(\frac{\Delta[PH_3]}{\Delta t}\right) = \left(\frac{1}{2}\right)\left(\frac{\Delta[P]}{\Delta t}\right) = \left(\frac{1}{3}\right)\left(\frac{\Delta[H_2]}{\Delta t}\right)$$

Recall that we include the minus sign in the first term because the concentration of a reactant decreases with time; thus, $\Delta[PH_3]/\Delta t$ is negative, but the reaction rate is positive.

PRACTICE PROBLEM 16-2: For the reaction

$$3BrO^-(aq) \rightarrow BrO_3^-(aq) + 2Br^-(aq)$$

express the rate in terms of the rate of change in the concentration of each species.

Answer:

$$\text{rate} = -\left(\frac{1}{3}\right)\left(\frac{\Delta[BrO^-]}{\Delta t}\right) = \frac{\Delta[BrO_3^-]}{\Delta t} = \left(\frac{1}{2}\right)\left(\frac{\Delta[Br^-]}{\Delta t}\right)$$

Table 16-1 gives the numerical values of $[N_2O_5]$, $[NO_2]$, and $[O_2]$ that correspond to Figure 16-1. An inspection of the data in Table 16-1 shows that the reaction rate decreases as $[N_2O_5]$ decreases. For example, the rate of decomposition of N_2O_5 over the first 10 min of the reaction is

$$\begin{aligned}\text{rate} &= -\frac{\Delta[N_2O_5]}{\Delta t}\\ &= -\frac{(0.92\times10^{-2} - 1.24\times10^{-2})\text{ M}}{(10-0)\text{ min}}\\ &= 3.2\times10^{-4}\text{ mol}\cdot\text{L}^{-1}\cdot\text{min}^{-1}\end{aligned}$$

whereas the rate over the period 10 min to 20 min is

$$\begin{aligned}\text{rate} &= -\frac{(0.68\times10^{-2} - 0.92\times10^{-2})\text{ M}}{(20-10)\text{ min}}\\ &= 2.4\times10^{-4}\text{ mol}\cdot\text{L}^{-1}\cdot\text{min}^{-1}\end{aligned}$$

Table 16-1 Concentration of $N_2O_5(g)$, $NO_2(g)$, and $O_2(g)$ as a function of time at 45°C for the reaction

$$N_2O_5(g) \rightarrow 2NO_2(g) + \tfrac{1}{2}O_2(g)$$

t/min	$[N_2O_5]$/M	$[NO_2]$/M	$[O_2]$/M
0	1.24×10^{-2}	0	0
10	0.92×10^{-2}	0.64×10^{-2}	0.16×10^{-2}
20	0.68×10^{-2}	1.12×10^{-2}	0.28×10^{-2}
30	0.50×10^{-2}	1.48×10^{-2}	0.37×10^{-2}
40	0.37×10^{-2}	1.74×10^{-2}	0.44×10^{-2}
50	0.27×10^{-2}	1.92×10^{-2}	0.48×10^{-2}
60	0.20×10^{-2}	2.08×10^{-2}	0.52×10^{-2}
70	0.14×10^{-2}	2.18×10^{-2}	0.55×10^{-2}
80	0.11×10^{-2}	2.26×10^{-2}	0.57×10^{-2}
90	0.08×10^{-2}	2.32×10^{-2}	0.58×10^{-2}
100	0.06×10^{-2}	2.36×10^{-2}	0.59×10^{-2}

EXAMPLE 16-3: Use the data in Table 16-1 to calculate the rate of production of NO_2 over the first 10 min of the reaction

$$N_2O_5(g) \rightarrow 2NO_2(g) + \tfrac{1}{2}O_2(g)$$

Solution: Because NO_2 is a product, the rate of production of NO_2 is given by $\Delta[NO_2]/\Delta t$. For the first 10 min of the reaction

$$\frac{\Delta[NO_2]}{\Delta t} = \frac{(0.64 \times 10^{-2} - 0)\ \text{M}}{(10 - 0)\ \text{min}} = 6.4 \times 10^{-4}\ \text{M}\cdot\text{min}^{-1}$$

Note that the rate of production of NO_2 over the first 10 min is twice as great as the rate of consumption of N_2O_5 over the same time period.

PRACTICE PROBLEM 16-3: Use the data in Table 16-1 to calculate the rate of production of $NO_2(g)$ over the time interval 30 to 40 min.

Answer: $2.6 \times 10^{-4}\ \text{M}\cdot\text{min}^{-1}$

It is important to realize that, given the initial concentrations in Table 16-1, the concentrations of $NO_2(g)$ and $O_2(g)$ can be calculated from the concentration of $N_2O_5(g)$ at any time. The following Example illustrates such a calculation.

EXAMPLE 16-4: If $[N_2O_5] = 0.28 \times 10^{-2}$ M at 50 min, calculate $[NO_2]$ and $[O_2]$.

Solution: The chemical equation is

$$N_2O_5(g) \rightarrow 2NO_2(g) + \tfrac{1}{2}O_2(g)$$

The number of moles per liter of $N_2O_5(g)$ that have reacted in 50 min is the difference between the initial concentration and the concentration at 50 min.

$$\begin{aligned}\text{mol}\cdot\text{L}^{-1}\ N_2O_5(g)\ \text{reacted} &= [N_2O_5]_0 - [N_2O_5] \\ &= 1.24 \times 10^{-2}\ \text{M} - 0.28 \times 10^{-2}\ \text{M} \\ &= 0.96 \times 10^{-2}\ \text{M}\end{aligned}$$

According to the chemical equation, 2 mol of $NO_2(g)$ are produced for each mole of $N_2O_5(g)$ that reacts, so, per liter, we have

$$\begin{aligned}\begin{pmatrix}\text{mol}\cdot\text{L}^{-1}\ NO_2(g) \\ \text{produced}\end{pmatrix} &= (0.96 \times 10^{-2}\ \text{mol}\cdot\text{L}^{-1}\ N_2O_5)\left(\frac{2\ \text{mol}\cdot\text{L}^{-1}\ NO_2}{1\ \text{mol}\cdot\text{L}^{-1}\ N_2O_5}\right) \\ &= 1.92 \times 10^{-2}\ \text{M}\end{aligned}$$

Similarly, for $O_2(g)$ we have

$$\begin{aligned}\text{mol}\cdot\text{L}^{-1}\ O_2(g)\ \text{produced} &= (0.96 \times 10^{-2}\ N_2O_5)\left(\frac{1/2\ \text{mol}\cdot\text{L}^{-1}\ O_2}{1\ \text{mol}\cdot\text{L}^{-1}\ N_2O_5}\right) \\ &= 0.48 \times 10^{-2}\ \text{M}\end{aligned}$$

PRACTICE PROBLEM 16-4: Using the initial concentrations and the fact that $[O_2] = 0.55 \times 10^{-2}$ M at 70 min in Table 16-1, calculate $[N_2O_5]$ and $[NO_2]$.

Answer: $[N_2O_5] = 1.4 \times 10^{-3}$ M; $[NO_2] = 0.022$ M

16-2. THE RATE LAW OF A REACTION CAN BE DETERMINED BY THE METHOD OF INITIAL RATES

We have noted that the rate of a reaction changes as the reactants are consumed and the products are formed. This change occurs because the rates of chemical reactions usually depend on the concentrations of one or more of the reactants. The mathematical equation that gives this dependence is called the **rate law** of the reaction. Many experiments show that the rate of a reaction usually is proportional to the concentrations of the reactants raised to small integer powers. Thus, for the case of the thermal decomposition of N_2O_5, we will assume that

$$\text{rate} \propto [N_2O_5]^x \tag{16-3}$$

We can write this proportionality as an equation by inserting a proportionality constant:

$$\text{rate} = k[N_2O_5]^x \tag{16-4}$$

The proportionality constant, k, in a rate law is called the **rate constant** of the reaction.

The value of x in Equation (16-4) must be determined *experimentally;* it is *not* necessarily the same as the balancing coefficient of N_2O_5 in the chemical equation. *There is no relation between the balancing coefficients in a chemical equation and the reaction rate law.* For example, the rate law for the reaction

$$2H_2O_2(aq) \longrightarrow 2H_2O(l) + O_2(g)$$

is

$$\text{rate} = k[H_2O_2]$$

whereas the rate law for the reaction

$$2NOBr(g) \longrightarrow 2NO(g) + Br_2(g)$$

is

$$\text{rate} = k[NOBr]^2$$

In practice, rate laws often are determined from data involving reaction rates at the early stage of reactions. Using the **method of initial rates,** we measure the rate of a reaction over an initial time interval that is short enough so that the concentrations of the reactants do not vary appreciably from their initial values. If we use zeros as subscripts to denote the initial values of the rate and of the various concentrations, then for any reaction described by an equation of the form

$$aA + bB + cC \longrightarrow \text{products}$$

we can write

$$(\text{rate})_0 = k[A]_0^x[B]_0^y[C]_0^z \tag{16-5}$$

PRACTICE PROBLEM 16-12: A possible mechanism for the reaction

$$Fe^{2+}(aq) + HNO_2(aq) + H^+(aq) \rightarrow Fe^{3+}(aq) + NO(g) + H_2O(l)$$

is the three-step mechanism:

$$HNO_2(aq) \xrightarrow{k_1} NO^+(aq) + OH^-(aq)$$

$$Fe^{2+}(aq) + NO^+(aq) \xrightarrow{k_2} Fe^{3+}(aq) + NO(g)$$

$$H^+(aq) + OH^-(aq) \xrightarrow{k_3} H_2O(l)$$

Identify any intermediate species and write the rate law for each step.

Answer: $NO^+(aq)$ and $OH^-(aq)$ are intermediate species. The rate laws are $k_1[HNO_2]$, $k_2[Fe^{2+}][NO^+]$, and $k_3[H^+][OH^-]$, respectively.

16-8. SOME REACTION MECHANISMS HAVE A RATE-DETERMINING STEP

Let's consider once again the $NO_2 + CO$ reaction,

$$NO_2(g) + CO(g) \rightarrow NO(g) + CO_2(g)$$

We proposed in the previous section that this reaction has a two-step mechanism:

$$NO_2(g) + NO_2(g) \xrightarrow{k_1} NO_3(g) + NO(g) \qquad \text{rate}_1 = k_1[NO_2]^2$$

$$NO_3(g) + CO(g) \xrightarrow{k_2} NO_2(g) + CO_2(g) \qquad \text{rate}_2 = k_2[NO_3][CO]$$

It turns out that the first step is much slower than the second. Because the reaction proceeds through both steps, the slow step acts as a bottleneck. In general, the rate of the overall reaction can be no faster than the rate of the slowest step. In this case, the rate of the overall reaction, therefore, will be given by the rate of the first step alone. The overall reaction rate for the $NO_2 + CO$ reaction will thus be

$$\text{rate} = k_1[NO_2]^2$$

which, in fact, is the experimentally observed rate law. In effect, the CO molecules have to wait around for NO_3 molecules to be produced. Once formed, these are consumed very rapidly by reaction with CO.

If one step in a reaction mechanism is much slower than any of the other steps, then that step effectively controls the overall reaction rate and is called the **rate-determining step.** Not all reaction mechanisms have a rate-determining step, but when one does occur, the overall reaction rate is limited by the rate of the rate-determining step.

EXAMPLE 16-13: In Example 16-12 we claimed that the reaction

$$2NO_2(g) + F_2(g) \rightarrow 2NO_2F(g)$$

proceeds by a two-step mechanism. If the rate of the first step is much slower than that of the second step, then deduce the rate law of the overall reaction.

Solution: The first step is a rate-determining step; hence the rate law for the overall reaction is the rate law of the first step,

$$\text{rate} = k_1[NO_2][F_2]$$

PRACTICE PROBLEM 16-13: Another possible mechanism for the reaction in Practice Problem 16-12 is

$$Fe^{2+}(aq) + HNO_2(aq) \xrightarrow{k_1} Fe^{3+}(aq) + OH^-(aq) + NO(g)$$

$$H^+(aq) + OH^-(aq) \xrightarrow{k_2} H_2O(l)$$

If the first step occurs much more slowly than the second step, then what would be the rate law of the overall reaction?

Answer: first order in $[Fe^{2+}]$, first order in $[HNO_2]$, and second order overall

16-9. REACTANTS MUST SURMOUNT AN ENERGY BARRIER TO REACT

Because an elementary process occurs in a single step, it is the simplest type of a reaction to treat theoretically. One approach is called the **collision theory of reaction rates.** Its basic postulate is that two species must collide with each other in order to react. In 1 L of a mixture of the gases A and B at 1.0 atm, the collision frequency between A and B molecules is about $10^{31}\ s^{-1}$. If every collision led to a reaction, then the initial reaction rate would be about

$$\left(\frac{10^{31}s^{-1}}{1L}\right)\left(\frac{1}{6.022\times10^{23}\ mol^{-1}}\right) \simeq 10^7\ mol\cdot L^{-1}\cdot s^{-1}$$

In fact, only a few reactions occur at this very high rate; most occur at a much lower rate. The inescapable conclusion is that most collisions do not lead to a reaction. Unless two conditions are met, the colliding molecules simply bounce off each other unchanged.

The first and more obvious condition is that the molecules must collide with sufficient energy either to break or to rearrange bonds. The other condition is that the molecules must collide in some preferred relative orientation. Figure 16-8 illustrates this idea for the collision of a F_2 molecule with a NO_2 molecule. A reactive collision will occur only if

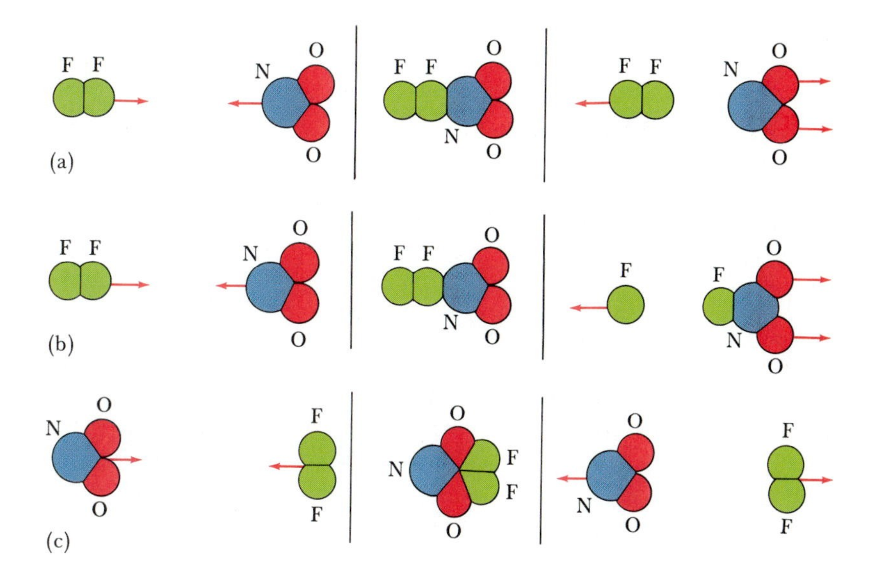

Figure 16-8 Molecular view of the reaction

$$NO_2(g) + F_2(g) \rightarrow FNO_2(g) + F(g)$$

(a) Nonreactive collision. Molecules bounce off one another without reacting because the kinetic energy of the colliding particles is less than the energy required to break the F_2 bond and form the N—F bond. (b) Reactive collision. Molecules collide with sufficient kinetic energy and react because they also have the correct orientation for reaction. (c) Nonreactive collision. Molecules collide with sufficient kinetic energy but do not react because they do not have the correct orientation for reaction.

the two molecules have sufficient energy and if, in addition, the F_2 molecule strikes the nitrogen atom of the NO_2 molecule. Even if the two molecules collide very energetically, collisions with any other relative orientation will not lead to a reaction.

These two requirements are the basis of the collision theory of reaction rates. To incorporate them, we write the reaction rate for the elementary process

$$A + B \rightarrow C + D$$

as

$$\text{rate} = \begin{pmatrix}\text{collision}\\ \text{frequency}\end{pmatrix} \begin{pmatrix}\text{fraction of}\\ \text{collisions with}\\ \text{the required}\\ \text{energy}\end{pmatrix} \begin{pmatrix}\text{fraction of collisions}\\ \text{in which molecules}\\ \text{have the required}\\ \text{relative orientation}\end{pmatrix} \qquad (16\text{-}21)$$

Let's look at each of the terms in parentheses individually. The two major factors that affect the collision frequency are concentration and temperature. The greater the concentration of the reactant molecules, the greater the frequency of collisions. For two reactant species, A and B, the collision frequency is proportional to the product of their concentrations, or

$$\text{collision frequency} \propto [A][B] \qquad (16\text{-}22)$$

The collision frequency also increases with temperature because the root-mean-square molecular speed increases with temperature [see Equation (7-28)]. The faster the molecules are moving, the more frequently they collide.

Now let's consider the second term in parentheses. Figure 16-9 shows a plot of the number of collisions per second of the A and B molecules versus the kinetic energy of the colliding molecules. Note that the fraction of collisions with a kinetic energy in excess of some fixed energy, such as E_a in the figure, increases with increasing temperature. The quantity E_a in Figure 16-9 represents the minimum energy necessary to cause a reaction between the colliding molecules and is called the **activation energy** of the process. Molecules that collide with a relative ki-

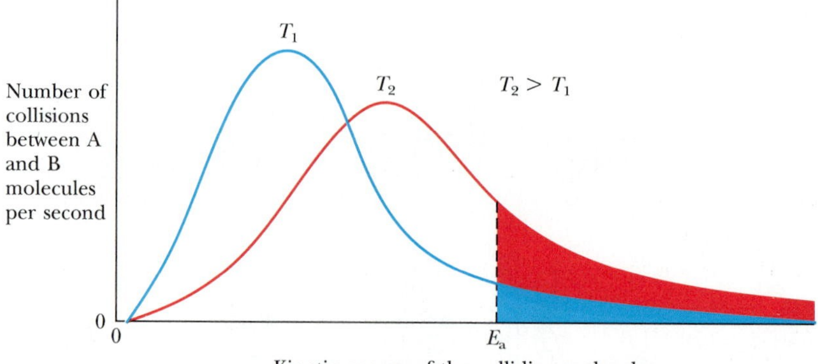

Figure 16-9 Plots of the number of collisions between A and B molecules per unit time versus the kinetic energy of the colliding molecules at two temperatures, T_1 and T_2 ($T_2 > T_1$). Note that the area under the curve beyond E_a is greater for the T_2 curve than for the T_1 curve. The number of collisions with a kinetic energy E_a or greater increases with increasing temperature.

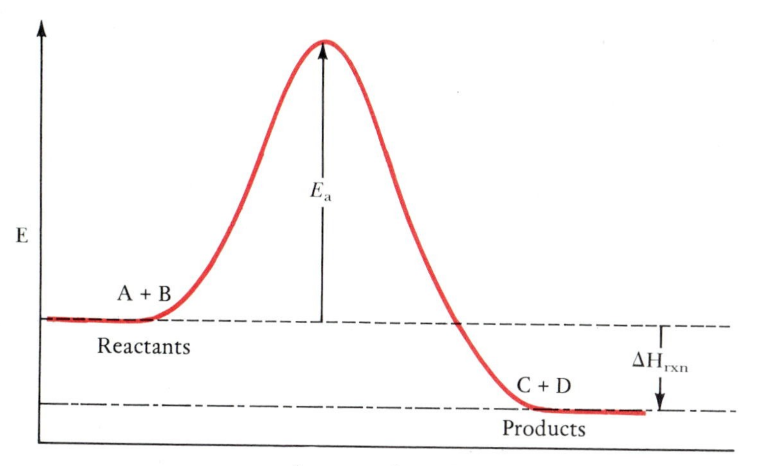

Figure 16-10 The energy of the molecules as the reaction $A + B \rightarrow C + D$ proceeds. The value of E_a has no relationship to the value of the overall energy change of the reaction, ΔH_{rxn}.

netic energy less than E_a simply bounce off each other. A colliding pair that does have the required energy E_a can react, provided they have the required relative orientation.

Let's consider the final term in Equation (16-21). As we have seen, two molecules can react only if they have the correct relative orientation. As we saw in Figure 16-8, this depends on the shapes of the colliding molecules.

In Equation (16-21) then, only the collision frequency depends on the concentrations of A and B. If we substitute Equation (16-22) into (16-21), we see that for an elementary process involving two species A and B

$$\text{rate} = k[\text{A}][\text{B}] \tag{16-23}$$

where the rate constant k depends on the other factors present in Equation (16-21); consequently, it depends on temperature and the shapes of the molecules. Equation (16-23) gives us the rate law for an elementary process directly from its chemical equation. As we stated in Section 16-7, this is true, of course, only for an elementary process.

Figure 16-10 illustrates the activation energy for the elementary reaction $A + B \rightarrow C + D$. The vertical axis in Figure 16-10 represents the energy, E, of the reactants and products and the horizontal axis is a schematic representation of the reaction pathway, beginning with the reactants A and B and ending up with the products C and D. The horizontal axis is labeled "progress of reaction" and represents how far the reaction has proceeded from reactants to products. Figure 16-10 is called an **activation energy diagram.** For the reaction to take place, the molecules must overcome an energy barrier of height E_a; they must collide with sufficient energy to go over the activation energy "hump."

Note for the reaction depicted in Figure 16-10 that the products have a lower energy than the reactants, so $\Delta H^\circ_{rxn} < 0$; therefore, Figure 16-10 represents an exothermic reaction (see Section 8-3). Also note that the activation energy in going from products to reactants is as follows: $E_a + |\Delta H^\circ_{rxn}|$, where $|\Delta H^\circ_{rxn}|$ denotes the absolute magnitude of ΔH°_{rxn}, that is, the value of ΔH°_{rxn} without its negative sign. (remember that ΔH°_{rxn} is a negative quantity in Figure 16-10).

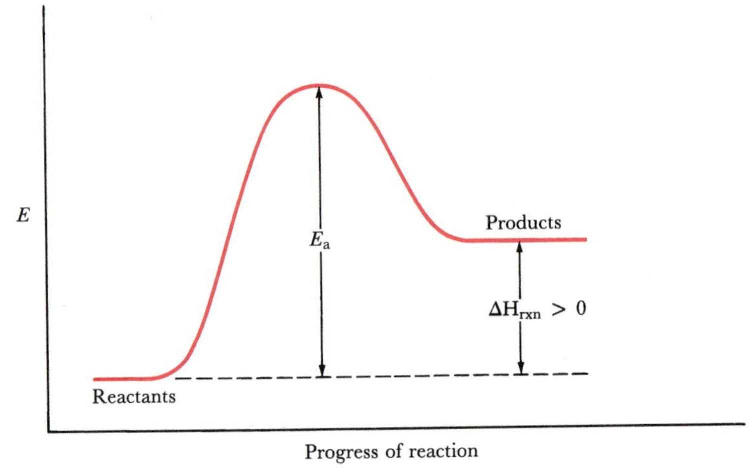

Figure 16-11 Activation energy diagram for an endothermic reaction.

EXAMPLE 16-14: Sketch an activation energy diagram for an endothermic reaction. Determine the activation energy in both the forward and reverse directions.

Solution: In an endothermic reaction, the energy of the reactants is lower than that of the products, so the activation energy diagram looks like the curve in Figure 16-11. The activation energy in the forward direction is E_a, but the activation energy in the reverse reaction is $E_a - \Delta H^\circ_{rxn}$.

PRACTICE PROBLEM 16-14: If the value of ΔH°_{rxn} for a reaction is -80 kJ and its activation energy is 50 kJ, then what is the value of the activation energy in the reverse direction?

Answer: 130 kJ

The maximum energy in an activation energy diagram like that in Figures 16-10 and 16-11 has a useful physical interpretation. To develop this idea, let's consider the reaction between chloromethane, $CH_3Cl(aq)$, and a hydroxide ion, $OH^-(aq)$, to produce methyl alcohol, $CH_3OH(aq)$, and $Cl^-(aq)$:

$$CH_3Cl(aq) + OH^-(aq) \rightarrow CH_3OH(aq) + Cl^-(aq)$$

This reaction has been determined to take place in one step, and the rate law has been determined experimentally to be

$$\text{rate} = k[CH_3Cl][OH^-]$$

We can view this reaction by the following pathway:

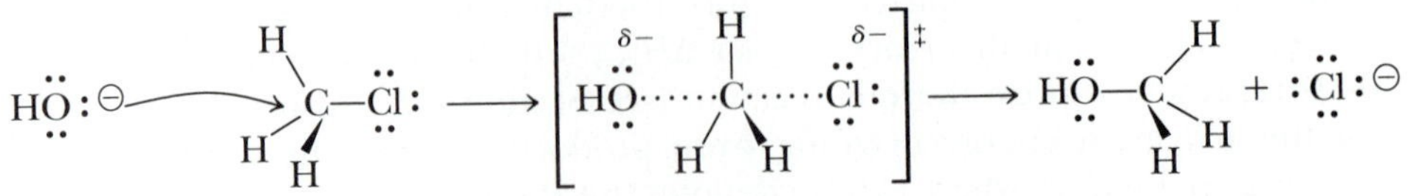

$$HO:^{\ominus} + H_3C{-}Cl: \longrightarrow \left[\overset{\delta-}{HO}\cdots CH_3 \cdots \overset{\delta-}{Cl}: \right]^{\ddagger} \longrightarrow HO{-}CH_3 + :Cl:^{\ominus}$$

As the hydroxide ion approaches the chloromethane molecule, a carbon-oxygen bond begins to form, while the carbon-chlorine bond begins to break. The species shown in square brackets with a ‡ super-

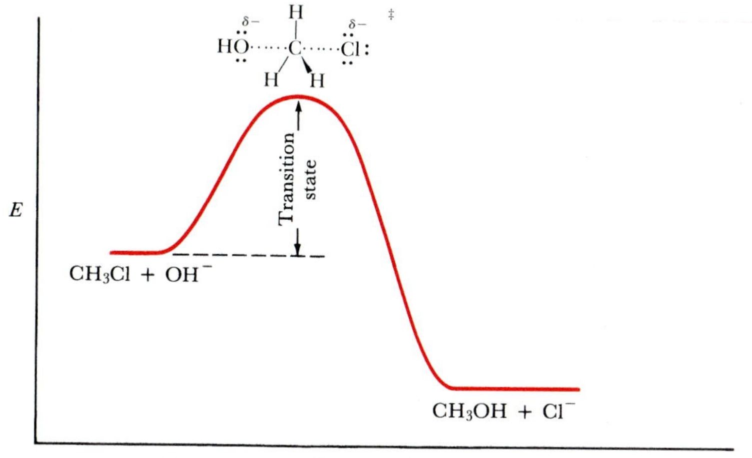

Figure 16-12 Activation energy diagram for the replacement of a chlorine atom by a hydroxyl group in chloromethane. There is only one transition state, which involves both reactant molecules.

script is called an **activated complex.** An activated complex represents the state with the least amount of additional energy necessary to pass from reactants to products. It is an intrinsically unstable species because it sits at the maximum in the activation energy diagram, as shown in Figure 16-12. An activated complex cannot be isolated or detected by any ordinary means. In this regard, an activated complex is completely different from an intermediate species, which may have a lifetime millions of times longer than an activated complex and can often be detected experimentally, or even isolated. Although activated complexes cannot be isolated, they are very useful in describing reaction mechanisms.

16-10. THE ARRHENIUS EQUATION DESCRIBES THE TEMPERATURE DEPENDENCE OF A REACTION RATE CONSTANT

In almost all cases, an increase in temperature will increase the rate of a reaction. We cook foods in order to increase the rates of the biochemical reactions that take place. In this section we study the effect of temperature on reaction rate constants.

Let's go back to Equation (16-21), which expresses the rate of a reaction as the product of three factors: collision frequency, the fraction of collisions with sufficient kinetic energy, and the fraction of collisions in which the colliding pair of molecules has the appropriate relative orientation. We have seen that two of these terms depend on the temperature. We can use the kinetic theory of gases to evaluate each of the terms. The final result for the reaction rate constant, k, can be expressed as

$$k = pze^{-E_a/RT} \qquad (16\text{-}24)$$

In this equation, p, called a **steric factor,** is the fraction of collisions with the correct relative orientations; z is the collision frequency; E_a is the activation energy; R is the molar gas constant ($8.314\ \text{J}\cdot\text{mol}^{-1}\cdot\text{K}^{-1}$); and T is the Kelvin temperature.

The factor $e^{-E_a/RT}$ is the fraction of collisions with energy greater than E_a. This factor varies strongly with temperature, but the quantities

p and z do not vary appreciably. We will denote their product by A and consider it to be a constant. Thus, Equation (16-24) can be written as

$$k = Ae^{-E_a/RT} \tag{16-25}$$

where A is a constant that is proportional to the collision frequency. Equation (16-25), which was first proposed in 1889 by the Swedish chemist Svante Arrhenius, is now called the **Arrhenius equation.** It expresses the temperature dependence of a reaction rate constant.

If we take the natural logarithm of both sides of Equation (16-25), then we obtain

$$\ln k = \ln A - \frac{E_a}{RT} \tag{16-26}$$

Equation (16-26) enables us to compare the values of rate constants at different temperatures. We can write Equation (16-26) for k_2 at temperature T_2 and for k_1 at temperature T_1.

$$\ln k_2 = -\left(\frac{E_a}{R}\right)\left(\frac{1}{T_2}\right) + \ln A$$

$$\ln k_1 = -\left(\frac{E_a}{R}\right)\left(\frac{1}{T_1}\right) + \ln A$$

Now subtract the second equation from the first:

$$\ln k_2 - \ln k_1 = -\frac{E_a}{R}\left(\frac{1}{T_2} - \frac{1}{T_1}\right) = -\frac{E_a}{R}\left(\frac{T_1 - T_2}{T_1 T_2}\right)$$

Using the property of logarithms that $\ln k_2 - \ln k_1 = \ln (k_2/k_1)$, we obtain

$$\ln\left(\frac{k_2}{k_1}\right) = \frac{E_a}{R}\left(\frac{T_2 - T_1}{T_1 T_2}\right) \tag{16-27}$$

We can use this equation to calculate an activation energy if the rate constant is known at two different temperatures. For example, consider the reaction whose equation is

$$NO_2(g) \rightarrow NO(g) + \tfrac{1}{2}O_2(g)$$

Let's determine the activation energy, given that $k = 0.71\ M^{-1}\cdot s^{-1}$ at 604 K and $1.81\ M^{-1}\cdot s^{-1}$ at 627 K. We let $k_1 = 0.71\ M^{-1}\cdot s^{-1}$, $T_1 =$ 604 K, $k_2 = 1.81\ M^{-1}\cdot s^{-1}$, and $T_2 = 627$ K. If we substitute these values into Equation (16-27), then we have

$$\ln\left(\frac{1.81\ M^{-1}\cdot s^{-1}}{0.71\ M^{-1}\cdot s^{-1}}\right) = 0.936$$

$$= \left(\frac{E_a}{8.314\ J\cdot K^{-1}\cdot mol^{-1}}\right)\left(\frac{627\ K - 604\ K}{(627\ K)(604\ K)}\right)$$

$$= \frac{E_a}{1.37 \times 10^5\ J\cdot mol^{-1}}$$

or, solving for E_a,

$$E_a = (1.37 \times 10^5\ J\cdot mol^{-1})(0.936) = 1.28 \times 10^5\ J\cdot mol^{-1}$$
$$= 128\ kJ\cdot mol^{-1}$$

The Arrhenius equation also can be used to calculate the value of a rate constant at one temperature if its value is known at some other temperature and if the value of E_a is known.

EXAMPLE 16-15: The activation energy for the reaction

$$2NO_2(g) + F_2(g) \rightarrow 2NO_2F(g)$$

is $E_a = 43.5\ kJ\cdot mol^{-1}$. Estimate the increase in the rate constant of the reaction for an increase in temperature from 300 K to 310 K.

Solution: We use Equation (16-27):

$$\ln\left(\frac{k_2}{k_1}\right) = \frac{E_a}{R}\left(\frac{T_2 - T_1}{T_1 T_2}\right)$$

Inserting the quantities $E_a = 43.5\ kJ\cdot mol^{-1}$, $T_1 = 300$ K, and $T_2 = 310$ K into this equation yields

$$\ln\left(\frac{k_2}{k_1}\right) = \left(\frac{43.5 \times 10^3\ J\cdot mol^{-1}}{8.314\ J\cdot K^{-1}\cdot mol^{-1}}\right)\left(\frac{310\ K - 300\ K}{(310\ K)(300\ K)}\right)$$

Thus,

$$\frac{k_2}{k_1} = e^{0.563} = 1.8$$

Thus, the reaction rate constant increases by about a factor of 2 for the 10-K temperature increase.

PRACTICE PROBLEM 16-15: The major reaction that occurs when an egg is boiled in water is the denaturation of the egg protein. Given that this is a first-order reaction with $E_a \simeq 42$ kJ, calculate how much time is required to cook an egg at 92°C (the boiling point of water at 8000 ft elevation) to the same extent as an egg cooked for 3.0 minutes at 100°C (the boiling point of pure water at sea level).

Answer: 4.0 min

16-11. A CATALYST IS A SUBSTANCE THAT INCREASES THE REACTION RATE BUT IS NOT CONSUMED IN THE REACTION

The rates of many reactions are increased by catalysts. A **catalyst** is a reaction facilitator that acts by providing a different and faster reaction mechanism than is possible in the absence of the catalyst. For example, the reaction rate law for the reaction

$$\underset{\text{an alkene}}{(H)_2C{=}C(CH_3)_2}(aq) + H_2O(l) \xrightarrow{H^+(aq)} \underset{\text{an alcohol}}{H_3C{-}C(CH_3)(OH){-}CH_3}\ (aq)$$

is

$$\text{rate} = k[\text{alkene}][H^+]$$

The solvated hydrogen ion, $H^+(aq)$, does not appear as a reactant in this reaction, but nevertheless the reaction rate is proportional to $[H^+]$.

The $H^+(aq)$ ion presumably facilitates the reaction by attaching to the carbon atom that is bonded to the two hydrogen atoms. Thus, a plausible mechanism for this reaction is as follows:

$$H^+(aq) + (H)_2C{=}C(CH_3)_2 \xrightarrow{\text{slow}} H_3C{-}C^{\oplus}(CH_3)_2$$

intermediate

The intermediate reacts rapidly with water to form the alcohol:

$$H_3C{-}C^{\oplus}(CH_3)_2 + H_2O \xrightarrow{\text{fast}} H_3C{-}C(CH_3)(OH){-}CH_3 + H^+(aq)$$

Note that $H^+(aq)$ is regenerated in the second step, so it is not consumed by the reaction.

The catalyst $H^+(aq)$ acts by providing a new reaction pathway with a lower activation energy, and thus a larger rate constant. Because the rate of a reaction is proportional to the rate constant, the fact that the rate constant is larger means that the reaction goes faster.

The role of a catalyst is illustrated in Figure 16-13. Note that the catalyzed reaction pathway has a lower activation energy than that of the uncatalyzed reaction. This lower activation energy implies not only that it is easier to go from reactants to products but also that it is easier to go from products to reactants. Therefore, a catalyst increases the rates of both the forward and reverse reactions. Thus, a catalyst does not affect the final amounts of reactants and products. In effect, a catalyst helps to get the job done faster, but the final result is the same.

Consider the reaction of aqueous cerium(IV) ions, $Ce^{4+}(aq)$, with aqueous thallium(I) ions, $Tl^+(aq)$:

$$2Ce^{4+}(aq) + Tl^+(aq) \rightarrow 2Ce^{3+}(aq) + Tl^{3+}(aq)$$

This reactions occurs very slowly and its rate law is

$$\text{rate} = k[Tl^+][Ce^{4+}]^2$$

The low reaction rate is thought to be a consequence of the requirement that the reactive event, that is, the simultaneous transfer of two

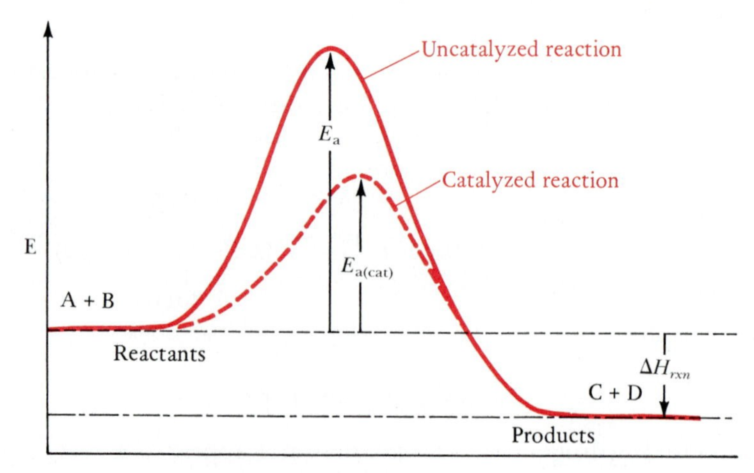

Figure 16-13 A comparison of the activation energies for the uncatalyzed, E_a, and catalyzed, E_a (cat), reaction $A + B \rightarrow C + D$. The catalyst lowers the activation energy barrier to the reaction and thereby increases the reaction rate.

electrons from a $Tl^+(aq)$ ion to two different $Ce^{4+}(aq)$ ions, requires that two $Ce^{4+}(aq)$ ions be present simultaneously near a $Tl^+(aq)$ ion. Three-body encounters are only about 10^{-5} times as likely to occur as two-body encounters; for this reason, the reaction rate is low. The $2Ce^{4+}(aq) + Tl^+(aq)$ reaction is catalyzed by $Mn^{2+}(aq)$. The catalytic action of Mn^{2+} has been attributed to the availability of the Mn^{3+} and Mn^{4+} oxidation states, which provide a new reaction pathway involving a three-step sequence of two-body elementary processes:

$$Ce^{4+}(aq) + Mn^{2+}(aq) \xrightarrow{\text{slow}} Mn^{3+}(aq) + Ce^{3+}(aq) \quad \text{(rate-determining)}$$

$$Ce^{4+}(aq) + Mn^{3+}(aq) \xrightarrow{\text{fast}} Mn^{4+}(aq) + Ce^{3+}(aq)$$

$$Tl^{+}(aq) + Mn^{4+}(aq) \xrightarrow{\text{fast}} Mn^{2+}(aq) + Tl^{3+}(aq) \quad \left(\begin{array}{c}\text{two electrons}\\ \text{transferred}\end{array}\right)$$

The sum of these three equations corresponds to the overall reaction stoichiometry. The rate law for the $Mn^{2+}(aq)$-catalyzed reaction is determined by the slowest step in the mechanism; thus, the rate law is

$$\text{rate} = k_{cat}[Ce^{4+}][Mn^{2+}]$$

where k_{cat} is the rate constant for the catalyzed reaction. Note that the rate law for the catalyzed reaction is different from that for the uncatalyzed reaction. Different mechanisms usually (but not always) give rise to different rate laws.

EXAMPLE 16-16: Suppose that a reaction is catalyzed by two different catalysts and that the activation energies of the two catalyzed reactions are $30.0\ kJ\cdot mol^{-1}$ and $50.0\ kJ\cdot mol^{-1}$ respectively. Use Equation (16-26) to calculate the ratio of the rate constants for the two catalyzed reactions, assuming that A in Equation 16-26 is the same for both.

Solution: We write Equation (16-26) for one of the catalyzed reactions as

$$\begin{aligned}\ln k_1 &= -\frac{E_1}{RT} + \ln A_1 \\ &= -\frac{30.0 \times 10^3\ J\cdot mol^{-1}}{(8.314\ J\cdot K^{-1}\cdot mol^{-1})(300\ K)} + \ln A_1 \\ &= -12.03 + \ln A_1\end{aligned}$$

and for the other catalyzed reaction

$$\begin{aligned}\ln k_2 &= -\frac{50.0 \times 10^3\ J\cdot mol^{-1}}{(8.314\ J\cdot K^{-1}\cdot mol^{-1})(300\ K)} + \ln A_2 \\ &= -20.05 + \ln A_2\end{aligned}$$

If we subtract the second equation from the first, and assume $\ln A_2$ is equal to $\ln A_1$, then we have

$$\ln k_1 - \ln k_2 = \ln\left(\frac{k_1}{k_2}\right) = 8.02$$

or

$$\frac{k_1}{k_2} = e^{8.02} = 3.04 \times 10^3$$

Thus, the rate of the reaction is about 3000 larger with one catalyst rather than the other.

PRACTICE PROBLEM 16-16: Calculate the ratio of the rate constants for a catalyzed reaction and for an uncatalyzed reaction at 25°C if the activation energy is lowered by 5.0 $kJ \cdot mol^{-1}$ by the introduction of a catalyst (assume that the Arrhenius equation A factors are the same).

Answer: 7.5

Figure 16-14 When the catalytic effect of platinum metal on the combustion of hydrogen was discovered in the early 1800s, the process was used to produce cigar lighters, which became very fashionable.

In each of the two catalyzed reactions that we have discussed so far, the catalyst is in the same phase as the reaction mixture. We call such a catalyst a **homogeneous catalyst. Heterogeneous catalysts** are in a different phase than the reaction mixture. Many industrial chemical reactions are catalyzed by metal surfaces, which are heterogeneous catalysts (Figure 16-14). For example, platinum and palladium are used as surface catalysts for a variety of reactions, such as the hydrogenation of double bonds:

$$H_2C{=}CH_2(g) + H_2(g) \xrightarrow{Pt(s)} H{-}\underset{H}{\overset{H}{C}}{-}\underset{H}{\overset{H}{C}}{-}H(g)$$

The first step in this hydrogenation involves the adsorption of hydrogen onto the platinum surface; this step is followed by dissociation of the adsorbed H_2 into adsorbed hydrogen atoms.

$$H_2(\text{surface}) \rightarrow 2H(\text{surface})$$

The adsorbed hydrogen atoms can move around on the platinum surface and eventually react stepwise with adsorbed ethylene, H_2CCH_2, to form ethane, H_3CCH_3:

$$H_2C{=}CH_2(\text{surface}) + H(\text{surface}) \rightarrow H_2\dot{C}{-}CH_3(\text{surface})$$
$$H_2\dot{C}{-}CH_3(\text{surface}) + H(\text{surface}) \rightarrow H_3C{-}CH_3(g)$$

The ethane produced does not interact strongly with the platinum surface and thus leaves the surface immediately after it is formed.

Platinum metal also catalyzes a wide variety of oxygenation reactions, including the oxidation of SO_2 to SO_3 in the production of sulfuric acid (Chapter 5). The first step is the adsorption of O_2 onto the platinum surface. The adsorbed O_2 dissociates into O atoms to form a surface layer of reactive O atoms. The final step involves the rapid reaction of SO_2 with surface O atoms (Figure 16-15)

$$SO_2 + O(\text{surface}) \rightarrow SO_3$$

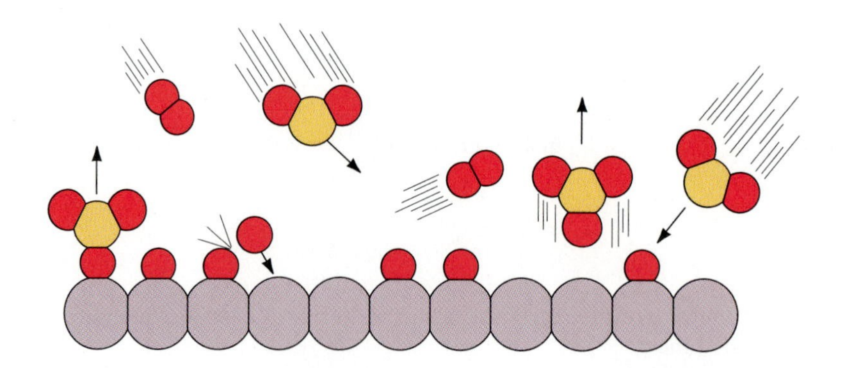

Figure 16-15 Heterogeneous (contact) catalysis by a platinum metal surface. The catalyzed reaction is

$$2SO_2(g) + O_2(g) \xrightarrow{Pt(s)} 2SO_3(g)$$

The platinum surface catalyzes the reaction by causing the dissociation of adsorbed O_2 molecules into O atoms. The surface-bound O atoms then react with SO_2 to give SO_3.

Enzymes are proteins that catalyze chemical reactions in living systems (proteins are discussed in Section 30-4) (Figure 16-16). Without enzymes, most biochemical reactions are too slow to be of any consequence. Cells contain thousands of different enzymes. The absence or an insufficient quantity of an enzyme can lead to serious physiological disorders. Albinism, cystinosis, and phenylketonuria are three of many examples.

A remarkable property of enzymes is their extraordinary catalytic specificity. An enzyme usually catalyzes a single chemical reaction or a set of closely related reactions. The exact manner in which many enzymes function is still a topic of research, but one simple picture of enzyme activity is given by the **lock-and-key theory** (Figure 16-17). In this picture, the enzyme binds the **substrate,** which is the substance that is reacting, to its surface. The substrate molecule is bound in such a way that the substrate is susceptible to chemical attack. The binding site on the enzyme is such that it can bind only one substrate or a closely related one. The particular shape of and the nature and location of atoms at the binding site account for the extraordinary specificity of enzymes.

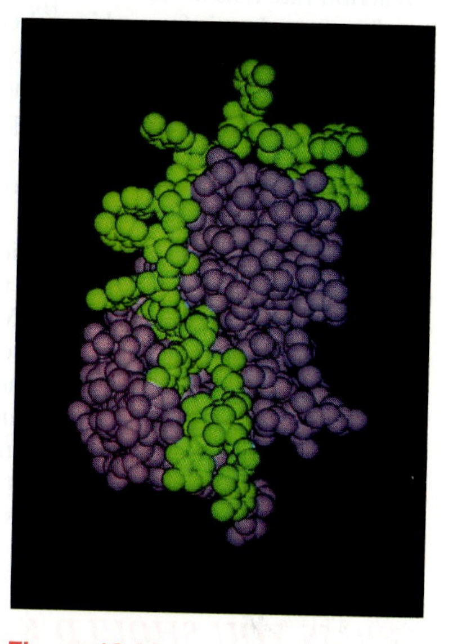

Figure 16-16 A computer-generated molecular model of the enzyme ribonuclease.

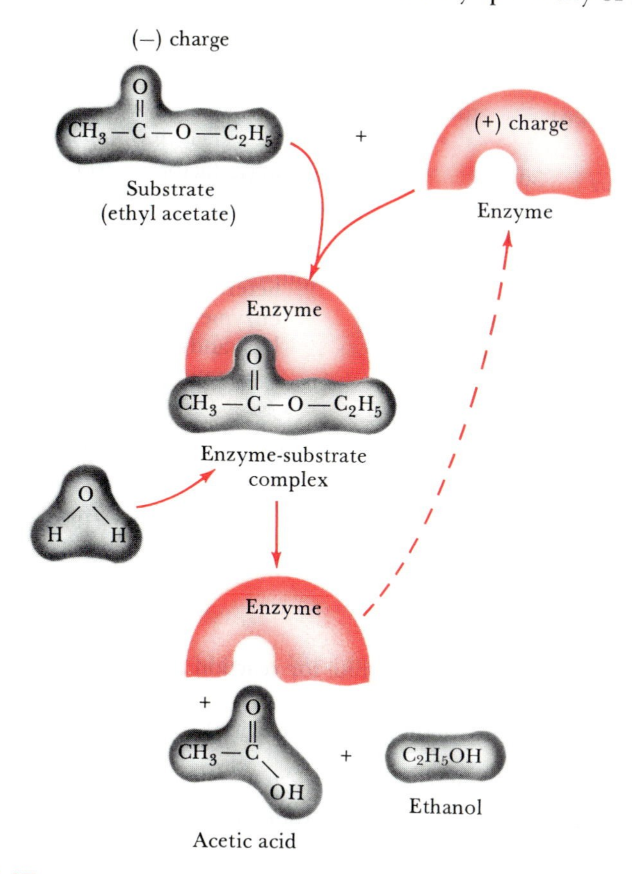

Figure 16-17 The lock-and-key theory of enzyme-catalyzed reactions. The reaction in this case is the hydrolysis of ethyl acetate:

$$CH_3COOC_2H_5 + H_2O \rightarrow C_2H_5OH + HC_2H_3O_2$$

The shape of the binding site of the enzyme and the charges there are perfectly suited to accommodate the substrate, ethyl acetate. Because of the specific way the enzyme binds the ethyl acetate, the appropriate bond is exposed and readily attacked by a water molecule. By holding the substrate in this way, the enzyme speeds up the reaction rate.

Find the average rate of disappearance of $H^+(aq)$ for the time interval between each measurement. What is the average rate of disappearance of CH_3OH for the same time intervals? The average rate of appearance of CH_3Cl?

INITIAL RATES

16-11. Sulfuryl chloride decomposes according to the equation

$$SO_2Cl_2(g) \rightarrow SO_2(g) + Cl_2(g)$$

Using the following initial-rate data, determine the order of the reaction with respect to SO_2Cl_2:

$[SO_2Cl_2]_0$/mol·L^{-1}	(rate)$_0$/mol·L^{-1}·s^{-1}
0.10	2.2×10^{-6}
0.20	4.4×10^{-6}
0.30	6.6×10^{-6}
0.40	8.8×10^{-6}

16-12. Nitrosyl bromide decomposes according to the equation

$$2NOBr(g) \rightarrow 2NO(g) + Br_2(g)$$

Using the following initial-rate data, determine the order of the reaction with respect to $NOBr(g)$:

$[NOBr]_0$/mol·L^{-1}	(rate)$_0$/mol·L^{-1}·s^{-1}
0.20	0.80
0.40	3.20
0.60	7.20
0.80	12.80

16-13. The reaction

$$C_2H_5Cl(g) \rightarrow C_2H_4(g) + HCl(g)$$

was studied at 300 K, and the following data were collected:

Run	Initial concentration, $[C_2H_5Cl]_0$/M	Initial rate of formation of C_2H_4/M·s^{-1}
1	0.33	2.40×10^{-30}
2	0.55	4.00×10^{-30}
3	0.90	6.54×10^{-30}

Determine the rate law and the rate constant for the reaction.

16-14. The reaction

$$2C_5H_6(g) \rightarrow C_{10}H_{12}(g)$$

was studied at 373 K, and the following data were collected:

Run	Initial pressure, $P_{C_5H_6}$/torr	Initial rate of formation of $C_{10}H_{12}$/torr·s^{-1}
1	200	5.76
2	314	14.2
3	576	47.8

Determine the rate law in terms of pressure rather than concentration, and calculate the rate constant for the reaction.

16-15. The reaction

$$2NOCl(g) \rightarrow 2NO(g) + Cl_2(g)$$

was studied at 400 K, and the following data were collected:

Run	Initial concentration, $[NOCl]_0$/M	Initial rate of formation of NO/M·s^{-1}
1	0.25	1.75×10^{-6}
2	0.42	4.94×10^{-6}
3	0.65	1.18×10^{-5}

Determine the rate law and the rate constant for the reaction.

16-16. The following data were obtained for the decomposition of N_2O_3:

$$N_2O_3(g) \rightarrow NO(g) + NO_2(g)$$

Initial pressure, $P_{N_2O_3}$/torr	Initial rate of formation of NO_2/torr·s^{-1}
0.91	5.5
1.4	8.4
2.1	13

Determine the rate law for the reaction, expressed in terms of $P_{N_2O_3}$ rather than $[N_2O_3]$. Calculate the rate constant for the reaction.

16-17. Consider the reaction

$$Cr(H_2O)_6^{3+}(aq) + SCN^-(aq) \rightarrow Cr(H_2O)_5SCN^{2+}(aq) + H_2O(l)$$

for which the following initial-rate data were obtained at 25°C:

$[Cr(H_2O)_6^{3+}]_0$/M	$[SCN^-]_0$/M	$(rate)_0$/M·s^{-1}
1.0×10^{-4}	0.10	2.0×10^{-11}
1.0×10^{-3}	0.10	2.0×10^{-10}
1.5×10^{-3}	0.20	6.0×10^{-10}
1.5×10^{-3}	0.50	1.5×10^{-9}

Determine the rate law and the rate constant for the reaction.

16-18. The reaction

$$CoBr(NH_3)_5^{2+}(aq) + OH^-(aq) \rightarrow Co(NH_3)_5OH^{2+}(aq) + Br^-(aq)$$

was studied at 25°C, and the following initial-rate data were collected:

$[CoBr(NH_3)_5^{2+}]_0$/M	$[OH^-]_0$/M	$(rate)_0$/M·s^{-1}
0.030	0.030	1.37×10^{-3}
0.060	0.030	2.74×10^{-3}
0.030	0.090	4.11×10^{-3}
0.090	0.090	1.23×10^{-2}

Determine the rate law, the overall order of the rate law, and the value of the rate constant for the reaction.

16-19. Given the following initial-rate data at 300 K for the reaction

$$2NO_2(g) + O_3(g) \rightarrow N_2O_5(g) + O_2(g)$$

determine the reaction rate law and the value of the rate constant:

$[NO_2]_0$/M	$[O_3]_0$/M	$(rate)_0$/M·s^{-1}
0.65	0.80	2.61×10^4
1.10	0.80	4.40×10^4
1.70	1.55	1.32×10^5

16-20. Given the following initial-rate data for the reaction

$$CH_3COCH_3(aq) + Br_2(aq) \xrightarrow{H^+(aq)} CH_3COCH_2Br(aq) + H^+(aq) + Br^-(aq)$$

determine the reaction rate law and the value of the rate constant:

$[CH_3COCH_3]_0$/M	$[Br_2]_0$/M	$[H^+]_0$/M	$(rate)_0$/M·s^{-1}
1.00	1.00	1.00	4.0×10^{-3}
1.74	1.00	1.00	7.0×10^{-3}
1.74	1.40	1.00	9.8×10^{-3}
1.00	1.40	2.00	11.3×10^{-3}

FIRST-ORDER REACTIONS

16-21. The reaction

$$SO_2Cl_2(g) \rightarrow SO_2(g) + Cl_2(g)$$

is first order with a rate constant of 2.2×10^{-5} s^{-1} at 320°C. What fraction of a sample of SO_2Cl_2 will remain if it is heated for 5.0 h at 320°C?

16-22. The rate constant for the first-order reaction

cyclopropane (g) $\rightarrow CH_3CH{=}CH_2(g)$

cyclopropane   propene

at 500°C is 5.5×10^{-4} s^{-1}. Calculate the half-life of cyclopropane at 500°C. Given an initial cyclopropane concentration of 1.00×10^{-3} M at 500°C, calculate the concentration of cyclopropane that remains after 2.0 h.

16-23. Azomethane, $CH_3N_2CH_3$, decomposes according to the equation

$$CH_3N_2CH_3(g) \rightarrow CH_3CH_3(g) + N_2(g)$$

Given that the decomposition is a first-order process with $k = 4.0 \times 10^{-4}$ s^{-1} at 300°C, calculate the fraction of azomethane that remains after 1.0 h.

16-24. Methyl iodide, CH_3I, decomposes according to the equation

$$2CH_3I(g) \rightarrow C_2H_6(g) + I_2(g)$$

Given that the decomposition is a first-order process with $k = 1.5 \times 10^{-4}$ s^{-1} at 300°C, calculate the fraction of methyl iodide that remains after one millisecond.

16-25. Peroxydisulfate ion, $S_2O_8^{2-}$, decomposes in aqueous solution according to the equation

$$S_2O_8^{2-}(aq) + H_2O(l) \rightarrow 2SO_4^{2-}(aq) + \tfrac{1}{2}O_2(g) + 2H^+(aq)$$

Given the following data from an experiment with $[S_2O_8^{2-}]_0 = 0.100$ M in a solution with $[H^+]$ fixed at

17 CHEMICAL EQUILIBRIUM

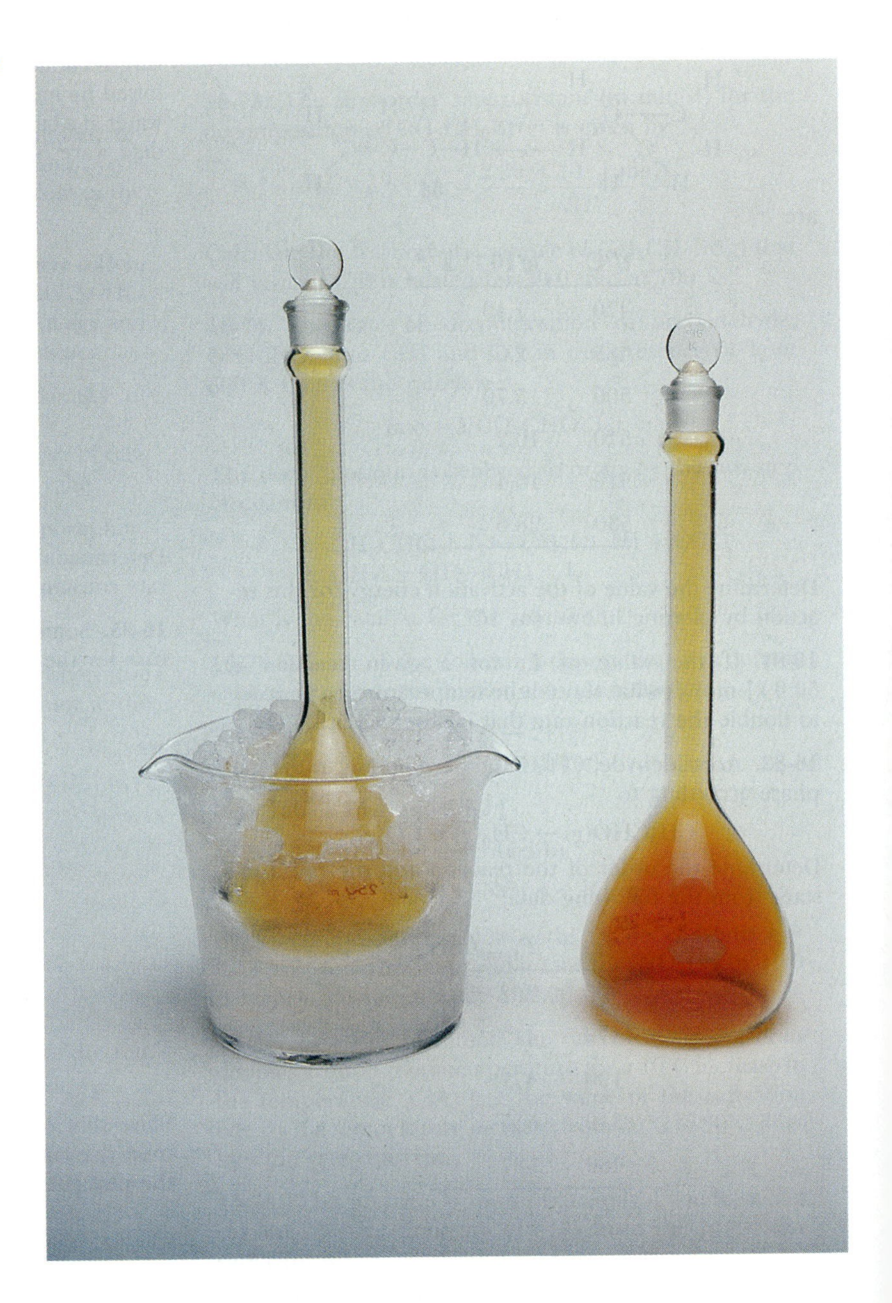

Effect of temperature on the reaction equilibrium

$$\underset{\text{colorless}}{N_2O_4(g)} \rightleftharpoons \underset{\text{brown}}{2NO_2(g)}$$

An increase in temperature from 0°C (ice water) to 25°C converts some of the N_2O_4 to NO_2 and results in a darker color for the reaction mixture.

So far we have tacitly assumed that chemical reactions go in only one direction, forward from reactants to products. Generally, however, a reaction also proceeds in the reverse direction, from products to reactants. As the concentrations of products build up, the rate of the reverse reaction can become significant; and it continues to increase until it equals the rate of the forward reaction. When that happens, the reaction appears to have stopped. No more reactant disappears, and no more product appears. Reactant molecules continue to react to produce product molecules, and product molecules continue to react to produce reactant molecules, but the concentrations of reactants and products no longer change. The system has reached a state of equilibrium.

If we prepare a mixture of reactants and products, then in what direction does the reaction proceed toward equilibrium and what are the final concentrations? If conditions then change, disturbing equilibrium, in which direction does the reaction shift? In this chapter, we show how to answer these questions. We show that a chemical reaction at a given temperature is characterized by a quantity called an equilibrium constant. By studying reaction systems at equilibrium, we determine how to maximize the amount of a desired product.

17-1. A CHEMICAL EQUILIBRIUM IS A DYNAMIC EQUILIBRIUM

Consider the chemical reaction described by the equation

$$\underset{\text{colorless}}{N_2O_4(g)} \rightleftharpoons \underset{\text{brown}}{2NO_2(g)} \qquad (17\text{-}1)$$

in which the colorless gas dinitrogen tetroxide, $N_2O_4(g)$, dissociates into the reddish brown gas nitrogen dioxide, $NO_2(g)$. The reaction described by Equation (17-1), like all chemical reactions, is really two opposing reactions. The **forward reaction** is the dissociation of N_2O_4 molecules into NO_2 molecules, and the **reverse reaction** is the association of NO_2 molecules into N_2O_4 molecules. We have recognized the existence of both a forward reaction and reverse reaction by separating the two sides of Equation (17-1) by *two* arrows pointing in opposite directions.

Suppose that we start with only $N_2O_4(g)$. Initially, the reaction mixture is colorless. As N_2O_4 molecules dissociate into NO_2 molecules, the mixture becomes reddish brown. As the concentration of NO_2 increases, more and more NO_2 molecules collide and associate back into N_2O_4 molecules. Thus, the reverse rate of the process described by Equation (17-1) increases with time. As N_2O_4 molecules dissociate, there are fewer remaining, so the forward rate decreases. Eventually, the forward rate and the reverse rate become equal and a state of **chemical equilibrium** exists (Figure 17-1). At equilibrium, the concentrations of N_2O_4 and NO_2 no longer change with time. A chemical equilibrium is a **dynamic equilibrium** because N_2O_4 molecules are still dissociating into NO_2 molecules and NO_2 molecules are still associating into N_2O_4 molecules. The rates of these two processes are exactly the same, however, so there is no net change in the concentrations of N_2O_4 and NO_2.

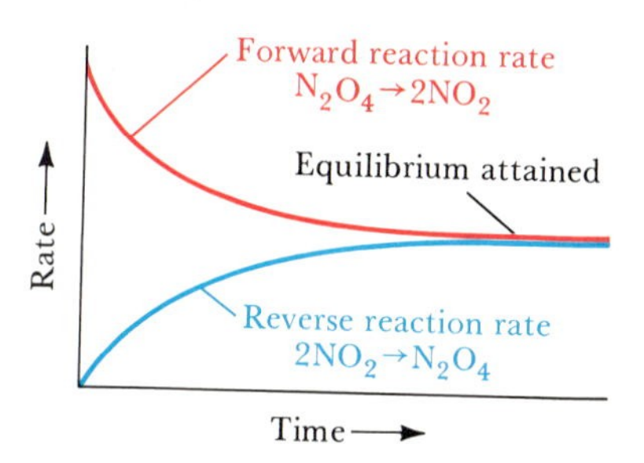

Figure 17-1 Dinitrogen tetroxide, $N_2O_4(g)$, injected into an empty vessel will dissociate according to

$$N_2O_4(g) \rightleftharpoons 2NO_2(g)$$

As the $N_2O_4(g)$ dissociates, its concentration decreases, so the rate of the forward reaction decreases (red curve). As the concentration of $NO_2(g)$ builds up, the rate of the reverse reaction increases (blue line). Eventually the two rates become equal, and a state of equilibrium results. The concentrations of $N_2O_4(g)$ and $NO_2(g)$ no longer change with time.

(c) decrease in P_{H_2} (d) increase in P_{CH_4}
(e) addition of C(s)

17-38. For the chemical equilibrium

$$Ni(s) + 4CO(g) \rightleftharpoons Ni(CO)_4(g) \qquad \Delta H^\circ_{rxn} < 0$$

predict the way in which the equilibrium will shift in response to each of the following changes in conditions (if the equilibrium is unaffected by the change, then write *no change*):

(a) increase in temperature
(b) increase in reaction volume
(c) removal of $Ni(CO)_4(g)$
(d) addition of Ni(s)

17-39. For the chemical equilibrium

$$2SO_2(g) + O_2(g) \rightleftharpoons 2SO_3(g) \qquad \Delta H^\circ_{rxn} = -198 \text{ kJ}$$

predict the direction in which the equilibrium will shift in response to each of the following changes in conditions:

(a) increase in temperature
(b) increase in reaction volume
(c) decrease in $[O_2]$
(d) increase in $[SO_2]$

17-40. For the chemical equilibrium

$$N_2(aq) \rightleftharpoons N_2(g) \qquad \Delta H^\circ_{rxn} > 0$$

in which direction will the equilibrium shift in response to the following changes in conditions?

(a) increase in temperature
(b) increase in volume over the solution
(c) addition of $H_2O(l)$
(d) addition of $N_2(g)$

17-41. Several key reactions in coal gasification are

1. the synthesis gas reaction:

$$C(s) + H_2O(g) \rightleftharpoons CO(g) + H_2(g) \qquad \Delta H^\circ_{rxn} = +131 \text{ kJ}$$

2. the water-gas-shift reaction:

$$CO(g) + H_2O(g) \rightleftharpoons CO_2(g) + H_2(g) \qquad \Delta H^\circ_{rxn} = -41 \text{ kJ}$$

3. the catalytic methanation reaction:

$$CO(g) + 3H_2(g) \rightleftharpoons H_2O(g) + CH_4(g) \quad \Delta H^\circ_{rxn} = -206 \text{ kJ}$$

(a) Write the equilibrium-constant expressions in terms of concentrations, K_c, for each of these equations.
(b) Predict the direction in which each equilibrium shifts in response to (i) an increase in temperature and (ii) a decrease in reaction volume.

17-42. An important modern chemical problem is the liquefaction of coal, because it is still relatively abundant whereas oil is a dwindling resource. The first step is heating the coal with steam to produce synthesis gas:

$$C(s) + H_2O(g) \rightleftharpoons CO(g) + H_2(g) \qquad \Delta H^\circ_{rxn} = 131 \text{ kJ}$$

Carbon monoxide can be hydrogenated to form the important chemical, methyl alcohol:

$$CO(g) + 2H_2(g) \rightleftharpoons CH_3OH(g) \qquad \Delta H^\circ_{rxn} = -128 \text{ kJ}$$

Use Le Châtelier's principle to suggest conditions that maximize the yield of CH_3OH from $CO(g)$ and $H_2(g)$.

QUANTITATIVE APPLICATION OF LE CHÂTELIER'S PRINCIPLE

17-43. At 320 K an equilibrium mixture of $NO_2(g)$ and $N_2O_4(g)$ has partial pressures of 393 torr and 292 torr, respectively. A quantity of $NO_2(g)$ is injected into the mixture, and the total pressure jumps to 812 torr. Calculate the new partial pressures after equilibrium is reestablished. The appropriate chemical equation is

$$N_2O_4(g) \rightleftharpoons 2NO_2(g)$$

17-44. An equilibrium mixture of $PCl_5(g)$, $PCl_3(g)$, and $Cl_2(g)$ has partial pressures of 217 torr, 13.2 torr, and 13.2 torr, respectively. A quantity of $Cl_2(g)$ is injected into the mixture, and the total pressure jumps to 263 torr. Calculate the new partial pressures after equilibrium is reestablished. The appropriate chemical equation is

$$PCl_3(g) + Cl_2(g) \rightleftharpoons PCl_5(g)$$

17-45. Ammonium hydrogen sulfide decomposes according to

$$NH_4HS(s) \rightleftharpoons NH_3(g) + H_2S(g)$$

In a certain experiment, $NH_4HS(s)$ is placed in a sealed 1.00-L container at 25°C and the total equilibrium pressure is observed to be 0.664 atm, with a small amount of $NH_4HS(s)$ remaining. A quantity of $NH_3(g)$ is injected into the container and the total pressure jumps to 0.906 atm. Calculate the total pressure after equilibrium is reestablished.

17-46. Dinitrogen tetroxide decomposes according to

$$N_2O_4(g) \rightleftharpoons 2NO_2(g)$$

In a certain experiment, $N_2O_4(g)$ at an initial pressure of 0.554 atm is introduced into an empty reaction container; after equilibrium is established, the total pressure is 0.770 atm. A quantity of $NO_2(g)$ is injected into the container and the total pressure jumps to 1.124 atm. Calculate the total pressure after equilibrium is reestablished.

17-47. Consider the reaction described by

$$CO_2(g) + H_2(g) \rightleftharpoons CO(g) + H_2O(g)$$

An equilibrium mixture of these gases has the partial pressures $P_{CO} = 512$ torr, $P_{H_2O} = 77$ torr, $P_{H_2} = 192$ torr, and $P_{CO_2} = 384$ torr. If the volume of the reaction container is doubled, then what will the new values of the partial pressures be?

17-48. Ammonium bromide decomposes according to

$$NH_4Br(s) \rightleftharpoons NH_3(g) + HBr(g)$$

Some $NH_4Br(s)$ is introduced into an empty reaction container; after equilibrium is established, the total pressure is 26.4 torr. Calculate the total pressure if the volume of the reaction container is halved.

REACTION QUOTIENT CALCULATIONS

17-49. At 900 K the equilibrium constant for the equation

$$2SO_2(g) + O_2(g) \rightleftharpoons 2SO_3(g)$$

is 13 M^{-1}. If we mix the following concentrations of the three gases, predict in which direction the reaction will proceed toward equilibrium:

Part	$[SO_2]$/M	$[O_2]$/M	$[SO_3]$/M
(a)	0.40	0.20	0.10
(b)	0.05	0.10	0.30

17-50. Suppose that $H_2(g)$ and $CH_4(g)$ are brought into contact with C(s) at 500°C with $P_{H_2} = 0.20$ atm and $P_{CH_4} = 3.0$ atm. Is the reaction described by the equation

$$C(s) + 2H_2(g) \rightleftharpoons CH_4(g) \qquad K_p = 2.69 \times 10^3 \text{ atm}^{-1}$$

at equilibrium under these conditions? If not, in what direction will the reaction proceed to attain equilibrium?

17-51. Suppose we have a mixture of the gases H_2, CO_2, CO, and H_2O at 1260 K, with $P_{H_2} = 0.55$ atm, $P_{CO_2} = 0.20$ atm, $P_{CO} = 1.25$ atm, and $P_{H_2O} = 0.10$ atm. Is the reaction described by the equation

$$H_2(g) + CO_2(g) \rightleftharpoons CO(g) + H_2O(g) \qquad K_p = 1.59$$

at equilibrium under these conditions? If not, in what direction will the reaction proceed to attain equilibrium?

17-52. Suppose $S_2(g)$ and $CS_2(g)$ are brought into contact with solid carbon at 900 K with $P_{S_2} = 1.78$ atm and $P_{CS_2} = 0.794$ atm. Is the reaction described by the equation

$$S_2(g) + C(s) \rightleftharpoons CS_2(g) \qquad K_p = 9.40$$

at equilibrium under these conditions? If not, in what direction will the reaction proceed to attain equilibrium?

17-53. The equilibrium constant for the chemical equation

$$2SO_2(g) + O_2(g) \rightleftharpoons 2SO_3(g)$$

is $K_p = 0.14$ atm^{-1} at 900 K. Suppose the reaction system is prepared at 900 K with the initial pressures $P_{O_2} = 0.50$ atm, $P_{SO_2} = 0.30$ atm, and $P_{SO_3} = 0.20$ atm.

(a) Calculate the value of Q for the reaction with these pressures.

(b) Indicate the direction in which the reaction proceeds toward equilibrium

17-54. Given that $K_p = 2.25 \times 10^4$ atm^{-2} at 25°C for the equation

$$2H_2(g) + CO(g) \rightleftharpoons CH_3OH(g)$$

predict the direction in which a reaction mixture for which $P_{CH_3OH} = 10.0$ atm, $P_{H_2} = 0.010$ atm, and $P_{CO} = 0.0050$ atm proceeds to attain equilibrium.

ADDITIONAL PROBLEMS

17-55. The value of K_p for the chemical equation

$$CuSO_4{\cdot}4NH_3(s) \rightleftharpoons CuSO_4{\cdot}2NH_3(s) + 2NH_3(g)$$

is 6.66×10^{-3} atm^2 at 20°C. Calculate the equilibrium pressure of ammonia at 20°C.

17-56. The equilibrium constant for the chemical equation

$$N_2(g) + 3H_2(g) \rightleftharpoons 2NH_3(g)$$

is $K_p = 0.10$ atm $^{-2}$ at 227°C. Calculate the value of K_c for the reaction at 227°C.

17-57. At 500°C hydrogen iodide decomposes according to

$$2HI(g) \rightleftharpoons H_2(g) + I_2(g)$$

For HI heated to 500°C in a 1.00-L reaction vessel, chemical analysis gave the following concentrations at equilibrium: $[H_2] = 0.42$ M, $[I_2] = 0.42$ M, and $[HI] = 3.52$ M. If an additional mole of HI is introduced into the reaction vessel, what are the equilibrium concentrations after the new equilibrium has been reached?

17-58. The equilibrium constant for the methanol synthesis equation

$$2H_2(g) + CO(g) \rightleftharpoons CH_3OH(g)$$

is $K_p = 2.25 \times 10^4$ atm^{-2} at 25°C.

(a) Calculate the value of P_{CH_3OH} at equilibrium when $P_{H_2} = 0.020$ atm and $P_{CO} = 0.010$ atm.
(b) Given that at equilibrium $P_{total} = 10.0$ atm and $P_{H_2} = 0.020$ atm, calculate P_{CO} and P_{CH_3OH}.

17-59. Given

$$SO_2(g) + NO_2(g) \rightleftharpoons SO_3(g) + NO(g) \qquad \Delta H^\circ_{rxn} = -42 \text{ kJ}$$

Complete the following table:

Change	Effect on equilibrium
(a) decrease in total volume	
(b) increase in temperature	
(c) increase in partial pressure of $NO_2(g)$	
(d) decrease in partial pressure of products	

17-60. According to Table 17-1, $K_c = 0.20$ M at 100°C for the chemical equation

$$N_2O_4(g) \rightleftharpoons 2NO_2(g)$$

Calculate K_p at the same temperature.

17-61. Tin can be prepared by heating SnO_2 ore with hydrogen gas:

$$SnO_2(s) + 2H_2(g) \rightleftharpoons Sn(s) + 2H_2O(g)$$

When the reactants are heated to 500°C in a closed vessel, $[H_2O] = [H_2] = 0.25$ M at equilibrium. If more hydrogen is added so that its initial concentration becomes 0.50 M, what are the concentrations of H_2 and H_2O when equilibrium is restored?

17-62. The equilibrium constant for the chemical equation

$$N_2(g) + 3H_2(g) \rightleftharpoons 2NH_3(g)$$

is $K_p = 0.10\ \text{atm}^{-2}$ at 227°C.

(a) Given that at equilibrium $P_{N_2} = 1.00$ atm and $P_{H_2} =$ 3.00 atm, calculate P_{NH_3} at equilibrium.
(b) Given that at equilibrium the total pressure is 2.00 atm and also that the mole fraction of H_2, X_{H_2}, is 0.20, calculate X_{NH_3}. (Note that $X_{N_2} + X_{H_2} + X_{NH_3} = 1$.)

17-63. The decomposition of ammonium carbamate, NH_2COONH_4, takes place according to

$$NH_2COONH_4(s) \rightleftharpoons 2NH_3(g) + CO_2(g)$$

Show that if all the NH_3 and CO_2 result from the decomposition of ammonium carbamate, then $K_p = (4/27)P^3$, where P is the total pressure at equilibrium.

17-64. The equilibrium constant for the chemical equation

$$SO_2(g) + NO_2(g) \rightleftharpoons SO_3(g) + NO(g)$$

is 3.0. Calculate the number of moles of NO_2 that must be added to 2.4 mol of SO_2 in order to form 1.2 mol of SO_3 at equilibrium.

17-65. Show that, for a reaction involving gaseous products and/or reactants,

$$K_p = K_c(RT)^{\Delta n}$$

where Δn is the number of moles of gaseous products minus the number of moles of gaseous reactants in the chemical equation as written.

17-66. Diatomic chlorine dissociates to chlorine atoms at elevated temperatures. For example, $K_p = 0.570$ atm at 2000°C. Calculate the fraction of chlorine molecules that are dissociated at 2000°C.

17-67. The value of the equilibrium constant for the equation

$$H_2(g) + I_2(g) \rightleftharpoons 2HI(g)$$

is $K = 85$ at 553 K.

(a) Is it possible at 553 K to have an equilibrium reaction mixture for which $P_{HI} = P_{H_2} = P_{I_2}$?
(b) Suppose a 5.0-g sample of HI(g) is heated to 553 K in a 2.00-L vessel. Calculate the composition of the equilibrium reaction mixture.

17-68. The value of the equilibrium constant for the equation

$$H_2(g) + I_2(g) \rightleftharpoons 2HI(g)$$

is $K = 85$ at 280°C. Calculate the value of K at 280°C for

$$HI(g) \rightleftharpoons \tfrac{1}{2}H_2(g) + \tfrac{1}{2}I_2(g)$$

17-69. The equilibrium constant at 823 K for

$$MgCl_2(s) + \tfrac{1}{2}O_2(g) \rightleftharpoons MgO(s) + Cl_2(g)$$

is $K_p = 1.75\ \text{atm}^{1/2}$. Suppose that 50 g of $MgCl_2(s)$ is placed in a reaction vessel with 2.00 L of oxygen at 25°C and 1.00 atm and that the reaction vessel is sealed and heated to 823 K until equilibrium is attained. Calculate P_{Cl_2} and P_{O_2} at equilibrium.

17-70. Consider the reaction equilibrium

$$COCl_2(g) \rightleftharpoons CO(g) + Cl_2(g)$$

If 2.00 mol of $COCl_2(g)$ is introduced into a 10.0-L flask at 1000°C, calculate the equilibrium concentrations of all species at this temperature. At 1000°C, $K_c = 0.329$ M.

17-71. Osmium dioxide occurs either as a black powder or as brown crystals. The density of the black powder form is 7.7 g·mL^{-1} and the density of the brown crystalline form is 11.4 g·mL^{-1}. Which is the more stable form at high pressure?

17-72. Ammonia, $NH_3(g)$, and hydrogen chloride, HCl(g), with ammonia in excess, are injected into a reaction vessel maintained at 300°C. A white powder of ammonium chloride, $NH_4Cl(s)$, is observed to form according to

$$NH_3(g) + HCl(g) \rightleftharpoons NH_4Cl(s)$$

with $K_p = 17.6\ \text{atm}^{-2}$ at 300°C. When the system comes to equilibrium, the total pressure is 2.740 atm. Calculate the partial pressure of each gas at equilibrium.

17-73. For the reaction described by

$$SO_2(g) + \tfrac{1}{2}O_2(g) \rightleftharpoons SO_3(g)$$

$K_p = 1.8 \times 10^{12}\ \text{atm}^{-1/2}$ at 310 K. Calculate the value of K_c for

$$2SO_2(g) + O_2(g) \rightleftharpoons 2SO_3(g)$$

17-74. Phosphorus pentachloride decomposes according to

$$PCl_5(g) \rightleftharpoons PCl_3(g) + Cl_2(g)$$

with $K_p = 1.78$ atm at 250°C. How many moles of $PCl_5(g)$ must be added to a 1.00-L container at 250°C to obtain a concentration of 0.100 M of $PCl_3(g)$?

17-75. An equilibrium mixture of $CO(g)$, $Cl_2(g)$, and $COCl_2(g)$ has partial pressures $P_{CO} = P_{Cl_2} = 1.08$ atm and $P_{COCl_2} = 0.142$ atm. A quantity of $CO(g)$ is suddenly injected into the reaction vessel and the total pressure jumps to 3.27 atm. Calculate the total pressure after equilibrium is reestablished. The relevant chemical equation is

$$CO(g) + Cl_2(g) \rightleftharpoons COCl_2(g)$$

17-76. The value of K associated with

$$H_2(g) + I_2(g) \rightleftharpoons 2HI(g)$$

is $K = 54.4$ at 355°C. What percentage of $I_2(g)$ will be converted to $HI(g)$ if 0.200 mol each of $H_2(g)$ and $I_2(g)$ are mixed and allowed to come to equilibrium at 355°C in a 1.00-L container?

17-77. Referring to Problem 17-76, calculate the percentage of $I_2(g)$ that will be converted to $HI(g)$ at equilibrium at 355°C if 2.00 mol of $I_2(g)$ are mixed with 2.00 mol of $H_2(g)$.

17-78. Antimony pentachloride, $SbCl_5(s)$, decomposes to antimony trichloride, $SbCl_3(s)$, and chlorine, $Cl_2(g)$, according to

$$SbCl_5(s) \rightleftharpoons SbCl_3(s) + Cl_2(g)$$

A 0.50-mol sample of $SbCl_5(s)$ is put into a closed 1.00-L container and heated to 250°C. At equilibrium the mole fraction of $Cl_2(g)$ is found to be 0.428. Calculate the value of K_p.

17-79. Calculate the partial pressures in the equilibrium gas mixture that results when 26.1 torr of $CO_2(g)$ and 26.1 torr of $H_2(g)$ are mixed at 1000°C. The relevant chemical equation is

$$CO_2(g) + H_2(g) \rightleftharpoons CO(g) + H_2O(g) \qquad K = 0.719$$

17-80. Calculate the equilibrium partial pressures in Problem 17-79 if the volume of the reaction container is doubled.

17-81. Ammonium chloride decomposes according to

$$NH_4Cl(s) \rightleftharpoons NH_3(g) + HCl(g)$$

with $K_p = 5.67 \times 10^{-2}$ atm^2 at 300°C. Calculate the partial pressure of each gas and the number of grams of $NH_4Cl(s)$ produced if equal molar quantities of $NH_3(g)$ and $HCl(g)$ at an initial total pressure of 8.76 atm are injected into a 2.00-L container at 300°C.

17-82. Sulfuryl chloride, $SO_2Cl_2(g)$, decomposes to $SO_2(g)$ and $Cl_2(g)$ at 100°C according to

$$SO_2Cl_2(g) \rightleftharpoons SO_2(g) + Cl_2(g)$$

A 6.175-g sample of $SO_2Cl_2(g)$ is placed in an evacuated 1.00-L container at 100°C, and the total pressure at equilibrium is found to be 2.38 atm. Calculate the partial pressures of $SO_2Cl_2(g)$, $SO_2(g)$, and $Cl_2(g)$ and the value of K_p.

17-83. An equilibrium mixture contains 0.20 mol of hydrogen gas, 0.80 mol of carbon dioxide, 0.10 mol of carbon monoxide, and 0.40 mol of water vapor in a 1.00-L container. How many moles of carbon dioxide would have to be added at constant temperature and volume to increase the amount of carbon monoxide to 0.20 mol? The equation for the reaction is

$$CO(g) + H_2O(g) \rightleftharpoons CO_2(g) + H_2(g)$$

17-84. Consider the methanation reaction

$$CO(g) + 3H_2(g) \rightleftharpoons CH_4(g) + H_2O(g)$$

It was found that 0.613 mol of $CO(g)$, 0.387 mol of $H_2(g)$, 0.387 mol of $CH_4(g)$, and 0.387 mol of $H_2O(g)$ were present in an equilibrium mixture in a 1.00-L container. All the water vapor was removed, and the system allowed to come to equilibrium again. Calculate the concentration of all gases in the new equilibrium system. (You must solve the resulting equation by trial and error.)

17-85. The decomposition of ammonium hydrogen sulfide is an endothermic reaction. The equation for the reaction is

$$NH_4HS(s) \rightleftharpoons NH_3(g) + H_2S(g)$$

A 5.2589-g sample of solid ammonium hydrogen sulfide is placed in an evacuated 3.00-L container at 25°C. After equilibrium was established, the total pressure inside the vessel is 0.659 atm. Some solid ammonium hydrogen sulfide remains in the flask.

(a) What is the value of the equilibrium constant, K_p?
(b) What percentage of the solid placed in the flask has reacted?

17-86. Given:

$$2BrCl(g) \rightleftharpoons Cl_2(g) + Br_2(g) \qquad K_p = 0.45$$
$$2IBr(g) \rightleftharpoons Br_2(g) + I_2(g) \qquad K_p = 21.0$$

Determine K_p for the reaction

$$BrCl(g) + \tfrac{1}{2}I_2(g) \rightleftharpoons IBr(g) + \tfrac{1}{2}Cl_2(g)$$

17-87. Acetic acid in the vapor phase is in equilibrium with its dimer

$$2HC_2H_3O_2(g) \rightleftharpoons (HC_2H_3O_2)_2(g)$$
$$K_p = 3.72 \text{ atm}^{-1} \text{ at } 100°C$$

If the total pressure is 1.50 atm, calculate the partial pressure of the dimer.

17-88. Dinitrogen tetraoxide decomposes according to

$$N_2O_4(g) \rightleftharpoons 2NO_2(g)$$

Pure $N_2O_4(g)$ was placed in an empty container at 127°C at a pressure of 0.0438 atm. After the system reached equilibrium, the total pressure was 0.0743 atm. Calculate the value of K_p for this equation.

The Ozone Hole

Stratospheric *ozone*, O_3, is formed by the action of short-wavelength ultraviolet (uv) radiation in sunlight, which dissociates oxygen molecules into oxygen atoms:

$$O_2(g) \xrightarrow{uv} 2O(g)$$

The oxygen atoms produced react with other O_2 molecules to form ozone:

$$O(g) + O_2(g) + M(g) \rightarrow O_3(g) + M(g)$$

where $M(g)$ indicates that some other molecule (e.g., another O_2 or N_2 molecule) or atom must be present to carry away the energy released when the new oxygen–oxygen bond in ozone is formed. As noted in Interchapter C, stratospheric oxygen and ozone protect us from uv light by absorbing over 99 percent of all of the sunlight with wavelengths below 310 nm. Solar uv radiation is dangerous, because it can cause blindness and skin cancers.

The possibility that stratospheric ozone can be depleted by trace gases arising from human sources was first recognized in the early 1960s by H. S. Johnston of the University of California, Berkeley. The main culprits, chlorofluorocarbons, or CFCs, were identified in the middle 1970s by F. S. Rowland and M. Molina at the University of California, Irvine.

The CFCs were developed in response to the pressing need for nontoxic and nonflammable substances to replace sulfur dioxide and ammonia in refrigerators and air conditioners. Some of the common CFCs are listed in Table H-1. The commercial designation of a CFC is read as follows: the units digit is the number of F atoms, the tens digit is the number of H atoms plus 1, and the hundreds digit is the number of carbon atoms minus 1. Thus CFC-113 has 3 fluorine atoms, 0 hydrogen atoms, and 2 carbon atoms. The commercial CFCs are ideally suited to the task for which they were designed: they are nonflammable, noncorrosive, nontoxic, and odorless; and their vapor pressures and heats of vaporization make them truly superior refrigerants. The excellent chemical properties of CFCs led to their extensive use as propellants in aerosol spray cans, as cleaning agents for electrical components, and as foaming agents in the manufacture of plastics.

The remarkable chemical inertness of CFCs (at least those without hydrogen atoms) gives them a lifetime of the order of 100 years in the earth's atmosphere. Once in the atmosphere, CFCs move benignly in air currents to heights of 25 to 40 km. At these elevations the CFCs are above the protective ozone layer and the trouble starts. Chlorine atoms are photochemically dissociated from CFCs by short-wavelength uv radiation; for example,

$$CF_2Cl_2 \xrightarrow{uv} CF_2Cl + Cl$$

The resulting chlorine atoms decompose ozone via the reactions

$$Cl + O_3 \rightarrow ClO + O_2 \qquad \text{(H-1)}$$

$$ClO + O \rightarrow Cl + O_2 \qquad \text{(H-2)}$$

Note that reaction (2) regenerates the Cl atom consumed in reaction (1), which can then return to reaction (1) to destroy still another ozone molecule. The atmospheric lifetime of the chlorine atoms is about 1–2 years, during which time, on average,

Table H-1 Some common chlorofluorocarbons

Chemical name	Formula	Commercial name
trichlorofluoromethane	CCl_3F	CFC-11 (Freon-11)
dichlorodifluoromethane	CCl_2F_2	CFC-12 (Freon-12)
chlorodifluoromethane	$CHClF_2$	CFC-22 (Freon-22)
1,1,2-trichloro-1,2,2-trifluoroethane	$Cl_2FCCClF_2$	CFC-113 (Freon-113)

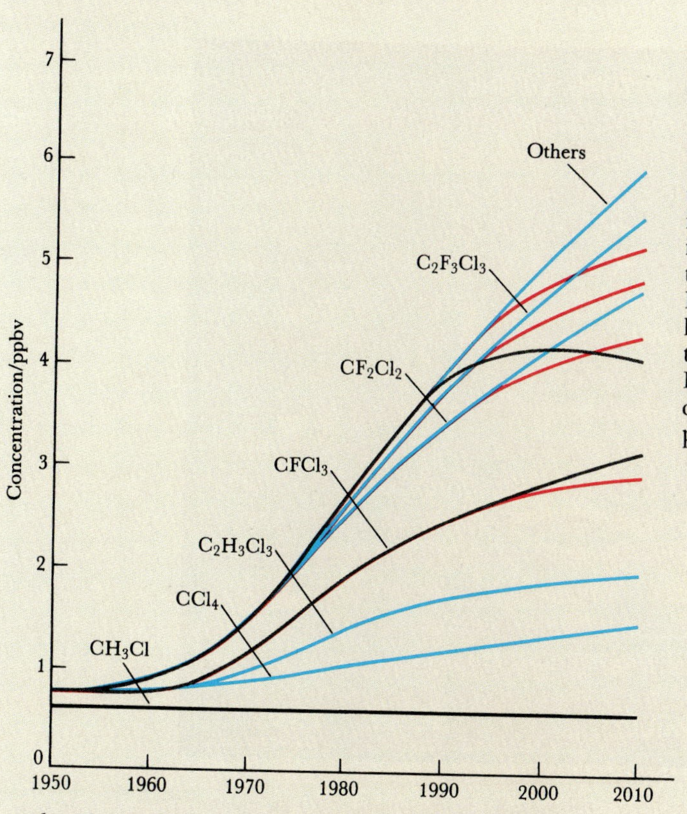

Figure H-1 The cumulative concentration of organochlorine in the atmosphere has grown substantially since the 1960s as a result of the introduction of chlorofluorocarbons for a wide variety of commercial uses. The flat curve (*black*) of the only organochlorine compound with an important natural source, methyl chloride (CH_3Cl), indicates the probable level of such substances in the atmosphere at the beginning of the century. Future levels include those projected from current levels of use (*blue*), those that will result if the terms of the United Nations Environment Programme protocol of 1987 are met (*red*), and the cumulative level that would result from a complete phase-out over the next decade (*gray*).

each Cl atom produced destroys about 100,000 ozone molecules. The annual emission of CFCs into the atmosphere at present is about 1 million tons, which corresponds to an ozone loss roughly 10^5 times as great. The buildup of CFCs in the atmosphere is shown in Figure H-1.

The depletion of O_3 over the South Pole during the polar spring is shown in Figures H-2 and H-3,

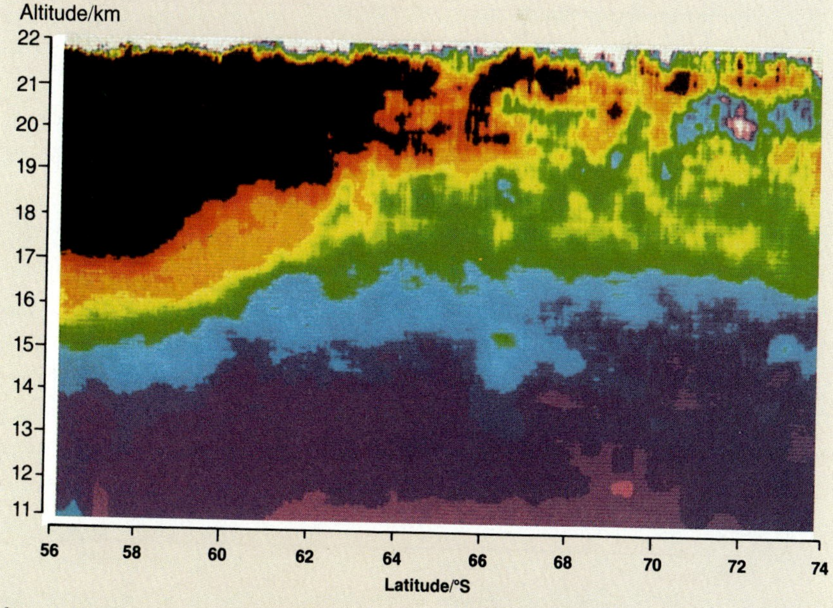

Figure H-2 An airborne lidar (lasar radar) system was used to measure ozone concentrations above Antarctica. The computer plot shows ozone levels on September 26, 1987. The black areas in the upper left corner represent ozone concentrations of more than 2.5 ppm by volume outside the ozone hole. As the plane flew further south, much lower amounts of ozone were recorded between 15 and 22 km altitude. *Orange* represents ozone concentrations of about 2.0 ppm; *yellow,* about 1.5 ppm; *green,* about 1.0 ppm; and *blue* 0.5 ppm. (From *Chemical and Engineering News,* May 30, 1988.)

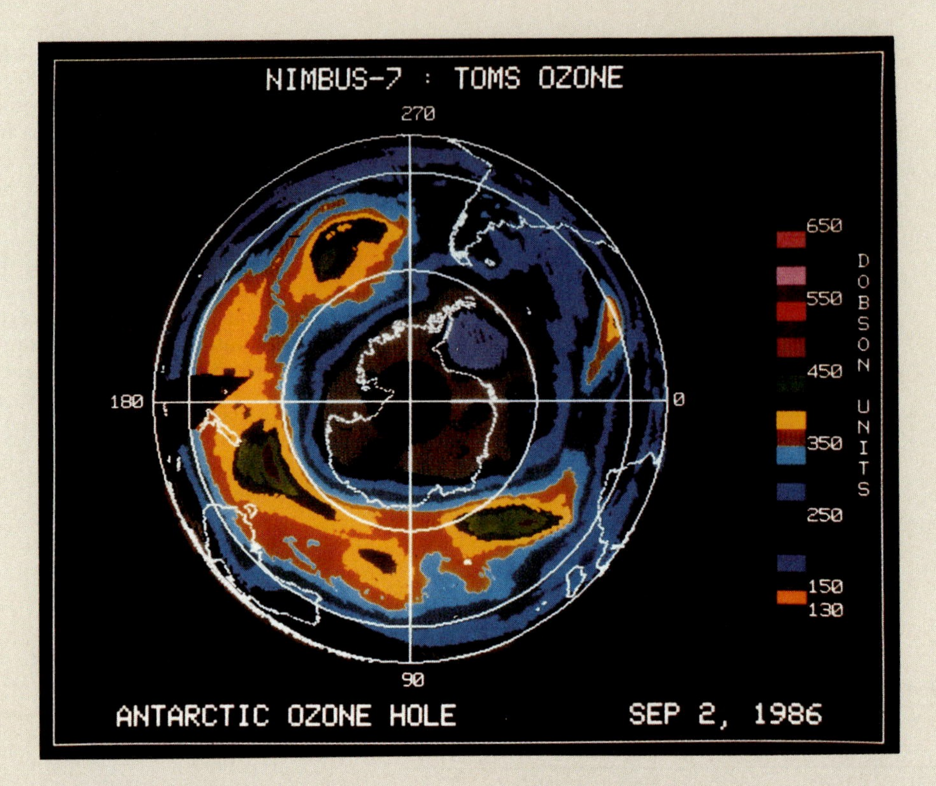

Figure H-3 A map of the ozone over Antarctica obtained from the Nimbus 7 satellite. The black, white, and purple regions surrounding the pole (at the crosshairs) show greatly diminished ozone, now generally acknowledged to be caused by chlorofluorocarbons (CFCs). This "ozone hole" permits dangerous ultraviolet light from the sun to reach the earth.

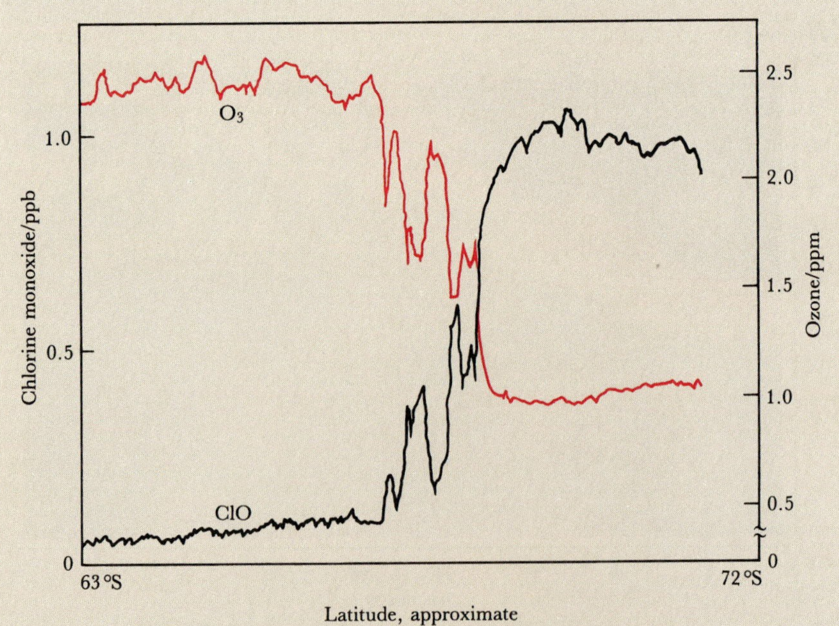

Figure H-4 Instruments aboard NASA's ER-2 research airplane measured concentrations of chlorine monoxide and ozone simultaneously as the plane flew from Punta Arenas, Chile (53°S), to 72°S. The data shown here were collected on September 16, 1987. As the plane entered the ozone hole, concentrations of chlorine monoxide increased to about 500 times normal levels, while those of ozone plummeted.

(Adapted from *Chemical and Engineering News,* May 1988.)

and the strong correlation between O_3 depletion and ClO levels is shown in Figure H-4.

There is little doubt that some CFCs pose a major threat to the ozone layer and that the need to find suitable substitutes is acute. A clue to a possible solution of the problem lies in the observation that hydrogen-containing CFCs, such as CH_3CCl_3 and $CHClF_2$, have relatively short atmospheric lifetimes of 6 to 7 years, which is not long enough for most of them to reach the stratosphere. The short lifetimes of hydrogen-containing CFCs is a result of their ability to react with atmospheric hydroxyl radicals, OH,

$$CHClF_2 + OH \rightarrow H_2O + CClF_2$$

followed by the decomposition of the reactive CFC radicals.

At 1 atm pressure, the atmosphere contains about 10^6 hydroxyl radicals per cubic centimeter. These hydroxyl radicals function as a chemical "detergent" in that they react readily with a wide variety of species, including hydrogen-containing CFCs, and thereby convert them to more benign substances. The replacement of non-hydrogen-containing CFCs, such as CCl_3F and CCl_2F_2, with hydrogen-containing compounds, such as CH_3CCl_3 and $CHClF_2$, could be used to slow the depletion of the ozone layer. Nonetheless, the problem will remain for a long time, because of the long atmospheric lifetimes of non-hydrogen-containing CFCs.

said to be a Brønsted-Lowry acid because it donates a proton to the solvent $H_2O(l)$ to produce a hydronium ion. The H_2O functions as a Brønsted-Lowry base in accepting a proton from $HCl(aq)$.

A reaction involving the transfer of a proton from one molecule to another is called a **proton-transfer reaction** or a **protonation reaction. Acid-base reactions** are proton-transfer reactions.

18-2. IN AN AQUEOUS SOLUTION THE PRODUCT OF THE ION CONCENTRATIONS $[H_3O^+]$ AND $[OH^-]$ IS A CONSTANT

Pure water contains a small number of hydronium ions and hydroxide ions, $H_3O^+(aq)$ and $OH^-(aq)$, respectively, which arise from the equilibrium

$$H_2O(l) + H_2O(l) \rightleftharpoons H_3O^+(aq) + OH^-(aq) \qquad (18\text{-}1)$$

In this reaction water molecules transfer protons to other water molecules. Note that water acts both as an acid (proton donor) and as a base (proton acceptor). The reaction described by Equation (18-1) is called an **autoprotonation reaction.**

The equilibrium-constant expression for Equation (18-1) is

$$K_w = [H_3O^+][OH^-] \qquad (18\text{-}2)$$

where the subscript w refers to water. Note that the concentration of water does not appear in the K_w expression because the value of $[H_2O]$ is effectively constant in aqueous solutions (Section 17-3). The equilibrium constant K_w is called the **ion-product constant of water.** At 25°C the experimental value of K_w is

$$K_w = [H_3O^+][OH^-] = 1.00 \times 10^{-14}\ M^2 \qquad (18\text{-}3)$$

This small value of K_w means that in pure water the concentrations of $H_3O^+(aq)$ and $OH^-(aq)$ are low; that is, the equilibrium represented by Equation (18-1) lies far to the left. From the stoichiometry of Equation (18-1) we note that if we start with pure water, then $H_3O^+(aq)$ and $OH^-(aq)$ are produced on a one-for-one basis. Therefore, *in pure water* we have the equality

$$[H_3O^+] = [OH^-]$$

Using this equation to eliminate $[OH^-]$ in Equation (18-3) yields

$$[H_3O^+]^2 = 1.00 \times 10^{-14}\ M^2$$

Taking the square root of both sides yields

$$[H_3O^+] = 1.00 \times 10^{-7}\ M$$

Because $[H_3O^+] = [OH^-]$, we conclude that

$$[OH^-] = 1.00 \times 10^{-7}\ M$$

Thus, both $[H_3O^+]$ and $[OH^-]$ are equal to 1.00×10^{-7} M in pure water at 25°C. Although $[H_3O^+] = [OH^-]$ for pure water, this expression is not necessarily true when substances are dissolved in water.

A **neutral** aqueous solution is defined as one in which

$$[H_3O^+] = [OH^-] \qquad \text{(neutral solution)}$$

An **acidic** aqueous solution is defined as one in which

$$[H_3O^+] > [OH^-] \qquad \text{(acidic solution)}$$

A **basic** aqueous solution is defined as one in which

$$[OH^-] > [H_3O^+] \qquad \text{(basic solution)}$$

18-3. STRONG ACIDS AND STRONG BASES ARE COMPLETELY DISSOCIATED IN AQUEOUS SOLUTIONS

We noted in Chapter 10 that aqueous solutions of electrolytes conduct an electric current. This electrical conductance is proportional to the number of ions that are available to conduct a current. Conductivity measurements on dilute HCl(*aq*) solutions show that HCl in water is completely dissociated into $H_3O^+(aq)$ and $Cl^-(aq)$. There are essentially no undissociated HCl molecules in aqueous solution. Acids that are completely dissociated are referred to as **strong acids.** The term *strong* refers to the ability of such acids to donate protons to water molecules. Strong acids transfer all their dissociable protons to water molecules.

EXAMPLE 18-1: Calculate $[H_3O^+]$, $[Cl^-]$, and $[OH^-]$ in a 0.15 M aqueous solution of HCl(*aq*).

Solution: Because HCl is a strong acid in water, it is completely dissociated, so $[H_3O^+] = 0.15$ M and $[Cl^-] = 0.15$ M. The corresponding value of $[OH^-]$ in this solution can be computed from the K_w expression:

$$K_w = [H_3O^+][OH^-] = 1.00 \times 10^{-14}\ \text{M}^2$$

$$[OH^-] = \frac{1.00 \times 10^{-14}\ \text{M}^2}{[H_3O^+]} = \frac{1.00 \times 10^{-14}\ \text{M}^2}{0.15\ \text{M}} = 6.7 \times 10^{-14}\ \text{M}$$

Because $[H_3O^+] \gg [OH^-]$, the solution is strongly acidic. Note that we ignored the small contribution to $[H_3O^+]$ arising from the dissociation of water, which is very small compared with 0.15 M.

PRACTICE PROBLEM 18-1: Calculate $[H_3O^+]$, $[NO_3^-]$, and $[OH^-]$ in a 0.600 M aqueous solution of $HNO_3(aq)$.

Answer: $[H_3O^+] = [NO_3^-] = 0.600$ M; $[OH^-] = 1.67 \times 10^{-14}$ M

Conductivity measurements also show that sodium hydroxide in water is completely dissociated; that is, it exists as $Na^+(aq)$ and $OH^-(aq)$:

$$NaOH(s) \xrightarrow[H_2O(l)]{} Na^+(aq) + OH^-(aq)$$

There is essentially no undissociated NaOH present in aqueous solution. Sodium hydroxide is a base because $OH^-(aq)$ is a proton acceptor:

$$H_3O^+(aq) + OH^-(aq) \rightarrow 2H_2O(l)$$

Completely dissociated bases are referred to as **strong bases.**

EXAMPLE 18-2: Calculate $[OH^-]$, $[Na^+]$, and $[H_3O^+]$ in a 0.15 M aqueous solution of NaOH(*aq*).

Solution: Because NaOH is a strong base in water, it is completely dissociated, so $[OH^-] = 0.15$ M and $[Na^+] = 0.15$ M. The value of $[H_3O^+]$ can be computed from the K_w expression:

$$[H_3O^+] = \frac{1.00 \times 10^{-14}\ M^2}{[OH^-]} = \frac{1.00 \times 10^{-14}\ M^2}{0.15\ M} = 6.7 \times 10^{-14}\ M$$

Because $[OH^-] \gg [H_3O^+]$, the solution is strongly basic.

PRACTICE PROBLEM 18-2: Calculate $[Ba^{2+}]$, $[OH^-]$, and $[H_3O^+]$ in a 0.250 M aqueous solution of barium hydroxide.

Answer: $[Ba^{2+}] = 0.250$ M; $[OH^-] = 0.500$ M; $[H_3O^+] = 2.00 \times 10^{-14}$ M

Table 18-1 Strong acids and strong bases in water

Formula	Name
Strong acids	
$HClO_4$	perchloric acid
HNO_3	nitric acid
H_2SO_4	sulfuric acid*
HCl	hydrochloric acid
HBr	hydrobromic acid
HI	hydroiodic acid
Strong bases	
LiOH	lithium hydroxide
NaOH	sodium hydroxide
KOH	potassium hydroxide
RbOH	rubidium hydroxide
CsOH	cesium hydroxide
TlOH	thallium(I) hydroxide
$Ca(OH)_2$	calcium hydroxide
$Sr(OH)_2$	strontium hydroxide
$Ba(OH)_2$	barium hydroxide

*First proton only.

Table 18-2 Hydrogen halide molar bond enthalpies

HX	Molar bond enthalpy/kJ·mol^{-1}
HF	569
HCl	431
HBr	368
HI	297

There are only a few strong acids and bases in water. Most acids and bases, when dissolved in water, are only partly dissociated into their constituent ions. Acids that are incompletely dissociated are called **weak acids,** and bases that are incompletely dissociated are called **weak bases.**

Table 18-1 lists common strong acids and bases. (Note that these are the first two classes of strong electrolytes given on page 356.) You should memorize their formulas, because this information is essential in working problems in acid-base chemistry. Note that three of the six strong acids are halogen acids (HCl, HBr, and HI) and that five of the nine strong bases are alkali metal hydroxides (LiOH, NaOH, KOH, RbOH, and CsOH) and another three of the strong bases are alkaline earth metal hydroxides [$Ca(OH)_2$, $Sr(OH)_2$, and $Ba(OH)_2$]. Unlike the other halogen acids, HF(*aq*) is a weak acid because it has a much stronger bond than the H—X bonds in the other halogen acids (Table 18-2).

18-4. ALMOST ALL ORGANIC ACIDS ARE WEAK ACIDS

The most common organic acids are the **carboxylic acids,** which have the general formula RCOOH, where R is a hydrogen atom or an **alkyl group** such as methyl (CH_3—) and ethyl (CH_3CH_2—). The —COOH group is called the **carboxyl group.**

The two simplest carboxylic acids are formic acid (the major irritant in the bite of ants) and acetic acid (familiar in vinegar, which is a 5 percent aqueous solution of acetic acid):

$$\mathrm{H{-}C({=}\ddot{O}{:}){-}\ddot{\underset{\cdot\cdot}{O}}{-}H} \qquad \mathrm{CH_3{-}C({=}\ddot{O}{:}){-}\ddot{\underset{\cdot\cdot}{O}}{-}H}$$

formic acid, HCOOH   acetic acid, CH_3COOH

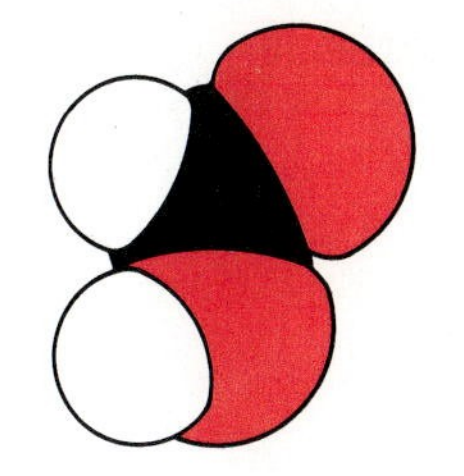

formic acid

The carboxyl group produces hydronium ions in water, for example,

$$\mathrm{CH_3{-}C({=}\ddot{O}{:}){-}\ddot{\underset{\cdot\cdot}{O}}{-}H}(aq) + H_2O(l) \rightleftharpoons H_3O^+(aq) + \mathrm{CH_3{-}C({=}\ddot{O}{:}){-}\ddot{\underset{\cdot\cdot}{O}}{:}^{\ominus}}(aq)$$

acetic acid   acetate ion

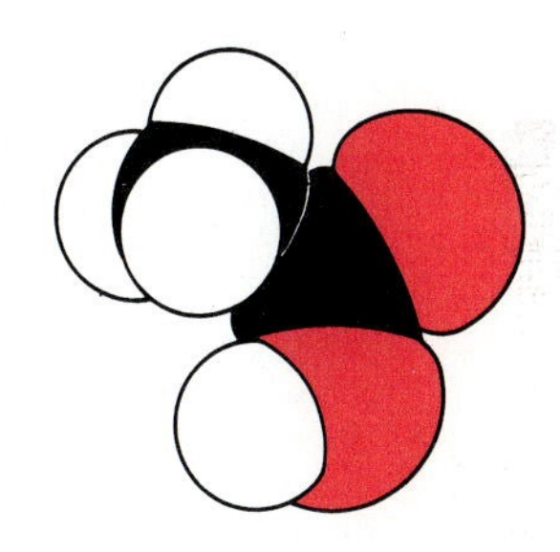

acetic acid

Note that the hydrogen atom in the carboxyl group is the acidic hydrogen atom. Organic acids also react with bases such as sodium hydroxide to produce salts and water:

$$\mathrm{H{-}C({=}\ddot{O}{:}){-}\ddot{\underset{\cdot\cdot}{O}}{-}H}(aq) + NaOH(aq) \longrightarrow Na^+(aq) + \mathrm{H{-}C({=}\ddot{O}{:}){-}\ddot{\underset{\cdot\cdot}{O}}{:}^{\ominus}}(aq) + H_2O(l)$$

formic acid   sodium formate

Up to this point we have written the formula of acetic acid and the acetate ion as $HC_2H_3O_2$ and $C_2H_3O_2^-$, respectively. Now that we know that acetic acid is a carboxylic acid, we will usually write these formulas as CH_3COOH and CH_3COO^-, respectively. Similarly, we will write the formulas of formic acid and its ion as HCOOH and $HCOO^-$, rather than as HCH_2O and CH_2O^-.

EXAMPLE 18-3: Complete and balance the following chemical equation:

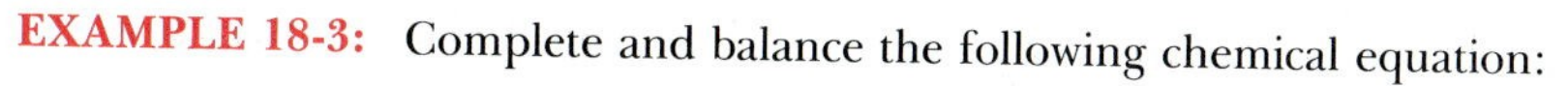

$$CH_3COOH(aq) + Ca(OH)_2(aq) \rightarrow$$

and name the salt produced in the reaction.

Solution: Each acetic acid molecule contributes one hydrogen ion, so 2 mol of acetic acid are required to neutralize completely 1 mol of calcium hydroxide. The balanced equation is

$$2CH_3COOH(aq) + Ca(OH)_2(aq) \rightarrow Ca(CH_3COO)_2(aq) + 2H_2O(l)$$

The product is a salt of calcium hydroxide and acetic acid. To name the anion of the salt, we change the *-ic* ending of the acid to *-ate* and drop the word *acid.* Thus, we have, in this case, calcium acetate.

PRACTICE PROBLEM 18-3: Oxalic acid occurs naturally in many plants, such as spinach and rhubarb. The Lewis formula of oxalic acid is

$$\mathrm{H{-}\ddot{\underset{\cdot\cdot}{O}}{-}C({=}\ddot{O}{:}){-}C({=}\ddot{O}{:}){-}\ddot{\underset{\cdot\cdot}{O}}{-}H}$$

oxalic acid

Because an oxalic acid molecule has two carboxylic groups, it is called a dicarboxylic acid. Complete and balance the equation

$$HOOCCOOH(aq) + KOH(aq) \rightarrow$$

Answer: $HOOCCOOH(aq) + 2KOH(aq) \rightarrow K(OOCCOO)_2(aq) + 2H_2O(l)$

The anion formed by a carboxylic acid is called a **carboxylate ion** and has the general formula $RCOO^-$. The carboxylate ion may be described by the two resonance formulas

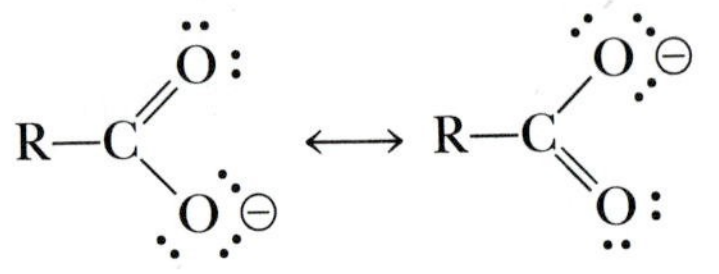

R—C(=Ö:)(—Ö:⊖) ⟷ R—C(—Ö:⊖)(=Ö:)

whose resonance hybrid is

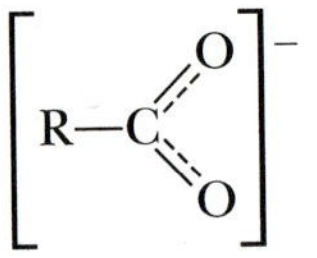

$[R—C(O)(O)]^-$

The Lewis formula of the hybrid shows that the negative charge is distributed equally between the two oxygen atoms. The delocalization of the negative charge over the two oxygen atoms confers an added degree of stability to a carboxylate ion. The two carbon-oxygen bonds in formic acid have different lengths, but in the formate ion the two carbon-oxygen bond lengths are identical and are intermediate between those of single and double carbon-oxygen bonds:

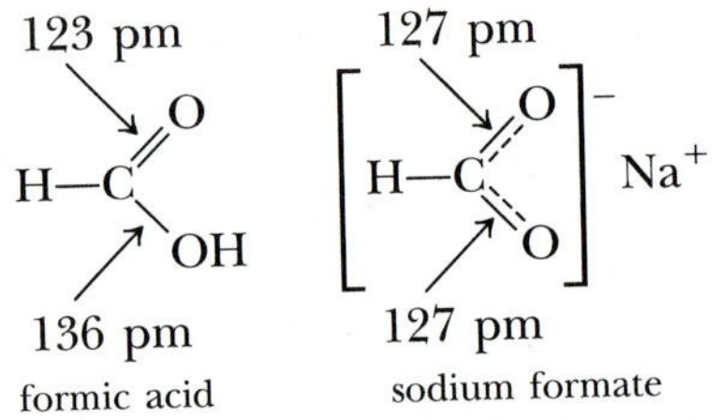

123 pm → H—C(=O)—OH ← 136 pm
formic acid

127 pm → $[H—C(O)(O)]^-$ Na^+ ← 127 pm
sodium formate

18-5. *pH* IS A MEASURE OF THE ACIDITY OF AN AQUEOUS SOLUTION

You will find throughout your study of chemistry that the rates of many chemical reactions depend on the concentration of $H_3O^+(aq)$ in the reaction mixture. This effect may be so even if $H_3O^+(aq)$ is not one of the reactants or products. Thus, the addition of a small amount of $H_3O^+(aq)$ can dramatically alter the rates of many reactions (Section 16-10).

As we shall see, concentrations of $H_3O^+(aq)$ often lie in the range from 1 M to 10^{-14} M. Such a wide range of concentrations makes it awkward to plot these values on ordinary graphs. Because of this, it is convenient to use a logarithmic scale and to define a quantity called **pH** as

$$pH = -\log [H_3O^+] \qquad (18\text{-}4)$$

Note that here we use a common logarithm, that is, a logarithm to the base 10. (The properties of logarithms are reviewed in Appendix A.) Equation (18-4) enables us to calculate the pH corresponding to various values of $[H_3O^+]$.

EXAMPLE 18-4: Calculate the pH of a solution that has a $H_3O^+(aq)$ concentration of 5.0×10^{-10} M.

Solution: If you have a hand calculator, enter 5.0×10^{-10}. Now press the "log" key and then the "change-sign" (or "+/−") key to get

$$\text{pH} = -\log (5.0 \times 10^{-10}) = 9.30$$

If you use a table of logarithms instead of a calculator, you need the relation

$$\log ab = \log a + \log b$$

Then, because $\log 5.0 = 0.70$,

$$\begin{aligned}\log (5.0 \times 10^{-10}) &= \log 5.0 + \log 10^{-10} \\ &= 0.70 - 10 = -9.30\end{aligned}$$

or pH = 9.30.

PRACTICE PROBLEM 18-4: Calculate the pH of a solution that has an $H_3O^+(aq)$ concentration of 4.82×10^{-2} M.

Answer: 1.32

EXAMPLE 18-5: Calculate the pH of an aqueous solution at 25°C that has been prepared by dissolving 0.26 g of calcium hydroxide in water and diluting the solution to a final volume of 0.500 L.

Solution: We first calculate the number of moles of $Ca(OH)_2$. The formula mass of $Ca(OH)_2$ is 74.1, so the number of moles is

$$(0.26 \text{ g Ca(OH)}_2)\left(\frac{1 \text{ mol Ca(OH)}_2}{74.1 \text{ g Ca(OH)}_2}\right) = 3.5 \times 10^{-3} \text{ mol}$$

The molarity of the solution is the number of moles per liter of solution:

$$\text{molarity} = \frac{\text{moles of solute}}{\text{liters of solution}} = \frac{3.5 \times 10^{-3} \text{ mol}}{0.500 \text{ L}} = 7.0 \times 10^{-3} \text{ M}$$

Calcium hydroxide is a strong base and yields two $OH^-(aq)$ per mole of $Ca(OH)_2(aq)$. Therefore, the molarity of the $OH^-(aq)$ is

$$[OH^-] = (2)(7.0 \times 10^{-3} \text{ M}) = 1.4 \times 10^{-2} \text{ M}$$

The value of $[H_3O^+]$ is calculated by using the ion-product constant of water:

$$[H_3O^+] = \frac{1.00 \times 10^{-14} \text{ M}^2}{[OH^-]} = \frac{1.00 \times 10^{-14} \text{ M}^2}{1.4 \times 10^{-2} \text{ M}} = 7.1 \times 10^{-13} \text{ M}$$

The pH of the solution is

$$\text{pH} = -\log[\text{H}_3\text{O}^+] = -\log(7.1 \times 10^{-13}) = 12.15$$

PRACTICE PROBLEM 18-5: Calculate the pH of a solution that results when 20.0 mL of 6.00 M NaOH(*aq*) is diluted with water to a final volume of 75.0 mL.

Answer: 14.20

We can solve problems like Example 18-5 a little more quickly if we first introduce a new quantity called pOH. We take the logarithm of Equation (18-3) to obtain

$$\log([\text{H}_3\text{O}^+][\text{OH}^-]) = \log(1.00 \times 10^{-14}) = -14.00$$

We use the fact that

$$\log([\text{H}_3\text{O}^+][\text{OH}^-]) = \log[\text{H}_3\text{O}^+] + \log[\text{OH}^-]$$

and then multiply by -1 to obtain

$$-\log[\text{H}_3\text{O}^+] - \log[\text{OH}^-] = 14.00$$

If we define a new quantity called **pOH** as

$$\text{pOH} = -\log[\text{OH}^-] \qquad (18\text{-}5)$$

then we can say that, at 25°C,

$$\text{pH} + \text{pOH} = 14.00 \qquad (18\text{-}6)$$

Equations (18-5) and (18-6) are often useful in solving problems involving basic solutions. In Example 18-5 above we found that $[\text{OH}^-] = 1.4 \times 10^{-2}$ M. We substitute this value in Equation (18-5) to get pOH = 1.85, so the pH is 14.00 − 1.85 = 12.15.

At 25°C, pure water has a hydronium ion concentration of $[\text{H}_3\text{O}^+] = 1.00 \times 10^{-7}$ M; therefore, the pH of a neutral aqueous solution at 25°C is

$$\text{pH} = -\log[\text{H}_3\text{O}^+] = -\log[1.00 \times 10^{-7}] = 7.00$$

We note that at 25°C acidic solutions have $[\text{H}_3\text{O}^+]$ values greater than 1.00×10^{-7} M. Thus, acidic solutions have pH values less than 7.00. Basic solutions have a $[\text{H}_3\text{O}^+]$ less than 1.00×10^{-7} M and pH values greater than 7.00. The pH scale is shown schematically as

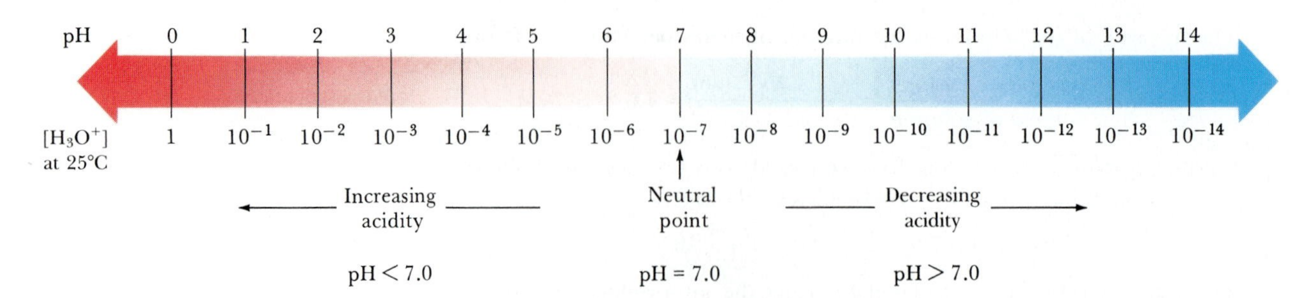

Note that a change in pH of one unit corresponds to a 10-fold change in $[\text{H}_3\text{O}^+]$. The pH values of various aqueous solutions are given in Figure 18-2.

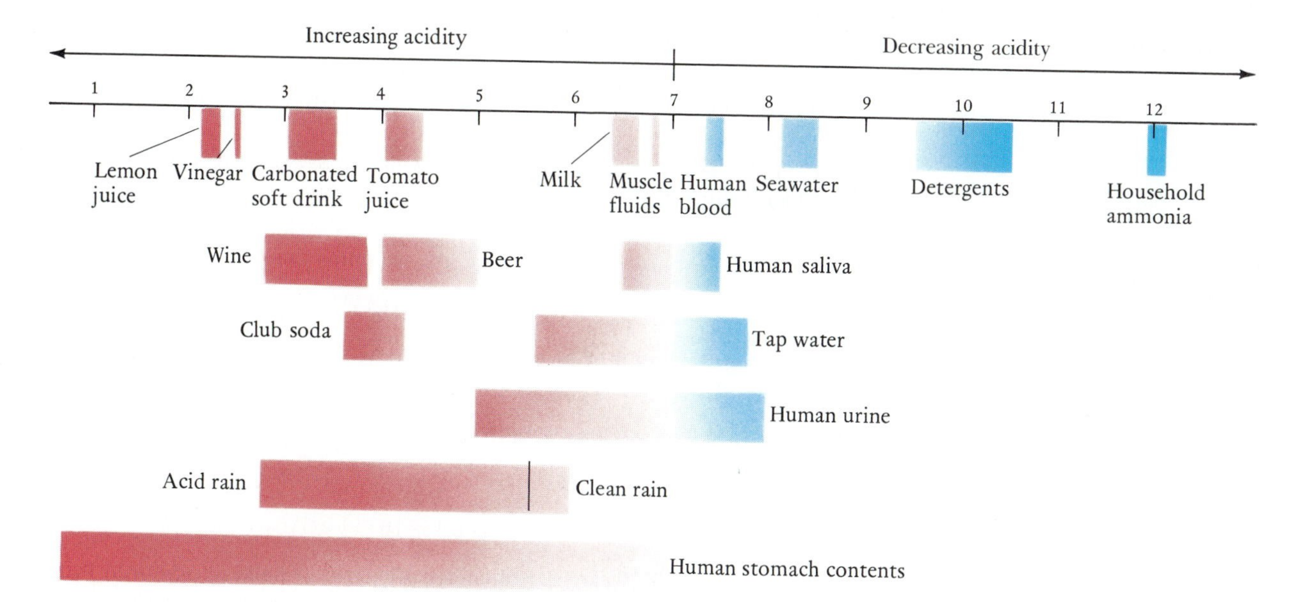

Figure 18-2 The range of pH values for common aqueous solutions.

EXAMPLE 18-6: Compare the pHs of two solutions for which $[H_3O^+] = 5.46 \times 10^{-5}$ M and $[H_3O^+] = 5.46 \times 10^{-3}$ M.

Solution: The two solutions have pH values of $pH = -\log(5.46 \times 10^{-5}) = 4.26$ and $pH = -\log(5.46 \times 10^{-3}) = 2.26$. Notice that these differ by 2 pH units (4.26 − 2.26), whereas the hydronium ion concentrations differ by a factor of 100. The pH scale is a logarithmic scale, just like the Richter scale of earthquake intensities. One unit on each scale corresponds to a factor of 10.

PRACTICE PROBLEM 18-6: Calculate the difference in pH values of two solutions for which the $OH^-(aq)$ concentrations differ by a factor of 10,000.

Answer: 4 pH units

The pH of a solution is conveniently measured in the laboratory with a **pH meter,** an electronic device that responds to the $[H_3O^+]$ of a solution (Figure 18-3). The meter scale or digital readout of the device is set up to display pH directly.

Figure 18-3 A pH meter. The electrodes are placed in the solution to measure the pH, which is displayed in digital form on the meter.

Examples 18-4 and 18-5 illustrate how to calculate pH from $[H_3O^+]$. It is often necessary to do the inverse calculation, that is, to calculate $[H_3O^+]$ from the pH, as shown in the following Example.

EXAMPLE 18-7: The pH of milk is about 6.5. Calculate the value of $[H_3O^+]$ for milk.

Solution: From the definition of pH, we know that

$$pH = -\log [H_3O^+]$$

From the definition of logarithms, we know that if $y = \log x$, then $x = 10^y$. Thus,

$$[H_3O^+] = 10^{-pH}$$

The pH of milk is 6.5, so

$$[H_3O^+] = 10^{-6.5}$$

The quantity $10^{-6.5}$ can be evaluated easily on your hand calculator by using the inverse logarithm operation (the "10^x" key on some calculators):

$$[H_3O^+] = 10^{-6.5} = 3 \times 10^{-7}\ M$$

If you are using a table of logarithms instead of a hand calculator, then first you must write $10^{-6.5}$ as

$$10^{-6.5} = 10^{0.5} \times 10^{-7}$$

The antilogarithm of 0.5 is 3, so once again we find that

$$[H_3O^+] = 3 \times 10^{-7}\ M$$

PRACTICE PROBLEM 18-7: The pH of a household ammonia solution is about 12. Calculate the concentration of $OH^-(aq)$ in household ammonia.

Answer: 1×10^{-2} M

18-6. WEAK ACIDS AND WEAK BASES ARE DISSOCIATED ONLY PARTIALLY IN WATER

If 0.10 mol of hydrogen chloride gas is dissolved in enough water to make 1.00 L of aqueous solution, then the observed pH of the resulting solution is 1.00. However, if 0.10 mol of hydrogen fluoride gas is dissolved in enough water to make 1.00 L of aqueous solution, then the observed pH of the resulting solution is 2.10. In each case we calculate the value of $[H_3O^+]$ from the pH by using the relation (see Example 18-7)

$$[H_3O^+] = 10^{-pH} \tag{18-7}$$

The values that we obtain using Equation (18-7) are

0.10 M HCl(*aq*)	0.10 M HF(*aq*)
pH = 1.00	pH = 2.10
$[H_3O^+] = 10^{-1.00} = 0.10$ M	$[H_3O^+] = 10^{-2.10} = 7.9 \times 10^{-3}$ M

Comparison of these two values of $[H_3O^+]$ shows that, unlike hydrochloric acid, hydrofluoric acid is only partially dissociated in water. The **percentage of dissociation** (% diss) of the hydrofluoric acid in the 0.10 M solution is only

$$\left(\frac{[H_3O^+]}{[HF]_0}\right) \times 100 = \left(\frac{0.0079\ M}{0.10\ M}\right) \times 100 = 7.9\%$$

where the subscript 0 denotes the stoichiometric concentration. In contrast, hydrochloric acid is completely dissociated:

$$\left(\frac{[H_3O^+]}{[HCl]_0}\right) \times 100 = \left(\frac{0.10\ M}{0.10\ M}\right) \times 100 = 100\%$$

By **stoichiometric concentration,** we mean the concentration at which a solution is prepared. For example, a $HF(aq)$ solution of a stoichiometric concentration of 0.100 M means that the solution was prepared by adding 0.100 moles of HF to enough water to make exactly one liter of solution. The label on a bottle of such a solution would read "0.100 M hydrofluoric acid" or "0.100 M HF." Because hydrofluoric acid only partially dissociates in water, however, the actual concentration of undissociated HF molecules at equilibrium is somewhat less than 0.100 M.

At any given stoichiometric concentration, a weak electrolyte such as $HF(aq)$ will be a poorer conductor than a strong electrolyte such as $HCl(aq)$, because fewer ions are available in $HF(aq)$ to conduct a current. The ratio of the electrical conductivity of a 1.0 M $HF(aq)$ solution to that of a 1.0 M $HCl(aq)$ solution is about 0.08, in accord with their relative amounts of dissociation. Figure 18-4 illustrates the difference between the reactions of a strong acid and a weak acid on magnesium.

Let's consider a weak base in aqueous solution. Ammonia is a base in aqueous solution because it accepts a proton from water according to

$$NH_3(aq) + H_2O(l) \rightleftharpoons NH_4^+(aq) + OH^-(aq)$$

Ammonia is said to be a weak base because the position of the above equilibrium does not lie far to the right; in other words, not many of the ammonia molecules are protonated.

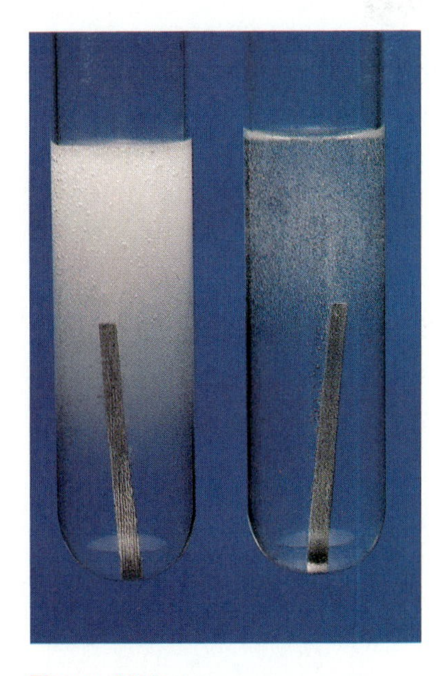

Figure 18-4 The reactions of a strong acid and of a weak acid with magnesium. Note that the strong acid produces a much more vigorous reaction.

EXAMPLE 18-8: The pH of a 0.20 M $NH_3(aq)$ solution is 11.27. Calculate the percentage of ammonia molecules that are protonated in this solution.

Solution: The equation for the reaction is

$$NH_3(aq) + H_2O(l) \rightleftharpoons NH_4^+(aq) + OH^-(aq)$$

The percentage of ammonia molecules that are protonated is

$$\%\ \text{protonated} = \left(\frac{[NH_4^+]}{[NH_3]_0}\right) \times 100$$

Because $[NH_4^+] = [OH^-]$, we have

$$\%\ \text{protonated} = \left(\frac{[OH^-]}{[NH_3]_0}\right) \times 100$$

We need to determine $[OH^-]$. The pH is 11.27, so

$$pOH = 14.00 - 11.27 = 2.73$$

The value of $[OH^-]$ can be calculated from the pOH:

$$[OH^-] = 10^{-pOH} = 10^{-2.73} = 1.9 \times 10^{-3}\ M$$

The percentage of ammonia molecules that are protonated in a 0.20 M $NH_3(aq)$ solution is

$$\%\ \text{protonated} = \left(\frac{1.9 \times 10^{-3}\ M}{0.20\ M}\right) \times 100 = 0.95\%$$

Thus, fewer than 1 percent of the ammonia molecules are protonated.

PRACTICE PROBLEM 18-8: The pH of a 0.400 M solution of formic acid, $HCOOH(aq)$, is 2.08 and that of a 0.400 M solution of hydrocyanic acid, $HCN(aq)$, is 4.86. Compare the percentages of acid molecules that are dissociated in the two solutions.

Answer: HCOOH, 2.08%; HCN, 0.00345%

18-7. ACIDS WITH LARGE VALUES OF K_a ARE STRONGER THAN ACIDS WITH SMALLER VALUES OF K_a

The equilibrium-constant expression for an **acid-dissociation reaction** is written according to the law of concentration action. Let's consider the weak acid acetic acid, $CH_3COOH(aq)$, which produces $H_3O^+(aq)$ by the protonation of water

$$CH_3COOH(aq) + H_2O(l) \rightarrow H_3O^+(aq) + CH_3COO^-(aq) \quad (18\text{-}8)$$

The equilibrium-constant expression for this equation is

$$K_a = \frac{[H_3O^+][CH_3COO^-]}{[CH_3COOH]} \quad (18\text{-}9)$$

where the subscript a on K reminds us that K_a is an **acid-dissociation constant.** Note that the $H_2O(l)$ concentration does not appear in the K_a expression. The experimental value of K_a at 25°C for acetic acid is 1.74×10^{-5} M.

How do we calculate the pH of a 0.050 M acetic acid solution? The small value of K_a of acetic acid means that only a small fraction of the acetic acid molecules dissociate. Most of them remain in the undissociated form as $CH_3COOH(aq)$, so we expect the equilibrium concentration of undissociated acetic acid not to differ significantly from the stoichiometric concentration, 0.050 M. We know that all the acetic acid either remains as $CH_3COOH(aq)$ or dissociates to $CH_3COO^-(aq)$, so

$$0.050\ M = [CH_3COOH) + [CH_3COO^-] \quad (18\text{-}10)$$

This equation expresses the fact that all the "CH_3COO" is in the form of either undissociated acetic acid, $CH_3COOH(aq)$, or acetate ion,

$CH_3COO^-(aq)$. A condition such as Equation (18-10), which accounts for all of a certain type of unit, is called a **material balance condition.**

The dissociation of acetic acid is one source of $H_3O^+(aq)$. Another source of $H_3O^+(aq)$ is the reaction represented by Equation (18-1):

$$H_2O(l) + H_2O(l) \rightleftharpoons H_3O^+(aq) + OH^-(aq) \qquad (18\text{-}11)$$

for which

$$[H_3O^+][OH^-] = K_w = 1.00 \times 10^{-14}\ M^2 \qquad (18\text{-}12)$$

at 25°C. It would appear at first sight that we must consider both Equations (18-8) and (18-11), along with their associated equilibrium-constant expressions. However, because K_a is so much larger than K_w, we can neglect the contribution that the autoprotonation reaction of water makes to the overall concentration of $H_3O^+(aq)$. We can use only Equation (18-8) and write

$$[H_3O^+] = [CH_3COO^-] \qquad (18\text{-}13)$$

Next, we set up a concentration table for the dissociation reaction:

Concentration	$CH_3COOH(aq)$	+	$H_2O(l)$	$\rightleftharpoons$	$H_3O^+(aq)$	+	$CH_3COO^-(aq)$
initial	0.050 M		—		≃0		0
equilibrium	0.050 M − $[CH_3COO^-]$		—		$[H_3O^+]$		$[CH_3COO^-]$
equilibrium (substituting $[H_3O^+]$ for $[CH_3COO^-]$)	0.050 M − $[H_3O^+]$		—		$[H_3O^+]$		$[H_3O^+]$

Note that we have set the initial concentration of $H_3O^+(aq)$ equal to 0 because we are neglecting the proton contribution of the reaction described by Equation (18-11). We substitute the values from the concentration table into Equation (18-9) to get

$$1.74 \times 10^{-5}\ M = \frac{[H_3O^+]^2}{0.050\ M - [H_3O^+]} \qquad (18\text{-}14)$$

Equation (18-14) can be written in the standard form of a quadratic equation:

$$[H_3O^+]^2 + (1.74 \times 10^{-5}\ M)[H_3O^+] - 8.70 \times 10^{-7}\ M^2 = 0$$

The two solutions to this equation are

$$[H_3O^+] = 9.24 \times 10^{-4}\ M \text{ and } -1.86 \times 10^{-3}\ M$$

We reject the physically unacceptable negative concentration, so

$$[H_3O^+] = 9.24 \times 10^{-4}\ M$$

The pH of a 0.050 M acetic acid solution is

$$pH = -\log[H_3O^+] = -\log[9.24 \times 10^{-4}] = 3.03$$

The concentration of the other species in the solution can also be obtained. From Equation (18-13), we have

$$[CH_3COO^-] = [H_3O^+] = 9.24 \times 10^{-4}\ M$$

and from Equation (18-10), we have

$$[CH_3COOH] = 0.050\text{ M} - [CH_3COO^-]$$
$$= 0.050\text{ M} - 9.24 \times 10^{-4}\text{ M} = 0.049\text{ M}$$

Note that the equilibrium concentration of undissociated acetic acid is indeed very close to its stoichiometric concentration. We can calculate $[OH^-]$ from Equation (18-12)

$$[OH^-] = \frac{1.00 \times 10^{-14}\text{ M}^2}{[H_3O^+]} = \frac{1.00 \times 10^{-14}\text{ M}^2}{9.24 \times 10^{-4}\text{ M}} = 1.08 \times 10^{-11}\text{ M}$$

The percentage of dissociation of acetic acid in a 0.050 M solution of $CH_3COOH(aq)$ is

$$\%\text{ diss} = \left(\frac{[H_3O^+]}{[CH_3COOH]_0}\right) \times 100 = \left(\frac{9.24 \times 10^{-4}\text{ M}}{0.050\text{ M}}\right) \times 100 = 1.8\%$$

Thus, we see that over 98 percent of the acetic acid in a 0.050 M solution remains undissociated.

Figure 18-5 shows how the percentage of dissociation of acetic acid increases as the acetic acid concentration decreases. This effect can be understood in terms of Le Châtelier's principle (Section 17-7). The addition of water increases the volume available to the reaction. Thus, the reaction equilibrium shifts to the side with the greater number of moles of solute species; that is, the equilibrium shifts in the direction

$$CH_3COOH(aq) + H_2O(l) \rightarrow H_3O^+(aq) + CH_3COO^-(aq)$$

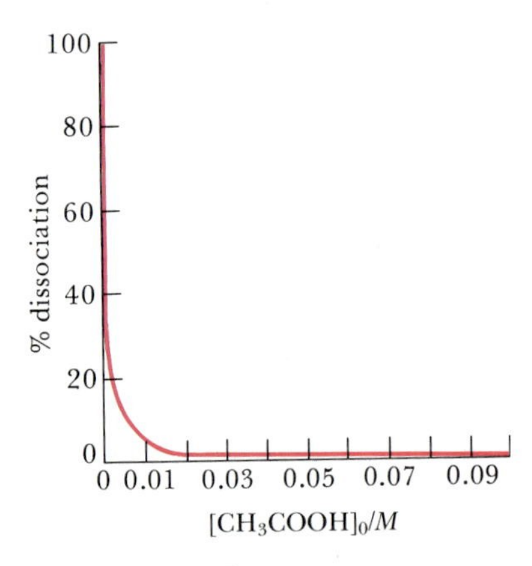

Figure 18-5 The percent dissociation of acetic acid as a function of the stoichiometric (initial) concentration of acid, $[CH_3COOH]_0$. The calculations are carried out by solving the quadratic equation

$$\frac{[H_3O^+]^2}{[CH_3COOH]_0 - [H_3O^+]} = 1.74 \times 10^{-5}\text{ M}$$

for $[H_3O^+]$, for various values of $[CH_3COOH]_0$. The percent dissociation is then computed by using

$$\%\text{ dissociation} = \left(\frac{[H_3O^+]}{[CH_3COOH]_0}\right) \times 100$$

The procedure works down to $[CH_3COOH]_0 = 10^{-5}$ M; below this concentration the contribution to $[H_3O^+]$ from the autoprotonation reaction of water (Equation 18-1) must be taken into account.

EXAMPLE 18-9: Calculate the pH and the concentrations of all the species in a 0.050 M aqueous solution of chloroacetic acid, $ClCH_2COOH(aq)$. The value of K_a for chloroacetic acid is 1.35×10^{-3} M.

Solution: The relevant equation for the dissociation reaction is

$$ClCH_2COOH(aq) + H_2O(l) \rightleftharpoons H_3O^+(aq) + ClCH_2COO^-(aq) \qquad K_a = 1.35 \times 10^{-3}\text{ M}$$

The only other source of $H_3O^+(aq)$ is

$$H_2O(l) + H_2O(l) \rightleftharpoons H_3O^+(aq) + OH^-(aq) \qquad K_w = 1.00 \times 10^{-14}\text{ M}^2$$

Because $K_a \gg K_w$, we can neglect this source of $H_3O^+(aq)$ in setting up a concentration table.

Concentration	$ClCH_2COOH(aq)$ +	$H_2O(l)$	$\rightleftharpoons H_3O^+(aq)$ +	$ClCH_2COO^-(aq)$
initial	0.050 M	—	≈0	0
equilibrium	$0.050\text{ M} - [ClCH_2COO^-]$	—	$[H_3O^+]$	$[ClCH_2COO^-]$
equilibrium (substituting $[H_3O^+]$ for $[ClCH_2COO^-]$)	$0.050\text{ M} - [H_3O^+]$	—	$[H_3O^+]$	$[H_3O^+]$

The equilibrium-constant expression is

$$K_a = \frac{[H_3O^+][ClCH_2COO^-]}{[ClCH_2COOH]} = 1.35 \times 10^{-3}\ M$$

Substituting the appropriate entries from the table gives

$$\frac{[H_3O^+]^2}{0.050\ M - [H_3O^+]} = 1.35 \times 10^{-3}\ M$$

We then write this equation in the standard form of a quadratic equation:

$$[H_3O^+]^2 + (1.35 \times 10^{-3}\ M)[H_3O^+] - 6.75 \times 10^{-5}\ M^2 = 0$$

The two solutions to this equation are $[H_3O^+] = 7.57 \times 10^{-3}$ M and -8.92×10^{-3} M. We reject the physically unacceptable negative concentration, so the pH of the solution is

$$pH = -\log [H_3O^+] = -\log (7.57 \times 10^{-3}) = 2.12$$

The concentration of $ClCH_2COO^-(aq)$ is given by

$$[ClCH_2COO^-] = [H_3O^+] = 7.57 \times 10^{-3}\ M$$

The concentration of $ClCH_2COOH(aq)$ is given by

$$\begin{aligned}[ClCH_2COOH] &= 0.050\ M - [ClCH_2COO^-] \\ &= 0.050\ M - 7.57 \times 10^{-3}\ M = 0.042\ M\end{aligned}$$

And the concentration of $OH^-(aq)$ is given by

$$[OH^-] = \frac{1.00 \times 10^{-14}\ M^2}{[H_3O^+]} = \frac{1.00 \times 10^{-14}\ M^2}{7.57 \times 10^{-3}\ M} = 1.32 \times 10^{-12}\ M$$

Note that the pH of this 0.050 M $ClCH_2COOH(aq)$ solution is almost an entire pH unit less than that of the 0.050 M $CH_3COOH(aq)$ solution calculated above. In other words, 0.050 M $ClCH_2COOH(aq)$ is about 10 times more acidic than $CH_3COOH(aq)$. The percentage of dissociation of 0.050 M $ClCH_2COOH(aq)$ is

$$\begin{aligned}\%\ \text{diss} &= \left(\frac{[H_3O^+]}{[ClCH_2COOH]_0}\right) \times 100 \\ &= \left(\frac{7.57 \times 10^{-3}\ M}{0.050\ M}\right) \times 100 = 15.1\%\end{aligned}$$

compared with only 1.8% for 0.050 M $CH_3COOH(aq)$.

PRACTICE PROBLEM 18-9: Calculate the pH and the concentrations of all the species in a 0.250 M aqueous solution of benzoic acid, $C_6H_5COOH(aq)$. The value of K_a for benzoic acid is 6.46×10^{-5} M.

Answer: $[H_3O^+] = [C_6H_5COO^-] = 3.99 \times 10^{-3}$ M; $[C_6H_5COOH] = 0.246$ M; $[OH^-] = 2.51 \times 10^{-12}$ M; pH = 2.40

For a given stoichiometric concentration, the percentage of dissociation of an acid depends on the value of K_a. The larger the value of K_a, the stronger the acid. The small value of K_a reflects the fact that acetic acid is only slightly dissociated in aqueous solution. Table 18-3 gives the K_a values for a number of weak acids.

Table 18-3 **Values of K_a and pK_a for weak acids in water at 25°C**

Acid	Formula	K_a/M	pK_a
acetic	CH_3COOH	1.74×10^{-5}	4.76
benzoic	C_6H_5COOH	6.46×10^{-5}	4.19
chloroacetic	$ClCH_2COOH$	1.35×10^{-3}	2.87
chlorous	$HClO_2$	1.15×10^{-2}	1.94
cyanic	HCNO	2.19×10^{-4}	3.66
dichloroacetic	$Cl_2CHCOOH$	1.38×10^{-3}	2.86
fluoroacetic	FCH_2COOH	2.75×10^{-3}	2.56
formic	HCOOH	1.78×10^{-4}	3.75
hydrozoic	HN_3	1.91×10^{-5}	4.72
hydrocyanic	HCN	4.79×10^{-10}	9.32
hydrofluoric	HF	6.76×10^{-4}	3.17
iodic	HIO_3	0.157	0.80
lactic	$CH_3CHOHCOOH$	1.41×10^{-4}	3.85
nitrous	HNO_2	4.47×10^{-4}	3.35
phenol	C_6H_5OH	1.0×10^{-10}	10.0
propionic	CH_3CH_2COOH	1.35×10^{-5}	4.87

Because K_a values for aqueous acids range over many powers of 10 (Table 18-3), it is convenient to define a quantity **pK_a** as

$$pK_a = -\log K_a \tag{18-15}$$

Note the similarity in the definitions of pH (Equation 18-4) and pK_a. The pK_a values at 25°C for several weak acids in water are given in Table 18-3. From these pK_a values, we see that the stronger the acid, the smaller the value of pK_a.

EXAMPLE 18-10: Given that $K_a = 1.74 \times 10^{-5}$ M for acetic acid in water at 25°C, calculate pK_a.

Solution: Using Equation (18-15), we have

$$\begin{aligned} pK_a &= -\log K_a \\ &= -\log (1.74 \times 10^{-5}) \\ &= 4.76 \end{aligned}$$

in agreement with the value in Table 18-3.

PRACTICE PROBLEM 18-10: Given that the value of pK_a for iodic acid, $HIO_3(aq)$, is 0.804 at 25°C, calculate the value of K_a.

Answer: 0.157 M

18-8. THE METHOD OF SUCCESSIVE APPROXIMATIONS IS OFTEN USED IN SOLVING ACID-BASE EQUILIBRIUM PROBLEMS

The use of the quadratic formula to solve Equation (18-14) is effective but tedious. An alternate, and usually faster, method of solution of a quadratic equation is the **method of successive approximations.** Let's reconsider Equation (18-14):

$$1.74 \times 10^{-5}\ \mathrm{M} = \frac{[\mathrm{H_3O^+}]^2}{0.050\ \mathrm{M} - [\mathrm{H_3O^+}]} \tag{18-14}$$

Because acetic acid is a weak acid (as indicated by the value of K_a), we expect that the percentage of dissociation of the acid is small. Furthermore, because essentially all the $H_3O^+(aq)$ arises from dissociation of the acid, we expect that $[H_3O^+]$ will be small relative to the initial concentration of the acid. We can express this algebraically as

$$0.050\ \mathrm{M} - [\mathrm{H_3O^+}] \simeq 0.050\ \mathrm{M}$$

Using this approximation in Equation (18-14) yields for $[H_3O^+]$

$$[\mathrm{H_3O^+}]^2 \simeq (0.050\ \mathrm{M})(1.74 \times 10^{-5}\ \mathrm{M})$$

or

$$[\mathrm{H_3O^+}] \simeq \sqrt{(0.050\ \mathrm{M})(1.74 \times 10^{-5}\ \mathrm{M})} = 9.33 \times 10^{-4}\ \mathrm{M}$$

Note that this first approximation is close to the value of 9.24×10^{-4} M obtained by the solution of the full quadratic equation in the previous section.

We can now use this approximate value of $[H_3O^+]$ in the denominator of the right-hand side of Equation (18-14) to obtain a second, more accurate, approximation; that is, we take

$$1.74 \times 10^{-5}\ \mathrm{M} \simeq \frac{[\mathrm{H_3O^+}]^2}{0.050\ \mathrm{M} - 9.33 \times 10^{-4}\ \mathrm{M}}$$

Thus, we calculate that

$$[\mathrm{H_3O^+}] \simeq \sqrt{(0.0491\ \mathrm{M})(1.74 \times 10^{-5}\ \mathrm{M})}$$
$$= 9.24 \times 10^{-4}\ \mathrm{M}$$

which is the same as the exact solution. If we repeat the approximation procedure, using 9.24×10^{-4} M in the denominator, then we find no further change in the value of $[H_3O^+]$ to three significant figures. Therefore, we have found that

$$1.74 \times 10^{-5}\ \mathrm{M} = \frac{(9.24 \times 10^{-4}\ \mathrm{M})^2}{0.050\ \mathrm{M} - 9.24 \times 10^{-4}\ \mathrm{M}}$$

Identical values of $[H_3O^+]$ in the numerator and the denominator means that the original equation (Equation 18-14) is satisifed. In other words, an unchanged value of $[H_3O^+]$ on successive approximations means that we have found the correct solution.

This method of successive approximations is generally much faster and easier than the use of the full quadratic equation, particularly if you use a hand calculator. We will use the method of successive approximations often to solve equilibrium problems.

Compounds in which the hydrogen atoms of ammonia are substituted by hydrocarbon groups are called **amines.** Other organic bases are

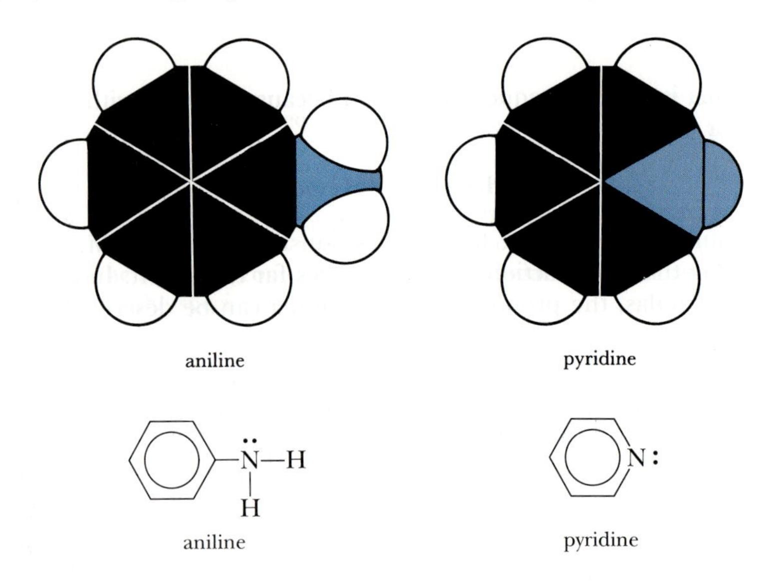

Notice that the nitrogen atom in each of these bases has a lone pair of electrons. Pyridine, for example, is basic in aqueous solution because of the reaction

$$\mathrm{C_5H_5N\!:}(aq) + \mathrm{H_2O}(l) \rightleftharpoons \mathrm{C_5H_5\overset{\oplus}{N}{-}H}(aq) + \mathrm{OH^-}(aq)$$

pyridine — pyridinium ion

Table 18-4 Values of K_b and pK_b for weak bases in water at 25°C

Base	Formula	K_b/M	pK_b
ammonia	$\mathrm{H{-}\ddot{N}{-}H}$ H	1.75×10^{-5}	4.76
methylamine	$\mathrm{H{-}\ddot{N}{-}CH_3}$ H	4.59×10^{-4}	3.34
dimethylamine	$\mathrm{H_3C{-}\ddot{N}{-}CH_3}$ H	5.81×10^{-4}	3.24
trimethylamine	$\mathrm{H_3C{-}\ddot{N}{-}CH_3}$ $\mathrm{CH_3}$	6.11×10^{-5}	4.21
ethylamine	$\mathrm{CH_3CH_2{-}\ddot{N}{-}H}$ H	4.27×10^{-4}	3.37
hydroxylamine	$\mathrm{H{-}\ddot{N}{-}OH}$ H	1.07×10^{-8}	7.97
aniline	$\mathrm{C_6H_5{-}\ddot{N}{-}H}$ H	4.17×10^{-10}	9.38
pyridine	$\mathrm{C_5H_5N\!:}$	1.46×10^{-9}	8.84

If we abbreviate pyridine by Py and the pyridinium ion by PyH^+, then the equilibrium-constant expression for this reaction is

$$K_b = \frac{[PyH^+][OH^-]}{[Py]}$$

The subscript b on K indicates that the reaction is a reaction of a weak base with water, that is, a base-protonation reaction. Thus, K_b is called the **base-protonation constant.** The values of K_b and pK_b for various weak bases are given in Table 18-4. By analogy to pK_a, $\mathbf{p}\boldsymbol{K}_\mathbf{b}$ is defined as

$$pK_b = -\log K_b \qquad (18\text{-}16)$$

The smaller the value of pK_b, the stronger the base.

EXAMPLE 18-12: Calculate the pH and the concentrations of all the species in a 0.755 M hydroxylamine, $HONH_2(aq)$, solution.

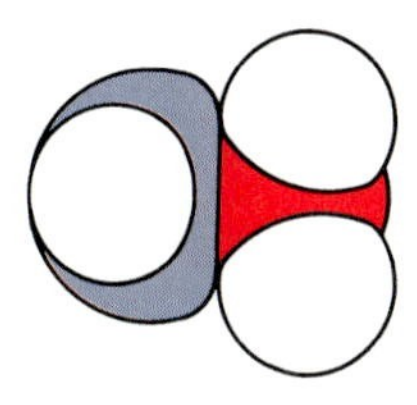

hydroxylamine

Solution: The equation for the base-protonation equilibrium is

$$HONH_2(aq) + H_2O(l) \rightleftharpoons HONH_3^+(aq) + OH^-(aq) \qquad K_b = 1.07 \times 10^{-8}\ M$$

where the value of K_b is from Table 18-4. The only other source of $OH^-(aq)$ is from the reaction described by Equation (18-1); but, because $K_b \gg K_w$, we can ignore this source of $OH^-(aq)$ and set up the following concentration table.

Concentration	$HONH_2(aq)$	+	$H_2O(l)$	⇌	$HONH_3^+(aq)$	+	$OH^-(aq)$
initial	0.755 M		—		0		≈0
equilibrium	0.775 M − $[HONH_3^+]$		—		$[HONH_3^+]$		$[OH^-]$
equilibrium (substituting $[OH^-]$ for $[HONH_3^+]$)	0.755 M − $[OH^-]$		—		$[OH^-]$		$[OH^-]$

The equilibrium-constant expression is

$$K_b = \frac{[HONH_3^+][OH^-]}{[HONH_2]} = \frac{[OH^-]^2}{0.755\ M - [OH^-]} = 1.07 \times 10^{-8}\ M$$

The method of successive approximations gives $[OH^-] = 8.99 \times 10^{-5}$ M, so the pOH is

$$pOH = -\log[OH^-] = 4.05$$

and the pH is

$$pH = 14.00 - pOH = 14.00 - 4.05 = 9.95$$

The concentration of $HONH_3^+(aq)$ is given by

$$[HONH_3^+] = [OH^-] = 8.99 \times 10^{-5}\ M$$

The concentration of $HONH_2(aq)$ is given by

$$[HONH_2] = 0.755\ M - [HONH_3^+]$$
$$= 0.755\ M - 8.99 \times 10^{-5}\ M = 0.755\ M$$

completely with added acid or added base and therefore is resistant to changes in pH.

Because this buffer solution contains both $CH_3COOH(aq)$ and $CH_3COO^-(aq)$, we have to consider the two equilibria

$$CH_3COOH(aq) + H_2O(l) \rightleftharpoons H_3O^+(aq) + CH_3COO^-(aq) \qquad K_a = 1.74 \times 10^{-5}\ M$$

$$CH_3COO^-(aq) + H_2O(l) \rightleftharpoons OH^-(aq) + CH_3COOH(aq) \qquad K_b = 5.75 \times 10^{-10}\ M$$

Because the equilibrium constants for both of these equations are small, both equilibria lie far to the left and neither $[CH_3COOH]$ nor $[CH_3COO^-]$ will differ significantly from its stoichiometric concentration. Consequently, we can write

$$[CH_3COOH] \simeq [CH_3COOH]_0$$

$$[CH_3COO^-] \simeq [CH_3COO^-]_0$$

where the zero subscript emphasizes that these are stoichiometric concentrations.

Now let's calculate the pH of a buffer solution that is 0.10 M in $CH_3COOH(aq)$ and 0.15 M in $CH_3COO^-(aq)$. The conjugate acid-base equilibrium involving these two species is

$$CH_3COOH(aq) + H_2O(l) \rightleftharpoons H_3O^+(aq) + CH_3COO^-(aq) \qquad K_a = 1.74 \times 10^{-5}\ M$$

If we neglect the source of $H_3O^+(aq)$ from the autoprotonation reaction of water—Equation (18-1)—then we can write the concentration table associated with this equation as

Concentration	$CH_3COOH(aq)$	$+ H_2O(l) \rightleftharpoons$	$H_3O^+(aq)$	$+ CH_3COO^-(aq)$
initial	0.10 M	—	≈ 0	0.15 M
equilibrium	$0.10\ M - [H_3O^+]$	—	$[H_3O^+]$	$0.15\ M + [H_3O^+]$

Substituting these values into the equilibrium-constant expression gives

$$\frac{[H_3O^+](0.15\ M + [H_3O^+])}{0.10\ M - [H_3O^+]} = 1.74 \times 10^{-5}\ M$$

We can solve this equation for $[H_3O^+]$ by using the quadratic equation, but because $CH_3COOH(aq)$ is a weak acid, we expect that $[H_3O^+]$ will be small. Therefore, we neglect $[H_3O^+]$ with respect to both 0.10 M and 0.15 M, and write

$$[H_3O^+] \simeq \frac{(0.10\ M)(1.74 \times 10^{-5}\ M)}{0.15\ M} = 1.16 \times 10^{-5}\ M$$

This small value for $[H_3O^+]$ justifies our approximation of neglecting $[H_3O^+]$ with respect to 0.10 M and 0.15 M. The pH of the solution is given by

$$pH = -\log[H_3O^+] = -\log(1.16 \times 10^{-5}) = 4.94$$

Generally, a buffer solution consists of a weak acid $HB(aq)$ and its conjugate base $B^-(aq)$. The two equilibria that we must consider are

$$HB(aq) + H_2O(l) \rightleftharpoons H_3O^+(aq) + B^-(aq)$$

and

$$B^-(aq) + H_2O(l) \rightleftharpoons OH^-(aq) + HB(aq)$$

If both K_a and K_b are small (less than 10^{-3} M, for example), then we can write

$$[HB] \simeq [HB]_0 \qquad (19\text{-}3a)$$

$$[B^-] \simeq [B^-]_0 \qquad (19\text{-}3b)$$

The equilibrium-constant expression involving the acid dissociation is

$$K_a = \frac{[H_3O^+][B^-]}{[HB]} \qquad (19\text{-}4)$$

If we substitute both parts of Equation (19-3) into Equation (19-4), then we obtain

$$K_a = \frac{[H_3O^+][B^-]_0}{[HB]_0}$$

Taking the logarithm of both sides yields

$$\log K_a = \log \frac{[H_3O^+][B^-]_0}{[HB]_0}$$

Using the fact that $\log ab = \log a + \log b$, we can rewrite this equation in the form

$$\log K_a = \log[H_3O^+] + \log\left(\frac{[B^-]_0}{[HB]_0}\right)$$

Finally, multiplying by -1 and using $pH = -\log[H_3O^+]$ and $pK_a = -\log K_a$, we get

$$pH = pK_a + \log\left(\frac{[B^-]_0}{[HB]_0}\right) \qquad (19\text{-}5)$$

Equation (19-5) is known as the **Henderson-Hasselbalch equation.** It enables us to use the stoichiometric concentrations directly, without the need for a stepwise analysis of the equilibrium. The Henderson-Hasselbalch equation is used extensively in biochemistry. Whenever Equation (19-5) is applied, two general conditions must be met: both K_a and K_b for the conjugate acid-base pair must be small (that is, less than about 10^{-3} M) and the ratio $[B^-]_0/[HB]_0$ must be between roughly 0.1 and 10.

Let's calculate the pH of a buffer that is 0.25 M in formic acid, $HCOOH(aq)$, and 0.15 M in potassium formate, $KHCOO(aq)$. Because potassium formate is a soluble salt consisting of potassium ions and formate ions, the solution will be 0.25 M in $HCOOH(aq)$ and 0.15 M in $HCOO^-(aq)$. We see from Table 18-5 that $K_a = 1.78 \times 10^{-4}$ M and $K_b = 5.62 \times 10^{-11}$ M. Because both K_a and K_b are small and the ratio of initial concentrations of base to acid is 0.6, we can use the Henderson-Hasselbalch equation:

$$pH \simeq pK_a + \log\left(\frac{0.15\text{ M}}{0.25\text{ M}}\right) = 3.75 - 0.22 = 3.53$$

Table 19-2 The pK_{ai} values and color changes of various indicators

Indicator	pK_a	pH range of color change	Color change
tropeolin 00	2.0	1.3–3.2	red to yellow
methyl orange	3.4	3.1–4.4	red to orange
methyl red	4.9	4.4–6.2	red to yellow
bromcresol purple	6.3	5.2–6.8	yellow to purple
bromthymol blue	7.1	6.2–7.6	yellow to blue
cresol red	8.2	7.2–8.8	yellow to red
thymol blue	8.9	8.0–9.6	yellow to blue
phenolphthalein	9.4	8.0–10.0	colorless to red
thymolphthalein	10.0	9.4–10.6	colorless to blue
alizarin yellow R	11.2	10.0–12.0	yellow to violet

blue. The value of pK_{ai} for bromthymol blue is 7.1, so the pH of the green solution is about 7. Generally the pH range over which an indicator changes color is approximately equal to pK_{ai} ± 1. Table 19-2 lists the values of the pK_a's and the color changes of a number of indicators.

Because of the intense color of indicators, only a very small concentration is necessary to produce a visible color. Thus, its contribution to the total acidity of the solution is negligible. Several indicators, together with their colors at various pH values, are shown in Figure 19-6. (Note the relation between Table 19-2 and Figure 19-6.) Using the indicators in Figure 19-6, we can estimate the pH of an aqueous solution to within about 0.5 pH unit. Of course, the solution must be colorless initially, otherwise the color change of the indicator may be obscured.

EXAMPLE 19-3: Estimate the pH of a colorless aqueous solution that turns blue when bromcresol green is added and yellow when bromthymol blue is added.

Solution: From Figure 19-6, we see that bromcresol green is blue at pH > 5 and that bromthymol blue is yellow at pH < 6. Therefore, the pH of the solution is between 5 and 6.

PRACTICE PROBLEM 19-3: Estimate the pH of a colorless aqueous solution that turns yellow when thymol blue is added and pink when methyl red is added.

Answer: Between 3 and 4

Indicators are used in titration solutions to signal the completion of the acid-base reaction. The point at which the indicator changes color is called the **end point** of the titration. It is important to distinguish between equivalence point and end point. The equivalence point is the point at which stoichiometrically equivalent amounts of acid and base

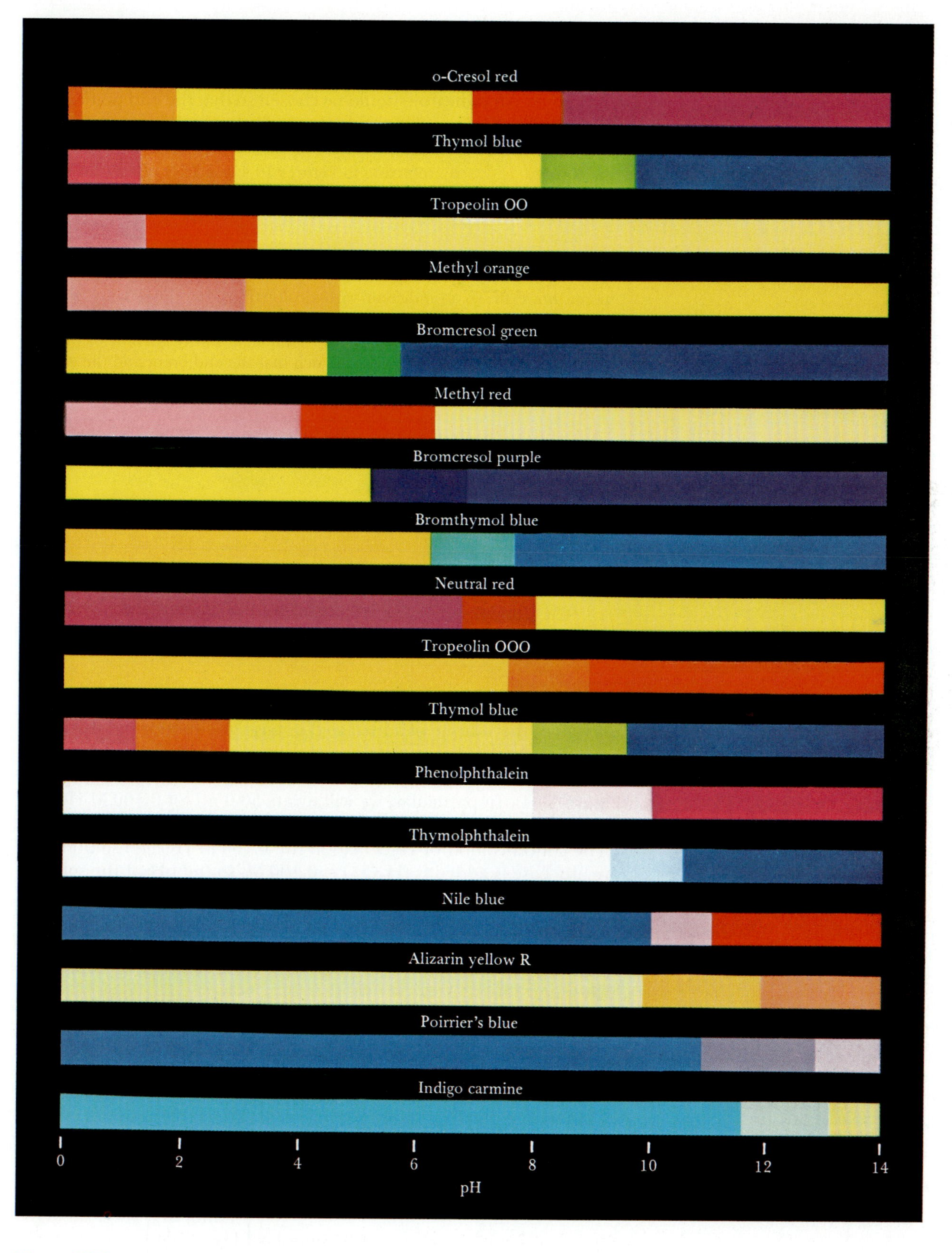

Figure 19-6 The colors of various indicators at different pH values.

have been brought together. The acid-base reaction is complete at the equivalence point. The end point is the point at which the indicator changes color. The end point is the experimental estimate of the equivalence point. An indicator should be chosen so that the end point corresponds as closely as possible with the equivalence point.

EXAMPLE 19-4: By referring to Figures 19-4 and 19-6, choose an indicator to signal the equivalence point shown in Figure 19-4.

Solution: From Figure 19-4, we note that the equivalence point occurs at pH = 7.0. The titration curve is very steep in the vicinity of the equivalence point, however, so an indicator with a color transition range lying between pH = 5 and pH = 9 would be suitable. Referring to Figure 19-6, we see that there are several possible choices, for example, bromthymol blue or phenolphthalein. We choose our indicator such that the end point and the equivalence point are the same within the required accuracy of the titration.

PRACTICE PROBLEM 19-4: By referring to Figures 19-9 and 19-6, choose an indicator (other than methyl red) to signal the equivalence point shown in Figure 19-9.

Answer: bromcresol green

19-4. THE pH CHANGES ABRUPTLY AT THE EQUIVALENCE POINT OF THE TITRATION OF A STRONG ACID WITH A STRONG BASE

Titration curves are obtained experimentally by measuring the pH as a function of the volume of added acid or base. To understand the origin of experimental titration curves and to interpret them, in this section we will calculate the titration curve of a strong acid with a strong base (Figure 19-4). In particular, we will titrate 50.0 mL of 0.100 M $HCl(aq)$ with 0.100 M $NaOH(aq)$.

The concentration of $H_3O^+(aq)$ in the 0.100 M $HCl(aq)$ solution before any base is added is $[H_3O^+] = 0.100$ M, so the pH of the solution initially is 1.00. After 10.0 mL of the 0.100 M $NaOH(aq)$ is added to the 50.0 mL of the initially 0.100 M $HCl(aq)$ solution, the total volume of the resulting solution is 50.0 mL + 10.0 mL = 60.0 mL. The $OH^-(aq)$ from the $NaOH(aq)$ reacts with the $H_3O^+(aq)$ from the $HCl(aq)$ to produce water:

$$H_3O^+(aq) + OH^-(aq) \rightarrow 2H_2O(l)$$

Thus, when 10.0 mL of 0.100 M $NaOH(aq)$ is added to 50.0 mL of 0.100 M $HCl(aq)$, the number of millimoles of $H_3O^+(aq)$ that reacts is equal to the number of millimoles of $OH^-(aq)$ added:

$$\begin{pmatrix}\text{mmol of}\\ OH^-(aq)\\ \text{added}\end{pmatrix} = (0.100\text{ M})(10.0\text{ mL})$$

$$= 1.00\text{ mmol} = \begin{pmatrix}\text{mmol of}\\ H_3O^+(aq)\\ \text{reacted}\end{pmatrix}$$

The total number of millimoles of $H_3O^+(aq)$ initially present in 50.0 mL of 0.100 M HCl(aq) is

$$\begin{pmatrix}\text{initial}\\ \text{mmol of}\\ H_3O^+(aq)\end{pmatrix} = MV = (0.100\ \text{M})(50.0\ \text{mL}) = 5.00\ \text{mmol}$$

The concentration of $H_3O^+(aq)$ that remains after the addition of 10.0 mL of 0.100 M NaOH is equal to the number of millimoles of $H_3O^+(aq)$ that remains unreacted divided by the total volume in milliliters of resulting solution. The number of millimoles of $H_3O^+(aq)$ unreacted is

$$\begin{pmatrix}\text{mmol of}\\ H_3O^+(aq)\\ \text{unreacted}\end{pmatrix} = \begin{pmatrix}\text{initial}\\ \text{mmol of}\\ H_3O^+(aq)\end{pmatrix} - \begin{pmatrix}\text{mmol of}\\ H_3O^+(aq)\\ \text{reacted}\end{pmatrix}$$

$$= 5.00\ \text{mmol} - 1.00\ \text{mmol}$$

$$= 4.00\ \text{mmol}$$

and the molarity of unreacted $H_3O^+(aq)$ is

$$\text{molarity} = \frac{4.00\ \text{mmol}}{60.0\ \text{mL}} = 6.67 \times 10^{-2}\ \text{M}$$

Therefore, the pH of the solution after the addition of 10.0 mL of NaOH(aq) is

$$\text{pH} = -\log\,[H_3O^+] = -\log\,(6.67 \times 10^{-2}) = 1.18$$

Proceeding in an analogous fashion, we can calculate the pH of the resulting solution after the addition of the volumes of NaOH(aq) (Table 19-3). Beyond the equivalence point, essentially all the $H_3O^+(aq)$ has reacted, and we simply have a diluted solution of NaOH(aq).

EXAMPLE 19-5: Calculate the pH of a solution obtained by adding 60.0 mL of 0.100 M NaOH(aq) to 50.0 mL of 0.100 M HCl(aq).

Solution: The total number of millimoles of $H_3O^+(aq)$ in 50.0 mL of 0.100 M HCl(aq) is

$$\begin{pmatrix}\text{mmol of}\\ H_3O^+(aq)\end{pmatrix} = MV = (0.100\ \text{M})(50.0\ \text{mL}) = 5.00\ \text{mmol}$$

The total number of millimoles of $OH^-(aq)$ in 60.0 mL of 0.100 M NaOH(aq) is

$$\begin{pmatrix}\text{mmol of}\\ OH^-(aq)\\ \text{added}\end{pmatrix} = MV = (0.100\ \text{M})(60.0\ \text{mL}) = 6.00\ \text{mmol}$$

Note that the number of millimoles of NaOH(aq) added (6.00 mmol) exceeds the number of millimoles of $H_3O^+(aq)$ present initially (5.00 mmol).

As usual, we set up a concentration table:

Concentration	$CH_3COOH(aq)$	+	$H_2O(l) \rightleftharpoons$	$H_3O^+(aq)$	+	$CH_3COO^-(aq)$
initial	0.100 M		—	$\simeq 0$		0
equilibrium	$0.100\ M - [H_3O^+]$		—	$[H_3O^+]$		$[H_3O^+]$

where we have neglected the contribution to $[H_3O^+]$ from the autoprotonation of water ($K_a \gg K_w$). Thus, Equation (19-11) becomes

$$\frac{[H_3O^+]^2}{0.100\ M - [H_3O^+]} = 1.74 \times 10^{-5}\ M \qquad (19\text{-}12)$$

Solving Equation (19-12) by the method of successive approximations, or by using the quadratic formula, yields $[H_3O^+] = 1.31 \times 10^{-3}$ M. Thus, we have

$$pH = -\log(1.31 \times 10^{-3}) = 2.88$$

in agreement with Figure 19-7.

To calculate the pH at the equivalence point, we must consider the principal species in the solution at the equivalence point, where the number of moles of NaOH(aq) added is equal to the number of moles of $CH_3COOH(aq)$ initially present. The equation for the reaction is

$$CH_3COOH(aq) + \cancel{Na^+(aq)} + OH^-(aq) \rightarrow \cancel{Na^+(aq)} + CH_3COO^-(aq) + H_2O(l) \qquad (19\text{-}13)$$

It is important to realize that the reaction described by Equation (19-13) goes essentially to completion even though $CH_3COOH(aq)$ is a weak acid. We can see this quantitatively by calculating the equilibrium constant of Equation (19-13). The net ionic equation for the titration of $CH_3COOH(aq)$ with NaOH(aq) is

$$CH_3COOH(aq) + OH^-(aq) \rightleftharpoons CH_3COO^-(aq) + H_2O(l) \qquad (19\text{-}14)$$

Equation (19-14) is simply the reverse of the base-protonation reaction for $CH_3COO^-(aq)$, so $K = (1/K_b) = 1.74 \times 10^9\ M^{-1}$ (see Equation 19-2). This large value of K indicates that the equilibrium described by Equation (19-14) lies far to the right and that $CH_3COOH(aq)$ reacts essentially completely when a strong base is added to the solution.

Because the reaction described by Equation (19-14) goes essentially to completion, the solution at the equivalence point of the titration consists of a $NaCH_3COO(aq)$ solution. To calculate the concentration of this $NaCH_3COO(aq)$ solution, we use the fact that we started with 50.0 mL of 0.100 M $CH_3COOH(aq)$. The initial number of millimoles of $CH_3COOH(aq)$ is

$$\begin{pmatrix}\text{initial} \\ \text{mmol of} \\ CH_3COOH(aq)\end{pmatrix} = (0.100\ M)(50.0\ mL) = 5.00\ mmol$$

According to Equation (19-13)

$$\begin{pmatrix}\text{mmol of } NaCH_3COO(aq) \\ \text{at the equivalence point}\end{pmatrix} = \begin{pmatrix}\text{initial mmol of} \\ CH_3COOH(aq)\end{pmatrix} = 5.00\ mmol$$

The total volume of the solution at the equivalence point is 100.0 mL, that is, 50.0 mL of $CH_3COOH(aq)$ plus 50.0 mL of added $NaOH(aq)$, so the stoichiometric concentration of the $NaCH_3COO(aq)$ solution at the equivalence point is

$$[CH_3COO^-]_0 = \frac{5.00 \text{ mmol}}{100.0 \text{ mL}} = 0.0500 \text{ M}$$

We emphasize here that 0.0500 M is the *stoichiometric* concentration of $NaCH_3COO(aq)$. The actual concentration of $CH_3COO^-(aq)$ will be slightly less than 0.0500 M because of the reaction described by

$$CH_3COO^-(aq) + H_2O(l) \rightleftharpoons OH^-(aq) + CH_3COOH(aq) \qquad K_b = 5.75 \times 10^{-10} \text{ M} \tag{19-15}$$

We calculated the pH of a solution of $NaCH_3COO(aq)$ in Section 18-11. In fact, Example 18-16 involves the calculation of the pH of a 0.0500 M $NaCH_3COO(aq)$ solution, giving a value of pH = 8.73. Thus, we see that the pH at the equivalence point in Figure 19-7 is 8.73.

EXAMPLE 19-6: Which indicator would you use to signal the equivalence point of the titration of 50.0 mL of 0.100 M $CH_3COOH(aq)$ with 0.100 M $NaOH(aq)$? The pH at the equivalence point is 8.73.

Solution: Referring to Figure 19-6, we see that phenolphthalein is a suitable indicator for the titration.

PRACTICE PROBLEM 19-6: Compare the use of bromthymol blue and phenolphthalein as indicators for the titration presented in Example 19-6.

Answer: Bromthymol blue changes color at too low a pH; phenolphthalein is a good choice.

19-6. *pH* = *pK*$_a$ AT THE MIDPOINT IN THE TITRATION OF A WEAK ACID WITH A STRONG BASE

We have calculated the pH at two points (the initial point and the equivalence point) on the titration curve of a weak acid with a strong base. Let's calculate the pH at the **midpoint,** which is the point at which one half of the acid has reacted with $NaOH(aq)$. The midpoint for the titration of 50.0 mL of 0.100 M $CH_3COOH(aq)$ with 0.100 M $NaOH(aq)$ occurs when 25.0 mL of $NaOH(aq)$ has been added. We can calculate the stoichiometric concentrations of $CH_3COOH(aq)$ and $CH_3COO^-(aq)$ at the midpoint. The total number of millimoles of $CH_3COOH(aq)$ present initially is

$$\begin{pmatrix} \text{initial} \\ \text{mmol of} \\ CH_3COOH(aq) \end{pmatrix} = MV = (0.100 \text{ M})(50.0 \text{ mL}) = 5.00 \text{ mmol}$$

The number of millimoles of NaOH(aq) added is half of 5.00 mmol, or 2.50 mmol, which corresponds to a volume

$$V = \frac{n}{M} = \frac{2.50 \text{ mmol}}{0.100 \text{ M}} = 25.0 \text{ mL}$$

Because the neutralization reaction (Equation 19-6) goes essentially to completion, all the $OH^-(aq)$ reacts with $CH_3COOH(aq)$. Furthermore, the number of millimoles of $CH_3COO^-(aq)$ produced is equal to the number of millimoles of $OH^-(aq)$ added, so

$$\begin{pmatrix} \text{mmol of} \\ CH_3COO^-(aq) \\ \text{produced} \end{pmatrix} = \text{mmol of } OH^-(aq) \text{ added} = 2.50 \text{ mmol}$$

The number of millimoles of $CH_3COOH(aq)$ unreacted is

$$\begin{pmatrix} \text{mmol of} \\ CH_3COOH(aq) \\ \text{unreacted} \end{pmatrix} = \begin{pmatrix} \text{initial} \\ \text{mmol of} \\ CH_3COOH(aq) \end{pmatrix} - \begin{pmatrix} \text{mmol of} \\ CH_3COO^-(aq) \\ \text{produced} \end{pmatrix}$$
$$= 5.00 \text{ mmol} - 2.50 \text{ mmol} = 2.50 \text{ mmol}$$

Thus, we find that the corresponding concentrations of both $CH_3COOH(aq)$ and $CH_3COO^-(aq)$ are

$$[CH_3COOH]_0 = [CH_3COO^-]_0 = \frac{2.50 \text{ mmol}}{75.0 \text{ mL}} = 0.0333 \text{ M} \qquad (19\text{-}16)$$

at the midpoint.

The actual concentrations of $CH_3COOH(aq)$ and $CH_3COO^-(aq)$ will differ slightly from their stoichiometric concentrations because of the reactions described by Equations (19-10) and (19-15). Because the concentrations of both $CH_3COOH(aq)$ and $CH_3COO^-(aq)$ are large compared with those of the other species in solution, they are called the **principal species.** We must use an equilibrium-constant expression that involves these principal species. Either Equation (19-10) or Equation (19-15) may be used because both equations involve these principal species. If we use Equation (19-10), we have

$$\frac{[H_3O^+][CH_3COO^-]}{[CH_3COOH]} = K_a$$

We set up a concentration table:

Concentration	$CH_3COOH(aq)$	+ $H_2O(l)$ ⇌	$H_3O^+(aq)$ +	$CH_3COO^-(aq)$
initial	0.0333 M	—	≈ 0	0.0333 M
equilibrium	0.0333 M − $[H_3O^+]$	—	$[H_3O^+]$	0.0333 M + $[H_3O^+]$

Substituting the equilibrium concentrations into Equation (19-11) gives

$$\frac{[H_3O^+](0.0333 \text{ M} + [H_3O^+])}{0.0333 \text{ M} - [H_3O^+]} = 1.74 \times 10^{-5} \text{ M}$$

Because acetic acid is a weak acid, we assume as a first approximation that $[H_3O^+]$ is small relative to 0.0333 M. With this approximation, we have

$$[H_3O^+] \simeq K_a = 1.74 \times 10^{-5}\ M$$

Because $[H_3O^+] \ll 0.0333$ M, our first approximation is justified, and the pH is given by

$$pH = pK_a = 4.76$$

Note that pH at the midpoint of the titration of a weak acid by a strong base is pK_a. Although we have shown this only for the titration of acetic acid with sodium hydroxide, this is a valid result if $[H_3O^+]$ is negligible compared with $[HB]_0$ and $[B^-]_0$, or equivalently, when K_a and K_b are much smaller than $[HB]_0$ and $[B^-]_0$. In such cases (which are very common), we can write

$$pH = pK_a \qquad \text{(midpoint)} \qquad (19\text{-}17)$$

We used Equation (19-10) in the above calculation because the solution contains the two principal species, $CH_3COOH(aq)$ and $CH_3COO^-(aq)$. Equation (19-15) also contains these two species, and the following Example shows that we also can use Equation (19-15).

EXAMPLE 19-7: Use Equation (19-15) (instead of Equation 19-10) to calculate the pH at the midpoint of the titration discussed above.

Solution: We set up a concentration table using Equation (19-15):

Concentration	$CH_3COO^-(aq)$	+ $H_2O(l)$ ⇌	$CH_3COOH(aq)$	+ $OH^-(aq)$
initial	0.0333 M	—	0.0333 M	$\simeq 0$
equilibrium	0.0333 M − $[OH^-]$	—	0.0333 M + $[OH^-]$	$[OH^-]$

The equilibrium-constant expression is

$$\frac{[OH^-](0.0333\ M + [OH^-])}{0.0333\ M - [OH^-]} = K_b = 5.75 \times 10^{-10}\ M$$

Because $CH_3COO^-(aq)$ is a weak base, we assume as a first approximation that $[OH^-]$ is small relative to 0.0333 M. With this approximation, we have

$$[OH^-] \simeq K_b = 5.75 \times 10^{-10}\ M$$

Because $[OH^-] \ll 0.0333$ M, our first approximation is justified, and

$$pOH = pK_b = 9.24$$

or pH = 4.76, as before.

PRACTICE PROBLEM 19-7: Calculate the pH at the midpoint for the titration of 50.0 mL of 0.100 M $HCNO(aq)$ with 0.250 M $NaOH(aq)$.

Answer: 3.66

The calculation of the entire titration curve of a weak acid with a strong base is quite lengthy and for the most part is left to the problems. The calculations of the pH at the initial point, the midpoint (pH = pK_a), and the equivalence point are almost sufficient to sketch the titration curve in Figure 19-7. All we really need is the pH at a point a little beyond the equivalence point. The following Example illustrates such a calculation.

EXAMPLE 19-8: Calculate the pH of the solution that results when 60.0 mL of 0.100 M NaOH(*aq*) is added to 50.0 mL of 0.100 M acetic acid.

Solution: The number of millimoles of NaOH added is

$$\begin{pmatrix}\text{mmol of}\\ \text{NaOH}(aq)\\ \text{added}\end{pmatrix} = (0.100\ \text{M})(60.0\ \text{mL}) = 6.00\ \text{mmol}$$

There are only 5.00 mmol of $CH_3COOH(aq)$ in the initial solution, so we have added an excess of NaOH(*aq*). The number of millimoles of unreacted NaOH(*aq*) is

$$\begin{pmatrix}\text{mmol of}\\ \text{NaOH}(aq)\\ \text{unreacted}\end{pmatrix} = \begin{pmatrix}\text{total mmol}\\ \text{of NaOH}(aq)\\ \text{added}\end{pmatrix} - \begin{pmatrix}\text{total mmol}\\ \text{of CH}_3\text{COOH}(aq)\\ \text{available}\end{pmatrix}$$

$$= 6.00\ \text{mmol} - 5.00\ \text{mmol} = 1.00\ \text{mmol}$$

The stoichiometric unreacted (excess) concentration of $OH^-(aq)$ in the final solution is

$$[\text{OH}^-]_0 = \frac{1.00\ \text{mmol}}{110.0\ \text{mL}} = 9.09 \times 10^{-3}\ \text{M}$$

The stoichiometric concentration of acetate ion in the final solution is

$$[\text{CH}_3\text{COO}^-]_0 = \frac{5.00\ \text{mmol}}{110.0\ \text{mL}} = 4.55 \times 10^{-2}\ \text{M}$$

The titration solution beyond the equivalence point consists of a mixture of sodium acetate and sodium hydroxide. The principal species are $CH_3COO^-(aq)$ and $OH^-(aq)$, so we will use a chemical equation and its corresponding equilibrium-constant expression that involves both of these species. A chemical equation that contains both species is Equation (19-15). We set up a concentration table, letting x denote the $OH^-(aq)$ formed in the reaction described by Equation (19-15); in other words, x is the concentration of $[OH^-]$ formed *in addition* to the stoichiometric excess of added NaOH(*aq*).

Concentration	**$CH_3COO^-(aq)$ +**	**$H_2O(l)$ ⇌**	**$CH_3COOH(aq)$ +**	**$OH^-(aq)$**
initial	0.0455 M	—	0	0.00909 M
equilibrium	0.0455 M − x	—	x	0.00909 M + x

Substituting these values into the equilibrium-constant expression gives

$$\frac{x(0.00909\ \text{M} + x)}{0.0455\ \text{M} - x} = 5.75 \times 10^{-10}\ \text{M}$$

If we assume that x is small compared with both 0.00909 M and 0.0455 M, then we obtain

$$x \simeq \frac{(0.0455\ \text{M})(5.75 \times 10^{-10}\ \text{M})}{0.00909\ \text{M}} = 2.88 \times 10^{-9}\ \text{M}$$

which indeed *is* small compared with both 0.00909 M and 0.0455 M. Thus, $[OH^-] = 0.00909$ M and pH = 11.96.

We can obtain this result more easily by the following argument. Because K_b is very small, the equilibrium in Equation (19-15) lies far to the left. The equilibrium is driven even further to the left (Le Châtelier's principle) by the presence of excess $OH^-(aq)$ in the solution at a stoichiometric concentration of $[OH^-]_0 = 9.09 \times 10^{-3}$ M. Thus, the reaction described by Equation (19-15) makes a negligible contribution to the total concentration of $OH^-(aq)$, and we have

$$[OH^-] \simeq [OH^-]_0 = 9.09 \times 10^{-3} \text{ M} \tag{19-18}$$

The pOH of the solution is

$$\text{pOH} = -\log[OH^-] = -\log(9.09 \times 10^{-3}) = 2.04$$

and the pH of the solution is

$$\text{pH} = 14.00 - \text{pOH} = 14.00 - 2.04 = 11.96$$

in agreement with the point at 60.0 mL NaOH(*aq*) in Figure 19-7.

PRACTICE PROBLEM 19-8: Calculate the pH of the solution that results when 55.0 mL of 0.100 M NaOH(*aq*) is added to 25.0 mL of 0.200 M $CH_3COOH(aq)$.

Answer: 11.8

As a final note, Figure 19-8 shows the concentrations of undissociated acetic acid and acetate ion plotted against the volume of added base. Note that $[CH_3COOH]$ decreases steadily, becoming essentially 0 at and beyond the equivalence point. Also, $[CH_3COO^-]$ increases steadily up to the equivalence point and then decreases steadily beyond it. No more $CH_3COO^-(aq)$ is produced once the equivalence point is reached, and it becomes diluted as base is added.

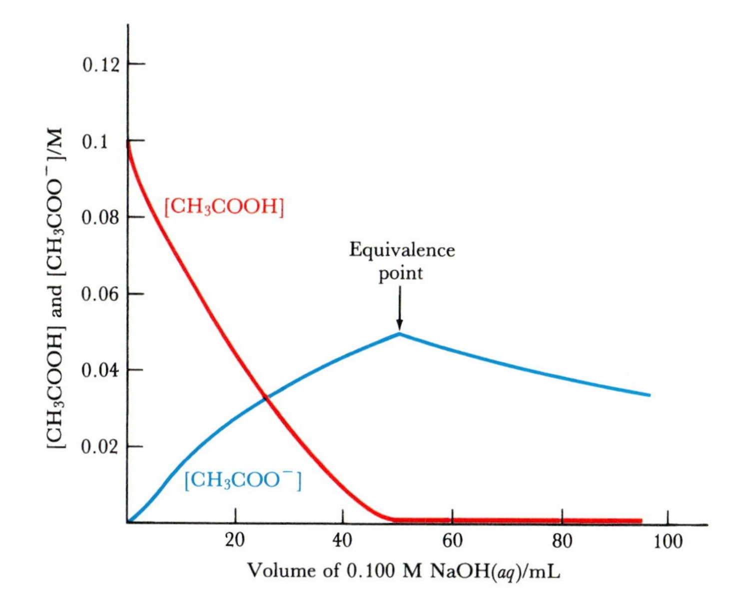

Figure 19-8 The concentration of undissociated acetic acid, $[CH_3COOH]$, and acetate ion, $[CH_3COO^-]$, versus the volume of 0.100 M NaOH(*aq*) used in the titration of 50.0 mL of 0.100 M $CH_3COOH(aq)$.

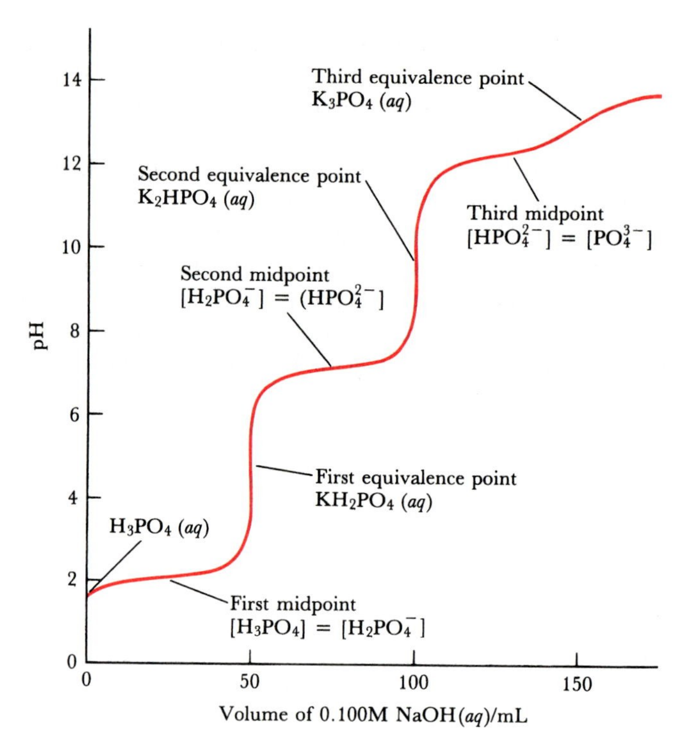

Figure 19-11 Titration curve for the titration of 0.100 M $H_3PO_4(aq)$ with 0.100 M NaOH(*aq*). Note the three equivalence points. The third equivalence point is not sharply defined because $HPO_4^{2-}(aq)$ is a very weak acid, so $PO_4^{3-}(aq)$ can compete with $H_2O(l)$ for protons. Note that the curve has three distinct sections: 0 to 50 mL, 50 mL to 100 mL, and 100 mL to 150 mL. In each section the behavior of pH versus volume of base is analogous to a titration curve for a monoprotic acid.

and

$$[H_3O^+] = [H_2PO_4^-] + 2[HPO_4^{2-}] + 3[PO_4^{3-}] + [OH^-]$$

or

$$0.023\ \text{M} \simeq 0.023\ \text{M} + 2(6.2 \times 10^{-8}\ \text{M}) + 3(1.2 \times 10^{-18}\ \text{M}) + (4.3 \times 10^{-13}\ \text{M})$$

$$\simeq 0.023\ \text{M}$$

Thus, we see that the assumptions we have used are valid. Calculations involving other polyprotic acids are carried out in a similar manner.

The titration curve for 0.10 M $H_3PO_4(aq)$ titrated with KOH(*aq*) is shown in Figure 19-11. Note that there are three equivalence points, because $H_3PO_4(aq)$ has three acidic protons.

EXAMPLE 19-11: Calculate the pH of a 0.10 M $H_2SO_4(aq)$ solution.

Solution: The fact that the first dissociation of $H_2SO_4(aq)$ is strong (Table 19-4) means that 0.10 M $H_2SO_4(aq)$ has stoichiometric concentrations of 0.10 M in $H_3O^+(aq)$ and 0.10 M in $HSO_4^-(aq)$:

$$\underset{0.10\ \text{M}}{H_2SO_4(aq)} + H_2O(l) \rightarrow \underset{0.10\ \text{M}}{H_3O^+(aq)} + \underset{0.10\ \text{M}}{HSO_4^-(aq)}$$

Because of the second dissociation, however, we have the equilibrium

$$HSO_4^-(aq) + H_2O(l) \rightleftharpoons H_3O^+(aq) + SO_4^{2-}(aq)$$

with

$$K_{a2} = \frac{[H_3O^+][SO_4^{2-}]}{[HSO_4^-]} = 0.010\ M$$

Because of this equilibrium reaction, both $[H_3O^+]_0$ and $[HSO_4^-]_0$ will be slightly different from 0.100 M.

We set up a concentration table. If we let x be the additional $[H_3O^+]$ due to the dissociation of $HSO_4^-(aq)$, then

Concentration	$HSO_4^-(aq)$	+	$H_2O(l)$	$\rightleftharpoons$	$H_3O^+(aq)$	+	$SO_4^{2-}(aq)$
initial	0.10 M		—		0.10 M		0
equilibrium	$0.10\ M - x$		—		$0.10\ M + x$		x

Substituting these values into the equilibrium-constant expression gives

$$\frac{(0.10\ M + x)x}{0.10\ M - x} = 0.010\ M \qquad (19\text{-}28)$$

If we assume that x is small compared with 0.10 M, then as a first approximation

$$x_1 \simeq \frac{(0.10\ M)(0.010\ M)}{0.10\ M} = 0.010\ M$$

Our result is 10 percent of 0.10 M. To obtain a better approximation, we substitute $x = 0.010$ M into the $0.10\ M + x$ and $0.10\ M - x$ terms in Equation (19-28) to obtain a second approximation.

$$x_2 \simeq \frac{(0.10\ M - 0.010\ M)(0.010\ M)}{0.10\ M + 0.010\ M} = 0.00818\ M$$

If each successive value of x is cycled through the equation several times more, we obtain successively 0.00849 M, 0.00846 M, 0.00844 M, and 0.00844 M. Thus,

$$[H_3O^+] = 0.10\ M + 0.00844\ M \simeq 0.11\ M$$

and the pH is 0.96.

PRACTICE PROBLEM 19-11: Calculate the pH and the concentrations of all the species in a 0.50 M aqueous solution of oxalic acid.

Answer: pH = 0.85; $[H_3O^+] = [HC_2O_4^-] = 0.14$ M; $[H_2C_2O_4] = 0.36$ M; $[C_2O_4^{2-}] = 5.4 \times 10^{-5}$ M

19-57. Suppose that you wish to determine whether a solution of unknown composition is buffered. Explain how you could do this with only two pH measurements.

19-58. The principal reaction when a salt composed of an acidic cation and a basic anion, for example, NH_4CH_3COO, is dissolved in water is of the type

$$NH_4^+(aq) + CH_3COO^-(aq) \rightleftharpoons NH_3(aq) + CH_3COOH(aq)$$

(a) Show that the value of the equilibrium constant for this reaction is given by

$$K \simeq \frac{K_{a,NH_4^+}}{K_{a,CH_3COOH}}$$

(b) Given the above stoichiometry, show that the $[H_3O^+]$ of the solution is equal to

$$[H_3O^+] \simeq (K_{a,NH_4^+}K_{a,CH_3COOH})^{1/2}$$

Note that $[H_3O^+]$ and thus the pH of the solution are independent of the concentration of the salt.

(c) Is an $NH_4CH_3COO(aq)$ solution a buffer? Explain.

19-59. A 1.20-g sample of an unknown acid is dissolved in water and titrated with 0.150 M $NaOH(aq)$ to the equivalence point. The volume of base required is 69.0 mL. Calculate the molecular mass of the acid. The titration curve shows only one dissociable proton per molecule.

19-60. Calculate the number of grams of $NaHCO_3(s)$ required to neutralize 2.00 g of citric acid in water. Citric acid is a triprotic acid, $H_3C_6H_5O_7$.

19-61. Determine how many of the hydrogen atoms in citric acid, $C_6H_8O_7$, are acidic if it requires 147.2 mL of 0.135 M $NaOH(aq)$ solution to titrate a 1.270-g sample of citric acid.

19-62. Calculate the pH of the solution that results when 2.00 g of $Mg(OH)_2(s)$ is dissolved in 850 mL of 0.160 M $HCl(aq)$.

19-63. Calculate $[H_2C_2O_4]$, $[HC_2O_4^-]$, and $[C_2O_4^{2-}]$ in a 0.125 M $H_2C_2O_4$ solution that is buffered at a pH of 5.00 ($pK_{a1} = 1.27$, $pK_{a2} = 4.27$).

19-64. A solution is 0.0500 M in $HCl(aq)$ and 0.060 M in $CH_3COOH(aq)$. Calculate the pH of the resulting solution if 30.0 mL of 0.120 M $NaOH(aq)$ is added to 50.0 mL of the original solution.

19-65. A commonly used buffer in biological experiments is a phosphate buffer containing NaH_2PO_4 and Na_2HPO_4. Estimate the pH of an aqueous solution that is

(a) 0.050 M NaH_2PO_4 and 0.050 M Na_2HPO_4

(b) 0.050 M NaH_2PO_4 and 0.10 M Na_2HPO_4

(c) 0.10 M NaH_2PO_4 and 0.050 M Na_2HPO_4

The relevant equation is

$$H_2PO_4^-(aq) + H_2O(l) \rightleftharpoons H_3O^+(aq) + HPO_4^{2-}(aq) \qquad K_a = 6.2 \times 10^{-8}\ M$$

19-66. Calculate the pH of a buffer solution obtained by dissolving 10.0 g of $KH_2PO_4(s)$ and 20.0 g of $Na_2HPO_4(s)$ in water and then diluting to 1.00 L. The relevant equation is

$$H_2PO_4^-(aq) + H_2O(l) \rightleftharpoons H_3O^+(aq) + HPO_4^{2-}(aq)$$

with $pK_a = 7.21$.

19-67. Suppose you are performing an experiment during which the pH must be maintained at 3.70. What would be an appropriate buffer to use? (See Table 18-5.)

19-68. Suppose you are performing an experiment during which the pH must be maintained at 5.16. What would be an appropriate buffer to use? (See Table 18-5.)

19-69. Vinegar is an aqueous solution of acetic acid. A 21.0-mL sample of vinegar requires 38.5 mL of 0.400 M $NaOH(aq)$ to neutralize the $CH_3COOH(aq)$. Given that the density of the vinegar is $1.060\ g\cdot mL^{-1}$, calculate the mass percentage of acetic acid in the vinegar.

19-70. A 2.00-g sample of acetylsalicylic acid, better known as aspirin, is dissolved in 100 mL of water and titrated with 0.200 M $NaOH(aq)$ to the equivalence point. The volume of base required is 55.5 mL. Calculate the molecular mass of the acetylsalicylic acid, which has one acidic proton per molecule.

19-71. Calculate the pH at the equivalence point in the titration of 50.0 mL of 0.125 M $NH_3(aq)$ with 0.175 M $HCl(aq)$.

19-72. Calculate the pH at the equivalence point in the titration of 17.5 mL of 0.098 M pyridine with 0.117 M $HI(aq)$.

19-73. Calculate the pH and the concentrations of all the species in 0.0250 M $H_2Se(aq)$.

19-74. Calculate the pH and the concentrations of all the species in 0.0250 M $H_2SO_4(aq)$.

19-75. Write out the "C_2O_4" balance condition and the electroneutrality condition in 0.200 M $H_2C_2O_4(aq)$.

19-76. Write out the "N_2H_4" balance condition and the electroneutrality condition in 0.200 M $N_2H_4(aq)$.

19-77. A buffer is prepared such that $[CH_3COOH]_0 = [CH_3COO^-]_0 = 0.050$ M. Calculate the volume of 0.10 M NaOH that can be added to 100 mL of the solution before its buffering capacity is lost. Assume the buffer capacity is lost when the ratio $[\text{base}]_0/[\text{acid}]_0$ is less than 0.1 or greater than 10.

19-78. A buffer solution is prepared such that $[H_3PO_4]_0 = [H_2PO_4^-]_0 = 0.20$ M. Calculate the volume of 0.010 M HCl that can be added to 200 mL of the solution before its buffering capacity is lost. Assume the buffer capacity is lost when the ratio $[\text{base}]_0/[\text{acid}]_0$ is less than 0.1 or greater than 10.

19-79. Calculate the mass of $NH_4Cl(s)$ that must be added to 1.00 L of 0.200 M $NH_3(aq)$ solution to obtain a solution of pH 9.50. Assume no change in volume.

19-80. Calculate the mass of $NaOH(s)$ that must be added to 500.0 mL of 0.120 M CH_3COOH to yield a solution of pH = 4.52. Assume no change in volume.

19-81. Describe how you would prepare a buffer solution of pH 9.2 starting with 0.10 M $NH_3(aq)$ and 0.10 M $HCl(aq)$.

19-82. How many grams of $NaNO_2(s)$ must be added to 300 mL of 0.200 M $HNO_2(aq)$ to give a pH of 3.70? Assume no volume change when the salt is added.

19-83. A 25.0-gram sample of $NH_4Cl(s)$ is added to 300 mL of 0.500 M $NH_3(aq)$. What is the pH? Assume no volume change when the salt is added.

19-84. What is the pH of the solution that results from the addition of 25.0 mL of 0.200 M $KOH(aq)$ to 50.0 mL of 0.150 M $HNO_2(aq)$?

19-85. A buffer solution consists of 1.00 M each of $HNO_2(aq)$ and $NaNO_2(aq)$. Calculate the change in pH if 50.0 mL of 0.650 M $HCl(aq)$ is added to a liter of the buffer solution.

19-86. Calculate the pH and the concentration of all the species in a 0.100 M $H_2S(aq)$ solution.

Natural Waters

Water is the solvent of life; the human body is about three-fourths water by mass. More chemical and biochemical reactions take place in water than in all other solvents combined. In the absence of water, no known form of life is possible.

The distribution of water on earth is shown in Figure I-1. About three-fourths of the earth's surface is covered with water. The total amount of water on earth is estimated to be 1.4×10^9 km^3, of which 97.4 percent is seawater and 2.6 percent is fresh water. Of the fresh water, four-fifths is located in ice caps and glaciers and about one-fifth is in relatively inaccessible groundwater. Less than 1 percent of the fresh water (0.014 percent of the total) is located in lakes, soils, rivers, biota, and the atmosphere.

We can classify water according to the amount of dissolved minerals it contains; seawater averages 3.5 percent by mass of dissolved minerals (Table I-1). About 75 elements have been detected in seawater, but only 9 species (Table I-2) constitute over 99.9 percent of the mass of the various substances dissolved in seawater. Sodium ions plus chloride ions constitute about 86 percent by mass of the dissolved species in seawater.

Most of the ionic constituents of seawater enter the ocean in the form of superheated (320°C), mineral-rich water that originates deep within the earth. This water escapes from vent holes in the ocean floor (Figure I-2). Manganese nodules form spontaneously in the vicinity of the vents. The nodules are porous, roughly spherical chunks of metal oxides ranging from 2 to 10 cm in diameter. In addition to manganese, the nodules contain iron, cobalt, nickel, copper, zinc, chromium, vanadium, tungsten, and lead. Manganese nodules are a potentially rich source of scarce metals, such as cobalt and chromium. Because they occur at great depths in the ocean, however, the nodules cannot be recovered economically at present.

Table I-1 Classification of water by amount of dissolved minerals

Type of water	Quantity of dissolved minerals/% by mass
fresh	0–0.1
brackish	0.1–1
salty	1–10
brine	>10

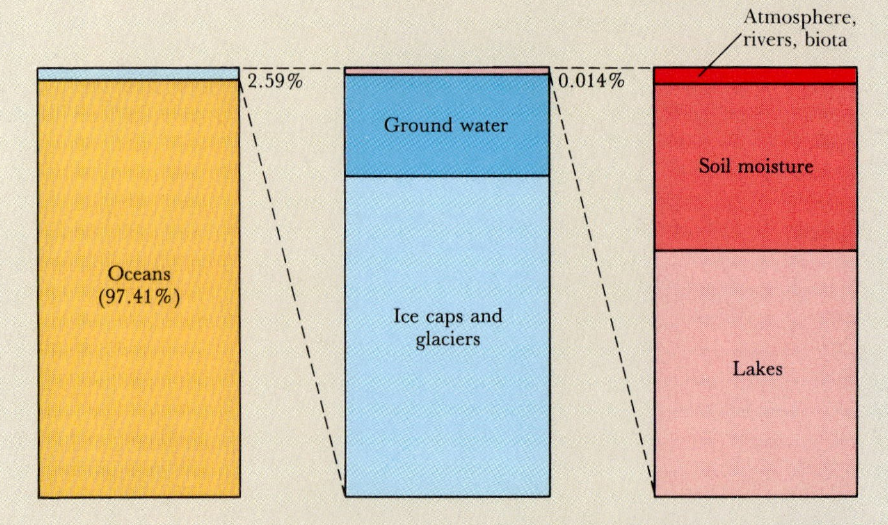

Figure I-1 The distribution of water on the earth is highly uneven. Most of it (97.41 percent) is in the oceans (*yellow*); only a small fraction (2.59 percent) is on the land (*blue*). Even most of the water on land is largely unavailable, because it is sequestered in the form of ice and snow or as groundwater; only a very small amount (0.014 percent) of the earth's total water is readily available to human beings and other organisms (*red*). The total available freshwater supply is about 2.0×10^5 km^3. (Adapted from *Scientific American*, September 1989.)

Table I-2 Concentrations of the principal ionic constituents of seawater and fresh water (average values)

Ion	Seawater concentration/mM	Fresh water concentration/mM
Cl^-	550	0.22
Na^+	460	0.27
SO_4^{2-}	28	0.12
Mg^{2+}	54	0.34
Ca^{2+}	10	0.38
K^+	10	0.06
HCO_3^-	2.3	0.96
Br^-	0.00083	0.03
CO_3^{2-}	0.0003	0.83

Temperature, oxygen concentration, carbon dioxide concentration, phosphate concentration (as HPO_4^{2-}), nitrate concentration, and pH play key roles in the chemistry and biochemistry of the oceans. The temperature of ocean water ranges from a high of about 32°C near the surface in some regions to a low of about −2°C near an ice shelf. The average surface temperature is around 22°C and decreases to about 2°C at a depth of 2 km; below 2 km the ocean temperature is fairly constant at 2°C. The pH of ocean water is remarkably constant at a value of 8.2. The ocean pH is maintained at 8.2 by the buffering action of ocean sediments containing carbonates and phosphates. The buffer capacity of the oceans makes them immune to acid rain, in contrast to many bodies of fresh water.

Variations in the concentration of O_2, NO_3^-, and HPO_4^{2-} with depth are shown in Figure I-3. All these constituents contribute to the development of plants and marine organisms such as phytoplankton, which constitute the primary sources in the ocean food chain for higher forms of marine life. The concentration of oxygen near the surface of the ocean is high relative to that of nitrate and hydrogen phosphate (Figure I-3), because of the dissolution of oxygen from the air and its production in photosynthesis by phytoplankton. The value of $[O_2]$ initially decreases with depth down to about 1.2 km, because of the consumption of oxygen during the decomposition of animal and plant matter. Below 1.2 km the concentration of $O_2(aq)$ increases again because of the absence of plant and animal matter.

Phosphates and nitrates rise to the surface from the nutrient-rich deep ocean water and are depleted near the surface as they are utilized in the production of marine life. The growth of phyto-

Figure I-2 A vent in the ocean floor through which hot, mineral-rich water enters the ocean from sources deep within the earth's mantle. These minerals are the major source of dissolved solids in the oceans. The structure showing at the bottom of the photo is the bathyscope from which the picture was taken.

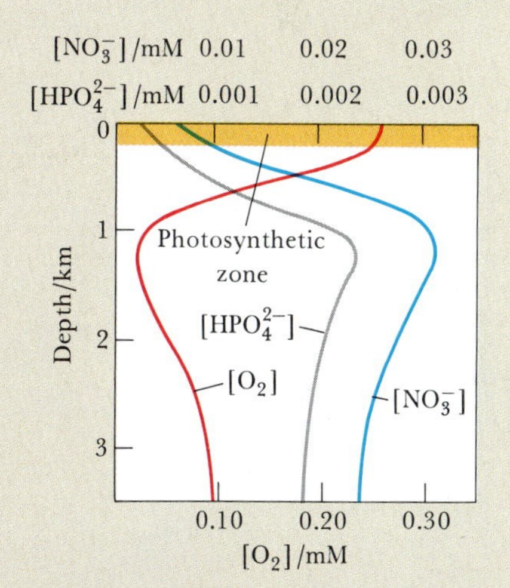

Figure I-3 Variation of $[O_2]$, $[HPO_4^{2-}]$, and $[NO_3^-]$ with ocean depth. Note that in the first kilometer the oxygen concentration decreases while the concentrations of nitrate and hydrogen phosphate increase. The value of the ratio $[NO_3^-]/[HPO_4^{2-}]$ is about 15, equal to the nitrogen-to-phosphorus ratio in phytoplankton, which constitute the first stage of the ocean food chain.

plankton in the oceans is limited by the amount of available phosphate, nitrate, and trace mineral nutrients. The carbon nutrient source is carbon dioxide from the atmosphere. At certain places in the ocean, the nutrient-rich deep water rises to the surface in large quantities as a result of prevailing winds that blow away the warmer and thus lighter surface water. These regions are especially rich in animal life and constitute the great ocean fishing areas. Examples are regions off the coasts of Newfoundland, Chile, and Peru.

Seawater constitutes an enormous source of useful chemicals, but at the present time only four chemicals are obtained from seawater on a commercial scale: pure water, sodium chloride, bromine, and magnesium hydroxide.

Pure water is obtained from seawater on a commercial scale by distillation and reverse osmosis techniques. The process of removing dissolved solids from seawater is called *desalination.* Economical methods of desalination are of paramount importance in the arid regions of the world.

About 40 million metric tons (1 metric ton = 1000 kg = 2200 lb) of sodium chloride is obtained from seawater each year. The process involves filtering the seawater to remove particulate matter, then allowing the filtered water to evaporate from storage ponds until the NaCl crystallizes from solution (Figure I-4).

Bromine is present in seawater as bromide ion at a concentration of 8.3×10^{-4} M; some oil-well brines have concentrations of $Br^-(aq)$ 10-fold higher than those of seawater. The economical recovery of bromine from seawater and brines depends on the fact that elemental bromine, Br_2, is a volatile liquid. To recover bromine from seawater, the pH is decreased from 8.2 to 3.5 by the addition of sulfuric acid, and the bromide ion is then converted to bromine by oxidation with chlorine:

$$2Br^-(aq) + Cl_2(g) \rightarrow Br_2(l) + 2Cl^-(aq)$$

The elemental bromine, Br_2, is swept out of the seawater as the gas by a stream of air passed through the solution.

Magnesium ion is present in seawater at a concentration of 0.054 M. It is separated from seawater by the addition of lime, $CaO(s)$, which precipitates the $Mg^{2+}(aq)$ as the hydroxide, $Mg(OH)_2(s)$, which is much less soluble in water than $Ca(OH)_2$:

$$Mg^{2+}(aq) + CaO(s) + H_2O(l) \rightarrow Mg(OH)_2(s) + Ca^{2+}(aq)$$

The $Mg(OH)_2(s)$ is collected by filtration and converted to magnesium chloride with hydrochloric

Figure I-4 Evaporation ponds are used to obtain salt from brines. The vivid coloration of the ponds is caused by the natural growth of certain algae and bacteria in the brine as it gradually becomes more concentrated.

acid. The magnesium chloride is recovered by evaporation of water from the solution until the $MgCl_2(s)$ crystallizes.

Fresh water that has been in contact with the earth for some time contains a variety of anions and cations (Table I-2). Hard water contains appreciable amounts of divalent cations, primarily Ca^{2+}, Mg^{2+}, and Fe^{2+}. The major anions in hard water are HCO_3^- and SO_4^{2-}. Bicarbonate ions occur in groundwater as a result of the interaction of water containing dissolved CO_2 with limestone ($CaCO_3$):

$$CaCO_3(s) + H_2O(l) + CO_2(aq) \rightleftharpoons Ca^{2+}(aq) + 2HCO_3^-(aq)$$

The reverse of this reaction results in the formation of stalagmites and stalactites in limestone caves. Magnesium ions in groundwater are the result of a similar reaction between $CO_2(aq)$ and dolomite, a mineral containing $CaCO_3$ and $MgCO_3$. Sulfate ions in fresh water result from the dissolution of calcium, magnesium, and iron sulfates.

The divalent cations in hard water form precipitates (scum) in the presence of soaps (Interchapter J). Hard water is classified as either temporary or permanent, depending on which anions it contains. *Temporary hard water* contains $HCO_3^-(aq)$ along with $Ca^{2+}(aq)$ and/or $Mg^{2+}(aq)$. When temporary hard water is heated, calcium carbonate or magnesium carbonate precipitates (thereby softening the water), as a result of the reaction

$$M^{2+}(aq) + 2HCO_3^-(aq) \rightarrow MCO_3(s) + H_2O(l) + CO_2(g)$$

where M^{2+} stands for either divalent metal ion. The metal carbonates that precipitate form deposits, called *boiler scale,* in boilers, hot-water pipes, and tea kettles (Figure I-5). Such deposits in water heaters act as heat insulators, and thereby increase the cost of heating water. Excessive deposits produce clogged pipes. In *permanent hard water* the primary anion is sulfate. Both calcium sulfate and magnesium sulfate are soluble in hot water and are not precipitated by heating. Thus, such water cannot be softened by heating.

Hard water can be converted by chemical means to soft water by removing the divalent cations. The softening process involves the use of water-insoluble ion-exchange resins that bind divalent metal ions and release sodium ions:

$$2[\underbrace{R{-}SO_3^-}_{\text{resin group}}Na^+] + M^{2+}(aq) \rightleftharpoons [(R{-}SO_3^-)_2M^{2+}] + 2Na^+(aq)$$

This cation-exchange reaction is used in home water-softening systems. Note that the divalent ions, $M^{2+}(aq)$, in the hard water are replaced with sodium ions, $Na^+(aq)$, which do not cause the precipitation of scum in the presence of soap. The ion-exchange resin can be reactivated by running a concentrated salt solution (NaCl) through the ion-exchange system, replacing Ca^{2+} and Mg^{2+} with Na^+ ions and thus reversing the cation-exchange reaction.

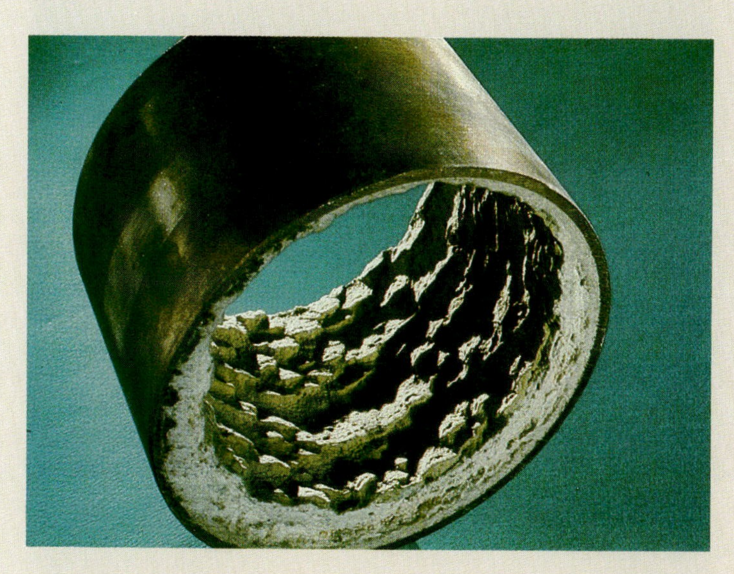

Figure I-5 Boiler scale consists of deposits of divalent metal carbonates such as $CaCO_3(s)$ and $MgCO_3(s)$.

20 SOLUBILITY AND PRECIPITATION REACTIONS

Precipitation of $Ag^+(aq)$ as $AgCl(s)$ by addition of $HCl(aq)$ to a solution containing $Ag^+(aq)$.

Chemists often are faced with the problem of determining which species are present in a sample of an unknown material. Analysis of this kind is called qualitative analysis, and many general chemistry laboratory courses have several experiments dealing with the techniques of qualitative analysis. Many of the procedures of qualitative analysis involve the formation and the separation of precipitates. We learned in Chapter 6 that many substances are insoluble or have a low solubility in water. In this chapter we treat solubility within the framework of chemical equilibria; this approach enables us to calculate the solubility of a solid not only in pure water but also in solutions of acids, bases, and salts. The type of calculations that we discuss in this chapter have many practical applications. For example, many metals are obtained from their ores by a series of reactions involving precipitates; the geological formation of minerals and rocks is governed by the relative solubilities of various substances; and the ultrapurification of materials used in computers and electronic devices often involves precipitation reactions.

20-1. THE LAW OF CONCENTRATION ACTION GOVERNS THE EQUILIBRIUM BETWEEN AN IONIC SOLID AND ITS CONSTITUENT IONS IN SOLUTION

In Chapter 6 we presented a few rules that enable us to determine which salts are soluble in water and which are not. We defined the **solubility** of a substance as the quantity of that substance that is dissolved in a saturated solution. We chose to call any substance whose solubility is less than 0.01 M insoluble. In this section we will discuss solubility more quantitatively. In particular, we will cast solubility in terms of chemical equilibria, using the idea of an equilibrium constant.

Consider the equilibrium between solid silver bromate and its constituent ions in water:

$$AgBrO_3(s) \rightleftharpoons Ag^+(aq) + BrO_3^-(aq) \qquad (20\text{-}1)$$

Application of the law of concentration action to Equation (20-1) yields the equilibrium-constant expression for this chemical equation:

$$K_{sp} = [Ag^+][BrO_3^-]$$

The subscript sp stands for solubility product, and K_{sp} is called the **solubility-product constant.** Note that $AgBrO_3(s)$ does not appear in the K_{sp} expression; as we discussed in Chapter 17, a pure solid does not appear in an equilibrium-constant expression because its concentration does not vary.

The experimental value of K_{sp} at 25°C for Equation (20-1) is

$$K_{sp} = [Ag^+][BrO_3^-] = 5.8 \times 10^{-5}\ M^2 \qquad (20\text{-}2)$$

Equation (20-2) states that if $AgBrO_3(s)$ is in equilibrium with an aqueous solution of $AgBrO_3(aq)$ at 25°C, then the product of the concentrations of $Ag^+(aq)$ and $BrO_3^-(aq)$ at equilibrium must equal $5.8 \times 10^{-5}\ M^2$. This relation holds irrespective of what other, if any, species are present in the solution.

A K_{sp} expression can be used to calculate the solubility of a solid. For example, suppose that excess $AgBrO_3(s)$ is in contact with water at 25°C. Then at equilibrium

$$[Ag^+][BrO_3^-] = 5.8 \times 10^{-5}\ M^2$$

From the reaction stoichiometry, Equation (20-1), we have

$$[Ag^+] = [BrO_3^-]$$

because each formula unit of $AgBrO_3$ that dissolves produces one $Ag^+(aq)$ ion and one $BrO_3^-(aq)$ ion and because $AgBrO_3(s)$ is the only source of $Ag^+(aq)$ and $BrO_3^-(aq)$. If we denote the solubility of $AgBrO_3(s)$ in units of molarity by s, then

$$s = \text{solubility of } AgBrO_3(s) \text{ in water} = [Ag^+] = [BrO_3^-]$$

because the concentration of either $Ag^+(aq)$ or $BrO_3^-(aq)$ is equal to the number of moles of dissolved salt per liter of solution. Thus, from the K_{sp} expression, we have

$$K_{sp} = 5.8 \times 10^{-5}\ M^2 = [Ag^+][BrO_3^-] = s^2$$

and

$$s = (5.8 \times 10^{-5}\ M^2)^{1/2} = 7.6 \times 10^{-3}\ M$$

The formula mass of $AgBrO_3(s)$ is 235.8, so the number of grams of $AgBrO_3(s)$ that dissolves in 1.00 L of solution at 25°C is

$$(7.6 \times 10^{-3}\ mol \cdot L^{-1})\left(\frac{235.8\ g\ AgBrO_3}{1\ mol\ AgBrO_3}\right) = 1.8\ g \cdot L^{-1}$$

EXAMPLE 20-1: The value of K_{sp} for $BaCrO_4(s)$ in equilibrium with an aqueous solution of its constituent ions at 25°C is $1.2 \times 10^{-10}\ M^2$. Write the chemical equation that represents the solubility equilibrium for $BaCrO_4(s)$ and calculate its solubility in water (in $g \cdot L^{-1}$) at 25°C.

Solution: The solubility equilibrium is

$$BaCrO_4(s) \rightleftharpoons Ba^{2+}(aq) + CrO_4^{2-}(aq)$$

The K_{sp} expression is

$$K_{sp} = [Ba^{2+}][CrO_4^{2-}] = 1.2 \times 10^{-10}\ M^2$$

If $BaCrO_4(s)$ is equilibrated with pure water, then, from the reaction stoichiometry, we have at equilibrium

$$[Ba^{2+}] = [CrO_4^{2-}] = s$$

where s is the solubility of $BaCrO_4(s)$ in pure water. Thus,

$$K_{sp} = s^2 = 1.2 \times 10^{-10}\ M^2$$

and

$$s = (1.2 \times 10^{-10}\ M^2)^{1/2} = 1.1 \times 10^{-5}\ M$$

The solubility in grams per liter is given by

$$s = (1.1 \times 10^{-5}\ M)\left(\frac{253.3\ g\ BaCrO_4}{1\ mol\ BaCrO_4}\right) = 2.8 \times 10^{-3}\ g \cdot L^{-1}$$

PRACTICE PROBLEM 20-1: Calculate the solubility of $TlBrO_3(s)$ in water at 25°C given that $K_{sp} = 1.7 \times 10^{-4}\ M^2$.

Answer: $4.3\ g{\cdot}L^{-1}$

We saw that when excess $AgBrO_3(s)$ is in equilibrium with pure water, we have $[Ag^+] = [BrO_3^-]$, because each $AgBrO_3$ unit that dissolves yields one $Ag^+(aq)$ and one $BrO_3^-(aq)$ and there is no other source of $Ag^+(aq)$ and $BrO_3^-(aq)$. Now consider the problem of calculating the solubility in water of copper(II) iodate, $Cu(IO_3)_2(s)$, which yields *two* $IO_3^-(aq)$ and one $Cu^{2+}(aq)$ for each $Cu(IO_3)_2$ unit that dissolves.

The chemical equation that represents the solubility equilibrium of $Cu(IO_3)_2(s)$ in water is

$$Cu(IO_3)_2(s) \rightleftharpoons Cu^{2+}(aq) + 2IO_3^-(aq)$$

According to the law of concentration action, the K_{sp} expression for this equation is

$$K_{sp} = [Cu^{2+}][IO_3^-]^2$$

The experimental value of K_{sp} at 25°C is $7.4 \times 10^{-8}\ M^3$, so we have at 25°C

$$K_{sp} = [Cu^{2+}][IO_3^-]^2 = 7.4 \times 10^{-8}\ M^3 \qquad (20\text{-}3)$$

Note that it is the *square* of the concentration of $IO_3^-(aq)$ that appears in the K_{sp} expression for $Cu(IO_3)_2(s)$ because each formula unit of $Cu(IO_3)_2$ that dissolves produces two iodate ions. Thus, when $Cu(IO_3)_2(s)$ is in equilibrium with its constituent ions in water, the concentration of iodate ion is twice as great as the concentration of copper(II) ion, if there is no other source of $Cu^{2+}(aq)$ and $IO_3^-(aq)$:

$$[IO_3^-] = 2[Cu^{2+}]$$

The solubility of $Cu(IO_3)_2(s)$ in pure water is equal to $[Cu^{2+}]$ because each mole of $Cu(IO_3)_2$ that dissolves yields one mole of $Cu^{2+}(aq)$. If we denote the solubility of $Cu(IO_3)_2(s)$ in pure water by s, then

$$s = \text{solubility of } Cu(IO_3)_2(s) \text{ in water} = [Cu^{2+}] = \frac{[IO_3^-]}{2}$$

It follows then that $[IO_3^-] = 2s$. Combining these results with the K_{sp} expression, Equation (20-3), yields

$$7.4 \times 10^{-8}\ M^3 = [Cu^{2+}][IO_3^-]^2 = (s)(2s)^2 = 4s^3$$

therefore,

$$s = \left(\frac{7.4 \times 10^{-8}\ M^3}{4}\right)^{1/3} = 2.6 \times 10^{-3}\ M$$

Note that $[Cu^{2+}] = s = 2.6 \times 10^{-3}\ M$ and also that $[IO_3^-] = 2s = 5.2 \times 10^{-3}\ M$. The solubility of $Cu(IO_3)_2(s)$ in grams per liter is

$$s = (2.6 \times 10^{-3}\ M)\left(\frac{413.3\ g\ Cu(IO_3)_2}{1\ mol\ Cu(IO_3)_2}\right) = 1.1\ g{\cdot}L^{-1}$$

Generally for a salt with the formula A_xB_y, the solubility-product constant is given by

$$K_{sp} = [A]^x[B]^y \quad (20\text{-}4)$$

and the solubility is given by

$$s = \frac{[A]}{x} = \frac{[B]}{y} \quad (20\text{-}5)$$

Various solubility-product constants are given in Table 20-1.

EXAMPLE 20-2: The solubility-product constant for silver chromate in equilibrium with its constituent ions in water at 25°C is $1.1 \times 10^{-12}\ M^3$. Calculate the value of $[Ag^+]$ that results when pure water is saturated with $Ag_2CrO_4(s)$.

Solution: The $Ag_2CrO_4(s)$ solubility equilibrium is

$$Ag_2CrO_4(s) \rightleftharpoons 2Ag^+(aq) + CrO_4^{2-}(aq)$$

and the solubility-product expression is

$$K_{sp} = [Ag^+]^2[CrO_4^{2-}] = 1.1 \times 10^{-12}\ M^3$$

Each Ag_2CrO_4 unit that dissolves yields two $Ag^+(aq)$ and one $CrO_4^{2-}(aq)$; thus,

$$s = \frac{[Ag^+]}{2} = [CrO_4^{2-}]$$

Substitution of this result into the K_{sp} expression for $Ag_2CrO_4(s)$ yields

$$K_{sp} = (2s)^2(s) = 1.1 \times 10^{-12}\ M^3$$

so

$$s^3 = \frac{1.1 \times 10^{-12}\ M^3}{4}$$

Solving for s yields

$$s = 6.5 \times 10^{-5}\ M$$

The value of $[Ag^+]$ is

$$[Ag^+] = 2s = 1.3 \times 10^{-4}\ M$$

PRACTICE PROBLEM 20-2: Calculate the solubility (in $g \cdot L^{-1}$) of mercury(I) chloride, Hg_2Cl_2, in water at 25°C given that $K_{sp} = 1.3 \times 10^{-18}\ M^3$.

Answer: $3.25 \times 10^{-4}\ g \cdot L^{-1}$

EXAMPLE 20-3: The solubility of $PbCl_2$ in water at 25°C is $4.41\ g \cdot L^{-1}$. Calculate the value of K_{sp} of $PbCl_2$.

Solution: The solubility of $PbCl_2$ in moles per liter is

$$s = (4.41\ g \cdot L^{-1})\left(\frac{1\ mol\ PbCl_2}{278.1\ g\ PbCl_2}\right) = 1.59 \times 10^{-2}\ M$$

Table 20-1 Solubility-product constants for various salts in water at 25°C

Bromates	K_{sp}
$AgBrO_3$	5.8×10^{-5} M^2
$Pb(BrO_3)_2$	7.9×10^{-6} M^3
$TlBrO_3$	1.7×10^{-4} M^2

Bromides	K_{sp}
$AgBr$	5.0×10^{-13} M^2
$CuBr$	5.3×10^{-9} M^2
Hg_2Br_2*	5.6×10^{-23} M^3
$HgBr_2$	1.3×10^{-19} M^3
$PbBr_2$	4.0×10^{-5} M^3
$TlBr$	3.4×10^{-6} M^2

Carbonates	K_{sp}
Ag_2CO_3	8.1×10^{-12} M^3
$BaCO_3$	5.1×10^{-9} M^2
$CaCO_3$	2.8×10^{-9} M^2
$CdCO_3$	1.8×10^{-14} M^2
$CoCO_3$	1.0×10^{-10} M^2
$CuCO_3$	1.4×10^{-10} M^2
$FeCO_3$	3.2×10^{-11} M^2
$MgCO_3$	3.5×10^{-8} M^2
$MnCO_3$	1.8×10^{-11} M^2
$NiCO_3$	1.3×10^{-7} M^2
$PbCO_3$	7.4×10^{-14} M^2
$SrCO_3$	1.1×10^{-10} M^2
$ZnCO_3$	1.4×10^{-11} M^2

Chlorides	K_{sp}
$AgCl$	1.8×10^{-10} M^2
$CuCl$	1.2×10^{-6} M^2
Hg_2Cl_2*	1.3×10^{-18} M^3
$PbCl_2$	1.6×10^{-5} M^3
$TlCl$	1.7×10^{-4} M^2

Chromates	K_{sp}
Ag_2CrO_4	1.1×10^{-12} M^3
$BaCrO_4$	1.2×10^{-10} M^2
$CuCrO_4$	3.6×10^{-6} M^2
Hg_2CrO_4	2.0×10^{-9} M^2
$PbCrO_4$	2.8×10^{-13} M^2
Tl_2CrO_4	9.8×10^{-15} M^3

Cyanides	K_{sp}
$AgCN$	2.2×10^{-16} M^2
$Hg_2(CN)_2$	5×10^{-40} M^3
$Zn(CN)_2$	3×10^{-16} M^3

Fluorides	K_{sp}
BaF_2	1.0×10^{-6} M^3
CaF_2	5.3×10^{-9} M^3
MgF_2	6.5×10^{-9} M^3
PbI_2	7.7×10^{-8} M^3
SrF_2	2.5×10^{-9} M^3

Hydroxides	K_{sp}
$Al(OH)_3$	1.3×10^{-33} M^4
$Ca(OH)_2$	5.5×10^{-6} M^3
$Cd(OH)_2$	2.5×10^{-14} M^3
$Co(OH)_2$	1.3×10^{-15} M^3
$Cr(OH)_3$	6.3×10^{-31} M^4
$Cu(OH)_2$	2.2×10^{-20} M^3
$Fe(OH)_2$	8.0×10^{-16} M^3
$Fe(OH)_3$	1.0×10^{-38} M^4
$Mg(OH)_2$	1.8×10^{-11} M^3
$Ni(OH)_2$	2.0×10^{-15} M^3
$Pb(OH)_2$	1.2×10^{-15} M^3
$Sn(OH)_2$	1.4×10^{-28} M^3
$Zn(OH)_2$	1.0×10^{-15} M^3

Iodates	K_{sp}
$AgIO_3$	3.1×10^{-8} M^2
$Ba(IO_3)_2$	1.5×10^{-9} M^3
$Ca(IO_3)_2$	7.1×10^{-7} M^3
$Cd(IO_3)_2$	2.3×10^{-8} M^3
$Cu(IO_3)_2$	7.4×10^{-8} M^3
$Pb(IO_3)_2$	2.5×10^{-13} M^3
$TlIO_3$	3.1×10^{-6} M^2
$Zn(IO_3)_2$	3.9×10^{-6} M^3

Iodides	K_{sp}
AgI	8.3×10^{-17} M^2
CuI	1.1×10^{-12} M^2
Hg_2I_2*	4.5×10^{-29} M^3
PbI_2	7.1×10^{-9} M^3
TlI	6.5×10^{-8} M^2

Oxalates	K_{sp}
$Ag_2C_2O_4$	3.4×10^{-11} M^3
CaC_2O_4	4×10^{-9} M^2
MgC_2O_4	7×10^{-7} M^2
SrC_2O_4	4×10^{-7} M^2

Sulfates	K_{sp}
Ag_2SO_4	1.4×10^{-5} M^3
$BaSO_4$	1.1×10^{-10} M^2
$CaSO_4$	9.1×10^{-6} M^2
Hg_2SO_4	7.4×10^{-7} M^2
$PbSO_4$	1.6×10^{-8} M^2
$SrSO_4$	3.2×10^{-7} M^2

Sulfides	K_{sp}
Ag_2S	8×10^{-51} M^3
CdS	8.0×10^{-27} M^2
CoS	5×10^{-22} M^2
CuS	6.3×10^{-36} M^2
FeS	6.3×10^{-18} M^2
HgS	4×10^{-53} M^2
MnS	2.5×10^{-13} M^2
NiS	1.3×10^{-25} M^2
PbS	8.0×10^{-28} M^2
SnS	1.0×10^{-25} M^2
Tl_2S	6×10^{-22} M^3
ZnS	1.6×10^{-24} M^2

Thiocyanates	K_{sp}
$AgSCN$	1.1×10^{-12} M^2
$Cu(SCN)_2$	4.0×10^{-14} M^3
$Hg_2(SCN)_2$	3.0×10^{-20} M^3
$Hg(SCN)_2$	2.8×10^{-20} M^3
$TlSCN$	1.6×10^{-4} M^2

*Recall that Hg(I) exists as $Hg_2^{2+}(aq)$ in aqueous solution.

Using the K_{sp} values from Table 20-1, we have that

$$[Mn^{2+}][S^{2-}] = 2.5 \times 10^{-13}\ M^2$$
$$[Pb^{2+}][S^{2-}] = 8.0 \times 10^{-28}\ M^2$$

Therefore, the concentrations of $Mn^{2+}(aq)$ and $Pb^{2+}(aq)$ are given by

$$[Mn^{2+}] = \frac{2.5 \times 10^{-13}\ M^2}{[S^{2-}]} = \frac{2.5 \times 10^{-13}\ M^2}{1.1 \times 10^{-13}\ M} = 2.3\ M$$

$$[Pb^{2+}] = \frac{8.0 \times 10^{-28}\ M^2}{[S^{2-}]} = \frac{8.0 \times 10^{-28}\ M^2}{1.1 \times 10^{-13}\ M} = 7.3 \times 10^{-15}\ M$$

Because one formula unit of $Mn^{2+}(aq)$ occurs in solution for each formula unit of $MnS(s)$ that dissolves, $[Mn^{2+}]$ is equal to the solubility of $MnS(s)$. Similarly, $[Pb^{2+}]$ is equal to the solubility of $PbS(s)$. Thus, at pH = 4.0, $MnS(s)$ is soluble (2.3 M), whereas $PbS(s)$ is insoluble (7.3×10^{-15} M).

At pH = 7.0,

$$[S^{2-}] = \frac{1.1 \times 10^{-21}\ M^3}{(1.0 \times 10^{-7}\ M)^2} = 1.1 \times 10^{-7}\ M$$

so

$$[Mn^{2+}] = 2.3 \times 10^{-6}\ M$$
$$[Pb^{2+}] = 7.3 \times 10^{-21}\ M$$

These results show that both $MnS(s)$ and $PbS(s)$ are insoluble at a pH of 7.0.

PRACTICE PROBLEM 20-10: Calculate the solubilities at 25°C of $CdS(s)$ and $CuS(s)$ in a saturated aqueous solution of H_2S at pH = 0.50.

Answer: $[Cd^{2+}] = 7.3 \times 10^{-7}$ M and $[Cu^{2+}] = 5.7 \times 10^{-16}$ M. Both salts are insoluble at pH = 0.50.

20-8. SOME METAL CATIONS CAN BE SEPARATED FROM A MIXTURE BY THE FORMATION OF AN INSOLUBLE HYDROXIDE OF ONE OF THEM

Just as we can separate certain metal ions in aqueous solution by selective precipitation as metal sulfides, we can also selectively precipitate certain metal ions as hydroxides. For example, consider the hydroxide of $Zn^{2+}(aq)$, for which

$$Zn(OH)_2(s) \rightleftharpoons Zn^{2+}(aq) + 2OH^-(aq)$$

and

$$K_{sp} = 1.0 \times 10^{-15}\ M^3 = [Zn^{2+}][OH^-]^2$$

The solubility of $Zn(OH)_2(s)$ in water can be calculated from the K_{sp} expression:

$$s = [Zn^{2+}] = \frac{1.0 \times 10^{-15}\ M^3}{[OH^-]^2} \qquad (20\text{-}22)$$

Table 20-3 **Solubility of $Zn(OH)_2(s)$ in water at 25°C at various pH values**

pH	$[H_3O^+]$/M	$[H_3O^+]^2/M^2$	$[Zn^{2+}]$/M
6.5	3.2×10^{-7}	1.0×10^{-13}	1.0
6.8	1.6×10^{-7}	2.5×10^{-14}	0.25
7.0	1.0×10^{-7}	1.0×10^{-14}	0.10
7.5	3.2×10^{-8}	1.0×10^{-15}	0.010
8.0	1.0×10^{-8}	1.0×10^{-16}	0.0010
8.5	3.2×10^{-9}	1.0×10^{-17}	0.00010

The concentration of $OH^-(aq)$ can be related to $[H_3O^+]$ by using the ion product constant expression for water:

$$[OH^-] = \frac{K_w}{[H_3O^+]} = \frac{1.00 \times 10^{-14}\ M^2}{[H_3O^+]}$$

Substitution of this equation into Equation (20-22) yields

$$\begin{aligned} s = [Zn^{2+}] &= \frac{1.0 \times 10^{-15}\ M^3}{(1.00 \times 10^{-14}\ M^2)^2}[H_3O^+]^2 \\ &= (1.0 \times 10^{13}\ M^{-1})[H_3O^+]^2 \end{aligned} \quad (20\text{-}23)$$

From this expression we can calculate the solubility of $Zn(OH)_2(s)$, that is, $[Zn^{2+}]$, at various pH values, as shown in Table 20-3. These results are plotted in Figure 20-5, together with the analogous results for the solubility of $Fe(OH)_3(s)$. Note that $Fe^{3+}(aq)$ can be separated from $Zn^{2+}(aq)$ by adjusting the pH of the solution to about 5 with an acetic acid-acetate buffer. At pH $\simeq$ 5, the $Fe(OH)_3(s)$ precipitates and the $Zn^{2+}(aq)$ remains in solution.

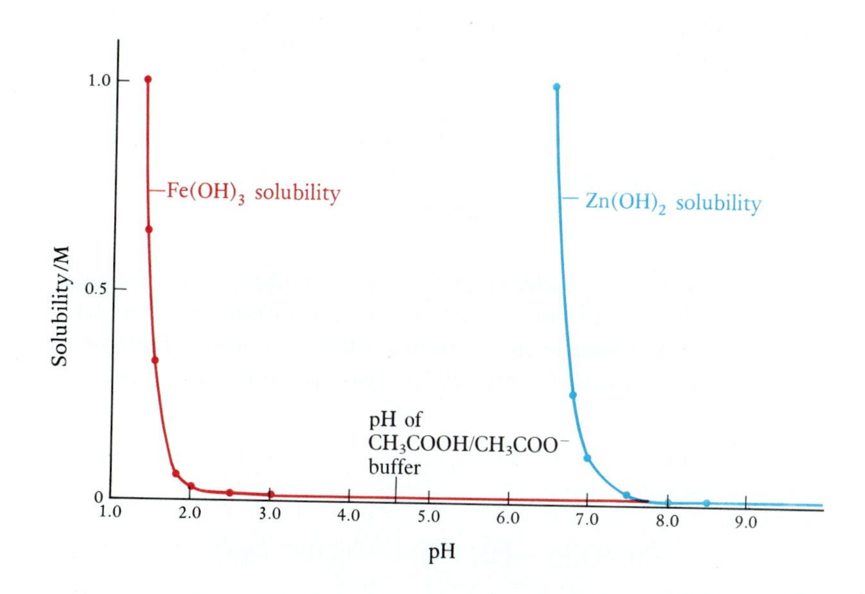

Figure 20-5 Solubilities of $Fe(OH)_3(s)$ and $Zn(OH)_2(s)$ as a function of pH. Note that a much lower pH is required to dissolve $Fe(OH)_3(s)$ than to dissolve $Zn(OH)_2(s)$. Therefore, at pH = 4.8, for example, $Fe(OH)_3(s)$ precipitates and $Zn^{2+}(aq)$ remains in solution. The $Fe(OH)_3(s)$ can be filtered off, thereby separating $Fe^{3+}(aq)$ from $Zn^{2+}(aq)$.

EXAMPLE 20-11: Calculate the solubilities of $Zn(OH)_2(s)$ and $Fe(OH)_3(s)$ in an aqueous solution buffered at pH = 6.8.

Solution: The solubility of $Zn(OH)_2(s)$ in a solution buffered at pH = 6.8 can be calculated from Equation (20-23):

$$s = [Zn^{2+}] = (1.0 \times 10^{13}\ M^{-1})[H_3O^+]^2$$

At pH = 6.8, we have $[H_3O^+] = 10^{-pH} = 10^{-6.8} = 1.58 \times 10^{-7}$ M; thus,

$$s = (1.0 \times 10^{13}\ M^{-1})(1.58 \times 10^{-7}\ M)^2 = 0.25\ M$$

The solubility of $Fe(OH)_3(s)$($K_{sp} = 1.0 \times 10^{-38}\ M^4$ from Table 20-1) is computed in a manner analogous to that just described for $Zn(OH)_2(s)$. Thus, we write

$$Fe(OH)_3(s) \rightleftharpoons Fe^{3+}(aq) + 3OH^-(aq)$$

and

$$K_{sp} = [Fe^{3+}][OH^-]^3 = 1.0 \times 10^{-38}\ M^4$$

The solubility of $Fe(OH)_3(s)$ is equal to $Fe^{3+}(aq)$, so

$$s = [Fe^{3+}] = \frac{1.0 \times 10^{-38}\ M^4}{[OH^-]^3} = \frac{1.0 \times 10^{-38}\ M^4[H_3O^+]^3}{K_w^3}$$

The final form here is obtained by using the relation $[OH^-] = K_w/[H_3O^+]$. Using the fact that $K_w = 1.0 \times 10^{-14}\ M^2$, we have

$$s = [Fe^{3+}] = \frac{1.0 \times 10^{-38}\ M^4[H_3O^+]^3}{(1.0 \times 10^{-14}\ M^2)^3} = (1.0 \times 10^4\ M^{-2})[H_3O^+]^3$$

At $[H_3O^+] = 1.58 \times 10^{-7}$ M, we calculate

$$s = [Fe^{3+}] = (1.0 \times 10^4\ M^{-2})(1.58 \times 10^{-7}\ M)^3 = 3.9 \times 10^{-17}\ M$$

At pH = 6.8, $Fe(OH)_3(s)$ is insoluble and $Zn(OH)_2$ is soluble (Figure 20-5).

PRACTICE PROBLEM 20-11: Calculate the solubilities at 25°C of $Cd(OH)_2(s)$ and $Cu(OH)_2(s)$ in aqueous solution as a function of pH. In what pH range can the two hydroxides be separated?

Answer: $[Cd^{2+}] = (2.5 \times 10^{14}\ M^{-1})[H_3O^+]^2$ and $[Cu^{2+}] = (2.2 \times 10^8\ M^{-1})[H_3O^+]^2$. At pH = 7.0 ($[H_3O^+] = 10^{-7}$ M), $Cd(OH)_2(s)$ is very soluble but $Cu(OH)_2(s)$ is not.

Figure 20-6 Aluminum hydroxide occurs as a white, flocculent precipitate that is used to clarify water.

20-9. AMPHOTERIC METAL HYDROXIDES DISSOLVE IN BOTH HIGHLY ACIDIC AND HIGHLY BASIC SOLUTIONS

Many metal oxides and hydroxides that are insoluble in neutral aqueous solutions dissolve in both acidic and basic solutions. Such hydroxides are called **amphoteric metal hydroxides.** Aluminum hydroxide, $Al(OH)_3(s)$, is an example (Figure 20-6). The equations for the relevant reactions are

$$Al(OH)_3(s) + 3H_3O^+(aq) \rightleftharpoons Al^{3+}(aq) + 6H_2O(l) \qquad (20\text{-}24)$$

and

$$Al(OH)_3(s) + OH^-(aq) \rightleftharpoons Al(OH)_4^-(aq) \qquad (20\text{-}25)$$

Table 20-4 **Equilibrium constants for amphoteric metal hydroxides in water at 25°C**

Reaction	K
$Al(OH)_3(s) + OH^-(aq) \rightleftharpoons Al(OH)_4^-(aq)$	40
$Pb(OH)_2(s) + OH^-(aq) \rightleftharpoons Pb(OH)_3^-(aq)$	0.08
$Zn(OH)_2(s) + 2OH^-(aq) \rightleftharpoons Zn(OH)_4^{2-}(aq)$	0.05 M^{-1}
$Cr(OH)_3(s) + OH^-(aq) \rightleftharpoons Cr(OH)_4^-(aq)$	0.04
$Sn(OH)_2(s) + OH^-(aq) \rightleftharpoons Sn(OH)_3^-(aq)$	0.01

In acidic solutions, $Al(OH)_3(s)$ dissolves because of a reaction that is similar to an acid-base neutralization reaction; and in basic solutions, it dissolves because of the formation of a soluble hydroxy complex ion, $Al(OH)_4^-(aq)$. The total solubility of $Al(OH)_3$ at any pH is given by

$$s = [Al^{3+}] + [Al(OH)_4^-] \qquad (20\text{-}26)$$

The value of $[Al^{3+}]$ can be obtained from the solubility product constant expression of $Al(OH)_3(s)$

$$K_{sp} = [Al^{3+}][OH^-]^3 = 1.3 \times 10^{-33}\ M^4 \qquad (20\text{-}27)$$

and that of $[Al(OH)_4^-]$ from the equilibrium-constant expression for the reaction described by Equation (20-25)

$$\frac{[Al(OH)_4^-]}{[OH^-]} = K = 40 \qquad (20\text{-}28)$$

where the value of K is given in Table 20-4. At a pH of 12.0, $[OH^-] = 1.0 \times 10^{-2}$ M, so Equation (20-27) gives

$$[Al^{3+}] = \frac{1.3 \times 10^{-33}\ M^4}{[OH^-]^3} = \frac{1.3 \times 10^{-33}\ M^4}{(1.0 \times 10^{-2}\ M)^3} = 1.3 \times 10^{-27}\ M$$

and Equation (20-28) gives

$$[Al(OH)_4^-] = 40[OH^-] = (40)(1.0 \times 10^{-2}\ M) = 0.40\ M$$

Therefore, the solubility of $Al(OH)_3(s)$ at pH = 12.0 is

$$s = [Al^{3+}] + [Al(OH)_4^-] = 1.3 \times 10^{-27}\ M + 0.40\ M \simeq 0.40\ M$$

If we computed the solubility of $Al(OH)_3(s)$ at a pH = 12.0 without considering the reactions described by Equation (20-25), then our result would be in error by a factor of about 10^{26}.

At lower values of pH (from 4.0 to 10.0), the reactions described by Equations (20-24) and (20-25) yield very little $Al^{3+}(aq)$ *or* $Al(OH)_4^-(aq)$. For example, at pH = 9.0, $[OH^-] = 1.0 \times 10^{-5}$ M, so Equation (20-27) gives

$$[Al^{3+}] = \frac{1.3 \times 10^{-33}\ M^4}{[OH^-]^3} = \frac{1.3 \times 10^{-33}\ M^4}{(1.0 \times 10^{-5}\ M)^3} = 1.3 \times 10^{-18}\ M$$

and Equation (20-28) gives

$$[Al(OH)_4^-] = (40)(1.0 \times 10^{-5}\ M) = 4.0 \times 10^{-4}\ M$$

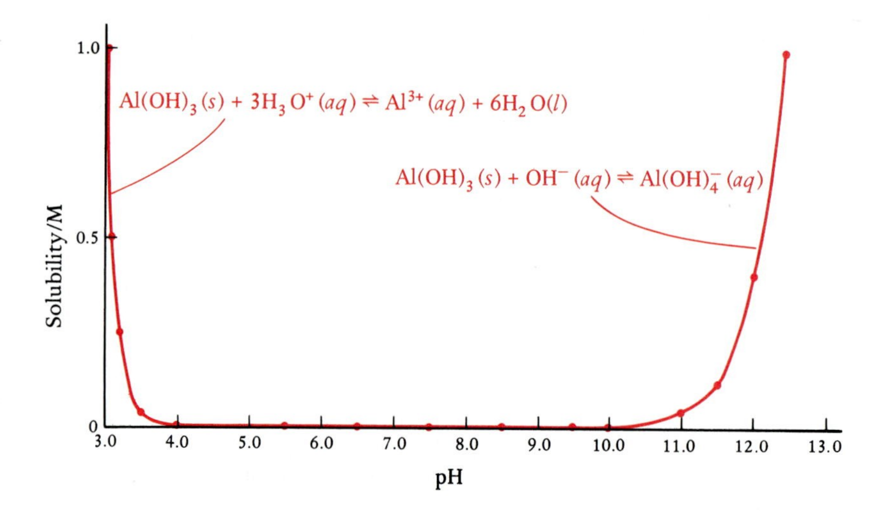

Figure 20-7 The solubility of $Al(OH)_3(s)$ as a function of pH. The amphoteric nature of $Al(OH)_3$ is clearly shown by its solubility in both highly acidic and highly basic solutions. Note that $Al(OH)_3(s)$ is essentially insoluble over the pH range 4 to 10.

For lower values of pH, however, Equation (20-24) gives significant values of $Al^{3+}(aq)$. For example, at pH = 3.0,

$$[Al^{3+}] = \frac{1.3 \times 10^{-33}\ M^4}{(1.0 \times 10^{-11}\ M)^3} = 1.3\ M$$

whereas Equation (20-28) gives

$$[Al(OH)_4^-] = (40)(1.0 \times 10^{-11}\ M) = 4.0 \times 10^{-10}\ M$$

Thus, we see that $Al(OH)_3(s)$ dissolves in acidic solutions and in basic solutions, but not in neutral solutions. The amphoteric behavior of $Al(OH)_3(s)$ is illustrated in Figure 20-7. Some other examples of amphoteric metal hydroxides are given in Table 20-4.

EXAMPLE 20-12: Use the equilibrium-constant data for zinc hydroxide in Table 20-4 to calculate its solubility in a solution buffered at pH = 7.0, 10.0, and 14.0.

Solution: Two equilibrium expressions that we can use are

$$Zn(OH)_2(s) \rightleftharpoons Zn^{2+}(aq) + 2OH^-(aq) \qquad K_{sp} = 1.0 \times 10^{-15}\ M^3$$

and

$$Zn(OH)_2(s) + 2OH^-(aq) \rightleftharpoons Zn(OH)_4^{2-}(aq) \qquad K = 0.05\ M^{-1}$$

and the two equilibrium-constant expressions are

$$[Zn^{2+}][OH^-]^2 = 1.0 \times 10^{-15}\ M^3$$

and

$$[Zn(OH)_4^{2-}] = (0.05\ M^{-1})[OH^-]^2$$

The solubility of $Zn(OH)_2(s)$ at any pH is given by

$$s = [Zn^{2+}] + [Zn(OH)_4^{2-}]$$

At pH = 7.0, $[OH^-] = 1.0 \times 10^{-7}$ M, so

$$[Zn^{2+}] = \frac{1.0 \times 10^{-15}\ M^3}{(1.0 \times 10^{-7}\ M)^2} = 0.10\ M$$

and

$$[Zn(OH)_4^{2-}] = (0.05\ M^{-1})(1.0 \times 10^{-7}\ M)^2 = 5.0 \times 10^{-16}\ M$$

The total solubility of $Zn(OH)_2(s)$ is given by

$$s = [Zn^{2+}] + [Zn(OH)_4^{2-}] = 0.10\ M + 5.0 \times 10^{-16}\ M = 0.10\ M$$

in agreement with Figure 20-5.

At pH = 10.0, $[OH^-] = 1.0 \times 10^{-4}$ M, so

$$[Zn^{2+}] = \frac{1.0 \times 10^{-15}\ M^3}{(1.0 \times 10^{-4}\ M)^2} = 1.0 \times 10^{-7}\ M$$

and

$$[Zn(OH)_4^{2-}] = (0.05\ M^{-1})(1.0 \times 10^{-4}\ M)^2 = 5.0 \times 10^{-10}\ M$$

Thus,

$$s = [Zn^{2+}] + [Zn(OH)_4^{2-}] = 1.0 \times 10^{-7}\ M + 5.0 \times 10^{-10}\ M$$
$$= 1.0 \times 10^{-7}\ M$$

At pH = 14.0, $[OH^-] = 1.0$ M, so

$$[Zn^{2+}] = \frac{1.0 \times 10^{-15}\ M^3}{(1.0\ M)^2} = 1.0 \times 10^{-15}\ M$$

and

$$[Zn(OH)_4^{2-}] = (0.05\ M^{-1})(1.0\ M)^2 = 0.05\ M$$

Notice that the solubility of $Zn(OH)_2(s)$ increases with increasing pH, but not as dramatically as in the case of $Al(OH)_3(s)$; the value of K for the complexation reaction given in Table 20-4 is not as large.

PRACTICE PROBLEM 20-12: Calculate the solubility of $Pb(OH)_2(s)$ in water at 25°C as a function of pH over the range pH = 6.0 to pH = 14.0. Use increments of two pH units.

Answer:

pH	6.0	8.0	10.0	12.0	14.0
s/M	1.2	0.0012	8×10^{-6}	8×10^{-4}	0.08

20-10. QUALITATIVE ANALYSIS IS THE IDENTIFICATION OF THE SPECIES PRESENT IN A SAMPLE

An essential feature of a qualitative analysis scheme for a large group of ions is the *successive* removal of subgroups of the ions by precipitation reactions. It is essential to carry out the separation steps in a *systematic* fashion; otherwise ions that are presumed to have been removed may interfere with subsequent steps in the analytical scheme. If you are in doubt about how the test results look with certain mixtures of ions, then

ble to separate the $Ca^{2+}(aq)$ and the $Ba^{2+}(aq)$ by selectively precipitating the $Ba^{2+}(aq)$ with $SO_4^{2-}(aq)$? Use a criterion of precipitating 99.99 percent of the $Ba^{2+}(aq)$ without causing $Ca^{2+}(aq)$ to precipitate.

20-43. A solution contains 0.050 M $Ca^{2+}(aq)$ and 0.025 M $Ag^{+}(aq)$. Can 99 percent of either ion be precipitated by adding $SO_4^{2-}(aq)$, without precipitating the other metal ion? What will be the concentration of $Ca^{2+}(aq)$ when $Ag_2SO_4(s)$ begins to precipitate?

20-44. A solution contains 0.0100 M $Pb^{2+}(aq)$ and 0.0100 M $Sr^{2+}(aq)$. Can 99 percent of either ion be precipitated by adding $SO_4^{2-}(aq)$, without precipitating the other metal ion? What will be the concentration of $Pb^{2+}(aq)$ when $SrSO_4(s)$ begins to precipitate?

SEPARATION OF CATIONS AS HYDROXIDES AND SULFIDES

20-45. Calculate the solubility of $Cr(OH)_3(s)$ and $Ni(OH)_2(s)$ in an aqueous solution buffered at pH = 5.0. Can $Cr(OH)_3$ be separated from $Ni(OH)_2$ at this pH?

20-46. Calculate the solubility of $Cu(OH)_2(s)$ and $Zn(OH)_2(s)$ in an aqueous solution buffered at pH = 4.0. Can $Cu(OH)_2$ be separated from $Zn(OH)_2$ at this pH?

20-47. Calculate the solubility of $CuS(s)$ in a solution buffered at pH = 2.0 and saturated with hydrogen sulfide so that $[H_2S] = 0.10$ M.

20-48. Calculate the solubility of $SnS(s)$ in a solution buffered at pH = 2.0 and saturated with hydrogen sulfide so that $[H_2S] = 0.10$ M.

20-49. What must the pH of a buffered solution saturated with H_2S ($[H_2S] = 0.10$ M) be in order to precipitate PbS leaving $[Pb^{2+}] = 1 \times 10^{-6}$ M, without precipitating any MnS? The original solution is 0.025 M in both $Pb^{2+}(aq)$ and $Mn^{2+}(aq)$.

20-50. Iron(II) sulfide is used as the pigment in black paint. A sample of $FeS(s)$ is suspected of containing lead(II) sulfide, which can cause lead poisoning if ingested. Suggest a scheme based on pH for separating FeS from PbS.

20-51. Use the equilibrium-constant data in Table 20-4 to estimate the solubility of tin(II) hydroxide in a solution buffered at pH = 13.0.

20-52. Use the equilibrium-constant data in Table 20-4 to estimate the solubility of lead(II) hydroxide in a solution buffered at pH = 13.0.

ADDITIONAL PROBLEMS

20-53. One treatment for poisoning by soluble lead compounds is to give $MgSO_4(aq)$ or $Na_2SO_4(aq)$ as soon as possible. Explain in chemical terms why this procedure is effective.

20-54. Calculate the solubility of HgS and CdS at pH = 3.0 and 6.0 for aqueous solutions that are saturated with H_2S ($[H_2S] = 0.10$ M).

20-55. It is observed that a precipitate forms when a 2.0 M $NaOH(aq)$ solution is added dropwise to a 0.10 M $Pb(NO_3)_2(aq)$ solution and that, on further addition of $NaOH(aq)$, the precipitate dissolves. Explain these observations using balanced chemical equations.

20-56. Insoluble $Pb(OH)_2$ and $Sn(OH)_2$ are formed when sodium hydroxide is added to a solution containing $Pb^{2+}(aq)$ and $Sn^{2+}(aq)$. At what pH can $Pb(OH)_2$ be separated from $Sn(OH)_2$? Assume that an effective separation requires a maximum concentration of the less soluble hydroxide of 1×10^{-6} M.

20-57. Oxalic acid and soluble oxalates can cause death if swallowed. The recommended treatment for oxalic acid or oxalate poisoning is to give, as soon as possible, a glassful of limewater (saturated solution of calcium hydroxide) or a 1 percent calcium chloride solution, followed by inducing vomiting several times. Then give 15 to 30 g of Epsom salt ($MgSO_4$) in water and do not induce vomiting. Explain in chemical terms why this procedure is effective.

20-58. A solution 0.30 M in $H_3O^{+}(aq)$ containing $Mn^{2+}(aq)$, $Cd^{2+}(aq)$, and $Fe^{2+}(aq)$ all at 0.010 M was saturated with $H_2S(g)$ at 25°C. Calculate the equilibrium concentrations of $Mn^{2+}(aq)$, $Cd^{2+}(aq)$, and $Fe^{2+}(aq)$. Assume that the solution is continuously saturated with H_2S and that the pH remains constant.

20-59. It is observed that a precipitate forms when 2.0 M $KOH(aq)$ solution is added dropwise to a 0.20 M $Zn(ClO_4)_2(aq)$ solution and that, on further addition of $KOH(aq)$, the precipitate dissolves. Explain these observations using balanced chemical equations.

20-60. A deposit of limestone is analyzed for its calcium and magnesium content. A sample is dissolved, and then the calcium and magnesium are precipitated as $Ca(OH)_2$ and $Mg(OH)_2$. At what pH can $Ca(OH)_2$ be separated from $Mg(OH)_2$? Assume that an effective separation requires a maximum concentration of the less soluble hydroxide of 1×10^{-6} M.

20-61. A 2.000-g sample of a salt deposit was dissolved in an aqueous solution. A solution of $AgNO_3$ was added to precipitate all the chloride ions as AgCl. The precipitate was filtered, dried, and weighed. The amount of AgCl obtained was 4.188 g. Calculate the mass percentage of chloride ion in the sample.

20-62. Consider the following chemical equation:

$$PbCrO_4(s) + 3OH^-(aq) \rightleftharpoons Pb(OH)_3^-(aq) + CrO_4^{2-}(aq)$$

Predict whether the solubility of $PbCrO_4(s)$ is increased, is decreased, or remains unchanged by

(a) a decrease in the concentration of $OH^-(aq)$
(b) an increase in the amount of $PbCrO_4(s)$
(c) an increase in the pH of the solution
(d) dissolution of $Na_2CrO_4(s)$
(e) addition of $H_2O(l)$ to the system
(f) addition of $HClO_4(aq)$

20-63. Given the equation

$$Ag^+(aq) + 2NH_3(aq) \rightleftharpoons Ag(NH_3)_2^+(aq) \qquad K_{comp} = 2.0 \times 10^7\ M^{-2}$$

determine the final concentration of $NH_3(aq)$ that is required to dissolve 250 mg of $AgCl(s)$ in 100 mL of solution.

20-64. Given the following data at 25°C

solubility of $I_2(s)$ in $H_2O(l)$:	0.00132 M
solubility of $I_2(s)$ in 0.1000 M $KI(aq)$:	0.05135 M

calculate the equilibrium constants for the following set of equations:

$$I_2(s) \rightleftharpoons I_2(aq)$$
$$I_2(s) + I^-(aq) \rightleftharpoons I_3^-(aq)$$
$$I_2(aq) + I^-(aq) \rightleftharpoons I_3^-(aq)$$

20-65. Use the K_{sp} data in Table 20-1 to calculate the equilibrium constants for the following set of equations:

(1) $Ag_2CrO_4(s) + 2Br^-(aq) \rightleftharpoons 2AgBr(s) + CrO_4^{2-}(aq)$
(2) $PbCO_3(s) + Ca^{2+}(aq) \rightleftharpoons CaCO_3(s) + Pb^{2+}(aq)$

20-66. Calculate the pH at which $Ca(OH)_2(s)$ will begin to precipitate from a solution that is 2.0×10^{-2} M in $Ca^{2+}(aq)$ at 25°C.

20-67. Suppose we have a solution containing $Pb^{2+}(aq)$ and $NO_3^-(aq)$. A solution of $NaCl(aq)$ is added slowly until no further precipitation occurs. The precipitate is collected by filtration, dried, and weighed. A total of 12.79 g of $PbCl_2(s)$ is obtained from 200.0 mL of the original solution. Calculate the mass of $Pb(NO_3)_2$ present and the molarity of the solution.

20-68. Excess $HgI_2(s)$ was equilibrated with a solution of 0.10 M in $KI(aq)$. Calculate the solubility of $HgI_2(s)$ in this solution given

$$HgI_2(s) \rightleftharpoons Hg^{2+}(aq) + 2I^-(aq) \qquad K_{sp} = 2.0 \times 10^{-28}\ M^3$$
$$HgI_2(s) + 2I^-(aq) \rightleftharpoons HgI_4^{2-}(aq) \qquad K_{comp} = 0.79\ M^{-1}$$

20-69. Calculate the solubility of silver acetate ($K_{sp} = 4.0 \times 10^{-3}\ M^2$ at 25°C) in solutions buffered at pH = 2.0, 4.0, 6.0, 8.0, and 10.0.

20-70. Calculate the value of the equilibrium constant for the reaction

$$Ag_2SO_4(s) + 2Br^-(aq) \rightleftharpoons 2AgBr(s) + SO_4^{2-}(aq)$$

20-71. Given that $K_{comp} = 0.05\ M^{-1}$ for

$$Zn(OH)_2(s) + 2OH^-(aq) \rightleftharpoons Zn(OH)_4^{2-}(aq)$$

calculate the solubility of $Zn(OH)_2(s)$ in a solution buffered at pH = 12.0.

20-72. Calculate the value of the equilibrium constant for the reaction

$$Ag_2SO_4(s) + Ca^{2+}(aq) \rightleftharpoons CaSO_4(s) + 2Ag^+(aq)$$

Calculate $[Ag^+]$, $[Ca^{2+}]$, and $[SO_4^{2-}]$ when excess $CaSO_4(s)$ is equilibrated with 0.100 M $AgNO_3(aq)$.

20-73. In each of the following cases, the two solutions indicated are mixed. In each case for which a precipitate forms on mixing, write the complete equation and the net ionic equation. If no precipitate forms, then write "no reaction." Use the solubility rules and assume that all solutions before mixing are 0.20 M and that equal volumes of the two solutions are mixed.

(a) $Hg_2(ClO_4)_2(aq) + NaBr(aq) \rightarrow$
(b) $Fe(ClO_4)_3(aq) \rightarrow NaOH(aq) \rightarrow$
(c) $Pb(NO_3)_2(aq) + LiIO_3(aq) \rightarrow$
(d) $H_2SO_4(aq) + Pb(NO_3)_2(aq) \rightarrow$

20-74. In each of the following cases, the two solutions indicated are mixed. In each case for which a precipitate forms on mixing, write the complete equation and the net ionic equation. If no precipitate forms, then write "no reaction." Use the solubility rules and assume that all solutions before mixing are 0.20 M and that equal volumes of the two solutions are mixed.

(a) $Hg_2(NO_3)_2(aq) + KCl(aq) \rightarrow$
(b) $Zn(ClO_4)_2(aq) + Na_2S(aq) \rightarrow$
(c) $CaCl_2(aq) + Na_2CO_3(aq) \rightarrow$
(d) $Cu(ClO_4)_2(aq) + LiOH(aq) \rightarrow$

20-75. The equilibrium constant for the equation

$$AgCl(s) + 2S_2O_3^{2-}(aq) \rightleftharpoons Ag(S_2O_3)_2^{3-}(aq) + Cl^-(aq)$$

is 5.20×10^3 at 25°C. Calculate the solubility of $AgCl(s)$ in a solution whose *equilibrium* concentration of $S_2O_3^{2-}(aq)$ is 0.010 M.

20-76. Copper(I) ions in aqueous solution react with NH_3 according to

$$Cu^+(aq) + 2NH_3(aq) \rightleftharpoons Cu(NH_3)_2^+(aq) \qquad K = 6.3 \times 10^{10}\ M^{-2}$$

Calculate the solubility of $CuBr(s)$ in a solution in which the equilibrium concentration of $NH_3(aq)$ is 0.15 M.

20-77. Consider the chemical equilibrium

$$AgBr(s) + 2S_2O_3^{2-}(aq) \rightleftharpoons Ag(S_2O_3)_2^{3-}(aq) + Br^-(aq)$$

Predict whether the solubility of AgBr(*s*) is increased, decreased, or unchanged by

(a) an increase in the concentration of $Na_2S_2O_3(aq)$
(b) a decrease in the amount of AgBr(*s*)
(c) dissolution of NaBr(*s*)
(d) dissolution of $NaNO_3(s)$

20-78. Consider the chemical equilibrium

$$PbI_2(s) + 3OH^-(aq) \rightleftharpoons Pb(OH)_3^-(aq) + 2I^-(aq)$$

Predict whether the solubility of $PbI_2(s)$ is increased, decreased, or unaffected by

(a) an increase in the concentration of $OH^-(aq)$
(b) a decrease in the amount of $PbI_2(s)$
(c) a decrease in the concentration of $I^-(aq)$

20-79. Predict which of the following compounds are soluble in water:

(a) $(NH_4)_2CO_3$ (b) $Ag_2C_2O_4$
(c) $PbSO_4$ (d) CuO

20-80. Predict which of the following compounds are soluble in water:

(a) K_2CO_3 (b) $SnSO_4$
(c) $CaCl_2$ (c) ZnS

20-81. Which of the following compounds are soluble in water?

(a) $Zn(ClO_4)_2(s)$ (b) $AgBrO_3(s)$
(c) $CdSO_4(s)$ (d) $Fe(OH)_2(s)$
(e) $Mn(NO_3)_2(s)$

20-82. Which of the following compounds are soluble in water?

(a) $CaSO_4(s)$ (b) $AgNO_2(s)$
(c) $Cu(CH_3COO)_2(s)$ (d) $NiI_2(s)$
(e) $Fe(NO_3)_3(s)$

20-83. The equilibrium constant for the equation

$$Al(OH)_3(s) + OH^-(aq) \rightleftharpoons Al(OH)_4^-(aq)$$

is $K = 40$ at 25°C. Calculate the solubility of $Al(OH)_3(s)$ in a solution buffered at pH = 12.0 at 25°C.

20-84. The equilibrium constant for the equation

$$Zn(OH)_2(s) + 2OH^-(aq) \rightleftharpoons Zn(OH)_4^{2-}(aq)$$

is $K = 0.050\ M^{-1}$. Calculate the solubility of $Zn(OH)_2(s)$ in a 0.10 M NaOH(*aq*) solution.

20-85. The solubility product of zinc hydroxide is the following: $1.0 \times 10^{-15}\ M^3$ at 25°C. Calculate the pH of a saturated $Zn(OH)_2(aq)$ solution at 25°C.

20-86. Given that the pH of a saturated $Ca(OH)_2(aq)$ solution is 12.45, calculate the solubility of $Ca(OH)_2(s)$ in water at 25°C.

20-87. Calculate the solubility of $Cu(OH)_2(s)$ in aqueous solution buffered at pH = 7.0.

20-88. Calculate the solubility of $Cd(OH)_2(s)$ in an aqueous solution buffered at pH = 9.0.

20-89. Use Le Châtelier's principle to predict the effect on the solubility of

(a) ZnS(*s*) when $HNO_3(aq)$ is added to a saturated ZnS(*aq*) solution
(b) AgI(*s*) when $NH_3(g)$ is added to a saturated AgI(*aq*) solution

20-90. For the equilibrium

$$ZnS(s) + 2H_3O^+(aq) \rightleftharpoons Zn^{2+}(aq) + H_2S(aq) + 2H_2O(l)$$

predict the direction of shift in response to each of the following changes in conditions. (If the equilibrium is unaffected by the change, then write "no change.")

(a) bubbling in HCl(*g*)
(b) diluting the solution
(c) increasing the pH of the solution

20-91. Mercury(I) chloride is sparingly soluble in aqueous solutions. What is the solubility (in $mol\cdot L^{-1}$) of mercury(I) chloride

(a) in 1.5 M KCl(*aq*)?
(b) in 1.5 M mercury(I) nitrate aqueous solution?
(c) in pure water?

20-92. Silver chromate is sparingly soluble in aqueous solutions. What is the solubility (in $mol\cdot L^{-1}$) of silver chromate

(a) in 1.5 M potassium chromate aqueous solution?
(b) in 1.5 M silver nitrate aqueous solution?
(c) in pure water?

20-93. What is the equilibrium chloride ion concentration in a solution made by mixing 50.0 mL of 1.00 M sodium chloride with 50.0 mL of 1.00 M mercury(I) nitrate?

20-94. What is the equilibrium chromate ion concentration in a solution made by mixing 200.0 mL of 0.200 M silver nitrate with 200.0 mL of 0.100 M potassium chromate?

20-95. What is the equilibrium mercury(I) ion concentration in a solution made by mixing 100.0 mL of 0.200 M mercury(I) nitrate with 150.0 mL of 0.100 M aluminum chloride? What fraction of the mercury(I) ion is not precipitated?

20-96. What is the equilibrium silver ion concentration in a solution made by mixing 500 mL of 0.200 M silver nitrate with 1200 mL of 0.100 M potassium chloride? What fraction of the silver ion is not precipitated?

20-97. Determine the molar solubility of silver iodide in 14.0 M aqueous ammonia. Use the criterion that soluble means at least 0.10 moles of the salt dissolve per liter of solution to determine whether silver iodide is soluble in aqueous ammonia.

20-98. Some oil-well brines contain iodide ion. One particular brine sample had 6.5 mg iodide ion per liter. If equal volumes of this brine sample are mixed with a solution that is 0.0010 M each in lead(II) ion and silver(I) ion, which metal ion will precipitate as an iodide?

20-99. Which is more soluble in pure water, silver chloride or silver chromate?

Soaps and Detergents

Natural soaps are sodium salts of fatty acids, which are organic acids containing long hydrocarbon chains. A typical example is sodium stearate, $C_{17}H_{35}COO^-Na^+$:

$$\underbrace{CH_3CH_2CH_2CH_2CH_2CH_2CH_2CH_2CH_2CH_2CH_2CH_2CH_2CH_2CH_2CH_2CH_2}_{\text{hydrocarbon portion}}\overbrace{COO^-}^{\text{carboxyl portion}}Na^+$$

Soap is effective as a cleaning agent because the hydrocarbon portion of the molecule has a strong affinity for grease and oils, whereas the charged (anionic) portion has a strong affinity for water. The anion of a soap molecule can be represented schematically as

hydrocarbon portion — anion portion

When soap molecules that are dissolved in water come into contact with grease, the hydrocarbon portions stick into the grease, leaving the anion portions at the grease-water interface. The penetration of the grease by the hydrocarbon portion of soap molecules is followed by a remarkable phenomenon—the formation of *micelles* (Figure J-1), which are small, spherical grease-soap droplets that are soluble in water because of the polar groups on their surface. Micelles do not combine into larger drops because their surfaces are all negatively charged. The micelles encapsulate small grease particles and are subsequently rinsed away, leaving a clean region behind.

Detergents constitute a broad class of cleaning agents composed of a wide range of ingredients (Table J-1). The key ingredients in detergents are *surfactants*, a term derived from the phrase "*surface-active agent*." Water has an unusually high surface tension, because the water molecules at the surface are, on average, stabilized by hydrogen bonding to other water molecules only half as much as are

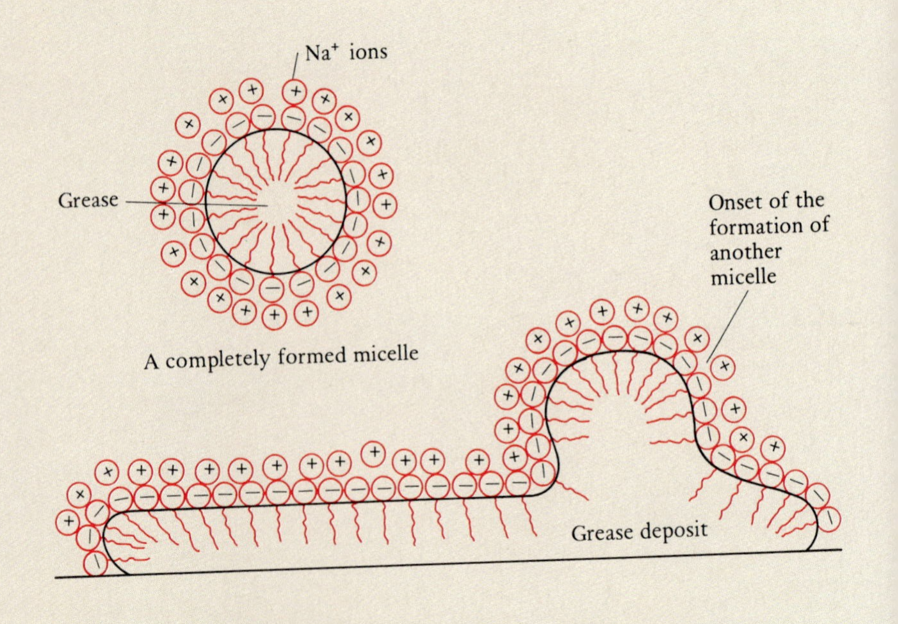

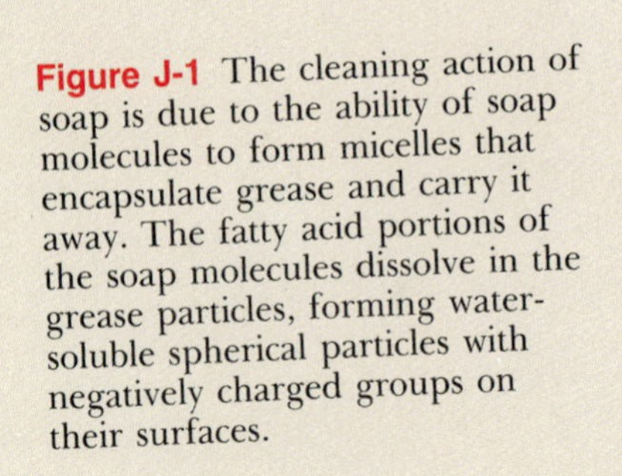

Figure J-1 The cleaning action of soap is due to the ability of soap molecules to form micelles that encapsulate grease and carry it away. The fatty acid portions of the soap molecules dissolve in the grease particles, forming water-soluble spherical particles with negatively charged groups on their surfaces.

Table J-1 Components of detergents and their functions

Component (percent)	Function	Example
surfactants (0–20%)	lower surface tension of water	alkyl benzene sulfonates, $RC_6H_4SO_3^-Na^+$
builders (0–50%)	bind divalent metal ions such as Ca^{2+} and Mg^{2+} and thereby soften the water	sodium triphosphate, $Na_5P_3O_{10}$ sodium citrate (see text)
ion exchangers (0–50%)	remove Ca^{2+} from water by exchanging two Na^+ for each Ca^{2+}	zeolites, e.g., $Na_2O \cdot Al_2O_3 \cdot 2SiO_2 \cdot 4.5H_2O$
alkalis (0–35%)	provide alkalinity to water, which aids in cleaning by maintaining surfactants in anionic form	Na_2CO_3, Na_4SiO_4
bleaches (dry) (0–5%)	remove stains by an oxidative reaction that releases hydrogen peroxide	peroxides, e.g., sodium perborate, $NaBO_3 \cdot 4H_2O$
enzymes (0–2%)	decompose proteinaceous materials	wide variety used
optical brighteners (0–0.25%)	convert uv light into visible light by absorption and reemission	stilbene derivatives, $RC_6H_4CH{=}CH_2$
fragrances (small)	provide a pleasant odor	wide variety used
fillers (to balance to 100%)	act as bulking agents	sodium sulfate

water molecules in the interior of the liquid. Surfactants lower the surface tension of water by concentrating in the surface layer of the water and thereby eliminating the relatively high-energy surface water molecules. The reduction in surface tension flattens water droplets (Figure J-2) and produces the well-known wetting ("sheeting") action of surfactants. Surfactant molecules invariably have *amphipathic* structures; that is, the molecules have an oil-soluble end and a water-soluble end analogous to soaps, which are also surfactants (see Figure J-1). The water-soluble end sticks into the water, and the oil-soluble end sits on the surface of the liquid.

Synthetic surfactants were first prepared around 2500 BC by the Sumerians, who lived in the region now occupied by Iran and Iraq. The Sumarian surfactant was a soap formed from animal fat and plant ash and consisted of salts of carboxylic (organic) acids. In 1988 the U.S. annual production of detergents was around 40 billion pounds.

Surfactants are classified as *anionic, cationic, nonionic,* or *amphoteric,* depending on the net charge on the organic portion of the surfactant. Most soaps involve alkali metal salts of carboxylic acids ("car-

Figure J-2 Surfactants (surface-active agents) lower the surface tension of water by concentrating in the surface layer. The decrease in surface tension causes the droplet of water to flatten, as shown here. The two water drops have the same volume but the lower drop contains a small amount of surfactant.

Large quantities of surfactants were used to break up the 1989 Alaskan oil spill from the tanker Exxon Valdez. This photograph shows the surfactants in action.

boxylates") and are thus anionic surfactants. The long-hydrocarbon carboxylates, such as sodium stearate, are derived from tallow and other natural fats. Use of soaps today is confined primarily to bath bar soap. Household and other detergents use more effective surfactants. Fatty alcohol sulfates (e.g., $RCH_2OSO_3Na^+$, where R is a hydrocarbon group) are used in hair shampoos, hand dishwashing liquids, and mild laundry products, because they have minimal effects on skin and they are inexpensive. Household laundry and dishwashing products contain linear-alkyl-chain benzene sulfonates for example,

R—C_6H_4—SO_3Na

Cationic surfactants used in detergents are quarternary ("fourfold") ammonium salts of the type $R_4N^+Cl^-$. The R_4N^+ cations bind to the negative charges on fabrics and thereby function as fabric softeners. This property also makes these compounds useful as antistatic agents in various laundry products. Many $R_4N^+Cl^-$ salts also act as germicides.

Nonionic surfactants are dominated by various polyethers (i.e., molecules that contain C—O—C bonds), which have superb wetting and dispersing properties. These compounds are liquids at room temperature, and their major use is in liquid laundry detergents.

Although surfactants are the stars in detergents, many other substances play important supporting roles, most of which are summarized in Table J-1. For example, *builders* are used to soften hard water by binding divalent metal ions, such as $Ca^{2+}(aq)$ and Mg^{2+}. Builders inhibit the precipitation of divalent metal salts of surfactant anions. Sodium poly-

Soap bubbles formed by surfactant action.

phosphates, such as the tripolyphosphate $Na_5P_3O_{10}$,

$$^{-}O-\underset{\underset{O}{\|}}{\overset{\overset{O^-}{|}}{P}}-O-\underset{\underset{O}{\|}}{\overset{\overset{O^-}{|}}{P}}-O-\underset{\underset{O}{\|}}{\overset{\overset{O^-}{|}}{P}}-O^-$$

the tripolyphosphate ion, $P_3O_{10}^{5-}$

are very effective at bonding Ca^{2+} and Mg^{2+}, for example, in the form $CaP_3O_{10}^{3-}(aq)$ and $MgP_3O_{10}^{3-}(aq)$, respectively. Polyphosphates have the major disadvantage of promoting the rapid growth of algae, which can cause *eutrophication* (oxygen depletion as a result of algae growth) of bodies of water into which the polyphosphates have been discharged. For this reason polyphosphates have been replaced in most detergents by sodium citrate and cation-absorbing zeolites. The citrate ion,

$$\begin{array}{c} CH_2COO^- \\ | \\ HOCCOO^- \\ | \\ CH_2COO^- \end{array}$$

the citrate ion, $C_6H_5O_7^{3-}$

binds divalent ions in a manner analogous to tripolyphosphate, forming complexes of the type $MC_6H_5O_7^-(aq)$.

Some detergents contain dry bleaches, which also are sold separately. On heating in water, dry bleaches release hydrogen peroxide, H_2O_2, which is the active bleaching agent. The most extensively used dry bleaching agent is sodium perborate tetrahydrate, $NaBO_3{\cdot}4H_2O(s)$, or better, $Na_2B_2(O_2)_2(OH)_4{\cdot}6H_2O(s)$. Sodium perborate contains anions of the type

$$\left[\begin{array}{ccccc} HO & & O-O & & OH \\ & B & & B & \\ HO & & O-O & & OH \end{array}\right]^{2-}$$

$B_2(O_2)_2(OH)_4^{2-}$

which have two peroxide bonds (—O—O—) per formula unit. In water at about 60°C, the $[B_2(O_2)_2(OH)_4]^{2-}$ ions decompose to yield hydrogen peroxide, which removes stains by oxidative attack.

Optical brighteners really do make whites and colors look brighter. These additives absorb ultraviolet light and then emit part of the absorbed light energy in the visible region, thereby giving the fabric a brighter appearance. The most common optical brighteners are derivatives of stilbene, $C_6H_5CH{=}CH_2$.

Enzyme additives provide detergents with a completely different mode of cleaning action than that provided by surfactants and bleaches. Enzyme cleaning action involves the removal of protein-containing stains by enzymatic hydrolysis of peptide bonds in the protein polymer chain:

$$H_2O + R-\underset{\underset{O}{\|}}{C}-\underset{\underset{H}{|}}{N}-R' \xrightarrow{\text{enzyme}} R-\underset{\underset{O}{\|}}{C}-OH + H-\underset{\underset{H}{|}}{N}-R'$$

(the arrow labelled "amide bond" points to the C—N bond)

The variety and range of function of modern detergents are truly remarkable. Detergents are an excellent example of how a clear understanding of the underlying chemistry can lead directly to the development of useful and effective new products.

21 OXIDATION-REDUCTION REACTIONS

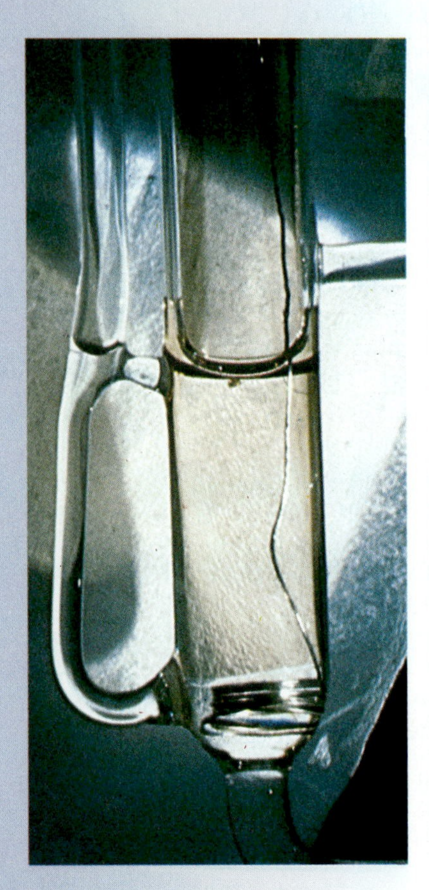
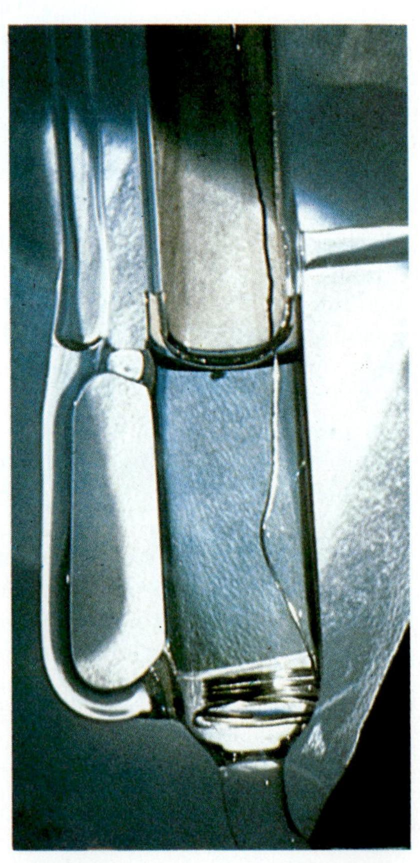
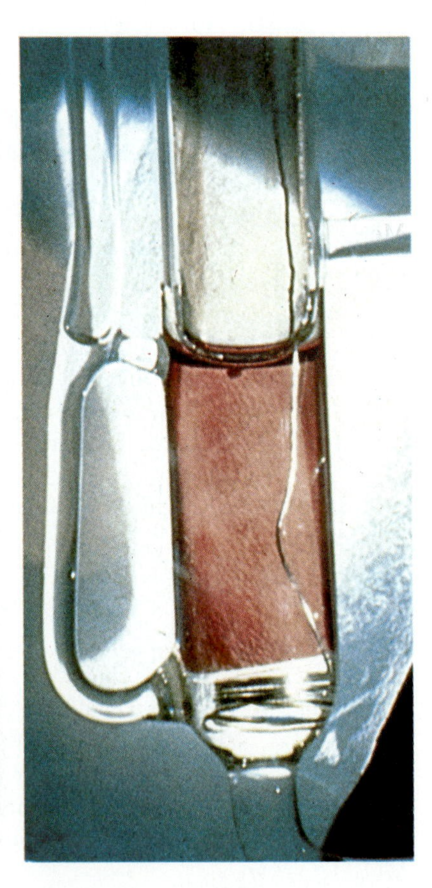

The complex ion RuL_3^{2+}, where L is an organic species, is attached to an electronically conductive polymer deposited on SnO_2, an arrangement that permits redox reactions to occur when electrons are supplied or withdrawn. The oxidation states of the ruthenium in the three cases are pale orange, +2; blue, 0; and cherry red, −4. *(Courtesy of Professor Elliot Morris, Colorado State University.)*

All chemical reactions can be assigned to one of two classes: reactions in which electrons are transferred from one reactant to another and reactions in which electrons are not transferred. We learned in Chapter 4 that reactions in which electrons are transferred from one reactant to another are called oxidation-reduction (redox) reactions or electron-transfer reactions. Many redox reactions are of major technological importance. Some examples are corrosion of metals and alloys, electroplating, batteries, and combustion reactions.

Recall from Chapter 4 that we use oxidation states to determine whether a reaction involves electron transfer and, if so, which species is oxidized and which species is reduced. Much of this chapter is devoted to the method of half reactions for balancing equations for oxidation-reduction reactions. The final two sections of the chapter discuss redox reactions used in chemical analyses and the corrosion of metals by redox reactions. A thorough understanding of redox reactions in terms of oxidation and reduction half reactions is an essential prerequisite to our study of electrochemistry in the next chapter.

21-1. AN OXIDATION STATE CAN BE ASSIGNED TO EACH ATOM IN A CHEMICAL SPECIES

Methods of balancing equations for oxidation-reduction reactions involve the ability to determine whether the reaction is a redox reaction. To do this we must be able to assign oxidation states to atoms in a chemical species. Oxidation states are assigned according to a set of rules, which originate from a consideration of the number of valence electrons and the electronegativities of the various elements in a species. For monatomic ions only, the assigned oxidation state is the actual charge on the ion. The procedure of assigning oxidation states to elements in a chemical species containing two or more atoms is described by the set of rules discussed in Section 4-1, which you should review. The rules given in Section 4-1 do not cover all possible cases. A more general method, based on Lewis electron-dot formulas, is used for cases not covered by the rules. Before we proceed with this method, recall that the assignment of oxidation states is, in essence, an electron-bookkeeping device. Assigned oxidation states in polyatomic species do not represent the actual charges on the individual atoms in the species.

In the Lewis-formula method, we assign oxidation states by the following four steps:

1. Write the Lewis formula for the molecule or ion.

2. Assign all the electrons in each bond to the more electronegative atom (Table 21-1) in the bond. If the two atoms have identical electronegativities, then divide the bonding electrons equally between them.

3. Add up the total number of valence electrons assigned to each atom in step 2.

Table 21-1 Electronegativities of selected elements

H 2.1	C 2.5	N 3.0	O 3.5	F 4.0
	Si 1.8	P 2.1	S 2.5	Cl 3.0
		As 2.0	Se 2.4	Br 2.8
				I 2.5

Note that for the series

$$\underset{4.0}{F} > \underset{3.5}{O} > \underset{3.0}{N \simeq Cl} > \underset{2.5}{C \simeq S} > \underset{2.1}{H \simeq P}$$

the electronegativities decrease by about 0.5 units between each inequality sign.

4. Assign an oxidation state to each *main-group element* in the species according to the formula:

$$\begin{pmatrix}\text{oxidation}\\ \text{state}\end{pmatrix} = \begin{pmatrix}\text{group number of}\\ \text{the element in}\\ \text{the periodic table}\end{pmatrix} - \begin{pmatrix}\text{total number of}\\ \text{valence electrons}\\ \text{assigned to}\\ \text{the element}\end{pmatrix} \qquad (21\text{-}1)$$

Recall that the group number of a main-group element (i.e., 1 through 8) is also the number of valence electrons in a free (nonbonded) atom. According to Equation (21-1), we subtract from that number the number of valence electrons assigned to the bonded atom in the species. The shared (bonding) electrons are credited to the more electronegative atom of the two bonded atoms.

The Lewis-formula method for the assignment of oxidation states applies to all the main-group elements, but generally it is used only when the simpler rules for assigning oxidation states given in Section 4-1 fail. The Lewis-formula method is best illustrated by examples.

EXAMPLE 21-1: Assign oxidation states to each atom in the following molecules:

(a) PCl_3 phosphorus trichloride

(b) CSe_2 carbon diselenide

Solution:

(a) The Lewis formula for phosphorus trichloride is

$$:\ddot{\underset{\cdot\cdot}{Cl}}-\ddot{\underset{|}{P}}-\ddot{\underset{\cdot\cdot}{Cl}}:$$
$$:\underset{\cdot\cdot}{Cl}:$$

Because chlorine is more electronegative than phosphorus (Table 21-1), we assign the electrons in each covalent bond to the chlorine atoms:

$$:\ddot{\underset{\cdot\cdot}{Cl}}: \quad \ddot{P} \quad :\ddot{\underset{\cdot\cdot}{Cl}}:$$
$$:\ddot{\underset{\cdot\cdot}{Cl}}:$$

Therefore,

the oxidation state of P in $PCl_3 = 5 - 2 = +3$, and

the oxidation state of Cl in $PCl_3 = 7 - 8 = -1$

(b) The Lewis formula for carbon diselenide is

$$\ddot{\underset{\cdot\cdot}{Se}}=C=\ddot{\underset{\cdot\cdot}{Se}}$$

Because carbon is more electronegative than selenium, we assign the electrons in each covalent bond to the carbon atoms:

$$\ddot{\underset{\cdot\cdot}{Se}} \qquad ::C:: \qquad \ddot{\underset{\cdot\cdot}{Se}}$$

Therefore,

the oxidation state of C in $CSe_2 = 4 - 8 = -4$, and
the oxidation state of Se in $CSe_2 = 6 - 4 = +2$

PRACTICE PROBLEM 21-1: Use the Lewis-formula method to assign oxidation states to the elements in the following compounds:

(a) HCN hydrogen cyanide

(b) SNF sulfur nitrogen fluoride

Answer: (a) H(+1), C(+2), N(−3); (b) S(+2), N(−1), F(−1)

When a table of electronegativities of elements is not available, in many cases we can look for analogies with other elements in the periodic table. The name or formula of the compound can also provide clues. This procedure is illustrated in the following Example.

EXAMPLE 21-2: Working by analogy with other elements in the periodic table and also using the clues provided by the names, assign oxidation states to each of the elements in the following compounds:
(a) As_2S_5 arsenic pentasulfide
(b) In_2Se_3 indium selenide
The more electronegative element usually is listed second in the formula and in binary compounds is given the *-ide* suffix.

Solution: (a) The name *arsenic pentasulfide* tells us that sulfur is more electronegative than arsenic, so the sulfur is assigned a negative oxidation state. Arsenic is below nitrogen, and sulfur is below oxygen in the periodic table; thus, an analogous compound is dinitrogen pentoxide, N_2O_5. Therefore, we assign sulfur an oxidation state of -2 (analogous to -2 for oxygen in oxides). To determine the oxidation state, x, of arsenic, we use the fact that the compound is electrically neutral

$$2x + 5(-2) = 0$$
$$x = +5$$

(b) Note that Se is below S in the periodic table and recall that the oxidation state of sulfur in many metal sulfides is -2. By analogy with sulfur, we assign an oxidation state of -2 to selenium. Thus, the oxidation state, x, of indium is

$$2x + 3(-2) = 0$$
$$x = +3$$

Note that indium is below aluminum in the periodic table, and aluminum has a characteristic oxidation state of $+3$ in its compounds.

PRACTICE PROBLEM 21-2: Using the procedure illustrated in Example 21-2, determine the oxidation states of the elements in the following compounds:
(a) GaAs gallium arsenide (b) Sb_2Te_3 antimony telluride

Answer: (a) As(-3), Ga($+3$); (b) Te(-2), Sb($+3$)

With some practice (which you can get by doing Problems 21-1 through 21-10), you will find that assigning oxidation states to elements in species is straightforward.

21-2. OXIDATION-REDUCTION REACTIONS INVOLVE THE TRANSFER OF ELECTRONS FROM ONE REACTANT TO ANOTHER

As you learned in Chapter 4, in an **oxidation-reduction (redox) reaction,** the **reducing agent** (the **electron donor**) donates electrons to the **oxidizing agent** (the **electron acceptor**). The reducing agent loses electrons and is **oxidized** (**oxidation** denotes a loss of electrons), whereas the oxidizing agent gains electrons and is **reduced** (**reduction** denotes a gain of electrons).

Consider the oxidation-reduction reaction between zinc metal and $Cu^{2+}(aq)$ in aqueous solution (see Figure 4-17):

$$Zn(s) + Cu^{2+}(aq) \rightarrow Zn^{2+}(aq) + Cu(s)$$

We say the $Cu^{2+}(aq)$ is reduced to $Cu(s)$ because the process involves a decrease (reduction) in the oxidation state of copper (from +2 to 0):

$$Cu^{2+}(aq) + 2e^- \rightarrow Cu(s) \qquad \text{(reduction)}$$

We say the $Zn(s)$ is oxidized because the process involves an increase in the oxidation state of zinc (from 0 to +2):

$$Zn(s) \rightarrow Zn^{2+}(aq) + 2e^- \qquad \text{(oxidation)}$$

An essential feature of oxidation-reduction reactions is that in one reactant the oxidation state of an element increases, and in another reactant the oxidation state of an element decreases. Thus, oxidation-reduction reactions involve a simultaneous oxidation and reduction. The simultaneous changes in oxidation states in oxidation-reduction reactions are the result of the transfer of electrons from one reactant to another; therefore, oxidation-reduction reactions are also called **electron-transfer reactions.**

EXAMPLE 21-3: In the following chemical equation, identify the atom that is oxidized, the atom that is reduced, the oxidizing agent, and the reducing agent:

$$MnO_2(s) + 4HCl(aq) \rightarrow MnCl_2(aq) + Cl_2(g) + 2H_2O(l)$$

Solution: The oxidation state of Mn is +4 in MnO_2 and +2 in $MnCl_2$. The oxidation state of Cl is −1 in HCl and 0 in Cl_2. Therefore, Cl is oxidized and Mn is reduced in this reaction. The reactant that contains the atom that is reduced is MnO_2, so MnO_2 is the oxidizing agent. The reactant that contains the atom that is oxidized is HCl, so HCl is the reducing agent. Notice that two electrons are transferred in this reaction: one manganese atom accepts two electrons and each of two chlorine atoms donates one electron. This reaction is used on a laboratory scale to prepare chlorine gas for use in chemical reactions.

PRACTICE PROBLEM 21-3: In the following chemical equation, identify the reducing agent, the oxidizing agent, the species reduced, and the species oxidized:

$$O_2(aq) + 6I^-(aq) + 2H_2O(l) \rightarrow 2I_3^-(aq) + 4OH^-(aq)$$

Answer: O_2 is the oxidizing agent and the species reduced; I^- is the reducing agent and the species oxidized.

The redox reactions between organic compounds and strong oxidizing agents such as $KMnO_4$ can be very vigorous, as illustrated in Figure 21-1. Table 21-2 summarizes the terms used to describe oxidation-reduction reactions.

Table 21-2 Summary of oxidation-reduction reactions

The reducing agent:
contains the atom that is oxidized
contains the atom whose oxidation state increases
is the electron donor
The oxidizing agent:
contains the atom that is reduced
contains the atom whose oxidation state decreases
is the electron acceptor

21-3. ELECTRON-TRANSFER REACTIONS CAN BE SEPARATED INTO TWO HALF REACTIONS

The electron-transfer reaction described by the equation

$$Zn(s) + Cu^{2+}(aq) \rightarrow Zn^{2+}(aq) + Cu(s) \qquad (21\text{-}2)$$

can be separated into two **half reaction** equations, one representing the oxidation process and the other representing the reduction process

$$Zn(s) \rightarrow Zn^{2+}(aq) + 2e^- \qquad \text{(oxidation)} \qquad (21\text{-}3)$$

$$Cu^{2+}(aq) + 2e^- \rightarrow Cu(s) \qquad \text{(reduction)} \qquad (21\text{-}4)$$

If we add Equations (21-3) and (21-4), then we obtain Equation (21-2). The half reaction equation in which electrons appear on the right-hand side (Equation 21-3) is called the **oxidation half reaction** (recall that oxidation is a *loss* of electrons). The half reaction equation in which electrons appear on the left-hand side (Equation 21-4) is called the **reduction half reaction** (recall that reduction is a *gain* of electrons). The oxidation half reaction supplies electrons to the reduction half reaction (Figure 21-1).

(a)

(b)

Figure 21-1 (a) The redox reaction between the oxidizing agent $KMnO_4$ (purple-black solid) and glycerin, $HOCH_2CHOHCH_2OH$, (clear liquid in eyedropper) is very vigorous as shown in (b). The primary reaction products are $MnO_2(s)$, $CO_2(g)$, $H_2O(g)$, and $K_2O(s)$. Organic compounds should not be brought into contact with strong oxidizing agents, except under carefully controlled conditions—otherwise serious injury could result.

EXAMPLE 21-4: Identify the oxidizing and reducing agents and the oxidation and reduction half reactions in the reaction described by the equation

$$Tl^+(aq) + 2Ce^{4+}(aq) \rightarrow 2Ce^{3+}(aq) + Tl^{3+}(aq)$$

This reaction is used in the standard method for determining the concentration of thallium(I) in analytical chemistry.

Solution: The oxidation state of thallium increases from +1 in Tl^+ to +3 in Tl^{3+}. Thus, Tl^+ is oxidized and acts as the reducing agent (electron donor). The oxidation state of cerium decreases from +4 in Ce^{4+} to +3 in Ce^{3+}. Thus, Ce^{4+} is reduced and acts as the oxidizing agent (electron acceptor).

We identify the two half reactions by writing the equations for the oxidation and the reduction reactions separately:

$$Tl^+(aq) \rightarrow Tl^{3+}(aq) + 2e^- \qquad \text{(oxidation)}$$

$$Ce^{4+}(aq) + e^- \rightarrow Ce^{3+}(aq) \qquad \text{(reduction)}$$

Note that Tl^+ is a two-electron reducing agent, whereas Ce^{4+} is a one-electron oxidizing agent. Two moles of Ce^{4+} are required to oxidize 1 mol of Tl^+. Thus, the complete balanced equation is obtained by multiplying the cerium half reaction equation by 2 and adding the result to the thallium half reaction equation:

$$Tl^+(aq) + 2Ce^{4+}(aq) \rightarrow Tl^{3+}(aq) + 2Ce^{3+}(aq)$$

A redox reaction, in which the same species is oxidized and reduced, is called a **disproportionation reaction.** The following practice problem involves a disproportionation reaction.

PRACTICE PROBLEM 21-4: Identify the oxidation and the reduction half reactions and the reducing and oxidizing agents in the reaction described by the equation:

$$2Cu^+(aq) \rightarrow Cu(s) + Cu^{2+}(aq)$$

Answer:

$$Cu^+ \rightarrow Cu^{2+} + e^- \quad \text{(oxidation)}$$
$$Cu^+ + e^- \rightarrow Cu \quad \text{(reduction)}$$

Cu^+ is both the reducing agent and the oxidizing agent in this reaction.

(a)

(b)

Figure 21-2 Chlorine in aqueous solution oxidizes iron metal to $Fe^{3+}(aq)$, which is yellow in the presence of $Cl^-(aq)$ as a result of the formation of the yellow ion $FeCl^{2+}(aq)$. (a) Before addition of iron powder to $Cl_2(aq)$. (b) After addition of iron powder and occurrence of the redox reaction.

21-4. EQUATIONS FOR OXIDATION-REDUCTION REACTIONS CAN BE BALANCED BY BALANCING EACH HALF REACTION SEPARATELY

Consider the equation for the reaction between iron metal and aqueous chlorine (Figure 21-2):

$$Fe(s) + Cl_2(aq) \rightarrow Fe^{3+}(aq) + Cl^-(aq) \qquad \text{(not balanced)}$$

This equation as it stands is not balanced. If we write

$$Fe(s) + Cl_2(aq) \rightarrow Fe^{3+}(aq) + 2Cl^-(aq) \qquad \text{(not balanced)}$$

then the equation is balanced with respect to the elements but *not* with respect to charge. The net charge on the left-hand side is zero, whereas the net charge on the right-hand side is $+3 + 2(-1) = +1$. The balanced equation

$$2Fe(s) + 3Cl_2(aq) \rightarrow 2Fe^{3+}(aq) + 6Cl^-(aq)$$

has the same number of atoms of each type on both sides and the same net charge (zero in this case) on both sides. Attempting to balance equations like this one by guessing the balancing coefficients can be time consuming and frustrating. We need a systematic procedure to balance redox equations.

In Chapter 4 we discussed a way to balance redox reactions by the oxidation-state method. In this chapter we develop a different method of balancing redox reactions, the **method of half reactions.** This method can be used to balance even the most complicated equation in a straightforward and systematic way. We first apply the method of half reactions to equations for reactions that take place in acidic solution, and then we discuss the method's application to equations occurring in basic solution.

Let us illustrate the method of half reactions by balancing the equation involving the oxidation of $Fe^{2+}(aq)$ by dichromate, a reaction used in the determination of the iron content of ores (Figure 21-3):

$$Fe^{2+}(aq) + Cr_2O_7^{2-}(aq) \xrightarrow{H^+(aq)} Fe^{3+}(aq) + Cr^{3+}(aq) \qquad (21\text{-}5)$$

The $H^+(aq)$ over the arrow in the chemical equation indicates that the reaction occurs in acidic solution.

To balance equations for redox reactions in acidic solution by the method of half reactions, we use the following sequence of steps:

1. *Identify the species oxidized and reduced and then separate the equation into two equations representing the oxidation half reaction and the reduction half reaction.*

The oxidation state of iron increases from +2 to +3, and the oxidation state of chromium decreases from +6 (in $Cr_2O_7^{2-}$) to +3. Thus, the two half reactions are

$$Fe^{2+} \rightarrow Fe^{3+} \quad \text{(oxidation)}$$
$$Cr_2O_7^{2-} \rightarrow Cr^{3+} \quad \text{(reduction)}$$

2. *Balance the equation for each half reaction with respect to all elements other than oxygen and hydrogen.*

The equation for the iron half reaction is already balanced with respect to iron (one Fe on each side). We balance the equation for the chromium half reaction with respect to chromium by placing a 2 in front of Cr^{3+}:

$$Fe^{2+} \rightarrow Fe^{3+}$$
$$Cr_2O_7^{2-} \rightarrow 2Cr^{3+}$$

3. *Balance each half reaction equation with respect to oxygen. To accomplish this step, add the appropriate number of H_2O to the side deficient in oxygen atoms.*

Only the chromium half reaction involves oxygen. There are seven oxygen atoms on the left and none on the right. Therefore, we balance with respect to oxygen by adding seven H_2O to the right-hand side of the equation for the chromium half reaction:

$$Fe^{2+} \rightarrow Fe^{3+}$$
$$Cr_2O_7^{2-} \rightarrow 2Cr^{3+} + 7H_2O$$

4. *Balance each half reaction with respect to hydrogen by adding the appropriate number of H^+ to the side deficient in hydrogen atoms.*

Only the chromium half reaction involves hydrogen. There are 14 hydrogens on the right and none on the left. Therefore, we balance the hydrogen by adding 14 H^+ to the left-hand side of the equation for the chromium half reaction:

$$Fe^{2+} \rightarrow Fe^{3+}$$
$$14H^+ + Cr_2O_7^{2-} \rightarrow 2Cr^{3+} + 7H_2O$$

The two half reaction equations are now balanced with respect to atoms, but they are not balanced with respect to charge.

5. *Balance each half reaction equation with respect to charge by adding the appropriate number of electrons to the side with the excess positive charge.*

The equation for the iron half reaction has a charge of +2 on the left and +3 on the right. Thus, we balance the charge by adding one electron to the right-hand side:

$$Fe^{2+} \rightarrow Fe^{3+} + e^- \quad \text{(oxidation)}$$

The equation for the chromium half reaction has a net charge of +12 [=14(+1) + (−2)] on the left and +6 [=2(+3)] on the right. Thus, we balance the charge by adding six electrons to the left-hand side:

$$14H^+ + Cr_2O_7^{2-} + 6e^- \rightarrow 2Cr^{3+} + 7H_2O \quad \text{(reduction)}$$

The two half reaction equations are now balanced. Note that the iron half reaction donates electrons (electrons on the right-hand

Figure 21-3 Addition of a 0.10 M solution of the oxidizing agent potassium dichromate, $K_2Cr_2O_7(aq)$ (orange solution), to a solution containing 0.10 M iron(II) sulfate, $FeSO_4(aq)$ (pale green), and 0.10 M $H_2SO_4(aq)$.) The iron is oxidized from $Fe^{2+}(aq)$ to $Fe^{3+}(aq)$, and the chromium is reduced from Cr(VI) in $Cr_2O_7^{2-}$ to the green ion $Cr^{3+}(aq)$.

side) and the chromium half reaction accepts electrons (electrons on the left-hand side).

6. *Make the number of electrons supplied by the oxidation half reaction equal to the number of electrons consumed by the reduction half reaction (conservation of electrons).*

   The iron half reaction supplies *one* electron for each Fe^{2+} that is oxidized to Fe^{3+}, and the chromium half reaction consumes *six* electrons for each $Cr_2O_7^{2-}$ that is reduced to Cr^{3+}. Therefore, we multiply the equation for the iron half reaction through by 6:

$$6Fe^{2+} \rightarrow 6Fe^{3+} + 6e^-$$
$$14H^+ + Cr_2O_7^{2-} + 6e^- \rightarrow 2Cr^{3+} + 7H_2O$$

   This change balances the number of electrons supplied and consumed.

7. *Obtain the complete balanced equation by adding the two balanced half reaction equations and canceling or combining any like terms.*

   Adding the equations for the two half reactions and canceling the $6e^-$ terms that appear on both sides yield

$$6Fe^{2+} \rightarrow 6Fe^{3+} + 6e^-$$
$$\underline{14H^+ + Cr_2O_7^{2-} + 6e^- \rightarrow 2Cr^{3+} + 7H_2O}$$
$$6Fe^{2+} + 14H^+ + Cr_2O_7^{2-} \rightarrow 6Fe^{3+} + 2Cr^{3+} + 7H_2O$$

   Note that the electrons cancel. No electrons ever appear in the complete balanced equation, because electrons are conserved. This fact serves as a check on your results. You should also check that the equation is balanced with respect to each element and with respect to charge. As a final step, we rewrite the balanced equation with phases indicated:

$$6Fe^{2+}(aq) + 14H^+(aq) + Cr_2O_7^{2-}(aq) \rightarrow 6Fe^{3+}(aq) + 2Cr^{3+}(aq) + 7H_2O(l)$$

Although the method of half reactions involves numerous steps, it is actually simple to use and, with a little practice, becomes straightforward. Also, even though it may seem arbitrary, the use of H_2O and H^+ to balance the half reactions (steps 3 and 4) is logical because in acidic aqueous solution these species are always present in appreciable concentrations and therefore are readily available to participate in chemical reactions.

EXAMPLE 21-5: Balance the following equation:

$$Fe^{2+}(aq) + O_2(g) \xrightarrow{H^+(aq)} Fe^{3+}(aq) + H_2O(l)$$

The reaction described by this equation occurs in the air oxidation of aqueous solutions containing $Fe^{2+}(aq)$, such as $FeSO_4(aq)$.

Solution: The oxidation state of iron changes from +2 (in Fe^{2+}) to +3 (in Fe^{3+}) and that of oxygen changes from 0 (in O_2) to −2 (in H_2O). Thus, the two half reactions are

$$Fe^{2+} \rightarrow Fe^{3+} \qquad \text{(oxidation)}$$
$$O_2 \rightarrow H_2O \qquad \text{(reduction)}$$

Let's balance the equation for each half reaction in turn. The oxidation half reaction equation is balanced with respect to iron. To balance it with respect to charge, we add one electron to the right-hand side:

$$Fe^{2+} \rightarrow Fe^{3+} + e^{-}$$

To balance the reduction half-reaction equation with respect to oxygen, we place a 2 in front of H_2O on the right-hand side:

$$O_2 \rightarrow 2H_2O$$

To balance it with respect to hydrogen, we add $4H^+$ to the left-hand side:

$$4H^+ + O_2 \rightarrow 2H_2O$$

Now we add $4e^-$ to the left-hand side to balance it with respect to charge:

$$4H^+ + O_2 + 4e^- \rightarrow 2H_2O \qquad \text{(reduction)}$$

The oxidation half reaction as written supplies *one* electron and the reduction half reaction as written consumes *four* electrons. If we multiply the oxidation half reaction equation by 4, then both half reaction equations will involve four electrons:

$$4Fe^{2+} \rightarrow 4Fe^{3+} + 4e^{-}$$
$$4H^+ + O_2 + 4e^- \rightarrow 2H_2O$$

Addition of these two half reaction equations yields

$$4Fe^{2+} + 4H^+ + O_2 \rightarrow 4Fe^{3+} + 2H_2O$$

Finally, we indicate the phases and write

$$4Fe^{2+}(aq) + 4H^+(aq) + O_2(g) \rightarrow 4Fe^{3+}(aq) + 2H_2O(l)$$

Note that this equation is balanced with respect to each element and with respect to charge.

PRACTICE PROBLEM 21-5: Balance the following redox equation:

$$Br^-(aq) + MnO_4^-(aq) \xrightarrow{H^+(aq)} BrO_3^-(aq) + Mn^{2+}(aq)$$

which is used to prepare bromates (BrO_3^-) from bromides.

Answer: $5Br^-(aq) + 18H^+(aq) + 6MnO_4^-(aq) \rightarrow 5BrO_3^-(aq) + 6Mn^{2+}(aq) + 9H_2O(l)$

The study of the redox chemistry of species is facilitated by considering the possible half reactions involving a particular species. Thus, it is useful to know how to balance half reactions without reference to a complete balanced equation. The following Example illustrates the balancing of an equation for a single half reaction in acidic solution.

EXAMPLE 21-6: Hydrogen peroxide is used extensively as an oxidizing agent in acidic aqueous solution. For example, H_2O_2 is used as a bleaching agent for human hair and to kill bacteria (antiseptic agent). Write a balanced half reaction for H_2O_2 acting as an oxidizing agent in acidic aqueous solution.

PRACTICE PROBLEM 21-7: Balance the following equation in basic aqueous solution:

$$Fe(OH)_2(s) + O_2(g) \xrightarrow{OH^-(aq)} Fe(OH)_3(s)$$

Answer: $4Fe(OH)_2(s) + 2H_2O(l) + O_2(g) \rightarrow 4Fe(OH)_3(s)$

21-6. OXIDATION-REDUCTION REACTIONS ARE USED IN CHEMICAL ANALYSES

Figure 21-4 Titration of a solution containing $Fe^{2+}(aq)$ with a solution containing the oxidizing agent permanganate, $MnO_4^-(aq)$ (purple). The $MnO_4^-(aq)$ oxidizes $Fe^{2+}(aq)$ to $Fe^{3+}(aq)$ and is itself reduced to $Mn^{2+}(aq)$.

As we learned in Section 6-7, oxidation-reduction reactions are used extensively in chemical analysis. To further illustrate this application of redox reactions, we consider the analytical determination of iron(II) using a solution of potassium permanganate of known concentration. Suppose that a 3.532-g sample of iron ore is dissolved in $H_2SO_4(aq)$ and suppose that we reduce any iron(III) present to Fe^{2+} by adding powdered zinc to the solution. Now consider the titration of the resulting filtered $Fe^{2+}(aq)$ solution with the oxidizing agent potassium permanganate, $KMnO_4(aq)$ (Figure 21-4). The balanced equation for the redox reaction is

$$5Fe^{2+}(aq) + MnO_4^-(aq) + 8H^+(aq) \rightarrow 5Fe^{3+}(aq) + Mn^{2+}(aq) + 4H_2O(l)$$

The equilibrium constant of this equation at 25°C is very large ($K = 3 \times 10^{62}\ M^{-8}$), so essentially all the added $KMnO_4(aq)$ oxidizes the $Fe^{2+}(aq)$ to $Fe^{3+}(aq)$. Such a reaction is said to be a **quantitative reaction,** because it goes essentially to completion, that is, essentially all the Fe^{2+} is converted to Fe^{3+} by the permanganate.

If 34.58 mL of 0.1108 M $KMnO_4(aq)$ are required to oxidize all the $Fe^{2+}(aq)$, then what is the mass percentage of iron in the ore? The number of millimoles of $KMnO_4(aq)$ required to oxidize the Fe^{2+} to Fe^{3+} is

$$\text{mmol of } KMnO_4 = MV = (0.1108\ \text{mol}\cdot\text{L}^{-1})(34.58\ \text{mL})$$
$$= 3.831\ \text{mmol}$$

From the balanced equation, we see that 5 mmol of $Fe^{2+}(aq)$ are oxidized for each millimole of $KMnO_4(aq)$ added, so

$$\text{mmol of } Fe^{2+} = (3.831\ \text{mmol } KMnO_4)\left(\frac{5\ \text{mmol } Fe^{2+}}{1\ \text{mmol } KMnO_4}\right)$$
$$= 19.16\ \text{mmol} = 0.01916\ \text{mol}$$

The atomic mass of iron is 55.85, so

$$\left(\begin{matrix}\text{grams}\\ \text{of Fe}\end{matrix}\right) = \left(\begin{matrix}\text{grams}\\ \text{of } Fe^{2+}\end{matrix}\right) = (0.01916\ \text{mol})\left(\frac{55.85\ \text{g Fe}}{1\ \text{mol Fe}}\right)$$
$$= 1.070\ \text{g}$$

The mass percentage of iron in the ore sample is

$$\text{mass \% Fe} = \left(\frac{1.070\ \text{g}}{3.523\ \text{g}}\right) \times 100 = 30.37\%$$

EXAMPLE 21-8: The concentration of ozone in a sample of air can be determined by reacting the sample with a buffered solution of potassium iodide, KI(*aq*). The $O_3(g)$ oxidizes $I^-(aq)$ to $I_3^-(aq)$ according to

$$O_3(g) + 3I^-(aq) + 2H^+(aq) \rightarrow O_2(g) + I_3^-(aq) + H_2O(l)$$

The concentration of the $I_3^-(aq)$ formed in the reaction is determined by titration with a sodium thiosulfate solution of known concentration

$$\underset{\text{thiosulfate}}{2S_2O_3^{2-}(aq)} + I_3^-(aq) \rightarrow \underset{\text{tetrathionate}}{S_4O_6^{2-}(aq)} + 3I^-(aq)$$

Suppose that a 50.00-mL sample of KI(*aq*) has reacted with a 43.15-g sample of air. If 34.56 mL of 0.002475 M $Na_2S_2O_3(aq)$ is required to titrate the $I_3^-(aq)$ produced, calculate the mass percentage of ozone in the mixture.

Solution: The number of millimoles of $Na_2S_2O_3(aq)$ required is

$$\text{mmol of } S_2O_3^{2-} = (0.002475\ \text{mol}\cdot\text{L}^{-1})(34.56\ \text{mL}) = 8.554 \times 10^{-2}\ \text{mmol}$$

The number of millimoles of $I_3^-(aq)$ reduced by the $S_2O_3^{2-}(aq)$ is

$$\begin{aligned}\text{mmol of } I_3^-(aq) &= (8.554 \times 10^{-2}\ \text{mmol } S_2O_3^{2-})\left(\frac{1\ \text{mmol } I_3^-}{2\ \text{mmol } S_2O_3^{2-}}\right)\\ &= 4.277 \times 10^{-2}\ \text{mmol}\end{aligned}$$

According to the stoichiometry of the equation for the O_3 plus I^- reaction,

$$\text{mmol of } O_3 = \text{mmol of } I_3^- = 4.277 \times 10^{-2}\ \text{mmol}$$

The mass of ozone is

$$\begin{aligned}\left(\begin{matrix}\text{grams}\\ \text{of } O_3\end{matrix}\right) &= (4.277 \times 10^{-2}\ \text{mmol})\left(\frac{1\ \text{mol}}{1000\ \text{mmol}}\right)\left(\frac{48.00\ \text{g } O_3}{1\ \text{mol } O_3}\right)\\ &= 2.053 \times 10^{-3}\ \text{g}\end{aligned}$$

and the mass percentage of ozone in the air sample is

$$\begin{aligned}\text{mass \% } O_3 &= \left(\frac{\text{mass of ozone}}{\text{mass of sample}}\right) \times 100 = \left(\frac{2.053 \times 10^{-3}\ \text{g}}{43.15\ \text{g}}\right) \times 100\\ &= 4.758 \times 10^{-3}\ \%\end{aligned}$$

Solutions containing thiosulfate, $S_2O_3^{2-}(aq)$, are used extensively in analytical chemistry to determine the concentration of $I_3^-(aq)$ or $I_2(aq)$ in a solution (Figure 21-5).

Figure 21-5 The reaction of $Na_2S_2O_3(aq)$ (in the presence of starch) with $I_2(aq)$. Iodine combines with starch to form the blue starch-iodine complex. When the iodine is reduced by sodium thiosulfate, the blue color disappears.

PRACTICE PROBLEM 21-8: The amount of ethanol in air exhaled from the lungs can be determined using a breath analyzer or "breathalyzer." This procedure is used by law enforcement agencies to determine the level of intoxication of drivers of motor vehicles suspected of driving while intoxicated. The exhaled air is bubbled through a solution containing the orange-yellow compound potassium dichromate, $K_2Cr_2O_7$, dissolved in aqueous sulfuric acid. The ethanol, C_2H_5OH, is oxidized by the dichromate to acetic acid, CH_3COOH, and the orange-yellow dichromate is reduced to $Cr^{3+}(aq)$, which is green. The amount of C_2H_5OH present is directly proportional to the decrease in the concentration of $Cr_2O_7^{2-}(aq)$.
(a) Balance the redox equation

$$C_2H_5OH(aq) + Cr_2O_7^{2-}(aq) \xrightarrow{H^+(aq)} CH_3COOH(aq) + Cr^{3+}(aq)$$

(b) Calculate the number of milligrams of ethanol required to decrease the concentration of 10.0 mL of a 0.0100 M $K_2Cr_2O_7(aq)$ solution by 10 percent.

Answer: (a)

$$3C_2H_5OH(aq) + 2Cr_2O_7^{2-}(aq) + 16H^+(aq) \rightarrow 3CH_3COOH(aq) + 4Cr^{3+}(aq) + 11H_2O(l)$$

(b) 0.69 mg

Redox reactions are used extensively in analytical chemistry to determine unknown concentrations of reducing and oxidizing agents. These reactions are especially useful in analytical chemistry, because we can choose a reactant that makes the reaction quantitative—that is, we choose a reactant that gives the resulting redox reaction a very large equilibrium constant. The methods of selecting such a reactant are based on the principles of electrochemistry described in Chapter 23.

21-7. BILLIONS OF DOLLARS ARE SPENT EACH YEAR TO PROTECT METALS FROM CORROSION

We are all familiar with corrosion, the best-known example of which is the rusting of iron and steel. Brown rust (Figure 21-6) is iron(III) oxide, and the corrosion of iron proceeds by air oxidation of the iron:

$$4Fe(s) + 3O_2(g) \xrightarrow{H_2O(l)} \underset{\text{rust}}{2Fe_2O_3(s)}$$

(a)

(b)

Figure 21-6 Scanning electron photomicrographs of iron oxides. (a) The angular crystals are green rust, Fe(II) oxides. (b) The column-shaped and rounded crystals are black rust, Fe_3O_4, and brown rust, Fe_2O_3, respectively. Because this type of microscope does not use optical wavelengths, the different rust colors are not actually seen in such pictures.

Most metals when exposed to air develop an oxide film. In some cases this film is very thin and protects the metal, and the metal maintains its luster. Examples are nickel and chromium. However, depending on the metal, the humidity, the acidity, and the presence of certain anions, corrosion can completely destroy a metal. For example, anions such as Cl^-, which is present in sea spray and on roads treated with rock salt, promote corrosion through the formation of chloro complexes and by increasing the electrical conductivity of the solution causing the corrosion. Certain gaseous species, such as the oxides of sulfur and nitrogen, combine with water to form acids that attack metals.

Corrosion is a major problem costing billions of dollars annually for replacements for the corroded parts. Research is directed toward understanding corrosion mechanisms, because a detailed understanding of these mechanisms can provide important clues to methods for preventing the process.

Corrosion of metals involves redox reactions between different sections of the same piece of metal or between two dissimilar metals in electrical contact with each other. One metal piece acts as the reducing agent, and the other provides the conducting surface on which reduction occurs. For example, iron in contact with air and moisture corrodes according to the mechanism sketched in Figure 21-7. The oxidation step is described by the equation

$$2Fe(s) + 4OH^-(aq) \rightarrow 2Fe(OH)_2(s) + 4e^-$$

and the reduction step is

$$2H_2O(l) + O_2(g) + 4e^- \rightarrow 4OH^-(aq)$$

where the $O_2(g)$ comes from the air. The iron(II) hydroxide formed is rapidly air-oxidized in the presence of water to iron(III) hydroxide:

$$4Fe(OH)_2(s) + O_2(g) + 2H_2O(l) \rightarrow 4Fe(OH)_3(s)$$

which in turn converts spontaneously to iron(III) oxide (rust):

$$2Fe(OH)_3(s) \rightarrow Fe_2O_3{\cdot}3H_2O(s)$$

The corrosion of aluminum in air is not so pronounced as that of iron, because the Al_2O_3 film that forms is tough, adherent, and impervious to oxygen. The same is true for chromium and nickel.

The simplest method of corrosion prevention is to provide a protective layer of paint or of a corrosion-resistant metal, such as chromium

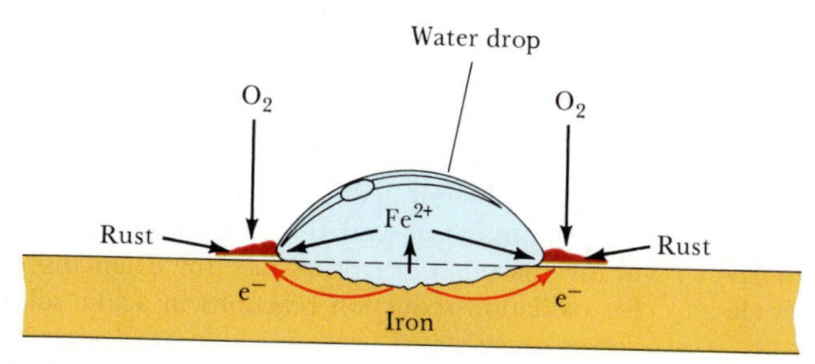

Figure 21-7 Corrosion of iron. A drop of water on an iron surface can act as a corrosion center. The iron is oxidized by oxygen from the air. Moisture is necessary for corrosion because the mechanism involves the formation of dissolved Fe(II) ions. Salts promote the corrosion by enabling a larger current flow between the active regions.

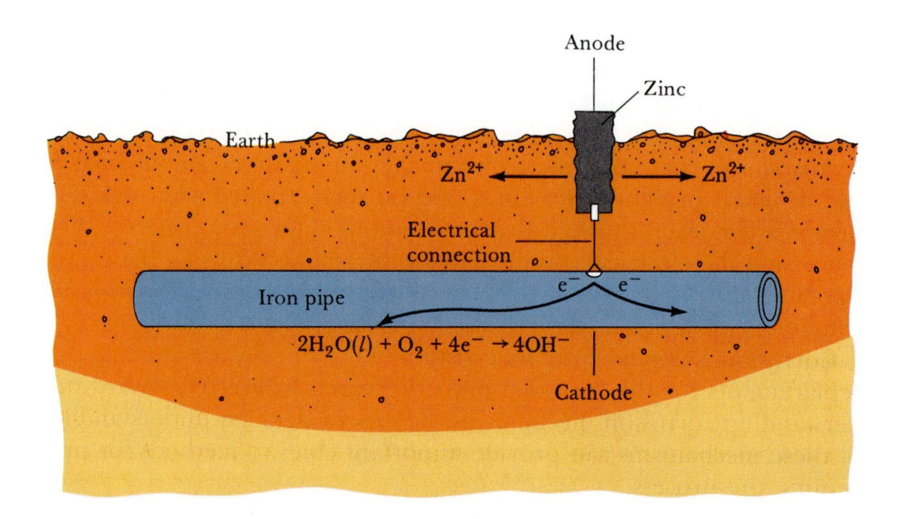

Figure 21-8 Protection of an iron pipe from corrosion with sacrificial zinc metal. Zinc is a stronger reducing agent than iron and thus is preferentially oxidized. The electrons produced in the oxidation flow to the iron pipe, on the surface of which O_2 is reduced to hydroxide ion. The net process is

$$2Zn(s) + O_2(aq) + 2H_2O(l) \rightarrow 2Zn(OH)_2(s)$$

and the iron remains intact.

or nickel. The weakness of such methods is that any scratch or crack in the protective layer exposes the metal surface. The exposed surface, even though small in area, can act as an electron donor in conjunction with other exposed metal parts, which act as electron acceptors. This combination then leads to corrosion of the metal under the no-longer-protective layer.

Another anticorrosion technique uses a replaceable sacrificial metal, which is a piece of metal electrically connected to a less active metal (Figure 21-8). The more active metal is the stronger reducing agent and is thus preferentially oxidized; oxygen is reduced on the surface of the less active metal. Sacrificial metals are used to protect water pipes and ship propellers. This method is also the basis of **galvanization,** in which iron is protected from corrosion either by a zinc coating, a process used in the manufacture of automobile bodies, or by impregnation of zinc in the steel body parts. A crack in the zinc coating does not affect the corrosion protection provided, because the zinc coating functions like the zinc metal shown in Figure 21-8. Note that the less active metal (iron) promotes the corrosion of the more active metal (zinc) by providing a metal surface for reduction of oxygen.

SUMMARY

To determine whether a reaction involves electron transfer, we first assign oxidation states to each element on both sides of the equation. Oxidation states are assigned using the rules described in Section 4-1, or by the use of Lewis formulas and electronegativities, or by analogy. In an electron-transfer reaction, the oxidation state of one element increases and the oxidation state of another element decreases.

Electron-transfer reactions can be separated into two half reactions: the oxidation half reaction (represented by the equation with electrons on the right) and the reduction half reaction (represented by the equation with electrons on the left). The oxidation half reaction supplies electrons to the reduction half reaction. The equations for electron-transfer reactions can be balanced by a systematic procedure once the oxidation half reaction and the reduction half reaction have been identified by the assignment of oxidation states to the atoms involved in the reaction. The procedure for balancing equations for oxidation-reduction reactions in acidic solutions involves seven steps:

1. Separate the equation into an oxidation half reaction equation and a reduction half reaction equation.
2. Balance each half reaction equation with respect to all elements other than oxygen and hydrogen.

3. Balance each half reaction equation with respect to oxygen by adding the appropriate number of H_2O to the side deficient in oxygen.
4. Balance each half reaction equation with respect to hydrogen by adding the appropriate number of H^+ to the side deficient in hydrogen.
5. Balance each half reaction equation with respect to charge by adding the appropriate number of electrons to the side with the excess positive charge.
6. Multiply each half reaction equation by an integer that makes the number of electrons supplied by the oxidation half reaction equation equal to the number of electrons accepted by the reduction half reaction equation.
7. Obtain the complete balanced equation by adding the two half reaction equations and canceling any like terms.

To balance the equation for a redox reaction that takes place in basic solution, steps 3 and 4 are changed to balance each half reaction equation with respect to hydrogen by adding one H_2O to the side with excess oxygen and twice as many OH^- to the other side. If the equation involves only excess OH^- for O and H, then balance the equation using OH^-.

Corrosion of metals and metal alloys usually involves air oxidation of the metal. Anticorrosion methods involve either protective coatings or the use of a more reactive metal that is preferentially oxidized (corroded).

TERMS YOU SHOULD KNOW

oxidation-reduction reaction 755
redox reaction 755
reducing agent 755
electron donor 755
oxidizing agent 755
electron acceptor 755
oxidized 755
oxidation 755
reduced 756
reduction 756
electron-transfer reaction 756
half reaction 757
oxidation half reaction 757
reduction half reaction 757
disproportionation reaction 757
method of half reactions 758
quantitative reaction 764
corrosion 767
galvanization 768

EQUATIONS YOU SHOULD KNOW HOW TO USE

$$\begin{pmatrix}\text{oxidation}\\\text{state}\end{pmatrix} = \begin{pmatrix}\text{group number of}\\\text{the element in the}\\\text{periodic table}\end{pmatrix} - \begin{pmatrix}\text{total number of}\\\text{valence electrons}\\\text{assigned to the}\\\text{element}\end{pmatrix} \quad (23\text{-}1) \quad \begin{pmatrix}\text{Lewis-formula method}\\\text{of assigning}\\\text{oxidation states}\end{pmatrix}$$

PROBLEMS

OXIDATION STATES

21-1. Assign oxidation states to the oxygen atoms in

(a) $O_2(g)$ (b) $KO_2(s)$
(c) $Na_2O_2(s)$ (d) $OF_2(g)$

21-2. Assign oxidation states to the oxygen atoms in

(a) $HO_2(g)$ (b) $S_2O_8^{2-}(aq)$ $[O_3S{-}O{-}O{-}SO_3]^{2-}$
(c) $O_3^-(aq)$ (d) $H_3C{-}O{-}O{-}CH_3(l)$

21-3. Assign oxidation states to the atoms in

(a) $ClO^-(aq)$ (b) $ClO_2^-(aq)$
(c) $ClO_3^-(aq)$ (d) $ClO_4^-(aq)$

21-4. Assign oxidation states to the atoms in

(a) $Cl_2O(g)$ (b) $ClO_2(g)$
(c) $Cl_2O_7(l)$ (d) $Cl_2O_5(l)$

21-5. Assign oxidation states to the atoms in

(a) $HCHO(g)$ (b) $CH_3OH(l)$
(c) $CH_3C(=\ddot{O}:)CH_3(l)$ (d) $CH_3COOH(l)$

21-6. Assign oxidation states to the atoms in

(a) $CS_2(l)$ (b) $CH_3S{-}SCH_3(l)$
(c) $HCONH_2(l)$ (d) $H_2NC(=\ddot{S}:)NH_2(s)$

21-7. Assign oxidation states to the atoms in

(a) $KMnO_4(s)$ (b) $MnO_4^{2-}(aq)$
(c) $MnO_2(s)$ (d) $Mn(ClO_4)_3(s)$

21-8. Assign oxidation states to the atoms in

(a) $Fe_2O_3{\cdot}3H_2O(s)$ (b) $Fe_3O_4(s)$
(c) $KCN(s)$ (d) $KCNO(s)$

21-9. Assign oxidation states to the atoms in

(a) $MoSe_2(s)$ (b) $SiC(s)$
(c) $GaAs(s)$ (d) $K_2S_2O_3(s)$

21-10. Assign oxidation states to the atoms in

(a) $GeH_4(g)$ (b) $XeOF_4(s)$
(c) $XeO_2F(s)$ (d) $XeO_4^{2-}(aq)$

OXIDIZING AGENTS AND REDUCING AGENTS

21-11. Identify the oxidizing and reducing agents in the equation

$$I_2(s) + 2Na_2S_2O_3(aq) \rightarrow 2NaI(aq) + Na_2S_4O_6(aq)$$

21-12. Sodium sulfide is manufactured by reacting sodium sulfate with carbon in the form of coke:

$$Na_2SO_4(s) + 4C(s) \rightarrow Na_2S(s) + 4CO(g)$$

Identify the oxidizing and reducing agents in this reaction.

21-13. Sodium nitrite, an important chemical in the dye industry, is manufactured by the reaction between sodium nitrate and lead:

$$NaNO_3(aq) + Pb(s) \rightarrow NaNO_2(aq) + PbO(s)$$

Identify the oxidizing and reducing agents in this equation.

21-14. Sodium chlorite, an industrial bleaching agent, is prepared as shown by the equation

$$4NaOH(aq) + Ca(OH)_2(aq) + C(s) + 4ClO_2(g) \rightarrow 4NaClO_2(aq) + CaCO_3(s) + 3H_2O(l)$$

Identify the oxidizing and reducing agents in this equation.

21-15. Identify the oxidizing and reducing agents and the oxidation and reduction half reaction equations in the following equations:

(a) $2Fe^{3+}(aq) + 2I^-(aq) \rightarrow 2Fe^{2+}(aq) + I_2(s)$
(b) $2Ti^{2+}(aq) + Co^{2+}(aq) \rightarrow 2Ti^{3+}(aq) + Co(s)$

21-16. Identify the oxidizing and reducing agents and the oxidation and reduction half reaction equations in the following equations:

(a) $H_2S(aq) + ClO^-(aq) \rightarrow S(s) + Cl^-(aq) + H_2O(l)$
(b) $In^+(aq) + 2Fe^{3+}(aq) \rightarrow 2Fe^{2+}(aq) + In^{3+}(aq)$

21-17. Potassium superoxide, KO_2, is a strong oxidizing agent. Explain why.

21-18. Lithium aluminum hydride, $LiAlH_4$, is a strong reducing agent. Explain why.

BALANCING OXIDATION-REDUCTION EQUATIONS

21-19. Balance the following equations for reactions that occur in acidic solution:

(a) $MnO(s) + PbO_2(s) \rightarrow MnO_4^-(aq) + Pb^{2+}(aq)$
(b) $As_2S_5(s) + NO_3^-(aq) \rightarrow H_3AsO_4(aq) + HSO_4^-(aq) + NO_2(g)$

For each of these reactions, identify the

electron donor	reducing agent
electron acceptor	species oxidized
oxidizing agent	species reduced

21-20. Balance the following equations for reactions that occur in acidic solution:

(a) $ZnS(s) + NO_3^-(aq) \rightarrow Zn^{2+}(aq) + S(s) + NO(g)$
(b) $MnO_4^-(aq) + HNO_2(aq) \rightarrow NO_3^-(aq) + Mn^{2+}(aq)$

For each of these reactions, identify the

electron donor	reducing agent
electron acceptor	species oxidized
oxidizing agent	species reduced

21-21. Complete and balance the following equations:

(a) $NH_4^+(aq) + NO_3^-(aq) \rightarrow N_2O(g)$ (acidic)
(b) $Fe(s) + O_2(g) \rightarrow Fe_2O_3 \cdot 3H_2O(s)$ (basic)

21-22. Complete and balance the following equations:

(a) $CoCl_2(s) + Na_2O_2(aq) \rightarrow Co(OH)_3(s) + Cl^-(aq) + Na^+(aq)$ (basic)
(b) $C_2O_4^{2-}(aq) + MnO_2(s) \rightarrow Mn^{2+}(aq) + CO_2(g)$ (acidic)

21-23. Complete and balance the following equations:

(a) $Fe(OH)_2(s) + O_2(g) \rightarrow Fe(OH)_3(s)$ (basic)
(b) $Cu(s) + NO_3^-(aq) \rightarrow Cu^{2+}(aq) + NO(g)$ (acidic)

21-24. Complete and balance the following equations:

(a) $Cr_2O_7^{2-}(aq) + I^-(aq) \rightarrow Cr^{3+}(aq) + I_2(s)$ (acidic)
(b) $CuS(s) + NO_3^-(aq) \rightarrow Cu^{2+}(aq) + S(s) + NO(g)$ (acidic)

21-25. Use the method of half reactions to balance the following equations:

(a) $IO_4^-(aq) + I^-(aq) \rightarrow IO_3^-(aq) + I_3^-(aq)$ (basic)
(b) $H_2MoO_4(aq) + Cr^{2+}(aq) \rightarrow Mo(s) + Cr^{3+}(aq)$ (acidic)

21-26. Use the method of half reactions to balance the following equations:

(a) $N_2H_4(aq) + Cu(OH)_2(s) \rightarrow N_2(g) + Cu(s)$ (basic)
(b) $H_3AsO_3(aq) + I_2(aq) \rightarrow H_3AsO_4(aq) + I^-(aq)$ (acidic)

21-27. Use the method of half reactions to balance the following equations:

(a) $CrO_4^{2-}(aq) + Cl^-(aq) \rightarrow Cr^{3+}(aq) + ClO_2^-(aq)$ (acidic)

(b) $Cu^{2+}(aq) + S_2O_3^{2-}(aq) \rightarrow Cu^+(aq) + S_4O_6^{2-}(aq)$ (acidic)

21-28. Use the method of half reactions to balance the following equations:

(a) $Co(OH)_2(s) + SO_3^{2-}(aq) \rightarrow SO_4^{2-}(aq) + Co(s)$ (basic)

(b) $IO_3^-(aq) + I^-(aq) \rightarrow I_3^-(aq)$ (acidic)

21-29. For the strong of heart: balance

$$CrI_3(s) + Cl_2(g) \rightarrow CrO_4^{2-}(aq) + IO_4^-(aq) + Cl^-(aq) \quad \text{(basic)}$$

21-30. For the strong of heart: balance

$$C_2H_5OH(aq) + I_3^-(aq) \rightarrow CO_2(g) + CHO_2^-(aq) + CHI_3(aq) + I^-(aq) \quad \text{(acidic)}$$

21-31. Balance the following equations for half reactions that occur in acid solution:

(a) $Mo^{3+}(aq) \rightarrow MoO_2^{2+}(aq)$
(b) $P_4(s) \rightarrow H_3PO_4(aq)$
(c) $S_2O_8^{2-}(aq) \rightarrow HSO_4^-(aq)$

21-32. Balance the following equations for half reactions that occur in acidic solution:

(a) $H_2BO_3^-(aq) \rightarrow BH_4^-(aq)$
(b) $ClO_3^-(aq) \rightarrow Cl_2(g)$
(c) $Cl_2(g) \rightarrow HClO(aq)$

21-33. Balance the following half reaction equations:

(a) $WO_3(s) \rightarrow W_2O_5(s)$ (acidic)
(b) $U^{4+}(aq) \rightarrow UO_2^+(aq)$ (acidic)
(c) $Zn(s) \rightarrow Zn(OH)_4^{2-}(aq)$ (basic)

21-34. Balance the following half reaction equations:

(a) $OsO_4(s) \rightarrow Os(s)$ (acidic)
(b) $S(s) \rightarrow SO_3^{2-}(aq)$ (basic)
(c) $Sn(s) \rightarrow HSnO_2^-(aq)$ (basic)

21-35. Balance the following equations for half reactions that occur in basic solution:

(a) $SO_3^{2-}(aq) \rightarrow S_2O_4^{2-}(aq)$
(b) $Cu(OH)_2(s) \rightarrow Cu_2O(s)$
(c) $AgO(s) \rightarrow Ag_2O(s)$

21-36. Balance the following equations for half reactions:

(a) $Au(CN)_2^-(aq) \rightarrow Au(s) + CN^-(aq)$ (acidic)
(b) $MnO_4^-(aq) \rightarrow MnO_2(s)$ (acidic)
(c) $Cr(OH)_3(s) \rightarrow CrO_4^{2-}(aq)$ (basic)

CALCULATIONS INVOLVING OXIDATION-REDUCTION EQUATIONS

21-37. The quantity of antimony in a sample can be determined by an oxidation-reduction titration with an oxidizing agent. A 9.62-g sample of stibnite, an ore of antimony, is dissolved in hot, concentrated HCl(*aq*) and passed over a reducing agent so that all the antimony is in the form Sb^{3+}. The $Sb^{3+}(aq)$ is completely oxidized by 43.7 mL of a 0.125 M solution of $KBrO_3$. The unbalanced equation for the reaction is

$$BrO_3^-(aq) + Sb^{3+}(aq) \rightarrow Br^-(aq) + Sb^{5+}(aq)$$

Calculate the amount of antimony in the sample and its percentage in the ore.

21-38. An ore is to be analyzed for its iron content by an oxidation-reduction titration with permanganate ion. A 4.23-g sample of the ore is dissolved in hydrochloric acid and passed over a reducing agent so that all the iron is in the form Fe^{2+}. The $Fe^{2+}(aq)$ is completely oxidized by 31.6 mL of a 0.0512 M solution of $KMnO_4$. The unbalanced equation for the reaction is

$$KMnO_4(aq) + HCl(aq) + FeCl_2(aq) \rightarrow MnCl_2(aq) + FeCl_3(aq) + H_2O(l) + KCl(aq)$$

Calculate the amount of iron in the sample and its mass percentage in the ore.

21-39. A rock sample is to be assayed for its tin content by an oxidation-reduction titration with $I_3^-(aq)$. A 10.0-g sample of the rock is crushed, dissolved in sulfuric acid, and passed over a reducing agent so that all the tin is in the form Sn^{2+}. The $Sn^{2+}(aq)$ is completely oxidized by 34.6 mL of a 0.556 M solution of NaI_3. The unbalanced equation for the reaction is

$$I_3^-(aq) + Sn^{2+}(aq) \rightarrow Sn^{4+}(aq) + I^-(aq)$$

Calculate the amount of tin in the sample and its mass percentage in the rock.

21-40. Sodium chlorite, $NaClO_2(s)$, is a powerful but stable oxidizing agent used in the paper industry, especially for the final whitening of paper. Sodium chlorite is capable of bleaching materials containing cellulose without oxidizing the cellulose. Sodium chlorite is made by the reaction.

$$NaOH(aq) + Ca(OH)_2(s) + C(s) + ClO_2(g) \rightarrow NaClO_2(aq) + CaCO_3(s)$$

Balance the equation for the reaction and calculate the number of kilograms of $ClO_2(g)$ required to make 1.00 metric ton of $NaClO_2$.

21-41. The amount of $I_3^-(aq)$ in a solution can be determined by titration with a solution containing a known concentration of $S_2O_3^{2-}(aq)$ (thiosulfate ion). The determination is based on the unbalanced equation

$$I_3^-(aq) + S_2O_3^{2-}(aq) \rightarrow I^-(aq) + S_4O_6^{2-}(aq)$$

22 ENTROPY, GIBBS FREE ENERGY, AND CHEMICAL REACTIVITY

The dispersal of ink throughout the entire volume of water is an example of a spontaneous process.

It is natural to ask why some substances react with each other and others do not. Why do some reactions occur spontaneously (i.e., on their own) and others require a continuous input of energy from an external source? In this chapter we discover the condition that must be met for a reaction to be spontaneous—a condition called the criterion for reaction spontaneity. The observation that all highly exothermic reactions are spontaneous led the French chemist Marcelin Berthelot to put forth, in the 1860s, the hypothesis that all spontaneous reactions are exothermic. Berthelot's criterion of spontaneity for a chemical reaction was based on the sign of the enthalpy change, ΔH_{rxn}, for the reaction. According to Berthelot, if ΔH_{rxn} is negative, then a reaction is spontaneous.

Berthelot's criterion of reaction spontaneity was shown to be incorrect, however; and it was superseded by the criterion of reaction spontaneity developed by the American thermodynamicist J. Willard Gibbs. Gibbs showed that reaction spontaneity is not just a matter of energy changes. Another property, called entropy, also must be considered when determining whether a reaction is spontaneous. We will show in this chapter that entropy is a measure of the randomness or disorder of a system and that the spontaneity of a chemical reaction is governed by both the energy change and the entropy change. We will discuss how the entropy concept is based on a fundamental principle called the second law of thermodynamics. We will show that the Gibbs criterion of reaction spontaneity is given by the sign of a new quantity, called the Gibbs free energy change, which depends on both energy and entropy changes for the reaction. In fact, it is equal to the maximum amount of energy, in the form of work, that can be extracted from a reaction system to perform tasks, such as lifting a weight or powering an electric motor.

22-1. NOT ALL SPONTANEOUS REACTIONS EVOLVE ENERGY

In Chapter 8 we discussed enthalpy changes for chemical reactions. Recall that an exothermic chemical reaction is one in which energy is evolved as heat. Exothermic reactions go energetically downhill in the sense that the total enthalpy of the products is less than the total enthalpy of the reactants (Figure 22-1). For an exothermic reaction, the **enthalpy change, ΔH_{rxn},** is negative. For an exothermic reaction run at standard conditions (all species at 1 atm pressure, all solutes at 1 M), we have

$$\Delta H^\circ_{rxn} = H^\circ(\text{products}) - H^\circ(\text{reactants}) < 0$$

where the superscript degree signs denote *standard* values for the thermodynamic quantities. Thus, ΔH°_{rxn} is the **standard enthalpy change** for the reaction.

In Chapter 8 we learned to calculate ΔH°_{rxn} from the values in a table of heats of formation (Table 8-1). Recall that the value of ΔH°_{rxn} is approximately equal to the value of the standard energy change, ΔU°_{rxn},

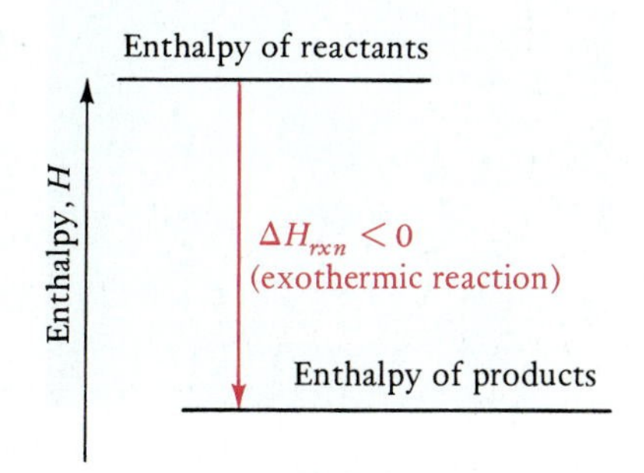

Figure 22-1 In an exothermic reaction, the total enthalpy of the products is less than the total enthalpy of the reactants.

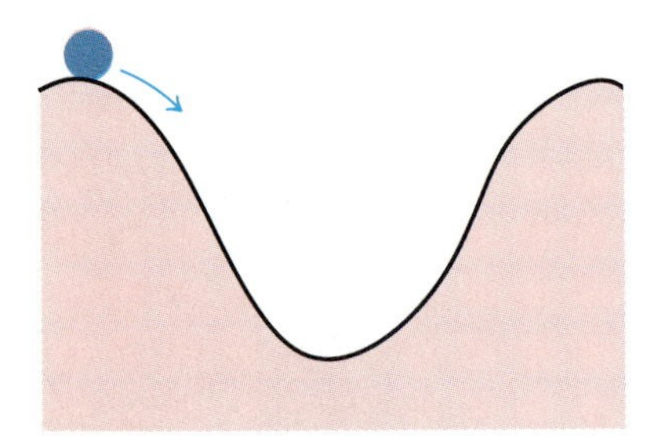

(a) Initial state

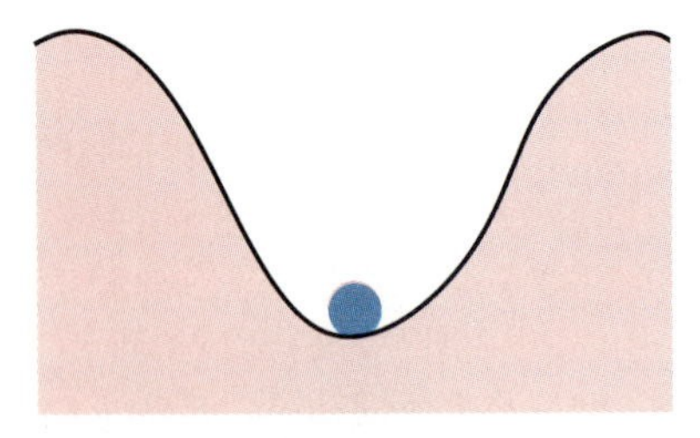

(b) Final state

Figure 22-2 A ball at the top of a hill will roll down the hill and eventually come to rest. The potential energy of the ball at the bottom of the hill is less than that at the top. The ball spontaneously goes from a state of high potential energy to a state of low potential energy.

Figure 22-3 The combustion of methane, CH_4, in air to yield gaseous carbon dioxide, CO_2, and water, H_2O, is an example of a spontaneous, exothermic process.

for a reaction. Also, recall that $\Delta H^\circ_{rxn} \simeq \Delta H_{rxn}$, because enthalpy is relatively independent of pressure and concentration.

A process that takes place without the input of energy from an external source is said to be **spontaneous.** The natural tendency of simple mechanical systems is to undergo processes that lead to a decrease in energy of the system. For example, water flows spontaneously (i.e., without any energy input) downhill. Water at the bottom of a waterfall has a lower potential energy than water at the top. To get water back to the top of the waterfall, we have to use a pump, which requires energy input. Thus, the flow of water uphill is not a spontaneous process. As another example, if we release the ball shown at the top of the hill in Figure 22-2, then it spontaneously rolls down the hill and eventually comes to rest at the bottom of the valley. The ball at the lowest point of the valley has the lowest possible potential energy for this system.

We have all observed a wide variety of spontaneous chemical processes. For example, natural gas, CH_4, once ignited, burns spontaneously in air to yield carbon dioxide and water (Figure 22-3):

$$CH_4(g) + 2O_2(g) \rightarrow CO_2(g) + 2H_2O(g) \qquad \Delta H^\circ_{rxn} = -802\ \text{kJ}$$

Iron, on exposure to air and moisture, spontaneously rusts:

$$4Fe(s) + 3O_2(g) \xrightarrow[H_2O(l)]{} 2Fe_2O_3(s) \qquad \Delta H^\circ_{rxn} = -1648\ \text{kJ}$$

Zinc metal reacts spontaneously with 1.0 M hydrochloric acid to yield hydrogen gas and aqueous zinc chloride:

$$Zn(s) + 2HCl(aq) \rightarrow H_2(g) + ZnCl_2(aq) \qquad \Delta H^\circ_{rxn} = -150\ \text{kJ}$$

Furthermore, these reactions do not occur spontaneously in the reverse direction.

The three chemical reactions we have considered so far have been highly exothermic; and, by analogy with mechanical processes (such as a ball rolling downhill) that move spontaneously to lower energy states, we predict (correctly) that they occur spontaneously. Many spontaneous processes, however, are not exothermic. Let's consider two gases, such as $N_2(g)$ and $I_2(g)$, occupying two separate containers, as shown in Figure 22-4. We all know that, if allowed to, the gases will mix spontaneously. Furthermore, simple gaseous mixtures do not separate spontaneously. It would be a potentially disastrous occurrence if the air in a room suddenly separated so that part of the room contained pure oxygen and the rest contained nitrogen. Although the mixing of two gases is a spontaneous process, it turns out that $\Delta H^\circ \simeq 0$ for such a process. Surely it is not the energy change that drives the gas-mixing process.

Not only do some spontaneous processes have $\Delta H^\circ \simeq 0$, but others are even endothermic. The energy of the products is *greater* than that of the reactants. For example, ordinary ice at any temperature greater than 0°C spontaneously melts:

$$H_2O(s) \rightarrow H_2O(l) \qquad \Delta H^\circ_{fusion} = +6.0\ \text{kJ}$$

and table salt spontaneously dissolves in water at 20°C:

$$NaCl(s) \xrightarrow[H_2O(l)]{} Na^+(aq) + Cl^-(aq) \qquad \Delta H^\circ_{solution} = +6.4\ \text{kJ}$$

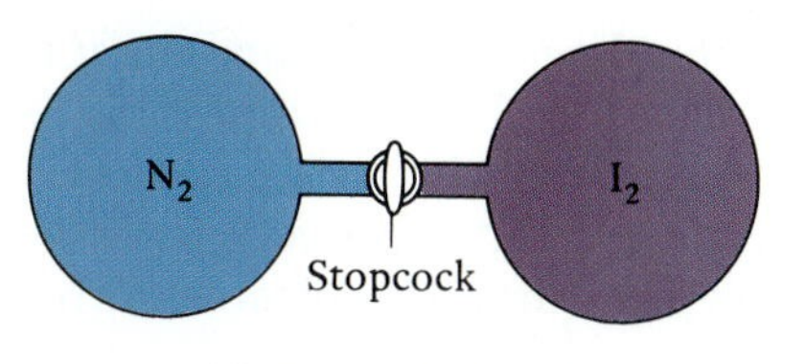

(a) Stopcock closed (b) After stopcock is opened

Figure 22-4 A simple example of a spontaneous process with $\Delta H^\circ_{rxn} \simeq 0$ is the mixing of two gases.

An especially interesting example of a spontaneous endothermic reaction is the reaction of solid barium hydroxide, $Ba(OH)_2(s)$, and solid ammonium nitrate, $NH_4NO_3(s)$:

$$Ba(OH)_2(s) + 2NH_4NO_3(s) \rightarrow Ba(NO_3)_2(s) + 2H_2O(l) + 2NH_3(aq)$$

The energy absorbed from the reaction system is sufficient to drop the temperature of the reaction system from 20°C (room temperature) to well below the freezing point of water (Figure 22-5), as evidenced by the freezing of the water placed between the reaction flask and the block of wood. Mixtures of chemicals such as these are used in commercially available **cold packs** for the first-aid treatment of bruises and strains.

That spontaneous processes can occur with negative, zero, or positive values of ΔH clearly indicates that heat evolution (that is, $\Delta H < 0$) is *not* a general criterion of spontaneity for a process. The spontaneity of a process is not determined solely by the enthalpy change. There is an additional factor involved—the entropy change for the process.

(a)

(b)

Figure 22-5 (a) Samples of ammonium nitrate (left) and barium hydroxide (right). The water in the eyedropper is placed on the wood block, the chemicals are mixed in the flask, and the flask is then placed on the wood block over the water. (b) The endothermic reaction between $Ba(OH)_2(s)$ and $NH_4NO_3(s)$ produces a large temperature drop that causes the water to freeze under the flask. The ice bonds the flask to the wood block.

22-2. THE SECOND LAW OF THERMODYNAMICS PLACES AN ADDITIONAL RESTRICTION ON ENERGY TRANSFERS

The only restriction placed on energy transfers by the first law of thermodynamics (Chapter 8) is the conservation of energy. We know from everyday experience, however, that there are other restrictions on energy transfers. For example, energy as heat always flows spontaneously from a region of higher temperature to a region of lower temperature. A piece of paper, once ignited, burns spontaneously in oxygen, but the reverse process, that is, the spontaneous recombination of the combustion products to the piece of paper and oxygen gas, has never been observed in nature. Nor will any incantation cause a scrambled egg to reassemble itself. Innumerable other naturally occurring processes are also unidirectional. Spontaneous, unidirectional processes often are referred to as **irreversible processes.**

In thermodynamics, we always distinguish between the **system,** that part of the universe where the change of interest occurs, and its **surroundings,** which is the rest of the universe. Thermodynamics states that when an actual process occurs, the system and its surroundings cannot both be restored exactly to their original states. If, for example, we restore the system to its initial state by doing work on the system, then it is impossible to restore the surroundings to its original state; if we restore the surroundings to its original state, then it is impossible to restore the system to its original state. All naturally occurring processes are in this sense **irreversible.** Nature is never exactly the same today as it was yesterday. When a process occurs in nature, the universe is irreversibly changed.

The second law of thermodynamics and the concept of entropy, which is embodied in the second law, arose from a definitive analysis of the operation of heat engines (especially steam engines) by the French engineer Sadi Carnot, who published his results in 1824. It is interesting to note that Sadi's father, Lazare Carnot, in 1803 published the definitive work on the operation of purely mechanical engines, that is, engines based on the first law of thermodynamics, which involve, for example, pulleys, gears, and inclined planes.

Heat engines, of which steam engines and internal combustion engines (e.g., automobile engines) are prime examples, are engines designed to convert heat into work. It was well known in the early part of the nineteenth century that all heat engines must be operated at a temperature higher than that of their surroundings in order to function. What Sadi Carnot showed, in effect, was why this is so. In addition, he showed why it was not possible to convert *all* the energy input as heat into energy output as work, and also how the ratio of the work output to the heat input of the engine depends on the temperature difference between the system (the engine) and the surroundings (usually the atmosphere).

Sadi Carnot's work led directly to the discovery of a new thermodynamic state function (Section 8-2)—called **entropy** and given the symbol S—that is closely associated with the transfer of energy as heat. Carnot's key result is most compactly expressed in the equation

$$\Delta S_{sys} \geq \frac{q_{sys}}{T_{sys}} \qquad (22\text{-}1)$$

Equation (22-1) is a mathematical statement of the **second law of thermodynamics;** it tells us that if energy is transferred either to or from a system as heat, q_{sys}, at the Kelvin temperature, T_{sys}, then the **entropy change** of the system, ΔS_{sys}, is greater than or equal to q_{sys}/T_{sys}. The equality sign applies if the process is reversible. The heat transfer process is said to be **reversible** if there is no heat generation due to friction and the heat is transferred at a uniform temperature, $T_{sys} = T_{surroundings}$. The inequality sign applies if the process is irreversible. Note from Equation (22-1) that ΔS_{sys} is greater than zero when q_{sys} is positive (heat in), and ΔS_{sys} is less than zero when q_{sys} is negative (heat out). Also note from Equation (22-1) that the units of entropy are joules per kelvin ($J{\cdot}K^{-1}$).

The second law of thermodynamics also states (as can be shown using Equation 22-1) that the *total* entropy change for spontaneous processes must always be positive. By total entropy change, we mean not only the entropy of the system that we are studying but also that of its surroundings as well. Because the most general example of a system and its surroundings includes the entire universe and because all naturally occurring processes are spontaneous, the second law of thermodynamics tells us that the entropy of the universe always increases when a natural process occurs. Unlike energy, entropy is not conserved.

Although in any process energy is neither created nor destroyed, the use of energy inevitably leads to a lowering of the "quality" of the energy. In other words, the energy has a decreased capacity to perform further work. The energy, although conserved, is said to be *degraded.* The extent of the degradation of energy in a process is given quantitatively by the increase in the total entropy (for the system plus its surroundings) that occurs when the process takes place.

The concept of entropy also can be developed in molecular terms. In the next few sections, we will see how the entropy concept gives us insight into chemical reactivity.

22-3. ENTROPY IS A MEASURE OF THE AMOUNT OF DISORDER OR RANDOMNESS IN A SYSTEM

On a molecular level the entropy of a substance is a quantitative measure of the amount of disorder in a substance. The disorder is of two types: **positional disorder,** which refers to the distribution of the particles in space, and **thermal disorder,** which refers to the distribution of the available energy among the particles. Any process that produces a more random distribution of the particles in space gives rise to an increase in the total entropy of the substance. So does any constant-pressure process that increases the temperature of the particles.

In Figure 22-6, the molar entropy of oxygen is plotted against temperature from 0 K to 300 K. The entropy is zero at 0 K. At this temperature (absolute zero), there is no positional disorder, because the O_2 molecules in the crystal are perfectly arrayed in the lattice. Nor is there any thermal disorder, because all the molecules are in the lowest possible energy state. Entropy is associated with disorder—no disorder, no entropy. The statement that the entropy of a perfect crystalline substance is zero at absolute zero is known as the **third law of thermodynamics.**

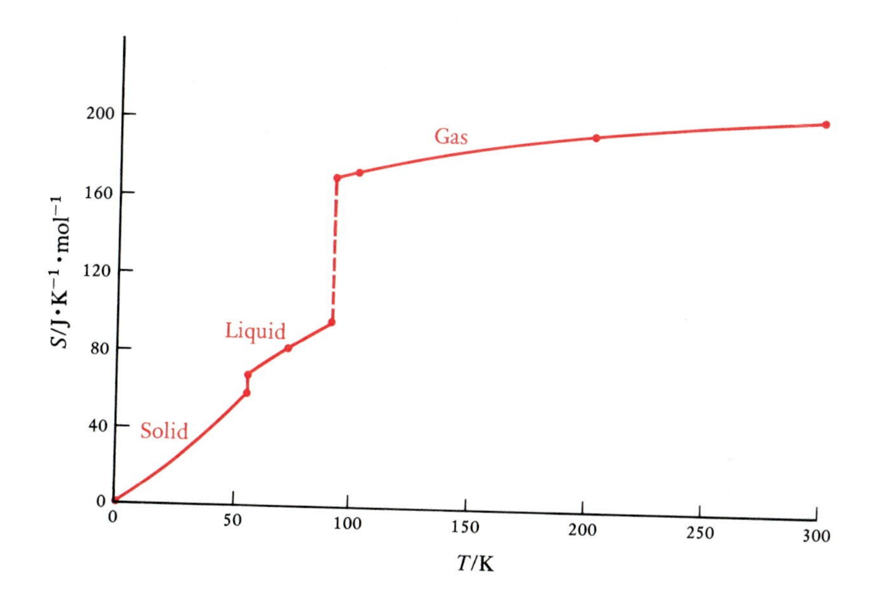

Figure 22-6 The molar entropy of oxygen as a function of temperature at 1 atm. Note that the entropy of each phase increases smoothly with increasing temperature. This gradual increase in entropy with increase in temperature is a result of the increase in thermal disorder. The jumps in entropy that occur when the solid melts and the liquid vaporizes are the result of an increase in positional disorder of the O_2 molecules.

As the temperature is increased from 0 K (Figure 22-6), the oxygen molecules begin to vibrate more freely about their lattice sites. The increase in temperature causes an increase in thermal disorder. When the temperature of a substance held at constant pressure increases, the entropy of the substance increases. We can now see how the picture of entropy as disorder agrees with our statement of the second law of thermodynamics: when energy as heat is added to a system, the entropy of the system always increases. The increase in entropy that arises from an increase in temperature is associated with the greater amount of energy that must be distributed among the molecules of the substance. Another way of approaching this concept is to think of the energy levels associated with molecular orbitals that we developed in Chapter 13. Molecules, like atoms, are restricted to discrete, quantized energy levels. The greater the amount of energy stored in a substance, the greater the number of ways in which the energy can be distributed and thus the greater the thermal disorder and the greater the entropy.

Figure 22-6 shows that relatively large increases in entropy occur when solid oxygen is converted to liquid oxygen (melting) and when liquid oxygen is converted to gaseous oxygen (vaporization). The reasons for these entropy increases are discussed in the next section. Figure 22-6 also shows that at fixed pressure the entropy of liquid and gaseous oxygen, like that for solid oxygen, increases as the temperature increases. And, as the temperature increases, the motions of the liquid and the gaseous oxygen molecules become more vigorous. In other words, average molecular speed increases with temperature (Section 7-11). The increased molecular speed means an increase in thermal disorder, which is the molecular basis for the increase in entropy.

Although not shown in Figure 22-6, the entropy of a substance can also increase at fixed temperature, as a result of an increase in positional disorder. For example, when the volume of a gas is increased at a fixed temperature, the molar entropy of the gas increases, because the gas molecules have a larger available volume in which to move and thus have a higher degree of positional disorder. Similarly, the molar entropy of a solute increases when additional solvent is added to the solution, because the solute particles have more room in which to move and thus have a higher degree of positional disorder.

EXAMPLE 22-1: Predict the sign of the entropy change, ΔS_{sys}, for the following processes:

(a) $H_2O(g)$ (1 atm, 25°C) $\rightarrow$ $H_2O(g)$ (0.01 atm, 25°C)

(b) the dissolution of sodium chloride in water

(c) $NaCl(aq)$ (4.8 M, 20°C) $\rightarrow$ $NaCl(aq)$ (1.0 M, 20°C)

(d) $C_6H_6(g)$ (10 torr, 25°C) $\rightarrow$ C_6H_6 (adsorbed on a metal surface at 25°C)

Solution: (a) A gas at a lower pressure, and thus with greater volume, is more disordered than a gas at a higher pressure at the same temperature; therefore,

$$\Delta S_{gas} \text{ (constant-temperature expansion)} > 0$$

(b) Upon dissolving, the sodium and chloride ions are dispersed throughout the solvent and are much more disordered than in the crystal, so

$$\Delta S_{solid} \text{ (dissolution)} > 0$$

(c) Dilution with solvent increases the positional disorder of the solute, so

$$\Delta S_{\text{solute}}\ (\text{dilution}) > 0$$

(d) The adsorption of a gas onto a surface produces a decrease in the entropy of adsorbed molecules because of the decreased freedom of movement (see Figure 22-7), so

$$\Delta S_{\text{benzene}}\ (\text{adsorption}) < 0$$

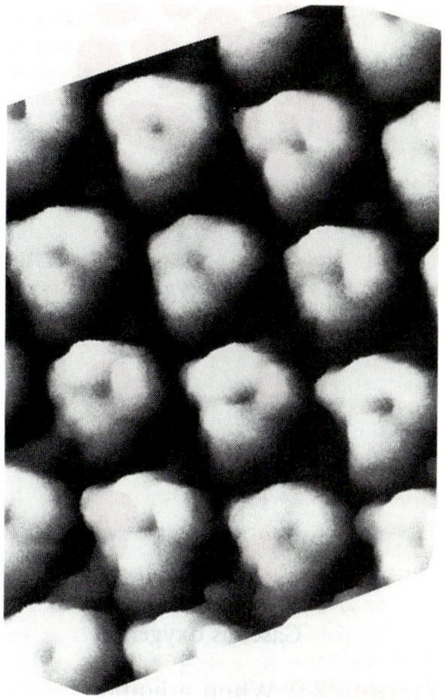

PRACTICE PROBLEM 22-1: Predict the sign of the entropy change, ΔS_{sys}, for the following processes:
(a) $H_2O(l) \rightarrow H_2O(s)$ (0°C, 1 atm)
(b) $Cu(s)$ (1 atm, 500°C) $\rightarrow$ $Cu(s)$ (1 atm, 25°C)
(c) $4Al(s) + 3O_2(g) \rightarrow 2Al_2O_3(s)$ (25°C, 1 atm)

Answer: (a) $\Delta S_{\text{sys}} < 0$; (b) $\Delta S_{\text{sys}} < 0$; (c) $\Delta S_{\text{sys}} < 0$

22-4. THERE IS AN INCREASE IN ENTROPY ON MELTING AND VAPORIZATION

Figure 22-6 shows that the entropy of a particular phase increases smoothly with increasing temperature. At a phase transition, however, there is a jump in entropy. Let's investigate the origin of the two entropy jumps in Figure 22-6. When solid oxygen melts to liquid oxygen (at 54.4 K), the ordered lattice of the solid breaks down into a liquid, where the molecules no longer are confined to lattice sites. Each molecule moves throughout the liquid volume, so there is an increase in the positional disorder of the oxygen (Figure 22-8). This increase in positional disorder leads to an increase in entropy. Because all the entropy increase occurs at the melting point, which is a fixed temperature, ΔS of melting (fusion) appears as a vertical jump in Figure 22-6.

We can use Equation (22-1) to calculate the change in the entropy of one mole of a substance upon fusion, called the **molar entropy of fusion** and written ΔS_{fus}. Because melting (as well as vaporization) can be carried out under essentially reversible conditions, we use the equality sign in Equation (22-1), together with the relation $q = \Delta H_{\text{fus}}$, to obtain

$$\Delta S_{\text{fus}} = \frac{\Delta H_{\text{fus}}}{T_{\text{m}}} \qquad (22\text{-}2)$$

Figure 22-7 Scanning tunneling micrograph of benzene molecules adsorbed on the surface of a rhodium crystal. The hexagonal shape of the benzene molecules is revealed in this remarkable micrograph.

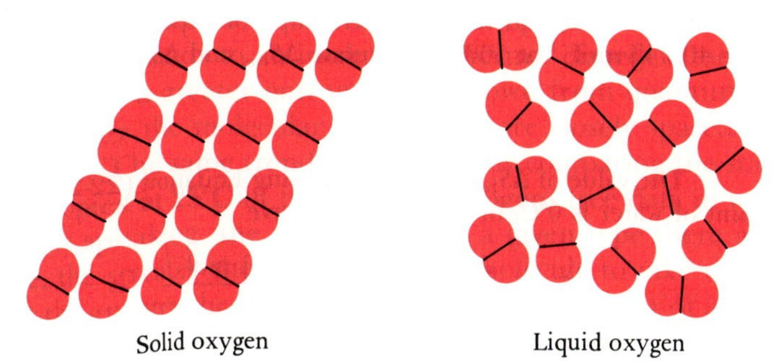

Figure 22-8 When a solid melts, the molecules become free to move throughout the volume of the liquid. This increased positional disorder in the liquid implies that, under the same conditions, the entropy of a liquid is always greater than the entropy of the solid.

so the Gibbs criteria predict that two gases mix spontaneously and that a drop of ink becomes uniformly dispersed throughout a volume of water. Furthermore, note that $\Delta G_{rxn} > 0$ for the reverse reactions, because $\Delta S < 0$ for the reverse reactions. Therefore, the Gibbs criteria predict that we will never see the reverse processes occur spontaneously.

Highly exothermic processes, like most combustion reactions, have large negative values of ΔH_{rxn}. In such cases ΔG_{rxn} is negative irrespective of the sign of ΔS_{rxn}, because in such cases the $\Delta H_{rxn} < 0$ term in Equation (22-7) dominates the $-T\,\Delta S_{rxn}$ term, even when $\Delta S_{rxn} < 0$. For combustion reactions, or other highly exothermic reactions, ΔH_{rxn} is an adequate criterion of reaction spontaneity. It was in his generalization of this observation to all reactions that Berthelot made his mistake.

In general, a system will change spontaneously in such a way that its Gibbs free energy is minimized. The Gibbs free energy of a substance consists of an enthalpy term and an entropy term. The enthalpy term is fairly independent of pressure (for a gaseous species) or concentration (for a species in solution), but as we showed in Section 22-3, the entropy term depends strongly on these quantities. In any spontaneous process, the composition of a system will change through the consumption of some species and the formation of others, thereby changing their pressures or concentrations so that the Gibbs free energy of the system is minimized.

In Chapter 17 we presented another set of criteria that governed the direction in which a chemical reaction proceeds. These criteria involved the ratio of the reaction quotient Q to the equilibrium constant K. Recall that

1. If $Q/K > 1$ or $Q > K$, then the reaction proceeds spontaneously from right to left.
2. If $Q/K < 1$ or $Q < K$, then the reaction proceeds spontaneously from left to right.
3. If $Q/K = 1$ or $Q = K$, then the reaction is at equilibrium.

Because the sign of ΔG_{rxn} also determines whether a reaction is spontaneous, we should expect a relation between ΔG_{rxn} and Q/K; the relation is

$$\Delta G_{rxn} = RT \ln\left(\frac{Q}{K}\right) \tag{22-8}$$

Note that because the units of Q and K are the same, the units cancel in the ratio Q/K. Also note from Equation (22-8) that

1. If $Q/K > 1$, then $\Delta G_{rxn} = RT \ln(>1) > 0$; that is, ΔG_{rxn} is positive.
2. If $Q/K < 1$, then $\Delta G_{rxn} = RT \ln(<1) < 0$; that is, ΔG_{rxn} is negative.
3. If $Q/K = 1$, then $\Delta G_{rxn} = RT \ln(1) = 0$; that is, ΔG_{rxn} is 0.

In applying the Gibbs criteria to a reaction, it is worth remembering that even when a reaction is spontaneous it may not occur at a detectable rate. *Spontaneous is not synonymous with immediate.* On the other hand, if $\Delta G_{rxn} > 0$, then the reaction will not occur under the prevailing conditions. The *no* of thermodynamics is emphatic; the *yes* of thermodynamics is actually a *maybe*. For a reaction to occur, it is *absolutely necessary* that $\Delta G_{rxn} < 0$, but a negative value of ΔG_{rxn} is not sufficient to guarantee that the reaction will occur at a detectable rate.

The following Example illustrates the use of Equation (22-8).

EXAMPLE 22-4: The reaction system

$$C(s) + H_2O(g) \rightleftharpoons CO(g) + H_2(g)$$

is prepared at 800°C with the initial concentrations $[CO]_0 = 0.500$ M, $[H_2]_0 = 4.25 \times 10^{-2}$ M, and $[H_2O]_0 = 0.150$ M. Given that $K_c = 7.99 \times 10^{-2}$ M at 800°C, calculate ΔG_{rxn} for this reaction system and indicate in which direction the reaction will proceed spontaneously.

Solution: The value of Q for this reaction system is given by

$$Q_c = \frac{[CO]_0[H_2]_0}{[H_2O]_0} = \frac{(0.500\ \text{M})(4.25 \times 10^{-2}\ \text{M})}{0.150\ \text{M}} = 0.142\ \text{M}$$

The value of Q_c/K_c is $(0.142\ \text{M}/7.99 \times 10^{-2}\ \text{M}) = 1.78$, so

$$\begin{aligned}\Delta G_{rxn} &= RT \ln\left(\frac{Q_c}{K_c}\right)\\ &= (8.314\ \text{J}\cdot\text{K}^{-1})(1073\ \text{K}) \ln (1.78)\\ &= +5.14\ \text{kJ}\end{aligned}$$

The positive value of ΔG_{rxn}, which results from the fact that $Q_c > K_c$, implies that the reaction system will evolve in such a way that the concentrations of $CO(g)$ and $H_2(g)$ will decrease and that of $H_2O(g)$ will increase. That is, the reaction

$$C(s) + H_2O(g,\ 0.150\ \text{M}) \rightleftharpoons CO(g,\ 0.500\ \text{M}) + H_2(g,\ 0.0425\ \text{M})$$

proceeds spontaneously from right to left.

Note that the units of R in Equation (22-8) are $8.314\ \text{J}\cdot\text{K}^{-1}$ rather than $8.314\ \text{J}\cdot\text{K}^{-1}\cdot\text{mol}^{-1}$. This difference results from the cancellation of the mol^{-1} unit in the derivation of this equation. The balancing coefficients in a chemical equation are the relative *numbers of moles* of products and reactants, but the exponents on the concentration (or pressure) factors in the equilibrium constant expression are unitless. The mol factor that arises in this process cancels the mol^{-1} unit on R.

PRACTICE PROBLEM 22-4: Suppose we add 5.0 mL of 0.10 M $AgNO_3(aq)$ to 100 mL of water that is in equilibrium with $AgCl(s)$ at 25°C. Calculate ΔG_{rxn} for the reaction

$$Ag^+(aq) + Cl^-(aq) \rightarrow AgCl(s)$$

immediately after adding the $AgNO_3(aq)$. Take $K_{sp} = 1.8 \times 10^{-10}\ \text{M}^2$ for $AgCl(s)$. Is your result consistent with the common-ion effect discussed in Chapter 20?

Answer: $\Delta G_{rxn} = -14.6$ kJ; yes

At this point you may think that the introduction of ΔG_{rxn} has not given us anything new, because we can determine the direction in which a reaction will proceed spontaneously by simply using the value of Q/K. But what if we are not given the value of the equilibrium con-

and the enthalpy release is

$$\text{enthalpy release} = H(\text{bond})_P = (1\ \text{mol})H(\text{C—Cl}) + (1\ \text{mol})H(\text{H—Cl})$$
$$= 331\ \text{kJ} + 431\ \text{kJ} = 762\ \text{kJ}$$

The heat of reaction is given by

$$\Delta H^\circ_{rxn} = H(\text{bond})_R - H(\text{bond})_P$$
$$= 657\ \text{kJ} - 762\ \text{kJ} = -105\ \text{kJ}$$

EXAMPLE 22-11: Chemical reactions can be used to produce flames for heating. The nonnuclear reaction with the highest attainable flame temperature (approximately 6000°C, about the surface temperature of the sun) is that between hydrogen and fluorine:

$$\text{H—H}(g) + \text{F—F}(g) \rightarrow 2\text{H—F}(g)$$

Use the molar bond enthalpies in Table 22-2 to estimate ΔH°_{rxn} for this equation.

Solution: The reaction involves the rupture of one hydrogen-hydrogen and one fluorine-fluorine bond and the formation of two hydrogen-fluorine bonds. Thus,

and $$\Delta H^\circ_{rxn} \simeq H(\text{H—H}) + H(\text{F—F}) - 2H(\text{H—F})$$

$$\Delta H^\circ_{rxn} \simeq (1\ \text{mol})(435\ \text{kJ}\cdot\text{mol}^{-1}) + (1\ \text{mol})(155\ \text{kJ}\cdot\text{mol}^{-1})$$
$$- (2\ \text{mol})(565\ \text{kJ}\cdot\text{mol}^{-1})$$
$$\simeq -540\ \text{kJ}$$

The value of ΔH°_{rxn} we would obtain using data from Table 22-1 is -542.2 kJ, a value in agreement with what we have found here using bond energies.

PRACTICE PROBLEM 22-11: Hydrazine, N_2H_4, and its derivatives are used as rocket fuels. Use the molar bond enthalpies in Table 22-2 to calculate the molar enthalpy of formation of $N_2H_4(g)$.

Answer: $\Delta H^\circ_f \simeq 96\ \text{kJ}\cdot\text{mol}^{-1}$

Although the value of ΔH°_{rxn} is determined primarily by the difference in bond energies of the reactants and the products, the attractive forces *between* chemical species in the liquid and in solid phases (Chapter 14) can also make significant contributions to the value of ΔH°_{rxn}. As an example, consider the vaporization of liquid water at 25°C. Using Table 22-1, we calculate

$$H_2O(l) \rightarrow H_2O(g) \qquad \Delta H^\circ_{rxn} = +44.0\ \text{kJ}$$

Note that no internal oxygen-hydrogen bonds are broken in this process. In the vaporization of water, it is the attractive forces *between* the water molecules that must be overcome. For this reason, the calculation of ΔH°_{rxn} values from bond enthalpies is restricted to gas-phase reactions. It should also be noted that the bond energies given in Table 22-2 are *average* values; thus, calculated values of ΔH°_{rxn} based on bond energies are only approximations to actual experimental values.

SUMMARY

Not all chemical reactions that evolve energy are spontaneous. Reaction spontaneity is determined by both energy and entropy changes. In general, a chemical equilibrium involves a compromise between minimization of the energy and maximization of the entropy. In any naturally occurring process, the second law of thermodynamics requires that the total entropy (system + surroundings) of the system must increase.

Entropy is a measure of the disorder, or randomness, of a system. The entropy of a compound increases with increasing temperature. Both fusion and vaporization processes lead to an increase in the entropy of a compound, because molecules in a liquid are more disordered than the same molecules in the solid phase and molecules in a gas are more disordered than the same molecules in the liquid phase.

The entropy of a perfect crystalline substance is zero at 0 K (third law of thermodynamics). This property of crystals is the basis of the absolute entropy scale. The entropy of a compound at $T > 0$ K is always positive and increases with increasing temperature. The standard entropy change for a chemical reaction, ΔS°_{rxn}, is expressed in terms of the standard entropies of the products minus the standard entropies of the reactants.

The Gibbs criterion for reaction spontaneity is that the Gibbs energy change for the reaction, ΔG_{rxn}, must be less than zero. The value of ΔG_{rxn} depends on the values of ΔH_{rxn} and ΔS_{rxn}. For a reaction with $\Delta G_{rxn} < 0$, the value of ΔG_{rxn} equals the maximum amount of work that can be obtained from the reaction under the stated conditions. The value of ΔG_{rxn} is related to the value of Q/K.

The criteria for reaction spontaneity are summarized as follows:

Reaction type	Value of (Q/K)	Gibbs free energy change
spontaneous reaction	<1	$\Delta G_{rxn} < 0$
nonspontaneous reaction	>1	$\Delta G_{rxn} > 0$
reaction at equilibrium	=1	$\Delta G_{rxn} = 0$

Spontaneous is not synonymous with immediate. The fact that $\Delta G_{rxn} < 0$ is not sufficient to ensure that a reaction proceeds toward equilibrium at a detectable rate.

The standard Gibbs free energy change, ΔG°_{rxn}, is equal to the value of ΔG_{rxn} when all products and reactants are at standard conditions (all phases at 1 atm, all solutes at 1 M). The value of ΔG°_{rxn} can be calculated from a table of standard molar Gibbs free energies of formation ΔG°_f (Table 22-1 and Appendix F). The value of ΔG°_{rxn} can be used to calculate the value of the equilibrium constant for the equation (Equation 22-10).

The van't Hoff equation describes the change in the value of an equilibrium constant with temperature, and the Clapeyron-Clausius equation describes the change in equilibrium vapor pressure with temperature. The value of ΔH°_{rxn} for a particular reaction is determined primarily by the difference in the bond energies of the reactant and the product molecules. The breaking of chemical bonds in the reactant molecules consumes energy, and the formation of chemical bonds in the product molecules releases energy. The difference between these values is ΔH°_{rxn}.

TERMS THAT YOU SHOULD KNOW

enthalpy change, ΔH_{rxn} 775
standard enthalpy change, ΔH°_{rxn} 775
spontaneous process 776
cold pack 777
irreversible process 777
system 777
surroundings 777
irreversible 778
entropy, S 778
second law of thermodynamics 778
entropy change, ΔS 778
reversible process 778
positional disorder 779
thermal disorder 779
third law of thermodynamics 779
molar entropy of fusion, ΔS_{fus} 781
molar entropy of vaporization, ΔS_{vap} 781
absolute molar entropy 783
standard conditions 783
standard molar entropy, S° 783
standard entropy change for a reaction, ΔS°_{rxn} 785
entropy-driven reaction 786
entropy-favored reaction 786
energy-favored reaction 786
Gibbs free energy, G 787
Gibbs free energy change, ΔG_{rxn} 787
Gibbs criteria for reaction spontaneity 787
standard Gibbs free energy change, ΔG°_{rxn} 790
standard molar Gibbs free energy of formation, ΔG°_f 793
van't Hoff equation 796
Fischer-Tropsch synthesis 796
Clapeyron-Clausius equation 798
molar bond enthalpy 800
bond enthalpies 800
bond energies 800

22-43. Use the data in Appendix F to calculate ΔG°_{rxn} and ΔH°_{rxn} at 25°C for the equation

$$2HCl(g) + F_2(g) \rightleftharpoons 2HF(g) + Cl_2(g)$$

Calculate the equilibrium constant at 25°C for the equation.

22-44. Use the data in Appendix F to calculate ΔG°_{rxn} and ΔH°_{rxn} at 25°C for the equation

$$Fe_3O_4(s) + 2C(s, \text{graphite}) \rightarrow 3Fe(s) + 2CO_2(g)$$

Calculate the equilibrium constant at 25°C for the equation.

22-45. The reaction described by the equation

$$2SO_2(g) + O_2(g) \rightleftharpoons 2SO_3(g)$$

is an important reaction in the manufacture of sulfuric acid. Use the data in Appendix F to calculate the values of ΔG°_{rxn} and ΔH°_{rxn} at 25°C for the reaction. Calculate the equilibrium constant for the reaction at 25°C.

22-46. Use the data in Appendix F to calculate ΔG°_{rxn} and ΔH°_{rxn} at 25°C for the equation

$$H_2(g) + I_2(g) \rightleftharpoons 2HI(g)$$

Calculate the equilibrium constant for the equation at 25°C.

22-47. Use the data in Appendix F to calculate the value of ΔG°_{rxn}, ΔH°_{rxn}, and ΔS°_{rxn} at 25°C for the reaction described by the equation

$$H_2(g) + CO_2(g) \rightleftharpoons H_2O(g) + CO(g)$$

What drives the reaction and in what direction at standard conditions?

22-48. Use the data in Appendix F for $CH_4(g)$, $Cl_2(g)$, and $HCl(g)$ to calculate ΔG°_f and ΔH°_f at 25°C for $CCl_4(l)$, given that $\Delta G^\circ_{rxn} = -395.7$ kJ and $\Delta H^\circ_{rxn} = -429.8$ kJ for the equation

$$CH_4(g) + 4Cl_2(g) \rightleftharpoons CCl_4(l) + 4HCl(g)$$

Compare your results for $CCl_4(l)$ with those given in Table 22-1.

22-49. Calculate the maximum amount of work that can be obtained from the combustion of 1.00 mol of ethane, $C_2H_6(g)$, at 25°C and standard conditions.

22-50. Calculate the maximum amount of work that can be obtained from the combustion of 1.00 mol of methane, $CH_4(g)$, at 25°C and standard conditions.

TEMPERATURE DEPENDENCE OF EQUILIBRIUM CONSTANTS

22-51. For the equation $N_2(g) + O_2(g) \rightleftharpoons 2NO(g)$, use the following data to calculate ΔH°_{rxn}:

T/K	$K_p/10^{-4}$
2000	4.08
2100	6.86
2200	11.0
2300	16.9
2400	25.1

Note that $-\Delta H^\circ_{rxn}/R$ is the slope of the ln K_p vs. $1/T$ plot.

22-52. For the dissociation of $Br_2(g)$ into $2Br(g)$, use the following data to calculate ΔH°_{rxn}:

t/°C	$K_p/10^{-3}$ atm
850	0.600
900	1.45
950	3.26
1000	6.88

22-53. For the equation

$$H_2(g) + CO_2(g) \rightleftharpoons CO(g) + H_2O(g)$$

use the following data to calculate ΔH°_{rxn}:

t/°C	K
600	0.39
700	0.64
800	0.96
900	1.34
1000	1.77

22-54. For the equation $2SO_2(g) + O_2(g) \rightleftharpoons 2SO_3(g)$, use the following data to calculate ΔH°_{rxn}:

T/K	K_p/atm^{-1}
900	43.1
1000	3.46
1100	0.44
1170	0.13

22-55. Use the data in Appendix F to calculate the value of ΔH°_{rxn} for the equation

$$PCl_3(g) + Cl_2(g) \rightleftharpoons PCl_5(g)$$

Given that $K_p = 0.562$ atm^{-1} at 250°C, calculate the value of K_p at 400°C.

22-56. Use the data in Appendix F to calculate the value of ΔH°_{rxn} for the equation

$$H_2(g) + I_2(g) \rightleftharpoons 2HI(g)$$

Given that $K = 58.0$ at 400°C, calculate K at 500°C.

CLAPEYRON-CLAUSIUS EQUATION

22-57. Acetone, a widely used solvent (as nail-polish remover, for example), has a normal boiling point of 56.2°C and a molar enthalpy of vaporization of 31.97 $kJ\cdot mol^{-1}$. Calculate the equilibrium vapor pressure of acetone at 20.0°C.

22-58. Diethyl ether is a volatile liquid whose vapor is highly combustible. The equilibrium vapor pressure over ether at 20.0°C is 380 torr. Calculate the vapor pressure over ether when it is stored in the refrigerator at 4.0°C ($\Delta H_{vap} = 29.1\ kJ\cdot mol^{-1}$).

22-59. The heat of vaporization of benzene (C_6H_6) is 32.3 $kJ\cdot mol^{-1}$. Given that the vapor pressure of benzene is 387 torr at 60.0°C, calculate the normal boiling point of benzene.

22-60. Mercury is an ideal substance to use in manometers and in studying the effect of pressure on the volume of gases. Its surface is fairly inert and few gases are soluble in mercury. We now consider whether the partial pressure of mercury vapor contributes significantly to the pressure of the gas above mercury. Using the data in Table 14-1, calculate the vapor pressure of mercury at 25°C and at 100°C.

22-61. Carbon tetrachloride, CCl_4, has a vapor pressure of 92.68 torr at 23.50°C and 221.6 torr at 45.00°C. Calculate ΔH_{vap} for CCl_4.

22-62. The vapor pressure of bromine is 133 torr at 20.0°C and 48.10 torr at 0.00°C. Calculate ΔH_{vap} for bromine.

22-63. The molar heat of vaporization of lead is 178 $kJ\cdot mol^{-1}$. Calculate the ratio of the vapor pressure of lead at 1300°C to that at 500°C.

22-64. The molar heat of vaporization of NaCl is 180 $kJ\cdot mol^{-1}$. Calculate the ratio of the vapor pressure of NaCl at 1100°C to that at 900°C.

BOND ENTHALPIES

22-65. The enthalpy change for the equation

$$ClF_3(g) \rightarrow Cl(g) + 3F(g)$$

is 514 kJ. Calculate the average chlorine-fluorine bond energy in ClF_3.

22-66. The enthalpy change for the equation

$$OF_2(g) \rightarrow O(g) + 2F(g)$$

is 368 kJ. Calculate the average oxygen-fluorine bond energy of OF_2.

22-67. Use the bond enthalpy data in Table 22-2 to estimate ΔH°_{rxn} for the equation

$$CCl_4(g) + 2F_2(g) \rightarrow CF_4(g) + 2Cl_2(g)$$

22-68. Use the bond enthalpy data in Table 22-2 to estimate ΔH°_{rxn} for the reaction

$$CH_3OH(g) + F_2(g) \rightarrow FCH_2OH(g) + HF(g)$$

The bonding in CH_3OH and FCH_2OH is

```
   O—H         O—H
   |           |
H—C—H       H—C—H
   |           |
   H           F
```

22-69. The formation of water from oxygen and hydrogen involves the reaction

$$2H_2(g) + O_2(g) \rightarrow 2H_2O(g)$$

Use the bond energies given in Table 22-2 and the ΔH°_f value for $H_2O(g)$ given in Table 22-1 to calculate the oxygen-oxygen bond energy in $O_2(g)$.

22-70. The formation of ammonia from hydrogen and nitrogen involves the reaction

$$N_2(g) + 3H_2(g) \rightarrow 2NH_3(g)$$

Use the bond energies given in Table 22-2 and the ΔH°_f value for $NH_3(g)$ given in Table 22-1 to calculate the bond energy in $N_2(g)$.

22-71. Given that

$$\Delta H^\circ_f[H(g)] = 218\ kJ\cdot mol^{-1}$$

$$\Delta H^\circ_f[C(g)] = 709\ kJ\cdot mol^{-1}$$

$$\Delta H^\circ_f[CH_4(g)] = -74.86\ kJ\cdot mol^{-1}$$

calculate the average carbon-hydrogen bond energy in CH_4.

22-72. Given that

$$\Delta H^\circ_f[Cl(g)] = 128\ kJ\cdot mol^{-1}$$

$$\Delta H^\circ_f[C(g)] = 709\ kJ\cdot mol^{-1}$$

$$\Delta H^\circ_f[CCl_4(g)] = -103\ kJ\cdot mol^{-1}$$

calculate the average carbon-chlorine bond energy in CCl_4.

22-73. Calculate ΔS_{fus} and ΔS_{vap} for the alkali metals:

Metal	T_m/K	ΔH_{fus}/kJ·mol^{-1}	T_b/K	ΔH_{vap}/kJ·mol^{-1}
Li	454	2.99	1615	134.7
Na	371	2.60	1156	89.6
K	336	2.33	1033	77.1
Rb	312	2.34	956	69
Cs	302	2.10	942	66

22-74. Suppose that you see an advertisement for a catalyst that decomposes water into hydrogen and oxygen at room temperature. Would you be skeptical of this claim? Explain.

22-75. Is it possible to have a reaction for which K is either zero or infinite? Explain in terms of ΔG°_{rxn}.

22-76. Given the following possibilities for ΔG°_{rxn}, what can you say in each case about the magnitude of the equilibrium constant for the reaction?

(a) $\Delta G^\circ_{rxn} > 0$ (b) $\Delta G^\circ_{rxn} = 0$ (c) $\Delta G^\circ_{rxn} < 0$

22-77. Hydrogen peroxide can be prepared in several ways. One method is the reaction between hydrogen and oxygen:

$$H_2(g) + O_2(g) \rightleftharpoons H_2O_2(l)$$

Another method is the reaction between water and oxygen:

$$2H_2O(l) + O_2(g) \rightleftharpoons 2H_2O_2(l)$$

Calculate the value of ΔG°_{rxn} for both reactions. Predict which method requires less energy under standard conditions.

22-78. Given the following Gibbs free energies at 25°C

Substance	ΔG°_f/kJ·mol^{-1}
$Ag^+(aq)$	77.1
$Cl^-(aq)$	−131.2
$AgCl(s)$	−109.7
$Br^-(aq)$	−102.8
$AgBr(s)$	−96.8

calculate the solubility-product constant of (a) AgCl and (b) AgBr.

22-79. Plot the data in Problem 22-51 and show that $\ln K$ versus $1/T$ is a straight line. Evaluate ΔH°_{rxn} from this plot.

22-80. Discuss the possible effects of a catalyst on the value of ΔG°_{rxn}.

22-81. Calculate the value (in kilojoules) of the change in ΔG°_{rxn} that corresponds to a 10-fold change in K at 25°C.

22-82. Glucose is a primary fuel in the production of energy in biological systems. Given that $\Delta G^\circ_f = -916$ kJ·mol^{-1} for glucose, calculate the maximum amount of work that can be obtained from the complete combustion of 1.00 mol of glucose under standard conditions:

$$C_6H_{12}O_6(s) + 6O_2(g) \rightleftharpoons 6CO_2(g) + 6H_2O(l)$$

22-83. The solubility of gases in water decreases with increasing temperature. What does this tell you about the heats of solution of gases?

22-84. Estimate the heat of solution of AgCl(s) in water from the following data:

t/°C	K_{sp}/M^2
50.0	13.2×10^{-10}
100.0	2.15×10^{-8}

22-85. The variation of the Henry's law constant with temperature for the dissolution of CO_2 in water is

t/°C	K/atm·M^{-1}
0	13.2
25	29.4

Calculate ΔH°_{rxn} for the process

$$CO_2(aq) \rightleftharpoons CO_2(g)$$

22-86. The vapor pressure of water above mixtures of $CuCl_2 \cdot H_2O(s)$ and $CuCl_2 \cdot 2H_2O(s)$ is 3.72 torr at 18.0°C and 91.2 torr at 60.0°C. Calculate ΔH°_{rxn} for the equilibrium

$$CuCl_2 \cdot 2H_2O(s) \rightleftharpoons CuCl_2 \cdot H_2O(s) + H_2O(g)$$

22-87. Given the following data, calculate ΔH°_{rxn}, ΔS°_{rxn}, and ΔG°_{rxn} at 298 K for the equilibrium

$$Mg(s) + 2HCl(aq) \rightleftharpoons H_2(g) + MgCl_2(aq)$$

Species	ΔH°_f/kJ·mol^{-1}	S°/J·K^{-1}·mol^{-1}
$H_2(g)$	0	130.6
$HCl(aq)$	−167.2	56.5
$Mg(s)$	0	32.6
$MgCl_2(aq)$	−801.2	−25.1

22-88. Given that $\Delta G^\circ_f = 3.142$ kJ·mol^{-1} for $Br_2(g)$ at 25°C, calculate the vapor pressure of bromine at 25°C.

22-89. Ethylamine undergoes an endothermic dissociation producing ethylene and ammonia:

$$H_3C{-}CH_2{-}NH_2(g) \longrightarrow H_2C{=}CH_2(g) + H{-}\underset{H}{N}{-}H(g)$$

Using the data in Table 22-2, calculate ΔH°_{rxn}.

22-90. Consider the dissociation of water:

$$H_2O(l) \rightleftharpoons H^+(aq) + OH^-(aq)$$

(a) Determine ΔH°_{rxn}, given that the ΔH°_f values for $H^+(aq)$ and $OH^-(aq)$ are 0 and $-230.0\ kJ\cdot mol^{-1}$, respectively.

22-91. Consider the dissolution of silver iodide in water. The reaction is:

	$AgI(s)$	$\rightleftharpoons$ $Ag^+(aq)$	+ $I^-(aq)$
$\Delta H^\circ_f/kJ\cdot mol^{-1}$	−62.4	105.9	−55.9

(a) Determine the value of ΔH°_{rxn} for the reaction.
(b) Given $K_{sp} = 8.3 \times 10^{-17}\ M^2$ at 25°C, calculate the value of K_{sp} at 35°C. Assume ΔH°_{rxn} is constant over this temperature range.

22-92. Write a balanced chemical equation for the *formation* of one mol of each of the substances from the elements in their standard states at 25°C.

(a) $(NH_4)_2Cr_2O_7(s)$ (b) $KIO_3(s)$
(c) $C_2H_5OH(l)$ (d) $Ca_3(PO_4)_2(s)$

22-93. Given the following data:

Equation	$\Delta H^\circ_{rxn}/kJ$	$\Delta G^\circ_{rxn}/kJ$
$V(s) + \frac{1}{2}O_2 \rightleftharpoons VO(s)$	−103.2	−96.6
$V(s) + O_2(g) \rightleftharpoons VO_2(s)$	−171.5	12.3
$2V(s) + \frac{3}{2}O_2(g) \rightleftharpoons V_2O_3(s)$	−291.3	−272.3
$2V(s) + \frac{5}{2}O_2(g) \rightleftharpoons V_2O_5(s)$	−370.6	−339.3

(a) Determine ΔG°_{rxn} for the reaction described by the equation

$$V_2O_3(s) \rightleftharpoons V_2O_5(s) + VO(s) \qquad \text{(not balanced)}$$

Note that this reaction involves the self oxidation-reduction of $V_2O_3(s)$ and is called a disproportionation reaction.

(b) Which of the above reactions proceeds with the greatest decrease in entropy at 25°C?

22-94. Use the following equations to calculate the value of ΔG°_f for the HBr(*g*) at 25°C.

Equation	$\Delta G^\circ/kJ$
$Br_2(l) \rightleftharpoons Br_2(g)$	3.14
$HBr(g) \rightleftharpoons H(g) + Br(g)$	339.12
$H_2(g) \rightleftharpoons 2H(g)$	406.53
$Br_2(g) \rightleftharpoons 2Br(g)$	161.71

22-95. The thermal decomposition of ammonium chloride has the following stoichiometry:

$$NH_4Cl(s) \rightleftharpoons NH_3(g) + HCl(g)$$

Given the ΔG°_f values $NH_4Cl(s) = -202.97\ kJ\cdot mol^{-1}$, $NH_3(g) = -16.64\ kJ\cdot mol^{-1}$, and $HCl(g) = -95.3\ kJ\cdot mol^{-1}$:

(a) Calculate the value of ΔG°_{rxn} for the reaction at 25°C.
(b) What is the value of the equilibrium constant for the reaction at 25°C?
(c) Determine the equilibrium and partial pressure of HCl above a one-gram sample of NH_4Cl in a 1.0-L container at 25°C.

22-96. Which of the following metal oxides can be reduced by using carbon in a wood fire (temperature is about 600°C), assuming standard conditions? Assume that the reactions are

$$2Fe_2O_3(s) + 3C(\text{graphite}) \rightarrow 4Fe(s) + 3CO_2(g)$$
$$2CuO(s) + C(\text{graphite}) \rightarrow 2Cu(s) + CO_2(g)$$

These reactions presumably led to the first preparations of the elements iron and copper from their ores.

22-97. The value of ΔG°_f for HCl(*g*) is $-95.3\ kJ\cdot mol^{-1}$ at 25°C.

(a) What is the value of ΔG for the formation of HCl(*g*) at 298 K if the partial pressures are $P_{H_2} = 3.5$ atm, $P_{Cl_2} = 1.5$ atm, and $P_{HCl} = 0.31$ atm.
(b) Is the process *more* or *less* favorable under these conditions than under standard state conditions?

22-98. Use Equation (22-1) to show that $\Delta S_{tot} = \Delta S_{sys} + \Delta S_{surr} \geq 0$ for the transfer of energy as heat q from the system to the surroundings or the reverse. (Hint: Note that $q_{sys} = -q_{surr}$.)

23 ELECTROCHEMISTRY

An electrochemical plant used to produce high-purity copper metal from impure copper ores. Gleaming sheets of pure copper, in the process of being removed from one of the production units, are visible in the center of the photograph.

In the last chapter we used the Gibbs free energy to predict whether a reaction is spontaneous. We showed that if the free energy change is greater than zero, then energy must be supplied to drive the reaction. For an oxidation-reduction reaction, that energy can be provided by an electric current. Conversely, when an oxidation-reduction reaction is spontaneous, we can use it to obtain electricity directly from chemicals, without the need for heat engines driven by combustion reactions. These two processes are the subject of electrochemistry.

Electrochemical principles explain the batteries that power our watches and hand calculators. They enable physicians to guide the rhythm of a heartbeat and to heal broken bones. Our senses and brains function through electrical pulses, controlled in part by electrochemical processes. Chemists rely on electrochemical sensors to measure pH, to perform potentiometric titrations, and to monitor a wide variety of ionic species and enzymes, even in the presence of numerous other species. Electrochemistry is also used to combat corrosion and to protect the environment.

In this chapter we show how to calculate the current needed to produce a given amount of reaction product. We also show how to predict the voltage produced in an electrochemical reaction and the work that can be obtained. We will apply these results to a variety of commercial processes.

23-1. CHEMICAL REACTIONS CAN OCCUR AS A RESULT OF THE PASSAGE OF AN ELECTRIC CURRENT THROUGH A SOLUTION

The science of electrochemistry began in 1791 when Luigi Galvani, an Italian scientist, showed that the contraction of a frog's leg produces an electric current. While other scientists of that time were invoking mysterious forces to explain Galvani's "animal electricity," his countryman, Alessandro Volta, was using experiments to show that the phenomenon could be explained in purely physical and chemical terms. Volta constructed a device that produced an electric current of considerable power. In this **voltaic pile** (Figure 23-1), alternate disks of dissimilar metals, such as zinc and copper, are separated by damp cloths soaked in salt water. The electric current is produced by a chemical reaction between the substances in the pile. The reaction occurs when an electrical connection is made between the top and the bottom of the pile.

To study the chemical reaction in a voltaic pile, we need a measure of how strongly the electric current produced is driven through a wire. **Voltage** (V), named after Volta, is the electrical energy (U) per unit of charge (Z):

$$V = U/Z \qquad (23\text{-}1)$$

In mechanical systems, force is the mechanical energy per unit of distance. Voltage plays a role in electrical systems analogous to force in mechanical systems. For this reason, a source of voltage arising from a chemical reaction is often called the **electromotive force,** or **emf.** The

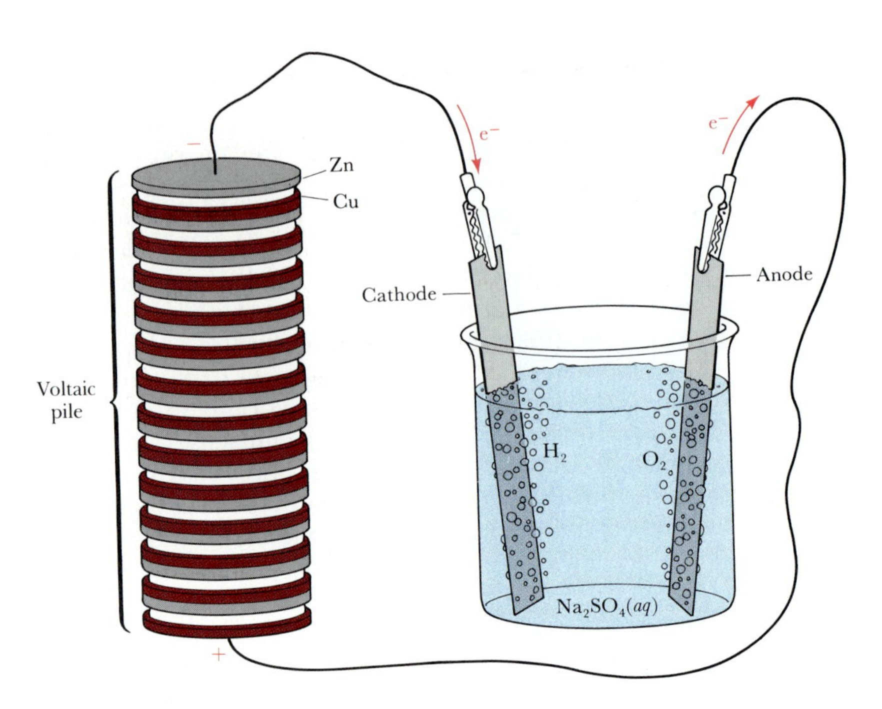

Figure 23-1 Electrochemical decomposition of water containing sodium sulfate, $Na_2SO_4(aq)$. The gas evolved from the platinum strip connected to the top zinc plate of the voltaic pile is hydrogen. The gas evolved from the platinum strip connected to the bottom copper plate of the voltaic pile is oxygen. The current through the $Na_2SO_4(aq)$ solution is carried by $Na^+(aq)$ and $SO_4^{2-}(aq)$ ions.

SI unit of voltage is a volt, also named in honor of Volta, and denoted by V. A **volt** is defined as one joule per coulomb, $1\ V = 1\ J{\cdot}C^{-1}$. Recall that the SI unit of electrical charge is the **coulomb;** the charge on an electron is 1.602×10^{-19} C.

Electric currents are measured in **amperes** (A), and one ampere is the flow of one coulomb (C) of charge per second. In an equation, we have

$$\text{current} = \frac{\text{charge}}{\text{time}}$$

The total charge that flows is

$$\text{charge} = \text{current} \times \text{time}$$

Using SI units, we have

$$\text{coulombs} = \text{amperes} \times \text{seconds}$$

Representing charge (in coulombs) by Z, current (in amperes) by I, and time (in seconds) by t, we have

$$Z = It \tag{23-2}$$

EXAMPLE 23-1: If a current of 1.50 A flows for 5.00 min, then what quantity of charge has flowed?

Solution: The total charge is equal to the current multiplied by time, so

$$\begin{aligned}\text{charge} &= (1.5\ \text{A})(5.00\ \text{min}) \\ &= (1.50\ \text{C}{\cdot}\text{s}^{-1})(5.00\ \text{min})(60\ \text{s}{\cdot}\text{min}^{-1}) = 450\ \text{C}\end{aligned}$$

PRACTICE PROBLEM 23-1: An electron's charge is 1.60×10^{-19} C. Calculate the number of electrons that move through a cross section of a metal wire in 1.00 s when the current is 1.00 A.

Answer: 6.25×10^{18} electrons

The voltage developed by a voltaic pile depends on the nature of the pair of metals used to construct the pile and also is greater the larger the number of sets of metal disks. Voltaic piles were used by the English scientist Humphrey Davy to discover several reactive metals (Na, K, Mg, Ca, Sr, and Ba) in the early 1800s. Davy passed the current from a voltaic pile through molten salts of these metals. The metals were deposited on metal strips attached by wires to the ends of the voltaic pile (Figure 23-1).

23-2. AN ELECTROCHEMICAL CELL PRODUCES ELECTRICITY DIRECTLY FROM A CHEMICAL REACTION

In this section we consider the use of a spontaneous chemical reaction ($\Delta G_{rxn} < 0$) to produce an electric current. The voltaic pile was the first example of such a device.

Recall that electrons are transferred from one substance to another in oxidation-reduction (electron-transfer) reactions (Chapter 21). Figure 23-2 shows that when a zinc rod is immersed in an aqueous solution of copper sulfate, the following reaction occurs spontaneously:

$$\mathrm{Zn}(s) + \mathrm{Cu}^{2+}(aq) \rightarrow \mathrm{Cu}(s) + \mathrm{Zn}^{2+}(aq)$$

It is possible to use redox reactions to produce an electric current. The basic idea is to keep the reactants [$\mathrm{Zn}(s)$ and $\mathrm{Cu}^{2+}(aq)$ in the equation above] and the products [$\mathrm{Cu}(s)$ and $\mathrm{Zn}^{2+}(aq)$] separated physically in such a way that the electrons are transferred from the reducing agent to the substance being reduced after passing through an external circuit. A setup in which an electric current is obtained from a chemical reaction is called an **electrochemical cell** or a **voltaic cell.** The simple elec-

(a)

(b)

Figure 23-2 (a) When a zinc rod is placed in a copper sulfate solution, zinc replaces the copper in solution and (b) elemental copper forms.

$Zn^{2+}(aq)$ and $Cu^{2+}(aq)$? (See Example 23-5 for any data necessary to carry out the calculations.)

Answer: $E_{final} = 0$; $[Zn^{2+}] = 0.20$ M and $[Cu^{2+}] = 1.3 \times 10^{-38}$ M

Substitution of $E^\circ_{rxn} = \left(\frac{0.0592\text{ V}}{n}\right) \log K$ from Equation (23-11) into Equation (23-10) yields an especially useful form of the Nernst equation:

$$E_{rxn} = E^\circ_{rxn} - \left(\frac{0.0592\text{ V}}{n}\right) \log Q \qquad (23\text{-}13)$$

where the voltages E_{rxn} and E°_{rxn} are in volts and all values are at 25.0°C. Equation (23-13) tells us that the cell voltage differs from the standard cell voltage when the reaction quotient is not equal to 1.00. If $Q < 1$, then $E_{rxn} > E^\circ_{rxn}$; if $Q > 1$, then $E_{rxn} < E^\circ_{rxn}$.

When E°_{rxn} is known, Equation (23-13) enables us to calculate the cell voltage for any concentrations of the reactants and products of the species involved in the net cell reaction. Conversely, when E_{rxn} is measured for a cell with known concentrations of products and reactants, we can use Equation (23-13) to calculate E°_{rxn}.

EXAMPLE 23-6: The measured voltage at 25.0°C of a cell in which the reaction described by the equation

$$Zn(s) + Cu^{2+}(aq,\ 1.00\text{ M}) \rightleftharpoons Cu(s) + Zn^{2+}(aq,\ 0.100\text{ M})$$

occurs at the concentrations shown is 1.13 V. Calculate E°_{rxn} for the cell equation.

Solution: From Equation (23-13) with $n = 2$, we obtain

$$E_{rxn} = E^\circ_{rxn} - \left(\frac{0.0592\text{ V}}{n}\right) \log Q = E^\circ_{rxn} - \left(\frac{0.0592\text{ V}}{2}\right) \log \frac{[Zn^{2+}]}{[Cu^{2+}]}$$

[Note that $Zn(s)$ and $Cu(s)$ do not appear in the Q expression because they are both solids.] Therefore,

$$1.13\text{ V} = E^\circ_{rxn} - \left(\frac{0.0592\text{ V}}{2}\right) \log \left(\frac{0.100\text{ M}}{1.00\text{ M}}\right)$$

from which we calculate

$$E^\circ_{rxn} = 1.13\text{ V} + (0.0296\text{ V}) \log 0.100 = 1.10\text{ V}$$

Note that $E^\circ_{rxn} < E_{rxn}$ because $Q < 1$. A value of E°_{rxn} can be used in Equation (23-12) to calculate the value of the equilibrium constant for the cell equation. (Example 23-5.)

Application of the Nernst equation at 25°C to the cell equation for other values of Q yields the graph shown in Figure 23-7. Figure 23-7 shows that when the value of E° is known for a cell reaction we can use Equation (23-10) to calculate the value of the cell voltage, E, for any given set of values of the cell reactant and product concentrations.

PRACTICE PROBLEM 23-6: Use the Nernst equation to calculate the voltage of the zinc-copper cell at 25°C when $[Zn^{2+}] = 1.00$ M and $[Cu^{2+}] = 0.010$ M (see Example 23-6).

Answer: $E_{rxn} = 1.04$ V

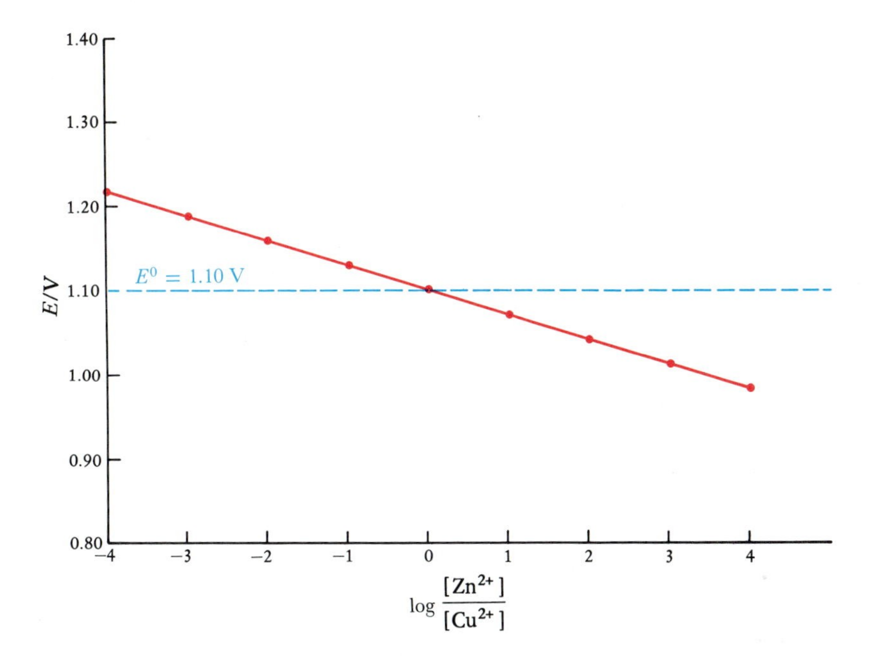

Figure 23-7 Plot of E (in volts) versus log Q for the cell equation

$$Zn(s) + Cu^{2+}(aq) \rightleftharpoons Cu(s) + Zn^{2+}(aq)$$

The plot is based on the Nernst equation applied to the above cell equation, that is,

$$E = 1.10\ V - (0.030\ V) \log \frac{[Zn^{2+}]}{[Cu^{2+}]}$$

A 10-fold increase in the value of $Q = [Zn^{2+}]/[Cu^{2+}]$ changes the cell voltage by −0.030 V. The plot illustrates the linear dependence of cell voltage on log Q.

23-5. E° VALUES CAN BE ASSIGNED TO HALF REACTION EQUATIONS

It is not possible to measure the voltage of a single electrode; only the difference in voltage between two electrodes can be measured. However, if we agree to choose a numerical value for the standard voltage of a particular electrode, then we can assign voltages to equations for half reactions.

In this section we see how the voltages assigned to half reactions are used to calculate $E^\circ_{rxn} = E^\circ_{cell}$ values for complete cell-reaction equations. Our convention is to set the voltage of the **hydrogen electrode** under standard conditions ($[H^+] = 1.00$ M, $P_{H_2} = 1.00$ atm) equal to zero; that is, we assign the value $E^\circ = 0$ (at all temperatures) to the electrode half reaction

$$2H^+(aq,\ 1\ M) + 2e^- \rightarrow H_2(g,\ 1\ atm) \quad E^\circ_{red} = 0 \text{ (by convention)} \tag{23-14}$$

Equation (23-14) assigns a **standard reduction voltage** denoted E°_{red} to a half reaction written as a reduction, that is, written with electrons on the left. A complete cell involves two half reactions: a reduction half reaction, with a standard reduction voltage E°_{red}, and an oxidation half reaction, with a standard oxidation voltage E°_{ox}. An oxidation half reaction is a half reaction written as an oxidation, that is, written with electrons on the right. We now write an equation for the voltage of a complete cell, which involves a reduction half reaction and an oxidation half reaction

$$E^\circ_{rxn} = E^\circ_{red} + E^\circ_{ox} \tag{23-15}$$

Equation (23-15) is used together with Equation (23-14) to set up a table of **standard reduction voltages,** which also are called **standard reduction potentials.** Because, for a particular half reaction, oxidation is the reverse of reduction, we have

$$E^\circ_{red} = -E^\circ_{ox} \quad \text{(same half reaction)} \tag{23-16}$$

Thus, given that for the reduction half reaction equation

$$Zn^{2+}(aq,\ 1\ M) + 2e^- \rightarrow Zn(s) \qquad E^\circ_{red} = -0.76\ V$$

we have for the corresponding oxidation half reaction equation

$$Zn(s) \rightarrow Zn^{2+}(aq,\ 1\ M) + 2e^- \qquad E^\circ_{ox} = +0.76\ V$$

To see how standard reduction voltages are obtained, consider the cell in Figure 23-5:

$$Zn(s)|ZnCl_2(aq)||HCl(aq)|H_2(g)|Pt(s)$$

The experimental value for the standard cell voltage at 25°C is $E^\circ_{cell} = 0.76\ V = E^\circ_{red} + E^\circ_{ox}$. The right-hand electrode involves the pair H^+/H_2, and the left-hand electrode involves the pair Zn/Zn^{2+}. Thus,

$$E^\circ_{cell} = E^\circ_{red}[H^+/H_2] + E^\circ_{ox}[Zn/Zn^{2+}] = 0.76\ V \qquad (23\text{-}17)$$

Table 23-1 Standard reduction voltages at 25.0°C for aqueous solutions

Electrode half reaction	E°_{red}/V
Acidic solutions	
$F_2(g) + 2e^- \rightarrow 2F^-(aq)$	+2.87
$O_3(g) + 2H^+(aq) + 2e^- \rightarrow O_2(g) + H_2O(l)$	+2.07
$Co^{3+}(aq) + e^- \rightarrow Co^{2+}(aq)$	+1.81
$Cl_2(g) + 2e^- \rightarrow 2Cl^-(aq)$	+1.36
$O_2(g) + 4H^+(aq) + 4e^- \rightarrow 2H_2O(l)$	+1.23
$Ag^+(aq) + e^- \rightarrow Ag(s)$	+0.80
$Cu^+(aq) + e^- \rightarrow Cu(s)$	+0.52
$Cu^{2+}(aq) + 2e^- \rightarrow Cu(s)$	+0.34
$AgCl(s) + e^- \rightarrow Ag(s) + Cl^-(aq)$	+0.22
$Cu^{2+}(aq) + e^- \rightarrow Cu^+(aq)$	+0.15
$2H^+(aq) + 2e^- \rightarrow H_2(g)$	+0.0
$Pb^{2+}(aq) + 2e^- \rightarrow Pb(s)$	−0.13
$V^{3+}(aq) + e^- \rightarrow V^{2+}(aq)$	−0.24
$Zn^{2+}(aq) + 2e^- \rightarrow Zn(s)$	−0.76
$Al^{3+}(aq) + 3e^- \rightarrow Al(s)$	−1.66
$H_2(g) + 2e^- \rightarrow 2H^-(aq)$	−2.25
$Mg^{2+}(aq) + 2e^- \rightarrow Mg(s)$	−2.36
$Na^+(aq) + e^- \rightarrow Na(s)$	−2.71
$Li^+(aq) + e^- \rightarrow Li(s)$	−3.05
Basic solutions	
$O_2(g) + 2H_2O(l) + 4e^- \rightarrow 4OH^-(aq)$	+0.40
$Cu(OH)_2(s) + 2e^- \rightarrow Cu(s) + 2OH^-(aq)$	−0.22
$O_2(g) + e^- \rightarrow O_2^-(aq)$	−0.56
$2H_2O(l) + 2e^- \rightarrow H_2(g) + 2OH^-(aq)$	−0.83
$2SO_3^{2-}(aq) + 2H_2O(l) + 2e^- \rightarrow S_2O_4^{2-}(aq) + 4OH^-(aq)$	−1.12

where the brackets enclose the oxidized and reduced species for each electrode. Note that we indicate the process as a reduction reaction in $E^\circ_{red}[H^+/H_2]$ and as an oxidation reaction in $E^\circ_{ox}[Zn/Zn^{2+}]$. Substitution of $E^\circ_{red}[H^+/H_2] = 0$ into Equation (23-17) yields

$$E^\circ_{cell} = 0 + E^\circ_{ox}[Zn/Zn^{2+}] = 0.76\ V$$

or

$$E^\circ_{ox}[Zn/Zn^{2+}] = -E^\circ_{red}[Zn^{2+}/Zn] = +0.76\ V$$

Thus, the standard reduction voltage for the electrode half reaction

$$Zn^{2+}(aq,\ 1\ M) + 2e^- \rightarrow Zn(s)$$

is $E^\circ_{red}[Zn/Zn^{2+}] = -0.76\ V$ at 25°C.

Once we know the standard reduction voltage of an electrode relative to the hydrogen electrode, we can use that value to determine the standard reduction voltage of other electrodes. Tables of E°_{red} values, such as Table 23-1, are used to calculate E°_{cell} values for cells. These E°_{cell} values can, in turn, be used to calculate equilibrium constants for chemical equations. The procedure for using standard reduction voltages to calculate a standard cell voltage is illustrated in the following Example.

A more complete listing of E°_{red} values appears in Appendix I. You will find it useful in solving the problems at the end of this chapter.

EXAMPLE 23-7: The standard voltage of the cell

$$Zn(s)|ZnSO_4(aq)||CuSO_4(aq)|Cu(s)$$

at 25°C is $E^\circ_{cell} = 1.10\ V$. Given that $E^\circ_{red} = -0.76\ V$ for the electrode half reaction

$$Zn^{2+}(aq,\ 1\ M) + 2e^- \rightarrow Zn(s)$$

calculate E°_{red} at 25°C for the electrode half reaction

$$Cu^{2+}(aq,\ 1\ M) + 2e^- \rightarrow Cu(s)$$

Solution: We assume that oxidation takes place at the left-hand electrode, so the half reactions are

$Zn(s) \rightarrow Zn^{2+}(aq) + 2e^-$ (oxidation, left-hand electrode, $E^\circ_{ox}[Zn/Zn^{2+}]$)
$Cu^{2+}(aq) + 2e^- \rightarrow Cu(s)$ (reduction, right-hand electrode, $E^\circ_{red}[Cu^{2+}/Cu]$)

Therefore,

$$E^\circ_{cell} = E^\circ_{red}[Cu^{2+}/Cu] + E^\circ_{ox}[Zn/Zn^{2+}] = 1.10\ V$$

Because

$$E^\circ_{ox}[Zn/Zn^{2+}] = -E^\circ_{red}[Zn^{2+}/Zn] = +0.76\ V$$

we have

$$E^\circ_{cell} = E^\circ_{red}[Cu^{2+}/Cu] + 0.76\ V = 1.10\ V$$

Thus,

$$E^\circ_{red}[Cu^{2+}/Cu] = 1.10\ V - 0.76\ V = +0.34\ V$$

The standard reduction voltage of the $Cu^{2+}(aq)|Cu(s)$ electrode is $E^\circ_{red} = +0.34\ V$.

Thus, we see that the standard reduction voltage of an electrode can be obtained from the standard cell voltage of a cell for which the standard reduction voltage of the other electrode is known.

PRACTICE PROBLEM 23-7: Use data from Table 23-1 to calculate the value of E°_{cell} at 25°C for an electrochemical cell with the net cell reaction

$$2H_2(g) + O_2(g) \rightleftharpoons 2H_2O(l)$$

Electrochemical cells using this reaction are the on-board power sources for the U.S. space shuttle vehicles.

Answer: 1.23 V

increasing strength of oxidizing agents ↑	E°_{red}/V	increasing strength of reducing agents ↓
$F_2(g) + 2e^- \rightarrow 2F^-(aq)$	+2.87	
·	·	
·	·	
·	·	
·	·	
$H_2(g) + 2e^- \rightarrow 2H^+(aq)$	0	
·	·	
·	·	
·	·	
·	·	
$Li^+(aq) + e^- \rightarrow Li(s)$	−3.05	

In the arrangement of E°_{red} values used in Table 23-1, more positive E°_{red} values for half reaction equations indicate more powerful oxidizing agents. Thus, we can represent Table 23-1 in schematic form as shown in the margin. Recall that the more positive the E°_{red} value for a half reaction, the stronger is the oxidizing agent (electron acceptor) in the half reaction. The more negative the E°_{red} value for a half reaction, the stronger is the reducing agent (electron donor) in the half reaction. Thus, fluorine is the strongest oxidizing agent (most positive E°_{red} value) and lithium is the strongest reducing agent (most negative E°_{red} value) in Table 23-1. The arrangement of half reactions in order of the standard reduction voltages is called the **electromotive force series** (or **emf series**).

Table 23-1 and Appendix I contain a tremendous amount of information on the chemistry of species in aqueous solutions. The E°_{red} values are used to predict many of the reactions that can occur between oxidizing and reducing agents.

EXAMPLE 23-8: As an example of the application of the data in Table 23-1, let's consider the question of whether $Co^{3+}(aq)$, which is a fairly strong oxidizing agent [$Co^{3+}(aq) + e^- \rightarrow Co^{2+}(aq)$] is capable of oxidizing water to oxygen gas in acidic aqueous solution under standard conditions at 25°C.

Solution: The oxidation of water to oxygen gas by $Co^{3+}(aq)$ under standard conditions ($Q = 1$) will be a spontaneous process if $E^\circ_{rxn} = E^\circ_{ox} + E^\circ_{red}$ is greater than zero, because when $E^\circ_{rxn} > 0$, the value of $\Delta G^\circ_{rxn} = -nFE^\circ_{rxn}$ is less than zero. The two half reactions involved are

$$Co^{3+}(aq) + e^- \rightarrow Co^{2+}(aq) \qquad E^\circ_{red} = +1.81\ V$$

$$2H_2O(l) \rightarrow O_2(g) + 4H^+(aq) + 4e^- \qquad E^\circ_{ox} = -E^\circ_{red} = -1.23\ V$$

The E°_{red} value for the two half reactions were obtained from Table 23-1. The value of E°_{rxn} is $E^\circ_{rxn} = E^\circ_{red}[Co^{3+}/Co^{2+}] + E^\circ_{ox}[H_2O/O_2]$

$$E^\circ_{rxn} = 1.81\ V - 1.23\ V = +0.58\ V > 0$$

Note that we do *not* multiply E°_{red} by 4, because the magnitude of a cell voltage or half-cell voltage is independent of the quantity of material involved. The positive value of E°_{rxn} means that $Co^{3+}(aq)$ is capable of oxidizing water at 25°C under standard conditions. The rate of oxidation of water by $Co^{3+}(aq)$ is fairly rapid at 25°C, and $Co^{3+}(aq)$ cannot persist at appreciable concentrations in water. Spontaneity of the reaction under nonstandard conditions is determined by the value of E_{rxn} rather than E°_{rxn}. The value of

E_{rxn} can be calculated using the Nernst equation (23-13), if the numerical value of Q for the balanced chemical equation

$$4Co^{3+}(aq) + 2H_2O(l) \rightleftharpoons 4Co^{2+}(aq) + O_2(g) + 4H^+(aq)$$

$$Q = \frac{[Co^{2+}]^4 P_{O_2} [H^+]^4}{[Co^{3+}]^4}$$

is known.

PRACTICE PROBLEM 23-8: Assume that a solution, which is initially 0.10 M in $Co^{3+}(aq)$ and buffered at pH = 1.00, is allowed to attain equilibrium in a beaker open to the atmosphere. Take the equilibrium value of P_{O_2} to be 0.20 atm, and compute the values of $[Co^{3+}]$ and $[Co^{2+}]$ at equilibrium at 25°C. (See Example 23-8 for the data necessary to solve this problem. Hint: First calculate the value of the equilibrium constant K using the value of E°_{rxn}.)

Answer: $P_{O_2} = 0.20$ atm (fixed by the atmosphere); $[Co^{2+}] = 0.10$ M; $[Co^{3+}] = 1.1 \times 10^{-12}$ M

EXAMPLE 23-9: The $V^{2+}(aq)$ ion is a moderately strong reducing agent $[V^{2+}(aq) \rightarrow V^{3+}(aq) + e^-]$ in acidic aqueous solution. Is $V^{2+}(aq)$ capable of liberating $H_2(g)$ from an aqueous solution under standard conditions at 25°C?

Solution: To liberate $H_2(g)$ from water, $V^{2+}(aq)$ must act as a reducing agent. Therefore, the equation for the reaction is

$$2V^{2+}(aq) + 2H^+(aq) \rightleftharpoons 2V^{3+}(aq) + H_2(g)$$

The value of E°_{rxn} for this equation is calculated using data from Table 23-1:

$$\begin{aligned} E^\circ_{rxn} &= E^\circ_{red}[H^+/H_2] + E^\circ_{ox}[V^{2+}/V^{3+}] \\ &= 0 + (+0.24\ \text{V}) = +0.24\ \text{V} > 0 \end{aligned}$$

Because E°_{rxn} is positive, $V^{2+}(aq)$ is capable of liberating $H_2(g)$ from an acidic, aqueous solution when $Q = 1$ at 25°C.

PRACTICE PROBLEM 23-9: Assuming standard conditions at 25°C, determine the voltage ranges of E°_{red} values for which (a) reducing agents cannot liberate $H_2(g)$ from water, and (b) oxidizing agents cannot liberate $O_2(g)$ from water in acidic solutions. (These voltage ranges define a region of stability for reducing agents and oxidizing agents in water at 25°C.)

Answer: (a)For $E^\circ_{red} > 0$ V, the reducing agent in the half reaction is stable [cannot reduce $H^+(aq)$ to $H_2(g)$]. (b) For $E^\circ_{red} < 1.23$ V, the oxidizing agent in the half reaction is stable [cannot oxidize $H_2O(l)$ to $O_2(g)$].

23-6. ELECTROCHEMICAL CELLS ARE USED TO DETERMINE CONCENTRATIONS OF IONS

We have shown in Section 23-4 how the Nernst equation is used to calculate the voltage of an electrochemical cell when the concentrations of the species involved in the cell reaction are known. Conversely, a measured cell voltage can be used to determine the concentration of a species in solution.

Consider the cell reaction given by

$$\mathrm{Zn}(s) + \mathrm{Cu}^{2+}(aq) \rightleftharpoons \mathrm{Cu}(s) + \mathrm{Zn}^{2+}(aq) \tag{23-18}$$

Application of the Nernst equation (23-13) to Equation (23-18) at 25.0°C yields

$$E_{\mathrm{rxn}} = E^{\circ}_{\mathrm{rxn}} - \left(\frac{0.0592\ \mathrm{V}}{2}\right)\log\frac{[\mathrm{Zn}^{2+}]}{[\mathrm{Cu}^{2+}]} \tag{23-19}$$

We know from Example 23-7 that $E^{\circ}_{\mathrm{cell}} = 1.10$ V at 25°C for Equation (23-18), so

$$E_{\mathrm{rxn}} = 1.10\ \mathrm{V} - (0.0296\ \mathrm{V})\log\frac{[\mathrm{Zn}^{2+}]}{[\mathrm{Cu}^{2+}]} \tag{23-20}$$

If we measure E_{rxn} at a known value of, say, $[\mathrm{Cu}^{2+}]$, then we can use Equation (23-20) to calculate the value of $[\mathrm{Zn}^{2+}]$ in the solution containing $\mathrm{Zn}^{2+}(aq)$. For example, suppose that when $[\mathrm{Cu}^{2+}] = 0.10$ M, we find that $E_{\mathrm{rxn}} = 1.20$ V. Substitution of these values for E and $[\mathrm{Cu}^{2+}]$ into Equation (23-20) yields

$$1.20\ \mathrm{V} = 1.10\ \mathrm{V} - (0.0296\ \mathrm{V})\log\frac{[\mathrm{Zn}^{2+}]}{0.10\ \mathrm{M}}$$

from which we calculate

$$\log\frac{[\mathrm{Zn}^{2+}]}{0.10\ \mathrm{M}} = \frac{1.10\ \mathrm{V} - 1.20\ \mathrm{V}}{0.0296\ \mathrm{V}} = -3.38$$

or

$$[\mathrm{Zn}^{2+}] = (0.10\ \mathrm{M})(10^{-3.38}) = 4.2 \times 10^{-5}\ \mathrm{M}$$

Electrochemical cells are used extensively in analytical chemistry to determine the concentrations of ions in solution. For example, this principle is used routinely in electroanalytical devices to determine pH, as illustrated in the following Example.

EXAMPLE 23-10: The measured voltage of the cell with the net cell equation

$$\mathrm{H_2}(g) + 2\mathrm{AgCl}(s) \rightarrow 2\mathrm{H}^+(aq) + 2\mathrm{Cl}^-(aq,\ 1.0\ \mathrm{M}) + 2\mathrm{Ag}(s)$$

is +0.34 V at 25°C when the pressure of $\mathrm{H_2}(g)$ is 1.00 atm. Use these results to calculate the pH of the solution.

Solution: The equations for the half reactions are

$$\mathrm{H_2}(g) \rightarrow 2\mathrm{H}^+(aq) + 2\mathrm{e}^-$$
$$2\mathrm{AgCl}(s) + 2\mathrm{e}^- \rightarrow 2\mathrm{Ag}(s) + 2\mathrm{Cl}^-(aq)$$

Application of the Nernst equation to the net cell equation (for which $n = 2$) yields

$$E_{\mathrm{rxn}} = E^{\circ}_{\mathrm{rxn}} - \left(\frac{0.0592\ \mathrm{V}}{2}\right)\log\left(\frac{[\mathrm{H}^+]^2[\mathrm{Cl}^-]^2}{P_{\mathrm{H_2}}}\right)$$

The value of E°_{rxn} is calculated from the E°_{rxn} values for the half reactions (Table 23-1) by using Equation (23-15):

$$E^{\circ}_{\mathrm{rxn}} = E^{\circ}_{\mathrm{red}}[\mathrm{AgCl/Ag}] + E^{\circ}_{\mathrm{ox}}[\mathrm{H_2/H^+}]$$
$$= +0.22\ \mathrm{V} + 0 = +0.22\ \mathrm{V}$$

Substitution of the known values of E_{rxn}, E°_{rxn}, $[Cl^-]$, and P_{H_2} into the Nernst equation yields

$$+0.34\ V = +0.22\ V - (0.0296\ V)\log\left(\frac{[H^+]^2(1.0)^2}{1.00}\right)$$

Solving for $pH = -\log[H^+]$ gives

$$pH = -\log[H^+] = \frac{+0.34\ V - 0.22\ V}{2(0.0296\ V)} = +2.0$$

This Example shows how cells can be used to measure pH values. Such measurements are made routinely using pH meters and hydrogen-ion sensitive glass electrodes that have a voltage dependence on $[H^+]$, like the hydrogen gas electrode in this Example, but involve a different chemical mechanism for detecting the concentration of $H^+(aq)$.

PRACTICE PROBLEM 23-10: Calculate the number of millivolts change in the voltage of the cell described in Example 23-10 per tenfold change in the hydrogen-ion concentration.

Answer: 59.2 mV

23-7. THE ELECTRICAL ENERGY RELEASED FROM AN ELECTROCHEMICAL CELL CAN DO USEFUL WORK

We have shown that the voltage of an electrochemical cell is a measure of the driving force of the cell reaction toward equilibrium. A spontaneous cell reaction creates a voltage difference between the electrodes, which can be used to produce an electric current. Electrochemical cells therefore provide a mechanism for converting chemical energy into electrical energy. By calculating the Gibbs free energy change for the reaction, we can determine how much electrical energy can be supplied to an external device, such as an electric motor.

Equation (23-5) relates the value of ΔG_{rxn} directly to the cell voltage:

$$\Delta G_{rxn} = -nFE_{rxn}$$

For a spontaneous reaction, ΔG_{rxn} is negative. From Chapter 22, we know that ΔG_{rxn} is then equal in value to the maximum amount of electrical energy that can be obtained from the reaction.

EXAMPLE 23-11: The reaction in an electrochemical cell is described by the equation

$$2H_2(g) + O_2(g) \rightleftharpoons 2H_2O(l)$$

At 25°C, with $P_{H_2} = P_{O_2} = 1.00$ atm, the cell voltage is 1.23 V. Determine whether the cell reaction is spontaneous; and, if it is, find the maximum amount of electrical energy it can provide.

Solution: The value of ΔG_{rxn} can be calculated using Equation (23-5), but we first must determine the value of n for the reaction. In the reaction, 2 mol of oxygen *atoms* (1 mol of O_2) are reduced from an oxidation state of zero to an oxidation state of −2. Thus, the reaction requires 4 mol of electrons per mole of O_2, so $n = 4$ mol. Therefore,

$$\begin{aligned}\Delta G_{rxn} &= -nFE_{rxn} \\ &= -(4\ \text{mol})(96{,}500\ \text{C}\cdot\text{mol}^{-1})(1.23\ \text{V}) \\ &= -4.75 \times 10^5\ \text{C}\cdot\text{V} = -4.75 \times 10^5\ \text{J} = -475\ \text{kJ}\end{aligned}$$

The negative value of ΔG_{rxn} indicates that the reaction is spontaneous, which we already know, because mixtures of hydrogen and oxygen gas can react explosively. We conclude that the reaction of 2 mol of $H_2(g)$ at 1 atm with 1 mol of $O_2(g)$ at 1 atm can provide a maximum of 4.75×10^5 J to an external device.

PRACTICE PROBLEM 23-11: Calculate the value of ΔG_{rxn} for the reaction described in Example 23-11 when $P_{H_2} = 50$ atm and $P_{O_2} = 50$ atm.

Answer: $E_{rxn} = 1.31$ V; $\Delta G_{rxn} = -506$ kJ

The maximum amount of energy that we calculated in Example 23-11 is an ideal value. In practice, we would obtain less than this amount because part of the available energy is lost as heat. This loss is a consequence of the second law of thermodynamics.

If ΔG_{rxn} is positive, then its value is the minimum energy that must be *supplied* in order to make the reaction occur. Example 23-11 shows that $\Delta G_{rxn} = +4.75 \times 10^5$ J for the decomposition of water into $H_2(g)$ and $O_2(g)$ at 1 atm. Thus, we must supply at least 4.75×10^5 J to decompose 2 mol of water. This energy could be supplied by an electric current at an appropriate voltage, as is done in electrolysis.

Figure 23-8 Michael Faraday (1791–1867), English chemist and physicist, discovered the laws of electrolysis and electromagnetic induction. He also played a major role in introducing the concept of electric and magnetic fields in terms of lines of force. When Sir Humphrey Davy, who discovered several elements and a variety of important chemical processes, was asked to name his most important discovery, he replied, "Michael Faraday."

23-8. ELECTROLYSIS IS DESCRIBED QUANTITATIVELY BY FARADAY'S LAWS

Michael Faraday (Figure 23-8), an English experimentalist with few peers, was trained by Humphrey Davy. Faraday used the voltaic pile (Section 23-1) to investigate the effect of the passage of an electric current through various electrolyte solutions. Faraday's primary observation was that the passage of an electric current through a solution causes the occurrence of chemical reactions *that would not otherwise occur.* The process by which a chemical reaction is *made* to occur by the passage of an electric current through the solution is called **electrolysis.** As noted in Section 23-2, the current enters and leaves the solution via electrodes. Faraday carried out an extensive series of experiments to determine the amount of electricity required to deposit measured quantities of metals on electrodes as a result of the passage of current through his electrolysis apparatus. The electrodes in Faraday's experiments were electrochemically inert in the sense that the electrode itself (e.g., Pt) was unchanged by the passage of current.

Faraday discovered that the metal ions of many salts are deposited as pure metal when an electric current is passed through aqueous solutions of their salts. For example, pure silver is deposited from a solution of $AgNO_3(aq)$ and pure copper is deposited from $Cu(NO_3)_2(aq)$. The reactions that take place at the electrode are

$$Ag^+(aq) + e^- \rightarrow Ag(s)$$
$$Cu^{2+}(aq) + 2e^- \rightarrow Cu(s)$$

respectively. Note that supplying 1 mol of electrons deposits 1 mol of silver from $Ag^+(aq)$, but 2 mol of electrons is needed to deposit 1 mol of copper from $Cu^{2+}(aq)$ (Figure 23-9).

The number of electrons supplied to drive the reaction can be controlled by regulating the electric current through the solution, which, in turn, is controlled by the applied voltage. Consider the passage of an electric current through a solution of $AgNO_3(aq)$. The charge that flows through the solution is directly proportional to the number of electrons that participate in the electrochemical reaction

$$Ag^+(aq) + e^- \rightarrow Ag(s)$$

The number of electrons consumed by Ag^+ ions is directly proportional to the total charge that flows through the solution.

Suppose that a current of 0.850 A flows through the solution for 20.0 min. The total charge that passes through the solution is (Equation 23-2)

$$\text{charge} = (0.850\ \text{C}\cdot\text{s}^{-1})(20.0\ \text{min})(60\ \text{s}\cdot\text{min}^{-1}) = 1020\ \text{C}$$

The number of moles of electrons that correspond to a given amount of charge that flows through the solution is equal to the ratio of the charge that flows to Faraday's constant (the charge on a mole of electrons)

$$\text{mol of electrons} = \frac{1020\ \text{C}}{9.65 \times 10^4\ \text{C}\cdot\text{mol}^{-1}} = 1.06 \times 10^{-2}\ \text{mol}$$

Figure 23-9 Electrodeposition of silver [from $AgNO_3(aq)$, left] and copper [from $Cu(NO_3)_2(aq)$, right]. The same quantity of electricity flows through the two solutions because they are placed in series. The number of moles of silver deposited after a given time is twice as great as the number of moles of copper deposited, because the reduction of 1 mol of Ag^+ requires 1 mol of electrons, whereas the reduction of 1 mol of Cu^{2+} requires 2 mol of electrons.

Because one electron deposits one atom of silver, we have

$$\text{mol of Ag deposited} = \text{mol of electrons} = 1.06 \times 10^{-2}\ \text{mol}$$

The number of grams of silver deposited by the passage of 0.850 A through a $AgNO_3(aq)$ solution for 20.0 min is

$$\text{mass of Ag deposited} = (1.06 \times 10^{-2}\ \text{mol})\left(\frac{107.9\ \text{g}}{1\ \text{mol}}\right) = 1.14\ \text{g}$$

These results for silver deposition are an illustration of **Faraday's laws of electrolysis:**

First law: The extent to which an electrochemical reaction occurs depends solely on the quantity of electricity that is passed through a solution.

Second law: The mass of a substance that is produced as a result of the passage of a given quantity of electricity is directly proportional to the molar mass of the substance divided by the number of electrons consumed or produced per formula unit.

We can write Faraday's second law as an equation. First, we denote the molar mass of the metal or gas produced by M and divide by the number of electrons, n, required to produce one formula unit of the substance. Then we have

$$\left(\begin{array}{c}\text{mass produced}\\ \text{per mole of}\\ \text{electrons used in}\\ \text{the electrolysis}\end{array}\right) = \frac{M}{n} \qquad (23\text{-}21)$$

Some examples of M and n are

Process	$M/\text{g}\cdot\text{mol}^{-1}$	n	M/n
$Cu^{2+}(aq) + 2e^- \rightarrow \underline{Cu}(s)$	63.55	2	31.78
$Ag^+(aq) + e^- \rightarrow \underline{Ag}(s)$	107.9	1	107.9
$2H_2O(l) \rightarrow \underline{O_2}(g) + 4H^+(aq) + 4e^-$	32.00	4	8.00

Preparation of hydrogen by electrolysis of an aqueous sulfuric acid solution. Hydrogen gas is liberated at the cathode, and oxygen gas is liberated at the anode. Note, as is required by the reaction stoichiometry, that the volume of $H_2(g)$ liberated is twice as great as that of $O_2(g)$:

$$2H_2O(l) \xrightarrow{\text{electrolysis}} 2H_2(g) + O_2(g)$$

Also note that $n = 4$ for the equation.

The underlined species in the table is the species produced at the cathode (for reduction) or the anode (for oxidation) in the electrolysis. The number of moles of electrons used in the electrolysis is equal to the total charge, Z, that is passed through the solution divided by Faraday's constant, F. Thus,

$$\begin{pmatrix}\text{moles of electrons} \\ \text{used in the} \\ \text{electrolysis}\end{pmatrix} = \frac{Z}{F} = \frac{It}{F} \qquad (23\text{-}22)$$

where we have used Equation (23-2), that is, $Z = It$. So, we can express Faraday's laws quantitatively by the equation

$$m = \begin{pmatrix}\text{mass deposited} \\ \text{as metal or} \\ \text{evolved as gas}\end{pmatrix} = \begin{pmatrix}\text{moles of} \\ \text{electrons} \\ \text{used in the} \\ \text{electrolysis}\end{pmatrix}\begin{pmatrix}\text{mass deposited as} \\ \text{metal or evolved} \\ \text{as gas per mole} \\ \text{of electrons used} \\ \text{in the electrolysis}\end{pmatrix}$$

Thus,

$$m = \left(\frac{It}{F}\right)\left(\frac{M}{n}\right) \qquad (23\text{-}23)$$

Problems 23-55 to 23-64 involve the use of Equation 23-23.

23-9. MANY CHEMICALS ARE PREPARED ON AN INDUSTRIAL SCALE BY ELECTROLYSIS

The alkali and alkaline earth metals are prepared on a commercial scale by electrolysis. All the sodium hydroxide (about 20 billion pounds annually) and most of the chlorine produced in the United States (about 20 billion pounds annually) are made by the **chlor-alkali process,** which involves the electrolysis of concentrated aqueous solutions of sodium chloride (Figures 23-10 and 23-11). The overall electrolysis reaction is described by the equation

$$2NaCl(aq) + 2H_2O(l) \xrightarrow{\text{electrolysis}} 2NaOH(aq) + Cl_2(g) + H_2(g)$$

The corresponding net ionic equation is

$$2Cl^-(aq) + 2H_2O(l) \rightarrow 2OH^-(aq) + Cl_2(g) + H_2(g)$$

The two half reaction equations are the reduction of water at the cathode

$$2H_2O(l) + 2e^- \rightarrow 2OH^-(aq) + H_2(g) \qquad \text{(cathode)}$$

and the oxidation of chloride ions at the anode

$$2Cl^-(aq) \rightarrow Cl_2(g) + 2e^- \qquad \text{(anode)}$$

The electrolysis must be carried out in an apparatus in which the sodium hydroxide and hydrogen produced at the cathode and the chlorine produced at the anode are separated by a special membrane. If they were not separated, then the sodium hydroxide and chlorine would react to form sodium hypochlorite:

$$2NaOH(aq) + Cl_2(g) \rightarrow NaClO(aq) + NaCl(aq) + H_2O(l)$$

In fact, sodium hypochlorite, which is a commonly used bleaching agent, is prepared in this manner. Household bleach is an aqueous solution that is about 5 percent by mass of NaClO.

Figure 23-10 A chlor-alkali plant in which chlorine gas and an aqueous solution of sodium hydroxide are obtained by electrolysis from an aqueous solution of sodium chloride.

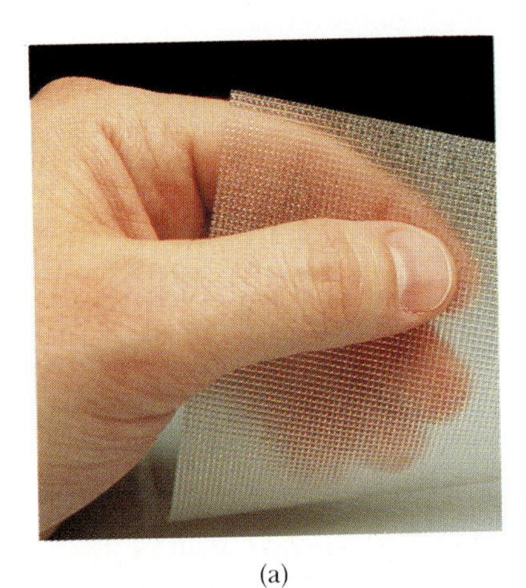

(a)

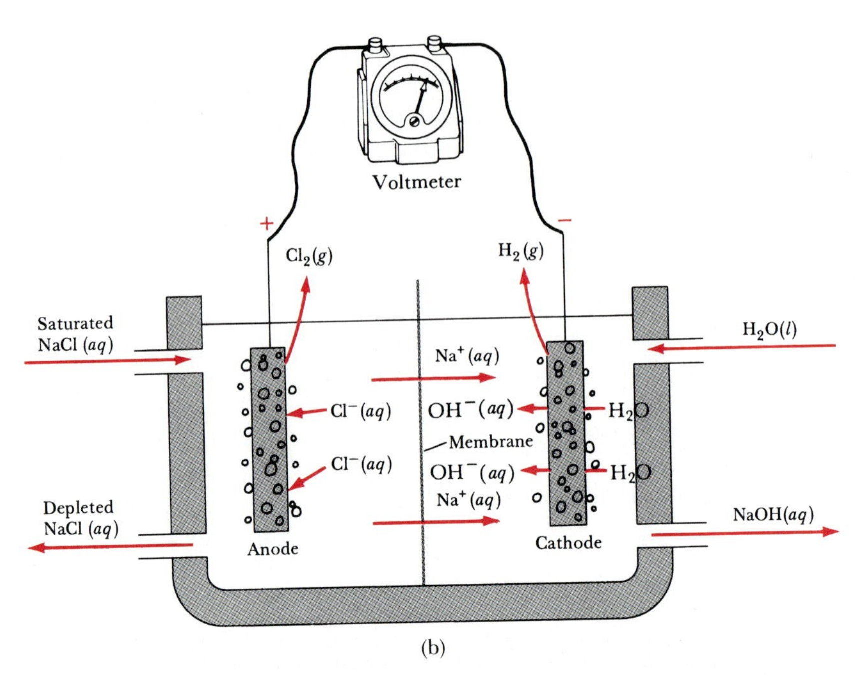

(b)

Figure 23-11 A chlor-alkali membrane cell. (a) A Nafion separator membrane. The membrane is a polymeric material with negatively charged groups ($—SO_3^-$) that permit the passage of Na^+ from $—SO_3^-$ group to $—SO_3^-$ group in the membrane. The membrane is supported by a Teflon grid for additional mechanical strength. (b) A schematic view of the cell. Hydrogen gas and aqueous sodium hydroxide solution are produced at the cathode (−) and chlorine gas is produced at the anode (+). The migration of $Na^+(aq)$ through the membrane maintains equal numbers of positive and negative charges in the separate cell solutions and also carries the current through the cell solutions.

Figure 23-11 shows a schematic drawing of a chlor-alkali membrane cell. The anode and cathode compartments are separated by a Nafion membrane, which has a high internal negative charge. The Nafion membrane excludes the negatively charged $Cl^-(aq)$ ions but allows the positively charged $Na^+(aq)$ ions to pass through. A saturated sodium chloride solution enters the anode compartment, where the $Cl^-(aq)$ is oxidized to $Cl_2(g)$. The excess $Na^+(aq)$ migrates through the membrane to the cathode, where water is reduced to $H_2(g)$ and $OH^-(aq)$ according to the cathode half reaction.

Over 5 million tons of 99.7 percent pure aluminum metal are produced by electrolysis each year in the United States. Most aluminum metal is produced by the electrolytic **Hall process,** which was patented by Charles Hall in 1889, when he was 26 years old. Hall conceived his process for the electrochemical production of aluminum while a student at Oberlin College in Ohio. Hall faced two major challenges in devising his process. The first challenge was that aluminum metal cannot be produced by the electrolysis of an aqueous solution containing $Al^{3+}(aq)$, because $H^+(aq)$ is reduced at an applied voltage much lower (about 1.66 V lower, based on the E°_{red} values; Table 23-1) than the voltage required to reduce $Al^{3+}(aq)$ at the cathode. The second challenge was that the aluminum ores melt at temperatures above 2000°C, which exceeds the boiling point of aluminum metal, thus ruling out electrolysis of the molten ore. Hall's brilliant solution to these problems was to use a molten salt electrolyte consisting of powdered aluminum oxide dissolved in the mineral cryolite, Na_3AlF_6. Aluminum oxide dissolves in cryolite to form a conductive solution with a melting point low enough to allow the operation of the cell at temperatures below 1000°C.

The electrolysis apparatus used today is essentially the same as that used in Hall's original process. (Figure 23-12 shows a portion of an electrochemical aluminum production facility.) The electrolysis is carried out at about 980°C, at which temperature aluminum is a liquid and can be siphoned off from the cathode compartment. Electrical contact to the molten aluminum cathode is made through a carbon shell that constitutes the bottom of the electrode compartment. The consumable anodes are composed of a petroleum coke that is obtained by heating to dryness the heavy petroleum fraction remaining from petroleum refining. The equation for the overall electrochemical reaction is

$$2Al_2O_3(soln) + 3C(s) \rightarrow 4Al(l) + 3CO_2(g)$$

Aluminum, which is a fairly reactive metal, protects itself from corrosion by spontaneously forming a thin, tough, adherent layer of aluminum oxide, Al_2O_3. The oxide coating is what gives aluminum metal its dull cast. Electrolysis is used to protect other reactive metals that do not self-protect as aluminum does. By electrolysis, a thin layer of a relatively nonreactive metal such as nickel, chromium, tin, silver, or gold is deposited on the surface of the reactive metal. The production of a layer of protective metal by electrolysis is called **electroplating.** For example, gold electroplating is used to coat base metals with a thin layer of gold that functions both as a decorative and a protective coating. Electroplating has innumerable industrial applications ranging from the manufacture of heavy machinery to the production of microcircuits.

Figure 23-12 The Hall process on an industrial scale. Shown here is an Alcoa potline (198 pots) at the Massena, New York, plant.

SUMMARY

An electrochemical cell provides the means for obtaining electricity from an electron-transfer reaction. The cell consists of a pair of metal electrodes in contact with an electrolyte solution. The dependence of the cell voltage on the concentrations of the reactants and products of the cell reaction is described quantitatively by the Nernst equation.

The standard cell voltage, E°_{rxn}, is the voltage of the cell when the reaction quotient $Q = 1$. The equilibrium constant of the cell reaction can be calculated from the value of E°_{rxn}. Electrode reactions can be arranged in a series of decreasing electrode reduction voltages. The assignment of standard reduction voltages to electrode reactions is achieved by setting $E^\circ_{red} = 0$ for the hydrogen gas electrode reaction. The E°_{red} values for electrode reactions given in Table 23-1 and Appendix I can be used to calculate E°_{rxn} values for reactions and to predict the thermodynamic stability of oxidizing agents and reducing agents. Electrochemical cells are used to determine the concentration of various ions in a solution.

Electrolysis is the process by which a chemical reaction is driven uphill on the Gibbs free energy scale by the application of a voltage across electrodes placed in a solution. The extent to which the electrochemical reaction occurs is proportional to the current that flows through the solution, according to Faraday's laws of electrolysis.

The maximum possible amount of energy that can be obtained from an electrochemical cell for the performance of useful work is given by $\Delta G_{rxn} = -nFE_{rxn}$, where E_{rxn} is the cell voltage, F is the faraday (96,500 coulombs per mole of electrons), and n is the number of electrons that cancel out when the two half reactions are combined to yield the net balanced cell reaction.

23-47. An electrochemical cell is set up so that the reaction described by the equation

$$Zn(s) + Cu^{2+}(aq) \rightleftharpoons Zn^{2+}(aq) + Cu(s)$$

occurs. At 25°C the measured cell voltage is 1.05 V. Calculate the value of ΔG_{rxn} for the reaction.

23-48. An electrochemical cell is set up so that the reaction described by the equation

$$H_2O_2(aq) + Fe(s) + 2H^+(aq) \rightleftharpoons Fe^{2+}(aq) + 2H_2O(l)$$

occurs. At 25°C the measured cell voltage is 2.03 V. Calculate the value of ΔG_{rxn}.

23-49. An electrochemical cell is set up so that the reaction described by the equation

$$2NO_3^-(aq) + 4H^+(aq) + Cu(s) \rightleftharpoons 2NO_2(g) + 2H_2O(l) + Cu^{2+}(aq)$$

occurs. The standard cell voltage is $E^\circ_{rxn} = 0.65$ V. Calculate the value of ΔG°_{rxn}.

23-50. An electrochemical cell is set up so that the reaction described by the equation

$$Cr_2O_7^{2-}(aq) + 14H^+(aq) + 6Fe^{2+}(aq) \rightleftharpoons 2Cr^{3+}(aq) + 6Fe^{3+}(aq) + 7H_2O(l)$$

occurs. At 25°C, the standard cell voltage is 0.56 V. Calculate the value of ΔG°_{rxn}.

23-51. Use the data in Appendix I to calculate the ΔG°_{rxn} values for the following equations:

(a) $2Ag(s) + F_2(g) \rightleftharpoons 2Ag^+(aq) + 2F^-(aq)$
(b) $\frac{1}{2}H_2(g) + Fe^{3+}(aq) \rightleftharpoons Fe^{2+}(aq) + H^+(aq)$

23-52. Use the data in Appendix I to calculate the ΔG°_{rxn} values for the following equations:

(a) $Zn(s) + Cu^{2+}(aq) \rightleftharpoons Zn^{2+}(aq) + Cu(s)$
(b) $Ag(s) + Fe^{3+}(aq) \rightleftharpoons Fe^{2+}(aq) + Ag^+(aq)$

23-53. For the following electrochemical cell

$$Zn(s)|Zn^{2+}(aq, 0.010\text{ M})||Cd^{2+}(aq, 0.050\text{ M})|Cd(s)$$

use the data in Appendix I to calculate the values at 25°C of E°_{rxn}, ΔG°_{rxn}, ΔG_{rxn}, and E_{rxn} for the cell equation. What is the equation for the cell reaction?

23-54. The standard voltage of the following cell at 25°C is $E^\circ_{rxn} = 1.08$ V:

$$Co(s)|Co^{2+}(aq, 0.0155\text{ M})||Ag^+(aq, 1.50\text{ M})|Ag(s)$$

Calculate ΔG°_{rxn}, ΔG_{rxn}, and E_{rxn}. What is the equation for the cell reaction?

ELECTROLYSIS

23-55. How long will it take an electric current of 1.25 A to deposit all the copper from 500 mL of 0.150 M $CuSO_4(aq)$?

23-56. How much silver is deposited if an electric current of 0.150 A flows through a silver nitrate solution for 20.0 min?

23-57. Cesium metal is produced by the electrolysis of molten cesium cyanide. Calculate how much Cs(*s*) is deposited from CsCN(*l*) in 30 min by a current of 500 mA.

23-58. Beryllium occurs naturally in the form of beryl. The metal is produced from its ore by electrolysis after the ore has been converted to the oxide and then to the chloride. How much Be(*s*) is deposited from a $BeCl_2$ melt by a current of 5.0 A that flows for 1.0 h?

23-59. Fluorine is manufactured by the electrolysis of HF dissolved in molten KF. The equation is

$$2HF(KF) \rightarrow H_2(g) + F_2(g)$$

The KF acts as a solvent for HF and as the conductor of electricity. A commercial cell for producing fluorine operates at a current of 1500 A. How much F_2 can be produced per 24 h? Why isn't the electrolysis of liquid HF alone used?

23-60. Suppose that it is planned to electrodeposit 200 mg of gold onto the surface of a steel object via the process

$$Au(CN)_2^-(aq) + e^- \rightarrow Au(s) + 2CN^-(aq)$$

If the electric current in the circuit is set at 30 mA, for how long should the current be passed?

23-61. Gallium is produced by the electrolysis of a solution obtained by dissolving gallium oxide in concentrated NaOH(*aq*). Calculate the amount of Ga(*s*) deposited from a Ga(III) solution by a current of 0.50 A that flows for 30 min.

23-62. Sodium metal is produced commercially by the electrolysis of molten sodium chloride. How much sodium will be produced if a current of 1.00×10^3 A is passed through NaCl(*l*) for 8.0 h? How many liters of chlorine at 10 atm and 20°C will be produced?

23-63. Hydrogen, $H_2(g)$, and oxygen, $O_2(g)$, can be produced by the electrolysis of water:

$$2H_2O(l) \xrightarrow{\text{electrolysis}} 2H_2(g) + O_2(g)$$

Calculate the volume of $O_2(g)$ produced at 25°C and 1.00 atm when a current of 30.35 A is passed through a $K_2SO_4(aq)$ solution for 2.00 h.

23-64. Bauxite, the principal source of aluminum oxide, contains about 55 percent Al_2O_3 by mass. How much bauxite is required to produce the 5 million tons of aluminum metal produced each year by electrolysis?

ADDITIONAL PROBLEMS

23-65. Write the balanced chemical equation for the cell reaction in the following electrochemical cell:

$Pt(s)|MnO_4^-(aq), Mn^{2+}(aq), H^+(aq)||IO_3^-(aq), I^-(aq), H^+(aq)|Pt(s)$

23-66. A rechargeable silver oxide cell, involving a $Ag_2O_2(s)|Ag(s)$ cathode, in a LCD (liquid crystal display) calculator is estimated to last 1000 h while drawing a current of only 0.10 mA. What mass of silver will be produced over the lifetime of the cell?

23-67. The cell diagram for the Edison cell, which is used extensively as a car and truck battery in Europe, is

$Fe(s)|Fe(OH)_2(s)|NaOH(aq)|NiOOH(s), Ni(OH)_2(s)|steel$

where the steel electrode is nonreactive and the comma between $NiOOH(s)$ and $Ni(OH)_2(s)$ denotes a heterogeneous mixture of the two solids. Determine the equation for the cell reaction.

23-68. The cell diagram for the lead-acid cell which is used in automobile and truck batteries is

$Pb(s)|PbSO_4(s)|H_2SO_4(aq)|PbO_2(s), PbSO_4(s)|Pb(s)$

where the comma between $PbO_2(s)$ and $PbSO_4(s)$ denotes a heterogeneous mixture of the two solids and the right-hand lead electrode is nonreactive.

(a) Determine the equation for the net cell reaction
(b) Use the data in Table 23-1 to calculate E°_{rxn}
(c) Calculate ΔG°_{rxn}
(d) How many lead-acid cells in a 12V car battery?

23-69. Many metals can be refined electrolytically. The impure metal is used as the anode, and the cathode is made of the pure metal. The electrodes are placed in an electrolyte containing a salt of the metal being refined. When an electric current is passed between these electrodes, the metal leaves the impure anode and is deposited in a pure form on the cathode. How many ampere-hours of electricity are required to refine electrolytically 1 metric ton of copper? (Use $n = 2$.) An ampere-hour is an ampere times an hour.

23-70. Write a balanced equation for the cell reaction in the following electrochemical cell and calculate the cell voltage, E_{cell}:

$Pt(s)|H_2(g, 0.50 \text{ atm})|H_2SO_4(aq, 1.00 \text{ M})|PbSO_4(s)|Pb(s)$

23-71. Two electrolytic cells are placed in series. One cell contains a solution of $AgClO_4(aq)$ and the other cell contains a solution of $Cd(ClO_4)_2(aq)$. An electric current is passed through the two cells until 0.876 g of Ag is deposited. How much Cd will be deposited?

23-72. Explain why $F_2(g)$ cannot be prepared by electrolysis of $NaF(aq)$.

23-73. Given that $E^\circ_{rxn} = +0.728$ V at 25°C for the cell

$Ag(s)|AgBr(s)|Br^-(aq)||Ag^+(aq)|Ag(s)$

write the cell equation and determine the solubility-product constant of $AgBr(s)$ in water at 25°C.

23-74. The Weston standard cell is given by

$Cd(Hg)|CdSO_4(aq, satd)|Hg_2SO_4(s)|Hg(l)$
(12.5% Cd)

Write the equation that occurs in the cell. Ten Weston standard cells that use a saturated $CdSO_4(aq)$ solution are maintained at the U.S. Bureau of Standards as the official unit of voltage. The voltage of each cell is virtually constant at 1.01857 V. Explain why the voltage remains constant.

23-75. Suppose a zinc rod is dipped into a 1.0 M $CuSO_4(aq)$ solution containing a copper rod and the system is allowed to stand for several hours. What do you predict for the voltage measured between the $Zn(s)$ and $Cu(s)$ rods?

23-76. A battery that operates at −50°C was developed for the exploration of the moon and Mars. The electrodes are magnesium metal/magnesium chloride and silver chloride/silver. The electrolyte is potassium thiocyanate, KSCN, in liquid ammonia. Draw the cell diagram and write the equation for the reaction for the cell.

23-77. Suppose the leads of an electrochemical cell are connected together external to the cell and the cell is allowed to come to equilibrium. What will be the value of the cell voltage at equilibrium?

23-78. A battery that operates at 500°C was developed for the exploration of Venus. The electrodes are a magnesium metal anode and a mixture of copper(I) and copper(II) oxides in contact with an inert steel cathode. The electrolyte is a mixture of LiCl and KCl, which is melted to activate the cell. The MgO that is produced is sparingly soluble in the molten salt mixture and precipitates. Draw the cell diagram and write the equation for the reaction for the cell.

23-79. Electrolysis can be used to determine atomic masses. A current of 0.600 A deposits 2.42 g of a certain metal in exactly 1 h. Calculate the atomic mass of the metal if $n = 1$. What is the metal?

23-80. From 1882 to 1895 home electricity was provided as direct current rather than as alternating current, as is now the case. Thomas Edison invented a meter to measure the amount of electricity used by a consumer. A small amount of current was diverted to an electrolysis cell that consisted of zinc electrodes in a zinc sulfate solution. Once a month the cathode was removed, washed, dried, and weighed. The bill was figured in ampere-hours (Problem 23-69). In 1888 Boston Edison Company had 800 chemical meters in service. In one case, in one 30-day period, 65 g of zinc was deposited on the cathode. The meter used 11 percent of the current into the house. How many coulombs were used in the month? Calculate the current used in ampere-hours.

same mass an an electron but a positive charge. The symbol for a positron in nuclear equations is ${}^{0}_{+1}e$. Two examples of positron emission are

$$ {}^{38}_{19}K \rightarrow {}^{38}_{18}Ar + {}^{0}_{+1}e $$

$$ {}^{120}_{51}Sb \rightarrow {}^{120}_{50}Sn + {}^{0}_{+1}e $$

The emission of a positron can be viewed as a result of the conversion of a proton to a neutron in the nucleus. The process can be represented as

$$ {}^{1}_{1}p \rightarrow {}^{1}_{0}n + {}^{0}_{+1}e $$

Positrons exist for only a very short time. They combine with electrons in about 10^{-9} s, and the resulting energy produced appears in the form of γ-ray photons. The reverse process is shown in Figure 24-2.

Figure 24-2 An electron and a positron annihilate each other to produce two γ-ray photons. The reverse process also occurs; that is, a high-energy γ-ray photon can transform spontaneously into an electron and a positron, as seen in this photograph from a bubble chamber. Because it is uncharged, the photon leaves no track, but the electron and positron leave visible spiral tracks. An externally applied magnetic field causes the tracks to be spirals, and the opposite charges on the electron and positron cause them to spiral in opposite directions.

Another type of nuclear transformation is called **electron capture.** In an electron-capture process, one of the innermost electrons of an atom is absorbed by the nucleus. The process involves the conversion of a proton to a neutron and the emission of an X-ray photon. (Free electrons do not exist in nuclei.) An example of an electron-capture process is

$$ {}^{195}_{79}Au + {}^{0}_{-1}e \rightarrow {}^{195}_{78}Pt $$

Positron emission and electron capture lead to the same result: they both decrease Z by one and leave A unchanged. Electron capture usually occurs in nuclei with Z greater than 80, whereas positron emission usually occurs in nuclei with values of Z less than about 30. Both processes are observed in elements with Z values in the range $30 < Z < 80$. Table 24-1 summarizes the various types of nuclear decay processes.

When a radioactive nucleus emits a particle and transforms to another nucleus, we say that it *decays* to that nucleus. Thus, the expression **radioactive decay** refers to a process in which one nucleus is converted into another.

Table 24-1 Various particles emitted in radioactive processes

Emission	Symbol	Change in nucleus: Mass number	Change in nucleus: Atomic number	Example
α	${}^{4}_{2}He$	decreases by 4	decreases by 2	${}^{238}_{92}U \rightarrow {}^{234}_{90}Th + {}^{4}_{2}He$
β	${}^{0}_{-1}e$	no change	increases by 1	${}^{14}_{6}C \rightarrow {}^{14}_{7}N + {}^{0}_{-1}e$
γ	${}^{0}_{0}\gamma$	no change	no change	${}^{16}_{7}N \rightarrow {}^{16}_{8}O + {}^{0}_{-1}e + \gamma$
positron	${}^{0}_{+1}e$	no change	decreases by 1	${}^{38}_{19}K \rightarrow {}^{38}_{18}Ar + {}^{0}_{+1}e$
electron capture	EC	no change	decreases by 1	${}^{195}_{79}Au + {}^{0}_{-1}e \rightarrow {}^{195}_{78}Pt$
neutron	${}^{1}_{0}n$	decreases by 1	no change	${}^{137}_{53}I \rightarrow {}^{136}_{53}I + {}^{1}_{0}n$

EXAMPLE 24-2: Fill in the missing symbols in the following nuclear equations:
(a) ${}^{214}_{82}Pb \rightarrow {}^{214}_{83}Bi + ?$ (b) ${}^{11}_{6}C \rightarrow {}^{0}_{+1}e + ?$
(c) $? \rightarrow {}^{0}_{-1}e + {}^{97}_{41}Nb$

Solution: (a) The missing particle has a charge of -1 ($82 = 83 - 1$), and A does not change; thus, the missing particle is a β-particle, ${}^{0}_{-1}e$.
(b) The missing particle has $Z = 5$ and $A = 11$, so it is ${}^{11}_{5}B$.
(c) The missing particle has $Z = 40$ and $A = 97$. The element that has $Z = 40$ is zirconium, so the nucleus that decays is ${}^{97}_{40}Zr$.

PRACTICE PROBLEM 24-2: Phosphorus-32 (a β-emitter) is used extensively as a **radiotracer,** that is, a radioactive isotope used to assist in mapping chemical and biochemical reaction pathways. Write a balanced nuclear equation to describe the decay of phosphorus-32.

Answer: ${}^{32}_{15}P \rightarrow {}^{0}_{-1}e + {}^{32}_{16}S$

24-3. EMISSIONS FROM RADIOACTIVE SUBSTANCES CAN BE DETECTED BY SEVERAL MEANS

There are several methods available to detect emissions from radioactive nuclei. The most common method is based on Becquerel's original observation that the emissions from radioactive substances affect photographic film in a manner similar to light. The difference is that the film can remain covered with a thin layer of opaque plastic that keeps out the visible light but permits the passage of the emissions from radioactive materials. This technique is used in the **film badges** worn by people who work with radioactive substances. The extent of darkening of the developed film provides a quantitative measure of the degree of exposure of the film badge to radiation.

Gamma-ray and β-particle emissions can be detected rapidly and conveniently with a **Geiger counter** (Figure 24-3). The Geiger counter was developed in 1908 by Hans Geiger, a colleague of Ernest Rutherford. A high voltage is continuously applied to electrodes in contact with the gas (Figure 24-4), producing an electric current. When β-

Figure 24-3 A Geiger counter (see also Figure 24-4).

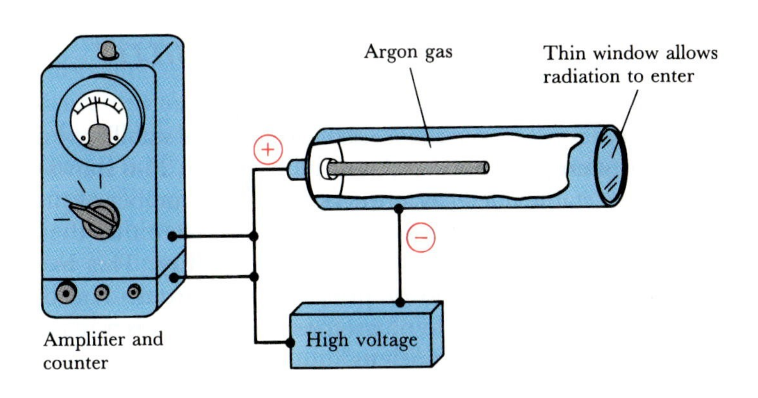

Figure 24-4 Diagram of a Geiger counter. Radiation that enters the thin window ionizes the argon gas in the chamber, and the ions produced carry current between the negative and positive electrodes. This current is detected by the amplifier and counter.

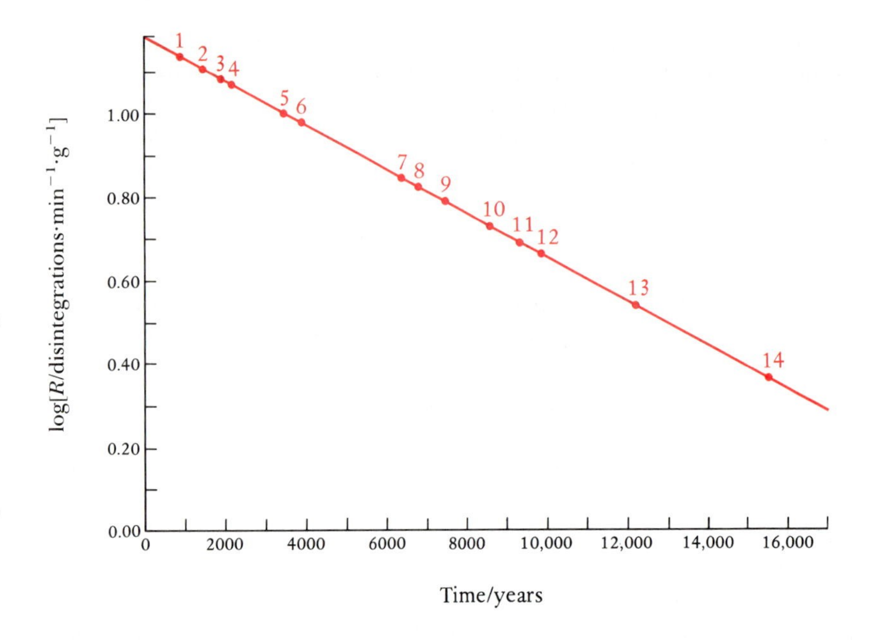

Figure 24-9 Plot of the logarithm of R, the number of disintegrations per minute per gram of carbon, versus time for various samples dated using the carbon-14 method. As Equation (24-6) indicates, this plot is a straight line. The numbers on the curve designate archaeological objects that have been dated by the carbon-14 method: (1) Charcoal from earliest Polynesian culture in Hawaii (946 ± 180 years). (2) Wooden lintels from a Mayan site in Tikal, Guatemala (1503 ± 110 years). (3) Linen wrappings from the Dead Sea Scrolls (1917 ± 200 years). (4) Wood from a coffin from the Egyptian Ptolemaic period (2190 ± 450 years). (5) Samples of oak from an ancient cooking place at Killeens, County Cork (3506 ± 230 years). (6) Charcoal sample from Stonehenge (3798 ± 275 years). (7) Charcoal from a tree destroyed by the explosion of Mount Mazama, the explosion that formed Crater Lake in Oregon (6453 ± 250 years). (8) Land-snail shells found at Jarmo, Iraq (6707 ± 320 years). (9) Charcoal from an archaeological site near Beer-Sheba, Israel (7240 ± 520 years). (10) Burned animal bones found near a site inhabited by humans in Palli Aike Cave in southern Chile (8639 ± 450 years). (11) Woven rope sandals found in Fork Rock Cave, Oregon (9053 ± 350 years). (12) Buried bison bone from Folsom Man site near Lubbock, Texas (9883 ± 350 years). (13) Glacial wood found near Skunk River, Iowa, (12,200 ± 500 years). (14) Charcoal from the Lascaux cave in France, which contains many cave paintings (15,516 ± 900 years).

PRACTICE PROBLEM 24-6: Recent (1988) measurements on samples from the Shroud of Turin give a carbon-14 content that is 0.928 times that for a contemporary carbon sample of biological origin. Estimate the age of the Shroud of Turin.

Answer: t = 618 years, or about 1372 A.D.

Radiocarbon dating has been used to date many archaeological objects. Figure 24-9 shows a plot of log R versus t and the ages of artifacts that have been determined by the carbon-14 method.

24-8. RADIOISOTOPES CAN BE PRODUCED IN THE LABORATORY

We have learned that numerous radioactive decay reactions occur spontaneously in nature, changing one element into another and emitting particles in the process. Such a reaction, in which one element is converted to another, is called **transmutation.** In 1919, the New Zealand-born physicist Ernest Rutherford discovered he could carry out transmutations in the laboratory. Working in the Cavendish Laboratories in England, he bombarded nitrogen with a beam of α-particles and was able to detect the reaction

$$^{14}_{7}\text{N} + {}^{4}_{2}\text{He} \rightarrow {}^{17}_{8}\text{O} + {}^{1}_{1}\text{H}$$

where $^{1}_{1}\text{H}$ denotes a proton. Rutherford's experiment was the first laboratory synthesis of a nucleus. Following Rutherford's lead, hundreds of different transmutations have been achieved. The products of many of these reactions are radioactive isotopes that are not found in nature and therefore are called **artificial radioisotopes.** If these isotopes ever did exist in nature, they have long since disappeared, because their half-lives are so much shorter than the age of the earth.

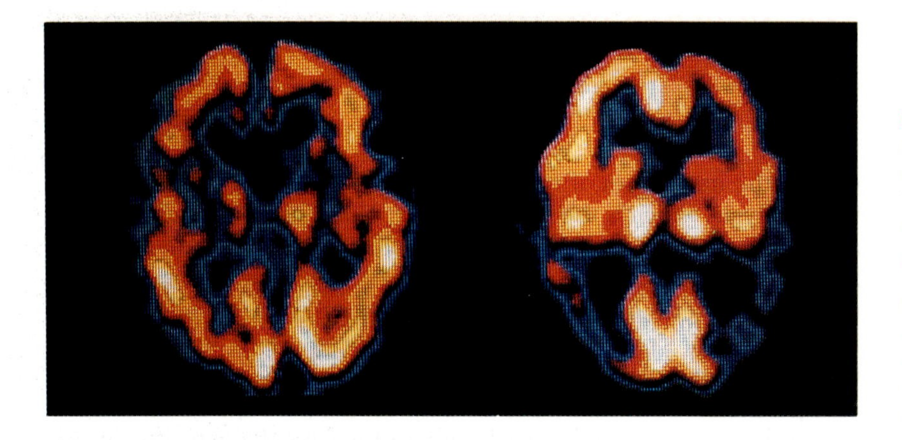

Figure 24-10 A molecule labeled with the radioisotope iodine-123 is used to study cerebral blood flow. The γ-ray photons emitted by iodine-123 are detected by probes placed around the patient's head. In the photos shown, a normal brain (left) and the brain of a patient with Alzheimer's disease (right) are compared. The distribution of ^{123}I is radically different in the two cases.

Many of the radioisotopes produced have applications in numerous areas, including medicine, agriculture, and oil exploration. For example, iodine-131 is used to measure the activity of the thyroid gland, and iodine-123 is used to monitor brain activity (Figure 24-10). A few of the many other radioisotopes that are used in medicine are sodium-24 (to follow blood circulation), technetium-99 (for brain, liver, and spleen tests), and phosphorus-32 (for treatment of leukemia).

Before the development of nuclear science, uranium lay at the end of the periodic table. Since the 1940s, elements with $Z = 93$ to 107 and $Z = 109$ have been synthesized. Many of these elements were produced at the Lawrence Radiation Laboratory in Berkeley, California, as indicated by the names californium (98), berkelium (97), americium (95), and Lawrencium (103) (Figure 24-11). Many of these **transuranium elements** can be made in commercial quantities. For example, americium-241, an α- and β-emitter with a half-life of 432 years, is used in smoke detectors (Figure 24-12).

Figure 24-11 The elements beyond uranium in the periodic table are called transuranium elements. With the exception of recently discovered traces of plutonium, these elements do not occur in nature but have been created in the laboratory. Shown here is the first visible sample of americium produced, with a mass of only a few micrograms.

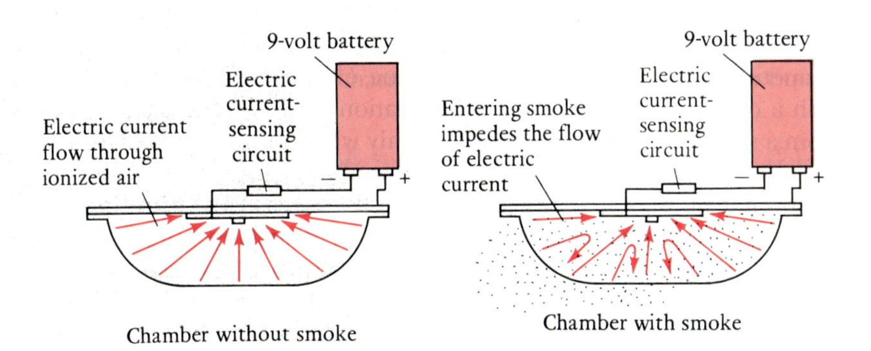

Figure 24-12 Diagram of a home smoke detector. A small quantity of americium-241 ($t_{1/2} = 432$ years) ionizes the air in the ionization chamber. An electric voltage is applied across the ionization chamber, and the ions in the air conduct an electric current, which is constantly monitored electronically. When smoke particles pass through the ionization chamber, they impede the flow of electricity, which is detected by electronic circuitry. This reduction then triggers a horn alarm. A weak battery also produces a decreased current flow which in some smoke detectors is signaled by an audible "beep."

Figure 24-20 The Princeton University Tokomak Fusion Reactor, which is designed to study nuclear fusion as a controlled energy source.

The control of fusion reactions for the generation of electricity is one of the most exciting, important, and difficult technological problems of our time. The best possibilities seem to be the reactions

$$^{2}_{1}\text{H} + ^{2}_{1}\text{H} \rightarrow ^{3}_{2}\text{He} + ^{1}_{0}\text{n}$$

and

$$^{2}_{1}\text{H} + ^{2}_{1}\text{H} \rightarrow ^{3}_{1}\text{H} + ^{1}_{1}\text{H}$$

and

$$^{1}_{1}\text{H} + ^{2}_{1}\text{H} \rightarrow ^{3}_{2}\text{He}$$

Because hydrogen and deuterium are plentiful, the successful development of a fusion reactor would totally change the world we live in by allowing all countries to have an essentially inexhaustible supply of energy. The technological problems that bar the way are staggering, however. In addition to being heated to millions of degrees, the hydrogen and deuterium must be contained long enough for the reaction to occur. Of course, there is no structural material that can withstand such temperatures, but present research is aimed at confining the nuclei by using very strong magnetic and electric fields. The enormous temperatures required also can be achieved with high-power laser beams. Controlled fusion is the subject of intensive research (Figure 24-20). Some progress appears to have been made, but a fusion reactor capable of net energy production remains to be developed.

24-16. EXPOSURE TO RADIATION DAMAGES CELLS, TISSUES, AND GENES

Example 24-4 notes that γ-radiation from cobalt-60 can be used to destroy cancerous cells. Actually, γ-radiation destroys healthy cells as well, but it is more destructive to cancerous cells because they grow and divide more rapidly than normal cells and so are more vulnerable. The effect of various kinds of radiation on living systems has received much study and has resulted in the imposition of legal limits on exposure to radiation of various types.

The standard measure of the activity level of a radioactive substance is its **specific activity,** which is the number of nuclei that disintegrate per second per gram of the radioactive isotope. The specific activity of a radioisotope is related to its half-life, $t_{1/2}$, and its atomic mass M by the equation (see Problem 24-81)

$$\text{specific activity} = \left(\frac{4.17 \times 10^{23}\ \text{disintegrations}\cdot\text{g}^{-1}}{Mt_{1/2}}\right) \quad (24\text{-}10)$$

Because specific activity is defined as the number of disintegrations per second per gram, the value of $t_{1/2}$ in Equation (24-10) must be in seconds. As an example, let's calculate the specific activity of radium-226, which has a half-life of 1600 years. We first convert the half-life to seconds:

$$(1600\ \text{yr})(365\ \text{day}\cdot\text{yr}^{-1})(24\ \text{h}\cdot\text{day}^{-1})(60\ \text{min}\cdot\text{h}^{-1})(60\ \text{s}\cdot\text{min}^{-1}) = 5.05 \times 10^{10}\ \text{s}$$

The specific activity of radium-226 is thus

$$\text{specific activity} = \frac{4.17 \times 10^{23}\ \text{disintegrations}\cdot\text{g}^{-1}}{(226)(5.05 \times 10^{10}\ \text{s})} = 3.7 \times 10^{10}\ \text{disintegrations}\cdot\text{s}^{-1}\cdot\text{g}^{-1}$$

The quantity 3.7×10^{10} disintegrations·s^{-1} is called a **curie,** Ci, after Marie Curie, one of the pioneers in research in radioactivity (Figure 24-21). She received a Nobel Prize in physics in 1903 for her research on radioactivity, and a Nobel Prize in chemistry in 1911 for her research on radiochemistry. She is one of only four scientists to earn two Nobel Prizes, and she is the only scientist to earn a Nobel Prize in two different sciences.

Table 24-5 lists the specific activities of a number of important radioactive isotopes. The SI unit of the rate of radioactive decay is a **becquerel** (Bq), which is defined as one disintegration per second: 1 Bq ≈ 27 pCi (picocuries).

Figure 24-21 Marie Curie with her daughter Irène. Both were pioneers in nuclear chemical research. Marie and her husband, Pierre, discovered radium and several other radioactive elements in the early 1900s. Irène and her husband, Frédéric Joliot, made the first artificial radioisotope, a process which has led to numerous applications in chemistry, biology, and medicine. Marie shared the 1903 Nobel Prize in physics with her husband and Henri Becquerel, and she received the Nobel Prize in chemistry singly in 1911. Irène Joliot-Curie shared the 1935 Nobel Prize in chemistry with her husband.

Table 24-5 Specific activities of important radioactive isotopes

Isotope	Half-life	Specific activity*/ disintegrations·$\text{s}^{-1}\cdot\text{g}^{-1}$	Specific activity/ $\text{Ci}\cdot\text{g}^{-1}$
radium-226	1600 years	3.7×10^{10}	1.00
uranium-238	4.47×10^{9} years	1.3×10^{4}	3.5×10^{-7}
plutonium-239	2.41×10^{4} years	2.3×10^{9}	6.2×10^{-2}
iodine-131	8.04 days	4.6×10^{15}	1.2×10^{5}
cobalt-60	5.27 years	4.2×10^{13}	1.1×10^{3}
strontium-90	29 years	5.1×10^{12}	1.4×10^{2}
cesium-137	30.2 years	3.2×10^{12}	87
carbon-14	5730 years	1.7×10^{11}	4.6

*The SI unit of the rate of radioactive decay is a becquerel (Bq), which is equal to one disintegration per second.

EXAMPLE 24-9: In reading about the use of uranium-238 in radiodating rocks, you may have wondered how it is possible to determine a half-life as long as 4.47×10^9 years. The answer lies in our ability to count individual radioactive decay events and in the fact that Avogadro's number is enormous. There are 6.022×10^{23} uranium-238 nuclei in a sample that contains 238 g of uranium-238. Calculate the number of uranium-238 nuclei that disintegrate in 10 s in a sample that contains 2.0 mg of uranium-238.

Solution: Using Equation (24-10), we find that the specific activity of uranium-238, which has a half-life of 4.47×10^9 years, is

$$\text{specific activity} = \frac{4.17 \times 10^{23}\ \text{disintegrations}\cdot\text{g}^{-1}}{(238)(4.47 \times 10^9\ \text{years})(3.15 \times 10^7\ \text{s}\cdot\text{year}^{-1})}$$
$$= 1.3 \times 10^4\ \text{disintegrations}\cdot\text{s}^{-1}\cdot\text{g}^{-1}$$

Thus, the number of disintegrations in 2.0 mg in 10 s is

$$(1.2 \times 10^4\ \text{disintegrations}\cdot\text{s}^{-1}\cdot\text{g}^{-1})(2.0 \times 10^{-3}\ \text{g})(10\ \text{s}) = 240\ \text{disintegrations}$$

Consequently, about 240 uranium-238 nuclei disintegrate in a 2.0-mg sample in only 10 s. It is easy with a modern radiocounting apparatus to measure such a large number of events in a 10-s interval. The half-life of uranium-238 can be measured by counting the number of disintegrations in a known mass of the radioisotope and then carrying out the reverse of the calculation just given with $t_{1/2}$ as the unknown, which is calculated from the measured specific activity.

PRACTICE PROBLEM 24-9: The half-life for iodine-131, which is used extensively in nuclear medicine (Table 24-5), is 8.04 days. Iodine-131 decays by β-particle and γ-ray emission. Calculate the specific activity and the total number of β-particles emitted per second by a sample that contains 2.00 mg of ^{131}I.

Answer: specific activity = 4.58×10^{15} disintegrations$\cdot g^{-1}\cdot s^{-1}$ and 9.16×10^{12} disintegrations per second for 2.00 mg of iodine-131

The damage that is produced by radiation depends on more than just the specific activity. As the radiation passes through tissue, it ionizes molecules and breaks chemical bonds, leaving behind a trail of molecular damage. The extent of the damage produced depends on the energy and type of radiation. The different types of radiation vary in their ability to penetrate matter. Alpha particles can be stopped by a sheet of paper or by the skin, but ingested α-emitters are especially dangerous, because of their high energy and ionizing ability. Beta particles penetrate deeper than α-particles, but β-particles of moderate energy are stopped by about 1 cm of water. Gamma rays are a very penetrating form of radiation, and walls of lead bricks are required to stop them. Neutrons are the most penetrating type of radiation and can cause extensive biological damage deep within the human body. The quantitative assessment of the dangers of low dosage levels of radiation is highly controversial and involves extensive political as well as scientific questions. Only further research can provide the definitive answers.

24-17. RADON IS A MAJOR HEALTH HAZARD

The existence of radioactive radon gas in the air in homes in the United States was brought sharply into focus in 1984 when an engineer at the Limerock nuclear power plant in Pennsylvania repeatedly triggered the plant's radioactivity detectors. The source of the radioactivity was found to be the engineer's home, which registered 2.7×10^3 picocuries per liter ($pCi{\cdot}L^{-1}$) of air. This radioactivity level is 675 times the Environmental Protection Agency's recommended maximum level of 4 $pCi{\cdot}L^{-1}$ of air. The radioactivity level of 4 $pCi{\cdot}L^{-1}$ corresponds to about nine particle emissions per minute per liter of air. **Radon** is a naturally occurring, radioactive noble gas formed in the radioactive decay of radium-226, which in turn arises from the radioactive decay of uranium ores. The normal background (i.e., natural) level of radon in outdoor air is about 0.2 $pCi{\cdot}L^{-1}$, and the average indoor level is about 2 $pCi{\cdot}L^{-1}$. Over 50 percent of natural background radiation results from radon decay. Indoor levels of radon are generally higher than outdoor levels, because radon enters a house through the ground and indoor airflow is much more restricted than outdoor airflow.

Three radon isotopes are produced in the decay of uranium ores—radon-219, radon-220, and radon-222. The half-life of ^{219}Ra and ^{220}Ra are of the order of a few seconds; hence, these two radioisotopes decay before they can leave the ground. In contrast, ^{222}Ra has a half-life of 3.8 days and is an α-emitter. The decay products of radon-222 are called radon daughters or progeny. Two of the ^{222}Ra progeny are polonium-218 and polonium-214, which are α-emitters with half-lives of 3.1 min and 2×10^{-4} s, respectively. Radon is a noble gas and is thus chemically inert and diffuses out of the ground in which it is formed. Similarly, inhaled radon is generally exhaled unless it decays to ^{218}Po and ^{214}Po while in the lungs. These polonium isotopes are fairly chemically reactive and bind to lung tissue. They emit α-particles, which damage cells and trigger the development of lung cancers. Although α-particles have a fairly short penetration range of about 70 μm, this distance is roughly twice the thickness of cell walls in the lungs. The U.S. Environmental Protection Agency (EPA) estimates that radon causes 5000 to 20,000 deaths from lung cancer annually in the United States. Radon is now recognized as a major cause of lung cancers, comparable in significance to smoking cigarettes, and especially deadly in combination. The presence of radon can be detected using commercially available test kits.

A home radon detector.

Provided the structure itself is not built from radioactive mine tailings or highly radioactive materials such as uranium-rich granites, it is usually not difficult to reduce radon levels within the structure. The basic procedure is to seal off the points of entry of radon gas into the structure. Common avenues of entry are cracks in the basement walls and floor, openings around pipes, and so forth. It is also advisable to avoid a tightly sealed structure that has a low level of outside airflow into the structure.

We can set the solubility, s, equal to $[Li^+]$ or $[F^-]$ and write

$$K_{sp} = s^2 = 3.8 \times 10^{-3}\ M^2$$

or

$$\begin{aligned} s &= 0.062\ M \\ &= (0.062\ mol{\cdot}L^{-1})\left(\frac{25.94\ g\ LiF}{1\ mol\ LiF}\right) \\ &= 1.61\ g{\cdot}L^{-1} \end{aligned}$$

PRACTICE PROBLEM 25-2: The solubility-product constant of potassium periodate in water is $8.3 \times 10^{-4}\ M^2$ at 25°C. Calculate the solubility of $KIO_4(s)$ in grams per liter.

Answer: $6.6\ g{\cdot}L^{-1}$

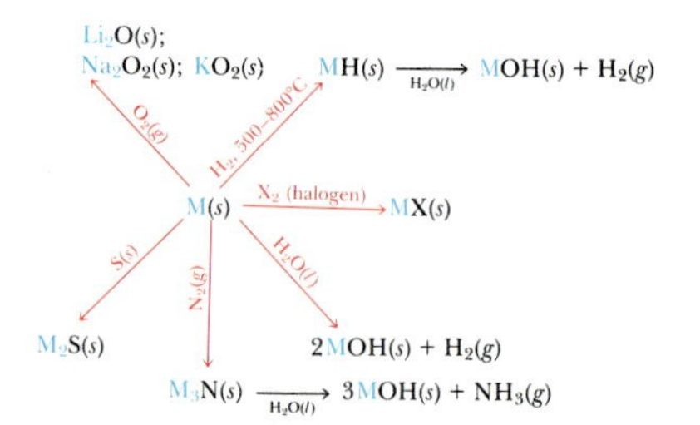

Figure 25-10 Representative reactions of the Group 1 metals.

The reactions of alkali metals are summarized in Figure 25-10; and other important alkali metal compounds and their uses are listed in Table 25-9.

Table 25-9 Commercially important alkali metal compounds and their uses

Compound	Uses
lithium aluminum hydride, $LiAlH_4(s)$	production of many pharmaceuticals and perfumes
lithium borohydride, $LiBH_4(s)$	organic synthesis
lithium carbonate, $Li_2CO_3(s)$	to treat schizophrenia
sodium hydrogen carbonate (sodium bicarbonate), $NaHCO_3(s)$	manufacture of effervescent salts and beverages, baking powder, gold plating
sodium carbonate, $Na_2CO_3(s)$	manufacture of glass, pulp and paper, soaps and detergents
sodium hydroxide, $NaOH(s)$	production of rayon, cellulose, oven cleaner
sodium sulfate decahydrate (Glauber's salt), $Na_2SO_4{\cdot}10H_2O(s)$	solar heating storage, air conditioning
sodium cyanide, $NaCN(s)$	extraction of gold and silver from ores; electroplating solutions; fumigant for fruit trees
potassium carbonate (potash), $K_2CO_3(s)$	manufacture of special glass for optical instruments, soft soaps
potassium nitrate, $KNO_3(s)$	pyrotechnics, explosives, matches; tobacco treatment
dipotassium hydrogen phosphate, $K_2HPO_4(s)$	buffering agent

25-4. THE CHEMICAL PROPERTIES OF THE ALKALINE EARTH METALS

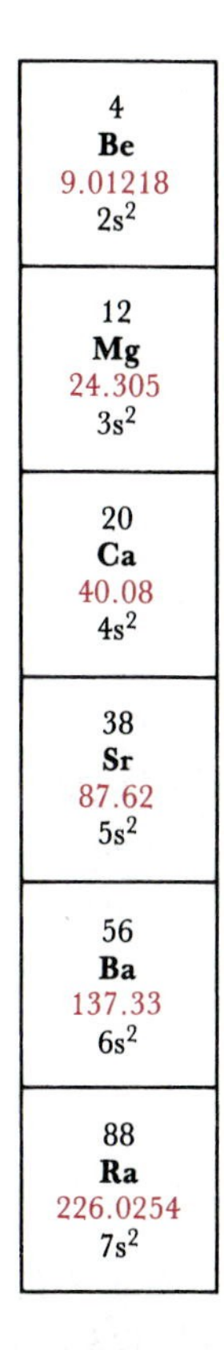

The Group 2 elements (beryllium, magnesium, calcium, strontium, barium, and radium) are called the **alkaline earth metals** (Figure 25-11). They are not as reactive as the Group 1 metals, but they are much too reactive to be found in the free state in nature. They have a ground-state electron configuration [noble gas]ns^2 and attain a noble-gas electron configuration by the loss of the two electrons from the outermost s orbital.

$$M\{[\text{noble gas}]ns^2\} \rightarrow M^{2+}[\text{noble gas}] + 2e^-$$

Beryllium is a relatively rare element but occurs as localized surface deposits of the mineral *beryl* (Figure 25-12). Essentially unlimited quantities of magnesium are readily available in seawater, where $Mg^{2+}(aq)$ occurs at a concentration of 0.054 M. Calcium, strontium, and barium rank 5th, 18th, and 19th in abundance in the earth's crust, occurring primarily as carbonates and sulfates (Table 25-10). All isotopes of radium are radioactive, with the longest-lived one (Ra-226) having a half-life of 1600 years.

The chemistry of the Group 2 elements involves primarily the metals and the +2 ions. With few exceptions, the reactivity of the Group 2 elements increases from beryllium to barium. As in all the s-block and p-block groups, the first member of the family differs in several respects from the other members of the family. The anomalous properties of beryllium are attributed to the very small ionic radius of Be^{2+}. The radius of Be^{2+} is similar to that of Al^{3+}, and beryllium(II) has some chemical properties like those of aluminum(III).

The atomic and physical properties of the Group 2 elements are given in Table 25-11. The periodic trends in the atomic properties of the Group 2 elements are shown clearly in the data in Table 25-11, except for radium, which in some cases appears anomalous. As we go down the group, the ionization energy and the electronegativity decrease, whereas the atomic radii and ionic radii increase. These trends are a direct consequence of the increase in size of the atoms and ions with increase in atomic number.

Figure 25-12 The mineral beryl, $Be_3Al_2Si_6O_{18}$, occurs in light-green hexagonal prisms. Beryl is the chief source of beryllium and is also used as a gem.

Figure 25-11 The Group 2 elements. Top row: beryllium, magnesium, and calcium. Bottom row: strontium and barium.

Table 25-10 Major sources and uses of the alkaline earth metals

Metal	Sources	Uses
beryllium	beryllium aluminum silicates, including *beryl*, $Be_3Al_2Si_6O_{18}$	lightweight alloys (improves corrosion resistance and resistance to fatigue and temperature changes); gyroscopes; nuclear reactors (absorbs neutrons)
magnesium	*dolomite*, $CaMg(CO_3)_2$; carbonates and silicates; seawater and well brines	alloys for airplanes; flashbulbs; pyrotechnics; corrosion protection for metals
calcium	*limestone*, $CaCO_3$; *gypsum*, $CaSO_4{\cdot}2H_2O$; *apatite*, $Ca_{10}(OH)_2(PO_4)_6$ (major constituent of tooth enamel)	alloys; production of chromium and other metals
strontium	*celestite*, $SrSO_4$; *strontianite*, $SrCO_3$	alloys
barium	*witherite*, $BaCO_3$; *barite*, $BaSO_4$	lubricant on rotors of anodes in vacuum X-ray tubes; spark-plug alloys
radium	*pitchblende* and *carnotite* ores	skin cancer treatments

25-4A. The Small Size of Be^{2+} Makes the Chemistry of Beryllium Different from That of the Other Group 2 Metals

Beryllium metal is steel gray, light, very hard, and high-melting. Like lithium, it differs significantly in its chemistry from the other elements in its group. Because of the small size of the Be^{2+} ion, all beryllium(II) compounds involve appreciable covalent bonding, and there are no crystalline compounds or solutions involving Be^{2+} as such. The other Group 2 metals have larger sizes and lower ionization energies, making them more electropositive than beryllium. As a consequence, the ionic

Table 25-11 Properties of the Group 2 elements

Element	Number of naturally occurring isotopes	Metal radius/pm	Ionic radius of M^{2+}/pm	Sum of first and second ionization energies of M(g)/ $kJ{\cdot}mol^{-1}$
beryllium	1	110	31	2656
magnesium	3	160	65	2187
calcium	6	190	99	1734
strontium	4	210	113	1608
barium	7	220	135	1462
radium	4	225	150	~1480

nature of the compounds of the alkaline earth metals increases as one moves down a group in the periodic table. Beryllium and its compounds are highly toxic, especially when their dusts are inhaled. The free element is prepared on a commercial scale by electrolysis of the halides and by reduction of BeF_2 with magnesium.

Beryllium metal is fairly unreactive at room temperature. Hot (400°C) beryllium metal reacts with oxygen to form the oxide, $BeO(s)$; with nitrogen to form the nitride, $Be_3N_2(s)$; and with halogens to form the halides, BeX_2. Some of the more common reactions of beryllium are diagramed in Figure 25-13.

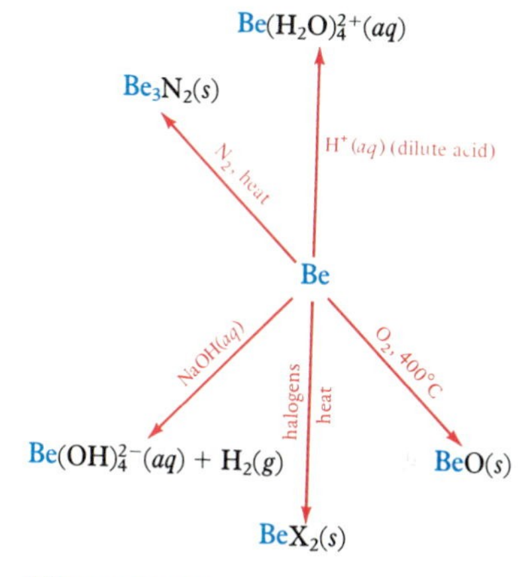

Figure 25-13 Representative reactions of beryllium.

25-4B. Magnesium, Calcium, Strontium, and Barium Form Ionic Compounds Involving M^{2+} Ions

Magnesium, calcium, strontium, and barium are prepared by high-temperature electrolysis of the molten chloride; for example,

$$CaCl_2(l) \xrightarrow[\text{high T}]{\text{electrolysis}} Ca(l) + Cl_2(g)$$

The metals magnesium, calcium, strontium, barium, and radium are silvery white in appearance when freshly cut but tarnish readily in air as a result of the formation of the metal oxide on the surface. The free metals have limited commercial use (Table 25-10). The metals are highly electropositive and readily form M^{2+} ions. The alkaline earth $M^{2+}(aq)$ ions are neutral (in an acid-base sense) in aqueous solution.

Beryllium and magnesium react slowly with water at ordinary temperatures, although hot magnesium reacts violently with water. The other alkaline earth metals react more rapidly with water, but the rates of these reactions are still much slower than those for the alkali metals.

The alkaline earth metals burn in oxygen to form oxides having the formula MO, which are ionic solids. Magnesium is used as an incendiary in warfare because of its vigorous reaction with oxygen. It burns even more rapidly when sprayed with water and reacts with carbon dioxide via the reaction

$$2Mg(s) + CO_2(g) \rightarrow 2MgO(s) + C(s)$$

Table 25-11 ***Continued***

Pauling electronegativity	Melting point/°C	Boiling point/°C	Density at 25°C/ g·cm^{-3}	ΔH_{fus}/ kJ·mol^{-1}	ΔH_{vap}/ kJ·mol^{-1}
1.5	1278	2970	1.85	9.8	140
1.2	651	1107	1.74	9.2	95
1.0	845	1487	1.55	9.1	80
1.0	769	1384	2.54	8.2	76
0.9	725	1740	3.51	7.5	65
0.9	~700	~1740	6.0	8.0	110

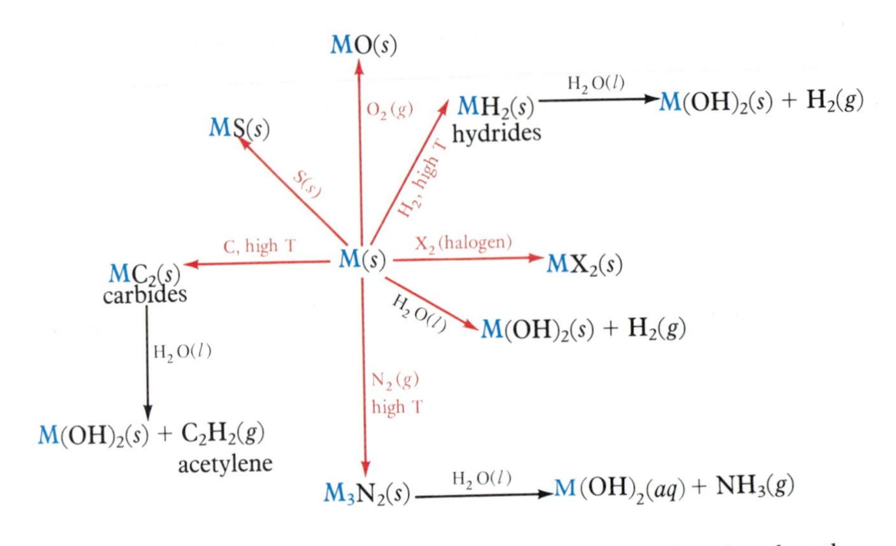

Figure 25-14 Representative reactions of the Group 2 metals.

Covering burning magnesium with sand slows the combustion, but the molten magnesium reacts with silicon dioxide (the principal component of sand) to form magnesium oxide:

$$2Mg(l) + SiO_2(s) \rightarrow 2MgO(s) + Si(s)$$

Magnesium wool, which has a large surface area, is used in flashbulbs. The brilliant flash is produced by the reaction of magnesium with oxygen.

Like the alkali metals, the alkaline earth metals show an increasing tendency to form peroxides with increasing size. Strontium peroxide, SrO_2, is formed at high oxygen pressure; and barium peroxide, BaO_2, forms readily in air at 500°C.

Except for beryllium, the alkaline earth metals react vigorously with dilute acids:

$$Mg(s) + 2HCl(aq) \rightarrow MgCl_2(aq) + H_2(g)$$

Beryllium reacts slowly with dilute acids.

The alkaline earth metals magnesium, calcium, strontium, and barium react with most of the nonmetals to form ionic binary compounds. Some common reactions are summarized in Figure 25-14.

Figure 25-15 Stalactites and stalagmites are produced when calcium carbonate precipitates from ground water. Shown here is the Powerhouse Cave in West Virginia.

25-4C. Many Alkaline Earth Metal Compounds Are Important Commercially

Magnesium sulfate heptahydrate, $MgSO_4 \cdot 7H_2O$, known as **Epsom salt,** is a cathartic, or purgative. The name Epsom comes from the place where the compound was first discovered in 1695, a natural spring in Epsom, England. Magnesium hydroxide is only slightly soluble in water, and suspensions of it are sold as the antacid **milk of magnesia.**

Calcium is an essential constituent of bones and teeth, plants, egg shells, the shells of marine organisms, and limestone (Figure 25-15). The ion Ca^{2+} plays a major role in muscle contraction, vision, and nerve excitation. Calcium oxide, or **quicklime,** is the sixth-ranked industrial chemical; over 30 billion pounds are produced annually in the United States. It is made by heating limestone:

$$CaCO_3(s) \rightarrow CaO(s) + CO_2(g)$$

Table 25-12 Important compounds of the Group 2 elements

Compound	Uses
beryllium oxide, $BeO(s)$	nuclear reactor fuel moderator; electrical insulator
magnesium oxide, $MgO(s)$	talcum powder; component of fire bricks; optical instruments
magnesium perchlorate, $Mg(ClO_4)_2(s)$	desiccant
calcium hydrogen sulfite, $Ca(HSO_3)_2(s)$	germicide, preservative, disinfectant; beer manufacture
calcium carbonate, $CaCO_3(s)$	antacid in wine-making; manufacture of pharmaceuticals
calcium chloride, $CaCl_2(s)$	de-icer on roads; to keep dust down on dirt roads; fire extinguishers
calcium hypochlorite, $Ca(OCl)_2(s)$	bleaching powder; sugar refining; algicide
strontium nitrate, $Sr(NO_3)_2(s)$	signal flares (red)
strontium sulfide, $SrS(s)$	luminous paints
barium carbonate, $BaCO_3(s)$	rat poison
barium nitrate, $Ba(NO_3)_2(s)$	pyrotechnics (green flame); signal flares

Figure 25-16 A red signal flare. The red color comes from light emitted by electronically excited strontium atoms.

Calcium oxide is mixed with water to form calcium hydroxide, also called **slaked lime,** which is used to make cement, mortar, and plaster. **Plaster of Paris** is $CaSO_4{\cdot}\frac{1}{2}H_2O$, which combines with water to form gypsum:

$$\underset{\text{plaster of Paris}}{CaSO_4{\cdot}\tfrac{1}{2}H_2O(s)} + \tfrac{3}{2}H_2O(l) \rightarrow \underset{\text{gypsum}}{CaSO_4{\cdot}2H_2O(s)}$$

Asbestos is a calcium magnesium silicate, a naturally occurring mineral, with the approximate composition $CaMg_3(SiO_3)_4$. It can resist very high temperatures, but because small asbestos fibers are a confirmed carcinogen, it is being phased out as a construction material.

Strontium salts produce a brilliant red flame and are used in signal flares and fireworks (Figure 25-16). The radioactive isotope strontium-90, which is produced in atomic bomb explosions, is a major health hazard because it behaves like calcium and is incorporated into bone, thereby causing various cancers.

Some commercially useful compounds of the Group 2 elements are listed in Table 25-12.

25-5. THE CHEMICAL PROPERTIES OF THE GROUP 3 ELEMENTS

5 **B** 10.81 $2s^22p^1$
13 **Al** 26.98154 $3s^23p^1$
31 **Ga** 69.72 $4s^24p^1$
49 **In** 114.82 $5s^25p^1$
81 **Tl** 204.37 $6s^26p^1$

The Group 3 elements are boron, aluminum, gallium, indium, and thallium. Boron is a semimetal, and the other members of the series are metals, with the metallic character of the elements increasing as we descend the group in the periodic table. The electron configuration of the members of the group is [noble gas]ns^2np^1, so the common oxidation states of the Group 3 elements are 0 and +3. The increasing ten-

Table 25-13 Properties of the Group 3 elements

Element	Number of naturally occurring isotopes	Metal radius/pm	Ionic radius of M^{3+}/pm	Sum of the first three ionization energies of M(g)/ $kJ \cdot mol^{-1}$
boron	2	85	20	6886
aluminum	1	125	50	5137
gallium	2	130	62	5520
indium	2	155	81	5063
thallium	2	190	95*	5415

*Ionic radius of Tl^{+}: 144 pm.

dency on descending a group column to have an oxidation state that is two less than the maximum possible value first appears in Group 3, where In^{+} and Tl^{+} are significant oxidation states of indium and thallium, respectively.

The chemistry of boron, the first member of the group, differs in many respects from that of the rest of the group. Boron has a metallic luster but behaves more like the semimetal silicon than like the metal aluminum.

As in Groups 1 and 2, atomic and ionic radii and density increase on descending the group (Table 25-13). The enthalpies of fusion (except that for thallium) and vaporization and the boiling point also decrease on descending the group (Table 25-13). These trends are a direct consequence of the increase in size and mass with increase in atomic number.

The major sources and commercial uses of the Group 3 elements are given in Table 25-14.

Table 25-14 Major sources and uses of the Group 3 metals

Metal	Sources	Uses
boron	*kernite,* $Na_2B_4O_7 \cdot 4H_2O$ *borax,* $Na_2B_4O_7 \cdot 10H_2O$	in instruments used for absorbing and detecting neutrons; hardening agent in alloys
aluminum	*bauxite,* AlO(OH); clays	in aircraft and rockets, utensils, electrical conductors, photography, explosives, fireworks, paint, building decoration, and telescope mirrors
gallium	trace impurity in bauxite and zinc and copper minerals	high-temperature heat-transfer fluid
indium	by-product in lead and zinc production	low-melting alloys in safety devices, sprinklers
thallium	by-product from production of other metals	no significant commercial uses

Table 25-13 *Continued*

Pauling electro-negativity	Melting point/°C	Boiling point/°C	Density at 25°C/ $g \cdot cm^{-3}$	ΔH_{fus}/ $kJ \cdot mol^{-1}$	ΔH_{vap}/ $kJ \cdot mol^{-1}$
1.9	2180	~3650	2.35	23.6	505
1.5	660	2467	2.70	10.5	291
1.6	30	2250	5.90	5.6	270
1.7	157	2070	7.30	3.3	232
1.8	304	1457	11.85	4.3	166

25-5A. The Bonding in Boron Compounds Is Covalent

Boron(III) is always covalently bonded; boron forms no simple cations of the type B^{3+}. For example, the boron trihalides (BX_3) are trigonal planar molecules with X—B—X bond angles of 120°, as predicted by VSEPR theory.

The boron trihalides react with water to form boric acid, $B(OH)_3$, and the hydrohalic acid. For example,

$$BCl_3(g) + 3H_2O(l) \rightarrow B(OH)_3(s) + 3H^+(aq) + 3Cl^-(aq)$$

Boric acid is usually made by adding hydrochloric acid or sulfuric acid to **borax,** $Na_2B_4O_7 \cdot 10H_2O(s)$:

$$Na_2B_4O_7 \cdot 10H_2O(s) + 2HCl(aq) \rightarrow 4B(OH)_3(aq) + 5H_2O(l) + 2NaCl(aq)$$

Boric acid is a moderately soluble monoprotic weak acid in water; its formula is usually written as $B(OH)_3$ rather than H_3BO_3 or $HBO(OH)_2$, because it acts as a Lewis acid by accepting a hydroxide ion rather than by donating a proton (Section 18-12):

$$B(OH)_3(aq) + H_2O(l) \rightleftharpoons B(OH)_4^-(aq) + H^+(aq) \qquad pK_a = 9.23 \text{ at } 25°C$$

Aqueous solutions of boric acid are used in mouth and eye washes. Borax, which is found in large deposits in certain desert regions of California, was known and used thousands of years ago to glaze pottery (Figure 25-17).

The principal oxide of boron, $B_2O_3(s)$, is obtained by heating boric acid:

$$2B(OH)_3(s) \rightarrow B_2O_3(s) + 3H_2O(g)$$

Boron trioxide, commonly known as boric oxide, is a colorless, vitreous substance that is extremely difficult to crystallize. It is fused with $SiO_2(s)$ (sand) and $Na_2CO_3(s)$ to make heat-resistant glassware such as Pyrex and as a fire-resistant additive for paints.

Figure 25-17 Aerial view of a mine in Boron, California, where massive deposits of borax are located.

EXAMPLE 25-3: Calculate the pH of a 0.10 M $B(OH)_3(aq)$ solution at 25°C.

Solution: The chemical equation is

$$B(OH)_3(aq) + H_2O(l) \rightleftharpoons H^+(aq) + B(OH)_4^-(aq) \qquad pK_a = 9.3$$

deposits of **bauxite,** AlO(OH) (Figure 25-21), which is the chief source of aluminum. The bauxite is first converted to aluminum oxide, $Al_2O_3(s)$ which is electrolyzed to produce aluminum.

Figure 25-21 Bauxite, a reddish brown ore, is the principal source of aluminum. Bauxite is heated with coke to produce aluminum oxide, a white powder (the white pellets shown). The oxide is further refined through electrolysis to produce aluminum.

Originally the Al_2O_3 was dissolved in molten *cryolite,* Na_3AlF_6, because the Al_2O_3-Na_3AlF_6 mixture melts at a relatively low temperature, but insufficient naturally occurring quantities of cryolite exist for world production of aluminum, so it is now manufactured by reacting $Al_2O_3(s)$ with HF and NaOH in a lead vessel according to

$$12HF(g) + Al_2O_3(s) + 6NaOH(s) \rightarrow \underset{\text{cryolite}}{2Na_3AlF_6(s)} + 9H_2O(g)$$

The cryolite is used to obtain aluminum metal by the **Hall process** (Chapter 23), in which a molten mixture of cryolite, together with CaF_2 and NaF, is electrolyzed at 800–1000°C. The other Group 3 metals are also obtained by electrolysis of the appropriate molten halide salt or by electrolysis of aqueous solutions of their salts.

Aluminum is a light, soft metal that resists corrosion by forming a tough, adherent protective layer of the oxide, Al_2O_3. Structural alloys of aluminum for aircraft and automobiles contain silicon, copper, magnesium, and other metals, which increase the strength and stiffness of the aluminum.

Gallium, indium, and thallium are soft, silvery white metals. Gallium melts at a temperature lower than body temperature and has the widest liquid range (2220°C) of any known substance. Indium is soft enough to find use as a metallic O-ring material in metal high-vacuum fittings.

25-5D. The Group 3 Oxides Become Increasingly Basic on Descending the Group

The Group 3 metals aluminum, gallium, indium, and thallium exhibit two important trends that are also found to various degrees in Groups 4, 5, 6, and 7. Two trends are found in Group 3 metals:

1. As mass increases, a higher oxidation state is less stable than a lower oxidation state. For the Group 3 metals with greater mass, the stability of the M(III) state is reduced relative to that of the M(I) state. Although the trivalent state is important for all Group 3 metals, Tl(I) is also an important oxidation state in the chemistry of thallium.
2. As mass increases, the metallic character increases for identical oxidation states. This increase is illustrated by an increase in the basicity of the oxides.

Aluminum and gallium are amphoteric; they dissolve in both strong aqueous acids and strong aqueous bases:

$$2Al(s) + 6H^+(aq) \rightarrow 2Al^{3+}(aq) + 3H_2(g)$$

$$2Al(s) + 6H_2O(l) + 2OH^-(aq) \rightarrow \underset{\text{aluminate ion}}{2Al(OH)_4^-(aq)} + 3H_2(g)$$

The reaction of aluminum with concentrated aqueous sodium hydroxide is used in the commercial drain cleaner Drāno. The heat melts the grease and the gas evolved in the reaction agitates the solid materials blocking the drain.

Figure 25-22 The elements indium (small shiny piece at the left) and thallium (sliced rod showing lustrous metal. The tarnish is due to reaction with air).

The hydroxides and oxides of aluminum and gallium are also amphoteric. For example,

$$Al(OH)_3(s) + OH^-(aq) \rightarrow Al(OH)_4^-(aq)$$
$$Al(OH)_3(s) + 3H^+(aq) \rightarrow Al^{3+}(aq) + 3H_2O(l)$$

Indium and thallium (Figure 25-22) react with aqueous solutions of strong acids, but they are unaffected by strong bases. The oxides and hydroxides of indium and thallium are not amphoteric but basic.

Compounds of the Group 3 metals exhibit both ionic and covalent bonding; however, ionic bonding is somewhat favored. All four metals react with halogens to form compounds with the empirical formula MX_3. The MX_3 fluorides are ionic, whereas the chlorides, bromides, and iodides are low-melting compounds that are dimeric in the vapor state. Note the similarity of the halide-bridge structure of the dimer Al_2Cl_6 shown in Figure 25-23 to that of diborane (Figure 25-19).

The salts $LiAlH_4$ and $LiGaH_4$, which contain the tetrahedral hydride ions MH_4^-, can be prepared from the respective halides and lithium hydride, LiH. For example,

$$4LiH(soln) + AlCl_3(soln) \xrightarrow{\text{ether}} LiAlH_4(soln) + 3LiCl(s)$$

These hydrides are useful reducing agents in numerous aprotic solvents but are violently decomposed by water and occasionally ignite or explode on contact with air.

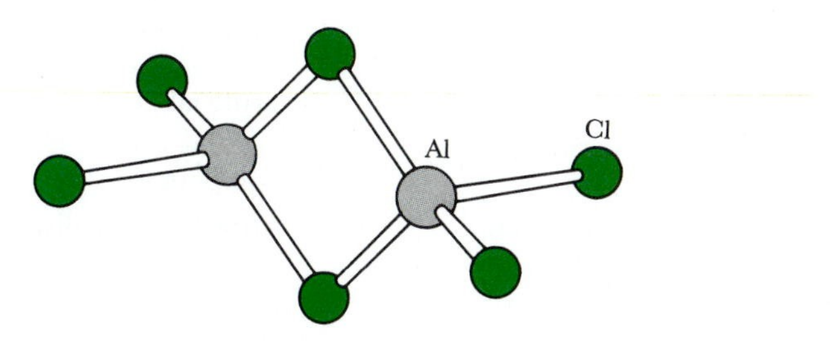

Figure 25-23 The structure of $Al_2Cl_6(g)$. Compare with the structure of diborane (Figure 25-19).

EXAMPLE 25-4: Sodium borohydride, $NaBH_4$, and lithium aluminum hydride, $LiAlH_4$, react with aldehydes or ketones to produce alcohols

$$\underset{\text{formaldehyde}}{H_2CO} \xrightarrow{NaBH_4} \underset{\text{methanol}}{CH_3OH}$$

Determine the oxidation number of the carbon atom in formaldehyde and methanol and thus show that formaldehyde has been reduced to methanol.

Solution: According to the rules given in Section 4-1, we assign the hydrogen atoms an oxidation state of +1 and the oxygen atom an oxidation state of −2. The oxidation state of the carbon atom in each molecule is given by

formaldehyde $\quad 2(+1) + (-2) + x = 0, \quad \text{or } x = 0$

methanol $\quad 4(+1) + (-2) + x = 0, \quad \text{or } x = -2$

Thus, the oxidation state of the carbon atom has been changed from 0 to −2, so we say that the formaldehyde has been reduced to methanol.

PRACTICE PROBLEM 25-4: Unlike sodium borohydride, which reduces only aldehydes and ketones, lithium aluminum hydride rapidly reduces carboxylic acids as well; for example,

$$\underset{\text{formic acid}}{HCOOH} \xrightarrow[H_2O]{LiAlH_4} \underset{\text{methanol}}{CH_3OH} \qquad \text{(not balanced)}$$

Show, using oxidation state changes, that the formic acid has been reduced to methanol.

Answer: The oxidation state of carbon goes from +2 to −2.

Table 25-15 lists some commercially important Group 3 compounds.

Table 25-15 Important compounds of the Group 3 elements

Compound	Uses
boron trioxide, $B_2O_3(s)$	heat-resistant glassware (Pyrex); fire retardant
boron carbide, $B_4C(s)$	abrasive; wear-resistant tools
boron nitride, $BN(s)$	lubricant; refractory, nose cone windows; cutting tools
aluminum ammonium sulfate, $Al(NH_4)(SO_4)_2(s)$	purification of drinking water; soil acidification
aluminum oxide (alumina), $Al_2O_3(s)$	manufacture of abrasives, refractories, ceramics, spark plugs; artificial gems
aluminum borohydride, $Al(BH_4)_3(s)$	reducing agent; rocket fuel component
aluminum hydroxychloride, $AlOHCl_2(s)$	antiperspirant; disinfectant
gallium arsenide, $GaAs(s)$	semiconductors for use in transistors and solar cells
thallium(I) sulfate, $Tl_2SO_4(s)$	rat and ant poison

Figure 25-24 The Group 4 elements. From left to right: carbon (as graphite), silicon, germanium, tin, lead.

25-6. THE CHEMICAL PROPERTIES OF THE GROUP 4 ELEMENTS

The Group 4 elements—carbon, silicon, germanium, tin, and lead—are among the most abundant elements in nature and the most important in industry (Figure 25-24). Carbon is widely distributed in nature both as the free element and in compounds. The great majority of naturally occurring carbon occurs in coal, petroleum, limestone ($CaCO_3$), *dolomite* [$MgCa(CO_3)_2$], and a few other deposits. Carbon is also a principal element in all living matter, and the study of its compounds forms the vast fields of organic chemistry and biochemistry.

6 C 12.011 $2s^22p^2$
14 Si 28.0855 $3s^23p^2$
32 Ge 72.59 $4s^24p^2$
50 Sn 118.69 $5s^25p^2$
82 Pb 207.2 $6s^26p^2$

Silicon constitutes 28 percent of the mass of the earth's mantle and is the second most abundant element in the mantle, exceeded only by oxygen. Silicon does not occur as the free element in nature; it occurs primarily as the oxide and in numerous silicates. Most sands are primarily silicon dioxide, and many rocks and minerals are silicate materials. Germanium and tin rank in the range 40th to 50th in elemental abundance. The presence of small amounts of germanium in coal deposits serves as a commercial source of that element. The most important source of tin is the mineral **cassiterite,** SnO_2, from which tin is obtained by reduction with coke. Lead is the most abundant of the heavy metals, its most important ore being **galena,** PbS. The principal sources and commercial uses of the Group 4 elements are given in Table 25-16. Table 25-17 lists some of the properties of the Group 4 elements.

These elements provide the best example of a group in which the first member has properties different from those of the rest. Carbon is decidedly nonmetallic, and the other members of the group become increasingly metallic as atomic number increases. The properties of the remaining members vary more smoothly from silicon to lead: silicon and germanium are semimetals; tin and lead are metals. As in Group 3, there is a tendency for the elements with higher atomic numbers to exhibit an oxidation state of two less than the maximum of +4. The common oxidation state of silicon and germanium is +4, but lead and, to some extent, tin have a common oxidation state of +2.

Table 25-16 Sources and uses of the Group 4 elements

Element	Principal sources	Uses
carbon	coal and petroleum	fuels; production of iron, furnace linings, electrodes (coke); lubricant, fibers, pencils, airframe structures (graphite); decolorizer, air purification, catalyst (activated charcoal); rubber and printing inks (carbon black); drill bits, abrasives (diamond)
silicon	*quartzite* or sand (SiO_2)	steel alloys; silicones; semiconductor in integrated circuits, rectifiers, transistors, solar batteries
germanium	coal ash; by-product of zinc refining	solid-state electronic devices; alloying agent
tin	*cassiterite* (SnO_2)	food packaging; tin plate; pewter; bronze; soft solder
lead	*galena* (PbS)	storage batteries; solder and low-melting alloys; type metals; ballast; lead shot; cable covering

25-6A. Diamond and Graphite Are Allotropic Forms of Carbon

Figure 25-25 Synthetic diamonds.

As discussed in Section 14-10, solid carbon displays two important allotropic forms, diamond and graphite. Recall that diamond has an extended, covalently bonded tetrahedral structure and that graphite has a layered structure. Because diamond is the hardest naturally occurring substance known, it is extensively used as an abrasive and cutting material where a very high resistance to wear is required. Graphite, however, is the stable form of carbon at ordinary temperatures and pressures. It exists as a solid at a higher temperature than any other material known (4100 K).

To produce diamonds (Figure 25-25), it is necessary to subject graphite to a pressure and temperature that lie above the graphite-diamond

Table 25-17 Properties of the Group 4 elements

Element	Number of naturally occurring isotopes	Atomic radius/pm	Ionization energy/MJ·mol^{-1} First	Second	Third	Fourth
carbon	3	70	1.09	2.35	4.62	6.22
silicon	3	110	0.786	1.57	3.23	4.36
germanium	5	125	0.761	1.53	3.30	4.41
tin	10	145	0.708	1.41	2.94	3.93
lead	4	180	0.715	1.45	3.08	4.08

equilibrium line in Figure 25-26. For example, at 300 K, a pressure greater than 15,000 atm is required. Note that high pressure favors the solid form with the higher density (smaller molar volume). However, even at 300 K and 15,000 atm, the rate of conversion of graphite to diamond is extremely slow. The first successful synthesis of diamonds from graphite was carried out at General Electric Laboratories in 1955. At 2500 K and 150,000 atm, essentially complete conversion of graphite to diamond occurs in a few minutes. A rapid decrease in the pressure and temperature traps the carbon in the diamond form.

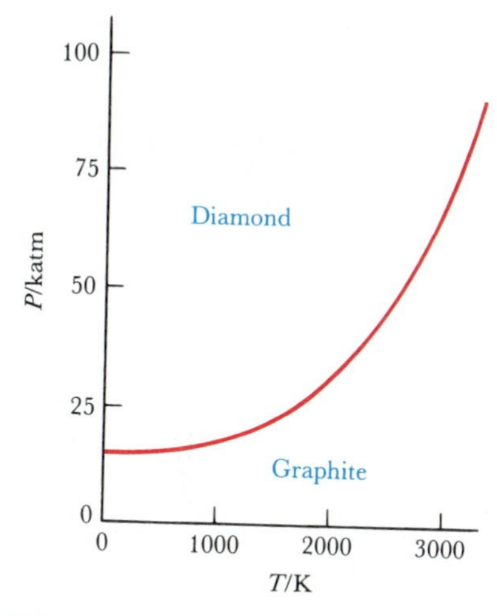

Figure 25-26 Graphite-diamond phase diagram. Above the curve, diamond is the stable form of solid carbon; below the curve, graphite is the stable form. At 25°C the equilibrium transition pressure for the conversion of graphite to diamond is 15,000 atm. Diamond has a higher density than graphite, so diamond is the stable form of C(*s*) at high pressures. Diamonds at 1 atm are unstable with respect to graphite, but the rate of conversion of diamonds to graphite at 25°C is completely negligible.

EXAMPLE 25-5: Use Le Châtelier's principle and the data given in the caption to Figure 25-26 to predict whether graphite or diamond is the stable allotropic form of carbon at high pressure.

Solution: Consider the equation

$$\mathrm{C}(\textit{diamond}) \rightleftharpoons \mathrm{C}(\textit{graphite})$$

The density of graphite is 2.2 $g \cdot cm^{-3}$, so its specific volume is 0.45 $cm^3 \cdot {}^{-1}$; for diamond, the density is 3.5 $g \cdot cm^{-3}$ and its specific volume is 0.29 $cm^3 \cdot g^{-1}$. Le Châtelier's principle says that the form having the smaller specific volume is favored, so we predict (correctly) that diamond is the stable form at high pressure (see Figure 25-26).

PRACTICE PROBLEM 25-5: Calcite and aragonite are two different crystal modifications of calcium carbonate, $CaCO_3(s)$. The densities of the two forms are

$$\text{calcite, } d = 2.71\ g \cdot cm^{-3}\text{; aragonite, } d = 2.94\ g \cdot cm^{-3}$$

Which form is more stable at high pressures? The calcite-aragonite phase transition has been proposed as a geobarometer, that is, as an indicator of the minimum pressure to which the $CaCO_3(s)$ was subjected.

Answer: aragonite

Table 25-17 ***Continued***

Pauling electro-negativity	Melting point/°C	Boiling point/°C	Density at 20°C/ $g \cdot cm^{-3}$	ΔH_{fus}/ $kJ \cdot mol^{-1}$	ΔH_{vap}/ $kJ \cdot mol^{-1}$
2.5	Sublimes at ~3900		2.27 (graphite) 3.51 (diamond)	105 (ΔH_{sub})	—
1.8	1410	~3000	2.33	50.2	359
1.8	940	2850	5.32	36.8	328
1.8	232	2620	7.28 (white)	6.99	296
1.8	328	1750	11.34	4.77	179

Figure 25-27 The flask on the left contains nitrogen dioxide, $NO_2(g)$, and the dish in front contains activated carbon, a black powder. The flask on the right shows the result of adding the activated carbon to the flask containing $NO_2(g)$. Note that the $NO_2(g)$ has been completely absorbed.

Another important form of carbon is **activated carbon,** an amorphous form characterized by its ability to absorb large quantities of gases (Figure 25-27). Activated carbon can be produced by heating wood or other carbonaceous material to 800–900°C with steam or carbon dioxide. This process results in a porous product with a honeycomb structure. The ability of activated carbon to absorb large quantities of gases is due to its very large surface area, which averages about 1000 square meters per gram.

25-6B. Carbon Ranks Second Among the Elements in the Number of Compounds Formed

Carbon forms more compounds than any other element except hydrogen. Most of these compounds are classified as organic compounds, which are discussed in Chapter 29. Although the classification of compounds into inorganic compounds and organic compounds is artificial, we will discuss a few of the important "inorganic" compounds of carbon here.

Binary compounds in which carbon is combined with less electronegative elements are called carbides. One of the most important carbides is calcium acetylide, which more commonly is called calcium carbide. It is produced industrially by the reaction of lime (CaO) and coke:

$$CaO(s) + 3C(s) \xrightarrow{2000°C} CaC_2(s) + CO(g)$$

Calcium carbide is a gray-black, hard solid with a melting point over 2000°C. Calcium carbide reacts with water to produce acetylene:

$$CaC_2(s) + 2H_2O(l) \rightarrow C_2H_2(g) + Ca(OH)_2(s)$$

At one time, this reaction represented one of the major sources of acetylene for the chemical industry and for oxyacetylene welding, but its use has declined because of the high cost of the energy used in the production of the calcium carbide. Acetylene is now produced from petroleum and natural gas.

EXAMPLE 25-6: Cyanogen, $C_2N_2(g)$, is a very poisonous gas that can be made by mixing warm concentrated solutions of potassium cyanide and copper(II) sulfate:

$$2Cu^{2+}(aq) + 4CN^-(aq) \rightarrow 2CuCN(s) + C_2N_2(g)$$

Write the Lewis formula of cyanogen.

Solution: There are $(2 \times 4) + (2 \times 5) = 18$ valence electrons. If we place the two carbon atoms in central positions, then

$$:N{\equiv}C{-}C{\equiv}N:$$

PRACTICE PROBLEM 25-6: Finely powdered calcium carbide will react with nitrogen at high temperature and at about two atmospheres pressure to form calcium cyanamide, $CaCN_2(s)$, according to

$$CaC_2(s) + N_2(g) \rightarrow CaCN_2(s) + C(s)$$

Write the Lewis formula for a cyanamide ion.

Answer: $^{\ominus}\ddot{\underset{\cdot\cdot}{N}}{=}C{=}\ddot{\underset{\cdot\cdot}{N}}^{\ominus}$

One other industrially important carbide is silicon carbide, SiC, also known as **carborundum.** Carborundum is one of the hardest known materials and is used as an abrasive for cutting metals and polishing glass. Its structure is similar to the cubic crystal structure of diamond.

Carbon has a number of oxides, but only two of them are particularly stable. When carbon is burned in a limited amount of oxygen, carbon monoxide predominates. When an excess of oxygen is used, carbon dioxide results. Carbon monoxide is an odorless, colorless, tasteless gas that burns in oxygen to produce carbon dioxide. It is highly poisonous because it binds to hemoglobin much more strongly than oxygen does and thereby blocks the oxygen-carrying ability of hemoglobin. It is used as a fuel and as a reducing agent in metallurgy.

Carbon dioxide is an odorless, colorless gas. Over 50 percent of the carbon dioxide produced industrially is used as a refrigerant, either as a liquid or as a solid (dry ice), and about 25 percent is used to carbonate soft drinks. The phase diagram of carbon dioxide is discussed in Section 14-8.

Carbon forms several sulfides, but only one of them, carbon disulfide, CS_2, is stable at room temperature. Carbon disulfide is a colorless, poisonous, flammable liquid. The purified liquid has a sweet, pleasing odor, but the commonly occurring commercial and reagent grades have an extremely disagreeable odor that is due to organic impurities. Large quantities of CS_2 are used in the manufacture of rayon, carbon tetrachloride, and cellophane and as a solvent for a number of substances.

Carbon also forms several important nitrogen-containing compounds. Hydrogen cyanide, HCN, is a colorless, extremely poisonous gas that dissolves in water to form the very weak acid, hydrocyanic acid ($pK_a = 9.32$ at 25°C). Salts of hydrocyanic acid, called cyanides, are prepared by direct neutralization. Sodium cyanide, NaCN, is used in the extraction of gold and silver from their ores (Chapter 28) and in the electroplating industry.

Figure 25-28 Carbon forms binary compounds with the halogens. Here we see carbon tetrachloride, $CCl_4(l)$, a clear liquid, carbon tetrabromide, $CBr_4(s)$, a white solid, and carbon tetraiodide, $CI_4(s)$, a red crystalline solid.

dihalides, SnX_2, and tetrahalides, SnX_4, which are also covalent. Lead reacts with the halogens to form dihalides, PbX_2, which have well-defined ionic character.

Table 25-19 lists some of the important compounds of germanium, tin, and lead.

Table 25-19 Important compounds of germanium, tin, and lead

Compound	Uses
germanium dioxide, $GeO_2(s)$	infrared-transmitting glass; transistors and diodes
tin(II) chloride, $SnCl_2(s)$	reducing agent in dye manufacture; tin galvanizing; soldering flux
tin(II) fluoride, $SnF_2(s)$	toothpaste additive
lead(II) oxide, $PbO(s)$	glazing pottery and ceramics; lead glass
lead dioxide, $PbO_2(s)$	oxidizing agent; matches; lead-acid storage batteries; pyrotechnics
lead azide, $Pb(N_3)_2(s)$	detonating agent (primer)

TERMS YOU SHOULD KNOW

PROBLEMS

PERIODIC PROPERTIES

25-1. Without using your text, sketch a periodic table and locate the *s*-block elements, *p*-block elements, *d*-block elements, and *f*-block elements.

25-2. How many elements are there in each *d*-transition series and in each *f*-transition series?

25-3. Why are the lanthanide series and the actinide series sometimes called the inner transition metal series?

25-4. List the key distinctions between metals, semimetals, and nonmetals.

25-5. Determine for each of the following elements whether they are metals, nonmetals, or semimetals.

(a) Li (b) Se (c) B (d) Cs
(e) La (f) Mg (g) As (h) Si

25-6. Using only a periodic table, write the ground-state electron configurations of selenium, cesium, antimony, and indium.

25-7. Classify the following elements as *s*-block, *p*-block, *d*-block, or *f*-block elements.

(a) Ba (b) W (c) I (d) Te
(e) Ga (f) Ar (g) Th (h) Sm

25-8. Based on what you know from the periodic table, which of the following compounds would you expect not to exist?

(a) NCl_5 (b) NaO (c) Mn_2O_7 (d) CCl_6

25-9. A certain element, X, forms the following compounds: X_2O_3; XH_3; and XF_3. What are the possible group numbers for this element?

25-10. Sketch a periodic table, indicating the trend of electronegativities from highest to lowest. Do the same for atomic radii, metallic character, and ionization energy.

25-11. Determine for each of the following properties of elements whether the indicated property generally increases or decreases on moving from left to right across a row of the periodic table.

(a) metallic character (b) electronegativity
(c) atomic size (d) acidity of oxides
(e) ionization energy

25-12. Explain why the enthalpies of vaporization of the noble gases increase with increasing atomic number.

25-13. What is one of the most important differences between the bonding involving elements in the second row and bonding involving elements in the third row of the periodic table?

25-14. Write chemical equations that show the amphoteric property of gallium oxide.

HYDROGEN

25-15. Using the data in Table 25-5, calculate the atomic mass of naturally occurring hydrogen.

25-16. Suggest how tritium can be used to study the movement of ground water.

25-17. Calculate the number of disintegrations per second in a sample consisting of one micromole of tritium.

25-18. A sample of water containing a trace amount of T_2O, tritium oxide, is found to have 1.12×10^4 disintegrations per second. What will be its activity after 50 years?

25-19. Complete and balance the following equations:

(a) $Fe_2O_3(s) + H_2(g) \xrightarrow{\text{high T}}$
(b) $LiH(s) + H_2O(l) \rightarrow$
(c) $Mg(s) + H_2(g) \rightarrow$
(d) $K(s) + H_2(g) \rightarrow$

25-20. Complete and balance the following equations:

(a) $Zn(s) + HBr(aq) \rightarrow$
(b) $C(s) + H_2O(g) \xrightarrow[1000°C]{Fe}$
(c) $D_2(g) + N_2(g) \xrightarrow{Fe/Mo}$
(d) $Li(s) + D_2O(l) \rightarrow$

25-21. Lithium metal is often cleaned by treatment with ethanol, C_2H_5OH. By analogy with the reaction of Li(*s*) with water, complete and balance the following equation:

$$Li(s) + CH_3CH_2OH(l) \rightarrow$$

25-22. Hydrogen is the most exothermic fuel on a mass basis. Suggest some reasons why hydrogen is not a widely used fuel.

25-23. How many grams of zinc are required to generate 500 mL of hydrogen at 20°C and 740 torr by the reaction between zinc and hydrochloric acid?

25-24. Calculate the relative masses of hydrogen produced by the reaction of hydrochloric acid with 100 g of iron and 100 g of zinc.

25-25. What volume of hydrogen at 250°C and 10.0 atm is required to reduce 2.50 metric tons of tungsten(VI) oxide to the metal?

25-26. Calculate the enthalpy of combustion of one gram of hydrogen.

25-27. Given that $\Delta G^\circ_{rxn} = -191$ kJ at 25°C for the reaction

$$H_2(g) + Cl_2(g) \rightleftharpoons 2HCl(g)$$

calculate the maximum voltage that can be obtained from a fuel cell utilizing this reaction with each gas at 1.00 atm pressure.

25-28. Given that $\Delta G^\circ_{rxn} = -237$ kJ at 25°C for the reaction

$$H_2(g) + \tfrac{1}{2}O_2(g) \rightleftharpoons H_2O(l)$$

calculate the minimum voltage required to decompose water electrolytically at standard conditions.

THE ALKALI METALS

25-29. Why must the alkali metals be stored under kerosene?

25-30. Explain why the reactivities of the alkali metals increase with atomic number.

25-31. How is sodium metal produced commercially?

25-32. Complete and balance the following equations:

(a) $Na(s) + H_2O(l) \rightarrow$
(b) $K(s) + Br_2(l) \rightarrow$
(c) $Li(s) + N_2(g) \rightarrow$
(d) $Na(s) + H_2(g) \xrightarrow{600°C}$
(e) $NaH(s) + H_2O(l) \rightarrow$

25-33. Complete and balance the following equations. Assume an excess of oxygen.

(a) $Li(s) + O_2(g) \rightarrow$
(b) $Na(s) + O_2(g) \rightarrow$
(c) $K(s) + O_2(g) \rightarrow$
(d) $Cs(s) + O_2(g) \rightarrow$

25-34. The sodium-sulfur battery has been extensively studied as a potential power source for electric-powered vehicles. This high-temperature battery uses the elements sodium and sulfur in molten form. Write the anode and cathode half reactions and the net cell reaction on discharge.

25-35. Explain why sodium metal cannot be prepared by the electrolysis of an $NaCl(aq)$ solution.

25-36. Solutions of sodium metal in liquid ammonia decompose in the presence of a rusty nail, liberating hydrogen gas and forming a white precipitate. Postulate a balanced chemical equation to explain these observations. Formulate your answer by analogy with the reaction between sodium metal and water.

25-37. Sodium peroxide is prepared by first oxidizing sodium to Na_2O in a limited supply of $O_2(g)$ and then reacting this further to give Na_2O_2. Why can't the peroxides of potassium, rubidium, and cesium be prepared in this manner?

25-38. Use molecular orbital theory (Section 13-3) to argue that the superoxide ion is paramagnetic and that the peroxide ion is diamagnetic (not paramagnetic).

25-39. Sodium hydride is used to extract titanium from $TiCl_4$ according to

$$TiCl_4(g) + 4NaH(s) \xrightarrow{400^\circ C} Ti(s) + 4NaCl(s) + 2H_2(g)$$

How many grams of NaH are required to react with 1 kg of $TiCl_4$?

25-40. Sodium hydride reacts with SO_2 to produce sodium dithionite, $Na_2S_2O_4$, according to

$$2SO_2(l) + 2NaH(s) \rightarrow Na_2S_2O_4(s) + H_2(g)$$

How many grams of sodium dithionite can be obtained from 100 g of NaH?

25-41. Lithium peroxide is used as an oxygen source for self-contained breathing apparatus on space capsules. The relevant reaction is

$$2Li_2O_2(s) + 2CO_2(g) \rightarrow 2Li_2CO_3(s) + O_2(g)$$

What volume of oxygen (at 37°C and 1.0 atm) can be obtained from 454 g Li_2O_2?

25-42. Use the data in Table 25-8 to calculate the values of ΔS_{vap} for the alkali metals.

THE ALKALINE EARTH METALS

25-43. Complete and balance the following equations:

(a) $Ca(s) + H_2(g) \xrightarrow{500^\circ C}$
(b) $Mg(s) + N_2(g) \xrightarrow{500^\circ C}$
(c) $Sr(s) + S(s) \xrightarrow{500^\circ C}$
(d) $Ba(s) + O_2(g) \xrightarrow{500^\circ C}$

25-44. Complete and balance the following equations:

(a) $Ca(s) + H_2O(l) \rightarrow$
(b) $Sr_3N_2(s) + H_2O(l) \rightarrow$
(c) $CaC_2(s) + H_2O(l) \rightarrow$
(d) $Ca(s) + C(s) \xrightarrow{500^\circ C}$

25-45. Complete and balance the following equations:

(a) $Be(s) + HCl(aq) \rightarrow$
(b) $Be(s) + NaOH(aq) \rightarrow$
(c) $Be(s) + N_2(g) \xrightarrow{500^\circ C}$
(d) $Be(s) + O_2(g) \xrightarrow{400^\circ C}$

25-46. Burning magnesium, which can occur in automobile fires, should not be attacked with either water or carbon dioxide extinguishers. Why not?

25-47. Magnesium hydroxide is only slightly soluble in water, but a suspension of magnesium hydroxide (milk of magnesia) in water is used as an antacid.

(a) Write a balanced chemical equation for the neutralization of stomach acid [$HCl(aq)$] by milk of magnesia. (b) Given that stomach acid is about 0.10 M $HCl(aq)$, calculate the number of milligrams of $Mg(OH)_2(s)$ required to neutralize 1.0 mL of stomach acid.

25-48. An old industrial preparation of hydrogen peroxide involves the reaction of oxygen with barium oxide at 500°C to form barium peroxide, followed by the treatment of the peroxide with aqueous acid. Write balanced chemical equations for the process.

25-49. Suggest a method for the preparation of magnesium chloride from magnesium carbonate.

25-50. Suggest a method for the preparation of calcium nitrate from calcium carbonate.

25-51. (a) Use VSEPR theory to predict the structure of beryllium chloride, $BeCl_2$.

(b) Use hybrid orbitals to describe the bonding in $BeCl_2$.

25-52. (a) Use VSEPR theory to predict the shape of the tetrafluoroberyllate(II) ion BeF_4^{2-}.

(b) Use hybrid orbitals to describe the bonding in BeF_4^{2-}.

25-53. The solubilities (in grams per 100 mL of solution) of the alkaline earth hydroxides in water at 20°C are

$Mg(OH)_2$	9×10^{-4}	$Sr(OH)_2$	0.93
$Ca(OH)_2$	0.18	$Ba(OH)_2$	5.8

Calculate the pH of a saturated solution in each case.

25-54. Both $BaCO_3$ and $BaSO_4$ are insoluble in basic solution. In acidic solution, $BaCO_3$ dissolves but $BaSO_4$ does not. Explain.

THE GROUP 3 ELEMENTS

25-55. Do the acidities of the oxides of the Group 3 elements increase or decrease upon descending the group in the periodic table?

25-56. Write chemical equations describing the amphoteric nature of $Al(OH)_3(s)$ and $Ga(OH)_3(s)$.

25-57. The formula for boric acid is often written as $B(OH)_3$ rather than H_3BO_3. A good reason for doing this is because boric acid acts not as a Brønsted-Lowry acid but as a Lewis acid. Write a chemical equation describing the Lewis acid property of an aqueous solution of boric acid.

25-58. Diborane is often prepared by the reaction of boron trifluoride with sodium borohydride, with sodium tetrafluoroborate(III) as the other product. Write a balanced chemical equation for this reaction.

25-59. How many valence electrons are there in (a) $B_3H_8^-$; (b) $B_{10}H_{10}^{2-}$?

25-60. Use VSEPR theory to predict the structures of the following species.

(a) AlF_6^{3-} (b) $Al(OH)_4^-$ (c) $AlOF$

25-61. Use VSEPR theory to predict the structures of the following species:

(a) $GaCl_3$ (b) GaF_2^+ (c) $GaBr_4^-$

25-62. Use VSEPR theory to predict the shape of the tetrafluoroborate(III) ion, BF_4^-. Describe the bonding in BF_4^- using hybrid orbitals.

25-63. The structure of tetraborane, B_4H_{10}, is shown in Figure 25-18. Describe the bonding in terms of hybrid orbitals.

25-64. Describe the bonding in $B_3H_8^-$ in terms of hybrid orbitals (see Figure 25-18).

25-65. The pK_a values for the Group 3 $M^{3+}(aq)$ ions are as follows:

Ion	pK_a at 25°C
$Al^{3+}(aq)$	4.95
$Ga^{3+}(aq)$	2.62
$In^{3+}(aq)$	3.70
$Tl^{3+}(aq)$	1.14

Calculate the pH of a 0.10 M aqueous solution of each of these ions.

25-66. A certain borane is found to be 81.10% boron and 18.90% hydrogen. What is the simplest formula of this borane? Referring to Figure 25-18, identify which borane it is.

THE GROUP 4 ELEMENTS

25-67. Complete and balance the following equations:

(a) $Al_2O_3(s) + C(s) \xrightarrow[\text{furnace}]{\text{electric}}$

(b) $CaC_2(s) + D_2O(l) \xrightarrow{\text{heat}}$

(c) $PbS(s) + O_2(g) \xrightarrow{\text{heat}}$

25-68. Do the acidities of the Group 4 oxides increase or decrease with increasing atomic number?

25-69. Given that graphite is converted to diamond under high pressures, use Le Châtelier's principle to deduce the relative densities of the two substances.

25-70. Explain on a molecular level why diamond is an extremely hard substance and graphite is slippery. Give applications of these properties.

25-71. Explain on a molecular level why diamond is a poor conductor of electricity and graphite is a good conductor.

25-72. Clean rainwater (as opposed to acid rain) is slightly acidic, with a pH of about 5.6. Explain the acidity of clean rainwater.

25-73. Give two examples (not mentioned in the text) of elements that can be added to make *n*-type and *p*-type silicon semiconductors, respectively.

25-74. Determine the oxidation states of the Group 4 elements in the following compounds.

(a) SiC (b) CS_2 (c) $NaCN$ (d) $Si_6O_{18}^{12-}$

25-75. Determine the oxidation state(s) of the Group 4 elements in the following compounds.

(a) PbO_2
(b) Pb_3O_4 ("red lead")
(c) Al_4C_3
(d) CaC_2

25-76. Draw structures for the following ions.

(a) $Si_2O_7^{6-}$ (b) $Si_3O_{10}^{8-}$ (c) $Si_6O_{18}^{12-}$

25-77. Write the chemical equation that describes the etching of glass by hydrofluoric acid. (One of the products is H_2SiF_6.)

25-78. Use VSEPR theory to predict the shapes of

(a) $GeBr_4$ (b) $SnCl_2$ (c) GeG_6^{2-} (d) SiH_3^-

25-79. Use hybrid orbitals to describe the bonding in the silanes, SiH_4 and Si_2H_6.

25-80. Write chemical equations describing how tin and lead are obtained from their ores.

25-81. Explain how photochromic glass works.

25-82. Many fine museum pieces and organ pipes made of tin have been ruined because their temperatures were allowed to drop below 13°C for appreciable periods of time. Explain.

26 THE CHEMISTRY OF THE MAIN-GROUP ELEMENTS II

Elemental sulfur occurs in large underground deposits along the Gulf Coast of the United States. The sulfur shown here was mined by the Frasch process and is awaiting shipment at Pennzoil's terminal at Galveston, Texas.

In this chapter we continue our discussion of the chemistry of the main-group elements. We shall observe the same periodic trends we studied in Chapter 25 for Groups 1 through 4, both across the periodic table and within a group. The metallic character of the elements decreases from Group 1 to Group 7, so in this chapter we discuss mostly semimetals and nonmetals. We also expect the oxides of these elements to be more acidic than those of the elements in Groups 1 through 4. These and other conclusions based on the basic principles discussed earlier will enable us to understand more of the diversity of chemical properties and behavior.

GROUPS 5 THROUGH 8

26-1. THE CHEMICAL PROPERTIES OF THE GROUP 5 ELEMENTS

The Group 5 elements are nitrogen, phosphorus, arsenic, antimony, and bismuth (Figure 26-1). As you might expect from periodic trends, nitrogen and phosphorus are nonmetals, arsenic and antimony are semimetals, and bismuth is a metal. Group 5 nicely illustrates the trend from acidic to basic oxides on descending a group in the periodic table. The oxides of nitrogen, phosphorus, and arsenic are acidic, those of antimony are amphoteric, and bismuth(III) oxide, Bi_2O_3, is basic. There is also an increase in stability of the lower oxidation state with increasing atomic number. Thus, Bi_2O_3 is the only stable oxide of bismuth, whereas the other members of the group also have oxides of the type M_2O_5. Examination of compounds of the Group 5 family gives us an opportunity to note that the nonmetals of the fourth row—arsenic, selenium, and bromine—favor an oxidation state two less than their maximum. Thus, although AsF_3, $AsCl_3$, $AsBr_3$, and AsI_3 exist, the only known arsenic(V) halide is AsF_5.

Nitrogen makes up almost 80 percent of the earth's atmosphere, from which it can be obtained by fractional distillation of liquefied air. This method exploits the difference in the boiling points of nitrogen

7 **N** 14.0067 $2s^22p^3$
15 **P** 30.97376 $3s^23p^3$
33 **As** 74.9216 $4s^24p^3$
51 **Sb** 121.75 $5s^25p^3$
83 **Bi** 208.9804 $6s^26p^3$

Figure 26-1 The Group 5 elements. Back row from left to right: liquid nitrogen, phosphorus, and arsenic. Front row: antimony and bismuth.

sodium nitrite, both stable compounds, to water. Even so, the solution must be heated carefully to avoid an explosion.

Nitrogen fixation is the conversion of the free element to its compounds, and nitrogen fixation by microorganisms is an important source of plant nutrients. The most common of these nitrogen-fixing bacteria belong to the species *Rhizobium*. They invade the roots of leguminous plants such as alfalfa, clover, beans, and peas and form nodules on the roots of these legumes. The relationship between *Rhizobium* and the host plant is symbiotic (mutually beneficial) (Figure 26-3). The plant produces carbohydrates through photosynthesis, and *Rhizobium* uses the carbohydrate as fuel for fixing the nitrogen, which is incorporated into plant protein. Alfalfa is the most potent nitrogen-fixer, followed by clover, soybeans, other beans, peas, and peanuts. In modern agriculture, crops are rotated, that is, plantings of a nonleguminous crop and a leguminous crop are alternated from time to time on one piece of land. The leguminous crop is either harvested, leaving behind nitrogen-rich roots, or plowed into the soil, adding both nitrogen and organic matter. A plowed-back crop of alfalfa may add as much as 400 pounds of fixed nitrogen to the soil per acre.

Figure 26-3 Nitrogen-fixing nodules on the roots of a leguminous plant, soybeans. The nodules contain *Rhizobium,* a soil bacterium that converts atmospheric elemental nitrogen to water-soluble nitrogen compounds.

26-1B. Most Nitrogen Is Converted to Ammonia by the Haber Process

Because of its importance to agriculture, nitrogen fixation was a major application of modern chemistry; but because of the great strength of the nitrogen-nitrogen triple bond, it was also a major challenge. This challenge was met in the early 1900s by the **Haber process,** in which nitrogen reacts directly with hydrogen at high pressure and high temperature in the presence of an $Fe_2O_3(s)/Co_2O_3(s)$ catalyst to form ammonia

$$N_2(g) + 3H_2(g) \xrightarrow[500°\text{C}]{300\text{ atm}} 2NH_3(g)$$

Over 30 billion pounds of ammonia are produced annually in the United States by the Haber process. Rated in terms of pounds produced per year, ammonia is the fifth-ranked industrial chemical in the United States.

Ammonia is a colorless gas with a sharp, irritating odor. It is the effective agent in some forms of "smelling salts." Household ammonia is an aqueous solution of ammonia and a detergent. Ammonia was the first complex molecule to be identified in interstellar space. It occurs in galactic dust clouds in the Milky Way and, in solid form, in the rings of Saturn.

Unlike nitrogen, which is sparingly soluble in water, ammonia is very soluble. Over 700 mL of ammonia at 0°C and 1 atm will dissolve in 1 mL of water. The solubility of ammonia in water can be demonstrated nicely by the **fountain effect.** This effect can be observed with the simple laboratory setup shown in Figure 26-4. A dry flask is filled with anhydrous ("dry") ammonia gas at atmospheric pressure. When just a few drops of water are squirted into the flask from a syringe, some of the ammonia dissolves in the drops. The pressure of the gaseous ammonia falls below that of the atmosphere, and water is forced from the beaker, up the vertical glass tubing into the flask, producing a spectacular fountain.

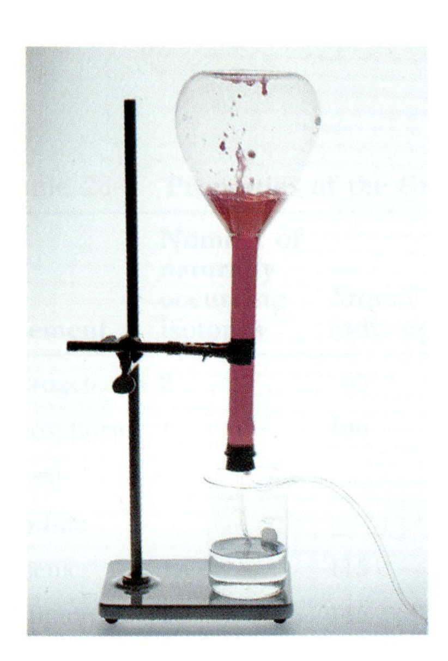

Figure 26-4 An ammonia fountain demonstrates the large solubility of ammonia in water.

Ammonia reacts with various acids to form ammonium compounds, which contain the ion NH_4^+. For example,

$$NH_3(aq) + HCl(aq) \rightarrow NH_4Cl(aq)$$

Many ammonium compounds are important commercially, particularly as fertilizers. Ammonia is rich in nitrogen, binds to many components of soil, and is easily converted to usable plant food. Concentrated aqueous solutions of ammonia or pure liquid ammonia can be sprayed directly into the soil (Figure 26-5). The increased growth of plants when fertilized by ammonia is spectacular. Liquid ammonia is toxic and injurious to living tissue, however, and must be handled carefully.

For some purposes, it is more convenient to use a solid fertilizer instead of ammonia solutions. For example, ammonia combines directly with sulfuric acid to produce ammonium sulfate:

$$2NH_3(aq) + H_2SO_4(aq) \rightarrow (NH_4)_2SO_4(aq)$$

Ammonium sulfate is the most important solid fertilizer in the world. The annual U.S. production of ammonium sulfate is over 4 billion pounds.

Figure 26-5 Spraying ammonia into the soil. Liquid ammonia, called anhydrous ammonia, is used extensively as a fertilizer because it is cheap, high in nitrogen, and easy to apply.

26-1C. Nitric Acid Is Produced by the Ostwald Process

About half of all the ammonia produced is converted to nitric acid by the **Ostwald process.** The first step in this process is the conversion of ammonia to nitrogen oxide:

$$4NH_3(g) + 5O_2(g) \xrightarrow[826^\circ C]{Pt} 4NO(g) + 6H_2O(g)$$

The second step in the Ostwald process involves the oxidation of nitrogen oxide to nitrogen dioxide by reaction with oxygen:

$$2NO(g) + O_2(g) \rightarrow 2NO_2(g)$$

In the final step, the $NO_2(g)$ is dissolved in water to yield nitric acid:

$$3NO_2(g) + H_2O(l) \rightarrow 2HNO_3(aq) + NO(g)$$

The NO(g) evolved is recycled back to the second step.

Laboratory grade nitric acid is approximately 70 percent HNO_3 by mass, with a density of 1.42 $g\cdot mL^{-1}$ and a concentration of 16 M. The U.S. annual production of nitric acid is almost 16 billion pounds, which makes it the 12th-ranked industrial chemical. Nitric acid is the least expensive potent oxidizing agent and is used in a large number of important chemical processes, including the production of explosives such as trinitrotoluene (TNT), nitroglycerine, and nitrocellulose (gun cotton). It is also used in etching and photoengraving processes to produce grooves in metal surfaces.

EXAMPLE 26-2: Verify that laboratory grade nitric acid, which is 70 percent HNO_3 by mass and has a density of 1.42 $g\cdot mL^{-1}$, is approximately 16 M.

Solution: Consider a 1-L sample, which has a mass of

$$\begin{pmatrix}\text{mass of}\\ \text{1-L sample}\end{pmatrix} = (1.42\ g\cdot mL^{-1})\left(\frac{1000\ mL}{1\ L}\right) = 1420\ g\cdot L^{-1}$$

Of this sample, 70 percent by mass is HNO_3, or

$$\text{mass of } HNO_3 = (1420\text{ g})(0.70) = 994\text{ g}$$

The number of moles of NHO_3 in our sample is

$$\text{mol of } HNO_3 = (994\text{ g } HNO_3)\left(\frac{1\text{ mol } HNO_3}{63.02\text{ g } HNO_3}\right) = 15.8\text{ mol}$$

Because we have taken a 1-L sample, the molarity is 15.8 M.

PRACTICE PROBLEM 26-2: Describe how you would prepare 1.00 L of 6.0 M $HNO_3(aq)$ from laboratory grade nitric acid.

Answer: Add 380 mL of laboratory grade nitric acid to enough water to make 1.00 L of solution.

26-1D. Nitrogen Forms Several Important Compounds with Hydrogen and Oxygen

The most important nitrogen-hydrogen compounds after ammonia are hydrazine, N_2H_4, and hydrazoic acid, HN_3. Hydrazine is a colorless, fuming, reactive liquid. It is produced by the **Raschig synthesis,** in which ammonia is reacted with hypochlorite ion (household bleach is sodium hypochlorite in water) in basic solution:

$$2NH_3(aq) + ClO^-(aq) \xrightarrow{OH^-(aq)} N_2H_4(aq) + H_2O(l) + Cl^-(aq)$$

(Household bleach should *never* be mixed with household ammonia because extremely toxic and explosive chloramines, such as H_2NCl and $HNCl_2$, are produced as by-products.) The reaction of hydrazine with oxygen,

$$N_2H_4(l) + O_2(g) \rightarrow N_2(g) + 2H_2O(g)$$

is accompanied by the release of a large amount of energy, and hydrazine and some of its derivatives are used as rocket fuels.

EXAMPLE 26-3: Use the data in Table 8-2 to calculate the value of ΔH°_{rxn} for the reaction of hydrazine with oxygen at 25°C.

Solution: The equation for the reaction at 25°C is

$$N_2H_4(l) + O_2(g) \rightarrow N_2(g) + 2H_2O(l)$$

We calculate ΔH°_{rxn} from

$$\begin{aligned}\Delta H^\circ_{rxn} &= \Delta H^\circ_f[N_2(g)] + 2\Delta H^\circ_f[H_2O(l)] - \Delta H^\circ_f[N_2H_4(l)] - \Delta H^\circ_f[O_2(g)] \\ &= (1\text{ mol})(0) = (2\text{ mol})(-285.8\text{ kJ}\cdot\text{mol}^{-1}) - (1\text{ mol})(50.6\text{ kJ}\cdot\text{mol}^{-1}) \\ &\qquad - (1\text{ mol})(0) \\ &= -622.2\text{ kJ}\end{aligned}$$

PRACTICE PROBLEM 26-3: A reaction that can be used for the quantitative estimation of hydrazine is described by the (unbalanced) equation in acidic solution

$$MnO_4^-(aq) + N_2H_4(aq) \rightarrow Mn^{2+}(aq) + N_2(g)$$

Balance this equation.

Answer:

$$4MnO_4^-(aq) + 5N_2H_4(aq) + 12H^+(aq) \rightarrow 4Mn^{2+}(aq) + 5N_2(g) + 16H_2O(l)$$

Nitrous acid, HNO_2, is an important oxyacid of nitrogen. Salts of nitrous acid are called nitrites. Sodium nitrite, $NaNO_2$, is used as a meat preservative. The nitrite ion combines with the hemoglobin in meat to produce a deep red color. The main problem with the extensive use of nitrites in foods is that the nitrite ion reacts with amines in the body's gastric juices to produce compounds called nitrosamines [such as $(CH_3)_2NNO$, dimethylnitrosamine], which are carcinogenic.

The reaction of nitrous acid with hydrazine in acidic solution yields hydrazoic acid:

$$N_2H_4(aq) + HNO_2(aq) \rightarrow HN_3(aq) + 2H_2O(l)$$

Hydrazoic acid is a colorless, toxic liquid and a dangerous explosive. In aqueous solution, HN_3 is a weak acid, with $pK_a = 4.72$ at 25°C. Its lead and mercury salts, $Pb(N_3)_2$ and $Hg(N_3)_2$, which are called **azides,** are used in detonation caps; both compounds are dangerously explosive. Sodium azide, NaN_3, is used as the gas source in automobile air bags, which inflate rapidly on impact.

Nitrogen forms a number of oxides, with nitrogen assuming oxidation states of +1 through +5 (Table 26-3). Dinitrogen oxide (nitrous oxide), also known as laughing gas, was once used as a general anesthetic, but its primary use now is as an aerosol and canned whipped cream propellant. Dinitrogen oxide can be produced by a cautious thermal decomposition of NH_4NO_3:

$$NH_4NO_3(s) \rightarrow N_2O(g) + 2H_2O(l)$$

Table 26-3 The oxides of nitrogen

Formula	Systematic name	Description
$N_2O(g)$	dinitrogen oxide (nitrous oxide)	colorless, rather unreactive gas
$NO(g)$	nitrogen oxide (nitric oxide)	colorless, paramagnetic, reactive gas
$N_2O_3(g)$	dinitrogen trioxide	dark blue solid (b.p. 3.5°C); dissociates in gas phase to NO and NO_2
$NO_2(g)$	nitrogen dioxide	brown, paramagnetic, reactive gas; dimerizes reversibly to N_2O_4
$N_2O_4(g)$	dinitrogen tetroxide	colorless gas (b.p. 21°C); dissociates reversibly to NO_2
$N_2O_5(s)$	dinitrogen pentoxide	colorless, ionic solid; unstable as a gas

Nitrogen oxide is produced in the oxidation of copper by 6M nitric acid:

$$3Cu(s) + 8HNO_3(aq) \rightarrow 3Cu(NO_3)_2(aq) + 2NO(g) + 4H_2O(l)$$

Although NO(*g*) is colorless, this reaction produces a brown gas if it is run in a vessel that is open to the atmosphere. The brown gas results from the rapid production of nitrogen dioxide by the reaction.

$$\underset{\text{colorless}}{2NO(g)} + O_2(g) \rightarrow \underset{\text{brown}}{2NO_2(g)}$$

In the gas phase, nitrogen dioxide dimerizes to form dinitrogen tetroxide:

$$2NO_2(g) \rightleftharpoons N_2O_4(g) \qquad \Delta H^\circ_{rxn} = -57.2\ \text{kJ}$$

Because this reaction is exothermic, an increase in temperature results in the formation of more $NO_2(g)$ and, hence, a more reddish brown mixture (Figure 26-6). Some commercially important nitrogen-containing compounds are given in Table 26-4.

Table 26-4 Some important compounds of nitrogen

Compound	Uses
ammonia, $NH_3(g)$	fertilizers; manufacture of nitric acid, explosives; synthetic fibers; refrigerant
nitric acid, $HNO_3(l)$	manufacture of fertilizers, explosives, lacquers, synthetic fabrics, drugs, and dyes; oxidizing agent; metallurgy; ore flotation
ammonium nitrate, $NH_4NO_3(s)$	fertilizer; explosives; herbicides and insecticides; solid rocket propellant
sodium cyanide, NaCN(*s*)	extraction of gold and silver from their ores; insecticide; fumigant; manufacture of dyes and pigments

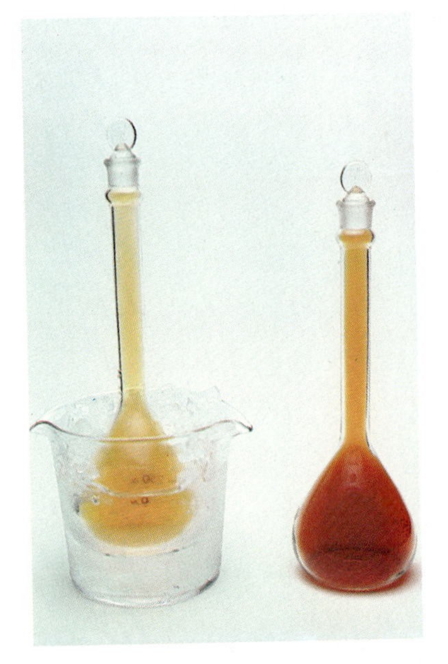

Figure 26-6 An increase in temperature from 0°C (ice water) to 25°C favors conversion of N_2O_4 to NO_2, a brown gas.

26-1E. There Are Two Principal Allotropes of Solid Phosphorus

The most important allotropic forms of solid phosphorus are **white phosphorus** and **red phosphorus** (Figure 26-7). White phosphorus is a white, transparent, waxy, crystalline solid that often appears pale yellow because of impurities. It is insoluble in water and alcohol but soluble in carbon disulfide. A characteristic property of white phosphorus is its high chemical reactivity. It ignites spontaneously in air at about 25°C. White phosphorus should never be allowed to come into contact with the skin because body temperature (37°C) is sufficient to ignite it spontaneously. Phosphorus burns are extremely painful and slow to heal. In addition, white phosphorus is very poisonous. White phosphorus should always be kept under water and handled with forceps.

When white phosphorus is heated above 250°C in the absence of air, a form called red phosphorus is produced. Red phosphorus is a red to violet powder that is less reactive than white phosphorus. The chemical reactions that the red form undergoes are the same as those of the

Figure 26-7 Red and white phosphorus. White phosphorus, one of the principal allotropes of solid phosphorus, is very reactive and must be handled with care because it produces severe burns when it comes in contact with skin. White phosphorus is usually stored under water, but the sample shown here has a yellowish cast as a result of surface reactions with air. Red phosphorus is much less reactive than white phosphorus and does not require special handling.

white form, but they generally occur only at higher temperatures. The toxicity of red phosphorus is much lower than that of white phosphorus. Whereas red phosphorus consists of large, random aggregates of phosphorus atoms, white phosphorus consists of tetrahedral P_4 molecules (Figure 26-8).

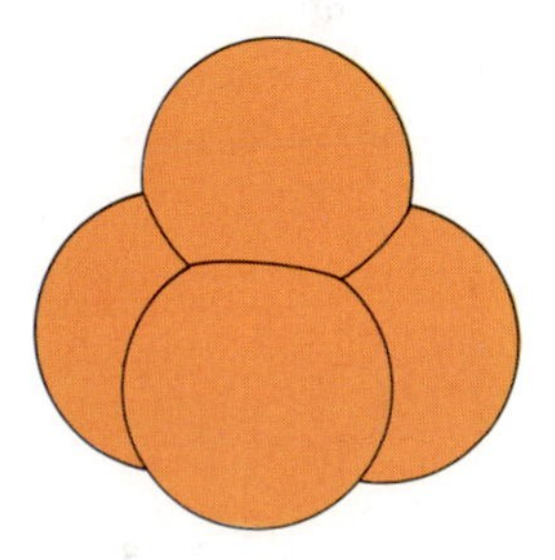

Figure 26-8 White phosphorus consists of tetrahedral P_4 molecules.

Most of the phosphorus that is produced is used to make phosphoric acid or other phosphorus compounds. Elemental phosphorus, however, is used in the manufacture of fireworks, matches, rat poisons, incendiary shells, smoke bombs, and tracer bullets.

Phosphorus is not found as the free element in nature. The principal sources are the minerals calcium phosphate, hydroxyapatite, fluorapatite, and chlorapatite (Figure 26-9). These ores collectively are called phosphate rock. Vast phosphate rock deposits occur in Russia, in Morocco, and in Florida, Tennessee, and Idaho. Although some phosphate rock is used to make elemental phosphorus, most phosphate rock is used in the production of fertilizers. Phosphorus is a required nutrient of all plants, and phosphorus compounds have long been used as a fertilizer. In spite of its great abundance, phosphate rock cannot be used as a fertilizer because, as the name implies, it is insoluble in water.

Figure 26-9 The apatite minerals. Left to right: hydroxyapatite, $Ca_{10}(OH)_2(PO_4)_6$; fluorapatite, $Ca_{10}F_2(PO_4)_6$; and chlorapatite, $Ca_{10}Cl_2(PO_4)_6$.

Consequently, plants are not able to assimilate the phosphorus from phosphate rock. To produce a water-soluble source of phosphorus, phosphate rock is reacted with sulfuric acid to produce a water-soluble product called **superphosphate,** $Ca(H_2PO_4)_2$, one of the world's most important fertilizers.

26-1F. The Oxides of Phosphorus Are Acid Anhydrides

White phosphorus reacts directly with oxygen to produce the oxides P_4O_6 and P_4O_{10}. With excess phosphorus present, P_4O_6 is formed:

$$\underset{\text{excess}}{P_4(s)} + 3O_2(g) \rightarrow P_4O_6(s)$$

With excess oxygen present, P_4O_{10} is formed:

$$P_4(s) + \underset{\text{excess}}{5O_2(g)} \rightarrow P_4O_{10}(s)$$

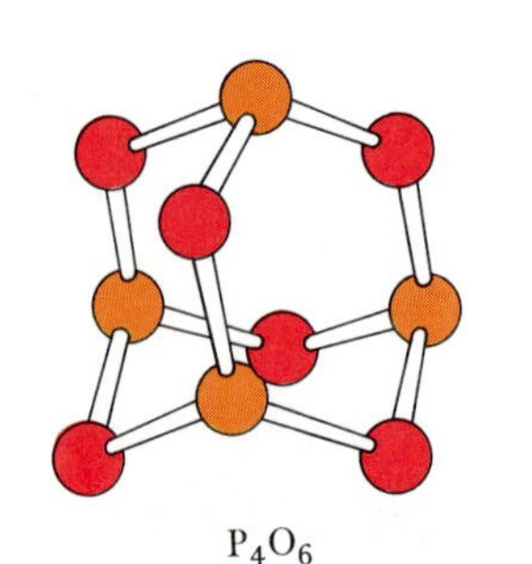

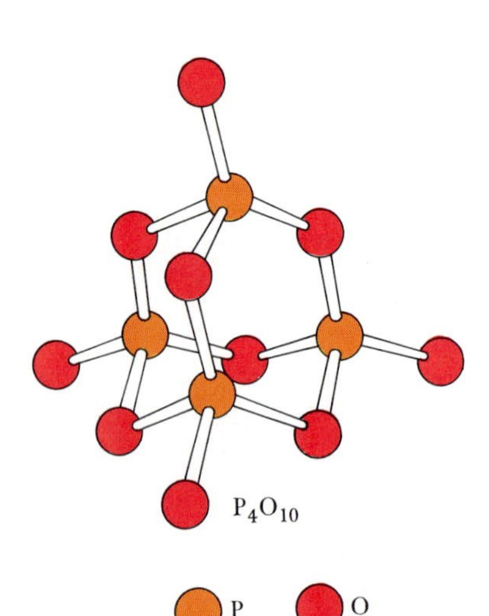

Figure 26-10 Structure of P_4O_6 and P_4O_{10}. (a) The P_4O_6 molecule can be viewed as arising from the tetrahedral P_4 molecule when an oxygen atom is inserted between each pair of adjacent phosphorus atoms. (b) The P_4O_{10} molecule can be viewed as arising from P_4O_6 when an oxygen atom is attached to each of the four phosphorus atoms. Note that there are no phosphorus-phosphorus bonds in either P_4O_6 or P_4O_{10}.

Before the actual molecular formulas of these phosphorus oxides were known, the empirical formulas $P_2O_3(s)$ and $P_2O_5(s)$ were used. Consequently, $P_4O_6(s)$ and $P_4O_{10}(s)$ are still commonly called phosphorus trioxide and phosphorus pentoxide.

It is interesting to compare the structures of P_4O_6 and P_4O_{10} (Figure 26-10). The structure of P_4O_6 is obtained from that of P_4 by inserting an oxygen atom between each pair of adjacent phosphorus atoms; there are six edges on a tetrahedron, and thus a total of six oxygen atoms are required. The structure of P_4O_{10} is obtained from that of P_4O_6 by attaching an additional oxygen atom to each of the four phosphorus atoms.

The phosphorus oxides P_4O_6 and P_4O_{10} react with water to form the phosphorus oxyacids: phosphorous acid, H_3PO_3, and phosphoric acid, H_3PO_4:

$$P_4O_6(s) + 6H_2O(l) \rightarrow 4H_3PO_3(aq)$$
$$P_4O_{10}(s) + 6H_2O(l) \rightarrow 4H_3PO_4(aq)$$

Phosphorus pentoxide is a powerful dehydrating agent capable of removing water from concentrated sulfuric acid, which is itself a strong dehydrating agent.

$$P_4O_{10}(s) + 6H_2SO_4(l) \rightarrow 4H_3PO_4(l) + 6SO_3(g)$$

Hypophosphorous acid, H_3PO_2, is a third oxyacid of phosphorus. The structures of phosphoric acid, phosphorous acid, and hypophosphorous acid are shown in Figure 26-11.

Table 26-5 **The pK_a values in water at 25°C of the phosphorus oxyacids**

Acid	pK_{a1}	pK_{a2}	pK_{a3}
phosphoric acid, H_3PO_4	2.15	7.21	12.36
phosphorous acid, $H_2(HPO_3)$	1.20	6.59	none
hypophosphorous acid, $H(H_2PO_2)$	1.2	none	none

The hydrogen atoms attached to the phosphorus atom in the oxyacids of phosphorus are not dissociable in aqueous solutions. Thus, phosphoric acid, H_3PO_4, is triprotic; phosphorous acid, $H_2(HPO_3)$, is diprotic; and hypophosphorous acid, $H(H_2PO_2)$, is monoprotic. The pK_a values at 25°C are given in Table 26-5.

EXAMPLE 26-4: Calculate the pH of a H_3PO_4-$H_2PO_4^-$ buffer that is 0.100 M in $H_3PO_4(aq)$ and 0.126 M in $KH_2PO_4(aq)$.

Solution: To calculate the pH of a buffer, we use the Henderson-Hasselbalch equation (Equation 19-5),

$$pH = pK_a + \log\frac{[\text{conjugate base}]_0}{[\text{conjugate acid}]_0}$$

In our case, the acid is $H_3PO_4(aq)$ and its conjugate base is $H_2PO_4^-(aq)$, so

$$pH = 2.2 + \log\left(\frac{[H_2PO_4^-]}{[H_3PO_4]}\right)$$

$$= 2.2 + \log\left(\frac{0.126\text{ M}}{0.100\text{ M}}\right) = 2.3$$

PRACTICE PROBLEM 26-4: Calculate (a) the change in pH if 10.0 mL of 0.150 M HCl(aq) is added to 100 mL of the buffer in Example 26-4.

Answer: (a) The pH decreases to 2.2.

Almost 23 billion pounds of phosphoric acid are produced annually in the United States alone, which makes it the 7th-ranked industrial chemical. Commercial phosphoric acid is sold as an 85 percent by mass (85 g of H_3PO_4 to 15 g of H_2O) solution, equivalent to 15 M. It is used extensively in the production of soft drinks, and many of its salts are used in the food industry. For example, the monosodium salt, NaH_2PO_4, is used in a variety of foods to control acidity; and calcium dihydrogen phosphate, $Ca(H_2PO_4)_2$, is the acidic ingredient in **baking powder.** The evolution of carbon dioxide that takes place when baking powder is heated can be represented as

$$\underbrace{Ca(H_2PO_4)_2(s) + 2NaHCO_3(s)}_{\text{baking powder}} \xrightarrow{300°C} 2CO_2(g) + 2H_2O(g) + CaHPO_4(s) + Na_2HPO_4(s)$$

The slowly evolving $CO_2(g)$ gets trapped in small gas pockets and thereby causes the cake or bread to rise.

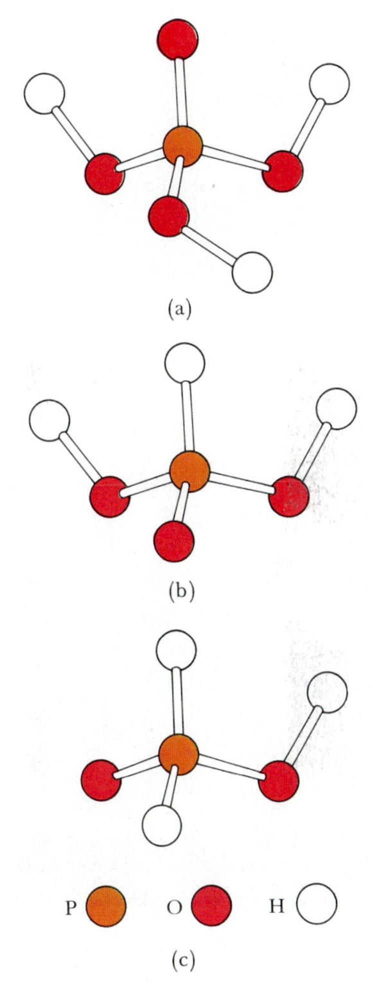

Figure 26-11 The structure of (a) phosphoric acid, (b) phosphorous acid, and (c) hypophosphorous acid. Note that all three hydrogen atoms of phosphoric acid are attached to oxygen atoms. One of the hydrogen atoms in phosphorous acid is attached directly to the phosphorus atom, and two of the hydrogen atoms in hypophosphorous acid are attached to the phosphorus atom. Only those hydrogen atoms attached to oxygen atoms are dissociable, so phosphoric acid is triprotic, phophorous acid is diprotic, and hypophosphorous acid is monoprotic.

26-1G. Phosphorus Forms a Number of Binary Compounds

Phosphorus reacts directly with reactive metals, such as sodium and calcium, to form phosphides. A typical reaction is

$$12Na(s) + P_4(s) \rightarrow 4Na_3P(s)$$

Note the similarity of phosphides to nitrides. Most metal phosphides react vigorously with water to produce phosphine, $PH_3(g)$:

$$Ca_3P_2(s) + 6H_2O(l) \rightarrow 2PH_3(g) + 3Ca(OH)_2(aq)$$

Figure 26-12 The reaction between phosphorus and bromine.

Phosphine has a trigonal pyramidal structure with an H—P—H bond angle of 93.7°. It is a colorless, extremely toxic gas with an offensive odor like that of rotten fish. Unlike ammonia, phosphine does not act as a base toward water, and few phosphonium (PH_4^+) salts are stable. Phosphine can also be prepared by the reaction of white phosphorus with a strong base.

Phosphorus reacts directly with chlorine and bromine to form halides of the form PX_3 and PX_5 (Figure 26-12). If an excess of phosphorus is used, then the trihalide is favored. If an excess of halide is used, then the pentahalide is favored.

When phosphorus is heated with sulfur, the yellow crystalline compound tetraphosphorus trisulfide, P_4S_3, is formed. Matches that can be ignited by striking on any rough surface contain a tip composed of the yellow P_4S_3 on top of a red portion that contains lead dioxide, PbO_2, together with antimony sulfide, Sb_2S_3 (Figure 26-13). Friction causes the P_4S_3 to ignite in air, and the heat produced then initiates a reaction between antimony sulfide and lead dioxide, which produces a flame.

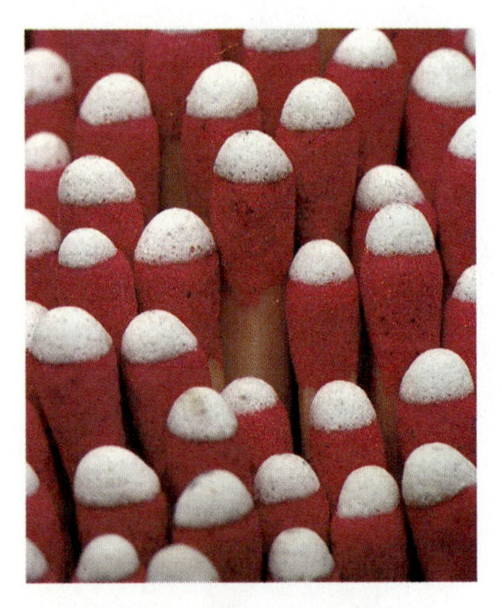

Figure 26-13 Phosphorus in the form of P_4S_3 is one of the principal components of "strike-anywhere" matches.

Safety matches consist of a mixture of potassium chlorate and antimony sulfide. The match is ignited by striking it on a special rough surface composed of a mixture of red phosphorus, glue, and abrasive. The red phosphorus is ignited by friction and in turn ignites the reaction mixture in the match head.

26-1H. Many Phosphorus Compounds Are Important Biologically

The energy requirements for many biochemical reactions are supplied by a substance called **adenosine triphosphate,** or simply **ATP** (Figure 26-14). The chain of three phosphate groups in ATP makes it an energy-rich molecule. Under physiological conditions, the reaction of 1 mol of ATP with water to produce adenosine diphosphate (ADP) and

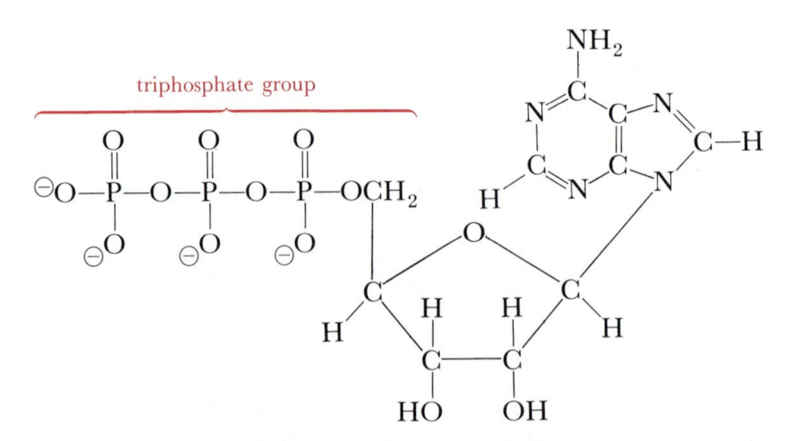

Figure 26-14 The Lewis formula of ATP. ADP is similar but has two phosphate groups joined together instead of three.

a hydrogen phosphate ion releases 31 kJ of energy. This energy is used by all living species to drive biochemical reactions. We can represent the reaction of ATP with water schematically by

$$ATP(aq) + H_2O(l) \rightarrow ADP(aq) + HPO_4^{2-}(aq)$$

Adenosine diphosphate is converted into ATP by the biochemical oxidation of food molecules, and then the ATP is available to supply energy for muscular activity, synthesis of proteins and other biochemical molecules, production of nerve signals, and other biological activity. In other words, ATP is a biological fuel. The formation and utilization of ATP occur on the average within about 1 min of each other. The amount of ATP used by the human body is truly remarkable: at rest over a 24-h period about 40 kg of ATP is utilized. For strenuous exercise, the rate of utilization of ATP can reach 5 kg in 10 min.

Many organic phosphates are potent insecticides that are also highly toxic to humans. These insecticides act by blocking the transmission of electrical signals in the respiratory system, thereby causing paralysis and death by suffocation. Fortunately, such poisons do not last for long in the environment because they are destroyed over a period of several days by reaction with water. An important example of an organophosphorus insecticide is malathion, which has been used to combat the Mediterranean fruit fly infestation in California. Malathion is toxic to humans, but only at fairly large doses. An enzyme in human gastric juice decomposes malathion (insects lack this enzyme); thus, malathion is most toxic to humans when it is absorbed directly into the bloodstream, as, for example, when it comes into contact with a cut in the skin.

Some commercially important compounds of phosphorus are given in Table 26-6.

26-11. Arsenic and Antimony Are Semimetals; Bismuth Is the Only Group 5 Metal

Neither arsenic nor antimony is particularly abundant. Common ores of arsenic are the sulfide minerals *realgar* (As_4S_4) and *orpiment* (As_2S_3),

Table 26-6 Some important compounds of phosphorus

Compound	Uses
phosphorus(V) sulfide, $P_2S_5(g)$	safety matches; oil additive
phosphorus(V) oxide, $P_4O_{10}(s)$	dehydrating agent
phosphoric acid, $H_3PO_4(l)$	fertilizers; soaps and detergents; soft drinks; soil stabilizer
sodium phosphates(s)	synthetic detergents; water softeners; leavening agents
calcium phosphates, $CaHPO_4(s)$ and $Ca(H_2PO_4)_2(s)$	fertilizers; poultry and animal feeds

Figure 26-15 The minerals orpiment (left), stibnite (center), and realgar (right).

found in Turkey, Russia, Eastern Europe, and Nevada. The most important ore of antimony is *stibnite* (Sb_2S_3), which is found in China, South Africa, Mexico, and Bolivia (Figure 26-15).

The sulfides of arsenic are converted to the oxides by roasting in air:

$$2As_2S_3(s) + 9O_2(g) \rightarrow As_4O_6(s) + 6SO_2(g)$$
$$As_4S_4(s) + 7O_2(g) \rightarrow As_4O_6(s) + 4SO_2(g)$$

The oxide is reduced to the element with carbon or hydrogen:

$$As_4O_6(s) + 3C(s) \rightarrow 3CO_2(g) + 4As(l)$$

Like other fourth-row posttransition elements, arsenic has a tendency to favor an oxidation state of two less than the maximum for the group. Thus, although PCl_5 and $SbCl_5$ are stable species, $AsCl_5$ decomposes above $-50°C$. Furthermore, antimony trioxide can be prepared by burning antimony in oxygen, but the pentoxide cannot be produced this way. The relative instability of the $+5$ oxidation state of arsenic means that As_4O_{10} and H_3AsO_4 are strong oxidizing agents.

In accord with the fact that the character of the oxides shifts from acidic to basic going down any one group in the periodic table, the oxides of arsenic are acidic:

$$As_4O_6 + 12NaOH(aq) \rightarrow \underset{\text{sodium arsenite}}{4Na_3AsO_3(aq)} + 6H_2O(l)$$

$$As_4O_{10}(s) + 12NaOH(aq) \rightarrow \underset{\text{sodium arsenate}}{4Na_3AsO_4(aq)} + 6H_2O(l)$$

and the oxides of antimony are amphoteric:

$$Sb_2O_3(s) + 6HCl(aq) \rightarrow 2SbCl_3(s) + 3H_2O(l)$$
$$Sb_2O_3(s) + 6NaOH(aq) \rightarrow \underset{\text{sodium antimonite}}{2Na_3SbO_3(aq)} + 3H_2O(l)$$

and bismuth(III) oxide is basic:

$$Bi_2O_3(s) + 6HCl(aq) \rightarrow 2BiCl_3(s) + 3H_2O(l)$$

Table 26-7 Some compounds of arsenic, antimony, and bismuth

Compound	Uses
lead(II) arsenate, $Pb_3(AsO_4)_2(s)$	insecticide
sodium arsenite, $NaAsO_2(s)$	arsenical soap; antiseptic; herbicide; insecticide; fungicide
arsenic(III) oxide, $As_4O_6(s)$	manufacture of glass; insecticide; rodenticide; wood preservative
antimony trichloride, $SbCl_3(s)$	fireproofing textiles
antimony sulfide, $Sb_2S_3(s)$	vermilion or yellow pigment; pyrotechnics; ruby glass
bismuth oxychloride, $BiOCl(s)$	face powder; artificial pearls
bismuth subnitrate, $BiONO_3(s)$	cosmetics; enamel flux

Although arsenic compounds are generally poisonous, trace amounts of arsenic are essential to the growth of red blood cells in bone marrow. The average healthy human body contains about 7 mg of arsenic.

Antimony is distinctly more metallic in character than arsenic; and numerous alloys contain antimony, which acts to prevent corrosion and to increase the resistance of the alloys to fracture as a result of thermal shock. Some compounds of arsenic and antimony and their uses are given in Table 26-7.

The principal compounds of bismuth contain bismuth(III), an oxidation state that is two less than the group number. Bismuth is a grey-white metal with a red tinge whose common source is the sulfide ore *bismuthinite*, Bi_2S_3.

The oxide Bi_2O_3 is soluble in strongly acidic aqueous solutions (Figure 26-16). The bismuthyl ion, $BiO^+(aq)$, and the bismuthate ion, $BiO_3^-(aq)$, are important in the aqueous-solution chemistry of bismuth. The bismuthyl ion forms insoluble compounds such as BiOCl and BiO(OH), whereas BiO_3^- is a powerful oxidizing agent. Some bismuth compounds and their major uses are given in Table 26-7.

Figure 26-16 Bismuth metal and bismuth(III) oxide, Bi_2O_3.

26-2. THE CHEMICAL PROPERTIES OF THE GROUP 6 ELEMENTS

8 **O** 15.9994 $2s^22p^4$
16 **S** 32.06 $3s^23p^4$
34 **Se** 78.96 $4s^24p^4$
52 **Te** 127.60 $5s^25p^4$
84 **Po** (209) $6s^26p^4$

The Group 6 elements progress as expected from nonmetallic to metallic properties with increasing atomic number (Figure 26-17). Oxygen, sulfur, and selenium are nonmetals, tellurium is a semimetal, and polonium is a semimetal with some metallic properties. As in other groups, there is a significant difference between the chemical properties of the first member and the other members. Oxygen is limited to two bonds (e.g., H_2O) or occasionally three bonds (e.g., H_3O^+), whereas the other members of the group may use *d* orbitals to form compounds such as SF_4 and TeF_6. As in Groups 4 and 5, there is a decrease in thermal stability of the binary hydrogen compounds in going from H_2S to H_2Po. In addition, as atomic number increases, there is an increasing stability of the oxidation state +4, which is two less than the group number, 6.

Oxygen is the most abundant element on earth and the third most abundant element in the universe, ranking behind hydrogen and helium. Most rocks contain a large amount of combined oxygen. For example, sand is predominantly silicon dioxide (SiO_2) and consists of more than 50 percent oxygen by mass. Almost 90 percent of the mass of the oceans and two thirds of the mass of the human body are oxygen. Air is 21 percent oxygen by volume. We can live weeks without food, days without water, but only minutes without oxygen.

Sulfur is widely distributed in nature but not usually in sufficient concentration to merit commercial mining. The two most important sources of sulfur are elemental sulfur from large salt domes offshore along the Gulf of Mexico and hydrogen sulfide from natural gas and petroleum refining. Selenium and tellurium are relatively rare ele-

Figure 26-17 The Group 6 elements. From left to right: oxygen, sulfur, selenium, and tellurium.

Table 26-8 Sources and uses of the Group 6 elements

Element	Principal sources	Uses
oxygen	fractional distillation of liquid air	blast furnaces; steel production; production of methyl alcohol, acetylene, ethylene oxide, etc.; rocket propellant; sewage treatment; breathing apparatus; many other uses
sulfur	native from underground deposits (Frasch process); natural gas and petroleum by-product	manufacture of sulfuric acid; paper manufacture; rubber vulcanization; drugs and pharmaceuticals, dyes; fungicides; insecticides
selenium	anode muds of the electrolyte refining of copper and lead	photocells; exposure meters; solar cells; xerography
tellurium	anode muds of the electrolyte refining of copper and lead	alloys to improve machinability of copper and stainless steels

ments. Both are found in association with metal sulfide ores and are obtained commercially as by-products of the refining of copper and lead (Table 26-8). Polonium has no stable isotopes, although minute quantities of polonium-210 occur in uranium ores.

Table 26-9 presents some atomic and physical properties of the Group 6 elements. The usual trends with increasing atomic number are evident.

26-2A. Over 35 Billion Pounds of Oxygen Are Produced Annually in the United States

Oxygen is an odorless, tasteless gas that exists as a diatomic molecule, O_2. Although colorless in the gaseous state, both liquid and solid oxygen are pale blue.

Table 26-9 Properties of the Group 6 elements

Element	Number of naturally occurring isotopes	Atomic radius/pm	Ionic radius/pm	Pauling atomic electro-negativity	Melting point/°C	Boiling point/°C	ΔH_{fus}/ $kJ\cdot mol^{-1}$	ΔH_{vap}/ $kJ\cdot mol^{-1}$	Density at 20°C/ $g\cdot cm^{-3}$
oxygen	3	60	140	3.5	−218.8	−183.0	0.443	6.82	1.33×10^{-3}
sulfur	4	100	184	2.5	119	445	1.72	8.37	1.96
selenium	6	115	196	2.4	221	685	5.44	—	4.79
tellurium	7	140	221	2.1	450	1009	17.5	50.6	6.24
polonium	0	190	—	2.0	254 ± 5	962 ± 2	—	—	9.4

Industrially, oxygen is produced by the **fractional distillation** of liquid air. The pure oxygen thus obtained is compressed in steel cylinders to a pressure of about 150 atm.

A frequently used method for preparing oxygen in the laboratory is the thermal decomposition of potassium chlorate, $KClO_3$. The chemical equation for the reaction is

$$2KClO_3(s) \xrightarrow{MnO_2(s)} 2KCl(s) + 3O_2(g)$$

This reaction requires a temperature of about 400°C; but if a small amount of the catalyst manganese dioxide (MnO_2) is added, the reaction occurs rapidly at 250°C. An alternate method for the laboratory preparation of oxygen is to add sodium peroxide (Na_2O_2) to water:

$$2Na_2O_2(s) + 2H_2O(l) \rightarrow 4NaOH(aq) + O_2(g)$$

This rapid and convenient reaction does not require heat. Oxygen also can be prepared by the electrolysis of water (Section 25-2C):

$$2H_2O(l) \xrightarrow{\text{electrolysis}} 2H_2(g) + O_2(g)$$

Almost 38 billion pounds of oxygen are produced annually in the United States, making it the third most important industrial chemical. The major commercial use of oxygen is in the blast furnaces used to manufacture steel. Oxygen is also used in hospitals, in oxyhydrogen and oxyacetylene torches for welding metals, and to facilitate breathing at high altitudes and under water. Tremendous quantities of oxygen are used directly from the air as a reactant in the combustion of fossil fuels, which supply 93 percent of the energy consumed in the United States.

26-2B. Oxygen in the Earth's Atmosphere Is Produced by Photosynthesis

Most of the oxygen in the atmosphere is the result of **photosynthesis,** the process by which green plants combine $CO_2(g)$ and $H_2O(g)$ into carbohydrates and $O_2(g)$ under the influence of visible light (Figure 26-18). The carbohydrates appear in the plants as starch, cellulose, and sugars. The reaction is described schematically as

$$CO_2(g) + H_2O(g) \xrightarrow{\text{visible light}} \text{carbohydrate} + O_2(g)$$

When the carbohydrate is glucose, we have

$$6CO_2(g) + 6H_2O(l) \rightarrow C_6H_{12}O_6(aq) + 6O_2(g) \qquad \Delta G^\circ_{rxn} = +2870 \text{ kJ at } 25°C$$

The reaction is driven up the Gibbs free energy "hill" ($\Delta G^\circ_{rxn} > 0$) by the energy obtained from sunlight.

In one year, more than 10^{10} metric tons of carbon is incorporated into carbohydrates by photosynthesis. In the hundreds of millions of years during which plant life has existed on earth, photosynthesis has produced much more oxygen than the amount now present in the atmosphere. Most of the oxygen released by plants is consumed by oxidation reactions involving both living and nonliving entities rather than remaining in the atmosphere.

Figure 26-18 Oxygen is produced by photosynthesis. An underwater plant produces oxygen gas, which appears as gas bubbles around the leaves.

26-2C. Oxygen Reacts Directly with Most Other Elements

Oxygen is very reactive and is the second most electronegative element. It reacts directly with all the other elements except the halogens, the noble gases, and some of the less reactive metals to form a wide variety of compounds. Only fluorine reacts with more elements than does oxygen. Compounds containing oxygen constitute 30 of the top 50 industrial chemicals (Appendix J).

Oxygen forms oxides with many elements. Most metals react rather slowly with oxygen at ordinary temperatures but react more rapidly as the temperature is increased. For example, iron, in the form of steel wool, burns vigorously in pure oxygen but does not burn in air. All hydrocarbons burn in oxygen to give carbon dioxide and water, including gasoline, which is a mixture of hydrocarbons. A combustion reaction with which we are all familiar is the burning of a candle. The wax in a candle is composed of long-chain hydrocarbons, such as $C_{20}H_{42}$. The molten wax rises up the wick to the combustion zone the way ink rises in a piece of blotting paper.

26-2D. Some Metals React with Oxygen to Yield Peroxides

Although most metals yield oxides when they react with oxygen, some of the more reactive metals, such as sodium, potassium, and rubidium, yield peroxides and superoxides. Peroxides are compounds in which the negative ion is the peroxide ion, O_2^{2-}, and superoxides are compounds in which the negative ion is the superoxide ion, O_2^-. For example,

$$2Na(s) + O_2(g) \rightarrow \underset{\text{sodium peroxide}}{Na_2O_2(s)}$$

$$K(s) + O_2(g) \rightarrow \underset{\text{potassium superoxide}}{KO_2(s)}$$

One of the most important peroxides is hydrogen peroxide, H_2O_2, a colorless, syrupy liquid that explodes violently when heated. Hydrogen peroxide is a strong oxidizing agent that oxidizes a wide variety of organic substances. Dilute aqueous solutions of hydrogen peroxide are fairly safe to use. A 3 percent aqueous solution is sold in drugstores and used as a mild antiseptic and as a bleach. More concentrated solutions (30 percent) of hydrogen peroxide are used industrially as a bleaching agent for hair, flour, textile fibers, fats, and oils, in the artificial aging of wines and liquor, and for control of pollution in sewage effluents (Table 26-10).

Table 26-10 Commercially available aqueous solutions of hydrogen peroxide

$[H_2O_2]$/%	Use
3	antiseptic
6	hair bleach
30	industrial and laboratory oxidizing agent
85	potent oxidizing agent

26-2E. Ozone Is a Potent Oxidizing Agent

When a spark is passed through oxygen, some of the oxygen is converted to ozone, O_3:

$$3O_2(g) \rightarrow 2O_3(g)$$

Ozone is a light blue gas at room temperature. It has a sharp, characteristic odor that is frequently apparent after electrical storms and near high-voltage generators and copying machines. Liquid ozone (boiling point, −112°C) is a deep blue, explosive liquid (Figure 26-19). Ozone is so reactive that it cannot be transported safely but must be generated as needed. Relatively unreactive metals such as silver and mercury, which do not react with oxygen, react with ozone to form oxides. Ozone is used as a bleaching agent and is being considered as a replacement for chlorine in water treatment because of the environmental problem involving chlorinated hydrocarbons.

(a)

(b)

Figure 26-19 (a) Although gaseous oxygen is colorless, liquid oxygen is pale blue. (b) Solid ozone is dark blue.

EXAMPLE 26-5: The quantity of ozone in a sample can be determined by means of the quantitative reaction of ozone with iodide in acidic solution. The unbalanced equation is

$$I^-(aq) + O_3(g) \rightarrow O_2(g) + I_3^-(aq)$$

Balance this equation.

Solution: The equations for the two half reactions are

$$I^- \rightarrow I_3^-$$
$$O_3 \rightarrow O_2$$

Following the steps given in Section 21-4, we get the two balanced equations for the half reactions:

$$3I^- \rightarrow I_3^- + 2e^-$$
$$2H^+ + O_3 + 2e^- \rightarrow H_2O + O_2$$

Add the two balanced equations for the half reactions and include the phases to obtain

$$O_3(g) + 3I^-(aq) + 2H^+(aq) \rightarrow H_2O(l) + O_2(g) + I_3^-(aq)$$

PRACTICE PROBLEM 26-5: Ozone oxidizes acidified iron(II) solutions to iron(III) solutions, with the color changing from green to yellow in the process. Given that $O_2(g)$ is one of the products, complete and balance the equation for the reaction.

Answer:

$$2H^+(aq) + 2Fe^{2+}(aq) + O_3(g) \rightarrow 2Fe^{3+}(aq) + O_2(g) + H_2O(l)$$

Ozone is produced in the stratosphere by the reactions

$$O_2(g) \xrightarrow{h\nu} 2O(g)$$
$$O(g) + O_2(g) \rightarrow O_3(g)$$

Stratospheric ozone screens out ultraviolet radiation in the wavelength region 240 nm to 310 nm via the reaction

$$O_3(g) \xrightarrow{h\nu} O_2(g) + O(g)$$

and thereby protects life on earth from the destructive effects of this radiation (Interchapter C).

26-2F. Sulfur Exists as Rings of Eight Sulfur Atoms

Sulfur is a yellow, tasteless, odorless solid that is often found in nature as the free element (Frontispiece). Sulfur is essentially insoluble in water but dissolves readily in carbon disulfide, CS_2 (Figure 26-20). It does not react with dilute acids or bases, but it does react with many metals at elevated temperatures to form metal sulfides.

Prior to 1900, most of the world's supply of sulfur came from Sicily, where sulfur occurs at the surfaces around hot springs and volcanoes. In the early 1900s, however, large subsurface deposits of sulfur were found along the Gulf Coast of the United States. The sulfur occurs in limestone caves, more than 1000 feet beneath layers of rock, clay, and quicksand. The recovery of the sulfur from these deposits posed a great technological problem that was solved by the engineer Herman Frasch. The **Frasch process** uses an arrangement of three concentric pipes (diameters of 1, 3, and 6 inches) placed in a bore hole that penetrates to the base of the sulfur-bearing calcite ($CaCO_3$) rock formation (Figure 26-21a). Pressurized hot water (180°C) is forced down the space between the 6-inch and 3-inch pipes to melt the sulfur (melting point, 119°C). The molten sulfur, which is twice as dense as water, sinks to the bottom of the deposit and then is forced up the space between the

Figure 26-20 Sulfur is insoluble in water (left) but is soluble in carbon disulfide, CS_2, (right).

Hot compressed air
Liquid sulfur
180°C water at
7 atm pressure
Calcite
Sulfur-bearing
calcite
Liquid
sulfur

(a)

(b)

Figure 26-21 The Frasch process for sulfur extraction. (a) Three concentric pipes are sunk into sulfur-bearing calcite rock. Water at 180°C and 7 atm is forced down the outermost pipe to melt the sulfur. Hot compressed air is forced down the innermost pipe and mixes with the molten sulfur to form a foam of water, air, and sulfur. The mixture rises to the surface through the middle pipe. The sulfur, when dried, has a purity of 99.5 percent.
(b) Sulfur with hot water and air can be seen surfacing from Culberson Mine in West Texas.

3-inch and 1-inch pipes as a foam by the action of compressed air injected through the innermost pipe (Figure 26-21b). The molten sulfur rises to the surface, where it is pumped into tank cars for shipment or into a storage area. About 40 percent of the U.S. annual sulfur production of over 11 million metric tons is obtained by the Frasch process from the region around the Gulf of Mexico in Louisiana and Texas.

About 50 percent of U.S. and most Canadian sulfur is produced by the **Claus process,** in which sulfur is obtained from the hydrogen sulfide that occurs in some natural gas deposits and from that produced when sulfur is removed from petroleum. In the Claus process, hydrogen sulfide is burned in air to produce sulfur dioxide, which is then reacted with additional hydrogen sulfide to produce sulfur:

$$2H_2S(g) + 3O_2(g) \rightarrow 2SO_2(g) + 2H_2O(g)$$

$$SO_2(g) + 2H_2S(g) \rightarrow 3S(s) + 2H_2O(g)$$

These reactions also are thought to be responsible for the surface deposits of sulfur around hot springs and volcanoes.

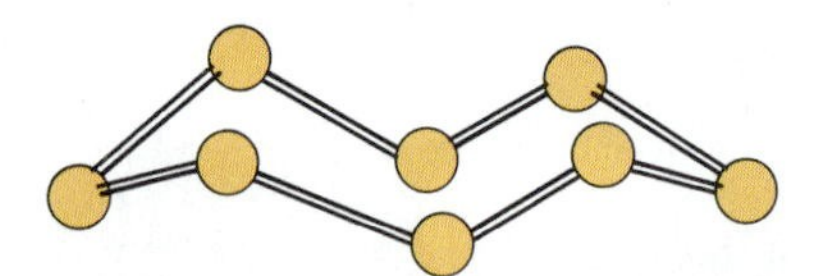

Figure 26-22 Under most conditions, sulfur exists as eight-membered rings, S_8. The ring is not flat but puckered in such a way that four of the atoms lie in one plane and four lie in another plane.

Below 96°C, sulfur exists as yellow, transparent, rhombic crystals. If rhombic sulfur is heated above 96°C, then it becomes opaque and the crystals expand into monoclinic crystals. Monoclinic sulfur is the stable form from 96°C to the melting point. The molecular units of the rhombic form are rings containing eight sulfur atoms, S_8 (Figure 26-22). The molecular units of monoclinic sulfur are also S_8 rings, but the rings themselves are arranged differently in the crystal.

Monoclinic sulfur melts at 119°C to a thin, pale yellow liquid consisting of S_8 rings. Upon heating to about 150°C there is little change; but beyond 150°C, the liquid sulfur begins to thicken and turns reddish brown. By 200°C, the liquid is so thick it hardly pours (Figure 26-23).

(a)

(b)

(c)

Figure 26-23 Molten sulfur at various temperatures. (a) Sulfur melts at 119°C to a thin, yellow liquid. (b) By 200°C the liquid turns reddish brown and is so thick that it hardly pours. (c) Upon further heating, the liquid becomes fluid once again and pours easily.

The molecular explanation for this behavior is simple. At about 150°C, thermal agitation causes the S_8 rings to begin to break apart and form chains of sulfur atoms:

```
   S—S
  /   \
 S     S
 |     |                 S    S    S    S·
 S     S  →  ·S  /  \  /  \  /  \  /
  \   /          S    S    S    S
   S—S
```

These free radical chains can then join together (polymerize) to form longer chains, which become entangled in one another and cause the liquid to thicken. Above 250°C, the liquid begins to flow more easily because the thermal agitation is sufficient to begin to break the chains of sulfur atoms. At the boiling point (445°C), liquid sulfur pours freely and the vapor molecules consist mostly of S_8 rings.

If liquid sulfur at about 200°C is placed quickly in cold water (this process is called **quenching**), then a rubbery substance known as plastic sulfur is formed (Figure 26-24). The material is rubbery because the long, coiled chains of sulfur atoms can straighten out some if they are pulled. As plastic sulfur cools, it slowly becomes hard again as it rearranges itself into the rhombic form.

Figure 26-24 If liquid sulfur at about 200°C is cooled quickly by pouring it into cold water, a rubbery substance called plastic sulfur is formed.

26-2G. Sulfuric Acid Is the Leading Industrial Chemical

Sulfur, which constitutes only 0.05 percent of the earth's crust, is not one of the most prevalent elements. Yet it is one of the most commercially important ones because it is the starting material for the most important industrial chemical, sulfuric acid. Most sulfuric acid is made by the **contact process.** First the sulfur is burned in oxygen to produce sulfur dioxide:

$$S(s) + O_2(g) \rightarrow SO_2(g)$$

The sulfur dioxide is then converted to sulfur trioxide in the presence of the catalyst vanadium pentoxide:

$$2SO_2(g) + O_2(g) \xrightarrow{V_2O_5(s)} 2SO_3(g)$$

The sulfur trioxide is absorbed into nearly pure liquid sulfuric acid to form **fuming sulfuric acid (oleum):**

$$H_2SO_4(l) + SO_3(g) \rightarrow \underset{\text{oleum}}{H_2S_2O_7} \ (35\% \ in \ H_2SO_4)$$

The oleum is added to water or aqueous sulfuric acid to produce the desired final concentration of aqueous sulfuric acid. Sulfur trioxide cannot be absorbed directly in water because the acid mist of H_2SO_4 that forms is very difficult to condense.

Almost 90 billion pounds of sulfuric acid are produced annually in the United States. Commercial-grade sulfuric acid is one of the least expensive chemicals, costing less than 10 cents per pound in bulk quantities. Very large quantities of sulfuric acid are used in the production of fertilizers and numerous industrial chemicals, the petroleum industry, metallurgical processes, synthetic fiber production, and paints, pigments, and explosives manufacture.

others were named helium (sun), neon (new), krypton (hidden), and xenon (stranger). Helium was named after the Greek word for sun (*helios*) because its presence in the sun had been determined earlier by spectroscopic methods. For their work in discovering and characterizing an entire new family of elements, Rayleigh received the 1904 Nobel Prize in physics and Ramsay received the 1904 Nobel Prize in chemistry.

All the noble gases are colorless, odorless, and relatively inert. Despite the fact that helium is denser and hence has less lifting power than hydrogen, it is used in lighter-than-air craft because it is nonflammable. Helium is also used in welding to provide an inert atmosphere around the welding flame and thus reduce corrosion of the heated metal. Neon is used in neon signs, which are essentially discharge tubes filled with a single noble gas or a noble-gas mixture. When placed in a discharge tube, neon emits an orange-yellow glow that penetrates fog very well. Argon, the most plentiful and least expensive noble gas, often is used in fluorescent and incandescent light bulbs because it does not react with the discharge electrodes or the hot filament. Krypton and xenon are scarce and costly, which limits their application, although they are used in lasers, flashtubes for high-speed photography, and automobile-engine timing lights (Figure 26-38).

Figure 26-38 Xenon flash lamps.

26-4B. Xenon Forms Compounds with Fluorine and Oxygen

Prior to 1962 most chemists believed, and all chemistry textbooks proclaimed, that the noble gases did not form any chemical compounds. In fact, the gases helium through xenon were called the inert gases, indicating that they did not undergo any chemical reactions.

In 1962, Neil Bartlett, then of the University of British Columbia, was working with the extremely strong oxidizing agent platinum hexafluoride, PtF_6, which oxidizes $O_2(g)$ to produce the ionic compound $O_2^+PtF_6^-$:

$$O_2(g) + PtF_6(s) \rightarrow O_2^+PtF_6^-$$

Bartlett realized that the ionization energy of $O_2(g)$ ($1171\ kJ{\cdot}mol^{-1}$) is about the same as the ionization energy of $Xe(g)$ ($1176\ kJ{\cdot}mol^{-1}$), so he reasoned that xenon might react with PtF_6 in an analogous manner. When he mixed xenon and PtF_6 in a reaction chamber, he obtained a definite chemical reaction that at the time was thought to be the formation of $Xe^+PtF_6^-$. It has since been found that the product of the reaction is more complex than $Xe^+PtF_6^-$, but nevertheless Bartlett showed that xenon will react with a strong oxidizing agent under the right conditions of temperature and pressure. Bartlett's discovery prompted other research groups to investigate reactions of xenon, and within a year or so several other compounds of xenon were synthesized.

Figure 26-39 Xenon tetrafluoride crystals. Xenon tetrafluoride was first prepared in 1962 by the direct combination of $Xe(g)$ and $F_2(g)$ at 6 atm and 400°C.

Three xenon fluorides can be prepared by the direct combination of xenon and fluorine in a nickel vessel:

$$Xe(g) + F_2(g) \rightleftharpoons XeF_2(s)$$
$$XeF_2(s) + F_2(g) \rightleftharpoons XeF_4(s)$$
$$XeF_4(s) + F_2(g) \rightleftharpoons XeF_6(s)$$

As these three equations indicate, the reaction of a mixture of xenon and fluorine yields a mixture of XeF_2, XeF_4, and XeF_6 (Figure 26-39). The chief difficulty is the separation of the products. A favorable yield of XeF_2 can be obtained by using a large excess of xenon. Xenon difluoride forms large, colorless crystals that melt at 130°C. It is a linear molecule, as predicted by VSEPR theory (AX_2E_3). Xenon difluoride is soluble in water and evidently exists as XeF_2 molecules in solution. Xenon tetrafluoride can be obtained in quantitative yield by reacting a 1:5 mixture of Xe and F_2 at 400°C and 6 atm in a nickel vessel. Xenon tetrafluoride forms colorless crystals that melt at 177°C. The molecule is square planar, as predicted by VSEPR theory (AX_4E_2).

Xenon forms chemical bonds with the most electronegative elements, fluorine and oxygen, and exhibits oxidation states of +2, +4, +6, and +8. Xenon, having the greatest atomic size of any of the nonradioactive noble gases, has the smallest ionization energy. Hence, except for radon, xenon is the most "reactive" noble gas, and we expect the reactivity of the noble gases to decrease from xenon to helium. The only known molecule containing krypton is KrF_2, and no isolable compounds of argon have yet been reported.

TERMS YOU SHOULD KNOW

phosphate rock 928
nitrogen fixation 930
Haber process 930
fountain effect 930
Ostwald process 931
Raschig synthesis 932
azide 933
allotrope 934
white phosphorous 934
red phosphorous 934
superphosphate 936
baking powder 937
adenine triphosphate (ATP) 938
fractional distillation 944
photosynthesis 944
Frasch process 947
Claus process 948
quenching 949
contact process 949
oleum (fuming sulfuric acid) 949
mercaptan 952
thio- 953
halogen 955
bifluoride ion 958
tincture of iodine 962

AN EQUATION YOU SHOULD KNOW HOW TO USE

$pK_a = 8 - 5.5\, n_O$ (26-1) (approximate value of the pK_a of an oxyacid)

PROBLEMS

GROUP 5 ELEMENTS

26-1. Complete and balance the following equations:

(a) $P_4(s) + \underset{\text{excess}}{O_2(g)} \rightarrow$
(b) $P_4O_6(s) + H_2O(l) \rightarrow$
(c) $P_4O_{10}(s) + H_2O(l) \rightarrow$

26-2. Complete and balance the following equations:

(a) $NH_4NO_3(s) \xrightarrow[\text{heat}]{}$
(b) $N_2O_5(s) + H_2O(l) \rightarrow$
(c) $NH_4NO_2(aq) \xrightarrow[\text{heat}]{}$

26-3. How could you prepare $ND_3(g)$, using D_2O as a source of deuterium?

26-4. Why do solutions of nitric acid often have a brown-yellow color?

26-5. Why are there no nitrogen pentahalides?

26-6. Write Lewis formulas for the following species (include resonance forms where appropriate).

(a) N_2O_4 (b) N_2O_3 (c) CN_2^{2-}
(d) H_2N_2 (e) N_3^-

26-7. Write Lewis formulas for the following species (include resonance forms where appropriate).

(a) $HONH_2$ (b) NO_3^- (c) NO_2^-
(d) N_2O (e) NO_2

26-8. How could you prepare PD_3, using D_2O as a source of deuterium?

26-9. Write chemical equations for the reactions of phosphorus pentoxide with sulfuric acid and with nitric acid.

26-10. Phosphine dissolves in liquid ammonia to give $NH_4^+PH_2^-$. Use VSEPR theory to predict the shape of PH_2^-.

26-11. Use VSEPR theory to predict the shapes of

(a) $POCl_3$ (b) PO_4^{3-} (c) PCl_6^-

26-12. When P_4O_{10} and P_4S_{10} are heated in the appropriate proportions above 400°C, $P_4O_6S_4$, a colorless, hygroscopic crystalline substance, is obtained. Using Figure 26-10 as a guide, predict the structure of $P_4O_6S_4$. What about the structure of $P_4O_4S_6$?

26-13. Write chemical equations describing the acidic characters of As_4O_6 and As_4O_{10}, the amphoteric character of Sb_2O_3, and the basic character of Bi_2O_3.

26-14. For the equilibrium

$$\underset{\text{brown}}{2NO_2(g)} \rightleftharpoons \underset{\text{colorless}}{N_2O_4(g)} \qquad \Delta H^\circ_{rxn} = -57.2 \text{ kJ}$$

does the reaction mixture become increasingly colored with increasing or decreasing temperature?

THE GROUP 6 ELEMENTS

26-15. What is the source of most of the oxygen in the earth's atmosphere?

26-16. Give two methods used to produce small quantities of oxygen in the laboratory.

26-17. What is the heat-producing reaction of an oxyacetylene torch?

26-18. It has been determined that in the oxidation of H_2O_2 in aqueous solution by $MnO_4^-(aq)$, $Ce^{4+}(aq)$, and other strong oxidizing agents, the $O_2(g)$ produced comes entirely from the H_2O_2 and not from water. How could this be determined?

26-19. Describe what happens at various stages when sulfur (initially in the rhombic form) is heated slowly from 90°C to 450°C.

26-20. Describe, using balanced chemical equations, the contact process for the manufacture of sulfuric acid.

26-21. The gas inside some tennis balls is about 50 percent SF_6 and 50 percent air. Such balls retain their bounce longer than balls charged solely with air. Suggest an explanation based on molecular size for this observation.

26-22. There are two known isomers of S_2F_2 with significantly different sulfur-sulfur bond lengths. Propose structures for the two isomers and write the Lewis formulas.

26-23. Draw a Lewis formula for disulfuric acid, $H_2S_2O_7$ (there is an S—O—S linkage).

26-24. Draw a Lewis formula for peroxomonosulfuric acid, H_2SO_5. This acid is also known as Caro's acid.

26-25. The compounds SF_4 and SF_6 are both very unstable with respect to reaction with water, as shown by the following ΔG°_{rxn} values.

(a) $SF_4(g) + 2H_2O(l) \rightarrow$
$$SO_2(aq) + 4HF(aq) \qquad \Delta G^\circ_{rxn} = -282 \text{ kJ}$$

(b) $SF_6(g) + 4H_2O(l) \rightarrow$
$$2H^+(aq) + SO_4^{2-}(aq) + 6HF(aq) \qquad \Delta G^\circ_{rxn} = -472 \text{ kJ}$$

Although $SF_4(g)$ reacts rapidly with water, $SF_6(g)$ does not, being inert even to hot $NaOH(aq)$ or $HNO_3(aq)$. Consider the bonding in the two molecules and offer an explanation for the observed difference in reactivities.

26-26. Given the thermodynamic data, $\Delta G_f^\circ[SF_4(g)] = -731.3 \text{ kJ·mol}^{-1}$ and $\Delta G_f^\circ[SF_6(g)] = -1105.3 \text{ kJ·mol}^{-1}$ at 25°C, calculate the equilibrium constant of the reaction

$$SF_4(g) + F_2(g) \rightleftharpoons SF_6(g)$$

Given that $\Delta H^\circ_{rxn} = -434.1$ kJ at 25°C, is the production of SF_6 more favored at high or low temperatures?

26-27. A 35.0-mL sample of $I_3^-(aq)$ requires 28.5 mL of 0.150 M $Na_2S_2O_3(aq)$ to react with all the $I_3^-(aq)$. Calculate the concentration of $I_3^-(aq)$ in the sample.

26-28. The concentration of ozone in oxygen-ozone mixtures can be determined by passing the gas mixture into a buffered $KI(aq)$ solution. The $O_3(g)$ oxidizes $I^-(aq)$ to $I_3^-(aq)$:

$$O_3(g) + 3I^-(aq) + H_2O(l) \rightarrow O_2(g) + I_3^-(aq) + 2OH^-(aq)$$

The concentration of $I_3^-(aq)$ formed is then determined by titration with $Na_2S_2O_3(aq)$. Given that 22.50 mL of 0.0100 M $Na_2S_2O_3(aq)$ are required to titrate the $I_3^-(aq)$ in a 50.0-mL sample of $KI(aq)$ that was equilibrated with a $O_2 + O_3$ sample, calculate the number of moles of O_3 in the sample.

THE HALOGENS

26-29. Balance the following equations:

(a) $NaCl(aq) + H_2SO_4(aq) + MnO_2(s) \rightarrow$
$$Na_2SO_4(aq) + MnCl_2(aq) + H_2O(l) + Cl_2(g)$$

(b) $NaIO_3(aq) + NaHSO_3(aq) \rightarrow$
$$I_2(s) + Na_2SO_4(aq) + H_2SO_4(aq) + H_2O(l)$$

(c) $Br_2(l) + NaOH(aq) \rightarrow$
$$NaBr(aq) + NaBrO_3(aq) + H_2O(l)$$

26-30. Chlorine oxidizes iodine to iodic acid in water. Balance the following equation:

$$Cl_2(g) + I_2(s) + H_2O(l) \rightarrow HCl(aq) + HIO_3(aq)$$

26-31. Iodine is oxidized to iodic acid by concentrated nitric acid. Balance the following equation:

$$I_2(s) + HNO_3(aq) \rightarrow HIO_3(aq) + NO(g) + H_2O(l)$$

26-32. Name the following oxyacids:

(a) $HBrO_2$ (b) HIO (c) $HBrO_4$ (d) HIO_3

26-33. Name the following oxyacids:

(a) HNO_2 (b) H_2SO_3 (c) H_3PO_2
(d) H_3PO_3 (e) $H_2N_2O_2$

and the following salts:

(a) K_2SO_3 (b) $Ca(NO_2)_2$
(c) KIO_2 (d) $Mg(BrO)_2$

26-34. Why is the heat evolved per mole in the neutralization reaction of $HCl(aq)$ by $KOH(aq)$ the same as that for $HBr(aq)$ by $KOH(aq)$? Why is the heat evolved per mole much less for the reaction $HF(aq)$ plus $KOH(aq)$?

26-35. Suggest an explanation for why, in contrast to NH_3, NF_3 is not at all basic.

26-36. What is the oxidation state of the oxygen atom in HOF, an unstable substance that decomposes to HF and O_2?

26-37. When perchloric acid is dehydrated by P_4O_{10}, a colorless, unstable oily liquid, Cl_2O_7, is produced. Use VSEPR theory to describe the shape of Cl_2O_7 (there is a Cl—O—Cl bond).

26-38. Use VSEPR theory to predict the shapes of the following interhalogen cations.

(a) ClF_2^+ (b) ClF_4^+ (c) ClF_6^+

26-39. Determine the oxidation state of each halogen in the following compounds.

(a) IF_5 (b) $NaClO$ (c) $KBrO_3$
(d) ClF (e) $NaIO_3$

26-40. The acid $HF(aq)$ differs from the other hydrohalic acids in that it is a weak acid (25°C data):

$$HF(aq) + H_2O(l) \rightleftharpoons H_3O^+(aq) + F^-(aq) \qquad pK_a = 3.17$$

and in that the ion $HF_2^-(aq)$ forms readily:

$$HF(aq) + F^-(aq) \rightleftharpoons HF_2^-(aq) \qquad K = 5.1\ M^{-1}$$

Suppose we have a solution with a stoichiometric concentration of $HF(aq)$ of 0.10 M that is buffered at pH = 3.00. Calculate the concentrations of $F^-(aq)$, $HF(aq)$, and $HF_2^-(aq)$ in the solution.

26-41. Given that $\Delta G_f^\circ[I_2(aq)] = 16.40\ kJ\cdot mol^{-1}$, $\Delta G_f^\circ[I^-(aq)] = -51.57\ kJ\cdot mol^{-1}$, and $\Delta G_f^\circ[I_3^-(aq)] = -51.40\ kJ\cdot mol^{-1}$ at 25°C, calculate the equilibrium constant for the reaction

$$I_2(aq) + I^-(aq) \rightleftharpoons I_3^-(aq)$$

26-42. Given that $\Delta G_f^\circ[ICl(aq)] = -17.1\ kJ\cdot mol^{-1}$, $\Delta G_f^\circ[Cl^-(aq)] = -131.23\ kJ\cdot mol^{-1}$, and $\Delta G_f^\circ[ICl_2^-(aq)] = -161.0\ kJ\cdot mol^{-1}$ at 25°C, calculate the equilibrium constant for the reaction

$$ICl(aq) + Cl^-(aq) \rightleftharpoons ICl_2^-(aq)$$

26-43. Iodine pentoxide is a reagent for the quantitative determination of carbon monoxide. The reaction is

$$5CO(g) + I_2O_5(s) \rightarrow I_2(s) + 5CO_2(g)$$

The iodine produced is dissolved in $KI(aq)$ and then determined by reaction with $Na_2S_2O_3$:

$$2S_2O_3^{2-}(aq) + I_3^-(aq) \rightarrow 3I^-(aq) + S_4O_6^{2-}(aq)$$

Calculate the moles of CO required to produce sufficient $I_3^-(aq)$ to react completely with the $S_2O_3^{2-}(aq)$ in 10.0 mL of 0.0350 M $Na_2S_2O_3(aq)$.

26-44. The solubility of $I_2(s)$ in water at 25°C is 0.0013 M and

$$I_2(aq) + I^-(aq) \rightleftharpoons I_3^-(aq) \qquad K = 700\ M^{-1}$$

at 25°C. Calculate the solubility of $I_2(s)$ in a solution that is initially 0.10 M in $KI(aq)$.

THE NOBLE GASES

26-45. Discuss how the noble gases were discovered by Lord Rayleigh.

26-46. What is the source of $He(g)$ in natural gas deposits?

26-47. Sketch an experimental setup for removing O_2, H_2O, and CO_2 from air. How could you remove the remaining N_2?

26-48. Nitrogen is a relatively inert gas. Suggest an experiment to distinguish between nitrogen and argon.

26-49. When Bartlett prepared $O_2^+PtF_6^-$ in 1962, what reasoning did he use to conjecture that it might be possible to prepare $Xe^+PtF_6^-$?

26-50. Use the data in Table 26-18 to calculate the values of ΔS_{vap} and ΔS_{fus} for the noble gases. Compare your results with the values of ΔS_{vap} from Trouton's rule (see Problem 14-44).

26-51. Why do both van der Waals constants, *a* and *b*, increase with increasing atomic number for the noble gases?

26-52. Why does ΔH_{vap} increase with increasing atomic number for the noble gases?

Chemistry and Photography

A **photograph** is a permanent, visible image formed on a light-sensitive surface by the action of light (Figure L-1). The essential chemical process is the darkening of white, crystalline silver salts such as silver chloride, AgCl(*s*), when exposed to sunlight. The underlying chemical reaction is the photochemical reduction of the silver ion Ag^+ of the salt to elemental silver Ag(*s*). As more and more silver ions are reduced, the resultant silver atoms aggregate into clusters of a sufficient size to scatter light, and hence appear as small, dark specks.

The first known application of the photoreduction of silver ions seems to have been as early as 1727, when the German physician J. H. Schulze spread a paste of chalk and silver nitrate, $AgNO_3(s)$, over a flat, solid surface and, using stencils, produced a darkened image upon exposing the surface to sunlight. Unfortunately, Schulze's images could not be viewed in direct light because the unexposed silver salts were immediately reduced, thus darkening the entire surface. It was not until 1819 that the British chemist John Herschel discovered the fixing properties of sodium thiosulfate (hypo), $Na_2S_2O_3$, thus paving the way for permanent pictorial reproductions. It was Herschel who gave us the terms "photography,"

(a)

(b)

Figure L-1 (a) A key resting on silver halide crystals is exposed to light. In (b) the key is removed, showing the exposed and the unexposed silver halide crystals.

"negative," and "fixing," and who later suggested using glass as the flat surface to support the layer of silver salt, and hence the term "photographic plate." Photography became available to everyone when George Eastman introduced film in the form of a flexible roll of celluloid in 1889.

There are essentially five steps in producing a photograph:

1. Preparing the light-sensitive surface
2. Producing a latent image by exposing the surface to light
3. Developing the latent image to produce a negative
4. Making the image permanent by fixing it
5. Producing a positive print from the negative by development and fixation

Let's look at each of these steps in more detail.

In modern photographic films, the light-sensitive material is a silver halide, AgX(*s*), or a mixture of silver halides (the photographic industry uses over 3 million pounds of silver per year). The silver halide is produced in the form of very small crystals, or "grains," of diameters less than a millionth of a meter, by controlled precipitation through the reaction of silver nitrate with an alkali halide; for example,

$$AgNO_3(aq) + NaBr(aq) \rightarrow NaNO_3(aq) + AgBr(s)$$

The type of halide that is used depends upon the particular film, but silver bromide, AgBr, is the most common. Silver iodide, AgI, and mixtures involving silver iodide are used for especially fast film. Silver halide crystals are sensitive only to light with wavelengths shorter than about 500 nm, which is in the blue region of the spectrum; they are said to be blue sensitive, but green and red insensitive. To achieve sensitivity in the green (500–600 nm) and red (600–700 nm) regions, the crystals are washed in dye solutions, so that the dye molecules absorb onto the surfaces of the crystals. These dye molecules then absorb radiation in the green and red regions and transfer their energy to the crystal. Sensitization to green and red as well as blue is important even in black-and-white photography, because the different colors are represented as various shades of gray. The sensitized silver halide grains are then uniformly dispersed in some gelatinous substance such as gelatin, which is then coated onto a support such as plastic (celluloid), paper, or glass. This dispersion of silver halide crystals is called an **emulsion.**

When the emulsion is exposed to light, photons fall upon the silver halide grains and interact with the halide ions. The halide ions lose electrons, which migrate to the surface of the crystal and reduce silver ions:

$$X^-(s) + h\nu \rightarrow X(s) + e^-$$
$$Ag^+(surface) + e^- \rightarrow Ag(surface)$$

The reduced silver atoms at the surfaces of the grains aggregate into small clusters containing 10 to 100 atoms. The distribution of these little clusters varies over the surface of the emulsion according to the pattern and intensity of the exposure, forming what is called a **latent image.**

The latent image is now intensified by a **developer,** which is a mild reducing agent that preferentially reduces the silver ions in the grains that have been exposed. The grains which have not been exposed develop slowly or not at all. A common developer is hydroquinone, and the chemical reaction of the development process is

$$2AgX(s) + HO{-}C_6H_4{-}OH(aq) + 2OH^-(aq) \rightarrow$$

silver halide   hydroquinone

$$2Ag(s) + O{=}C_6H_4{=}O(aq) + 2H_2O(l) + 2X^-$$

quinone

Once development has reached the desired point, the developer reaction is quenched by transferring the film from the mildly basic developer solution to an acidic "stop" bath, which is commonly dilute acetic acid.

The actual mechanism by which the developer reduces all the silver ions in a grain that contains just one or at most a few clusters of silver atoms on its surface is not well understood. But when you realize that a grain of AgBr with a diameter of 10^{-6} m contains about 10^{11} silver ions, and that the surface clusters of silver atoms contain only about 100 silver atoms, you see that there is an amplification by a factor of 10^9. This tremendous amplification is the primary reason for the preference of silver halides over other photosensitive materials.

Now the film is **fixed** by removing all the remaining (unexposed) silver halide; otherwise it would be reduced when the negative is exposed to sunlight, resulting in a totally black negative. The remaining silver halide is commonly removed with an aqueous solution of sodium thiosulfate, $Na_2S_2O_3(aq)$ (hypo), which reacts with the silver halide according to

$$AgX(s) + 2Na_2S_2O_3(aq) \rightarrow Na_3[Ag(S_2O_3)_2](aq) + NaX(aq)$$

The products in this reaction are water soluble, and so the hypo solution effectively solubilizes the water-insoluble silver halide, which is then simply washed off the negative. The result is a permanent negative that can be exposed to light with no adverse effect. It is called a negative because the areas that correspond to the greatest exposure appear the darkest. A positive print is obtained by passing light through the negative onto printing paper, which is specially impregnated paper which is developed and fixed by a process similar to that used to produce the negative.

Up to this point we have discussed only black-and-white photography. To understand how color photography works, we must first discuss the nature of colored light. Experiments with colored light have established that all the colors that can be recognized by the human eye can be composed from three fundamental or **primary colors.** One of the primary colors may be chosen to be magenta, which absorbs green light but transmits red and blue. A good example of a magenta substance is the $Ti(H_2O)_6^{3+}(aq)$ ion (Figure L-2). A solution of $Ti(H_2O)_6^{3+}(aq)$ absorbs green light but permits blue and red light to pass through it, thus producing a magenta color (Figure L-3). Magenta is thus said to **complementary** to green (Figure L-4). As Figure L-4 shows, the other two primary colors are yellow, which absorbs blue but transmits green and red; and cyan (greenish blue), which absorbs red but transmits blue and green. A combination of yellow with magenta appears red; yellow with cyan appears green; and magenta with cyan appears blue. A combination of yellow, cyan, and magenta pro-

Figure L-2 A solution of $Ti(H_2O)_6^{3+}(aq)$ appears magenta because it absorbs green light and transmits red and blue light (Figure L-3).

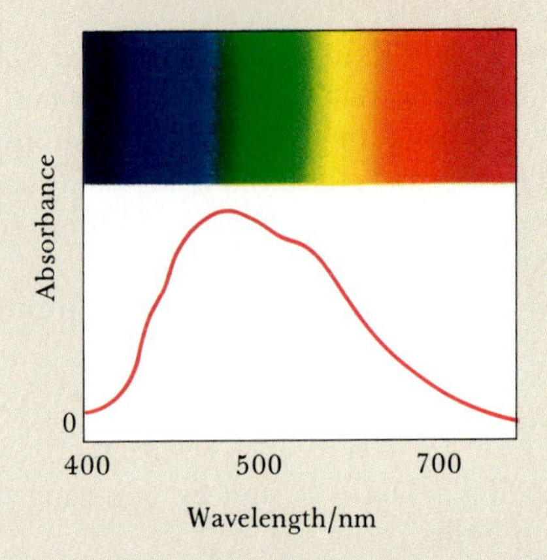

Figure L-3 Absorption spectrum of the magenta ion, $Ti(H_2O)_6^{3+}(aq)$. The absorption in the green region of the spectrum causes solutions of $Ti(H_2O)_6^{3+}(aq)$ and colorless anions to appear magenta.

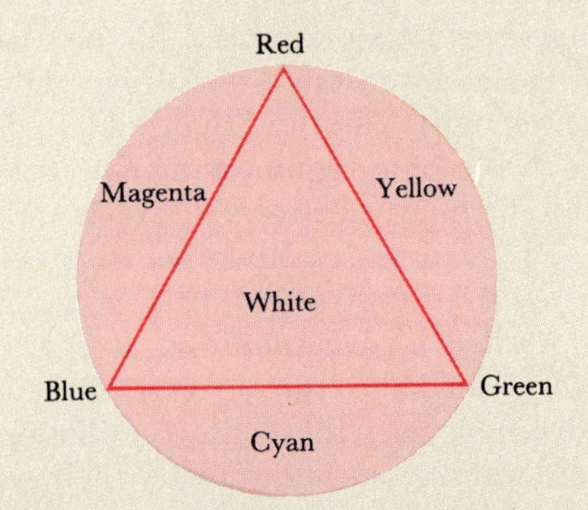

Figure L-4 A triangular illustration of the composition of the three primary colors and their complementary colors. For example, magenta consists of a mixture of red and blue light and is complementary to green. Thus, a magenta-colored substance (Figure L-2) transmits red and blue light but absorbs green light (Figure L-3).

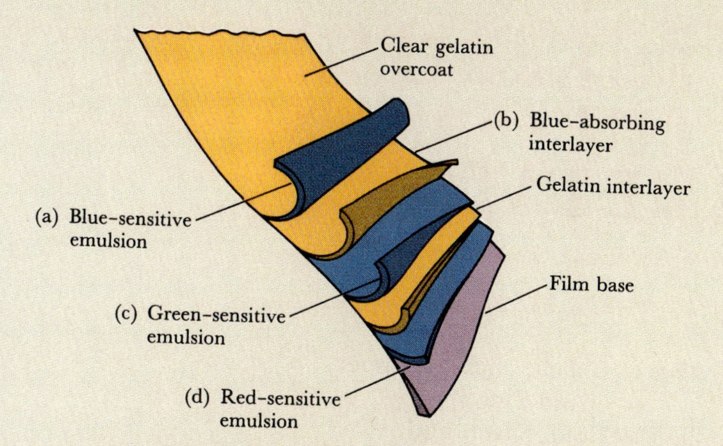

Figure L-5 The various layers of silver halide emulsions (blue) in color photographic film.

duces a range of grays through black, depending upon the relative intensities of the three colors.

As in black-and-white film, it is the exposure of silver halide salts that ultimately results in the formation of a latent image. There are various types of color film, such as Kodachrome and Ektachrome, but most color film consists of a number of layers of separated silver halide emulsions, each being made sensitive to a particular primary color by the inclusion of a dye (Figure L-5). Silver halides are most sensitive to blue and violet light, so the uppermost silver halide emission, (a) in Figure L-5, is exposed by blue and violet light. The next layer, (b), contains a yellow dye, which during exposure prevents blue and violet light from penetrating to the lower layers. Layer (c) contains a magenta-colored dye, which absorbs green light and sensitizes the silver halide grains to that region of the spectrum. Layer (d) contains a cyan dye, which absorbs red light and therefore sensitizes the silver halide to red light. Accordingly, when such a film is exposed to light, the blue-sensitive emulsion is exposed by blue light, the green-sensitive emulsion is exposed by green light, and the red-sensitive emulsion is exposed by red light. The resulting latent images in each emulsion are developed separately with special developers (a cyan developer, a yellow developer, and a magenta developer) to produce a negative, from which a color positive can be obtained by printing onto a similar emulsion package.

27 THE CHEMISTRY OF THE TRANSITION METALS

The 3*d* transition-metal series. Top row from left to right: scandium, titanium, vanadium, chromium, and manganese. Bottom row: iron, cobalt, nickel, copper, and zinc.

In this chapter we conclude our systematic study of the elements with a discussion of the *d* transition metals, or the ***d*-block elements.** They are called the 3*d* transition-metal series, the 4*d* transition-metal series, and the 5*d* transition-metal series, and so on, to indicate the subshell that is being filled within each series.

Transition-metal alloys are the structural backbone of modern civilization. Human development itself was marked by progress to the Bronze Age and then to the Iron Age. The Industrial Revolution was powered by steam engines made from steels. Today we rely on countless exotic alloys developed to meet specialized requirements as different as those of space shuttles, personal computers, and ordinary house paint. Table 27-1 describes the major properties, sources, and uses of the 3*d* transition metals.

The ten members of each *d* transition-metal series correspond to the ability of a *d* subshell to hold a maximum of 10 electrons. Figure 27-1 gives the ground-state outer electron configurations of the 3*d*, 4*d*, and 5*d* series. Note that the filling of the *d* orbitals is not perfectly regular in all cases. For example, the ground-state outer electron configuration of chromium is $4s^1 3d^5$ rather than $4s^2 3d^4$ and that of copper is $4s^1 3d^{10}$ rather than $4s^2 3d^9$.

We will not discuss the chemistry of all the transition metals in this chapter but will focus only on the first transition-metal series. Just as the first member in the main-group families differs significantly from the others, the chemistry of the first transition-metal series also differs

Table 27-1 Properties of the 3*d* transition metals

Element	Density/g·cm^{-3}	Melting point/°C	Principal sources	Uses
scandium	3.0	1541	*thortveitite,* $(Sc,Y)_2Si_2O_7$	no major industrial uses
titanium	4.5	1660	*rutile,* TiO_2	high-temperature, lightweight steel alloys; TiO_2 in white paints
vanadium	6.0	1890	*vanadinite,* $(PbO)_9(V_2O_5)_3PbCl_2$	vanadium steels (rust-resistant)
chromium	7.2	1860	*chromite,* $FeCr_2O_4$	stainless steels; chrome plating
manganese	7.4	1244	*pyrolusite,* MnO_2 *manganosite,* MnO nodules on ocean floor	alloys
iron	7.9	1535	*hematite,* Fe_2O_3 *magnetite,* Fe_3O_4	steels
cobalt	8.9	1490	*cobaltite,* $CoS_2 \cdot CoAs_2$ *linnaeite,* Co_3S_4	alloys; cobalt-60 radiology
nickel	8.9	1455	*pentlandite,* $(Fe,Ni)_9S_8$ *pyrrhotite,* $F_{0.8}S$	nickel plating; coins; magnets; catalysts
copper	9.0	1083	*chalcopyrite,* $CuFeS_2$ *chalcocite,* Cu_2S *malachite,* $Cu_2(CO_3)(OH)_2$	bronzes; brass; coins; electric conductors
zinc	7.1	420	*zinc blende,* ZnS *smithsonite,* $ZnCO_3$	galvanizing; bronze; brass; dry cells

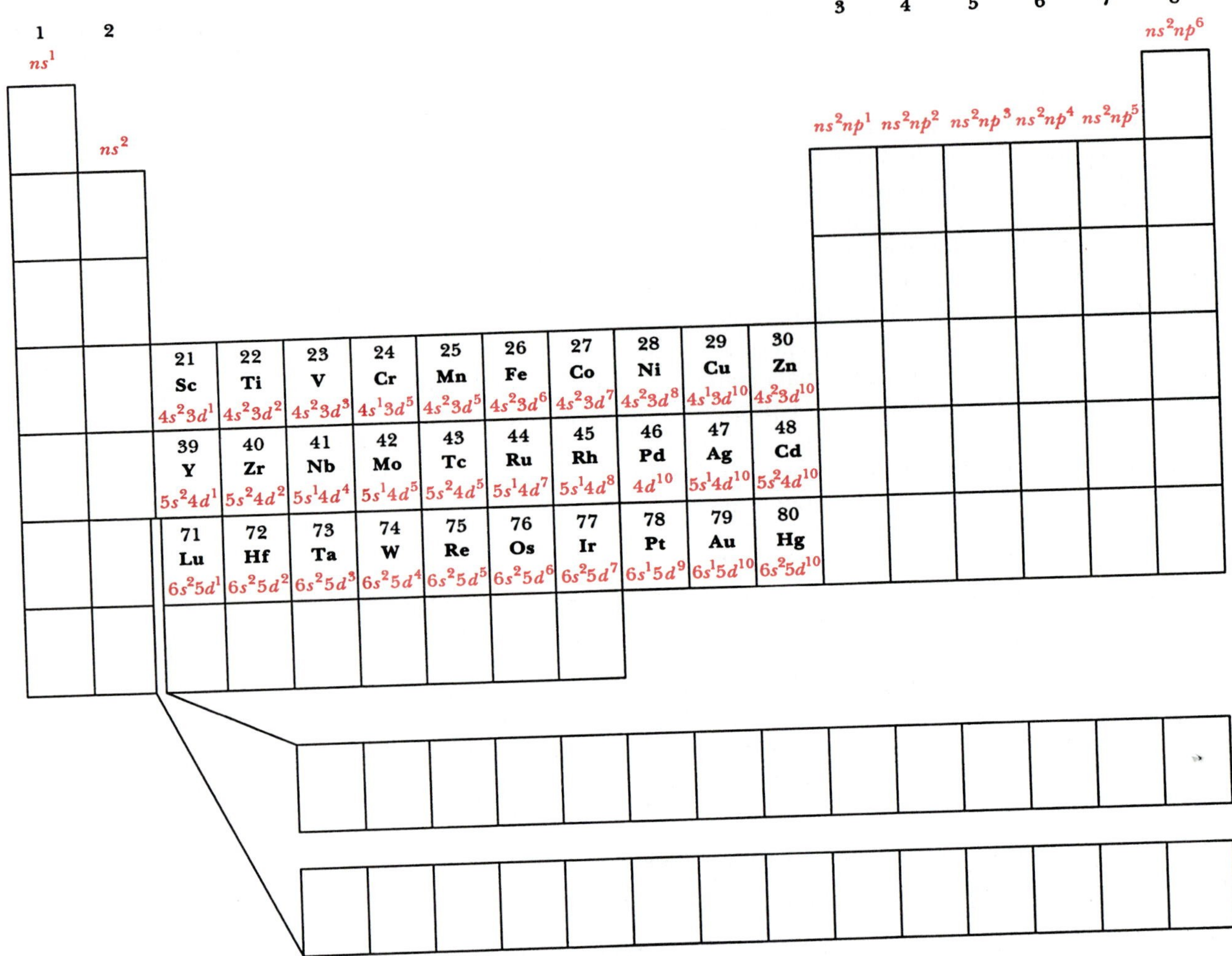

Figure 27-1 The ground-state outer electron configurations of the 3*d*, 4*d*, and 5*d* transition-metal series.

appreciably from that of the other series. In particular, the aqueous-solution chemistry of the first transition-metal series is simpler than that of the heavier transition metals.

The transition metals also have a rich and fascinating chemistry involving numerous ligands and different oxidation states. We will describe the coordination chemistry of the transition metals in the next chapter.

27-1. THE MAXIMUM OXIDATION STATES OF SCANDIUM THROUGH MANGANESE ARE EQUAL TO THE TOTAL NUMBER OF 4s AND 3d ELECTRONS

The chemistry of even the first transition-metal series is especially rich because of the several oxidation states available to many of the metals (Table 27-2). Despite the differences in the chemistry of the transition metals, certain trends do exist:

1. For scandium through manganese, the highest oxidation state is equal to the total number of 4*s* plus 3*d* electrons, and this oxidation state is achieved primarily in oxygen compounds or in fluorides and

chlorides. Furthermore, the stability of the highest oxidation state decreases from scandium to manganese. We have

Sc_2O_3 a stable oxide

TiO_2 a stable, common ore of titanium

V_2O_5 mild oxidizing agent

CrO_4^{2-} strong oxidizing agent

MnO_4^- strong oxidizing agent

2. Except for scandium and titanium, all the 3*d* transition metals form divalent ions in aqueous solution.
3. For a given metal, the oxides become more acidic with increasing oxidation state.
4. For a given metal, the halides become more covalent with increasing oxidation state and react more readily with water.

We can see these trends as we briefly discuss the chemical properties of the metals in the first transition-metal series.

Table 27-2 The common oxidation states of the 3*d* transition metals

Element	Common oxidation states
scandium	+3
titanium	+4
vanadium	+2, +3, +4, +5
chromium	+2, +3, +6
manganese	+2, +4, +7
iron	+2, +3
cobalt	+2, +3
nickel	+2
copper	+1, +2
zinc	+2

Scandium

The ground-state electron configuration of scandium is $[Ar]3d^14s^2$, and it is somewhat similar to aluminum in its chemical properties. Scandium has a +3 oxidation state in almost all its compounds. It forms a very stable oxide, Sc_2O_3, and forms halides with the formula ScX_3. The addition of base to $Sc^{3+}(aq)$ produces a white, gelatinous precipitate with the formula $Sc_2O_3 \cdot nH_2O$. Like $Al(OH)_3$, this hydrated Sc_2O_3 is amphoteric. Scandium and its compounds have little technological importance at present.

Titanium

The ground-state electron configuration of titanium is $[Ar]4s^23d^2$. Its most common and stable oxidation state by far is +4, as in the compounds TiO_2 and $TiCl_4$, which are covalently bonded.

Pure titanium is a lustrous, white metal and is the second most abundant transition metal (Figure 27-2). It is used to make lightweight alloys

Figure 27-2 Titanium has a relatively low density, high strength, and excellent corrosion resistance. It is also easily machined.

that are stable at high temperatures and are used in missiles and high-performance aircraft. Titanium is as strong as most steels but 50 percent lighter. It is 60 percent heavier than aluminum but twice as strong. In addition, it has excellent resistance to corrosion.

Pure titanium is difficult to prepare because the metal is very reactive at high temperatures. The most important ore of titanium is **rutile,** which is primarily TiO_2 (Figure 27-3). Pure titanium metal is produced by first converting TiO_2 to $TiCl_4$—by heating TiO_2 to red heat in the presence of carbon and chlorine. The $TiCl_4$, which is a colorless, covalent liquid, is reduced to the metal by reacting it with magnesium in an inert atmosphere of argon. Most titanium is used in the production of titanium steels, but TiO_2, which is white when pure, is used as the white pigment in many paints. Titanium tetrachloride is used to make smoke screens; when it is sprayed into the air, it reacts with moisture to produce a dense and persistent white cloud of TiO_2. Titanium dioxide is the only transition-metal compound ranked (44th) in the top 50 industrial chemicals (Appendix J). Commercially important titanium compounds are given in Table 27-3.

Figure 27-3 Rutile, an ore of titanium.

Table 27-3 Important compounds of titanium, vanadium, chromium, and manganese

Compound	Uses
titanium dioxide, $TiO_2(s)$	ceramic colorant; white paints and lacquers; inks and plastics; gemstones
titanium tetrachloride, $TiCl_4(l)$	smoke screens; iridescent glass; artificial pearls
vanadium pentoxide, $V_2O_5(s)$	production of sulfuric acid; manufacture of yellow glass
vanadyl sulfate, $VOSO_4(s)$	blue and green colored glasses and glazes on pottery
chromium(IV) oxide, $CrO_2(s)$	constituent of magnetic recording tapes for better resolution and high-frequency response
chromium(III) oxide, $Cr_2O_3(s)$	constituent of abrasives, refractory materials, and semiconductors; green pigment, especially for coloring glass
sodium dichromate, $Na_2Cr_2O_7(s)$	leather tanning; textile manufacture; metal corrosion inhibitor
manganese(IV) oxide, $MnO_2(s)$	manufacture of manganese steel; alkaline batteries and dry cells; printing and dyeing textiles; pigment in brick industry
manganese(II) sulfate, $MnSO_4(s)$	dyeing; red glazes on porcelain; fertilizers for vines, tobacco
potassium permanganate, $KMnO_4(s)$	oxidizing agent; medical disinfectant (bladder infection); water and air purification

EXAMPLE 27-1: Write the chemical equation for the reaction of $TiCl_4(l)$ with water vapor.

Solution: We represent the water vapor by $H_2O(g)$ and write

$$TiCl_4(l) + 2H_2O(g) \rightarrow TiO_2(s) + 4HCl(g)$$

PRACTICE PROBLEM 27-1: Write the chemical equations for (a) the conversion of rutile $\{TiO_2(s)\}$ to $TiCl_4(g)$ in the presence of carbon and chlorine at high temperature to produce $TiCl_4(g)$ and $CO(g)$ and (b) the reaction of $TiCl_4(g)$ with magnesium metal to produce metallic titanium.

Answer:
(a) $TiO_2(s) + 2Cl_2(g) + 2C(s) \rightarrow TiCl_4(g) + 2CO(g)$
(b) $TiCl_4(g) + 2Mg(s) \rightarrow Ti(s) + 2MgCl_2(s)$

Vanadium

The ground-state electron configuration of vanadium is $[Ar]4s^2 3d^3$. Its maximum oxidation state is +5; the +2, +3, and +4 oxidation states are common, with the +2 state being the least common. Vanadium(V) oxide is obtained when vanadium is burned in excess oxygen. It is used as a catalyst in several industrial processes, including the oxidation of SO_2 to SO_3, which is one step in the production of sulfuric acid by the contact process (Chapter 26).

Vanadium pentoxide is amphoteric. It dissolves in concentrated bases such as $NaOH(aq)$ to produce a colorless solution in which the principal species above pH = 13 is believed to be $VO_4^{3-}(aq)$. As the pH is lowered, the solution turns orange, and at pH = 2, a precipitate occurs, which redissolves at a lower pH to give a pale-yellow solution, in which the principal species is believed to be $VO_2^+(aq)$ (Figure 27-4).

Except for V_2O_5, the compounds of vanadium have limited commercial importance, but vanadium itself is used in alloy steels, particularly **ferrovanadium.**

Figure 27-4 Solutions of V_2O_5 at various pH values. Left, V_2O_5 dissolved in $NaOH(aq)$ at pH = 13, where the principal species is $VO_4^{3-}(aq)$. Center left, V_2O_5 dissolved in hydrochloric acid at pH = 0 where the principal species is the vanadyl ion $VO_2^+(aq)$. Center right, V_2O_5 dissolved in hydrochloric acid at pH = 2 with the formation of a precipitate. Right, V_2O_5 dissolved in hydrochloric acid at pH = 4.

Figure 27-5 Left, $Na_2CrO_4(aq)$ in $NaOH(aq)$ at pH = 8, where CrO_4^{2-} is the principal species. Center, a solution from the left with pH adjusted to 4, where the principal species is $Cr_2O_7^{2-}(aq)$. Right, a solution from the center with pH adjusted to 0, where the principal species is $H_2CrO_4(aq)$.

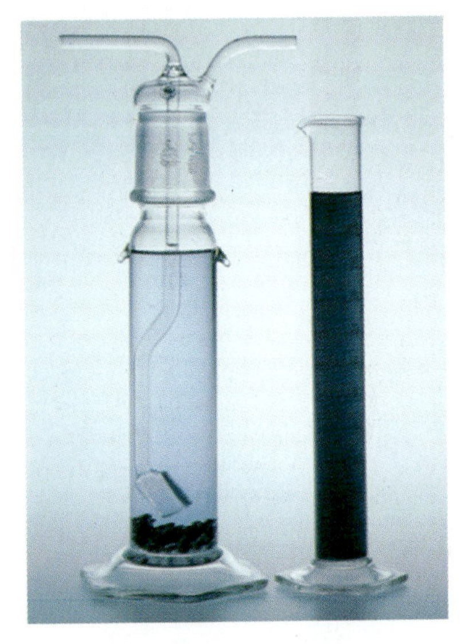

Figure 27-6 (Left) A chromous bubbler, which is used to remove traces of oxygen from unreactive gases such as nitrogen or argon. The blue color of the solution is due to $Cr^{2+}(aq)$, and the solid at the bottom of the tube is a zinc amalgam (Zn + Hg). The gas to be deoxygenated enters the bubbler at the left inlet tube, bubbles up through the solution of $Cr^{2+}(aq)$, and exits from the right tube. A glass membrane at the bottom of the inlet tube breaks up the entering gas stream into small bubbles. (Right) The green solution in the graduated cylinder is $Cr_2(SO_4)_3(aq)$, which is used to prepare the chromous bubbler. The green $Cr^{3+}(aq)$ is reduced to the blue $Cr^{2+}(aq)$ by the zinc amalgam in the bubbler.

27-2. THE +6 OXIDATION STATE OF CHROMIUM AND THE +7 OXIDATION STATE OF MANGANESE ARE STRONGLY OXIDIZING

Chromium

Chromium, with the ground-state electron configuration $[Ar]4s^1 3d^5$, has a maximum oxidation state of +6, although the +2 and +3 states are common. Whereas the +4 oxidation state for titanium and the +5 state of vanadium are only mildly oxidizing, the +6 oxidation state of chromium is strongly oxidizing. Indeed, the dichromate ion in acidic solution is a strong oxidizing agent (Figure 27-5):

$$14H^+(aq) + Cr_2O_7^{2-}(aq) + 6e^- \rightarrow 2Cr^{3+}(aq) + 7H_2O(l) \qquad E^\circ_{red} = 1.33\ \text{V}$$

Recall from our discussion of the electromotive force series in Section 23-5 that the larger the positive value of E°_{red}, the stronger the agent's oxidizing power. In contrast to chromium(VI) compounds, the chromium(II) ion is a fairly strong reducing agent:

$$Cr^{3+}(aq) + e^- \rightarrow Cr^{2+}(aq) \qquad E^\circ_{red} = -0.41\ \text{V}$$

A **chromous bubbler** is a freshly prepared solution of $Cr^{2+}(aq)$ that is used to remove traces of oxygen from gases (Figure 27-6). The chromous bubbler solution is prepared by reacting zinc metal with chromium(III) nitrate to form chromium(II):

$$Zn(s) + 2Cr^{3+}(aq) \rightarrow 2Cr^{2+}(aq) + Zn^{2+}(aq)$$

The resulting chromium(II) nitrate solution reduces oxygen to water as it is bubbled through the solution:

$$4Cr^{2+}(aq) + O_2(g) + 4H^+(aq) \rightarrow 4Cr^{3+}(aq) + 2H_2O(l)$$

The equilibrium constant for this reaction is very large, making it essentially a quantitative reaction.

Manganese

The highest oxidation state of manganese is +7, which is best known in the strongly oxidizing permanganate ion, MnO_4^-:

Acidic solution

$$MnO_4^-(aq) + 8H^+(aq) + 5e^- \rightarrow Mn^{2+}(aq) + 4H_2O(l) \qquad E^\circ_{red} = +1.507 \text{ V}$$

Basic solution

$$MnO_4^-(aq) + 2H_2O(l) + 3e^- \rightarrow MnO_2(s) + 4OH^-(aq) \qquad E^\circ_{red} = +1.23 \text{ V}$$

The most important permanganate salt is potassium permanganate, $KMnO_4$, which is used as an oxidizing agent in industry and medicine, as well as in many general chemistry laboratories. Freshly prepared solutions of potassium permanganate are deep purple but turn brown on long standing, because permanganate ion oxidizes water to oxygen and is thereby reduced to MnO_2 (Figure 27-7). The net equation is

$$\underset{\text{purple}}{4MnO_4^-(aq)} + 2H_2O(l) \rightarrow \underset{\text{brown}}{4MnO_2(s)} + 3O_2(g) + 4OH^-(aq)$$

The reaction is catalyzed by $MnO_2(s)$ and thus is **autocatalytic.**

Figure 27-7 Potassium permanganate, $KMnO_4$, in water is a strong oxidizing agent. (Left) Freshly prepared $KMnO_4(aq)$ is purple. (Right) As the solution stands, brown $MnO_2(s)$ precipitates as a result of the reduction of $KMnO_4(aq)$ by $H_2O(l)$.

EXAMPLE 27-2: Potassium permanganate is used as an oxidizing agent in analytical chemistry. A 3.68-g sample of stibnite (Sb_2S_3), an ore of antimony, was dissolved in acid and the resulting Sb(III) was oxidized to Sb(V) by titrating with 38.6 mL of 0.125 M $KMnO_4(aq)$ solution. Determine the mass percentage of Sb_2S_3 in the ore.

Solution: We do not need to know the details of the species involved in the antimony oxidation, so the two half reactions are

$$\text{Sb(III)} \rightarrow \text{Sb(V)} + 2e^-$$

$$MnO_4^-(aq) + 8H^+(aq) + 5e^- \rightarrow Mn^{2+}(aq) + 4H_2O(l)$$

The overall balanced equation is

$$5\text{Sb(III)} + 2MnO_4^-(aq) + 16H^+(aq) \rightarrow 5\text{Sb(V)} + 2Mn^{2+}(aq) + 8H_2O(l)$$

The number of millimoles of MnO_4^- is given by

$$\text{mmol of } MnO_4^- = (38.6 \text{ mL})(0.125 \text{ M}) = 4.825 \text{ mmol}$$

The number of millimoles of Sb(III) is

$$\text{mmol of Sb(III)} = (4.825 \text{ mmol } MnO_4^-)\left(\frac{5 \text{ mol Sb(III)}}{2 \text{ mol } MnO_4^-}\right) = 12.06 \text{ mmol}$$

so the number of millimoles of Sb_2S_3 originally present in the sample of ore is

$$\text{mmol of } Sb_2S_3 = (12.06 \text{ mmol})\left(\frac{1 \text{ mol } Sb_2S_3}{2 \text{ mol Sb(III)}}\right) = 6.03 \text{ mmol}$$

The mass of Sb_2S_3 is

$$\text{mass of } Sb_2S_3 = (6.03 \text{ mmol } Sb_2S_3)\left(\frac{339.8 \text{ g } Sb_2S_3}{1000 \text{ mmol } Sb_2S_3}\right) = 2.05 \text{ g}$$

The mass percentage of Sb_2S_3 in the ore is

$$\text{mass }\% = \left(\frac{\text{mass } Sb_2S_3}{\text{mass of ore}}\right) \times 100 = \left(\frac{2.05 \text{ g}}{3.68 \text{ g}}\right) \times 100 = 55.7\%$$

PRACTICE PROBLEM 27-2: Permanganate ion reacts quantitatively with oxalate ion in acidic solution to give manganese(II) and carbon dioxide. Complete and balance the equation for the reaction.

Answer:

$$2MnO_4^-(aq) + 5C_2O_4^{2-}(aq) + 16H^+(aq) \rightarrow 2Mn^{2+}(aq) + 10CO_2(g) + 8H_2O(l)$$

Manganese(II) forms soluble salts with most common anions. For the +3 and +4 oxidation states, the most important compounds are the oxides Mn_2O_3 and MnO_2. Manganese dioxide as the mineral *pyrolusite* is an important ore of manganese and is the source of most manganese compounds. Commercially important manganese compounds are given in Table 27-3.

27-3. IRON IS PRODUCED IN A BLAST FURNACE

Iron

Starting with iron, we no longer associate the highest oxidation state with the total number of 4*s* and 3*d* electrons. Although the highest known oxidation state of iron is +6, it is very rare. Only the +2 and +3 oxidation states of iron are common.

Iron is found in nature as *hematite*, Fe_2O_3; *magnetite*, Fe_3O_4; *siderite*, $FeCO_3$; and *iron pyrite*, FeS_2, called **fool's gold** (Figure 27-8). It is the

Figure 27-8 Iron ores. Clockwise from the left: magnetite, Fe_3O_4; siderite, $FeCO_3$; iron pyrite, FeS_2; and hematite, Fe_2O_3.

Table 27-4 Important iron compounds

Compound	Uses
iron(III) chloride, $FeCl_3(s)$	treatment of wastewater; etching for engraving copper for printed circuitry; feed additive
iron(III) oxide, $Fe_2O_3(s)$	metallurgy; paint and rubber pigment; memory cores for computers; magnetic tapes; polishing agent for glass, precious metals, and diamonds
iron(II) sulfate, $FeSO_4(s)$	flour enrichment; wood preservative; water and sewage treatment; manufacture of ink
iron(III) sulfate, $Fe_2(SO_4)_3(s)$	soil conditioner; disinfectant; etching aluminum; wastewater treatment

most abundant transition metal—constituting 4.7 percent by mass of the earth's crust, the cheapest metal, and, in the form of steel, the most useful. Pure iron is a silvery white, soft metal that rusts rapidly in moist air. We have already discussed this familiar example of corrosion in Section 21-7. Iron is little used as the pure element but is strengthened greatly by the addition of small amounts of carbon and of various other transition metals. Its principal use is in the production of steel, but some other uses are given in Table 27-4.

Iron is also a vital constituent of mammalian cells. It occurs in a number of proteins that transport oxygen from the lungs to muscles and other tissues, where the oxygen is involved in the cellular oxidation reactions essential to respiration and metabolism. It has been estimated that an adult human being contains about 5 g of iron.

Millions of tons of iron are produced annually in the United States by the reaction of Fe_2O_3 with coke; this reaction is carried out in a blast furnace. About 100 ft high and 25 ft wide, the modern **blast furnace** produces about 5000 tons of iron daily (Figure 27-9). A mixture of iron ore, coke, and limestone ($CaCO_3$) is loaded into the top, and preheated compressed air and oxygen are blown in near the bottom. The reaction of the coke and the oxygen to produce carbon dioxide gives off a great deal of heat, and the temperature in the lower region of a blast furnace is around 1900°C. As the CO_2 rises, it reacts with more coke to produce hot carbon monoxide, which reduces the iron ore to iron. The molten iron metal is denser than the other substances and drops to the bottom, where it can be drained off to form ingots of what is called **pig iron.**

The function of the limestone is to remove the sand and gravel that normally contaminate the iron ore. The intense heat decomposes the limestone to $CaO(s)$ and $CO_2(g)$. The $CaO(s)$ combines with the sand and gravel (both of which are primarily silicon dioxide) to form molten calcium silicate, $CaSiO_3(l)$.

$$CaO(s) + SiO_2(s) \rightarrow \underset{\text{slag}}{CaSiO_3(l)}$$

The molten calcium silicate, called **slag,** floats on top of the molten iron and is drained off periodically. It is used in building materials, such as

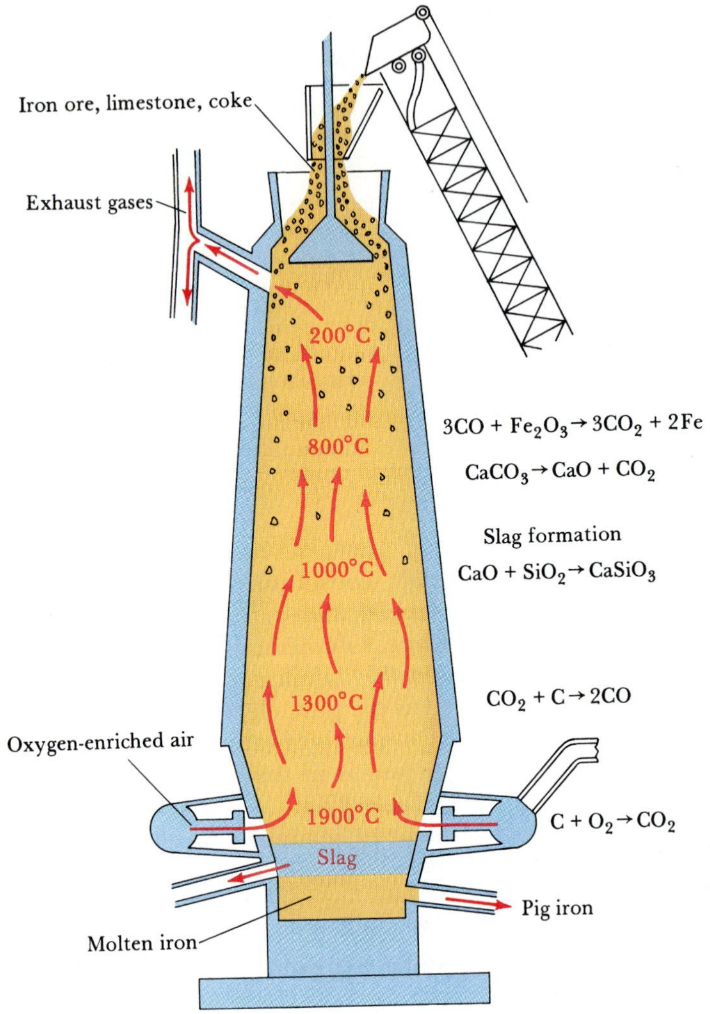

Figure 27-9 Iron is produced in a blast furnace. A typical blast furnace like the one shown in this diagram runs continuously and consumes about 120 railroad cars of iron ore, 50 railroad cars of coke, and 40 railroad cars of limestone per day. The 5000 tons of iron produced requires about 75 railroad cars to transport it.

cement and concrete aggregate, rock-wool insulation, and cinder block, and as railroad ballast.

Pig iron contains about 4 or 5 percent carbon together with lesser amounts of silicon, manganese, phosphorus, and sulfur. It is brittle, difficult to weld, and not strong enough for structural applications. To be useful, pig iron must be converted to steel, which is an alloy of iron with small but controlled amounts of other metals and between 0.1 and 1.5 percent carbon. Steel is made from pig iron in several different processes, all of which use oxygen to oxidize most of the impurities. One such process is the **basic oxygen process,** in which hot, pure O_2 gas is blown through molten pig iron (Figure 27-10). The oxidation of carbon and phosphorus is complete in less than 1 h. The desired carbon content of the steel is then achieved by adding high-carbon steel alloy.

There are two types of steels: carbon steels and alloy steels. Both types contain carbon, but carbon steels contain essentially no other metals besides iron. About 90 percent of all steel produced is carbon steel. Carbon steel that contains less than 0.2 percent carbon is called **mild steel.** Mild steels are malleable and ductile and are used where load-bearing ability is not a consideration. **Medium steels,** which contain 0.2

Figure 27-10 Molten iron being charged into a basic oxygen furnace. Most steel is produced by a process called the basic oxygen process. A typical basic oxygen furnace is charged with about 200 tons of molten pig iron, 100 tons of scrap iron, and 20 tons of limestone (to form a slag). A stream of hot oxygen is blown through the molten mixture, and the oxidized impurities are blown out of the iron. High-quality steel is produced in an hour or less.

to 0.6 percent carbon, are used for such structural materials as beams and girders and for railroad equipment. **High-carbon steels** contain 0.8 to 1.5 percent carbon and are used to make drill bits, knives, and other tools in which hardness is important.

Alloy steels contain other metals in small amounts. Different metals give different properties to steels. The alloy steels called **stainless steels** contain high percentages of chromium and nickel. Stainless steels resist corrosion and are used for cutlery and hospital equipment. The most common stainless steel contains 18 percent chromium and 8 percent nickel.

27-4. THE +2 OXIDATION STATE IS THE MOST IMPORTANT OXIDATION STATE FOR COBALT, NICKEL, COPPER, AND ZINC

As we go from iron to zinc, there is an increasing prominence of the +2 oxidation state. Most compounds of cobalt and almost all compounds of nickel and zinc involve the metal in the +2 oxidation state. Only copper, which has an important +1 oxidation state, and cobalt, which has an important +3 oxidation state, have extensive chemistries not involving the +2 state.

Cobalt

Cobalt is a fairly rare element and is usually found associated with nickel in nature. It is a hard, bluish-white metal that is used in the manufacture of high-temperature alloys such as Alnico, an aluminum-nickel-cobalt alloy used in permanent magnets. The pure metal is relatively unreactive and dissolves only slowly in dilute mineral acids. When cobalt is burned in oxygen, a mixture of CoO and Co_3O_4 is obtained. Blue "cobalt" glass contains a small amount of CoO(*s*) (Figure 27-11). Most simple cobalt salts involve Co(II). The species $Co^{3+}(aq)$ is a strong oxidizing agent:

$$Co^{3+}(aq) + e^- \rightarrow Co^{2+}(aq) \qquad E^\circ_{red} = +1.84 \text{ V}$$

and oxidizes $H_2O(l)$. Commercially important cobalt compounds are given in Table 27-5.

Figure 27-11 Cobalt glass. The characteristic blue color is due to the presence of small amounts of CoO(*s*).

Nickel

Nickel occurs in a variety of sulfide ores, the most important deposit being found in the Sudbury basin of Ontario, Canada. The metal is obtained by roasting the ore to obtain NiO and then reducing with hydrogen or carbon. Very pure nickel is obtained by electrolysis. Nickel is a silvery metal that takes a beautiful high polish, which is protected by a spontaneously formed transparent layer of the oxide NiO. It is used in a number of magnetic alloys and in the alloy Monel, which is used to handle fluorine and other reactive fluorine compounds.

Nickel is more reactive than cobalt and dissolves readily in dilute acids. The aqueous solution chemistry of nickel involves primarily the species $Ni^{2+}(aq)$. Important nickel compounds are given in Table 27-5.

Copper

Copper is only slightly less abundant than nickel, but deposits of the free metal are very rare. It generally occurs as various sulfides. Most copper-containing deposits have a copper content of less than 1 percent, although some richer deposits have up to 4 percent copper (Figure 27-12). Copper ores contain other metals and semimetals such as selenium and tellurium, which are important by-products of copper production. Important copper minerals are *chalcocite,* Cu_2S; *chalcopyrite,* $CuFeS_2$; and *malachite,* $CuCO_3 \cdot Cu(OH)_2$. The total world output of copper metal is about 10 million metric tons, with U.S. production accounting for over 10 percent of the total. The extraction of copper from its ores is a multistep process, the final step being purification by electrolysis.

Table 27-5 Important compounds of cobalt and nickel

Compound	Uses
cobalt(II) oxide, CoO(*s*)	glass and ceramic coloring and decolorization
cobalt(II) phosphate, $Co_3(PO_4)_2(s)$	lavender pigment in paints and ceramics
cobalt(II) sulfate, $CoSO_4(s)$	storage batteries; agent in ceramics, enamels, glazes to prevent discoloring
nickel(II) cyanide, $Ni(CN)_2(s)$	metallurgy; electroplating
nickel(II) chloride, $NiCl_2(s)$	nickelplating; absorbent for NH_3 in gas masks

Copper is a reddish, soft, ductile metal that takes on a bright metallic luster. Its important use is as an electrical conductor. The only metal that is a better conductor is silver, but the price of silver precludes its widespread use. Although copper is fairly unreactive, its surface turns green after long exposure to the atmosphere. The green patina (Figure 27-13) is due to the surface formation of copper hydroxocarbonate and hydroxosulfate.

Brass, an alloy of copper with zinc, and **bronze,** an alloy of copper with tin, are among the earliest known alloys. Bronze usually contains from 5 to 10 percent tin and is very resistant to corrosion. It is used for casting, marine equipment, fine arts work, and spark-resistant tools. Yellow brasses contain about 35 percent zinc and have good ductility and high strength. Brass is used for piping, valves, hose nozzles, marine equipment, and jewelry and in the fine arts.

Figure 27-12 An open pit copper mine located about 25 miles south of Tucson, Arizona. Over 90,000 tons of copper ore are mined per day at this site.

Copper does not replace hydrogen from dilute acids, but it reacts with oxidizing acids such as 6M nitric acid or hot concentrated sulfuric acid:

$$3Cu(s) + 8HNO_3(aq) \rightarrow 3Cu(NO_3)_2(aq) + 2NO(g) + 4H_2O(l)$$

$$Cu(s) + 2H_2SO_4(conc) \rightarrow CuSO_4(aq) + 2H_2O(l) + SO_2(g)$$

Commercially important copper compounds are given in Table 27-6.

Most compounds of copper involve Cu(II), many of which are blue or bluish green. The most common copper(II) salt is copper(II) sulfate pentahydrate, $CuSO_4 \cdot 5H_2O(s)$, which occurs as beautiful blue crystals

Table 27-6 Important compounds of copper and zinc

Compound	Uses
copper(II) arsenite, $Cu(AsO_2)_2(s)$ (Scheele's green)	pigment; wood preservative; insecticide, fungicide, rodenticide, mosquito control
copper(II) nitrate, $Cu(NO_3)_2(s)$	light-sensitive paper; insecticide for vines; wood preservative
copper(I) oxide, $Cu_2O(s)$	antifouling paints for wood and steel exposed to seawater; fungicide; porcelain red glaze; red glass
copper(II) sulfate pentahydrate $CuSO_4 \cdot 5H_2O(s)$	soil and feed additive; germicide; petroleum and rubber industries; laundry and metal-marking inks
zinc(II) carbonate, $ZnCO_3(s)$	white pigment; porcelain, pottery, and rubber manufacture; astringent and antiseptic
zinc(II) chloride, $ZnCl_2(s)$	deodorants; disinfectants; fireproofing and preserving wood; adhesives; dental cements; taxidermist fluid; artificial silk; parchment paper
zinc(II) oxide, $ZnO(s)$	ointment; pigment and mold inhibitor in paints; floor tile; cosmetics; color photography; dental cements; automobile tires

Figure 27-13 The green patina of the Statue of Liberty is due to the surface formation of copper hydroxocarbonate and hydroxosulfate compounds. A major restoration of the statue's surface ended in 1986.

(Figure 27-14). When the crystals are heated gently, the water of hydration is driven off to produce anhydrous copper(II) sulfate, $CuSO_4$, which is a white powder (Figure 27-14). When $NH_3(aq)$ is added to aqueous solutions containing $Cu^{2+}(aq)$, an intense dark blue color occurs as a result of the formation of a copper-ammonia complex ion (see Chapter 28):

$$[Cu(H_2O)_6]^{2+}(aq) + 4NH_3(aq) \rightarrow [Cu(NH_3)_4(H_2O)_2]^{2+}(aq) + 4H_2O(l)$$

A comparison of the colors of aqueous solutions containing $Cu(H_2O)_6^{2+}(aq)$ and $Cu(NH_3)_4(H_2O)_2^{2+}(aq)$ is shown in Figure 27-15.

Copper(I) salts are often colorless and only slightly soluble in water. The $Cu^+(aq)$ ion is unstable and undergoes a disproportionation reaction according to

$$2Cu^+(aq) \rightarrow Cu(s) + Cu^{2+}(aq) \qquad K = 1.2 \times 10^6\ M^{-1}$$

Figure 27-14 Copper(II) sulfate pentahydrate, $CuSO_4{\cdot}5H_2O(s)$, crystals are blue, and copper(II) sulfate, $CuSO_4$, is a white powder.

EXAMPLE 27-3: Use the data in Table 23-1 to verify the value of the equilibrium constant given above.

Solution: From Table 23-1, we have

$$Cu^+(aq) + e^- \rightarrow Cu(s) \qquad E^\circ_{red} = +0.52\ V$$
$$Cu^{2+}(aq) + 2e^- \rightarrow Cu(s) \qquad E^\circ_{red} = +0.34\ V$$

Multiply the first equation by 2 and add the reverse of the second to obtain

$$2Cu^+(aq) \rightarrow Cu(s) + Cu^{2+}(aq)$$

$$E^\circ_{rxn} = E^\circ_{red}[Cu^+/Cu] + E^\circ_{ox}[Cu/Cu^{2+}] = +0.52\ V + (-0.34\ V) = 0.18\ V$$

Equation (23-12) gives the relation between K and E°_{rxn}:

$$\log K = \frac{nE^\circ_{rxn}}{0.0592\ V} = \frac{(2)(0.18\ V)}{0.0592\ V} = 6.08$$

or

$$K = 1.2 \times 10^6\ M^{-1}$$

The large value of K implies that the equilibrium lies far to the right and that $Cu^+(aq)$ disproportionates to $Cu(s)$ and $Cu^{2+}(aq)$. However, the formation of a very insoluble Cu^+ salt can drive the disproportionation reaction equilibrium to the left. That is the case with $CuI(s)$, which is stable and does not undergo disproportionation to $Cu(s)$ and $CuI_2(aq)$.

Figure 27-15 A comparison of the color of aqueous solutions containing $Cu(H_2O)_6^{2+}(aq)$ and $Cu(NH_3)_4(H_2O)_2^{2+}(aq)$.

PRACTICE PROBLEM 27-3: A simple method for the quantitative determination of copper(II) in solution is to add an *excess* of $KI(aq)$ to a measured volume of the $Cu^{2+}(aq)$ solution. Instead of a precipitate of $CuI_2(s)$ as you might expect, the actual products are $CuI(s)$ and $I_3^-(aq)$:

$$2Cu^{2+}(aq) + 5I^-(aq) \rightarrow 2CuI(s) + I_3^-(aq)$$

The I_3^- is back-titrated with $Na_2S_2O_3(aq)$ according to

$$I_3^-(aq) + 2S_2O_3^{2-}(aq) \rightarrow 3I^-(aq) + S_4O_6^{2-}(aq)$$

Suppose that excess $KI(aq)$ is added to 25.0 mL of a $Cu^{2+}(aq)$ solution and that it requires 18.5 mL of 0.400 M $Na_2S_2O_3(aq)$ to titrate the liberated iodine $\{I_3^-(aq)\}$. Calculate the concentration of the $Cu^{2+}(aq)$ solution.

Answer: 0.296 M

Zinc

Zinc, which is widely distributed in nature, is about as abundant as copper. Its principal ores are *sphalerite* (or zinc blende), ZnS, and *smithsonite,* $ZnCO_3$, from which zinc is obtained by roasting and reduction of the resultant ZnO with carbon. Zinc is a shiny white metal with a bluish gray luster.

The 3*d* subshell of zinc is completely filled, so zinc behaves more like a Group 2 metal than like a transition metal. Metallic zinc is a strong reducing agent. It dissolves readily in dilute acids and combines with oxygen, sulfur, phosphorus, and the halogens upon being heated. The only important oxidation state of zinc is +2, and zinc(II) salts are colorless, unless color is imparted by the anion. Commercially important zinc(II) salts are given in Table 27-6.

Figure 27-16 Gold often occurs in the free state in nature. Here we see native gold on a quartz crystal.

27-5. GOLD, SILVER, AND MERCURY HAVE BEEN KNOWN SINCE ANCIENT TIMES

Gold

Gold is a very dense, soft, yellow metal with a high luster (Figure 27-16). It is found in nature as the free element and in tellurides. It occurs in veins and alluvial deposits and is often separated from rocks and other minerals by sluicing or panning. Over two thirds of the gold produced by the Western world comes from South Africa. In many mining operations, about 5 g of gold is recovered from 1 ton of rock.

Pure gold is soft and often alloyed to make it harder. The amount of gold in an alloy is expressed in karats: pure gold is 24 karat; coinage gold is 22 **karat,** or $(22/24) \times 100 = 92$ percent. White gold, which is used in jewelry, is usually an alloy of gold and nickel. Gold is very unreactive, so it has a remarkable resistance to corrosion. It is also an excellent conductor of electricity. In addition to its use in jewelry and as a world monetary standard, gold is also used in microelectronic devices (Figure 27-17). It is also used extensively in dentistry for tooth crowns.

Gold is extracted from ores by reaction with sodium cyanide, NaCN, and oxygen:

$$4Au(s) + 8CN^-(aq) + O_2(g) + 2H_2O(l) \rightarrow 4Au(CN)_2^-(aq) + 4OH^-(aq)$$

This oxidation-reduction reaction is driven by the formation of the very stable $Au(CN)_2^-(aq)$ complex ion. The gold is recovered from the $Au(CN)_2^-$ either by the replacement reaction:

$$2Au(CN)_2^-(aq) + Zn(dust) \rightarrow Zn(CN)_4^{2-}(aq) + 2Au(s)$$

or by electrolysis. Important gold compounds are listed in Table 27-7.

Figure 27-17 Gold is used in the production of printed circuits.

Silver

Silver is a lustrous, white metal whose ductility and malleability are exceeded only by those of gold and palladium. Pure silver also has the highest electrical conductivity of all metals. Most of the silver produced today is a by-product of the production of other metals such as copper,

Table 27-7 Important compounds of gold, silver, and mercury

Compound	Uses
gold(I) stannate, $Au_2SnO_2(s)$	manufacture of ruby glass, colored enamels, and porcelain
tetrachloroauric(III) acid, $HAuCl_4(s)$	photography; gold plating; gilding glass and porcelain
silver iodide, $AgI(s)$	dispersed in clouds to induce rain; photography (fast film)
silver nitrate, $AgNO_3(s)$	manufacture of mirrors; silverplating; hair-darkening agent; eye drops for newborn infants
silver bromide, $AgBr(s)$	photography; photosensitive lenses
mercury(I) chloride, $Hg_2Cl_2(s)$ (calomel)	fungicide; control of root maggots on cabbage and onions; calomel electrodes
mercury(II) chloride, $HgCl_2(s)$	preservative for wood and anatomical specimens; embalming agent; photographic intensifier
mercury(II) oxide, red, $HgO(s)$	marine paints; porcelain pigments; anode material in mercury batteries

lead, and zinc. Its uses include jewelry, silverware, high-capacity batteries, and coinage. However, about a third of the silver produced is used in photography in the form of silver halides (Interchapter L). Other important compounds appear in Table 27-7.

The silver halides illustrate the phenomenon that the first member of a family in the periodic table may differ some from the others. Although $AgCl(s)$, $AgBr(s)$, and $AgI(s)$ are insoluble in water, $AgF(s)$ is very soluble, having a solubility in water of $1800\ g{\cdot}L^{-1}$.

Figure 27-18 Cinnabar, $HgS(s)$, and mercury, $Hg(l)$.

Mercury

Like gold and silver, mercury has been known for thousands of years; it used to be called quicksilver because it is the only metal that is a liquid at 25°C and often forms small balls that move around rapidly on uneven surfaces. The principal ore of mercury is **cinnabar,** HgS, which was widely used in the ancient world as a vermilion pigment (Figure 27-18). The most extensive and richest deposits of cinnabar occur in the Almaden region of Spain, the world's largest producer of mercury. The metal is easily recovered from its ore by roasting:

$$HgS(s) + O_2(g) \rightarrow Hg(l) + SO_2(g)$$

and the mercury is purified by distillation.

Mercury is not very reactive. On being heated, it reacts with oxygen, sulfur, and the halogens, but not with nitrogen, phosphorus, hydrogen, or carbon. When mercury is heated in air to around 300°C, it reacts with oxygen to produce the bright orange-red mercury(II) oxide. When $HgO(s)$ is heated to about 400°C, it decomposes according to

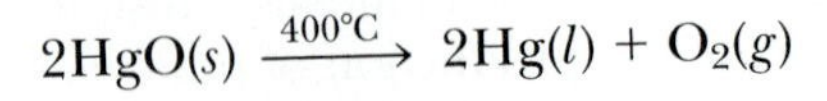

$$2HgO(s) \xrightarrow{400^{\circ}C} 2Hg(l) + O_2(g)$$

Like copper, mercury does not replace hydrogen from acids, but it does react with oxidizing acids such as 6M nitric acid or hot concentrated sulfuric acid:

$$3Hg(l) + 8HNO_3(aq) \rightarrow 3Hg(NO_3)_2(aq) + 2NO(g) + 4H_2O(l)$$
$$Hg(l) + 2H_2SO_4(aq) \rightarrow HgSO_4(aq) + SO_2(aq) + 2H_2O(l)$$

Mercury compounds occur as Hg(I) or Hg(II), with Hg(II) being more common. Except for the nitrate, acetate, and perchlorate salts, Hg(I) salts are insoluble. A notable feature of Hg(I) salts is that they consist of the diatomic ion Hg_2^{2+}. Many of the salts of Hg(I) and Hg(II) are covalently bonded. Important mercury compounds are listed in Table 27-7.

Mercury compounds are very poisonous and have been used in insecticides, fungicides, rodenticides, and disinfectants. In the last century, mercury compounds were used in the production of felt for hats; the felt workers suffered from a nervous disorder called "hatter's shakes," which led to the expression "mad as a hatter." The discharge of mercury-containing industrial wastes into rivers, lakes and oceans caused serious environmental problems, culminating in the Minamata disaster in Japan in 1952, when over 50 people died of mercury poisoning. The mercury effluent was converted to the organomercury compounds by certain sedimentary bacteria, entered the marine food chain, and became concentrated in fish, which was the main diet of the fishing village of Minamata. Since then, the disposal of mercury wastes has been regulated and mercury levels in food are constantly monitored. In 1972, more than 90 nations agreed on an international ban on the dumping of mercury wastes.

TERMS YOU SHOULD KNOW

d-block elements 975
rutile 978
ferrovanadium 979
chromous bubbler 980
autocatalytic 981
iron pyrite 982
fool's gold 982
blast furnace 983
pig iron 983
slag 983
basic oxygen process 984
mild steel 984
medium steel 984
high-carbon steel 985
stainless steel 985
brass 987
bronze 987
karat 989
cinnabar 990

PROBLEMS

27-1. Which metal has the highest melting point?

27-2. Which are the two densest metals?

27-3. Which is the most abundant transition metal?

27-4. Which are the only two 3*d* transition metals that do not readily form divalent ions in aqueous solution?

27-5. Describe how titanium metal is produced from its ore.

27-6. Give the highest oxidation states for Sc, Ti, V, Cr, and Mn.

27-7. Name a catalyst in the production of sulfuric acid by the contact process.

27-8. The chief ore of chromium is *chromite* ($FeCr_2O_4$). Chromium of high purity can be obtained from chromite by oxidizing chromium(III) to chromium(VI) in the form of sodium dichromate, $Na_2Cr_2O_7$, and then reducing it with carbon according to

$$Na_2Cr_2O_7(s) + 2C(s) \rightarrow Cr_2O_3(s) + Na_2CO_3(s) + CO(g)$$

The oxide is then reduced with aluminum by the thermite reaction:

$$Cr_2O_3(s) + 2Al(s) \rightarrow Al_2O_3(s) + 2Cr(s)$$

If an ore is 65.0 percent chromite, then how many grams of pure chromium can be obtained from 100 grams of ore?

27-9. A small quantity of $Na_2Cr_2O_7$ is often added to water stored in steel drums used in some passive home solar heating systems. The dichromate acts as a corrosion inhibitor by forming an impervious layer of $Cr_2O_3(s)$ on the iron surface. Write a balanced chemical equation for the process in which Cr_2O_3 is formed.

27-10. A chromous bubbler is used to remove traces of oxygen from various gases, for example, from tank nitrogen. The bubbler solution is prepared by reducing $Cr^{3+}(aq)$ to $Cr^{2+}(aq)$ with excess zinc metal. Chromium(II) rapidly reduces O_2 to water and forms chromium(III), which is then reduced back to Cr(II) by the zinc. Write balanced chemical equations for the various chemical reactions involved in the operation of the bubbler.

27-11. Why are solutions of potassium permanganate stored in dark bottles?

27-12. The reaction

$$MnO_2(s) + 4HCl(g) \rightarrow MnCl_2(aq) + Cl_2(g) + 2H_2O(l)$$

is often used to generate small quantities of $Cl_2(g)$ in the laboratory. How many grams of $Cl_2(g)$ can be generated by reacting 4.5 g of MnO_2 with an excess of $HCl(aq)$?

27-13. Use VSEPR theory to predict the shapes of

(a) $TiCl_4$ (b) VF_5 (c) CrO_4^{2-} (d) MnO_4^-

27-14. Describe the principal reactions that take place in a blast furnace.

27-15. Describe the basic oxygen process.

27-16. Suppose that an iron ore consists of 50 percent Fe_2O_3 and 50 percent SiO_2. How many metric tons of iron and slag will be produced from 10,000 metric tons of ore?

27-17. Why must solutions of $Co^{3+}(aq)$ be prepared freshly?

27-18. Which two elements are important by-products of copper production?

27-19. A copper ore consists of 2.65 percent chalcopyrite. How many tons of ore must be processed to obtain one metric ton of copper?

27-20. When copper reacts with nitric acid, $NO(g)$ is evolved. Although $NO(g)$ is colorless, it appears as though a brown-red gas is evolved. Explain.

27-21. What is the percentage of gold in 14-karat gold?

27-22. Copper metal reacts with concentrated nitric acid but not with concentrated hydrochloric acid.

(a) Explain these observations by using standard reduction voltages.
(b) Write balanced chemical equations for the two reactions that occur between copper and dilute and concentrated nitric acid.

27-23. Gold is recovered by leaching crushed ores with aqueous cyanide solutions. The resultant complex, $[Au(CN)_2]^-(aq)$, is reduced with zinc metal to produce gold. Write the chemical equations for the reactions that take place.

27-24. The vapor pressure of mercury varies with temperature as

P_{Hg}/torr	t/°C
1	126
10	184
40	229
100	261
400	323
760	357

Use the relation $\ln(P_2/P_1) = (\Delta H_{vap}/R)(T_1 - T_2)/T_1T_2$ or graph the data to estimate the vapor pressure of mercury at 25°C. The enthalpy of vaporization of mercury is $59.1\ kJ \cdot mol^{-1}$.

27-25. Vitamin B_{12} has the molecular formula, $C_{63}H_{90}O_{14}PCo$.

(a) Calculate the percentage by mass of cobalt in vitamin B_{12}.
(b) How much cobalt (in grams) is present in the U.S. recommended daily allowance for vitamin B_{12} of 6.0 mg per day?

27-26. Alloys are usually mixtures of metals. Give the elemental compositions of the common alloys: brass, bronze, common solder, and white gold. You may have to consult a handbook of chemistry in the library for the necessary information.

27-27. Ores containing as little as 0.25 percent copper are sometimes used to obtain copper metal. What mass of such an ore is needed to produce 91 metric tons of copper, the amount used in the Statue of Liberty?

28 TRANSITION-METAL COMPLEXES

Crystals of the chlorides of the transition metals: (*top row*) scandium through manganese; (*bottom row*) iron through zinc. Most compounds of the transition metals are colored.

Figure 28-5 Aqueous solutions of Ni(II) complex ions: $[Ni(H_2O)_6]^{2+}$ (*green*), $[Ni(NH_3)_6]^{2+}$ (*violet*), and $[Ni(CN)_4]^{2-}$ (*orange-yellow*).

general, light-induced electronic transitions of transition-metal species give rise to their color (Figure 28-5).

Not all complex ions are octahedral. For example, if we add $NaCN(aq)$ to a solution containing the blue-violet $[Ni(NH_3)_6]^{2+}(aq)$ ion, then cyanide ions displace the NH_3 ligands from the nickel(II) ion to form the bright yellow $[Ni(CN)_4]^{2-}(aq)$ ion (Figure 28-6):

$$\underset{\text{blue-violet}}{[Ni(NH_3)_6]^{2+}(aq)} + \underset{\text{colorless}}{4CN^-(aq)} \rightleftharpoons \underset{\text{yellow}}{[Ni(CN)_4]^{2-}(aq)} + \underset{\text{colorless}}{6NH_3(aq)}$$

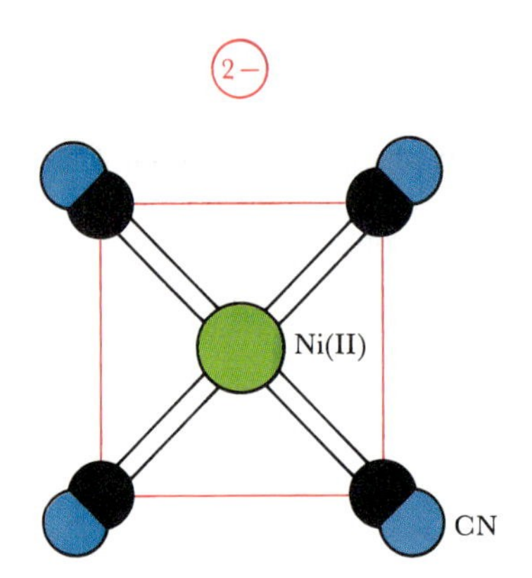

Figure 28-6 The $[Ni(CN)_4]^{2-}(aq)$ ion is square planar.

The structure of the complex ion $[Ni(CN)_4]^{2-}(aq)$ is *square planar;* the four CN^- ligands are arranged in a plane at the four corners of an imaginary square around the nickel(II) ion, and the cyanide ions are bonded to the Ni(II) ion through the carbon atoms of the cyanide ions.

Note that the overall charge on the $[Ni(CN)_4]^{2-}$ complex is -2, because the nickel(II) ion contributes a charge of $+2$ and the four CN^- ligands each contribute a charge of -1 $[+2 + 4(-1) = -2]$.

EXAMPLE 28-2: Determine the oxidation state of platinum in $[Pt(NH_3)_4Cl_2]^{2+}$.

Solution: The complex ion $[Pt(NH_3)_4Cl_2]^{2+}$ has two kinds of ligands around the central metal ion: four NH_3 ligands and two Cl^- ligands. The charge on each NH_3 ligand is 0; the charge on each Cl^- is -1; and the overall charge is $+2$. Denoting the charge on the Pt ion as x, we have

$$\underset{\text{charge on Pt}}{x} + \underbrace{4(0)}_{4NH_3} + \underbrace{2(-1)}_{2Cl^-} = \underset{\text{overall charge on ion}}{+2}$$

or

$$x + 2(-1) = +2$$
$$x = +4$$

Thus, the oxidation state of platinum in the complex ion $[Pt(NH_3)_4Cl_2]^{2+}$ is $+4$.

PRACTICE PROBLEM 28-2: Determine the oxidation state of the transition-metal ion in each of the following complex ions: (a) $[Fe(CN)_6]^{3-}$; (b) $[PtCl_6]^{2-}$; (c) $[Pt(NH_3)_3Cl_3]^+$.

Answer: (a) +3; (b) +4; (c) +4

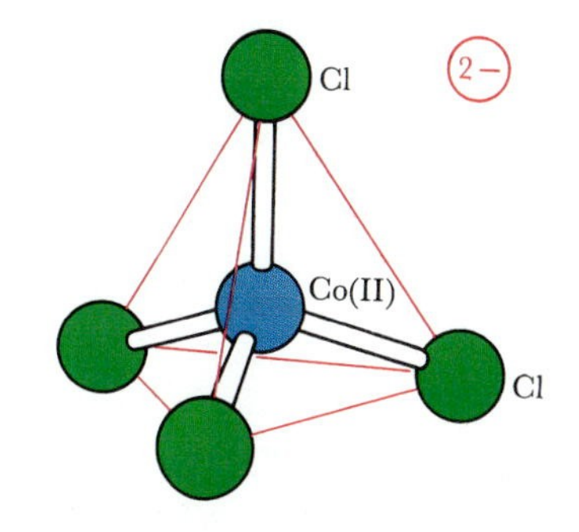

Figure 28-7 The blue $[CoCl_4]^{2-}(aq)$ ion is tetrahedral.

Devices that indicate humidity levels by changing color are based on ligand-substitution reactions. For example, consider

$$\underset{\text{pink}}{2[Co(H_2O)_6]Cl_2(s)} \rightleftharpoons \underset{\text{blue}}{Co[CoCl_4](s)} + 12H_2O(g)$$

The ion $[Co(H_2O)_6]^{2+}$ is pink, and $[CoCl_4]^{2-}$ is blue (Figure 28-7). When the humidity is high, $Co[CoCl_4]$ reacts with the water vapor in the air to form the pink ion, $[Co(H_2O)_6]^{2+}$. When the humidity is low, the equilibrium shifts from left to right, forming the blue ion, $[CoCl_4]^{2-}$. The ion $[CoCl_4]^{2-}$ is *tetrahedral,* with the cobalt(II) ion at the center of a tetrahedron formed by the four chloride ligands.

In many qualitative analysis schemes, AgCl(*s*) is separated from other insoluble chlorides by the addition of $NH_3(aq)$ to form the soluble salt $[Ag(NH_3)_2]Cl$, which contains the complex ion $[Ag(NH_3)_2]^+$:

$$AgCl(s) + 2NH_3(aq) \rightarrow [Ag(NH_3)_2]^+(aq) + Cl^-(aq)$$

In the complex ion $[Ag(NH_3)_2]^+$, the N—Ag—N atoms are arranged in a straight line; for this reason, the complex ion is usually referred to as *linear* (Figure 28-8).

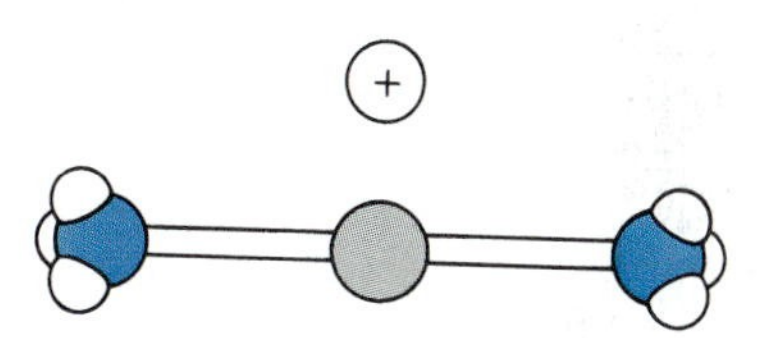

Figure 28-8 The two nitrogen atoms and the silver atom in the ion $[Ag(NH_3)_2]^+(aq)$ are in a straight line.

The most common structures for transition-metal complexes are octahedral, tetrahedral, square planar, and linear. Of these four geometries, octahedral is by far the most common. Examples of transition-metal complexes with these four geometries are given in Table 28-1.

Table 28-1 Examples of transition-metal complexes of various geometries

Octahedral*	Tetrahedral*	Square planar	Linear
$[Fe(H_2O)_6]^{2+}$	$[Zn(NH_3)_4]^{2+}$	$[Pt(CN)_4]^{2-}$	$[AgCl_2]^-$
$[Co(NO_2)_6]^{3-}$	$[CoCl_4]^{2-}$	$[AuCl_4]^-$	$[CuI_2]^-$
$[Ru(NH_3)_6]^{3+}$	$[FeCl_4]^-$	$[Rh(CN)_4]^{3-}$	$[Ag(CN)_2]^-$
$[Pt(NH_3)_6]^{4+}$	$[Ni(CO)_4]$	$[Pt(NH_3)_4]^{2+}$	$[AuCl_2]^-$
$[Cr(CO)_6]$	$[HgI_4]^{2-}$	$[Pd(ox)_2]^{2-}$	
$[Co(ox)_3]^{3-}$	$[Cd(en)_2]^{2+}$	$[Cu(NH_3)_4]^{2+}$	

*en, ethylenediamine; ox, oxalate.

28-3. TRANSITION-METAL COMPLEXES HAVE A SYSTEMATIC NOMENCLATURE

The wide variety and large number of possible transition-metal complexes make a systematic procedure for naming them essential. An example of a systematic name for a compound that contains a transition-metal complex ion is

$$[Ni(NH_3)_6](NO_3)_2 \qquad \text{hexaamminenickel(II) nitrate}$$

Let's analyze the name of this compound. As with any salt, the cation is named first. Thus, this name tells us that the compound consists of a hexaamminenickel(II) cation and nitrate anions. The Greek prefix *hexa-* denotes six, and *ammine-* denotes the ligand NH_3. Thus, the *hexaammine-* part of the name tells us that there are six NH_3 ligands in the cation. The Roman numeral II tells us that the nickel is in the +2 oxidation state. Because ammonia is a neutral molecule and nickel is in a +2 oxidation state, the charge on the complex cation is +2. Its formula is $[Ni(NH_3)_6]^{2+}$.

A simplified set of nomenclature rules for complexes is as follows:

Table 28-2 Names for common ligands

Ligand*	Name as ligand
F^-	fluoro
Cl^-	chloro
Br^-	bromo
I^-	iodo
$\underline{C}N^-$	cyano
$\underline{O}H^-$	hydroxo
$\underline{N}O_2^-$	nitro
$\underline{O}NO$	nitrito
$\underline{C}O$	carbonyl
$H_2\underline{O}$	aqua
$\underline{N}H_3$	ammine

*For ligands with two or more different atoms, the underlined atom is the one bonded to the metal.

1. *Name the cation first and then the anion:* for example, potassium tetracyanonickelate(II), $K_2[Ni(CN)_4]$.

2. *In any complex ion or neutral complex, name the ligands first and then the metal:* for example, hexaamminenickel(II) or tetracyanonickelate(II). If there is more than one type of ligands in the complex, then name them in alphabetical order: for example, diamminedichloroplatinum(II), $[Pt(NH_3)_2Cl_2]$.

3. *End the names of negative ligands in the letter* o, *but give neutral ligands the name of the ligand molecule.* Some common neutral ligands have special names, such as *aqua* for H_2O, *ammine* for NH_3, and *carbonyl* for CO. Table 28-2 lists the names of a number of ligands.

4. *Denote the number of ligands of a particular type by a Greek prefix, such as di-, tri-, tetra-, penta-, or hexa-. (Mono is not used.)*

5. *If the complex ion is a cation or a neutral atom, then use the ordinary name for the metal; if the complex ion is an anion, then end the name of the metal in -ate:* for example, tetrachlorocobaltate(II), $[CoCl_4]^{2-}$, where the suffix *-ate* on the metal name tells us that the complex ion is an anion. Table 28-3 lists a few exceptions to this rule.

6. *Denote the oxidation state of the metal by a Roman numeral or zero in parentheses following the name of the metal.*

Table 28-3 Exceptions to rule 5 for naming complexes

Metal	Name in complex anion	Complex anion	Name of complex anion
silver	argentate	$[AgCl_2]^-$	dichloroargentate(I)
gold	aurate	$[Au(CN)_4]^-$	tetracyanoaurate(III)
copper	cuprate	$[CuCl_4]^{2-}$	tetrachlorocuprate(II)
iron	ferrate	$[Fe(CN)_6]^{3-}$	hexacyanoferrate(III)

The application of these rules is illustrated in the following Example.

EXAMPLE 28-3: The water-soluble yellow-orange compound $Na_3[Co(NO_2)_6]$ is used in some qualitative analysis schemes to test for $K^+(aq)$. Almost all potassium salts are water soluble, but $K_2Na[Co(NO_2)_6]$ is only slightly soluble in water. The equation for the precipitation reaction is

$$Na^+(aq) + 2K^+(aq) + [Co(NO_2)_6]^{3-}(aq) \rightarrow K_2Na[Co(NO_2)_6](s)$$

Name the $[Co(NO_2)_6]^{3-}$ ion.

Solution: The oxidation state of cobalt in the complex ion is determined as follows: the overall charge on the ion is −3, and there are six nitrite ions in the complex, each with a charge of −1. Denoting the oxidation state of cobalt as x, we have

$$\underbrace{x}_{\text{Co}} + \underbrace{6(-1)}_{6NO_2^-} = \underbrace{-3}_{\text{net charge on complex ion}}$$

or $x = +3$. The ion is called hexanitrocobaltate(III), where the *-ate* ending tells us that the complex is an anion.

PRACTICE PROBLEM 28-3: Name the salt $[Ag(NH_3)_2]_3[Fe(CN)_6]$.

Answer: diamminesilver(I) hexacyanoferrate(III)

Other examples of the nomenclature of complexes are

$$K_2[Ni(CN)_4] \quad \overbrace{\text{potassium}}^{\text{cation}} \; \overbrace{\underbrace{\text{tetracyano}}_{\substack{4CN^- \\ \text{ligands}}}\underbrace{\text{nickelate(II)}}_{\substack{\text{Ni in } +2 \\ \text{oxidation} \\ \text{state}}}}^{\text{complex anion}}$$

and

$$[Co(H_2O)_4Cl_2]Cl \quad \overbrace{\underbrace{\text{tetraaqua}}_{\substack{4H_2O \\ \text{ligands}}}\underbrace{\text{dichloro}}_{\substack{2Cl^- \\ \text{ligands}}}\underbrace{\text{cobalt(III)}}_{\substack{\text{Co in } +3 \\ \text{oxidation} \\ \text{state}}}}^{\text{complex cation}} \; \overbrace{\text{chloride}}^{\text{anion}}$$

The rules for writing a chemical formula from the name of a complex follow from the nomenclature rules. For example, the formula for the compound named potassium hexacyanoferrate(II) is determined as follows:

1. The cation is potassium, K^+.
2. The complex anion contains six CN^- (hexacyano) ions and an iron atom. The oxidation state of the iron is +2, as indicated by the Roman numeral. The ending *-ate* tells us that the complex is an anion.
3. The charge on the complex anion is calculated by adding up the charges on the metal ion and the ligands:

$$\underbrace{(+2)}_{\text{Fe(II)}} + \underbrace{6(-1)}_{6CN^-} = \underbrace{-4}_{\text{net charge on complex}}$$

4. The formulas for complex ions and neutral complexes are enclosed in brackets, so we write the formula for the anion as $[Fe(CN)_6]^{4-}$. The formula for the salt is $K_4[Fe(CN)_6]$ because four K^+ ions are required to balance the −4 charge on the anion.

EXAMPLE 28-4: Give the formula for the compound hexamminecobalt(III) hexachlorocobaltate(III).

Solution: In this case both the cation and the anion are complex ions. The cation has six NH_3 ligands with zero charge and one cobalt in a +3 oxidation state. Therefore, the formula for the cation is

$$[Co(NH_3)_6]^{3+}$$

The anion has six Cl^- with a total ligand charge of −6 plus one cobalt in a +3 oxidation state. Therefore, the net charge on the complex anion is

$$6(-1) + (+3) = -3$$

and the formula is

$$[CoCl_6]^{3-}$$

The magnitudes of the charges on the cation and the anion are equal; thus, the ions appear in the formula for the salt on a one-to-one basis:

$$[Co(NH_3)_6][CoCl_6]$$

PRACTICE PROBLEM 28-4: Give the chemical formula for the compound hexaaquanickel(II) diaquatetrabromochromate(III).

Answer: $[Ni(H_2O)_6][Cr(H_2O)_2Br_4]_2$

$[Co(ox)_3]^{3-}$

(a)

$[Co(en)_3]^{3+}$

(b)

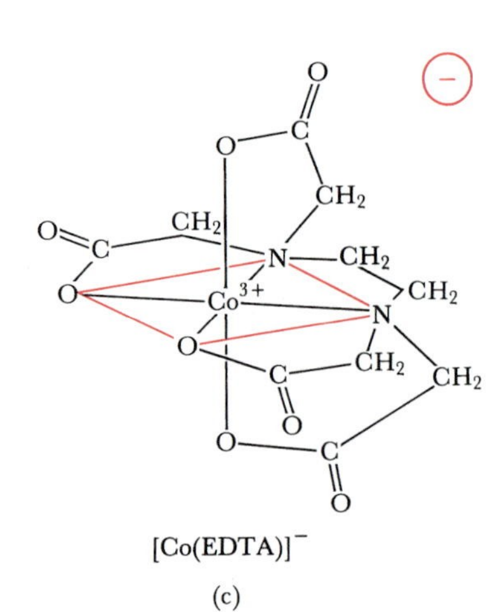

$[Co(EDTA)]^-$

(c)

Figure 28-9 Bidentate chelate complexes of cobalt(III), involving the ligands (a) oxalate, $C_2O_4^{2-}$, (b) ethylenediamine, en, and (c) ethylenediaminetetraacetate, $EDTA^{4-}$.

28-4. POLYDENTATE LIGANDS BIND TO MORE THAN ONE COORDINATION POSITION AROUND THE METAL ION

Certain ligands can bond to a central metal cation at more than one point of attachment, or **coordination position.** Examples of ligands that bond to two coordination positions are the oxalate ion (abbreviated ox) and ethylenediamine (abbreviated en):

ligating atoms oxalate ion (ox^{2-})

ligating atoms ethylenediamine (en)

The atoms of the ligand that attach to the metal ion are called **ligating atoms.** Two complexes involving these two ligands are shown in Figure 28-9a and b.

Ligands that attach to a metal ion at more than one coordination position are called **polydentate ligands** or **chelating ligands.** The resulting complex is called a **chelate,** which comes from the Greek word meaning claw. We can visualize the attachment of a chelating ligand as a species grasping a metal ion with molecular claws. A chelating ligand that attaches to two metal coordination positions is called **bidentate** (two teeth); one that attaches to three positions is **tridentate** (three teeth); and so on.

Ethylenediaminetetraacetate ion, $EDTA^{4-}$, is the best known example of a hexadentate (i.e., six-coordinate) ligand:

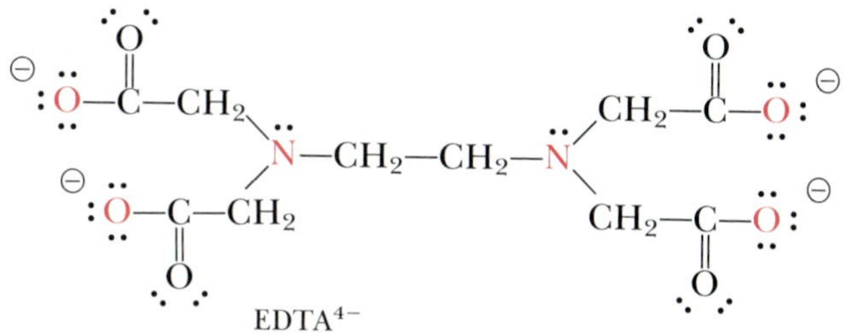

$$\begin{matrix} {}^{-}\!:\ddot{O}-\overset{\overset{\displaystyle O}{\|}}{C}-CH_2 & & & & CH_2-\overset{\overset{\displaystyle O}{\|}}{C}-\ddot{O}\!:^{-} \\ & \ddot{N}-CH_2-CH_2-\ddot{N} & & & \\ {}^{-}\!:\ddot{O}-\underset{\underset{\displaystyle O}{\|}}{C}-CH_2 & & & & CH_2-\underset{\underset{\displaystyle O}{\|}}{C}-\ddot{O}\!:^{-} \end{matrix}$$

$EDTA^{4-}$

The six ligating atoms are shown in color. Figure 28-9c shows the structure of the $[Co(EDTA)]^-$ complex ion. The complex ion $EDTA^{4-}$ binds strongly to a number of metal ions and has a great variety of uses. It is used as an antidote for poisoning by heavy metals such as lead and mercury; as a food preservative, because it complexes with and renders inactive metal ions that catalyze the reactions involved in the spoiling process; as an analytical reagent in the analysis of the hardness of water; in detergents, soaps, and shampoos; and to decontaminate radioactive surfaces.

The nomenclature for complex ions and molecules that contain polydentate ligands follows the rules listed in Section 28-3, with one additional rule:

7. *If the the ligand attached to the metal ion is a polydentate ligand, then enclose the ligand name in parentheses and use the prefix bis- for two ligands and tris- for three ligands:* for example, tris(ethylenediamine)cobalt(III), $[Co(H_2NCH_2CH_2NH_2)_3]^{3+}$. The parentheses are not used if the complex contains only one of the polydentate ligands of a particular type.

EXAMPLE 28-5: Give the chemical formula for ammonium tris(oxalato)ferrate(III).

Solution: The cation is the ammonium ion, NH_4^+. The anion is a complex ion with three (tris) oxalate ions, $C_2O_4^{2-}$, and an iron atom in the +3 oxidation state. The net charge on the complex anion is

$$\underbrace{(+3)}_{Fe(III)} + \underbrace{3(-2)}_{3C_2O_4^{2-}} = \underbrace{-3}_{\text{net charge}}$$

Therefore, the formula for the complex anion is $[Fe(C_2O_4)_3]^{3-}$ and the formula for ammonium tris(oxalato)ferrate(III) is $(NH_4)_3[Fe(C_2O_4)_3]$. Note that we do not say triammonium, because the number of ammonium ions is unambiguously fixed by the net charge of −3 on the complex anion. There are four ions per formula unit in $(NH_4)_3[Fe(C_2O_4)_3]$, three NH_4^+, and one $[Fe(C_2O_4)_3]^{3-}$.

PRACTICE PROBLEM 28-5: Name the following compounds: (a) $K_2[Fe(EDTA)]$; (b) $Na[Co(C_2O_4)_2(en)]$.

Answer: (a) potassium ethylenediaminetetraacetatoferrate(II); (b) sodium ethylenediaminebis(oxalato)cobaltate(III)

Table 28-4 Examples of nomenclature for transition-metal compounds

Compound	Name
$[Co(NH_3)_6]Cl_3$	hexaamminecobalt(III) chloride
$K[AuCl_4]$	potassium tetrachloroaurate(III)
$Cu_2[Fe(CN)_6]$	copper(II) hexacyanoferrate(II)
$[Pt(NH_3)_6]Cl_4$	hexaammineplatinum(IV) chloride
$[Cu(NH_3)_4(H_2O)_2]Cl_2$	tetraamminediaquacopper(II) chloride
$K[V(CO)_6]$	potassium hexacarbonylvanadate(−I)
$K_3[CoF_6]$	potassium hexafluorocobaltate(III)
$Cr(CO)_6$	hexacarbonylchromium(0)

The nomenclature of transition-metal complexes may appear cumbersome at first because of the length of the names, but with a little practice you will find it straightforward. Table 28-4 gives several additional examples of names of transition-metal complexes. You should try to name each one from the formula and write the formula from each name.

28-5. SOME OCTAHEDRAL AND SQUARE-PLANAR TRANSITION-METAL COMPLEXES CAN EXIST IN ISOMERIC FORMS

Consider the compound

$$[Pt(NH_3)_2Cl_2] \qquad \text{diamminedichloroplatinum(II)}$$

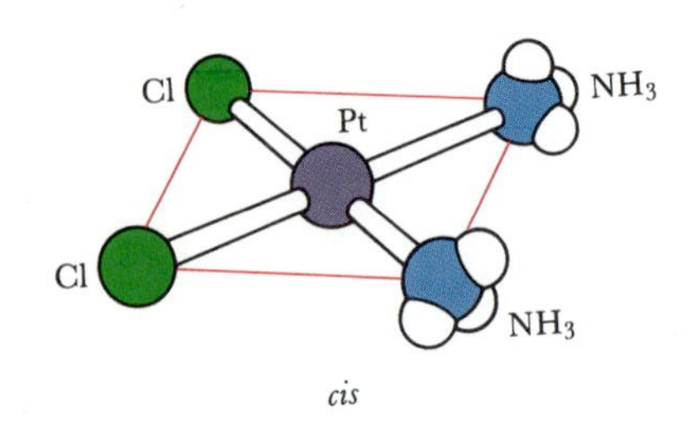

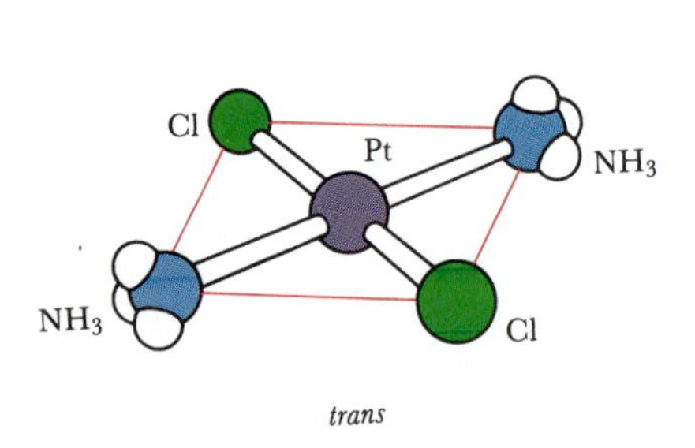

Figure 28-10 Cis and trans isomers of the square-planar complexes of diamminedichloroplatinum(II).

Platinum(II) complexes are invariably square planar. As shown in Figure 28-10, there are two possible arrangements of the four ligands around the central platinum(II) ion; these two complexes are **cis and trans isomers.** Recall from Section 13-11 that cis and trans compounds are **geometric isomers,** because they differ in the spatial arrangements of the constituent atoms. The designation *cis* ("on the same side") tells us that identical ligands are placed adjacent to each other (on the same side) in the structure. The designation *trans* ("opposite") tells us that the identical ligands are placed directly opposite each other in the structure. The cis and trans isomers of $[Pt(NH_3)_2Cl_2]$ are different compounds with different physical and chemical properties. For example, the *cis*-diamminedichloroplatinum(II) isomer is manufactured as the potent anticancer drug **cisplatin,** whereas the trans isomer does not exhibit anticancer activity. How the cis isomer destroys cancer cells is not fully understood, but the evidence suggests that it interferes with the cell duplication process by inserting into the DNA double helix (Chapter 30).

Cis and trans isomers are also found in certain octahedral complexes. Consider the octahedral ion tetraamminedichlorocobalt(III), $[Co(NH_3)_4Cl_2]^+$. The two Cl^- ligands can be placed in adjacent (cis) or

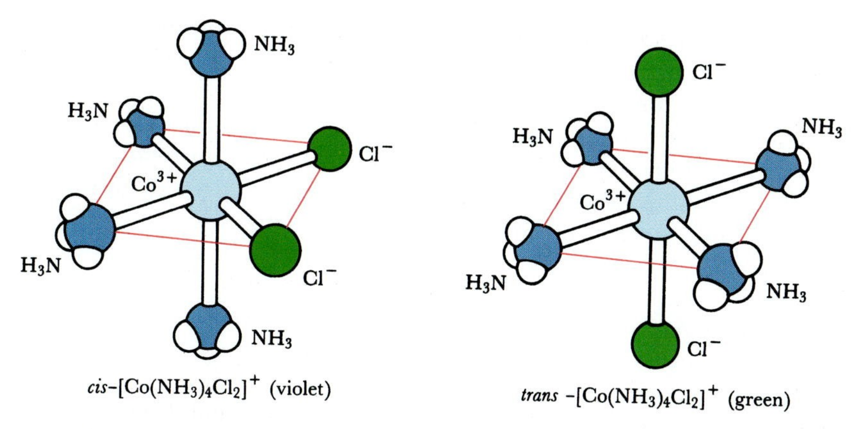

Figure 28-11 Cis and trans isomers of octahedral complexes.

opposite (trans) positions around the central cobalt(III) ion, as shown in Figure 28-11. Note that because the six coordination positions in an octahedral complex are equivalent, any other cis placement of the two Cl^- ligands around the cobalt(III) ion yields a structure identical to the cis structure shown in the figure, except for its orientation in space; this equivalency is also true for the trans placement of the two Cl^- ligands.

EXAMPLE 28-6: The compound $[Co(NH_3)_3Cl_3]$ exists in two isomeric forms. Draw the structures of the two isomers of this neutral complex.

Solution: The structures of the two isomers are shown in Figure 28-12. One compound is denoted *cis-cis* because each Cl^- is adjacent to the two others. The other compound is denoted *cis-trans* because one Cl^- is adjacent to the third chloride and one Cl^- is opposite it.

PRACTICE PROBLEM 28-6: Draw the structures of the various possible geometric isomers of $[CoCl_4BrI]^{3-}$.

Answer: There are two geometric isomers, one with the Br^- and I^- trans to each other and one with the Br^- and I^- cis to each other.

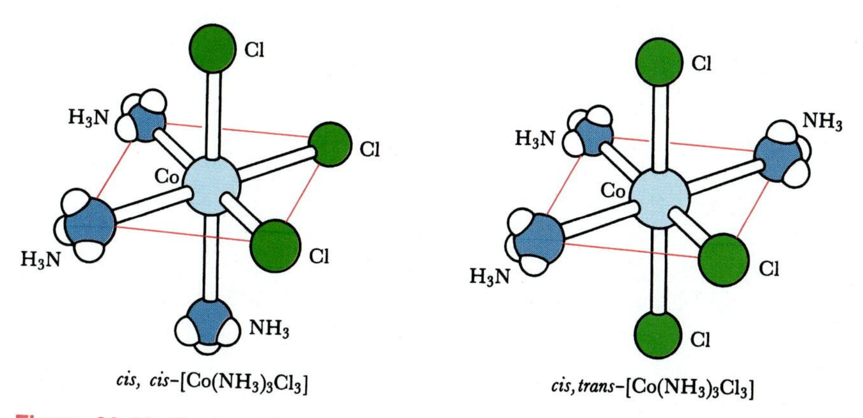

Figure 28-12 Cis,cis and cis,trans geometrical isomers of octahedral complexes.

Figure 28-13 The Swiss chemist Alfred Werner (1866–1919), who received the Nobel Prize in Chemistry in 1913 for his work on transition-metal compounds. He deduced the structures of numerous complexes solely from the number of isomers formed. His predictions were confirmed by X-ray crystallography.

The structures of transition-metal complexes were worked out by the Swiss chemist Alfred Werner in the late nineteenth and early twentieth centuries, without the aid of modern X-ray structure determination methods (Figure 28-13). In 1893, at the age of 26, he proposed a correct structural theory based on the number of different types of complexes, including isomers, that could be prepared for platinum(II), platinum(IV), and cobalt(III) amminechloro complexes. In 1913 Werner was awarded the Nobel Prize for his research in transition-metal chemistry.

This section completes our introduction to the structure and nomenclature of transition-metal complexes. In the next section we turn to a consideration of interactions between the central metal ion *d* orbitals and the coordinated ligands. We will show how the ligands perturb the various *d* orbitals in different ways, depending on the type and the spatial arrangement of the coordinated ligands.

28-6. THE FIVE d ORBITALS OF A TRANSITION-METAL ION IN AN OCTAHEDRAL COMPLEX ARE SPLIT INTO TWO GROUPS BY THE LIGANDS

The five *d* orbitals (Figure 28-2) in a gas-phase transition-metal atom or ion without any attached ligands all have the same energy. However, when six identical ligands are attached to the transition-metal ion to form an octahedral complex, the *d* orbitals on the metal ion are split into two sets, each set with a different energy (Figure 28-14). This outcome is called ***d*-orbital splitting.**

The difference in the energies of the two sets of *d* orbitals means that a *d* electron can be excited to a state of higher energy by absorbing light. We will show that this absorption of light by electrons in the *d* orbitals accounts for the colors of many coordination compounds. The lower-energy set of orbitals, called $\mathbf{t_{2g}}$ **orbitals,** consists of three orbitals (d_{xy}, d_{xz}, and d_{yz}), and the higher-energy set of orbitals, called $\mathbf{e_g}$ **orbitals,** consists of two orbitals ($d_{x^2-y^2}$ and d_{z^2}). The magnitude of the splitting of the *d* orbitals depends on both the central metal ion and the ligands. We will let Δ_o (where the subscript *o* stands for octahedral) be the energy difference between the t_{2g} and e_g orbitals. The t_{2g} orbitals can accommodate up to six electrons, and the e_g orbitals can accommodate up to four electrons.

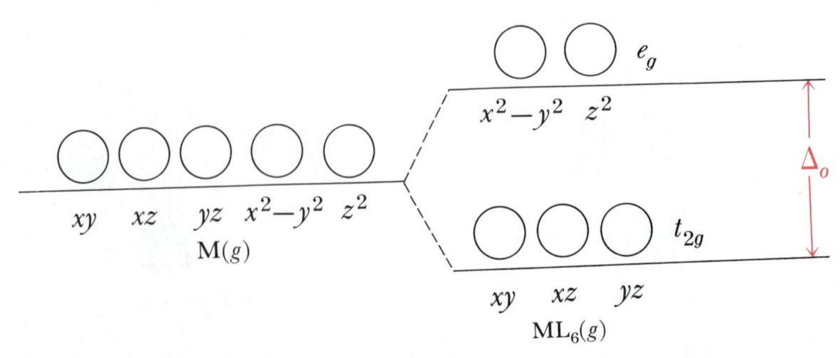

Figure 28-14 *d*-Orbital splitting pattern for a regular octahedral complex.

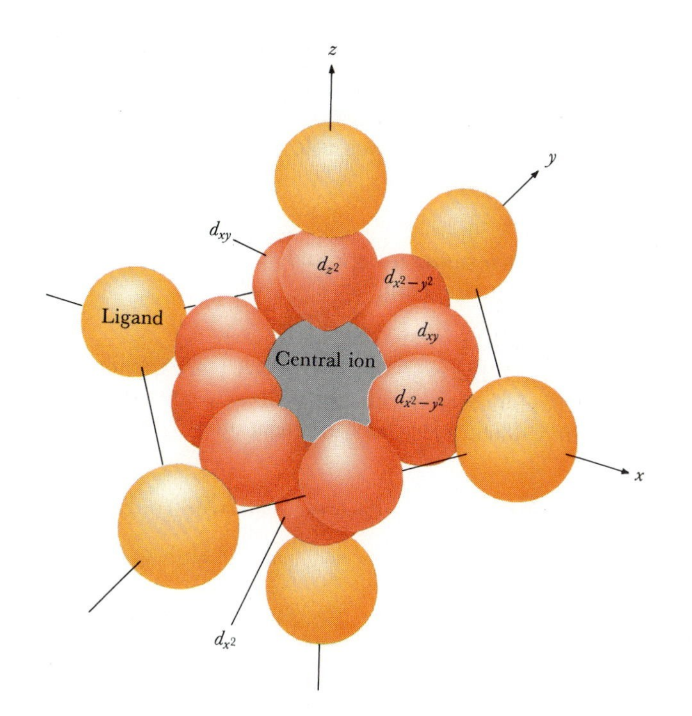

Figure 28-15 A regular octahedral complex, showing the orientation of the d_{z^2}, $d_{x^2-y^2}$, and d_{xy} orbitals relative to the ligands, which are brought in along the x, y, and z axes toward the central metal ion. For simplicity, the d_{xz} and d_{yz} orbitals are not shown. The d_{xy}, d_{xz}, and d_{yz} lobes point toward positions between the ligands, as shown here for d_{xy}.

The splitting of the d orbitals in an octahedral complex can be explained using a simple, electrostatic model. Consider the octahedral complex shown in Figure 28-15, where the ligands are located along the positive and negative x, y, and z axes. Each ligand possesses a lone pair of electrons that for the moment can be considered to be localized on its ligating atom and directed toward the transition-metal atom. Note that the d_{z^2} and $d_{x^2-y^2}$ orbitals (the e_g set) point *directly* toward these ligand lone pairs, whereas the d_{xy}, d_{xz}, and d_{yz} orbitals (the t_{2g} set) point *between* the ligand lone pairs. An electron placed in a d_{z^2} or a $d_{x^2-y^2}$ orbital will, therefore, experience a greater electrostatic repulsion (like charges repel) than one placed in a d_{xy}, d_{xz}, or d_{yz} orbital, because the e_g electron will have a *higher* probability of being close to the ligand lone pair electrons. Thus, electrons in the e_g set of orbitals will have a higher energy than those in the t_{2g} set. The d-orbital splitting pattern in an octahedral complex is thus seen to be a consequence of the positions of the ligands relative to the d orbitals.

To determine the ground-state d-orbital configuration of a transition metal in a complex, we place electrons in the t_{2g} and e_g orbitals in a manner similar to that used in Chapter 9 for atoms. Here also we must observe the restrictions of the Pauli exclusion principle (Section 9-17) and Hund's rule (Section 9-19). Thus, the orbitals of lower energy (the t_{2g} set) are occupied first, and as these orbitals are all of the same energy, each one must contain one electron before any one can contain two electrons. For example, an octahedral Cr(III) complex has three d electrons; and the d electron configuration would be $t_{2g}^3e_g^0$, or simply t_{2g}^3 (Figure 28-16). Note that Hund's rule requires that each t_{2g} orbital contain one electron and that all have the same spin.

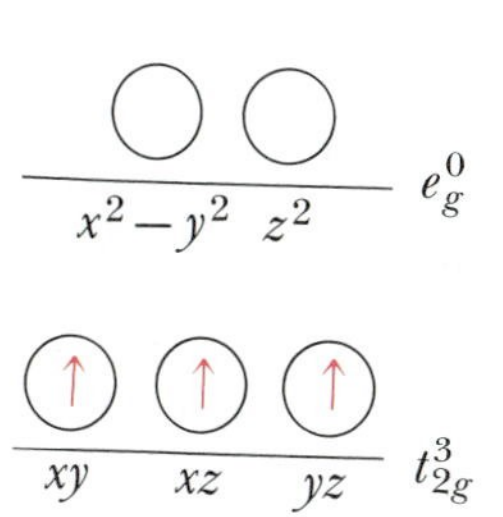

Figure 28-16 Ground-state d electron configuration of an octahedral d^3complex.

For an octahedral Ni(II) complex with eight d electrons, we can accommodate six of these eight electrons in the t_{2g} orbitals (spins paired). The remaining two electrons must occupy the higher-energy set (the e_g orbitals), giving the configuration $t_{2g}^6e_g^2$. By Hund's rule the two e_g orbitals have one electron each with parallel spins. This procedure can be used to obtain the ground-state *d* electron configurations for d^1–d^3 and d^8–d^{10} ions. For d^4–d^7 ions, on the other hand, two different *d* electron configurations are found to be possible. The reasons for this are explained in the next section.

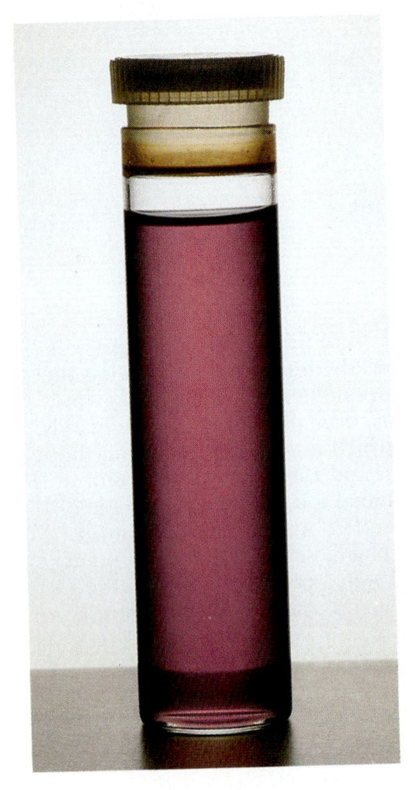

Figure 28-17 An aqueous solution of titanium(III) chloride. The d^1 ion $[Ti(H_2O)_6]^{3+}$ has a red-purple color.

EXAMPLE 28-7: Determine the ground-state *d* electron configuration of copper(II) in an octahedral complex.

Solution: Copper(II) has nine *d* electrons. We place the nine *d* electrons in the t_{2g} and e_g orbitals in accord with the Pauli exclusion principle and Hund's rule. The ground-state *d* electron configuration of copper(II) is $t_{2g}^6 e_g^3$.

PRACTICE PROBLEM 28-7: Give the *d* electron configuration of V(III) in an octahedral complex.

Answer: $t_{2g}^2 e_g^0$

The magnitude of the energy difference between the t_{2g} and e_g orbitals, denoted by Δ_o, depends on the central metal ion and the ligands. For most cases, the value of the energy difference, Δ_o (also called the **splitting energy**), corresponds to the values of photon energies in the visible region of the electromagnetic spectrum. In other words, the frequency of the radiation that is absorbed, which obeys the relation

$$E = h\nu = \Delta_o$$

often is in the visible region of the spectrum. Thus, many transition-metal complexes absorb light in the visible region as a result of *d* electron transitions and are colored.

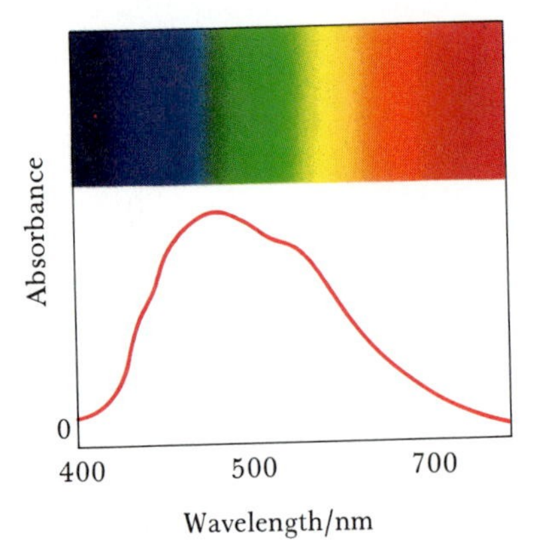

Figure 28-18 Absorption spectrum of the red-purple ion $[Ti(H_2O)_6]^{3+}$ in aqueous solution. The complex absorbs light in the blue, green, yellow, and orange regions, whereas most of the light in the red and purple regions passes through the sample and therefore is detected by the eye. Thus, the $[Ti(H_2O)_6]^{3+}(aq)$ complex is red-purple.

It is possible to understand the rich variety of colors of octahedral complex ions in terms of the magnitudes of Δ_o values and the electron occupancy of the t_{2g} and e_g orbitals. For example, consider the red-purple ion $[Ti(H_2O)_6]^{3+}(aq)$ (Figure 28-17). Reference to Figure 28-1 shows that titanium is the second member of the 3*d* transition series; thus, Ti(II) has two *d* electrons. The $[Ti(H_2O)_6]^{3+}$ complex, which contains Ti(III), has one fewer *d* electron than Ti(II); therefore, $[Ti(H_2O)_6]^{3+}$ is a d^1 ion. The ground-state and excited-state *d* electron configurations of the d^1 ion $[Ti(H_2O)_6]^{3+}$ are

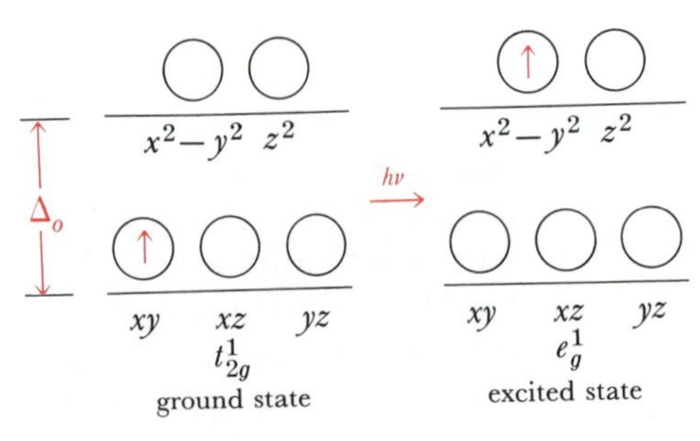

The absorption spectrum $[Ti(H_2O)_6]^{3+}(aq)$ in the visible region is shown in Figure 28-18. The absorption of a photon in the blue through

orange region excites the d electron from the lower-energy (t_{2g}) set of d orbitals to the higher-energy (e_g) set.

The $[Ti(H_2O)_6]^{3+}(aq)$ complex appears to be red-purple because the photons not absorbed by the complex have wavelengths corresponding to the red and purple regions of the spectrum. The colors of many familiar gemstones are due to the presence of transition-metal ions (Interchapter M).

28-7. d-ORBITAL ELECTRON CONFIGURATION IS THE KEY TO UNDERSTANDING MANY PROPERTIES OF THE d TRANSITION-METAL IONS

Not only the spectral properties, but also the magnetic properties of transition-metal complexes can be understood on the basis of the d electron configurations of the metal ion in the complex. Moving charges give rise to a magnetic field. Unpaired electrons act like tiny magnets as a result of their intrinsic spin. The magnetic fields from paired electrons, which have opposite spins, cancel out. Molecules with no unpaired electrons cannot be magnetized (in fact, they are repelled) by an external magnetic field and are called **diamagnetic.** Molecules with unpaired electrons can be magnetized by an external field, however, and are called **paramagnetic.** As we showed for oxygen in Section 13-3, paramagnetic molecules line up in an external magnetic field with the electron spins parallel to the applied field. Thus, a paramagnetic substance behaves like a collection of magnets and is drawn into an externally applied magnetic field (Figure 28-19). A diamagnetic substance is not drawn into an applied magnetic field and thus is easily distinguished from a paramagnetic substance. In some cases it is possible to determine the number of unpaired electrons by measuring the force with which the paramagnetic substance is drawn into the magnetic field.

Magnetic experiments have shown that the compound $K_4[Fe(CN)_6]$ is diamagnetic, whereas the compound $K_4[FeF_6]$ is paramagnetic. Furthermore, there are four unpaired electrons in $[FeF_6]^{4-}$. We can explain these observations in terms of d-orbital electron configurations. Let's consider the ion $[FeF_6]^{4-}$. Iron(II) has six d electrons; and, as noted in the previous section, d^6 systems have two possible d electron configurations. The two possibilities for the d electron configuration of an octahedral iron(II) complex ion are

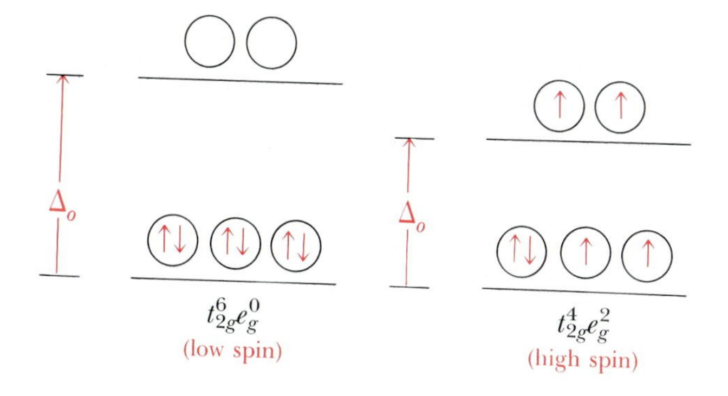

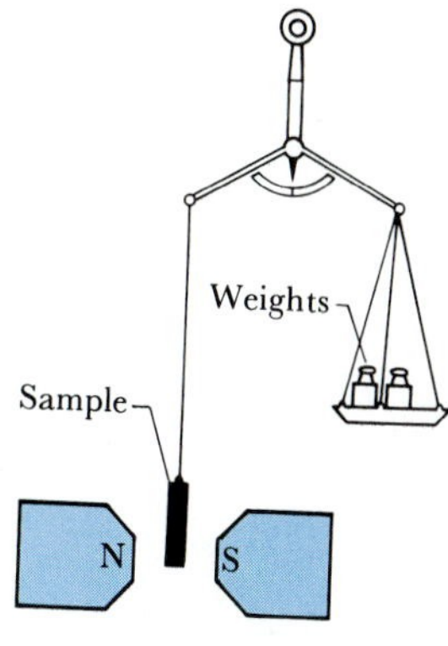

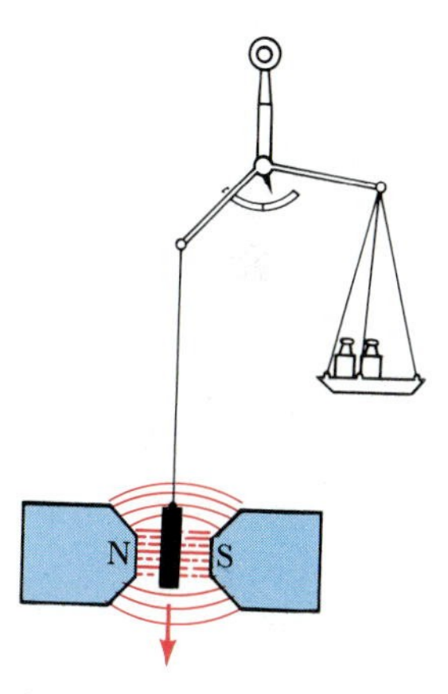

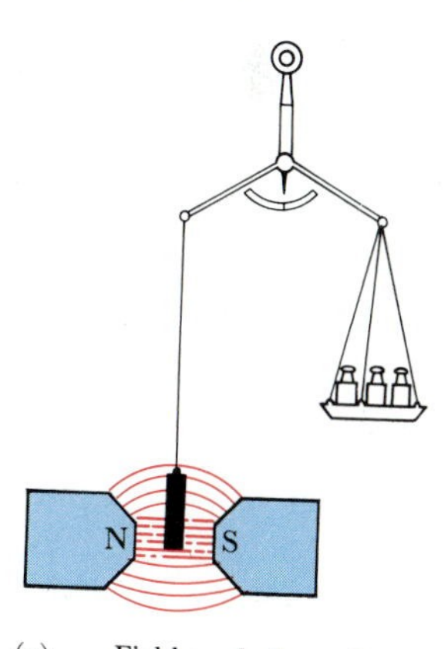

Figure 28-19 Attraction of a paramagnetic substance into a magnetic field. The magnetic attractive force on the sample makes it appear heavier. The number of unpaired electrons in the sample can be calculated from the apparent mass gain. Masses are added to the balance pan until balance is restored with the field on.

In one case the d electron configuration is $t_{2g}^6e_g^0$, and in the other case it is $t_{2g}^4e_g^2$. In the t_{2g}^6 configuration, the spins of all the electrons are paired; in the $t_{2g}^4e_g^2$ configuration, the electrons are in different orbitals with four unpaired electrons, in accord with Hund's rule. The $t_{2g}^6e_g^0$ configuration is said to be a **low-spin configuration,** and the $t_{2g}^4e_g^2$ configuration is said to be a **high-spin configuration.**

The value of Δ_o determines whether a d electron configuration is low spin or high spin. If Δ_o is small, then the d electrons occupy the e_g orbitals before they pair up in the t_{2g} orbitals. If Δ_o is large, then the d electrons fill the t_{2g} orbitals completely before occupying the higher-energy e_g orbitals. For example, the d^6 ion, $[FeF_6]^{4-}$, is high spin and has four unpaired electrons, as shown in the margin. This configuration occurs because Δ_o is small relative to the **electron-pairing energy,** which is the energy required to pair up the electrons in the t_{2g} orbitals. The d^6 ion, $[Fe(CN)_6]^{4-}$, on the other hand, is low spin and has no unpaired electrons, again, as shown in the margin. This configuration occurs because Δ_o is large relative to the energy required to pair up the electrons in the t_{2g} orbitals. In other words, it requires less energy to pair up two more electrons in the t_{2g} orbitals in $[Fe(CN)_6]^{4-}$ than to place the two electrons in the high-energy (large Δ_o) e_g orbitals (Figure 28-20). Thus, we see that $[FeF_6]^{4-}$ is paramagnetic, with four unpaired electrons, and that $[Fe(CN)_6]^{4-}$ is diamagnetic, with no unpaired electrons. The Δ_o values obtained from the spectra of the complexes show that the CN^- ligands interact much more strongly with iron(III) than do the F^- ligands; that is, as we note that $\Delta_o(CN^-) \gg \Delta_o(F^-)$. The increased d-orbital splitting energy for CN^- is sufficiently great to overcome the additional electron-electron repulsions that result from pairing up the electrons. A low-spin complex results whenever the energy difference between the t_{2g} and e_g orbitals is greater than the electron-pairing energy.

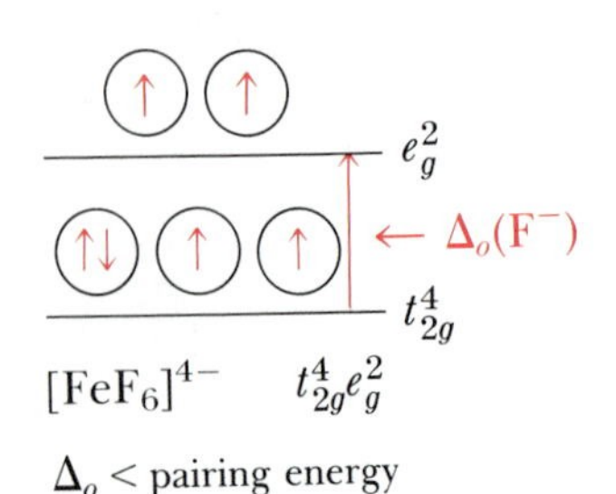

$[FeF_6]^{4-}$ $t_{2g}^4e_g^2$

$\Delta_o <$ pairing energy

e_g^0

$\leftarrow \Delta_o(CN^-)$

t_{2g}^6

$[Fe(CN)_6]^{4-}$ $t_{2g}^6e_g^0$

$\Delta_o >$ pairing energy

The various possible d electron configurations for octahedral transition-metal ions are shown in Figure 28-21. Note that for d^4, d^5, d^6, and d^7 octahedral complexes there are two possible d electron configurations. The high-spin configuration has the maximum possible number of unpaired d electrons and the low-spin configuration has the minimum possible number of unpaired d electrons.

- Low-spin complex:
  $\Delta_o >$ pairing energy

  High-spin complex:
  $\Delta_o <$ pairing energy

EXAMPLE 28-8: Give the d electron configuration of the low-spin complex $[Pt(NH_3)_6]^{4+}$.

Solution: Referring to Figure 28-1, we note that platinum is the eighth member of the $5d$ transition series. Therefore, platinum(II) is a d^8 ion. The platinum in $[Pt(NH_3)_6]^{4+}$ is platinum(IV); thus, platinum(IV) is a d^6 ion [two fewer d electrons than platinum(II)]. The d electron configuration of a low-spin d^6 ion is $t_{2g}^6e_g^0$ (Figure 28-21). The ion is diamagnetic because it has no unpaired electrons.

e_g^0

x^2-y^2 z^2

t_{2g}^6

xy xz yz

Figure 28-20 The low spin d^6 electron configuration.

PRACTICE PROBLEM 28-8: Predict which of the following ions can form both low-spin and high-spin complexes depending on the ligands: (a) Cr(III); (b) Mn(II); (c) Cu(III).

Answer: (b)

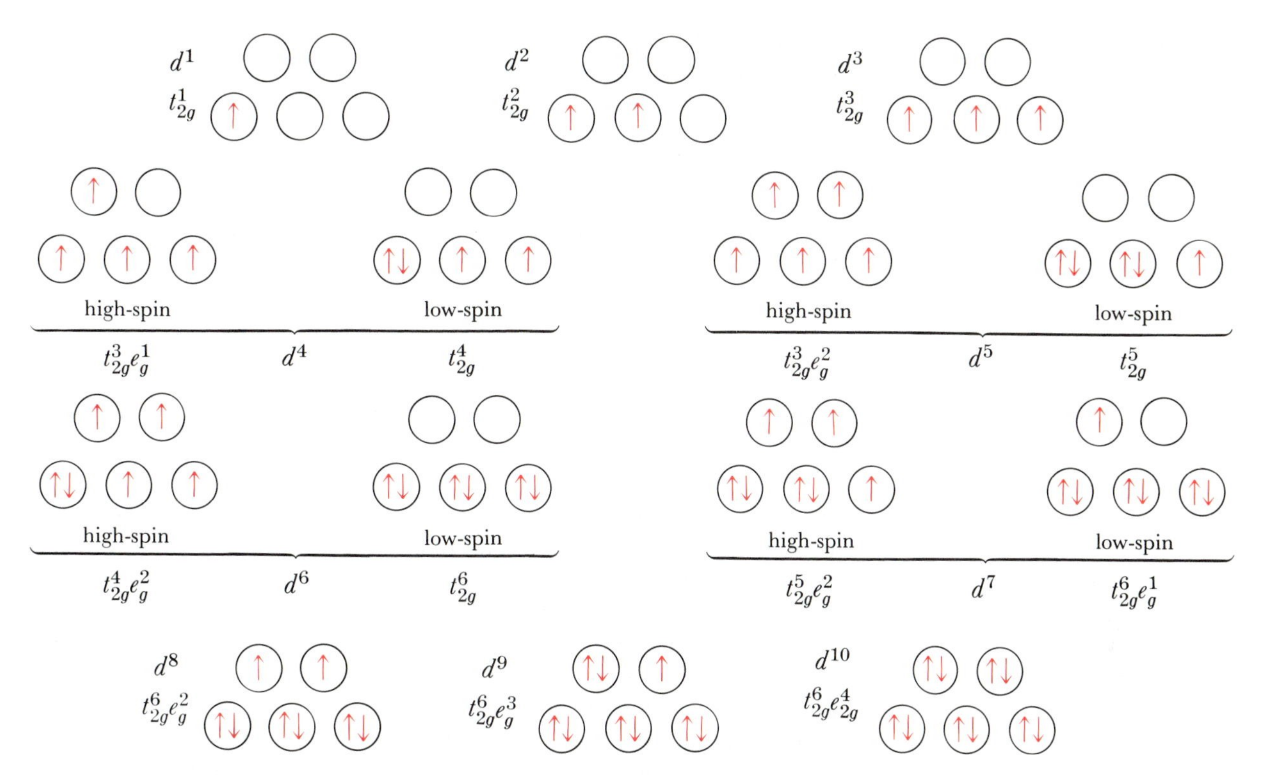

Figure 28-21 Possible ground-state d electron configurations for octahedral d^x ions, where x is the total number of outer d electrons in the transition metal. Note that for d^4, d^5, d^6, and d^7 ions there are two possibilities: the high-spin and the low-spin configurations.

A change in the nature of the coordination around a transition-metal ion can trigger a conversion from a high-spin to a low-spin d electron configuration. A particularly interesting case of such a conversion occurs in hemoglobin. This iron-containing protein is responsible for the color of red blood cells, which are about 35 percent hemoglobin by mass. Hemoglobin transports O_2 from the lungs to the various cells of the body, where it is used in the oxidation of glucose, and transports CO_2, a waste product of the oxidation processes (in the form of HCO_3^-), back to the lungs for elimination. A hemoglobin molecule contains four so-called heme groups, each of which acts as a pentadentate (five-coordinate) ligand to an Fe^{2+} ion (Figure 28-22). The vacant, sixth coordinate position about the Fe^{2+} ion is taken up either by an oxygen molecule, or a hydrogen carbonate ion (in CO_2 transport), or a water molecule. The hemoglobin-oxygen complex, which is called oxyhemoglobin, is bright red and has a low-spin d electron configuration. When oxygen is lost, the bluish-red high-spin deoxyhemoglobin forms.

$$\text{Hmb}\cdot 4\text{O}_2 \rightleftharpoons \text{Hmb} + 4\text{O}_2$$

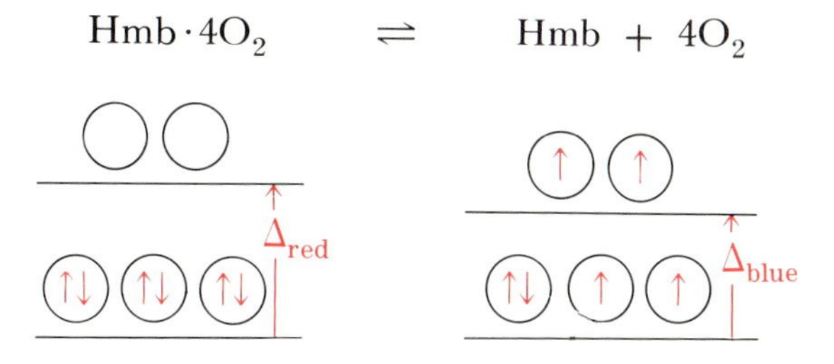

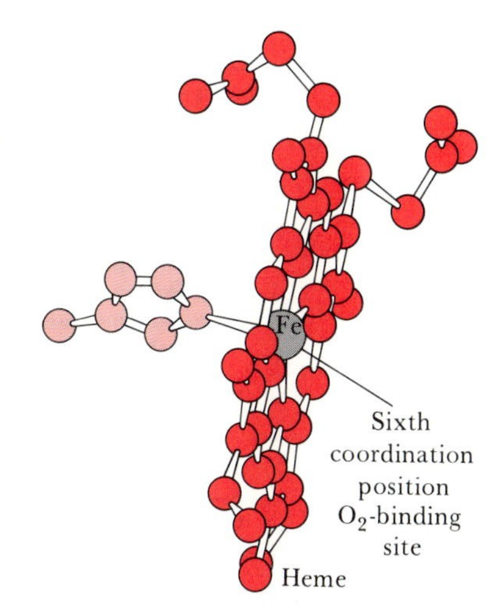

Figure 28-22 Model of the oxygen-binding site in a heme portion of hemoglobin. The iron atom is already attached to five atoms. The O_2 molecule attaches to the vacant sixth coordination position of the iron atom.

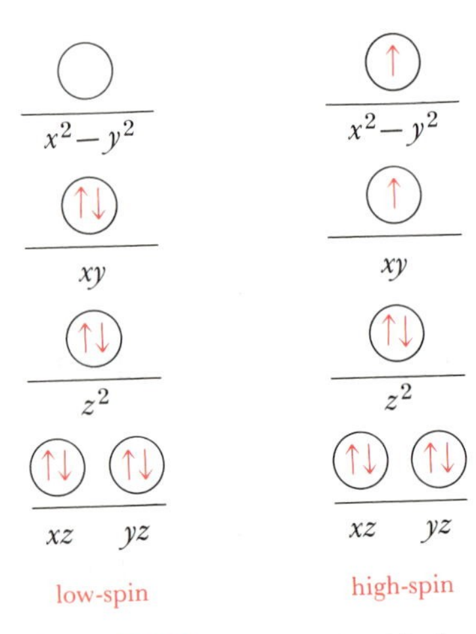

Figure 28-25 Possible d-orbital splitting patterns in Example 28-10.

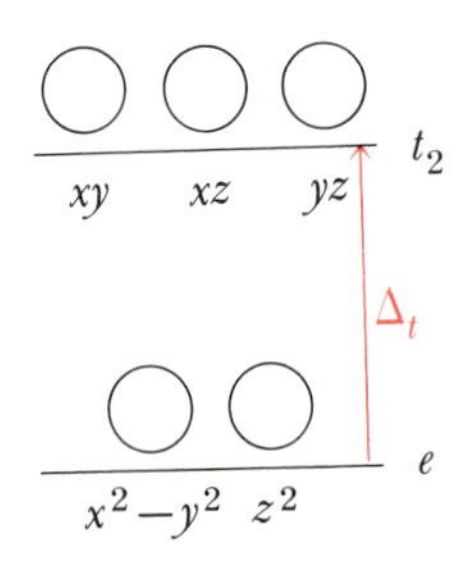

Figure 28-26 The d-orbital splitting pattern for a tetrahedral complex. Note that the pattern is the reverse of that for an octahedral complex. The difference in splitting patterns for tetrahedral and octahedral complexes is a direct consequence of the different placement of the ligands relative to the five d orbitals.

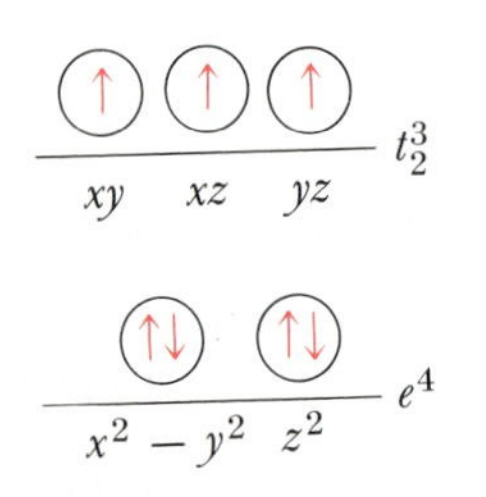

Figure 28-27 The d electron configuration of $[CoCl_4]^{2-}$ (see Example 28-11).

The actual configuration is the one with no unpaired electrons (low spin). The $d_{x^2-y^2}$ orbital is so high in energy relative to the d_{xy} orbital that d^8 square-planar complexes are all low spin.

PRACTICE PROBLEM 28-10: The $Ni^{2+}(aq)$ ion forms a square-planar complex with $CN^-(aq)$. Give the formula of the complex and its ground-state d electron configuration.

Answer: $[Ni(CN)_4]^{2-}$; $d_{xz}^2 d_{yz}^2 d_{z^2}^2 d_{xy}^2 d_{x^2-y^2}^0$

The d-orbital splitting pattern in a tetrahedral complex is shown in Figure 28-26. The two d orbitals that make up the lower set are called the ***e* orbitals,** and the three d orbitals that make up the upper set are called the **t_2 orbitals.** The d-orbital splitting energy for a tetrahedral complex is denoted by $\boldsymbol{\Delta_t}$, where t stands for tetrahedral. The value of Δ_t is only about half the value of Δ_o (the d-orbital splitting energy for an octahedral complex) because none of the d-orbital lobes in tetrahedral complexes point directly at the ligands, and because only four instead of six ligands are present. A consequence of the relatively small value of Δ_t is that no low-spin tetrahedral complexes exist.

EXAMPLE 28-11: Give the d-electron configuration in the tetrahedral complex $[CoCl_4]^{2-}$.

Solution: The oxidation state of cobalt in the complex is Co(II) because there are four Cl^- ligands and the net charge on the complex is −2. Cobalt is the seventh member of the 3d transition series; thus, cobalt(II) is a d^7 ion. Referring to Figure 28-26, we have, for the d-electron configuration of a tetrahedral d^7 ion, $e^4t_2^3$ (Figure 28-27).

PRACTICE PROBLEM 28-11: Suggest an explanation for the observation that tetrahedral complexes are much less common than octahedral complexes.

Answer: The value of Δ_t is only about one half that of Δ_o for the same ligands.

The d-orbital splitting patterns for octahedral, tetrahedral, and square-planar complexes are summarized in Figure 28-28.

The magnetic behavior of transition-metal complexes can, in some cases, be used to determine the structure of a complex. For example, experiments show that the compound $K_2[Ni(CN)_4]$ is diamagnetic and that the compound $K_2[NiBr_4]$ is paramagnetic. Using this information, we can predict the structures of the $[NiBr_4]^{2-}$ and $[Ni(CN)_4]^{2-}$ ions. Both of these complexes contain nickel(II), a d^8 ion. Because there are four ligands, the two structural possibilities are tetrahedral and square planar. The d^8electron distributions for tetrahedral and square-planar complexes are given in Figure 28-29. A tetrahedral d^8 ion has two unpaired d electrons and is therefore paramagnetic. A square-planar d^8 ion has no unpaired d electrons and is therefore diamagnetic. Thus, $[NiBr_4]^{2-}$, which is paramagnetic, must have a tetrahedral structure and $[Ni(CN)_4]^{2-}$, which is diamagnetic, must have a square-planar structure.

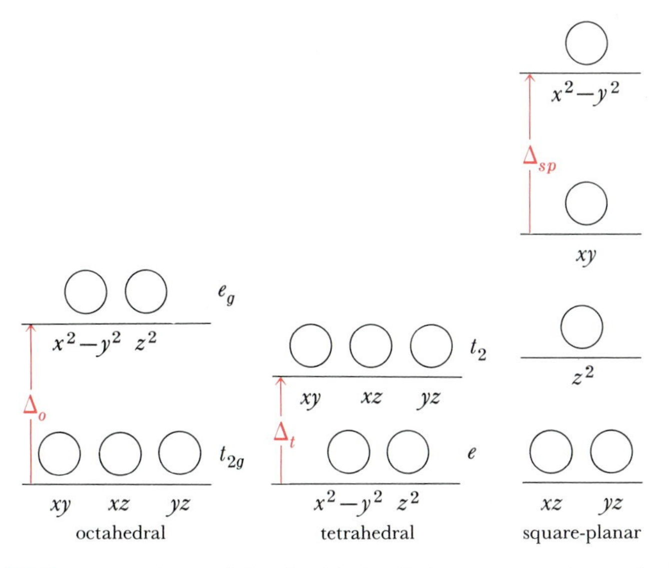

Figure 28-28 Comparison of the d-orbital splitting patterns in octahedral, tetrahedral, and square-planar complexes. The relative magnitudes of the d-orbital splitting energy for a particular metal ion with a particular ligand are $\Delta_{sp} \simeq \Delta_o \simeq 2\Delta_t$.

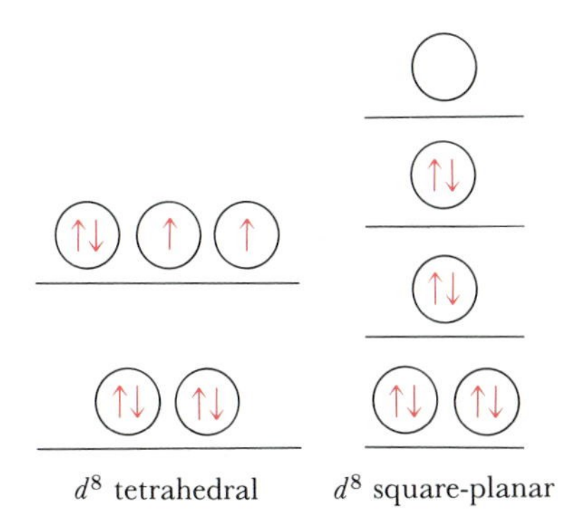

Figure 28-29 The d^8 electron configurations for tetrahedral and square-planar complexes.

28-10. TRANSITION-METAL COMPLEXES ARE CLASSIFIED AS EITHER INERT OR LABILE

Many of the chemical properties of complex ions can be understood or predicted from the d electron configurations. In this section we consider the relative rates at which complex ions undergo ligand-substitution reactions. An **inert complex** is one that only slowly exchanges its ligands with other available ligands. A **labile complex** is one that rapidly exchanges its ligands with other available ligands.

As a dramatic example of the contrasting ligand exchange rates of labile and inert complexes, consider the ligand-substitution reactions described by the equations

$$(1)\quad [Co(NH_3)_6]^{2+}(aq) + 6H_3O^+(aq) \rightleftharpoons [Co(H_2O)_6]^{2+}(aq) + 6NH_4^+(aq)$$

$$(2)\quad [Co(NH_3)_6]^{3+}(aq) + 6H_3O^+(aq) \rightleftharpoons [Co(H_2O)_6]^{3+}(aq) + 6NH_4^+(aq)$$

The equilibrium for both reactions lies far to the right. The equilibrium constants at 25°C are $K = 10^{25}$ and $K = 10^{51}$, respectively. Nonetheless, the reaction described by Equation (1), which involves the d^7 $[Co(NH_3)_6]^{2+}(aq)$ complex, attains equilibrium in 1 M $H^+(aq)$ in about 10 s, whereas the reaction described by Equation (2), which involves the d^6 $[Co(NH_3)_6]^{3+}(aq)$ complex, requires over a month to reach equilibrium at 25°C under the same conditions. In other words, a difference of only one d electron causes the reaction rates to differ by a factor of over 350,000.

The American chemist Henry Taube was the first to note that $t_{2g}^3e_g^0$, $t_{2g}^4e_g^0$, $t_{2g}^5e_g^0$, and $t_{2g}^6e_g^0$ octahedral complexes are inert and that all other octahedral complexes are labile (Figure 28-30). Thus, $[Co(NH_3)_6]^{2+}(aq)$, an octahedral high-spin d^7 ion ($t_{2g}^5e_g^2$), is labile; and $[Co(NH_3)_6]^{3+}(aq)$, an octahedral, low-spin d^6 ion ($t_{2g}^6e_g^0$), is inert.

Figure 28-30 Henry Taube, an American chemist, received the Nobel Prize in Chemistry in 1983 for his pioneering research on the rates and mechanisms of transition-metal coordination compounds.

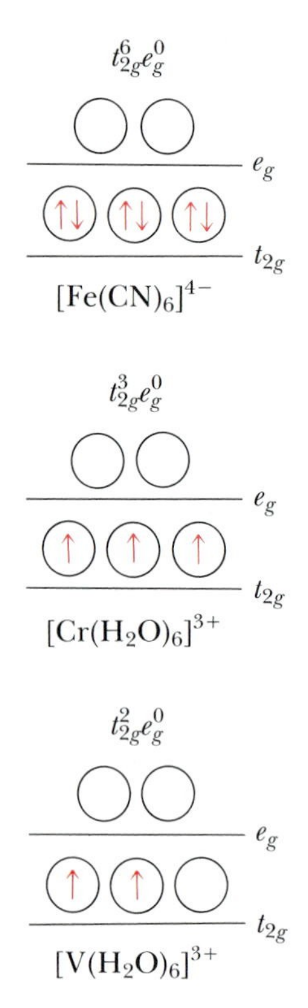

Figure 28-31 The *d*-electron configurations for the three complexes in Example 28-12.

EXAMPLE 28-12: Predict whether the following complex ions are labile or inert:

$$[Fe(CN)_6]^{4-},\ [Cr(H_2O)_6]^{3+},\ \text{and}\ [V(H_2O)_6]^{3+}$$

Solution: The *d* electron configurations of the three complexes are given in Figure 28-31. The d^6 low-spin ion $[Fe(CN)_6]^{4-}$ is $t_{2g}^6e_g^0$ and thus inert; the d^3 ion $[Cr(H_2O)_6]^{3+}$ is $t_{2g}^3e_g^0$ and thus inert; the d^2 ion $[V(H_2O)_6]^{3+}$ is $t_{2g}^2e_g^0$ and thus labile. These three predictions are confirmed by experimental results.

PRACTICE PROBLEM 28-12: In aqueous solutions with chloride ion at 1.0 M and in large excess, the half-lives for the reactant complex in the ligand-substitution reactions

(a) $\underset{\text{high spin}}{[Fe(H_2O)_6]^{3+}}(aq) + Cl^-(aq) \rightarrow [Fe(H_2O)_5Cl]^{2+}(aq) + H_2O(l)$

(b) $[Cr(H_2O)_6]^{3+}(aq) + Cl^-(aq) \rightarrow [Cr(H_2O)_5Cl]^{2+}(aq) + H_2O(l)$

are 0.085 and 3.5×10^5 s, respectively, at 25°C. Determine whether these data are consistent with Taube's rule for the relative rates of ligand-substitution reactions.

Answer: Data are consistent with Taube's rule; (a) $t_{2g}^3e_g^2$ and (b) $t_{2g}^3e_g^0$

Although **Taube's rule** can be given a theoretical foundation, we treat it here only as a useful empirical rule. If you pursue your studies in chemistry, you will learn about the explanation for Taube's rule and many other aspects of the rich and fascinating chemistry of the transition metals.

SUMMARY

There are three *d* transition-metal series (3*d*, 4*d*, and 5*d*), each having 10 members. The keys to understanding the chemistry of the *d* transition-metal series are the electron occupancy of the five *d* orbitals of the metal ion and the influence of the ligands on the relative energies and splitting patterns of the orbitals. Ligands are anions or neutral molecules that bind to metal ions or neutral atoms to form distinct chemical species called complexes, complex ions, or neutral complexes. Chelating or polydentate ligands are ligands that attach to two or more coordination positions on the metal ion. A transition-metal ion with *x d* electrons is called a d^x ion.

The structure of a complex with identical ligands may be octahedral (the most common), tetrahedral, square planar, or linear. Certain octahedral and square-planar complexes exist as cis and trans isomers. The *d*-orbital splitting pattern is different for each geometry. The determining factor of the splitting of the *d* orbitals is the placement of the ligands in the complex ion relative to the positions of the *d* orbitals. The splitting of the *d* orbitals gives rise to the possibility of low-spin and high-spin complexes for octahedral d^4, d^5, d^6, and d^7 ions. The *d*-orbital splittings are denoted as Δ_o (octahedral), Δ_{sp} (square planar), and Δ_t (tetrahedral).

A paramagnetic complex ion has unpaired electrons. Magnetic measurements can be used to detect the presence of unpaired electrons in a complex ion.

Transition-metal complexes are classified as either labile or inert, depending on whether they undergo ligand-substitution reactions rapidly or slowly. Inert octahedral complexes are those with the *d* electron configurations: $t_{2g}^3e_g^0$; $t_{2g}^4e_g^0$; $t_{2g}^5e_g^0$; and $t_{2g}^6e_g^0$.

TERMS YOU SHOULD KNOW

d_{xy}, d_{xz}, d_{yz}, $d_{x^2-y^2}$, d_{z^2} orbitals 995
d^x ion 996
complex ion 997
ligand 997
coordination number 997
coordination complex 997
neutral complex 997
ligand-substitution reaction 997

coordination position 1002
ligating atoms 1002
polydentate ligand 1002
chelating ligand 1002
chelate 1002
bidentate 1002
tridentate 1002
cis and trans isomers 1004
geometric isomers 1004
cisplatin 1004
d-orbital splitting 1006
t_{2g} and e_g orbitals 1006
Δ_o, splitting energy in an octahedral complex 1008
diamagnetic 1009
paramagnetic 1009
low-spin configuration 1010
high-spin configuration 1010
electron-pairing energy 1010
spectrochemical series 1012
Δ_{sp}, splitting energy in square-planar complex 1013
e and t_2 orbitals 1014
Δ_t, splitting energy in tetrahedral complex 1014
inert complex 1015
labile complex 1015
Taube's rule 1016

PROBLEMS

ELECTRON CONFIGURATIONS AND OXIDATION STATES

28-1. Write the ground-state electron configuration for

(a) Mn(II) (b) V(III)
(c) Ru(II) (d) Pt(IV)

28-2. Write the ground-state electron configuration for

(a) Co(III) (b) Ti(IV)
(c) Au(III) (d) Cu(I)

28-3. How many outer-shell *d* electrons are there in each of the following transition-metal ions?

(a) Ag(I) (b) Pd(IV)
(c) Ir(III) (d) Co(II)

28-4. How many outer-shell *d* electrons are there in each of the following transition-metal ions?

(a) Re(III) (b) Sc(III)
(c) Ru(IV) (d) Hg(II)

28-5. Give three examples of

(a) M(III) d^6 ions (b) M(IV) d^4 ions
(c) M(I) d^{10} ions

28-6. Give three examples of

(a) M(II) d^3 ions (b) M(I) d^8 ions
(c) M(IV) d^0 ions

28-7. Give the oxidation state of the metal in

(a) $[Os(NH_3)_4Cl_2]^+$ (b) $[CoCl_6]^{3-}$
(c) $[Fe(CN)_6]^{4-}$ (d) $[Nb(NO_2)_6]^{3-}$

28-8. Give the oxidation state of the metal in

(a) $[Ir(H_2O)_6]^{3+}$ (b) $[Co(NH_3)_3(CO)_3]^{3+}$
(c) $[CuCl_4]^{2-}$ (d) $[Ni(CN)_4]^{2-}$

28-9. Give the oxidation state of the metal in

(a) $[Cd(CN)_4]^{2-}$ (b) $[Pt(NH_3)_6]^{2+}$
(c) $[Pt(NH_3)_4Cl_2]$ (d) $[RhBr_6]^{3-}$

28-10. Give the oxidation state of the metal in

(a) $[Mo(CO)_4Cl_2]^+$ (b) $[Ta(NO_2)_3Cl_3]^{3-}$
(c) $[Co(CN)_6]^{3-}$ (d) $[Ni(CO)_4]$

IONS FROM COMPLEX SALTS

28-11. Name the major species present when the following compounds are dissolved in water and calculate the number of moles of each species present if 1.0 mol is dissolved:

(a) $K_3[Fe(CN)_6]$ (b) $[Ir(NH_3)_6](NO_3)_3$
(c) $[Pt(NH_3)_4Cl_2]Cl_2$ (d) $[Ru(NH_3)_6]Br_3$

28-12. Name the major species present when the following compounds are dissolved in water and calculate the number of moles of each species present if 1.0 mol is dissolved:

(a) $[Cr(NH_3)_6]Br_3$ (b) $[Pt(NH_3)_3Cl_3]Cl$
(c) $[Mo(H_2O)_6]Br_3$ (d) $K_4[Cr(CN)_6]$

28-13. Some of the first complexes discovered by Werner in the 1890s had the empirical formulas given below. Also given are the number of chloride ions per formula unit precipitated by the addition of $Ag^+(aq)$. Explain these observations.

Empirical formula	Number of Cl^- per formula unit precipitated by $Ag^+(aq)$
$PtCl_4 \cdot 6NH_3$	4
$PtCl_4 \cdot 5NH_3$	3
$PtCl_4 \cdot 4NH_3$	2
$PtCl_4 \cdot 3NH_3$	1
$PtCl_4 \cdot 2NH_3$	0

28-14. Some of the first complexes discovered by Werner in the 1890s had the empirical formulas given below. Also given are the number of chloride ions per formula unit precipitated by the addition of $Ag^+(aq)$. Explain these observations.

Empirical formula	Number of Cl^- per formula unit precipitated by $Ag^+(aq)$
$PtCl_2 \cdot 4NH_3$	2
$PtCl_2 \cdot 3NH_3$	1
$PtCl_2 \cdot 2NH_3$	0

CHEMICAL FORMULAS AND NAMES

28-15. Give the systematic name for

(a) $K_3[Cr(CN)_6]$ (b) $[Cr(H_2O)_5Cl](ClO_4)_2$
(c) $[Co(CO)_4Cl_2]ClO_4$ (d) $[Pt(NH_3)_4Br_2]Cl_2$

28-16. Give the systematic name for

(a) $K_3[Fe(CN)_6]$ (b) $[Ni(CO)_4]$
(c) $[Ru(H_2O)_6]Cl_3$ (d) $Na[Al(OH)_4]$

28-17. Give the systematic name for

(a) $[NH_4)_3[Co(NO_2)_6]$ (b) $[Ir(NH_3)_4Br_2]Br$
(c) $K_2[CuCl_4]$ (d) $[Ru(CO)_5]$

28-18. Give the systematic name for

(a) $Na[Au(CN)_4]$ (b) $[Cr(H_2O)_6]Cl_3$
(c) $Na_3[Co(CN)_6]$ (d) $[Cu(NH_3)_6]Cl_2$

28-19. Give the chemical formula for

(a) sodium pentacyanocarbonylferrate(II)
(b) ammonium *trans*-dichlorodiiodoaurate(III)
(c) potassium hexacyanocobaltate(III)
(d) calcium hexanitritocobaltate(III)

28-20. Give the chemical formula for

(a) sodium bromochlorodicyanonickelate(II)
(b) rubidium tetranitritocobaltate(II)
(c) potassium hexachlorovanadate(III)
(d) pentaamminechlorochromium(III) acetate

28-21. Give the chemical formula for

(a) triamminechloroplatinum(II) nitrate
(b) sodium tetrafluorocuprate(II)
(c) lithium hexanitrocobaltate(II)
(d) barium hexacyanoferrate(II)

28-22. Give the chemical formula for

(a) chlorohydroxobis(ethylenediamine)cobalt(III) nitrate
(b) bis(ethylenediamine)oxalatocadmium(II)
(c) lithium dinitrobis(oxalato)platinate(IV)
(d) bis(ethylenediamine)oxalatovanadium(III) acetate

GEOMETRIC ISOMERS

28-23. Draw all the geometric isomers for

(a) $[Co(en)_2Br_2]$ (b) $[RuCl_2Br_2(NO_2)_2]^{3-}$

28-24. Draw all the geometric isomers for

(a) $[Pd(C_2O_4)_2I_2]^{2-}$ (b) $[PtCl_3Br_3]^{2-}$

28-25. Draw the structure for

(a) *trans*-dichlorodibromoplatinum(IV) (square planar)
(b) potassium *trans*-dichlorodiodoaurate(III) (square planar)
(c) *cis,cis*-triamminetrichlorocobalt(III)
(d) *cis,trans*-triamminetrichloroplatinum(IV) chloride

28-26. Indicate whether each of the following complexes has geometric isomers:

(a) $[Cr(NH_3)_4Cl_2]^+$ (b) $[Cr(NH_3)_5Cl]^{2+}$
(c) $[Co(NH_3)_2Cl_2]^{2-}$ (tetrahedral)
(d) $[Pt(NH_3)_2Cl_2]$ (square planar)

HIGH-SPIN AND LOW-SPIN COMPLEXES

28-27. Write the *d*-orbital electron configurations for the following octahedral complex ions:

(a) an Nb(III) complex
(b) an Mo(II) complex if Δ_o is greater than the electron-pairing energy
(c) an Mn(II) complex if Δ_o is less than the electron-pairing energy
(d) an Au(I) complex
(e) an Ir(III) complex if Δ_o is greater than the electron-pairing energy

28-28. Write the *d*-orbital electron configurations for the following octahedral complex ions:

(a) a high-spin Ni(II) complex
(b) a high-spin Mn(II) complex
(c) a low-spin Fe(III) complex
(d) a Ti(IV) complex
(e) a Ni(II) complex

28-29. Classify the following complex ions as high spin or low spin:

(a) $[Fe(CN)_6]^{4-}$ (no unpaired electrons)
(b) $[Fe(CN)_6]^{3-}$ (one unpaired electron)
(c) $[Co(NH_3)_6]^{2+}$ (three unpaired electrons)
(d) $[CoF_6]^{3-}$ (four unpaired electrons)
(e) $[Mn(H_2O)_6]^{2+}$ (five unpaired electrons)

28-30. Classify the following complex ions as high spin or low spin:

(a) $[Mn(NH_3)_6]^{3+}$ (two unpaired electrons)
(b) $[Rh(CN)_6]^{3-}$ (no unpaired electrons)
(c) $[Co(C_2O_4)_3]^{4-}$ (three unpaired electrons)
(d) $[IrBr_6]^{4-}$ (three unpaired electrons)
(e) $[Ru(NH_3)_6]^{3+}$ (one unpaired electron)

PARAMAGNETISM IN COMPLEX IONS

28-31. Predict the number of unpaired electrons in

(a) $[VCl_6]^{3-}$ (b) $[CoCl_4]^{2-}$ (tetrahedral)
(c) $[Cr(CO)_6]$ (d) $[Cr(CN)_6]^{4-}$

28-32. Predict the number of unpaired electrons in

(a) $[Pd(NO_2)_4]^{2-}$ (square planar) (b) $[Rh(NH_3)_6]^{3+}$
(c) $[Ir(H_2O)_6]^{3+}$ (d) $[FeF_6]^{3-}$

28-33. Indicate whether each of the following complexes is paramagnetic:

(a) $[Co(en)_3]^{3+}$ (low spin)
(b) $[Fe(CN)_6]^{4-}$
(c) $[NiF_4]^{2-}$ (tetrahedral)
(d) $[CoBr_4]^{2-}$ (tetrahedral)

28-34. Indicate whether each of the following complexes is paramagnetic:

(a) $[Cu(NH_3)_6]^{2+}$ (b) $[CrF_6]^{3-}$
(c) $[CoCl_4]^{2-}$ (tetrahedral) (d) $[Zn(H_2O)_6]^{2+}$

28-35. The complex $[NiF_4]^{2-}$ is paramagnetic, but $[Ni(CN)_4]^{2-}$ is diamagnetic. Explain the difference.

28-36. The complex $[Fe(H_2O)_6]^{2+}$ is paramagnetic, whereas $[Fe(CN)_6]^{4-}$ is diamagnetic. Explain the difference.

INERT AND LABILE COMPLEXES

28-37. Of the following complexes, which would be expected to be inert to ligand substitution?

(a) $[Ti(H_2O)_6]^{3+}$ (b) $[VF_6]^{3-}$
(c) $[Cr(NO_2)_6]^{3-}$ (d) $[CuCl_6]^{4-}$

28-38. Of the following complexes, which would be expected to be inert to ligand substitution?

(a) $[Mo(NO_2)_6]^{2-}$ (b) $[Fe(CN)_6]^{4-}$
(c) $[Fe(CN)_6]^{3-}$ (d) $[Rh(NH_3)_6]^{3+}$

ADDITIONAL PROBLEMS

28-39. Give the chemical formula for

(a) hexanitrocobaltate(III)
(b) *trans*-dichlorobis(ethylenediamine)platinum(IV)
(c) pentacyanocarbonylferrate(II)
(d) *trans*-dichlorodiiodoaurate(III)

28-40. Write the formulas for

(a) hexaaquanickel(II) perchlorate
(b) triamminetrichloroplatinum(IV) bromide
(c) potassium chloropentacyanoferrate(III)
(d) strontium hexacyanoferrate(II)

28-41. Silver nitrate was added to solutions of the following octahedral complexes and AgCl was precipitated immediately in the mole ratios indicated:

Formula of the complex	(mol AgCl/mol complex)
$CoCl_3(NH_3)_6$	3
$CoCl_3(NH_3)_5$	2
$CoCl_3(NH_3)_4$ (purple)	1
$CoCl_3(NH_3)_4$ (green)	1

(a) Draw the structures expected for each of these complexes.
(b) Explain the fact that $CoCl_3(NH_3)_4$ can be purple or green but that both forms give 1 mol of AgCl per mol complex.

28-42. Arrange the following complexes in order of increasing values of Δ_o:

$$[Cr(H_2O)_6]^{3+} \quad [Co(NH_3)_6]^{3+} \quad [CrF_6]^{3-} \quad [Cr(CN)_6]^{3-} \quad [Ru(CN)_6]^{3-}$$

28-43. Comparing $[Co(CN)_6]^{3-}$ with $[CoCl_6]^{4-}$, indicate whether each of the following statements is true or false:

(a) $[Co(CN)_6]^{3-}$ has more d electrons than $[CoCl_6]^{4-}$.
(b) $[Co(CN)_6]^{3-}$ has the same number of d electrons as $[CoCl_6]^{4-}$.
(c) $[Co(CN)_6]^{3-}$ is paramagnetic, whereas $[CoCl_6]^{4-}$ is diamagnetic.
(d) $[Co(CN)_6]^{3-}$ is diamagnetic, whereas $[CoCl_6]^{4-}$ is paramagnetic.

28-44. Which of the following ions is inert to ligand substitution?

(a) $[Cr(C_2O_4)_3]^{3-}$ (b) $[Ti(H_2O)_6]^{3+}$
(c) $[Zn(H_2O)_6]^{2+}$ (d) $[Co(C_2O_4)_3]^{4-}$

28-45. How many unpaired electrons would you predict for the following complex ions?

(a) $[NiCl_4]^{2-}$ (tetrahedral)
(b) $[CoCl_4]^{2-}$ (tetrahedral)
(c) $[Co(CO)_6]^{3+}$
(d) $[Fe(CN)_6]^{3-}$

28-46. Why do you think most complexes of Zn(II) are colorless?

28-47. Excess $Pb_2[Fe(CN)_6](s)$ was equilibrated at 25°C with an aqueous solution of NaI(aq). The equilibrium concentrations of $I^-(aq)$ and $[Fe(CN)_6]^{4-}(aq)$ were found by chemical analysis to be 0.57 M and 0.11 M, respectively. Estimate the value of K_{sp} of $Pb_2[Fe(CN)_6](s)$. See Table 20-1 for the value of K_{sp} of $PbI_2(s)$.

28-48. Explain, using balanced chemical equations, the following observations:

(a) Dilute solutions of $CuSO_4(aq)$ are blue but become green on addition of 6 M HCl.
(b) Several commercially available rust removers contain sodium oxalate.
(c) Solid mercury(II) oxide is soluble in excess 2 M KI(aq) solution.
(d) Dilute aqueous solutions of iron(III) nitrate are yellow in color. The yellow color is removed by addition of excess 2 M HNO_3 but not by excess 2 M HCl.

28-49. Which of the following complexes is labile?

(a) $[Cr(CN)_6]^{4-}$ (b) $[Pt(NH_3)_6]^{4+}$
(c) $[Cu(NH_3)_6]^{2+}$ (d) $[W(NH_3)_6]^{2+}$

Natural gas consists primarily of methane, with small quantities of ethane and propane (Figure 29-2). The Lewis formulas of these molecules are

$$\begin{array}{ccc}&\mathrm{H}&\\&|&\\\mathrm{H}-&\mathrm{C}&-\mathrm{H}\\&|&\\&\mathrm{H}&\end{array}\qquad\begin{array}{cccc}&\mathrm{H}&\mathrm{H}&\\&|&|&\\\mathrm{H}-&\mathrm{C}-&\mathrm{C}&-\mathrm{H}\\&|&|&\\&\mathrm{H}&\mathrm{H}&\end{array}\qquad\begin{array}{ccccc}&\mathrm{H}&\mathrm{H}&\mathrm{H}&\\&|&|&|&\\\mathrm{H}-&\mathrm{C}-&\mathrm{C}-&\mathrm{C}&-\mathrm{H}\\&|&|&|&\\&\mathrm{H}&\mathrm{H}&\mathrm{H}&\end{array}$$

CH_4, methane $\qquad$ C_2H_6, ethane $\qquad$ C_3H_8, propane

Figure 29-2 Natural gas plant in West Virginia.

The formulas of ethane and propane are often written as CH_3CH_3 and $CH_3CH_2CH_3$, respectively. Such formulas, which can be thought of as abbreviations of the above Lewis formulas, are called **condensed structural formulas.**

In alkanes, the arrangement of the four bonds about each atom is tetrahedral. All the H—C—H, H—C—C, and C—C—C bond angles are 109.5°. In addition, the H—C bond lengths are all equal (110 pm) and the C—C bond lengths are all equal (154 pm). Each of the four bonds from a carbon atom is connected to a different atom. It is not possible to bond any additional atoms directly to carbon atoms in alkanes. The bonding about each carbon atom is said to be **saturated,** so alkanes are called **saturated hydrocarbons.**

The fourth member of the alkane series—butane, C_4H_{10}—is interesting because there are two different types of butane molecules (Figure 29-3). The Lewis formulas for the two forms of butane are

$$\begin{array}{cccccc}&\mathrm{H}&\mathrm{H}&\mathrm{H}&\mathrm{H}&\\&|&|&|&|&\\\mathrm{H}-&\mathrm{C}-&\mathrm{C}-&\mathrm{C}-&\mathrm{C}&-\mathrm{H}\\&|&|&|&|&\\&\mathrm{H}&\mathrm{H}&\mathrm{H}&\mathrm{H}&\end{array}\quad\text{and}\quad\begin{array}{ccccc}&\mathrm{H}&\mathrm{H}&\mathrm{H}&\\&|&|&|&\\\mathrm{H}-&\mathrm{C}-&\mathrm{C}-&\mathrm{C}&-\mathrm{H}\\&|&|&|&\\&\mathrm{H}&|&\mathrm{H}&\\&\mathrm{H}-&\mathrm{C}&-\mathrm{H}&\\&&|&&\\&&\mathrm{H}&&\end{array}$$

n-butane $\qquad$ isobutane

and their condensed structural formulas are

$$CH_3CH_2CH_2CH_3 \quad \text{and} \quad \begin{array}{c}CH_3\underset{|}{C}HCH_3\\ \ \ CH_3\end{array}$$

n-butane $\qquad$ isobutane

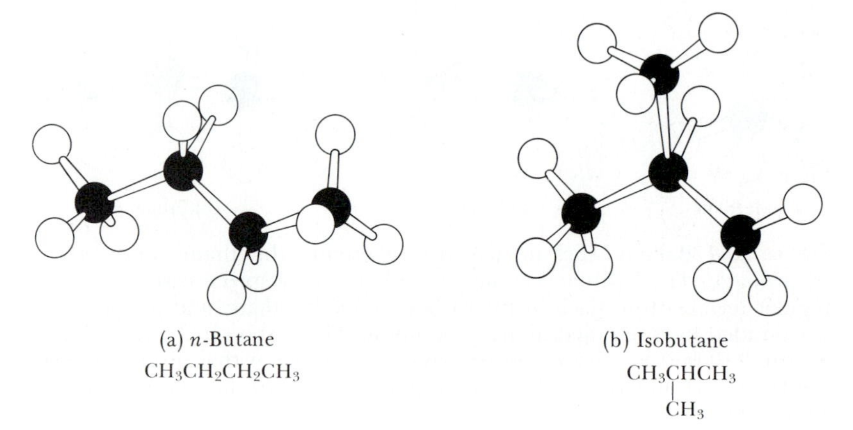

Figure 29-3 Ball-and-stick models of (a) *n*-butane and (b) isobutane. As in ethane and propane, the bonds from each carbon atom are tetrahedrally oriented.

The straight-chain molecule, that is, the one not containing any branches, is called normal butane, written *n*-butane; and the branched molecule is called isobutane. It is important to realize that *n*-butane and isobutane are different compounds, with different physical and chemical properties. For example, the boiling point of *n*-butane is −0.5°C and that of isobutane is −10.2°C. Compounds that have the same molecular formula but different structures are called **structural isomers.**

The bonding in alkanes can be conveniently described in terms of sp^3 hybrid orbitals on the carbon atoms (Section 13-7). Because the carbon-carbon bond in ethane is a σ bond (Figure 13-19), one $—CH_3$ group in ethane can rotate relative to the other. All carbon-carbon single bonds are formed in the same way as those in ethane, so all carbon-carbon single bonds are σ bonds. Consequently, rotation can occur about carbon-carbon single bonds. Because there is rotation about carbon-carbon single bonds, the formula

$$\begin{array}{l} CH_3CH_2CH_2 \\ \qquad\quad\;\;| \\ \qquad\quad\;\; CH_3 \end{array}$$

does *not* represent a third isomer of butane. It simply represents four carbon atoms joined in a chain and is just a somewhat misleading way of writing the structural formula for *n*-butane.

After butane, the alkanes are assigned systematic names that indicate the number of carbon atoms in the molecule. For example, C_5H_{12} is called pentane. The prefix *pent-* indicates that there are five carbon atoms, and the ending *-ane* denotes an alkane. The names of the first 10 straight-chain alkanes are given in Table 29-1. The first four *n*-alkanes are gases at room temperature (20°C), whereas *n*-pentane through *n*-decane are liquids at room temperature. Higher alkanes are waxy solids at room temperature. The boiling points of the *n*-alkanes increase with

Table 29-1 The names, melting points, and boiling points of the first 10 straight-chain alkanes

Name	Molecular formula*	Melting point/°C	Boiling point/°C (at atmospheric pressure)	
methane	CH_4	−183	−162	gases at 20°C
ethane	C_2H_6	−172	−89	
propane	C_3H_8	−188	−42	
n-butane	C_4H_{10}	−135	−0.5	
n-pentane	C_5H_{12}	−130	36	liquids at 20°C
n-hexane	C_6H_{14}	−95	69	
n-heptane	C_7H_{16}	−91	98	
n-octane	C_8H_{18}	−57	126	
n-nonane	C_9H_{20}	−54	151	
n-decane	$C_{10}H_{22}$	−30	174	

*The molecular formulas for the alkanes fit the general formula C_nH_{2n+2}, where n represents the number of carbon atoms in the molecule.

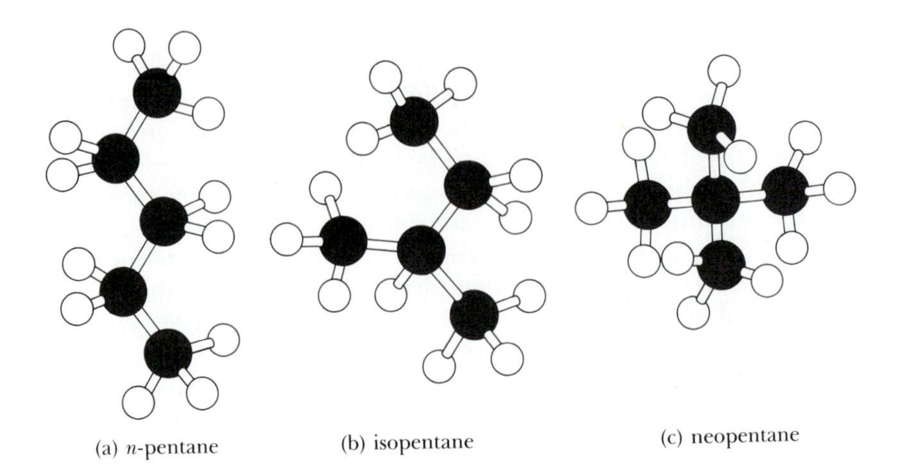

Figure 29-4 Ball-and-stick models of the three structural isomers of pentane: (a) *n*-pentane, (b) isopentane, and (c) neopentane.

molecular mass, because the total van der Waals force between molecules increases with the size of the molecules (Section 14-4).

The number of structural isomers increases very rapidly as the number of carbon atoms in an alkane increases. For example, there are three isomers of pentane (Figure 29-4):

$$\underset{n\text{-pentane}}{CH_3CH_2CH_2CH_2CH_3} \qquad \underset{\text{isopentane}}{CH_3\overset{\overset{CH_3}{|}}{C}HCH_2CH_3} \qquad \underset{\text{neopentane}}{CH_3\underset{\underset{CH_3}{|}}{\overset{\overset{CH_3}{|}}{C}}CH_3}$$

(Note that all three isomers have the molecular formula C_5H_{12}.) Hexane has 5 structural isomers, heptane has 9, octane has 18, nonane has 35, and decane has 75.

All the normal straight-chain alkanes through $C_{33}H_{68}$, as well as many branched-chain hydrocarbons, have been isolated from petroleum. (Table 29-2 lists common hydrocarbons obtained from petroleum.) A few alkanes occur elsewhere in nature. For example, the skin of an apple contains the C_{27} and C_{29} *n*-alkanes. These waxes are responsible for the waxy feel of an apple when it is polished. Long-chain saturated hydrocarbons often form part of the protective coating on leaves and fruits. Similar hydrocarbons are found in beeswax (Figure 29-5). Apparently the major function of these waxes in fruits is to retard water loss.

Figure 29-5 Bees produce high-formula-mass hydrocarbons, which are the major components of beeswax.

29-2. ALKANES ARE NOT VERY REACTIVE

Alkanes can be heated for long periods in strong acids or strong bases with no appreciable reaction. Nor do they react with oxidizing agents such as $KMnO_4$ or reducing agents such as NaH. We can write these nonreactions of the alkanes as

$$\left.\begin{array}{l} \text{alkane + strong acid (e.g., } H_2SO_4) \rightarrow \\ \text{alkane + strong base (e.g., KOH)} \rightarrow \\ \text{alkane + oxidizing agent (e.g., } KMnO_4) \rightarrow \\ \text{alkane + reducing agent (e.g., NaH)} \rightarrow \end{array}\right\} \text{ no reaction}$$

Table 29-2 Common petroleum products

Product	Alkanes present	Boiling range/°C
natural gas	C_1	−162
liquefied petroleum gas (LPG), propane, butane	C_3–C_4	−42 to 0
petroleum ether (solvent)	C_5–C_7	30 to 98
gasoline	C_5–C_{10}	36 to 175
kerosene, jet fuel	C_{10}–C_{18}	175 to 275
diesel fuel	C_{12}–C_{20}	190 to 330
fuel oil	C_{14}–C_{22}	230 to 360
lubricating oil	C_{20}–C_{30}	above 350
mineral oil (refined)	C_{20}–C_{30}	above 350
		Melting range/°C
petroleum jelly	C_{22}–C_{40}	40 to 60
paraffin	C_{25}–C_{50}	50 to 65

The alkanes do burn in oxygen, however, to produce carbon dioxide and water. These combustion reactions are exothermic and constitute the basis for the use of hydrocarbons as heating fuels. For example, the value of ΔH°_{rxn} for the combustion of propane

$$C_3H_8(g) + 5O_2(g) \rightarrow 3CO_2(g) + 4H_2O(l)$$

is −2220 kJ at 25°C.

The alkanes also can react with F_2, Cl_2, and Br_2, although a mixture of an alkane and chlorine will remain unreacted indefinitely in the dark. If such a mixture is heated or exposed to the ultraviolet radiation in sunlight, however, a reaction occurs in which one or more of the hydrogen atoms in the alkane are replaced by chlorine atoms. Such a reaction is called a **substitution reaction,** because the alkane hydrogen atoms are substituted by chlorine atoms. For example,

$$CH_4(g) + Cl_2(g) \xrightarrow{\text{uv}} CH_3Cl(g) + HCl(g)$$

(where uv denotes ultraviolet radiation). The product in this case can be considered to be a derivative of methane and therefore is named chloromethane. The function of the ultraviolet radiation is to break the bond in the Cl_2 molecule and produce free chlorine atoms. The chlorine atoms are free radicals—hence, highly reactive—and react with methane.

As the concentration of CH_3Cl builds up during the reaction, it reacts further with chlorine atoms to produce dichloromethane, CH_2Cl_2. The dichloromethane can react even further to produce trichloromethane, $CHCl_3$, and so on. Thus, these **free-radical reactions** lead to more than a single product. By varying the relative concentrations of CH_4 and Cl_2, however, we can favor one product over another.

EXAMPLE 29-1: Calculate the value of ΔG°_{rxn} for the chlorination of methane to give chloromethane.

Solution: The equation for the reaction is

$$CH_4(g) + Cl_2(g) \rightarrow CH_3Cl(g) + HCl(g)$$

Using the data from Table 22-1, we have

$$\begin{aligned}\Delta G^\circ_{rxn} &= \Delta G^\circ_f[HCl(g)] + \Delta G^\circ_f[CH_3Cl(g)] - \Delta G^\circ_f[CH_4(g)] - \Delta G^\circ_f[Cl_2(g)] \\ &= (1\ \text{mol})(-95.30\ \text{kJ}\cdot\text{mol}^{-1}) + (1\ \text{mol})(-57.40\ \text{kJ}\cdot\text{mol}^{-1}) \\ &\qquad - (1\ \text{mol})(-50.75\ \text{kJ}\cdot\text{mol}^{-1}) - (1\ \text{mol})(0) \\ &= -101.95\ \text{kJ}\end{aligned}$$

Thus, under standard conditions (1 atm), the formation of chloromethane from methane and chlorine is a spontaneous process.

PRACTICE PROBLEM 29-1: If a mixture of methane and chlorine is heated to temperatures above 300°C, then chlorination will occur. Given the large, negative value of ΔG°_{rxn} (Example 29-1), why does the thermal chlorination of methane not occur at room temperature?

Answer: The activation energy is very large (see Section 16-9).

An alkane in which hydrogen atoms are replaced by halogen atoms is called an **alkyl halide** or a **haloalkane.** Haloalkanes are named by prefixing the stem of the name of the attached halogen atoms to the name of the alkane. For example, the molecule (for simplicity, we will not usually include lone pairs in Lewis formulas when discussing organic compounds)

```
    H  H
    |  |
 H—C—C—Cl
    |  |
    H  H
```

is called chloroethane. Chloroethane is also called ethyl chloride; it is used as a topical anesthetic and acts by evaporative cooling.

If two hydrogen atoms in ethane are replaced by chlorine atoms, then there are two distinct products:

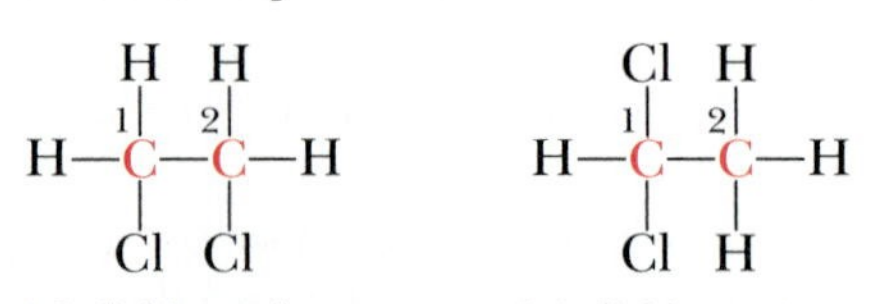

1,2-dichloroethane 1,1-dichloroethane

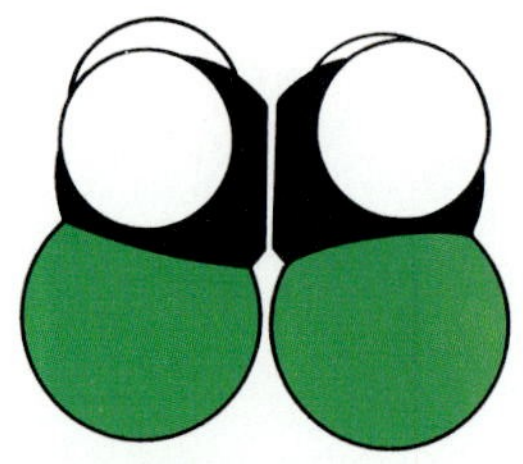

1,2-dichloroethane

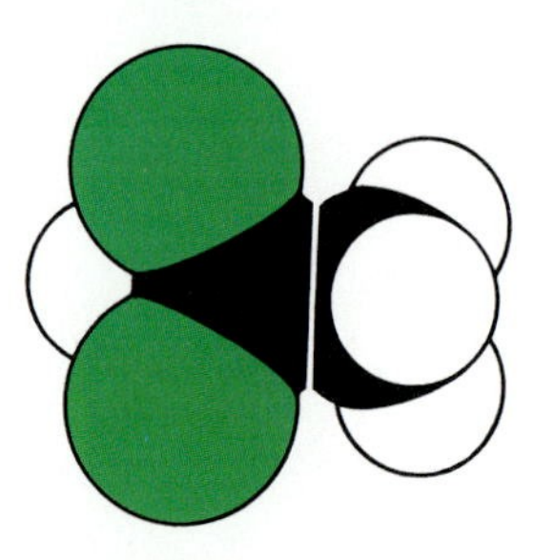

1,1-dichloroethane

These compounds provide another example of structural isomerism. We must distinguish between these two dichloroethanes. We can do this by numbering the carbon atoms along the alkane chain and designating which carbon atoms have attached chlorine atoms. Here, the dichloroethane shown on the left is called 1,2-dichloroethane and the one shown on the right is called 1,1-dichloroethane. Atoms or groups of atoms that replace a hydrogen atom bonded to a carbon atom are called **substituents.** Thus, the two chlorine atoms are the substituents in the **disubstituted** ethanes shown here.

EXAMPLE 29-2: Complete and balance the following equations. If there is no reaction, then write N.R.

(a) $C_4H_{10}(g) + O_2(g) \rightarrow$

(b) $C_2H_6(g) + Cl_2(g) \xrightarrow[\text{at room temperature}]{\text{dark}}$

(c) $C_3H_8(g) + Cl_2(g) \xrightarrow{\text{uv}}$

Solution: (a) The equation for the combustion of butane is

$$C_4H_{10}(g) + \tfrac{13}{2}O_2(g) \rightarrow 4CO_2(g) + 5H_2O(g)$$

(b) Chlorine and ethane do not react in the dark:

$$C_2H_6(g) + Cl_2(g) \xrightarrow{\text{dark}} \text{N.R.}$$

(c) Under ultraviolet radiation, we have

$$C_3H_8(g) + Cl_2(g) \xrightarrow{\text{uv}} C_3H_7Cl(g) + HCl(g)$$

Because this is a free-radical reaction, dichloropropane and polychloropropanes also are formed.

PRACTICE PROBLEM 29-2: Use the data in Table 8-2 to calculate ΔH°_{rxn} at 25°C for the combustion of one mole of hexane, $C_6H_{14}(l)$.

Answer: −4162.8 kJ

29-3. ALKANES AND SUBSTITUTED ALKANES CAN BE NAMED SYSTEMATICALLY ACCORDING TO IUPAC RULES

Structural isomerism leads to an enormous number of alkanes and substituted alkanes. Consequently, it is necessary to have a systematic method of naming alkanes and their derivatives simply and unambiguously. A system of nomenclature for organic molecules has been devised and is used by chemists throughout the world. This system, which has been recommended by the International Union of Pure and Applied Chemistry (IUPAC), makes the structure apparent from the name of the compound.

The **IUPAC nomenclature** rules for alkanes and their derivatives are as follows:

1. For straight-chain alkanes, use the names in Table 29-1 *without* the *n* prefix. Thus, the straight-chain alkane containing eight carbon atoms is called octane.
2. To name a branched or a substituted alkane, first identify the longest chain (the main chain) of consecutive carbon atoms in the molecule. Name this main chain according to rule 1. For example, the

main chain in the following molecule has five carbon atoms (shown in color):

$$CH_3-\overset{\overset{Cl}{|}}{C}H-\overset{\overset{CH_3}{|}}{C}H-CH_2-CH_3$$

Because there are five carbon atoms in the main chain, this substituted alkane is named as a substituted pentane, even though the molecule has six carbon atoms in all.

3. Number the carbon atoms in the main chain consecutively, starting at the end that gives the *lowest* numbers to the carbon atoms that have attached groups. For our substituted pentane, we have

$$\overset{1}{CH_3}-\overset{\overset{Cl}{2|}}{C}H-\overset{\overset{CH_3}{3|}}{C}H-\overset{4}{CH_2}-\overset{5}{CH_3}$$

We number the carbon atoms from left to right so that the attached groups are on the lowest-numbered carbon atoms, 2 and 3, in this case. Note that if we were to number the chain from right to left, the carbon atoms with attached groups would be 3 and 4.

4. Name the groups attached to the main chain according to Table 29-3 and indicate their position along the chain by showing the number of the carbon atom to which they are attached. The substituted alkane we are using as our example is 2-chloro-3-methylpentane. Punctuation is important in assigning IUPAC names. Numbers are separated from letters by hyphens, and the name is written as one word.

5. When two or more different groups are attached to the main chain, list them in alphabetical order. For example, as we just saw,

$$CH_3-\overset{\overset{Cl}{|}}{C}H-\overset{\overset{CH_3}{|}}{C}H-CH_2-CH_3$$

is called 2-chloro-3-methylpentane, whereas

$$CH_3-\overset{\overset{CH_3}{|}}{C}H-\overset{\overset{Cl}{|}}{C}H-CH_2-CH_3$$

is called 3-chloro-2-methylpentane.

6. When two or more identical groups are attached to the main chain, use prefixes such as *di-*, *tri-*, or *tetra-*. For example,

$$\overset{1}{CH_3}-\overset{\overset{CH_3}{2|}}{C}H-\overset{\overset{CH_3}{3|}}{C}H-\overset{4}{CH_2}-\overset{5}{CH_3}$$

is 2,3-dimethylpentane. Note that the numbers are separated by commas. Every attached group must be named and numbered, even if two identical groups are attached to the same carbon atom. For example, the IUPAC name for

$$\overset{1}{CH_3}-\underset{\underset{CH_3}{|}}{\overset{\overset{CH_3}{2|}}{C}}-\overset{\overset{CH_3}{3|}}{C}H-\overset{4}{CH_2}-\overset{5}{CH_3}$$

is 2,2,3,-trimethylpentane.

Table 29-3 Common substituent groups

Group	Name*
$-CH_3$	methyl
$-CH_2CH_3$	ethyl
$-CH_2CH_2CH_3$	propyl
$CH_3\underset{\mid}{C}HCH_3$	isopropyl
$-F$	fluoro
$-Cl$	chloro
$-Br$	bromo
$-I$	iodo
$-NH_2$	amino
$-NO_2$	nitro

*Groups that are derived from alkanes are called **alkyl groups.** The first four groups here are alkyl groups; they are named by dropping the *-ane* ending from the name of the alkane and adding *-yl*.

The assignment of IUPAC names is best learned by example.

EXAMPLE 29-3: Assign an IUPAC name to neopentane.

Solution: The structural formula for neopentane is

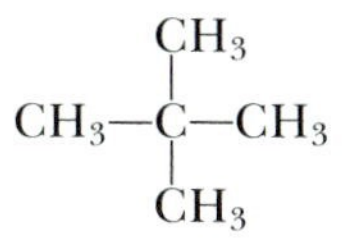

$$\begin{array}{c} CH_3 \\ | \\ CH_3-C-CH_3 \\ | \\ CH_3 \end{array}$$

Its main chain is

$$\begin{array}{c} \;^{1} \quad\; ^{2}| \quad ^{3} \\ CH_3-C-CH_3 \\ | \end{array}$$

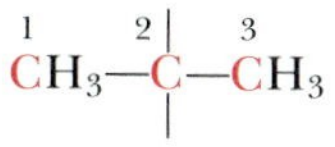

so we name it as a derivative of propane. The IUPAC name is 2,2-dimethylpropane.

PRACTICE PROBLEM 29-3: Using information from Table 29-3, draw a structural formula for 2-chloro-2-methylbutane.

Answer:

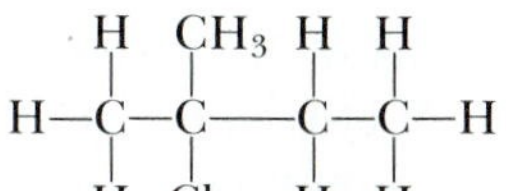

$$\begin{array}{ccccccccc} & & H & & CH_3 & & H & & H \\ & & | & & | & & | & & | \\ H & - & C & - & C & - & C & - & C & - H \\ & & | & & | & & | & & | \\ & & H & & Cl & & H & & H \end{array}$$

EXAMPLE 29-4: Write the IUPAC name for

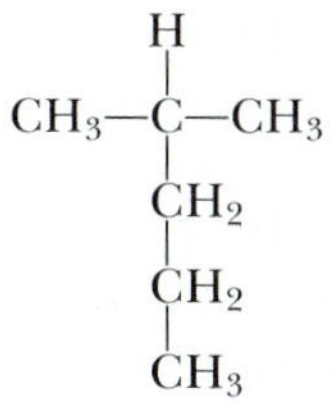

$$\begin{array}{c} H \\ | \\ CH_3-C-CH_3 \\ | \\ CH_2 \\ | \\ CH_2 \\ | \\ CH_3 \end{array}$$

Solution: If you think that the main chain consists of three (left to right) or four (top to bottom) carbon atoms, then you are making an error made by many beginners. The main chain consists of *five* carbon atoms:

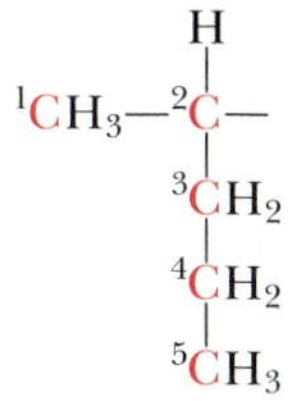

$$\begin{array}{c} H \\ | \\ {}^{1}CH_3-{}^{2}C- \\ | \\ {}^{3}CH_2 \\ | \\ {}^{4}CH_2 \\ | \\ {}^{5}CH_3 \end{array}$$

The IUPAC name is 2-methylpentane.

PRACTICE PROBLEM 29-4: Using information from Table 29-3, draw structural formulas for (a) 1-chloro-2-methylbutane and (b) 2,4-dimethyl-3-nitropentane.

Answer: **(a)** $H_2C—CH—CH_2CH_3$ (Cl on C_1, CH_3 on C_2) **(b)** $CH_3CHCHCHCH_3$ (CH_3 on C_2 and C_4, NO_2 on C_3)

29-4. HYDROCARBONS THAT CONTAIN DOUBLE BONDS ARE CALLED ALKENES

All the hydrocarbons that we have discussed so far are saturated hydrocarbons; that is, each carbon atom is bonded to four other atoms. There is another class of hydrocarbons called **unsaturated hydrocarbons,** in which not all the carbon atoms are bonded to four other atoms. These molecules necessarily contain double or triple bonds. The double bonds and triple bonds serve as functional groups. Unsaturated hydrocarbons that contain one or more double bonds are called **alkenes.** The simplest alkene, C_2H_4, is called ethene—or, more commonly, ethylene:

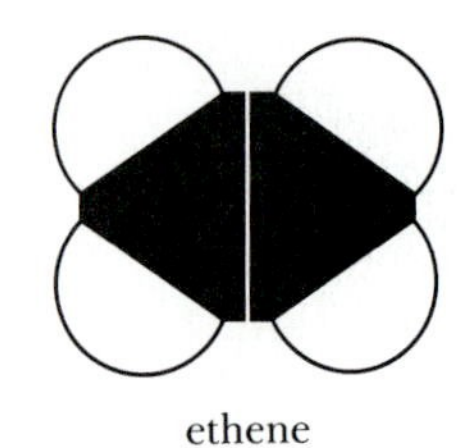

ethene

$$H_2C{=}CH_2$$

Ethylene is a colorless gas with a sweet odor and taste. It is highly flammable, and mixtures of ethylene and oxygen are highly explosive. Ethylene ranks third in U.S. annual production; 35 billion pounds of ethylene were produced in the United Sates in 1989. This is 150 pounds for every man, woman, and child. Ethylene is the starting material for about 40 percent of all organic substances produced commercially. It is used to make polyethylene, ethylene dichloride, vinyl chloride and polyvinyl chloride (PVC), vinyl acetate and polyvinyl acetate, styrene, polyesters, refrigerants, and anesthetics. Ethylene is also used in orchard spray to accelerate the ripening of fruit.

We learned in Section 13-10 that the bonding in ethene can be described by sp^2 orbitals on each carbon atom (Figure 13-25). The double bond consists of a σ bond and a π bond. The σ bond results from the combination of two sp^2 orbitals, one from each carbon atom; and the π bond results from the combination of two p orbitals, also one from each carbon atom. The π orbital maintains the σ-bond framework in a planar shape and prevents rotation about the double bond. Consequently, all six atoms in an ethene molecule lie in one plane, and there are cis and trans isomers of 1,2-dichloroethene (see Section 13-11):

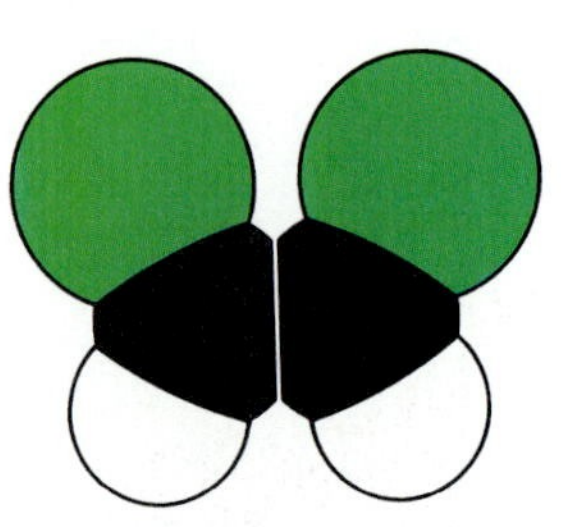

cis-1,2-dichloroethene

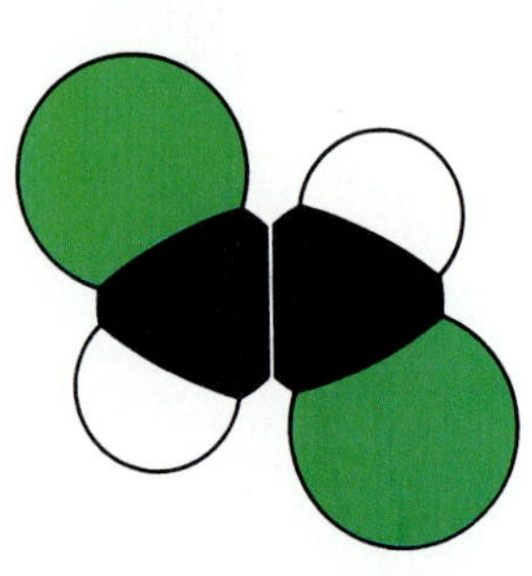

trans-1,2-dichloroethene

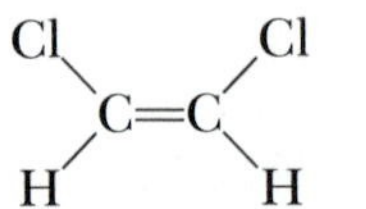

cis-1,2-dichloroethene

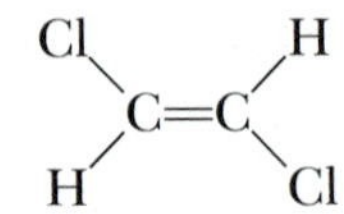

trans-1,2-dichloroethene

The IUPAC nomenclature for alkenes uses the longest chain of consecutive carbon atoms *containing the double bond* to denote the parent compound. The parent compound is named by replacing the *-ane* of the corresponding alkane with *-ene* and using a number to designate the carbon atom preceding the double bond. Thus, we have

$$\mathrm{H_2C{=}CH_2} \qquad \mathrm{H_2C{=}CH{-}CH_3}$$

ethene propene

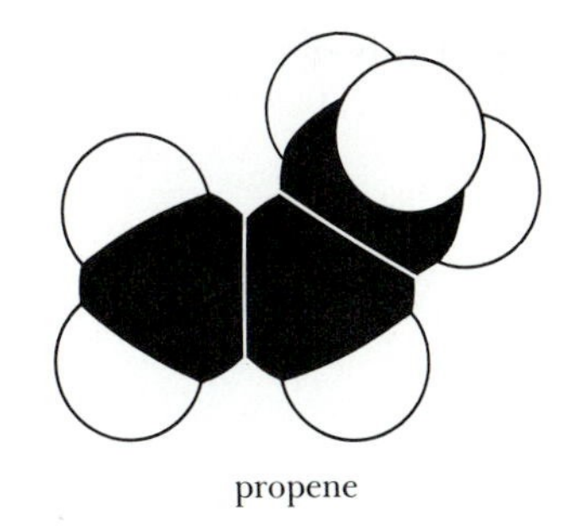

propene

There are two possible positions for the double bond in butene, so we have

$$\overset{1}{\mathrm{C}}\mathrm{H_2}{=}\overset{2}{\mathrm{C}}\mathrm{H}{-}\overset{3}{\mathrm{C}}\mathrm{H_2}\overset{4}{\mathrm{C}}\mathrm{H_3} \qquad \overset{1}{\mathrm{C}}\mathrm{H_3}{-}\overset{2}{\mathrm{C}}\mathrm{H}{=}\overset{3}{\mathrm{C}}\mathrm{H}{-}\overset{4}{\mathrm{C}}\mathrm{H_3}$$

1-butene 2-butene

The planar C=C portion of each of these molecules is colored red. These structures indicate that the name 2-butene is ambiguous because of cis-trans isomerism. The cis-trans isomers of 2-butene are

$$\mathrm{H_3C(H)C{=}C(H)CH_3} \qquad \mathrm{H(H_3C)C{=}C(H)CH_3}$$

cis-2-butene (m.p. −139°C) *trans*-2-butene (m.p. −106°C)

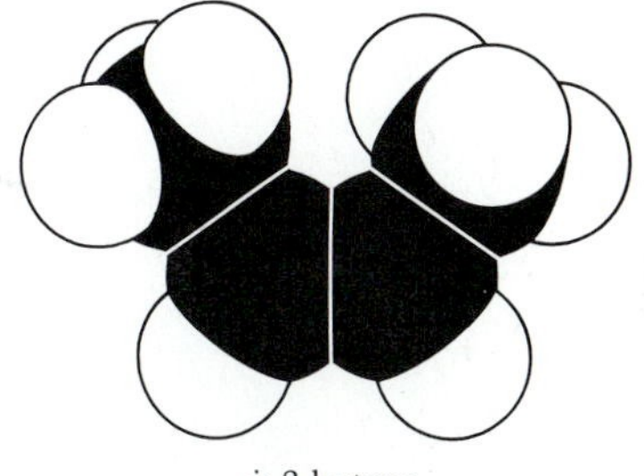

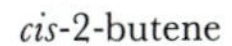

cis-2-butene

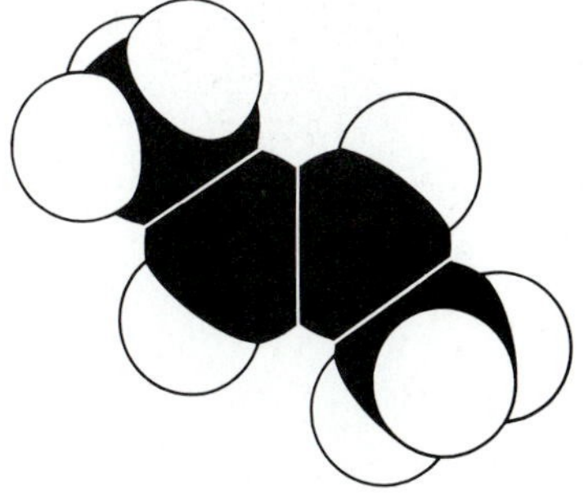

trans-2-butene

EXAMPLE 29-5: Discuss the shape of a propene molecule, whose structural formula is

$$\mathrm{H_2C{=}CH{-}CH_3}$$

Solution: The $\rangle C{=}C\langle$ portion of propene is planar, so all three carbon atoms and the three hydrogen atoms attached to the double-bonded carbon atoms lie in one plane, as the red region shows:

$$\mathrm{H_2C{=}CH{-}CH_3}$$

The methyl group can rotate about the carbon-carbon single bond.

PRACTICE PROBLEM 29-5: Describe the bonding in propene.

Answer: It is appropriate to use sp^3 orbitals to describe the bonding about the $-CH_3$ carbon atom and sp^2 orbitals for the carbon atoms in the double bond. The double bond can be described in terms of a σ bond $[C(sp^2) + C(sp^2)]$ and a π bond $[C(2p) + C(2p)]$.

EXAMPLE 29-6: Name the compound whose Lewis formula is

$$\begin{array}{c} CH_3\quad H \\ \diagdown\ \diagup \\ H\qquad\quad C \\ \diagdown\qquad\diagup\quad\diagdown \\ C{=}C\qquad CH_2CH_3 \\ \diagup\qquad\quad\diagdown \\ CH_3\qquad\quad H \end{array}$$

Solution: The longest chain containing the double bond consists of six carbon atoms, so the parent compound is a hexene. In particular, it is a 2-hexene because the double bond occurs after the second carbon atom in the chain. The configuration of the molecule is trans because the carbon atoms lie on opposite sides of the double bond. In addition, there is a methyl group attached to the fourth carbon atom, so the name of the compound is 4-methyl-*trans*-2-hexene.

PRACTICE PROBLEM 29-6: Write the Lewis formula of 2,2-dichloro-*cis*-3-hexene.

Answer:

$$\begin{array}{l} \qquad\quad\ \ Cl\qquad H \\ \qquad\quad\ \ |\qquad\ \diagup \\ CH_3{-}C{-}C \\ \qquad\quad\ \ |\qquad\ \ \diagdown\!\!\diagdown \\ \qquad\quad\ \ Cl\qquad\ \ C{-}H \\ \qquad\qquad\qquad\diagup \\ \qquad\qquad\quad CH_2 \\ \qquad\qquad\quad\ | \\ \qquad\qquad\quad CH_3 \end{array}$$

29-5. ALKENES UNDERGO ADDITION REACTIONS AS WELL AS COMBUSTION REACTIONS AND SUBSTITUTION REACTIONS

Alkenes are more reactive than alkanes because the carbon-carbon double bond provides a reactive center in the molecule. In a sense, the double bond has "extra" electrons available for reaction. So, besides the combustion and substitution reactions that alkanes undergo, alkenes undergo **addition reactions.** Examples of addition reactions are

1. Addition of hydrogen, called **hydrogenation:**

$$\begin{array}{c} H_3C\qquad H \\ \diagdown\quad\diagup \\ C{=}C \\ \diagup\quad\diagdown \\ H\qquad\ H \end{array}(g) + H_2(g) \xrightarrow[\text{high P, high T}]{\text{catalyst}} CH_3{-}\begin{array}{c} H \\ | \\ C \\ | \\ H \end{array}{-}\begin{array}{c} H \\ | \\ C \\ | \\ H \end{array}{-}H(g)$$

This reaction requires a catalyst and high pressure and temperature. Usually powdered nickel or platinum is used as the catalyst. "Hydrogenated vegetable oils" are made by hydrogenating the double bonds in the molecules that constitute vegetable oils. This hydrogenation makes vegetable oils solid at room temperature.

2. Addition of chlorine or bromine:

$$H_3C(H)C{=}CH_2(g) + Br_2(l) \rightarrow CH_3{-}CHBr{-}CH_2Br(l)$$

This reaction can be carried out either with pure chlorine or bromine or by dissolving the halogen in some solvent, such as carbon tetrachloride. The addition reaction with bromine is a useful qualitative test for the presence of double bonds. A solution of bromine in carbon tetrachloride is red, whereas alkenes and bromoalkanes are usually colorless. As the bromine adds to the double bond, the red color disappears, a result providing a simple test for the presence of double bonds.

3. Addition of hydrogen chloride:

$$H_3C(H)C{=}CH_2(g) + HCl(g) \rightarrow CH_3{-}CHCl{-}CH_3(g) \quad \text{(sole product)}$$

$$H_3C(H)C{=}CH_2(g) + HCl(g) \rightarrow CH_3{-}CH_2{-}CH_2Cl(g) \quad \text{(none produced)}$$

Although two different products might seem possible in this reaction, only one is found. There is a simple rule for determining which product is produced: **Markovnikov's rule** states that, when HX adds to an alkene, the hydrogen atom attaches to the carbon atom in the double bond already bearing the larger number of hydrogen atoms. More succinctly, the hydrogen-rich gets hydrogen-richer.

4. Addition of water. In the presence of acid, which catalyzes the reaction, water adds to the more reactive alkenes:

$$H_3C(H)C{=}CH_2(g) + HOH(l) \xrightarrow{\text{acid}} CH_3{-}CH(OH){-}CH_2(H)(l) \quad \text{(sole product)}$$

$$H_3C(H)C{=}CH_2(g) + HOH(l) \xrightarrow{\text{acid}} CH_3{-}CH(H){-}CH_2(OH)(l) \quad \text{(none produced)}$$

Note that the addition of water to an alkene obeys Markovnikov's rule. Simply picture the water as H—OH.

EXAMPLE 29-7: Use Markovnikov's rule to predict the product of the reaction

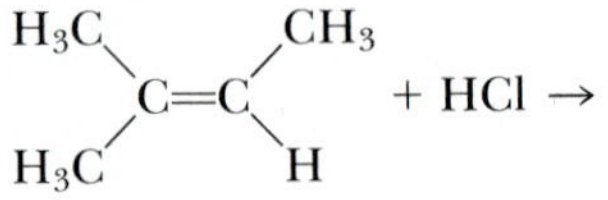

$$(H_3C)_2C{=}C(CH_3)H + HCl \rightarrow$$

Solution: According to Markovnikov's rule, the H of HCl will end up on the carbon atom of the double bond that already has the greater number of hydrogen atoms. This rule gives the sole product

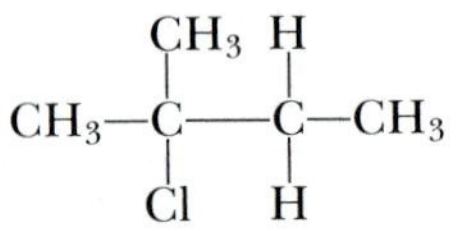

$$CH_3{-}C(CH_3)(Cl){-}CH(H){-}CH_3$$

PRACTICE PROBLEM 29-7: Write structural formulas for and assign IUPAC names to the products of the following reactions:
(a) addition of Br_2 to *cis*-2-butene
(b) addition of HBr to 1-butene

Answer: (a) 2,3-dibromobutane; (b) 2-bromobutane

29-6. HYDROCARBONS THAT CONTAIN A TRIPLE BOND ARE CALLED ALKYNES

Alkynes are hydrocarbons that contain at least one carbon-carbon triple bond. The simplest alkyne is acetylene, C_2H_2,

$$H{-}C{\equiv}C{-}H$$

Acetylene is the common name of C_2H_2. Its IUPAC name is ethyne; this name is formed by replacing the *-ane* ending of the parent alkane with the ending *-yne,* which is characteristic of alkynes (see Problem 29-27). We learned in Section 13-12 that acetylene is a linear molecule whose bonding can be described in terms of *sp* hybrid orbitals on the carbon atoms (Figure 13-28). Acetylene is a colorless gas, which is produced primarily from petroleum. It is used in oxyacetylene torches, which produce relatively high temperatures. Acetylene also is used as the starting material for a number of plastics.

Acetylene can also be produced by the reaction of calcium carbide and water:

$$CaC_2(s) + 2H_2O(l) \rightarrow Ca(OH)_2(s) + C_2H_2(g)$$

This reaction is used by spelunkers as a light source in caves (Figure 29-6). Acetylene is produced by allowing $H_2O(l)$ to drop slowly onto $CaC_2(s)$ in a canister. The $C_2H_2(g)$ pressure builds up, leaks out of the canister through a nozzle, and is burned in air:

$$2C_2H_2(g) + 5O_2(air) \rightarrow 4CO_2(g) + 2H_2O(g)$$

(a)

(b)

Figure 29-6 (a) Chip Clark, spelunker, with a calcium carbide lamp on his helmet. (b) The reaction of calcium carbide with water yields acetylene gas and calcium hydroxide. The acetylene gas burns in air and is used to provide light in lamps on hats used by spelunkers.

Besides the combustion reaction that all hydrocarbons undergo, acetylene and related hydrocarbons undergo the same type of addition reactions as alkenes do.

EXAMPLE 29-8: Predict the product when HCl reacts with C_2H_2.

Solution: The reaction can be broken down into two steps. The first step is

$$\mathrm{H{-}C{\equiv}C{-}H + HCl \rightarrow \underset{H}{\overset{Cl}{}}\!\!>C{=}C<\!\!\underset{H}{\overset{H}{}}}$$

The second step is

$$\mathrm{\underset{H}{\overset{Cl}{}}\!\!>C{=}C<\!\!\underset{H}{\overset{H}{}} + HCl \rightarrow H{-}\underset{Cl}{\overset{Cl}{C}}{-}\underset{H}{\overset{H}{C}}{-}H}$$

Note that we have used Markovnikov's rule to predict the product in the second step. The sole product is 1,1-dichloroethane.

PRACTICE PROBLEM 29-8: Predict and name the product formed when 2 mol of HBr react with propyne.

Answer: 2,2-dibromopropane

29-7. BENZENE BELONGS TO A CLASS OF HYDROCARBONS CALLED AROMATIC HYDROCARBONS

Benzene is a colorless, carcinogenic, flammable liquid with a characteristic odor. It belongs to the class of hydrocarbons called **aromatic hydrocarbons.** These hydrocarbons have relatively stable rings as a result of π-electron delocalization (Section 13-13). Benzene and many other aromatic hydrocarbons are obtained from petroleum and coal tar. It is ranked 16th in U.S. production, with almost 12 billion pounds produced annually.

We noted in Sections 13-13 and 11-6 that benzene is a planar, perfectly hexagonal molecule and that the two principal resonance forms are

The resonance hybrid is

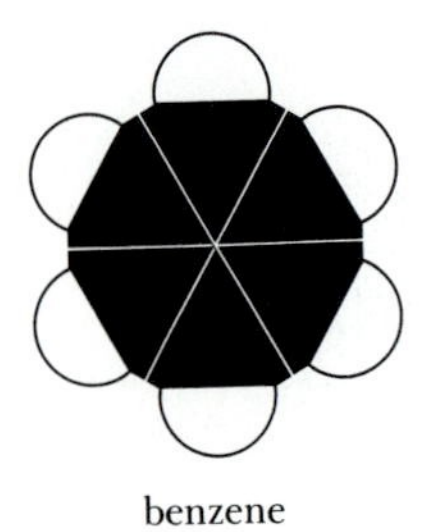

benzene

or, more compactly,

where a hydrogen atom is understood to be bonded to the carbon atom at each vertex in the benzene ring. Although each resonance form of benzene shows localized double bonds, the resonance hybrid does not. Benzene does not have localized double bonds and does not react as an unsaturated hydrocarbon. In fact, the π-electron delocalization causes the ring to be so stable that most of the reactions that benzene undergoes are substitution reactions in which the hydrogen atoms on the ring are replaced by other atoms or groups. For example, the usual reaction of benzene with bromine is a substitution reaction:

$$C_6H_6(l) + Br_2(l) \xrightarrow{FeBr_3} C_6H_5Br(l) + HBr(g)$$

The fact that only one monobromobenzene has ever been isolated indicates that all the hydrogen atoms in benzene are equivalent, as is confirmed by X-ray and spectroscopic data.

Some other types of substitution reactions that benzene undergoes are

Nitration

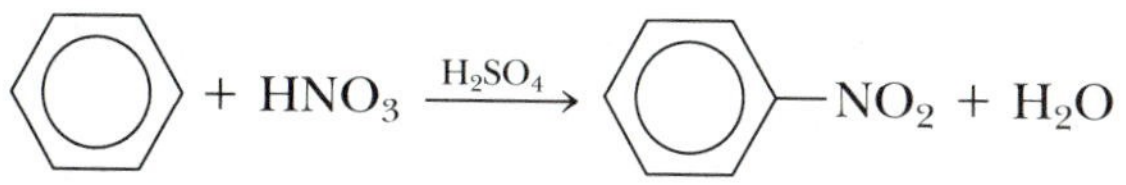

$$C_6H_6 + HNO_3 \xrightarrow{H_2SO_4} C_6H_5{-}NO_2 + H_2O$$

nitrobenzene

Sulfonation

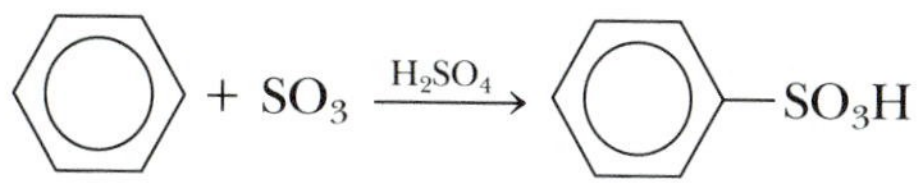

$$C_6H_6 + SO_3 \xrightarrow{H_2SO_4} C_6H_5{-}SO_3H$$

benzenesulfonic acid

Alkylation

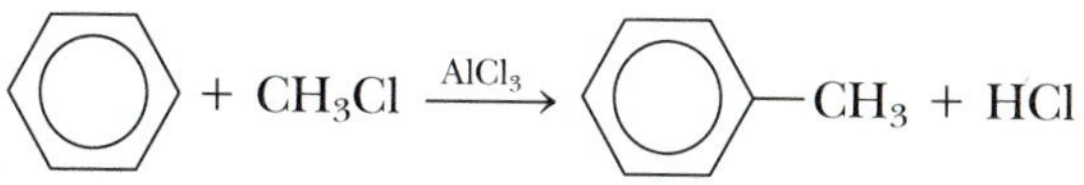

$$C_6H_6 + CH_3Cl \xrightarrow{AlCl_3} C_6H_5{-}CH_3 + HCl$$

methylbenzene (toluene)

EXAMPLE 29-9: Draw structural formulas for all the isomers of dibromobenzene.

Solution: Because the benzene ring is a regular hexagon, there are three isomers of dibromobenzene:

Br, Br (I) — Br, Br (II) — Br, Br (III)

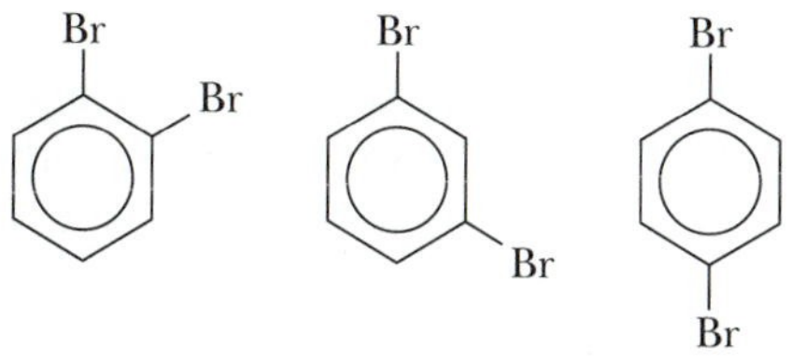

I II III

We can name these compounds by numbering the carbon atoms sequentially:

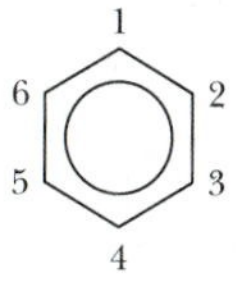

1 2 3 4 5 6

Therefore, we have

compound I	1,2-dibromobenzene
compound II	1,3-dibromobenzene
compound III	1,4-dibromobenzene

One of the substituents on the ring is always placed at the number 1 position.

PRACTICE PROBLEM 29-9: Name the following compounds:

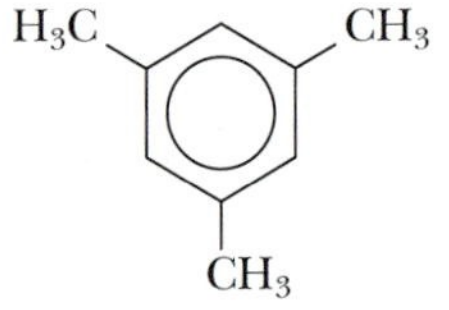

H_3C CH_3 CH_3 (a)

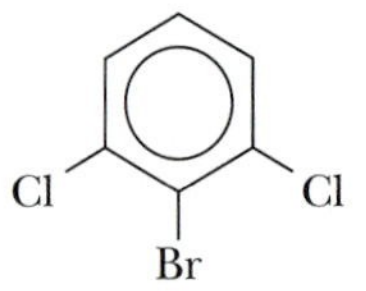

Cl Cl Br (b)

Answer: (a) 1,3,5-trimethylbenzene; (b) 1,3-dichloro-2-bromobenzene

There is a less systematic but more common way to designate the positions of the bromine atoms in the three disubstituted benzenes in Example 29-9. Substituents at the 1,2 positions are designated **ortho-** (*o*-), those at the 1,3 positions are designated **meta-** (*m*-), and those at the 1,4 positions are designated **para-** (*p*-). Thus, we write

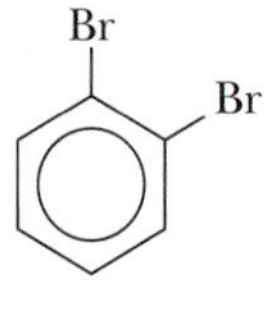

o-dibromobenzene (*ortho*-dibromobenzene)

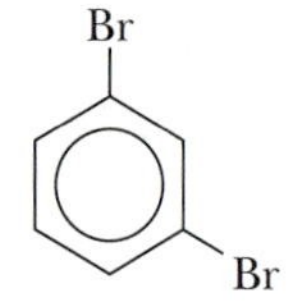

m-dibromobenzene (*meta*-dibromobenzene)

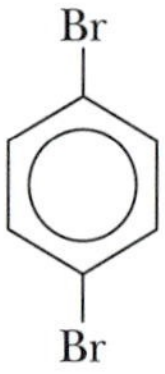

p-dibromobenzene (*para*-dibromobenzene)

Benzene is used in the manufacture of medicinal chemicals, dyes, plastics, varnishes, lacquers, linoleum, and many other products. Important derivatives of benzene are vanillin, styrene, and aspirin.

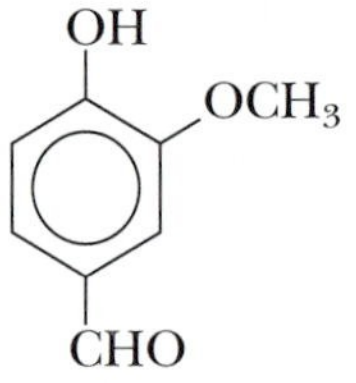

OH OCH_3 CHO vanillin

$CH{=}CH_2$ styrene

CH_3 O_2N NO_2 NO_2 2,4,6-trinitrotoluene (TNT)

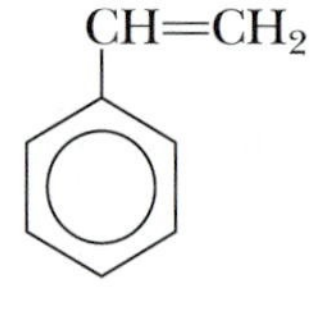

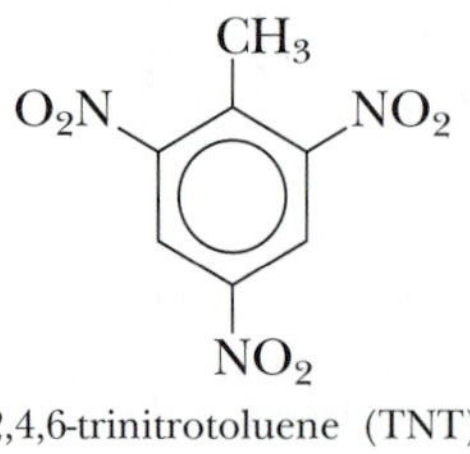

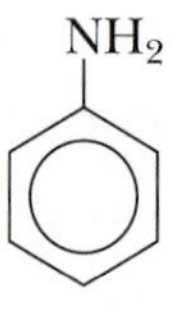

NH_2 aniline

O $O{-}C{-}CH_3$ COOH acetylsalicylic acid (aspirin)

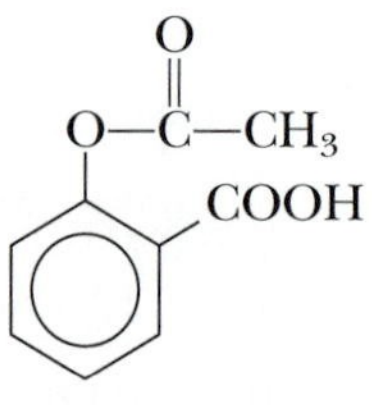

29-8. ALCOHOLS ARE ORGANIC COMPOUNDS THAT CONTAIN AN —OH GROUP

The class of organic compounds called **alcohols** is characterized by an —OH group attached to a hydrocarbon chain. Important common alcohols are (Figure 29-7)

CH_3OH	CH_3H_2OH	CH_3CHCH_3 (with OH on the middle C)
methanol (methyl alcohol)	ethanol (ethyl alcohol)	2-propanol (isopropyl alcohol)

These alcohols have both common names (given here in parentheses) and IUPAC names. The common names are formed by naming the alkyl group to which the —OH is attached and adding the word alcohol. The IUPAC name is formed by replacing the *-e* from the end of the alkane name of the longest chain of carbon atoms containing the —OH group with the suffix *-ol*. Thus, we have methanol, ethanol, and propanol. If the —OH group can be attached at more than one position, then its position is denoted by a number, as in

$\overset{3}{C}H_3\overset{2}{C}H_2\overset{1}{C}H_2OH$	$\overset{3}{C}H_3\overset{2}{C}H\overset{1}{C}H_3$ (with OH on C-2)
1-propanol	2-propanol

Methanol sometimes is called wood alcohol because it can be produced by heating wood in the absence of oxygen. Over 7 billion pounds of methanol are produced annually in the United States by the reaction

$$CO(g) + 2H_2(g) \xrightarrow[\text{high P, high T}]{\text{catalyst}} CH_3OH(l)$$

Methanol is highly toxic and can cause blindness and death if taken internally. During Prohibition in the United States, many people died

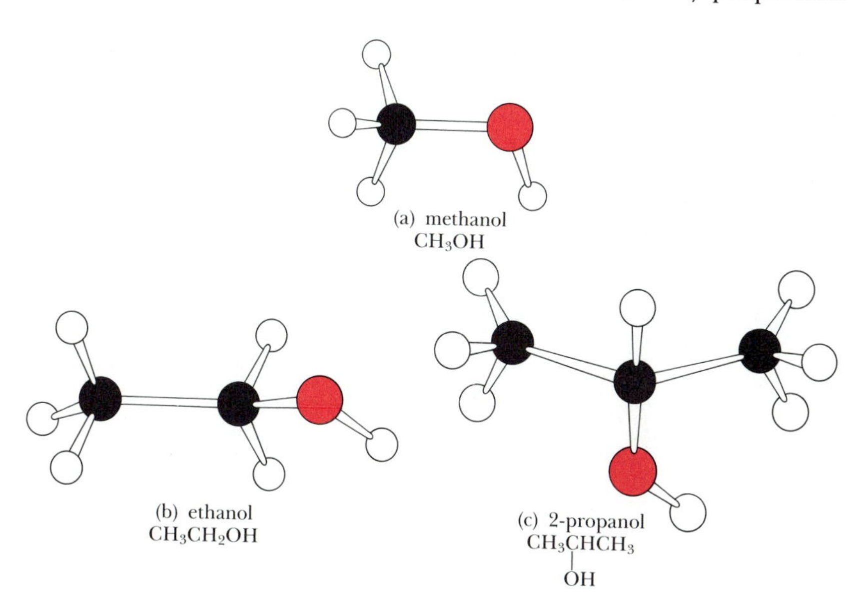

Figure 29-7 Ball-and-stick models of common alcohols: (a) methanol, (b) ethanol, and (c) 2-propanol.

or became seriously ill from drinking methanol, either because they were not aware of the difference between methanol and ethanol or because the alcohol they purchased contained methanol as a major impurity.

Ethanol is an ingredient in all fermented beverages. Various sugars and the starch in potatoes, grains, and similar substances can be used to produce ethanol by fermentation, a process in which yeast is used to convert sugar to ethanol. Although ethanol may be best known as an ingredient in alcoholic beverages, it is an important industrial chemical as well. An aqueous solution of 70 percent by volume of 2-propanol (isopropyl alcohol) is sold as rubbing alcohol.

It is sometimes convenient to view alcohols as derivatives of water, in which one of the hydrogen atoms is replaced by an alkyl group. This approach is most useful when the hydrocarbon portion of the alcohol is relatively small. The low-molecular-mass alcohols form hydrogen bonds and mix completely (i.e., are miscible) with water in all proportions. As the hydrocarbon portion becomes larger, however, the solubility of alcohols in water decreases and eventually approaches the low solubility of the comparable hydrocarbon. In addition, the low-molecular-mass alcohols have much higher boiling points than their comparable hydrocarbons. For example, whereas ethane (molecular mass 30) has a boiling point of −89°C, the boiling point of methyl alcohol (molecular mass 32) is 65°C. The relatively high boiling points of the low-molecular-mass alcohols are due to extensive hydrogen bonding in the liquids.

Alcohols undergo reactions analogous to those of water. For example, alcohols react with alkali metals to form **alkoxides** and liberate hydrogen:

$$2CH_3OH(l) + 2Na(s) \rightarrow \underset{\text{sodium methoxide}}{2NaOCH_3(alc)} + H_2(g)$$

Alkoxides are strong Lewis bases, that is, electron-pair donors (Section 18-12).

Table 29-4 Boiling points and solubilities in water of several pairs of hydrocarbons and alcohols of similar molecular masses

Name	Formula	Molecular mass	Boiling point/°C	Solubility in water/ g per 100 g H_2O
ethane	CH_3CH_3	30	−89	insoluble
methanol	CH_3OH	32	65	soluble in all proportions
propane	$CH_3CH_2CH_3$	44	−42	insoluble
ethanol	CH_3CH_2OH	46	78	soluble in all proportions
butane	$CH_3CH_2CH_2CH_3$	58	−0.5	insoluble
1-propanol	$CH_3CH_2CH_2OH$	60	97	soluble in all proportions
pentane	$CH_3CH_2CH_2CH_2CH_3$	72	36	insoluble
1-butanol	$CH_3CH_2CH_2CH_2OH$	74	117	9
hexane	$CH_3CH_2CH_2CH_2CH_2CH_3$	86	69	insoluble
1-pentanol	$CH_3CH_2CH_2CH_2CH_2OH$	88	138	2.7
heptane	$CH_3CH_2CH_2CH_2CH_2CH_2CH_3$	100	98	insoluble
1-hexanol	$CH_3CH_2CH_2CH_2CH_2CH_2OH$	102	157	0.6

Table 29-4 lists the boiling points and the solubilities in water of a number of alcohols. Note how the solubility decreases as the relative size of the alkyl portion of the molecule increases. Also note that for pairs of hydrocarbons and alcohols of comparable molecular mass, the alcohols have considerably higher boiling points.

Alcohols can be converted to alkenes by acid-catalyzed dehydration. For example,

$$CH_3CH_2OH(l) \xrightarrow[200°C]{H_2SO_4} H_2C{=}CH_2(g) + H_2O(l)$$

Many of the reactions that alcohols undergo depend on the number of carbon atoms that are bonded to the carbon atom bearing the —OH group. For this reason we define **primary, secondary,** and **tertiary alcohols.** A primary alcohol has one carbon atom bonded to the carbon atom bearing the —OH group; a secondary alcohol has two such carbon atoms; and a tertiary alcohol has three:

$CH_3CH_2CH_2OH$	$CH_3-CH(OH)-CH_3$	$CH_3-C(CH_3)(OH)-CH_3$
propanol	2-propanol	2-methyl-2-propanol (tertiary butyl alcohol)
a primary alcohol	a secondary alcohol	a tertiary alcohol

EXAMPLE 29-10: Classify the following alcohols as primary, secondary, or tertiary: (a) 2-methyl-2-butanol; (b) 3-pentanol.

Solution: (a) The structural formula of 2-methyl-2-butanol is

$$CH_3-C(CH_3)(OH)-CH_2CH_3$$

Consequently, 2-methyl-2-butanol is a tertiary alcohol.

(b) The structural formula of 3-pentanol is

$$CH_3CH_2CH(OH)CH_2CH_3$$

Consequently, 3-pentanol is a secondary alcohol.

PRACTICE PROBLEM 29-10: How many primary, secondary, and tertiary alcohols are there with the formula $C_5H_{11}OH$?

Answer: 4,3,1

29-9. ALDEHYDES AND KETONES CONTAIN A CARBON-OXYGEN DOUBLE BOND

Alcohols can be oxidized to produce aldehydes and ketones. **Aldehydes** are compounds that have the general formula RCHO

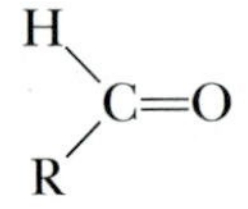

$$\begin{matrix} H \diagdown \\ \quad C{=}O \\ R \diagup \end{matrix}$$

where R is a hydrogen atom or an alkyl group. The —CHO group, called the **aldehyde group,** is characteristic of aldehydes. The two simplest aldehydes are

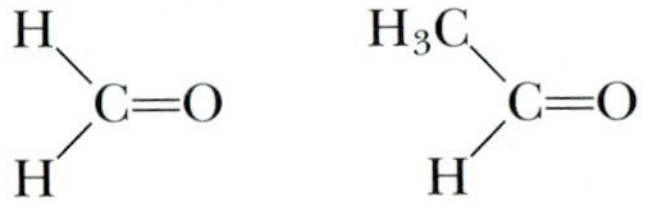

$$\begin{matrix} H \diagdown \\ \quad C{=}O \\ H \diagup \end{matrix} \qquad\qquad \begin{matrix} H_3C \diagdown \\ \quad C{=}O \\ H \diagup \end{matrix}$$

methanal (formaldehyde) $HCHO$ ethanal (acetaldehyde) CH_3CHO

Formaldehyde is a gas with an offensive and characteristic odor. It is one of the irritants in photochemical smog. Over 6 billion pounds of formaldehyde are manufactured annually in the United States. Most of this goes into the production of various plastic materials. A solution of formaldehyde in water, called formalin, preserves biological specimens. However, this use is being phased out because formaldehyde is a carcinogen.

Acetaldehyde is a colorless, flammable liquid with a pungent, fruity odor. It is used in the manufacture of perfumes, flavors, plastics, synthetic rubbers, dyes, and a variety of other organic chemicals.

Aldehydes are obtained from the oxidation of primary alcohols. A commonly used oxidizing agent for the oxidation of primary alcohols is $K_2Cr_2O_7$ dissolved in an aqueous acidic solution:

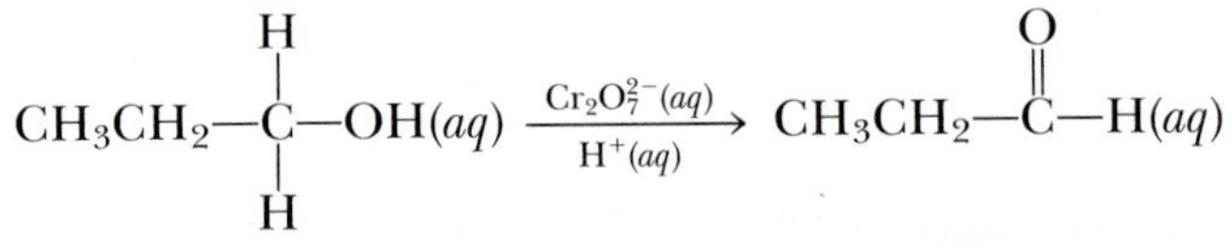

$$CH_3CH_2{-}\underset{H}{\overset{H}{\underset{|}{\overset{|}{C}}}}{-}OH(aq) \xrightarrow[H^+(aq)]{Cr_2O_7^{2-}(aq)} CH_3CH_2{-}\overset{O}{\overset{\|}{C}}{-}H(aq)$$

1-propanol   propanal

Aldehydes are readily oxidized to carboxylic acids (see Section 29-11), so it is necessary to stop the oxidation of a primary alcohol at the aldehyde stage in order to prevent the aldehyde from being oxidized further to the carboxylic acid. One way to do this is to remove the aldehyde as soon as it is formed by distilling it from the reaction mixture. Aldehydes do not hydrogen-bond and consequently have lower boiling points than the corresponding primary alcohols. The systematic nomenclature for aldehydes is developed in Problem 29-43.

Many law enforcement agencies use a device called a breathalyzer in the field to estimate the ethanol content of the blood of someone who is suspected of having been drinking alcoholic beverages. Breathalyzers function by analyzing the extent to which the oxidation of ethanol produces either acetaldehyde or acetic acid. One type of breathalyzer consists of a glass tube packed with an inert matrix that is impregnated with a potassium dichromate-sulfuric acid mixture. When alcohol vapor is exhaled into the tube, the ethanol is oxidized by the potassium dichromate. The potassium dichromate, which is reddish orange, is reduced to Cr^{3+}, which is green. The greater the concentration of alcohol vapor in the breath, the further down the glass tube the change from a reddish orange to a green color will occur. By suitable calibration, the length of the region experiencing a color change can be directly related to the concentration of ethanol vapor in the breath. This concentration in turn can be related to the concentration of ethanol in the blood because, as blood passes through the pulmonary arteries, an equilibrium is established between the ethanol in the blood and the ethanol in the lungs.

Secondary alcohols can be oxidized to compounds called **ketones,** whose general formula is

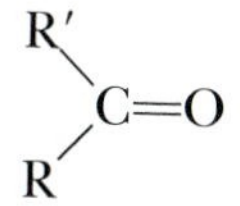

$$\begin{matrix} R' \\ & \diagdown \\ & & C{=}O \\ & \diagup \\ R \end{matrix}$$

where R and R′ are alkyl groups and the $\rangle C{=}O$ group is called a **carbonyl group.** The simplest ketone—acetone, $(CH_3)_2CO$—is prepared from the oxidation of 2-propanol:

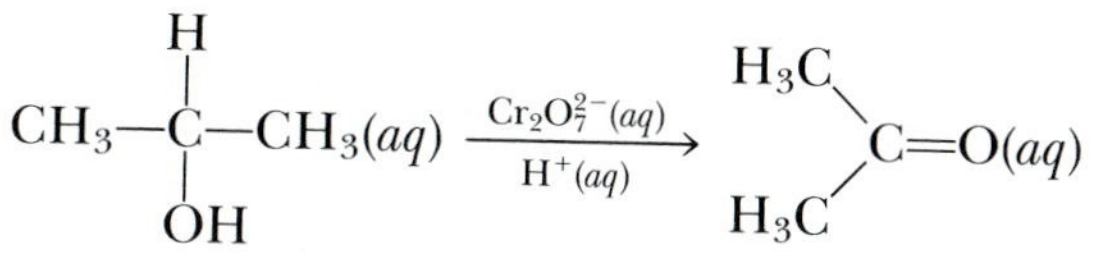

$$\underset{\text{2-propanol}}{CH_3{-}\underset{\underset{OH}{|}}{\overset{\overset{H}{|}}{C}}{-}CH_3(aq)} \xrightarrow[H^+(aq)]{Cr_2O_7^{2-}(aq)} \underset{\text{propanone (acetone)}}{(H_3C)_2C{=}O(aq)}$$

The systematic nomenclature for ketones is developed in Problem 29-47.

Acetone is a colorless, volatile liquid with a sweet odor. Over 2 billion pounds of acetone are produced annually in the United States. It is used extensively as a solvent because it dissolves many organic compounds yet is completely miscible with water. It dissolves fats, oils, waxes, rubber, lacquers, varnishes, rubber cements, and coatings ranging from nail polish to exterior enamel paints.

Tertiary alcohols are resistant to oxidation. The —OH-bearing carbon atom is already bonded to three other carbon atoms and so cannot form a double bond with an oxygen atom:

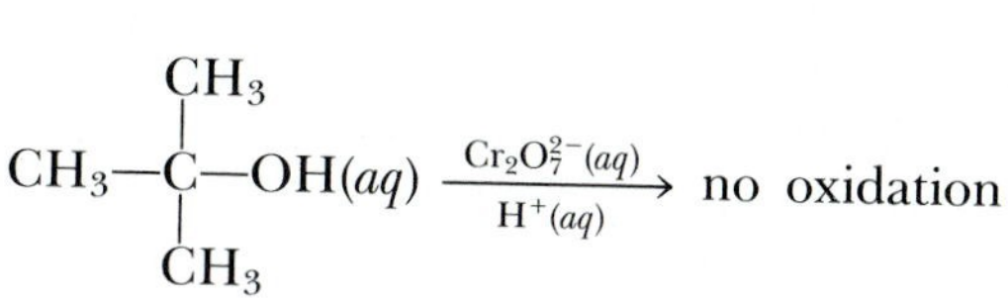

$$\underset{\substack{\text{tertiary alcohol:} \\ \text{2-methyl-2-propanol}}}{CH_3{-}\underset{\underset{CH_3}{|}}{\overset{\overset{CH_3}{|}}{C}}{-}OH(aq)} \xrightarrow[H^+(aq)]{Cr_2O_7^{2-}(aq)} \text{no oxidation}$$

EXAMPLE 29-11: Which alcohol would you use to produce $CH_3CH_2\underset{\underset{O}{\|}}{C}CH_3$?

Solution: The product here is a ketone, so we must oxidize a secondary alcohol. The secondary alcohol should have a methyl group and an ethyl group attached to the —OH-bearing carbon atom. Thus, we have

$$\underset{\text{2-butanol}}{CH_3CH_2-\underset{\underset{OH}{|}}{\overset{\overset{H}{|}}{C}}-CH_3(aq)} \xrightarrow[H^+(aq)]{Cr_2O_7^{2-}(aq)} \underset{\substack{\text{butanone} \\ \text{(methyl ethyl ketone)}}}{CH_3CH_2-\underset{\underset{O}{\|}}{C}-CH_3(aq)}$$

PRACTICE PROBLEM 29-11: Aldehydes and ketones can be reduced to alcohols by a reducing agent such as sodium borohydride, $NaBH_4$. Which aldehyde or ketone would you use to produce 2-propanol?

Answer: acetone

29-10. AMINES ARE ORGANIC DERIVATIVES OF AMMONIA

Amines are derivatives of ammonia in which one or more of the hydrogen atoms in NH_3 are replaced by hydrocarbon groups. Alkyl amines are frequently named by combining the names of the attached alkyl groups with the word *amine.* For example,

$$\underset{\text{methylamine}}{CH_3NH_2} \qquad \underset{\text{dimethylamine}}{(CH_3)_2NH} \qquad \underset{\text{dimethylethylamine}}{CH_3-\underset{\underset{CH_3}{|}}{N}-CH_2CH_3}$$

It is sometimes useful to classify amines according to the number of organic groups attached to the nitrogen atom. Thus, of the three amines shown here, methylamine is called a primary amine (with one attached group); dimethylamine is a secondary amine (with two attached groups); and dimethylethylamine is called a tertiary amine (with three attached groups).

The resemblance between amines and ammonia is much stronger than that between alcohols and water, but in both cases the similarity becomes less as the hydrocarbon portions of the molecules become large. Primary and secondary amines have at least one hydrogen atom bonded to the central nitrogen atom and so form hydrogen bonds in the liquid state. Thus, the boiling points of primary and secondary amines are higher than those of hydrocarbons of comparable molecular mass. The boiling points of tertiary amines, which have no hydrogen atom attached to the nitrogen atom, are closer to those of the comparable hydrocarbons; for example,

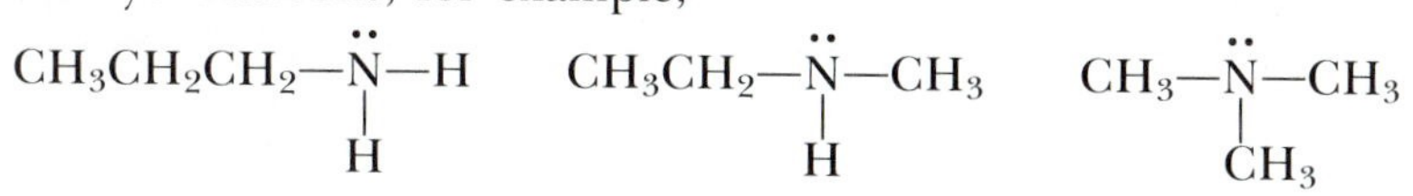

$$CH_3CH_2CH_2\text{—}\ddot{N}(H)\text{—}H \qquad CH_3CH_2\text{—}\ddot{N}(H)\text{—}CH_3 \qquad CH_3\text{—}\ddot{N}(CH_3)\text{—}CH_3$$

	propylamine	ethylmethylamine	trimethylamine
boiling point	49°C	36°C	2.9°C

Notice that the molecular mass of each of these amines is 59. The boiling point of butane (molecular mass 58) is −0.5°C.

Even tertiary amines can form hydrogen bonds with water molecules, so the solubilities of amines in water are similar to those of alcohols of the same molecular mass. Amines having up to three or four carbon atoms are completely miscible with water, but the solubility decreases rapidly with an increasing number of carbon atoms.

Like ammonia, amines are weak bases (Table 18-4) and react with acids to form crystalline, nonvolatile salts that are very water soluble. For example,

$$\underset{\text{dimethylamine}}{(CH_3)_2NH(aq)} + HBr(aq) \rightarrow \underset{\text{dimethylammonium ion}}{(CH_3)_2NH_2^+(aq)} + \underset{\text{bromide ion}}{Br^-(aq)}$$

Amines also behave as Lewis bases by reacting with alkyl halides, as in

$$\underset{\text{trimethylamine}}{(CH_3)_3N(aq)} + CH_3Cl(g) \rightarrow \underset{\text{tetramethylammonium ion}}{(CH_3)_4N^+(aq)} + \underset{\text{chloride ion}}{Cl^-(aq)}$$

The salts of amines are generally more water soluble than the amines themselves.

Many alkyl amines have strong odors, like that of decaying fish. The common names of some diamines (i.e., molecules containing two amino groups) suggest even worse odors:

$$\underset{\text{putrescine}}{H_2NCH_2CH_2CH_2CH_2NH_2} \qquad \underset{\text{cadaverine}}{H_2NCH_2CH_2CH_2CH_2CH_2NH_2}$$

29-11. THE REACTION OF A CARBOXYLIC ACID WITH AN ALCOHOL PRODUCES AN ESTER

One important use of aldehydes is in the production of **carboxylic acids** (see Chapter 18):

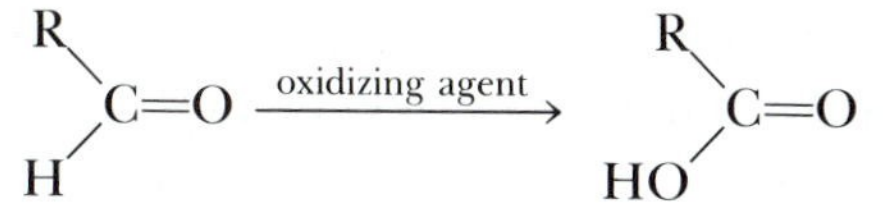

$$\underset{\text{aldehyde}}{R(H)C{=}O} \xrightarrow{\text{oxidizing agent}} \underset{\text{carboxylic acid}}{R(HO)C{=}O}$$

For example,

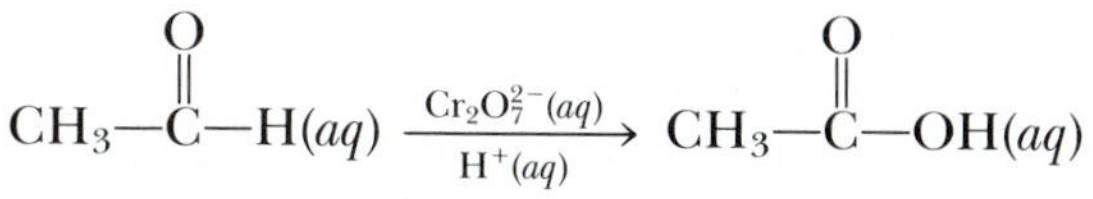

$$\underset{\substack{\text{ethanal}\\ \text{(acetaldehyde)}}}{CH_3\text{—}\overset{O}{\overset{\|}{C}}\text{—}H(aq)} \xrightarrow[H^+(aq)]{Cr_2O_7^{2-}(aq)} \underset{\substack{\text{ethanoic acid}\\ \text{(acetic acid)}}}{CH_3\text{—}\overset{O}{\overset{\|}{C}}\text{—}OH(aq)}$$

Note that the oxidizing agent effectively inserts an oxygen atom between the carbon atom and the hydrogen atom in the aldehyde group. We can also obtain carboxylic acids by the direct oxidation of primary alcohols. In the oxidation, the aldehyde occurs as an intermediate species:

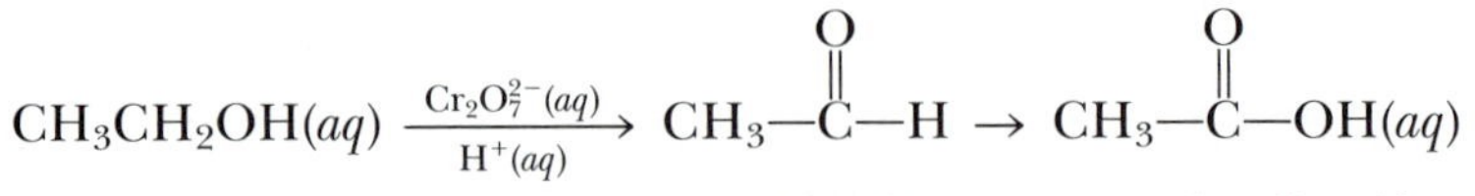

$$\underset{\text{a primary alcohol}}{CH_3CH_2OH(aq)} \xrightarrow[H^+(aq)]{Cr_2O_7^{2-}(aq)} \underset{\text{an aldehyde}}{CH_3{-}\overset{\overset{\displaystyle O}{\|}}{C}{-}H} \rightarrow \underset{\text{a carboxylic acid}}{CH_3{-}\overset{\overset{\displaystyle O}{\|}}{C}{-}OH(aq)}$$

The two simplest carboxylic organic acids are formic acid (methanoic acid) and acetic acid (ethanoic acid) (see Section 18-4):

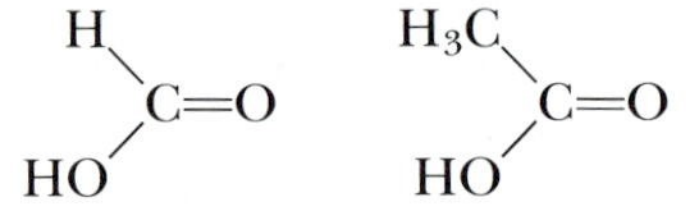

H(HO)C=O	H_3C(HO)C=O
methanoic acid (formic acid) HCOOH	ethanoic acid (acetic acid) CH_3COOH

Recall from Section 18-4 that —COOH is called a **carboxyl group.** The systematic nomenclature of carboxylic acids is developed in Problem 29-53.

Formic acid is a colorless, fuming liquid that is very soluble in water. Acetic acid is a clear, colorless liquid with a pungent odor. The pure compound is called glacial acetic acid because it freezes at 18°C. Acetic acid is ranked 34th among industrial chemicals in terms of annual U.S. production. It is used in the manufacture of plastics, pharmaceuticals, dyes, insecticides, photographic chemicals, acetates, and many other organic chemicals.

Carboxylic acids react with alcohols in the presence of an acid catalyst; the equation for the reaction is:

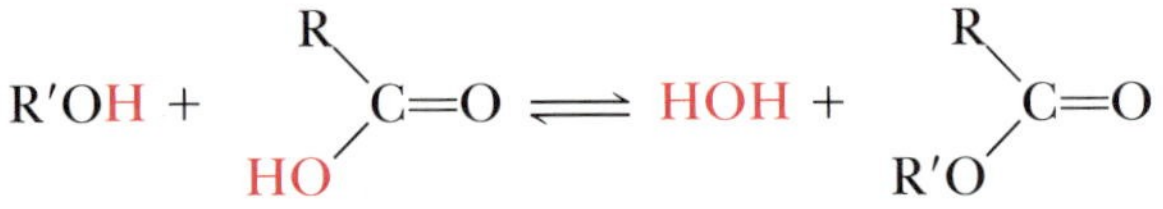

$$R'OH + R(HO)C{=}O \rightleftharpoons HOH + R(R'O)C{=}O$$

where R and R′ represent possibly different alkyl groups. The coloring used in this equation emphasizes that the water is formed from the —OH group of the acid and the hydrogen atom of the alcohol. This reaction is quite unlike an acid-base neutralization reaction, in which the water is formed by the reaction between a hydroxide ion from the base and a hydronium ion from the acid. Alcohols do not react like bases. The similarity in their chemical formulas, for example, CH_3OH versus NaOH, is superficial.

The product of the reaction between an organic acid and an alcohol is called an **ester.** As the carboxylic acid-alcohol reaction indicates, the general formula for an ester is

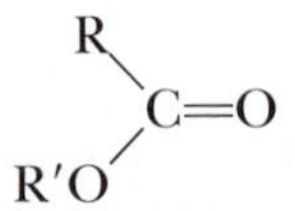

$$R(R'O)C{=}O$$

where the R′ group comes from the alcohol. The pleasant odors of flowers and fruits are due to esters. Table 29-5 lists some naturally occurring esters and their odors. Esters are named by first naming the alkyl group from the alcohol and then designating the acid, with its *-ic* ending changed to *-ate*. For example,

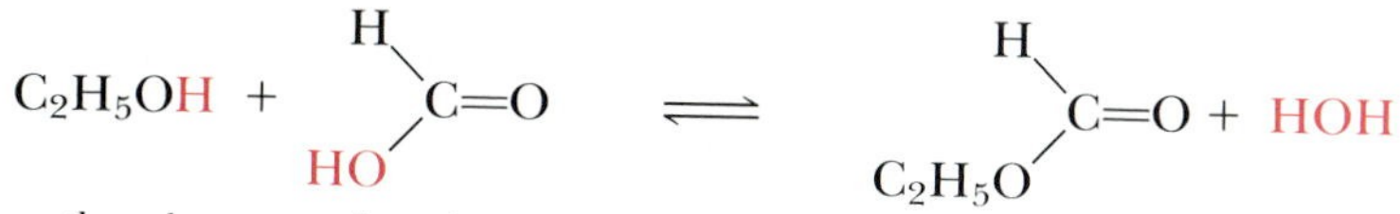

$$C_2H_5OH + \underset{\text{HO}}{\overset{\text{H}}{\diagdown}} C{=}O \rightleftharpoons \underset{C_2H_5O}{\overset{\text{H}}{\diagdown}} C{=}O + HOH$$

ethanol; methanoic acid (formic acid); ethyl methanoate (ethyl formate)

Such reactions are called **esterification reactions.**

We have written the above equations with double arrows to emphasize that esterification reactions are highly reversible. The equilibrium constants for esterification reactions are usually not large, so appreciable quantities of both reactants and products are present at equilibrium. You will learn more efficient methods of preparing esters when you take a course in organic chemistry.

Table 29-5 Various esters and their odors

Name	Odor
ethyl formate	rum
pentyl acetate	bananas
octyl acetate	oranges
methyl butyrate	apples
ethyl butyrate	pineapples
pentyl butyrate	apricots

EXAMPLE 29-12: Complete and balance the equation for the reaction between ethanol and acetic acid. Name the product.

Solution: The reaction between an organic acid and an alcohol yields an ester, so in this case we have

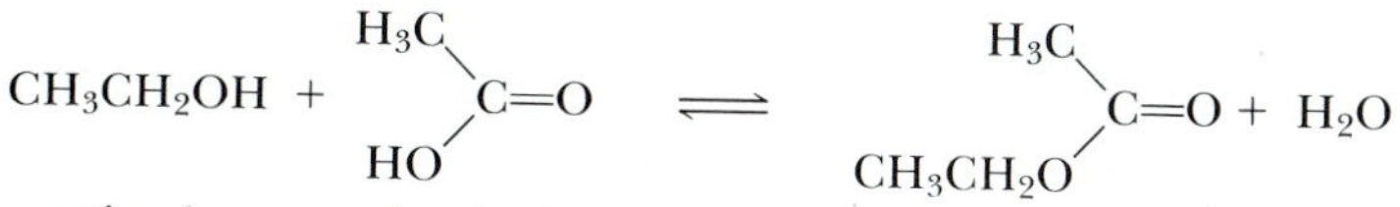

$$CH_3CH_2OH + \underset{\text{HO}}{\overset{H_3C}{\diagdown}} C{=}O \rightleftharpoons \underset{CH_3CH_2O}{\overset{H_3C}{\diagdown}} C{=}O + H_2O$$

ethanol; ethanoic acid (acetic acid); ethyl ethanoate (ethyl acetate)

PRACTICE PROBLEM 29-12: Fats and oils are called triglycerides, which are triesters of a trialcohol, glycerol (shown in the margin), with long hydrocarbon-chain carboxylic acids such as stearic acid, $CH_3(CH_2)_{16}COOH$. Write the Lewis formula for the triester of glycerol and stearic acid.

Answer: See the margin.

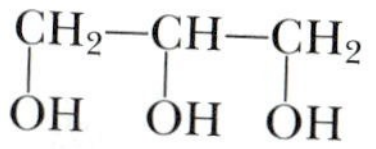

$$\underset{OH}{CH_2}-\underset{OH}{CH}-\underset{OH}{CH_2}$$

glycerol

$$\begin{array}{ccccc} CH_2 & - & CH_2 & - & CH_2 \\ | & & | & & | \\ O & & O & & O \\ | & & | & & | \\ O{=}C & & O{=}C & & O{=}C \\ | & & | & & | \\ (CH_2)_{16} & & (CH_2)_{16} & & (CH_2)_{16} \\ | & & | & & | \\ CH_3 & & CH_3 & & CH_3 \end{array}$$

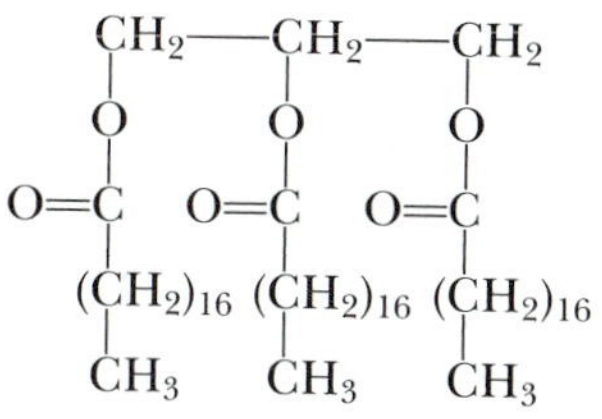

As Practice Problem 29-12 says, fats and oils are **triglycerides,** that is, triesters of glycerol and long-hydrocarbon-chain carboxylic acids called **fatty acids.** Because the fatty acids have long hydrocarbon chains, they are insoluble in water but soluble in organic solvents such as diethyl ether, chloroform, and acetone.

Complete hydrolysis of a fat or an oil yields three fatty acid molecules for every molecule of glycerol:

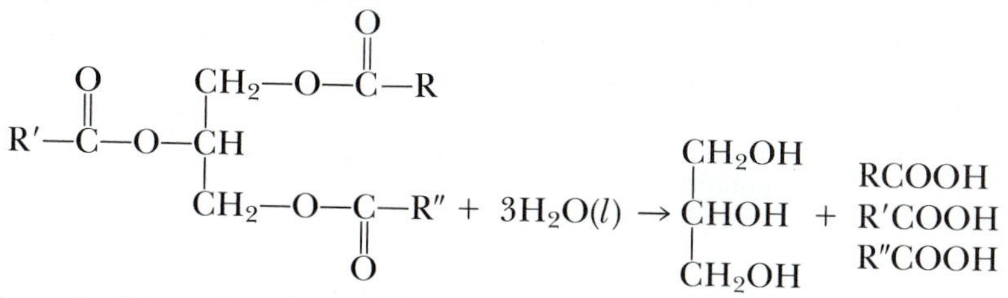

$$\begin{array}{l} \qquad\qquad\quad CH_2-O-\overset{O}{\overset{\|}{C}}-R \\ R'-\overset{O}{\overset{\|}{C}}-O-CH \\ \qquad\qquad\quad CH_2-O-\underset{O}{\underset{\|}{C}}-R'' \end{array} + 3H_2O(l) \rightarrow \begin{array}{c} CH_2OH \\ | \\ CHOH \\ | \\ CH_2OH \end{array} + \begin{array}{c} RCOOH \\ R'COOH \\ R''COOH \end{array}$$

where R, R′, and R″ are hydrocarbon chains. Of the 50 or more fatty acids that have been obtained from fats and oils, the three most abundant are

$CH_3(CH_2)_{14}COOH$ (palmitic acid) $\quad CH_3(CH_2)_{16}COOH$ (stearic acid) $\quad CH_3(CH_2)_7CH{=}CH(CH_2)_7COOH$ (oleic acid)

Note that palmitic acid and stearic acid have saturated hydrocarbon chains, whereas oleic acid has an unsaturated hydrocarbon chain. Triglycerides rich in saturated fatty acids are generally solids at room temperature and are called **fats.** Beef tallow is about 45 percent by mass saturated fatty acids. Triglycerides rich in unsaturated fatty acids are generally liquids at room temperature and are called **oils.** Vegetable oils such as corn oil, soybean oil, and olive oil are all over 85 percent by mass unsaturated fatty acids.

For reasons of convenience and dietary preference, oils often are converted to fats by hydrogenating some of the double bonds in the unsaturated triglyceride. The products are sold as shortening (Crisco and Spry), margarine, and other food products. Saturated fats are alleged to be a dietary factor in atherosclerosis; therefore, for health reasons, vegetable oils are becoming increasingly popular.

29-12. THE EMPIRICAL FORMULAS OF MANY ORGANIC COMPOUNDS CAN BE DETERMINED BY COMBUSTION ANALYSIS

Many organic compounds consist of only carbon and hydrogen or of carbon, hydrogen, and oxygen. When these compounds are burned in an excess of O_2, all the carbon in the original sample ends up in CO_2 and all the hydrogen ends up in H_2O. These facts are the basis of the determination of the percentage composition of such compounds by **combustion analysis,** the analysis of the products obtained when an organic compound is burned in an excess of oxygen.

After the sample is burned in excess oxygen, the resulting gaseous water and CO_2 are passed through chambers containing different substances, as shown in Figure 29-8. The water is absorbed in the magnesium perchlorate chamber by the reaction

$$\underset{\text{anhydrous}}{Mg(ClO_4)_2(s)} + 6H_2O(g) \rightarrow \underset{\text{hydrated}}{Mg(ClO_4)_2\cdot 6H_2O(s)}$$

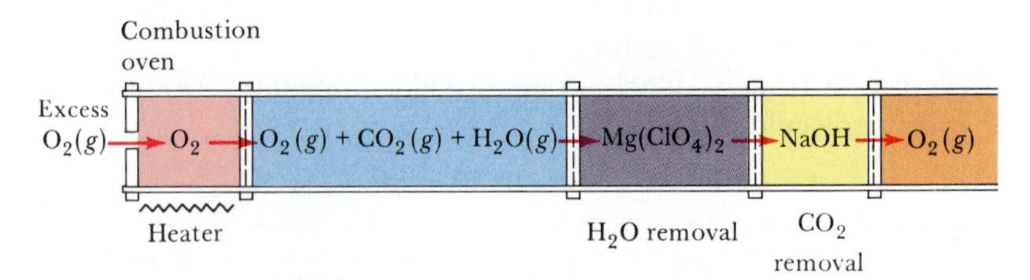

Figure 29-8 A schematic illustration of the removal of H_2O and CO_2 from combustion gases. Anhydrous magnesium perchlorate removes the water vapor:

$$Mg(ClO_4)_2(s) + 6H_2O(g) \rightarrow Mg(ClO_4)_2\cdot 6H_2O(s)$$

and then sodium hydroxide removes the CO_2:

$$NaOH(s) + CO_2(g) \rightarrow NaHCO_3(s)$$

Excess O_2 gas is used as a carrier gas to "sweep" all the H_2O and CO_2 from the combustion zone (left-hand compartment).

After passing through the magnesium perchlorate chamber, the carbon dioxide reacts with the sodium hydroxide in the next chamber according to

$$NaOH(s) + CO_2(g) \rightarrow NaHCO_3(s)$$

The masses of water and carbon dioxide formed in the combustion reaction are determined by measuring mass increases in the magnesium perchlorate and sodium hydroxide chambers. The masses of CO_2 and H_2O produced in the combustion reaction are converted readily to the stoichiometrically equivalent masses of carbon and hydrogen:

$$\text{mass of C} = \left(\begin{matrix}\text{mass of}\\ CO_2\text{ formed}\end{matrix}\right)\left(\begin{matrix}\text{fraction of the}\\ \text{mass of } CO_2\\ \text{due to C}\end{matrix}\right)$$

$$= \left(\begin{matrix}\text{mass of}\\ CO_2\text{ formed}\end{matrix}\right)\left(\frac{\text{atomic mass of C}}{\text{formula mass of } CO_2}\right)$$

and

$$\text{mass of H} = \left(\begin{matrix}\text{mass of}\\ H_2O\text{ formed}\end{matrix}\right)\left(\begin{matrix}\text{fraction of the}\\ \text{mass of } H_2O\\ \text{due to H}\end{matrix}\right)$$

$$= \left(\begin{matrix}\text{mass of}\\ H_2O\text{ formed}\end{matrix}\right)\left(\frac{2 \times \text{atomic mass of H}}{\text{formula mass of } H_2O}\right)$$

The mass percentages of carbon and hydrogen in the original sample are then calculated as follows:

$$\text{mass percentage of C} = \left(\frac{\text{mass of C}}{\text{mass of sample}}\right) \times 100$$

$$\text{mass percentage of H} = \left(\frac{\text{mass of H}}{\text{mass of sample}}\right) \times 100$$

If the original sample contains only carbon and hydrogen, then the mass percentages of C and H should sum to 100 percent, within experimental error. If the sample also contains oxygen, then the mass percentage of oxygen is determined by difference:

$$\left(\begin{matrix}\text{mass percentage}\\ \text{of O}\end{matrix}\right) = 100 - \left(\begin{matrix}\text{mass percentage}\\ \text{of C}\end{matrix}\right) - \left(\begin{matrix}\text{mass percentage}\\ \text{of H}\end{matrix}\right)$$

Example 29-13 outlines a calculation involving a combustion analysis.

EXAMPLE 29-13: A 1.250-g sample of the compound responsible for the odor of cloves, which is known to contain only the elements C, H, and O, is burned in a combustion analysis apparatus. The mass of CO_2 produced is 3.350 g, and the mass of water produced is 0.8232 g. (a) Determine the percentage composition of the sample. (b) Determine the molecular formula of the compound, given that its molecular mass is 164.

Solution: (a) The masses of carbon and hydrogen in the original sample are

$$\text{mass of C} = (3.350\text{ g } CO_2)\left(\frac{12.01\text{ g C}}{44.01\text{ g } CO_2}\right) = 0.9142\text{ g C}$$

$$\text{mass of H} = (0.8232\text{ g } H_2O)\left(\frac{2 \times 1.008\text{ g H}}{18.02\text{ g } H_2O}\right) = 0.09210\text{ g H}$$

The mass percentages of C and H are

$$\text{mass percentage of C} = \left(\frac{0.9142\text{ g}}{1.250\text{ g}}\right) \times 100 = 73.14\%$$

$$\text{mass percentage of H} = \left(\frac{0.09210\text{ g}}{1.250\text{ g}}\right) \times 100 = 7.37\%$$

The mass percentage of oxygen is obtained by difference:

$$\text{mass percentage of O} = 100.00 - 73.14 - 7.37 = 19.49\%$$

(b) To determine the molecular formula, first we calculate the empirical formula from the mass percentages, as described in Section 5-3. As usual, we consider a 100-g sample and write

$$73.14\text{ g C} \approx 7.37\text{ g H} \approx 19.49\text{ g O}$$

$$6.090\text{ mol C} \approx 7.312\text{ mol H} \approx 1.218\text{ mol O}$$

$$5.00\text{ mol C} \approx 6.00\text{ mol H} \approx 1.00\text{ mol O}$$

Thus, the empirical formula is C_5H_6O. The molecular mass is known to be 164, so the molecular formula is $C_{10}H_{12}O_2$.

PRACTICE PROBLEM 29-13: A 2.475-g sample of vitamin C is burned in a combustion analysis apparatus. The mass of CO_2 produced is found to be 3.710 g and the mass of H_2O produced is found to 1.013 g. (a) Assuming that vitamin C consists of carbon, hydrogen, and oxygen, determine its empirical formula. (b) A solution containing 18.61 g of vitamin C in 100 g of water has a freezing point of −1.97°C. Determine the molecular formula of vitamin C.

Answer: (a) $C_3H_4O_3$; (b) $C_6H_8O_6$

The molecular formula of a compound does not tell us in what arrangement the atoms are bonded together in the molecules of the compound. For example, the two compounds ethanol and dimethyl ether both have the molecular formula C_2H_6O, but their molecular structures are very different as shown in the margin.

```
   H  H
   |  |
H—C—C—OH
   |  |
   H  H
```

ethanol

```
   H     H
   |     |
H—C—O—C—H
   |     |
   H     H
```

dimethyl ether

Chemists use a variety of techniques for determining the structures of molecules. Many of these involve spectroscopy, which is based on the interaction of electromagnetic radiation with molecules to produce a characteristic spectrum. Spectroscopic methods are also routinely used in the identification of unknown substances, a problem that chemists face frequently. Fortunately, there are powerful analytical methods available for determining the chemical composition of unknown substances. Many of these analytical methods establish the identity of chemical compounds by showing that certain characteristic properties of the unknown compound are identical to those of a known compound. Proton magnetic resonance, discussed in Interchapter F, is one of the most powerful analytical methods used in organic chemistry.

SUMMARY

Organic chemistry is the chemistry of compounds that contain carbon atoms. The simplest organic compounds are hydrocarbons, which are compounds that consist of only hydrogen and carbon. Hydrocarbons are classified into alkanes, alkenes, alkynes, and aromatic hydrocarbons. Alkanes have only single bonds, alkenes have one or more carbon-carbon double bonds, alkynes have one or more carbon-carbon triple bonds, and aromatic hydrocarbons have rings with delocalized electrons. Alkanes, which are also called saturated hydrocarbons, are relatively unreactive and undergo primarily combustion and substitution reactions. Alkenes and alkynes, which are called unsaturated hydrocarbons, undergo not only the same reactions as saturated hydrocarbons but also addition reactions, in which small molecules, such as H_2, Cl_2, Br_2, HCl, and H_2O, add to the double or triple bonds.

Other classes of organic compounds are alcohols (ROH), aldehydes (RCHO), ketones (RCOR′), carboxylic acids (RCOOH), and esters (RCOOR′). Compounds in each class undergo characteristic reactions. For example, primary alcohols can be oxidized to aldehydes and secondary alcohols can be oxidized to ketones; aldehydes can be oxidized to carboxylic acids; and carboxylic acids and alcohols react to form esters.

Because of the great diversity of organic compounds, a systematic nomenclature is necessary. An internationally recognized system (IUPAC) has been developed that allows each organic compound to be named unambiguously.

TERMS YOU SHOULD KNOW

organic compound 1025
hydrocarbon 1025
alkane 1025
condensed structural formula 1026
saturated 1026
saturated hydrocarbon 1026
structural isomer 1027
substitution reaction 1029
free-radical reaction 1029
alkyl halide (haloalkane) 1030
substituent 1030
disubstituted 1030
IUPAC nomenclature 1031
alkyl group 1032
unsaturated hydrocarbon 1034
alkene 1034
addition reaction 1036
hydrogenation 1036
Markovnikov's rule 1037
alkyne 1038
aromatic hydrocarbon 1040
ortho- 1042
meta- 1042
para- 1042
alcohol 1043
alkoxide 1044
primary alcohol 1045
secondary alcohol 1045
tertiary alcohol 1045
aldehyde 1046
aldehyde group 1046
ketone 1047
carbonyl group 1047
amine 1048
carboxylic acid 1049
carboxyl group 1050
ester 1050
esterification reaction 1051
triglyceride 1051
fatty acid 1051
fat 1052
oil 1052
combustion analysis 1052

PROBLEMS

ALKANES

29-1. Complete and balance the following equations (if no reaction occurs, then write N.R.):

(a) $C_5H_{12}(g) + O_2(g) \rightarrow$

(b) $C_2H_6(g) + Cl_2(g) \xrightarrow{\text{dark}}$

(c) $C_4H_{10}(g) + H_2SO_4(aq) \rightarrow$

(d) $CH_4(g) + Cl_2(g) \xrightarrow{\text{uv}}$

29-2. Complete and balance the following equations (if no reaction occurs, then write N.R.):

(a) $C_2H_6(g) + KOH(aq) \rightarrow$
(b) $C_6H_{14}(l) + O_2(g) \rightarrow$
(c) $C_3H_8(g) + KMnO_4(aq) \rightarrow$
(d) $C_5H_{12}(g) + HCl(aq) \rightarrow$

29-3. Which of the following pairs of molecules are identical and which are different?

(a) $ClCH_2CH_2CH_3$ and $CH_3CH_2CH_2Cl$

(b) $CH_3CH_2\underset{\underset{\displaystyle CH_3}{|}}{C}HCl$ and $CH_3\underset{\underset{\displaystyle Cl}{|}}{C}HCH_2CH_3$

29-4. Which of the following pairs of molecules are identical and which are different?

(a) $CH_3CH_2\underset{\underset{\displaystyle Cl}{|}}{C}HCH_3$ and $CH_3\underset{\underset{\displaystyle Cl}{|}}{C}HCH_2CH_3$

(b) $H_2\overset{\overset{\displaystyle Cl}{|}}{C}CH_2CH_3$ and $CH_3CH_2CH_2Cl$

29-5. If you substitute a chlorine atom for a hydrogen atom in *n*-hexane, how many possible isomers do you get? Give IUPAC names for all these isomers.

29-6. Suppose that neopentane reacts with chlorine under uv irradiation such that two hydrogen atoms are replaced by two chlorine atoms. Write the Lewis formulas for the possible isomers and name each isomer.

29-7. Use the data in Table 8-2 to calculate the heat of combustion per gram of butane.

29-8. Use the data in Appendix F to calculate ΔH°_{rxn} at 25°C for the reaction of chlorine with methane to form chloromethane.

29-9. One class of saturated hydrocarbons has the general formula C_nH_{2n}. Given that these hydrocarbons are called cycloalkanes, write the Lewis formula of cyclopropane (C_3H_6) and cyclobutane (C_4H_8).

29-10. Write the Lewis formula of cyclopentane, 1,2-dichlorocyclobutane, and 1-methyl-2-chlorocyclopropane (see Problem 29-9).

IUPAC NOMENCLATURE

29-11. Give IUPAC names for the following compounds:

(a) $CH_3-\underset{\displaystyle Br}{\underset{|}{CH}}-\underset{\displaystyle Cl}{\underset{|}{CH}}-CH_3$

(b) $CH_3-\overset{\displaystyle CH_3}{\overset{|}{\underset{\displaystyle CH_3}{\underset{|}{C}}}}-CH_3$

29-12. Give IUPAC names for the following compounds:

(a) $CH_3-CH_2-\overset{\displaystyle Cl}{\overset{|}{CH}}-CH_2-CH_3$

(b) $CH_3-\overset{\displaystyle NO_2}{\overset{|}{CH}}-\overset{\displaystyle CH_3}{\overset{|}{\underset{\displaystyle CH_3}{\underset{|}{C}}}}-CH_3$

29-13. Explain why the following names are incorrect and give a correct IUPAC name in each case:

(a) 4-methylpentane
(b) 2-ethylbutane
(c) 2-propylhexane
(d) 2-dimethylpropane

29-14. Explain why the following names are incorrect and give a correct IUPAC name in each case:

(a) 2,3-dichloropropane
(b) 3-bromo-2-ethylpropane
(c) 1,1,3-trimethylpropane
(d) 2-dimethylbutane

29-15. Write the Lewis formula for and assign an IUPAC name to each of the following compounds:

(a) $CH_3CH(CH_3)CH_2CH_3$
(b) CH_2ClCH_2Br
(c) $CH_3CCl_2CCl_3$
(d) $CH_3C(CH_3)_2CH_3$

29-16. Write the Lewis formula for and assign an IUPAC name to each of the following compounds:

(a) $CH_3(CH_2)_4CH_3$
(b) $(CH_3)_2CHCH_2CH_2CH(CH_3)_2$
(c) $(CH_3)_4C$
(d) $CH_3CHClCCl_2CCl_3$

29-17. Write the structural formula for each of the following alkanes:

(a) 2,3-dimethylbutane
(b) 2,2,3-trimethylbutane
(c) 3,3-dimethyl-4-ethylhexane
(d) 4-isopropyloctane

29-18. Write the structural formula for each of the following trichloroalkanes:

(a) 1,1,2-trichlorobutane
(b) 1,1,1-trichloroethane
(c) 1,2,3-trichloropentane
(d) 2,2,4-trichlorohexane

29-19. Write the Lewis formulas for the following:

(a) 2,3-dimethylbutane
(b) 2-amino-3-methylbutane
(c) 3-chloro-3-ethylpentane
(d) 1,2,3-trichloropropane

29-20. Write the Lewis formula for each of the following:

(a) 2-nitropropane
(b) 3-aminopentane
(c) 1,1,2,2-tetrachlorobutane
(d) 3-methyl-2-chlorohexane

ALKENES

29-21. Name the product of the following reactions.

(a) addition of HCl to propene
(b) addition of HBr to 2-butene
(c) addition of HCl to 1-butene
(d) addition of HBr to 2-pentene

29-22. Write the structural formula for and assign an IUPAC name to the product when each of the following compounds reacts with 1 mol of bromine:

(a) 1-butene
(b) 2-butene

29-23. Complete the following equations and name the products:

(a) $H_2C{=}CHCH_2CH_3(g) + HCl(g) \rightarrow$

(b) $H_2C{=}CHCH_3(g) + Cl_2(g) \rightarrow$

29-24. Complete and balance the following equations:

(a) $(H_3C)_2C{=}CHCH_3(l) + Br_2(l) \rightarrow$

(b) $(H_3C)_2C{=}CHCH_3(l) + HCl(g) \rightarrow$

29-25. In each case, determine which unsaturated hydrocarbon reacts with what reagent to form the given product:

(a) $CH_3{-}C(CH_3)(OH){-}CH_2CH_3$

(b) $CH_3{-}CH(OH){-}CH_3$

(c) $CH_3CH{=}CHCH_3$

29-26. In each case, determine which unsaturated hydrocarbon reacts with what reagent to form the given product:

(a) $CH_3CHBrCHBrCH_3$

(b) $CH_3CCl_2CCl_2CH_3$

(c) $(CH_3)_3COH$

(d) $CH_3CH(CH_3)CHBrCH_2Br$

ALKYNES

29-27. Assign IUPAC names to the following alkynes:

(a) $CH_3{-}C{\equiv}CH$

(b) $CH_3{-}C{\equiv}C{-}CH_3$

(c) $CH_3{-}C{\equiv}C{-}C(CH_3)_2{-}CH_2CH_3$

29-28. Write a structural formula for each of the following alkynes:

(a) 2-pentyne
(b) 3-hexyne
(c) 5-ethyl-3-octyne
(d) 2,2-dimethyl-3-hexyne

29-29. Predict the product in the addition of 2 mol of HBr to 1 mol of propyne.

29-30. Predict the product in the addition of 2 mol of HBr to 1 mol of 1-butyne.

29-31. Complete and balance the following equations (assume complete saturation of the triple bonds):

(a) $CH_3C{\equiv}CH(g) + HCl(g) \rightarrow$
(b) $CH_3C{\equiv}CH(g) + Br_2(l) \rightarrow$

29-32. Complete and balance the following equations (assume complete saturation of the triple bonds):

(a) $CH_3C{\equiv}CCH_3(g) + HCl(g) \rightarrow$

(b) $CH_3C{\equiv}CCH_3(g) + H_2(g) \xrightarrow{Ni(s)}$

BENZENE

29-33. Write the structural formula for each of the following benzene derivatives:

(a) ethylbenzene
(b) 1,3,5-trichlorobenzene
(c) 2-chloro-1-methylbenzene
(d) 1-chloro-3-bromobenzene

29-34. Name the following benzene derivatives:

(a) benzene ring with NO_2 and Cl (meta)

(b) benzene ring with six CH_3 groups (H_3C, CH_3)

(c) benzene ring with Cl and Cl (para)

(d) benzene ring with Cl and Cl (ortho)

ALCOHOLS

29-35. Assign IUPAC names to each of the following alcohols:

(a) $CH_3CH(CH_3)OH$

(b) $CH_3CH_2C(CH_3)_2CH_2OH$

(c) $ClCH_2CH(OH)CH_2Cl$

(d) $CH_3CH_2C(CH_3)(H)OH$

29-36. Write a structural formula for each of the following alcohols:

(a) 2,3-dimethyl-1-butanol

(b) 2-methyl-4-hexanol

(c) 2-chloro-1-hexanol

(d) 1,2-dichloro-3-pentanol

29-37. Classify the following alcohols as primary, secondary, or tertiary:

(a) CH_3CH_2OH (b) $CH_3\underset{OH}{\overset{H}{C}}CH_3$

(c) $CH_3CH_2\underset{OH}{\overset{CH_3}{C}}H$ (d) $CH_3\underset{CH_3}{\overset{CH_3}{C}}OH$

29-38. Classify the following alcohols as primary, secondary, or tertiary:

(a) $CH_3CH_2\underset{CH_3}{\overset{CH_3}{C}}OH$ (b) $CH_3CH_2\overset{OH}{C}HCH_3$

(c) $CH_3CH_2CH_2CH_2OH$ (d) $CH_3\underset{CH_3}{\overset{CH_3}{C}}CH_2OH$

29-39. Complete and balance the following equations:

(a) $CH_3CH_2OH(l) + Na(s) \rightarrow$

(b) $CH_3CH_2CH_2OH(l) + Na(s) \rightarrow$

29-40. Metal alkoxides (see Problem 28-39) are bases that are comparable in strength to sodium hydroxide. Write an equation that illustrates the basic character of alkoxides.

29-41. What are the products of

(a) $CH_3CHOHCH_2CH_3(l) \xrightarrow[H^+(aq)]{CrO_7^{2-}(aq)}$

(b) $CH_3CH_2OH(l) + NaH(s) \rightarrow$

29-42. What are the products of

(a) $CH_3CH_2CH_2OH(l) \xrightarrow[200°C]{H_2SO_4(conc)}$

(b) $CH_3CH_2OH(l) \xrightarrow[H^+(aq)]{Cr_2O_7^{2-}(aq)}$

ALDEHYDES AND KETONES

29-43. The IUPAC nomenclature for aldehydes uses the longest chain of consecutive carbon atoms *containing the —CHO group* as the parent compound. The aldehyde is named by dropping the terminal *-e* and adding *-al* to the name of the corresponding alkane (compare with the nomenclature for alcohols). For example, the IUPAC name for acetaldehyde, CH_3CHO, is ethanal. Assign IUPAC names to the following aldehydes:

(a) $CH_3CH_2CH_2CHO$ (b) $CH_3\overset{CH_3}{C}HCH_2CHO$

(c) CH_3CH_2CHO (d) $CH_3\overset{CH_3}{C}H\underset{CH_3}{C}HCH_2CHO$

29-44. Using the IUPAC nomenclature for aldehydes presented in Problem 29-43, write the structural formula for each of the following aldehydes:

(a) propanal

(b) 2-methylpentanal

(c) 4-methylpentanal

(d) 3,3-dimethylhexanal

29-45. Determine which alcohol you would use to produce each of the following ketones:

(a) diethyl ketone

(b) methyl propyl ketone

(c) ethyl propyl ketone

29-46. Determine which alcohol you would use to produce each of the following aldehydes:

(a) ethanal

(b) 2-methylpropanal

(c) 2,2-dimethylpropanal

29-47. The IUPAC nomenclature for ketones uses the longest chain of consecutive atoms containing the carbonyl group as the parent compound. The ketone is named by dropping the terminal *-e* and adding *-one* to the name of the corresponding alkane, with a preceding numeral to indicate which carbon atom in the chain is the carbonyl carbon atom (if necessary). For example, the IUPAC name of

$$CH_3-\underset{\underset{O}{\|}}{C}-CH_2CH_2CH_3$$

is 2-pentanone. Assign IUPAC names to the following ketones

(a) $CH_3-\underset{\underset{O}{\|}}{C}-CH_3$ (b) $CH_3CH_2-\underset{\underset{O}{\|}}{C}-CH_2CH_3$

(c) $CH_3CH_2-\underset{\underset{O}{\|}}{C}-CH_3$ (d) $CH_3-\underset{\underset{O}{\|}}{C}-CH_2CH_3$

29-48. Write Lewis formulas for

(a) 2-methyl-3-pentanone

(b) 3,3-dimethyl-2-hexanone

(c) butanone

AMINES

29-49. In the IUPAC nomenclature of amines, the longest chain of carbon atoms that has an attached $-NH_2$ group (the amino group) is considered to be the parent and the $-NH_2$ group is considered to be a substituent, like $-Cl$ (the chloro group), for example. If more than one alkyl group is attached to the nitrogen atom, then the alkyl group containing the longest chain of carbon atoms is taken as the parent, and the other substituents on the nitrogen atom are named as alkyl groups preceded by *N*- to indicate that they are bonded to the amine nitrogen atom. For example,

$$\mathrm{CH_3CH_2\underset{\displaystyle CH_3}{\underset{|}{C}}H{-}CH_2{-}\underset{\displaystyle H}{\underset{|}{N}}{-}CH_3}$$

is named *N*-methyl-1-amino-2-methylbutane. Name the following amines:

(a) $\mathrm{CH_3{-}\underset{\displaystyle CH_3}{\underset{|}{C}}H{-}CH_2{-}\underset{\displaystyle H}{\underset{|}{N}}{-}CH_2CH_3}$

(b) $\mathrm{CH_3CH_2{-}\underset{\displaystyle CH_3}{\underset{|}{C}}H{-}\overset{\displaystyle CH_3}{\overset{|}{C}}H{-}\underset{\displaystyle H}{\underset{|}{N}}{-}CH_3}$

29-50. Write the Lewis formula of 3-amino-1-propanol and *N,N*-dimethyl-2-aminopentane (see Problem 29-49).

29-51. Complete and balance the following equations:

(a) $C_2H_5NH_2(aq) + HBr(aq) \rightarrow$

(b) $(CH_3)_2NH(aq) + H_2SO_4(aq) \rightarrow$

(c) 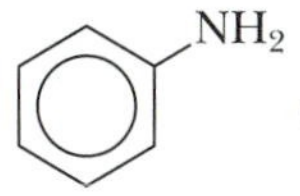 $C_6H_5NH_2$ (benzene ring with NH_2) $(aq) + HCl(aq) \rightarrow$

29-52. Complete and balance the following equations:

(a) $H_2NCH_2CH_2NH_2(aq) + HI(aq) \rightarrow$

(b) $CH_3NH_2(aq) + H_2O(l) \rightarrow$

(c) $H_2NCH_2CH(NH_2)_2(aq) + HCl(aq) \rightarrow$

CARBOXYLIC ACIDS

29-53. The IUPAC nomenclature for carboxylic acids uses the longest chain of consecutive carbon atoms *containing the —COOH group* as the parent compound. The parent compound is named by changing the ending *-e* in the name of the corresponding alkane to *-oic acid.* For example, the IUPAC name for formic acid is methanoic acid and that for acetic acid is ethanoic acid. Write the structural formula for each of the following carboxylic acids:

(a) propanoic acid
(b) 2-methylpropanoic acid
(c) 3,3-dimethylbutanoic acid
(d) 3-methylpentanoic acid

29-54. Using the IUPAC nomenclature for carboxylic acids presented in Problem 29-53, name the following compounds:

(a) $\mathrm{CH_3\overset{\displaystyle Cl}{\overset{|}{C}}HCH_2COOH}$

(b) $\mathrm{CH_3\overset{\displaystyle CH_3}{\overset{|}{\underset{\displaystyle CH_3}{\underset{|}{C}}}}COOH}$

(c) $\mathrm{CH_3\overset{\displaystyle Cl}{\overset{|}{\underset{\displaystyle H}{\underset{|}{C}}}}{-}\overset{\displaystyle CH_3}{\overset{|}{\underset{\displaystyle H}{\underset{|}{C}}}}CH_2COOH}$

(d) $\mathrm{Cl\overset{\displaystyle Cl}{\overset{|}{\underset{\displaystyle Cl}{\underset{|}{C}}}}{-}\overset{\displaystyle Cl}{\overset{|}{\underset{\displaystyle Cl}{\underset{|}{C}}}}COOH}$

29-55. Complete and balance the following equations:

(a) $HCOOH(aq) + NaOH(aq) \rightarrow$

(b) $HCOOH(aq) + CH_3OH(aq) \xrightarrow{H^+(aq)}$

(c) $HCOOH(aq) + Ca(OH)_2(aq) \rightarrow$

29-56. Complete and balance the following equations:

(a) $CH_3CH_2COOH(aq) + NH_3(aq) \rightarrow$

(b) $CH_3CH_2COOH(aq) + CH_3OH(aq) \xrightarrow{H^+(aq)}$

(c) $CH_3CH_2COOH(aq) + CH_3CH_2OH(aq) \xrightarrow{H^+(aq)}$

29-57. To form the IUPAC name of anions of carboxylic acids, the ending *-ic acid* is replaced by *-ate*. For example, the IUPAC name for sodium acetate is sodium ethanoate. Complete and balance the following equations and name the salt in each case:

(a) $CH_3CH_2COOH(aq) + KOH(aq) \rightarrow$

(b) $\mathrm{CH_3\underset{\displaystyle CH_3}{\underset{|}{C}}HCOOH}(aq) + KOH(aq) \rightarrow$

(c) $Cl_2CHCOOH(aq) + Ca(OH)_2(aq) \rightarrow$

29-58. Write the structural formula for each of the following salts (see Problem 29-57):

(a) sodium 2-chloropropanoate
(b) rubidium methanoate
(c) strontium 2,2-dimethylpropanoate
(d) lanthanum ethanoate

29-59. Complete and balance the following equations and name the products:

(a) 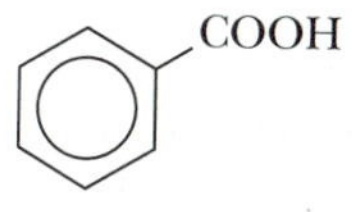 C_6H_5COOH (benzene ring with COOH; benzoic acid) $+ CH_3CH_2OH \xrightarrow{H^+(aq)}$

benzoic acid

(b) HOOC—COOH + $CH_3CH_2CH_2OH \xrightarrow{H^+(aq)}$
oxalic acid

(c) $CH_3COOH + CH_3CHOHCH_3 \xrightarrow{H^+(aq)}$

29-60. Complete and balance the following equations and name the products:

(a) COOH + KOH →
|
COOH
oxalic acid

(b) $CH_3(CH_2)_{16}COOH$ + NaOH →
stearic acid

(c) [benzene ring with two adjacent COOH groups] + KOH →
phthalic acid

COMBUSTION ANALYSIS

29-61. Combustion analysis of a 1.000-g sample of a compound known to contain only carbon, hydrogen, and oxygen gave 1.500 g of CO_2 and 0.409 g of H_2O. Determine the empirical formula of the compound.

29-62. Combustion analysis of a 1.000-g sample of a compound known to contain only carbon, hydrogen, and iron gave 2.367 g of CO_2 and 0.4835 g of H_2O. Determine the empirical formula of the compound.

29-63. Diethyl ether, often called simply ether, is a common solvent that contains carbon, hydrogen, and oxygen. A 1.23-g sample was burned under controlled conditions to produce 2.92 g of CO_2 and 1.49 g of H_2O. Determine the empirical formula of diethyl ether.

29-64. Butylated hydroxytoluene, BHT, a food preservative, contains carbon, hydrogen, and oxygen. A sample of 15.42 mg of BHT was burned in a stream of oxygen and yielded 46.20 mg CO_2 and 15.13 mg H_2O. Calculate the empirical formula of BHT.

29-65. Pyridine is recovered from coke-oven gases and is used extensively in the chemical industry, in particular, in the synthesis of vitamins and drugs. Pyridine contains carbon, hydrogen, and nitrogen. A 0.546-g sample was burned to produce 1.518 g of CO_2 and 0.311 g of H_2O. Determine the empirical formula of pyridine.

29-66. One of the additives to gasoline to prevent knocking was found to contain lead, carbon, and hydrogen. A 5.83-g sample was burned in an apparatus like that in Figure 29-8, and 6.34 g of CO_2 and 3.26 g of H_2O were produced. Determine the empirical formula of this additive.

ADDITIONAL PROBLEMS

29-67. In enumerating the monochlorosubstituted isomers of propane, we listed 1-chloropropane and 2-chloropropane but not 3-chloropropane. Why not?

29-68. A cylinder is labeled "PENTANE." When the gas inside the cylinder is monochlorinated, five isomers of formula $C_5H_{11}Cl$ result. Was the gas pure *n*-pentane, pure isopentane, pure neopentane, or a mixture of two or all three of these?

29-69. Consider the following structures:

(1) $(CH_3)_2CH—CH(CH_3)—CH_2—CH_3$ (drawn as: CH_3, CH_3 on CH; CH—CH(CH_3)—CH_2—CH_3)

(2) $CH_3CH_2—C(CH_3)—CH(CH_3)_2$ (drawn as: CH_3CH_2—C(CH_3)—CH_3, with CH bearing H_3C and CH_3)

(3) $CH_2(CH_3)—CH(CH_3)—CH(CH_2CH_3)—CH_3$

(4) $CH_2(CH_3)—CH(CH_3)—CH(CH_3)—CH_3$

Indicate which of these formulas represent(s)

(a) the same compound
(b) an isomer of octane
(c) a derivative of hexane
(d) the one with the most "methyl" groups

29-70. The boiling points and molecular masses of ethane, methylamine, and methanol are

Compound	Boiling point/°C	Molecular mass
CH_3CH_3	−89	30
CH_3NH_2	−6.3	31
CH_3OH	65	32

Explain why their boiling points are so different even though they have similar molecular masses.

29-71. The boiling points and molecular masses of 1-butanol, diethyl ether, and pentane are

Compound	Boiling point/°C	Molecular mass
$CH_3CH_2CH_2CH_2OH$	118	74
$CH_3CH_2OCH_2CH_3$	35	74
$CH_3CH_2CH_2CH_2CH_3$	36	72

Explain the relative values of these boiling points.

29-72. Alkoxides react with alkyl halides to produce ethers. For example,

$$Na^+CH_3O^- + CH_3Br \rightarrow CH_3OCH_3 + Na^+Br^-$$

The product here is called dimethyl ether. Ethers are compounds with the general formula ROR′, where R and R′ are alkyl groups. Devise a procedure to synthesize diethyl ether starting with ethylene and any inorganic substances. Diethyl ether was once used as an anesthetic.

29-73. An unknown gas was shown to have the molecular formula C_3H_6 from analysis of its combustion reaction. This gas reacts with HCl to give 2-chloropropane. What is the Lewis formula for the compound?

29-74. Animal fats and vegetable oils are esters of an alcohol called glycerol with long-chain acids called fatty acids. These esters are generally called glycerides. Write a Lewis formula for the glyceride product in the following reaction:

$$\underset{\text{palmitic acid}}{3CH_3(CH_2)_{14}COOH} + \underset{\text{glycerol}}{\begin{array}{c} CH_2OH \\ | \\ CHOH \\ | \\ CH_2OH \end{array}} \rightarrow 3H_2O + \text{a glyceride}$$

29-75. Gasoline is a mixture of hydrocarbons, primarily containing 5 to 10 carbon atoms. If we use octane as an example, calculate the volume of air at 20°C and 1 atm that is required to burn a tank of gasoline (say, 75 L). Air is 21 mol% oxygen. If ΔH°_{rxn} for the combustion of octane is $-48\ kJ\cdot g^{-1}$, calculate the energy produced when 75 L of octane is burned. Take the density of octane to be $0.80\ g\cdot mL^{-1}$.

29-76. Crude oil is traded by the barrel, and one barrel is equivalent to 42 U.S. gallons. Given that the average heat of combustion of hydrocarbons is about $-47\ kJ\cdot g^{-1}$, and that their average density is about $0.80\ g\cdot mL^{-1}$, calculate the available energy in one barrel of crude oil (1 gallon = 4 quarts and 1.06 quarts = 1 liter).

29-77. A component of natural gas was completely burned in O_2 to give CO_2 and H_2O. From the analysis of the products, it was determined that the compound is 81.71 percent carbon and 18.29 percent hydrogen by mass. What is its empirical formula? If 0.75 g of the gas occupies 386 mL at 0°C and 750 torr, then determine the molecular formula of the gas. Name the gas.

29-78. Methyl ketones such as acetone, a widely used solvent of commercial importance, can be detected by the iodoform (CHI_3) test, in which the ketone is reacted with hypoiodite in aqueous base:

$$CH_3-\underset{\underset{O}{\|}}{C}-CH_3(aq) + 3IO^-(aq) \rightarrow$$

$$CH_3-C\begin{array}{l} \nearrow O \\ \searrow O^- \end{array}(aq) + \underset{\text{yellow}}{CHI_3(s)} + 2OH^-(aq)$$

A 5.00-mL sample of acetone ($d = 0.792\ g\cdot mL^{-1}$) was allowed to react with excess hypoiodite under basic conditions. After filtration, washing, and drying, the iodoform product was found to weigh 15.6 g. Calculate the theoretical yield and the percentage yield (Chapter 5).

29-79. Write the structural formula for and assign an IUPAC name to the product when each of the following compounds reacts with 1-butene:

(a) Cl_2 (addition)
(b) HCl
(c) H_2O (acid catalyst)
(d) H_2 (platinum catalyst)

29-80. Write the structural formula for and assign an IUPAC name to the product when each of the following compounds reacts with 2-pentene:

(a) Cl_2 (addition)
(b) HCl
(c) H_2O (acid catalyst)
(d) H_2 (platinum catalyst)

29-81. Write a structural formula for the product when HCl is added to each of the following alkenes:

(a) 3-chloro-1-butene
(b) 1-bromo-2-butene
(c) 2-methyl-1-propene
(d) 1-chloro-1-propene

29-82. Write a structural formula for the alcohol formed when H_2O is added to each of the following alkenes:

(a) 3-methyl-1-butene
(b) 2-methyl-2-butene
(c) 2-pentene
(d) 2-butene

29-83. Aldehydes can be reduced to primary alcohols by the strong reducing agent sodium borohydride, $NaBH_4$. Determine which aldehyde you would use to produce each of the following primary alcohols:

(a) 1-propanol
(b) 2-methyl-1-propanol
(c) 2,2-dimethyl-1-butanol

29-84. Indicate whether each of the following alkenes shows cis-trans isomerism:

(a) $CH_2{=}CHCH_2CH_3$
(b) $CH_3CH{=}CHCH_2CH_3$
(c) $CH_2{=}\underset{\underset{CH_3}{|}}{C}CH_2CH_3$

29-85. Write the Lewis formulas of the eight isomeric amines with the formula $C_4H_{11}N$. Designate each one as a primary, secondary, or tertiary amine.

29-86. Which compound has the higher boiling point, 1-aminobutane or 1-butanol? Explain your answer.

30 SYNTHETIC AND NATURAL POLYMERS

The formation of nylon by a condensation polymerization reaction at the interface of two immiscible solvents. The water layer contains the diamine monomer $H_2N(CH_2)_6NH_2$ plus $NaOH(aq)$, and the upper hexane layer contains the compound $\mathrm{Cl\underset{\underset{\displaystyle O}{\|}}{C}(CH_2)_4\underset{\underset{\displaystyle O}{\|}}{C}Cl}$. The small molecule that splits out in the reaction is HCl. The polymer forms at the interface between the two solutions and is pulled out by the tweezers as a strand.

Some molecules contain so many atoms (up to tens of thousands) that understanding their structure would seem to be an impossible task. By recognizing that many of these macromolecules exhibit recurring structural motifs, however, chemists have come to understand how these molecules are constructed, and further, how to synthesize them. These molecules, called polymers, fall into two classes: natural and synthetic. Natural polymers include many of the biomolecules that are essential to life: proteins, nucleic acids, and carbohydrates among them. Synthetic polymers—developed in just the last 50 or so years—include plastics, synthetic rubbers, and synthetic fibers.

Enormous industries have been built around synthetic polymer chemistry, which have profoundly changed the quality of life in the modern world. It is estimated that about half of all industrial research chemists are involved in some aspect of polymer chemistry. Few of us have not heard of nylon, rayon, polyester, polyethylene, polystyrene, Teflon, Formica, and Saran, all of which are synthetic polymers. The technological impact of polymer chemistry is immense and continues to increase.

The fields of biochemistry, medicine, and molecular biology have been profoundly influenced by discoveries in polymer chemistry as well. In exploring the relationship between the three-dimensional structures of biomolecules and their biological function, biochemists have elucidated how nerve impulses travel and how enzymes catalyze biological reactions and have discovered the molecular mechanisms underlying many diseases. An understanding of polymers helped elucidate how DNA and RNA molecules store and transmit genetic information and direct the synthesis of proteins. The understanding of the structure and function of proteins stands as one of the greatest achievements of modern science and is still a highly active area of research.

Even though this chapter cannot possibly cover all the achievements of polymer chemistry, we hope that a brief introduction to a few synthetic and natural polymers will encourage you to pursue the subject further. We will first examine how a few synthetic polymers are constructed and how diverse properties arise from their structures. Then, as examples of natural polymers, we will briefly consider the structure and function of proteins and DNA.

30-1. POLYMERS ARE COMPOSED OF MANY MOLECULAR SUBUNITS JOINED END TO END

The simplest very large molecule, or **macromolecule,** is polyethylene. Polyethylene is formed by joining many ethylene molecules end to end. The repeated addition of small molecules to form a long, continuous chain is called **polymerization,** and the resulting chain is called a **polymer** (poly = many; mer = unit). The small molecules or units from which polymers are synthesized are called **monomers.**

The polymerization of the monomer ethylene can be initiated by a free radical, such as HO·, the hydroxyl radical. (Recall that a free radical is a species having one or more unpaired electrons.) The first step in the polymerization of ethylene is

$$HO\cdot + H_2C{=}CH_2 \rightarrow HOCH_2CH_2\cdot$$

The product is a free radical that can react with another ethylene molecule to give

$$HOCH_2CH_2\cdot + H_2C{=}CH_2 \rightarrow HOCH_2CH_2CH_2CH_2\cdot$$

The product of this step is also a free radical that can react with another ethylene molecule:

$$HOCH_2CH_2CH_2CH_2\cdot + H_2C{=}CH_2 \rightarrow HOCH_2CH_2CH_2CH_2CH_2CH_2\cdot$$

The product here is a reactive chain that can grow longer by the sequential addition of more ethylene molecules. The chain continues to grow until some **termination reaction,** such as the combination of two free radicals, occurs. The polyethylene molecules formed in this manner typically contain thousands of carbon atoms.

The polymerization of ethylene can be written schematically as

$$\underset{\text{monomer}}{nH_2C{=}CH_2} \rightarrow \underset{\text{polymer}}{-\!\!(CH_2CH_2)\!\!-_n}$$

The notation $-\!\!(CH_2CH_2)\!\!-_n$ means that the group enclosed in the parentheses is repeated n times; it also serves to identify the monomer unit. The free radical that initiates the polymerization reaction is not indicated, because n is large and thus the end group constitutes only a trivial fraction of the large polymer molecule. The precise number of monomer molecules incorporated into a polymer molecule is not important for typically large values of n. Polymer syntheses generally produce polymer molecules with a range of n values. The polymer properties are described in terms of the average value of n. It makes little difference whether a polyethylene molecule consists of 5000 or 5100 monomer units, for example.

Polyethylene is a tough, flexible plastic that is used in the manufacture of packaging films and sheets, wire and cable insulation, ice cube trays, refrigerator dishes, squeeze bottles, bags for foods and clothes, trash bags, and many other articles. Other well-known polymeric materials are made from other monomers. For example, Teflon is produced from the monomer tetrafluoroethylene:

$$n\,\underset{\text{tetrafluoroethylene}}{F_2C{=}CF_2} \rightarrow \underset{\text{Teflon}}{\left(-\overset{F}{\underset{F}{\overset{|}{\underset{|}{C}}}}-\overset{F}{\underset{F}{\overset{|}{\underset{|}{C}}}}-\right)_n}$$

Teflon is a tough, nonflammable, and exceptionally inert polymer with a slippery surface that is used for nonstick surfaces in pots and pans, and in electrical insulation, plastic pipes, and cryogenic bearings. Some other polymers that are produced from alkenes are polypropylene (indoor-outdoor carpeting, pipes, valves); polyvinyl chloride (PVC; pipes, floor tiles, records); Saran (food packaging, fibers); Orlon and Acrilan (fabrics); and polystyrene (Table 30-1).

Table 30-1 Common polymers prepared from substituted alkenes

Name	Structural unit	Name	Structural unit
polyethylene	$-(CH_2-CH_2)_n-$		
polypropylene	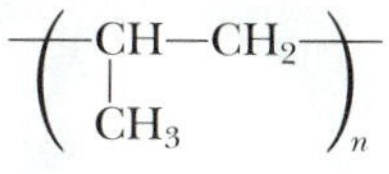 $-(CH(CH_3)-CH_2)_n-$	Teflon	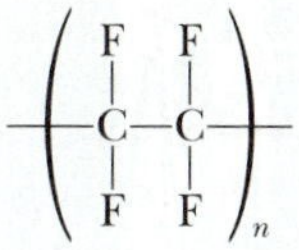 $-(CF_2-CF_2)_n-$
polyvinylchloride (PVC)	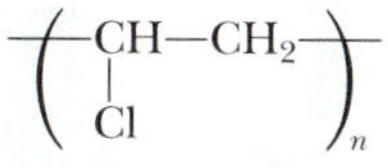 $-(CH(Cl)-CH_2)_n-$	Orlon, Acrilan	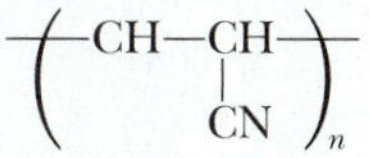 $-(CH-CH(CN))_n-$
Plexiglas	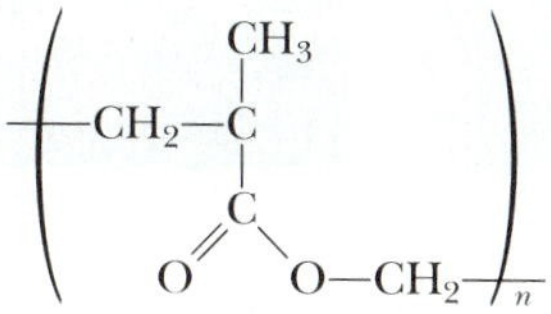 $-(CH_2-C(CH_3)(C(=O)O-CH_2))_n-$	polystyrene	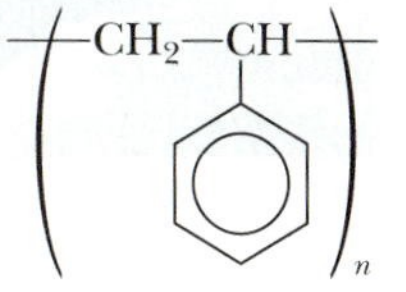 $-(CH_2-CH(C_6H_5))_n-$

30-2. NYLON AND DACRON ARE MADE BY CONDENSATION REACTIONS

The polymerization reaction of ethylene is called an **addition polymerization reaction** because it involves the direct addition of monomer molecules. Another type of polymerization reaction is a **condensation polymerization reaction.** In such a reaction, a small molecule, such as water, is split out as each monomer is added to the polymer chain. The polymerization product is called a **condensation polymer.** The synthesis of nylon is an example of a condensation polymerization reaction. Nylon is formed by the reaction of a diamino compound, such as 1,6-diaminohexane,

$$H-N(H)-CH_2CH_2CH_2CH_2CH_2CH_2-N(H)-H$$

and a dicarboxylic acid, such as adipic acid (hexanedioic acid)

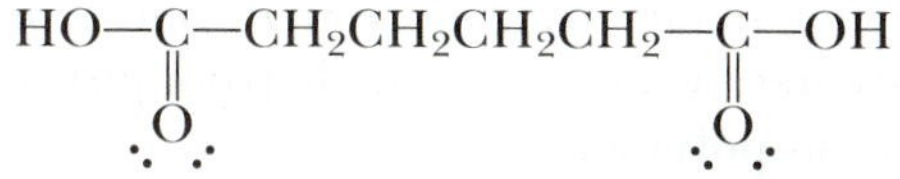

$$HO-C(=O)-CH_2CH_2CH_2CH_2-C(=O)-OH$$

These two molecules can be linked by the reaction

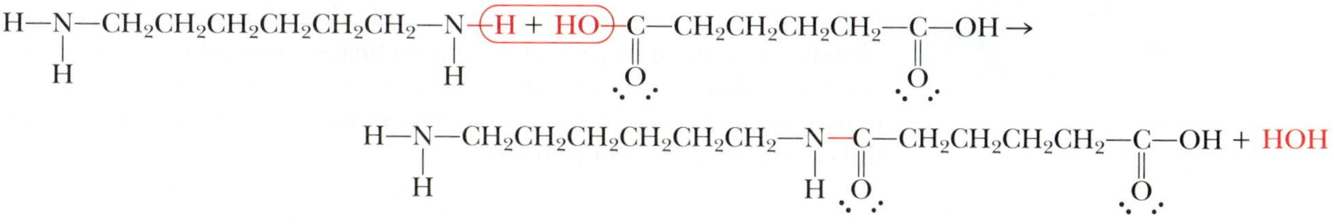

$$H-N(H)-CH_2CH_2CH_2CH_2CH_2CH_2-N(H)-H + HO-C(=O)-CH_2CH_2CH_2CH_2-C(=O)-OH \rightarrow$$

$$H-N(H)-CH_2CH_2CH_2CH_2CH_2CH_2-N(H)-C(=O)-CH_2CH_2CH_2CH_2-C(=O)-OH + HOH$$

The product in this reaction is called a **dimer** (di = two; mer = unit). Note that one end of the dimer is an amino group and the other end is a carboxyl group. The dimer can grow by the reaction of its amino end with a dicarboxylic acid monomer or by the reaction of its carboxyl end

and bisphenol A,

$$HO-C_6H_4-C(CH_3)_2-C_6H_4-OH$$

in the presence of pyridine, by splitting out two molecules of HCl per polymer unit. Determine the structural unit of Lexan.

Answer:

$$\left(-O-C_6H_4-C(CH_3)_2-C_6H_4-O-\overset{O}{\overset{\|}{C}}- \right)_n$$

Table 30-2 Annual reduction of some commercially important polymers

Polymer	Annual production/ millions of pounds
polyethylene	12400
polyvinylchloride	5700
polyester	4200
polypropylene	4000
polystyrene	3600
phenolics	2300
nylon	2300
polyesters	1000

Dacron, which is light and tough, is used to make clear films, skis, boat and aircraft components, surgical components, and permanent-press clothing. When used for clothing, it is usually blended in a roughly two-to-one ratio with cotton, because the resulting blended fibers are softer and pass moisture more readily than pure Dacron (Figure 30-3).

There are many other condensation polymers, including polycarbonates (used in safety helmets, lenses, electrical components, and photographic film), polyurethanes (insulation and furniture), and phenolics (brake linings and structural components). The essential feature in the synthesis of condensation polymers is the presence of reactive groups at both ends of the monomers (Table 30-2).

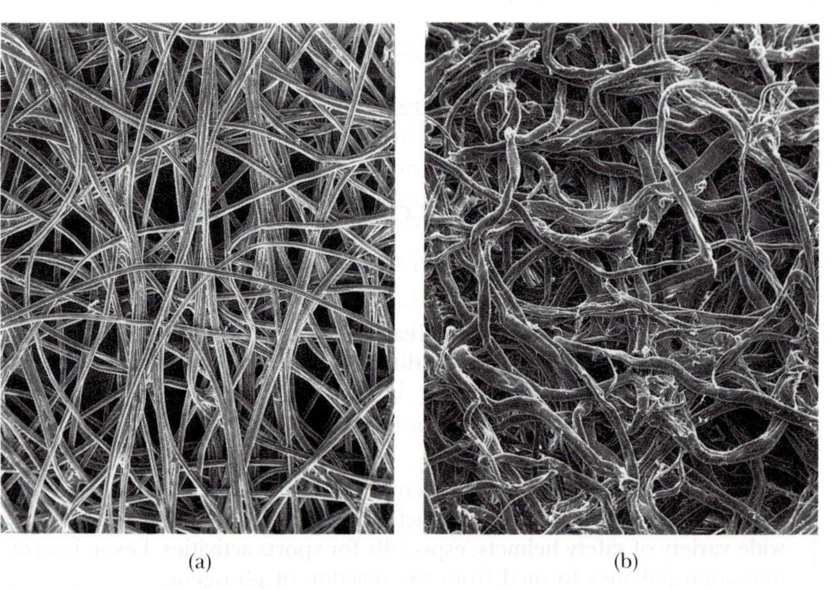

(a) (b)

Figure 30-3 (a) Synthetic Dacron fibers and (b) natural cotton fibers. A blend of these two fibers produces permanent-press clothing.

30-3. POLYMERS WITH CROSS-LINKED CHAINS ARE ELASTIC

Depending on the length of the polymer chains and on the temperature, a particular type of polymer may exist as a viscous liquid, a rubbery solid, a glass, or a partially crystalline solid. Generally, the longer the average length of the polymer chains, the less liquidlike the polymer at a given temperature. For chain lengths involving more than 10,000 bonds, liquid flow is negligible at normal temperatures but elasticlike deformations are possible.

The elasticity of polymers can be explained in terms of structure. The polymer chains in a sample are coiled and intertangled with one another. If the polymer is stretched, then the chains slowly untangle and the sample appears to flow. The relative movement of polymer chains that occurs when the sample is stretched can be decreased by connecting the polymer chains to one another through chemical bonds called **cross-links** (Figure 30-4a). When the cross-linked network is stretched, the coils become elongated (Figure 30-4b); but when the stress is released, the polymer network returns to its original coiled state. A cross-linked polymer that exhibits elastic behavior is called an **elastomer.** If the cross-links occur at average intervals of about 100 to 1000 bonds along the chain, then the polymer may be stretched to several times its unstretched length without breaking. The stretched polymer returns to its initial length when the force is released. The resistance of an elastomer to stretching can be increased by increasing the number of cross-links between chains. High elasticity is found in substances composed of long polymer chains joined by sparsely distributed cross-links, such as the polymer chains found in rubber bands.

Natural rubber is composed of chains of *cis*-1,4-isoprene units with an average chain length of 5000 isoprene units (Figure 30-5a). The major problem associated with natural rubber is tackiness, which causes particles of material to stick to the surface of the rubber. This problem was solved in 1839 by Charles Goodyear, who discovered that if sulfur is added to natural rubber and the mixture is heated (a process called **vulcanization**), then the rubber remains elastic but is much stronger and no longer tacky. The vulcanization of natural rubber involves the formation of —S—S— cross-links between the polyisoprene chains (Figure 30-5b).

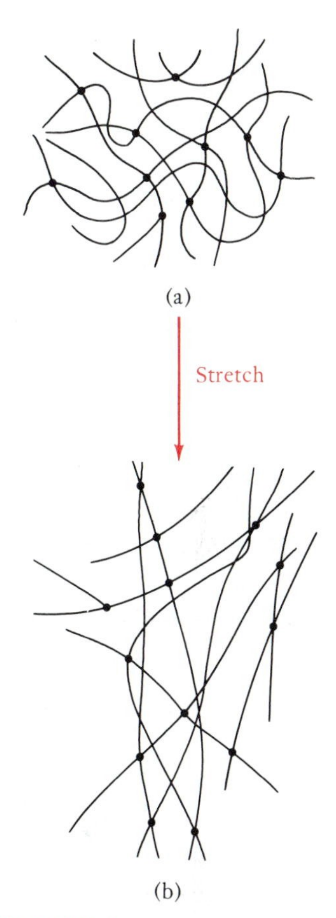

Figure 30-4 Schematic representation of a cross-linked polymer network. The cross-links between chains are represented by dots. (a) In the natural (unstretched) state, the polymer chains are coiled. (b) When the polymer is stretched, the chains become elongated.

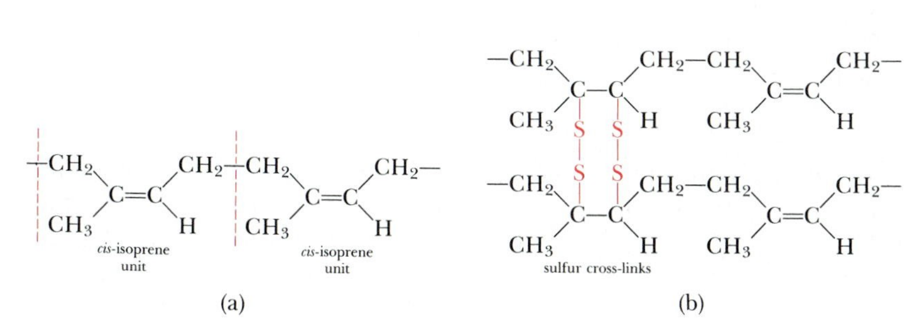

Figure 30-5 The vulcanization of rubber involves the formation of disulfide (—S—S—) cross-links between chains.

The U.S. sources of natural rubber were cut off by Japanese territorial expansions in 1941. As a result, research in the United States directed toward the production of synthetic rubber underwent a major expansion that culminated in the development of several varieties of synthetic rubber, including a product essentially identical to natural rubber. As with all rubbers, vulcanization is used to produce the cross-links that give rise to the desired degree of elasticity. The annual production of synthetic rubber in the United States exceeds 2 million metric tons.

EXAMPLE 30-2: An example of a synthetic rubber is polybutadiene, which is formed from the monomer butadiene, $H_2C{=}CHCH{=}CH_2$. Draw two polybutadiene chain segments that are cross-linked by vulcanization.

Solution: As in the case of the vulcanization of isoprene (Figure 30-5), the —S—S— cross-links result from the reaction of sulfur with two carbon-carbon double bonds in adjacent polymer chains:

```
—CH2         CH2CH2          CH2—
    \       /      \        /
     C — C          C = C
    / |  | \       /     \
   H  S  S  H     H       H
      |  |
   H  S  S  H     H       H
    \ |  | /       \     /
     C — C          C = C
    /       \      /     \
—CH2         CH2CH2       CH2—
```

PRACTICE PROBLEM 30-2: Chloroprene, $H_2C{=}C(Cl){-}CH{=}CH_2$, is polymerized to yield Neoprene, a rubber with excellent weather resistance and therefore used in automobile and high-quality garden hoses. Determine the polymer unit of Neoprene. (Hint: The resulting polymer is formed in an addition reaction and has one double bond per polymer unit.)

Answer:

```
 ⎛ —CH2          CH2— ⎞
 ⎜      \       /     ⎟
 ⎜       C  =  C      ⎟
 ⎜      /       \     ⎟
 ⎝    Cl         H    ⎠n
```

30-4. AMINO ACIDS ARE THE MONOMER UNITS OF POLYMERS CALLED PROTEINS

Proteins are natural polymers whose monomer units are **amino acids.** The word *protein* was coined in 1838 by the Swedish chemist Jöns Berzelius, drawing on the Greek word *proteios,* which means "of the first rank." As their name suggests, proteins are essential to life. Hemoglobin, which transports oxygen in the blood and hydrogen carbonate ions from cells, is a protein. Other globular (or roughly spherical) proteins act as catalysts (enzymes) in living things. The fibrous protein collagen provides the high tensile strength of skin and bone; other fibrous proteins include the antibodies that protect the body by combining with viruses, foreign bacteria, and cells from other organisms. All told, proteins constitute about 12 percent to 15 percent by mass of the human body.

The general formula of an amino acid is

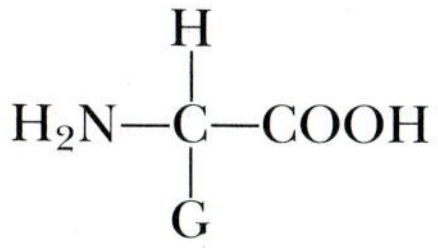

$$\begin{array}{c} \text{H} \\ | \\ \text{H}_2\text{N—C—COOH} \\ | \\ \text{G} \end{array}$$

These compounds are called amino acids because they contain both an amino group, $—NH_2$, and an acidic group, —COOH. Amino acids differ from one another only in the **side group** (denoted by G in the formula shown above) attached to the central carbon atom. A total of 20 amino acids are commonly found in proteins. At least some of these 20 amino acids are found in proteins at all levels of life, from the simplest bacteria to humans. Most natural proteins contain between 50 and 2000 of these monomer units, and the molecular mass of most protein polymer chains ranges from 550 to 220,000 $g \cdot mol^{-1}$.

Except for glycine, which is the simplest amino acid,

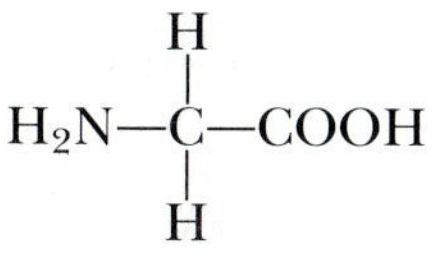

$$\begin{array}{c} \text{H} \\ | \\ \text{H}_2\text{N—C—COOH} \\ | \\ \text{H} \end{array}$$

glycine

all the amino acids have four different groups attached to the central carbon atom. For example, the structural formula for the amino acid alanine is

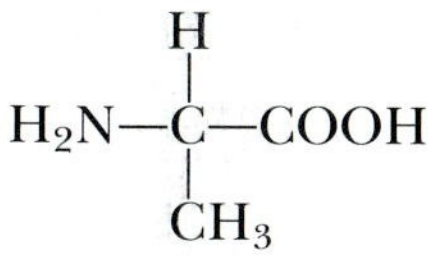

$$\begin{array}{c} \text{H} \\ | \\ \text{H}_2\text{N—C—COOH} \\ | \\ \text{CH}_3 \end{array}$$

alanine

Note that alanine has a methyl ($—CH_3$) side group. The side groups and the names of the corresponding amino acids are shown in Table 30-3.

The four bonds about the central carbon atom in an amino acid are tetrahedrally oriented, a geometry that can be represented as

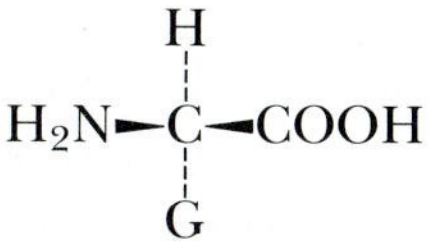

$$\begin{array}{c} \text{H} \\ \vdots \\ \text{H}_2\text{N}\blacktriangleright\text{C}\blacktriangleleft\text{COOH} \\ \vdots \\ \text{G} \end{array}$$

The dashed bonds indicate that —H and —G lie below the page; and the dark, wedge-shaped bonds indicate that the $—NH_2$ and —COOH groups lie above the page.

The amino acids display a type of isomerism that we have not studied so far in this text. They exist as **optical isomers,** which are nonsuperimposable isomers that are mirror images of each other. To get an idea of what a nonsuperimposable mirror image is, consider the mirror images

Table 30-3 The side groups and names of the 20 common amino acids of proteins

Side group	Amino acid
Nonpolar side groups	
$-H$	glycine (Gly)
$-CH_3$	alanine (Ala)
$-CHCH_3$ $\;\;|$ $\;\;CH_3$	valine (Val)
$-CH_2CHCH_3$ $\quad\;\;|$ $\quad\;\;CH_3$	leucine (Leu)
$-CHCH_2CH_3$ $\;\;|$ $\;\;CH_3$	isoleucine (Ile)
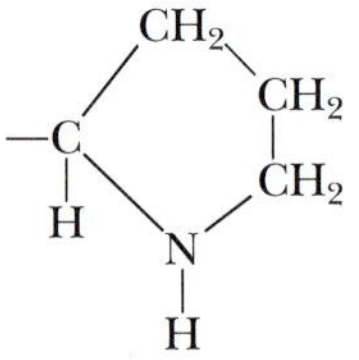 $-C(H)$ in ring with CH_2, CH_2, CH_2, $N-H$	proline (Pro)
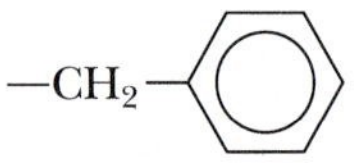 $-CH_2-C_6H_5$ (benzene ring)	phenylalanine (Phe)
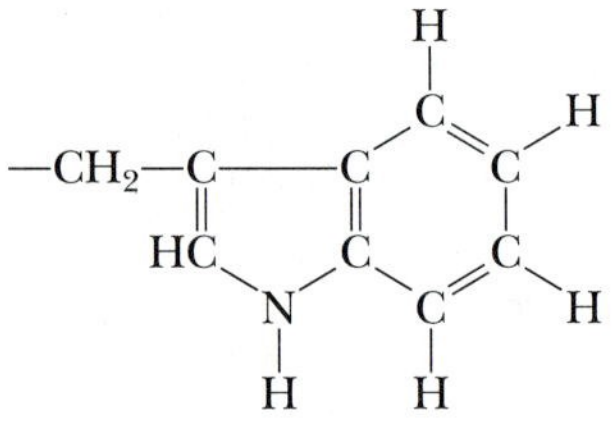 $-CH_2-C$ (indole ring: HC, $N-H$, $C-H$ ring carbons)	tryptophan (Trp)
$-CH_2CH_2SCH_3$	methionine (Met)
Uncharged polar side groups	
$-CH_2OH$	serine (Ser)
$-CHCH_3$ $\;\;|$ $\;\;OH$	threonine (Thr)
$-CH_2SH$	cysteine (Cys)
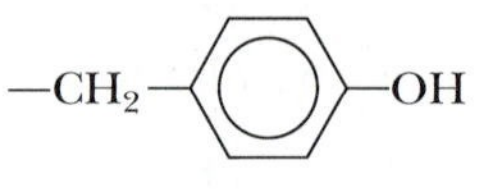 $-CH_2-C_6H_4-OH$ (benzene ring)	tyrosine (Tyr)
$-CH_2C(=O)-NH_2$	asparagine (Asn)
$-CH_2CH_2C(=O)-NH_2$	glutamine (Gln)
Acidic side groups	
$-CH_2C(=O)OH$	aspartic acid (Asp)
$-CH_2CH_2C(=O)OH$	glutamic acid (Glu)
Basic side groups	
$-CH_2CH_2CH_2CH_2NH_2$	lysine (Lys)
$-CH_2CH_2CH_2N(H)C(=NH)NH_2$	arginine (Arg)
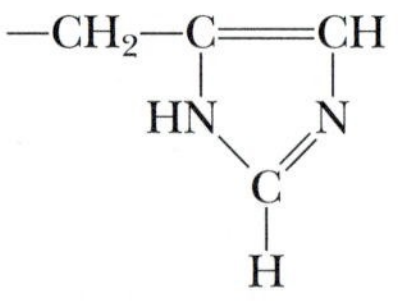 $-CH_2-C{=}CH$ (imidazole ring: HN, N, $C-H$)	histidine (His)

of the words MOM and DAD (Figure 30-6a). Whereas the mirror image of the word MOM is superimposable on the word MOM, the mirror image of the word DAD that is, ᗡAᗡ, is not superimposable on the word DAD. Similarly, your right hand is a mirror image of your left hand, and the two cannot be superimposed (Figure 30-6b). A molecule can exist in optically isomeric forms if the mirror image of the molecule cannot be superimposed onto itself. If we let the four groups attached to the central carbon atom in alanine be H, X, Y, and Z for simplicity, then we see

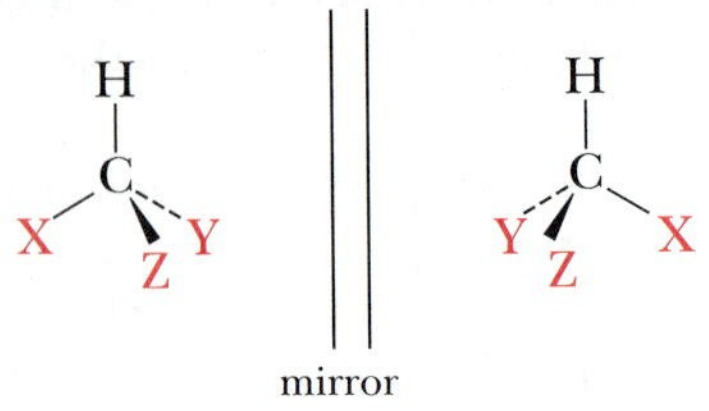

H, C, X, Y, Z | mirror | H, C, Y, Z, X

mirror

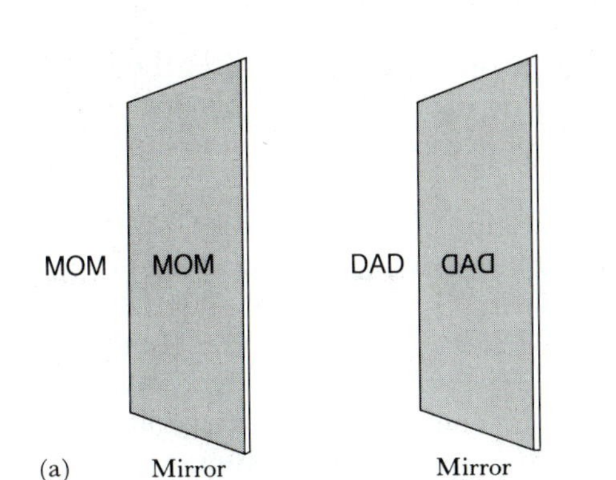

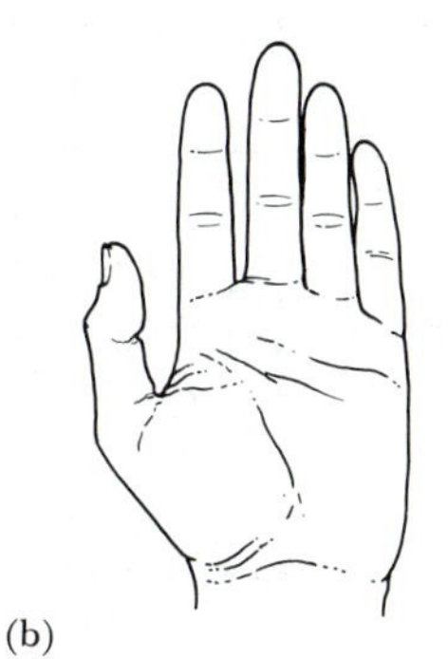

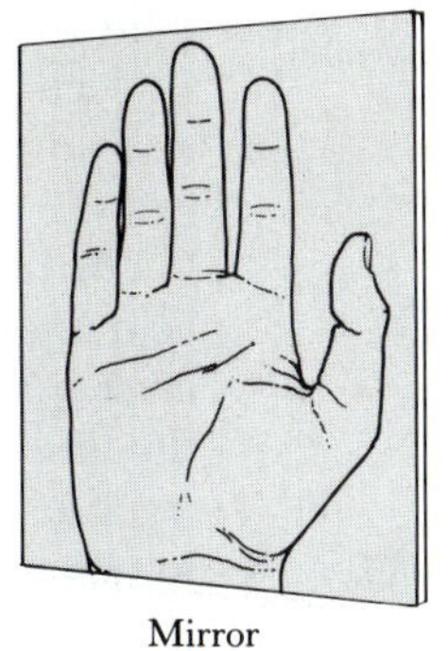

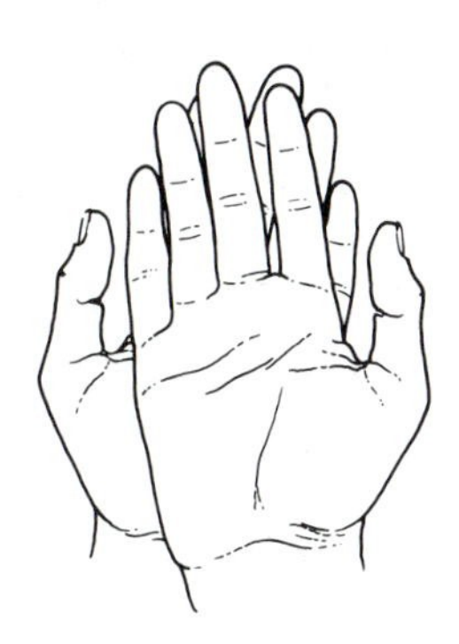

Figure 30-6 Mirror images may be superimposable or nonsuperimposable. (a) Mirror images of the words MOM and DAD. The mirror image of MOM is superimposable on the original, but the mirror image of DAD is not superimposable on the original. (b) Your two hands are an excellent example of nonsuperimposable mirror images. Your right hand is not superimposable on your left hand.

These two mirror images cannot be superimposed, just as a right hand cannot be superimposed onto a left hand. Two optical isomers are distinguished from each other by a D or L placed in front of the name of the amino acid. The D and L are derived from *dextro-* (right) and *levo-* (left).

$$\begin{array}{c} \text{H} \\ | \\ \text{H}_2\text{N} \blacktriangleleft \text{C} \blacktriangleright \text{COOH} \\ | \\ \text{CH}_3 \\ \text{D-alanine} \end{array} \qquad \begin{array}{c} \text{H} \\ | \\ \text{HOOC} \blacktriangleleft \text{C} \blacktriangleright \text{NH}_2 \\ | \\ \text{CH}_3 \\ \text{L-alanine} \end{array}$$

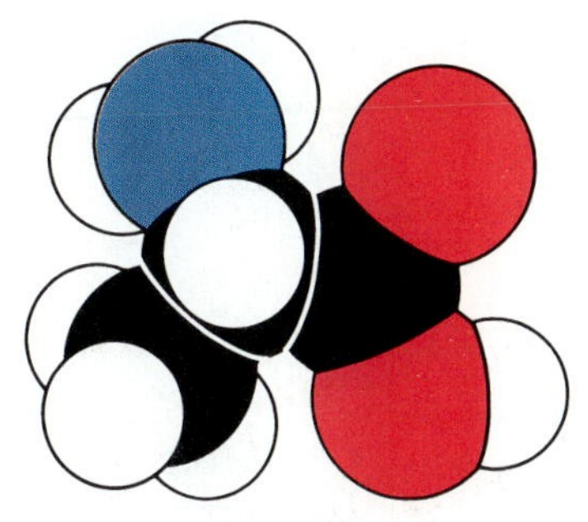

D-alanine

Optical isomers ordinarily display the same chemical properties; but, with very few exceptions, only the L isomers of the amino acids occur in biological systems. Biochemical reactions are exceptionally **stereospecific;** that is, they are exceptionally dependent on the shape of the reactants. Apparently life on earth originated from L amino acids; and once the process started, it continued to use only L isomers, which, unlike the D isomers, are recognized by enzymes.

L-alanine

30-5. PROTEINS ARE FORMED BY CONDENSATION REACTIONS OF AMINO ACIDS

Proteins are formed by condensation reactions similar to the reaction that results in the formation of nylon. The carboxyl group $\left(-\underset{\underset{\text{O}}{\|}}{\text{C}}-\text{O}-\text{H}\right)$ on one amino acid reacts with the amino group $\left(\text{H}-\underset{\underset{\text{H}}{|}}{\ddot{\text{N}}}-\right)$ on another, thereby forming a **peptide bond.** For example,

$$\text{H}_2\text{N}-\underset{\underset{\text{G}_1}{|}}{\overset{\overset{\text{H}}{|}}{\text{C}}}-\underset{\underset{\ddot{\text{O}}:}{\|}}{\text{C}}-\boxed{\text{OH} + \text{H}}-\underset{\underset{\text{H}}{|}}{\ddot{\text{N}}}-\underset{\underset{\text{G}_2}{|}}{\overset{\overset{\text{H}}{|}}{\text{C}}}-\text{COOH} \rightarrow \text{H}_2\text{N}-\underset{\underset{\text{G}_1}{|}}{\overset{\overset{\text{H}}{|}}{\text{C}}}-\underset{\underset{\ddot{\text{O}}:}{\|}}{\text{C}}\overset{\text{peptide bond (amide linkage)}}{-}\underset{\underset{\text{H}}{|}}{\ddot{\text{N}}}-\underset{\underset{\text{G}_2}{|}}{\overset{\overset{\text{H}}{|}}{\text{C}}}-\text{COOH} + \text{HOH}$$

The portion of the amino acid that remains in the chain after the water molecule is split out is called an **amino acid residue.** The product of the reaction is called a **dipeptide** because it contains *two* amino acid residues.

Further condensation reactions of a dipeptide with additional amino acid molecules produce a **polypeptide,** which is a polymer having amino acids as monomers. (Note that all proteins are polypeptides, but not all polypeptides are proteins. Proteins are naturally occurring polypeptides.) Polypeptides are thus long chains of amino acid residues joined together by peptide bonds. The chain to which the amino acid side groups are attached is called the **polypeptide backbone.** An example of a portion of a polypeptide is

$$
\begin{array}{cccccccccccccc}
 & & \mathrm{H} & & & & \mathrm{H} & & & & \mathrm{H} & & & \mathrm{H} \\
 & & | & & & & | & & & & | & & & | \\
- & \mathrm{N}- & \mathrm{C}- & \mathrm{C}- & \mathrm{N}- & & \mathrm{C}- & \mathrm{C}- & \mathrm{N}- & & \mathrm{C}- & \mathrm{C}- \mathrm{N}- & & \mathrm{C}-\mathrm{C}- \\
 & | & | & \| & | & & | & \| & | & & | & \| \quad | & & | \quad \| \\
 & \mathrm{H} & \mathrm{G_1} & \mathrm{O} & \mathrm{H} & & \mathrm{G_2} & \mathrm{O} & \mathrm{H} & & \mathrm{G_3} & \mathrm{O} \quad \mathrm{H} & & \mathrm{G_4} \quad \mathrm{O}
\end{array}
$$

a portion of a polypeptide

where the carbon atoms that are bonded to the amino acid side groups are shown in blue and the peptide bonds are shown in red.

As the number of amino acids in a polypeptide increases, it becomes unwieldy to write out the complete chemical formula of the polypeptide. For this reason, three-letter abbreviations are commonly used to designate the amino acids in a polypeptide chain (Table 30-3). For example, the amino acid alanine is designated Ala, whereas lysine is designated Lys.

EXAMPLE 30-3: Write an equation for the reaction between alanine and serine.

Solution: The side groups in alanine and serine are given in Table 30-3. The reaction between them can be written as

$$
\begin{array}{l}
\mathrm{H_2N{-}C(H)(CH_3){-}C({=}\ddot{O}){-}OH + H{-}\ddot{N}(H){-}C(H)(CH_2OH){-}COOH} \rightarrow \\
\qquad \mathrm{H_2N{-}C(H)(CH_3){-}C({=}\ddot{O}){-}\ddot{N}(H){-}C(H)(CH_2OH){-}COOH + HOH}
\end{array}
$$

This result is not the only one possible, however. A different dipeptide is formed when the carboxyl group on serine reacts with the amino group on alanine:

$$
\begin{array}{l}
\mathrm{H_2N{-}C(H)(HOCH_2){-}C({=}\ddot{O}){-}OH + H{-}\ddot{N}(H){-}C(H)(CH_3){-}COOH} \rightarrow \\
\qquad \mathrm{H_2N{-}C(H)(HOCH_2){-}C({=}\ddot{O}){-}\ddot{N}(H){-}C(H)(CH_3){-}COOH + HOH}
\end{array}
$$

Thus, we see that it is necessary to specify the order of the amino acids in a peptide.

Table 30-4 Number of amino acids in and formula mass of common proteins

Protein	Number of amino acids	Formula mass	Number of polypeptide chains
insulin (hormone)	51	5700	2
cobratoxin (snake toxin)	62	7000	1
myoglobin (carries oxygen in muscles)	153	16,900	1
keratin (wool protein)	204	21,000	1
actin (muscle protein)	410	46,000	1
hemoglobin (transports oxygen in bloodstream)	574	64,500	4
alcohol dehydrogenase (metabolism of ethanol)	748	80,000	2
γ-globulin (antibody)	1250	150,000	4
collagen (skin, tendons, cartilage)	3000	300,000	3

PRACTICE PROBLEM 30-3: Write out the chemical formula of the following tetrapeptide: (H_2N-end) Glu-Cys-Asp-Lys.

Answer:

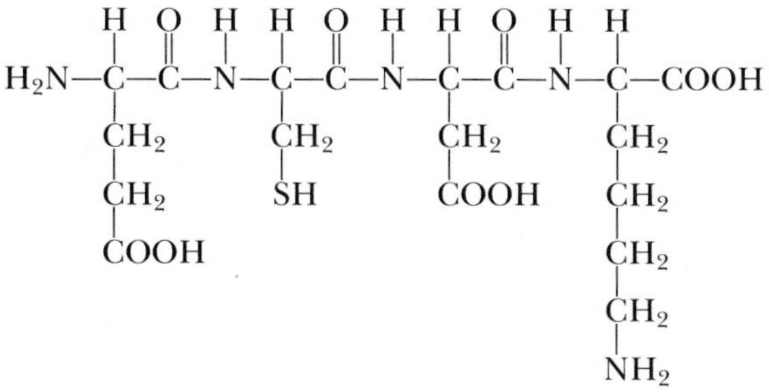

```
      H  O  H  H  O  H  H  O  H  H
      |  ‖  |  |  ‖  |  |  ‖  |  |
H2N—C—C—N—C—C—N—C—C—N—C—COOH
      |        |        |        |
     CH2      CH2      CH2      CH2
      |        |        |        |
     CH2      SH      COOH      CH2
      |                          |
     COOH                       CH2
                                 |
                                CH2
                                 |
                                NH2
```

Proteins are naturally occurring polypeptides. Each protein is characterized by a specific number and variety of amino acid units that occur in a specific order (sequence) along the polypeptide backbone. Table 30-4 lists some proteins and the number of amino acid units in each.

The sequence of the amino acid units in a polypeptide defines the **primary structure** of that polypeptide. The primary structure uniquely characterizes a protein. The primary structures of hundreds of proteins have been determined since the 1950s. Figure 30-7 shows the primary structure of the protein beef insulin, which is a polypeptide hormone that regulates carbohydrate metabolism. A deficiency of insulin in humans leads to *diabetes mellitus.*

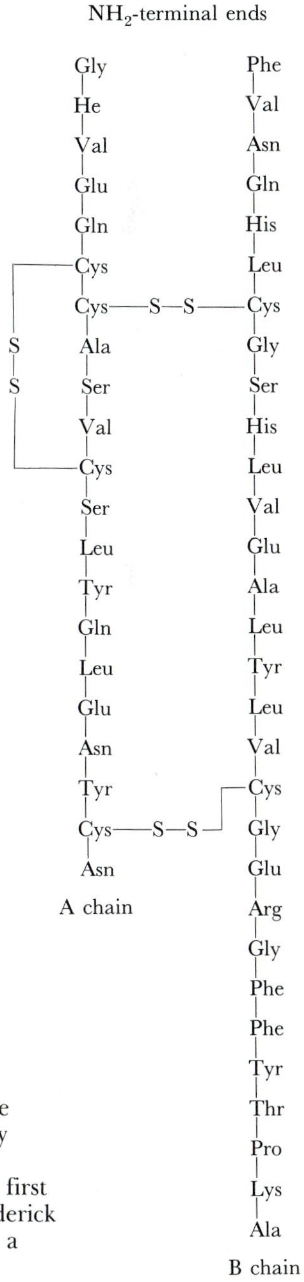

NH_2-terminal ends

A chain: Gly–Ile–Val–Glu–Gln–Cys–Cys–Ala–Ser–Val–Cys–Ser–Leu–Tyr–Gln–Leu–Glu–Asn–Tyr–Cys–Asn

B chain: Phe–Val–Asn–Gln–His–Leu–Cys–Gly–Ser–His–Leu–Val–Glu–Ala–Leu–Tyr–Leu–Val–Cys–Gly–Glu–Arg–Gly–Phe–Phe–Tyr–Thr–Pro–Lys–Ala

(Disulfide bridges: A-chain Cys(6)—S—S—Cys(11) within the A chain; A-chain Cys(7)—S—S—Cys(7) of the B chain; A-chain Cys(20)—S—S—Cys(19) of the B chain.)

Figure 30-7 The primary structure of the protein beef insulin. The amino acids are designated by standard three-letter abbreviations. The determination of the primary structure of a protein is like a complicated chemical jigsaw puzzle. The protein is hydrolyzed into shorter chains, which are separated and analyzed individually. The first primary structure determination was completed in 1953 by the British chemist Frederick Sanger, who received the 1958 Nobel Prize in chemistry for this work. He received a second Nobel Prize in 1980 for the sequencing of DNA.

30-6. THE SHAPE OF A PROTEIN MOLECULE IS CALLED ITS TERTIARY STRUCTURE

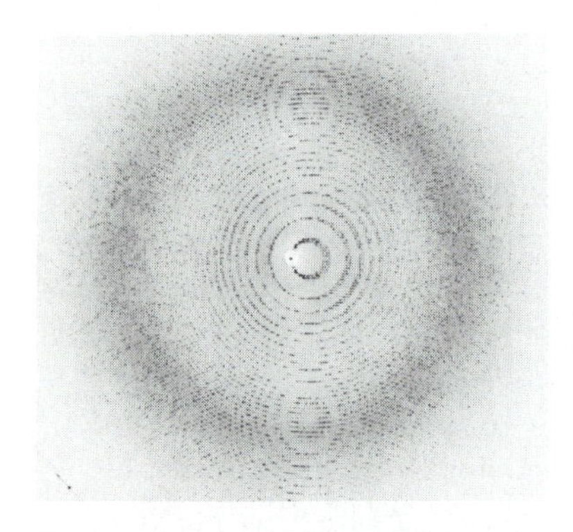

Figure 30-8 X-ray diffraction patterns like this one from a polio virus can be used to determine the structures of proteins and other biological molecules.

A key step in understanding how a particular protein functions is the determination of its shape. Because many proteins are extremely large molecules, this task is very complex. The definitive method for determining a protein's structure is X-ray crystallography. We saw in Chapter 14 that X-ray patterns can be used to determine the arrangement of atoms in crystalline solids. The X-ray patterns obtained from proteins, however, are more difficult to analyze and interpret because so many atoms are involved (Figure 30-8).

In the 1950s two American chemists, Linus Pauling and Robert B. Corey, were able to interpret X-ray patterns of proteins to show that many proteins have regions in which the chain twists into a helix (which is the shape of a spiral staircase). Pauling and Corey called the helix an ***α*-helix** (Figure 30-9). The helical shape results from the formation of hydrogen bonds between peptide linkages in the peptide chain. Individually, these hydrogen bonds are relatively weak, but collectively they combine to bend the protein chain into the *α*-helix. This coiled, helical

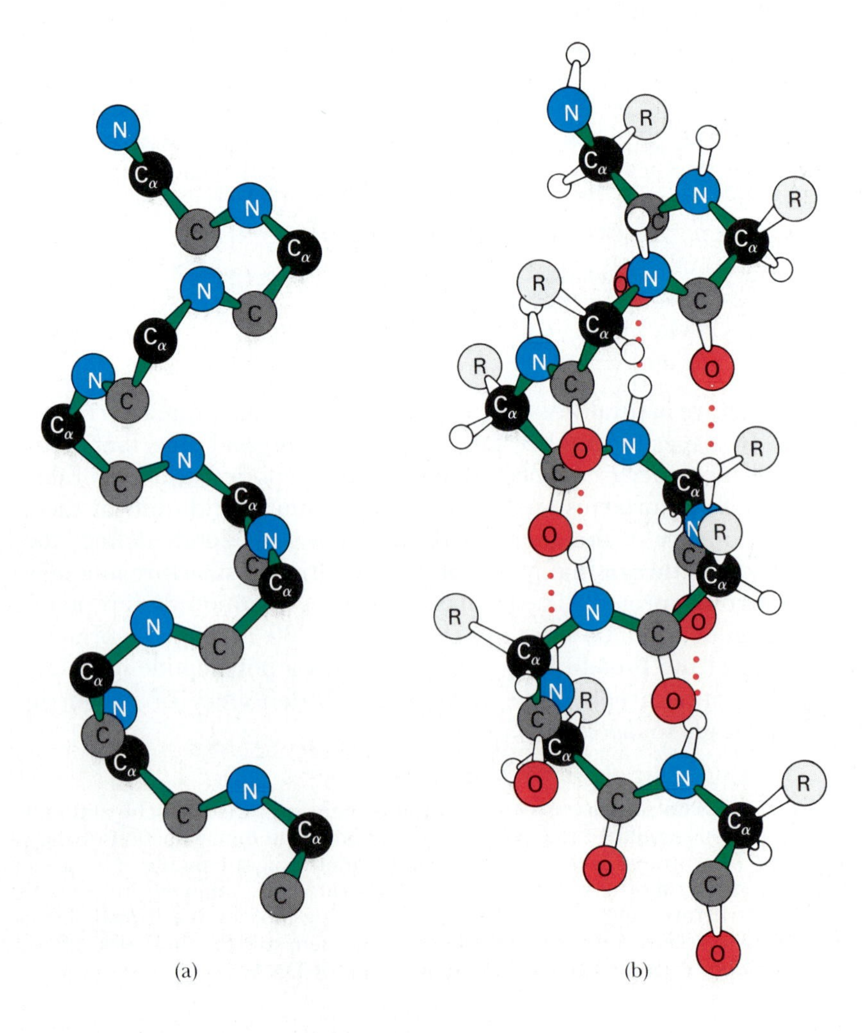

Figure 30-9 A segment of an *α*-helical region along a polypeptide chain. The chain is held in a helical shape by hydrogen bonds (dotted lines). The bond is formed between a hydrogen atom in one peptide bond and an oxygen atom in the fourth peptide bond further along the polypeptide chain; C refers to the carbon atoms to which the side groups, R, are attached. Part (a) shows just the backbone of the chain, whereas (b) shows the groups attached to the backbone and the hydrogen bonds.

shape in different regions of a protein chain is called **secondary structure.**

The shape of a protein molecule in water results from a complicated interplay between the amino acid side groups along the protein chain and the solvent, water. This interplay causes the protein to coil, fold, and bend into a three-dimensional shape called the **tertiary structure.** The tertiary structure of a protein is obtained from X-ray analysis. Because proteins play crucial roles in nearly all biological processes, biochemists are challenged to determine how amino acid sequences specify the conformations of proteins.

30-7. DNA IS A DOUBLE HELIX

The final class of biological polymers, or **biopolymers,** that we study in this chapter are the **polynucleotides.** The two most important polynucleotides are **DNA** (deoxyribonucleic acid) and **RNA** (ribonucleic acid). DNA occurs in the nuclei of cells and the genome of viruses and is the principal component of chromosomes (Figure 30-10). Genetic information that is passed from one generation to another is stored in DNA molecules. The discovery in 1953 of just how this is done has led to a revolution in biology that is as profound and far-reaching as the harnessing of nuclear energy in the 1940s and 1950s. The study of heredity, reproduction, and aging at the molecular level has produced an entirely new field of science—molecular biology—that has given birth to genetic engineering, with its awesome possibilities. In order to see how DNA can store and pass on information, we must examine its molecular structure.

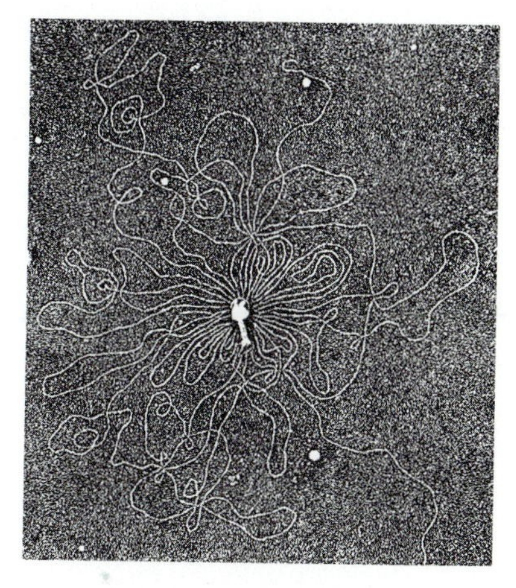

Figure 30-10 An electron micrograph of a virus particle that has burst and released strands of DNA. The long, cylindrical molecule is revealed beautifully in the photo.

DNA is a polynucleotide, by which we mean that it is a polymer made up of nucleotides. **Nucleotides,** the monomers of DNA and RNA, consist of three parts: a sugar (carbohydrate), a phosphate group, and a nitrogen-containing ring compound called a **base.** The sugar in DNA is 2-deoxyribose and that in RNA is ribose:

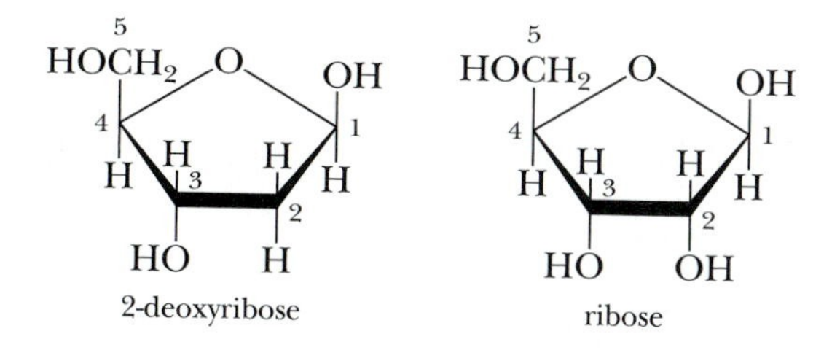

Carbon atoms, which are understood to constitute the vertices of these rings, are numbered 1 to 4 in the formulas. Notice that the difference between 2-deoxyribose and ribose is that 2-deoxyribose is lacking an oxygen atom at the number 2 carbon atom.

In both DNA and RNA, a phosphate group is attached to the number 5 carbon atom in the sugar as shown in the margin. The group labeled X is —OH in ribose and —H in 2-deoxyribose.

The five bases that occur in DNA and RNA are given in the margin of the next page. The bases are bonded to the ribose or deoxyribose rings by condensation reactions involving the hydrogen atoms shown in red on the bases in the margin and the —OH group on the number 1 carbon atom in the ribose and deoxyribose rings. For example, if thymine or adenine is the base, we have

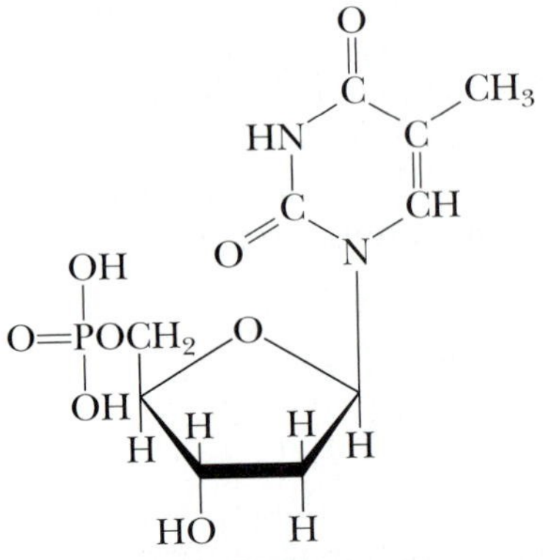

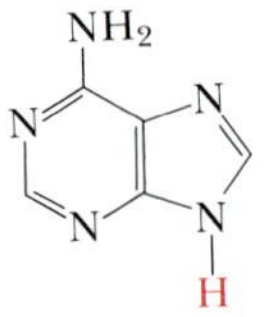

adenine

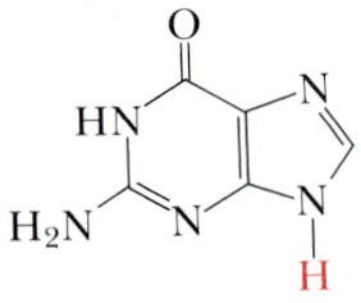

guanine

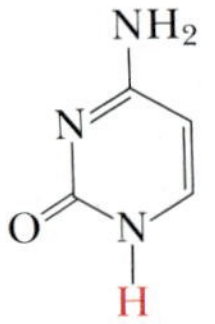

cytosine

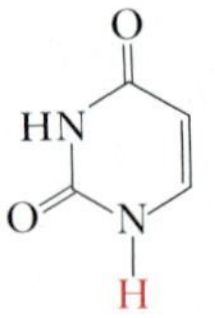

uracil

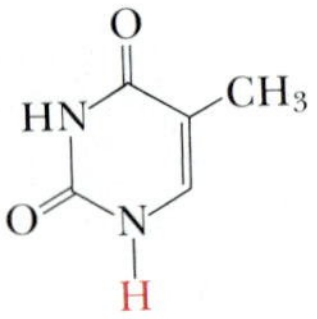

thymine

deoxythymidine 5-phosphate

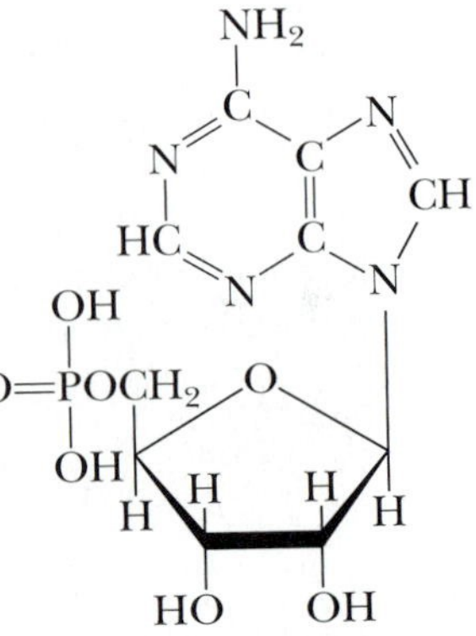

adenosine 5-phosphate

Both of these molecules are nucleotides. Deoxythymidine 5-phosphate is one of four monomers of DNA; and adenosine 5-phosphate is one of four monomers of RNA. DNA contains only the four bases adenine (A), guanine (G), cytosine (C), and thymine (T); and RNA contains adenine (A), guanine (G), cytosine (C), and uracil (U).

Nucleotides can be joined together by a condensation reaction between the phosphate group of one nucleotide and the 3-hydroxyl group of another. The result is a polynucleotide, part of which might look like this:

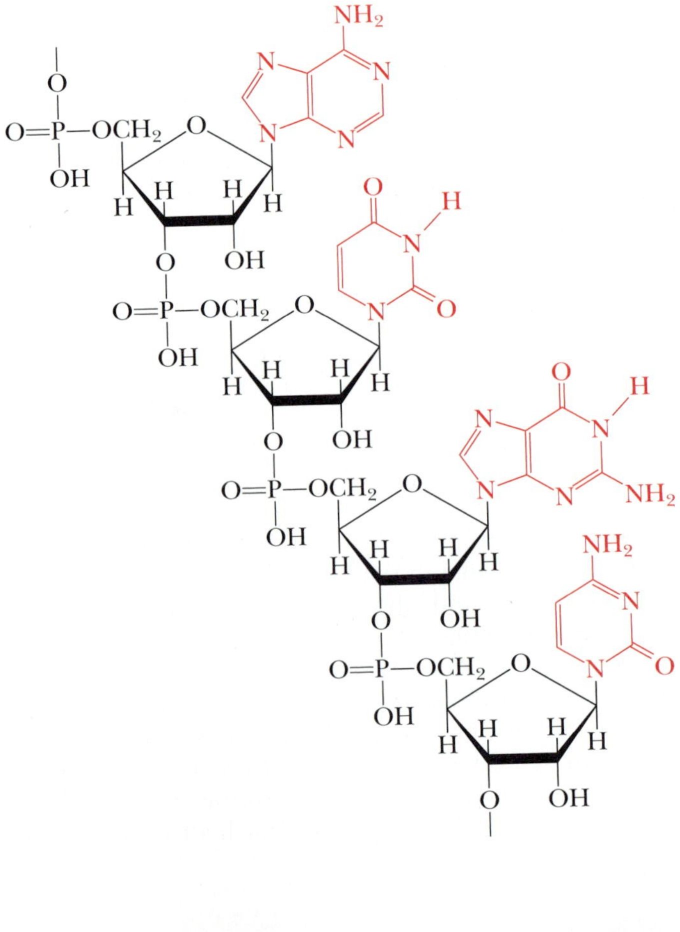

Figure 30-11 In the early 1950s, James Watson (left), who had recently received his Ph.D. in zoology from Indiana University, went to Cambridge University on a postdoctoral research fellowship. He and the British physicist Francis Crick (right) worked together on the molecular structure of DNA. In 1953 they proposed the double helix model of DNA, which explains elegantly how DNA can store and transmit genetic information. Their proposal is one of the most important scientific breakthroughs of modern times. Watson and Crick were awarded the Nobel Prize in physiology and medicine in 1962. The details of their discovery are given by Watson in his book *The Double Helix.*

Thus, we see that DNA and RNA consist of a **sugar-phosphate backbone** (shown in black) with bases attached at intervals (shown in red). Let's see now how a molecule like DNA can store and pass on genetic information.

The key to understanding how DNA works lies in its three-dimensional structure. In 1953, James Watson and Francis Crick (Figure 30-11) proposed that DNA consists of two polynucleotide chains intertwined in a **double helix** (Figure 30-12). Their proposal was based on two principal observations: X-ray data indicated that DNA is helical; and chemical analysis revealed that, regardless of the source of DNA, be it a simple bacterium or the higher vertebrates, the amount of guanine is always equal to the amount of cytosine and the amount of adenine is always equal to the amount of thymine.

Watson and Crick realized that the bases in DNA must somehow be paired. Working with molecular models, they discovered that adenine (A) and thymine (T) were of the right shape and size to form two hydrogen bonds:

CH_3 O···H—N
N—H···N
sugar sugar
1.1 nm
T A

Similarly, guanine (G) and cytosine (C) were found to form three hydrogen bonds:

N—H···O
N···H—N
O···H—N
sugar sugar
1.1 nm
C G

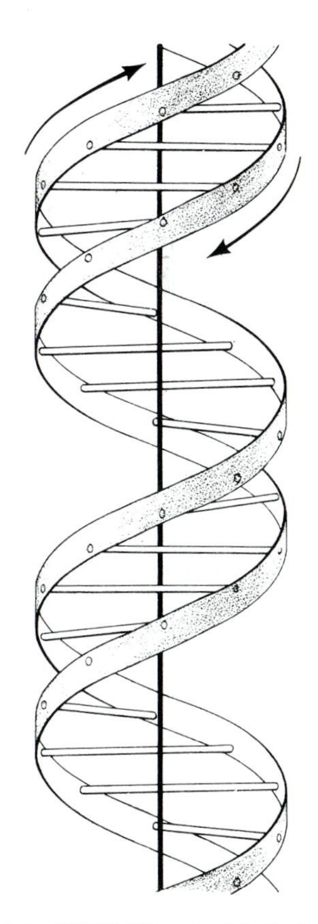

Figure 30-12 The double helix structure of DNA consists of two polynucleotide strands twisted about each other.

Notice that both the T—A and the C—G base pairs encompass a distance of 1.1 nm, thus allowing the two strands of the double helix to be evenly separated by 1.1 nm. Other possible base pairs, such as C—C, T—T, and C—T, encompass a narrower distance and A—A, G—G, and A—G, a wider one. Others that theoretically would be the right size (A—C and G—T) cannot pair because their atoms are not in suitable positions to form hydrogen bonds:

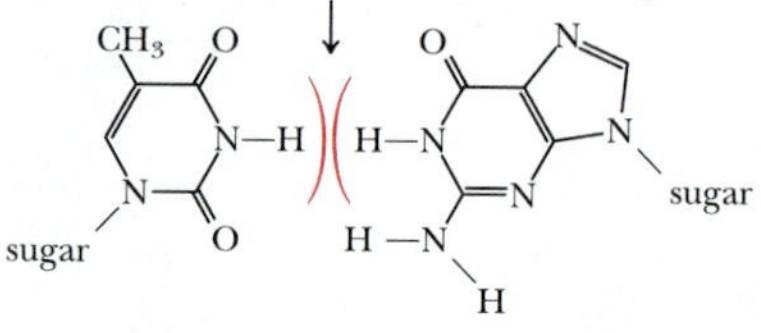

unfavorable for hydrogen bonding

T G

Thus, only A—T and G—C base pairs form, and this restricted base pairing accounts for the structure of DNA. The two strands of the double helix are said to be **complementary.**

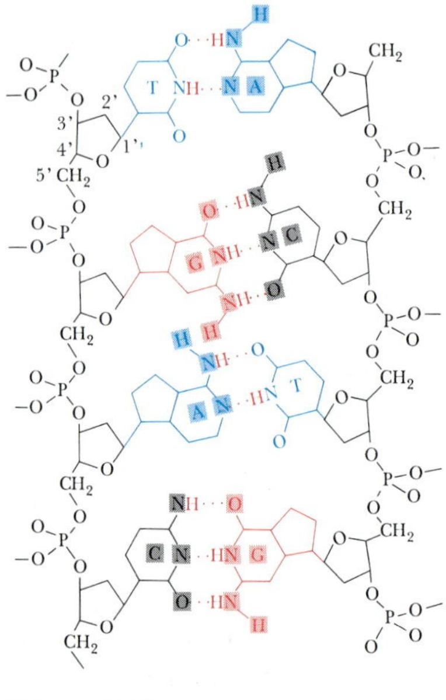

EXAMPLE 30-4: If the base sequence along a portion of one strand of a double helix is . . . AGCCTCG . . . , what must be the corresponding sequence on the other strand?

Solution: The two sequences must be complementary to each other, meaning that a T and an A must be opposite each other and a G and a C must be opposite each other. Thus, the other sequence must have the base sequence . . . TCGGAGC. . . .

PRACTICE PROBLEM 30-4: Determine the total number of hydrogen bonds between the seven base pairs in the double helix segment described in Example 30-4.

Answer: 19

Figure 30-13 Hydrogen bonds between complementary base pairs hold the two strands of DNA together in a double helix configuration.

The two strands of the DNA double helix are held together electrostatically by the hydrogen bonds between complementary bases (Figure 30-13).

30-8. DNA CAN DUPLICATE ITSELF

Hydrogen bonds, unlike covalent bonds, are weak enough to allow the double helix to uncoil into two separate strands at moderate temperatures (~40°C). Each strand can then act as a template for building a complementary strand, and the result is two double helices that are identical to the first. In this way genetic information is transmitted. Thus, the Watson-Crick model for DNA explains not only DNA structure but also DNA replication.

Living systems differ from one another by the myriad biochemical processes characteristic of each system. Almost all these biochemical reactions are controlled by enzymes, and many of them involve other proteins as well. In a sense, each living system is a reflection of its various proteins. What we mean by genetic information is the information that calls for the production of all the proteins characteristic of a given organism. This is the information stored in DNA.

Researchers discovered in the 1950s that each series of three bases along a DNA segment represents a code for binding to a particular amino acid during protein syntheses. For example, the triplet AAA is a DNA code for phenylalanine, and TTC is a DNA code for lysine. Thus, the segment AAATTC in DNA would give rise to a segment Phe-Lys in a protein. Because these code words are represented by a sequence of three bases, the code is called the **triplet code.**

Because the nature and the order of the bases are equivalent to genetic information, the bases are in the interior of the double helix for protection. A **gene** is a segment along a DNA molecule that codes the synthesis of one polypeptide. DNA can have a molecular mass of over 100,000,000.

The chemical reactions that are involved in transcribing the base sequence along a DNA strand into a protein molecule are complicated but fairly well understood. They involve several types of RNA and numerous enzymes and other proteins. If you go on to take a course in biochemistry or biology, you will study the DNA-protein pathway.

SUMMARY

Polymers are long, chainlike molecules that are formed by the bonding together of relatively small molecules called monomers. Monomers are the subunits, or links, of the molecular chains. Synthetic polymers of amazing diversity have been made by chemists in the last 50 years. All plastics, rubbers, synthetic fibers, and many other common materials are synthetic polymers.

Proteins are naturally occurring polymers whose monomers are the 20 common amino acids found in all living species. The order of the amino acids in a protein chain is called the primary structure of the chain and specifies the protein uniquely. A protein chain can be pictured as a polypeptide backbone with amino acid side groups attached at intervals. The side groups interact with each other and with solvent to effect a three-dimensional shape that is characteristic of the protein and is called its tertiary structure. The function and efficacy of a protein are extraordinarily sensitive to its tertiary structure.

Another class of biopolymers are polynucleotides. The polynucleotide DNA stores and transmits genetic information. The Watson and Crick double helical model pictures DNA as two complementary strands joined together by hydrogen bonds. This model explains how DNA can replicate itself and transmit genetic information from generation to generation. The genetic information is stored in DNA as a triplet code, defined by three sequential bases on the DNA strand. The codes spell out a sequence of amino acids during protein synthesis.

TERMS THAT YOU SHOULD KNOW

macromolecule 1063
polymerization 1063
polymer 1063
monomer 1063
termination reaction 1064
addition polymerization reaction 1065
condensation polymerization reaction 1065
condensation polymer 1065
dimer 1065
copolymer 1066
monopolymer 1066
cross-links 1069
elastomer 1069
vulcanization 1069
protein 1070
amino acid 1070
side group 1071

PROBLEMS

POLYMERS

30-1. What is the difference between an addition polymerization and a condensation polymerization? Give an example of each.

30-2. Sketch the equation for the reaction in which Orlon is produced. Is it an addition polymerization reaction or a condensation polymerization reaction?

30-3. Why do you think the nylon discussed in Section 30-2 is called nylon 66?

30-4. Polystyrene, which has many uses, including the manufacture of Styrofoam, has the formula

$$-\!\!\left(CH_2-\underset{\underset{C_6H_5}{|}}{CH}\right)\!\!-_n$$

Write the formula for the monomer.

OPTICAL ISOMERS

30-5. Indicate which of the following compounds can exist as optical isomers:

(a) CH_3Cl (b) $H_2N\underset{\underset{COOH}{|}}{C}HCH_2OH$

(c) $Cl-\overset{\overset{H}{|}}{\underset{\underset{Br}{|}}{C}}-COOH$ (d) $CH_3CH_2-\overset{\overset{CH_3}{|}}{\underset{\underset{Br}{|}}{Si}}-Cl$

30-6. Indicate which of the following compounds can exist as optical isomers:

(a) $Br-\overset{\overset{NH_2}{|}}{\underset{\underset{H}{|}}{C}}-COOH$ (b) CH_2Cl_2

(c) $(CH_3)_2SiCl_2$ (d) $F-\overset{\overset{Cl}{|}}{\underset{\underset{Br}{|}}{C}}-H$

POLYPEPTIDES

30-7. Write the equations for the reactions between tyrosine and valine.

30-8. Write the equations for the reactions between lysine and serine.

30-9. Draw the structural formulas for the two possible dipeptides that can be formed from the reaction between glycine and alanine.

30-10. Draw the structural formulas for the two possible dipeptides that can be formed from the reaction between valine and aspartic acid.

30-11. How many different tripeptides can be formed from two different amino acids? Draw the structural formula for each of your tripeptides.

30-12. How many different tripeptides can be formed from three different amino acids? Draw the structural formula for each of your tripeptides.

30-13. Draw the structural formula for the tripeptide Glu-Asp-Tyr.

30-14. Draw the structural formula for the tripeptide Glu-Val-Cys.

POLYNUCLEOTIDES

30-15. Draw the structural formula for the DNA triplet GAT.

30-16. Draw the structural formula for the DNA triplet ATC.

30-17. Draw the structural formula for the RNA triplet UCU.

30-18. Draw the structural formula for the RNA triplet CUG.

30-19. If the base sequence along a portion of one strand of a double helix is AAGTCTCGA, what must the corresponding sequence on the other strand be?

30-20. If the base sequence along a portion of one strand of a double helix is CATGGCTAA, what must the corresponding sequence on the other strand be?

30-21. Determine the complementary base sequence that corresponds to the following sequence of DNA bases:

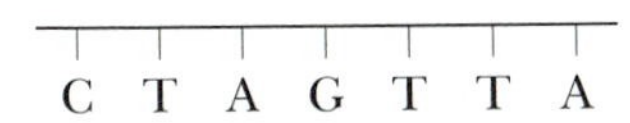

30-22. Determine the complementary base sequence that corresponds to the following sequence of DNA bases:

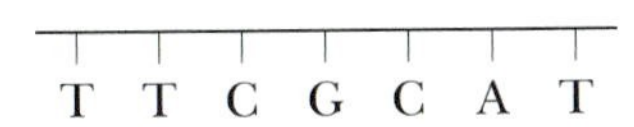

30-23. Suppose a segment along a double helix is

```
                              1
G  C  T  T  A  C  G

C  G  A  A  T  G  C
                              2
```

Draw the segments obtained when the DNA duplicates itself.

30-24. Suppose a segment along a double helix is

```
                              1
T  C  G  T  A  C  G

A  G  C  A  T  G  C
                              2
```

Draw the segments obtained when the DNA duplicates itself.

ENERGY AND BIOCHEMICAL REACTIONS

30-25. An important source of energy for cells is the combustion of glucose:

$$C_6H_{12}O_6(aq) + 6O_2(g) \rightarrow 6CO_2(g) + 6H_2O(l)$$
$$\Delta G^\circ_{rxn} = -2.87 \times 10^3 \text{ kJ at pH} = 7.0, 25^\circ\text{C}$$

Calculate the maximum amount of work that can be obtained from 1.0 g of glucose under standard conditions and pH = 7.0.

30-26. Plants synthesize carbohydrates from CO_2 and H_2O by the process of photosynthesis. For example,

$$6CO_2(g) + 6H_2O(l) \rightarrow \underset{\text{glucose}}{C_6H_{12}O_6(aq)} + 6O_2(g)$$

Calculate ΔG°_{rxn} for this reaction (see Problem 30-25). What is the equilibrium constant at 25°C for the chemical equation?

30-27. Given that at 25°C and pH = 7.0

$$\text{glucose}(aq) + 6O_2(g) \rightarrow 6CO_2(g) + 6H_2O(l) \qquad \Delta G^\circ_{rxn} = -2.87 \times 10^3 \text{ kJ}$$

$$\text{sucrose}(aq) + H_2O(l) \rightarrow \text{glucose}(aq) + \text{fructose}(aq) \qquad \Delta G^\circ_{rxn} = -29.3 \text{ kJ}$$

$$\text{fructose}(aq) \rightarrow \text{glucose}(aq) \qquad \Delta G^\circ_{rxn} = -1.6 \text{ kJ}$$

calculate the value of ΔG°_{rxn} for the combustion of sucrose. Calculate the maximum amount of work that can be obtained from the combustion of 1.0 g of sucrose under standard conditions.

30-28. An important biochemical reaction is the hydrolysis of adenosine triphosphate, ATP:

$$ATP(aq) + H_2O(l) \rightarrow ADP(aq) + HPO_4^{2-}(aq)$$
$$\Delta G^\circ_{rxn} = -29 \text{ kJ at pH} = 7.0 \text{ and } 25^\circ\text{C}$$

Calculate the value of ΔG°_{rxn} for the equation

$$\text{glucose}(aq) + 38HPO_4^{2-}(aq) + 38ADP(aq) + 6O_2(g) \rightarrow 6CO_2(g) + 44H_2O(l) + 38ATP(aq)$$

Calculate the equilibrium constant at 25°C for the equation. (See Problem 30-27.)

30-29. Much of the energy from the combustion of glucose is used to synthesize adenosine triphosphate, ATP. The hydrolysis of ATP provides the energy to drive unfavorable reactions, perform work, and carry out other functions of the cells. The equation is

$$ATP(aq) + H_2O(l) \rightarrow ADP(aq) + HPO_4^{2-}(aq)$$
$$\Delta G^\circ_{rxn} = -31 \text{ kJ (cellular conditions, } 37^\circ\text{C)}$$

Calculate the equilibrium constant at 37°C for this equation.

30-30. The hydrolysis of ATP provides the energy needed for the contraction of muscles. In a resting muscle, the concentration of ATP is 50 mM, the concentration of ADP is 0.5 mM, and the concentration of HPO_4^{2-} is 1.0 mM. Calculate the value of ΔG_{rxn} for the hydrolysis of ATP under these conditions. Take $\Delta G^\circ_{rxn} = -31$ kJ and a temperature of 37°C. (See Problem 30-29.)

30-31. Another important source of energy in biological systems is glycolysis, the process by which glucose is broken down to lactic acid:

$$C_6H_{12}O_6(aq) \rightarrow 2CH_3CHOHCOOH(aq)$$
$$\Delta G^\circ_{rxn} = -200 \text{ kJ at pH} = 7.0 \text{ and } 25^\circ\text{C}$$

Calculate the equilibrium constant at 25°C for the chemical equation.

30-32. The source of energy in human erythrocytes (red blood cells) is glycolysis. In erythrocytes the concentration of glucose is 5.0 mM and the concentration of lactic acid is 2.9 mM. Calculate the value of ΔG_{rxn} under these conditions at 25°C. See Problem 20-37 for the necessary data.

APPENDIX **A**

A Mathematical Review

A1. SCIENTIFIC NOTATION AND EXPONENTS

The numbers encountered in chemistry are often extremely large (such as Avogadro's number) or extremely small (such as the mass of an electron in kilograms). When working with such numbers, it is convenient to express them in **scientific notation,** where we write the number as a number between 1 and 10 multiplied by 10 raised to the appropriate power. For example, the number 171.3 is $1.713 \times 100 = 1.713 \times 10^2$ in scientific notation. Some other examples are

$$7320 = 7.32 \times 10^3$$
$$1{,}624{,}000 = 1.624 \times 10^6$$

The zeros in these numbers are not regarded as significant figures and are dropped in scientific notation. Notice that in each case the power of 10 is the number of places that the decimal point has been moved to the left:

7320. (3 places)   1624000. (6 places)

When numbers that are smaller than 1 are expressed in scientific notation, the 10 is raised to a negative power. For example, 0.614 becomes 6.14×10^{-1}. Recall that a negative exponent is governed by the relation

$$10^{-n} = \frac{1}{10^n} \qquad \text{(A1-1)}$$

Some other examples are

$$0.0005 = 5 \times 10^{-4}$$
$$0.000000000446 = 4.46 \times 10^{-10}$$

Notice that the power of 10 in each case is the number of places that the decimal point has been moved to the right:

0.0005 (4 places)   0.000000000446 (10 places)

It is necessary to be able to work with numbers in scientific notation. To add or subtract two or more numbers expressed in scientific notation, the power of 10 must be the same in both. For example, consider the sum

$$5.127 \times 10^4 + 1.073 \times 10^3$$

We rewrite the first number so that we have 10^3:

$$5.127 \times 10^4 = 51.27 \times 10^3$$

To change a number such as 51.27×10^3 to 5.127×10^4, we make the number in front one factor of 10 smaller; so we must make 10^3 one factor of 10 larger.

Note that we have changed the 10^4 factor to 10^3, so we must make the factor in front of 10^3 one power of 10 larger. Thus, we have

$$\begin{aligned} 5.127 \times 10^4 + 1.073 \times 10^3 &= (51.27 + 1.073) \times 10^3 \\ &= 52.34 \times 10^3 \\ &= 5.234 \times 10^4 \end{aligned}$$

Similarly, we have

$$(4.728 \times 10^{-6}) - (2.156 \times 10^{-7}) = (4.728 - 0.2156) \times 10^{-6} = 4.512 \times 10^{-6}$$

Note that in changing 2.156×10^{-7} to 0.2156×10^{-6}, we make 2.156 one factor of 10 smaller and 10^{-7} one factor of 10 larger.

When multiplying two numbers, we add the powers of 10 because of the relation

$$(10^x)(10^y) = 10^{x+y} \qquad \text{(A1-2)}$$

For example,

$$(5.00 \times 10^2)(4.00 \times 10^3) = (5.00)(4.00) \times 10^5 = 20.0 \times 10^5 = 2.00 \times 10^6$$

$$(3.014 \times 10^3)(8.217 \times 10^{-6}) = (3.014)(8.217) \times 10^{-3} = 24.77 \times 10^{-3} = 2.477 \times 10^{-2}$$

To divide, we subtract the power of 10 of the number in the denominator from the power of 10 of the number in the numerator because of the relation

$$\frac{10^x}{10^y} = 10^{x-y} \qquad \text{(A1-3)}$$

For example,

$$\frac{4.0 \times 10^{12}}{8.0 \times 10^{23}} = \left(\frac{4.0}{8.0}\right) \times 10^{12-23} = 0.50 \times 10^{-11} = 5.0 \times 10^{-12}$$

and

$$\frac{2.80 \times 10^{-4}}{4.73 \times 10^{-5}} = \left(\frac{2.80}{4.73}\right) \times 10^{-4+5} = 0.592 \times 10^1 = 5.92$$

To raise a number to a power, we use the fact that

$$(10^x)^n = 10^{nx} \qquad \text{(A1-4)}$$

For example,

$$(2.187 \times 10^2)^3 = (2.187)^3 \times 10^6 = 10.46 \times 10^6 = 1.046 \times 10^7$$

To take a root of a number, we use the relation

$$\sqrt[n]{10^x} = (10^x)^{1/n} = 10^{x/n} \qquad \text{(A1-5)}$$

Thus, the power of 10 must be written such that it is divisible by the root. For example,

$$\sqrt[3]{2.70 \times 10^{10}} = (2.70 \times 10^{10})^{1/3} = (27.0 \times 10^9)^{1/3} = (27.0)^{1/3} \times 10^3 = 3.00 \times 10^3$$

and

$$\sqrt{6.40 \times 10^5} = (6.40 \times 10^5)^{1/2} = (64.0 \times 10^4)^{1/2} = (64.0)^{1/2} \times 10^2 = 8.00 \times 10^2$$

A2. COMMON LOGARITHMS

You know that $100 = 10^2$, $1000 = 10^3$, and so on. You also know that

$$\sqrt{10} = 10^{1/2} = 10^{0.500} = 3.16$$

By taking the square root of both sides of

$$10^{0.500} = 3.16$$

we find that

$$\sqrt{10^{0.500}} = 10^{(1/2)(0.500)} = 10^{0.250} = \sqrt{3.16} = 1.78$$

Furthermore, because

$$(10^x)(10^y) = 10^{x+y}$$

we can write

$$10^{0.250} \times 10^{0.500} = 10^{0.750} = (3.16)(1.78) = 5.62$$

By continuing this process, we can express any number y as

$$y = 10^x \tag{A2-1}$$

The number x to which 10 must be raised to get y is called the **logarithm** of y and is written as

$$x = \log y \tag{A2-2}$$

Equations (A2-1) and (A2-2) are equivalent. For example, we have shown that

$$\begin{aligned} \log 1.78 &= 0.250 \\ \log 3.16 &= 0.500 \\ \log 5.62 &= 0.750 \\ \log 10.00 &= 1.000 \end{aligned}$$

Logarithms of other numbers may be obtained from tables or, more conveniently, from a hand calculator. If you use tables, you must always write the number y in standard notation. Thus, for example, you must write 782 as 7.82×10^2 and 0.000465 as 4.65×10^{-4}. To take the logarithm of such numbers, we use the fact that

$$\log ab = \log a + \log b \tag{A2-3}$$

Thus, we write

$$\begin{aligned} \log 782 &= \log (7.82 \times 10^2) = \log 7.82 + \log 10^2 \\ &= \log 7.82 + 2.000 \end{aligned}$$

Logarithm tables are set up such that the number a in $\log a$ is between 1 and 10 and the numbers in the tables are between 0 and 1. Thus, we find from any table of logarithms, for example, that

$$\log 4.12 = 0.6149$$

$$\log 8.37 = 0.9227$$

and so on. If we look up log 7.82, we find that it is equal to 0.8932. Therefore,

$$\begin{aligned} \log 782 = \log (7.82 \times 10^2) &= \log 7.82 + \log 10^2 \\ &= 0.8932 + 2.000 = 2.8932 \end{aligned}$$

If you use your calculator, you simply enter 782 and push a log key to get 2.8932 directly.

To find log 0.000465, we write

$$\begin{aligned} \log 0.000465 &= \log (4.65 \times 10^{-4}) \\ &= \log 4.65 + \log 10^{-4} \\ &= \log 4.65 - 4.000 \end{aligned}$$

We find $\log 4.65 = 0.6675$ from a table of logarithms, and so

$$\begin{aligned} \log 0.000465 &= 0.6675 - 4.000 \\ &= -3.3325 \end{aligned}$$

If you use your calculator, simply enter 0.000465 and push the LOG key to get −3.3325 directly. Although a hand calculator is much more convenient than a table of logarithms, you should be able to handle logarithms by either method.

Because logarithms are exponents ($y = 10^x$), they have certain special properties, such as

$$\log ab = \log a + \log b \tag{A2-3}$$

$$\log \frac{a}{b} = \log a - \log b \tag{A2-4}$$

$$\log a^n = n \log a \tag{A2-5}$$

$$\log \sqrt[n]{a} = \log a^{1/n} = \frac{1}{n} \log a \tag{A2-6}$$

If we let $a = 1$ in Equation (A2-4), then we have

$$\log \frac{1}{b} = \log 1 - \log b$$

or, because $\log 1 = 0$,

$$\log \frac{1}{b} = -\log b \tag{A2-7}$$

Thus, we change the sign of a logarithm by taking the reciprocal of its argument. Notice that because $\log 1 = 0$,

$$\log y > 0 \qquad \text{if } y > 1$$
$$\log y < 0 \qquad \text{if } y < 1$$

Up to this point we have found the value of x in

$$y = 10^x$$

when y is given. It is often necessary to find the value of y when x is given. Because x is called the logarithm of y, y is called the **antilogarithm** of x. For example, suppose that $x = 6.1303$ and we wish to find y. We write

$$y = 10^{6.1303} = 10^{0.1303} \times 10^6$$

From the logarithm table, we see that the number whose logarithm is 0.1303 is 1.35. Thus, we find that

$$10^{6.1303} = 1.35 \times 10^6$$

You can obtain this result directly from your calculator. On a calculator having INV and LOG keys, for example, enter 6.1303, press the INV key (for inverse) and then the LOG key.

To obtain the antilogarithm of y using logarithm tables, you must express y as

$$y = 10^a \times 10^n \tag{A2-8}$$

where n is a positive or negative integer and a is between 0 and 1. The quantity a is found in the table and the antilog of a is then read.

As another example, let's find the antilog of 1.9509. We write

$$y = 10^{1.9509} = 10^{0.9509} \times 10^1$$

We find the value 0.9509 in a log table and see that its antilog is 8.93. Thus, we have

$$y = 10^{1.9509} = 8.93 \times 10^1 = 89.3$$

You should be able to obtain this result directly from your calculator. If your calculator has a 10^x key, then you can obtain the antilog of 1.9509 by entering 1.9509 and pressing the 10^x key. This operation is equivalent to using the INV key followed by the LOG key.

In many problems, it is necessary to find the antilogarithm of negative numbers. For example, let's find the antilogarithm of -4.167, or the value of y in

$$y = 10^{-4.167}$$

Even though the exponent is negative, we still must express y in the form of Equation (A2-8). To do this, we write $-4.167 = 0.833 - 5.000$, so

$$y = 10^{0.833} \times 10^{-5}$$

Now we find 0.833 in a logarithm table and see that its antilogarithm is 6.81. Thus,

$$y = 6.81 \times 10^{-5}$$

You should be able to obtain this same result from your calculator by entering -4.167 and finding the inverse logarithm directly.

A3. NATURAL LOGARITHMS

The logarithms that we have just discussed in the previous section are called **common logarithms,** or **logarithms to the base 10.** The definitions of pH, pOH, and various pKs (as well as the Richter earthquake scale and the decibel sound scale, among others) are expressed in terms of common logarithms. Logarithms to another base arise naturally in calculus. The base is an irrational number called e:

$$e = 2.718281\ldots \tag{A3-1}$$

and the logarithms to this base, called **natural logarithms,** are designated by ln instead of log. Thus, we have

$$x = \ln y \tag{A3-2}$$

and its inverse

$$y = e^x \tag{A3-3}$$

Even if you have not taken a course in calculus, the functions $\ln y$ and e^x occur on all hand calculators nowadays. For example, by entering 2 and pushing the e^x key, you obtain

$$e^2 = 7.389056\ldots$$

By entering 2 followed by the change sign key and then the e^x key, you obtain

$$e^{-2} = 0.135335\ldots$$

Note that $e^{-2} = 1/e^2$, as you might expect. In fact, the mathematical properties of e^x and natural logarithms are similar to those of 10^x and $\log y$. For example,

$$\ln ab = \ln a + \ln b \tag{A3-4}$$

$$\ln \frac{a}{b} = \ln a - \ln b \tag{A3-5}$$

$$\ln a^n = n \ln a \tag{A3-6}$$

$$\ln \sqrt[n]{a} = \ln a^{1/n} = \frac{1}{n} \ln a \tag{A3-7}$$

and

$$e^{-x} = \frac{1}{e^x} \tag{A3-8}$$

$$e^a e^b = e^{a+b} \tag{A3-9}$$

$$\frac{e^a}{e^b} = e^{a-b} \tag{A3-10}$$

$$(e^x)^n = e^{nx}$$

Furthermore, from Equation (A3-10), with $a = b$ we get

$$e^0 = \frac{e^a}{e^a} = 1$$

so

$$\ln(1) = \ln e^0 = (0)\ln e = 0$$

Furthermore, from Equation (A3-5), with $a = 1$ we find that

$$\ln y > 0 \qquad \text{if } y > 1$$
$$\ln y < 0 \qquad \text{if } y < 1$$

just as in the case of common logarithms.

EXERCISE: Using your hand calculator, determine the following quantities:
(a) $e^{0.37}$ (b) $\ln(4.07)$
(c) $e^{-6.02}$ (d) $\ln(0.00965)$

Answer: (a) 1.45 (b) 1.404 (c) 2.430×10^{-3} (d) -4.640

EXERCISE: Given that (a) $\ln y = 3.065$ and (b) $\ln y = -0.605$, determine y.

Answer: (a) $y = e^{3.065} = 21.43$ and (b) $y = e^{-0.605} = 0.546$

A4. THE QUADRATIC FORMULA

The standard form for a quadratic equation in x is

$$ax^2 + bx + c = 0 \tag{A4-1}$$

where a, b, and c are consonants. The two solutions to the quadratic equation are

$$x = \frac{-b \pm \sqrt{b^2 - 4ac}}{2a} \tag{A4-2}$$

Equation (A4-2) is called the **quadratic formula** and is used to obtain the solutions to a quadratic equation expressed in the standard form.

For example, let's find the solutions to the quadratic equation

$$2x^2 - 3x - 1 = 0$$

In this case, $a = 2$, $b = -3$, and $c = -1$ and Equation (A4-2) gives

$$\begin{aligned} x &= \frac{3 \pm \sqrt{(-3)^2 - 4(2)(-1)}}{(2)(2)} \\ &= \frac{3 \pm 4.123}{4} \\ &= 1.781 \quad \text{and} \quad -0.2808 \end{aligned}$$

To use the quadratic formula, it is first necessary to put the quadratic equation in the standard form so that we know the values of the constants a, b, and c. For example, consider the problem of solving for x in the quadratic equation

$$\frac{x^2}{0.50 - x} = 0.040$$

To identify the constants a, b, and c, we must write this equation in the standard quadratic form. Multiplying both sides by $0.50 - x$ yields

$$\begin{aligned} x^2 &= (0.50 - x)0.040 \\ &= 0.020 - 0.040x \end{aligned}$$

Rearrangement to the standard quadratic form yields

$$x^2 + 0.040x - 0.020 = 0$$

Thus, $a = 1$, $b = 0.040$, and $c = -0.020$. Using Equation (A4-2), we have

$$x = \frac{-0.040 \pm \sqrt{(0.040)^2 - 4(1)(-0.020)}}{2(1)}$$

from which we compute

$$\begin{aligned} x &= \frac{-0.040 \pm \sqrt{0.0816}}{2} \\ &= \frac{-0.040 \pm 0.286}{2} \end{aligned}$$

Thus, the solutions for x are

$$x = \frac{-0.040 + 0.286}{2} = 0.123$$

and

$$x = \frac{-0.040 - 0.286}{2} = -0.163$$

If x represents, say, a concentration or gas pressure, then the only physically possible value is $+0.123$ because concentrations and pressures cannot have negative values.

A5. SUCCESSIVE APPROXIMATIONS

Many problems involving chemical equilibria lead to a quadratic equation of the form

$$\frac{x^2}{M_0 - x} = K \tag{A5-1}$$

where x is the concentration of a particular species, M_0 is an initial concentration, and K is an equilibrium constant. For example, the equation

$$\frac{[H_3O^+]^2}{0.100 - [H_3O^+]} = 2.19 \times 10^{-4}\ \text{M} \tag{A5-2}$$

arises if we wish to calculate $[H_3O^+]$ and the pH of a 0.100 M HCNO(*aq*) solution. If the value of K is small, then it is much more convenient to solve an equation like Equation (A5-1) by the **method of successive approximations** than by using the quadratic equation.

The first step in the method of successive approximations is to neglect the unknown in the denominator on the left-hand side of the equation. This step allows the unknown to be found by simply multiplying through by the initial concentration and taking the square root of both sides:

$$[H_3O^+]_1 \approx [(0.100\ \text{M})(2.19 \times 10^{-4}\ \text{M})]^{1/2} = 4.68 \times 10^{-3}\ \text{M} \tag{A5-3}$$

We have subscripted $[H_3O^+]$ with a 1 in this result because it represents a first approximation to $[H_3O^+]$. To obtain a second approximation, use the value of

$[H_3O^+]_1$ in the denominator of the left-hand side of Equation (A5-2), multiply both sides by the result in the denominator, and then take the square root:

$$[H_3O^+]_2 \simeq [(0.100\ M - 4.68 \times 10^{-3}\ M)(2.19 \times 10^{-4}\ M)]^{1/2}$$
$$= 4.57 \times 10^{-3}\ M$$

We now carry out the cycle, called an **iteration,** over again to obtain a third approximation:

$$[H_3O^+]_3 \simeq [(0.100\ M - 4.57 \times 10^{-3}\ M)(2.19\ 10^{-4}\ M)]^{1/2}$$
$$= 4.57 \times 10^{-3}\ M$$

Note that $[H_3O^+]_3 \simeq [H_3O^+]_2$; when this occurs, we say that the procedure has converged. After convergence is achieved, the same result will occur in any subsequent iteration, and the value obtained is the solution to the original equation because the equation is satisfied by the same value of $[H_3O^+]$ in the numerator as in the denominator.

The method of successive approximations is particularly convenient when you are using a hand calculator. Although it is usually necessary to carry out several iterations to obtain the solution, each cycle is easy to perform on a calculator, and the total effort involved usually is less than using the quadratic formula. In fact, as you use this method, think about the sequence of steps that you use on your calculator. Consider the equation

$$\frac{x^2}{0.250 - x} = 7.63 \times 10^{-4}$$

First, neglect x compared to 0.250, multiply through by 0.250, and take the square root to obtain

$$x_1 = 1.38 \times 10^{-2}$$

Now subtract x_1 from 0.250, multiply the result by 7.63×10^{-4}, and take the square root to obtain

$$x_2 = 1.34 \times 10^{-2}$$

One more iteration gives $x_3 = x_2$, and we say that the method has converged in three iterations.

Usually you will obtain convergence after only a few iterations. If you don't see the successive iterations approaching some value after just a few iterations, then probably it is more convenient to use the quadratic formula.

Here are some examples to practice with:

1. $\dfrac{x^2}{0.500 - x} = 1.07 \times 10^{-3}$ $(x_1 = 2.31 \times 10^{-2}, x_2 = 2.26 \times 10^{-2}, x_3 = 2.26 \times 10^{-2})$

2. $\dfrac{x^2}{0.0100 - x} = 6.80 \times 10^{-4}$ $(x_1 = 2.61 \times 10^{-3}, x_2 = 2.24 \times 10^{-3}, x_3 = 2.30 \times 10^{-3}, x_4 = 2.29 \times 10^{-3}, x_5 = 2.29 \times 10^{-3})$

3. $\dfrac{x_2}{0.150 - x} = 0.0360$ $(x_1 = 7.35 \times 10^{-2}, x_2 = 5.25 \times 10^{-2}, x_3 = 5.92 \times 10^{-2}, x_4 = 5.72 \times 10^{-2}, x_5 = 5.78 \times 10^{-2}, x_6 = 5.76 \times 10^{-2}, x_7 = 5.77 \times 10^{-2}, x_8 = 5.77 \times 10^{-2})$

Even in this last case, which requires eight iterations, the method of successive approximations is easier than using the quadratic formula.

A6. PLOTTING DATA

The human eye and brain are quite sensitive to recognizing straight lines, so it is always desirable to plot equations or experimental data such that a straight line is obtained. The mathematical equation for a straight line is of the form

$$y = mx + b \qquad \text{(A6-1)}$$

In this equation, m and b are constants: m is the **slope** of the line and b is its **intercept** with the y axis. The slope of a straight line is a measure of its steepness; it is defined as the ratio of its vertical rise to the corresponding horizontal distance.

Let's plot the two straight lines

$$\text{I} \quad y = x + 1$$
$$\text{II} \quad y = 2x - 2$$

We first make a table of values of x and y:

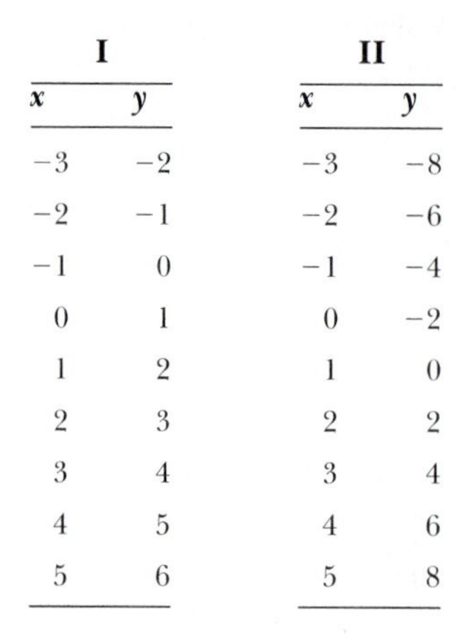

I			II	
x	y		x	y
−3	−2		−3	−8
−2	−1		−2	−6
−1	0		−1	−4
0	1		0	−2
1	2		1	0
2	3		2	2
3	4		3	4
4	5		4	6
5	6		5	8

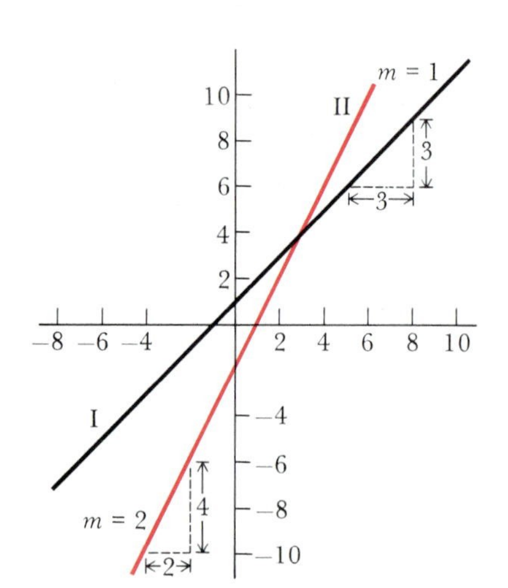

Figure A1 Plots of the equations (I) $y = x + 1$ and (II) $y = 2x - 2$.

These results are plotted in Figure A1. Note that curve I intersects the y axis at $y = 1$ ($b = 1$) and has a slope of 1 ($m = 1$). Curve II intersects the y axis at $y = -2$ ($b = -2$) and has a slope of 2 ($m = 2$).

Usually the equation to be plotted will not appear to be of the form of Equation (A6-1) at first. For example, consider the Boyle's law (Chapter 7) relation between the volume of a gas and its pressure:

$$V = \frac{c}{P} \qquad \text{constant temperature} \qquad \text{(A6-2)}$$

where c is a proportionality constant whose value depends on the temperature and the mass of a given sample. For example, for a 0.29-g sample of air, $c = 0.244$ L·atm at 25°C. Some results for such a sample are presented in Table A1, and the data in Table A1 are plotted as volume versus pressure in Figure A2.

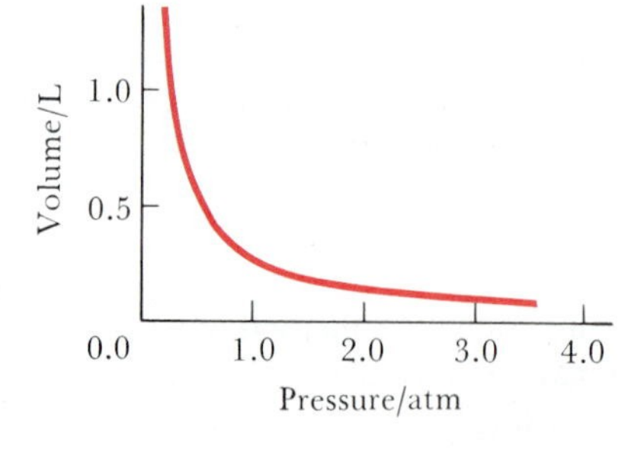

Figure A2 The volume of 0.29 g of air plotted versus pressure at 25°C. The data are given in Table A1. The curve in this figure is from the Boyle's law equation

$$V = \frac{0.244\ \text{L·atm}}{P}$$

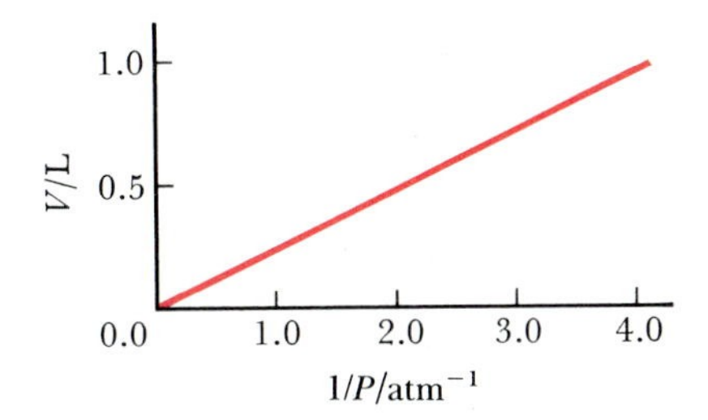

Figure A3 The volume of 0.29 g of air plotted against the reciprocal of the pressure ($1/P$) at 25°C. If we compare this curve with Figure A2, we see that a straight line results when V is plotted against $1/P$ instead of against P. Straight lines are much easier to work with than other curves, so it is usually desirable to plot equations and data in the form of a straight line.

It may appear at first sight that Equation (A6-2) is not of the form $y = mx$. However, if we let $V = y$ and $1/P = x$, then Equation (A6-2) becomes

$$y = cx$$

Thus, if we plot V versus $1/P$ instead of versus P, then we get a straight line. The data in Table A1 are plotted as V versus $1/P$ in Figure A3.

Table A1 Pressure-volume data for 0.29 g of air at 25°C

P/atm	V/L	$\frac{1}{P}$/atm^{-1}
0.26	0.938	3.85
0.41	0.595	2.44
0.83	0.294	1.20
1.20	0.203	0.83
2.10	0.116	0.48
2.63	0.093	0.38
3.14	0.078	0.32

EXAMPLE: Plot the equation

$$\ln P = -\frac{1640\ \text{K}}{T} + 10.560 \qquad \text{(A6-3)}$$

as a straight line. The quantity P is the pressure in units of torr, and T is the Kelvin temperature.

Solution: Comparing Equation (A6-3) to Equation (A6-1), we see that we can let

$$y = \ln P$$

$$x = \frac{1}{T}$$

which suggests that a straight line will result if we plot $\ln P$ versus $1/T$. Table A2 shows the numerical results and the following figure shows $\ln P$ plotted against $1/T$.

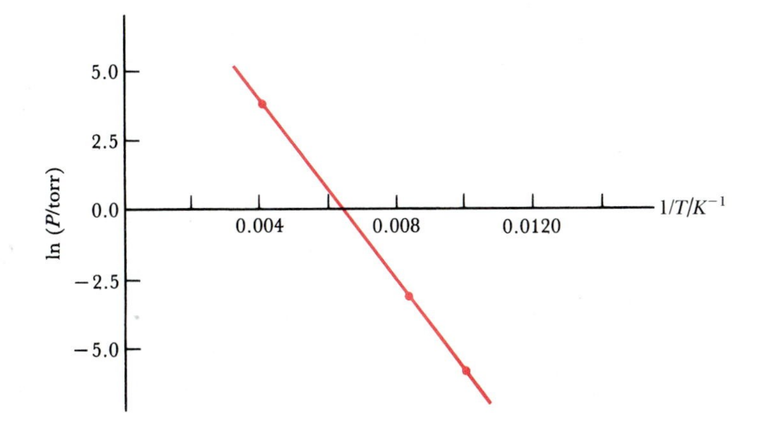

Table A2 Numerical results for plotting Equation (A6-3) as a straight line

T/K	$\frac{1}{T}/\frac{1}{K}$	$\ln P$
100	0.0100	−5.84
120	0.00833	−3.11
140	0.00714	−1.15
160	0.00625	+0.310
180	0.00556	1.45
200	0.00500	2.36
220	0.00456	3.11
240	0.00417	3.73

APPENDIX **B**

SI Units and Conversion Factors

Measurements and physical quantities in the sciences are expressed in the **metric system,** which is a system of units that was formalized by the French National Academy in 1790. There are several subsystems of units in the metric system; and in an international effort to achieve uniformity, the International System of Units (abbreviated SI from the French *Système Internationale d'Unites*) was adopted by the General Conference of Weights and Measures in 1960 as *the* recommended units for science and technology. The SI is constructed from the seven basic units given in Table B1. The first five units in Table B1 are used frequently in general chemistry. Each has a technical definition that serves to define the unit in an unambiguous, reproducible way, but here we simply relate the SI units to the English system.

In 1983, one meter was redefined as the distance that light travels through space in $^{1}/_{299,792,458}$ s.

1. Length: One meter is equivalent to 1.0936 yards, or to 39.370 in. Thus, a meter stick is about 3 in. longer than a yardstick.
2. Mass: One kilogram is equivalent to 2.2046 lb. The mass of a substance is determined by balancing it against a set of standard masses using a balance.
3. Temperature: The Kelvin temperature scale is related to the Celsius temperature scale. On the Celsius scale, the freezing point of water at 760 torr is set at 0°C and its boiling point at 760 torr is set at 100°C. The Kelvin and Celsius scales are related by the equation (Chapter 7)

$$\mathrm{T\ (in\ K)} = t\ (\mathrm{in\ °C}) + 273.15$$

   Recall that the freezing point of water is 32°F and its boiling point (at sea level) is 212°F on the Fahrenheit scale. The relation between the Celsius and Fahrenheit scales is given by

$$t\ (\mathrm{in\ °C}) = \frac{5}{9} t\ (\mathrm{in\ °F}) - 32 \qquad \text{(B-1)}$$

   Thus, for example, 50°F corresponds to 10°C and 86°F corresponds to 30°C. Note that the symbol for kelvin is K and not °K.
4. Amount of substance: One mole is the amount of substance that contains as many elementary entities as there are atoms in exactly 0.012 kg of carbon-12 (Chapter 5).

Table B1 **The seven SI basic units**

Physical quantity	Name of unit	Symbol
length	meter	m
mass	kilogram	kg
time	second	s
temperature	kelvin	K
amount of substance	mole	mol
electric current	ampere	A
luminous intensity	candela	cd

Table B2 **Prefixes used for multiples and fractions of SI units**

Prefix	Symbol	Multiple	Example
tera-	T	10^{12}	terawatt, 1 TW = 10^{12} W
giga-	G	10^{9}	gigavolt, 1 GV = 10^{9} V
mega-	M	10^{6}	megawatt, 1 MW = 10^{6} W
kilo-	k	10^{3}	kilometer, 1 km = 10^{3} m
deci-	d	10^{-1}	decimeter, 1 dm = 10^{-1} m
centi-	c	10^{-2}	centimeter, 1 cm = 10^{-2} m
milli-	m	10^{-3}	millisecond, 1 ms = 10^{-3} s
micro-	μ*	10^{-6}	microsecond, μs = 10^{-6} s
nano-	n	10^{-9}	nanosecond, 1 ns = 10^{-9} s
pico-	p	10^{-12}	picometer, 1 pm = 10^{-12} m
femto-	f	10^{-15}	femtometer, 1 fm = 10^{-15} m
atto-	a	10^{-18}	attomole, 1 amol = 10^{-18} mol

*This is the Greek letter mu, pronounced "mew."

Chemists can now measure processes that occur in one picosecond.

An important feature of the metric system and the SI is the use of prefixes to designate multiples of the basic units (Table B2).

The units of all quantities not listed in Table B1 involve combinations of the basic SI units and are called **derived units.** The derived units frequently used in general chemistry are given in Table B3. Many of these units may not be familiar to you unless you have had a course in physics. For example, the SI unit of force is a **newton** (N), which is defined as the force required to give a 1-kg body a speed of 1 $m{\cdot}s^{-1}$ when the force is applied for 1 s. The SI unit of pressure is the **pascal** (Pa). Pressure is force per area, and a pascal is defined as the pres-

Table B3 **Names and symbols for SI-derived units**

Quantity	Unit	Symbol	Definition
area	square meter	m^2	
volume	cubic meter	m^3	
density	kilogram per cubic meter	$kg{\cdot}m^{-3}$	
speed	meter per second	$m{\cdot}s^{-1}$	
frequency	hertz	Hz	s^{-1} (cycles per second)
force	newton	N	$kg{\cdot}m{\cdot}s^{-2}$
pressure	pascal	Pa	$N{\cdot}m^{-2} = kg{\cdot}m^{-1}{\cdot}s^{-2}$
energy	joule	J	$kg{\cdot}m^2{\cdot}s^{-2} = N{\cdot}m$
electric charge	coulomb	C	$A{\cdot}s$
electric potential difference	volt	V	$J{\cdot}A^{-1}{\cdot}s^{-1} = kg{\cdot}m^2{\cdot}s^{-3}{\cdot}A^{-1}$

sure produced by a force of 1 N acting on an area of 1 m^2. The SI unit of energy is the **joule** (J), which is the energy that a 1-kg mass has when it is traveling at a speed of 1 $m \cdot s^{-1}$. A joule is also the energy that a mass gains when it is acted upon by a force of 1 N through a distance of 1 m. Thus, we have $J = N \cdot m$.

Although the SI is gradually becoming the universally accepted system of units, there are a number of older units that are used frequently (Table B4). For example, volume is usually expressed in **liters** (L). A liter is defined as a cubic decimeter and is slightly larger than a quart, being equivalent to 1.0567 qt. The glassware in your laboratory is measured in milliliters (mL). One milliliter is equivalent to one cubic centimeter (cm^3).

The SI unit of pressure, the pascal, is rarely used in the United States. The most commonly used units of pressure are the **atmosphere** (atm) and the **torr** (Chapter 7).

We can use the relation between atmospheres and pascals to derive a relation between liter-atmospheres and joules. We use this relationship in Section 7-11. We start with

$$1 \text{ atm} = 101.32 \text{ kPa} = 1.0132 \times 10^5 \text{ Pa}$$

and multiply by L

$$1 \text{ L}\cdot\text{atm} = 1.0132 \times 10^5 \text{ L}\cdot\text{Pa}$$

Using the relations

$$\text{Pa} = \text{N}\cdot\text{m}^{-2} \qquad \text{J} = \text{N}\cdot\text{m} \qquad \text{L} = \text{dm}^3 = 10^{-3} \text{ m}^3$$

we obtain

$$\begin{aligned} 1 \text{ L}\cdot\text{atm} &= (1.0132 \times 10^5 \text{ N}\cdot\text{m}^{-2})(10^{-3} \text{ m}^3) \\ &= 101.32 \text{ N}\cdot\text{m} = 101.32 \text{ J} \end{aligned}$$

or writing this result as a unit conversion factor, we have

$$\frac{101.32 \text{ J}}{\text{L}\cdot\text{atm}} = 1$$

In particular, in Section 7-11 we need the relation (also see Problem 7-58)

$$0.08206 \text{ L}\cdot\text{atm} = (0.08206 \text{ L}\cdot\text{atm})(101.32 \text{ J}\cdot\text{L}^{-1}\cdot\text{atm}^{-1}) = 8.314 \text{ J}$$

The SI units and their conversion factors are given inside the back cover of this book.

Table B4 **Commonly used non-SI units**

Quantity	Unit	Symbol	SI definition
length	angstrom	Å	10^{-10} m
length	micron	μ	10^{-6} m = 1 μm
volume	liter	L	10^{-3} m^3
energy	calorie	cal	4.184 J
pressure	atmosphere	atm	101.325 kPa
pressure	torr	torr	133.322 Pa
pressure	bar	bar	10^5 Pa

APPENDIX C

Instructions for Building a Tetrahedron and an Octahedron

Tetrahedron: To construct a tetrahedron, trace out the following figure on a piece of light cardboard.

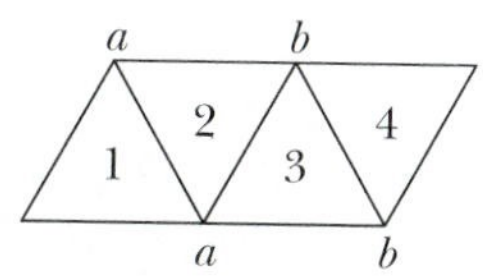

Bend face 1 upward about the line *aa*, and bend face 4 upward about the line *bb*:

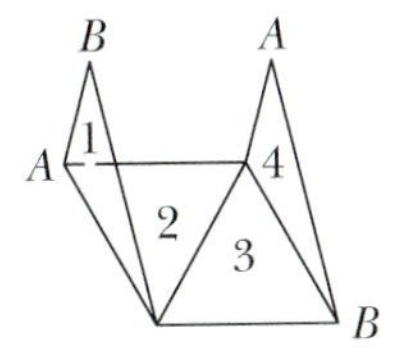

Now bend face 3 upward about the line between faces 2 and 3 and connect points *A* to *A* and *B* to *B* to get

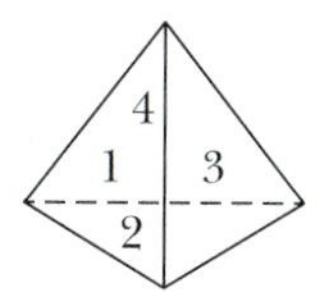

Octahedron: To construct an octahedron, trace out the following figure on a piece of light cardboard:

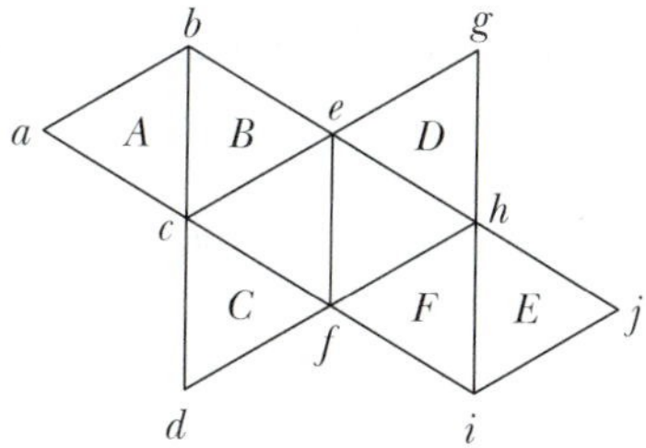

Bend face *A* upward about the line *bc* and face *C* upward about the line *cf*. Then bend face *B* upward about line *ce* and join points *a* and *d*. Lines *ac* and *cd* are now aligned and should be taped together. Now bend face *D* upward about the line *eh* and face *E* upward about the line *hi*. Then bend face *F* upward about the line *fh* and join points *g* and *j*. Lines *gh* and *hj* are now aligned and should be taped in place. Lastly, bend both sides upward about the line *ef* and tape the resulting octahedron together to get

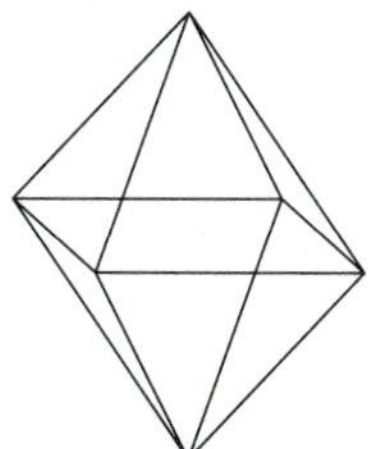

APPENDIX D

Symbols and Names of Elements with Atomic Numbers Greater Than 100

The elements with atomic numbers 93 and greater (called transuranium elements) are produced by particle accelerators such as cyclotrons; and traditionally, the research group that first produced an element assigned the name to that element. Conflict first arose when two groups claimed credit for the discovery of elements 104 and 105. For example, for element 105 a group in Berkeley, California, proposed the name rutherfordium, while another group, located in Dubna, USSR, proposed the name kurchatovium. Consequently, the International Union of Pure and Applied Chemistry (IUPAC) has recommended that newly discovered elements be assigned names that reflect their atomic numbers until the priority of their discovery has been determined. The symbols of these elements are formed according to their atomic numbers and consist of three letters, each being the first letter of the following roots:

0	nil	4	quad	8	oct
1	un	5	pent	9	enn
2	bi	6	hex		
3	tri	7	sept		

Thus, the symbols of the elements 104 through 110 are

104	Unq	106	Unh	108	Uno
105	Unp	107	Uns	109	Une
				110	Uun

To name an element, the three roots are joined in the order of the digits in the atomic number and the ending *ium* is added. Thus, for example, element 104 is called unnilquadium and element 109 is called unnilennium. For a further discussion, see the *Journal of Chemical Education* article by Mary V. Orna (*J. Chem. Educ.*, 59:123, 1982).

APPENDIX E

Table of Atomic Masses

Element	Symbol	Atomic number	Atomic mass*
actinium	Ac	89	(227)
aluminum	Al	13	26.98154
americium	Am	95	(243)
antimony	Sb	51	121.75
argon	Ar	18	39.948
arsenic	As	33	74.9216
astatine	At	85	(210)
barium	Ba	56	137.33
berkelium	Bk	97	(247)
beryllium	Be	4	9.01218
bismuth	Bi	83	208.9804
boron	B	5	10.81
bromine	Br	35	79.904
cadmium	Cd	48	112.41
calcium	Ca	20	40.08
californium	Cf	98	(251)
carbon	C	6	12.011
cerium	Ce	58	140.12
cesium	Cs	55	132.9054
chlorine	Cl	17	35.453
chromium	Cr	24	51.996
cobalt	Co	27	58.9332
copper	Cu	29	63.546
curium	Cm	96	(247)
dysprosium	Dy	66	162.50
einsteinium	Es	99	(252)
erbium	Er	68	167.26
europium	Eu	63	151.96
fermium	Fm	100	(257)
fluorine	F	9	18.998403
francium	Fr	87	(223)
gadolinium	Gd	64	157.25
gallium	Ga	31	69.72
germanium	Ge	32	72.59
gold	Au	79	196.9665
hafnium	Hf	72	178.49
helium	He	2	4.00260
holmium	Ho	67	164.9304
hydrogen	H	1	1.0079
indium	In	49	114.82
iodine	I	53	126.9045
iridium	Ir	77	192.22
iron	Fe	26	55.847
krypton	Kr	36	83.80
lanthanum	La	57	138.9055
lawrencium	Lr	103	(260)
lead	Pb	82	207.2
lithium	Li	3	6.941
lutetium	Lu	71	174.967

Element	Symbol	Atomic number	Atomic mass*
magnesium	Mg	12	24.305
manganese	Mn	25	54.9380
mendelevium	Md	101	(258)
mercury	Hg	80	200.59
molybdenum	Mo	42	95.94
neodymium	Nd	60	144.24
neon	Ne	10	20.179
neptunium	Np	93	(237)
nickel	Ni	28	58.70
niobium	Nb	41	92.9064
nitrogen	N	7	14.0067
nobelium	No	102	(259)
osmium	Os	76	190.2
oxygen	O	8	15.9994
palladium	Pd	46	106.4
phosphorus	P	15	30.97376
platinum	Pt	78	195.09
plutonium	Pu	94	(244)
polonium	Po	84	(209)
potassium	K	19	39.0983
praseodymium	Pr	59	140.9077
promethium	Pm	61	(145)
protactinium	Pa	91	(231)
radium	Ra	88	226.0254
radon	Rn	86	(222)
rhenium	Re	75	186.207
rhodium	Rh	45	102.9055
rubidium	Rb	37	85.4678
ruthenium	Ru	44	101.07
samarium	Sm	62	150.4
scandium	Sc	21	44.9559
selenium	Se	34	78.96
silicon	Si	14	28.0855
silver	Ag	47	107.868
sodium	Na	11	22.98977
strontium	Sr	38	87.62
sulfur	S	16	32.06
tantalum	Ta	73	180.9479
technetium	Tc	43	(98)
tellurium	Te	52	127.60
terbium	Tb	65	158.9254
thallium	Tl	81	204.37
thorium	Th	90	232.0381
thulium	Tm	69	168.9342
tin	Sn	50	118.69
titanium	Ti	22	47.90
tungsten	W	74	183.85
uranium	U	92	238.029
vanadium	V	23	50.9415
xenon	Xe	54	131.30
ytterbium	Yb	70	173.04
yttrium	Y	39	88.9059
zinc	Zn	30	65.38
zirconium	Zr	40	91.22

*A value given in parentheses denotes the mass number of the longest-lived isotope.

APPENDIX F

Standard Molar Entropies, Heats of Formation, Gibbs Free Energies of Formation, and Heat Capacities of Various Substances at 25°C and 1 atm

Substance	$S°$/J·K^{-1}·mol^{-1}	$\Delta H_f°$/kJ·mol^{-1}	$\Delta G_f°$/kJ·mol^{-1}	C_p/J·mol^{-1}·K^{-1}
aluminum				
$Al(s)$	28.3	0	0	24.35
$Al_2O_3(s)$	50.9	−1676	−1582	79.04
argon				
$Ar(g)$	154.7	0	0	20.786
barium				
$Ba(s)$	62.8	0	0	28.07
$BaCO_3(s)$	112.2	−1216	−1138	85.35
$BaO(s)$	70.2	−553.5	−525.1	47.78
$Ba^{2+}(aq)$	9.6	−537.64	−560.77	—
bromine				
$Br(g)$	174.9	111.9	82.43	20.786
$Br_2(g)$	245.4	30.91	3.14	36.02
$Br_2(l)$	152.2	0	0	75.689
$Br^-(aq)$	82.4	−121.55	−103.96	—
calcium				
$Ca(s)$	41.4	0	0	25.31
$CaC_2(s)$	69.9	−59.8	−64.8	62.76
$CaCO_3(s)$	92.9	−1207	−1129	81.88
$CaO(s)$	39.7	−635.1	−604.0	42.80
$CaSO_4(s)$	106.7	−1434.11	−1321.79	99.66
$Ca^{2+}(aq)$	−53.1	−542.83	−553.58	—
carbon				
$C(s,\text{diamond})$	2.377	1.90	2.900	6.113
$C(s,\text{graphite})$	5.74	0	0	8.527
$CH_4(g)$	186.2	−74.86	−50.75	35.309
$C_2H_2(g)$	200.8	226.7	209.2	43.93
$C_2H_4(g)$	219.6	52.28	68.12	43.56
$C_2H_6(g)$	229.5	−84.68	−32.89	52.63
$C_3H_8(g)$	269.9	−103.8	−23.49	
$C_6H_6(l)$	172.8	49.03	124.5	
$CH_3OH(l)$	126.9	−238.7	−166.3	81.6
$CH_3OH(g)$	239.81	−200.86	−161.96	43.89
$C_2H_5OH(l)$	160.8	−277.7	−174.8	111.46
$CH_3Cl(g)$	234.8	−80.83	−57.40	40.75
$CH_3Cl(l)$	145.3	−102	−51.5	
$CH_2Cl_2(g)$	270.4	−92.47	−65.90	50.96
$CH_2Cl_2(l)$	178.1	−121	−67.4	100.0
$CHCl_3(g)$	294.9	−103.1	−70.37	65.69
$CHCl_3(l)$	202.26	−134.5	−73.72	113.8
$CCl_4(g)$	308.7	−103.0	−60.63	83.30
$CCl_4(l)$	215.4	−135.4	−65.27	131.75
$CO(g)$	197.8	−110.5	−137.2	29.142
$CO_2(g)$	213.6	−393.5	−394.5	37.11

Substance	$S°/J \cdot K^{-1} \cdot mol^{-1}$	$\Delta H_f°/kJ \cdot mol^{-1}$	$\Delta G_f°/kJ \cdot mol^{-1}$	$C_p/J \cdot mol^{-1} \cdot K^{-1}$
chlorine				
$Cl(g)$	165.1	121.7	105.7	21.840
$Cl_2(g)$	222.9	0	0	33.907
$Cl^-(aq)$	56.5	−167.159	−131.228	—
copper				
$Cu(s)$	33.2	0	0	24.435
$CuO(s)$	43.6	−157	−130	42.30
$Cu_2O(s)$	93.1	−169	−146	63.64
$Cu^{2+}(aq)$	−99.6	64.77	65.49	—
fluorine				
$F(g)$	158.6	78.99	61.92	22.744
$F_2(g)$	203.7	0	0	31.30
$F^-(aq)$	−13.8	−332.63	−278.79	—
helium				
$He(g)$	126.0	0	0	20.786
hydrogen				
$H(g)$	114.6	218.0	203.3	20.784
$H_2(g)$	130.6	0	0	28.824
$H_2O(g)$	188.7	−241.8	−228.6	33.577
$H_2O(l)$	69.9	−285.8	−237.2	75.291
$H_2O_2(l)$	110.0	−187.8	−120.4	89.91
$HF(g)$	173.6	−271.1	−273	29.133
$HCl(g)$	186.8	−92.31	−95.30	29.12
$HBr(g)$	198.6	−36.4	−53.43	29.142
$HI(g)$	206.4	26.1	1.7	29.158
$H_2S(g)$	205.7	−20.1	−33.0	34.23
$H^+(aq)$	0	0	0	—
$OH^-(aq)$	−10.75	−230.0	−157.3	—
iodine				
$I(g)$	180.7	106.8	70.23	20.786
$I_2(g)$	260.6	62.4	19.36	36.90
$I_2(s)$	116.5	0	0	54.438
iron				
$Fe(s)$	27.3	0	0	25.10
$Fe_2O_3(s)$	87.9	−824.2	−742.2	103.85
$Fe_3O_4(s)$	146.3	−1118	−1015	143.43
krypton				
$K(g)$	164.0	0	0	20.786
magnesium				
$Mg(s)$	32.68	0	0	24.89
$MgO(s)$	26.94	−601.70	−569.43	37.15
$MgCO_3(s)$	65.7	−1095.8	−1012.1	75.52
$Mg^{2+}(aq)$	−138.1	−466.85	−454.8	—
neon				
$Ne(g)$	146.2	0	0	20.786
nitrogen				
$N(g)$	153.2	472.6	455.5	20.786
$N_2(g)$	191.5	0	0	29.125
$NH_3(g)$	192.5	−46.19	−16.64	35.06
$N_2H_4(l)$	121.0	50.6	149	98.87
$NO(g)$	210.6	90.37	86.69	29.844
$NO_2(g)$	240.4	33.85	51.84	37.30
$N_2O(g)$	220.2	81.55	103.6	38.45
$N_2O_4(g)$	304.3	9.66	98.29	77.28
$N_2O_4(l)$	209.2	19.50	97.54	142.7
$N_2O_5(s)$	113.1	−41.8	134.2	143.1
$NOCl(g)$	262	51.9	66.1	44.69
$NH_3(aq)$	111.3	−80.29	−26.50	—
$NH_4^+(aq)$	113.4	−132.51	−79.31	—
$NO_3^-(aq)$	146.4	−205.0	−108.74	—

Substance	$S°/J \cdot K^{-1} \cdot mol^{-1}$	$\Delta H_f°/kJ \cdot mol^{-1}$	$\Delta G_f°/kJ \cdot mol^{-1}$	$C_p/J \cdot mol^{-1} \cdot K^{-1}$
oxygen				
$O(g)$	161.0	247.5	230.1	21.912
$O_2(g)$	205.0	0	0	29.355
$O_3(g)$	238.8	142	163.4	39.20
phosphorus				
$P(s,\text{white})$	41.1	0	0	23.840
$P(s,\text{red})$	22.8	−18.4	−12.6	21.21
$P_4O_{10}(s)$	228.9	−2984	−2698	211.71
$POCl_3(g)$	325.3	−558.5	−513.0	84.96
$POCl_3(l)$	222	−597.0	−520.9	138.78
$PCl_3(g)$	311.7	−306.4	−286.3	71.84
$PCl_5(g)$	354.5	−375.0	−305.0	112.80
$PH_3(g)$	210.1	5.4	13.1	37.11
potassium				
$KCl(s)$	82.59	−436.747	−409.14	51.30
$KClO_3(s)$	143.1	−397.73	−296.25	100.25
$K^+(aq)$	102.5	−252.38	−283.27	—
silver				
$Ag(s)$	42.6	0	0	25.351
$AgBr(s)$	107.1	−100.4	−96.90	52.38
$AgCl(s)$	96.2	−127.1	−109.8	50.79
$Ag^+(aq)$	72.68	105.579	77.107	—
sodium				
$Na(g)$	153.6	107.1	77.30	20.786
$Na(s)$	51.3	0	0	28.24
$NaHCO_3(s)$	102	−947.7	−851.9	87.61
$Na_2CO_3(s)$	138.8	−1131.1	−1048.2	112.30
$Na_2O(s)$	75.06	−418.0	−379.1	69.12
$NaCl(s)$	72.13	−411.2	−384.0	50.50
$NaBr(s)$	86.82	−361.4	−349.3	51.83
$NaI(s)$	98.49	−287.8	−284.6	52.09
$Na^+(aq)$	59.0	−240.12	−261.905	—
sulfur				
$S(s,\text{rhombic})$	31.8	0	0	22.64
$S(s,\text{monoclinic})$	32.6	0.30	0.10	—
$SO_2(g)$	248.4	−296.8	−300.2	39.87
$SO_3(g)$	256.3	−395.7	−371.1	50.67
$SF_6(g)$	291.7	−1209.3	−1105.2	97.28
$SO_4^{2-}(aq)$	20.1	−909.27	−744.53	—
xenon				
$Xe(g)$	169.6	0	0	20.786
zinc				
$Zn(s)$	41.63	0	0	25.40
$ZnO(s)$	43.64	−348.3	−318.3	40.25
$ZnS(s)$	57.7	−206.0	−210.3	46.0
$Zn^{2+}(aq)$	−112.1	−153.89	−147.06	—

APPENDIX G

Acid Dissociation Constants of Weak Acids and Base Protonation Constants of Weak Bases in Water at 25°C

Name	Formula*	K_a/M	pK_a
acetic acid	$CH_3COO\underline{H}$	1.74×10^{-5}	4.76
arsenic acid	$\underline{H}O{-}As(=O)(O\underline{H}){-}O\underline{H}$	6.0×10^{-3} 1.10×10^{-7} 4.0×10^{-12}	2.22 6.96 11.40
benzene-1,2,3-tricarboxylic acid (hemimetallic acid)	$C_6H_3(COO\underline{H})_3$	1.32×10^{-3} 1.78×10^{-5} 7.4×10^{-8}	2.88 4.75 7.13
benzoic acid	$C_6H_5{-}COO\underline{H}$	6.46×10^{-5}	4.19
bromoacetic acid	$BrCH_2COO\underline{H}$	1.25×10^{-3}	2.90
butanoic acid	$CH_3CH_2CH_2COO\underline{H}$	1.52×10^{-5}	4.82
carbonic acid	$\underline{H}O{-}C(=O){-}O\underline{H}$	4.45×10^{-7} 4.69×10^{-11}	6.35 10.33
chloroacetic acid	$ClCH_2COO\underline{H}$	1.35×10^{-3}	2.87
chlorous acid	$O{=}Cl{-}O\underline{H}$	1.15×10^{-2}	1.94
cyanic acid	$\underline{H}CNO$	2.19×10^{-4}	3.66
dichloroacetic acid	$Cl_2CHCOO\underline{H}$	5.5×10^{-2}	1.26
formic acid	$HCOO\underline{H}$	1.78×10^{-4}	3.75
hydrazoic acid (hydrogen azide)	$\underline{H}N_3$	1.91×10^{-5}	4.72
hydrocyanic acid	$\underline{H}CN$	4.79×10^{-10}	9.32
hydrofluoric acid	$\underline{H}F$	6.76×10^{-4}	3.17
hydrosulfuric acid	$\underline{H_2}S$	9.12×10^{-8} 1.2×10^{-13}	7.04 12.92
hydrothiocyanic acid	$\underline{H}SCN$	0.13	0.89
hypobromous acid	$\underline{H}OBr$	2.3×10^{-9}	8.64
hypochlorous acid	$\underline{H}OCl$	3.0×10^{-8}	7.52
hypoiodus acid	$\underline{H}OI$	2.3×10^{-11}	10.64
iodic acid	$\underline{H}IO_3$	0.157	0.80
lactic acid	$\underline{H}C_3H_5O_3$	1.41×10^{-4}	3.85
nitrous acid	$\underline{H}NO_2$	4.47×10^{-4}	3.35
oxalic acid	$\underline{H}OOC{-}COO\underline{H}$	5.37×10^{-2} 5.37×10^{-5}	1.27 4.27

Name	Formula*	K_a/M	pK_a
phenol	$C_6H_5{-}O\underline{H}$	1.0×10^{-10}	10.00
phosphoric acid	$\underline{H}O{-}P(=O)(O\underline{H}){-}O\underline{H}$	7.08×10^{-3} 6.17×10^{-8} 4.37×10^{-13}	2.15 7.21 12.36
phosphorous acid	$\underline{H}O{-}P(=O)(O\underline{H}){-}H$	6.3×10^{-2} 2.57×10^{-7}	1.20 6.59
propanoic acid	$CH_3CH_2COO\underline{H}$	1.35×10^{-5}	4.87
sulfuric acid	$\underline{H}_2SO_4$	— 1.0×10^{-2}	— 2.00
sulfurous acid	$\underline{H}_2SO_3$	1.54×10^{-2} 6.6×10^{-8}	1.81 7.18
thiosulfuric acid	$\underline{H}O{-}S(=O)(=S){-}O\underline{H}$	0.3 3×10^{-2}	0.52 1.52

Name	Formula	Protonated form	K_b/M	pK_b
aniline	$C_6H_5{-}NH_2$	$C_6H_5{-}NH_3^+$	4.17×10^{-10}	9.38
2-aminoethanol (ethanolamine)	$HOCH_2CH_2NH_2$	$HOCH_2CH_2NH_3^+$	3.15×10^{-5}	4.50
ammonia	NH_3	NH_4^+	1.75×10^{-5}	4.76
benzylamine	$C_6H_5{-}CH_2NH_2$	$C_6H_5{-}CH_2NH_3^+$	2.24×10^{-5}	4.65
butylamine	$CH_3CH_2CH_2CH_2NH_2$	$CH_3CH_2CH_2CH_2NH_3^+$	4.37×10^{-4}	3.36
cyclohexylamine	$C_6H_{11}{-}NH_2$	$C_6H_{11}{-}NH_3^+$	4.37×10^{-4}	3.36
diethylamine	$(C_2H_5)_2NH$	$(C_2H_5)_2NH_2^+$	8.51×10^{-4}	3.07
dimethylamine	$(CH_3)_2NH$	$(CH_3)_2NH_2^+$	5.81×10^{-4}	3.24
ethylamine	$CH_3CH_2NH_2$	$CH_3CH_2NH_3^+$	4.27×10^{-4}	3.37
hydroxylamine	$HONH_2$	$HONH_3^+$	1.07×10^{-8}	7.97
methylamine	CH_3NH_2	$CH_3NH_3^+$	4.59×10^{-4}	3.34
piperidine	$C_5H_{10}NH$	$C_5H_{10}NH_2^+$	1.33×10^{-3}	2.88
propylamine	$CH_3CH_2CH_2NH_2$	$CH_3CH_2CH_2NH_3^+$	3.68×10^{-4}	3.43
pyridine	C_5H_5N	$C_5H_5NH^+$	1.46×10^{-9}	8.84
trimethylamine	$(CH_3)_3N$	$(CH_3)_3NH^+$	6.11×10^{-5}	4.21

*Acidic protons are underlined.

APPENDIX H

Solubility-Product Constants in Water at 25°C

Bromates	K_{sp}
$AgBrO_3$	5.5×10^{-5} M^2
$Pb(BrO_3)_2$	7.9×10^{-6} M^3
$TlBrO_3$	1.7×10^{-4} M^2

Bromides	K_{sp}
AgBr	5.0×10^{-13} M^2
CuBr	5.3×10^{-9} M^2
$Hg_2Br_2^*$	5.6×10^{-23} M^3
$HgBr_2$	1.3×10^{-19} M^3
$PbBr_2$	4.0×10^{-5} M^3
TlBr	3.4×10^{-6} M^2

Carbonates	K_{sp}
Ag_2CO_3	8.1×10^{-12} M^3
$BaCO_3$	5.1×10^{-9} M^2
$CaCO_3$	2.8×10^{-9} M^2
$CdCO_3$	1.8×10^{-14} M^2
$CoCO_3$	1.0×10^{-10} M^2
$CuCO_3$	1.4×10^{-10} M^2
$FeCO_3$	3.2×10^{-11} M^2
$MgCO_3$	3.5×10^{-8} M^2
$MnCO_3$	1.8×10^{-11} M^2
$NiCO_3$	1.3×10^{-7} M^2
$PbCO_3$	7.4×10^{-14} M^2
$SrCO_3$	1.1×10^{-10} M^2
$ZnCO_3$	1.4×10^{-11} M^2

Chlorides	K_{sp}
AgCl	1.8×10^{-10} M^2
CuCl	1.2×10^{-6} M^2
$Hg_2Cl_2^*$	1.3×10^{-18} M^3
$PbCl_2$	1.6×10^{-5} M^3
TlCl	1.7×10^{-4} M^2

Chromates	K_{sp}
Ag_2CrO_4	1.1×10^{-12} M^3
$BaCrO_4$	1.2×10^{-10} M^2
$CuCrO_4$	3.6×10^{-6} M^2
Hg_2CrO_4	2.0×10^{-9} M^2
$PbCrO_4$	2.8×10^{-13} M^2
Tl_2CrO_4	9.8×10^{-15} M^3

Cyanides	K_{sp}
AgCN	2.2×10^{-16} M^2
$Hg_2(CN)_2$	5×10^{-40} M^3
$Zn(CN)_2$	3×10^{-16} M^3

Fluorides	K_{sp}
BaF_2	1.0×10^{-6} M^3
CaF_2	5.3×10^{-9} M^3
MgF_2	6.5×10^{-9} M^3
PbF_2	7.7×10^{-8} M^3
SrF_2	2.5×10^{-9} M^3

Hydroxides	K_{sp}
$Al(OH)_3$	1.3×10^{-33} M^4
$Ca(OH)_2$	5.5×10^{-6} M^3
$Cd(OH)_2$	2.5×10^{-14} M^3
$Co(OH)_2$	1.3×10^{-15} M^3
$Cr(OH)_3$	6.3×10^{-31} M^4
$Cu(OH)_2$	2.2×10^{-20} M^3
$Fe(OH)_2$	8.0×10^{-16} M^3
$Fe(OH)_3$	1.0×10^{-38} M^4
$Mg(OH)_2$	1.8×10^{-11} M^3
$Ni(OH)_2$	2.0×10^{-15} M^3
$Pb(OH)_2$	1.2×10^{-15} M^3
$Sn(OH)_2$	1.4×10^{-28} M^3
$Zn(OH)_2$	1.0×10^{-15} M^3

Iodates	K_{sp}
$AgIO_3$	3.1×10^{-8} M^2
$Ba(IO_3)_2$	1.5×10^{-9} M^3
$Ca(IO_3)_2$	7.1×10^{-7} M^3
$Cd(IO_3)_2$	2.3×10^{-8} M^3
$Pb(IO_3)_2$	2.5×10^{-13} M^3
$TlIO_3$	3.1×10^{-6} M^2
$Zn(IO_3)_2$	3.9×10^{-6} M^3

Iodides	K_{sp}
AgI	8.3×10^{-17} M^2
CuI	1.1×10^{-12} M^2
$Hg_2I_2^*$	4.5×10^{-29} M^3
PbI_2	7.1×10^{-9} M^3
TlI	6.5×10^{-8} M^2

Oxalates	K_{sp}
$Ag_2C_2O_4$	3.4×10^{-11} M^3
CaC_2O_4	4×10^{-9} M^2
MgC_2O_4	7×10^{-7} M^2
SrC_2O_4	4×10^{-7} M^2

Sulfates	K_{sp}
Ag_2SO_4	1.4×10^{-5} M^3
$BaSO_4$	1.1×10^{-10} M^2
$CaSO_4$	9.1×10^{-6} M^2
Hg_2SO_4*	7.4×10^{-7} M^2
$PbSO_4$	1.6×10^{-8} M^2
$SrSO_4$	3.2×10^{-7} M^2

Sulfides	K_{sp}
Ag_2S	8×10^{-51} M^3
CdS	8.0×10^{-27} M^2
CoS	5×10^{-22} M^2
CuS	6.3×10^{-36} M^2
FeS	6.3×10^{-18} M^2
HgS	4×10^{-53} M^2
MnS	2.5×10^{-13} M^2
NiS	1.3×10^{-25} M^2
PbS	8.0×10^{-28} M^2
SnS	1.0×10^{-25} M^2
Tl_2S	6×10^{-22} M^3
ZnS	1.6×10^{-24} M^2

Thiocyanates	K_{sp}
AgSCN	1.1×10^{-12} M^2
$Cu(SCN)_2$	4.0×10^{-14} M^3
$Hg_2SCN)_2$*	3.0×10^{-20} M^3
$Hg(SCN)_2$	2.8×10^{-20} M^3
TlSCN	1.6×10^{-4} M^2

*Recall that the mercury(I) ion exists in aqueous solution as the dimer $Hg_2^{2+}(aq)$.

APPENDIX I

Standard Reduction Voltages for Aqueous Solutions at 25°C

Acidic solutions

Elements	Half reaction	$E°$/volts
aluminum	$Al^{3+}(aq) + 3e^- \rightleftharpoons Al(s)$	−1.66
barium	$Ba^{2+}(aq) + 2e^- \rightleftharpoons Ba(s)$	−2.912
beryllium	$Be^{2+}(aq) + 2e^- \rightleftharpoons Be(s)$	−1.85
bromine	$BrO_3^-(aq) + 6H^+(aq) + 5e^- \rightleftharpoons \frac{1}{2}Br_2(l) + 3H_2O(l)$	1.52
	$Br_2(l) + 2e^- \rightleftharpoons 2Br^-(aq)$	1.065
	$Br_3^- + 2e^- \rightleftharpoons 3Br^-(aq)$	1.051
cadmium	$Cd^{2+}(aq) + 2e^- \rightleftharpoons Cd(s)$	−0.402
calcium	$Ca^{2+}(aq) + 2e^- \rightleftharpoons Ca(s)$	−2.868
	$CaSO_4(s) + 2e^- \rightleftharpoons Ca(s) + SO_4^{2-}(aq)$	−2.936
cerium	$Ce^{3+}(aq) + 3e^- \rightleftharpoons Ce(s)$	−2.335
	$Ce^{4+}(aq) + e^- \rightarrow Ce^{3+}(aq)$	1.65
cesium	$Cs^+(aq) + e^- \rightleftharpoons Cs(s)$	−2.923
chlorine	$HClO(aq) + H^+(aq) + e^- \rightleftharpoons \frac{1}{2}Cl_2(g) + H_2O(l)$	1.63
	$ClO_3^-(aq) + 6H^+(aq) + 5e^- \rightleftharpoons \frac{1}{2}Cl_2(g) + 3H_2O(l)$	1.47
	$Cl_2(aq) + 2e^- \rightleftharpoons 2Cl^-(aq)$	1.395
	$Cl_2(g) + 2e^- \rightleftharpoons 2Cl^-(aq)$	1.358
	$ClO_4^-(aq) + 2H^+(aq) + 2e^- \rightleftharpoons ClO_3^-(aq) + H_2O(l)$	1.19
chromium	$Cr_2O_7^{2-}(aq) + 14H^+(aq) + 6e^- \rightleftharpoons 2Cr^{3+}(aq) + 7H_2O(l)$	1.33
	$Cr^{3+}(aq) + e^- \rightleftharpoons Cr^{2+}(aq)$	−0.41
	$Cr^{3+}(aq) + 3e^- \rightleftharpoons Cr(s)$	−0.74
	$Cr^{2+}(aq) + 2e^- \rightleftharpoons Cr(s)$	−0.91
cobalt	$Co^{3+}(aq) + e^- \rightleftharpoons Co^{2+}(aq)$	1.81
	$Co(NH_3)_6^{3+}(aq) + e^- \rightleftharpoons Co(NH_3)_6^{2+}(aq)$	0.1
	$Co^{2+}(aq) + 2e^- \rightleftharpoons Co(s)$	−0.277
copper	$Cu^+(aq) + e^- \rightleftharpoons Cu(s)$	0.518
	$Cu^{2+}(aq) + 2e^- \rightleftharpoons Cu(s)$	0.337
	$Cu^{2+}(aq) + e^- \rightleftharpoons Cu^+(aq)$	0.159
	$CuCl(s) + e^- \rightleftharpoons Cu(s) + Cl^-(aq)$	0.137
	$CuI(s) + e^- \rightleftharpoons Cu(s) + I^-(aq)$	−0.185
	$Cu(CN)_2^-(aq) + e^- \rightleftharpoons Cu(s) + 2CN^-(aq)$	−0.429
	$CuCN(s) + e^- \rightleftharpoons Cu(s) + CN^-(aq)$	−0.639
fluorine	$F_2(g) + 2e^- \rightleftharpoons 2F^-(aq)$	2.87
gadolinium	$Gd^{3+}(aq) + 3e^- \rightleftharpoons Gd(s)$	−2.40
gallium	$Ga^{3+}(aq) + 3e^- \rightleftharpoons Ga(s)$	−0.560
gold	$Au^+(aq) + e^- \rightleftharpoons Au(s)$	1.692
	$Au^{3+}(aq) + 2e^- \rightleftharpoons Au^+(aq)$	1.401
	$AuCl_2^-(aq) + e^- \rightleftharpoons Au(s) + 2Cl^-(aq)$	1.154
	$AuCl_4^-(aq) + 2e^- \rightleftharpoons AuCl_2^-(aq) + 2Cl^-(aq)$	0.926
hydrogen	$2H^+(aq) + 2e^- \rightleftharpoons H_2(g)$	0.000
	$H_2(g) + 2e^- \rightarrow H^-(aq)$	−2.25
iodine	$IO_3^-(aq) + 6H^+(aq) + 5e^- \rightleftharpoons \frac{1}{2}I_2(s) + 3H_2O(l)$	1.195
	$I_2(s) + 2e^- \rightleftharpoons 2I^-(aq)$	0.536
	$I_3^-(aq) + 2e^- \rightleftharpoons 3I^-(aq)$	0.536

Acidic solutions

Elements	Half reaction	$E°$/volts
iron	$Fe^{3+}(aq) + e^- \rightleftharpoons Fe^{2+}(aq)$	0.770
	$Fe(CN)_6^{3-}(aq) + e^- \rightleftharpoons Fe(CN)_6^{4-}(aq)$	0.356
	$Fe^{2+}(aq) + 2e^- \rightleftharpoons Fe(s)$	−0.440
lanthanum	$La^{3+}(aq) + 3e^- \rightleftharpoons La(s)$	−2.52
lead	$PbO_2(s) + 4H^+(aq) + SO_4^{2-}(aq) + 2e^- \rightleftharpoons PbSO_4(s) + 2H_2O(l)$	1.685
	$PbO_2(s) + 4H^+(aq) + 2e^- \rightleftharpoons Pb^{2+}(aq) + 2H_2O(l)$	1.455
	$Pb^{2+}(aq) + 2e^- \rightleftharpoons Pb(s)$	−0.126
	$PbF_2(s) + 2e^- \rightleftharpoons Pb(s) + 2F^-(aq)$	−0.350
	$PbSO_4(s) + 2e^- \rightleftharpoons Pb(s) + SO_4^{2-}(aq)$	−0.355
lithium	$Li^+(aq) + e^- \rightleftharpoons Li(s)$	−3.045
magnesium	$Mg^{2+}(aq) + 2e^- \rightleftharpoons Mg(s)$	−2.36
manganese	$MnO_4^-(aq) + 4H^+(aq) + 3e^- \rightleftharpoons MnO_2(s) + 2H_2O(l)$	1.695
	$Mn^{3+}(aq) + e^- \rightleftharpoons Mn^{2+}(aq)$	1.542
	$MnO_4^-(aq) + 8H^+(aq) + 5e^- \rightleftharpoons Mn^{2+}(aq) + 4H_2O(l)$	1.51
	$MnO_2(s) + 4H^+(aq) + 2e^- \rightleftharpoons Mn^{2+}(aq) + 2H_2O(l)$	1.229
	$MnO_4^-(aq) + e^- \rightleftharpoons MnO_4^{2-}(aq)$	0.558
	$Mn^{2+}(aq) + 2e^- \rightleftharpoons Mn(s)$	−1.182
mercury	$2Hg^{2+}(aq) + 2e^- \rightleftharpoons Hg_2^{2+}(aq)$	0.908
	$Hg^{2+}(aq) + 2e^- \rightleftharpoons Hg(l)$	0.854
	$Hg_2^{2+}(aq) + 2e^- \rightleftharpoons 2Hg(l)$	0.792
	$Hg_2SO_4(s) + 2e^- \rightleftharpoons 2Hg(l) + SO_4^{2-}(aq)$	0.614
	$Hg_2Cl_2(s) + 2e^- \rightleftharpoons 2Hg(l) + 2Cl^-(aq)$	0.268
	$Hg_2Br_2(s) + 2e^- \rightleftharpoons 2Hg(l) + 2Br^-(aq)$	0.140
nickel	$Ni(OH)_3(s) + 3H^+(aq) + e^- \rightleftharpoons Ni^{2+}(aq) + 3H_2O(l)$	2.08
	$Ni^{2+}(aq) + 2e^- \rightleftharpoons Ni(s)$	−0.231
nitrogen	$N_2O(g) + 2H^+(aq) + 2e^- \rightleftharpoons N_2(g) + H_2O(l)$	1.77
	$2NO(aq) + 2H^+(aq) + 2e^- \rightleftharpoons N_2O(g) + H_2O(l)$	1.59
	$HNO_2(aq) + H^+(aq) + e^- \rightleftharpoons NO(g) + H_2O(l)$	1.00
	$NO_3^-(aq) + 4H^+(aq) + 3e^- \rightleftharpoons NO(g) + 2H_2O(l)$	0.96
	$NO_3^-(aq) + 3H^+(aq) + 2e^- \rightleftharpoons HNO_2(aq) + H_2O(l)$	0.94
	$NO_3^-(aq) + 2H^+(aq) + e^- \rightleftharpoons \frac{1}{2}N_2O_4(g) + H_2O(l)$	0.80
oxygen	$O_3(g) + 2H^+(aq) + 2e^- \rightleftharpoons O_2(g) + H_2O(l)$	2.07
	$\frac{1}{2}O_2(g) + 2H^+(aq) + 2e^- \rightleftharpoons H_2O(l)$	1.229
	$O_2(g) + 2H^+(aq) + 2e^- \rightleftharpoons H_2O_2(l)$	0.682
palladium	$Pd^{2+}(aq) + 2e^- \rightleftharpoons Pd(s)$	0.915
phosphorus	$H_3PO_4(aq) + 2H^+(aq) + 2e^- \rightleftharpoons H_3PO_3(aq) + H_2O(l)$	−0.276
	$H_3PO_3(aq) + 2H^+(aq) + 2e^- \rightleftharpoons H_3PO_2(aq) + H_2O(l)$	−0.50
platinum	$Pt^{2+}(aq) + 2e^- \rightleftharpoons Pt(s)$	1.188
potassium	$K^+(aq) + e^- \rightleftharpoons K(s)$	−2.925
rubidium	$Rb^+(aq) + e^- \rightleftharpoons Rb(s)$	−2.924
scandium	$Sc^{3+}(aq) + 3e^- \rightleftharpoons Sc(s)$	−2.08
silver	$AgO(s) + H^+(aq) + e^- \rightleftharpoons \frac{1}{2}Ag_2O(s) + \frac{1}{2}H_2O(l)$	1.40
	$Ag^+(aq) + e^- \rightleftharpoons Ag(s)$	0.799
	$AgCl(s) + e^- \rightleftharpoons Ag(s) + Cl^-(aq)$	0.222
	$AgBr(s) + e^- \rightleftharpoons Ag(s) + Br^-(aq)$	0.071
	$Ag(S_2O_3)_2^{3-}(aq) + e^- \rightleftharpoons Ag(s) + 2S_2O_3^{2-}(aq)$	0.017
	$AgI(s) + e^- \rightleftharpoons Ag(s) + I^-(aq)$	−0.152
sodium	$Na^+(aq) + e^- \rightleftharpoons Na(s)$	−2.713
strontium	$Sr^{2+}(aq) + 2e^- \rightleftharpoons Sr(s)$	−2.886
sulfur	$S_2O_8^{2-}(aq) + 2e^- \rightleftharpoons 2SO_4^{2-}(aq)$	2.01
	$S_2O_6^{2-}(aq) + 4H^+(aq) + 2e^- \rightleftharpoons 2H_2SO_3(aq)$	0.57
	$2H_2SO_3(aq) + 2H^+(aq) + 4e^- \rightleftharpoons S_2O_3^{2-}(aq) + 3H_2O(l)$	0.40
	$S_4O_6^{2-}(aq) + 2e^- \rightleftharpoons 2S_2O_3^{2-}(aq)$	0.08
thallium	$Tl^{3+}(aq) + 2e^- \rightleftharpoons Tl^+(aq)$	1.28
	$Tl^+(aq) + e^- \rightleftharpoons Tl(s)$	−0.336
	$TlCl(s) + e^- \rightleftharpoons Tl(s) + Cl^-(aq)$	−0.557

Acidic solutions

Elements	Half reaction	$E°$/volts
tin	$Sn^{4+}(aq) + 2e^- \rightleftharpoons Sn^{2+}(aq)$	0.154
	$Sn(OH)_4(s) + 4H^+(aq) + 4e^- \rightleftharpoons Sn(s) + 4H_2O(l)$	−0.008
	$Sn(OH)_2(s) + 2H^+(aq) + 2e^- \rightleftharpoons Sn(s) + 2H_2O(l)$	−0.091
	$Sn^{2+}(aq) + 2e^- \rightleftharpoons Sn(s)$	−0.136
	$Sn(OH)_4(s) + H^+(aq) + 2e^- \rightleftharpoons Sn(OH)_3^-(aq) + H_2O(l)$	−0.349
vanadium	$VO_2^+(aq) + 2H^+(aq) + e^- \rightleftharpoons VO^{2+}(aq) + H_2O(l)$	1.000
	$VO^{2+}(aq) + 2H^+(aq) + e^- \rightleftharpoons V^{3+}(aq) + H_2O(l)$	0.337
	$V^{3+}(aq) + e^- \rightleftharpoons V^{2+}(aq)$	−0.24
	$V^{2+}(aq) + 2e^- \rightleftharpoons V(s)$	−1.18
zinc	$ZnO(s) + 2H^+(aq) + 2e^- \rightleftharpoons Zn(s) + H_2O(l)$	−0.439
	$Zn^{2+}(aq) + 2e^- \rightleftharpoons Zn(s)$	−0.764
	$Zn(NH_3)_4^{2+}(aq) + 2e^- \rightleftharpoons Zn(s) + 4NH_3(aq)$	−1.04
	$ZnCO_3(s) + 2e^- \rightleftharpoons Zn(s) + CO_3^{2-}(aq)$	−1.06
	$ZnS(s) + 2e^- \rightleftharpoons Zn(s) + S^{2-}(aq)$	−1.405

SOURCE: L. G. Sillén and A. E. Martell, *Stability Constants of Metal-Ion Complexes,* London: The Chemical Society, Special Publications no. 17 and 25, 1964 and 1971; G. Milazzo and S. Caroli, *Table of Standard Electrode Potentials* (New York: Wiley, 1978); T. Mussini, P. Longhi, and S. Rondinini, *Pure Appl. Chem.*, **57,** 169 (1985).

Basic solutions

Elements	Half reaction	$E°$/volts
aluminum	$Al(OH)_4^-(aq) + 3e^- \rightleftharpoons Al(s) + 4OH^-(aq)$	−2.33
bromine	$BrO^-(aq) + H_2O(l) + 2e^- \rightleftharpoons Br^-(aq) + 2OH^-(aq)$	0.76
	$BrO_3^-(aq) + 3H_2O(l) + 6e^- \rightleftharpoons Br^-(aq) + 6OH^-(aq)$	0.61
chlorine	$ClO^-(aq) + H_2O(l) + 2e^- \rightleftharpoons Cl^-(aq) + 2OH^-(aq)$	0.89
chromium	$CrO_4^{2-}(aq) + 4H_2O(l) + 3e^- \rightleftharpoons Cr(OH)_3(s) + 5OH^-(aq)$	−0.13
cobalt	$Co(OH)_2(s) + 2e^- \rightleftharpoons Co(s) + 2OH^-(aq)$	−0.73
copper	$Cu(OH)_2(s) + 2e^- \rightleftharpoons Cu(s) + 2OH^-(aq)$	−0.222
hydrogen	$H_2O(l) + e^- \rightleftharpoons \frac{1}{2}H_2(g) + OH^-(aq)$	−0.828
indium	$In(OH)_3(s) + 3e^- \rightleftharpoons In(s) + 3OH^-(aq)$	−1.00
iodine	$IO_3^-(aq) + 3H_2O(l) + 6e^- \rightleftharpoons I^-(aq) + 6OH^-$	0.26
iron	$Fe(OH)_3(s) + e^- \rightarrow Fe(OH)_2(s) + OH^-(aq)$	−0.56
lead	$3PbO_2(s) + 2H_2O(l) + 4e^- \rightleftharpoons Pb_3O_4(s) + 4OH^-(aq)$	0.295
	$Pb_3O_4(s) + H_2O(l) + 2e^- \rightleftharpoons 3PbO(s) + 2OH^-(aq)$	0.249
magnesium	$Mg(OH)^+(aq) + 2e^- \rightleftharpoons Mg(s) + OH^-(aq)$	−2.440
	$Mg(OH)_2(s) + 2e^- \rightleftharpoons Mg(s) + 2OH^-(aq)$	−2.690
manganese	$Mn(OH)_2(s) + 2e^- \rightleftharpoons Mn(s) + 2OH^-(aq)$	−1.55
mercury	$HgO(s) + H_2O(l) + 2e^- \rightleftharpoons Hg(l) + 2OH^-(aq)$	0.10
nickel	$Ni(OH)_2(s) + 2e^- \rightleftharpoons Ni(s) + 2OH^-(aq)$	−0.72
oxygen	$O_3(g) + H_2O(l) + 2e^- \rightleftharpoons O_2(g) + 2OH^-(aq)$	1.24
	$O_2(g) + 2H_2O(l) + 4e^- \rightleftharpoons 4OH^-(aq)$	0.40
sulfur	$2SO_3^{2-}(aq) + 3H_2O(l) + 4e^- \rightleftharpoons S_2O_3^{2-}(aq) + 6OH^-(aq)$	−0.58
	$SO_3^{2-}(aq) + 3H_2O(l) + 4e^- \rightleftharpoons S(s) + 6OH^-(aq)$	−0.66
	$SO_4^{2-}(aq) + H_2O(l) + 2e^- \rightleftharpoons SO_3^{2-}(aq) + 2OH^-(aq)$	−0.93
	$2SO_3^{2-}(aq) + 2H_2O(l) + 2e^- \rightleftharpoons S_2O_4^{2-} + 4OH^-(aq)$	−1.12
zinc	$Zn(OH)_3^-(aq) + 2e^- \rightleftharpoons Zn(s) + 3OH^-(aq)$	−1.183
	$Zn(OH)_4^{2-}(aq) + 2e^- \rightleftharpoons Zn(s) + 4OH^-(aq)$	−1.214
	$Zn(OH)_2(s) + 2e^- \rightleftharpoons Zn(s) + 2OH^-(aq)$	−1.245
	$ZnS(s) + 2e^- \rightleftharpoons Zn(s) + S^{2-}(aq)$	−1.405

SOURCE: L. G. Sillén and A. E. Martell, *Stability Constants of Metal-Ion Complexes,* London: The Chemical Society, Special Publications no. 17 and 25, 1964 and 1971; G. Milazzo and S. Caroli, *Table of Standard Electrode Potentials* (New York: Wiley, 1978); T. Mussini, P. Longhi, and S. Rondinini, *Pure Appl. Chem.* **57:** 169 (1985).

APPENDIX J

Top 50 Chemicals (1989)*

Rank	Name	Formula	Output/in billions of pounds
1	sulfuric acid	H_2SO_4	86.80
2	nitrogen	N_2	53.77
3	oxygen	O_2	37.75
4	ethylene (ethene)	C_2H_4	34.95
5	ammonia	NH_3	33.76
6	lime	CaO	32.99
7	phosphoric acid	H_3PO_4	23.12
8	chlorine	Cl_2	22.32
9	sodium hydroxide	NaOH	22.15
10	propylene	$CH_3CH{=}CH_2$	20.23
11	sodium carbonate	Na_2CO_3	19.79
12	nitric acid	HNO_3	15.98
13	urea	$(NH_2)_2CO$	15.47
14	ammonium nitrate	NH_4NO_3	14.11
15	ethylene dichloride	$ClCH_2CH_2Cl$	13.68
16	benzene	C_6H_6	11.68
17	carbon dioxide	CO_2	10.83
18	vinyl chloride	$CH_2{=}CHCl$	9.96
19	ethylbenzene	$C_6H_5C_2H_5$	9.22
20	terephthalic acid	$C_6H_4(COOH)_2$	8.31
21	styrene	$C_6H_5CH{=}CH_2$	8.13
22	methanol	CH_3OH	7.14
23	formaldehyde	HCHO	6.37
24	toluene	$C_6H_5CH_3$	5.84
25	xylene	$C_6H_4(CH_3)_2$	5.80
26	ethylene glycol	CH_2OHCH_2OH	5.50
27	*p*-xylene	$C_6H_4(CH_3)_2$	5.49
28	ethylene oxide	$CH_2{-}CH_2$ (bridged by O)	5.32
29	hydrochloric acid	HCl	5.26
30	methyl *tert*-butyl ether	$CH_3OC(CH_3)_3$	4.98

Rank	Name	Formula	Output/in billions of pounds
31	ammonium sulfate	$(NH_4)_2SO_4$	4.71
32	cumene	$C_6H_5CH(CH_3)_2$	4.54
33	phenol	C_6H_5OH	3.89
34	acetic acid	CH_3COOH	3.83
35	potash	K_2CO_3	3.35
36	propylene oxide	$CH_2{-}CH{-}CH_3$ (epoxide ring through O)	3.20
37	butadiene	$CH_2{=}CHCH{=}CH_2$	3.09
38	carbon black	C	2.91
39	acrylonitrile	$CH_2{=}CHCN$	2.61
40	acetone	$(CH_3)_2CO$	2.50
41	vinyl acetate	$CH_2{=}CH(OOCCH_3)$	2.47
42	cyclohexane	C_6H_{12}	2.39
43	aluminum sulfate	$Al_2(SO_4)_3$	2.35
44	titanium dioxide	TiO_2	2.22
45	calcium chloride	$CaCl_2$	1.92
46	sodium silicate	Na_2SiO_3	1.75
47	adipic acid	$HOOC(CH_2)_4COOH$	1.64
48	sodium sulfate	Na_2SO_4	1.60
49	isopropyl alcohol	$CH_3CHOHCH_3$	1.43
50	caprolactam	O=C ring: H_2C, CH_2, H_2C, CH_2, $H_2C{-}CH_2$	1.31

Chem. Eng. News, June 18, 1990.

APPENDIX K

Answers to the Odd-Numbered Problems

CHAPTER 1

1-1. $d < a < e < c < b = f$
1-3. 4.19×10^{-30} m^3
1-5. 13.6 g·cm^{-3}
1-7. (a) three (b) four (c) four (d) two (e) no uncertainty
1-9. (a) 2 (b) 2.08×10^4 (c) 2.8 (d) 3.4×10^{22}
1-11. (a) 1.06 qt (b) 2.99×10^8 m·s^{-1} (c) 1.987 cal·K^{-1}·mol^{-1}
1-13. 9.46×10^{15} m; 5.88×10^{12} mi
1-15. The 2-L bottle is the better buy.
1-21. 20°C; 293 K
1-23. 8.65 lb
1-25. 0.374 oz·in^{-3} to 0.427 oz·in^{-3}
1-27. 170 cm
1-29. 382 mL
1-31. It takes sound 8.71×10^5 times as long as light.
1-33. No two snowflakes have yet been observed to be the same.
1-35. 1.4 g
1-37. 172 mL
1-39. 473 mL milk, 15 mL baking soda, 180°C
1-41. 40°C; 313 K
1-43. Substitute Equation (1-2) into Equation (1-1) and collect like terms.
1-45. 1.60 g·mL^{-1}
1-47. The 440-yd race is 2 m longer.
1-49. 0.700 g·mL^{-1}
1-51. 22.8 mi·h^{-1}
1-53. 6170°F
1-55. 1.36×10^5 L; 1.36×10^5 kg
1-57. (a) 834 lb (b) 840 cm^3
1-59. 0.116 mm
1-61. 8.3 min

CHAPTER 2

2-1. (a) Se (b) In (c) Mn (d) Tm (e) Hg (f) Kr (g) Pd (h) Tl (i) U (j) W
2-3. (a) germanium (b) scandium (c) iridium (d) cesium (e) strontium (f) americium (g) molybdenum (h) indium (i) plutonium (j) xenon
2-5. 59.0% Na; 41.0% O
2-7. 80.0% Cu; 20% S
2-9. 60.0% K; 18.4% C; 21.5% N
2-11. (a) lithium sulfide (b) barium oxide (c) magnesium phosphide (d) cesium bromide
2-13. (a) silicon carbide (b) gallium phosphide (c) aluminum oxide (d) beryllium chloride
2-15. (a) chlorine trifluoride and chlorine pentafluoride (b) sulfur tetrafluoride and sulfur hexafluoride (c) krypton difluoride and krypton tetrafluoride (d) bromine oxide and bromine dioxide
2-17. (a) 79.90 (b) 159.70 (c) 181.88 (d) 283.88
2-19. (a) 121.93 (b) 197.9 (c) 120.91 (d) 389.91
2-21. 38.35% Cl; 61.65% F
2-23. 42.10% C; 6.479% H; 51.42% O
2-25. 63.34% Xe; 1.267 g Xe
2-27. (a) 53 protons, 53 electrons, 78 neutrons (b) 27 protons, 27 electrons, 33 neutrons (c) 19 protons, 19 electrons, 24 neutrons (d) 49 protons, 49 electrons, 64 neutrons
2-29. $^{14}_{6}$C, 6, 8, 14; $^{241}_{95}$Am, 95, 146, 241; $^{123}_{53}$I, 53, 70, 123; $^{18}_{8}$O, 8, 10, 18
2-31. $^{67}_{31}$Ga, 31, 36, 67; $^{15}_{7}$N, 7, 8, 15; $^{58}_{27}$Co, 27, 31, 58; $^{133}_{54}$Xe, 54, 79, 133
2-33. 1.0080
2-35. 20.18
2-37. 50.65% bromine-79; 49.35% bromine-81
2-39. 0.36%
2-41. (a) 54 (b) 54 (c) 36 (d) 10
2-43. (a) 54 (b) 78 (c) 24 (d) 18
2-45. (a) Ca^{2+}, Cl^-, S^{2-} (b) Rb^+, Sr^{2+}, Br^-, Se^{2-} (c) Na^+, Mg^{2+}, F^-, O^{2-} (d) Cs^+, Ba^{2+}, Te^{2-}
2-47. (a) 17.002 (b) 19.018 (c) 140.97196 (d) 172.786
2-49. Use a magnet to remove the iron filings.
2-51. See Section 2-12.
2-53. To convert the vapor from the distillation chamber into a liquid.
2-55. To sweep the sample through the column.
2-57. The original separation involved colored constituents. *Chroma* comes from the Greek word for color.
2-59. See Section 2-12.
2-61. The more dense gold settles at the center of the swirled pan while the sand particles are swirled out of the pan.
2-63. See Section 2-12.
2-65. 0.0840 H; 1.333 O
2-67. baking soda in refrigerator and gas masks
2-69. (a) 53 protons in I and 53 protons in I^- (b) 78 neutrons in I and 78 neutrons in I^- (c) 53 electrons in I and 54 electrons in I^-

CHAPTER 3

3-1. (a) $2P(s) + 3Br_2(l) \rightarrow 2PBr_3(l)$
(b) $2H_2O_2(l) \rightarrow 2H_2O(l) + O_2(g)$
(c) $4CoO(s) + O_2(g) \rightarrow 2Co_2O_3(s)$
(d) $PCl_5(s) + 4H_2O(l) \rightarrow H_3PO_4(l) + 5HCl(g)$
3-3. (a) $CaH_2(s) + 2H_2O(l) \rightarrow Ca(OH)_2(aq) + 2H_2(g)$
(b) $CaCO_3(s) + 2HCl(aq) \rightarrow CaCl_2(aq) + CO_2(g) + H_2O(l)$
(c) $C_6H_{12}O_2(aq) + 8O_2(g) \rightarrow 6CO_2(g) + 6H_2O(l)$
(d) $2Li(s) + 2CO_2(g) + 2H_2O(g) \rightarrow 2LiHCO_3(s) + H_2(g)$
3-5. (a) $2Na(s) + S(s) \rightarrow Na_2S(s)$
(b) $Ca(s) + Br_2(l) \rightarrow CaBr_2(s)$
(c) $2Ba(s) + O_2(g) \rightarrow 2BaO(s)$
(d) $2SO_2(g) + O_2(g) \rightarrow 2SO_3(g)$
(e) $3Mg(s) + N_2(g) \rightarrow Mg_3N_2(s)$
3-7. (a) $NaH(s) + H_2O(l) \rightarrow NaOH(aq) + H_2(g)$
sodium hydride, water, sodium hydroxide, hydrogen
(b) $2SO_2(g) + O_2(g) \rightarrow 2SO_3(g)$
sulfur dioxide, oxygen, sulfur trioxide
(c) $H_2S(g) + 2LiOH(aq) \rightarrow Li_2S(aq) + 2H_2O(l)$
hydrogen sulfide, lithium hydroxide, lithium sulfide, water
(d) $ZnO(s) + CO(g) \rightarrow Zn(s) + CO_2(g)$
zinc oxide, carbon monoxide, zinc, carbon dioxide
3-9. (a) $2Na(s) + I_2(s) \rightarrow 2NaI(s)$
sodium iodide
(b) $Sr(s) + H_2(g) \rightarrow SrH_2(s)$
strontium hydride
(c) $3Ca(s) + N_2(g) \rightarrow Ca_3N_2(s)$
calcium nitride
(d) $2Mg(s) + O_2(g) \rightarrow 2MgO(s)$
magnesium oxide
3-11. (a) solid (b) NaAt (c) white (d) At_2 (e) black
3-13. Tl-main group (5) metal, Eu-inner transition metal, Xe-main group (8) nonmetal, Hf-transition metal, Ru-transition metal, Am-inner transition metal, B-main group (3) semimetal
3-15. (a) $2Ra(s) + O_2(g) \rightarrow 2RaO(s)$
(b) $Ra(s) + Cl_2(g) \rightarrow RaCl_2(s)$
(c) $Ra(s) + 2HCl(g) \rightarrow RaCl_2(s) + H_2(g)$
(d) $Ra(s) + H_2(g) \rightarrow RaH_2(s)$
(e) $Ra(s) + S(s) \rightarrow RaS(s)$
3-17. Sn, Sb, Ar
3-19. (a) semimetal (b) metal (c) metal (d) semimetal (e) metal
3-21. (a) yes, Xe (b) no (c) yes, Ar (d) yes, Ar (e) yes, Ne (f) no
3-23. (a) $Mg^{2+}S^{2-}$ magnesium sulfide
(b) $Al^{3+}P^{3-}$ aluminum phosphide
(c) $Ba^{2+}F^{-}$ barium fluoride
(d) $Ga^{3+}O^{2-}$ gallium oxide
3-25. (a) Ga_2Se_3 (b) AlP (c) KI (d) SrF_2
3-27. (a) Li_3N (b) Ga_2Te_3 (c) Ba_3N_2 (d) $MgBr_2$
3-29. (a) Fe_2O_3 (b) CdS (c) RuF_3 (d) Tl_2S
3-31. (a) copper(I) iodide (b) mercury(I) bromide (c) cobalt(II) fluoride (d) iron(II) oxide
3-33. (a) CoP (b) MnO_2 (c) V_2O_5 (d) $TiCl_4$
3-35. (a) $2K(s) + 2H_2O(l) \rightarrow 2KOH(s) + H_2(g)$
(b) $KH(s) + H_2O(l) \rightarrow KOH(s) + H_2(g)$
(c) $SiO_2(s) + 3C(s) \rightarrow SiC(s) + 2CO(g)$
(d) $SiO_2(s) + 4HF(g) \rightarrow SiF_4(g) + 2H_2O(l)$
(e) $2P(s) + 3Cl_2(g) \rightarrow 2PCl_3(l)$
3-37. $Na_3P(s) + 3H_2O(l) \rightarrow PH_3(g) + 3NaOH(aq)$
3-39. We would predict that LiF would dissolve the least in water because the first member of a group does not behave chemically as the other members of the group.
3-41. (a) SnF_4 (b) HgS (c) Co_2O_3 (d) CrP
3-43. (a) $2Rb(s) + H_2(g) \rightarrow 2RbH(s)$
rubidium hydride
(b) $6Li(s) + N_2(g) \rightarrow 2Li_3N(s)$
lithium nitride
(c) $Sr(s) + Cl_2(g) \rightarrow SrCl_2(s)$
strontium chloride
(d) $2Sc(s) + 3S(s) \rightarrow Sc_2S_3(s)$
scandium sulfide
3-45. (a) chromium(II) chloride (b) iron(III) iodide (c) mercury(II) chloride (d) gold(I) sulfide
3-47. (a) $Ba(s) + I_2(s) \rightarrow BaI_2(s)$
barium iodide
(b) $4Al(s) + 3O_2(g) \rightarrow 2Al_2O_3(s)$
aluminum oxide
(c) $2Ga(s) + 3F_2(g) \rightarrow 2GaF_3(s)$
gallium fluoride
(d) $2K(s) + Br_2(l) \rightarrow 2KBr(s)$
potassium bromide

CHAPTER 4

4-1. (a) +2 (b) +3 (c) +5 (d) 8/3
4-3. (a) −1 (b) +4 (c) +5 (d) +3
4-5. (a) +4 (b) +1 (c) +5 (d) +3
4-7. (a) 0 (b) −4 (c) −2 (d) +2
4-9. (a) +3 (b) +3 (c) +3 (d) +3
4-11. (a) +4 (b) +2 (c) +5 (d) +3
4-13. (a) +6 (b) +6 (c) +3 (d) +7
4-15. (a) $Na_2SO_4(s) + 4C(s) \rightarrow Na_2S(s) + 4CO(g)$
The oxidizing agent is $Na_2SO_4(s)$ and the reducing agent is $C(s)$.
(b) $I_2O_5(s) + 5CO(g) \rightarrow 5CO_2(g) + I_2(g)$
The oxidizing agent is $I_2O_5(s)$ and the reducing agent is $CO(g)$.
4-17. (a) $3NO_2(g) + H_2O(l) \rightarrow 2HNO_3(aq) + NO(g)$
The electron donor is NO_2 and the electron acceptor is NO_2.
(b) $CaH_2(s) + 2H_2O(l) \rightarrow 2H_2(g) + Ca(OH)_2(aq)$
The electron donor is CaH_2 and the electron acceptor is H_2O.

4-19. (a) $4NaOH(aq) + Ca(OH)_2(aq) + C(s) + 4ClO_2(g) \rightarrow 4NaClO_2(aq) + CaCO_3(s) + 3H_2O(l)$
The species reduced is ClO_2 and the species oxidized is C.
(b) $2I_2(s) + N_2H_4(aq) \rightarrow 4HI(aq) + N_2(g)$
The species reduced is I_2 and the species oxidized is N_2H_4.
4-21. (a) Calcium is oxidized and chlorine is reduced. (b) Aluminum is oxidized and oxygen is reduced. (c) Rubidium is oxidized and bromine is reduced. (d) Sodium is oxidized and sulfur is reduced.
4-23. (a) two electrons (b) six electrons (c) one electron (d) two electrons
4-25. (a) decomposition (b) combination (c) single-replacement (d) double-replacement
4-27. decomposition; already balanced
(b) combination: $4Fe(s) + 3O_2(g) \rightarrow 2Fe_2O_3(s)$
(c) single-replacement: $2Al(s) + Mn_2O_3(s) \rightarrow 2Mn(s) + Al_2O_3(s)$
(d) double-replacement: $2AgNO_3(aq) + H_2SO_4(aq) \rightarrow Ag_2SO_4(s) + 2HNO_3(aq)$
(e) double-replacement: $Ca(OH)_2(aq) + 2HBr(aq) \rightarrow CaBr_2(aq) + 2H_2O(l)$
(f) single-replacement: $Cd(s) + 2HCl(aq) \rightarrow CdCl_2(aq) + H_2(g)$
4-29. (a) $3Mg(s) + N_2(g) \rightarrow Mg_3N_2(s)$
(b) $H_2(g) + S(s) \rightarrow H_2S(g)$
(c) $2K(s) + Br_2(l) \rightarrow 2KBr(s)$
(d) $4Al(s) + 3O_2(g) \rightarrow 2Al_2O_3(s)$
(e) $MgO(s) + SO_2(g) \rightarrow MgSO_3(s)$
4-31. (a) $Zn(s) + 2HBr(aq) \rightarrow ZnBr_2(aq) + H_2(g)$
(b) $2Al(s) + Fe_2O_3(s) \rightarrow 2Fe(s) + Al_2O_3(s)$
(c) $Pb(s) + Cu(NO_3)_2(aq) \rightarrow Cu(s) + Pb(NO_3)_2(aq)$
(d) $Br_2(l) + 2NaI(aq) \rightarrow 2NaBr(aq) + I_2(s)$
4-33. (a) $2H^+(aq) + S^{2-}(aq) \rightarrow H_2S(g)$
(b) $Pb^{2+}(aq) + S^{2-}(aq) \rightarrow PbS(s)$
(c) $H^+(aq) + OH^-(aq) \rightarrow H_2O(l)$
(d) $Na_2O(s) + 2H^+(aq) \rightarrow 2Na^+(aq) + H_2O(l)$
(e) $NH_3(aq) + H^+(aq) \rightarrow NH_4^+(aq)$
4-35. (a) $Fe(NO_3)_3(aq) + 3NaOH(aq) \rightarrow Fe(OH)_3(s) + 3NaNO_3(aq)$
$Fe^{3+}(aq) + 3OH^-(aq) \rightarrow Fe(OH)_3(s)$
(b) $Zn(ClO_4)_2(aq) + K_2S(aq) \rightarrow ZnS(s) + 2KClO_4(aq)$
$Zn^{2+}(aq) + S^{2-}(aq) \rightarrow ZnS(s)$
(c) $Pb(NO_3)_2(aq) + 2KOH(aq) \rightarrow Pb(OH)_2(s) + 2KNO_3(aq)$
$Pb^{2+}(aq) + 2OH^-(aq) \rightarrow Pb(OH)_2(s)$
(d) $Zn(NO_3)_2(aq) + Na_2CO_3(aq) \rightarrow ZnCO_3(s) + 2NaNO_3(aq)$
$Zn^{2+}(aq) + CO_3^{2-}(aq) \rightarrow ZnCO_3(s)$
(e) $Cu(ClO_4)_2(aq) + Na_2CO_3(aq) \rightarrow CuCO_3(s) + 2NaClO_4(aq)$
$Cu^{2+}(aq) + CO_3^{2-}(aq) \rightarrow CuCO_3(s)$
4-37. (a) acidic (b) acidic (c) basic (d) acidic (e) basic
4-39. (a) $2HClO_3(aq) + Ba(OH)_2(aq) \rightarrow \underset{\text{barium chlorate}}{Ba(ClO_3)_2(aq)} + 2H_2O(l)$
$2H^+(aq) + 2OH^-(aq) \rightarrow 2H_2O(l)$
(b) $HC_2H_3O_2(aq) + KOH(aq) \rightarrow \underset{\text{potassium acetate}}{KC_2H_3O_2(aq)} + H_2O(l)$
$HC_2H_3O_2(aq) + OH^-(aq) \rightarrow H_2O(l) + C_2H_3O_2^-(aq)$
(c) $2HI(aq) + Mg(OH)_2(s) \rightarrow \underset{\text{magnesium iodide}}{MgI_2(aq)} + 2H_2O(l)$
$Mg(OH)_2(s) + 2H^+(aq) \rightarrow Mg^{2+}(aq) + 2H_2O(l)$
(d) $H_2SO_4(aq) + 2RbOH(aq) \rightarrow \underset{\text{rubidium sulfate}}{Rb_2SO_4(aq)} + 2H_2O(l)$
$2H^+(aq) + 2OH^-(aq) \rightarrow 2H_2O(l)$
4-41. (a) calcium cyanide (b) silver perchlorate (c) potassium permanganate (d) strontium chromate
4-43. (a) ammonium sulfate (b) ammonium phosphate (c) calcium phosphate (d) potassium phosphate
4-45. (a) $Na_2S_2O_3$ (b) $KHCO_3$ (c) $NaClO$ (d) $CaSO_3$
4-47. (a) Na_2SO_3 (b) K_3PO_4 (c) Ag_2SO_4 (d) NH_4NO_3
4-49. (a) mercury(I) chloride (b) chromium(III) nitrate (c) cobalt(II) bromide (d) copper(II) carbonate
4-51. (a) Cr_2O_3 (b) $Sn(OH)_2$ (c) $Cu(C_2H_3O_2)_2$ (d) $Co_2(SO_4)_3$
4-53. (a) sulfurous acid (b) bromic acid (c) hypophosphorous acid (d) periodic acid
4-55. (a) potassium hypobromite (b) calcium hydrogen phosphite (c) lead(II) chlorite (d) nickel(II) perchlorate
4-57. $2Ca_3(PO_4)_2(s) + 6SiO_2(s) + 10C(s) \rightarrow CaSiO_3(l) + 10CO(g) + P_4(g)$
4-59. (a) +4 (b) +2 (c) +4 (d) +2 (e) 0 (f) −1 (g) −2 (h) −4/3
4-61. (a) $HCl(aq) + KCN(aq) \rightarrow HCN(g) + KCl(aq)$
(b) $2K(s) + 2H_2O(l) \rightarrow H_2(g) + 2KOH(aq)$
(c) $2H_2O_2(aq) \rightarrow O_2(g) + 2H_2O(l)$
(d) $H_2(g) + Br_2(l) \rightarrow 2HBr(g)$
4-63. $2Pb(l) + O_2(g) \rightarrow 2PbO(s)$
$Ag(l) + O_2(g) \rightarrow$ no reaction
4-65. $HgS(s) + O_2(g) \xrightarrow{\text{heat}} Hg(g) + SO_2(g) \xrightarrow{\text{cool}} Hg(l) + SO_2(g)$
4-67. Na, Fe, Sn, Au
4-69. (a) $2Na(s) + H_2(g) \rightarrow 2NaH(s)$
(b) $2Al(s) + 3S(s) \rightarrow Al_2S_3(s)$
(c) $H_2O(g) + C(s) \rightarrow CO(g) + H_2(g)$
(d) $C(s) + 2H_2(g) \rightarrow CH_4(g)$
(e) $PCl_3(l) + Cl_2(g) \rightarrow PCl_5(s)$
4-71. $2CuO(s) + C(s) \xrightarrow{\text{high T}} 2Cu(s) + CO_2(g)$
$SnO_2(s) + C(s) \xrightarrow{\text{high T}} Sn(s) + CO_2(g)$
$2Fe_2O_3(s) + 3C(s) \xrightarrow{\text{high T}} 4Fe(s) + 3CO_2(g)$
4-73. (a) $4PH_3(g) + 8O_2(g) \rightarrow P_4O_{10}(s) + 6H_2O(l)$
(b) $2Cr^{3+}(aq) + 10OH^-(aq) + 3H_2O_2(aq) \rightarrow 2CrO_4^{2-}(aq) + 8H_2O(l)$
(c) $4Sn^{2+}(aq) + 2HNO_3(aq) + 8H^+(aq) \rightarrow N_2O(g) + 4Sn^{4+}(aq) + 5H_2O(l)$
(d) $3F_2(g) + 3H_2O(l) \rightarrow 6HF(g) + O_3(g)$
4-75. (a) +5 (b) +4 (c) +7 (d) +6

CHAPTER 5

5-1. (a) 1.55 mol (b) 0.0167 mol (c) 7.77 mol (d) 1.78×10^4 mol
5-3. (a) 3.03 mol (b) 0.219 mol (c) 3.20×10^{-4} mol (d) 9.368×10^{-3} mol

5-5. The mass of 1 mol of baseballs is 1.4×10^{-2} that of the mass of the earth.
5-7. (a) 7.308×10^{-23} g (b) 2.992×10^{-22} g (c) 1.843×10^{-22} g
5-9. (a) 1.855×10^{-20} g (b) 3.0×10^{-7} g (c) 2.7×10^{-17} g (d) 5.3×10^{-17} g
5-11. 2.77 mol; 1.67×10^{24} molecules; 5.01×10^{24} atoms
5-13. CaC_2
5-15. $CuCl_2$
5-17. $CoCl_2$
5-19. (a) Li_2N (b) Li_3N (c) LiN_3 (d) $CaCl_2$
5-21. 144.3; neodymium
5-23. Barium
5-25. C_3H_6O
5-27. $Na_6P_6O_{18}$
5-29. 36.3 g
5-31. 81.5 g
5-33. 599 mg
5-35. 1.34×10^5 kg
5-37. 9.41 metric ton
5-39. 3.650%
5-41. 39.1 metric ton
5-43. 23.8 kg
5-45. 29.0 g
5-47. 46.3%
5-49. 63.0%
5-51. 1280 g Na_2CS_3; 442 g Na_2CO_3; 225 g H_2O
5-53. 66.4%
5-55. 10.0%
5-57. 71.4% NaCl
5-59. 62.9% Zn; 37.1% Mg
5-61. 43.3% K_2SO_4; 56.7% $MnSO_4$
5-63. 6.15 g Al; 3.72 g Mg
5-65. 2.731×10^{26}
5-67. 1.57 kg
5-69. 7.82 g
5-71. 5.77 kg
5-73. 321 g
5-75. 62.4%
5-77. 48.2% Na_2SO_4; 51.8% $NaHSO_4$
5-79. MoO_2
5-81. 25.6 metric ton
5-83. 358 g
5-85. 8.6×10^{22} atom
5-87. (a) 7.12 $cm^3 \cdot mol^{-1}$ (b) 18.02 $cm^3 \cdot mol^{-1}$ (c) 3.41 $cm^3 \cdot mol^{-1}$ (d) 2.24×10^4 $cm^3 \cdot mol^{-1}$
5-89. 11.2 g
5-91. 21.7 g
5-93. (a) 1.42×10^{23} atoms (b) 2.1×10^{21} atoms (c) 3.72×10^{22} atoms (d) 5.16×10^{25} atoms

CHAPTER 6

6-1. 0.0250 M
6-3. 14.3 M
6-5. (a) 3.195×10^{-3} mol (b) 1.0×10^{-6} mol
6-7. 8.3 mL
6-9. Dissolve 42.8 g of sucrose in sufficient water to make 500 mL of the solution.
6-11. 18.2 M
6-13. 148 mL NaOH(*aq*); 1.79 g H_2
6-15. 27.7 mL
6-17. 55.5 g
6-19. 39.0 mL
6-21. 1060 g
6-23. (a) insoluble (b) soluble (c) soluble (d) soluble (e) insoluble
6-25. (a) insoluble (b) soluble (c) soluble (d) insoluble (e) insoluble
6-27. (a) $Cu^{2+}(aq) + S^{2-}(aq) \rightarrow CuS(s)$
(b) $Mg^{2+}(aq) + CO_3^{2-}(aq) \rightarrow MgCO_3(s)$
(c) $Ba^{2+}(aq) + SO_4^{2-}(aq) \rightarrow BaSO_4(s)$
(d) $Hg_2^{2+}(aq) + 2Cl^-(aq) \rightarrow Hg_2Cl_2(s)$
6-29. 0.171 M
6-31. (a) 2.8 μL (b) 8.33 mL
6-33. 0.114 M
6-35. 52.1% NaOH
6-37. 20.7%
6-39. 0.400 M
6-41. 0.459 g
6-43. 14.1 mL
6-45. 0.124 M $Sc^{3+}(aq)$; 0.372 M $NO_3^-(aq)$
6-47. 11.7 M
6-49. 33.3 g
6-51. 44.5 mL
6-53. 10.1 mL
6-55. 5.24 g
6-57. 23.2 g
6-59. 1000 g
6-61. 23.8% NaCl; 76.2% KBr
6-63. 184 g
6-65. +2
6-67. 2.57% N
6-69. 2.10 g
6-71. 5.14 g
6-73. 0.2300 M
6-75. 6.89%
6-77. 0.0310 M
6-79. 0.608%
6-81. 66 mL
6-83. 13.2 M
6-85. 2.77 g
6-87. Dilute 25 mL of the 1.0 M solution with water to make 500 mL.
6-89. 60.1

CHAPTER 7

7-1. (a) 5.7×10^4 torr and 76 bar (b) 0.76 atm and 770 mbar (c) 5.3×10^5 Pa and 530 kPa (d) 1.23×10^5 Pa and 1.21 atm
7-3. 1.63 cm at the surface
7-5. (a) 310 K (b) 293 K (c) 14 K (d) 472 K
7-7. 20.1 mL

7-9. 50 L
7-11. 21 L
7-13. 1.7 atm, 18 mL $H_2O(l)$
7-15. 1.0×10^{20} molecules at −200°C and 0.0010 atm; 2.7×10^{22} molecules at 0°C and 1 atm
7-17. 4.1×10^{12} molecules
7-19. 35.0 L at 0°C and 1 atm; 50.3 L at 120°C and 1 atm
7-21. 848 mL
7-23. 1.75 g
7-25. 1630
7-27. C_2H_4
7-29. C_2H_4
7-31. 0.610 atm $H_2(g)$; 1.38 atm $N_2(g)$
7-33. 222 torr
7-35. 5.1 L; 10 atm
7-37. 442 $m \cdot s^{-1}$
7-39. $v_{rmsf} = \sqrt{2} v_{rmsi}$
7-41. $^{238}UF_6 < {}^{235}UF_6 < NO_2 < CO_2 < O_2 < N_2 < H_2O$
7-43. 2.2×10^{19} m
7-45. 1.7×10^{10} $collisions \cdot s^{-1}$
7-47. 200 $mL \cdot h^{-1}$
7-49. 70.5
7-51. 16.5 atm
7-53. 2.49×10^{19} molecules
7-55. 56.2 μg
7-57. 17.7 $g \cdot m^{-3}$
7-59. 1.013×10^5 $N \cdot M^{-2}$
7-61. The plot of V versus $1/P$ is a straight line.
7-63. 1.2 L
7-65. 8600 gal
7-67. 62.8%
7-69. 21.2 mL
7-71. 0.57 atm
7-73. 1.4 lb $O_2(g)$
7-75. 1764 K
7-77. 32.9 mL
7-79. 287 L $CO_2(g)$; 144 L $O_2(g)$
7-81. 4.02 g
7-83. 91.8
7-85. 3.0 L
7-87. 238 mL
7-89. 466 torr $N_2(g)$; 317 torr $O_2(g)$; 783 torr total
7-91. 0.452 M
7-93. 43% NaH; 57% CaH_2

CHAPTER 8

8-1. 130 m
8-3. 4.3 m
8-5. 77 m^2
8-7. 804 $kJ \cdot mol^{-1}$
8-9. 89.3 $kJ \cdot mol^{-1}$
8-11. −5.3 kJ
8-13. −204 kJ
8-15. 145.4 kJ
8-17. 93.6 kJ
8-19. 229.27 kJ
8-21. −4.2 kJ
8-23. 180 kJ
8-25. −106 kJ
8-27. (a) −534.2 kJ; exothermic (b) −44.2 kJ; exothermic (c) −429.8 kJ; exothermic
8-29. (a) −29.67 $kJ \cdot g^{-1}$ (b) −51.87 $kJ \cdot g^{-1}$
8-31. −1249.1 $kJ \cdot mol^{-1}$
8-33. (a) 472.6 $kJ \cdot mol^{-1}$ (b) 79.0 $kJ \cdot mol^{-1}$ (c) 218.0 $kJ \cdot mol^{-1}$ (d) 121.7 $kJ \cdot mol^{-1}$, N_2 (*g*) has the greatest bond strength
8-35. 32.4 kJ
8-37. 113 $J.K^{-1} \cdot mol^{-1}$
8-39. 3.34×10^6 J
8-41. 21.6°C
8-43. 350°C
8-45. −56.2 $kJ \cdot mol^{-1}$
8-47. -1.48×10^3 kJ
8-49. 17.2 $kJ \cdot mol^{-1}$
8-51. −2040 $kJ \cdot mol^{-1}$
8-53. −827 $kJ \cdot mol^{-1}$
8-55. 8.6×10^6 $barrels \cdot d^{-1}$
8-57. 24,000 kJ
8-59. −1234 kJ
8-61. Petroleum liquids are easier to distribute to the site at which they are used because they can be pumped from the source in underground pipes. Petroleum liquids do not have to be stored on the site of use as does coal. Petroleum liquids can be pumped directly to the device used to burn the fuel rather than transported by the more cumbersome means that coal requires.
8-63. 310 g ice
8-65. TlCl
8-67. ZnSe
8-69. 120 mi
8-71. (a) −565 kJ (b) −540 kJ (c) −180 kJ
8-73. −1049 kJ
8-75. 1.8×10^5 kW
8-77. 9.0×10^{10} kJ. The conversion of mass to energy is around 10^{10} times greater than the energies per gram of ordinary chemical processes.
8-79. −88.2 kJ

CHAPTER 9

9-1. Be, Kr, Ne, and He
9-3. The plot of $\log(I_n)$ versus the number of electrons removed suggests that the five electrons are arranged in two shells with two electrons in an inner, tightly held shell and three in an outer shell.
9-5. Li· Na· K· Rb· Cs· Fr·
:F· :Cl· :Br· :I· :At·
9-7. Ar ·S· :S:$^{2-}$ Al^{3+} :Cl:$^-$
9-9. 4.74×10^{14} s^{-1}

9-11. 286 nm
9-13. 6 photons
9-15. 9.94×10^{-19} J. Electrons will be ejected from the surface of the gold.
9-17. 2.07×10^{-19} J
9-19. 3.97 pm
9-21. 99.0 pm
9-23. 656 nm
9-25. The electron makes a transition from the ground state to the $n = 4$ state and then from the $n = 4$ state to the $n = 2$ state.
9-27. $IE(He^+) = 5.25$ MJ·mol^{-1}; $IE(Li^{2+}) =$ 11.8 MJ·mol^{-1}; $IE(Be^{3+}) = 21.0$ MJ·mol^{-1}
9-29. (a), (c), and (e) are possible.
9-31. (a) $4p$ (b) $3d$ (c) $4d$ (d) $2s$
9-33. n must be at least 3. l must be at least 3.
9-35.

n	l	m_l	m_s
3	2	−2	$+\frac{1}{2}$
			$-\frac{1}{2}$
		−1	$+\frac{1}{2}$
			$-\frac{1}{2}$
		0	$+\frac{1}{2}$
			$-\frac{1}{2}$
		+1	$+\frac{1}{2}$
			$-\frac{1}{2}$
		+2	$+\frac{1}{2}$
			$-\frac{1}{2}$

9-37. s orbital—2 electrons; p orbital—6 electrons; d orbital—10 electrons; f orbital—14 electrons
9-39. In a d transition series the five d orbitals are being filled. A set of five d orbitals can hold up to 10 electrons.
9-41. (d) is allowed.
9-43. (a) $1s^22s^22p^63s^23p^2$ 14 electrons, silicon
(b) $1s^22s^22p^63s^23p^64s^13d^5$ 24 electrons, chromium
(c) $1s^22s^22p^63s^23p^64s^23d^{10}4p^2$ 32 electrons, germanium
(d) $1s^22s^22p^63s^23p^64s^23d^{10}4p^5$ 35 electrons, bromine
(e) $1s^22s^22p^1$ 5 electrons, boron
9-45. (a) Ti: $[Ar]4s^23d^2$ (b) K: $[Ar]4s^1$
(c) Fe: $[Ar]4s^23d^6$ (d) As: $[Ar]4s^23d^{10}4p^3$
9-47. (a) Ca: $[Ar]4s^2$ (b) Br: $[Ar]4s^23d^{10}4p^5$
(c) Ag: $[Kr]5s^14d^{10}$ (d) Zn: $[Ar]4s^23d^{10}$
9-49. (a) Groups 1 and 2 (b) Groups 3, 4, 5, 6, 7, 8 (c) the transition metals (d) the lanthanides and actinides
9-51. (a) Ge: $[Ar]4s^23d^{10}4p^2$; two unpaired electrons
(b) Se: $[Ar]4s^23d^{10}4p^4$; two unpaired electrons
(c) V: $[Ar]4s^23d^3$; three unpaired electrons
(d) Fe: $[Ar]4s^23d^6$; four unpaired electrons
9-53. (a) 1 electron $\cdot H \cdot^-$ helium
(b) 2 electrons $:\ddot{\underset{..}{O}}:^{2-}$ neon (c) 4 electrons $:\ddot{\underset{..}{C}}:^{4-}$ neon
(d) 2 electrons $:\ddot{\underset{..}{S}}:^{2-}$ argon
9-55. (a) 15 + 3 = 18 electrons: $1s^22s^22p^63s^23p^6$; [Ar]
(b) 35 + 1 = 36 electrons: $1s^22s^22p^63s^23p^64s^23d^{10}4p^6$; [Kr]
(c) 34 + 2 = 36 electrons: $1s^22s^22p^63s^23p^64s^23d^{10}4p^6$; [Kr]
(d) 56 − 2 = 54 electrons:
$1s^22s^22p^63s^23p^64s^23d^{10}4p^65s^24d^{10}5p^6$; [Xe]
9-57. (a) 2 (b) zero (c) zero (d) 1
9-59. (a) $O(g) + 2e^- \rightarrow O^{2-}(g)$
$[He]2s^22p^4 + 2e^- \rightarrow [He]2s^22p^6$ or [Ne]
(b) $Ca(g) + Sr^{2+}(g) \rightarrow Sr(g) + Ca^{2+}(g)$
$[Ar]4s^2 + [Kr] \rightarrow [Kr]5s^2 + [Ar]$
9-61. (a) $1s^12s^1$ (b) $2s^1$ (c) $1s^22s^22p^53s^1$ (d) $1s^22s^22p^53s^1$
9-63. (a) P > N (b) P > S (c) S > Ar (d) Kr > Ar
9-65. (a) Li < Na < Rb < Cs (b) P < Al < Mg < Na
(c) Ca < Sr < Ba
9-67. The energy of attraction between the nucleus and the outer electrons is greater the higher the value of the nuclear charge and is less the higher the value of the principal quantum number. These two effects oppose one another as we move down a group. The underlying electrons in the noble-gas-like core partially screen the nuclear charge, and the farther an electron (larger n) is from the nucleus the lower is the ionization energy.
9-69. 5×10^{19} photons
9-71. 6.02×10^{-20} J
9-73. A plot of $\nu_{n\rightarrow 1}$ versus $1/n^2$ is a straight line.
9-75. (a) 20 kJ (b) −710 kJ (c) −5.14 MJ
9-77. (a) $Z = 8, 34, 52, 84$ (b) $Z = 1, 3, 19, 37, 55, 87$
9-79. (a) 3 (b) 4 (c) 2
9-81. $2s^1$, $2s^12p_x^1$, $2s^12p_x^12p_y^1$, $2s^12p_x^12p_y^12p_z^1$, $2s^22p_x^12p_y^12p_z^1$, $2s^22p_x^22p_y^12p_z^1$, $2s^22p_x^22p_y^22p_z^1$, $2s^22p_x^22p_y^22p_z^2$
9-83. 3.97×10^5 m·s^{-1}
9-85. (a) 2 (b) 7 (c) 6 (d) 1 (e) 2
9-87. See the *Study Guide/Solutions Manual.*
9-89. See the *Study Guide/Solutions Manual.*
9-91. 1.89×10^{22} photons
9-93. A plot of kinetic energy versus frequency is a straight line. The slope of the line gives the value of h.

CHAPTER 10

10-1. (a) $Ca([Ar]4s^2) + 2F([He]2s^22p^5) \rightarrow Ca^{2+}([Ar]) + 2F^-([Ne]) \rightarrow CaF_2(g)$
(b) $Sr([Kr]5s^2) + 2Br([Ar]4s^23d^{10}4p^5) \rightarrow Sr^{2+}([Kr]) + 2Br^-([Kr]) \rightarrow SrBr_2(g)$
(c) $2Al([Ne]3s^23p^1) + 3O([He]2s^22p^4) \rightarrow 2Al^{3+}([Ne]) + 3O^{2-}([Ne]) \rightarrow Al_2O_3(g)$
10-3. (a) $3Li\cdot + \cdot\ddot{\underset{\cdot}{N}}\cdot \rightarrow 3Li^+ + :\ddot{\underset{..}{N}}:^{3-}(Li_3N)$
(b) $Na\cdot + H\cdot \rightarrow Na^+ + H:^-(NaH)$
(c) $\cdot\dot{Al}\cdot + 3:\ddot{\underset{..}{I}}\cdot \rightarrow Al^{3+} + 3:\ddot{\underset{..}{I}}:^-(AlI_3)$
10-5. (a) $Cr^{2+}([Ar]3d^4)$ (b) $Cu^{2+}([Ar]3d^9)$
(c) $Co^{3+}([Ar]3d^6)$ (d) $Mn^{2+}([Ar]3d^5)$
10-7. (a) Fe, Ru, Os (b) Zn, Cd, Hg (c) Sc, Y, Lu
(d) Mn, Tc, Re
10-9. (a) $Cd^{2+}([Kr]4d^{10})$ (b) $In^{3+}([Kr]4d^{10})$
(c) $Tl^{3+}([Xe]4f^{14}5d^{10})$ (d) $Zn^{2+}([Ar]3d^{10})$
10-11. (d), (e), and (f) are isoelectronic ions.
10-13. (a) Y_2S_3 (b) $LaBr_3$ (c) MgTe (d) Rb_3N
(e) Al_2Se_3 (f) CaO

10-15. (a) Cl^- (b) Ag^+ (c) Cu^+ (d) O^{2-}
10-17. Cl > Br > I > H
10-19. $-270\ kJ\cdot mol^{-1}$
10-21. $196\ kJ\cdot mol^{-1}$ (b) $-53\ kJ\cdot mol^{-1}$ (c) $428\ kJ\cdot mol^{-1}$
10-23. -4.32×10^{-18} J
10-25. $-329\ kJ\cdot mol^{-1}$
10-27. $-130\ kJ\cdot mol^{-1}$
10-29. $-523\ kJ\cdot mol^{-1}$
10-31. $-581\ kJ\cdot mol^{-1}$
10-33. $-299\ kJ\cdot mol^{-1}$
10-35. $-795\ kJ\cdot mol^{-1}$
10-37. N^{3-}, O^{2-}, Na^+, Mg^{2+}, Al^{3+}
10-39. The outer electrons in K^+ and Cu^+ are in the same shell, but the nuclear charge of Cu is greater than that of K.
10-41. $-348\ kJ\cdot mol^{-1}$
10-43. Four unpaired electrons in Fe^{2+}; no unpaired electrons in Zn^{2+}
10-45. The magnitude of the electron affinity decreases and the ionic size of X^- increases.
10-47. Ti^+, Zr^+, Hf^+
10-49. (a) strong (b) weak (c) non (d) non
10-51. Gaseous calcium atoms will readily reduce $Ca^{2+}(g)$ to $Ca^+(g)$ ($\Delta H^\circ_{rxn} = -550$ kJ). However, in aqueous solution Ca^{2+} binds much more strongly to water molecules because of its +2 charge than does Ca^+. The high solvation energy of Ca^{2+} stabilizes it relative to Ca^+ in aqueous solution and $Ca^+(aq)$ is unstable relative to $Ca^{2+}(aq)$ and Ca(*s*).
10-53. Fe(II): $1s^22s^22p^63s^23p^63d^6$
Fe(II): $1s^22s^22p^63s^23p^63d^5$
10-55. Cl^-
10-57. weak; strong; strong; strong
10-59. The fluoride ion is so small that the coulombic force between the fluoride ion and a metal ion is greater than between any other anion and the metal ion.
10-61. $-1540\ kJ\cdot mol^{-1}$
10-63. The Tl^+ ion and the K^+ ion have identical charges and similar ionic radii and thus $Tl^+(aq)$ moves around in human tissue much as $K^+(aq)$ does.

CHAPTER 11

11-1. (a) :C̤̈l—S̤̈—C̤̈l: (b) :C̤̈l—Ge—C̤̈l: with :C̤̈l: above and below Ge (c) :B̤̈r—Ä̤s—B̤̈r: with :B̤̈r: below As (d) H—P̤̈—H with H below P

11-3. H—Ö̤—Ö̤—H

11-5. (a) H—C—H with H above and below C (b) H—C—F̤̈: with H above and below C (c) H—C—N̈—H with H above and below C and H below N

11-7. (a) H—C≡C—H (b) H—N̈=N̈—H (c) :C̤̈l—C=Ö: with :C̤̈l: above C

11-9. H—C—Ö̤—H with Ö= above C

11-11. H—C=C—C̤̈l: with H above each C

11-13. :F̤̈—N̈—F̤̈: with :F̤̈: below N

11-15. NNO is more likely.

11-17. [H—C═O with O above C]$^{\ominus}$

11-19. [O═C═O with O above C]$^{2-}$

11-21. (a) ·Ö̤—N̈=Ö: ⟷ :Ö=N̈—Ö̤· odd electron
(b) $\ominus$ $\oplus$:C≡O:
(c) ·Ö̤—Ö̤—Ö̤: ⟷ :Ö̤—Ö̤—Ö̤· odd electron
(d) $\ominus$:Ö̤—Ö̤· ⟷ ·Ö̤—Ö̤: $\ominus$ odd electron

11-23. H—C—N̈—N̈=Ö· with H above and below C is a free radical.

11-25. (a) :C̤̈l—P—C̤̈l: with four more Cl on P, $\ominus$ on P (b) :Ï̤—Ï̤—Ï̤: with $\ominus$ on central I (c) :F̤̈—Si—F̤̈: with four more F on Si, $2-$ on Si

11-27. :F̤̈—Ï—F̤̈: with :F̤̈: below I; I with five F

11-29. :F̤̈—S̈—F̤̈: with :F̤̈: above and below S; :F̤̈—S—F̤̈: with four more F on S

11-31. 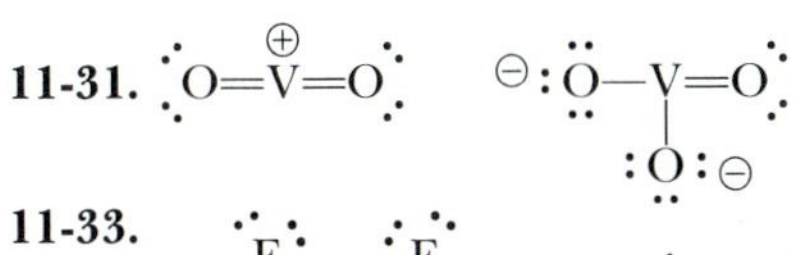

11-33.

:F: :F: (charge 2−)
:F—Ti—F:
:F: :F:

Br Br
Br—Ti—Br (charge ⊖)
:Br:

11-35.

δ+ δ−
:Br—Cl: or :Br—Cl: (arrow +⟶ above)

11-37. (a) δ− :F—N—F: δ− (3δ+ on N), :F: δ− below N (b) δ− :F—O—F: δ− (2δ+ on O)

(c) δ+ 2δ− δ+
:Br—O—Br:

11-39. (a) [O, O, S, O, O] (2−)

All four bonds are equivalent; they have the same bond lengths and the same bond energy.

(b) [O, O, P, O, O] (3−)

All four bonds are equivalent; they have the same bond lengths and the same bond energy.

(c)

H O
| ||
H—C—C==O (⊖)
|
H

The two C—O bonds are equivalent; they have the same bond lengths and the same bond energy.

11-41.

(a) :F: / :F—N(⊕)—F: / :F: (b) :F: / :F—Cl(⊕)—F: / :F: (c) H / H—P(⊕)—H / H

(d) F F / :F—As(⊖)—F: / F F (e) :F: / :F—Br(⊖)—F: / :F:

11-43. (a) ⊖:O:
⊖:O—Cl(2+)—O—H ⟷ O
||
O=Cl—O—H +
other resonance forms with an expanded valence shell

(b) O=N—O—H

(c) :O:⊖ / ⊖:O—I(3+)—O—H / :O:⊖ ⟷ O / || / O=I—O—H + / || / O

other resonance forms with an expanded valence shell

(d) ⊖:O—Br(⊕)—O—H ⟷ O=Br—O—H

11-45.
(a) :F—S—F: (b) :F: / :F—S—F: / :F:

(c)

F F F F
:F—S——S—F:
F F F F

(d) :F—S—S—F:

11-47.

⊖:O: :O:⊖
⊖:O—Cr(2+)—O—Cr(2+)—O:⊖ ⟷
⊖:O: :O:⊖

O O
|| ||
⊖:O—Cr—O—Cr—O:⊖ +
|| ||
O O

other resonance forms with expanded valence shells

11-49. (a) and (c)

11-51. FNNF

11-53.

H H H H H
| | | | |
(a) H—C—C—O—H (b) H—C—C—C—O—H
| | | | |
H H H H H

H H H
| | |
(c) H—C—C—C—H
| | |
H :O: H
|
H

11-55. H—O—O:⊖; sodium hydroperoxide; barium hydroperoxide

11-57.

(a) [S / || / S==C==S] (2−) (b) [O O / || || / O==C—C==O] (2−)

(c) [N≡C==S] (⊖) (d) ⊖:O—O· ⟷ ·O—O:⊖

11-59. (a) :Cl: bonded to a benzene ring (b) H—N—H bonded to a benzene ring

(c) O=C—O—H bonded to a benzene ring (d) :O—H bonded to a benzene ring

11-61. Ring of eight S atoms: S—S, :S: :S:, :S: :S:, S—S

11-63. (a) :Cl—Hg—Cl: (b) O=Ti=O

(c) UF_6: six F atoms bonded to U (d) O=Os=O with two further O atoms double-bonded to Os

11-65. Cl > S > Se > Sb > In

CHAPTER 12

12-1. (a), (b), and (d) have 90° angles.
12-3. (c) and (d) have 120° angles.
12-5. (a) bent (b) bent (c) linear (d) bent
12-7. (a) square planar (b) seesaw shaped (c) tetrahedral (d) seesaw shaped
12-9. (a) square pyramidal (b) trigonal bipyramidal (c) trigonal bipyramidal (d) square pyramidal
12-11. (a) octahedral, 1 (b) trigonal bipyramidal, 1,3 (c) square planar, 1 (d) trigonal planar, 3
12-13. (a) tetrahedral, 2 (b) octahedral, 1 (c) square planar, 1 (d) tetrahedral, 2
12-15. All AB compounds are linear. All AB_3 compounds are T-shaped. All AB_5 compounds are square pyramidal.
12-17. (a) trigonal pyramidal (b) tetrahedral (c) trigonal bipyramidal (d) tetrahedral
12-19. (a) trigonal planar (b) tetrahedral (c) linear (d) bent
12-21. (a) bent (b) square pyramidal (c) tetrahedral (d) trigonal pyramidal
12-23. (a) octahedral (b) trigonal bipyramidal (c) trigonal pyramidal (d) tetrahedral
12-25. NO_2^+ is a linear molecule with an O-N-O bond angle of 180°. NO_2^- is a bent molecule with a O-N-O bond angle of slightly less than 120°.
12-27. (a) linear, no dipole moment (b) trigonal bipyramidal, no dipole moment (c) seesaw shaped, dipole moment (d) bent, dipole moment
12-29. (a) trigonal planar, no dipole moment (b) bent, dipole moment (c) seesaw shaped, dipole moment (d) trigonal bipyramidal, no dipole moment
12-31. (a) tetrahedral, nonpolar (b) trigonal pyramidal, polar (c) square planar, nonpolar (d) seesaw shaped, polar
12-33. (a) $\delta-$:F—N—F: $\delta-$ with :F: $\delta-$ on N; N is $3\delta+$ (b) $\delta-$:F—O—F: $\delta-$; O is $2\delta+$

(c) :Br—O—Br: with $\delta+$, $2\delta-$, $\delta+$

12-35. (a) one possible arrangement (b) one possible arrangement (c) one possible arrangement (d) two isomers
12-37. (a) one possible arrangement (b) two isomers (c) two isomers
12-39. four isomers
12-41. (a) seesaw shaped (b) square planar (c) tetrahedral (d) tetrahedral
12-43. See Table 12-3.
12-45. (a) trigonal pyramidal (b) tetrahedral (c) seesaw shaped (d) octahedral
12-47. (a) bent (b) trigonal pyramidal (c) tetrahedral
12-49. (a) HF > HCl > HBr > HI (b) $NH_3 > PH_3 > AsH_3$ (c) $IF_3 > BrF_3 > ClF_3$ (d) $H_2O > H_2S > H_2Se > H_2Te$
12-51. XeF_4
12-53. (a) octahedral (b) linear (c) tetrahedral (d) tetrahedral
12-55. (a) trigonal bipyramidal (b) octahedral (c) octahedral (d) tetrahedral
12-57. (a) trigonal bipyramidal (b) trigonal pyramidal (c) tetrahedral
12-59. (a) tetrahedral (b) tetrahedral (c) tetrahedral
12-61. See the *Study Guide/Solutions Manual.*

CHAPTER 13

13-1. $(1\sigma)^2(1\sigma^*)^2(2\sigma)^2(2\sigma^*)^2$. The bond order is zero and so we predict that Be_2 does not exist.
13-3. The bond order of N_2 is 3 while the bond order of N_2^+ is 2½; therefore, the bond energy of N_2 is greater than that of N_2^+. The bond order of O_2 is 2 while the bond order of O_2^+ is 2½; therefore, the bond energy of O_2 is less than that of O_2^+.
13-5. C_2^{2-} has a larger bond energy and a shorter bond length than C_2
13-7. (a) 18 electrons
$(1\sigma)^2(1\sigma^*)^2(2\sigma)^2(2\sigma^*)^2(1\pi)^2(1\pi)^2(3\sigma)^2(1\pi^*)^2(1\pi^*)^2$
bond order = 1
(b) 11 electrons
$(1\sigma)^2(1\sigma^*)^2(2\sigma)^2(2\sigma^*)^2(1\pi)^2(1\pi)^3$
bond order = 1½
(c) 7 electrons
$(1\sigma)^2(1\sigma^*)^2(2\sigma)^2(2\sigma^*)^1$
bond order = ½
(d) 19 electrons
$(1\sigma)^2(1\sigma^*)^2(2\sigma)^2(2\sigma^*)^2(1\pi)^2(1\pi)^2(3\sigma)^2(1\pi^*)^2(1\pi^*)^2(3\sigma^*)^1$
bond order = ½

13-9. B_2^-. When the additional electron occupies a bonding orbital, a stronger net bonding will result.
13-11. B_2, C_2^{2+}, and F_2^{2+}
13-13. The three localized covalent bonds can be described as B(sp^2)-H(1s).
13-15. The four localized covalent bonds can be described as C(sp^3)-F(2p). The remaining 24 valence electrons are lone pairs on the fluorine atoms.
13-17. Three of the four σ bonds can be described as C(sp^3)-H(1s) and the other as C(sp^3)-Cl(3p). The remaining 6 valence electrons are lone pairs on the chlorine atom.
13-19. The three σ bonds can be described as O(sp^3)-H(1s). The lone electron pair occupies the remaining sp^3 orbital on the oxygen atom.
13-21. The three σ bonds can be described as N(sp^3)-F(2p). One lone pair occupies the remaining sp^3 orbital on the nitrogen atom and the remaining 18 valence electrons are lone pairs on the fluorine atoms.
13-23. The three σ bonds can be described as P(sp^3)-Cl(3p). One lone pair occupies the remaining sp^3 orbital on the phosphorus atom and the remaining 18 valence electrons are lone pairs on the chlorine atoms.
13-25. The four σ bonds can be described as O(sp^3)-H(1s), O(sp^3)-N(sp^3), and N(sp^3)-H(1s). Four of the valence electrons occupy the two remaining oxygen sp^3 orbitals, and two the remaining nitrogen sp^3 orbital.
13-27. The six σ bonds can be described as C(sp^3)-H(1s), C(sp^3)-N(sp^3), and N(sp^3)-H(1s). Two valence electrons occupy the remaining nitrogen sp^3 orbital.
13-29. The eight σ bonds can be described as C(sp^3)-H(1s) and C(sp^3)-O(sp^3). Four valence electrons occupy the two remaining oxygen sp^3 orbitals.
13-31. (a) five σ bonds and one π bond (b) nine σ bonds and two π bonds (c) seven σ bonds and one π bond (d) fourteen σ bonds and two π bonds
13-33. nine σ bonds and two π bonds
13-35. The σ bonds can be described as C(sp)-O(sp). Each of the two π bonds are formed by combining a carbon 2p orbital and an oxygen 2p orbital. One lone pair occupies the remaining carbon sp orbital and the other the remaining oxygen sp orbital.
13-37. The σ-bond framework is

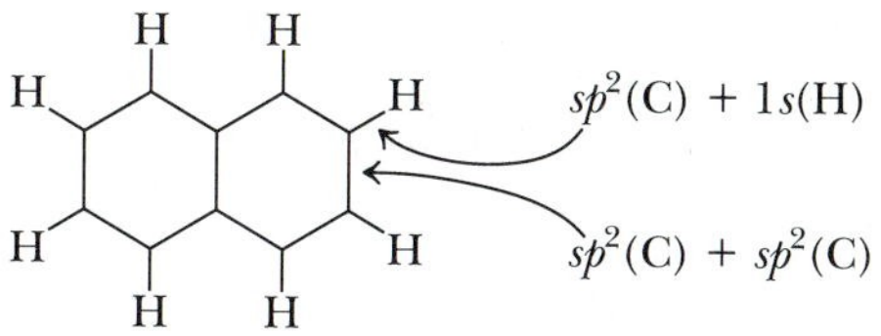

sp^2(C) + 1s(H)

sp^2(C) + sp^2(C)

The remaining 2p orbitals of each carbon atom combine to form ten π bond orbitals that are delocalized over the two rings. Ten valence electrons occupy five of the delocalized π bond orbitals.
13-39. The two σ bonds can be described as N(sp^2)-O(sp^2). Ten of the valence electrons occupy the remaining five sp^2 orbitals. The remaining four valence electrons occupy the π and π^{nb} orbitals. The total bond order is 3.
13-41. The five σ bonds can be described as N(sp^2)-H(1s), N(sp^2)-C(sp^2), C(sp^2)-O(sp^2), and C(sp^2)-H(1s). The remaining 2p orbitals form three delocalized π bond orbitals. Four valence electrons occupy the remaining two oxygen sp^2 orbitals and four occupy two of the delocalized π bond orbitals.
13-43. The four σ bonds formed by the -CH_3 carbon atom can be described as C(sp^3)-H(1s) and C(sp^3)-C(sp^2). Two σ bonds can be described as C(sp^2)-O(sp^2) and C(sp^2)-H(1s). The π bond orbital is formed by combining the remaining carbon 2p orbital with the remaining oxygen 2p orbital. Two valence electrons occupy the π bond orbital and four occupy the remaining two oxygen sp^2 orbitals.
13-45. The four σ bonds formed by the -CH_3 carbon atom can be described as C(sp^3)-H(1s) and C(sp^3)-C(sp). Two σ bonds can be described as C(sp)-C(sp) and C(sp)-H(1s). The two π bond orbitals are formed from the two remaining 2p orbitals on each carbon atom and are occupied by four valence electrons.
13-47. The two bonds can be described as Te(5p)-H(1s).
13-49. We predict that Cl_2 has the longer bond length.
13-51. CO has a shorter bond length and a greater bond energy than CO^+.
13-53. The bond length of nitrogen in the excited state is greater than that of nitrogen in the ground state.
13-55. (a) eight σ bonds and one π bond (b) six σ bonds and one π bond (c) eight σ bonds and two π bonds (d) eight σ bonds
13-57. There are cis-trans isomers of 1,2-dibromoethene.
13-59. The three σ bonds can be described as C(sp^2)-H(1s).
13-61. Six valence electrons occupy the three σ orbitals. Twelve valence electrons occupy the remaining six sp^2 orbitals as lone pairs. The π electron configuration is $(\pi)^2(\pi^{nb})^4$. The total bond order is 4. A total bond order of 4 is consistent with the Lewis formula.
13-63. Four valence electrons occupy the two σ orbitals. Eight valence electrons occupy the remaining four sp^2 orbitals as lone pairs. The π electron configuration is $(\pi)^4$. The total bond order is 4. A total bond order of 4 is consistent with the Lewis formula.

CHAPTER 14

14-1. 6870 kJ
14-3. 0.410 kJ
14-5. 418 J
14-7. 57.3 kJ

14-9. time to heat from 200 K to 234 K is 20.8 s; time to melt the mercury is 51.6 s; time to heat from 234 K to 630 K is 249 s; time to vaporize the mercury is 1330 s; time to heat from 630 K to 800 K is 79.2 s
14-11. 29.2 $kJ \cdot mol^{-1}$
14-13. Cl_2 is nonpolar. ClF is polar. NF_3 is polar. F_2 is nonpolar.
14-15. $T_b[He] < T_b[C_2H_6] < T_b[C_2H_5OH] < T_b[KBr]$
14-17. $\Delta H_{vap}[CH_4] < \Delta H_{vap}[C_2H_6] < \Delta H_{vap}[CH_3OH] < \Delta H_{vap}[C_2H_5OH]$
14-19. Liquid will be present.
14-21. We expect condensation to occur at around 60°C.
14-23. The dew point of the 20°C day is 15°C; the dew point of the 30°C day is 24°C.
14-25. 0.2 μJ
14-27. 36
14-29. The dielectric constant of *cis*-1,2-dichloroethene is the greater of the two.
14-31. (a) gas (b) solid (c) liquid (d) liquid
14-33. Solid oxygen does not melt under an applied pressure.
14-35. 2
14-37. 408.7 pm
14-39. 6.02×10^{23} $atom \cdot mol^{-1}$
14-41. 268.9 pm
14-43. It requires 0.334 kJ to melt 1 g of snow at 0°C.
14-45. 31.7 $kJ \cdot mol^{-1}$. Water is strongly hydrogen bonded, and so has strong specific intermolecular interactions. Therefore, Trouton's rule does not apply to water.
14-47. $\Delta H_{sub} = \Delta H_{fus} + \Delta H_{vap}$
14-49. Both form a diamondlike covalent crystal network.
14-51. The boiling point of water was used to determine the atmospheric pressure of Lhasa. A plot of atmospheric pressure versus altitude was used to determine the altitude of Lhasa.
14-53. 172 K and 10.7 torr
14-55. about 23°C
14-57. There are four formula units in a unit cell; therefore, the unit cell of CaO must be the NaCl type.
14-59. 3.995 $g \cdot cm^{-3}$
14-61. 9.26 $g \cdot cm^{-3}$
14-63. 1.28×10^{-3} $mol \cdot L^{-1}$
14-65. 0.38
14-67. There are three triple points. When heated from 40°C to 200°C, sulfur goes from rhombic to monoclinic to liquid to gas. Sublimation will occur below 10^{-5} atm.
14-69. The carbon-carbon bonds can be described as $C(sp^3)$-$C(sp^3)$.
14-71. 337.2 K and 1140 torr

CHAPTER 15

15-1. $X_{H_2O} = 0.911$; $X_{C_2H_5OH} = 0.0891$
15-3. Mix 431 g of acetone with 569 g of water.
15-5. 0.355
15-7. Dissolve 115 g of fromic acid in 1000 g of acetone.
15-9. 0.1026 m
15-11. (a) 2.0 m_c (b) 3.0 m_c (c) 1.0 m_c (d) 5.0 m_c
15-13. (a) 2 (b) 3 (c) 3
15-15. H_2SO_4
15-17. 46.9 torr; 0.2 torr
15-19. 23.64 torr; 0.12 torr
15-21. 51.3 torr; 7.9 torr
15-23. (a) 17.38 torr; 0.16 torr (b) 17.31 torr; 0.23 torr (c) 17.46 torr; 0.08 torr (d) 17.23 torr; 0.31 torr
15-25. (a) 1.09×10^{-3} atm (b) 2.10×10^{-3} atm (c) 3.05×10^{-3} atm
15-27. 9.78 m
15-29. 0.0228
15-31. 104.2°C
15-33. 81.9°C
15-35. 212.3°C
15-37. −3.10°C
15-39. 4.59°C
15-41. 3.18°C
15-43. 450
15-45. 10 m
15-47. 1.52 m_c. The colligative molality is only slightly greater than the molality indicating that $HgCl_2$ is essentially undissociated in water.
15-49. 6.4 atm
15-51. 5750
15-53. 13 L
15-55. (a) 412 torr (b) 0.466
15-57. 4.9×10^{-4} M
15-59. 0.069 M
15-61. 130 ft
15-63. 150,000
15-65. The boiling point of ethyl alcohol is 78°C and so its equilibrium vapor pressure is much greater than 1 atm at 100°C. Thus ethyl alcohol is much more readily lost by evaporation than a relatively high boiling liquid.
15-67. 0.115 m
15-69. 110°C
15-71. 0.952 m for methyl alcohol and 0.597 m for ethyl alcohol
15-73. (a) no net flow (b) no net flow (c) a net flow of water from the 0.10 M solution to the 0.50 M solution.
15-75. 4.78 M; 7.23 m. The molality at 0°C is the same as at 20°C.
15-77. (a) −0.64°C; 100.18°C (b) −0.81°C; 100.23°C (c) 171.2°C; 209.3°C
15-79. P_4
15-81. 2.26 M
15-83. 108.4°C
15-85. 18 M
15-87. The expression is derived from the definitions of the mole fraction and colligative molality. Take 1000 g of water.
15-89. 2.01

CHAPTER 16

16-1. rate = $-\frac{1}{2}\Delta[NOCl]/\Delta t$; rate = $\frac{1}{2}\Delta[NO]/\Delta t$; rate = $\Delta[Cl_2]/\Delta t$
16-3. (a) Measure the decrease in the yellow color due to $I_2(aq)$ as a function of time. (b) Measure the total pressure in the reaction vessel as a function of time.
16-5. rate = 7.2×10^{-2} M·s^{-1}
16-7. 7.5×10^{-4} atm·s^{-1}
16-9. 4.6×10^{-4} M·s^{-1}; 3.8×10^{-4} M·s^{-1}, 2.9×10^{-4} M·s^{-1}
16-11. first order; rate = $k[SO_2Cl_2]$
16-13. first order; rate = $(7.3 \times 10^{-30}\ s^{-1})[C_2H_5Cl]$
16-15. second order; rate = $(2.8 \times 10^{-5}\ M^{-1}\cdot s^{-1})[NOCl]^2$
16-17. rate = $(2.0 \times 10^{-6}\ M^{-1}\cdot s^{-1})[Cr(H_2O)_6^{3+}][SCN^-]$
16-19. rate = $(5.0 \times 10^4\ M^{-1}\cdot s^{-1})[NO_2][O_3]$
16-21. 0.67
16-23. 0.24
16-25. A plot of $\ln[S_2O_8^{2-}]$ versus time is a straight line indicating that the reaction is first order. rate = $(0.041\ min^{-1})[S_2O_8^{2-}]$
16-27. A plot of $\ln[C_2H_4O]$ versus time is a straight line indicating that the reaction is first order. rate = $(0.012\ min^{-1})[C_2H_4O]$
16-29. 8.7×10^{-3} M
16-31. A plot of $1/[NO_2]$ versus time is a straight line indicating that the reaction is second order. $k = 0.71\ M^{-1}\cdot s^{-1}$
16-33. 1.1×10^{-2} M^{-1}·s^{-1}
16-35. (a) rate = $k[N_2O][O]$ (b) rate = $k[O][O_3]$ (c) rate = $k[ClCO][Cl_2]$
16-37. The mechanism is consistent with the experimental rate equation.
16-39. 100 kJ·mol^{-1}
16-41. 3.06×10^{-2} s^{-1}
16-43. 0.017 h
16-45. A catalyst affects the rates of the forward and reverse reactions equally, so that the final concentrations are unaffected.
16-47. (a) $H^+(aq)$ and $Br^-(aq)$ (b) third order (c) 693 s
16-49. Run the reaction with all initial concentrations the same in vessels with different amounts of reaction vessel wall area in contact with the reaction mixture.
16-51. Except at very low pressure, the platinum surface will be completely covered with oxygen molecules. Thus, the value of $[O_2(surface)]$ will be essentially constant and the rate-determining step will be independent of the pressure of $O_2(g)$. The oxygen molecules react rapidly with SO_2 molecules except when the number of SO_2 molecules present is very small.
16-53. rate = $(1.0 \times 10^{-6}\ torr^{-1}\cdot s^{-1})P_{CO}^2$
16-55. (a) 3.6×10^{-4} M·s^{-1} (b) 1.3 mol·L^{-1}
16-57. 2.56×10^4 bacteria; 4.6×10^{-2} min^{-1}
16-59. 230 day
16-61. The formation of a covalent bond from two radicals does not involve any bond-breaking process, and so E_a is approximately zero.
16-63. rate = $(2.2 \times 10^{-5}\ s^{-1})P_{SO_2Cl_2}$
16-65. 0.80 M
16-67. 1.4×10^6 bacteria·mL^{-1}
16-69. 85 y
16-71. 110 kJ·mol^{-1}
16-73. 0.928
16-75. The dissociation of H_2
16-77. 51 kJ·mol^{-1}
16-79. Integrate the rate law $d[A]/dt = -k[A]^2$
16-81. 30°C
16-83. M^{-2}·s^{-1}
16-85. 100 kJ·mol^{-1}

CHAPTER 17

17-1. $[SbCl_3] = 0.126$ M; $[Cl_2] = 0.240$ M
17-3. (a) $K_c = [CO_2]/[CO]$; unitless (b) $K_c = [C_{10}H_{12}]/[C_5H_6]^2$; M^{-1} (c) $K_c = [NO_2]^4[O_2]/[N_2O_5]^2$; M^3
17-5. (a) $K_c = [SO_2][Cl_2]/[SO_2Cl_2]$; M (b) $K_c = [O_2]/[H_2O_2]^2$; M^{-1} (c) $K_c = 1/[H_2O]^3$; M^{-3}
17-7. (a) $K_p = P_{SO_2}P_{Cl_2}/P_{SO_2Cl_2}$; torr or atm (b) $K_p = P_{O_2}/P_{H_2O_2}^2$; torr^{-1} or atm^{-1} (c) $K_p = 1/P_{H_2O}^3$; torr^{-3} or atm^{-3}
17-9. 4.7×10^{-3} M
17-11. 6.7×10^{-3} atm^2
17-13. 50.3
17-15. 1.9 atm
17-17. 0.064 M
17-19. $[PCl_3] = 0.22$ M; $[Cl_2] = 0.22$ M; $[PCl_5] = 0.03$ M
17-21. 0.98 mol
17-23. 0.658 atm
17-25. $[Cl_2] = 0.05$ M; $[I_2] = 0.26$ M; $[ICl] = 0.34$ M
17-27. $P_{N_2O_4} = 0.28$ atm; $P_{NO_2} = 1.17$ atm
17-29. $P_{CO} = 3.00 \times 10^{-3}$ atm; $P_{CO_2} = 1.80$ atm
17-31. 36.9 atm^2
17-33. 8.11 atm
17-35. (a) from right to left (b) no change
17-37. (a) → (b) → (c) ← (d) ← (e) no change
17-39. (a) ← (b) ← (c) ← (d) →
17-41. (1) $K_c = [CO][H_2]/[H_2O]$; to the left (2) $K_c = [CO_2][H_2]/[CO][H_2O]$; no effect (3) $K_c = [H_2O][CH_4]/[CO][H_2]^3$; to the right
17-43. $P_{N_2O_4} = 340$ torr; $P_{NO_2} = 424$ torr
17-45. $P_{NH_3} = 0.475$ atm; $P_{H_2S} = 0.233$ atm
17-47. 0.535
17-49. (a) $Q_c = 0.31$ M^{-1}; $Q_c/K_c = 0.024$; → (b) $Q_c = 360$ M^{-1}; $Q_c/K_c = 28$; ←
17-51. $Q_p/K_p = 0.69$; from left to right
17-53. $Q_p/K_p = 6.4$ from right to left
17-55. 8.16×10^{-2} atm
17-57. $[HI] = 4.33$ M; $[H_2] = 0.52$ M; $[I_2] = 0.52$ M
17-59. (a) no change (b) ← (c) → (d) →
17-61. $[H_2O] = 0.38$ M; $[H_2] = 0.38$ M
17-63. See the *Study Guide/Solutions Manual.*
17-65. See the *Study Guide/Solutions Manual.*
17-67. (a) $Q_c = 1$; equilibrium is not possible (b) $[H_2] = 1.7 \times 10^{-3}$ M; $[HI] = 0.016$ M

17-69. $P_{Cl_2} = 2.24$ atm; $P_{O_2} = 1.64$ atm
17-71. the brown form
17-73. $1.6 \times 10^{26}\ M^{-1}$
17-75. $P_{COCl_2} = 0.236$ atm; $P_{CO} = 1.96$ atm; $P_{Cl_2} = 0.99$ atm; $P_{total} = 3.19$ atm
17-77. 78.7%
17-79. $P_{CO} = 12.0$ torr; $P_{H_2O} = 12.0$ torr; $P_{CO_2} = 14.1$ torr; $P_{H_2} = 14.1$ torr
17-81. 9.41 g
17-83. 3.3 mol
17-85. 39%
17-87. 0.985 atm

CHAPTER 18

18-1. $[H_3O^+] = 0.150$ M; $[ClO_4^-] = 0.150$ M; $[OH^-] = 6.67 \times 10^{-14}$ M
18-3. $[OH^-] = 1.81 \times 10^{-2}$ M; $[Tl^+] = 1.81 \times 10^{-2}$ M; $[H_3O^+] = 5.52 \times 10^{-13}$ M
18-5. 1.70; acidic
18-7. 1.10; acidic
18-9. 12.85
18-11. 1.6×10^{-7} M
18-13. 0.1 M
18-15. 3×10^{-7} M
18-17. 1.3×10^{-5} M
18-19. 1.7×10^{-5} M
18-21. 9×10^{-4}
18-23. $[H_3O^+] = 1.10 \times 10^{-3}$ M; $[C_6H_5COO^-] = 1.10 \times 10^{-3}$ M; $[C_6H_5COOH] = 0.019$ M; pH = 2.96
18-25. 1.6
18-27. 1.5
18-29. 1.8×10^{-5} M
18-31. 9.32
18-33. 11.75
18-35. (a) right to left (b) left to right (c) left to right (d) left to right
18-37. (a) right to left (b) not affected (c) right to left (d) left to right (e) right to left
18-39. (a) $C_6H_5COOH(aq)/C_6H_5COO^-(aq)$ and $H_2O(l)/H_3O^+(aq)$ (b) $CH_3NH_2(aq)/CH_3NH_3^+(aq)$ and $H_2O(l)/OH^-(aq)$ (c) $HCOOH(aq)/HCOO^-(aq)$ and $H_2O(l)/H_3O^+(aq)$
18-41. (a) $ClO^-(aq)$ (b) $NH_3(aq)$ (c) $N_3^-(aq)$ (d) $S^{2-}(aq)$
18-43. (a) acid; $CNO^-(aq)$ (b) base; $HOBr(aq)$ (c) acid; $ClO_3^-(aq)$ (d) acid; $CH_3NH_2(aq)$ (e) base; $ClNH_3^+(aq)$ (f) base; $HONH_3^+(aq)$
18-45. (a) 7.46×10^{-10} M (b) 1.48×10^{-11} M (c) 1.75×10^{-5} M (d) 1.58×10^{-7} M
18-47. (a) 0.490 (b) 3.21×10^{-6}
18-49. (a) acidic (b) acidic (c) basic (d) basic
18-51. (a) basic (b) neutral (c) basic (d) acidic
18-53. basic
18-55. 10.00
18-57. $[OH^-] = 3.02 \times 10^{-6}$ M; $[HCNO] = 3.02 \times 10^{-6}$ M; $[CNO^-] = 0.20$ M; $[H_3O^+] = 3.31 \times 10^{-9}$ M; pH = 8.48
18-59. 2.84
18-61. 4.47
18-63. 1.86
18-65. (a) Arrhenius acid and Brønsted-Lowry acid (b) Arrhenius acid, Brønsted-Lowry acid, and Lewis acid (c) Lewis base
18-67. (a) Lewis base (b) Lewis acid (c) Lewis base
18-69. 2.8×10^3
18-71. 2.64
18-73. 9.66×10^{-4} g per 100 mL of solution
18-75. 3.37
18-77. 6.94
18-79. The strongest acid is $NH_4^+(amm)$; the strongest base is $NH_2^-(amm)$
18-81. pH > 5.4
18-83. 1.9 g per 100 mL of solution
18-85. $[H_3O^+] = 5.08 \times 10^{-3}$ M; $[HCOO^-] = 5.08 \times 10^{-3}$ M; $[HCOOH] = 0.145$ M; $[OH^-] = 1.97 \times 10^{-12}$ M; pH = 2.29
18-87. $[OH^-] = 9.05 \times 10^{-3}$ M; $[(CH_3)_2NH_2^+] = 9.05 \times 10^{-3}$ M; $[(CH_3)_2NH] = 0.141$ M; $[H_3O^+] = 1.10 \times 10^{-12}$ M; pH = 11.96
18-89. 4.57%

CHAPTER 19

19-1. 4.76
19-3. 3.97
19-5. 5.06
19-7. (a) 0.33 (b) 0.03
19-9. pH = 4.66; change in pH = −0.11
19-11. 4.85
19-13. Nile blue or thymol blue
19-15. 4.5 ± 0.5
19-17. bromcresol green or methyl red
19-19. 3×10^{-5} M
19-21. (a) 12.35 (b) 13.11
19-23. (a) 100 mL (b) 45.0 mL
19-25. pH initially = 13.00; pH at equivalence point = 7.00; pH after addition of 100 mL of HCl(*aq*) = 1.48
19-27. 0.235 M
19-29. 9.48; thymolphthalein or phenolphthalein
19-31. (a) 2.73 (b) 4.16 (c) 4.76 (d) 8.88 (e) 11.59
19-33. (a) 9.17 (b) 5.34 (c) 3.78 (d) 3.15 (e) 2.51
19-35. 176
19-37. 50.0 mL
19-39. 90.1
19-41. 1.66
19-43. 340 mL
19-45. benzoic acid
19-47. See the *Study Guide/Solutions Manual.*
19-49. (a) 8.34 (b) $HCO_3^-(aq) + H_3O^+(aq) \rightleftharpoons H_2CO_3(aq) + H_2O(l)$ and $HCO_3^-(aq) + OH^-(aq) \rightleftharpoons CO_3^{2-}(aq) + H_2O(l)$
19-51. $3HCO_3^-(aq) + H_3C_6H_5O_7(aq) \rightleftharpoons 3CO_2(g) + C_6H_5O_7^{3-}(aq) + 3H_2O(l)$
19-53. 190 mL
19-55. 88.1
19-57. Measure the pH of the solution before and after dilution.
19-59. 116

19-61. three dissociable protons
19-63. $[C_2O_4^{2-}] = 0.105$ M; $[HC_2O_4^-] = 0.020$ M; $[H_2C_2O_4] = 3.7 \times 10^{-6}$ M
19-65. (a) 7.21 (b) 7.51 (c) 6.91
19-67. a solution of equal concentrations of HCOOH(*aq*) and NaHCOO(*aq*)
19-69. 4.15%
19-71. 5.19
19-73. $[H_3O^+] = 5.61 \times 10^{-3}$ M; $[HSe^-] = 5.61 \times 10^{-3}$ M; $[H_2Se] = 0.244$ M; $[Se^{2-}] = 1.0 \times 10^{-11}$ M; $[OH^-] = 1.78 \times 10^{-12}$ M; pH = 2.25
19-75. $0.200\ M = [H_2C_2O_4] + [HC_2O_4^-] + [C_2O_4^{2-}]$; $[H_3O^+] = [HC_2O_4^-] + 2[C_2O_4^{2-}] + [OH^-]$
19-77. 41 mL
19-79. 5.88 g
19-81. Add a volume of 0.10 M HCl(*aq*) that is equal to one half the volume of the original $NH_3(aq)$ solution.
19-83. 8.75
19-85. change in pH = 0.03

CHAPTER 20

20-1. (a) insoluble (b) soluble (c) soluble (d) soluble (e) insoluble
20-3. (a) insoluble (b) soluble (c) soluble (d) insoluble
20-5. (a) $CuCl_2(aq) + Na_2S(aq) \rightarrow CuS(s) + 2NaCl(aq)$
$Cu^{2+}(aq) + S^{2-}(aq) \rightarrow CuS(s)$
(b) $MgBr_2(aq) + K_2CO_3(aq) \rightarrow MgCO_3(s) + 2KBr(aq)$
$Mg^{2+}(aq) + CO_3^{2-}(aq) \rightarrow MgCO_3(s)$
(c) $BaCl_2(aq) + K_2SO_4(aq) \rightarrow BaSO_4(s) + 2KCl(aq)$
$Ba^{2+}(aq) + SO_4^{2-}(aq) \rightarrow BaSO_4(s)$
(d) $Hg_2(NO_3)_2(aq) + 2KCl(aq) \rightarrow Hg_2Cl_2(s) + 2KNO_3(aq)$
$Hg_2^{2+}(aq) + 2Cl^-(aq) \rightarrow Hg_2Cl_2(s)$
20-7. $1.7 \times 10^{-4}\ g \cdot L^{-1}$
20-9. $5.02 \times 10^{-13}\ M^2$
20-11. $9.6 \times 10^{-3}\ g \cdot L^{-1}$
20-13. $2.9 \times 10^{-3}\ M^2$
20-15. $1.4 \times 10^{-2}\ g \cdot L^{-1}$
20-17. $0.16\ g \cdot L^{-1}$
20-19. $4.9 \times 10^{-14}\ g \cdot L^{-1}$
20-21. $1.4 \times 10^{-8}\ g \cdot L^{-1}$
20-23. $0.21\ g \cdot L^{-1}$
20-25. $3.6\ g \cdot L^{-1}$
20-27. $0.42\ g \cdot L^{-1}$
20-29. (a) increased (b) decreased
20-31. $CaCO_3$; CaF_2; $PbSO_3$; $Fe(OH)_3$; ZnS
20-33. 1.8 M
20-35. $Q_{sp}/K_{sp} > 1$; AgCl(*s*) will precipitate.
20-37. 1.82×10^{-5} mol of $PbI_2(s)$ precipitated; $[Pb^{2+}] = 2.00$ M; $[NO_3^-] = 4.00$ M; $[Na^+] = 6.67 \times 10^{-4}$ M; $[I^-] = 6.0 \times 10^{-5}$ M
20-39. (a) 0.29 g (b) 1.8×10^{-10} M
20-41. Hg_2I_2 will precipitate. $[Hg_2^{2+}] = 6.4 \times 10^{-23}$ M. $Pb^{2+}(aq)$ can be separated from $Hg_2^{2+}(aq)$.
20-43. When $[SO_4^{2-}] = 0.022$ M, $Ag^+(aq)$ does not precipitate and 99% of $Ca^{2+}(aq)$ does precipitate.
20-45. Solubility of $Cr(OH)_3 = 6.3 \times 10^{-4}$ M and the solubility of $Ni(OH)_2 = 2.0 \times 10^3$ M. $Cr(OH)_3$ will precipitate while $Ni^{2+}(aq)$ remains in solution.
20-47. 5.7×10^{-19} M
20-49. At a pH of −0.07, essentially all the $Pb^{2+}(aq)$ precipitates as PbS and essentially all the $Mn^{2+}(aq)$ remains in solution.
20-51. 1×10^{-3} M
20-53. The $Pb^{2+}(aq)$ is removed from solution by the formation of insoluble $PbSO_4(s)$, which passes out of the body through the large intestine.
20-55. $Pb(NO_3)_2(aq) + 2NaOH(aq) \rightleftharpoons Pb(OH)_2(s) + 2NaNO_3(aq)$; $Pb(OH)_2(aq) + OH^-(aq) \rightleftharpoons Pb(OH)_3^-(aq)$
20-57. Calcium ions form an insoluble oxalate, $CaC_2O_4(s)$, which is removed by vomiting. The excess $Ca^{2+}(aq)$ is removed by adding $MgSO_4(aq)$ to form $CaSO_4(s)$, which is insoluble in water and in stomach acid.
20-59. $Zn(ClO_4)_2(aq) + 2KOH(aq) \rightleftharpoons Zn(OH)_2(s) + 2KClO_3(aq)$; $Zn(OH)_2(s) + 2OH^-(aq) \rightleftharpoons Zn(OH)_4^{2-}(aq)$
20-61. 51.75%
20-63. 0.29 M
20-65. (a) $4.4 \times 10^{12}\ M^{-1}$ (b) 2.6×10^{-5}
20-67. 0.230 M
20-69. at pH = 2.0, 1.5 M; at pH = 4.0, 0.16 M; at pH = 6.0, 0.065 M; at pH above 7, 0.063 M
20-71. 5.0×10^{-6} M
20-73. (a) $Hg_2(ClO_4)_2(aq) + 2NaBr(aq) \rightarrow 2NaClO_4(aq) + Hg_2Br_2(s)$
$Hg_2^{2+}(aq) + 2Br^-(aq) \rightarrow Hg_2Br_2(s)$
(b) $Fe(ClO_4)_2(aq) + 3NaOH(aq) \rightarrow 3NaClO_4(aq) + Fe(OH)_3(s)$
$Fe^{3+}(aq) + 3OH^-(aq) \rightarrow Fe(OH)_3(s)$
(c) $Pb(NO_3)_2(aq) + 2LiIO_3(aq) \rightarrow 2LiNO_3(aq) + Pb(IO_3)_2(s)$
$Pb^{2+}(aq) + 2IO_3^-(aq) \rightarrow Pb(IO_3)_2(s)$
(d) $H_2SO_4(aq) + Pb(NO_3)_2(aq) \rightarrow 2HNO_3(aq) + PbSO_4(s)$
$Pb^{2+}(aq) + SO_4^{2-}(aq) \rightarrow PbSO_4(s)$
20-75. 0.72 M
20-77. (a) increased (b) unchanged (c) decreased (d) unchanged
20-79. (a) soluble (b) insoluble (c) insoluble (d) insoluble
20-81. (a) soluble (b) insoluble (c) soluble (d) insoluble (e) soluble
20-83. 0.40 M
20-85. 9.10
20-87. 2.2×10^{-6} M
20-89. (a) increases (b) increases
20-91. (a) 5.8×10^{-19} M (b) 4.7×10^{-10} M (c) 6.9×10^{-7} M
20-93. 2.3×10^{-9} M
20-95. 4.1×10^{-14}
20-97. Silver iodide is not soluble in the aqueous ammonia solution.
20-99. $AgCrO_4$ is more soluble than AgCl in pure water.

CHAPTER 21

21-1. The oxidation states are (a) 0 (b) K = +1 and O = −½ (c) Na = +1 and O = −1 (d) F = −1 and O = +2

21-3. (a) +1 (b) +3 (c) +5 (d) +7
21-5. The oxidation states are (a) O = −2, C = 0, and H = +1 (b) O = −2, C = −2, and H = +1 (c) O = −2, C in $-CH_3$ = −3, C in CO = +2, and H = +1 (d) O = −2, C in $-CH_3$ = −3, C in −COOH = +3, and H = +1
21-7. (a) +7 (b) +6 (c) +4 (d) Cl = +7 and Mn = +3
21-9. (a) Se = −2 and Mo = +4 (b) C = −4 and Si = +4 (c) Ga = +3 and As = −3 (d) K = +1, O = −2, and S = +2
21-11. Iodine is reduced and I_2 acts as the oxidizing agent; sulfur is oxidized and $Na_2S_2O_3$ acts as the reducing agent.
21-13. Nitrogen is reduced and $NaNO_3$ acts as the oxidizing agent; lead is oxidized and Pb acts as the reducing agent.
21-15. (a) Iodine is oxidized and I^- acts as the reducing agent; iron is reduced and Fe^{3+} acts as the oxidizing agent.
(b) Titanium is oxidized and Ti^{2+} acts as the reducing agent; cobalt is reduced and Co^{2+} acts as the oxidizing agent.
21-17. The oxygen in KO_2 is in the unusual oxidation state of −½. The oxygen is easily reduced to its common oxidation state of −2.
21-19. (a) $2MnO(s) + 5PbO_2(s) + 8H^+(aq) \rightarrow 2MnO_4^-(aq) + 5Pb^{2+}(aq) + 4H_2O(l)$

electron donor, MnO; electron acceptor, PbO_2; species oxidized, Mn; species reduced, Pb; reducing agent, MnO; oxidizing agent, PbO_2.
(b) $As_2S_5(s) + 40NO_3^-(aq) + 35H^+(aq) \rightarrow 5HSO_4^-(aq) + 2H_3AsO_4(aq) + 40NO_2(g) + 12H_2O(l)$
electron donor, As_2S_5; electron acceptor, NO_3^-; species oxidized, S; species reduced, N; reducing agent, As_2S_5; oxidizing agent, NO_3^-.
21-21. (a) $NH_4^+(aq) + NO_3^-(aq) \rightarrow N_2O(g) + 2H_2O(l)$
(b) $4Fe(s) + 3O_2(g) + 6H_2O(l) \rightarrow 2Fe_2O_3{\cdot}3H_2O(s)$
21-23. (a) $4Fe(OH)_2(s) + O_2(g) + 2H_2O(l) \rightarrow 4Fe(OH)_3(s)$
(b) $3Cu(s) + 2NO_3^-(aq) + 8H^+(aq) \rightarrow 2NO(g) + 3Cu^{2+}(aq) + 4H_2O(l)$
21-25. (a) $IO_4^-(aq) + 3I^-(aq) + H_2O(l) \rightarrow IO_3^-(aq) + I_3^-(aq) + 2OH^-(aq)$
(b) $H_2MoO_4(aq) + 6Cr^{2+}(aq) + 6H^+(aq) \rightarrow Mo(s) + 6Cr^{3+}(aq) + 4H_2O(l)$
21-27. (a) $4CrO_4^{2-}(aq) + 3Cl^-(aq) + 20H^+(aq) \rightarrow 3ClO_2^-(aq) + 4Cr^{3+}(aq) + 10H_2O(l)$
(b) $2S_2O_3^{2-}(aq) + 2Cu^{2+}(aq) \rightarrow S_4O_6^{2-}(aq) + 2Cu^+(aq)$
21-29. $2CrI_3(s) + 27Cl_2(g) + 64OH^-(aq) \rightarrow 2CrO_4^{2-}(aq) + 6IO_4^-(aq) + 54Cl^-(aq) + 32H_2O(l)$
21-31. (a) $Mo^{3+}(aq) + 2H_2O(l) \rightarrow MoO_2^{2+}(aq) + 4H^+(aq) + 3e^-$
(b) $P_4(s) + 16H_2O(l) \rightarrow 4H_3PO_4(aq) + 20H^+ + 20e^-$
(c) $S_2O_8^{2-}(aq) + 2H^+(aq) + 2e^- \rightarrow 2HSO_4^-(aq)$
21-33. (a) $2WO_3(s) + 2H^+(aq) + 2e^- \rightarrow W_2O_5(s) + H_2O(l)$
(b) $U^{4+}(aq) + 2H_2O(l) \rightarrow UO_2^+(aq) + 4H^+(aq) + e^-$
(c) $Zn(s) + 4OH^-(aq) \rightarrow Zn(OH)_4^{2-}(aq) + 2e^-$
21-35. (a) $2SO_3^{2-}(aq) + 2H_2O(l) + 2e^- \rightarrow S_2O_4^{2-}(aq) + 4OH^-(aq)$
(b) $2Cu(OH)_2(s) + 2e^- \rightarrow Cu_2O(s) + 2OH^-(aq) + H_2O(l)$
(c) $2AgO(s) + H_2O(l) + 2e^- \rightarrow Ag_2O(s) + 2OH^-(aq)$
21-37. 20.8%
21-39. 22.8%
21-41. 0.400 M
21-43. 0.459 g
21-45. $4MnO_4^-(aq) + 2H_2O(l) \rightarrow 4MnO_2(s) + 3O_2(g) + 4OH^-(aq)$
21-47. (a) $Cr_2O_7^{2-}(aq) + 9I^-(aq) + 14H^+(aq) \rightarrow 2Cr^{3+}(aq) + 3I_3^-(aq) + 7H_2O(l)$
(b) $3IO_4^-(aq) + 33I^-(aq) + 24H^+(aq) \rightarrow 12I_3^-(aq) + 12H_2O(l)$
21-49. $4Ag^{2+}(aq) + 2H_2O(l) \rightarrow 4Ag^+(aq) + O_2(g) + 4H^+(aq)$
21-51. (a) $Cr_2O_7^{2-}(aq) + 8H^+(aq) + 3H_2O_2(aq) \rightarrow 2Cr^{3+}(aq) + 3O_2(g) + 7H_2O(l)$
(b) $2Cr^{3+}(aq) + Zn(s) \rightarrow 2Cr^{2+}(aq) + Zn^{2+}(aq)$
21-53. (a) $H_2O_2(aq) + 2H^+(aq) + 2e^- \rightarrow 2H_2O(l)$
(b) $H_2O_2(aq) \rightarrow O_2(g) + 2H^+(aq) + 2e^-$
The complete reaction is $2H_2O_2(aq) \rightarrow 2H_2O(l) + O_2(g)$
21-55. 0.018%
21-57. 1.82×10^{-4}%

CHAPTER 22

22-1. For CH_4, $\Delta S_{fus} = 10.3\ J{\cdot}K^{-1}{\cdot}mol^{-1}$ and $\Delta S_{vap} = 81.57\ J{\cdot}K^{-1}{\cdot}mol^{-1}$. For C_2H_6, $\Delta S_{fus} = 31.8\ J{\cdot}K^{-1}{\cdot}mol^{-1}$ and $\Delta S_{vap} = 84.78\ J{\cdot}K^{-1}{\cdot}mol^{-1}$. For C_3H_8, $\Delta S_{fus} = 38.5\ J{\cdot}K^{-1}{\cdot}mol^{-1}$ and $\Delta S_{vap} = 87.11\ J{\cdot}K^{-1}{\cdot}mol^{-1}$.
22-3. $\Delta S_{fus} = 18.11\ J{\cdot}K^{-1}{\cdot}mol^{-1}$ for CH_3OH; $31.64\ J{\cdot}K^{-1}{\cdot}mol^{-1}$ for C_2H_5OH; $35.32\ J{\cdot}K^{-1}{\cdot}mol^{-1}$ for C_3H_7OH. $\Delta S_{vap} = 111.1\ J{\cdot}K^{-1}{\cdot}mol^{-1}$ for CH_3OH; $115.0\ J{\cdot}K^{-1}{\cdot}mol^{-1}$ for C_2H_5OH; $117.6\ J{\cdot}K^{-1}{\cdot}mol^{-1}$ for C_3H_7OH
22-5. $\Delta S_{fus} = 12.7\ J{\cdot}K^{-1}{\cdot}mol^{-1}$; $\Delta S_{vap} = 88.0\ J{\cdot}K^{-1}{\cdot}mol^{-1}$. The hydrogen bonds in water make for a higher degree of order and the breaking of the hydrogen bonds produces a greater increase in disorder.
22-7. (a) $S°(H_2O) < S°(D_2O)$ (b) $S°$(ethylene oxide) < $S°$(ethanol) (c) $S°$(pyrrolidine) < $S°$(butyl amine)
22-9. $S°(CH_4) < S°(CH_3OH) < S°(CH_3Cl)$
22-11. $S°(Fe_2O_3) < S°(Fe_3O_4)$
22-13. Bromine molecules are much more restricted in movement in the liquid state than in the gaseous state. The positional disorder is greater in the gaseous state and thus the entropy of $Br_2(g)$ is greater than that of $Br_2(l)$.
22-15. (a) increase (b) increase (c) increase (d) decrease
22-17. $\Delta S°_{rxn}$ (c) = $\Delta S°_{rxn}$ (b) $\Delta < S°_{rxn}$ (a) $< \Delta S°_{rxn}$ (d)
22-19. (a) $-111.2\ J{\cdot}K^{-1}$ (b) $-332.1\ J{\cdot}K^{-1}$ (c) $134.0\ J{\cdot}K^{-1}$ (d) $-173.4\ J{\cdot}K^{-1}$
22-21. (a) $2.9\ J{\cdot}K^{-1}$ (b) $-189.2\ J{\cdot}K^{-1}$ (c) $-242.8\ J{\cdot}K^{-1}$ (d) $-111.8\ J{\cdot}K^{-1}$
22-23. Vaporization is entropy driven.
22-25. $\Delta S°_{rxn} = -429.6\ J{\cdot}K^{-1}$; $\Delta G°_{rxn} = -503$ kJ. The reaction $3C_2H_2(g) \rightarrow C_6H_6(l)$ is spontaneous.
22-27. $\Delta G°_{rxn} = -1314$ kJ. The reaction is spontaneous left to right at standard conditions. $\Delta G_{rxn} = -1281$ kJ. The reaction is spontaneous left to right under the conditions given.

22-29. $\Delta G_{rxn} = -50.1$ kJ. The reaction is spontaneous from left to right at the stated conditions.
22-31. $\Delta G^\circ_{rxn} = -36.5$ kJ. The reaction is spontaneous from left to right at standard conditions. $\Delta G_{rxn} = +15.6$ kJ. The reaction is spontaneous from right to left at the stated conditions.
22-33. $\Delta G^\circ_{rxn} = 19.1$ kJ. Nitric acid does not dissociate under standard conditions. $\Delta G_{rxn} = -38.0$ kJ. Nitric acid will dissociate under the stated conditions.
22-35. $\Delta G^\circ_{rxn} = 16.4$ kJ. Chloracetic acid does not dissociate under standard conditions. $\Delta G_{rxn} = -23.6$ kJ. Chloroacetic acid will dissociate under the stated conditions.
22-37. $\Delta G^\circ_{rxn} = 55.6$ kJ. It is not possible to prepare a solution that is 1.0 M in $Ag^+(aq)$ and $Cl^-(aq)$.
22-39. $\Delta G^\circ_{rxn} = -19.4$ kJ. The reaction is spontaneous from left to right at standard conditions. $\Delta G_{rxn} = -7.97$ kJ. The reaction is spontaneous from left to right at the stated conditions.
22-41. (a) $\Delta G^\circ_{rxn} = -29.1$ kJ; $K = 1.26 \times 10^5$ atm^{-3} (b) $\Delta G^\circ_{rxn} = 91.4$ kJ; $K = 9.53 \times 10^{-17}$ atm (c) $\Delta G^\circ_{rxn} = -142.2$ kJ; $K = 8.44 \times 10^{24}$ atm^{-2}
22-43. $\Delta G^\circ_{rxn} = -335$ kJ; $\Delta H^\circ_{rxn} = -357.6$ kJ; $K = 1.7 \times 10^{62}$
22-45. $\Delta G^\circ_{rxn} = -141.8$ kJ; $\Delta H^\circ_{rxn} = -197.8$ kJ; $K = 7.18 \times 10^{24}$ atm^{-1}
22-47. $\Delta G^\circ_{rxn} = 28.6$ kJ; $\Delta H^\circ_{rxn} = 41.2$ kJ; $\Delta S^\circ_{rxn} = 42.3$ J·K^{-1}. The reaction is spontaneous from right to left under standard conditions. The reaction is enthalpy driven.
22-49. $\Delta G^\circ_{rxn} = -1468$ kJ, which is the maximum work that can be obtained from the reaction at standard conditions.
22-51. 181 kJ
22-53. 34.6 kJ
22-55. 0.0167 atm^{-1}
22-57. 180 torr
22-59. 80.5°C
22-61. 31.8 kJ·mol^{-1}
22-63. 1.31×10^6
22-65. 171 kJ·mol^{-1}
22-67. −608 kJ
22-69. 502 kJ·mol^{-1}
22-71. 414 kJ·mol^{-1}
22-73. $\Delta S_{fus} = 6.59$ J·K^{-1}·mol^{-1} for Li; 7.01 J·K^{-1}·mol^{-1} for Na; 6.93 J·K^{-1}·mol^{-1} for K; 7.50 J·K^{-1}·mol^{-1} for Rb; 6.95 J·K^{-1}·mol^{-1} for Cs. $\Delta S_{vap} = 83.41$ J·K^{-1}·mol^{-1} for Li; 77.5 J·K^{-1}·mol^{-1} for Na; 74.6 J·K^{-1}·mol^{-1} for K; 72 J·K^{-1}·mol^{-1} for Rb; 70 J·K^{-1}·mol^{-1} for Cs
22-75. No. If K is infinite, then ΔG°_{rxn} is negative infinite and if K is negative infinite, ΔG°_{rxn} is infinite.
22-77. The reaction between hydrogen and oxygen is the more energy efficient because $\Delta G^\circ_{rxn}(1) = -120.4$ kJ is less than $\Delta G^\circ_{rxn}(2) = 116.8$ kJ.
22-79. A plot of ln K_p versus $1/T$ is a straight line. From the slope of the line $\Delta H^\circ_{rxn} = 181$ kJ.
22-81. A tenfold increase in K corresponds to a change of −5.71 kJ in ΔG°_{rxn}.
22-83. From the van't Hoff equation, ΔH°_{rxn} is negative.
22-85. 21.7 kJ
22-87. $\Delta H^\circ_{rxn} = -466.8$ kJ; $\Delta S^\circ_{rxn} = -40.1$ J·K^{-1}; $\Delta G^\circ_{rxn} = -454.9$ kJ
22-89. 49 kJ
22-91. $\Delta H^\circ_{rxn} = 112.4$ kJ; $K_{sp} = 3.6 \times 10^{-16}$ M^2
22-93. (a) 91.3 kJ (b) Equation (2) has the largest decrease in entropy.
22-95. (a) 91.0 kJ (b) 1.10×10^{-16} M^2 (c) $P_{HCl} = P_{NH_3} = 2.57 \times 10^{-7}$ atm; $P_{total} = 5.14 \times 10^{-7}$ atm
22-97. The reaction is more favorable under the conditions given.

CHAPTER 23

23-1. The reaction at the negative electrode is $Zn(s) \rightarrow Zn^{2+}(aq) + 2e^-$. The reaction at the positive electrode is $2Ag^+(aq) + 2e^- \rightarrow 2Ag(s)$. The cell diagram is $Zn(s)|ZnCl_2(aq)||AgNO_3(aq)|Ag(s)$
23-3. The reaction at the negative electrode is $V(s) \rightarrow V^{2+}(aq) + 2e^-$. The reaction at the positive electrode is $Cu^{2+}(aq) + 2e^- \rightarrow Cu(s)$. The cell diagram is $V(s)|VI_2(aq)||CuSO_4(aq)|Cu(s)$
23-5. The reaction at the left electrode is $Pb(s) + 2I^-(aq) \rightarrow PbI_2(s) + 2e^-$. The reaction at the right electrode is $2H^+(aq) + 2e^- \rightarrow H_2(g)$. The net cell reaction is $Pb(s) + 2HI(aq) \rightarrow PbI_2(s) + H_2(g)$.
23-7. The reaction at the left electrode is $In(s) \rightarrow In^{3+}(aq) + 3e^-$. The reaction at the right electrode is $Cd^{2+}(aq) + 2e^- \rightarrow Cd(s)$. The net cell reaction is $2In(s) + 3Cd^{2+}(aq) \rightarrow 2In^{3+}(aq) + 3Cd(s)$.
23-9. The reaction at the left electrode is $H_2(g) \rightarrow 2H^+(aq) + 2e^-$. The reaction at the right electrode is $Hg_2Cl_2(s) + 2e^- \rightarrow 2Hg(l) + 2Cl^-(aq)$. The net cell reaction is $H_2(g) + Hg_2Cl_2(s) \rightarrow 2Hg(l) + 2H^+(aq) + 2Cl^-(aq)$.
23-11. (a) no effect (b) decreases (c) increases (d) decreases
23-13. (a) no effect (b) increases (c) increases (d) decreases (e) decreases (f) increases
23-15. (a) increases (b) no effect (c) decreases (d) decreases
23-17. (a) 8 (b) 4
23-19. 7.6×10^{15}
23-21. $E^\circ_{rxn} = 0.270$ V; $K = 1.3 \times 10^9$
23-23. $E^\circ_{rxn} = 1.62$ V; $K = 1.2 \times 10^{82}$
23-25. 1.14 V
23-27. 8.3×10^{-5} M
23-29. 1.69
23-31. (a) −0.16 V (b) −0.03 V (c) 3.63 V
23-33. 1.39 V
23-35. $E_{rxn} = 0.27$ V. The reaction is spontaneous.
23-37. $E_{rxn} = 0.48$ V. $V^{2+}(aq)$ can liberate $H_2(g)$ under the given conditions.
23-39. 2.03 V
23-41. −0.58 V
23-43. $E_{rxn} = 0.82$ V. The reaction is spontaneous.
23-45. The value of E_{rxn} is positive for all values of $[Co^{2+}]$; the oxidation of water by $Co^{3+}(aq)$ is a spontaneous process.
23-47. −203 kJ

23-49. −130 kJ
23-51. (a) −400 kJ (b) −74 kJ
23-53. 0.38 V
23-55. 1.16×10^4 s
23-57. 1.24 g
23-59. 2.55×10^4 g. HF is a covalent compound and thus a poor conductor of electricity. Liquid HF is very corrosive and toxic.
23-61. 0.22 g
23-63. 13.8 L
23-65. $9H_2O(l) + 6Mn^{2+}(aq) + 5IO_3^-(aq) \rightarrow 18H^+(aq) + 6MnO_4^-(aq) + 5I^-(aq)$
23-67. $Fe(s) + 2NiOOH(s) + 2H_2O(l) \rightarrow 2Ni(OH)_2(s) + Fe(OH)_2(s)$
23-69. 8.44×10^5 A·h
23-71. 0.456 g
23-73. 5.0×10^{-13} M^2
23-75. zero
23-77. zero
23-79. 108; silver

CHAPTER 24

24-1. (a) $^{72}_{30}Zn \rightarrow {}^{0}_{-1}e + {}^{72}_{31}Ga$
(b) $^{230}_{92}U \rightarrow {}^{4}_{2}He + {}^{226}_{90}Th$
(c) $^{136}_{57}La \rightarrow {}^{0}_{+1}e + {}^{136}_{56}Ba$
(d) $^{14}_{7}N + {}^{1}_{0}n \rightarrow {}^{1}_{1}H + {}^{14}_{6}C$
24-3. (a) $^{25}_{12}Mg + {}^{4}_{2}He \rightarrow {}^{28}_{13}Al + {}^{1}_{1}H$
(b) $^{27}_{13}Al + {}^{1}_{0}n \rightarrow {}^{4}_{2}He + {}^{24}_{11}Na$
(c) $^{17}_{8}O + {}^{1}_{1}H \rightarrow {}^{4}_{2}He + {}^{14}_{7}N$
(d) $^{63}_{29}Cu + {}^{1}_{1}H \rightarrow {}^{63}_{30}Zn + {}^{1}_{0}n$
24-5. a positron-emitter
24-7. (a) positron-emitter (b) β-emitter (c) positron-emitter (d) β-emitter
24-9. (a)
24-11. (a), (b), and (d) are radioactive
24-13. 1.15×10^{-4}
24-15. 0.0136 mg
24-17. 0.63
24-19. 0.51
24-21. 5.3 days
24-23. 8660 years
24-25. 16,000 years
24-27. 5060 years
24-29. 2.44×10^9 years
24-31. 1.26×10^{-12} J·nucleon^{-1}
24-33. 1.37×10^{-12} J·nucleon^{-1}
24-35. 1.99 kg
24-37. 1.24×10^{20} Hz
24-39. -1.93×10^{13} J·mol^{-1}
24-41. 2 years
24-43. 9.5×10^7 Ci·g^{-1}
24-45. 1.5 μg
24-47. 4.6 μg
24-49. (a) $^{12}_{6}C + {}^{1}_{1}H \rightarrow {}^{13}_{7}N + \gamma$
(b) $^{13}_{7}N \rightarrow {}^{13}_{6}C + {}^{0}_{+1}e$
(c) $^{13}_{6}C + {}^{1}_{1}H \rightarrow {}^{14}_{7}N + \gamma$
(d) $^{14}_{7}N + {}^{1}_{1}H \rightarrow {}^{15}_{8}O + \gamma$
(e) $^{15}_{8}O \rightarrow {}^{15}_{7}N + {}^{0}_{+1}e$
(f) $^{15}_{7}N + {}^{1}_{1}H \rightarrow {}^{12}_{6}C + {}^{4}_{2}He$
The net equation is $4{}^{1}_{1}H \rightarrow {}^{4}_{2}He + 2{}^{0}_{+1}e + 3\gamma$
24-51. 0.0598
24-53. 0.46 disintegration·min^{-1}
24-55. 12.1 metric ton
24-57. 1.4×10^5 kg
24-59. 27.8 d
24-61. The source of the water molecules is from the —OH on the acid and the —H on the alcohol; thus the labeled oxygen will appear in the ester. Because of the acid-dissociation reaction of acetic acid, both oxygen atoms are labeled; thus the labeled oxygen will appear in both the water and ester.
24-63. 1.70×10^{12} J·mol^{-1}
24-65. 7.0×10^{11} J·mol^{-1}
24-67. -1.91×10^{10} J
24-69. To overcome the coulombic repulsions of the nuclei.
24-71. 5.00 L
24-73. 3.3×10^{-5} M
24-75. 1.7×10^{-8} M^2
24-77. 4000 disintegrations
24-79. 1.7×10^8 y

CHAPTER 25

25-1. See Figure 9-35.
25-3. The electrons in the *f* orbitals lie closer to the nucleus than do the *d* electrons in the transition metals.
25-5. (a) metal (b) nonmetal (c) semimetal (d) metal (e) metal (f) metal (g) semimetal (h) semimetal
25-7. (a) *s*-block (b) *d*-block (c) *p*-block (d) *p*-block (e) *p*-block (f) *p*-block (g) *f*-block (h) *f*-block
25-9. Group 3 or Group 5
25-11. (a) decrease (b) increase (c) decrease (d) increase (e) increase
25-13. The elements in the third row can use the 3*d* orbitals in addition to the 3*s* and 3*p* orbitals in forming bonds; thus they can form more than four bonds. The 3*d* orbitals of the elements in the first and second rows are of too high energy to be used in forming bonds.
25-15. 1.0079
25-17. 1.07×10^9 disintegration·s^{-1}·μmol^{-1}
25-19. (a) $Fe_2O_3(s) + 3H_2(g) \xrightarrow{\text{high T}} 2Fe(s) + 3H_2O(g)$
(b) $LiH(s) + H_2O(l) \rightarrow LiOH(aq) + H_2(g)$
(c) $Mg(s) + H_2(g) \rightarrow MgH_2(s)$
(d) $2K(s) + H_2(g) \rightarrow 2KH(s)$
25-21. $2Li(s) + 2CH_3CH_2OH(l) \rightarrow 2LiCH_3CH_2O(alc) + H_2(g)$
25-23. 1.32 g
25-25. 1.39×10^5 L
25-27. 0.990 V
25-29. They react with the oxygen and water vapor in the air, but not with kerosene.
25-31. The electrolysis of molten sodium chloride
25-33. (a) $4Li(s) + O_2(g) \rightarrow 2Li_2O(s)$

(b) $2Na(s) + O_2(g) \rightarrow Na_2O_2(s)$
(c) $K(s) + O_2(g) \rightarrow KO_2(s)$
(d) $Cs(s) + O_2(g) \rightarrow CsO_2(s)$

25-35. Any sodium metal produced will react with the water to form sodium hydroxide and hydrogen.

25-37. K, Rb, and Ce form superoxides in addition to oxides when oxidized.

25-39. 506 g

25-41. 126 L

25-43. (a) $Ca(s) + H_2(g) \xrightarrow{500°C} CaH_2(s)$
(b) $3Mg(s) + N_2(g) \xrightarrow{500°C} Mg_3N_2(s)$
(c) $Sr(s) + S(s) \xrightarrow{500°C} SrS(s)$
(d) $Ba(s) + O_2(g) \xrightarrow{500°C} BaO_2(s)$

25-45. (a) $Be(s) + 2HCl(aq) \rightarrow BeCl_2(aq) + H_2(g)$
(b) $Be(s) + NaOH(aq) + 2H_2O(l) \rightarrow Na_2Be(OH)_4(aq) + H_2(g)$
(c) $3Be(s) + N_2(g) \xrightarrow{500°C} Be_3N_2(s)$
(d) $2Be(s) + O_2(g) \xrightarrow{300°C} 2BeO(s)$

25-47. (a) $2HCl(aq) + Mg(OH)_2(s) \rightarrow MgCl_2(aq) + 2H_2O(l)$
(b) 2.9 mg

25-49. React magnesium chloride with an aqueous solution of hydrochloric acid followed by evaporation of the water and hydrogen chloride.

25-51. (a) linear (b) Each of the two bonds is formed by combining an *sp* hybrid orbital on Be with one 3*p* orbital on Cl.

25-53. 10.5; 12.69; 13.18; 13.83

25-55. They decrease upon descending the group.

25-57. $B(OH)_3(aq) + 2H_2O(l) \rightarrow B(OH)_4^-(aq) + H_3O^+(aq)$

25-59. (a) 18 (b) 42

25-61. (a) trigonal planar (b) linear (c) tetrahedral

25-63. See Section 25-5B.

25-65. 2.98; 1.83; 1.86; 1.25

25-67. (a) $Al_2O_3(s) + 3C(s) \xrightarrow[\text{furnace}]{\text{electric}} 2Al(s) + 3CO(g)$
(b) $CaC_2(s) + 2D_2O(l) \xrightarrow{heat} Ca(OD)_2(s) + C_2D_2(g)$
(c) $2PbS(s) + 3O_2(g) \rightarrow 2PbO(s) + 2SO_2(g)$

25-69. diamond

25-71. Diamond is essentially one large covalently bonded molecule and thus has no electrons available to conduct an electric current. Graphite has many delocalized electrons in the covalently bonded layers and thus has electrons available to conduct an electric current.

25-73. Al or Ga

25-75. (a) +4 (b) +8/3 (c) −4 (d) −1

25-77. $SiO_2(s) + 6HF(aq) \rightarrow H_2SiF_6(aq) + 2H_2O(l)$

25-79. Each Si-H bond can be described by combining an sp^3 hybrid orbital on Si with an 1*s* atomic orbital on one of the hydrogen atoms. Each Si-H bond can be described by combining an sp^3 hybrid orbital on Si with an 1*s* atomic orbital on one of the hydrogen atoms. The Si-Si bond can be described by combining an sp^3 hybrid orbital on each of the silicon atoms.

25-81. See Section 25-6E.

CHAPTER 26

26-1. (a) $P_4(s) + 5O_2(g) \rightarrow P_4O_{10}(s)$
excess
(b) $P_4O_6(s) + 6H_2O(l) \rightarrow 4H_3PO_3(aq)$
(c) $P_4O_{10}(s) + 6H_2O(l) \rightarrow 4H_3PO_4(aq)$

26-3. $Li_3N(s) + 3D_2O(l) \rightarrow 3LiOD(aq) + ND_3(g)$

26-5. Nitrogen can form only four covalent bonds using eight valence electrons.

26-7. (a) H—N(—H)—O—H

(b) $(^-O)(^-O)N{=}O$ plus other resonance forms

(c) $^-O{-}N{=}O \leftrightarrow O{=}N{-}O^-$

(d) $^-N{=}N^+{=}O \leftrightarrow N{\equiv}N^+{-}O^-$

(e) $^-O{-}N^+{=}O \leftrightarrow O{=}N^+{-}O^-$

26-9. $P_4O_{10}(s) + 6H_2SO_4(l) \rightarrow 4H_3PO_4(l) + 6SO_3(g)$
$P_4O_{10}(s) + 12HNO_3(l) \rightarrow 4H_3PO_4(l) + 6N_2O_5(s)$

26-11. (a) tetrahedral (b) tetrahedral (c) octahedral

26-13. $As_4O_6(s) + 12NaOH(aq) \rightarrow 4Na_3AsO_3(aq) + 6H_2O(l)$
$As_4O_{10}(s) + 20NaOH(aq) \rightarrow 4Na_3AsO_4(aq) + 10H_2O(l)$
$Sb_2O_3(s) + 6NaOH(aq) \rightarrow 2Na_3SbO_3(aq) + 3H_2O(l)$
$Sb_2O_3(s) + 6HCl(aq) \rightarrow 2SbCl_3(aq) + 3H_2O(l)$
$Bi_2O_3(s) + 6HCl(aq) \rightarrow 2BiCl_3(aq) + 3H_2O(l)$

26-15. photosynthesis in green plants

26-17. $2C_2H_2(g) + 5O_2(g) \rightarrow 4CO_2(g) + 2H_2O(g)$

26-19. See Section 26-2F.

26-21. Rate of effusion of SF_6 is much slower than that of the molecules in air.

26-23.

H—O—S(=O)(=O)—O—S(=O)(=O)—O—H plus other resonance forms

26-25. The sulfur atom in SF_6 is effectively hidden from the water molecules in the middle of an octahedron formed by the fluorine atoms. The sulfur atom in SF_4 is more open to the water molecules.

26-27. 0.0611 M

26-29. (a) $4NaCl(aq) + MnO_2(s) + 2H_2SO_4(aq) \rightarrow MnCl_2(aq) + Cl_2(g) + 2H_2O(l) + 2Na_2SO_4(aq)$
(b) $4NaIO_3(aq) + 10NaHSO_4(aq) \rightarrow 2I_2(s) + 3H_2SO_4(aq) + 7Na_2SO_4(aq) + 2H_2O(l)$
(c) $6Br_2(l) + 12NaOH(aq) \rightarrow 2NaBrO_3(aq) + 10NaBr(aq) + 6H_2O(l)$

26-31. $3I_3(s) + 10HNO_3(aq) \rightarrow 6HIO_3(aq) + 10NO(g) + 2H_2O(l)$

26-33. (a) nitrous acid (b) sulfurous acid (c) hypophosphorous acid (d) phosphorous acid (e) hyponitrous acid. (a) potassium sulfite (b) calcium nitrite (c) potassium iodite (d) magnesium hypobromite

26-35. The nitrogen atom in NH_3 has a partial negative charge; thus the lone pair on the nitrogen atom can be easily donated. The nitrogen atom in NF_3 has a partial positive charge; thus the lone pair on the nitrogen atom cannot be easily donated.
26-37. tetrahedral around each chlorine atom; bent around the oxygen atom
26-39. (a) −1 on F and +5 on I (b) +1 (c) +5 (d) −1 on F and +1 on Cl (e) +5
26-41. 700 M^{-1}
26-43. 8.75×10^{-4} mol
26-45. See Section 26-4A.
26-47. See Figure 26-36. The remaining nitrogen may be removed by the reaction with lithium.
26-49. The electronegativity of O_2 is about the same as that of Xe and so Xe may replace O_2 in $O_2^+PtF_6^-$.
26-51. The van der Waals constant b is proportional to the volume of the gas molecule, and the volume of noble gas molecules increases with increasing atomic number. The van der Waals constant a is related to the attraction between molecules. The London forces or van der Waals forces between molecules increase as the size of the molecules increases, and so the value of a increases with increasing atomic number.

CHAPTER 27

27-1. tungsten
27-3. iron
27-5. $TiO_2(s) + 2Cl_2(g) \rightarrow TiCl_4(l) + O_2(g)$
$TiCl_4(l) + Mg(l) \rightarrow Ti(s) + MgCl_2(l)$
27-7. V_2O_5
27-9. $2Cr_2O_7^{2-}(aq) + 4H^+(aq) \rightarrow 2Cr_2O_3(s) + 2H_2O(l) + 3O_2(g)$
27-11. It is decomposed to $MnO_2(s)$ in the presence of light.
27-13. (a) tetrahedral (b) bipyramidal (c) tetrahedral (d) tetrahedral
27-15. See Section 27-3.
27-17. The cobalt(III) ion is unstable and is readily reduced to cobalt(II).
27-19. 109 metric ton
27-21. 58%
27-23. $4Au(s) + 8CN^-(aq) + O_2(g) + 2H_2O(l) \rightarrow 4[Au(CN)_2]^-(aq) + 4OH^-(aq)$
$2[Au(CN)_2]^-(aq) + Zn(s) \rightarrow [Zn(CN)_4]^{2-}(aq) + 2Au(s)$
27-25. (a) 5.075% (b) 0.30 μg
27-27. 36,400 metric ton

CHAPTER 28

28-1. (a) $1s^22s^22p^63s^23p^63d^5$ or $[Ar]3d^5$
(b) $1s^22s^22p^63s^23p^63d^2$ or $[Ar]3d^2$
(c) $1s^22s^22p^63s^23p^63d^{10}4s^24p^64d^6$ or $[Kr]4d^6$
(d) $1s^22s^22p^63s^23p^63d^{10}4s^24p^64d^{10}4f^{14}5s^25p^65d^6$ or $[Xe]4f^{14}5d^6$
28-3. (a) 10 (b) 6 (c) 6 (d) 7
28-5. (a) Co(III), Rh(III), and Ir(III) (b) Fe(IV), Ru(IV), and Os(IV) (c) Cu(I), Ag(I), and Au(I)
28-7. (a) +3 (b) +3 (c) +2 (d) +3
28-9. (a) +2 (b) +2 (c) +2 (d) +3
28-11. (a) three moles of $K^+(aq)$ and one mole of $[Fe(CN)_6]^{3-}(aq)$; hexacyanoferrate(III) (b) one mole of $[Ir(NH_3)_6]^{3+}(aq)$ and three moles of $NO_3^-(aq)$; hexammineiridium(III) (c) one mole of $[Pt(NH_3)_4Cl_2]^{2+}(aq)$ and two moles of $Cl^-(aq)$; tetramminedichloroplatinum(IV) (d) one mole of $[Ru(NH_3)_6]^{3+}(aq)$ and three moles of $Br^-(aq)$; hexammineruthenium(III)
28-13. $[Pt(NH_3)_6]Cl_4$; $[Pt(NH_3)_5Cl]Cl_3$; $[Pt(NH_3)_4Cl_2]Cl_2$; $[Pt(NH_3)_3Cl_3]Cl$; $[Pt(NH_3)_2Cl_4]$
28-15. (a) potassium hexacyanochromate(III)
(b) pentaaquachlorochromium(III) perchlorate
(c) tetracarbonyldichlorocobalt(III) perchlorate
(d) tetraamminedibromoplatinum(IV) chloride
28-17. (a) ammonium hexanitrocobaltate(III)
(b) tetraamminedibromoiridium(III) bromide
(c) potassium tetrachlorocuprate(II)
(d) pentacarbonylruthenium(0)
28-19. (a) $Na_3[Fe(CN)_5CO]$ (b) *trans*-$NH_4[AuCl_2I_2]$ (c) $K_3[Co(CN)_6]$ (d) $Ca_3[Co(NO_2)_6]_2$
28-21. (a) $[Pt(NH_3)_3Cl]NO_3$ (b) $Na_2[CuF_4]$ (c) $Li_4[Co(NO_2)_6]$ (d) $Ba_2[Fe(CN)_6]$
28-23. (a) There are three isomers. (b) There are six isomers.
28-25.

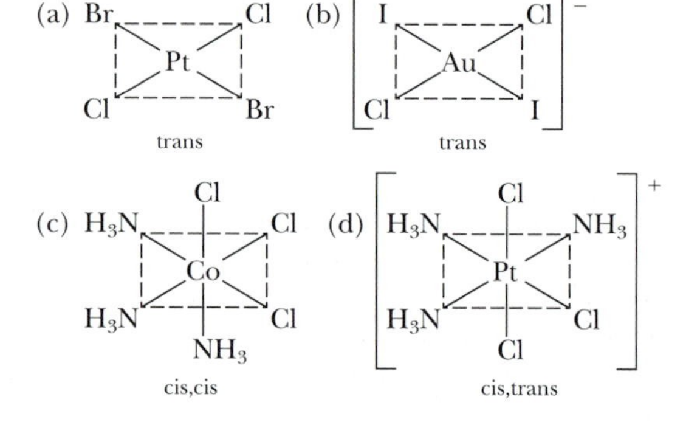

28-27. (a) t_{2g}^2 (b) t_{2g}^4 (c) $t_{2g}^3e_g^2$ (d) $t_{2g}^6e_g^4$ (e) $t_{2g}^6e_g^0$
28-29. (a) low spin (b) low spin (c) high spin (d) high spin (e) high spin
28-31. (a) 2 (b) 3 (c) 0 (d) 2
28-33. (a) diamagnetic (b) diamagnetic (c) paramagnetic (d) paramagnetic
28-35. Because $[NiF_4]^{2-}$ is paramagnetic, we predict that it is tetrahedral. Because $[Ni(CN)_4]^{2-}$ is diamagnetic, we predict that it is square planar.
28-37. (a) labile (b) labile (c) inert (d) labile
28-39. (a) $[Co(NO_2)_6]^{3-}$ (b) *trans*-$[PtCl_2(en)_2]^{2+}$ (c) $[Fe(CN)_5CO]^{3-}$ (d) *trans*-$[AuCl_2I_2]^-$

28-41.

$[Co(NH_3)_6]Cl_3$ $\quad$ $[Co(NH_3)_5Cl]Cl_2$

cis isomer $\quad$ trans isomer

$[Co(NH_3)_4Cl_2]Cl$

28-43. (a) false (b) false (c) false (d) true
28-45. (a) 2 (b) 3 (c) 0 (d) 1
28-47. $5.2 \times 10^{-17}\ M^3$
28-49. (c)

CHAPTER 29

29-1. (a) $C_5H_{12}(g) + 8O_2(g) \rightarrow 5CO_2(g) + 6H_2O(l)$

(b) $C_2H_6(g) + Cl_2(g) \xrightarrow{\text{dark}} \text{N.R.}$

(c) $C_4H_{10}(g) + H_2SO_4(aq) \rightarrow \text{N.R.}$

(d) $CH_4(g) + Cl_2(g) \xrightarrow{\text{UV}} CH_3Cl(g) + HCl(g)$
plus other chloromethanes such as $CH_2Cl_2(g)$

29-3. (a) identical (b) identical
29-5. The three isomers are 1-chlorohexane, 2-chlorohexane, and 3-chlorohexane.
29-7. $-49.50\ \text{kJ}\cdot\text{g}^{-1}$

29-9. cyclopropane $(CH_2)_3$; cyclobutane $(CH_2)_4$

29-11. (a) 2-bromo-3-chlorobutane or 2-chloro-3-bromobutane (b) 2,2-dimethylpropane
29-13. (a) It violates rule 2; 2-methylpentane. (b) It violates rule 2; 3-methylpentane. (c) It violates rule 2; 4-methyloctane. (d) It violates rule 6; 2,2-dimethylpropane.
29-15. (a) 2-methylbutane (b) 1-bromo-2-chloroethane or 2-bromo-1-chloroethane (c) 1,1,1,2,2-pentachloropropane (d) 2,2-dimethylpropane

29-17.

(a) $CH_3{-}CH(CH_3){-}CH(CH_3){-}CH_3$ (b) $CH_3{-}C(CH_3)_2{-}CH(CH_3){-}CH_3$

(c) $CH_3{-}CH_2{-}C(CH_3)_2{-}CH(CH_2CH_3){-}CH_2{-}CH_3$

(d) $CH_3{-}CH_2{-}CH_2{-}CH(CH(CH_3){-}CH_3){-}CH_2{-}CH_2{-}CH_2{-}CH_3$

29-19. (a) $H_3C{-}CH(CH_3){-}CH(CH_3){-}CH_3$ (b) $H_3C{-}CH(NH_2){-}CH(CH_3){-}CH_3$

(c) $H_3C{-}CH_2{-}CCl(CH_2CH_3){-}CH_2{-}CH_3$ (d) $H{-}CHCl{-}CHCl{-}CHCl{-}H$

29-21. (a) $H_2C{=}CH(CH_3) + HCl \longrightarrow H_3C{-}CHCl{-}CH_3$
2-chloropropane

(b) $H_3C(H)C{=}C(H)CH_3 + HBr \longrightarrow H_3C{-}CHBr{-}CH_2{-}CH_3$
2-bromobutane

(c) $H_2C{=}CH(CH_2CH_3) + HCl \longrightarrow H{-}CH_2{-}CHCl{-}CH_2{-}CH_2{-}H$
2-chlorobutane

(d) $H_3C(H)C{=}C(H)CH_2CH_3 + HBr \longrightarrow$

$H_3C{-}CH_2{-}CHBr{-}CH_2CH_3$
2-bromopentane

and $H_3C{-}CHBr{-}CH_2{-}CH_2CH_3$
2-bromopentane

29-23. (a) $H_2C{=}CH(CH_2CH_3) + HCl \longrightarrow$

$CH_3{-}CHCl{-}CH_2CH_3$
2-chlorobutane

(b) $H_2C{=}CH(CH_3) + Cl_2 \longrightarrow H{-}CHCl{-}CHCl{-}CH_3$
1,2-dichloropropane

29-25. (a) $(CH_3)_2C{=}CHCH_3$ or $CH_2{=}C(CH_3)CH_2CH_3$ and H_2O
(b) $CH_2{=}CHCH_3$ and H_2O
(c) $CH_3C{\equiv}CCH_3$ and H_2
29-27. (a) propyne (b) 2-butyne (c) 4,4-dimethyl-2-hexyne
29-29. 2,2-dibromopropane
29-31. (a) $CH_3CCl_2CH_3$ (b) $CH_3CBr_2CHBr_2$
29-33.

(a) benzene ring with CH_2CH_3 (b) benzene ring with Cl, Cl, Cl (1,3,5) (c) benzene ring with CH_3 and Cl (ortho) (d) benzene ring with Cl and Br (meta)

29-35. (a) 2-propanol (b) 2,2-dimethyl-1-butanol (c) 1,3-dichloro-2-propanol (d) 2-butanol
29-37. (a) primary (b) secondary (c) secondary (d) tertiary
29-39.

(a) $2CH_3CH_2OH(l) + 2Na(s) \rightarrow 2Na^+CH_3CH_2O^-(s) + H_2(g)$
(b) $2CH_3CH_2CH_2OH(l) + 2Na(s) \rightarrow 2Na^+CH_3CH_2CH_2O^-(s) + H_2(g)$

29-41.
(a) $CH_3CHOHCH_2CH_3(l) \xrightarrow{Cr_2O_7^{2-}} CH_3{-}\overset{O}{\overset{\|}{C}}{-}CH_2CH_3(l)$
(b) $CH_3CH_2OH(l) + NaH(s) \longrightarrow NaOCH_2CH_3(alc) + H_2(g)$
29-43. (a) butanal (b) 3-methylbutanal (c) propanal (d) 3,4-dimethylpentanal
29-45. (a) 3-pentanol (b) 2-pentanol (c) 3-hexanol
29-47. (a) propanone (b) 3-pentanone (c) butanone (d) butanone
29-49. (a) *N*-ethyl-1-amino-2-methylpropane (b) *N*-methyl-2-amino-3-methylpentane
29-51. (a) $C_2H_5NH_2(aq) + HBr(aq) \rightarrow C_2H_5NH_3Br(aq)$
(b) $2(CH_3)_2NH(aq) + H_2SO_4(aq) \rightarrow ((CH_3)_2NH_2)_2SO_4(aq)$

(c) $C_6H_5NH_2(aq) + HCl(aq) \longrightarrow C_6H_5NH_3Cl(aq)$

29-53. (a) CH_3CH_2COOH (b) $(CH_3)_2CHCOOH$
(c) $(CH_3)_3CCH_2COOH$
(d) $CH_3CH_2CH(CH_3)CH_2COOH$
29-55. (a) $HCOOH(aq) + NaOH(aq) \rightarrow NaHCOO(aq) + H_2O(l)$
(b) $HCOOH(aq) + CH_3OH(aq) \rightarrow CH_3OCHO(aq) + H_2O(l)$
(c) $2HCOOH(aq) + Ca(OH)_2(aq) \rightarrow Ca(HCOO)_2(aq) + 2H_2O(l)$
29-57. (a) potassium propanoate (b) potassium 2-methylpropanoate (c) calcium dichloroethanoate
29-59.
(a) $C_6H_5{-}COOH + HOCH_2CH_3 \rightarrow C_6H_5{-}COOCH_2CH_3 + H_2O$
(b) $HOOCCOOH + 2CH_3CH_2CH_2OH \rightarrow CH_3CH_2CH_2OOCCOOCH_2CH_2CH_3 + 2H_2O$
(c) $CH_3COOH + CH_3CHOHCH_3 \rightarrow CH_3COOCH(CH_3)_2 + H_2O$
29-61. $C_3H_4O_3$
29-63. $C_4H_{10}O$
29-65. C_5H_5N
29-67. They are the same molecule.
29-69. (a) 1 and 4 (b) 2 and 3 (c) 3 (d) 2
29-71. There is hydrogen bonding in 1-butanol but not in diethyl ether or pentane.
29-73. propene
29-75. 7.5×10^5 L of air; 2.9×10^6 kJ
29-77. C_3H_8, propane
29-79. (a) 1,2-dichlorobutane (b) 2-chlorobutane (c) 2-butanol (d) butane
29-81. (a) $CH_3CHClCHClCH_3$
(b) $BrCH_2CHClCH_2CH_3$ and $BrCH_2CH_2CHClCH_3$
(c) $CH_3CCl(CH_3)_2$ (d) $ClCH_2CHClCH_3$ and $Cl_2CHCH_2CH_3$
29-83. (a) propanal (b) 2-methylpropanal (c) 2,2-dimethylbutanal
29-85. The primary amines are $CH_3CH_2CH_2CH_2NH_2$, $(CH_3)_2CHCH_2NH_2$, $CH_3CH_2CH(CH_3)NH_2$, and $(CH_3)_3CNH_2$. The secondary amines are $CH_3CH_2CH_2NHCH_3$, $CH_3CH_2NHCH_2CH_3$, and $(CH_3)_2CHNHCH_3$. The tertiary amine is $CH_3CH_2N(CH_3)_2$.

CHAPTER 30

30-1. In addition to polymerization, monomers are joined to each other directly. In condensation polymerization, monomers are joined together with the formation of small molecules as joint products.
30-3. Both the dicarboxylic acid and diamine monomers contain six carbon atoms.
30-5. The molecules in (b), (c), and (d) are optical isomers.

30-7.

```
H2N—C(H)(CH2-C6H4-OH)—COOH + H2N—C(H)(CH(CH3)2)—COOH ⟶
H2N—C(H)(CH2-C6H4-OH)—C(=O)—N(H)—C(H)(CH(CH3)2)—COOH + H2O

or

H2N—C(H)(CH(CH3)2)—COOH + H2N—C(H)(CH2-C6H4-OH)—COOH ⟶
H2N—C(H)(CH(CH3)2)—C(=O)—N(H)—C(H)(CH2-C6H4-OH)—COOH + H2O
```

30-9.

```
H2N—C(H)(H)—C(=O)—N(H)—C(H)(CH3)—COOH    or    H2N—C(H)(CH3)—C(=O)—N(H)—C(H)(H)—COOH
       gly                 ala                          ala                 gly
```

30-11. $G_1G_1G_2$, $G_1G_2G_1$, $G_2G_1G_1$, $G_2G_2G_1$, $G_2G_1G_2$, $G_1G_2G_2$

30-13.

```
H2N—C(H)(CH2CH2COOH)—C(=O)—N(H)—C(H)(CH2COOH)—C(=O)—N(H)—C(H)(CH2-C6H4-OH)—COOH
```

30-15.

```
⊖O—P(=O)(O⊖)—OCH2 [deoxyribose–guanine (H2N, HN, C=O)]
  3'-O—P(=O)(O⊖)—O—CH2 [deoxyribose–adenine (NH2)]
  3'-O—P(=O)(O⊖)—O—CH2 [deoxyribose–thymine (HN, C=O, C—CH3)], 3'-OH
```

30-17.

```
O=P(O⊖)(O⊖)—OCH2 [ribose–uracil], 2'-OH
  3'-O—P(=O)(O⊖)—OCH2 [ribose–cytosine (NH2)], 2'-OH
  3'-O—P(=O)(O⊖)—OCH2 [ribose–uracil], OH OH
```

30-19. TTCAGAGCT

30-21. GATCAAT

30-23. The two strands come apart; complements to the two strands are formed to give two identical pairs of strands.

30-25. 15.9 kJ

30-27. $16.9\ kJ \cdot g^{-1}$

30-29. 1.7×10^5 M

30-31. 1.1×10^{35} M

PHOTO CREDITS

Laboratory photos by Chip Clark unless otherwise indicated; other photos are by the following:

p. 1, Dow Chemical; p. 2, California Water Service Company; p. 4, David Whillas/CSIRO Atmospheric Research, Australia; p. 6, Metropolitan Museum of Art, Purchase, Mr. and Mrs. Charles Wrightsman Gift, 1977 (1977.10); p. 7, Burndy Library; p. 8, (*bottom*) William Felger/Grant Heilman Photography; p. 10, (*bottom*) NASA; p. 14, Air Products and Chemicals, Inc.; p. 20, Echo Bay Mines; p. 28, Merck and Company, Inc.; p. 29, (*top*) INCO, (*bottom*) Schering-Plough; p. 30, Randall M. Feenstra and Joseph A. Stroscio/IBM; p. 35, (*bottom*) Division of Mineral Sciences, Smithsonian Institution; p. 36, Burndy Library; p. 37, The Science Museum, London; p. 44, (*top*) Cavendish Laboratory, (*bottom*) Deutsches Museum; p. 45, The Bettmann Archive; p. 48, Ontario Hydro; p. 53, Lawrence Berkeley Laboratory, U. of California; p. 56, Texasgulf; p. 66, City of Derby Museums and Art Gallery; p. 72, New York Public Library; p. 76, Alexander Boden; p. 78, American Chemical Society; p. 107, Grant Heilman/Grant Heilman; p. 114, Allied Chemical; p. 117, (*top*) W. H. Fahrenbach; p. 118, (*top*) TRW; p. 122, Ethyl Corp.; p. 142, Burndy Library; p. 156, Dennis Harding, Chevron Corporation; p. 161, NASA; p. 170, California Water Service Company; p. 195, Larry Lefever/Grant Heilman; p. 199, National Maritime Museum, Greenwich, England; p. 200, National Portrait Gallery; p. 202, American Institute of Physics; p. 216, Travis Amos; p. 226; U.S. Department of Energy; p. 238, NASA; p. 239, USGS; p. 242, NASA; p. 243, National Baseball Hall of Fame and Museum, Inc.; p. 246, U.S. Bureau of Reclamation; p. 253, Tenneco Inc.; p. 258, General Motors; p. 266, Travis Amos; p. 273, Parr Instrument Company; p. 277, (*top*) U.S. Air Force, p. 277, (*bottom*) Dennis Harding, Chevron Corporation; p. 279, Department of the Navy; p. 280, Peter Menzel; p. 281, Hydro-Québec; p. 290, Bellcore; p. 301, Bausch & Lomb; p. 302, Culver Pictures Inc.; p. 303, (*bottom right*) Yerkes Observatory; p. 306, American Institute of Physics, Center for History of Physics; p. 307, Education Development Center Inc., Newton Mass.; p. 308, (*left*) Susan Middleton, (*right*) Dr. Joan W. Nowicke, Smithsonian; p. 309, The Salvador Dali Museum, St. Petersburg, Florida; p. 310, (*top*) The Neils Bohr Institute; p. 316, University of Hamburg, from *From X-rays to Quarks,* by Emilio Segre, W. H. Freeman and Company, 1980; p. 318, A. K. Kleinschmidt/American Institute of Physics; p. 325, CERN/American Institute of Physics; p. 350, Lambda Physik; p. 353, IBM; p. 354, Will and Deni McIntyre/Photo Researchers; p. 365, Lawrence Berkeley Laboratory, U. of California; p. 375, The University of California Archives, the Bancroft Library; p. 409, Museum Boerhaave Leiden; p. 425, Argonne National Laboratory; p. 434, GE Research and Development Center; p. 438, Roger Ressmeyer/Starlight Photography; p. 445, Donald Clegg; p. 488, (*bottom*) John Shaw/Tom Stack & Associates; p. 498, (*top left*) Håkon Hope; p. 507, (*bottom right*) Corning Glass; p. 514, IBM; p. 515, IBM; p. 517, Argonne National Laboratory; p. 527, (*bottom*) John Lythgoe/Planet Earth Pictures; p. 529, (*bottom right*) Philip Saters/Planet Earth Pictures; p. 534, (*top*) Millipore, (*bottom left*) Millipore; p. 535, (*top*) Changmoon Park and William A. Goddard (Caltech) using Biograf software on the Evans and Sutherland PS 380 graphics systems, (*bottom*) "Biological membranes as bilayer couples," by M. Sheetz, R. Painter, and S. Singer, *J. of Cell Biology,* 1976, 70:193; p. 582, (*top*) Science Museum, London; p. 583, (*top*) Alex McPherson; p. 611, Smithsonian Institution; p. 613, General Electric; p. 619, Air Products and Chemicals, Inc.; p. 632, NASA; p. 634, Bill Wunsch/Denver Post; p. 711, (*bottom left*) Dudley Foster, Woods Hole Oceanographic Institute; p. 712, USGS; p. 713, Betz Laboratories, Inc.; p. 740, California Water Service Company; p. 748, Stephen Dalton/NHPA; p. 750, Wide World; p. 752, Elliot Morris, Colorado State University; p. 766, Rodney Cotterill and Flemming Kragh. The Technology University of Denmark; p. 776, (*bottom left*) Pacific Gas and Electric; p. 781, (*top right*) IBM Almaden Research Center; p. 784, Byron Crader/Ric Ergenbright Photography; p. 785, Yale University Library; p. 812, Manley Prim Photography, Inc.; p. 821, Travis Amos; p. 834, The Royal Institution; p. 837, Vulcan Materials Company; p. 838, E. I. du Pont de Nemours & Co.; p. 839, Alcoa; p. 849, GM; p. 850, Oak Ridge National Laboratory; p. 854, Lawrence Berkeley Laboratory, U. of California; p. 856, Bicron Corporation; p. 863, Lawrence Midgale/Photo Researchers; p. 866, Thomas A. Cahill, U. of California, Davis; p. 865, (*top*) Lawrence Berkeley Laboratory, U. of California, (*right*) Lawrence Berkeley Laboratory, U. of California; p. 871, Deutsches Museum; p. 873, (*bottom right*) Pacific Gas & Electric; p. 875, Naval Research Laboratory; p. 876, Princeton Plasma Physics Laboratory; p. 877, Archives of the Institute of Radium, from *From X-rays to Quarks,* by E. Segre, W. H. Freeman and Company, 1980; p. 879, Thomas A. Cahill, U. of California, Davis; p. 891, Lawrence Berkeley Laboratory, U. of California; p. 892, David Malin/Anglo-Australian Observatory; p. 895, (*top*) William Garnett, (*bottom*) Bill Tronca/Tom Stack & Associates; p. 899, (*bottom right*) Division of Mineral Sciences, Smithsonian Inst.; p. 905, U.S. Borax; p. 908, Alcoa; p. 912, General Electric; p. 917, Bellcore; p. 919, (*bottom right*) Steven Smale; p. 920, The Corning Museum of Glass, Corning, N.Y., Gift of The Ruth Bryan Strauss Memorial Foundation; p. 926, Pennzoil; p. 930, (*top left*) Runk/Schoenberger/Grant Heilman; p. 931, Grant Heilman/Grant Heilman; p. 944, Runk/Schoenberger/Grant Heilman; p. 946, (*bottom*) Ross Chapple; p. 947, (*bottom right*) Pennzoil; p. 958, The Corning Museum of Glass, Corning, N.Y.; p. 965, The Bettmann Archive; p. 966 (*top*) EG&G, Inc., Wellesley, Mass., (*bottom*) Argonne National Laboratory; p. 977, Martin Marietta; p. 984, Nippon Kohan K.K.; p. 985, Bethlehem Steel Corp.; p. 986, The Corning Museum of Glass, Corning, N.Y.; p. 987, (*top*) Bob Lynn, Cyprus Minerals, (*bottom*) Peter Arnold/Peter Arnold; p. 989, Hewlett Packard; p. 1006, University of Zurich; p. 1015, Stanford News Service; p. 1024, Pennzoil; p. 1026, Camille Vickers, Consolidated Natural Gas; p. 1028, Kenneth Lorenzen, U. of California, Davis; p. 1062, (*left*) Hagley Museum and Library; p. 1062, (*right*) Stanford News Service; p. 1067, General Electric Research and Development Center; p. 1068, Institute of Paper Chemistry; p. 1076, J. M. Hogle, Department of Molecular Biology, Research Institute of Scripps Clinic; p. 1077, A. K. Kleinschmidt, Elsevier; p. 1079, A. C. Barrington Brown, from *The Double Helix,* by J. D. Watson, Atheneum, N.Y., p. 215, copyright 1968.

INDEX

Physical Constants

Constant	*Symbol*	*Value*
atomic mass unit	amu	1.66056×10^{-27} kg
Avogadro's number	N	6.02205×10^{23} mol^{-1}
Bohr radius	a_0	5.292×10^{-11} m
Boltzmann constant	k	1.38066×10^{-23} J·K^{-1}
charge of a proton	e	1.60219×10^{-19} C
Faraday constant	F	96,485 C·mol^{-1}
gas constant	R	8.31441 J·K^{-1}·mol^{-1}
		0.08206 L·atm·K^{-1}·mol^{-1}
mass of an electron	m_e	9.10953×10^{-31} kg
		5.48580×10^{-4} amu
mass of a neutron	m_n	1.67495×10^{-27} kg
		1.00866 amu
mass of a proton	m_p	1.67265×10^{-27} kg
		1.00728 amu
Planck's constant	h	6.62618×10^{-34} J·s
speed of light	c	2.997925×10^{8} m·s^{-1}

SI Prefixes

Prefix	*Multiple*	*Symbol*	*Prefix*	*Multiple*	*Symbol*
tera	10^{12}	T	deci	10^{-1}	d
giga	10^{9}	G	centi	10^{-2}	c
mega	10^{6}	M	milli	10^{-3}	m
kilo	10^{3}	k	micro	10^{-6}	μ
			nano	10^{-9}	n
			pico	10^{-12}	p
			femto	10^{-15}	f
			atto	10^{-18}	a